Table G.1

SI Units and Prefixes[a]

SI Units	Quantity	Unit	SI Symbol	Formula
SI base units:	Length	meter	m	—
	Mass	kilogram	kg	—
	Time	second	s	—
	Temperature	kelvin	K	—
SI supplementary unit:	Plane angle	radian	rad	—
SI derived units:	Energy	joule	J	$N \cdot m$
	Force	newton	N	$kg \cdot m/s^2$
	Power	watt	W	J/s
	Pressure	pascal	Pa	N/m^2
	Work	joule	J	$N \cdot m$

SI prefixes	Multiplication Factor	Prefix	SI Symbol
	$1\ 000\ 000\ 000\ 000 = 10^{12}$	tera	T
	$1\ 000\ 000\ 000 = 10^{9}$	giga	G
	$1\ 000\ 000 = 10^{6}$	mega	M
	$1\ 000 = 10^{3}$	kilo	k
	$0.01 = 10^{-2}$	centi[b]	c
	$0.001 = 10^{-3}$	milli	m
	$0.000\ 001 = 10^{-6}$	micro	μ
	$0.000\ 000\ 001 = 10^{-9}$	nano	n
	$0.000\ 000\ 000\ 001 = 10^{-12}$	pico	p

[a]Source: ASTM Standard for Metric Practice E 380-97, 1997.

[b]To be avoided where possible.

Table G.2

Conversion Factors and Definitions

Fundamental Dimension	English Unit	Exact SI Value	Approximate SI Value
Length	1 in.	0.0254 m	—
Mass	1 lbm	0.453 592 37 kg	0.454 kg
Temperature	1°F	5/9 K	—

Definitions:

Acceleration of gravity: $g = 9.8066$ m/s^2 (= 32.174 ft/s^2)

Energy: Btu (British thermal unit) ≡ amount of energy required to raise the temperature of 1 lbm of water 1°F (1 Btu = 778.2 ft · lbf)
kilocalorie ≡ amount of energy required to raise the temperature of 1 kg of water 1 K(1 kcal = 4187 J)

Length: 1 mile = 5280 ft; 1 nautical mile = 6076.1 ft = 1852 m (exact)

Power: 1 horsepower ≡ 550 ft · lbf/s

Pressure: 1 bar ≡ 10^5 Pa

Temperature: degree Fahrenheit, $T_F = \frac{9}{5}T_C + 32$ (where T_C is degrees Celsius)
degree Rankine, $T_R = T_F + 459.67$
Kelvin, $T_K = T_C + 273.15$ (exact)

Viscosity: 1 Poise ≡ 0.1 kg/(m · s)
1 Stoke ≡ 0.0001 m^2/s

Volume: 1 gal ≡ 231 in.^{3} (1 ft^3 = 7.48 gal)

Useful Conversion Factors:

Length:
1 ft = 0.3048 m
1 in. = 25.4 mm

Mass:
1 lbm = 0.4536 kg
1 slug = 14.59 kg

Force:
1 lbf = 4.448 N
1 kgf = 9.807 N

Velocity:
1 ft/s = 0.3048 m/s
1 ft/s = 15/22 mph
1 mph = 0.447 m/s

Pressure:
1 psi = 6.895 kPa
1 lbf/ft^2 = 47.88 Pa
1 atm = 101.3 kPa
1 atm = 14.7 psi
1 in. Hg = 3.386 kPa
1 mm Hg = 133.3 Pa

Energy:
1 Btu = 1.055 kJ
1 ft · lbf = 1.356 J
1 cal = 4.187 J

Power:
1 hp = 745.7 W
1 ft · lbf/s = 1.356 W
1 Btu/hr = 0.2931 W

Area
1 ft^2 = 0.0929 m^2
1 acre = 4047 m^2

Volume:
1 ft^3 = 0.02832 m^3
1 gal (US) = 0.003785 m^3
1 gal (US) = 3.785 L

Volume flow rate:
1 ft^3/s = 0.02832 m^3/s
1 gpm = 6.309×10^{-5} m^3/s

Viscosity (dynamic)
1 lbf · s/ft^2 = 47.88 N · s/m^2
1 g/(cm · s) = 0.1 N · s/m^2
1 Poise = 0.1 N · s/m^2

Viscosity (kinematic)
1 ft^2/s = 0.0929 m^2/s
1 Stoke = 0.0001 m^2/s

Contents

Preface

Introduction

This text was written for an introductory course in fluid mechanics. Our approach to the subject, as in all previous editions, emphasizes the physical concepts of fluid mechanics and methods of analysis that begin from basic principles. The primary objective of this text is to help users develop an orderly approach to problem solving. Thus we always start from governing equations, state assumptions clearly, and try to relate mathematical results to corresponding physical behavior. We continue to emphasize the use of control volumes to maintain a practical problem-solving approach that is also theoretically inclusive.

Proven Problem-Solving Methodology

The Fox-McDonald-Pritchard solution methodology used in this text is illustrated in numerous Examples in each chapter. Solutions presented in the Examples have been prepared to illustrate good solution technique and to explain difficult points of theory. Examples are set apart in format from the text so that they are easy to identify and follow. Additional important information about the text and our procedures is given in the "Note to Student" in Section 1.1 of the printed text. We urge you to study this section carefully and to integrate the suggested procedures into your problem-solving and results-presentation approaches.

Goals and Advantages of Using This Text

Complete explanations presented in the text, together with numerous detailed Examples, make this book understandable for students, freeing the instructor to depart from conventional lecture teaching methods. Classroom time can be used to bring in outside material, expand on special topics (such as non-Newtonian flow, boundary-layer flow, lift and drag, or experimental methods), solve example problems, or explain difficult points of assigned homework problems. In addition, the 51 Example *Excel* workbooks are useful for presenting a variety of fluid mechanics phenomena, especially the effects produced when varying input parameters. Thus each class period can be used in the manner most appropriate to meet student needs.

When students finish the fluid mechanics course, we expect them to be able to apply the governing equations to a variety of problems, including those they have not encountered previously. We particularly emphasize physical concepts throughout to help students model the variety of phenomena that occur in real fluid flow situations. Although we collect, for convenience, useful equations at the end of most chapters, we stress that our philosophy is to minimize the use of so-called magic formulas and emphasize the systematic and fundamental approach to problem solving. By following this format, we believe students develop confidence in their ability to apply the material and to find that they can reason out solutions to rather challenging problems.

The book is well suited for independent study by students or practicing engineers. Its readability and clear examples help build confidence. Answer to Selected Problems are included, so students may check their own work.

Topical Coverage

The material has been selected carefully to include a broad range of topics suitable for a one- or two-semester course at the junior or senior level. We assume a background in rigid-body dynamics and mathematics through differential equations. A background in thermodynamics is desirable for studying compressible flow.

More advanced material, not typically covered in a first course, has been moved to the Web site (these sections are identified in the Table of Contents as being on the Web site). Advanced material is available to interested users of the book; available online, it does not interrupt the topic flow of the printed text.

Material in the printed text has been organized into broad topic areas:

- Introductory concepts, scope of fluid mechanics, and fluid statics (Chapters 1, 2, and 3)
- Development and application of control volume forms of basic equations (Chapter 4)
- Development and application of differential forms of basic equations (Chapters 5 and 6)
- Dimensional analysis and correlation of experimental data (Chapter 7)
- Applications for internal viscous incompressible flows (Chapter 8)
- Applications for external viscous incompressible flows (Chapter 9)
- Analysis of fluid machinery and system applications (Chapter 10)
- Analysis and applications of open-channel flows (Chapter 11)
- Analysis and applications of one- and two-dimensional compressible flows (Chapters 12 and 13)

Chapter 4 deals with analysis using both finite and differential control volumes. The Bernoulli equation is derived (in an optional subsection of Section 4.4) as an example application of the basic equations to a differential control volume. Being able to use the Bernoulli equation in Chapter 4 allows us to include more challenging problems dealing with the momentum equation for finite control volumes.

Another derivation of the Bernoulli equation is presented in Chapter 6, where it is obtained by integrating Euler's equation along a streamline. If an instructor chooses numerical values to calculate a quantitative answer. Solutions may be reproduced for classroom or library use, eliminating the labor of problem solving for the instructor who adopts the text.

The *Solutions Manual* is available online after the text is adopted. Visit the instructor section of the text's Web site at www.wiley.com/go/global/fox to request access to the password-protected online *Solutions Manual*.

- *Problem Key*: A list of all problems that are renumbered from the seventh edition of this title, to the eighth edition. There is no change to the actual solution to each of these problems.
- *PowerPoint Lecture Slides*: Lecture slides have been developed by Philip Pritchard, outlining the concepts in the book, and including appropriate illustrations and equations.
- *Image Gallery*: Illustrations from the text in a format appropriate to include in lecture presentations.

Additional Resources

- *A Brief Review of Microsoft Excel*: Prepared by Philip Pritchard and included on the book Web site as Appendix H, this resource will coach students in setting up and solving fluid mechanics problems using *Excel* spreadsheets. Visit www.wiley.com/go/global/fox to access it.
- *Excel Files*: These *Excel* Files and add-ins are for use with specific Examples from the text.
- *Additional Text Topics*: PDF files for these topics/sections are available only on the Web site. These topics are highlighted in the text's table of contents and in the chapters as being available on the Web site.
- *Answers to Selected Problems*: Answers to odd-numbered problems are listed at the end of the book as a useful aid for student self-study.
- *Videos*: Many worthwhile videos are available on the book Web site to demonstrate and clarify the basic principles of fluid mechanics. When it is appropriate to view these videos to aid in understanding concepts or phenomena, an icon appears in the margin of the printed text; the Web site provides links to both classic and new videos, and these are also listed in Appendix C.

Acknowledgments

We recognize that no single approach can satisfy all needs, and we are grateful to the many students and faculty whose comments have helped us improve on earlier editions of this book.

We wish to express our thanks to the contributors and reviewers of the *WileyPLUS* course:

Darrell W. Pepper, University of Nevada, Las Vegas
Brian P. Sangeorzan, Oakland University
Asghar Esmaeeli, Southern Illinois University, Carbondale
Andrew Gerhart, Lawrence Technological University
John Mitchell, University of Wisconsin, Madison
David Benson, Kettering University
Donald Fenton, Kansas State University
Alison Griffin, University of Central Florida

John Leylegian, Manhattan College
Mark Cummings, University of Idaho

We would also like to thank Bud Homsy for his assistance in getting permission from Stanford University, as well as the University of California, Santa Barbara, to license many of the videos we are making available to adopters of this edition, and we thank Gordon McCreight for his assistance with this process as well.

The following individuals are thanked for their invaluable contributions in developing interesting new problems for several chapters:

Kenneth W. Miller, St. Cloud State University
Darrell W. Pepper, University of Nevada, Las Vegas
Shizhi Qian, Old Dominion University
Thomas Shepard, University of Minnesota

The eighth edition was carefully reviewed in whole or part by:

John Abbitt, University of Florida
Soyoung Stephen Cha, University of Illinois, Chicago
Kangping Chen, Arizona State University
W. Scott Crawford, Stanford University
Timothy J. Fry, University of Dayton
James W. Leach, North Carolina State University
Jed E. Marquart, Ohio Northern University
Hans Mayer, California Polytechnic State University, San Luis Obispo
Karl R. Nelson, Colorado School of Mines
Siva Parameswaran, Texas Tech University
Brian P. Sangeorzan, Oakland University
Brian Savilonis, Worcester Polytechnic Institute
Hayley H. Shen, Clarkson University

We are extremely grateful for their comments and suggestions.

Finally, for this edition, we are very deeply indebted to John Leylegian, of Manhattan College, for his major contributions to this edition. He restructured Chapter 10 (and revised Appendix D), and he made significant contributions to changes in all the other chapters. He also took major responsibility for revising, updating, or replacing end-of-chapter problems for half of the chapters, as well as generating the corresponding parts of the solution manual. His expertise was essential for the revisions to Chapter 10.

We look forward to continued interactions with these and other colleagues who use the book.

Professor Pritchard appreciates the unstinting support of his wife, Penelope, who is keenly aware of all the hours that went into the effort of preparing this edition.

We welcome suggestions and/or criticisms from interested users of this book.

Philip J. Pritchard
August 2010

Professor Pritchard appreciates the unstinting support of his wife, Penelope, who is keenly aware of all the hours that went into the effort of preparing this edition.

We welcome suggestions and/or criticisms from interested users of this book.

Philip J. Pritchard
August 2010

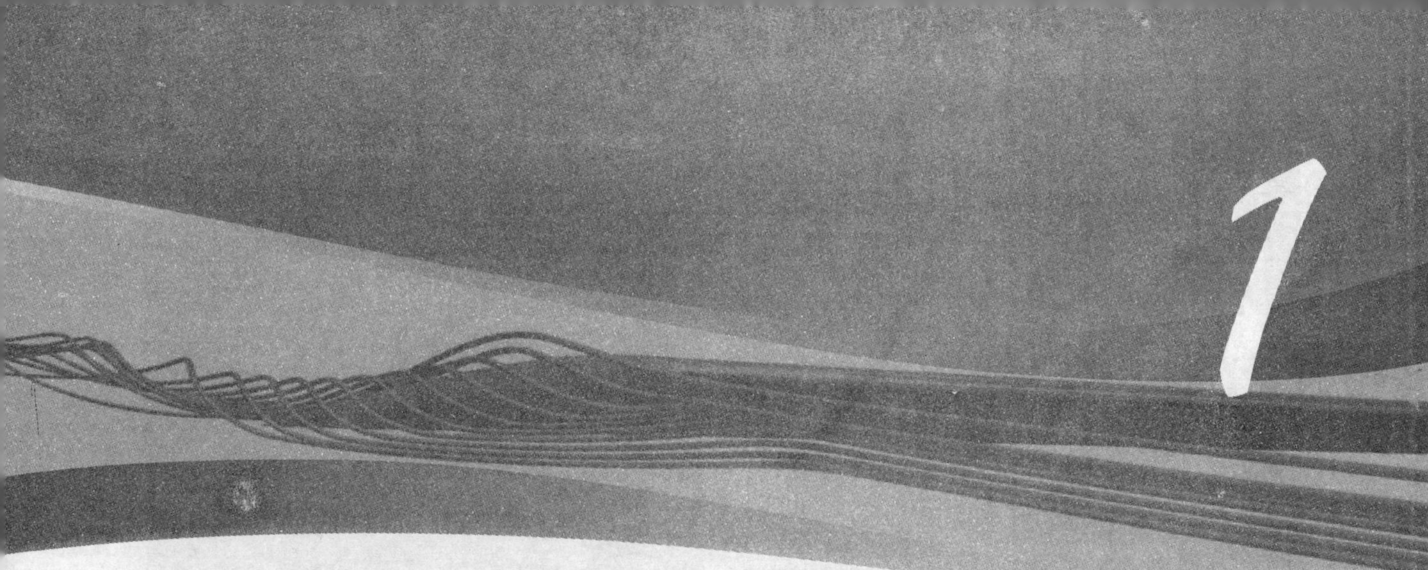

1

Introduction

Case Study in Energy and the Environment

Wind Power

At the beginning of each chapter we present a case study in the role of fluid mechanics in helping solve the energy crisis and in alleviating the environmental impact of our energy needs: the cases provide insight into the ongoing importance of the field of fluid mechanics. We have tried to present novel and original developments, not the kind of applications such as the ubiquitous wind farms. Please note that case studies represent a narrative; so each chapter's case study is not necessarily representative of the material in that chapter. Perhaps as a creative new engineer, you'll be able to create even better ways to extract renewable, nonpolluting forms of energy or invent something to make fluid-mechanics devices more efficient!

According to the July 16, 2009, edition of the *New York Times*, the global wind energy potential is much higher than previously estimated by both wind industry groups and government agencies. (Wind turbines are discussed in Chapter 10.) Using data from thousands of meteorological stations, the research indicates that the world's wind power potential is about 40 times greater than total current power consumption; previous studies had put that multiple at about seven times! In the lower 48 states, the potential from wind power is 16 times more

KiteGen's kites would fly at an altitude of about 1000 m and spin a power carousel on the ground. (Picture courtesy of Ben Shepard and Archer & Caldeira.)

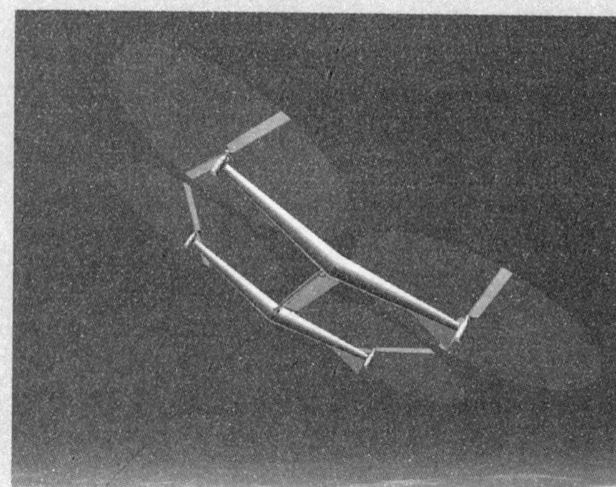

Sky Windpower's flying electric generators would fly at altitudes of about 10,000 m. (Picture courtesy of Ben Shepard and Archer & Caldeira.)

than total electricity demand in the United States, the researchers suggested, again much higher than a 2008 Department of Energy study that projected wind could supply a fifth of all electricity in the country by 2030. The findings indicate the validity of the often made claim that "the United States is the Saudi Arabia of wind." The new estimate is based the idea of deploying 2.5- to 3-megawatt (MW) wind turbines in rural areas that are neither frozen nor forested and also on shallow offshore locations, and it includes a conservative 20 percent estimate for capacity factor, which is a measure of how much energy a given turbine actually produces. It has been estimated that the total power from the wind that could conceivably be extracted is about 72 terawatts (TW, 72×10^{12} watts). Bearing in mind that the *total* power consumption by all humans was about 16 TW (as of 2006), it is clear that wind energy could supply all the world's needs for the foreseeable future!

One reason for the new estimate is due to the increasingly common use of very large turbines that rise to almost 100 m, where wind speeds are greater. Previous wind studies were based on the use of 50- to 80-m turbines. In addition, to reach even higher elevations (and hence wind speed), two approaches have been proposed. In a recent paper, Professor Archer at California State University and Professor Caldeira at the Carnegie Institution of Washington, Stanford, discussed some possibilities. One of these is a design of *KiteGen* (shown in the figure), consisting of tethered airfoils (kites) manipulated by a control unit and connected to a ground-based, carousel-shaped generator; the kites are maneuvered so that they drive the carousel, generating power, possibly as much as 100 MW. This approach would be best for the lowest few kilometers of the atmosphere. An approach using further increases in elevation is to generate electricity aloft and then transmit it to the surface with a tether. In the design proposed by *Sky Windpower*, four rotors are mounted on an airframe; the rotors both provide lift for the device and power electricity generation. The aircraft would lift themselves into place with supplied electricity to reach the desired altitude but would then generate up to 40 MW of power. Multiple arrays could be used for large-scale electricity generation. (Airfoils are discussed in Chapter 9.)

We shall examine some interesting developments in wind power in the *Case Studies in Energy and the Environment* in subsequent chapters.

We decided to title this textbook "Introduction to . . ." for the following reason: After studying the text, you will *not* be able to design the streamlining of a new car or an airplane, or design a new heart valve, or select the correct air extractors and ducting for a $100 million building; however, you *will* have developed a good understanding of the concepts behind all of these, and many other applications, and have made significant progress toward being ready to work on such state-of-the-art fluid mechanics projects.

To start toward this goal, in this chapter we cover some very basic topics: a case study, what fluid mechanics encompasses, the standard engineering definition of a fluid, and the basic equations and methods of analysis. Finally, we discuss some common engineering student pitfalls in areas such as unit systems and experimental analysis.

Note to Students 1.1

This is a student-oriented book: We believe it is quite comprehensive for an introductory text, and a student can successfully self-teach from it. However, most students will use the text in conjunction with one or two undergraduate courses. In either case, we recommend a thorough reading of the relevant chapters. In fact, a good approach is to read a chapter quickly once, then reread more carefully a second and even a third time, so that concepts develop a context and meaning. While students often find fluid mechanics quite challenging, we believe this approach, supplemented by your instructor's lectures that will hopefully amplify and expand upon the text material (if you are taking a course), will reveal fluid mechanics to be a fascinating and varied field of study.

Other sources of information on fluid mechanics are readily available. In addition to your professor, there are many other fluid mechanics texts and journals as well as the Internet (a recent Google search for "fluid mechanics" yielded 26.4 million links, including many with fluid mechanics calculators and animations!).

There are some prerequisites for reading this text. We assume you have already studied introductory thermodynamics, as well as statics, dynamics, and calculus; however, as needed, we will review some of this material.

It is our strong belief that one learns best by *doing*. This is true whether the subject under study is fluid mechanics, thermodynamics, or soccer. The fundamentals in any of these are few, and mastery of them comes through practice. *Thus it is extremely important that you solve problems.* The numerous problems included at the end of each chapter provide the opportunity to practice applying fundamentals to the solution of problems. Even though we provide for your convenience a summary of useful equations at the end of each chapter (except this one), you should avoid the temptation to adopt a so-called plug-and-chug approach to solving problems. Most of the problems are such that this approach simply will not work. In solving problems we strongly recommend that you proceed using the following logical steps:

1. State briefly and concisely (in your own words) the information given.
2. State the information to be found.
3. Draw a schematic of the system or control volume to be used in the analysis. Be sure to label the boundaries of the system or control volume and label appropriate coordinate directions.
4. Give the appropriate mathematical formulation of the *basic* laws that you consider necessary to solve the problem.
5. List the simplifying assumptions that you feel are appropriate in the problem.
6. Complete the analysis algebraically before substituting numerical values.
7. Substitute numerical values (using a consistent set of units) to obtain a numerical answer.
   a. Reference the source of values for any physical properties.
   b. Be sure the significant figures in the answer are consistent with the given data.
8. Check the answer and review the assumptions made in the solution to make sure they are reasonable.
9. Label the answer.

In your initial work this problem format may seem unnecessary and even long-winded. However, it is our experience that this approach to problem solving is ultimately the most efficient; it will also prepare you to be a successful professional, for which a major prerequisite is to be able to communicate information and the results of an analysis clearly and precisely. *This format is used in all Examples presented in this text*; answers to Examples are rounded to three significant figures.

Finally, *we strongly urge you to take advantage of the many Excel tools available for this book on the text Web site*, for use in solving problems. Many problems can be solved much more quickly using these tools; occasional problems can *only* be solved with the tools or with an equivalent computer application.

1.2 Scope of Fluid Mechanics

As the name implies, fluid mechanics is the study of fluids at rest or in motion. It has traditionally been applied in such areas as the design of canal, levee, and dam systems; the design of pumps, compressors, and piping and ducting used in the water and air conditioning systems of homes and businesses, as well as the piping systems needed in chemical plants; the aerodynamics of automobiles and sub- and supersonic airplanes; and the development of many different flow measurement devices such as gas pump meters.

While these are still extremely important areas (witness, for example, the current emphasis on automobile streamlining and the levee failures in New Orleans in 2005), fluid mechanics is truly a "high-tech" or "hot" discipline, and many exciting areas have developed in the last quarter-century. Some examples include environmental and energy issues (e.g., containing oil slicks, large-scale wind turbines, energy generation from ocean waves, the aerodynamics of large buildings, and the fluid mechanics of the atmosphere and ocean and of phenomena such as tornadoes, hurricanes, and tsunamis); biomechanics (e.g., artificial hearts and valves and other organs such as the liver; understanding of the fluid mechanics of blood, synovial fluid in the joints, the respiratory system, the circulatory system, and the urinary system); sport (design of bicycles and bicycle helmets, skis, and sprinting and swimming clothing, and the aerodynamics of the golf, tennis, and soccer ball); "smart fluids" (e.g., in automobile suspension systems to optimize motion under all terrain conditions, military uniforms containing a fluid layer that is "thin" until combat, when it can be "stiffened" to give the soldier strength and protection, and fluid lenses with humanlike properties for use in cameras and cell phones); and microfluids (e.g., for extremely precise administration of medications).

These are just a small sampling of the newer areas of fluid mechanics. They illustrate how the discipline is still highly relevant, and increasingly diverse, even though it may be thousands of years old.

1.3 Definition of a Fluid

CLASSIC VIDEO

Deformation of Continuous Media.

We already have a common-sense idea of when we are working with a fluid, as opposed to a solid: Fluids tend to flow when we interact with them (e.g., when you stir your morning coffee); solids tend to deform or bend (e.g., when you type on a keyboard, the springs under the keys compress). Engineers need a more formal and precise definition of a fluid: A *fluid* is a substance that deforms continuously under the application of a shear (tangential) stress no matter how small the shear stress may be. Because the fluid motion continues under the application of a shear stress, we can also define a fluid as any substance that cannot sustain a shear stress when at rest.

Hence liquids and gases (or vapors) are the forms, or phases, that fluids can take. We wish to distinguish these phases from the solid phase of matter. We can see the difference between solid and fluid behavior in Fig. 1.1. If we place a specimen of either substance between two plates (Fig. 1.1*a*) and then apply a shearing force F, each will initially deform (Fig. 1.1*b*); however, whereas a solid will then be at rest (assuming the force is not large enough to go beyond its elastic limit), a fluid will *continue* to deform (Fig. 1.1*c*, Fig. 1.1*d*, etc) as long as the force is applied. Note that a fluid in contact with a solid surface does not slip—it has the same velocity as that surface because of the *no-slip* condition, an experimental fact.

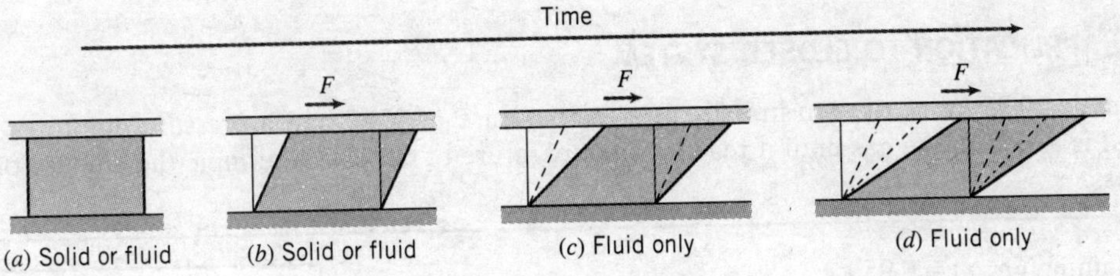

Fig. 1.1 Difference in behavior of a solid and a fluid due to a shear force.

The amount of deformation of the solid depends on the solid's modulus of rigidity G; in Chapter 2 we will learn that the *rate of deformation* of the fluid depends on the fluid's viscosity μ. We refer to solids as being *elastic* and fluids as being *viscous*. More informally, we say that solids exhibit "springiness." For example, when you drive over a pothole, the car bounces up and down due to the car suspension's metal coil springs compressing and expanding. On the other hand, fluids exhibit friction effects so that the suspension's shock absorbers (containing a fluid that is forced through a small opening as the car bounces) dissipate energy due to the fluid friction, which stops the bouncing after a few oscillations. If your shocks are "shot," the fluid they contained has leaked out so that there is almost no friction as the car bounces, and it bounces several times rather than quickly coming to rest. The idea that substances can be categorized as being either a solid or a liquid holds for most substances, but a number of substances exhibit both springiness and friction; they are *viscoelastic*. Many biological tissues are viscoelastic. For example, the synovial fluid in human knee joints lubricates those joints but also absorbs some of the shock occurring during walking or running. Note that the system of springs and shock absorbers comprising the car suspension is also viscoelastic, although the individual components are not. We will have more to say on this topic in Chapter 2.

CLASSIC VIDEO

Fundamentals—Boundary Layers.

Basic Equations 1.4

Analysis of any problem in fluid mechanics necessarily includes statement of the basic laws governing the fluid motion. The basic laws, which are applicable to any fluid, are:

1. The conservation of mass
2. Newton's second law of motion
3. The principle of angular momentum
4. The first law of thermodynamics
5. The second law of thermodynamics

Not all basic laws are always required to solve any one problem. On the other hand, in many problems it is necessary to bring into the analysis additional relations that describe the behavior of physical properties of fluids under given conditions.

For example, you probably recall studying properties of gases in basic physics or thermodynamics. The *ideal gas* equation of state

$$p = \rho RT \tag{1.1}$$

is a model that relates density to pressure and temperature for many gases under normal conditions. In Eq. 1.1, R is the gas constant. Values of R are given in Appendix A for several common gases; p and T in Eq. 1.1 are the absolute pressure and absolute temperature, respectively; ρ is density (mass per unit volume). Example 1.1 illustrates use of the ideal gas equation of state.

Example 1.1 FIRST LAW APPLICATION TO CLOSED SYSTEM

A piston-cylinder device contains 0.95 kg of oxygen initially at a temperature of 27°C and a pressure due to the weight of 150 kPa (abs). Heat is added to the gas until it reaches a temperature of 627°C. Determine the amount of heat added during the process.

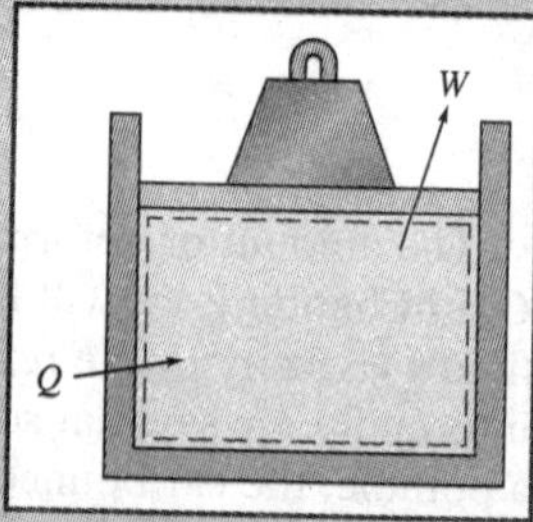

Given: Piston-cylinder containing O_2, $m = 0.95$ kg.

$$T_1 = 27°\text{C} \qquad T_2 = 627°\text{C}$$

Find: $Q_{1\to 2}$.

Solution: $p = \text{constant} = 150$ kPa (abs)
We are dealing with a system, $m = 0.95$ kg.

Governing equation: First law for the system, $Q_{12} - W_{12} = E_2 - E_1$

Assumptions: (1) $E = U$, since the system is stationary.
(2) Ideal gas with constant specific heats.

Under the above assumptions,

$$E_2 - E_1 = U_2 - U_1 = m(u_2 - u_1) = mc_v(T_2 - T_1)$$

The work done during the process is moving boundary work

$$W_{12} = \int_{V_1}^{V_2} p\,dV = p(V_2 - V_1)$$

For an ideal gas, $pV = mRT$. Hence $W_{12} = mR(T_2 - T_1)$. Then from the first law equation,

$$Q_{12} = E_2 - E_1 + W_{12} = mc_v(T_2 - T_1) + mR(T_2 - T_1)$$
$$Q_{12} = m(T_2 - T_1)(c_v + R)$$
$$Q_{12} = mc_p(T_2 - T_1) \quad \{R = c_p - c_v\}$$

From the Appendix, Table A.6, for O_2, $c_p = 909.4$ J/(kg · K). Solving for Q_{12}, we obtain

$$Q_{12} = 0.95\,\text{kg} \times 909 \frac{\text{J}}{\text{kg}\cdot\text{K}} \times 600\,\text{K} = 518\,\text{kJ} \longleftarrow Q_{12}$$

This problem:
- ✓ Was solved using the nine logical steps discussed earlier.
- ✓ Reviewed use of the ideal gas equation and the first law of thermodynamics for a system.

It is obvious that the basic laws with which we shall deal are the same as those used in mechanics and thermodynamics. Our task will be to formulate these laws in suitable forms to solve fluid flow problems and to apply them to a wide variety of situations.

We must emphasize that there are, as we shall see, many apparently simple problems in fluid mechanics that cannot be solved analytically. In such cases we must resort to more complicated numerical solutions and/or results of experimental tests.

1.5 Methods of Analysis

The first step in solving a problem is to define the system that you are attempting to analyze. In basic mechanics, we made extensive use of the *free-body diagram*. We will use a *system* or a *control volume*, depending on the problem being studied. These concepts are identical to the ones you used in thermodynamics (except you may have called them *closed system* and *open system*, respectively). We can use either one to get mathematical expressions for each of the basic laws. In thermodynamics they were mostly used to obtain expressions for conservation of mass and the first and second laws of thermodynamics; in our study of fluid mechanics, we will be most interested in conservation of mass and Newton's second law of motion. In thermodynamics our focus was energy; in fluid mechanics it will mainly be forces and motion. We must always be aware of whether we are using a system or a control volume approach because each leads to different mathematical expressions of these laws. At this point we review the definitions of systems and control volumes.

System and Control Volume

A *system* is defined as a fixed, identifiable quantity of mass; the system boundaries separate the system from the surroundings. The boundaries of the system may be fixed or movable; however, no mass crosses the system boundaries.

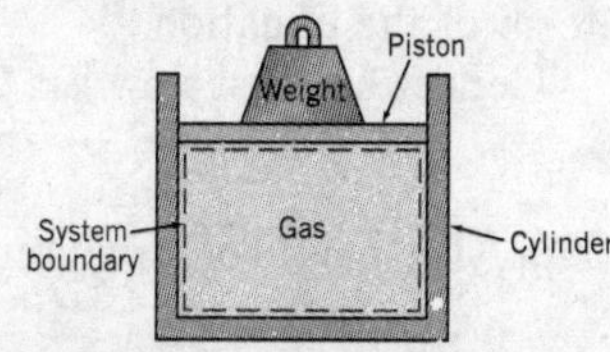

Fig. 1.2 Piston-cylinder assembly.

In the familiar piston-cylinder assembly from thermodynamics, Fig. 1.2, the gas in the cylinder is the system. If the gas is heated, the piston will lift the weight; the boundary of the system thus moves. Heat and work may cross the boundaries of the system, but the quantity of matter within the system boundaries remains fixed. No mass crosses the system boundaries.

In mechanics courses you used the free-body diagram (system approach) extensively. This was logical because you were dealing with an easily identifiable rigid body. However, in fluid mechanics we normally are concerned with the flow of fluids through devices such as compressors, turbines, pipelines, nozzles, and so on. In these cases it is difficult to focus attention on a fixed identifiable quantity of mass. It is much more convenient, for analysis, to focus attention on a volume in space through which the fluid flows. Consequently, we use the control volume approach.

A *control volume* is an arbitrary volume in space through which fluid flows. The geometric boundary of the control volume is called the control surface. The control surface may be real or imaginary; it may be at rest or in motion. Figure 1.3 shows flow through a pipe junction, with a control surface drawn on it. Note that some regions of the surface correspond to physical boundaries (the walls of the pipe) and others (at locations ①, ②, and ③) are parts of the surface that are imaginary (inlets or outlets). For the control volume defined by this surface, we could write equations for the basic laws and obtain results such as the flow rate at outlet ③ given the flow rates at inlet ①

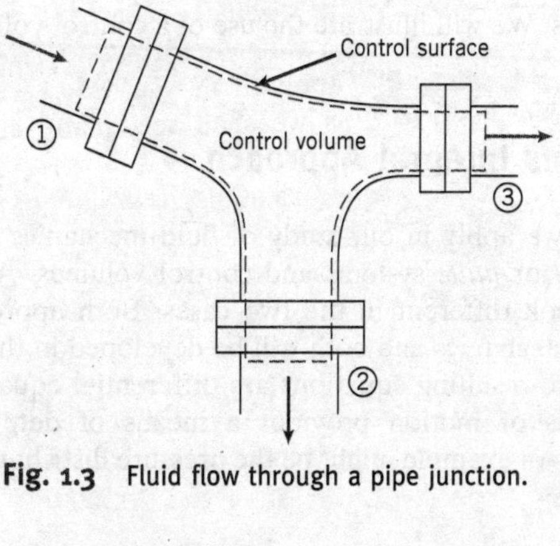

Fig. 1.3 Fluid flow through a pipe junction.

Example 1.2 MASS CONSERVATION APPLIED TO CONTROL VOLUME

A reducing water pipe section has an inlet diameter of 50 mm and exit diameter of 30 mm. If the steady inlet speed (averaged across the inlet area) is 2.5 m/s, find the exit speed.

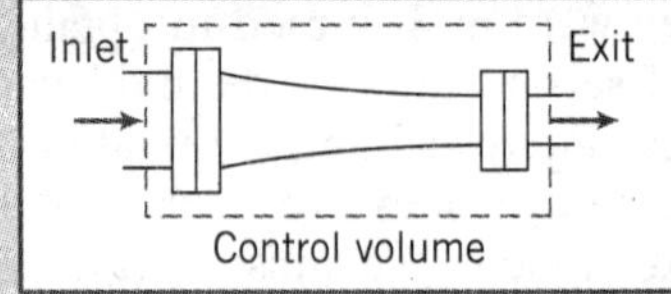

Given: Pipe, inlet $D_i = 50$ mm, exit $D_e = 30$ mm.
Inlet speed, $V_i = 2.5$ m/s.

Find: Exit speed, V_e.

Solution:

Assumption: Water is incompressible (density $\rho =$ constant).

The physical law we use here is the conservation of mass, which you learned in thermodynamics when studying turbines, boilers, and so on. You may have seen mass flow at an inlet or outlet expressed as either $\dot{m} = VA/v$ or $\dot{m} = \rho VA$ where V, A, v, and ρ are the speed, area, specific volume, and density, respectively. We will use the density form of the equation.

Hence the mass flow is:

$$\dot{m} = \rho VA$$

Applying mass conservation, from our study of thermodynamics,

$$\rho V_i A_i = \rho V_e A_e$$

(Note: $\rho_i = \rho_e = \rho$ by our first assumption.)
(Note: Even though we are already familiar with this equation from thermodynamics, we will derive it in Chapter 4.)

Solving for V_e,

$$V_e = V_i \frac{A_i}{A_e} = V_i \frac{\pi D_i^2/4}{\pi D_e^2/4} = V_i \left(\frac{D_i}{D_e}\right)^2$$

$$V_e = 2.7 \frac{\text{m}}{\text{s}} \left(\frac{50}{30}\right)^2 = 7.5 \frac{\text{m}}{\text{s}} \longleftarrow V_e$$

This problem:
- ✓ Was solved using the nine logical steps.
- ✓ Demonstrated use of a control volume and the mass conservation law.

and outlet ② (similar to a problem we will analyze in Example 4.1 in Chapter 4), the force required to hold the junction in place, and so on. It is always important to take care in selecting a control volume, as the choice has a big effect on the mathematical form of the basic laws. We will illustrate the use of a control volume with an example.

Differential versus Integral Approach

The basic laws that we apply in our study of fluid mechanics can be formulated in terms of *infinitesimal* or *finite* systems and control volumes. As you might suspect, the equations will look different in the two cases. Both approaches are important in the study of fluid mechanics and both will be developed in the course of our work.

In the first case the resulting equations are differential equations. Solution of the differential equations of motion provides a means of determining the detailed behavior of the flow. An example might be the pressure distribution on a wing surface.

Frequently the information sought does not require a detailed knowledge of the flow. We often are interested in the gross behavior of a device; in such cases it is more appropriate to use integral formulations of the basic laws. An example might be the overall lift a wing produces. Integral formulations, using finite systems or control volumes, usually are easier to treat analytically. The basic laws of mechanics and thermodynamics, formulated in terms of finite systems, are the basis for deriving the control volume equations in Chapter 4.

Methods of Description

Mechanics deals almost exclusively with systems; you have made extensive use of the basic equations applied to a fixed, identifiable quantity of mass. On the other hand, attempting to analyze thermodynamic devices, you often found it necessary to use a control volume (open system) analysis. Clearly, the type of analysis depends on the problem.

Where it is easy to keep track of identifiable elements of mass (e.g., in particle mechanics), we use a method of description that follows the particle. This sometimes is referred to as the *Lagrangian* method of description.

Consider, for example, the application of Newton's second law to a particle of fixed mass. Mathematically, we can write Newton's second law for a system of mass m as

$$\Sigma \vec{F} = m\vec{a} = m\frac{d\vec{V}}{dt} = m\frac{d^2\vec{r}}{dt^2} \tag{1.2}$$

In Eq. 1.2, $\Sigma\vec{F}$ is the sum of all external forces acting on the system, $\vec{a}$ is the acceleration of the center of mass of the system, $\vec{V}$ is the velocity of the center of mass of the system, and $\vec{r}$ is the position vector of the center of mass of the system relative to a fixed coordinate system.

Example 1.3 FREE FALL OF BALL IN AIR

The air resistance (drag force) on a 200 g ball in free flight is given by $F_D = 2 \times 10^{-4}\ V^2$, where F_D is in newtons and V is in meters per second. If the ball is dropped from rest 500 m above the ground, determine the speed at which it hits the ground. What percentage of the terminal speed is the result? (The *terminal speed* is the steady speed a falling body eventually attains.)

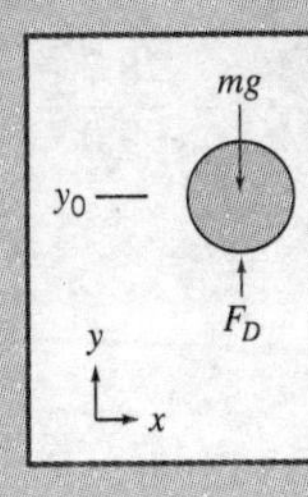

Given: Ball, $m = 0.2$ kg, released from rest at $y_0 = 500$ m.
Air resistance, $F_D = kV^2$, where $k = 2 \times 10^{-4}\ \text{N}\cdot\text{s}^2/\text{m}^2$.
Units: F_D(N), V(m/s).

Find: (a) Speed at which the ball hits the ground.
(b) Ratio of speed to terminal speed.

Solution:

Governing equation: $\Sigma\vec{F} = m\vec{a}$

Assumption: Neglect buoyancy force.

The motion of the ball is governed by the equation

$$\Sigma F_y = ma_y = m\frac{dV}{dt}$$

Since $V = V(y)$, we write $\Sigma F_y = m\dfrac{dV}{dy}\dfrac{dy}{dt} = mV\dfrac{dV}{dy}$ Then,

$$\Sigma F_y = F_D - mg = kV^2 - mg = mV\frac{dV}{dy}$$

Separating variables and integrating,

$$\int_{y_0}^{y} dy = \int_0^V \frac{mV dV}{kV^2 - mg}$$

$$y - y_0 = \left[\frac{m}{2k}\ln(kV^2 - mg)\right]_0^V = \frac{m}{2k}\ln\frac{kV^2 - mg}{-mg}$$

Taking antilogarithms, we obtain

$$kV^2 - mg = -mg\, e^{[(2k/m)(y - y_0)]}$$

Solving for V gives

$$V = \left\{\frac{mg}{k}\left(1 - e^{[(2k/m)(y - y_0)]}\right)\right\}^{1/2}$$

Substituting numerical values with $y = 0$ yields

$$V = \left\{0.2\,\text{kg} \times 9.81\frac{\text{m}}{\text{s}^2} \times \frac{\text{m}^2}{2 \times 10^{-4}\text{N}\cdot\text{s}^2} \times \frac{\text{N}\cdot\text{s}^2}{\text{kg}\cdot\text{m}}\left(1 - e^{[(2\times 2\times 10^{-4}/0.2)(-500)]}\right)\right\}^{1/2}$$

$V = 78.7\,\text{m/s}$ ← V

At terminal speed, $a_y = 0$ and $\Sigma F_y = 0 = kV_t^2 - mg$.

Then, $V_t = \left[\dfrac{mg}{k}\right]^{1/2} = \left[0.2\,\text{kg} \times 9.81\dfrac{\text{m}}{\text{s}^2} \times \dfrac{\text{m}^2}{2 \times 10^{-4}\text{N}\cdot\text{s}^2} \times \dfrac{\text{N}\cdot\text{s}^2}{\text{kg}\cdot\text{m}}\right]^{1/2}$

$= 99.0\ \text{m/s}$

The ratio of actual speed to terminal speed is

$\dfrac{V}{V_t} = \dfrac{78.7}{99.0} = 0.795$, or 79.5% ← $\dfrac{V}{V_t}$

This problem:

✓ Reviewed the methods used in particle mechanics.

✓ Introduced a variable aerodynamic drag force.

Try the *Excel* workbook for this Example for variations on this problem.

We could use this Lagrangian approach to analyze a fluid flow by assuming the fluid to be composed of a very large number of particles whose motion must be described. However, keeping track of the motion of each fluid particle would become a horrendous bookkeeping problem. Consequently, a particle description becomes unmanageable. Often we find it convenient to use a different type of description. Particularly with control volume analyses, it is convenient to use the field, or *Eulerian*, method of description, which focuses attention on the properties of a flow at a given point in space as a function of time. In the Eulerian method of description, the properties of a flow field are described as functions of space coordinates and time. We shall see in Chapter 2 that this method of description is a logical outgrowth of the assumption that fluids may be treated as continuous media.

Dimensions and Units 1.6

Engineering problems are solved to answer specific questions. It goes without saying that the answer must include units. In 1999, NASA's Mars Climate Observer crashed because the JPL engineers assumed that a measurement was in meters, but the supplying company's engineers had actually made the measurement in feet! Consequently, it is appropriate to present a brief review of dimensions and units. We say "review" because the topic is familiar from your earlier work in mechanics.

We refer to physical quantities such as length, time, mass, and temperature as *dimensions*. In terms of a particular system of dimensions, all measurable quantities are subdivided into two groups—*primary* quantities and *secondary* quantities. We refer to a small group of dimensions from which all others can be formed as primary quantities, for which we set up arbitrary scales of measure. Secondary quantities are those quantities whose dimensions are expressible in terms of the dimensions of the primary quantities.

Units are the arbitrary names (and magnitudes) assigned to the primary dimensions adopted as standards for measurement. For example, the primary dimension of length may be measured in units of meters, feet, yards, or miles. These units of length are related to each other through unit conversion factors (1 mile = 5280 feet = 1609 meters).

Systems of Dimensions

Any valid equation that relates physical quantities must be dimensionally homogeneous; each term in the equation must have the same dimensions. We recognize that Newton's second law ($\vec{F} \propto m\vec{a}$) relates the four dimensions, F, M, L, and t. Thus force and mass cannot both be selected as primary dimensions without introducing a constant of proportionality that has dimensions (and units).

Length and time are primary dimensions in all dimensional systems in common use. In some systems, mass is taken as a primary dimension. In others, force is selected as a primary dimension; a third system chooses both force and mass as primary dimensions. Thus we have three basic systems of dimensions, corresponding to the different ways of specifying the primary dimensions.

a. Mass $[M]$, length $[L]$, time $[t]$, temperature $[T]$

b. Force $[F]$, length $[L]$, time $[t]$, temperature $[T]$

c. Force $[F]$, mass $[M]$, length $[L]$, time $[t]$, temperature $[T]$

In system *a*, force $[F]$ is a secondary dimension and the constant of proportionality in Newton's second law is dimensionless. In system *b*, mass $[M]$ is a secondary dimension, and again the constant of proportionality in Newton's second law is dimensionless. In system *c*, both force $[F]$ and mass $[M]$ have been selected as primary dimensions. In this case the constant of proportionality, g_c (not to be confused with g, the acceleration of gravity!) in Newton's second law (written $\vec{F} = m\vec{a}/g_c$) is not dimensionless. The dimensions of g_c must in fact be $[ML/Ft^2]$ for the equation to be dimensionally homogeneous. The numerical value of the constant of proportionality depends on the units of measure chosen for each of the primary quantities.

Systems of Units

There is more than one way to select the unit of measure for each primary dimension. We shall present only the more common engineering systems of units for each of the basic systems of dimensions. Table 1.1 shows the basic units assigned to the primary dimensions for these systems. The units in parentheses are those assigned to that unit

Table 1.1

Common Unit Systems

System of Dimensions	Unit System	Force F	Mass M	Length L	Time t	Temperature T
a. MLtT	Système International d'Unités (SI)	(N)	kg	m	s	K
b. FLtT	British Gravitational (BG)	lbf	(slug)	ft	s	°R
c. FMLtT	English Engineering (EE)	lbf	lbm	ft	s	°R

system's secondary dimension. Following the table is a brief description of each of them.

a. MLtT

SI, which is the official abbreviation in all languages for the Système International d'Unités,[1] is an extension and refinement of the traditional metric system. More than 30 countries have declared it to be the only legally accepted system.

In the SI system of units, the unit of mass is the kilogram (kg), the unit of length is the meter (m), the unit of time is the second (s), and the unit of temperature is the kelvin (K). Force is a secondary dimension, and its unit, the newton (N), is defined from Newton's second law as

$$1\,\mathrm{N} \equiv 1\,\mathrm{kg} \cdot \mathrm{m/s^2}$$

In the Absolute Metric system of units, the unit of mass is the gram, the unit of length is the centimeter, the unit of time is the second, and the unit of temperature is the kelvin. Since force is a secondary dimension, the unit of force, the dyne, is defined in terms of Newton's second law as

$$1\,\mathrm{dyne} \equiv 1\,\mathrm{g} \cdot \mathrm{cm/s^2}$$

b. FLtT

CLASSIC VIDEO

Fluid Quantity and Flow.

In the British Gravitational system of units, the unit of force is the pound (lbf), the unit of length is the foot (ft), the unit of time is the second, and the unit of temperature is the degree Rankine (°R). Since mass is a secondary dimension, the unit of mass, the slug, is defined in terms of Newton's second law as

$$1\,\mathrm{slug} \equiv 1\,\mathrm{lbf} \cdot \mathrm{s^2/ft}$$

c. FMLtT

In the English Engineering system of units, the unit of force is the pound force (lbf), the unit of mass is the pound mass (lbm), the unit of length is the foot, the unit of time is the second, and the unit of temperature is the degree Rankine. Since both force and mass are chosen as primary dimensions, Newton's second law is written as

$$\vec{F} = \frac{m\vec{a}}{g_c}$$

[1] American Society for Testing and Materials, *ASTM Standard for Metric Practice*, E380-97. Conshohocken, PA: ASTM, 1997.

A force of one pound (1 lbf) is the force that gives a pound mass (1 lbm) an acceleration equal to the standard acceleration of gravity on Earth, 32.2 ft/s^2. From Newton's second law we see that

$$1\ \text{lbf} \equiv \frac{1\ \text{lbm} \times 32.2\ \text{ft/s}^2}{g_c}$$

or

$$g_c \equiv 32.2\ \text{ft} \cdot \text{lbm}/(\text{lbf} \cdot \text{s}^2)$$

The constant of proportionality, g_c, has both dimensions and units. The dimensions arose because we selected both force and mass as primary dimensions; the units (and the numerical value) are a consequence of our choices for the standards of measurement.

Since a force of 1 lbf accelerates 1 lbm at 32.2 ft/s^2, it would accelerate 32.2 lbm at 1 ft/s^2. A slug also is accelerated at 1 ft/s^2 by a force of 1 lbf. Therefore,

$$1\ \text{slug} \equiv 32.2\ \text{lbm}$$

Many textbooks and references use lb instead of lbf or lbm, leaving it up to the reader to determine from the context whether a force or mass is being referred to.

Preferred Systems of Units

In this text we shall use both the *SI* and the *British Gravitational* systems of units. In either case, the constant of proportionality in Newton's second law is dimensionless and has a value of unity. Consequently, Newton's second law is written as $\vec{F} = m\vec{a}$. In these systems, it follows that the gravitational force (the "weight"[2]) on an object of mass m is given by $W = mg$.

SI units and prefixes, together with other defined units and useful conversion factors, are summarized in Appendix G.

Example 1.4 USE OF UNITS

The label on a jar of peanut butter states its net weight is 510 g. Express its mass and weight in SI, BG, and EE units.

Given: Peanut butter "weight," $m = 510$ g.

Find: Mass and weight in SI, BG, and EE units.

Solution: This problem involves unit conversions and use of the equation relating weight and mass:

$$W = mg$$

The given "weight" is actually the mass because it is expressed in units of mass:

$$m_{\text{SI}} = 0.510\ \text{kg} \longleftarrow \underline{m_{\text{SI}}}$$

Using the conversions of Table G.2 (Appendix G),

$$m_{\text{EE}} = m_{\text{SI}}\left(\frac{1\ \text{lbm}}{0.454\ \text{kg}}\right) = 0.510\ \text{kg}\left(\frac{1\ \text{lbm}}{0.454\ \text{kg}}\right) = 1.12\ \text{lbm} \longleftarrow \underline{m_{\text{EE}}}$$

[2]Note that in the English Engineering system, the weight of an object is given by $W = mg/g_c$.

Using the fact that 1 slug = 32.2 lbm,

$$m_{BG} = m_{EE}\left(\frac{1\text{ slug}}{32.2\text{ lbm}}\right) = 1.12\text{ lbm}\left(\frac{1\text{ slug}}{32.2\text{ lbm}}\right)$$

$$= 0.0349\text{ slug} \longleftarrow \underline{m_{BG}}$$

To find the weight, we use

$$W = mg$$

In SI units, and using the definition of a newton,

$$W_{SI} = 0.510\text{ kg}\times 9.81\frac{\text{m}}{\text{s}^2} = 5.00\left(\frac{\text{kg}\cdot\text{m}}{\text{s}^2}\right)\left(\frac{\text{N}}{\text{kg}\cdot\text{m/s}^2}\right)$$

$$= 5.00\text{ N} \longleftarrow \underline{W_{SI}}$$

In BG units, and using the definition of a slug,

$$W_{BG} = 0.0349\text{ slug}\times 32.2\frac{\text{ft}}{\text{s}^2} = 1.12\frac{\text{slug}\cdot\text{ft}}{\text{s}^2}$$

$$= 1.12\left(\frac{\text{slug}\cdot\text{ft}}{\text{s}^2}\right)\left(\frac{\text{s}^2\cdot\text{lbf/ft}}{\text{slug}}\right) = 1.12\text{ lbf} \longleftarrow \underline{W_{BG}}$$

In EE units, we use the form $W = mg/g_c$, and using the definition of g_c,

$$W_{EE} = 1.12\text{ lbm}\times 32.2\frac{\text{ft}}{\text{s}^2}\times\frac{1}{g_c} = \frac{36.1}{g_c}\frac{\text{lbm}\cdot\text{ft}}{\text{s}^2}$$

$$= 36.1\left(\frac{\text{lbm}\cdot\text{ft}}{\text{s}^2}\right)\left(\frac{\text{lbf}\cdot\text{s}^2}{32.2\text{ ft}\cdot\text{lbm}}\right) = 1.12\text{ lbf} \longleftarrow \underline{W_{EE}}$$

This problem illustrates:
✓ Conversions from SI to BG and EE systems.
✓ Use of g_c in the EE system.

Notes: The student may feel this example involves a lot of unnecessary calculation details (e.g., a factor of 32.2 appears, then disappears), but it cannot be stressed enough that such steps should always be explicitly written out to minimize errors—if you do not write all steps and units down, it is just too easy, for example, to multiply by a conversion factor when you should be dividing by it. For the weights in SI, BG, and EE units, we could alternatively have looked up the conversion from newton to lbf.

Dimensional Consistency and "Engineering" Equations

In engineering, we strive to make equations and formulas have consistent dimensions. That is, each term in an equation, and obviously both sides of the equation, should be reducible to the same dimensions. For example, a very important equation we will derive later on is the Bernoulli equation

$$\frac{p_1}{\rho} + \frac{V_1^2}{2} + gz_1 = \frac{p_2}{\rho} + \frac{V_2^2}{2} + gz_2$$

which relates the pressure p, velocity V, and elevation z between points 1 and 2 along a streamline for a steady, frictionless incompressible flow (density ρ). This equation is dimensionally consistent because each term in the equation can be reduced to dimensions of L^2/t^2 (the pressure term dimensions are FL/M, but from Newton's law we find $F = M/Lt^2$, so $FL/M = ML^2/Mt^2 = L^2/t^2$).

Almost all equations you are likely to encounter will be dimensionally consistent. However, you should be alert to some still commonly used equations that are not; these are often "engineering" equations derived many years ago, or are empirical (based on experiment rather than theory), or are proprietary equations used in a particular industry or company. For example, civil engineers often use the semi-empirical Manning equation

$$V = \frac{R_h^{2/3}S_0^{1/2}}{n}$$

which gives the flow speed V in an open channel (such as a canal) as a function of the hydraulic radius R_h (which is a measure of the flow cross-section and contact surface area), the channel slope S_0, and a constant n (the Manning resistance coefficient). The value of this constant depends on the surface condition of the channel. For example, for a canal made from unfinished concrete, most references give $n \approx 0.014$. Unfortunately, the equation is dimensionally inconsistent! For the right side of the equation, R_h has dimensions L, and S_0 is dimensionless, so with a dimensionless constant n, we end up with dimensions of $L^{2/3}$; for the left side of the equation the dimensions must be L/t! A user of the equation is supposed to know that the values of n provided in most references will give correct results *only* if we ignore the dimensional inconsistency, always use R_h in meters, and interpret V to be in m/s! (The alert student will realize that this means that even though handbooks provide n values as just constants, they must have units of $\text{s/m}^{1/3}$.) Because the equation is dimensionally inconsistent, using the *same* value for n with R_h in ft does *not* give the correct value for V in ft/s.

A second type of problem is one in which the dimensions of an equation are consistent but use of units is not. The commonly used EER of an air conditioner is

$$EER = \frac{\text{cooling rate}}{\text{electrical input}}$$

which indicates how efficiently the AC works—a higher EER value indicates better performance. The equation *is* dimensionally consistent, with the EER being dimensionless (the cooling rate and electrical input are both measured in energy/time). However, it is *used*, in a sense, incorrectly, because the *units* traditionally used in it are not consistent. For example, a good EER value is 10, which would appear to imply you receive, say, 10 kW of cooling for each 1 kW of electrical power. In fact, an EER of 10 means you receive 10 *Btu/hr* of cooling for each 1 *W* of electrical power! Manufacturers, retailers, and customers all use the EER, in a sense, incorrectly in that they quote an EER of, say, 10, rather than the correct way, of 10 Btu/hr/W. (The EER, as used, is an everyday, inconsistent unit version of the coefficient of performance, COP, studied in thermodynamics.)

The two examples above illustrate the dangers in using certain equations. Almost all the equations encountered in this text will be dimensionally consistent, but you should be aware of the occasional troublesome equation you will encounter in your engineering studies.

As a final note on units, we stated earlier that we will use SI and BG units in this text. You will become very familiar with their use through using this text but should be aware that many of the units used, although they are scientifically and engineering-wise correct, are nevertheless not units you will use in everyday activities, and vice versa; we do not recommend asking your grocer to give you, say, 22 newtons, or 0.16 slugs, of potatoes; nor should you be expected to immediately know what, say, a motor oil viscosity of 5W20 means!

SI units and prefixes, other defined units, and useful conversions are given in Appendix G.

1.7 Analysis of Experimental Error

Most consumers are unaware of it but, as with most foodstuffs, soft drink containers are filled to plus or minus a certain amount, as allowed by law. Because it is difficult to precisely measure the filling of a container in a rapid production process, a 12-fl-oz container may actually contain 12.1, or 12.7, fl oz. The manufacturer is never supposed to supply less than the specified amount; and it will reduce profits if it is unnecessarily generous. Similarly, the supplier of components for the interior of a car must satisfy minimum and maximum dimensions (each component has what are called tolerances) so that the final appearance of the interior is visually appealing. Engineers performing experiments must measure not just data but also the uncertainties in their measurements. They must also somehow determine how these uncertainties affect the uncertainty in the final result.

All of these examples illustrate the importance of *experimental uncertainty*, that is, the study of uncertainties in measurements and their effect on overall results. There is always a trade-off in experimental work or in manufacturing: We can reduce the uncertainties to a desired level, but the smaller the uncertainty (the more precise the measurement or experiment), the more expensive the procedure will be. Furthermore, in a complex manufacture or experiment, it is not always easy to see which measurement uncertainty has the biggest influence on the final outcome.

Anyone involved in manufacturing, or in experimental work, should understand experimental uncertainties. Appendix F has details on this topic; there is a selection of problems on this topic at the end of this chapter.

1.8 Summary

In this chapter we introduced or reviewed a number of basic concepts and definitions, including:

- ✓ How fluids are defined, and the no-slip condition
- ✓ System/control volume concepts
- ✓ Lagrangian and Eulerian descriptions
- ✓ Units and dimensions (including SI, British Gravitational, and English Engineering systems)
- ✓ Experimental uncertainty

Case Study

"Fly Like a Bird"

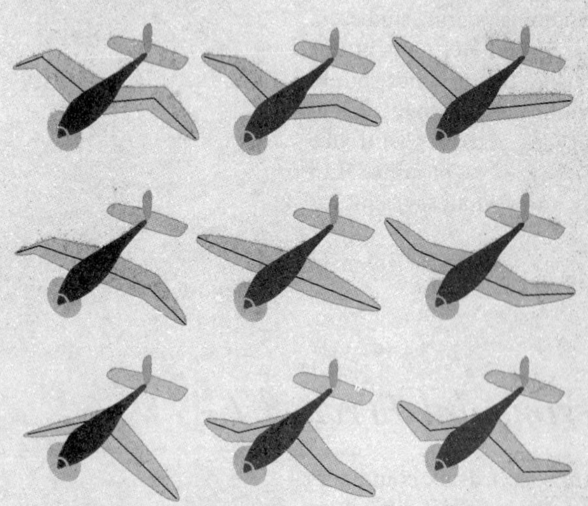

The airplane with various instantaneous wing shapes. (Courtesy of Dr. Rick Lind, University of Florida.)

At the end of each chapter, we present a case study: an interesting development in fluid mechanics chosen to illustrate that the field is constantly evolving.

No airplane, or airplane model, flies like a bird; aircraft all have fixed wings when in flight, whereas birds are (almost) constantly flapping away! One reason for this is that airplane and model wings must support relatively significant weight and are therefore thick and stiff; another reason is that we don't yet fully understand bird flight! Engineers at the University of Florida in Gainesville, led by researcher Rick Lind, have gone back to the drawing board and have developed a small surveillance aircraft (1/2 m wingspan, weight a total of 1 kg) that can change its wing shape during flight. While it is not true bird flight (the main propulsion is through a propeller), it is a radical departure from current airplane design. The airplane can change, for example, from an M shape wing configuration (very stable for gliding) to a W shape (for high maneuverability). It is amazingly dexterous: It can turn three rolls in less than a second (comparable to an F-15 fighter!), and its flight is sufficiently birdlike that it has attracted sparrows (friendly) and crows (unfriendly). Possible uses are in military surveillance, detection of biological agents in congested urban areas, and environmental studies in difficult airspaces such as forests.

Problems

Definition of a Fluid: Basic Equations

1.1 A number of common substances are

Tar	Sand
"Silly Putty"	Jello
Modeling clay	Toothpaste
Wax	Shaving cream

Some of these materials exhibit characteristics of both solid and fluid behavior under different conditions. Explain and give examples.

1.2 Give a word statement of each of the five basic conservation laws stated in Section 1.4, as they apply to a system.

Methods of Analysis

1.3 The barrel of a bicycle tire pump becomes quite warm during use. Explain the mechanisms responsible for the temperature increase.

1.4 Discuss the physics of skipping a stone across the water surface of a lake. Compare these mechanisms with a stone as it bounces after being thrown along a roadway.

1.5 Make a guess at the order of magnitude of the mass (e.g., 0.01, 0.1, 1.0, 10, 100, or 1000 kg) of standard air that is in a room 3 m by 3 m by 2.4 m, and then compute this mass in kg to see how close your estimate was.

1.6 A spherical tank of inside diameter 500 cm contains compressed oxygen at 7 MPa and 25°C. What is the mass of the oxygen?

1.7 Very small particles moving in fluids are known to experience a drag force proportional to speed. Consider a particle of net weight W dropped in a fluid. The particle experiences a drag force, $F_D = kV$, where V is the particle speed. Determine the time required for the particle to accelerate from rest to 95 percent of its terminal speed, V_t, in terms of k, W, and g.

1.8 Consider again the small particle of Problem 1.7. Express the distance required to reach 95 percent of its terminal speed in percent terms of g, k, and W.

1.9 A cylindrical tank must be designed to contain 5 kg of compressed nitrogen at a pressure of 200 atm (gage) and 20°C. The design constraints are that the length must be twice the diameter and the wall thickness must be 0.5 cm. What are the external dimensions?

1.10 In a combustion process, gasoline particles are to be dropped in air at 93°C. The particles must drop at least 25 cm in 1 s. Find the diameter d of droplets required for this. (The drag on these particles is given by $F_D = 3\pi\mu Vd$, where V is the particle speed and μ is the air viscosity. To solve this problem, use *Excel's Goal Seek*.)

1.11 For a small particle of styrofoam (16 kg/m^3) (spherical, with diameter $d = 0.3$ mm) falling in standard air at speed V, the drag is given by $F_D = 3\pi\mu Vd$, where μ is the air viscosity. Find the maximum speed starting from rest, and the time it takes to reach 95 percent of this speed. Plot the speed as a function of time.

1.12 In a pollution control experiment, minute solid particles (typical mass 5×10^{-11} kg) are dropped in air. The terminal speed of the particles is measured to be 5 cm. The drag of these particles is given by $F_D = kV$, where V is the instantaneous particle speed. Find the value of the constant k. Find the time required to reach 99 percent of terminal speed.

1.13 For Problem 1.12, find the distance the particles travel before reaching 99 percent of terminal speed. Plot the distance traveled as a function of time.

1.14 A sky diver with a mass of 70 kg jumps from an aircraft. The aerodynamic drag force acting on the sky diver is known to be $F_D = kV^2$, where $k = 0.25$ N·s^2/m^2. Determine the maximum speed of free fall for the sky diver and the speed reached after 100 m of fall. Plot the speed of the sky diver as a function of time and as a function of distance fallen.

1.15 For Problem 1.14, the initial horizontal speed of the sky diver is 70 m/s. As she falls, the k value for the vertical drag remains as before, but the value for horizontal motion is $k = 0.05$ N·s/m^2. Compute and plot the 2D trajectory of the sky diver.

1.16 The English perfected the longbow as a weapon after the Medieval period. In the hands of a skilled archer, the longbow was reputed to be accurate at ranges to 100 m or more. If the maximum altitude of an arrow is less than $h = 10$ m while traveling to a target 100 m away from the archer, and neglecting air resistance, estimate the speed and angle at which the arrow must leave the bow. Plot the required release speed and angle as a function of height h.

Dimensions and Units

1.17 For each quantity listed, indicate dimensions using mass as a primary dimension, and give typical SI and English units:
(a) Power
(b) Pressure
(c) Modulus of elasticity
(d) Angular velocity
(e) Energy
(f) Moment of a force
(g) Momentum
(h) Shear stress
(i) Strain
(j) Angular momentum

1.18 For each quantity listed, indicate dimensions using force as a primary dimension, and give typical SI and English units:
(a) Power
(b) Pressure
(c) Modulus of elasticity
(d) Angular velocity
(e) Energy
(f) Momentum
(g) Shear stress
(h) Specific heat

(i) Thermal expansion coefficient
(j) Angular momentum

1.19 Derive the following conversion factors:
(a) Convert a viscosity of 1 m^2/s to ft^2/s.
(b) Convert a power of 100 W to horsepower.
(c) Convert a specific energy of 1 kJ/kg to Btu/lbm.

1.20 Derive the following conversion factors:
(a) Convert a pressure of 1 psi to kPa.
(b) Convert a volume of 1 liter to gallons.
(c) Convert a viscosity of 1 $lbf \cdot s/ft^2$ to $N \cdot s/m^2$.

1.21 Derive the following conversion factors:
(a) Convert a specific heat of 4.18 kJ/kg · K to Btu/lbm · °R.
(b) Convert a speed of 30 m/s to mph.
(c) Convert a volume of 5.0 L to in^3.

1.22 Express the following in SI units:
(a) 5 acre · ft
(b) 150 in^3/s
(c) 3 gpm
(d) 3 mph/s

1.23 Express the following in SI units:
(a) 100 cfm (ft^3/min)
(b) 5 gal
(c) 65 mph
(d) 5.4 acres

1.24 Express the following in BG units:
(a) 50 m^2
(b) 250 cc
(c) 100 kW
(d) 5 kg/m^2

1.25 Express the following in BG units:
(a) 180 cc/min
(b) 300 kW · hr
(c) 50 $N \cdot s/m^2$
(d) 40 $m^2 \cdot hr$

1.26 While you're waiting for the ribs to cook, you muse about the propane tank of your barbecue. You're curious about the volume of propane versus the actual tank size. Find the liquid propane volume when full (the weight of the propane is specified on the tank). Compare this to the tank volume (take some measurements, and approximate the tank shape as a cylinder with a hemisphere on each end). Explain the discrepancy.

1.27 A farmer needs 4 cm of rain per week on his farm, with 10 hectares of crops. If there is a drought, how much water (L/min) will have to be supplied to maintain his crops?

1.28 Derive the following conversion factors:
(a) Convert a volume flow rate in cubic inches per minute to cubic millimeters per minute.
(b) Convert a volume flow rate in cubic meters per second to gallons per minute (gpm).
(c) Convert a volume flow rate in liters per minute to gpm.
(d) Convert a volume flow rate of air in standard cubic feet per minute (SCFM) to cubic meters per hour. A standard cubic foot of gas occupies one cubic foot at standard temperature and pressure ($T = 15$°C and $p = 101.3$ kPa absolute).

1.29 The density of mercury is given as 13,550 kg/m^3. Calculate the specific gravity and the specific volume in m^3/kg of the mercury. Calculate the specific weight in N/m^3 on Earth and on the moon. Acceleration of gravity on the moon is 1.67 m/s^2.

1.30 The kilogram force is commonly used in Europe as a unit of force. (1 kgf is the force exerted by a mass of 1 kg in standard gravity.) Moderate pressures, such as those for auto or truck tires, are conveniently expressed in units of kgf/cm^2. Convert 220 kPa to these units.

1.31 In Section 1.6 we learned that the Manning equation computes the flow speed V (m/s) in a canal made from unfinished concrete, given the hydraulic radius R_h (m), the channel slope S_0, and a Manning resistance coefficient constant value $n \approx 0.014$. For a canal with $R_h = 7.5$ m and a slope of 1/10, find the flow speed. Compare this result with that obtained using the same n value, but with R_h first converted to ft, with the answer assumed to be in ft/s. Finally, find the value of n if we wish to *correctly* use the equation for BG units (and compute V to check!).

1.32 From thermodynamics, we know that the coefficient of performance of an ideal air conditioner (COP_{ideal}) is given by

$$COP_{\text{ideal}} = \frac{T_L}{T_H - T_L}$$

where T_L and T_H are the room and outside temperatures (absolute). If an AC is to keep a room at 20°C when it is 40°C outside, find the COP_{ideal}. Convert to an *EER* value, and compare this to a typical Energy Star–compliant *EER* value.

1.33 The maximum theoretical flow rate (kg/s) through a supersonic nozzle is

$$\dot{m}_{\max} = 2.38 \frac{A_t p_0}{\sqrt{T_0}}$$

where A_t (m^2) is the nozzle throat area, p_0 (Pa) is the tank pressure, and T_0 (K) is the tank temperature. Is this equation dimensionally correct? If not, find the units of the 2.38 term.

1.34 The mean free path λ of a molecule of gas is the average distance it travels before collision with another molecule. It is given by

$$\lambda = C \frac{m}{\rho d^2}$$

where m and d are the molecule's mass and diameter, respectively, and ρ is the gas density. What are the dimensions of constant C for a dimensionally consistent equation?

1.35 In Chapter 9 we will study aerodynamics and learn that the drag force F_D on a body is given by

$$F_D = \frac{1}{2} \rho V^2 A C_D$$

Hence the drag depends on speed V, fluid density ρ, and body size (indicated by frontal area A) and shape (indicated by drag coefficient C_D). What are the dimensions of C_D?

1.36 A container weighs 15.5 N when empty. When filled with water at 32°C, the mass of the container and its contents is 36.5 kg. Find the weight of water in the container, and its volume in cubic meters, using data from Appendix A.

1.37 An important equation in the theory of vibrations is

$$m\frac{d^2x}{dt^2} + c\frac{dx}{dt} + kx = f(t)$$

where m (kg) is the mass and x (m) is the position at time t (s). For a dimensionally consistent equation, what are the dimensions of c, k, and f? What would be suitable units for c, k, and f in the SI and BG systems?

1.38 A parameter that is often used in describing pump performance is the specific speed, $N_{S_{cu}}$, given by

$$N_{S_{cu}} = \frac{N(\text{rpm})[Q(\text{gpm})]^{1/2}}{[H(\text{ft})]^{3/4}}$$

What are the units of specific speed? A particular pump has a specific speed of 2000. What will be the specific speed in SI units (angular velocity in rad/s)?

1.39 A particular pump has an "engineering" equation form of the performance characteristic equation given by $H\ (\text{m}) = 0.46 - 9.57 \times 10^{-7}\ [Q\ (\text{Lit/min})]^2$, relating the head H and flow rate Q. What are the units of the coefficients 1.5 and 4.5×10^{-5}? Derive an SI version of this equation.

Analysis of Experimental Error

1.40 Calculate the density of standard air in a laboratory from the ideal gas equation of state. Estimate the experimental uncertainty in the air density calculated for standard conditions (101.3 kPa and 15°C) if the uncertainty in measuring the barometer height is ±2.5 mm. of mercury and the uncertainty in measuring temperature is ±0.3°C.

1.41 Repeat the calculation of uncertainty described in Problem 1.40 for air in a hot air balloon. Assume the measured barometer height is 759 mm of mercury with an uncertainty of ±1 mm of mercury and the temperature is 60°C with an uncertainty of ±1°C. [Note that 759 mm of mercury corresponds to 101 kPa (abs).]

1.42 The mass of the standard American golf ball is 45.4 ± 0.3 g and its mean diameter is 43 ± 0.25 mm. Determine the density and specific gravity of the American golf ball. Estimate the uncertainties in the calculated values.

1.43 A can of pet food has the following internal dimensions: 102 mm height and 73 mm diameter (each ±1 mm at odds of 20 to 1). The label lists the mass of the contents as 397 g. Evaluate the magnitude and estimated uncertainty of the density of the pet food if the mass value is accurate to ±1 g at the same odds.

1.44 The mass flow rate in a water flow system determined by collecting the discharge over a timed interval is 0.2 kg/s. The scales used can be read to the nearest 0.05 kg and the stopwatch is accurate to 0.2 s. Estimate the precision with which the flow rate can be calculated for time intervals of (a) 10 s and (b) 1 min.

1.45 The mass flow rate of water in a tube is measured using a beaker to catch water during a timed interval. The nominal mass flow rate is 100 g/s. Assume that mass is measured using a balance with a least count of 1 g and a maximum capacity of 1 kg, and that the timer has a least count of 0.1 s. Estimate the time intervals and uncertainties in measured mass flow rate that would result from using 100, 500, and 1000 mL beakers. Would there be any advantage in using the largest beaker? Assume the tare mass of the empty 1000 mL beaker is 500 g.

1.46 The mass of the standard British golf ball is 45.9 ± 0.3 g and its mean diameter is 41.1 ± 0.3 mm. Determine the density and specific gravity of the British golf ball. Estimate the uncertainties in the calculated values.

1.47 The estimated dimensions of a soda can are $D = 66.0 \pm 0.5$ mm and $H = 110 \pm 0.5$ mm. Measure the mass of a full can and an empty can using a kitchen scale or postal scale. Estimate the volume of soda contained in the can. From your measurements estimate the depth to which the can is filled and the uncertainty in the estimate. Assume the value of SG = 1.055, as supplied by the bottler.

1.48 From Appendix A, the viscosity μ ($\text{N}\cdot\text{s/m}^2$) of water at temperature T (K) can be computed from $\mu = A10^{B/(T-C)}$, where $A = 2.414 \times 10^{-5}\ \text{N}\cdot\text{s/m}^2$, $B = 247.8$ K, and $C = 140$ K. Determine the viscosity of water at 30°C, and estimate its uncertainty if the uncertainty in temperature measurement is ±0.5°C.

1.49 Using the nominal dimensions of the soda can given in Problem 1.47, determine the precision with which the diameter and height must be measured to estimate the volume of the can within an uncertainty of ±0.5 percent.

1.50 An enthusiast magazine publishes data from its road tests on the lateral acceleration capability of cars. The measurements are made using a 46 m diameter skid pad. Assume the vehicle path deviates from the circle by ±0.6 m and that the vehicle speed is read from a fifth-wheel speed-measuring system to ±0.8 km/h. Estimate the experimental uncertainty in a reported lateral acceleration of 0.7 g. How would you improve the experimental procedure to reduce the uncertainty?

1.51 The height of a building may be estimated by measuring the horizontal distance to a point on the ground and the angle from this point to the top of the building. Assuming these measurements are $L = 30 \pm 0.15$ m and $\theta = 30 \pm 0.2°$, estimate the height H of the building and the uncertainty in the estimate. For the same building height and measurement uncertainties, use *Excel's Solver* to determine the angle (and the corresponding distance from the building) at which measurements should be made to minimize the uncertainty in estimated height. Evaluate and plot the optimum measurement angle as a function of building height for $15 \le H \le 300$ m.

1.52 An American golf ball is described in Problem 1.42 Assuming the measured mass and its uncertainty as given, determine the precision to which the diameter of the ball must be measured so the density of the ball may be estimated within an uncertainty of ±1 percent.

1.53 A syringe pump is to dispense liquid at a flow rate of 100 mL/min. The design for the piston drive is such that the uncertainty of the piston speed is 0.0025 cm/min, and the cylinder bore diameter has a maximum uncertainty of 0.00125 cm. Plot the uncertainty in the flow rate as a function of cylinder bore. Find the combination of piston speed and bore that minimizes the uncertainty in the flow rate.

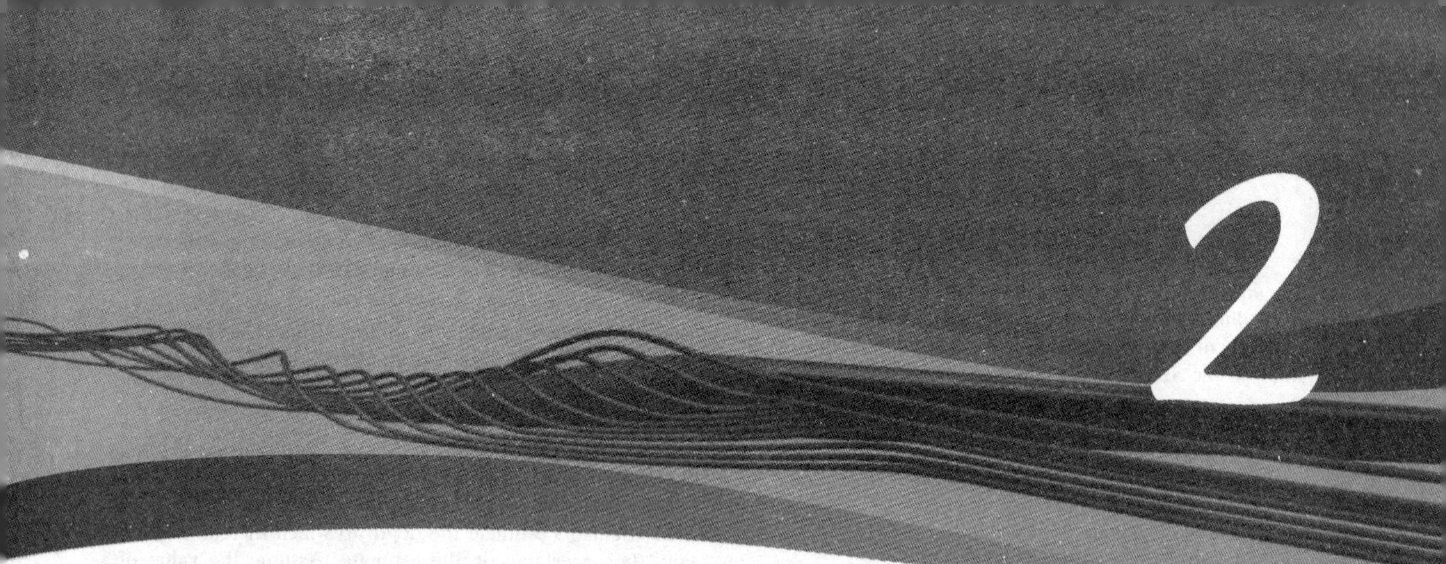

2 Fundamental Concepts

Case Study in Energy and the Environment

Ocean Power

We're not used to thinking of them this way, but the oceans are a huge repository of solar energy (and energy due to the moon's motion). The solar energy storage is initially thermal in nature, as the water surface is heated during the day. When the water cools overnight, thermal gradients are created that ultimately lead to ocean currents (as well as winds) containing huge amounts of energy. According to a 2009 U.S. Department of Energy study titled "Ocean Energy Technology," there are four types of ocean energy conversion: *wave energy, tidal energy, marine current energy*, and *ocean thermal energy conversion*.

The total power from waves believed to be available is about 2.7 TW, of which it is currently practical to extract 500 GW (500×10^9 W). Bear in mind that we mentioned in Chapter 1 that *total* power consumption by humans was about 16 TW (as of 2006), so at best wave power could supply about 3 percent of human needs using current technology. These devices work by either floating on the surface of the ocean or by being moored to the ocean floor. Many of these devices rely on buoyancy forces, which we will discuss in Chapter 3. For example, a device that floats on the surface may have joints hinged together that bend with the waves; this bending motion pumps fluid through turbines and creates electric power. Alternatively, stationary tethered devices use pressure fluctuations produced in long tubes from the waves swelling up and down; the bobbing motion drives a turbine. Wave energy is already reaching fairly advanced levels of development, with a number of companies being involved.

Tidal energy uses the 12-hr cycle due to the gravitational force of the moon; the difference in water height from low to high tide is an extractable form of potential energy. For example, water can be captured by using a barrier across an estuary during high tide and by forcing the water through a turbine during low tide. Alternatively, as shown in the figure, turbine systems could be mounted in such a way that they swing with the tide, extracting energy when the tide comes in and goes out. There are only about 20 locations on earth with tides sufficiently high to make tidal energy practical. The Bay of Fundy between Maine and Nova Scotia features the highest tides in the world, reaching 17 m. This area alone could produce up to 15 GW of power. The total wave energy power believed to be available is about 2.5 TW, of which, with current technology, it is practical to extract only about 65 GW.

Marine current energy is that due to ocean currents (which in turn are generated by solar heating and by the winds—ultimately solar in origin—as well as by the Earth's rotation). About 5 TW of power is believed to be available, of which it is currently practical to extract 450 GW; at best, this energy source will supply something less than 5 percent of total current needs. In the United States, it is most abundant off the coast of Florida in the flow known as the Gulf Stream. Kinetic energy can be captured from the Gulf Stream and other currents with submerged turbines that are very similar in appearance to miniature wind turbines. As with wind turbines, the continuous movement of the marine current moves the rotor blades to generate electric power. Turbines will be discussed in some detail in Chapter 10.

Ocean thermal energy conversion (OTEC), uses the ocean temperature difference between surface water and that at depths lower than 1000 m to extract energy. The temperature of ocean water at a depth of 1000 m is just above freezing; a temperature difference of as little as 20°C (293 K) can yield usable energy. (You can figure out the minimum surface

temperature required!) The warm surface water can be used as a heat source to evaporate a fluid such as ammonia, which can drive a turbine, and the deep water acts as a heat sink. Because of the temperatures involved, such devices will have a very low theoretical efficiency, but the amount of stored thermal energy is huge—about 200 TW of power!

Yet another form of ocean energy (ultimately traceable to solar energy) is that due to the variability of salinity due to water evaporation. When salty ocean water (brine) is separated from fresh water by a semipermeable membrane, a pressure gradient builds up across the membrane (osmotic pressure). We will learn in this text that a pressure gradient can be used as a driving force for energy generation. The exploitation of this energy is called *salinity gradient energy conversion*. This is a future technology with huge potential. There is about 1000 TW of energy available, or about 60 times total worldwide power usage!

We shall discuss some interesting developments in several of these energy conversion methods in *Case Studies in Energy and the Environment* in subsequent chapters.

In Chapter 1 we discussed in general terms what fluid mechanics is about, and described some of the approaches we will use in analyzing fluid mechanics problems. In this chapter we will be more specific in defining some important properties of fluids and ways in which flows can be described and characterized.

Fluid as a Continuum 2.1

We are all familiar with fluids—the most common being air and water—and we experience them as being "smooth," i.e., as being a continuous medium. Unless we use

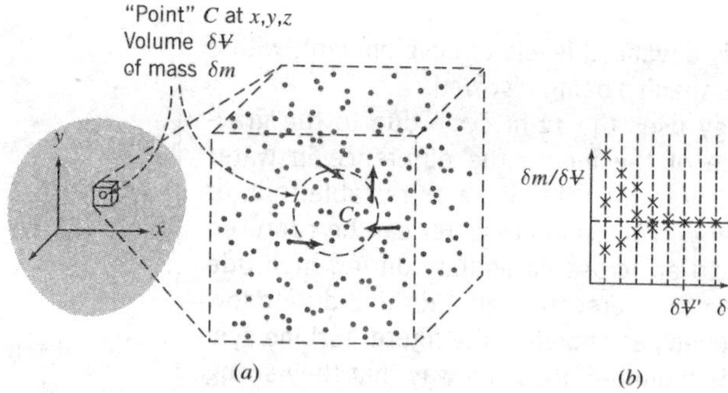

Fig. 2.1 Definition of density at a point.

VIDEO

Fluid as a Continuum.

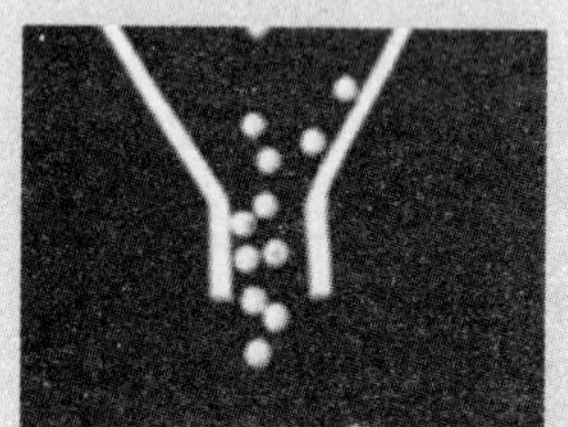

specialized equipment, we are not aware of the underlying molecular nature of fluids. This molecular structure is one in which the mass is *not* continuously distributed in space, but is concentrated in molecules that are separated by relatively large regions of empty space. The sketch in Fig. 2.1*a* shows a schematic representation of this. A region of space "filled" by a stationary fluid (e.g., air, treated as a single gas) looks like a continuous medium, but if we zoom in on a very small cube of it, we can see that we mostly have empty space, with gas molecules scattered around, moving at high speed (indicated by the gas temperature). Note that the size of the gas molecules is greatly exaggerated (they would be almost invisible even at this scale) and that we have placed velocity vectors only on a small sample. We wish to ask: What is the minimum volume, $\delta V'$, that a "point" C must be, so that we can talk about continuous fluid properties such as the density at a point? In other words, under what circumstances can a fluid be treated as a *continuum*, for which, by definition, properties vary smoothly from point to point? This is an important question because the concept of a continuum is the basis of classical fluid mechanics.

Consider how we determine the density at a point. Density is defined as mass per unit volume; in Fig. 2.1*a* the mass δm will be given by the instantaneous number of molecules in δV (and the mass of each molecule), so the average density in volume δV is given by $\rho = \delta m/\delta V$. We say "average" because the number of molecules in δV, and hence the density, fluctuates. For example, if the gas in Fig. 2.1*a* was air at standard temperature and pressure (STP[1]) and the volume δV was a sphere of diameter 0.01 μm, there might be 15 molecules in δV (as shown), but an instant later there might be 17 (three might enter while one leaves). Hence the density at "point" C randomly fluctuates in time, as shown in Fig. 2.1*b*. In this figure, each vertical dashed line represents a specific chosen volume, δV, and each data point represents the measured density at an instant. For very small volumes, the density varies greatly, but above a certain volume, $\delta V'$, the density becomes stable—the volume now encloses a huge number of molecules. For example, if $\delta V = 0.001$ mm^3 (about the size of a grain of sand), there will on average be 2.5×10^{13} molecules present. Hence we can conclude that air at STP (and other gases, and liquids) can be treated as a continuous medium as long as we consider a "point" to be no smaller than about this size; this is sufficiently precise for most engineering applications.

The concept of a continuum is the basis of classical fluid mechanics. The continuum assumption is valid in treating the behavior of fluids under normal conditions. It only breaks down when the mean free path of the molecules[2] becomes the same order of magnitude as the smallest significant characteristic dimension of the problem.

[1]STP for air are 15°C (288 K) and 101.3 kPa absolute, respectively.

[2]Approximately 6×10^{-8} m at STP (Standard Temperature and Pressure) for gas molecules that show ideal gas behavior [1].

This occurs in such specialized problems as rarefied gas flow (e.g., as encountered in flights into the upper reaches of the atmosphere). For these specialized cases (not covered in this text) we must abandon the concept of a continuum in favor of the microscopic and statistical points of view.

As a consequence of the continuum assumption, each fluid property is assumed to have a definite value at every point in space. Thus fluid properties such as density, temperature, velocity, and so on are considered to be continuous functions of position and time. For example, we now have a workable definition of density at a point,

$$\rho \equiv \lim_{\delta \mathcal{V} \to \delta \mathcal{V}'} \frac{\delta m}{\delta \mathcal{V}} \tag{2.1}$$

Since point C was arbitrary, the density at any other point in the fluid could be determined in the same manner. If density was measured simultaneously at an infinite number of points in the fluid, we would obtain an expression for the density distribution as a function of the space coordinates, $\rho = \rho(x, y, z)$, at the given instant.

The density at a point may also vary with time (as a result of work done on or by the fluid and/or heat transfer to the fluid). Thus the complete representation of density (the *field* representation) is given by

$$\rho = \rho(x, y, z, t) \tag{2.2}$$

Since density is a scalar quantity, requiring only the specification of a magnitude for a complete description, the field represented by Eq. 2.2 is a scalar field.

An alternative way of expressing the density of a substance (solid or fluid) is to compare it to an accepted reference value, typically the maximum density of water, ρ_{H_2O} (1000 kg/m^3 at 4°C (277 K)). Thus, the *specific gravity*, SG, of a substance is expressed as

$$SG = \frac{\rho}{\rho_{H_2O}} \tag{2.3}$$

For example, the SG of mercury is typically 13.6—mercury is 13.6 times as dense as water. Appendix A contains specific gravity data for selected engineering materials. The specific gravity of liquids is a function of temperature; for most liquids specific gravity decreases with increasing temperature.

The *specific weight*, γ, of a substance is another useful material property. It is defined as the weight of a substance per unit volume and given as

$$\gamma = \frac{mg}{\mathcal{V}} \rightarrow \gamma = \rho g \tag{2.4}$$

For example, the specific weight of water is approximately 9.81 kN/m^3.

Velocity Field 2.2

In the previous section we saw that the continuum assumption led directly to the notion of the density field. Other fluid properties also may be described by fields.

A very important property defined by a field is the velocity field, given by

$$\vec{V} = \vec{V}(x, y, z, t) \tag{2.5}$$

Velocity is a vector quantity, requiring a magnitude and direction for a complete description, so the velocity field (Eq. 2.5) is a vector field.

The velocity vector, $\vec{V}$, also can be written in terms of its three scalar components. Denoting the components in the x, y, and z directions by u, v, and w, then

$$\vec{V} = u\hat{i} + v\hat{j} + w\hat{k} \tag{2.6}$$

In general, each component, u, v, and w, will be a function of x, y, z, and t.

We need to be clear on what $\vec{V}(x, y, z, t)$ measures: It indicates the velocity of a fluid particle that is passing through the point x, y, z at time instant t, in the Eulerian sense. We can keep measuring the velocity at the same point or choose any other point x, y, z at the next time instant; the point x, y, z is *not* the ongoing position of an *individual* particle, but a point we choose to look at. (Hence x, y, and z are independent variables. In Chapter 5 we will discuss the *material derivative* of velocity, in which we *choose* $x = x_p(t), y = y_p(t)$, and $z = z_p(t)$, where $x_p(t)$, $y_p(t)$, $z_p(t)$ *is* the position of a specific particle.) We conclude that $\vec{V}(x, y, z, t)$ should be thought of as the velocity field of all particles, not just the velocity of an individual particle.

If properties at every point in a flow field do not change with time, the flow is termed *steady*. Stated mathematically, the definition of steady flow is

$$\frac{\partial \eta}{\partial t} = 0$$

where η represents any fluid property. Hence, for steady flow,

$$\frac{\partial \rho}{\partial t} = 0 \quad \text{or} \quad \rho = \rho(x, y, z)$$

and

$$\frac{\partial \vec{V}}{\partial t} = 0 \quad \text{or} \quad \vec{V} = \vec{V}(x, y, z)$$

In steady flow, any property may vary from point to point in the field, but all properties remain constant with time at every point.

One-, Two-, and Three-Dimensional Flows

A flow is classified as one-, two-, or three-dimensional depending on the number of space coordinates required to specify the velocity field.[3] Equation 2.5 indicates that the velocity field may be a function of three space coordinates and time. Such a flow field is termed *three-dimensional* (it is also *unsteady*) because the velocity at any point in the flow field depends on the three coordinates required to locate the point in space.

Although most flow fields are inherently three-dimensional, analysis based on fewer dimensions is frequently meaningful. Consider, for example, the steady flow through a long straight pipe that has a divergent section, as shown in Fig. 2.2. In this example, we are using cylindrical coordinates (r, θ, x). We will learn (in Chapter 8) that under certain circumstances (e.g., far from the entrance of the pipe and from the divergent section, where the flow can be quite complicated), the velocity distribution may be described by

$$u = u_{\text{max}}\left[1 - \left(\frac{r}{R}\right)^2\right] \tag{2.7}$$

This is shown on the left of Fig. 2.2. The velocity $u(r)$ is a function of only one coordinate, and so the flow is one-dimensional. On the other hand, in the diverging

[3]Some authors choose to classify a flow as one-, two-, or three-dimensional on the basis of the number of space coordinates required to specify *all* fluid properties. In this text, classification of flow fields will be based on the number of space coordinates required to specify the velocity field only.

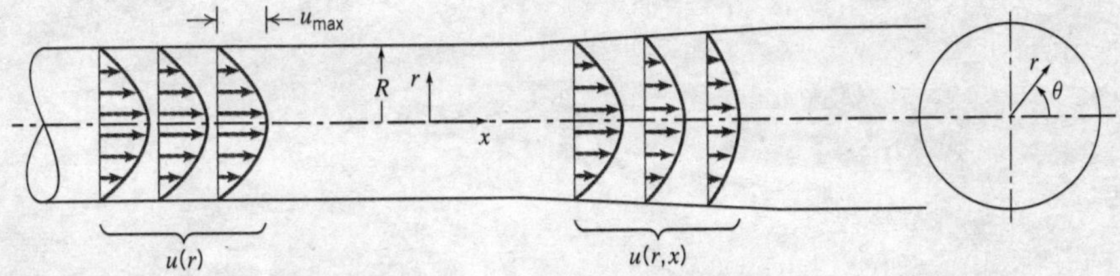

Fig. 2.2 Examples of one- and two-dimensional flows.

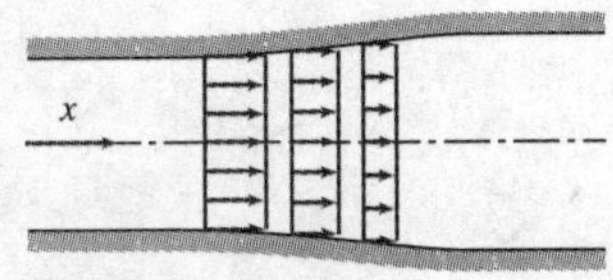

Fig. 2.3 Example of uniform flow at a section.

section, the velocity decreases in the x direction, and the flow becomes two-dimensional: $u = u(r, x)$.

As you might suspect, the complexity of analysis increases considerably with the number of dimensions of the flow field. For many problems encountered in engineering, a one-dimensional analysis is adequate to provide approximate solutions of engineering accuracy.

Since all fluids satisfying the continuum assumption must have zero relative velocity at a solid surface (to satisfy the no-slip condition), most flows are inherently two- or three-dimensional. To simplify the analysis it is often convenient to use the notion of *uniform flow* at a given cross section. In a flow that is uniform at a given cross section, the velocity is constant across any section normal to the flow. Under this assumption,[4] the two-dimensional flow of Fig. 2.2 is modeled as the flow shown in Fig. 2.3. In the flow of Fig. 2.3, the velocity field is a function of x alone, and thus the flow model is one-dimensional. (Other properties, such as density or pressure, also may be assumed uniform at a section, if appropriate.)

The term *uniform flow field* (as opposed to uniform flow at a cross section) is used to describe a flow in which the velocity is constant, i.e., independent of all space coordinates, throughout the entire flow field.

CLASSIC VIDEO

Flow Visualization.

Timelines, Pathlines, Streaklines, and Streamlines

Airplane and auto companies and college engineering laboratories, among others, frequently use wind tunnels to visualize flow fields [2]. For example, Fig. 2.4 shows a flow pattern for flow around a car mounted in a wind tunnel, generated by releasing smoke into the flow at five fixed upstream points. Flow patterns can be visualized using timelines, pathlines, streaklines, or streamlines.

If a number of adjacent fluid particles in a flow field are marked at a given instant, they form a line in the fluid at that instant; this line is called a *timeline*. Subsequent observations of the line may provide information about the flow field. For example, in discussing the behavior of a fluid under the action of a constant shear force (Section 1.2) timelines were introduced to demonstrate the deformation of a fluid at successive instants.

VIDEO

Streaklines.

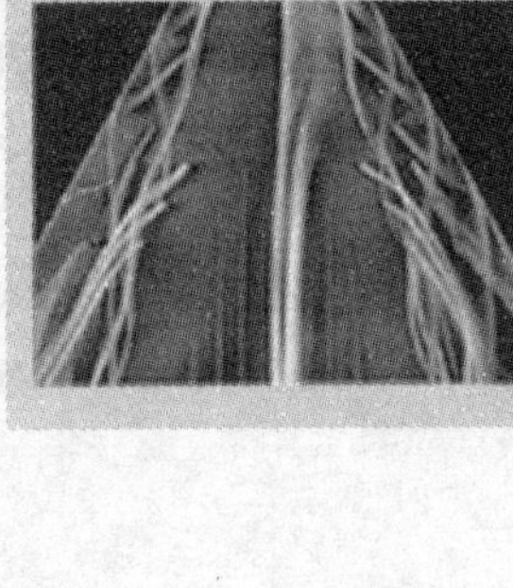

A *pathline* is the path or trajectory traced out by a moving fluid particle. To make a pathline visible, we might identify a fluid particle at a given instant, e.g., by the use of dye or smoke, and then take a long exposure photograph of its subsequent motion. The line traced out by the particle is a pathline. This approach might be used to study, for example, the trajectory of a contaminant leaving a smokestack.

On the other hand, we might choose to focus our attention on a fixed location in space and identify, again by the use of dye or smoke, all fluid particles passing through this point. After a short period of time we would have a number of identifiable fluid

[4]This may seem like an unrealistic simplification, but actually in many cases leads to useful results. Sweeping assumptions such as uniform flow at a cross section should always be reviewed carefully to be sure they provide a reasonable analytical model of the real flow.

Fig. 2.4 Streaklines over an automobile in a wind tunnel. (Courtesy of Andy Sacks/Getty Images, Inc.)

particles in the flow, all of which had, at some time, passed through one fixed location in space. The line joining these fluid particles is defined as a *streakline*.

Streamlines are lines drawn in the flow field so that at a given instant they are tangent to the direction of flow at every point in the flow field. Since the streamlines are tangent to the velocity vector at every point in the flow field, there can be no flow across a streamline. Streamlines are the most commonly used visualization technique. For example, they are used to study flow over an automobile in a computer simulation. The procedure used to obtain the equation for a streamline in two-dimensional flow is illustrated in Example 2.1.

VIDEO

Streamlines.

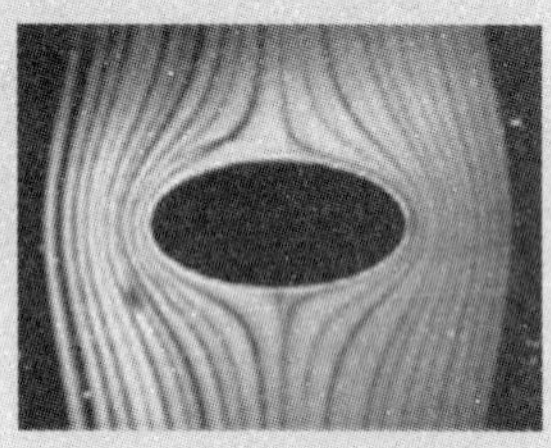

In steady flow, the velocity at each point in the flow field remains constant with time and, consequently, the streamline shapes do not vary from one instant to the next. This implies that a particle located on a given streamline will always move along the same streamline. Furthermore, consecutive particles passing through a fixed point in space will be on the same streamline and, subsequently, will remain on this streamline. Thus in a steady flow, pathlines, streaklines, and streamlines are identical lines in the flow field.

Figure 2.4 shows a photograph of five *streaklines* for flow over an automobile in a wind tunnel. A streakline is the line produced in a flow when all particles moving through a fixed point are marked in some way (e.g., using smoke). We can also define *streamlines*. These are lines drawn in the flow field so that *at a given instant* they are tangent to the direction of flow at every point in the flow field. Since the streamlines are tangent to the velocity vector at every point in the flow field, there is no flow across a streamline. *Pathlines* are as the name implies: They show, over time, the paths individual particles take (if you've seen time-lapse photos of nighttime traffic, you get the idea). Finally, *timelines* are created by marking a line in a flow and watching how it evolves over time.

We mentioned that Fig. 2.4 shows streaklines, but in fact the pattern shown also represents streamlines and pathlines! The steady pattern shown will exist as long as smoke is released from the five fixed points. If we were somehow to measure the velocity at all points at an instant, to generate streamlines, we'd get the same pattern; if we were instead to release only one smoke particle at each location, and video its motion over time, we'd see the particles follow the same curves. We conclude that for *steady* flow, streaklines, streamlines, and pathlines are identical.

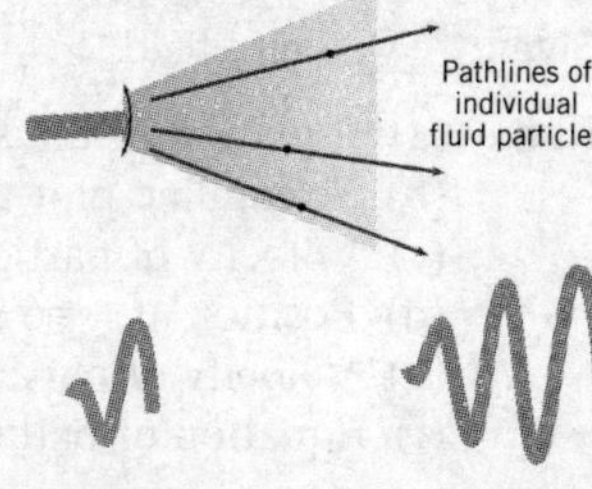

Fig. 2.5 Pathlines and streaklines for flow from the exit of an oscillating garden hose.

Things are quite different for *unsteady* flow. For unsteady flow, streaklines, streamlines, and pathlines will in general have differing shapes. For example, consider holding a garden hose and swinging it side to side as water exits at high speed, as shown in Fig. 2.5. We obtain a continuous sheet of water. If we consider individual water particles, we see that each particle, once ejected, follows a straight-line path (here, for simplicity, we ignore gravity): The pathlines are straight lines, as shown. On the other hand, if we start injecting dye into the water as it exits the hose, we will generate a streakline, and this takes the shape of an expanding sine wave, as shown.

Clearly, pathlines and streaklines do not coincide for this unsteady flow (we leave determination of streamlines to an exercise).

We can use the velocity field to derive the shapes of streaklines, pathlines, and streamlines. Starting with streamlines: Because the streamlines are parallel to the velocity vector, we can write (for 2D)

$$\left.\frac{dy}{dx}\right)_{\text{streamline}} = \frac{v(x,y)}{u(x,y)} \tag{2.8}$$

Note that streamlines are obtained at an instant in time; if the flow is unsteady, time t is held constant in Eq. 2.8. Solution of this equation gives the equation $y = y(x)$, with an undetermined integration constant, the value of which determines the particular streamline.

For pathlines (again considering 2D), we let $x = x_p(t)$ and $y = y_p(t)$, where $x_p(t)$ and $y_p(t)$ are the instantaneous coordinates of a specific particle. We then get

$$\left.\frac{dx}{dt}\right)_{\text{particle}} = u(x,y,t) \qquad \left.\frac{dy}{dt}\right)_{\text{particle}} = v(x,y,t) \tag{2.9}$$

The simultaneous solution of these equations gives the path of a particle in parametric form $x_p(t)$, $y_p(t)$.

The computation of streaklines is somewhat tricky. The first step is to compute the pathline of a particle (using Eqs. 2.9) that was released from the streak source point (coordinates x_0, y_0) at time t_0, in the form

$$x_{\text{particle}}(t) = x(t, x_0, y_0, t_0) \qquad y_{\text{particle}}(t) = y(t, x_0, y_0, t_0)$$

Then, instead of interpreting this as the position of a particle over time, we rewrite these equations as

$$x_{\text{streakline}}(t_0) = x(t, x_0, y_0, t_0) \qquad y_{\text{streakline}}(t_0) = y(t, x_0, y_0, t_0) \tag{2.10}$$

Equations 2.10 give the line generated (by time t) from a streak source at point (x_0, y_0). In these equations, t_0 (the release times of particles) is varied from 0 to t to show the instantaneous positions of all particles released up to time t!

Example 2.1 STREAMLINES AND PATHLINES IN TWO-DIMENSIONAL FLOW

A velocity field is given by $\vec{V} = Ax\hat{i} - Ay\hat{j}$; the units of velocity are m/s; x and y are given in meters; $A = 0.3\ \text{s}^{-1}$.

(a) Obtain an equation for the streamlines in the xy plane.
(b) Plot the streamline passing through the point $(x_0, y_0) = (2, 8)$.
(c) Determine the velocity of a particle at the point (2, 8).
(d) If the particle passing through the point (x_0, y_0) is marked at time $t = 0$, determine the location of the particle at time $t = 6$ s.
(e) What is the velocity of this particle at time $t = 6$ s?
(f) Show that the equation of the particle path (the pathline) is the same as the equation of the streamline.

Given: Velocity field, $\vec{V} = Ax\hat{i} - Ay\hat{j}$; x and y in meters; $A = 0.3\ \text{s}^{-1}$.

Find: (a) Equation of the streamlines in the xy plane.
(b) Streamline plot through point (2, 8).
(c) Velocity of particle at point (2, 8).
(d) Position at $t = 6$ s of particle located at (2, 8) at $t = 0$.
(e) Velocity of particle at position found in (d).
(f) Equation of pathline of particle located at (2, 8) at $t = 0$.

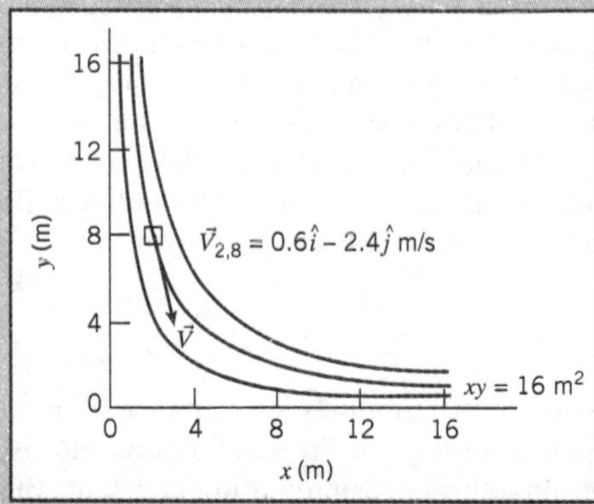

Solution:

(a) Streamlines are lines drawn in the flow field such that, at a given instant, they are tangent to the direction of flow at every point. Consequently,

$$\left.\frac{dy}{dx}\right)_{\text{streamline}} = \frac{v}{u} = \frac{-Ay}{Ax} = \frac{-y}{x}$$

Separating variables and integrating, we obtain

$$\int \frac{dy}{y} = -\int \frac{dx}{x}$$

or

$$\ln y = -\ln x + c_1$$

This can be written as $xy = c$ ←

(b) For the streamline passing through the point $(x_0, y_0) = (2, 8)$ the constant, c, has a value of 16 and the equation of the streamline through the point (2, 8) is

$$xy = x_0 y_0 = 16\ \text{m}^2 \leftarrow$$

The plot is as sketched above.

(c) The velocity field is $\vec{V} = Ax\hat{i} - Ay\hat{j}$. At the point (2, 8) the velocity is

$$\vec{V} = A(x\hat{i} - y\hat{j}) = 0.3\text{s}^{-1}(2\hat{i} - 8\hat{j})\text{m} = 0.6\hat{i} - 2.4\hat{j}\ \text{m/s} \leftarrow$$

(d) A particle moving in the flow field will have velocity given by

$$\vec{V} = Ax\hat{i} - Ay\hat{j}$$

Thus

$$u_p = \frac{dx}{dt} = Ax \quad \text{and} \quad v_p = \frac{dy}{dt} = -Ay$$

Separating variables and integrating (in each equation) gives

$$\int_{x_0}^{x} \frac{dx}{x} = \int_0^t A\,dt \quad \text{and} \quad \int_{y_0}^{y} \frac{dy}{y} = \int_0^t -A\,dt$$

Then

$$\ln\frac{x}{x_0} = At \quad \text{and} \quad \ln\frac{y}{y_0} = -At$$

or

$$x = x_0 e^{At} \quad \text{and} \quad y = y_0 e^{-At}$$

At $t = 6$ s,

$$x = 2\,\text{m}\,e^{(0.3)6} = 12.1\,\text{m} \quad \text{and} \quad y = 8\,\text{m}\,e^{-(0.3)6} = 1.32\,\text{m}$$

At $t = 6$ s, particle is at (12.1, 1.32) m ←

(e) At the point (12.1, 1.32) m,

$$\vec{V} = A(x\hat{i} - y\hat{j}) = 0.3\,\text{s}^{-1}(12.1\hat{i} - 1.32\hat{j})\text{m}$$
$$= 3.63\hat{i} - 0.396\hat{j}\,\text{m/s} \longleftarrow$$

(f) To determine the equation of the pathline, we use the parametric equations

$$x = x_0 e^{At} \quad \text{and} \quad y = y_0 e^{-At}$$

and eliminate t. Solving for e^{At} from both equations

$$e^{At} = \frac{y_0}{y} = \frac{x}{x_0}$$

Therefore $xy = x_0 y_0 = 16\ \text{m}^2 \longleftarrow$

Notes:

✓ This problem illustrates the method for computing streamlines and pathlines.

✓ Because this is a steady flow, the streamlines and pathlines have the same shape—in an unsteady flow this would not be true.

✓ When we follow a particle (the Lagrangian approach), its position (x, y) and velocity ($u_p = dx/dt$ and $v_p = dy/dt$) are functions of time, even though the flow is steady.

Stress Field 2.3

In our study of fluid mechanics, we will need to understand what kinds of forces act on fluid particles. Each fluid particle can experience: *surface forces* (pressure, friction) that are generated by contact with other particles or a solid surface; and *body forces* (such as gravity and electromagnetic) that are experienced throughout the particle.

The gravitational body force acting on an element of volume, $d\Psi$, is given by $\rho\vec{g}d\Psi$, where ρ is the density (mass per unit volume) and $\vec{g}$ is the local gravitational acceleration. Thus the gravitational body force per unit volume is $\rho\vec{g}$ and the gravitational body force per unit mass is $\vec{g}$.

Surface forces on a fluid particle lead to *stresses*. The concept of stress is useful for describing how forces acting on the boundaries of a medium (fluid or solid) are transmitted throughout the medium. You have probably seen stresses discussed in solid mechanics. For example, when you stand on a diving board, stresses are generated within the board. On the other hand, when a body moves through a fluid, stresses are developed within the fluid. The difference between a fluid and a solid is, as we've seen, that stresses in a fluid are mostly generated by motion rather than by deflection.

Imagine the surface of a fluid particle in contact with other fluid particles, and consider the contact force being generated between the particles. Consider a portion, $\delta\vec{A}$, of the surface at some point C. The orientation of $\delta\vec{A}$ is given by the unit vector, $\hat{n}$, shown in Fig. 2.6. The vector $\hat{n}$ is the outwardly drawn unit normal with respect to the particle.

The force, $\delta\vec{F}$, acting on $\delta\vec{A}$ may be resolved into two components, one normal to and the other tangent to the area. A *normal stress* σ_n and a *shear stress* τ_n are then defined as

$$\sigma_n = \lim_{\delta A_n \to 0} \frac{\delta F_n}{\delta A_n} \tag{2.11}$$

and

$$\tau_n = \lim_{\delta A_n \to 0} \frac{\delta F_t}{\delta A_n} \tag{2.12}$$

Subscript n on the stress is included as a reminder that the stresses are associated with the surface $\delta\vec{A}$ through C, having an outward normal in the $\hat{n}$ direction. The fluid is

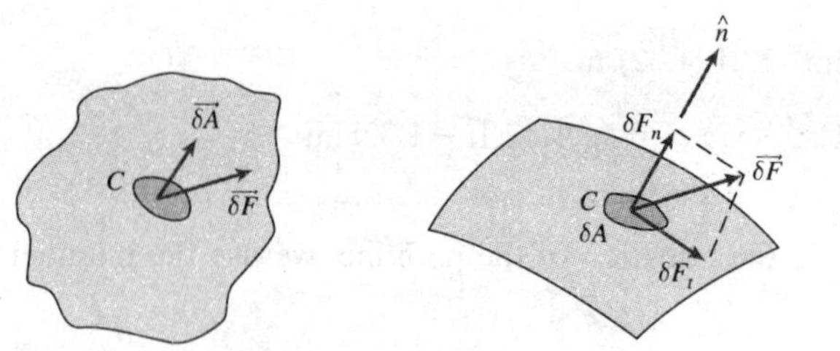

Fig. 2.6 The concept of stress in a continuum.

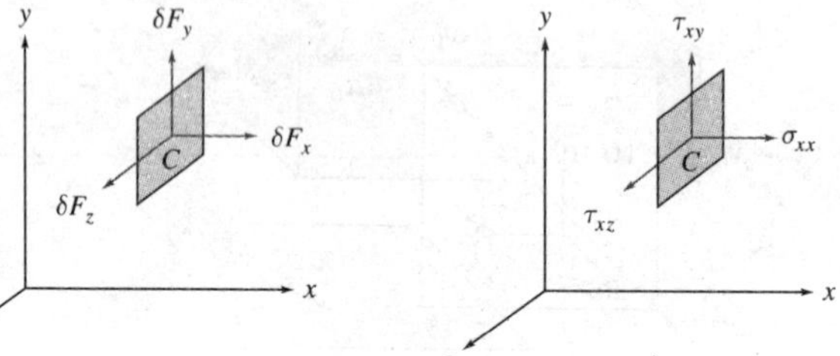

(*a*) Force components (*b*) Stress components

Fig. 2.7 Force and stress components on the element of area δA_x.

actually a continuum, so we could have imagined breaking it up any number of different ways into fluid particles around point *C*, and therefore obtained any number of different stresses at point *C*.

In dealing with vector quantities such as force, we usually consider components in an orthogonal coordinate system. In rectangular coordinates we might consider the stresses acting on planes whose outwardly drawn normals (again with respect to the material acted upon) are in the *x*, *y*, or *z* directions. In Fig. 2.7 we consider the stress on the element δA_x, whose outwardly drawn normal is in the *x* direction. The force, $\delta \vec{F}$, has been resolved into components along each of the coordinate directions. Dividing the magnitude of each force component by the area, δA_x, and taking the limit as δA_x approaches zero, we define the three stress components shown in Fig. 2.7*b*:

$$\sigma_{xx} = \lim_{\delta A_x \to 0} \frac{\delta F_x}{\delta A_x}$$

$$\tau_{xy} = \lim_{\delta A_x \to 0} \frac{\delta F_y}{\delta A_x} \qquad \tau_{xz} = \lim_{\delta A_x \to 0} \frac{\delta F_z}{\delta A_x} \tag{2.13}$$

We have used a double subscript notation to label the stresses. The *first* subscript (in this case, *x*) indicates the *plane* on which the stress acts (in this case, a surface perpendicular to the *x* axis). The *second* subscript indicates the *direction* in which the stress acts.

Consideration of area element δA_y would lead to the definitions of the stresses, σ_{yy}, τ_{yx}, and τ_{yz}; use of area element δA_z would similarly lead to the definitions of $\sigma_{zz}, \tau_{zx}, \tau_{zy}$.

Although we just looked at three orthogonal planes, an infinite number of planes can be passed through point *C*, resulting in an infinite number of stresses associated with planes through that point. Fortunately, the state of stress at a point can be described completely by specifying the stresses acting on *any* three mutually perpendicular planes through the point. The stress at a point is specified by the nine components

$$\begin{bmatrix} \sigma_{xx} & \tau_{xy} & \tau_{xz} \\ \tau_{yx} & \sigma_{yy} & \tau_{yz} \\ \tau_{zx} & \tau_{zy} & \sigma_{zz} \end{bmatrix}$$

where σ has been used to denote a normal stress, and τ to denote a shear stress. The notation for designating stress is shown in Fig. 2.8.

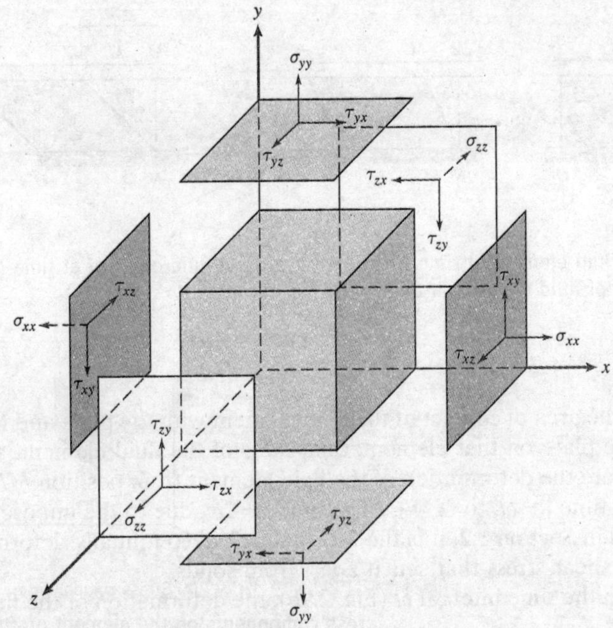

Fig. 2.8 Notation for stress.

Referring to the infinitesimal element shown in Fig. 2.8, we see that there are six planes (two x planes, two y planes, and two z planes) on which stresses may act. In order to designate the plane of interest, we could use terms like front and back, top and bottom, or left and right. However, it is more logical to name the planes in terms of the coordinate axes. The planes are named and denoted as positive or negative according to the direction of the outwardly drawn normal to the plane. Thus the top plane, for example, is a positive y plane and the back plane is a negative z plane.

It also is necessary to adopt a sign convention for stress. A stress component is positive when the direction of the stress component and the plane on which it acts are both positive or both negative. Thus $\tau_{yx} = 3.5\ \text{N/cm}^2$ represents a shear stress on a positive y plane in the positive x direction or a shear stress on a negative y plane in the negative x direction. In Fig. 2.8 all stresses have been drawn as positive stresses. Stress components are negative when the direction of the stress component and the plane on which it acts are of opposite sign.

Viscosity 2.4

Where do stresses come from? For a solid, stresses develop when the material is elastically deformed or strained; for a fluid, shear stresses arise due to viscous flow (we will discuss a fluid's normal stresses shortly). Hence we say solids are *elastic*, and fluids are *viscous* (and it's interesting to note that many biological tissues are *viscoelastic*, meaning they combine features of a solid and a fluid). For a fluid at rest, there will be no shear stresses. We will see that each fluid can be categorized by examining the relation between the applied shear stresses and the flow (specifically the rate of deformation) of the fluid.

Consider the behavior of a fluid element between the two infinite plates shown in Fig. 2.9*a*. The rectangular fluid element is initially at rest at time t. Let us now suppose a constant rightward force δF_x is applied to the upper plate so that it is dragged across the fluid at constant velocity δu. The relative shearing action of the infinite plates produces a shear stress, τ_{yx}, which acts on the fluid element and is given by

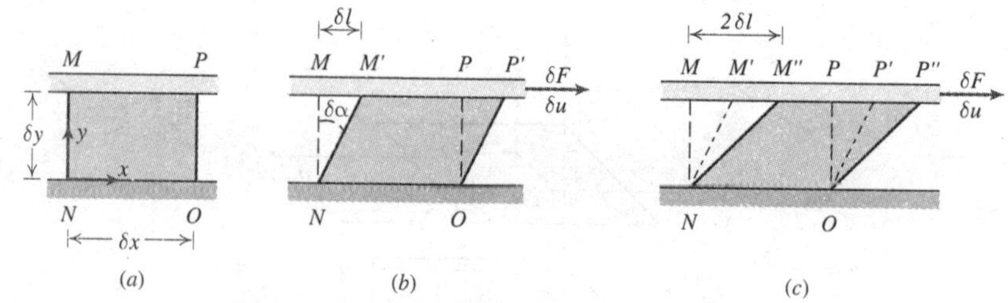

Fig. 2.9 (*a*) Fluid element at time t, (*b*) deformation of fluid element at time $t + \delta t$, and (*c*) deformation of fluid element at time $t + 2\delta t$.

$$\tau_{yx} = \lim_{\delta A_y \to 0} \frac{\delta F_x}{\delta A_y} = \frac{dF_x}{dA_y}$$

where δA_y is the area of contact of the fluid element with the plate and δF_x is the force exerted by the plate on that element. Snapshots of the fluid element, shown in Figs. 2.9*a–c*, illustrate the deformation of the fluid element from position $MNOP$ at time t, to $M'NOP'$ at time $t + \delta t$, to $M''NOP''$ at time $t + 2\delta t$, due to the imposed shear stress. As mentioned in Section 1.2, it is the fact that a fluid continually deforms in response to an applied shear stress that sets it apart from solids.

Focusing on the time interval δt (Fig. 2.9*b*), the deformation of the fluid is given by

$$\text{deformation rate} = \lim_{\delta t \to 0} \frac{\delta \alpha}{\delta t} = \frac{d\alpha}{dt}$$

We want to express $d\alpha/dt$ in terms of readily measurable quantities. This can be done easily. The distance, δl, between the points M and M' is given by

$$\delta l = \delta u \, \delta t$$

Alternatively, for small angles,

$$\delta l = \delta y \, \delta \alpha$$

Equating these two expressions for δl gives

$$\frac{\delta \alpha}{\delta t} = \frac{\delta u}{\delta y}$$

Taking the limits of both sides of the equality, we obtain

$$\frac{d\alpha}{dt} = \frac{du}{dy}$$

Thus, the fluid element of Fig. 2.9, when subjected to shear stress τ_{yx}, experiences a rate of deformation (*shear rate*) given by du/dy. We have established that any fluid that experiences a shear stress will flow (it will have a shear rate). What is the relation between shear stress and shear rate? Fluids in which shear stress is directly proportional to rate of deformation are *Newtonian fluids*. The term *non-Newtonian* is used to classify all fluids in which shear stress is not directly proportional to shear rate.

Newtonian Fluid

Most common fluids (the ones discussed in this text) such as water, air, and gasoline are Newtonian under normal conditions. If the fluid of Fig. 2.9 is Newtonian, then

$$\tau_{yx} \propto \frac{du}{dy} \tag{2.14}$$

We are familiar with the fact that some fluids resist motion more than others. For example, a container of SAE 30W oil is much harder to stir than one of water. Hence SAE 30W oil is much more viscous—it has a higher viscosity. (Note that a container of mercury is also harder to stir, but for a different reason.) The constant of proportionality in Eq. 2.14 is the *absolute* (or *dynamic*) *viscosity*, μ. Thus in terms of the coordinates of Fig. 2.9, Newton's law of viscosity is given for one-dimensional flow by

$$\tau_{yx} = \mu \frac{du}{dy} \tag{2.15}$$

Note that, since the dimensions of τ are $[F/L^2]$ and the dimensions of du/dy are $[1/t]$, μ has dimensions $[Ft/L^2]$. Since the dimensions of force, F, mass, M, length, L, and time, t, are related by Newton's second law of motion, the dimensions of μ can also be expressed as $[M/Lt]$. In the SI system the units of viscosity are kg/(m · s) or Pa · s (1 Pa · s = 1 N · s/m^2). The calculation of viscous shear stress is illustrated in Example 2.2.

In fluid mechanics the ratio of absolute viscosity, μ, to density, ρ, often arises. This ratio is given the name *kinematic viscosity* and is represented by the symbol ν. Since density has dimensions $[M/L^3]$, the dimensions of ν are $[L^2/t]$. In the SI system of units, the unit for ν is m^2/s.

Viscosity data for a number of common Newtonian fluids are given in Appendix A. Note that for gases, viscosity increases with temperature, whereas for liquids, viscosity decreases with increasing temperature.

Example 2.2 VISCOSITY AND SHEAR STRESS IN NEWTONIAN FLUID

An infinite plate is moved over a second plate on a layer of liquid as shown. For small gap width, d, we assume a linear velocity distribution in the liquid. The liquid viscosity is 0.0065 g/cm · s and its specific gravity is 0.88. Determine:

(a) The absolute viscosity of the liquid, in N · s/m^2.
(b) The kinematic viscosity of the liquid, in m^2/s.
(c) The shear stress on the upper plate, in N/m^2.
(d) The shear stress on the lower plate, in Pa.
(e) The direction of each shear stress calculated in parts (c) and (d).

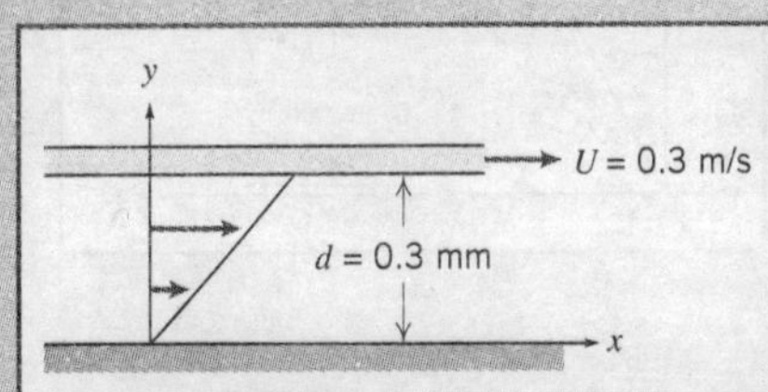

Given: Linear velocity profile in the liquid between infinite parallel plates as shown.

$$\mu = 0.0065 \text{ g/cm} \cdot \text{s}$$
$$SG = 0.88$$

Find: (a) μ in units of N · s/m^2.
(b) ν in units of m^2/s.
(c) τ on upper plate in units of N/m^2.
(d) τ on lower plate in units of Pa.
(e) Direction of stresses in parts (c) and (d).

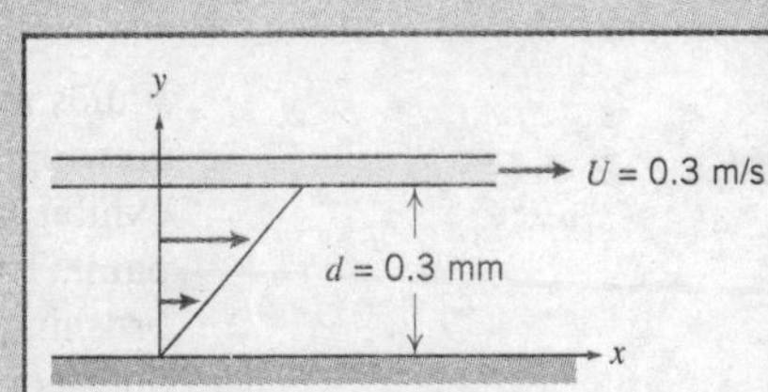

Solution:

Governing equation: $\tau_{yx} = \mu \dfrac{du}{dy}$ Definition: $\nu = \dfrac{\mu}{\rho}$

Assumptions: (1) Linear velocity distribution (given)
(2) Steady flow
(3) μ = constant

(a) $\mu = 0.0065 \dfrac{\text{g}}{\text{cm} \cdot \text{s}} \times \dfrac{\text{kg}}{1000 \text{ g}} \times \dfrac{9.81 \text{ m}}{\text{s}^2} \times \dfrac{100 \text{ cm}}{\text{m}} \times \dfrac{\text{s}^2}{9.81 \text{ m}}$

$\mu = 6.5 \times 10^{-4}$ N · s/m^2 ← μ

(b) $\nu = \dfrac{\mu}{\rho} = \dfrac{\mu}{SG\,\rho_{H_2O}}$

$= 6.5 \times 10^{-4} \dfrac{N \cdot s}{m^2} \times \dfrac{m^3}{(0.88)1000\ kg}$

$= 6.5 \times 10^{-4} \dfrac{kg \cdot m}{s^2} \times \dfrac{s}{m^2} \times \dfrac{m^3}{(0.88)1000\ kg}$

$\nu = 7.39 \times 10^{-7} m^2/s$ ← ν

(c) $\tau_{upper} = \tau_{yx,\,upper} = \mu \left.\dfrac{du}{dy}\right)_{y=d}$

Since u varies linearly with y,

$$\frac{du}{dy} = \frac{\Delta u}{\Delta y} = \frac{U - 0}{d - 0} = \frac{U}{d}$$

$$= 0.3 \frac{m}{s} \times \frac{1}{0.3\ mm} \times 1000 \frac{mm}{m} = 1000\ s^{-1}$$

$\tau_{upper} = \mu \dfrac{U}{d} = 6.5 \times 10^{-4} \dfrac{N \cdot s}{m^2} \times \dfrac{1000}{s} = 0.65\ N/m^2$ ← τ_{upper}

(d) $\tau_{lower} = \mu \dfrac{U}{d} = 0.65\ N/m^2 \times \dfrac{Pa \cdot m^2}{N} = 0.65\ Pa$ ← τ_{lower}

(e) Directions of shear stresses on upper and lower plates.

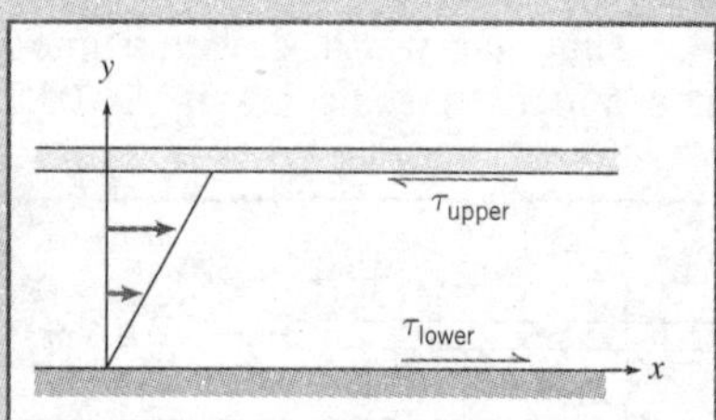

{The upper plate is a negative y surface; so positive τ_{yx} acts in the negative x direction.}

{The lower plate is a positive y surface; so positive τ_{yx} acts in the positive x direction.}

← (e)

Part (c) shows that the shear stress is:

- ✓ Constant across the gap for a linear velocity profile.
- ✓ Directly proportional to the speed of the upper plate (because of the linearity of Newtonian fluids).
- ✓ Inversely proportional to the gap between the plates.

Note that multiplying the shear stress by the plate area in such problems computes the force required to maintain the motion.

Non-Newtonian Fluids

Fluids in which shear stress is *not* directly proportional to deformation rate are non-Newtonian. Although we will not discuss these much in this text, many common fluids exhibit non-Newtonian behavior. Two familiar examples are toothpaste and Lucite[5] paint. The latter is very "thick" when in the can, but becomes "thin" when sheared by brushing. Toothpaste behaves as a "fluid" when squeezed from the tube. However, it does not run out by itself when the cap is removed. There is a threshold or yield stress below which toothpaste behaves as a solid. Strictly speaking, our definition of a fluid is valid only for materials that have zero yield stress. Non-Newtonian fluids commonly are classified as having time-independent or time-dependent behavior. Examples of time-independent behavior are shown in the rheological diagram of Fig. 2.10.

Numerous empirical equations have been proposed [3, 4] to model the observed relations between τ_{yx} and du/dy for time-independent fluids. They may be adequately

[5]Trademark, E. I. du Pont de Nemours & Company.

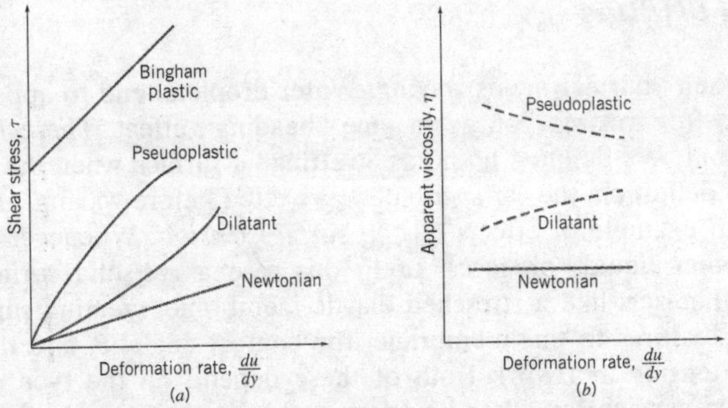

Fig. 2.10 (a) Shear stress, τ, and (b) apparent viscosity, η, as a function of deformation rate for one-dimensional flow of various non-Newtonian fluids.

represented for many engineering applications by the power law model, which for one-dimensional flow becomes

$$\tau_{yx} = k\left(\frac{du}{dy}\right)^n \tag{2.16}$$

where the exponent, n, is called the flow behavior index and the coefficient, k, the consistency index. This equation reduces to Newton's law of viscosity for $n = 1$ with $k = \mu$.

To ensure that τ_{yx} has the same sign as du/dy, Eq. 2.16 is rewritten in the form

$$\tau_{yx} = k\left|\frac{du}{dy}\right|^{n-1}\frac{du}{dy} = \eta\frac{du}{dy} \tag{2.17}$$

The term $\eta = k|du/dy|^{n-1}$ is referred to as the *apparent viscosity*. The idea behind Eq. 2.17 is that we end up with a viscosity η that is used in a formula that is the same form as Eq. 2.15, in which the Newtonian viscosity μ is used. The big difference is that while μ is constant (except for temperature effects), η depends on the shear rate. Most non-Newtonian fluids have apparent viscosities that are relatively high compared with the viscosity of water.

Fluids in which the apparent viscosity decreases with increasing deformation rate ($n < 1$) are called *pseudoplastic* (or shear thinning) fluids. Most non-Newtonian fluids fall into this group; examples include polymer solutions, colloidal suspensions, and paper pulp in water. If the apparent viscosity increases with increasing deformation rate ($n > 1$) the fluid is termed *dilatant* (or shear thickening). Suspensions of starch and of sand are examples of dilatant fluids. You can get an idea of the latter when you're on the beach—if you walk slowly (and hence generate a low shear rate) on very wet sand, you sink into it, but if you jog on it (generating a high shear rate), it's very firm.

A "fluid" that behaves as a solid until a minimum yield stress, τ_y, is exceeded and subsequently exhibits a linear relation between stress and rate of deformation is referred to as an ideal or *Bingham plastic*. The corresponding shear stress model is

$$\tau_{yx} = \tau_y + \mu_p\frac{du}{dy} \tag{2.18}$$

CLASSIC VIDEO

Rheological Behavior of Fluids.

Clay suspensions, drilling muds, and toothpaste are examples of substances exhibiting this behavior.

The study of non-Newtonian fluids is further complicated by the fact that the apparent viscosity may be time-dependent. *Thixotropic* fluids show a decrease in η with time under a constant applied shear stress; many paints are thixotropic. *Rheopectic* fluids show an increase in η with time. After deformation some fluids partially return to their original shape when the applied stress is released; such fluids are called *viscoelastic* (many biological fluids work this way).

2.5 Surface Tension

VIDEO

Molecular Interactions at the Interface.

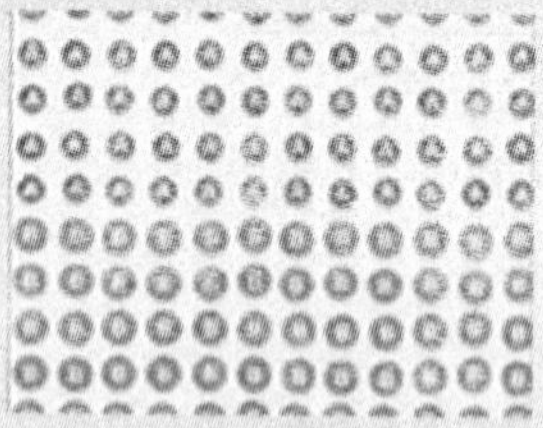

VIDEO

Shrinking Soap Film.

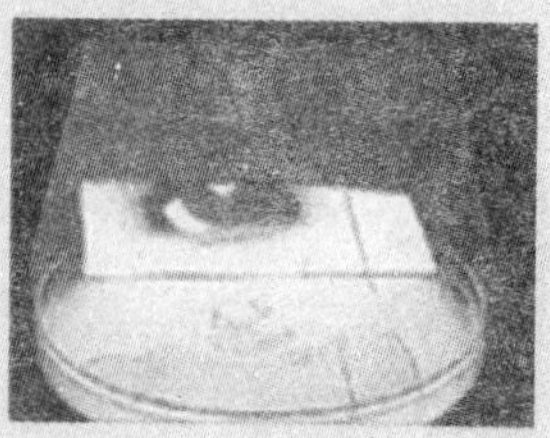

VIDEO

Soap Film Burst.

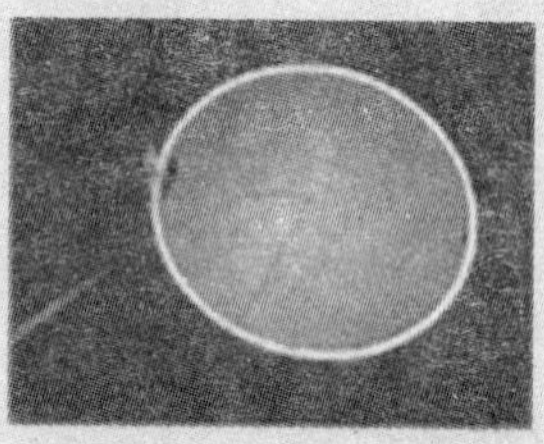

VIDEO

Wetted and Non-wetted Surfaces.

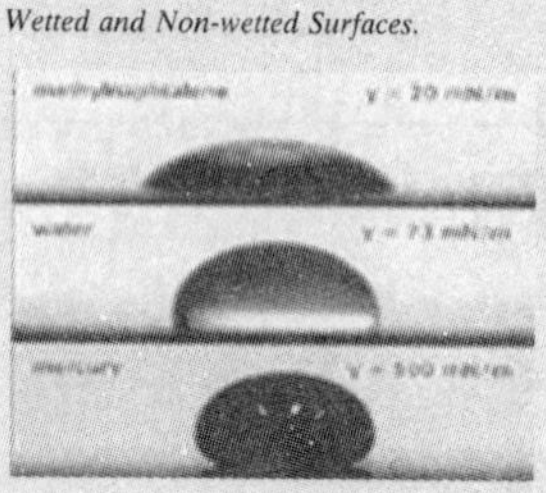

You can tell when your car needs waxing: Water droplets tend to appear somewhat flattened out. After waxing, you get a nice "beading" effect. These two cases are shown in Fig. 2.11. We define a liquid as "wetting" a surface when the *contact angle* $\theta < 90°$. By this definition, the car's surface was wetted before waxing, and not wetted after. This is an example of effects due to *surface tension*. Whenever a liquid is in contact with other liquids or gases, or in this case a gas/solid surface, an interface develops that acts like a stretched elastic membrane, creating surface tension. There are two features to this membrane: the contact angle, θ, and the magnitude of the surface tension, σ (N/m). Both of these depend on the type of liquid and the type of solid surface (or other liquid or gas) with which it shares an interface. In the car-waxing example, the contact angle changed from being smaller than 90° to larger than 90° because, in effect, the waxing changed the nature of the solid surface. Factors that affect the contact angle include the cleanliness of the surface and the purity of the liquid.

Other examples of surface tension effects arise when you are able to place a needle on a water surface and, similarly, when small water insects are able to walk on the surface of the water.

Appendix A contains data for surface tension and contact angle for common liquids in the presence of air and of water.

A force balance on a segment of interface shows that there is a pressure jump across the imagined elastic membrane whenever the interface is curved. For a water droplet in air, pressure in the water is higher than ambient; the same is true for a gas bubble in liquid. For a soap bubble in air, surface tension acts on both inside and outside interfaces between the soap film and air along the curved bubble surface. Surface tension also leads to the phenomena of capillary (i.e., very small wavelength) waves on a liquid surface [5], and capillary rise or depression, discussed below in Example 2.3.

In engineering, probably the most important effect of surface tension is the creation of a curved *meniscus* that appears in manometers or barometers, leading to a (usually unwanted) *capillary rise* (or depression), as shown in Fig. 2.12. This rise may be pronounced if the liquid is in a small-diameter tube or narrow gap, as shown in Example 2.3.

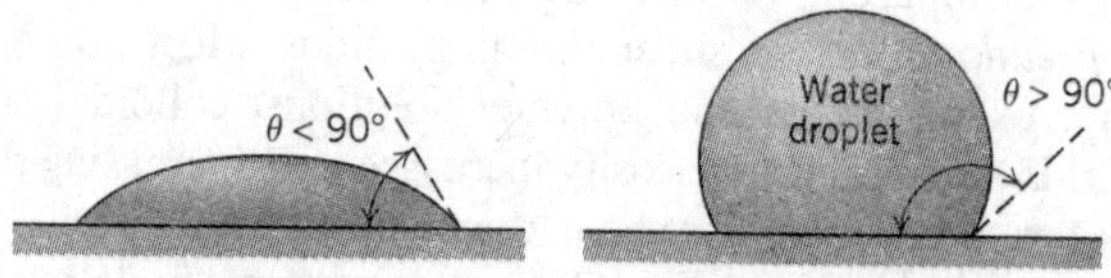

(*a*) A "wetted" surface (*b*) A nonwetted surface

Fig. 2.11 Surface tension effects on water droplets.

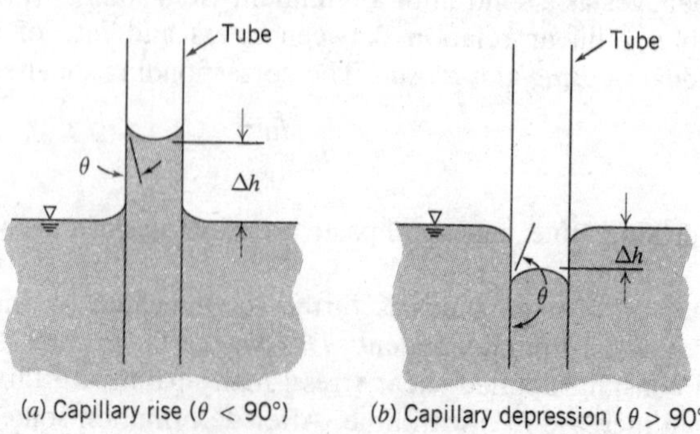

(*a*) Capillary rise ($\theta < 90°$) (*b*) Capillary depression ($\theta > 90°$)

Fig. 2.12 Capillary rise and capillary depression inside and outside a circular tube.

Example 2.3 ANALYSIS OF CAPILLARY EFFECT IN A TUBE

Create a graph showing the capillary rise or fall of a column of water or mercury, respectively, as a function of tube diameter D. Find the minimum diameter of each column required so that the height magnitude will be less than 1 mm.

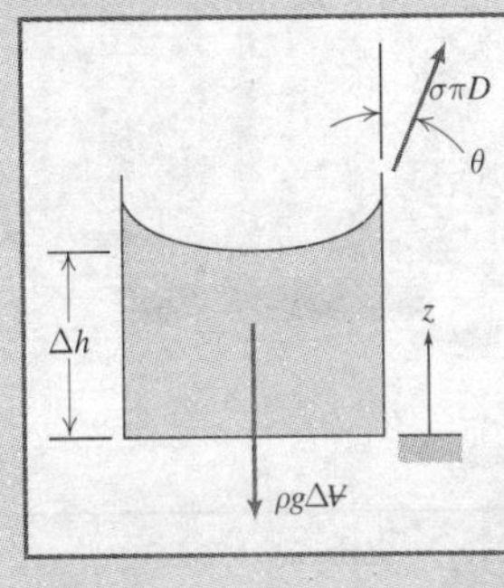

Given: Tube dipped in liquid as in Fig. 2.12.

Find: A general expression for Δh as a function of D.

Solution:
Apply free-body diagram analysis, and sum vertical forces.

Governing equation:

$$\sum F_z = 0$$

Assumptions: (1) Measure to middle of meniscus
(2) Neglect volume in meniscus region

Summing forces in the z direction:

$$\sum F_z = \sigma \pi D \cos\theta - \rho g \Delta V = 0 \tag{1}$$

If we neglect the volume in the meniscus region:

$$\Delta V \approx \frac{\pi D^2}{4} \Delta h$$

Substituting in Eq. (1) and solving for Δh gives

$$\Delta h = \frac{4\sigma \cos\theta}{\rho g D} \longleftarrow \Delta h$$

For water, $\sigma = 72.8$ mN/m and $\theta \approx 0°$, and for mercury, $\sigma = 484$ mN/m and $\theta = 140°$ (Table A.4). Plotting,

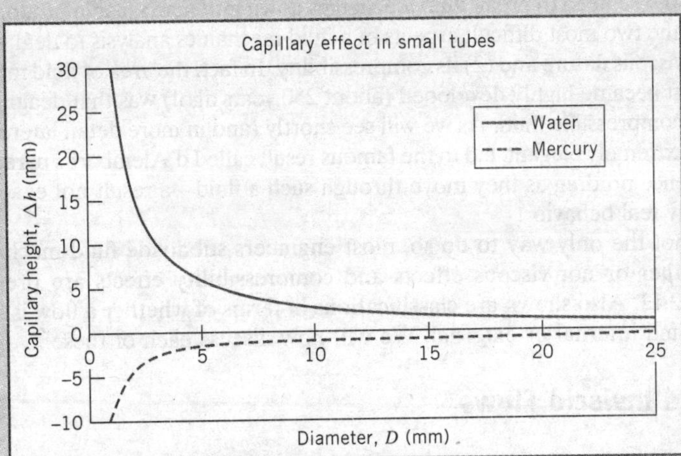

Using the above equation to compute $D_{\min}$ for $\Delta h = 1$ mm, we find for mercury and water

$$D_{M_{\min}} = 11.2 \text{ mm} \quad \text{and} \quad D_{W_{\min}} = 30 \text{ mm}$$

Notes:
- ✓ This problem reviewed use of the free-body diagram approach.
- ✓ It turns out that neglecting the volume in the meniscus region is only valid when Δh is large compared with D. However, in this problem we have the result that Δh is about 1 mm when D is 11.2 mm (or 30 mm); hence the results can only be very approximate.
- The graph and results were generated from the *Excel* workbook.

VIDEO

Capillary Rise.

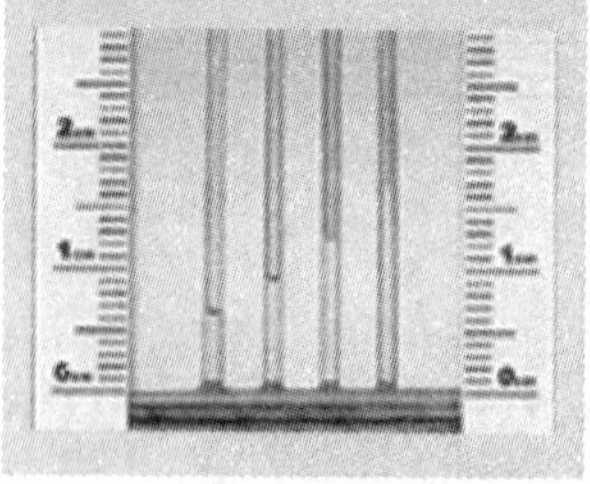

Folsom [6] shows that the simple analysis of Example 2.3 overpredicts the capillary effect and gives reasonable results only for tube diameters less than 2.54 mm. Over a diameter range $2.54 < D < 27.94$ mm, experimental data for the capillary rise with a water-air interface are correlated by the empirical expression $\Delta h = 0.400/e^{4.37D}$.

Manometer and barometer readings should be made at the level of the middle of the meniscus. This is away from the maximum effects of surface tension and thus nearest to the proper liquid level.

All surface tension data in Appendix A were measured for pure liquids in contact with clean vertical surfaces. Impurities in the liquid, dirt on the surface, or surface inclination can cause an indistinct meniscus; under such conditions it may be difficult to determine liquid level accurately. Liquid level is most distinct in a vertical tube. When inclined tubes are used to increase manometer sensitivity (see Section 3.3) it is important to make each reading at the same point on the meniscus and to avoid use of tubes inclined less than about 15° from horizontal.

CLASSIC VIDEO

Surface Tension in Fluid Mechanics.

Surfactant compounds reduce surface tension significantly (more than 40% with little change in other properties [7]) when added to water. They have wide commercial application: Most detergents contain surfactants to help water penetrate and lift soil from surfaces. Surfactants also have major industrial applications in catalysis, aerosols, and oil field recovery.

2.6 Description and Classification of Fluid Motions

In Chapter 1 and in this chapter, we have almost completed our brief introduction to some concepts and ideas that are often needed when studying fluid mechanics. Before beginning detailed analysis of fluid mechanics in the rest of this text, we will describe some interesting examples to illustrate a broad classification of fluid mechanics on the basis of important flow characteristics. Fluid mechanics is a huge discipline: It covers everything from the aerodynamics of a supersonic transport vehicle to the lubrication of human joints by sinovial fluid. We need to break fluid mechanics down into manageable proportions. It turns out that the two most difficult aspects of a fluid mechanics analysis to deal with are: (1) the fluid's viscous nature and (2) its compressibility. In fact, the area of fluid mechanics theory that first became highly developed (about 250 years ago!) was that dealing with a frictionless, incompressible fluid. As we will see shortly (and in more detail later on), this theory, while extremely elegant, led to the famous result called d'Alembert's paradox: All bodies experience no drag as they move through such a fluid—a result not exactly consistent with any real behavior!

Although not the only way to do so, most engineers subdivide fluid mechanics in terms of whether or not viscous effects and compressibility effects are present, as shown in Fig. 2.13. Also shown are classifications in terms of whether a flow is laminar or turbulent, and internal or external. We will now discuss each of these.

Viscous and Inviscid Flows

VIDEO

Examples of Flow over a Sphere.

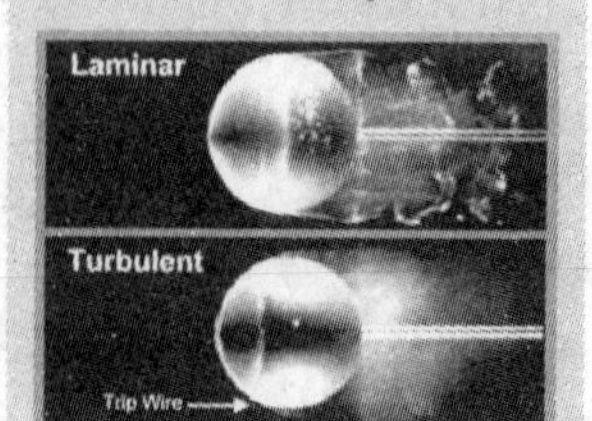

When you send a ball flying through the air (as in a game of baseball, soccer, or any number of other sports), in addition to gravity the ball experiences the aerodynamic drag of the air. The question arises: What is the nature of the drag force of the air on the ball? At first glance, we might conclude that it's due to friction of the air as it flows over the ball; a little more reflection might lead to the conclusion that because air has such a low viscosity, friction might not contribute much to the drag, and the drag might be due to the pressure build-up in front of the ball as it pushes the air out of the way. The question arises: Can we predict ahead of time the relative importance of the viscous force, and force due to the pressure build-up in front of the ball? Can we make similar predictions for *any* object, for example, an automobile, a submarine, a

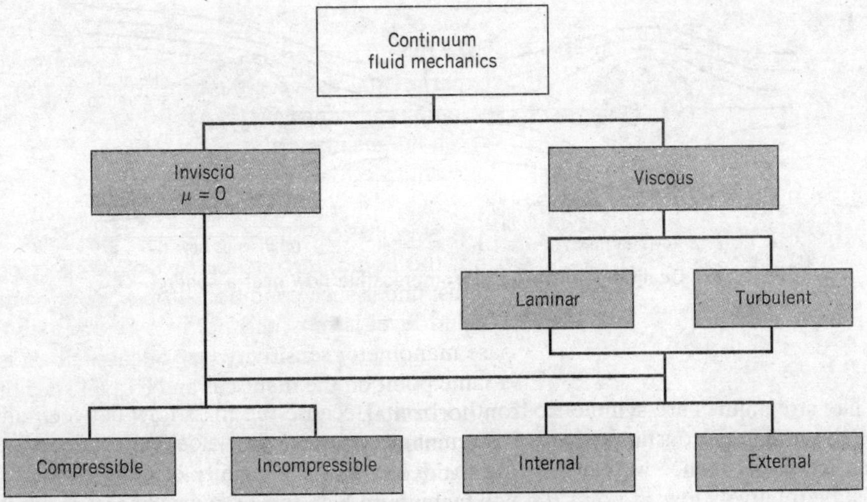

Fig. 2.13 Possible classification of continuum fluid mechanics.

VIDEO

Internal Laminar Flow in a Tube.

red blood cell, moving through *any* fluid, for example, air, water, blood plasma? The answer (which we'll discuss in much more detail in Chapter 7) is that we can! It turns out that we can estimate whether or not viscous forces, as opposed to pressure forces, are negligible by simply computing the Reynolds number

$$Re = \rho\frac{VL}{\mu}$$

where ρ and μ are the fluid density and viscosity, respectively, and V and L are the typical or "characteristic" velocity and size scale of the flow (in this example the ball velocity and diameter), respectively. If the Reynolds number is "large," viscous effects will be negligible (but will still have important consequences, as we'll soon see), at least in most of the flow; if the Reynolds number is small, viscous effects will be dominant. Finally, if the Reynolds number is neither large nor small, no general conclusions can be drawn.

VIDEO

The Space Shuttle: An External Turbulent Flow.

To illustrate this very powerful idea, consider two simple examples. First, the drag on your ball: Suppose you kick a soccer ball (diameter = 22.23 cm) so it moves at 97 km/h. The Reynolds number (using air properties from Table A.10) for this case is about 400,000—by any measure a large number; hence the drag on the soccer ball is almost entirely due to the pressure build-up in front of it. For our second example, consider a dust particle (modeled as a sphere of diameter 1 mm) falling under gravity at a terminal velocity of 1 cm/s: In this case $Re \approx 0.7$—a quite small number; hence the drag is mostly due to the friction of the air. Of course, in both of these examples, if we wish to *determine* the drag force, we would have to do substantially more analysis.

These examples illustrate an important point: A flow is considered to be friction dominated (or not) based not just on the fluid's viscosity, but on the complete flow system. In these examples, the airflow was low friction for the soccer ball, but was high friction for the dust particle.

CLASSIC VIDEO

Fundamentals—Boundary Layers.

Let's return for a moment to the idealized notion of frictionless flow, called *inviscid flow*. This is the branch shown on the left in Fig. 2.13. This branch encompasses most aerodynamics, and among other things explains, for example, why sub- and supersonic aircraft have differing shapes, how a wing generates lift, and so forth. If this theory is applied to the ball flying through the air (a flow that is also incompressible), it predicts streamlines (in coordinates attached to the sphere) as shown in Fig. 2.14*a*.

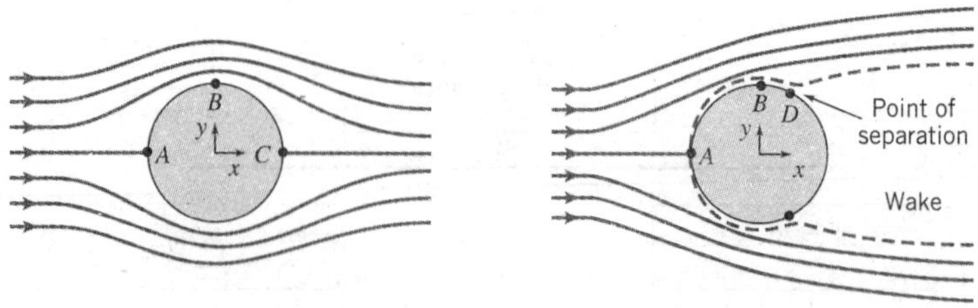

(a) Inviscid flow (b) Viscous flow

Fig. 2.14 Qualitative picture of incompressible flow over a sphere.

The streamlines are symmetric front-to-back. Because the mass flow between any two streamlines is constant, wherever streamlines open up, the velocity must decrease, and vice versa. Hence we can see that the velocity in the vicinity of points A and C must be relatively low; at point B it will be high. In fact, the air comes to rest at points A and C: They are *stagnation points*. It turns out that (as we'll learn in Chapter 6) the pressure in this flow is high wherever the velocity is low, and vice versa. Hence, points A and C have relatively large (and equal) pressures; point B will be a point of low pressure. In fact, the pressure distribution on the sphere is symmetric front-to-back, and there is no net drag force due to pressure. Because we're assuming inviscid flow, there can be no drag due to friction either. Hence we have d'Alembert's paradox of 1752: The ball experiences no drag!

VIDEO

Boundary Layer Flow.

This is obviously unrealistic. On the other hand, everything seems logically consistent: We established that Re for the sphere was very large (400,000), indicating friction is negligible. We then used inviscid flow theory to obtain our no-drag result. How can we reconcile this theory with reality? It took about 150 years after the paradox first appeared for the answer, obtained by Prandtl in 1904: The no-slip condition (Section 1.2) requires that the velocity everywhere on the surface of the sphere be zero (in sphere coordinates), but inviscid theory states that it's high at point B. Prandtl suggested that even though friction is negligible in general for high-Reynolds number flows, there will always be a thin *boundary layer*, in which friction is significant and across the width of which the velocity increases rapidly from zero (at the surface) to the value inviscid flow theory predicts (on the outer edge of the boundary layer). This is shown in Fig. 2.14b from point A to point B, and in more detail in Fig. 2.15.

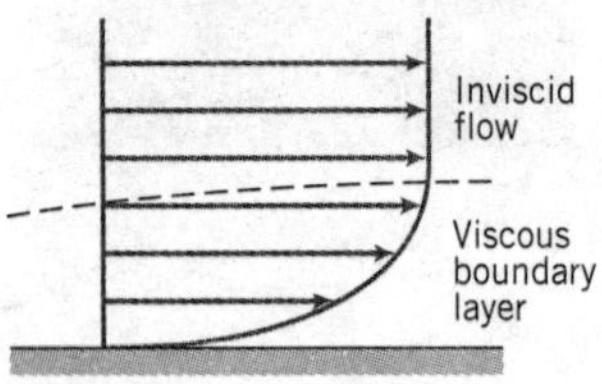

Fig. 2.15 Schematic of a boundary layer.

This boundary layer immediately allows us to reconcile theory and experiment: Once we have friction in a boundary layer we will have drag. However, this boundary layer has another important consequence: It often leads to bodies having a *wake*, as shown in Fig. 2.14b from point D onwards. Point D is a *separation point*, where fluid particles are pushed off the object and cause a wake to develop. Consider once again the original inviscid flow (Fig. 2.14a): As a particle moves along the surface from point B to C, it moves from low to high pressure. This *adverse pressure gradient* (a pressure change opposing fluid motion) causes the particles to slow down as they move along the rear of the sphere. If we now add to this the fact that the particles are moving in a boundary layer with friction that also slows down the fluid, the particles will eventually be brought to rest and then pushed off the sphere by the following particles, forming the wake. This is generally very bad news: It turns out that the wake will always be relatively low pressure, but the front of the sphere will still have relatively high pressure. Hence, the sphere will now have a quite large *pressure drag* (or *form drag*—so called because it's due to the shape of the object).

This description reconciles the inviscid flow no-drag result with the experimental result of significant drag on a sphere. It's interesting to note that although the boundary layer is necessary to explain the drag on the sphere, the drag is actually due mostly to the asymmetric pressure distribution created by the boundary layer separation—drag directly due to friction is still negligible!

We can also now begin to see how *streamlining* of a body works. The drag force in most aerodynamics is due to the low-pressure wake: If we can reduce or eliminate the wake, drag will be greatly reduced. If we consider once again why the separation occurred, we recall two features: Boundary layer friction slowed down the particles, but so did the adverse pressure gradient. The pressure increased very rapidly across the back half of the sphere in Fig. 2.14*a* because the streamlines opened up so rapidly. If we make the sphere teardrop shaped, as in Fig. 2.16, the streamlines open up gradually, and hence the pressure will increase slowly, to such an extent that fluid particles are not forced to separate from the object until they almost reach the end of the object, as shown. The wake is much smaller (and it turns out the pressure will not be as low as before), leading to much less pressure drag. The only negative aspect of this streamlining is that the total surface area on which friction occurs is larger, so drag due to friction will increase a little.

We should point out that none of this discussion applies to the example of a falling dust particle: This low-Reynolds number flow was viscous throughout—there is no inviscid region.

Finally, this discussion illustrates the very significant difference between inviscid flow ($\mu = 0$) and flows in which viscosity is negligible but not zero ($\mu \rightarrow 0$).

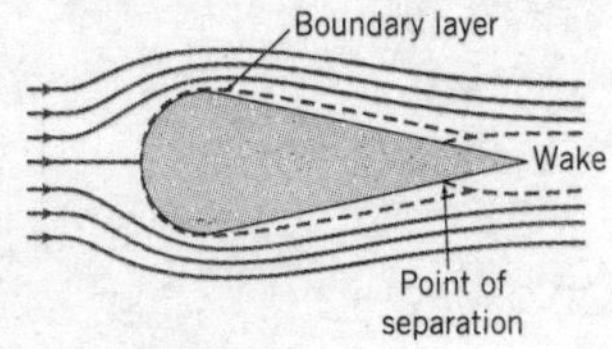

Fig. 2.16 Flow over a streamlined object.

VIDEO

Streamlined Flow over an Airfoil.

Laminar and Turbulent Flows

If you turn on a faucet (that doesn't have an aerator or other attachment) at a very low flow rate the water will flow out very smoothly—almost "glass-like." If you increase the flow rate, the water will exit in a churned-up, chaotic manner. These are examples of how a viscous flow can be laminar or turbulent, respectively. A *laminar* flow is one in which the fluid particles move in smooth layers, or laminas; a *turbulent* flow is one in which the fluid particles rapidly mix as they move along due to random three-dimensional velocity fluctuations. Typical examples of pathlines of each of these are illustrated in Fig. 2.17, which shows a one-dimensional flow. In most fluid mechanics problems—for example, flow of water in a pipe—turbulence is an unwanted but often unavoidable phenomenon, because it generates more resistance to flow; in other problems—for example, the flow of blood through blood vessels—it is desirable because the random mixing allows all of the blood cells to contact the walls of the blood vessels to exchange oxygen and other nutrients.

VIDEO

Streamlines around a Car.

The velocity of the laminar flow is simply u; the velocity of the turbulent flow is given by the mean velocity $\bar{u}$ plus the three components of randomly fluctuating velocity u', v', and w'.

CLASSIC VIDEO

Fluid Dynamics of Drag, I–IV.

Although many turbulent flows of interest are steady in the mean ($\bar{u}$ is not a function of time), the presence of the random, high-frequency velocity fluctuations makes the analysis of turbulent flows extremely difficult. In a one-dimensional laminar flow, the shear stress is related to the velocity gradient by the simple relation

$$\tau_{yx} = \mu \frac{du}{dy} \tag{2.15}$$

VIDEO

Laminar and Turbulent Flow.

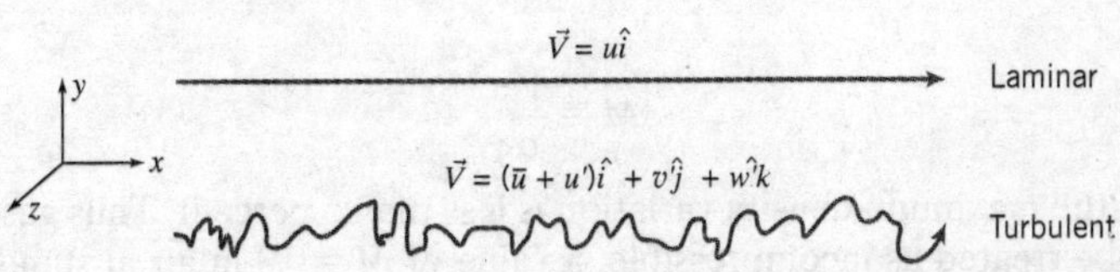

Fig. 2.17 Particle pathlines in one-dimensional laminar and turbulent flows.

For a turbulent flow in which the mean velocity field is one-dimensional, no such simple relation is valid. Random, three-dimensional velocity fluctuations (u', v', and w') transport momentum across the mean flow streamlines, increasing the effective shear stress. (This apparent stress is discussed in more detail in Chapter 8.) Consequently, in turbulent flow there is no universal relationship between the stress field and the mean-velocity field. Thus in turbulent flows we must rely heavily on semi-empirical theories and on experimental data.

Compressible and Incompressible Flows

Flows in which variations in density are negligible are termed *incompressible*; when density variations within a flow are not negligible, the flow is called *compressible*. The most common example of compressible flow concerns the flow of gases, while the flow of liquids may frequently be treated as incompressible.

For many liquids, density is only a weak function of temperature. At modest pressures, liquids may be considered incompressible. However, at high pressures, compressibility effects in liquids can be important. Pressure and density changes in liquids are related by the *bulk compressibility modulus*, or modulus of elasticity,

$$E_v \equiv \frac{dp}{(d\rho/\rho)} \tag{2.19}$$

If the bulk modulus is independent of temperature, then density is only a function of pressure (the fluid is *barotropic*). Bulk modulus data for some common liquids are given in Appendix A.

CLASSIC VIDEO

Cavitation.

Water hammer and cavitation are examples of the importance of compressibility effects in liquid flows. *Water hammer* is caused by acoustic waves propagating and reflecting in a confined liquid, for example, when a valve is closed abruptly. The resulting noise can be similar to "hammering" on the pipes, hence the term.

Cavitation occurs when vapor pockets form in a liquid flow because of local reductions in pressure (for example at the tip of a boat's propeller blades). Depending on the number and distribution of particles in the liquid to which very small pockets of undissolved gas or air may attach, the local pressure at the onset of cavitation may be at or below the vapor pressure of the liquid. These particles act as nucleation sites to initiate vaporization.

Vapor pressure of a liquid is the partial pressure of the vapor in contact with the saturated liquid at a given temperature. When pressure in a liquid is reduced to less than the vapor pressure, the liquid may change phase suddenly and "flash" to vapor.

The vapor pockets in a liquid flow may alter the geometry of the flow field substantially. When adjacent to a surface, the growth and collapse of vapor bubbles can cause serious damage by eroding the surface material.

Very pure liquids can sustain large negative pressures—as much as -6 MPa for distilled water—before the liquid "ruptures" and vaporization occurs. Undissolved air is invariably present near the free surface of water or seawater, so cavitation occurs where the local total pressure is quite close to the vapor pressure.

It turns out that gas flows with negligible heat transfer also may be considered incompressible provided that the flow speeds are small relative to the speed of sound; the ratio of the flow speed, V, to the local speed of sound, c, in the gas is defined as the Mach number,

$$M \equiv \frac{V}{c}$$

For $M < 0.3$, the maximum density variation is less than 5 percent. Thus gas flows with $M < 0.3$ can be treated as incompressible; a value of $M = 0.3$ in air at standard conditions corresponds to a speed of approximately 100 m/s. For example, although it might be a little counterintuitive, when you drive your car at 105 km/h the air flowing around it has negligible change in density. As we shall see in Chapter 12, the speed of sound in an ideal gas is given by $c = \sqrt{kRT}$, where k is the ratio of specific heats, R is the gas constant, and T is the absolute temperature. For air at STP, $k = 1.40$ and $R = 286.9$ J/kg · K. Values of k and R are supplied in Appendix A for several selected common gases at STP. In addition, Appendix A contains some useful data on atmospheric properties, such as temperature, at various elevations.

Compressible flows occur frequently in engineering applications. Common examples include compressed air systems used to power shop tools and dental drills, transmission of gases in pipelines at high pressure, and pneumatic or fluidic control and sensing systems. Compressibility effects are very important in the design of modern high-speed aircraft and missiles, power plants, fans, and compressors.

VIDEO

Compressible Flow: Shock Waves.

Internal and External Flows

Flows completely bounded by solid surfaces are called *internal* or *duct flows*. Flows over bodies immersed in an unbounded fluid are termed *external flows*. Both internal and external flows may be laminar or turbulent, compressible or incompressible.

We mentioned an example of internal flow when we discussed the flow out of a faucet—the flow in the pipe leading to the faucet is an internal flow. It turns out that we have a Reynolds number for pipe flows defined as $Re = \rho \overline{V} D/\mu$, where $\overline{V}$ is the average flow velocity and D is the pipe diameter (note that we do *not* use the pipe length!). This Reynolds number indicates whether a pipe flow will be laminar or turbulent. Flow will generally be laminar for $Re \leq 2300$ and turbulent for larger values: Flow in a pipe of constant diameter will be entirely laminar or entirely turbulent, depending on the value of the velocity $\overline{V}$. We will explore internal flows in detail in Chapter 8.

We already saw some examples of external flows when we discussed the flow over a sphere (Fig. 2.14*b*) and a streamlined object (Fig. 2.16). What we didn't mention was that these flows could be laminar or turbulent. In addition, we mentioned boundary layers (Fig. 2.15): It turns out these also can be laminar or turbulent. When we discuss these in detail (Chapter 9), we'll start with the simplest kind of boundary layer—that over a flat plate—and learn that just as we have a Reynolds number for the overall external flow that indicates the relative significance of viscous forces, there will also be a boundary-layer Reynolds number $Re_x = \rho U_\infty x/\mu$ where in this case the characteristic velocity U_∞ is the velocity immediately outside the boundary layer and the characteristic length x is the distance along the plate. Hence, at the leading edge of the plate $Re_x = 0$, and at the end of a plate of length L, it will be $Re_x = \rho U_\infty L/\mu$. The significance of this Reynolds number is that (as we'll learn) the boundary layer will be laminar for $Re_x \leq 5 \times 10^5$ and turbulent for larger values: A boundary layer will start out laminar, and if the plate is long enough the boundary layer will transition to become turbulent.

It is clear by now that computing a Reynolds number is often very informative for both internal and external flows. We will discuss this and other important *dimensionless groups* (such as the Mach number) in Chapter 7.

The internal flow through fluid machines is considered in Chapter 10. The principle of angular momentum is applied to develop fundamental equations for fluid machines. Pumps, fans, blowers, compressors, and propellers that add energy to fluid streams are considered, as are turbines and windmills that extract energy. The chapter features detailed discussion of operation of fluid systems.

The internal flow of liquids in which the duct does not flow full—where there is a free surface subject to a constant pressure—is termed *open-channel* flow. Common examples of open-channel flow include flow in rivers, irrigation ditches, and aqueducts. Open-channel flow will be treated in Chapter 11.

Both internal and external flows can be compressible or incompressible. Compressible flows can be divided into subsonic and supersonic regimes. We will study compressible flows in Chapters 12 and 13 and see among other things that *supersonic flows* ($M > 1$) will behave very differently than *subsonic flows* ($M < 1$). For example, supersonic flows can experience oblique and normal shocks, and can also behave in a counterintuitive way—e.g., a supersonic nozzle (a device to accelerate a flow) must be divergent (i.e., it has *increasing* cross-sectional area) in the direction of flow! We note here also that in a subsonic nozzle (which has a convergent cross-sectional area), the pressure of the flow at the exit plane will always be the ambient pressure; for a sonic flow, the exit pressure can be higher than ambient; and for a supersonic flow the exit pressure can be greater than, equal to, or less than the ambient pressure!

2.7 Summary and Useful Equations

In this chapter we have completed our review of some of the fundamental concepts we will utilize in our study of fluid mechanics. Some of these are:

- ✓ How to describe flows (timelines, pathlines, streamlines, streaklines).
- ✓ Forces (surface, body) and stresses (shear, normal).
- ✓ Types of fluids (Newtonian, non-Newtonian—dilatant, pseudoplastic, thixotropic, rheopectic, Bingham plastic) and viscosity (kinematic, dynamic, apparent).
- ✓ Types of flow (viscous/inviscid, laminar/turbulent, compressible/incompressible, internal/external).

We also briefly discussed some interesting phenomena, such as surface tension, boundary layers, wakes, and streamlining. Finally, we introduced two very useful dimensionless groups—the Reynolds number and the Mach number.

Note: Most of the Useful Equations in the table below have a number of constraints or limitations—*be sure to refer to their page numbers for details*!

Useful Equations

Definition of specific gravity:	$SG = \dfrac{\rho}{\rho_{H_2O}}$	**(2.3)**	Page 23
Definition of specific weight:	$\gamma = \dfrac{mg}{V} \rightarrow \gamma = \rho g$	**(2.4)**	Page 23
Definition of streamlines (2D):	$\left.\dfrac{dy}{dx}\right)_{\text{streamline}} = \dfrac{v(x,y)}{u(x,y)}$	**(2.8)**	Page 27
Definition of pathlines (2D):	$\left.\dfrac{dx}{dt}\right)_{\text{particle}} = u(x,y,t) \quad \left.\dfrac{dy}{dt}\right)_{\text{particle}} = v(x,y,t)$	**(2.9)**	Page 27
Definition of streaklines (2D):	$x_{\text{streakline}}(t_0) = x(t, x_0, y_0, t_0) \quad y_{\text{streakline}}(t_0) = y(t, x_0, y_0, t_0)$	**(2.10)**	Page 27
Newton's law of viscosity (1D flow):	$\tau_{yx} = \mu \dfrac{du}{dy}$	**(2.15)**	Page 33
Shear stress for a non-Newtonian fluid (1D flow):	$\tau_{yx} = k\left\|\dfrac{du}{dy}\right\|^{n-1}\dfrac{du}{dy} = \eta \dfrac{du}{dy}$	**(2.17)**	Page 35

Case Study

Fluid Mechanics and Your MP3 Player

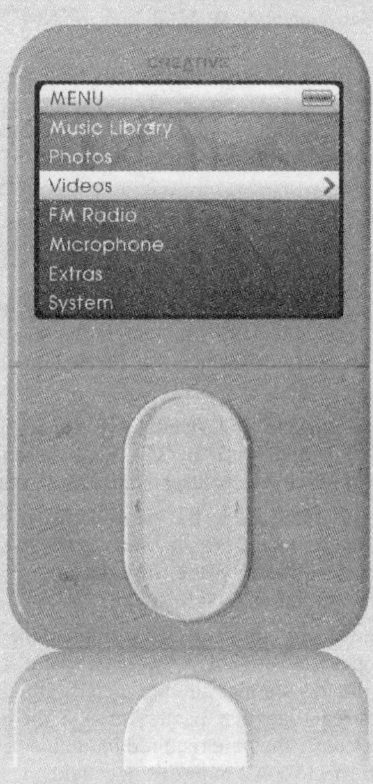

The MP3 player of one of the authors.

Some people have the impression that fluid mechanics is old- or low-tech: water flow in a household pipe, the fluid forces acting on a dam, and so on. While it's true that many concepts in fluid mechanics are hundreds of years old, there are still lots of exciting new areas of research and development. Everyone has heard of the relatively high-tech area of fluid mechanics called streamlining (of cars, aircraft, racing bikes, and racing swimsuits, to mention a few), but there are many others.

If you're a typical engineering student, there's a decent chance that while reading this chapter you're listening to music on your MP3 player; you can thank fluid mechanics for your ability to do this! The tiny hard disk drive (HDD) in many of these devices typically holds about 250 gigabytes (GB) of data, so the disk platter must have a huge density (greater than 40,000 tracks per cm); in addition, the read/write head must get very close to the platter as it transfers data (typically the head is about 0.05 μm above the platter surface—a human hair is about 100 μm). The platter also spins at something greater than 500 revolutions per second! Hence the bearings in which the spindle of the platter spins must have very low friction but also have virtually no play or looseness—otherwise, at worst, the head will crash into the platter or, at best, you won't be able to read the data (it will be too closely packed). Designing such a bearing presents quite a challenge. Until a few years ago, most hard drives used ball bearings (BBs), which are essentially just like those in the wheel of a bicycle; they work on the principle that a spindle can rotate if it is held by a ring of small spheres that are supported in a cage. The problems with BBs are that they have a lot of components; they are very difficult to build to the precision needed for the HDD; they are vulnerable to shock (if you drop an HDD with such a drive, you're likely to dent one of the spheres as it hits the spindle, destroying the bearing); and they are relatively noisy.

Hard-drive makers are increasingly moving to fluid dynamic bearings (FDBs). These are mechanically much simpler than BBs; they consist basically of the spindle directly mounted in the bearing opening, with only a specially formulated viscous lubricant (such as ester oil) in the gap of only a few microns. The spindle and/or bearing surfaces have a herringbone pattern of grooves to maintain the oil in place. These bearings are extremely durable (they can often survive a shock of 500 g!) and low noise; they will also allow rotation speeds in excess of 15,000 rpm in the future, making data access even faster than with current devices. FDBs have been used before, in devices such as gyroscopes, but making them at such a small scale is new. Some FDBs even use pressurized air as the lubrication fluid, but one of the problems with these is that they sometimes stop working when you take them on an airplane flight—the cabin pressure is insufficient to maintain the pressure the bearing needs!

In recent times the price and capacity of flash memory have improved so much that many MP3 players are switching to this technology from HDDs. Eventually, notebook and desktop PCs will also switch to flash memory, but at least for the next few years HDDs will be the primary storage medium. Your PC will still have vital fluid-mechanical components!

References

1. Vincenti, W. G., and C. H. Kruger Jr., *Introduction to Physical Gas Dynamics*. New York: Wiley, 1965.
2. Merzkirch, W., *Flow Visualization*, 2nd ed. New York: Academic Press, 1987.
3. Tanner, R. I., *Engineering Rheology*. Oxford: Clarendon Press, 1985.
4. Macosko, C. W., *Rheology: Principles, Measurements, and Applications*. New York: VCH Publishers, 1994.
5. Loh, W. H. T., "Theory of the Hydraulic Analogy for Steady and Unsteady Gas Dynamics," in *Modern Developments in Gas Dynamics*, W. H. T. Loh, ed. New York: Plenum, 1969.
6. Folsom, R. G., "Manometer Errors Due to Capillarity," *Instruments*, 9, 1, 1937, pp. 36–37.
7. Waugh, J. G., and G. W. Stubstad, *Hydroballistics Modeling*. San Diego: Naval Undersea Center, ca. 1972.

Problems

Velocity Field

2.1 For the velocity fields given below, determine:

a. whether the flow field is one-, two-, or three-dimensional, and why.

b. whether the flow is steady or unsteady, and why.

(The quantities a and b are constants.)

(1) $\vec{V} = [(ax+t)e^{by}]\hat{i}$ (2) $\vec{V} = (ax - by)\hat{i}$

(3) $\vec{V} = ax\hat{i} + [e^{bx}]\hat{j}$ (4) $\vec{V} = ax\hat{i} + bx^2\hat{j} + ax\hat{k}$

(5) $\vec{V} = ax\hat{i} + [e^{bt}]\hat{j}$ (6) $\vec{V} = ax\hat{i} + bx^2\hat{j} + ay\hat{k}$

(7) $\vec{V} = ax\hat{i} + [e^{bt}]\hat{j} + ay\hat{k}$ (8) $\vec{V} = ax\hat{i} + [e^{by}]\hat{j} + az\hat{k}$

2.2 For the velocity fields given below, determine:

a. whether the flow field is one-, two-, or three-dimensional, and why.

b. whether the flow is steady or unsteady, and why.

(The quantities a and b are constants.)

(1) $\vec{V} = [ay^2e^{-bt}]\hat{i}$ (2) $\vec{V} = ax^2\hat{i} + bx\hat{j} + c\hat{k}$

(3) $\vec{V} = axy\hat{i} - byt\hat{j}$ (4) $\vec{V} = ax\hat{i} - by\hat{j} + ct\hat{k}$

(5) $\vec{V} = [ae^{-bx}]\hat{i} + bt^2\hat{j}$ (6) $\vec{V} = a(x^2+y^2)^{1/2}(1/z^3)\hat{k}$

(7) $\vec{V} = (ax+t)\hat{i} - by^2\hat{j}$ (8) $\vec{V} = ax^2\hat{i} + bxz\hat{j} + cy\hat{k}$

2.3 A viscous liquid is sheared between two parallel disks; the upper disk rotates and the lower one is fixed. The velocity field between the disks is given by $\vec{V} = \hat{e}_\theta r\omega z/h$. (The origin of coordinates is located at the center of the lower disk; the upper disk is located at $z = h$.) What are the dimensions of this velocity field? Does this velocity field satisfy appropriate physical boundary conditions? What are they?

2.4 For the velocity field $\vec{V} = Ax^2y\hat{i} + Bxy^2\hat{j}$, where $A = 2\ \text{m}^{-2}\text{s}^{-1}$ and $B = 1\ \text{m}^{-2}\text{s}^{-1}$, and the coordinates are measured in meters, obtain an equation for the flow streamlines. Plot several streamlines in the first quadrant.

2.5 The velocity field $\vec{V} = Ax\hat{i} - Ay\hat{j}$, where $A = 2\ \text{s}^{-1}$, can be interpreted to represent flow in a corner. Find an equation for the flow streamlines. Explain the relevance of A. Plot several streamlines in the first quadrant, including the one that passes through the point $(x, y) = (0, 0)$.

2.6 A velocity field is specified as $\vec{V} = axy\hat{i} + by^2\hat{j}$, where $a = 2\ \text{m}^{-1}\text{s}^{-1}$, $b = -6\ \text{m}^{-1}\text{s}^{-1}$, and the coordinates are measured in meters. Is the flow field one-, two-, or three-dimensional? Why? Calculate the velocity components at the point (2, ½). Develop an equation for the streamline passing through this point. Plot several streamlines in the first quadrant including the one that passes through the point (2, ½).

2.7 A velocity field is given by $\vec{V} = ax\hat{i} - bty\hat{j}$, where $a = 1\ \text{s}^{-1}$ and $b = 1\ \text{s}^{-2}$. Find the equation of the streamlines at any time t. Plot several streamlines in the first quadrant at $t = 0$ s, $t = 1$ s, and $t = 20$ s.

2.8 A velocity field is given by $\vec{V} = ax^3\hat{i} + bxy^3\hat{j}$, where $a = 1\ \text{m}^{-2}\text{s}^{-1}$ and $b = 1\ \text{m}^{-3}\text{s}^{-1}$. Find the equation of the streamlines. Plot several streamlines in the first quadrant.

2.9 A flow is described by the velocity field $\vec{V} = (Ax + B)\hat{i} + (-Ay)\hat{j}$, where $A = 3$ m/s/m and $B = 6$ m/s. Plot a few streamlines in the xy plane, including the one that passes through the point $(x, y) = (0.3, 0.6)$.

2.10 The velocity for a steady, incompressible flow in the xy plane is given by $\vec{V} = \hat{i}A/x + \hat{j}Ay/x^2$, where $A = 2\ \text{m}^2/\text{s}$, and the coordinates are measured in meters. Obtain an equation for the streamline that passes through the point $(x, y) = (1, 3)$. Calculate the time required for a fluid particle to move from $x = 1$ m to $x = 2$ m in this flow field.

2.11 The flow field for an atmospheric flow is given by

$$\vec{V} = -\frac{My}{2\pi}\hat{i} + \frac{Mx}{2\pi}\hat{j}$$

where $M = 1\ \text{s}^{-1}$, and the x and y coordinates are the parallel to the local latitude and longitude. Plot the velocity magnitude along the x axis, along the y axis, and along the line $y = x$, and discuss the velocity direction with respect to these three axes. For each plot use a range x or y = 0 km to 1 km. Find the equation for the streamlines and sketch several of them. What does this flow field model?

2.12 The flow field for an atmospheric flow is given by

$$\vec{V} = -\frac{Ky}{2\pi(x^2+y^2)}\hat{i} + \frac{Kx}{2\pi(x^2+y^2)}\hat{j}$$

where $K = 10^5\ \text{m}^2/\text{s}$, and the x and y coordinates are parallel to the local latitude and longitude. Plot the velocity magnitude along the x axis, along the y axis, and along the line $y = x$, and discuss the velocity direction with respect to these three axes. For each plot use a range x or y = − 1 km to 1 km, excluding $|x|$ or $|y| < 100$ m. Find the equation for the streamlines and sketch several of them. What does this flow field model?

2.13 A flow field is given by

$$\vec{V} = -\frac{qx}{2\pi(x^2+y^2)}\hat{i} - \frac{qy}{2\pi(x^2+y^2)}\hat{j}$$

where $q = 5 \times 10^4$ m²/s. Plot the velocity magnitude along the x axis, along the y axis, and along the line $y = x$, and discuss the velocity direction with respect to these three axes. For each plot use a range x or $y = -1$ km to 1 km, excluding $|x|$ or $|y| < 100$ m. Find the equation for the streamlines and sketch several of them. What does this flow field model?

2.14 Beginning with the velocity field of Problem 2.5, show that the parametric equations for particle motion are given by $x_p = c_1 e^{At}$ and $y_p = c_2 e^{-At}$. Obtain the equation for the pathline of the particle located at the point $(x, y) = (2, 2)$ at the instant $t = 0$. Compare this pathline with the streamline through the same point.

2.15 A flow field is given by $\vec{V} = Ax\hat{i} + 2Ay\hat{j}$, where $A = 2\,\text{s}^{-1}$. Verify that the parametric equations for particle motion are given by $x_p = c_1 e^{At}$ and $y_p = c_2 e^{2At}$. Obtain the equation for the pathline of the particle located at the point $(x, y) = (2, 2)$ at the instant $t = 0$. Compare this pathline with the streamline through the same point.

2.16 A velocity field is given by $\vec{V} = ayt\hat{i} - bx\hat{j}$, where $a = 1\,\text{s}^{-2}$ and $b = 4\,\text{s}^{-1}$. Find the equation of the streamlines at any time t. Plot several streamlines at $t = 0$ s, $t = 1$ s, and $t = 20$ s.

2.17 Verify that $x_p = -a\sin(\omega t)$, $y_p = a\cos(\omega t)$ is the equation for the pathlines of particles for the flow field of Problem 2.12. Find the frequency of motion ω as a function of the amplitude of motion, a, and K. Verify that $x_p = -a\sin(\omega t)$, $y_p = a\cos(\omega t)$ is also the equation for the pathlines of particles for the flow field of Problem 2.11, except that ω is now a function of M. Plot typical pathlines for both flow fields and discuss the difference.

2.18 Air flows downward toward an infinitely wide horizontal flat plate. The velocity field is given by $\vec{V} = (ax\hat{i} - ay\hat{j})(2 + \cos\omega t)$, where $a = 5\,\text{s}^{-1}$, $\omega = 2\pi\,\text{s}^{-1}$, x and y (measured in meters) are horizontal and vertically upward, respectively, and t is in s. Obtain an algebraic equation for a streamline at $t = 0$. Plot the streamline that passes through point $(x, y) = (3, 3)$ at this instant. Will the streamline change with time? Explain briefly. Show the velocity vector on your plot at the same point and time. Is the velocity vector tangent to the streamline? Explain.

2.19 Consider the flow described by the velocity field $\vec{V} = A(1 + Bt)\hat{i} + Cty\hat{j}$, with $A = 1$ m/s, $B = 1\,\text{s}^{-1}$, and $C = 1\,\text{s}^{-2}$. Coordinates are measured in meters. Plot the pathline traced out by the particle that passes through the point (1, 1) at time $t = 0$. Compare with the streamlines plotted through the same point at the instants $t = 0$, 1, and 2 s.

2.20 Consider the flow described by the velocity field $\vec{V} = Bx(1 + At)\hat{i} + Cy\hat{j}$, with $A = 0.5\,\text{s}^{-1}$ and $B = C = 1\,\text{s}^{-1}$. Coordinates are measured in meters. Plot the pathline traced out by the particle that passes through the point (1, 1) at time $t = 0$. Compare with the streamlines plotted through the same point at the instants $t = 0$, 1, and 2 s.

2.21 Consider the flow field given in Eulerian description by the expression $\vec{V} = A\hat{i} - Bt\hat{j}$, where $A = 2$ m/s, $B = 2$ m/s², and the coordinates are measured in meters. Derive the Lagrangian position functions for the fluid particle that was located at the point $(x, y) = (1, 1)$ at the instant $t = 0$. Obtain an algebraic expression for the pathline followed by this particle. Plot the pathline and compare with the streamlines plotted through the same point at the instants $t = 0$, 1, and 2 s.

2.22 Consider the velocity field $V = ax\hat{i} + by(1 + ct)\hat{j}$, where $a = b = 2\,\text{s}^{-1}$ and $c = 0.4\,\text{s}^{-1}$. Coordinates are measured in meters. For the particle that passes through the point $(x, y) = (1, 1)$ at the instant $t = 0$, plot the pathline during the interval from $t = 0$ to 1.5 s. Compare this pathline with the streamlines plotted through the same point at the instants $t = 0$, 1, and 1.5 s.

2.23 Consider the flow field given in Eulerian descriptionby the expression $\vec{V} = ax\hat{i} + byt\hat{j}$, where $a = 0.2\,\text{s}^{-1}$, $b = 0.04\,\text{s}^{-2}$, and the coordinates are measured in meters. Derive the Lagrangian position functions for the fluid particle that was located at the point $(x, y) = (1, 1)$ at the instant $t = 0$. Obtain an algebraic expression for the pathline followed by this particle. Plot the pathline and compare with the streamlines plotted through the same point at the instants $t = 0$, 10, and 20 s.

2.24 A velocity field is given by $\vec{V} = axt\hat{i} - by\hat{j}$, where $a = 0.1\,\text{s}^{-2}$ and $b = 1\,\text{s}^{-1}$. For the particle that passes through the point $(x, y) = (1, 1)$ at instant $t = 0$ s, plot the pathline during the interval from $t = 0$ to $t = 3$ s. Compare with the streamlines plotted through the same point at the instants $t = 0$, 1, and 2 s.

2.25 Consider the flow field $\vec{V} = axt\hat{i} + b\hat{j}$, where $a = 0.1\,\text{s}^{-2}$ and $b = 4$ m/s. Coordinates are measured in meters. For the particle that passes through the point $(x, y) = (3, 1)$ at the instant $t = 0$, plot the pathline during the interval from $t = 0$ to 3 s. Compare this pathline with the streamlines plotted through the same point at the instants $t = 1$, 2, and 3 s.

2.26 Consider the garden hose of Fig. 2.5. Suppose the velocity field is given by $\vec{V} = u_0\hat{i} + v_0\sin[\omega(t - x/u_0)]\hat{j}$, where the x direction is horizontal and the origin is at the mean position of the hose, $u_0 = 10$ m/s, $v_0 = 2$ m/s, and $\omega = 5$ cycle/s. Find and plot on one graph the instantaneous streamlines that pass through the origin at $t = 0$ s, 0.05 s, 0.1 s, and 0.15 s. Also find and plot on one graph the pathlines of particles that left the origin at the same four times.

2.27 Using the data of Problem 2.26, find and plot the streakline shape produced after the first second of flow.

2.28 Consider the velocity field of Problem 2.20. Plot the streakline formed by particles that passed through the point (1, 1) during the interval from $t = 0$ to $t = 3$ s. Compare with the streamlines plotted through the same point at the instants $t = 0$, 1, and 2 s.

2.29 Streaklines are traced out by neutrally buoyant marker fluid injected into a flow field from a fixed point in space. A particle of the marker fluid that is at point (x, y) at time t must have passed through the injection point (x_0, y_0) at some earlier instant $t = \tau$. The time history of a marker particle may be found by solving the pathline equations for the initial conditions that $x = x_0$, $y = y_0$ when $t = \tau$. The present locations of particles on the streakline are obtained by setting τ equal to values in the range $0 \le \tau \le t$. Consider the flow field $\vec{V} = ax(1 + bt)\hat{i} + cy\hat{j}$, where $a = c = 1\,\text{s}^{-1}$ and $b = 0.2\,\text{s}^{-1}$. Coordinates are measured in meters. Plot the streakline that passes through the initial point $(x_0, y_0) = (1, 1)$, during the interval from $t = 0$ to $t = 3$ s. Compare with the streamline plotted through tne same point at the instants $t = 0$, 1, and 2 s.

2.30 Consider the flow field $\vec{V} = axt\hat{i} + b\hat{j}$, where $a = 1/4\ \text{s}^{-2}$ and $b = 1/3$ m/s. Coordinates are measured in meters. For the particle that passes through the point $(x, y) = (1, 2)$ at the instant $t = 0$, plot the pathline during the time interval from $t = 0$ to 3 s. Compare this pathline with the streakline through the same point at the instant $t = 3$ s.

2.31 A flow is described by velocity field $\vec{V} = ay^2\hat{i} + b\hat{j}$, where $a = 1\ \text{m}^{-1}\text{s}^{-1}$ and $b = 2$ m/s. Coordinates are measured in meters. Obtain the equation for the streamline passing through point (6, 6). At $t = 1$ s, what are the coordinates of the particle that passed through point (1, 4) at $t = 0$? At $t = 3$ s, what are the coordinates of the particle that passed through point $(-3, 0)$ 2 s earlier? Show that pathlines, streamlines, and streaklines for this flow coincide.

2.32 Tiny hydrogen bubbles are being used as tracers to visualize a flow. All the bubbles are generated at the origin $(x = 0, y = 0)$. The velocity field is unsteady and obeys the equations:

$$u = 1\ \text{m/s} \qquad v = 2\ \text{m/s} \qquad 0 \le t < 2\ \text{s}$$
$$u = 0 \qquad v = -1\ \text{m/s} \qquad 0 \le t \le 4\ \text{s}$$

Plot the pathlines of bubbles that leave the origin at $t = 0$, 1, 2, 3, and 4 s. Mark the locations of these five bubbles at $t = 4$ s. Use a dashed line to indicate the position of a streakline at $t = 4$ s.

2.33 A flow is described by velocity field $\vec{V} = ax\hat{i} + b\hat{j}$, where $a = 1/5\ \text{s}^{-1}$ and $b = 1$ m/s. Coordinates are measured in meters. Obtain the equation for the streamline passing through point (1, 1). At $t = 5$ s, what are the coordinates of the particle that initially (at $t = 0$) passed through point (1, 1)? What are its coordinates at $t = 10$ s? Plot the streamline and the initial, 5 s, and 10 s positions of the particle. What conclusions can you draw about the pathline, streamline, and streakline for this flow?

2.34 A flow is described by velocity field $\vec{V} = a\hat{i} + bx\hat{j}$, where $a = 2$ m/s and $b = 1\ \text{s}^{-1}$. Coordinates are measured in meters. Obtain the equation for the streamline passing through point (2, 5). At $t = 2$ s, what are the coordinates of the particle that passed through point (0, 4) at $t = 0$? At $t = 3$ s, what are the coordinates of the particle that passed through point (1, 4.25) 2 s earlier? What conclusions can you draw about the pathline, streamline, and streakline for this flow?

2.35 A flow is described by velocity field $\vec{V} = ay\hat{i} + bt\hat{j}$, where $a = 0.2\ \text{s}^{-1}$ and $b = 0.4\ \text{m/s}^2$. At $t = 2$ s, what are the coordinates of the particle that passed through point (1, 2) at $t = 0$? At $t = 3$ s, what are the coordinates of the particle that passed through point (1, 2) at $t = 2$ s? Plot the pathline and streakline through point (1, 2), and plot the streamlines through the same point at the instants $t = 0$, 1, 2, and 3 s.

2.36 A flow is described by velocity field $\vec{V} = at\hat{i} + b\hat{j}$, where $a = 0.4\ \text{m/s}^2$ and $b = 2$ m/s. At $t = 2$ s, what are the coordinates of the particle that passed through point (2, 1) at $t = 0$? At $t = 3$ s, what are the coordinates of the particle that passed through point (2, 1) at $t = 2$ s? Plot the pathline and streakline through point (2, 1) and compare with the streamlines through the same point at the instants $t = 0$, 1, and 2 s.

Viscosity

2.37 The variation with temperature of the viscosity of air is represented well by the empirical Sutherland correlation

$$\mu = \frac{bT^{1/2}}{1 + S/T}$$

Best-fit values of b and S are given in Appendix A. Develop an equation in SI units for kinematic viscosity versus temperature for air at atmospheric pressure. Assume ideal gas behavior. Check by using the equation to compute the kinematic viscosity of air at 0°C and at 100°C and comparing to the data in Appendix 10 (Table A.10); plot the kinematic viscosity for a temperature range of 0°C to 100°C, using the equation and the data in Table A.10.

2.38 The variation with temperature of the viscosity of air is correlated well by the empirical Sutherland equation

$$\mu = \frac{bT^{1/2}}{1 + S/T}$$

Best-fit values of b and S are given in Appendix A for use with SI units. Use these values to develop an equation for calculating air viscosity in British Gravitational units as a function of absolute temperature in degrees Rankine. Check your result using data from Appendix A.

2.39 Some experimental data for the viscosity of helium at 1 atm are

T, °C	0	100	200	300	400
μ, N·s/m² (× 10⁵)	1.86	2.31	2.72	3.11	3.46

Using the approach described in Appendix A.3, correlate these data to the empirical Sutherland equation

$$\mu = \frac{bT^{1/2}}{1 + S/T}$$

(where T is in kelvin) and obtain values for constants b and S.

2.40 The velocity distribution for laminar flow between parallel plates is given by

$$\frac{u}{u_{\max}} = 1 - \left(\frac{2y}{h}\right)^2$$

where h is the distance separating the plates and the origin is placed midway between the plates. Consider a flow of water at 15°C, with $u_{\max} = 0.10$ m/s and $h = 0.1$ mm. Calculate the shear stress on the upper plate and give its direction. Sketch the variation of shear stress across the channel.

2.41 The velocity distribution for laminar flow between parallel plates is given by

$$\frac{u}{u_{\max}} = 1 - \left(\frac{2y}{h}\right)^2$$

where h is the distance separating the plates and the origin is placed midway between the plates. Consider a flow of water at 15°C with maximum speed of 0.05 m/s and $h = 0.1$ mm.

Calculate the force on a 1 m^2 section of the lower plate and give its direction.

2.42 Explain how an ice skate interacts with the ice surface. What mechanism acts to reduce sliding friction between skate and ice?

2.43 Crude oil, with specific gravity SG = 0.85 and viscosity $\mu = 0.1\ \text{N}\cdot\text{s/m}^2$ flows steadily down a surface inclined $\theta = 45$ degrees below the horizontal in a film of thickness $h = 2.5$ mm. The velocity profile is given by

$$u = \frac{\rho g}{\mu}\left(hy - \frac{y^2}{2}\right)\sin\theta$$

(Coordinate x is along the surface and y is normal to the surface.) Plot the velocity profile. Determine the magnitude and direction of the shear stress that acts on the surface.

2.44 A female freestyle ice skater, weighing 450 N glides on one skate at speed $V = 6$ m/s. Her weight is supported by a thin film of liquid water melted from the ice by the pressure of the skate blade.. Assume the blade is $L = 0.3$ m long and $w = 3$ mm wide, and that the water film is $h = 0.0015$ mm, thick. Estimate the deceleration of the skater that results from viscous shear in the water film, if end effects are neglected.

2.45 A block weighing 45 N and having dimensions 250 mm on each edge is pulled up an inclined surface on which there is a film of SAE 10W oil at 37°C. If the speed of the block is 0.6 m/s and the oil film is 0.025 mm thick, find the force required to pull the block. Assume the velocity distribution in the oil film is linear. The surface is inclined at an angle of 25° from the horizontal.

2.46 A block of mass 10 kg and measuring 250 mm on each edge is pulled up an inclined surface on which there is a film of SAE 10W-30 oil at −1.1°C (the oil film is 0.025 mm thick). Find the steady speed of the block if it is released. If a force of 75 N is applied to pull the block up the incline, find the steady speed of the block. If the force is now applied to push the block down the incline, find the steady speed of the block. Assume the velocity distribution in the oil film is linear. The surface is inclined at an angle of 30° from the horizontal.

2.47 Tape is to be coated on both sides with glue by drawing it through a narrow gap. The tape is 0.38 mm thick and 25 mm wide. It is centered in the gap with a clearance of 0.3 mm on each side. The glue, of viscosity $\mu = 1\ \text{N}\cdot\text{s/m}^2$ completely fills the space between the tape and gap. If the tape can withstand a maximum tensile force of 110 N, determine the maximum gap region through which it can be pulled at a speed of 1 m/s.

2.48 A 73 mm diameter aluminum (SG = 2.64) piston of 100-mm length resides in a stationary 75 mm inner diameter steel tube lined with SAE 10W-30 oil at 25°C. A mass $m = 2$ kg is suspended from the free end of the piston. The piston is set into motion by cutting a support cord. What is the terminal velocity of mass m? Assume a linear velocity profile within the oil.

2.49 The piston in Problem 2.48 is traveling at terminal speed. The mass m now disconnects from the piston. Plot the piston speed vs. time. How long does it take the piston to come within 1 percent of its new terminal speed?

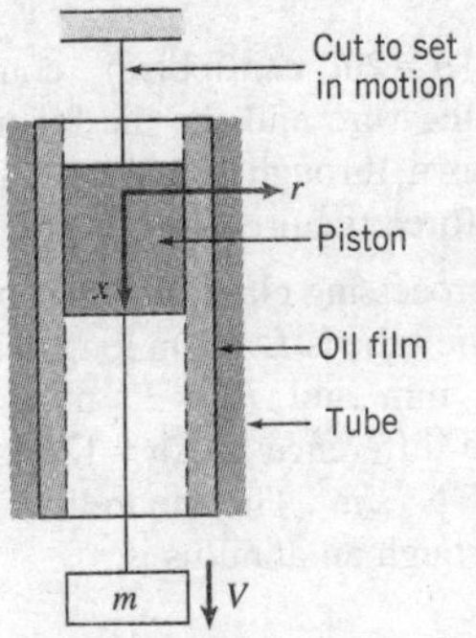

P2.48, P2.49

2.50 A block of mass M slides on a thin film of oil. The film thickness is h and the area of the block is A. When released, mass m exerts tension on the cord, causing the block to accelerate. Neglect friction in the pulley and air resistance. Develop an algebraic expression for the viscous force that acts on the block when it moves at speed V. Derive a differential equation for the block speed as a function of time. Obtain an expression for the block speed as a function of time. The mass $M = 5$ kg, $m = 1$ kg, $A = 25\ \text{cm}^2$, and $h = 0.5$ mm. If it takes 1 s for the speed to reach 1 m/s, find the oil viscosity μ. Plot the curve for $V(t)$.

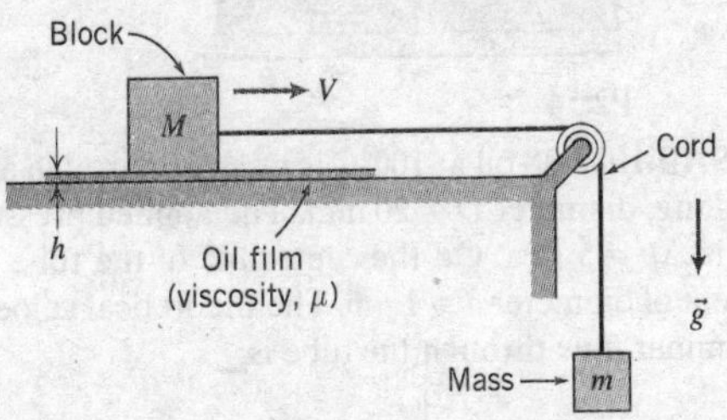

P2.50

2.51 A block 0.1 m square, with 5 kg mass, slides down a smooth incline, 30° below the horizontal, on a film of SAE 30 oil at 20°C that is 0.20 mm thick. If the block is released from rest at $t = 0$, what is its initial acceleration? Derive an expression for the speed of the block as a function of time. Plot the curve for $V(t)$. Find the speed after 0.1 s. If we want the mass to instead reach a speed of 0.3 m/s at this time, find the viscosity μ of the oil we would have to use.

2.52 A block that is a mm square slides across a flat plate on a thin film of oil. The oil has viscosity μ and the film is h mm thick. The block of mass M moves at steady speed U under the influence of constant force F. Indicate the magnitude and direction of the shear stresses on the bottom of the block and the plate. If the force is removed suddenly and the block begins to slow, sketch the resulting speed versus time curve for the block. Obtain an expression for the time required for the block to lose 95 percent of its initial speed.

2.53 Magnet wire is to be coated with varnish for insulation by drawing it through a circular die of 1.0 mm diameter. The wire diameter is 0.9 mm and it is centered in the die.

The varnish ($\mu = 20$ centipoise) completely fills the space between the wire and the die for a length of 50 mm. The wire is drawn through the die at a speed of 50 m/s. Determine the force required to pull the wire.

2.54 In a food-processing plant, honey is pumped through an annular tube. The tube is $L = 2$ m long, with inner and outer radii of $R_i = 5$ mm and $R_o = 25$ mm, respectively. The applied pressure difference is $\Delta p = 125$ kPa, and the honey viscosity is $\mu = 5\ \text{N}\cdot\text{s/m}^2$. The theoretical velocity profile for laminar flow through an annulus is:

$$u_z(r) = \frac{1}{4\mu}\left(\frac{\Delta p}{L}\right)\left[R_i^2 - r^2 - \frac{R_o^2 - R_i^2}{\ln\left(\frac{R_i}{R_o}\right)}\cdot \ln\left(\frac{r}{R_i}\right)\right]$$

Show that the no-slip condition is satisfied by this expression. Find the location at which the shear stress is zero. Find the viscous forces acting on the inner and outer surfaces, and compare these to the force $\Delta p \pi (R_o^2 - R_i^2)$. Explain.

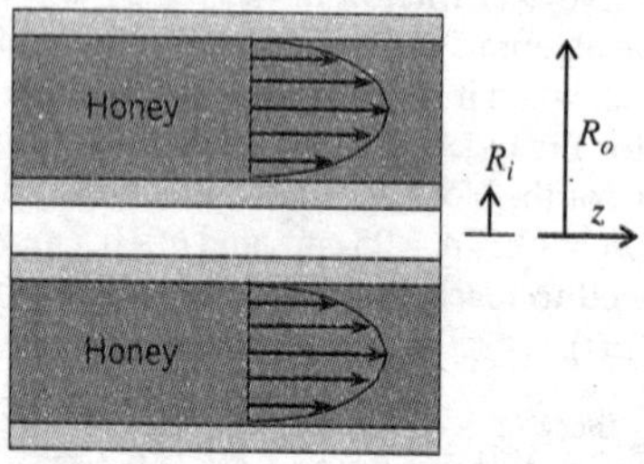

P2.54

2.55 SAE 10W-30 oil at 100°C is pumped through a tube $L =$ 10 m long, diameter $D = 20$ mm. The applied pressure difference is $\Delta p = 5$ kPa. On the centerline of the tube is a metal filament of diameter $d = 1\ \mu$m. The theoretical velocity profile for laminar flow through the tube is:

$$V(r) = \frac{1}{16\mu}\left(\frac{\Delta p}{L}\right)\left[d^2 - 4r^2 - \frac{D^2 - d^2}{\ln\left(\frac{d}{D}\right)}\cdot \ln\left(\frac{2r}{d}\right)\right]$$

Show that the no-slip condition is satisfied by this expression. Find the location at which the shear stress is zero, and the stress on the tube and on the filament. Plot the velocity distribution and the stress distribution. (For the stress curve, set an upper limit on stress of 5 Pa.) Discuss the results.

2.56 Fluids of viscosities $\mu_1 = 0.1\ \text{N}\cdot\text{s/m}^2$ and $\mu_2 = 0.15\ \text{N}\cdot\text{s/m}^2$ are contained between two plates (each plate is 1 m² in area).

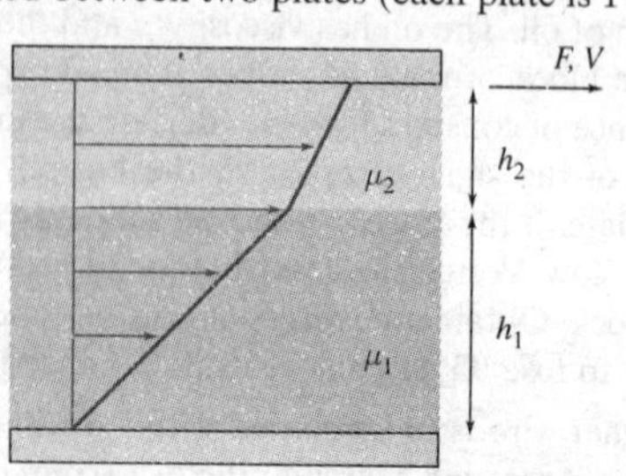

P2.56

The thicknesses are $h_1 = 0.5$ mm and $h_2 = 0.3$ mm, respectively. Find the force F to make the upper plate move at a speed of 1 m/s. What is the fluid velocity at the interface between the two fluids?

2.57 Fluids of viscosities $\mu_1 = 0.15\ \text{N}\cdot\text{s/m}^2$, $\mu_2 = 0.5\ \text{N}\cdot\text{s/m}^2$, and $\mu_3 = 0.2\ \text{N}\cdot\text{s/m}^2$ are contained between two plates (each plate is 1 m² in area). The thicknesses are $h_1 = 0.5$ mm, $h_2 = 0.25$ mm, and $h_3 = 0.2$ mm, respectively. Find the steady speed V of the upper plate and the velocities at the two interfaces due to a force $F = 100$ N. Plot the velocity distribution.

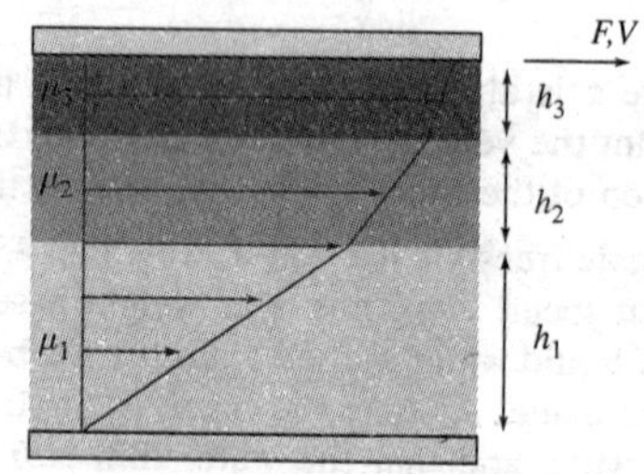

P2.57

2.58 A concentric cylinder viscometer may be formed by rotating the inner member of a pair of closely fitting cylinders. The annular gap is small so that a linear velocity profile will exist in the liquid sample. Consider a viscometer with an inner cylinder of 100 mm diameter and 200 mm height, and a clearance gap width of 0.001 in., filled with castor oil at 32°C. Determine the torque required to turn the inner cylinder at 400 rpm.

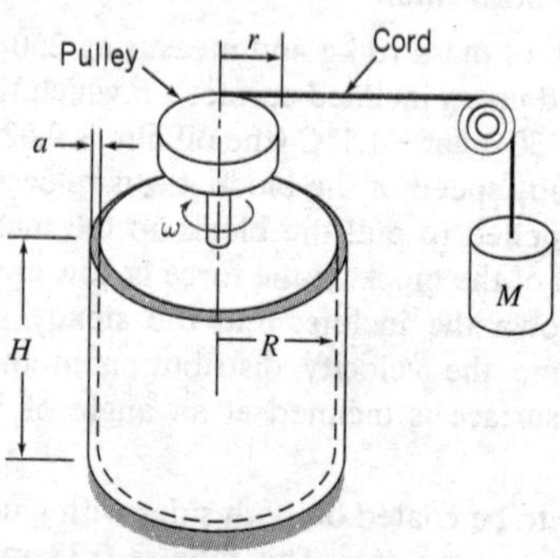

P2.58, P2.59, P2.60, P2.61

2.59 A concentric cylinder viscometer may be formed by rotating the inner member of a pair of closely fitting cylinders. For small clearances, a linear velocity profile may be assumed in the liquid filling the annular clearance gap. A viscometer has an inner cylinder of 75 mm diameter and 150 mm height, with a clearance gap width of 0.02 mm. A torque of 0.021 N · m is required to turn the inner cylinder at 100 rpm. Determine the viscosity of the liquid in the clearance gap of the viscometer.

2.60 A concentric cylinder viscometer is driven by a falling mass M connected by a cord and pulley to the inner cylinder, as shown. The liquid to be tested fills the annular gap of width a and height H. After a brief starting transient, the mass falls at constant speed V_m. Develop an algebraic expression for the viscosity of the liquid in the device in terms of M, g, V_m, r, R, a, and H. Evaluate the viscosity of the liquid using:

$M = 0.10$ kg	$r = 25$ mm
$R = 50$ mm	$a = 0.20$ mm
$H = 80$ mm	$V_m = 30$ mm/s

2.61 The viscometer of Problem 2.60 is being used to verify that the viscosity of a particular fluid is $\mu = 0.1\ \text{N}\cdot\text{s/m}^2$. Unfortunately the cord snaps during the experiment. How long will it take the cylinder to lose 99% of its speed? The moment of inertia of the cylinder/pulley system is $0.0273\ \text{kg}\cdot\text{m}^2$.

2.62 A shaft with outside diameter of 18 mm turns at 20 revolutions per second inside a stationary journal bearing 60 mm long. A thin film of oil 0.2 mm thick fills the concentric annulus between the shaft and journal. The torque needed to turn the shaft is 0.0036 N·m. Estimate the viscosity of the oil that fills the gap.

2.63 The thin outer cylinder (mass m_2 and radius R) of a small portable concentric cylinder viscometer is driven by a falling mass, m_1, attached to a cord. The inner cylinder is stationary. The clearance between the cylinders is a. Neglect bearing friction, air resistance, and the mass of liquid in the viscometer. Obtain an algebraic expression for the torque due to viscous shear that acts on the cylinder at angular speed ω. Derive and solve a differential equation for the angular speed of the outer cylinder as a function of time. Obtain an expression for the maximum angular speed of the cylinder.

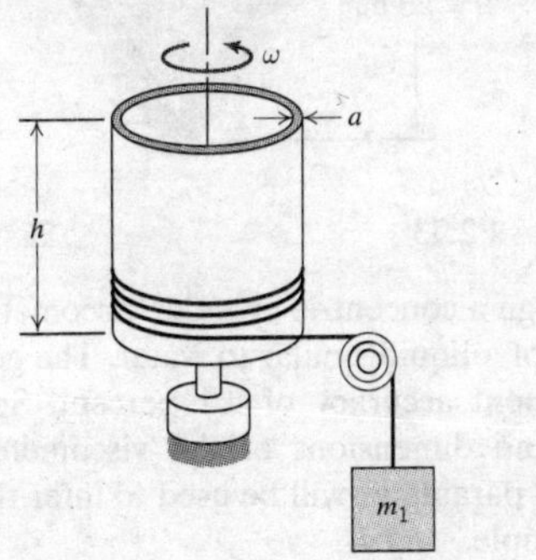

P2.63

2.64 A shock-free coupling for a low-power mechanical drive is to be made from a pair of concentric cylinders. The annular space between the cylinders is to be filled with oil. The drive must transmit power, $\mathcal{P} = 10$ W. Other dimensions and properties are as shown. Neglect any bearing friction and end effects. Assume the minimum practical gap clearance δ for the device is $\delta = 0.25$ mm. Dow manufactures silicone fluids with viscosities as high as 10^6 centipoise. Determine the viscosity that should be specified to satisfy the requirement for this device.

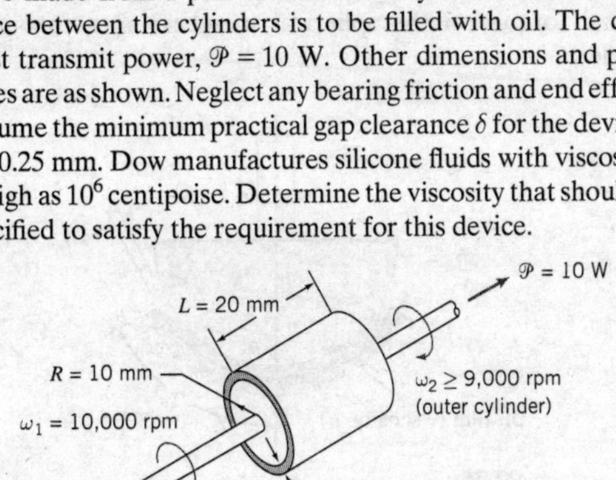

P2.64

2.65 A circular aluminum shaft mounted in a journal is shown. The symmetric clearance gap between the shaft and journal is filled with SAE 10W-30 oil at $T = 30°$C. The shaft is caused to turn by the attached mass and cord. Develop and solve a differential equation for the angular speed of the shaft as a function of time. Calculate the maximum angular speed of the shaft and the time required to reach 95 percent of this speed.

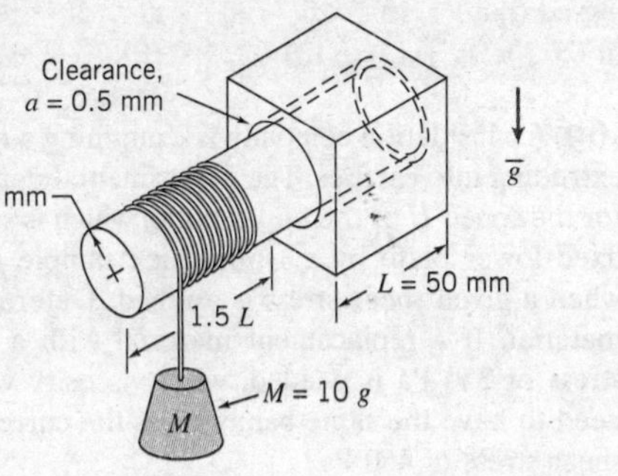

P2.65

2.66 A proposal has been made to use a pair of parallel disks to measure the viscosity of a liquid sample. The upper disk rotates at height h above the lower disk. The viscosity of the liquid in the gap is to be calculated from measurements of the torque needed to turn the upper disk steadily. Obtain an algebraic expression for the torque needed to turn the disk. Could we use this device to measure the viscosity of a non-Newtonian fluid? Explain.

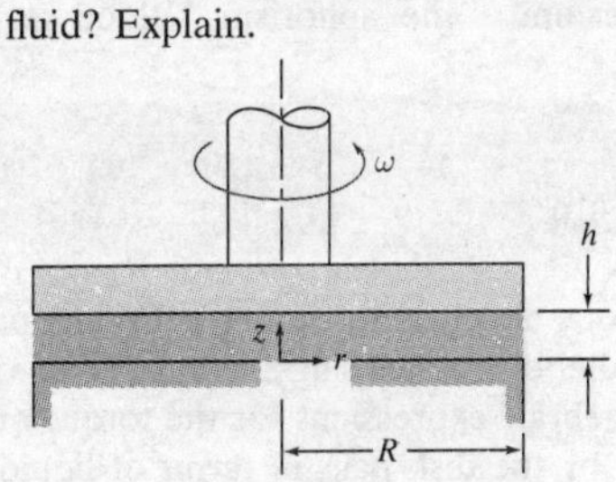

P2.66

2.67 The cone and plate viscometer shown is an instrument used frequently to characterize non-Newtonian fluids. It consists of a flat plate and a rotating cone with a very obtuse angle (typically θ is less than 0.5 degrees). The apex of the cone just touches the plate surface and the liquid to be tested fills the narrow gap formed by the cone and plate. Derive an expression for the shear rate in the liquid that fills the gap in terms of the geometry of the system. Evaluate the torque on the driven cone in terms of the shear stress and geometry of the system.

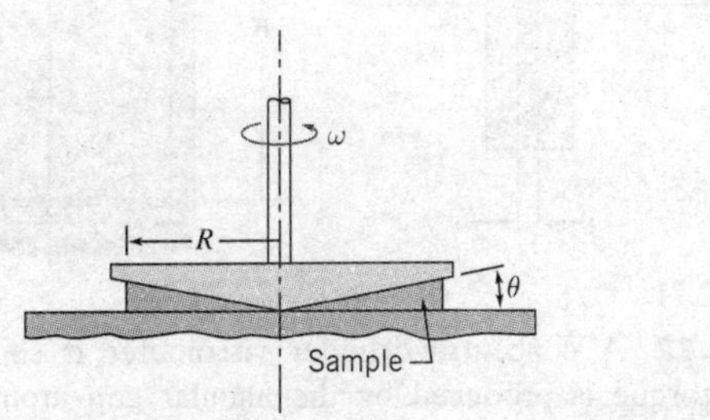

P2.67, P2.68

2.68 The viscometer of Problem 2.67 is used to measure the apparent viscosity of a fluid. The data below are obtained. What kind of non-Newtonian fluid is this? Find the values of k and n used in Eqs. 2.16 and 2.17 in defining the apparent viscosity of a fluid. (Assume θ is 0.5 degrees.) Predict the viscosity at 90 and 100 rpm.

Speed (rpm)	10	20	30	40	50	60	70	80
μ (N · s/m²)	0.121	0.139	0.153	0.159	0.172	0.172	0.183	0.185

2.69 An insulation company is examining a new material for extruding into cavities. The experimental data is given below for the speed U of the upper plate, which is separated from a fixed lower plate by a 1-mm-thick sample of the material, when a given shear stress is applied. Determine the type of material. If a replacement material with a minimum yield stress of 250 Pa is needed, what viscosity will the material need to have the same behavior as the current material at a shear stress of 450 Pa?

τ (Pa)	50	100	150	163	171	170	202	246	349	444
U (m/s)	0	0	0	0.005	0.01	0.025	0.05	0.1	0.2	0.3

2.70 A viscometer is used to measure the viscosity of a patient's blood. The deformation rate (shear rate)–shear stress data is shown below. Plot the apparent viscosity versus deformation rate. Find the value of k and n in Eq. 2.17, and from this examine the aphorism "Blood is thicker than water."

du/dy (s⁻¹)	5	10	25	50	100	200	300	400
τ (Pa)	0.0457	0.119	0.241	0.375	0.634	1.06	1.46	1.78

2.71 A viscous clutch is to be made from a pair of closely spaced parallel disks enclosing a thin layer of viscous liquid. Develop algebraic expressions for the torque and the power transmitted by the disk pair, in terms of liquid viscosity, μ, disk radius, R, disk spacing, a, and the angular speeds: ω_i of the input disk and ω_o of the output disk. Also develop expressions for the slip ratio, $s = \Delta\omega/\omega_i$, in terms of ω_i and the torque transmitted. Determine the efficiency, η, in terms of the slip ratio.

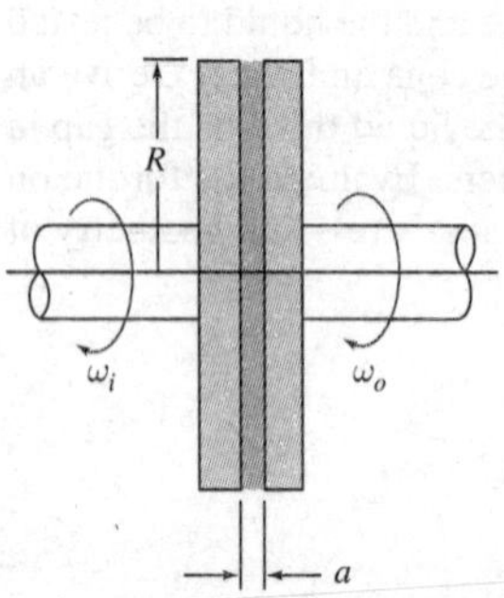

P2.71

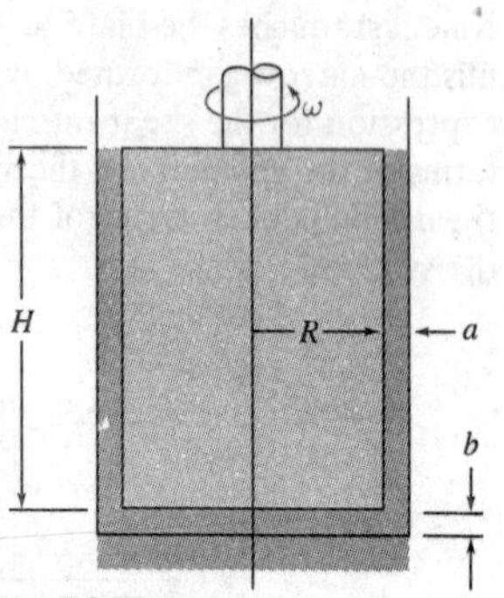

P2.72

2.72 A concentric-cylinder viscometer is shown. Viscous torque is produced by the annular gap around the inner cylinder. Additional viscous torque is produced by the flat bottom of the inner cylinder as it rotates above the flat bottom of the stationary outer cylinder. Obtain an algebraic expression for the viscous torque due to flow in the annular gap of width a. Obtain an algebraic expression for the viscous torque due to flow in the bottom clearance gap of height b. Prepare a plot showing the ratio, b/a, required to hold the bottom torque to 1 percent or less of the annulus torque, versus the other geometric variables. What are the design implications? What modifications to the design can you recommend?

2.73 A viscometer is built from a conical pointed shaft that turns in a conical bearing, as shown. The gap between shaft and bearing is filled with a sample of the test oil. Obtain an algebraic expression for the viscosity μ of the oil as a function of viscometer geometry (H, a, and θ), turning speed ω, and applied torque T. For the data given, find by referring to Figure A.2 in Appendix A, the type of oil for which the applied torque is 0.325 N · m. The oil is at 20°C. *Hint*: First obtain an expression for the shear stress on the surface of the conical shaft as a function of z.

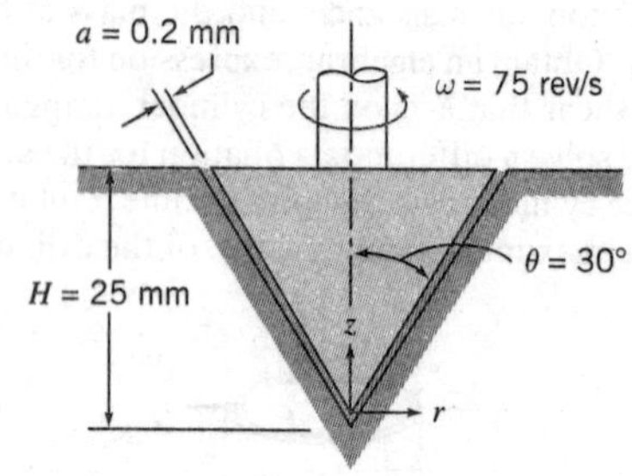

P2.73

2.74 Design a concentric-cylinder viscometer to measure the viscosity of a liquid similar to water. The goal is to achieve a measurement accuracy of ±1 percent. Specify the configuration and dimensions of the viscometer. Indicate what measured parameter will be used to infer the viscosity of the liquid sample.

2.75 A spherical thrust bearing is shown. The gap between the spherical member and the housing is of constant width h. Obtain and plot an algebraic expression for the nondimensional torque on the spherical member, as a function of angle α.

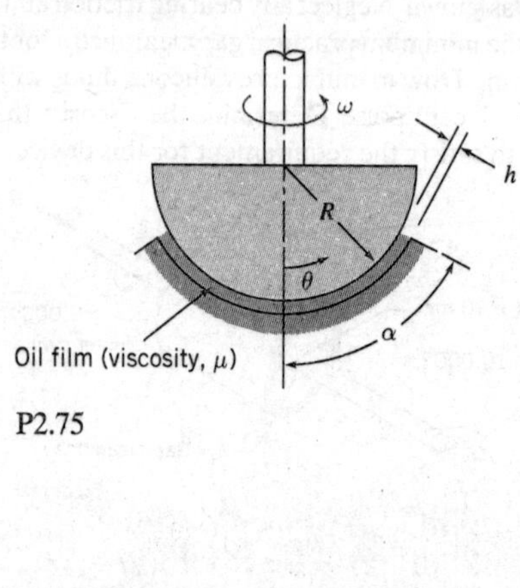

P2.75

2.76 A cross section of a rotating bearing is shown. The spherical member rotates with angular speed ω, a small distance, a, above the plane surface. The narrow gap is filled with viscous oil, having $\mu = 1250$ cp. Obtain an algebraic expression for the shear stress acting on the spherical member. Evaluate the maximum shear stress that acts on the spherical member for the conditions shown. (Is the maximum necessarily located at the maximum radius?) Develop an algebraic expression (in the form of an integral) for the total viscous shear torque that acts on the spherical member. Calculate the torque using the dimensions shown.

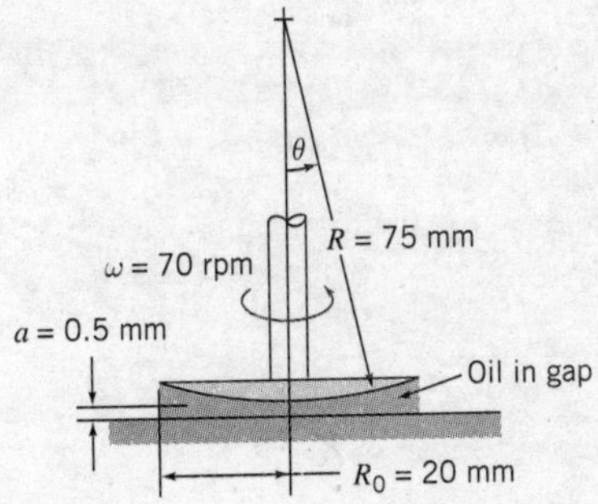

P 2.76

Surface Tension

2.77 Small gas bubbles form in soda when a bottle or can is opened. The average bubble diameter is about 0.1 mm. Estimate the pressure difference between the inside and outside of such a bubble.

2.78 You intend to gently place several steel needles on the free surface of the water in a large tank. The needles come in two lengths: Some are 5 cm long, and some are 10 cm long. Needles of each length are available with diameters of 1 mm, 2.5 mm, and 5 mm. Make a prediction as to which needles, if any, will float.

2.79 According to Folsom [6], the capillary rise Δh (mm) of a water-air interface in a tube is correlated by the following empirical expression:

$$\Delta h = Ae^{-b \cdot D}$$

where D (mm) is the tube diameter, $A = 0.400$, and $b = 4.37$. You do an experiment to measure Δh versus D and obtain:

D (mm)	2.5	5.0	7.5	10.0	12.5	15.0	17.5	20.0	22.5	25.0	27.5
Δh (mm)	5.8	4.6	2.25	1.48	1.30	0.83	0.43	0.25	0.15	0.1	0.08

What are the values of A and b that best fit this data using *Excel*'s *Trendline* feature? Do they agree with Folsom's values? How good is the data?

2.80 Slowly fill a glass with water to the maximum possible level. Observe the water level closely. Explain how it can be higher than the rim of the glass.

2.81 Plan an experiment to measure the surface tension of a liquid similar to water. If necessary, review the NCFMF video *Surface Tension* for ideas. Which method would be most suitable for use in an undergraduate laboratory? What experimental precision could be expected?

Description and Classification of Fluid Motions

2.82 Water usually is assumed to be incompressible when evaluating static pressure variations. Actually it is 100 times more compressible than steel. Assuming the bulk modulus of water is constant, compute the percentage change in density for water raised to a gage pressure of 10 MPa. Plot the percentage change in water density as a function of p/p_{atm} up to a pressure of 350 MPa which is the approximate pressure used for high-speed cutting jets of water to cut concrete and other composite materials. Would constant density be a reasonable assumption for engineering calculations for cutting jets?

2.83 The viscous boundary layer velocity profile shown in Fig. 2.15 can be approximated by a parabolic equation,

$$u(y) = a + b\left(\frac{y}{\delta}\right) + c\left(\frac{y}{\delta}\right)^2$$

The boundary condition is $u = U$ (the free stream velocity) at the boundary edge δ (where the viscous friction becomes zero). Find the values of a, b, and c.

2.84 The viscous boundary layer velocity profile shown in Fig. 2.15 can be approximated by a cubic equation,

$$u(y) = a + b\left(\frac{y}{\delta}\right) + c\left(\frac{y}{\delta}\right)^3$$

The boundary condition is $u = U$ (the free stream velocity) at the boundary edge δ (where the viscous friction becomes zero). Find the values of a, b, and c.

2.85 At what minimum speed (in km/hr) would an automobile have to travel for compressibility effects to be important? Assume the local air temperature is 15.5°C.

2.86 In a food industry process, carbon tetrachloride at 20°C flows through a tapered nozzle from an inlet diameter $D_{in} = 50$ mm to an outlet diameter of D_{out}. The area varies linearly with distance along the nozzle, and the exit area is one-fifth of the inlet area; the nozzle length is 250 mm. The flow rate is $Q = 2$ L/min. It is important for the process that the flow exits the nozzle as a turbulent flow. Does it? If so, at what point along the nozzle does the flow become turbulent?

2.87 What is the Reynolds number of water at 20°C flowing at 0.25 m/s through a 5 mm diameter tube? If the pipe is now heated, at what mean water temperature will the flow transition to turbulence? Assume the velocity of the flow remains constant.

2.88 A supersonic aircraft travels at 2700 km/hr at an altitude of 27 km. What is the Mach number of the aircraft? At what approximate distance measured from the leading edge of the aircraft's wing does the boundary layer change from laminar to turbulent?

2.89 SAE 30 oil at 100°C flows through a 12 mm diameter stainless-steel tube. What is the specific gravity and specific weight of the oil? If the oil discharged from the tube fills a 100-mL graduated cylinder in 9 seconds, is the flow laminar or turbulent?

2.90 A seaplane is flying at 160 km/h through air at 7°C. At what distance from the leading edge of the underside of the fuselage does the boundary layer transition to turbulence? How does this boundary layer transition change as the underside of the fuselage touches the water during landing? Assume the water temperature is also 7°C.

2.91 An airliner is cruising at an altitude of 5.5 km with a speed of 700 km/hr. As the airliner increases its altitude, it adjusts its speed so that the Mach number remains constant. Provide a sketch of speed vs. altitude. What is the speed of the airliner at an altitude of 8 km?

2.92 How does an airplane wing develop lift?

3 Fluid Statics

Case Study in Energy and the Environment

Wave Power: *Wavebob*

Humans have been interested in tapping the immense power of the ocean for centuries, but with fossil fuels (oil and gas) becoming depleted, the development of ocean energy technology is becoming important. Wave power in particular is attractive to a number of countries with access to a suitable resource. Geographically and commercially it's believed the richest wave energy resources currently are off the Atlantic coast of Europe (in particular near Ireland, the UK, and Portugal), the west coast of North America (from San Francisco to British Columbia), Hawaii, and New Zealand.

A family of devices called *point absorbers* is being developed by a number of companies. These are usually axisymmetric about a vertical axis, and by definition they are small compared to the wavelength of the waves that they are designed to exploit. The devices usually operate in a vertical mode, often referred to as *heave*; a surface-piercing float rises and falls with the passing waves and reacts against either the seabed or something attached to it. These devices ultimately depend on a buoyancy force, one of the topics of this chapter.

A company named *Wavebob Ltd.* has developed one of the simplest of these devices. This innovative eponymous device, as shown in the figure, is proving to be successful for extracting wave energy. The figure does not indicate the size of the device, but it is quite large; the upper champer has a diameter of 20 m. It looks like just another buoy floating on the surface, but underneath it is constantly harvesting energy. The lower component of the *Wavebob* is tethered to the ocean floor and so remains in its vertical location, while the section at the surface oscillates as the waves move over it. Hence the distance between the two components is constantly changing, with a significant force between them; work can thus be done on an electrical generator. The two components of the machinery contain electronic systems that can be controlled remotely or self-regulating, and these make the internal mechanism automatically react to changing ocean and wave conditions by retuning as needed, so that at all times the maximum amount of energy is harvested.

It has already been tested in the Atlantic Ocean off the coast of Ireland and is designed to have a 25-year life span and to be able to survive all but the very worst storms. Each *Wavebob* is expected to produce about 500 kW of power or more, sufficient electricity for over a thousand homes; it is intended to be part of a large array of such devices. It seems likely this device will become ubiquitous because it is relatively inexpensive, very low maintenance, and durable, and it takes up only a small area.

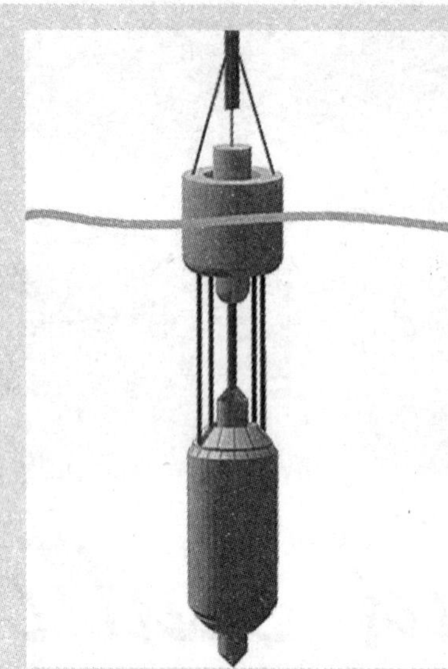

Schematic of *Wavebob* (Picture courtesy of Gráinne Byrne, *Wavebob* Ltd.)

In Chapter 1, we defined a fluid as any substance that flows (continuously deforms) when it experiences a shear stress; hence for a static fluid (or one undergoing "rigid-body" motion) only normal stress is present—in other words, pressure. We will study the topic of fluid statics (often called *hydrostatics*, even though it is not restricted to water) in this chapter.

Although fluid statics problems are the simplest kind of fluid mechanics problems, this is not the only reason we will study them. The pressure generated within a static fluid is an important phenomenon in many practical situations. Using the principles of hydrostatics, we can compute forces on submerged objects, develop instruments for measuring pressures, and deduce properties of the atmosphere and oceans. The principles of hydrostatics also may be used to determine the forces developed by hydraulic systems in applications such as industrial presses or automobile brakes.

In a static, homogeneous fluid, or in a fluid undergoing rigid-body motion, a fluid particle retains its identity for all time, and fluid elements do not deform. We may apply Newton's second law of motion to evaluate the forces acting on the particle.

3.1 The Basic Equation of Fluid Statics

The first objective of this chapter is to obtain an equation for computing the pressure field in a static fluid. We will deduce what we already know from everyday experience, that the pressure increases with depth. To do this, we apply Newton's second law to a differential fluid element of mass $dm = \rho\, d\mathcal{V}$, with sides dx, dy, and dz, as shown

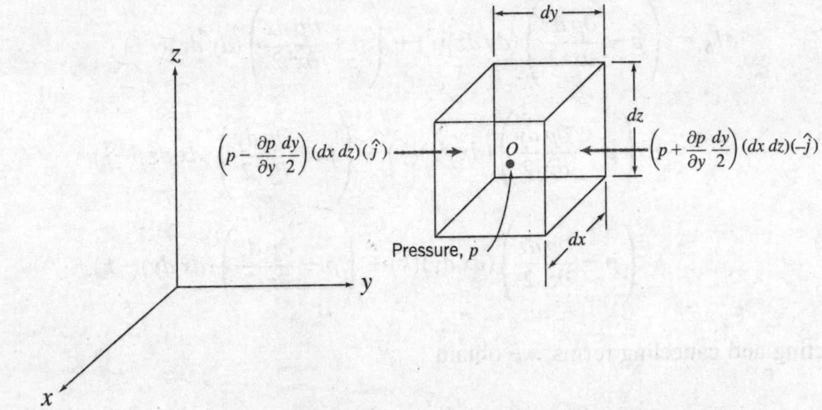

Fig. 3.1 Differential fluid element and pressure forces in the y direction.

in Fig. 3.1. The fluid element is stationary relative to the stationary rectangular coordinate system shown. (Fluids in rigid-body motion will be treated in Section 3.7 on the Web.)

From our previous discussion, recall that two general types of forces may be applied to a fluid: body forces and surface forces. The only body force that must be considered in most engineering problems is due to gravity. In some situations body forces caused by electric or magnetic fields might be present; they will not be considered in this text.

> CLASSIC VIDEO
> Magnetohydrodynamics.

For a differential fluid element, the body force is

$$d\vec{F}_B = \vec{g}\,dm = \vec{g}\rho\,d\mathcal{V}$$

where $\vec{g}$ is the local gravity vector, ρ is the density, and $d\mathcal{V}$ is the volume of the element. In Cartesian coordinates $d\mathcal{V} = dx\,dy\,dz$, so

$$d\vec{F}_B = \rho\vec{g}\,dx\,dy\,dz$$

In a static fluid there are no shear stresses, so the only surface force is the pressure force. Pressure is a scalar field, $p = p(x, y, z)$; in general we expect the pressure to vary with position within the fluid. The net pressure force that results from this variation can be found by summing the forces that act on the six faces of the fluid element.

Let the pressure be p at the center, O, of the element. To determine the pressure at each of the six faces of the element, we use a Taylor series expansion of the pressure about point O. The pressure at the left face of the differential element is

$$p_L = p + \frac{\partial p}{\partial y}(y_L - y) = p + \frac{\partial p}{\partial y}\left(-\frac{dy}{2}\right) = p - \frac{\partial p}{\partial y}\frac{dy}{2}$$

(Terms of higher order are omitted because they will vanish in the subsequent limiting process.) The pressure on the right face of the differential element is

$$p_R = p + \frac{\partial p}{\partial y}(y_R - y) = p + \frac{\partial p}{\partial y}\frac{dy}{2}$$

The pressure *forces* acting on the two y surfaces of the differential element are shown in Fig. 3.1. Each pressure force is a product of three factors. The first is the magnitude of the pressure. This magnitude is multiplied by the area of the face to give the magnitude of the pressure force, and a unit vector is introduced to indicate direction. Note also in Fig. 3.1 that the pressure force on each face acts *against* the face. A positive pressure corresponds to a *compressive* normal stress.

Pressure forces on the other faces of the element are obtained in the same way. Combining all such forces gives the net surface force acting on the element. Thus

$$d\vec{F}_S = \left(p - \frac{\partial p}{\partial x}\frac{dx}{2}\right)(dy\,dz)(\hat{i}) + \left(p + \frac{\partial p}{\partial x}\frac{dx}{2}\right)(dy\,dz)(-\hat{i})$$

$$+ \left(p - \frac{\partial p}{\partial y}\frac{dy}{2}\right)(dx\,dz)(\hat{j}) + \left(p + \frac{\partial p}{\partial y}\frac{dy}{2}\right)(dx\,dz)(-\hat{j})$$

$$+ \left(p - \frac{\partial p}{\partial z}\frac{dz}{2}\right)(dx\,dy)(\hat{k}) + \left(p + \frac{\partial p}{\partial z}\frac{dz}{2}\right)(dx\,dy)(-\hat{k})$$

Collecting and canceling terms, we obtain

$$d\vec{F}_S = -\left(\frac{\partial p}{\partial x}\hat{i} + \frac{\partial p}{\partial y}\hat{j} + \frac{\partial p}{\partial z}\hat{k}\right)dx\,dy\,dz \tag{3.1a}$$

The term in parentheses is called the gradient of the pressure or simply the pressure gradient and may be written grad p or ∇p. In rectangular coordinates

$$\text{grad}\, p \equiv \nabla p \equiv \left(\hat{i}\frac{\partial p}{\partial x} + \hat{j}\frac{\partial p}{\partial y} + \hat{k}\frac{\partial p}{\partial z}\right) \equiv \left(\hat{i}\frac{\partial}{\partial x} + \hat{j}\frac{\partial}{\partial y} + \hat{k}\frac{\partial}{\partial z}\right)p$$

The gradient can be viewed as a vector operator; taking the gradient of a scalar field gives a vector field. Using the gradient designation, Eq. 3.1a can be written as

$$d\vec{F}_S = -\text{grad}\, p\,(dx\,dy\,dz) = -\nabla p\,dx\,dy\,dz \tag{3.1b}$$

Physically the gradient of pressure is the negative of the surface force per unit volume due to pressure. Note that the pressure magnitude itself is not relevant in computing the net pressure force; instead what counts is the rate of change of pressure with distance, the *pressure gradient*. We shall encounter this term throughout our study of fluid mechanics.

We combine the formulations for surface and body forces that we have developed to obtain the total force acting on a fluid element. Thus

$$d\vec{F} = d\vec{F}_S + d\vec{F}_B = (-\nabla p + \rho\vec{g})\,dx\,dy\,dz = (-\nabla p + \rho\vec{g})\,d\forall$$

or on a per unit volume basis

$$\frac{d\vec{F}}{d\forall} = -\nabla p + \rho\vec{g} \tag{3.2}$$

For a fluid particle, Newton's second law gives $\vec{F} = \vec{a}\,dm = \vec{a}\rho d\forall$. For a static fluid, $\vec{a} = 0$. Thus

$$\frac{d\vec{F}}{d\forall} = \rho\vec{a} = 0$$

Substituting for $d\vec{F}/d\forall$ from Eq. 3.2, we obtain

$$-\nabla p + \rho\vec{g} = 0 \tag{3.3}$$

Let us review this equation briefly. The physical significance of each term is

$$\begin{array}{ccccc} -\nabla p & + & \rho\vec{g} & = 0 \\ \left\{\begin{array}{c}\text{net pressure force}\\ \text{per unit volume}\\ \text{at a point}\end{array}\right\} & + & \left\{\begin{array}{c}\text{body force per}\\ \text{unit volume}\\ \text{at a point}\end{array}\right\} & = 0 \end{array}$$

This is a vector equation, which means that it is equivalent to three component equations that must be satisfied individually. The component equations are

$$\left.\begin{aligned} -\frac{\partial p}{\partial x} + \rho g_x &= 0 \quad x \text{ direction} \\ -\frac{\partial p}{\partial y} + \rho g_y &= 0 \quad y \text{ direction} \\ -\frac{\partial p}{\partial z} + \rho g_z &= 0 \quad z \text{ direction} \end{aligned}\right\} \tag{3.4}$$

Equations 3.4 describe the pressure variation in each of the three coordinate directions in a static fluid. It is convenient to choose a coordinate system such that the gravity vector is aligned with one of the coordinate axes. If the coordinate system is chosen with the z axis directed vertically upward, as in Fig. 3.1, then $g_x = 0$, $g_y = 0$, and $g_z = -g$. Under these conditions, the component equations become

$$\frac{\partial p}{\partial x} = 0 \qquad \frac{\partial p}{\partial y} = 0 \qquad \frac{\partial p}{\partial z} = -\rho g \tag{3.5}$$

Equations 3.5 indicate that, under the assumptions made, the pressure is independent of coordinates x and y; it depends on z alone. Thus since p is a function of a single variable, a total derivative may be used instead of a partial derivative. With these simplifications, Eqs. 3.5 are finally reduced to

$$\frac{dp}{dz} = -\rho g \equiv -\gamma \tag{3.6}$$

Restrictions: (1) Static fluid.
(2) Gravity is the only body force.
(3) The z axis is vertical and upward.

In Eq. 3.6, γ is the specific weight of the fluid. This equation is the basic pressure-height relation of fluid statics. It is subject to the restrictions noted. Therefore it must be applied only where these restrictions are reasonable for the physical situation. To determine the pressure distribution in a static fluid, Eq. 3.6 may be integrated and appropriate boundary conditions applied.

Before considering specific applications of this equation, it is important to remember that pressure values must be stated with respect to a reference level. If the reference level is a vacuum, pressures are termed *absolute*, as shown in Fig. 3.2.

Most pressure gages indicate a pressure *difference*—the difference between the measured pressure and the ambient level (usually atmospheric pressure). Pressure levels measured with respect to atmospheric pressure are termed *gage* pressures. Thus

$$p_{\text{gage}} = p_{\text{absolute}} - p_{\text{atmosphere}}$$

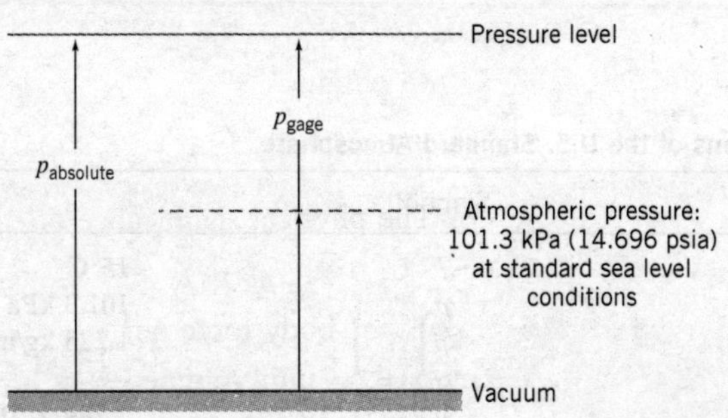

Fig. 3.2 Absolute and gage pressures, showing reference levels.

For example, a tire gage might indicate 207 kPa; the absolute pressure would be about 308 kPa. Absolute pressures must be used in all calculations with the ideal gas equation or other equations of state.

3.2 The Standard Atmosphere

Scientists and engineers sometimes need a numerical or analytical model of the Earth's atmosphere in order to simulate climate variations to study, for example, effects of global warming. There is no single standard model. An International Standard Atmosphere (ISA) has been defined by the International Civil Aviation Organization (ICAO); there is also a similar U.S. Standard Atmosphere.

The temperature profile of the U.S. Standard Atmosphere is shown in Fig. 3.3. Additional property values are tabulated as functions of elevation in Appendix A. Sea level conditions of the U.S. Standard Atmosphere are summarized in Table 3.1.

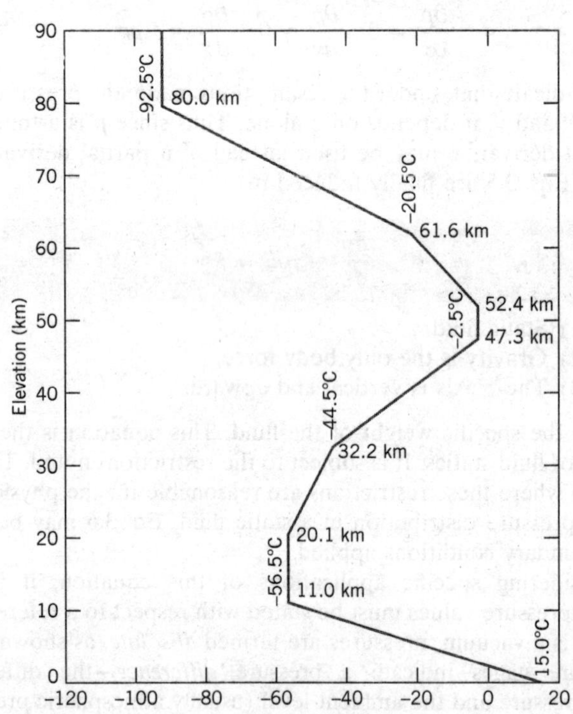

Fig. 3.3 Temperature variation with altitude in the U.S. Standard Atmosphere.

Table 3.1

Sea Level Conditions of the U.S. Standard Atmosphere

Property	Symbol	SI
Temperature	T	15°C
Pressure	p	101.3 kPa (abs)
Density	ρ	1.225 kg/m^3
Specific weight	γ	—
Viscosity	μ	1.789×10^{-5} kg/(m·s) (Pa·s)

Pressure Variation in a Static Fluid 3.3

We proved that pressure variation in any static fluid is described by the basic pressure-height relation

$$\frac{dp}{dz} = -\rho g \tag{3.6}$$

Although ρg may be defined as the specific weight, γ, it has been written as ρg in Eq. 3.6 to emphasize that *both* ρ and g must be considered variables. In order to integrate Eq. 3.6 to find the pressure distribution, we need information about variations in both ρ and g.

For most practical engineering situations, the variation in g is negligible. Only for a purpose such as computing very precisely the pressure change over a large elevation difference would the variation in g need to be included. Unless we state otherwise, we shall assume g to be constant with elevation at any given location.

Incompressible Liquids: Manometers

For an incompressible fluid, $\rho =$ constant. Then for constant gravity,

$$\frac{dp}{dz} = -\rho g = \text{constant}$$

To determine the pressure variation, we must integrate and apply appropriate boundary conditions. If the pressure at the reference level, z_0, is designated as p_0, then the pressure, p, at level z is found by integration:

$$\int_{p_0}^{p} dp = -\int_{z_0}^{z} \rho g\, dz$$

or

$$p - p_0 = -\rho g(z - z_0) = \rho g(z_0 - z)$$

For liquids, it is often convenient to take the origin of the coordinate system at the free surface (reference level) and to measure distances as positive downward from the free surface as in Fig. 3.4.

With h measured positive downward, we have

$$z_0 - z = h$$

and obtain

$$p - p_0 = \Delta p = \rho g h \tag{3.7}$$

Equation 3.7 indicates that the pressure difference between two points in a static incompressible fluid can be determined by measuring the elevation difference between the two points. Devices used for this purpose are called *manometers*.

Use of Eq. 3.7 for a manometer is illustrated in Example 3.1.

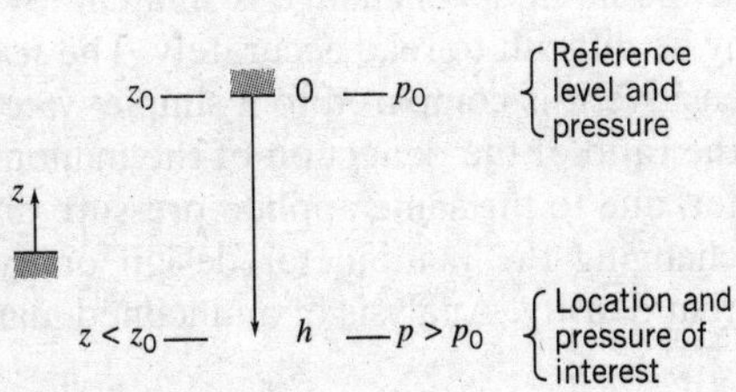

Fig. 3.4 Use of z and h coordinates.

Example 3.1 SYSTOLIC AND DIASTOLIC PRESSURE

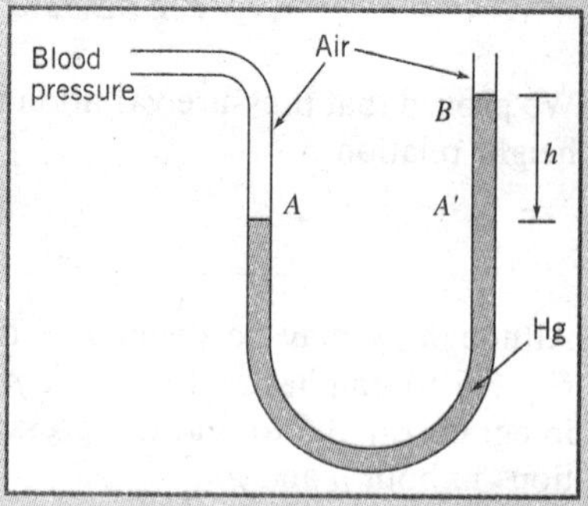

Normal blood pressure for a human is 120/80 mm Hg. By modeling a sphygmomanometer pressure gage as a U-tube manometer, convert these pressures to kPa.

Given: Gage pressures of 120 and 80 mm Hg.

Find: The corresponding pressures in kPa.

Solution:

Apply hydrostatic equation to points A, A', and B.

Governing equation:

$$p - p_0 = \Delta p = \rho g h \tag{3.7}$$

Assumptions: (1) Static fluid.
(2) Incompressible fluids.
(3) Neglect air density ($\ll$ Hg density).

Applying the governing equation between points A' and B (and p_B is atmospheric and therefore zero gage):

$$p_{A'} = p_B + \rho_{Hg} g h = SG_{Hg} \rho_{H_2O} g h$$

In addition, the pressure increases as we go downward from point A' to the bottom of the manometer, and decreases by an equal amount as we return up the left branch to point A. This means points A and A' have the same pressure, so we end up with

$$p_A = p_{A'} = SG_{Hg} \rho_{H_2O} g h$$

Substituting $SG_{Hg} = 13.6$ and $\rho_{H_2O} = 1000 \text{ kg/m}^3$ from Appendix A.1 yields for the systolic pressure ($h = 120$ mm Hg)

$$p_{systolic} = p_A = 13.6 \times 1000 \frac{\text{kg}}{\text{m}^3} \times 9.81 \frac{\text{m}}{\text{s}^2} \times 120 \frac{\text{m}}{100} \times \frac{\text{N} \cdot \text{s}^2}{\text{kg} \cdot \text{m}}$$

$$= 16{,}000 \frac{\text{N}}{\text{m}^2} = 16 \text{ kPa}$$

By a similar process, the diastolic pressure ($h = 80$ mm Hg) is

$$p_{diastolic} = 10.67 \text{ kPa} \longleftarrow p_{diastolic}$$

Notes:

- ✓ Two points at the same level in a continuous single fluid have the same pressure.
- ✓ In manometer problems we neglect change in pressure with depth for a gas: $\rho_{gas} \ll \rho_{liquid}$.
- ✓ This problem shows the conversion from mm Hg to psi, using Eq. 3.7: 120 mm Hg is equivalent to about 2.32 psi. More generally, 1 atm = 14.7 psi = 101 kPa = 760 mm Hg.

Manometers are simple and inexpensive devices used frequently for pressure measurements. Because the liquid level change is small at low pressure differential, a U-tube manometer may be difficult to read accurately. The *sensitivity* of a manometer is a measure of how sensitive it is compared to a simple water-filled U-tube manometer. Specifically, it is the ratio of the deflection of the manometer to that of a water-filled U-tube manometer, due to the same applied pressure difference Δp. Sensitivity can be increased by changing the manometer design or by using two immiscible liquids of slightly different density. Analysis of an inclined manometer is illustrated in Example 3.2.

Example 3.2 ANALYSIS OF INCLINED-TUBE MANOMETER

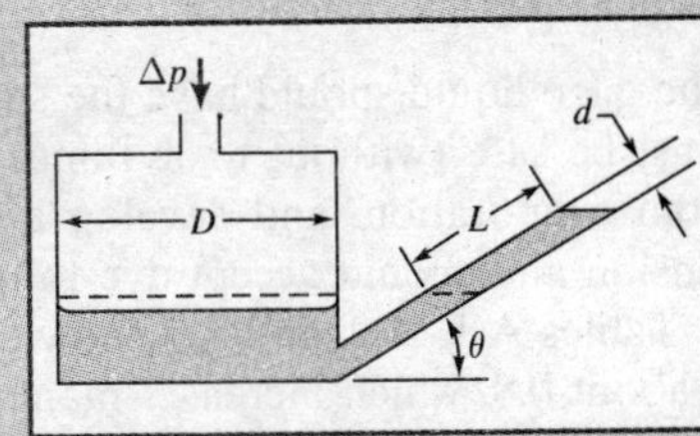

An inclined-tube reservoir manometer is constructed as shown. Derive a general expression for the liquid deflection, L, in the inclined tube, due to the applied pressure difference, Δp. Also obtain an expression for the manometer sensitivity, and discuss the effect on sensitivity of D, d, θ, and SG.

Given: Inclined-tube reservoir manometer.

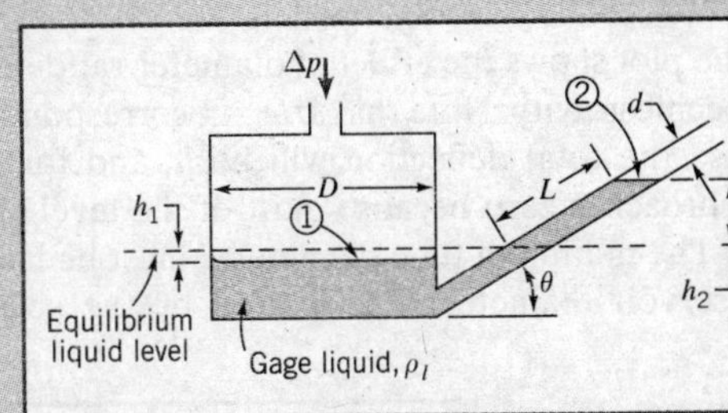

Find: Expression for L in terms of Δp.
General expression for manometer sensitivity.
Effect of parameter values on sensitivity.

Solution:
Use the equilibrium liquid level as a reference.

Governing equations: $p - p_0 = \Delta p = \rho g h \qquad SG = \dfrac{\rho}{\rho_{H_2O}}$

Assumptions: (1) Static fluid.
(2) Incompressible fluid.

Applying the governing equation between points *1* and *2*

$$p_1 - p_2 = \Delta p = \rho_l g(h_1 + h_2) \tag{1}$$

To eliminate h_1, we recognize that the *volume* of manometer liquid remains constant; the volume displaced from the reservoir must equal the volume that rises in the tube, so

$$\frac{\pi D^2}{4} h_1 = \frac{\pi d^2}{4} L \quad \text{or} \quad h_1 = L\left(\frac{d}{D}\right)^2$$

In addition, from the geometry of the manometer, $h_2 = L \sin\theta$. Substituting into Eq. 1 gives

$$\Delta p = \rho_l g \left[L\sin\theta + L\left(\frac{d}{D}\right)^2 \right] = \rho_l g L \left[\sin\theta + \left(\frac{d}{D}\right)^2 \right]$$

Thus

$$L = \frac{\Delta p}{\rho_l g \left[\sin\theta + \left(\dfrac{d}{D}\right)^2 \right]} \qquad L$$

To find the sensitivity of the manometer, we need to compare this to the deflection h a simple U-tube manometer, using water (density ρ), would experience,

$$h = \frac{\Delta p}{\rho g}$$

The sensitivity s is then

$$s = \frac{L}{h} = \frac{1}{SG_l \left[\sin\theta + \left(\dfrac{d}{D}\right)^2 \right]} \qquad s$$

where we have used $SG_l = \rho_l/\rho$. This result shows that to increase sensitivity, SG_l, $\sin\theta$, and d/D each should be made as small as possible. Thus the designer must choose a gage liquid and two geometric parameters to complete a design, as discussed below.

Gage Liquid

The gage liquid should have the smallest possible specific gravity to increase sensitivity. In addition, the gage liquid must be safe (without toxic fumes or flammability), be immiscible with the fluid being gaged, suffer minimal loss from evaporation, and develop a satisfactory meniscus. Thus the gage liquid should have relatively low surface tension and should accept dye to improve its visibility.

Tables A.1, A.2, and A.4 show that hydrocarbon liquids satisfy many of these criteria. The lowest specific gravity is about 0.8, which increases manometer sensitivity by 25 percent compared to water.

Diameter Ratio

The plot shows the effect of diameter ratio on sensitivity for a vertical reservoir manometer with gage liquid of unity specific gravity. Note that $d/D = 1$ corresponds to an ordinary U-tube manometer; its sensitivity is 0.5 because for this case the total deflection will be h, and for each side it will be $h/2$, so $L = h/2$. Sensitivity doubles to 1.0 as d/D approaches zero because most of the level change occurs in the measuring tube.

The minimum tube diameter d must be larger than about 6 mm to avoid excessive capillary effect. The maximum reservoir diameter D is limited by the size of the manometer. If D is set at 60 mm, so that d/D is 0.1, then

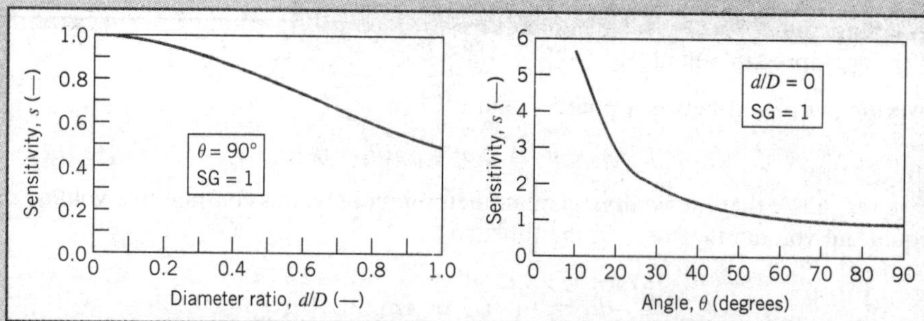

$(d/D)^2 = 0.01$, and the sensitivity increases to 0.99, very close to the maximum attainable value of 1.0.

Inclination Angle

The final plot shows the effect of inclination angle on sensitivity for $d/D = 0$. Sensitivity increases sharply as inclination angle is reduced below 30 degrees. A practical limit is reached at about 10 degrees: The meniscus becomes indistinct and the level hard to read for smaller angles.

Summary

Combining the best values ($SG = 0.8$, $d/D = 0.1$, and $\theta = 10$ degrees) gives a manometer sensitivity of 6.81. Physically this is the ratio of observed gage liquid deflection to equivalent water column height. Thus the deflection in the inclined tube is amplified 6.81 times compared to a vertical water column. With improved sensitivity, a small pressure difference can be read more accurately than with a water manometer, or a smaller pressure difference can be read with the same accuracy.

The graphs were generated from the *Excel* workbook for this Example. This workbook has more detailed graphs, showing sensitivity curves for a range of values of d/D and θ.

Students sometimes have trouble analyzing multiple-liquid manometer situations. The following rules of thumb are useful:

1. Any two points at the same elevation in a continuous region of the same liquid are at the same pressure.
2. Pressure increases as one goes *down* a liquid column (remember the pressure change on diving into a swimming pool).

To find the pressure difference Δp between two points separated by a series of fluids, we can use the following modification of Eq. 3.7:

$$\Delta p = g \sum_i \rho_i h_i \tag{3.8}$$

where ρ_i and h_i represent the densities and depths of the various fluids, respectively. Use care in applying signs to the depths h_i; they will be positive downwards, and negative upwards. Example 3.3 illustrates the use of a multiple-liquid manometer for measuring a pressure difference.

Example 3.3 MULTIPLE-LIQUID MANOMETER

Water flows through pipes A and B. Lubricating oil is in the upper portion of the inverted U. Mercury is in the bottom of the manometer bends. Determine the pressure difference, $p_A - p_B$, in units of kPa.

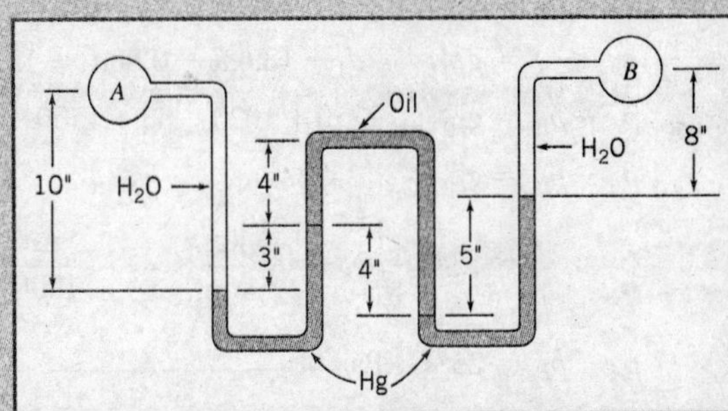

Given: Multiple-liquid manometer as shown.

Find: Pressure difference, $p_A - p_B$, in kPa.

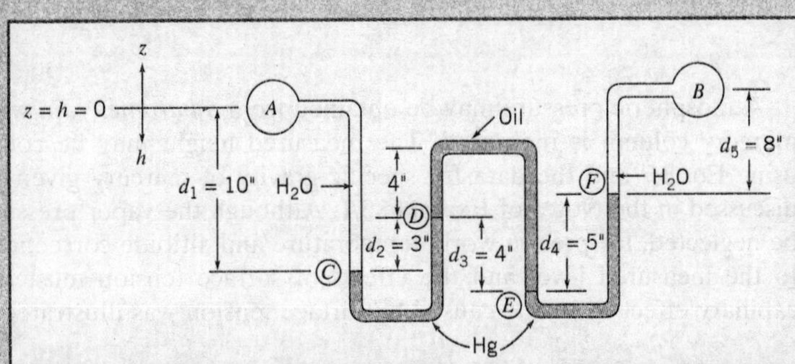

Solution:

Governing equations: $\Delta p = g \sum_i \rho_i h_i \qquad SG = \dfrac{\rho}{\rho_{H_2O}}$

Assumptions: (1) Static fluid.
(2) Incompressible fluid.

Applying the governing equation, working from point B to A

$$p_A - p_B = \Delta p = g(\rho_{H_2O} d_5 + \rho_{Hg} d_4 - \rho_{oil} d_3 + \rho_{Hg} d_2 - \rho_{H_2O} d_1) \tag{1}$$

This equation can also be derived by repeatedly using Eq. 3.7 in the following form:

$$p_2 - p_1 = \rho g (h_2 - h_1)$$

Beginning at point A and applying the equation between successive points along the manometer gives

$$p_C - p_A = +\rho_{H_2O} g d_1$$
$$p_D - p_C = -\rho_{Hg} g d_2$$
$$p_E - p_D = +\rho_{oil} g d_3$$
$$p_F - p_E = -\rho_{Hg} g d_4$$
$$p_B - p_F = -\rho_{H_2O} g d_5$$

Multiplying each equation by minus one and adding, we obtain Eq. (1)

$$p_A - p_B = (p_A - p_C) + (p_C - p_D) + (p_D - p_E) + (p_E - p_F) + (p_F - p_B)$$
$$= -\rho_{H_2O} g d_1 + \rho_{Hg} g d_2 - \rho_{oil} g d_3 + \rho_{Hg} g d_4 + \rho_{H_2O} g d_5$$

Substituting $\rho = SG\rho_{H_2O}$ with $SG_{Hg} = 13.6$ and $SG_{oil} = 0.88$ (Table A.2), yields

$$p_A - p_B = g(-\rho_{H_2O} d_1 + 13.6\rho_{H_2O} d_2 - 0.88\rho_{H_2O} d_3 + 13.6\rho_{H_2O} d_4 + \rho_{H_2O} d_5)$$
$$= g\rho_{H_2O}(-d_1 + 13.6 d_2 - 0.88 d_3 + 13.6 d_4 + d_5)$$
$$p_A - p_B = g\rho_{H_2O}(-250 + 1020 - 88 + 1700 + 200)\text{ mm}$$
$$p_A - p_B = g\rho_{H_2O} \times 2582\text{ mm}$$
$$= 9.81\,\frac{\text{m}}{\text{s}^2} \times 1000\,\frac{\text{kg}}{\text{m}^3} \times 2582\,\frac{\text{m}}{1000} \times \frac{\text{N}\cdot\text{s}^2}{\text{kg}\cdot\text{m}}$$
$$p_A - p_B = 25.33\text{ kPa} \longleftarrow p_A - p_B$$

This Example shows use of both Eq. 3.7 and Eq. 3.8. Use of either equation is a matter of personal preference.

Atmospheric pressure may be obtained from a *barometer*, in which the height of a mercury column is measured. The measured height may be converted to pressure using Eq. 3.7 and the data for specific gravity of mercury given in Appendix A, as discussed in the Notes of Example 3.1. Although the vapor pressure of mercury may be neglected, for precise work, temperature and altitude corrections must be applied to the measured level and the effects of surface tension must be considered. The capillary effect in a tube caused by surface tension was illustrated in Example 2.3.

Gases

In many practical engineering problems density will vary appreciably with altitude, and accurate results will require that this variation be accounted for. Pressure variation in a compressible fluid can be evaluated by integrating Eq. 3.6 if the density can be expressed as a function of p or z. Property information or an equation of state may be used to obtain the required relation for density. Several types of property variation may be analyzed. (See Example 3.4.)

The density of gases generally depends on pressure and temperature. The ideal gas equation of state,

$$p = \rho RT \tag{1.1}$$

where R is the gas constant (see Appendix A) and T the absolute temperature, accurately models the behavior of most gases under engineering conditions. However, the use of Eq. 1.1 introduces the gas temperature as an additional variable. Therefore, an additional assumption must be made about temperature variation before Eq. 3.6 can be integrated.

In the U.S. Standard Atmosphere the temperature decreases linearly with altitude up to an elevation of 11.0 km. For a linear temperature variation with altitude given by $T = T_0 - mz$, we obtain, from Eq. 3.6,

$$dp = -\rho g\, dz = -\frac{pg}{RT}dz = -\frac{pg}{R(T_0 - mz)}dz$$

Separating variables and integrating from $z = 0$ where $p = p_0$ to elevation z where the pressure is p gives

$$\int_{p_0}^{p} \frac{dp}{p} = -\int_0^z \frac{g\,dz}{R(T_0 - mz)}$$

Then

$$\ln\frac{p}{p_0} = \frac{g}{mR}\ln\left(\frac{T_0 - mz}{T_0}\right) = \frac{g}{mR}\ln\left(1 - \frac{mz}{T_0}\right)$$

and the pressure variation, in a gas whose temperature varies linearly with elevation, is given by

$$p = p_0\left(1 - \frac{mz}{T_0}\right)^{g/mR} = p_0\left(\frac{T}{T_0}\right)^{g/mR} \tag{3.9}$$

Example 3.4 PRESSURE AND DENSITY VARIATION IN THE ATMOSPHERE

The maximum power output capability of a gasoline or diesel engine decreases with altitude because the air density and hence the mass flow rate of air decrease. A truck leaves Denver (elevation 1610 m) on a day when the local temperature and barometric pressure are 27°C and 630 mm of mercury, respectively. It travels through Vail Pass (elevation 3230 mm), where the temperature is 17°C. Determine the local barometric pressure at Vail Pass and the percent change in density.

Given: Truck travels from Denver to Vail Pass.

Denver: $z = 1610$ m, $p = 630$ mm Hg, $T = 27°$C

Vail Pass: $z = 3230$ m, $T = 17°$C

Find: Atmospheric pressure at Vail Pass.
Percent change in air density between Denver and Vail.

Solution:

Governing equations: $\dfrac{dp}{dz} = -\rho g \qquad p = \rho RT$

Assumptions: (1) Static fluid.
(2) Air behaves as an ideal gas.

We shall consider four assumptions for property variations with altitude.

(a) If we assume temperature varies linearly with altitude, Eq. 3.9 gives

$$\frac{p}{p_0} = \left(\frac{T}{T_0}\right)^{g/mR}$$

Evaluating the constant m gives

$$m = \frac{T_0 - T}{z - z_0} = \frac{(27 - 17)°\text{C}}{(3230 - 1610)\text{ m}} = 6.17 \times 10^{-3}\,°\text{C/m}$$

and

$$\frac{g}{mR} = 9.81\frac{\text{m}}{\text{s}^2} \times \frac{1}{6.17\times10^{-3}} \times \frac{1}{286.9\,\text{J/kg}^\circ\text{K}} \times \frac{\text{N}\cdot\text{s}^2}{\text{kg}\cdot\text{m}} \times \frac{\text{J}}{\text{N}\cdot\text{m}} = 5.54$$

Thus

$$\frac{p}{p_0} = \left(\frac{T}{T_0}\right)^{g/mR} = \left(\frac{273+17}{273+27}\right)^{5.54} = (0.967)^{5.54} = 0.829$$

and

$$p = 0.829\,p_0 = (0.829)630\,\text{mm Hg} = 522.3\,\text{mm Hg} = 69.6\,\text{kPa} \longleftarrow \quad p$$

Note that temperature must be expressed as an absolute temperature in the ideal gas equation of state.

The percent change in density is given by

$$\frac{\rho-\rho_0}{\rho_0} = \frac{\rho}{\rho_0} - 1 = \frac{p}{p_0}\frac{T_0}{T} - 1 = \frac{0.829}{0.967} - 1 = -0.143 \quad \text{or} \quad -14.3\% \longleftarrow \quad \frac{\Delta\rho}{\rho_0}$$

(b) For ρ assumed constant ($=\rho_0$),

$$p = p_0 - \rho_0 g(z-z_0) = p_0 - \frac{p_0 g(z-z_0)}{RT_0} = p_0\left[1 - \frac{g(z-z_0)}{RT_0}\right]$$

$$p = 513.3\,\text{mm Hg} = 68.4\,\text{kPa} \quad \text{and} \quad \frac{\Delta\rho}{\rho_0} = 0 \longleftarrow \quad p, \frac{\Delta\rho}{\rho_0}$$

(c) If we assume the temperature is constant, then

$$dp = -\rho g\,dz = -\frac{p}{RT}g\,dz$$

and

$$\int_{p_0}^{p}\frac{dp}{p} = -\int_{z_0}^{z}\frac{g}{RT}dz$$

$$p = p_0 \exp\left[\frac{-g(z-z_0)}{RT}\right]$$

For $T = \text{constant} = T_0$,

$$p = 523.8\,\text{mm Hg} = 69.8\,\text{kPa} \quad \text{and} \quad \frac{\Delta\rho}{\rho_0} = -16.9\% \longleftarrow \quad p, \frac{\Delta\rho}{\rho_0}$$

(d) For an adiabatic atmosphere $p/\rho^k = \text{constant}$,

$$p = p_0\left(\frac{T}{T_0}\right)^{k/k-1} = 559.5\,\text{mm Hg} = 74.6\,\text{kPa} \quad \text{and} \quad \frac{\Delta\rho}{\rho_0} = -8.2\% \longleftarrow \quad p, \frac{\Delta\rho}{\rho_0}$$

We note that over the modest change in elevation the predicted pressure is not strongly dependent on the assumed property variation; values calculated under four different assumptions vary by a maximum of approximately 9 percent. There is considerably greater variation in the predicted percent change in density. The assumption of a linear temperature variation with altitude is the most reasonable assumption.

This Example shows use of the ideal gas equation with the basic pressure-height relation to obtain the change in pressure with height in the atmosphere under various atmospheric assumptions.

Hydraulic Systems 3.4

Hydraulic systems are characterized by very high pressures, so by comparison hydrostatic pressure variations often may be neglected. Automobile hydraulic brakes develop pressures up to 10 MPa aircraft and machinery hydraulic actuation systems frequently are designed for pressures up to 40 MPa and jacks use pressures to 70 MPa. Special-purpose laboratory test equipment is commercially available for use at pressures to 1000 MPa!

VIDEO

Hydraulic Force Amplification.

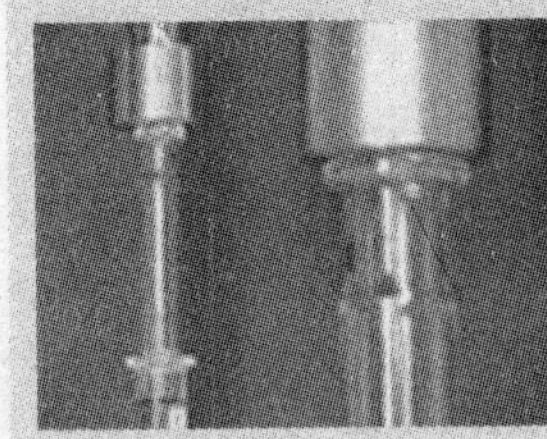

Although liquids are generally considered incompressible at ordinary pressures, density changes may be appreciable at high pressures. Bulk moduli of hydraulic fluids also may vary sharply at high pressures. In problems involving unsteady flow, both compressibility of the fluid and elasticity of the boundary structure (e.g., the pipe walls) must be considered. Analysis of problems such as water hammer noise and vibration in hydraulic systems, actuators, and shock absorbers quickly becomes complex and is beyond the scope of this book.

Hydrostatic Force on Submerged Surfaces 3.5

Now that we have determined how the pressure varies in a static fluid, we can examine the force on a surface submerged in a liquid.

In order to determine completely the resultant force acting on a submerged surface, we must specify:

1. The magnitude of the force.
2. The direction of the force.
3. The line of action of the force.

We shall consider both plane and curved submerged surfaces.

Hydrostatic Force on a Plane Submerged Surface

A plane submerged surface, on whose upper face we wish to determine the resultant hydrostatic force, is shown in Fig. 3.5. The coordinates are important: They have been chosen so that the surface lies in the xy plane, and the origin O is located at the intersection of the plane surface (or its extension) and the free surface. As well as

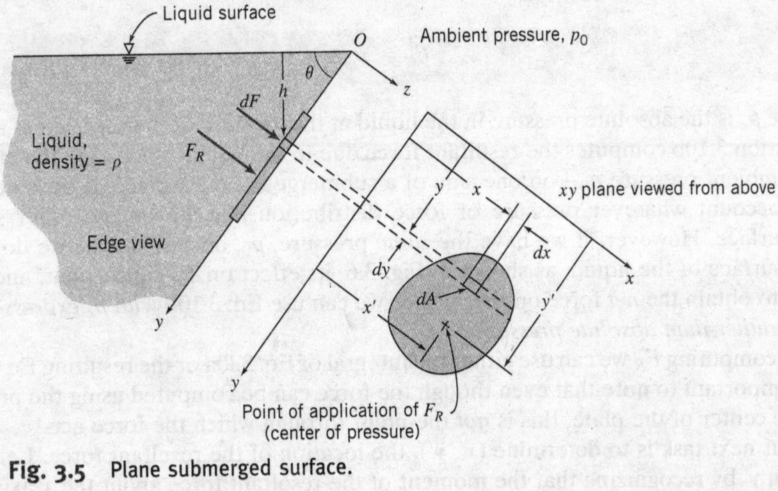

Fig. 3.5 Plane submerged surface.

the magnitude of the force F_R, we wish to locate the point (with coordinates x', y') through which it acts on the surface.

Since there are no shear stresses in a static fluid, the hydrostatic force on any element of the surface acts normal to the surface. The pressure force acting on an element $dA = dx\,dy$ of the upper surface is given by

$$dF = p\,dA$$

The *resultant* force acting on the surface is found by summing the contributions of the infinitesimal forces over the entire area.

Usually when we sum forces we must do so in a vectorial sense. However, in this case all of the infinitesimal forces are perpendicular to the plane, and hence so is the resultant force. Its magnitude is given by

$$F_R = \int_A p\,dA \tag{3.10a}$$

In order to evaluate the integral in Eq. 3.10a, both the pressure, p, and the element of area, dA, must be expressed in terms of the same variables.

We can use Eq. 3.7 to express the pressure p at depth h in the liquid as

$$p = p_0 + \rho g h$$

In this expression p_0 is the pressure at the free surface ($h = 0$).

In addition, we have, from the system geometry, $h = y \sin\theta$. Using this expression and the above expression for pressure in Eq. 3.10a,

$$F_R = \int_A p\,dA = \int_A (p_0 + \rho g h)dA = \int_A (p_0 + \rho g y \sin\theta)dA$$

$$F_R = p_0 \int_A dA + \rho g \sin\theta \int_A y\,dA = p_0 A + \rho g \sin\theta \int_A y\,dA$$

The integral is the first moment of the surface area about the x axis, which may be written

$$\int_A y\,dA = y_c A$$

where y_c is the y coordinate of the *centroid* of the area, A. Thus,

$$F_R = p_0 A + \rho g \sin\theta\, y_c A = (p_0 + \rho g h_c)A$$

or

$$F_R = p_c A \tag{3.10b}$$

where p_c is the absolute pressure in the liquid at the location of the centroid of area A. Equation 3.10b computes the resultant force due to the liquid—including the effect of the ambient pressure p_0—on one side of a submerged plane surface. It does not take into account whatever pressure or force distribution may be on the other side of the surface. However, if we have the *same* pressure, p_0, on this side as we do at the free surface of the liquid, as shown in Fig. 3.6, its effect on F_R cancels out, and if we wish to obtain the *net* force on the surface we can use Eq. 3.10b *with p_c expressed as a gage rather than absolute pressure.*

In computing F_R we can use either the integral of Eq. 3.10a or the resulting Eq. 3.10b. It is important to note that even though the force can be computed using the pressure at the center of the plate, this is *not* the point through which the force acts!

Our next task is to determine (x', y'), the location of the resultant force. Let's first obtain y' by recognizing that the moment of the resultant force about the x axis must

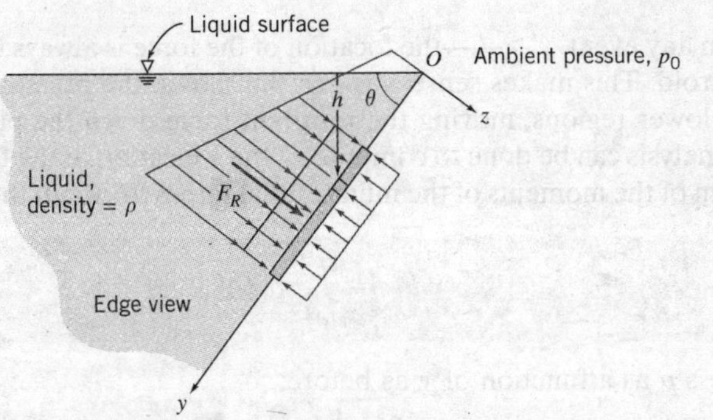

Fig. 3.6 Pressure distribution on plane submerged surface.

be equal to the moment due to the distributed pressure force. Taking the sum (i.e., integral) of the moments of the infinitesimal forces dF about the x axis we obtain

$$y'F_R = \int_A yp\, dA \tag{3.11a}$$

We can integrate by expressing p as a function of y as before:

$$y'F_R = \int_A yp\, dA = \int_A y(p_0 + \rho g h)\, dA = \int_A (p_0 y + \rho g y^2 \sin\theta)\, dA$$
$$= p_0 \int_A y\, dA + \rho g \sin\theta \int_A y^2 dA$$

The first integral is our familiar $y_c A$. The second integral, $\int_A y^2\, dA$, is the second moment of area about the x axis, I_{xx}. We can use the parallel axis theorem, $I_{xx} = I_{\hat{x}\hat{x}} + Ay_c^2$, to replace I_{xx} with the standard second moment of area, about the centroidal $\hat{x}$ axis. Using all of these, we find

$$y'F_R = p_0 y_c A + \rho g \sin\theta (I_{\hat{x}\hat{x}} + Ay_c^2) = y_c(p_0 + \rho g y_c \sin\theta)A + \rho g \sin\theta\, I_{\hat{x}\hat{x}}$$
$$= y_c(p_0 + \rho g h_c)A + \rho g \sin\theta I_{\hat{x}\hat{x}} = y_c F_R + \rho g \sin\theta I_{\hat{x}\hat{x}}$$

Finally, we obtain for y':

$$y' = y_c + \frac{\rho g \sin\theta\, I_{\hat{x}\hat{x}}}{F_R} \tag{3.11b}$$

Equation 3.11b is convenient for computing the location y' of the force on the submerged side of the surface when we include the ambient pressure p_0. If we have the same ambient pressure acting on the other side of the surface we can use Eq. 3.10b with p_0 neglected to compute the net force,

$$F_R = p_{c_{\text{gage}}} A = \rho g h_c A = \rho g y_c \sin\theta\, A$$

and Eq. 3.11b becomes for this case

$$y' = y_c + \frac{I_{\hat{x}\hat{x}}}{Ay_c} \tag{3.11c}$$

Equation 3.11a is the integral equation for computing the location y' of the resultant force; Eq. 3.11b is a useful algebraic form for computing y' when we are interested in the resultant force on the submerged side of the surface; Eq. 3.11c is for computing y' when we are interested in the net force for the case when the same p_0 acts at the free surface and on the other side of the submerged surface. For problems that have a pressure on the other side that is *not* p_0, we can either analyze each side of the surface separately or reduce the two pressure distributions to one net pressure distribution, in effect creating a system to be solved using Eq. 3.10b with p_c expressed as a gage pressure.

Note that in any event, $y' > y_c$—the location of the force is always below the level of the plate centroid. This makes sense—as Fig. 3.6 shows, the pressures will always be larger on the lower regions, moving the resultant force down the plate.

A similar analysis can be done to compute x', the x location of the force on the plate. Taking the sum of the moments of the infinitesimal forces dF about the y axis we obtain

$$x'F_R = \int_A xp\,dA \tag{3.12a}$$

We can express p as a function of y as before:

$$x'F_R = \int_A xp\,dA = \int_A x(p_0 + \rho gh)\,dA = \int_A (p_0 x + \rho gxy\sin\theta)\,dA$$
$$= p_0\int_A x\,dA + \rho g\sin\theta\int_A xy\,dA$$

The first integral is x_cA (where x_c is the distance of the centroid from y axis). The second integral is $\int_A xy\,dA = I_{xy}$. Using the parallel axis theorem, $I_{xy} = I_{\hat{x}\hat{y}} + Ax_c y_c$, we find

$$x'F_R = p_0x_cA + \rho g\sin\theta(I_{\hat{x}\hat{y}} + Ax_cy_c) = x_c(p_0 + \rho gy_c\sin\theta)A + \rho g\sin\theta\, I_{\hat{x}\hat{y}}$$
$$= x_c(p_0 + \rho gh_c)A + \rho g\sin\theta\, I_{\hat{x}\hat{y}} = x_cF_R + \rho g\sin\theta\, I_{\hat{x}\hat{y}}$$

Finally, we obtain for x':

$$x' = x_c + \frac{\rho g\sin\theta\, I_{\hat{x}\hat{y}}}{F_R} \tag{3.12b}$$

Equation 3.12b is convenient for computing x' when we include the ambient pressure p_0. If we have ambient pressure also acting on the other side of the surface we can again use Eq. 3.10b with p_0 neglected to compute the net force and Eq. 3.12b becomes for this case

$$x' = x_c + \frac{I_{\hat{x}\hat{y}}}{Ay_c} \tag{3.12c}$$

Equation 3.12a is the integral equation for computing the location x' of the resultant force; Eq. 3.12b can be used for computations when we are interested in the force on the submerged side only; Eq. 3.12c is useful when we have p_0 on the other side of the surface and we are interested in the net force.

In summary, Eqs. 3.10 through 3.12 constitute a complete set of equations for computing the magnitude and location of the force due to hydrostatic pressure on any submerged plane surface. The direction of the force will always be perpendicular to the plane.

We can now consider several examples using these equations. In Example 3.5 we use both the integral and algebraic sets of equations.

Example 3.5 RESULTANT FORCE ON INCLINED PLANE SUBMERGED SURFACE

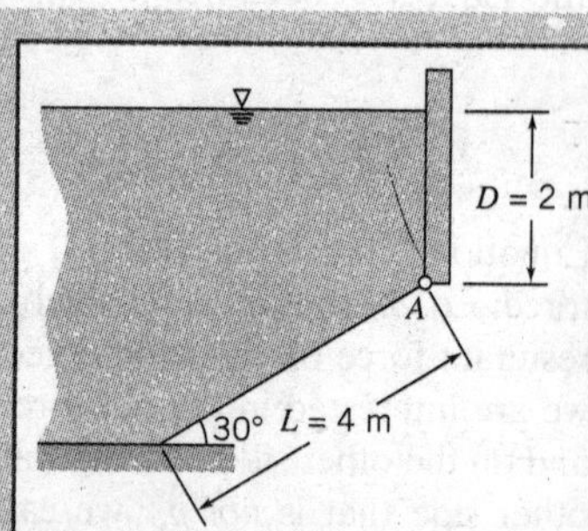

The inclined surface shown, hinged along edge A, is 5 m wide. Determine the resultant force, F_R, of the water and the air on the inclined surface.

Given: Rectangular gate, hinged along A, $w = 5$ m.

Find: Resultant force, F_R, of the water and the air on the gate.

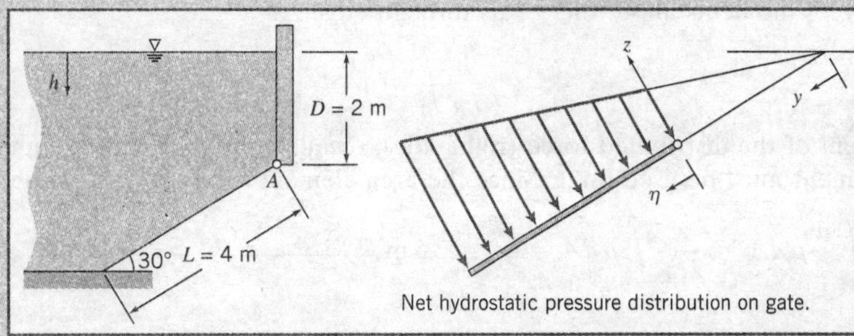

Net hydrostatic pressure distribution on gate.

Solution:

In order to completely determine F_R, we need to find (a) the magnitude and (b) the line of action of the force (the direction of the force is perpendicular to the surface). We will solve this problem by using (i) direct integration and (ii) the algebraic equations.

Direct Integration

Governing equations: $p = p_0 + \rho g h \quad F_R = \int_A p\, dA \quad \eta' F_R = \int_A \eta p\, dA \quad x' F_R = \int_A x p\, dA$

Because atmospheric pressure p_0 acts on both sides of the plate its effect cancels, and we can work in gage pressures ($p = \rho g h$). In addition, while we *could* integrate using the y variable, it will be more convenient here to define a variable η, as shown in the figure.

Using η to obtain expressions for h and dA, then

$$h = D + \eta \sin 30^\circ \qquad \text{and} \qquad dA = w\, d\eta$$

Applying these to the governing equation for the resultant force,

$$F_R = \int_A p\, dA = \int_0^L \rho g (D + \eta \sin 30^\circ) w\, d\eta$$

$$= \rho g w \left[D\eta + \frac{\eta^2}{2} \sin 30^\circ \right]_0^L = \rho g w \left[DL + \frac{L^2}{2} \sin 30^\circ \right]$$

$$= 999 \frac{\text{kg}}{\text{m}^3} \times 9.81 \frac{\text{m}}{\text{s}^2} \times 5\,\text{m} \left[2\,\text{m} \times 4\,\text{m} + \frac{16\,\text{m}^2}{2} \times \frac{1}{2} \right] \frac{\text{N} \cdot \text{s}^2}{\text{kg} \cdot \text{m}}$$

$$F_R = 588\,\text{kN} \qquad F_R$$

For the location of the force we compute η' (the distance from the top edge of the plate),

$$\eta' F_R = \int_A \eta p\, dA$$

Then

$$\eta' = \frac{1}{F_R} \int_A \eta p\, dA = \frac{1}{F_R} \int_0^L \eta p w\, d\eta = \frac{\rho g w}{F_R} \int_0^L \eta (D + \eta \sin 30^\circ)\, d\eta$$

$$= \frac{\rho g w}{F_R} \left[\frac{D\eta^2}{2} + \frac{\eta^3}{3} \sin 30^\circ \right]_0^L = \frac{\rho g w}{F_R} \left[\frac{DL^2}{2} + \frac{L^3}{3} \sin 30^\circ \right]$$

$$= 999 \frac{\text{kg}}{\text{m}^3} \times 9.8 \frac{\text{m}}{\text{s}^2} \times \frac{5\,\text{m}}{5.88 \times 10^5\,\text{N}} \left[\frac{2\,\text{m} \times 16\,\text{m}^2}{2} + \frac{64\,\text{m}^3}{3} \times \frac{1}{2} \right] \frac{\text{N} \cdot \text{s}^2}{\text{kg} \cdot \text{m}}$$

$$\eta' = 2.22\,\text{m} \quad \text{and} \quad y' = \frac{D}{\sin 30^\circ} + \eta' = \frac{2\,\text{m}}{\sin 30^\circ} + 2.22\,\text{m} = 6.22\,\text{m} \qquad y'$$

Also, from consideration of moments about the y axis through edge A,

$$x' = \frac{1}{F_R}\int_A xp\,dA$$

In calculating the moment of the distributed force (right side), recall, from your earlier courses in statics, that the centroid of the area element must be used for x. Since the area element is of constant width, then $x = w/2$, and

$$x' = \frac{1}{F_R}\int_A \frac{w}{2}p\,dA = \frac{w}{2F_R}\int_A p\,dA = \frac{w}{2} = 2.5\text{ m} \qquad x'$$

Algebraic Equations

In using the algebraic equations we need to take care in selecting the appropriate set. In this problem we have $p_0 = p_{atm}$ on both sides of the plate, so Eq. 3.10b with p_c as a gage pressure is used for the net force:

$$F_R = p_c A = \rho g h_c A = \rho g\left(D + \frac{L}{2}\sin 30^\circ\right)Lw$$

$$F_R = \rho g w\left[DL + \frac{L^2}{2}\sin 30^\circ\right]$$

This is the same expression as was obtained by direct integration.

The y coordinate of the center of pressure is given by Eq. 3.11c:

$$y' = y_c + \frac{I_{\hat{x}\hat{x}}}{Ay_c} \tag{3.11c}$$

For the inclined rectangular gate

$$y_c = \frac{D}{\sin 30^\circ} + \frac{L}{2} = \frac{2\text{ m}}{\sin 30^\circ} + \frac{4\text{ m}}{2} = 6\text{ m}$$

$$A = Lw = 4\text{ m} \times 5\text{ m} = 20\text{ m}^2$$

$$I_{\hat{x}\hat{x}} = \frac{1}{12}wL^3 = \frac{1}{12} \times 5\text{ m} \times (4\text{ m})^3 = 26.7\text{ m}^2$$

$$y' = y_c + \frac{I_{\hat{x}\hat{x}}}{Ay_c} = 6\text{ m} + 26.7\text{ m}^4 \times \frac{1}{20\text{ m}^2} \times \frac{1}{6\text{ m}} = 6.22\text{ m} \qquad y'$$

The x coordinate of the center of pressure is given by Eq. 3.12c:

$$x' = x_c + \frac{I_{\hat{x}\hat{y}}}{Ay_c} \tag{3.12c}$$

For the rectangular gate $I_{\hat{x}\hat{y}} = 0$ and $x' = x_c = 2.5$ m. x'

This Example shows
- ✓ Use of integral and algebraic equations.
- ✓ Use of the algebraic equations for computing the *net* force.

Example 3.6 FORCE ON VERTICAL PLANE SUBMERGED SURFACE WITH NONZERO GAGE PRESSURE AT FREE SURFACE

The door shown in the side of the tank is hinged along its bottom edge. A pressure of 4790 Pa (gage) is applied to the liquid free surface. Find the force, F_t, required to keep the door closed.

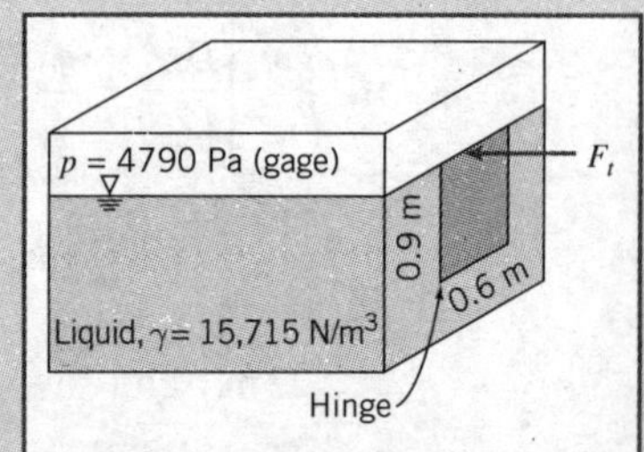

Given: Door as shown in the figure.

Find: Force required to keep door shut.

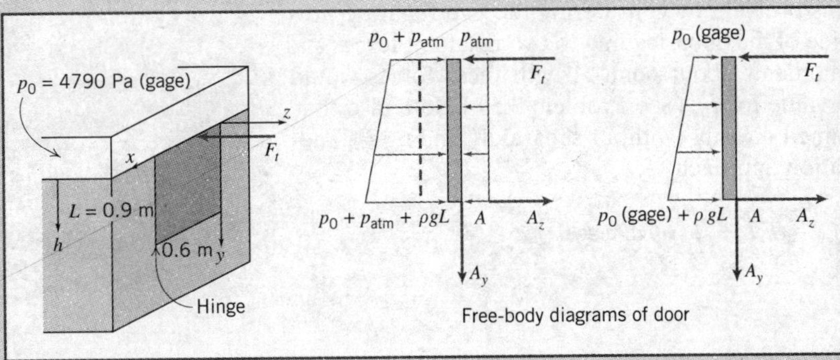

Free-body diagrams of door

Solution:

This problem requires a free-body diagram (FBD) of the door. The pressure distributions on the inside and outside of the door will lead to a net force (and its location) that will be included in the FBD. We need to be careful in choosing the equations for computing the resultant force and its location. We can either use absolute pressures (as on the left FBD) and compute two forces (one on each side) or gage pressures and compute one force (as on the right FBD). For simplicity we will use gage pressures. The right-hand FBD makes clear we should use Eqs. 3.10b and 3.11b, which were derived for problems in which we wish to include thc effects of an ambient pressure (p_0), or in other words, for problems when we have a nonzero gage pressure at the free surface. The components of force due to the hinge are A_y and A_z. The force F_t can be found by taking moments about A (the hinge).

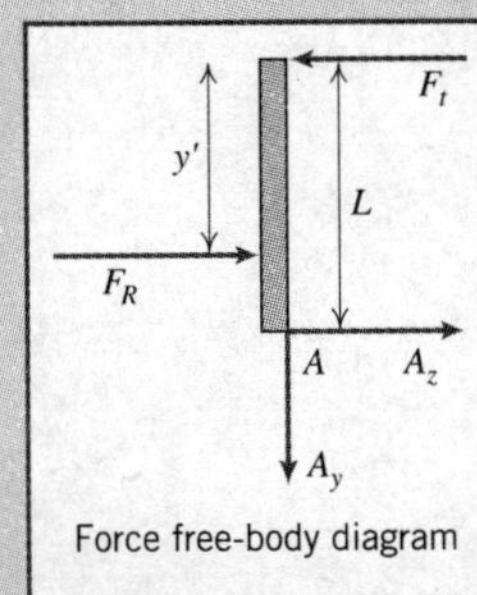

Force free-body diagram

Governing equations:

$$F_R = p_c A \qquad y' = y_c + \frac{\rho g \sin\theta \, I_{\hat{x}\hat{x}}}{F_R} \qquad \sum M_A = 0$$

The resultant force and its location are

$$F_R = (p_0 + \rho g h_c)A = \left(p_0 + \gamma\frac{L}{2}\right)bL \tag{1}$$

and

$$y' = y_c + \frac{\rho g \sin 90^\circ \, I_{\hat{x}\hat{x}}}{F_R} = \frac{L}{2} + \frac{\gamma b L^3/12}{\left(p_0 + \gamma\frac{L}{2}\right)bL} = \frac{L}{2} + \frac{\gamma L^2/12}{\left(p_0 + \gamma\frac{L}{2}\right)} \tag{2}$$

Taking moments about point A

$$\sum M_A = F_t L - F_R(L - y') = 0 \qquad \text{or} \qquad F_t = F_R\left(1 - \frac{y'}{L}\right)$$

Using Eqs. 1 and 2 in this equation we find

$$F_t = \left(p_0 + \gamma\frac{L}{2}\right)bL\left[1 - \frac{1}{2} - \frac{\gamma L^2/12}{\left(p_0 + \gamma\frac{L}{2}\right)}\right]$$

$$F_t = \left(p_0 + \gamma\frac{L}{2}\right)\frac{bL}{2} + \gamma\frac{bL^2}{12} = \frac{p_0 bL}{2} + \frac{\gamma b L^2}{6}$$

$$= 4790\,\frac{\text{N}}{\text{m}^2} \times 0.6\,\text{m} \times 0.9\,\text{m} \times \frac{1}{2} + 15{,}715\,\frac{\text{N}}{\text{m}^3} \times 0.6\,\text{m} \times 0.81\,\text{m}^2 \times \frac{1}{6} \tag{3}$$

$$F_t = 2566\,\text{N} \longleftarrow F_t$$

We could have solved this problem by considering the two separate pressure distributions on each side of the door, leading to two resultant forces and their locations. Summing moments about point A with these forces would also have yielded the same value for F_t. (See Problem 3.59.) Note also that Eq. 3 could have been obtained directly (without separately finding F_R and y') by using a direct integration approach:

$$\sum M_A = F_t L - \int_A y\, p\, dA = 0$$

This Example shows:

- ✓ Use of algebraic equations for nonzero gage pressure at the liquid free surface.
- ✓ Use of the moment equation from statics for computing the required applied force.

Hydrostatic Force on a Curved Submerged Surface

For curved surfaces, we will once again derive expressions for the resultant force by integrating the pressure distribution over the surface. However, unlike for the plane surface, we have a more complicated problem—the pressure force is normal to the surface at each point, but now the infinitesimal area elements point in varying directions because of the surface curvature. This means that instead of integrating over an element dA we need to integrate over vector element $d\vec{A}$. This will initially lead to a more complicated analysis, but we will see that a simple solution technique will be developed.

Consider the curved surface shown in Fig. 3.7. The pressure force acting on the element of area, $d\vec{A}$, is given by

$$d\vec{F} = -p\, d\vec{A}$$

where the minus sign indicates that the force acts on the area, in the direction opposite to the area normal. The resultant force is given by

$$\vec{F}_R = -\int_A p\, d\vec{A} \tag{3.13}$$

We can write

$$\vec{F}_R = \hat{i}F_{R_x} + \hat{j}F_{R_y} + \hat{k}F_{R_z}$$

where F_{Rx}, F_{Ry}, and F_{Rz} are the components of $\vec{F}_R$ in the positive x, y, and z directions, respectively.

To evaluate the component of the force in a given direction, we take the dot product of the force with the unit vector in the given direction. For example, taking the dot product of each side of Eq. 3.13 with unit vector $\hat{i}$ gives

$$F_{R_x} = \vec{F}_R \cdot \hat{i} = \int d\vec{F} \cdot \hat{i} = -\int_A p\, d\vec{A} \cdot \hat{i} = -\int_{A_x} p\, dA_x$$

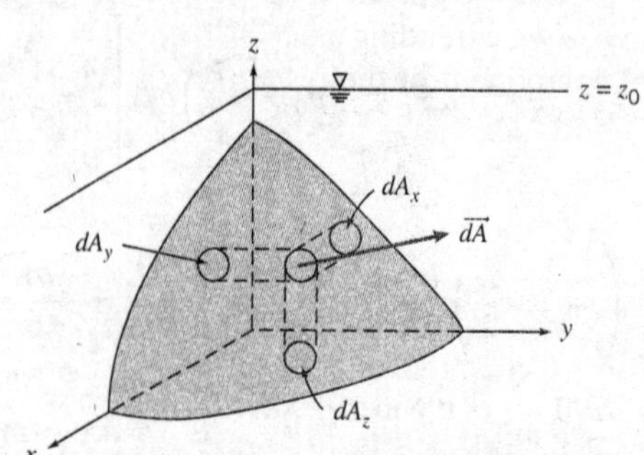

Fig. 3.7 Curved submerged surface.

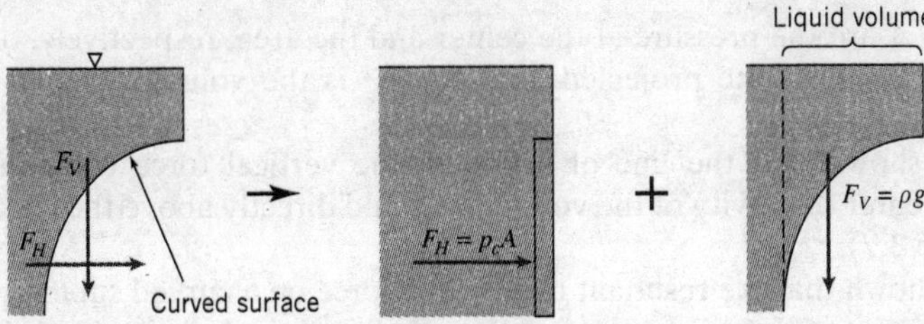

Fig. 3.8 Forces on curved submerged surface.

where dA_x is the projection of $d\vec{A}$ on a plane perpendicular to the x axis (see Fig. 3.7), and the minus sign indicates that the x component of the resultant force is in the negative x direction.

Since, in any problem, the direction of the force component can be determined by inspection, the use of vectors is not necessary. In general, the magnitude of the component of the resultant force in the l direction is given by

$$F_{R_l} = \int_{A_l} p\, dA_l \tag{3.14}$$

where dA_l is the projection of the area element dA on a plane perpendicular to the l direction. The line of action of each component of the resultant force is found by recognizing that the moment of the resultant force component about a given axis must be equal to the moment of the corresponding distributed force component about the same axis.

Equation 3.14 can be used for the horizontal forces F_{R_x} and F_{R_y}. We have the interesting result that *the horizontal force and its location are the same as for an imaginary vertical plane surface of the same projected area.* This is illustrated in Fig. 3.8, where we have called the horizontal force F_H.

Figure 3.8 also illustrates how we can compute the vertical component of force: With atmospheric pressure at the free surface and on the other side of the curved surface *the net vertical force will be equal to the weight of fluid directly above the surface.* This can be seen by applying Eq. 3.14 to determine the magnitude of the vertical component of the resultant force, obtaining

$$F_{R_z} = F_V = \int p\, dA_z$$

Since $p = \rho g h$,

$$F_V = \int \rho g h\, dA_z = \int \rho g\, d\forall$$

where $\rho g h\, dA_z = \rho g\, d\forall$ is the weight of a differential cylinder of liquid above the element of surface area, dA_z, extending a distance h from the curved surface to the free surface. The vertical component of the resultant force is obtained by integrating over the entire submerged surface. Thus

$$F_V = \int_{A_z} \rho g h\, dA_z = \int_{\forall} \rho g\, d\forall = \rho g \forall$$

In summary, for a curved surface we can use two simple formulas for computing the horizontal and vertical force components due to the fluid only (no ambient pressure),

$$F_H = p_c A \quad \text{and} \quad F_V = \rho g \forall \tag{3.15}$$

where p_c and A are the pressure at the center and the area, respectively, of a vertical plane surface of the same projected area, and $V\!\!\!\!-$ is the volume of fluid above the curved surface.

It can be shown that the line of action of the vertical force component passes through the center of gravity of the volume of liquid directly above the curved surface (see Example 3.7).

We have shown that the resultant hydrostatic force on a curved submerged surface is specified in terms of its components. We recall from our study of statics that the resultant of any force system can be represented by a force-couple system, i.e., the resultant force applied at a point and a couple about that point. If the force and the couple vectors are orthogonal (as is the case for a two-dimensional curved surface), the resultant can be represented as a pure force with a unique line of action. Otherwise the resultant may be represented as a "wrench," also having a unique line of action.

Example 3.7 FORCE COMPONENTS ON A CURVED SUBMERGED SURFACE

The gate shown is hinged at O and has constant width, $w = 5$ m. The equation of the surface is $x = y^2/a$, where $a = 4$ m. The depth of water to the right of the gate is $D = 4$ m. Find the magnitude of the force, F_a, applied as shown, required to maintain the gate in equilibrium if the weight of the gate is neglected.

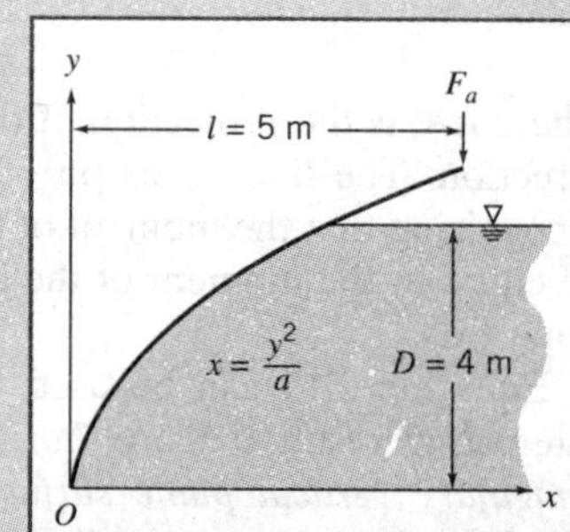

Given: Gate of constant width, $w = 5$ m.
Equation of surface in xy plane is $x = y^2/a$, where $a = 4$ m.
Water stands at depth $D = 4$ m to the right of the gate.
Force F_a is applied as shown, and weight of gate is to be neglected. (Note that for simplicity we do not show the reactions at O.)

Find: Force F_a required to maintain the gate in equilibrium.

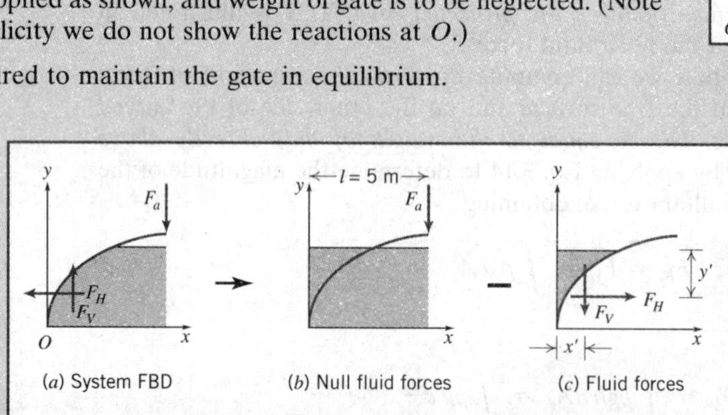

Solution:
We will take moments about point O after finding the magnitudes and locations of the horizontal and vertical forces due to the water. The free body diagram (FBD) of the system is shown above in part (a). Before proceeding we need to think about how we compute F_V, the vertical component of the fluid force—we have stated that it is equal (in magnitude and location) to the weight of fluid directly above the curved surface. However, we have no fluid directly above the gate, even though it is clear that the fluid does exert a vertical force! We need to do a "thought experiment" in which we imagine having a system with water on both sides of the gate (with null effect), minus a system with water directly above the gate (which generates fluid forces). This logic is demonstrated above: the system FBD(a) = the null FBD(b) − the fluid forces FBD(c). Thus the vertical and horizontal fluid forces on the system, FBD(a), are equal and opposite to those on FBD(c). In summary, the magnitude and location of the vertical fluid force F_V are given by the weight and location of the centroid of the fluid "above" the gate; the magnitude and location of the horizontal fluid force F_H are given by the magnitude and location of the force on an equivalent vertical flat plate.

Governing equations: $F_H = p_c A$ $\quad y' = y_c + \frac{I_{\hat{x}\hat{x}}}{Ay_c}$ $\quad F_V = \rho g V\!\!\!/$ $\quad x' =$ water center of gravity

For F_H, the centroid, area, and second moment of the equivalent vertical flat plate are, respectively, $y_c = h_c = D/2$, $A = Dw$, and $I_{\hat{x}\hat{x}} = wD^3/12$.

$$F_H = p_c A = \rho g h_c A$$
$$= \rho g \frac{D}{2} Dw = \rho g \frac{D^2}{2} w = 999 \frac{\text{kg}}{\text{m}^3} \times 9.81 \frac{\text{m}}{\text{s}^2} \times \frac{(4\,\text{m}^2)}{2} \times 5\,\text{m} \times \frac{\text{N}\cdot\text{s}^2}{\text{kg}\cdot\text{m}} \tag{1}$$
$$F_H = 392\,\text{kN}$$

and

$$y' = y_c + \frac{I_{\hat{x}\hat{x}}}{Ay_c}$$
$$= \frac{D}{2} + \frac{wD^3/12}{wDD/2} = \frac{D}{2} + \frac{D}{6}$$
$$y' = \frac{2}{3}D = \frac{2}{3} \times 4\,\text{m} = 2.67\,\text{m} \tag{2}$$

For F_V, we need to compute the weight of water "above" the gate. To do this we define a differential column of volume $(D - y)\,w\,dx$ and integrate

$$F_V = \rho g V\!\!\!/ = \rho g \int_0^{D^2/a} (D - y) w\, dx = \rho g w \int_0^{D^2/a} (D - \sqrt{a} x^{1/2}) dx$$
$$= \rho g w \left[Dx - \frac{2}{3}\sqrt{a} x^{3/2} \right]_0^{D^3/a} = \rho g w \left[\frac{D^3}{a} - \frac{2}{3}\sqrt{a}\frac{D^3}{a^{3/2}} \right] = \frac{\rho g w D^3}{3a}$$
$$F_V = 999 \frac{\text{kg}}{\text{m}^3} \times 9.81 \frac{\text{m}}{\text{s}^2} \times 5\,\text{m} \times \frac{(4)^3\text{m}^3}{3} \times \frac{1}{4\,\text{m}} \times \frac{\text{N}\cdot\text{s}^2}{\text{kg}\cdot\text{m}} = 261\,\text{kN} \tag{3}$$

The location x' of this force is given by the location of the center of gravity of the water "above" the gate. We recall from statics that this can be obtained by using the notion that the moment of F_V and the moment of the sum of the differential weights about the y axis must be equal, so

$$x'F_V = \rho g \int_0^{D^2/a} x(D - y) w\, dx = \rho g w \int_0^{D^2/a} (D - \sqrt{a} x^{3/2}) dx$$
$$x'F_V = \rho g w \left[\frac{D}{2}x^2 - \frac{2}{5}\sqrt{a} x^{5/2} \right]_0^{D^2/a} = \rho g w \left[\frac{D^5}{2a^2} - \frac{2}{5}\sqrt{a}\frac{D^5}{a^{5/2}} \right] = \frac{\rho g w D^5}{10a^2}$$
$$x' = \frac{\rho g w D^5}{10a^2 F_V} = \frac{3D^2}{10a} = \frac{3}{10} \times \frac{(4)^2\,\text{m}^2}{4\,\text{m}} = 1.2\,\text{m} \tag{4}$$

Now that we have determined the fluid forces, we can finally take moments about O (taking care to use the appropriate signs), using the results of Eqs. 1 through 4

$$\sum M_O = -lF_a + x'F_V + (D - y')F_H = 0$$
$$F_a = \frac{1}{l}[x'F_V + (D - y')F_H]$$
$$= \frac{1}{5\,\text{m}}[1.2\,\text{m} \times 261\,\text{kN} + (4 - 2.67)\,\text{m} \times 392\,\text{kN}]$$
$$F_a = 167\,\text{kN} \qquad \longleftarrow \qquad F_a$$

This Example shows:

- ✓ Use of vertical flat plate equations for the horizontal force, and fluid weight equations for the vertical force, on a curved surface.
- ✓ The use of "thought experiments" to convert a problem with fluid below a curved surface into an equivalent problem with fluid above.

*3.6 Buoyancy and Stability

If an object is immersed in a liquid, or floating on its surface, the net vertical force acting on it due to liquid pressure is termed *buoyancy*. Consider an object totally immersed in static liquid, as shown in Fig. 3.9.

The vertical force on the body due to hydrostatic pressure may be found most easily by considering cylindrical volume elements similar to the one shown in Fig. 3.9.

We recall that we can use Eq. 3.7 for computing the pressure p at depth h in a liquid,

$$p = p_0 + \rho g h$$

The net vertical pressure force on the element is then

$$dF_z = (p_0 + \rho g h_2)\, dA - (p_0 + \rho g h_1)\, dA = \rho g (h_2 - h_1)\, dA$$

But $(h_2 - h_1)dA = d\mathcal{V}$, the volume of the element. Thus

$$F_z = \int dF_z = \int_{\mathcal{V}} \rho g\, d\mathcal{V} = \rho g \mathcal{V}$$

where $\mathcal{V}$ is the volume of the object. Hence we conclude that for a submerged body *the buoyancy force of the fluid is equal to the weight of displaced fluid*,

$$F_{\text{buoyancy}} = \rho g \mathcal{V} \tag{3.16}$$

This relation reportedly was used by Archimedes in 220 B.C. to determine the gold content in the crown of King Hiero II. Consequently, it is often called "Archimedes' Principle." In more current technical applications, Eq. 3.16 is used to design displacement vessels, flotation gear, and submersibles.*

The submerged object need not be solid. Hydrogen bubbles, used to visualize streaklines and timelines in water (see Section 2.2), are positively buoyant; they rise slowly as they are swept along by the flow. Conversely, water droplets in oil are negatively buoyant and tend to sink.

Airships and balloons are termed "lighter-than-air" craft. The density of an ideal gas is proportional to molecular weight, so hydrogen and helium are less dense than air at the same temperature and pressure. Hydrogen ($M_m = 2$) is less dense than helium ($M_m = 4$), but extremely flammable, whereas helium is inert. Hydrogen has not been used commercially since the disastrous explosion of the German passenger airship *Hindenburg* in 1937. The use of buoyancy force to generate lift is illustrated in Example 3.8.

Equation 3.16 predicts the net vertical pressure force on a body that is totally submerged in a single liquid. In cases of partial immersion, a floating body displaces its own weight of the liquid in which it floats.

The line of action of the buoyancy force, which may be found using the methods of Section 3.5, acts through the centroid of the displaced volume. Since floating bodies

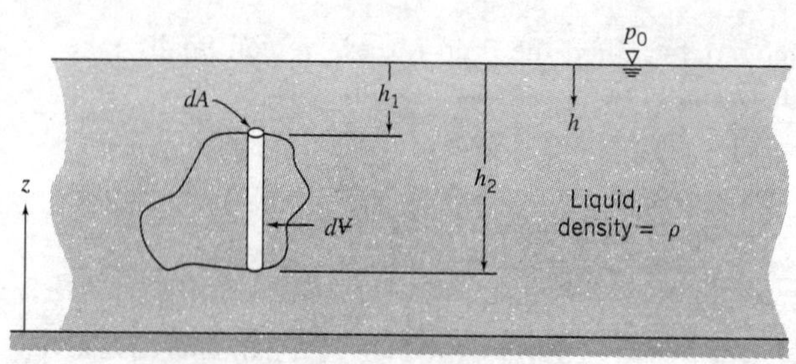

Fig. 3.9 Immersed body in static liquid.

*This section may be omitted without loss of continuity in the text material.

Example 3.8 BUOYANCY FORCE IN A HOT AIR BALLOON

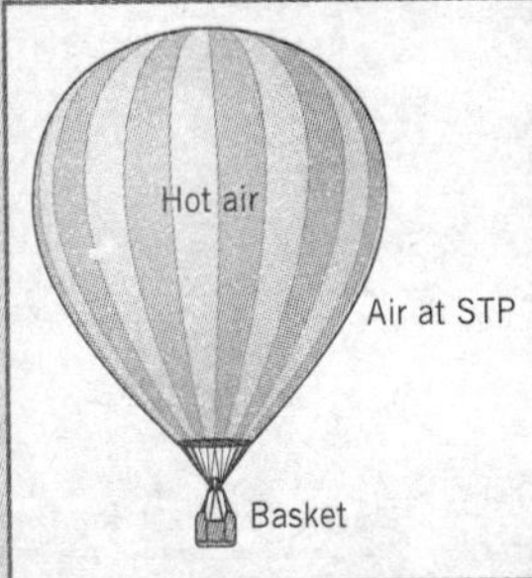

A hot air balloon (approximated as a sphere of diameter 15 m) is to lift a basket load of 2670 N. To what temperature must the air be heated in order to achieve liftoff?

Given: Atmosphere at STP, diameter of balloon $d = 15$ m, and load $W_{\text{load}} = 2670$ N.

Find: The hot air temperature to attain liftoff.

Solution:
Apply the buoyancy equation to determine the lift generated by atmosphere, and apply the vertical force equilibrium equation to obtain the hot air density. Then use the ideal gas equation to obtain the hot air temperature.

Governing equations:

$$F_{\text{buoyancy}} = \rho g \forall \qquad \sum F_y = 0 \qquad p = \rho R T$$

Assumptions: (1) Ideal gas.
(2) Atmospheric pressure throughout.

Summing vertical forces

$$\sum F_y = F_{\text{buoyancy}} - W_{\text{hot air}} - W_{\text{load}} = \rho_{\text{atm}} g \forall - \rho_{\text{hot air}} g \forall - W_{\text{load}} = 0$$

Rearranging and solving for $\rho_{\text{hot air}}$ (using data from Appendix A),

$$\rho_{\text{hot air}} = \rho_{\text{atm}} - \frac{W_{\text{load}}}{g\forall} = \rho_{\text{atm}} - \frac{6W_{\text{load}}}{\pi d^3 g}$$

$$= 1.227\frac{\text{kg}}{\text{m}^3} - 6 \times \frac{2670\,\text{N}}{\pi (15)^3\,\text{m}^3} \times \frac{\text{s}^2}{81\,\text{m}} \times \frac{\text{kg}\cdot\text{m}}{\text{N}\cdot\text{s}^2}$$

$$\rho_{\text{hot air}} = (1.227 - 0.154)\frac{\text{kg}}{\text{m}^3} = 1.073\frac{\text{kg}}{\text{m}^3}$$

Finally, to obtain the temperature of this hot air, we can use the ideal gas equation in the following form

$$\frac{p_{\text{hot air}}}{\rho_{\text{hot air}} R T_{\text{hot air}}} = \frac{p_{\text{atm}}}{\rho_{\text{atm}} R T_{\text{atm}}}$$

and with $p_{\text{hot air}} = p_{\text{atm}}$

$$T_{\text{hot air}} = T_{\text{atm}} \frac{\rho_{\text{atm}}}{\rho_{\text{hot air}}} = (273 + 15)^\circ\text{K} \times \frac{1.227}{1.073} = 329^\circ\text{K}$$

$$T_{\text{hot air}} = 56^\circ\text{C} \longleftarrow T_{\text{hot air}}$$

Notes:
- ✓ Absolute pressures and temperatures are always used in the ideal gas equation.
- ✓ This problem demonstrates that for lighter-than-air vehicles the buoyancy force exceeds the vehicle weight—that is, the weight of fluid (air) displaced exceeds the vehicle weight.

are in equilibrium under body and buoyancy forces, the location of the line of action of the buoyancy force determines stability, as shown in Fig. 3.10.

The weight of an object acts through its center of gravity, CG. In Fig. 3.10*a*, the lines of action of the buoyancy and the weight are offset in such a way as to produce a couple that tends to right the craft. In Fig. 3.10*b*, the couple tends to capsize the craft.

Ballast may be needed to achieve roll stability. Wooden warships carried stone ballast low in the hull to offset the weight of the heavy cannon on upper gun decks.

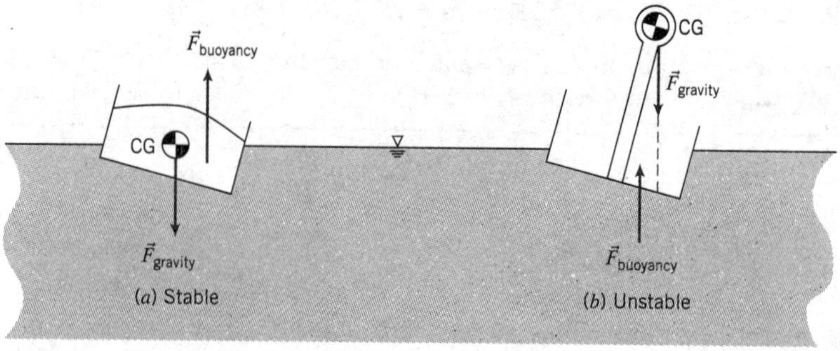

Fig. 3.10 Stability of floating bodies.

Modern ships can have stability problems as well: overloaded ferry boats have capsized when passengers all gathered on one side of the upper deck, shifting the CG laterally. In stacking containers high on the deck of a container ship, care is needed to avoid raising the center of gravity to a level that may result in the unstable condition depicted in Fig. 3.10*b*.

For a vessel with a relatively flat bottom, as shown in Fig. 3.10*a*, the restoring moment increases as roll angle becomes larger. At some angle, typically that at which the edge of the deck goes below water level, the restoring moment peaks and starts to decrease. The moment may become zero at some large roll angle, known as the angle of vanishing stability. The vessel may capsize if the roll exceeds this angle; then, if still intact, the vessel may find a new equilibrium state upside down.

The actual shape of the restoring moment curve depends on hull shape. A broad beam gives a large lateral shift in the line of action of the buoyancy force and thus a high restoring moment. High freeboard above the water line increases the angle at which the moment curve peaks, but may make the moment drop rapidly above this angle.

Sailing vessels are subjected to large lateral forces as wind engages the sails (a boat under sail in a brisk wind typically operates at a considerable roll angle). The lateral wind force must be counteracted by a heavily weighted keel extended below the hull bottom. In small sailboats, crew members may lean far over the side to add additional restoring moment to prevent capsizing [2].

Within broad limits, the buoyancy of a surface vessel is adjusted automatically as the vessel rides higher or lower in the water. However, craft that operate fully submerged must actively adjust buoyancy and gravity forces to remain neutrally buoyant. For submarines this is accomplished using tanks which are flooded to reduce excess buoyancy or blown out with compressed air to increase buoyancy [1]. Airships may vent gas to descend or drop ballast to rise. Buoyancy of a hot-air balloon is controlled by varying the air temperature within the balloon envelope.

For deep ocean dives use of compressed air becomes impractical because of the high pressures (the Pacific Ocean is over 10 km deep; seawater pressure at this depth is greater than 1000 atmospheres!). A liquid such as gasoline, which is buoyant in seawater, may be used to provide buoyancy. However, because gasoline is more compressible than water, its buoyancy decreases as the dive gets deeper. Therefore it is necessary to carry and drop ballast to achieve positive buoyancy for the return trip to the surface.

The most structurally efficient hull shape for airships and submarines has a circular cross-section. The buoyancy force passes through the center of the circle. Therefore, for roll stability the CG must be located below the hull centerline. Thus the crew compartment of an airship is placed beneath the hull to lower the CG.

3.7 Fluids in Rigid-Body Motion (on the Web)

3.8 Summary and Useful Equations

In this chapter we have reviewed the basic concepts of fluid statics. This included:

- ✓ Deriving the basic equation of fluid statics in vector form.
- ✓ Applying this equation to compute the pressure variation in a static fluid:
  - ○ Incompressible liquids: pressure increases uniformly with depth.
  - ○ Gases: pressure decreases nonuniformly with elevation (dependent on other thermodynamic properties).
- ✓ Study of:
  - ○ Gage and absolute pressure.
  - ○ Use of manometers and barometers.
- ✓ Analysis of the fluid force magnitude and location on submerged:
  - ○ Plane surfaces.
  - ○ Curved surfaces.
- ✓ *Derivation and use of Archimedes' Principle of Buoyancy.
- ✓ *Analysis of rigid-body fluid motion (on the Web).

Note: Most of the Useful Equations in the table below have a number of constraints or limitations—*be sure to refer to their page numbers for details*!

Useful Equations

Hydrostatic pressure variation:	$\frac{dp}{dz} = -\rho g \equiv -\gamma$	**(3.6)**	Page 59
Hydrostatic pressure variation (incompressible fluid):	$p - p_0 = \Delta p = \rho g h$	**(3.7)**	Page 61
Hydrostatic pressure variation (several incompressible fluids):	$\Delta p = g \sum_i \rho_i h_i$	**(3.8)**	Page 65
Hydrostatic force on submerged plane (integral form):	$F_R = \int_A p\,dA$	**(3.10a)**	Page 70
Hydrostatic force on submerged plane:	$F_R = p_c A$	**(3.10b)**	Page 70
Location y' of hydrostatic force on submerged plane (integral):	$y' F_R = \int_A yp\,dA$	**(3.11a)**	Page 71
Location y' of hydrostatic force on submerged plane (algebraic):	$y' = y_c + \frac{\rho g \sin\theta\, I_{\hat{x}\hat{x}}}{F_R}$	**(3.11b)**	Page 71
Location y' of hydrostatic force on submerged plane (p_0 neglected):	$y' = y_c + \frac{I_{\hat{x}\hat{x}}}{A y_c}$	**(3.11c)**	Page 71
Location x' of hydrostatic force on submerged plane (integral):	$x' F_R = \int_A x\,p\,dA$	**(3.12a)**	Page 72
Location x' of hydrostatic force on submerged plane (algebraic):	$x' = x_c + \frac{\rho g \sin\theta\, I_{\hat{x}\hat{y}}}{F_R}$	**(3.12b)**	Page 72
Location x' of hydrostatic force on submerged plane (p_0 neglected):	$x' = x_c + \frac{I_{\hat{x}\hat{y}}}{A y_c}$	**(3.12c)**	Page 72
Horizontal and vertical hydrostatic forces on curved submerged surface:	$F_H = p_c A$ and $F_V = \rho g \maltese$	**(3.15)**	Page 77
Buoyancy force on submerged object:	$F_{\text{buoyancy}} = \rho g \maltese$	**(3.16)**	Page 80

We have now concluded our introduction to the fundamental concepts of fluid mechanics, and the basic concepts of fluid statics. In the next chapter we will begin our study of fluids in motion.

*These topics apply to sections that may be omitted without loss of continuity in the text material.

Case Study

The Falkirk Wheel

The Falkirk Wheel. Courtesy of Alan Crawford/iStockphoto.

Hydrostatics, the study of fluids at rest, is an ancient discipline, so one might think there are no new or exciting applications still to be developed. The Falkirk wheel in Scotland is a dramatic demonstration that this is not the case; it is a novel replacement for a lock, a device for moving a boat from one water level to another. The wheel, which has a diameter of 35 m, consists of two sets of axe-shaped opposing arms (which take the shape of a Celtic-inspired, double-headed axe). Sitting in bearings in the ends of these arms are two water-filled caissons, or tanks, each with a capacity of 80,000 gal. The hydrostatics concept of Archimedes' principle, which we studied in this chapter, states that floating objects displace their own weight of water. Hence, the boat shown entering the lower caisson displaces water from the caisson weighing exactly the same as the boat itself. This means the entire wheel remains balanced at all times (both caissons always carry the same weight, whether containing boats or not), and so, despite its enormous mass, it rotates through 180° in less than four minutes while using very little power. The electric motors used for this use 22.5 kilowatts (kW) of power, so the energy used in four minutes is about 1.5 kilowatt-hours (kWh); even at current prices, this works out to be only a few cents worth of energy.

References

1. Burcher, R., and L. Rydill, *Concepts in Submarine Design*. Cambridge, UK: Cambridge University Press, 1994.

2. Marchaj, C. A., *Aero-Hydrodynamics of Sailing*, rev. ed. Camden, ME: International Marine Publishing, 1988.

Problems

3.1 Compressed nitrogen (63.5 kg) is stored in a spherical tank of diameter $D = 0.75$ m at a temperature of 25°C. What is the pressure inside the tank? If the maximum allowable stress in the tank is 210 MPa find the minimum theoretical wall thickness of the tank.

Standard Atmosphere

3.2 Because the pressure falls, water boils at a lower temperature with increasing altitude. Consequently, cake mixes and boiled eggs, among other foods, must be cooked different lengths of time. Determine the boiling temperature of water at 1000 and 2000 m elevation on a standard day, and compare with the sea-level value.

3.3 Ear "popping" is an unpleasant phenomenon sometimes experienced when a change in pressure occurs, for example in a fast-moving elevator or in an airplane. If you are in a two-seater airplane at 3000 m and a descent of 100 m causes your ears to "pop," what is the pressure change that your ears "pop" at, in millimeters of mercury? If the airplane now rises to 8000 m and again begins descending, how far will the airplane descend before your ears "pop" again? Assume a U.S. Standard Atmosphere.

3.4 When you are on a mountain face and boil water, you notice that the water temperature is 90°C. What is your approximate altitude? The next day, you are at a location where it boils at 85°C. How high did you climb between the two days? Assume a U.S. Atmosphere.

Pressure Variation in a Static Fluid

3.5 A 125-mL cube of solid oak is held submerged by a tether as shown. Calculate the actual force of the water on the bottom surface of the cube and the tension in the tether.

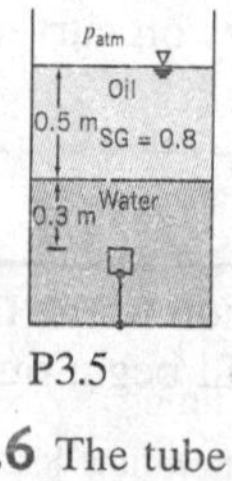

P3.5

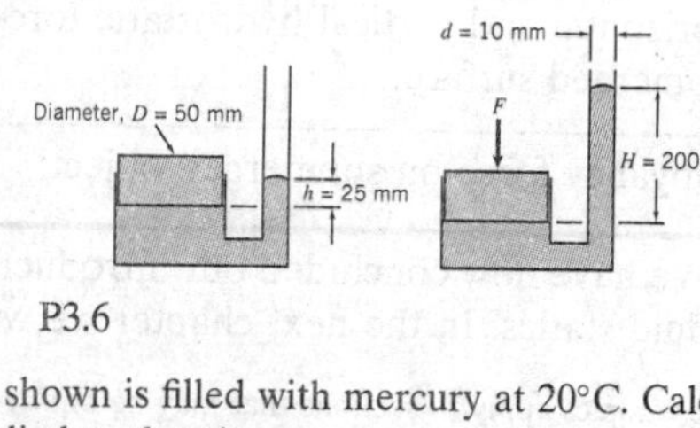

P3.6

3.6 The tube shown is filled with mercury at 20°C. Calculate the force applied to the piston.

3.7 The following pressure and temperature measurements were taken by a meteorological balloon rising through the lower atmosphere:

p (in 10^3 Pa)	101.4	100.8	100.2	99.6	99.0	98.4	97.8	97.2	96.6	96.0	95.4
T (in °C)	12.0	11.1	10.5	10.2	10.1	10.0	10.3	10.8	11.6	12.2	12.1

The initial values (top of table) correspond to ground level. Using the ideal gas law ($p = \rho RT$ with $R = 287\ \mathrm{m^2/(s^2 \cdot K)}$), compute and plot the variation of air density (kg/m³) with height.

3.8 A hollow metal cube with sides 100 mm floats at the interface between a layer of water and a layer of SAE 10W oil such that 10% of the cube is exposed to the oil. What is the pressure difference between the upper and lower horizontal surfaces? What is the average density of the cube?

3.9 Your pressure gage indicates that the pressure in your cold tires is 0.25 MPa (gage) on a mountain at an elevation of 3500 m. What is the absolute pressure? After you drive down to sea level, your tires have warmed to 25°C. What pressure does your gage now indicate? Assume a Standard Atmosphere.

3.10 An air bubble, 8 mm in diameter, is released from the regulator of a scuba diver swimming 30 m below the sea surface. (The water temperature is 30°C.) Estimate the diameter of the bubble just before it reaches the water surface.

3.11 A cube with 150 mm sides is suspended in a fluid by a wire. The top of the cube is horizontal and 203 mm. below the free surface. If the cube has a mass of 29 kg and the tension in the wire is $T = 226$ N, compute the fluid specific gravity, and from this determine the fluid. What are the gage pressures on the upper and lower surfaces?

3.12 Assuming the bulk modulus is constant for seawater, derive an expression for the density variation with depth, h, below the surface. Show that the result may be written

$$\rho \approx \rho_0 + bh$$

where ρ_0 is the density at the surface. Evaluate the constant b. Then, using the approximation, obtain an equation for the variation of pressure with depth below the surface. Determine the depth in meters at which the error in pressure predicted by the approximate solution is 0.01 percent.

3.13 Oceanographic research vessels have descended to 10 km below sea level. At these extreme depths, the compressibility of seawater can be significant. One may model the behavior of seawater by assuming that its bulk modulus remains constant. Using this assumption, evaluate the deviations in density and pressure compared with values computed using the incompressible assumption at a depth, h, of 10 km in seawater. Express your answers as a percentage. Plot the results over the range $0 \le h \le 11$ km.

3.14 An inverted cylindrical container is lowered slowly beneath the surface of a pool of water. Air trapped in the container is compressed isothermally as the hydrostatic pressure increases. Develop an expression for the water height, y, inside the container in terms of the container height, H, and depth of submersion, h. Plot y/H versus h/H.

3.15 You close the top of your straw with your thumb and lift the straw out of your glass containing Coke. Holding it vertically, the total length of the straw is 45 cm, but the Coke held in the straw is in the bottom 15 cm. What is the pressure in the straw just below your thumb? Ignore any surface tension effects.

3.16 A water tank filled with water to a depth of 5 m has in inspection cover (2.5 cm × 2.5 cm) at its base, held in place by a plastic bracket. The bracket can hold a load of 40 N. Is the bracket strong enough? If it is, what would the water depth have to be to cause the bracket to break?

3.17 A container with two circular vertical tubes of diameters $d_1 = 39.5$ mm and $d_2 = 12.7$ mm is partially filled with mercury. The equilibrium level of the liquid is shown in the left diagram. A cylindrical object made from solid brass is placed in the larger tube so that it floats, as shown in the right diagram. The object is $D = 37.5$ mm in diameter and $H = 76.2$ mm high. Calculate the pressure at the lower surface needed to float the object. Determine the new equilibrium level, h, of the mercury with the brass cylinder in place.

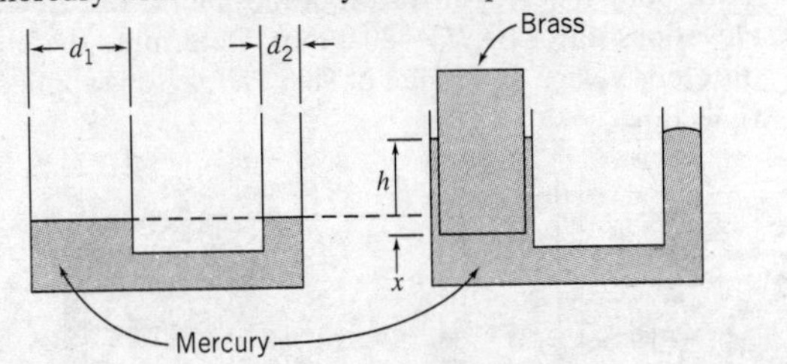

P3.17

3.18 A partitioned tank as shown contains water and mercury. What is the gage pressure in the air trapped in the left chamber? What pressure would the air on the left need to be pumped to in order to bring the water and mercury free surfaces level?

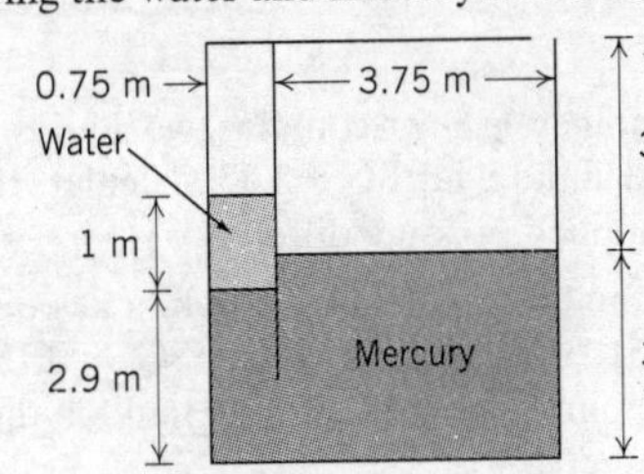

P3.18, P3.19

3.19 In the tank of Problem 3.18, if the opening to atmosphere on the right chamber is first sealed, what pressure would the air on the left now need to be pumped to in order to bring the water and mercury free surfaces level? (Assume the air trapped in the right chamber behaves isothermally.)

3.20 Consider the two-fluid manometer shown. Calculate the applied pressure difference.

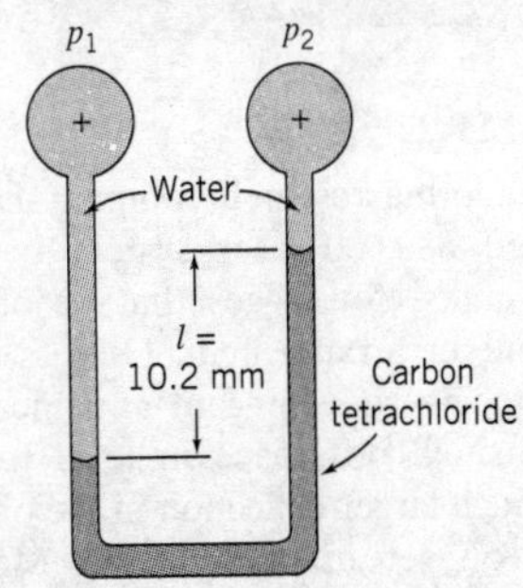

P3.20

3.21 A manometer is formed from glass tubing with uniform inside diameter, $D = 6.35$ mm, as shown. The U-tube is partially filled with water. Then $\forall = 3.25\ \text{cm}^3$ of Meriam red oil is added to the left side. Calculate the equilibrium height, H, when both legs of the U-tube are open to the atmosphere.

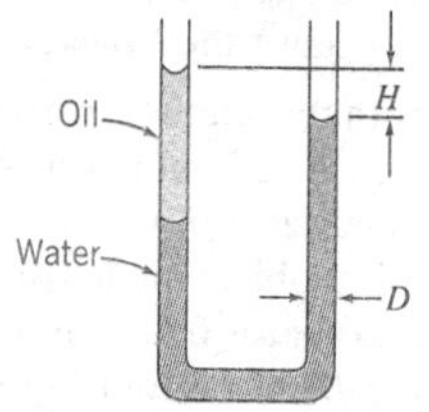

P3.21

3.22 The manometer shown contains water and kerosene. With both tubes open to the atmosphere, the free-surface elevations differ by $H_0 = 20.0$ mm. Determine the elevation difference when a pressure of 98.0 Pa (gage) is applied to the right tube.

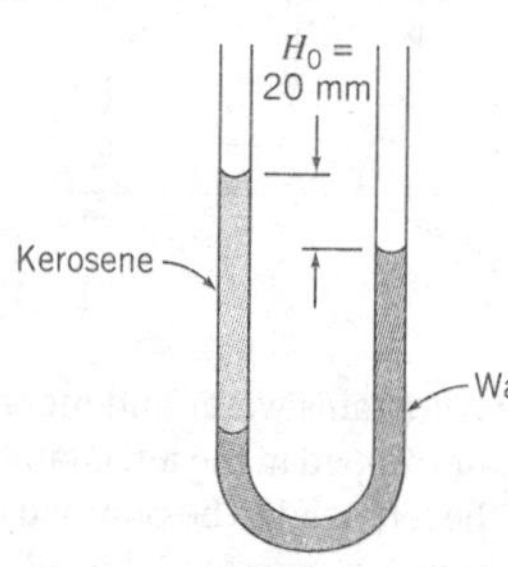

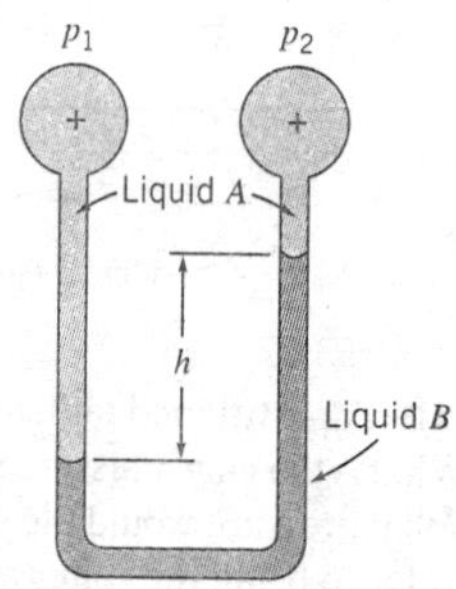

P3.22 P3.23

3.23 The manometer shown contains two liquids. Liquid A has SG = 0.88 and liquid B has SG = 2.95. Calculate the deflection, h, when the applied pressure difference is $p_1 - p_2 = 860$ Pa.

3.24 Determine the gage pressure in kPa at point a, if liquid A has SG = 1.20 and liquid B has SG = 0.75. The liquid surrounding point a is water, and the tank on the left is open to the atmosphere.

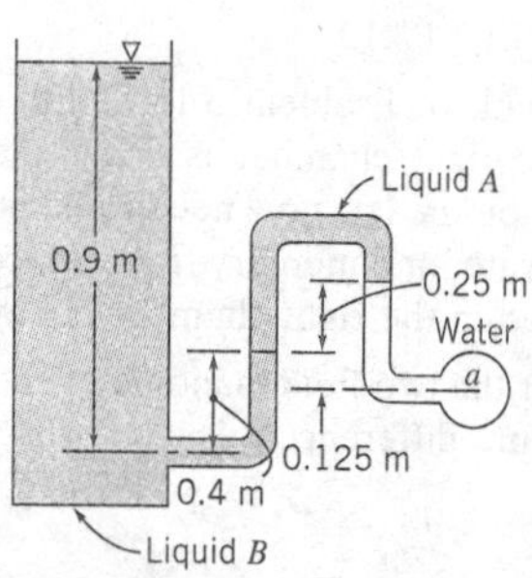

P3.24

3.25 An engineering research company is evaluating using a sophisticated \$80,000 laser system between two large water storage tanks. You suggest that the job can be done with a \$200 manometer arrangement. Oil less dense than water can be used to give a significant amplification of meniscus movement; a small difference in level between the tanks will cause a much larger deflection in the oil levels in the manometer. If you set up a rig using Meriam red oil as the manometer fluid, determine the amplification factor that will be seen in the rig.

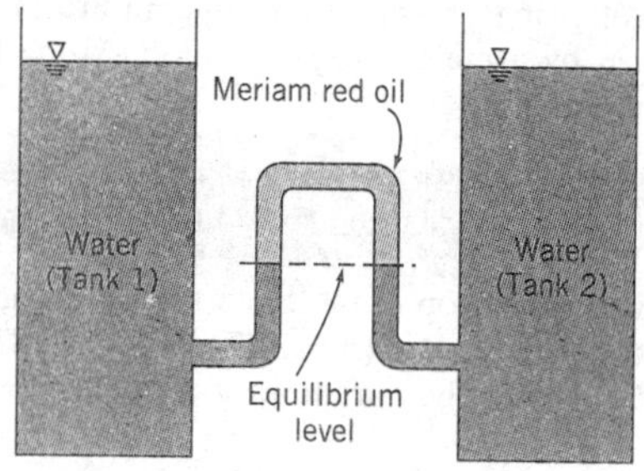

P3.25

3.26 Water flows downward along a pipe that is inclined at 30° below the horizontal, as shown. Pressure difference $p_A - p_B$ is due partly to gravity and partly to friction. Derive an algebraic expression for the pressure difference. Evaluate the pressure difference if $L = 1.5$ m and $h = 150$ mm.

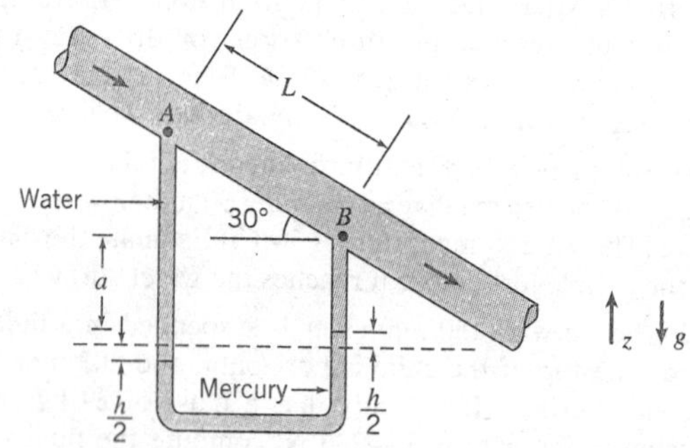

P3.26

3.27 Consider a tank containing mercury, water, benzene, and air as shown. Find the air pressure (gage). If an opening is made in the top of the tank, find the equilibrium level of the mercury in the manometer.

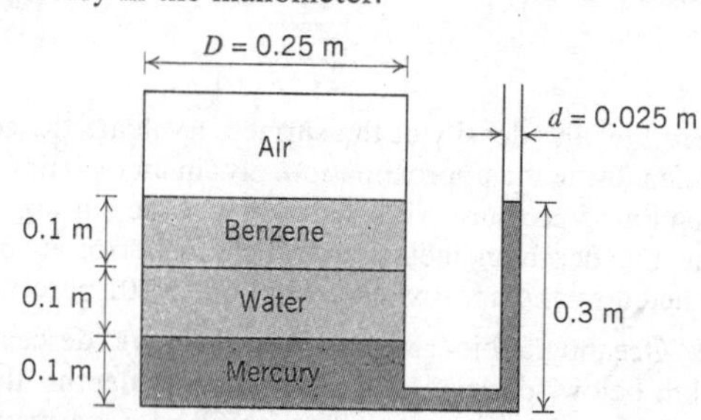

P3.27

3.28 A reservoir manometer has vertical tubes of diameter $D = 18$ mm and $d = 6$ mm. The manometer liquid is Meriam red oil. Develop an algebraic expression for liquid deflection L in the small tube when gage pressure Δp is applied to the reservoir. Evaluate the liquid deflection when the applied pressure is equivalent to 25 mm of water (gage).

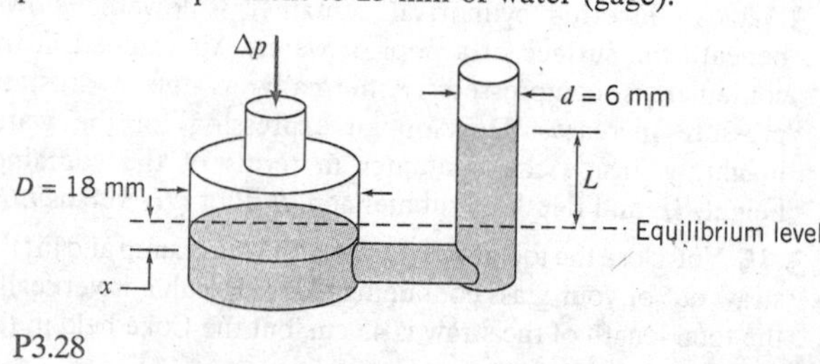

P3.28

3.29 A rectangular tank, open to the atmosphere, is filled with water to a depth of 2.5 m as shown. A U-tube manometer is connected to the tank at a location 0.7 m above the tank bottom. If the zero level of the Meriam blue manometer fluid is 0.2 m below the connection, determine the deflection l after the manometer is connected and all air has been removed from the connecting leg.

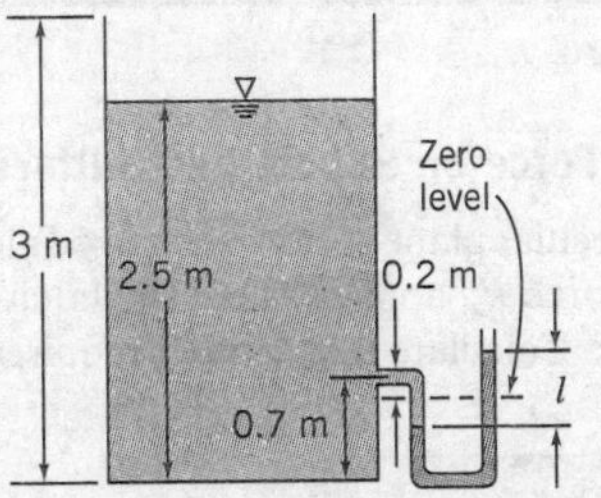

P3.29, P3.31, P3.37

3.30 A reservoir manometer is calibrated for use with a liquid of specific gravity 0.827. The reservoir diameter is 16 mm and the (vertical) tube diameter is 5 mm. Calculate the required distance between marks on the vertical scale for 25 mm. of water pressure difference.

3.31 The manometer fluid of Problem 3.29 is replaced with mercury (same zero level). The tank is sealed and the air pressure is increased to a gage pressure of 50.65 kPa. Determine the deflection l.

3.32 The inclined-tube manometer shown has $D = 96$ mm and $d = 8$ mm. Determine the angle, θ, required to provide a 5 : 1 increase in liquid deflection, L, compared with the total deflection in a regular U-tube manometer. Evaluate the sensitivity of this inclined-tube manometer.

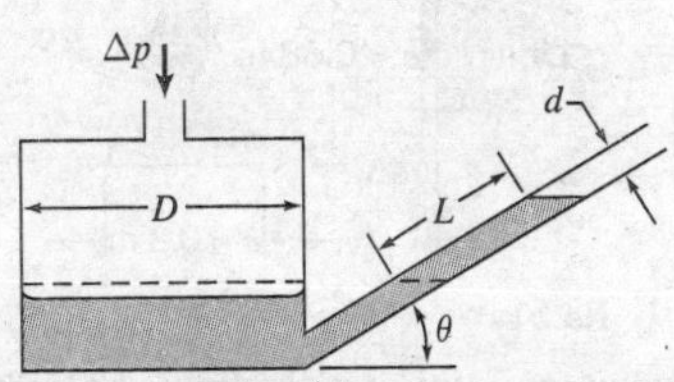

P3.32, P3.33

3.33 The inclined-tube manometer shown has $D = 76$ mm and $d = 8$ mm, and is filled with Meriam red oil. Compute the angle, θ, that will give a 15-cm oil deflection along the inclined tube for an applied pressure of 25 mm of water (gage). Determine the sensitivity of this manometer.

3.34 A barometer accidentally contains 165 mm of water on top of the mercury column (so there is also water vapor instead of a vacuum at the top of the barometer). On a day when the temperature is 21°C, the mercury column height is 720 mm (corrected for thermal expansion). Determine the barometric pressure in Pa. If the ambient temperature increased to 29°C and the barometric pressure did not change, would the mercury column be longer, be shorter, or remain the same length? Justify your answer.

3.35 A student wishes to design a manometer with better sensitivity than a water-filled U-tube of constant diameter. The student's concept involves using tubes with different diameters and two liquids, as shown. Evaluate the deflection h of this manometer, if the applied pressure difference is $\Delta p = 250$ N/m^2. Determine the sensitivity of this manometer. Plot the manometer sensitivity as a function of the diameter ratio d_2/d_1.

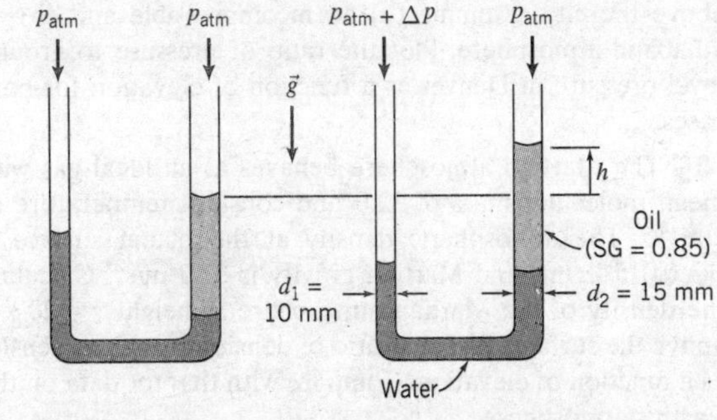

P3.35

3.36 A water column stands 50 mm high in a 2.5 mm diameter glass tube. What would be the column height if the surface tension were zero? What would be the column height in a 1.0-mm diameter tube?

3.37 If the tank of Problem 3.29 is sealed tightly and water drains slowly from the bottom of the tank, determine the deflection, l, after the system has attained equilibrium.

3.38 Consider a small-diameter open-ended tube inserted at the interface between two immiscible fluids of different densities. Derive an expression for the height difference Δh between the interface level inside and outside the tube in terms of tube diameter D, the two fluid densities ρ_1 and ρ_2, and the surface tension σ and angle θ for the two fluids' interface. If the two fluids are water and mercury, find the height difference if the tube diameter is 40 mils (1 mil = 0.0254 mm).

3.39 You have a manometer consisting of a tube that is 1.25 cm inner diameter (ID). On one side, the manometer leg contains mercury, 10 cc of an oil (SG = 1.4), and 3 cc of air as a bubble in the oil. The other leg contains only mercury. Both legs are open to the atmosphere and are in a static condition. An accident occurs in which 3 cc of the oil and the air bubble are removed from one leg. How much do the mercury height levels change?

3.40 Compare the height due to capillary action of water exposed to air in a circular tube of diameter $D = 0.5$ mm, and between two infinite vertical parallel plates of gap $a = 0.5$ mm.

3.41 Two vertical glass plates 300 mm × 300 mm are placed in an open tank containing water. At one end the gap between the plates is 0.1 mm, and at the other it is 2 mm. Plot the curve of water height between the plates from one end of the pair to the other.

3.42 Based on the atmospheric temperature data of the U.S. Standard Atmosphere of Fig. 3.3, compute and plot the pressure variation with altitude, and compare with the pressure data of Table A.3.

3.43 On a certain calm day, a mild inversion causes the atmospheric temperature to remain constant at 30°C between sea level and 5000-m altitude. Under these conditions, (a) calculate the elevation change for which a 3 percent reduction in air pressure occurs, (b) determine the change of elevation necessary to effect a 5 percent reduction in density, and (c) plot p_2/p_1 and ρ_2/ρ_1 as a function of Δz.

3.44 At ground level in Denver, Colorado, the atmospheric pressure and temperature are 83.2 kPa and 25°C. Calculate the pressure on Pike's Peak at an elevation of 2690 m above the city assuming (a) an incompressible and (b) an adiabatic atmosphere. Plot the ratio of pressure to ground level pressure in Denver as a function of elevation for both cases.

3.45 The Martian atmosphere behaves as an ideal gas with mean molecular mass of 32.0 and constant temperature of 200 K. The atmospheric density at the planet surface is $\rho = 0.015$ kg/m^3 and Martian gravity is 3.92 m/s^2. Calculate the density of the Martian atmosphere at height $z = 20$ km above the surface. Plot the ratio of density to surface density as a function of elevation. Compare with that for data on the Earth's atmosphere.

3.46 A door 1 m wide and 1.5 m high is located in a plane vertical wall of a water tank. The door is hinged along its upper edge, which is 1 m below the water surface. Atmospheric pressure acts on the outer surface of the door and at the water surface. (a) Determine the magnitude and line of action of the total resultant force from all fluids acting on the door. (b) If the water surface gage pressure is raised to 0.3 atm, what is the resultant force and where is its line of action? (c) Plot the ratios F/F_0 and y'/y_c for different values of the surface pressure ratio p_s/p_{atm}. (F_0 is the resultant force when $p_s = p_{\text{atm}}$.)

3.47 A door 1 m wide and 1.5 m high is located in a plane vertical wall of a water tank. The door is hinged along its upper edge, which is 1 m below the water surface. Atmospheric pressure acts on the outer surface of the door. (a) If the pressure at the water surface is atmospheric, what force must be applied at the lower edge of the door in order to keep the door from opening? (b) If the water surface gage pressure is raised to 0.5 atm, what force must be applied at the lower edge of the door to keep the door from opening? (c) Find the ratio F/F_0 as a function of the surface pressure ratio p_s/p_{atm}. (F_0 is the force required when $p_s = p_{\text{atm}}$.)

3.48 A hydropneumatic elevator consists of a piston-cylinder assembly to lift the elevator cab. Hydraulic oil, stored in an accumulator tank pressurized by air, is valved to the piston as needed to lift the elevator. When the elevator descends, oil is returned to the accumulator. Design the least expensive accumulator that can satisfy the system requirements. Assume the lift is 3 floors, the maximum load is 10 passengers, and the maximum system pressure is 800 kPa (gage). For column bending strength, the piston diameter must be at least 150 mm. The elevator cab and piston have a combined mass of 3000 kg, and are to be purchased. Perform the analysis needed to define, as a function of system operating pressure, the piston diameter, the accumulator volume and diameter, and the wall thickness. Discuss safety features that your company should specify for the complete elevator system. Would it be preferable to use a completely pneumatic design or a completely hydraulic design? Why?

3.49 Find the pressures at points A, B, and C, as shown in the figure, and in the two air cavities.

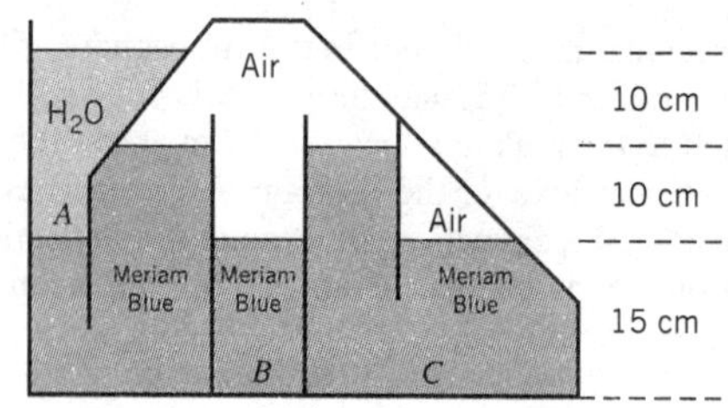

P3.49

Hydrostatic Force on Submerged Surfaces

3.50 Semicircular plane gate AB is hinged along B and held by horizontal force F_A applied at A. The liquid to the left of the gate is water. Calculate the force F_A required for equilibrium.

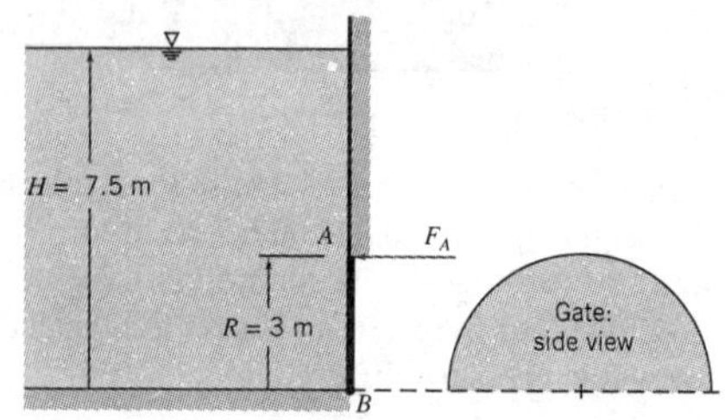

P3.50

3.51 A triangular access port must be provided in the side of a form containing liquid concrete. Using the coordinates and dimensions shown, determine the resultant force that acts on the port and its point of application.

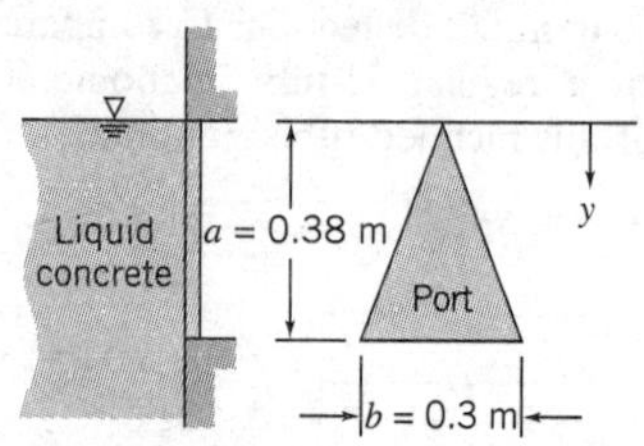

P3.51

3.52 A plane gate of uniform thickness holds back a depth of water as shown. Find the minimum weight needed to keep the gate closed.

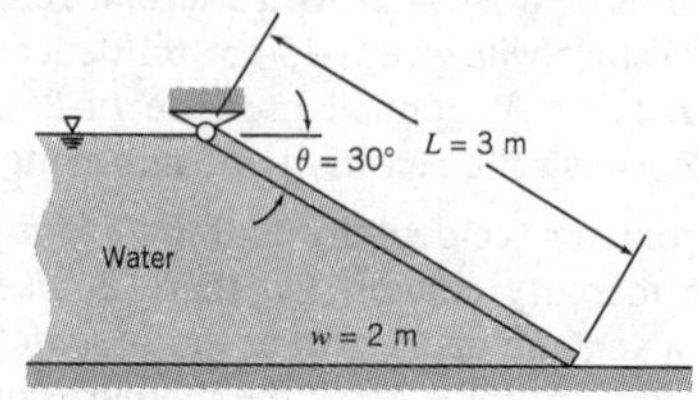

P3.52

3.53 Consider a semicylindrical trough of radius R and length L. Develop general expressions for the magnitude and line of action of the hydrostatic force on one end, if the trough is partially filled with water and open to atmosphere. Plot the results (in nondimensional form) over the range of water depth $0 \le d/R \le 1$.

3.54 A rectangular gate (width $w = 2$ m) is hinged as shown, with a stop on the lower edge. At what depth H will the gate tip?

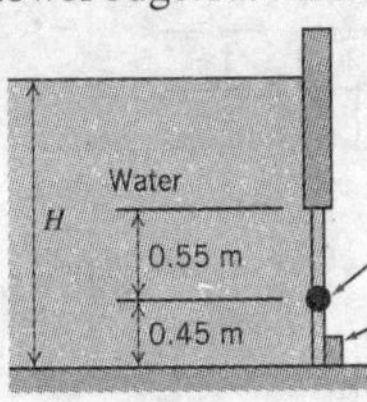

P3.54

3.55 For a mug of tea (65 mm diameter), imagine it cut symmetrically in half by a vertical plane. Find the force that each half experiences due to an 80-mm depth of tea.

3.56 Gates in the Poe Lock at Sault Ste. Marie, Michigan, close a channel $W = 34$ m wide, $L = 360$ m long, and $D = 10$ m deep. The geometry of one pair of gates is shown; each gate is hinged at the channel wall. When closed, the gate edges are forced together at the center of the channel by water pressure. Evaluate the force exerted by the water on gate A. Determine the magnitude and direction of the force components exerted by the gate on the hinge. (Neglect the weight of the gate.)

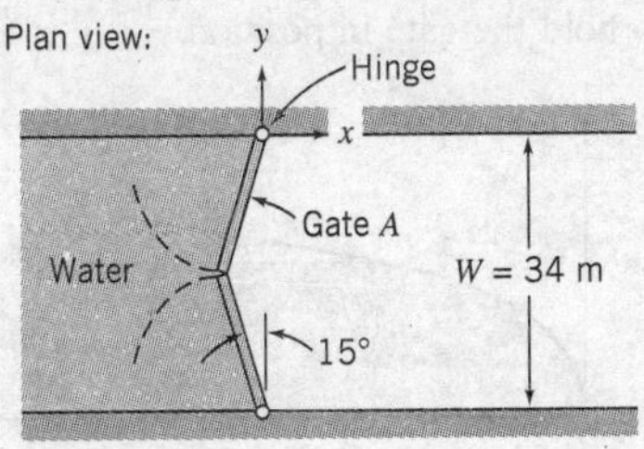

P3.56

3.57 A section of vertical wall is to be constructed from ready-mix concrete poured between forms. The wall is to be 3 m high, 0.25 m thick, and 5 m wide. Calculate the force exerted by the ready-mix concrete on each form. Determine the line of application of the force.

3.58 A window in the shape of an isosceles triangle and hinged at the top is placed in the vertical wall of a form that contains liquid concrete. Determine the minimum force that must be applied at point D to keep the window closed for the configuration of form and concrete shown. Plot the results over the range of concrete depth $0 \leq c \leq a$

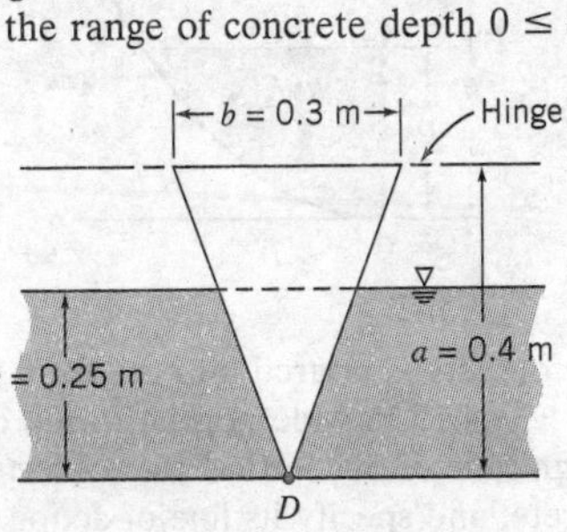

P3.58

3.59 Solve Example 3.6 again using the two separate pressures method. Consider the distributed force to be the sum of a force F_1 caused by the uniform gage pressure and a force F_2 caused by the liquid. Solve for these forces and their lines of action. Then sum moments about the hinge axis to calculate F_t.

3.60 A large open tank contains water and is connected to a 2 m diameter conduit as shown. A circular plug is used to seal the conduit. Determine the magnitude, direction, and location of the force of the water on the plug.

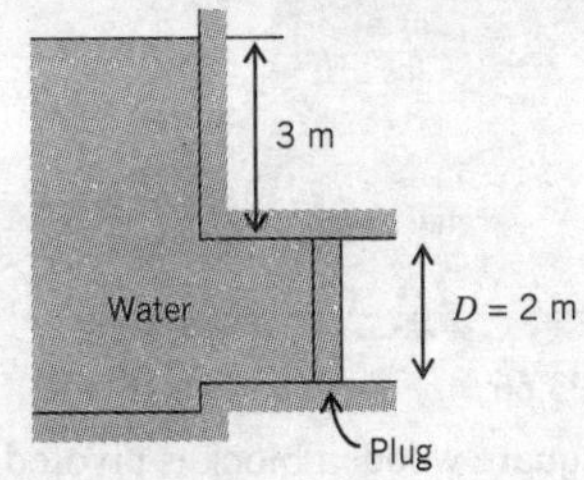

P3.60

3.61 What holds up a car on its rubber tires? Most people would tell you that it is the air pressure inside the tires. However, the air pressure is the same all around the hub (inner wheel), and the air pressure inside the tire therefore pushes down from the top as much as it pushes up from below, having no net effect on the hub. Resolve this paradox by explaining where the force is that keeps the car off the ground.

3.62 The circular access port in the side of a water standpipe has a diameter of 0.6 m and is held in place by eight bolts evenly spaced around the circumference. If the standpipe diameter is 7 m and the center of the port is located 12 m below the free surface of the water, determine (a) the total force on the port and (b) the appropriate bolt diameter.

3.63 As water rises on the left side of the rectangular gate, the gate will open automatically. At what depth above the hinge will this occur? Neglect the mass of the gate.

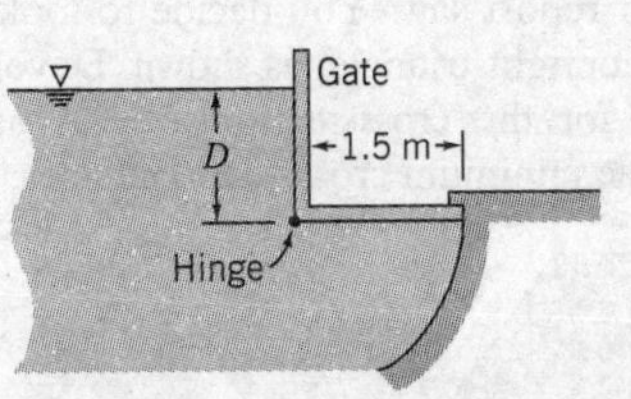

P3.63

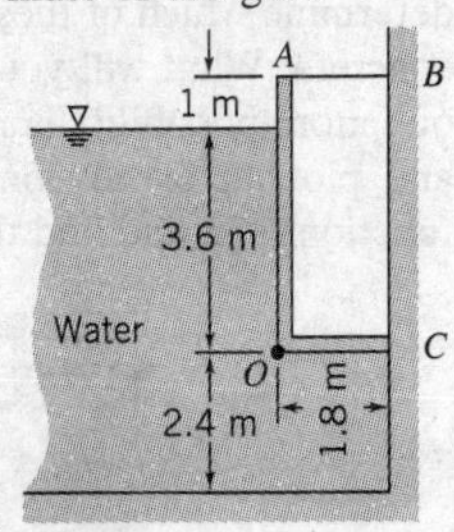

P3.64

3.64 The gate AOC shown is 1.8 m wide and is hinged along O. Neglecting the weight of the gate, determine the force in bar AB. The gate is sealed at C.

3.65 The gate shown is 3 m wide and for analysis can be considered massless. For what depth of water will this rectangular gate be in equilibrium as shown?

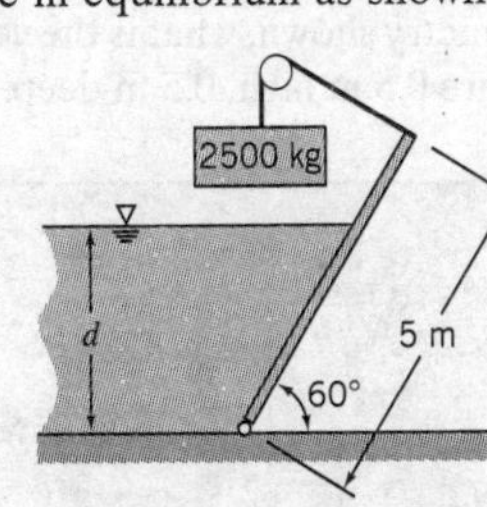

P3.65

3.66 The gate shown is hinged at H. The gate is 3 m wide normal to the plane of the diagram. Calculate the force required at A to hold the gate closed.

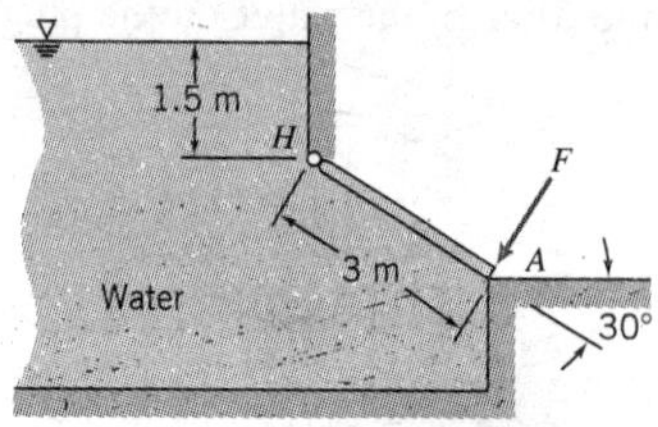

P3.66

3.67 A long, square wooden block is pivoted along one edge. The block is in equilibrium when immersed in water to the depth shown. Evaluate the specific gravity of the wood, if friction in the pivot is negligible.

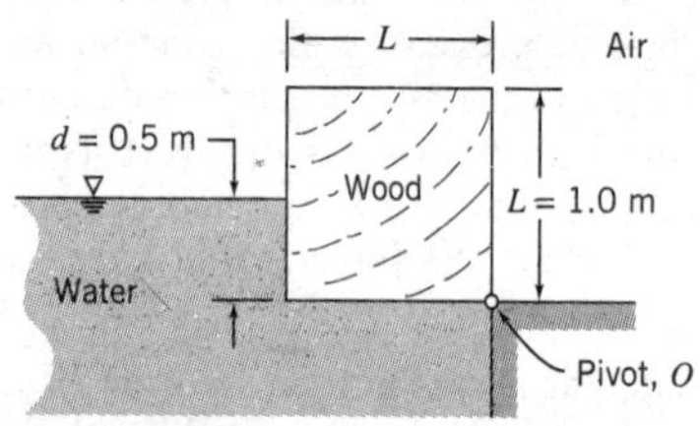

P3.67

3.68 A solid concrete dam is to be built to hold back a depth D of water. For ease of construction the walls of the dam must be planar. Your supervisor asks you to consider the following dam cross-sections: a rectangle, a right triangle with the hypotenuse in contact with the water, and a right triangle with the vertical in contact with the water. She wishes you to determine which of these would require the least amount of concrete. What will your report say? You decide to look at one more possibility: a nonright triangle, as shown. Develop and plot an expression for the cross-section area A as a function of a, and find the minimum cross-sectional area.

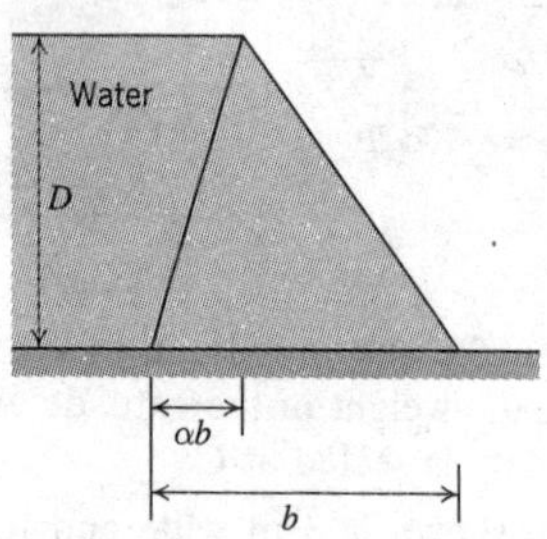

P3.68

3.69 For the geometry shown, what is the vertical force on the dam? The steps are 0.5 m high, 0.5 m deep, and 3 m wide.

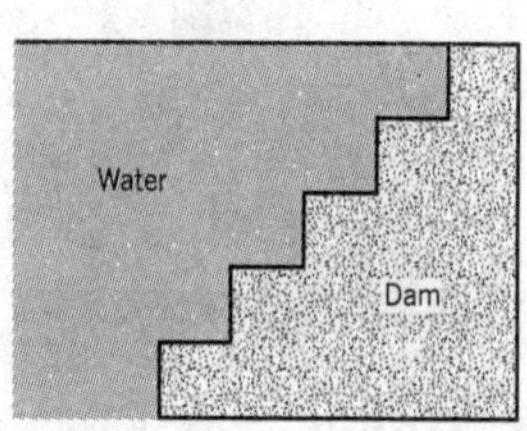

P3.69

3.70 For the dam shown, what is the vertical force of the water on the dam?

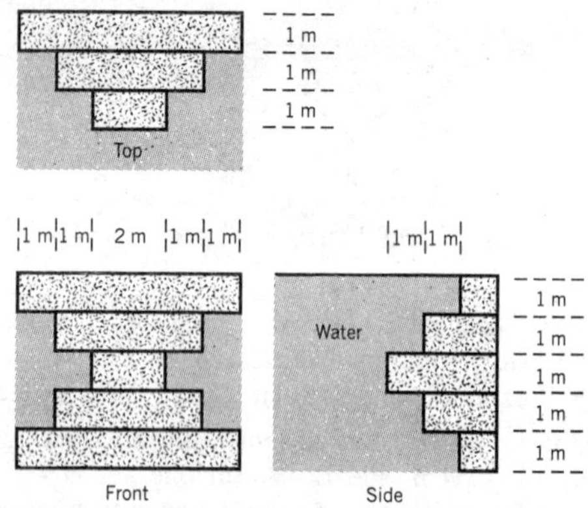

P3.70

3.71 The gate shown is 1.5 m wide and pivoted at O; $a = 1.0\ \text{m}^{-2}$, $D = 1.20$ m, and $H = 1.40$ m. Determine (a) the magnitude and moment of the vertical component of the force about O, and (b) the horizontal force that must be applied at point A to hold the gate in position.

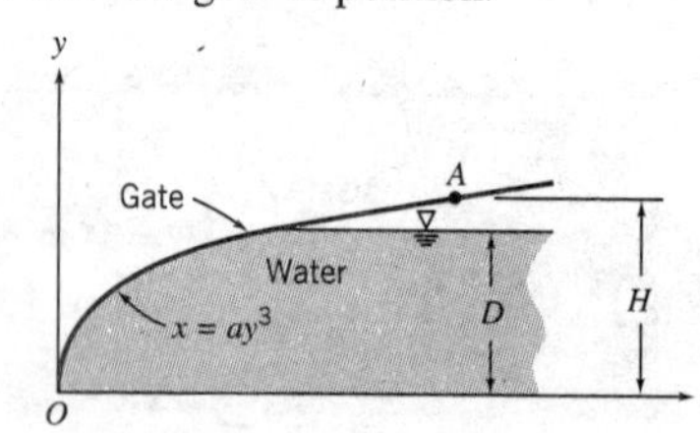

P3.71

3.72 The parabolic gate shown is 2 m wide and pivoted at O; $c = 0.25\ \text{m}^{-1}$, $D = 2$ m, and $H = 3$ m. Determine (a) the magnitude and line of action of the vertical force on the gate due to the water, (b) the horizontal force applied at A required to maintain the gate in equilibrium, and (c) the vertical force applied at A required to maintain the gate in equilibrium.

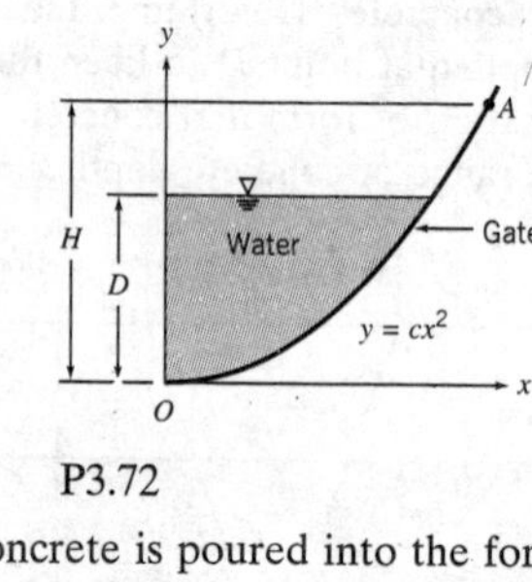

P3.72

3.73 Liquid concrete is poured into the form ($R = 0.313$ m). The form is $w = 4.25$ m wide normal to the diagram. Compute the magnitude of the vertical force exerted on the form by the concrete, and specify its line of action.

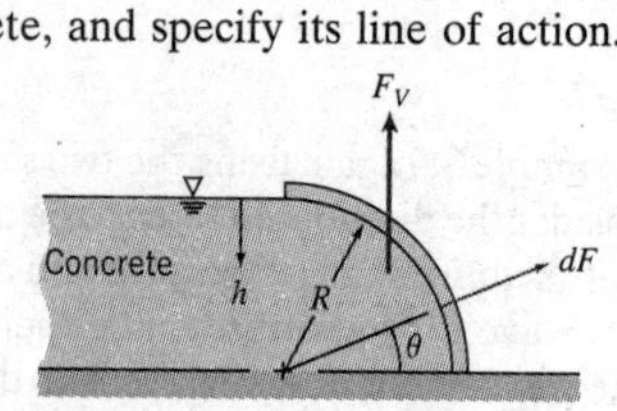

P3.73

3.74 An open tank is filled with water to the depth indicated. Atmospheric pressure acts on all outer surfaces of the tank. Determine the magnitude and line of action of the vertical component of the force of the water on the curved part of the tank bottom.

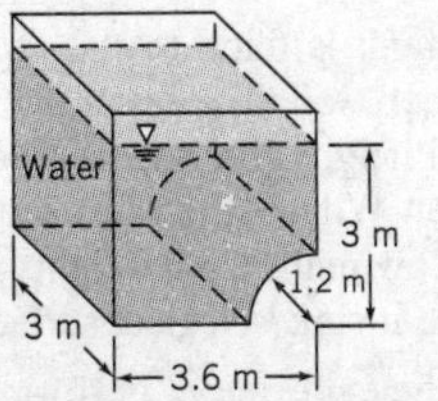

P3.74

3.75 A spillway gate formed in the shape of a circular arc is w m wide. Find the magnitude and line of action of the vertical component of the force due to all fluids acting on the gate.

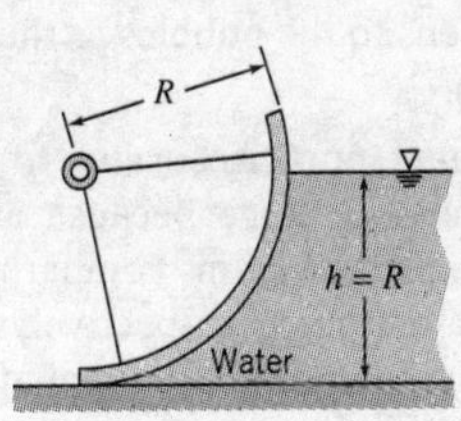

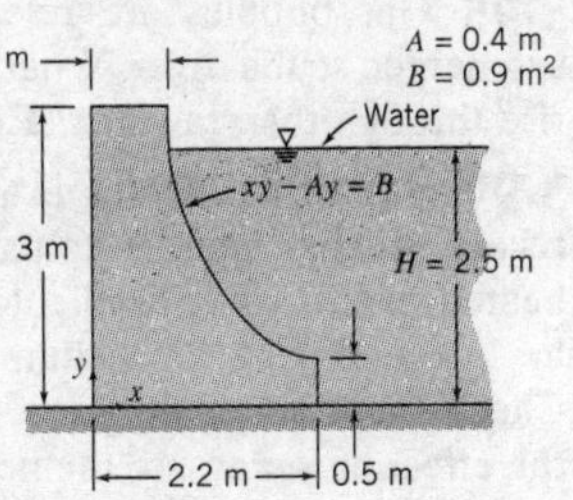

P3.75 P3.76

3.76 A dam is to be constructed using the cross-section shown. Assume the dam width is $w = 50$ m. For water height $H = 2.5$ m, calculate the magnitude and line of action of the vertical force of water on the dam face. Is it possible for water forces to overturn this dam? Under what circumstances will this happen?

3.77 A Tainter gate used to control water flow from the Uniontown Dam on the Ohio River is shown; the gate width is $w = 35$ m. Determine the magnitude, direction, and line of action of the force from the water acting on the gate.

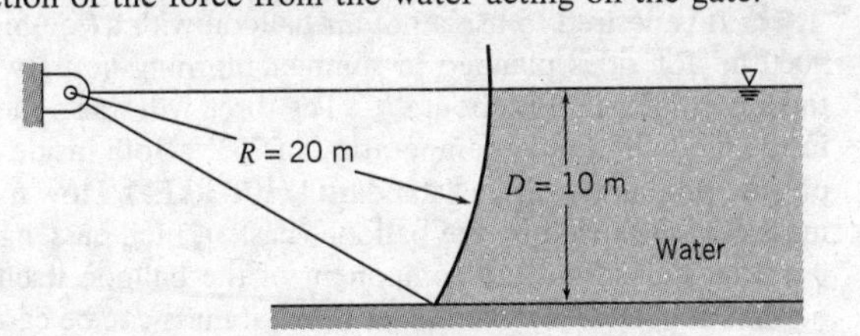

P3.77

3.78 A gate, in the shape of a quarter-cylinder, hinged at A and sealed at B, is 3 m wide. The bottom of the gate is 4.5 m below the water surface. Determine the force on the stop at B if the gate is made of concrete; $R = 3$ m.

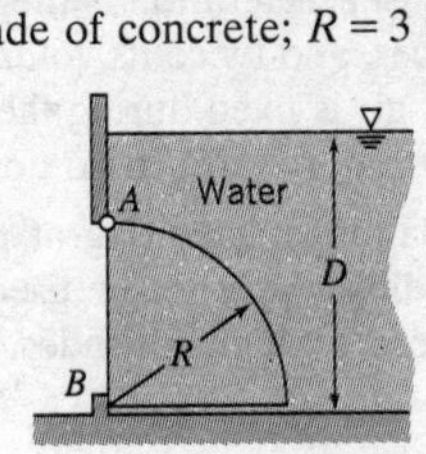

P3.78

3.79 Consider the cylindrical weir of diameter 3 m and length 6 m. If the fluid on the left has a specific gravity of 1.6, and on the right has a specific gravity of 0.8, find the magnitude and direction of the resultant force.

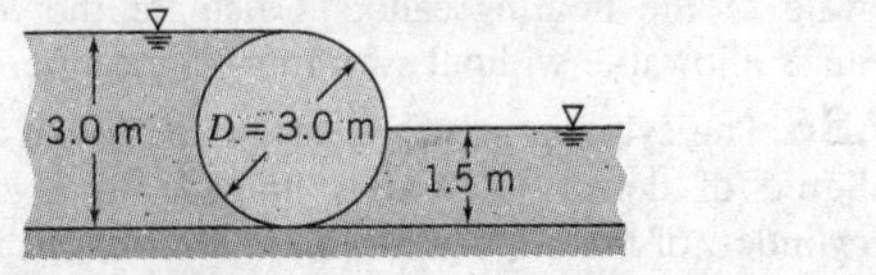

P3.79, P3.80

3.80 A cylindrical weir has a diameter of 3 m and a length of 6 m. Find the magnitude and direction of the resultant force acting on the weir from the water.

3.81 A cylindrical log of diameter D rests against the top of a dam. The water is level with the top of the log and the center of the log is level with the top of the dam. Obtain expressions for (a) the mass of the log per unit length and (b) the contact force per unit length between the log and dam.

3.82 A curved surface is formed as a quarter of a circular cylinder with $R = 0.750$ m as shown. The surface is $w = 3.55$ m wide. Water stands to the right of the curved surface to depth $H = 0.650$ m. Calculate the vertical hydrostatic force on the curved surface. Evaluate the line of action of this force. Find the magnitude and line of action of the horizontal force on the surface.

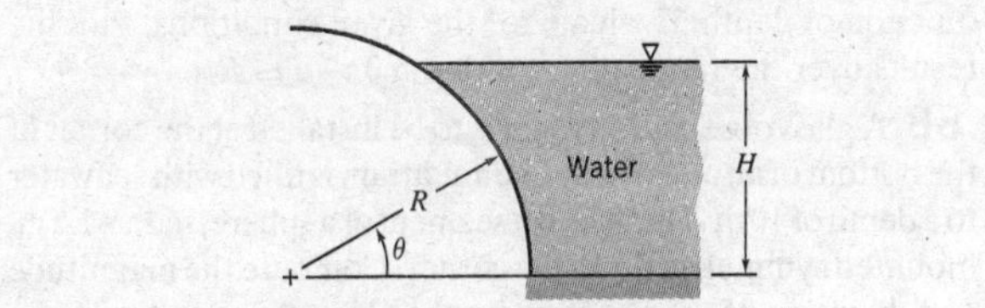

P3.82

Buoyancy and Stability

3.83 If you throw an anchor out of your canoe but the rope is too short for the anchor to rest on the bottom of the pond, will your canoe float higher, lower, or stay the same? Prove your answer.

3.84 A curved submerged surface, in the shape of a quarter cylinder with radius $R = 0.3$ m is shown. The form can withstand a maximum vertical load of 1.6 kN before breaking. The width is $w = 1.25$ m. Find the maximum depth H to which the form may be filled. Find the line of action of the vertical force for this condition. Plot the results over the range of concrete depth $0 \leq H \leq R$.

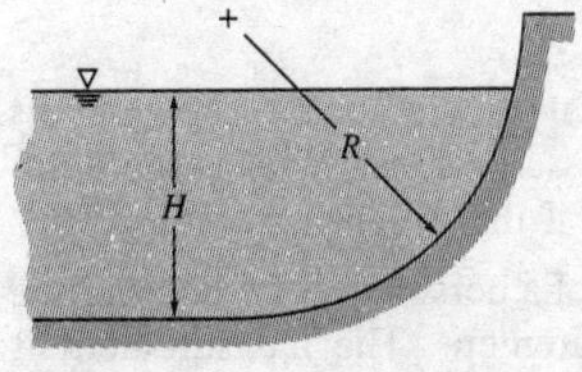

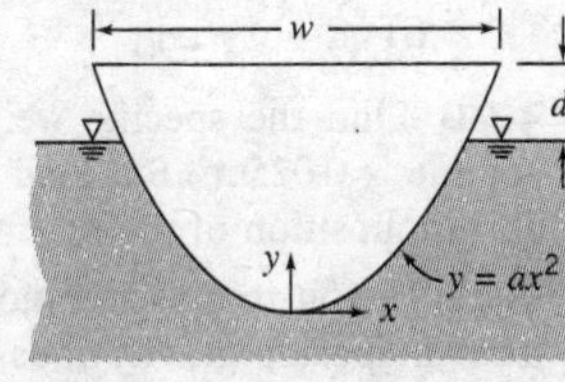

P3.84 P3.85

3.85 The cross-sectional shape of a canoe is modeled by the curve $y = ax^2$, where $a = 3.89\ \text{m}^{-1}$ and the coordinates are in feet. Assume the width of the canoe is constant at $w = 0.6$ m over its entire length $L = 5.25$ m. Set up a general algebraic expression relating the total mass of the canoe and its contents to distance d between the water surface and the gunwale of the floating canoe. Calculate the maximum total mass allowable without swamping the canoe.

3.86 The cylinder shown is supported by an incompressible liquid of density ρ, and is hinged along its length. The cylinder, of mass M, length L, and radius R, is immersed in liquid to depth H. Obtain a general expression for the cylinder specific gravity versus the ratio of liquid depth to cylinder radius, $\alpha = H/R$, needed to hold the cylinder in equilibrium for $0 \leq \alpha < 1$. Plot the results.

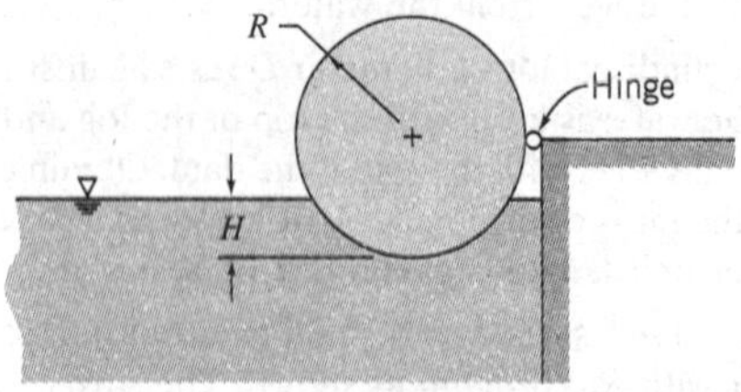

P3.86

3.87 A canoe is represented by a right semicircular cylinder, with $R = 0.35$ m and $L = 5.25$ m. The canoe floats in water that is $d = 0.245$ m deep. Set up a general algebraic expression for the total mass (canoe and contents) that can be floated, as a function of depth. Evaluate for the given conditions. Plot the results over the range of water depth $0 \leq d \leq R$.

3.88 A glass observation room is to be installed at the corner of the bottom of an aquarium. The aquarium is filled with seawater to a depth of 10 m. The glass is a segment of a sphere, radius 1.5 m, mounted symmetrically in the corner. Compute the magnitude and direction of the net force on the glass structure.

***3.89** A hydrometer is a specific gravity indicator, the value being indicated by the level at which the free surface intersects the stem when floating in a liquid. The 1.0 mark is the level when in distilled water. For the unit shown, the immersed volume in distilled water is 15 cm³. The stem is 6 mm in diameter. Find the distance, h, from the 1.0 mark to the surface when the hydrometer is placed in a nitric acid solution of specific gravity 1.5.

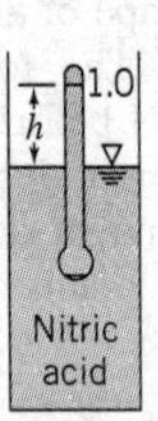

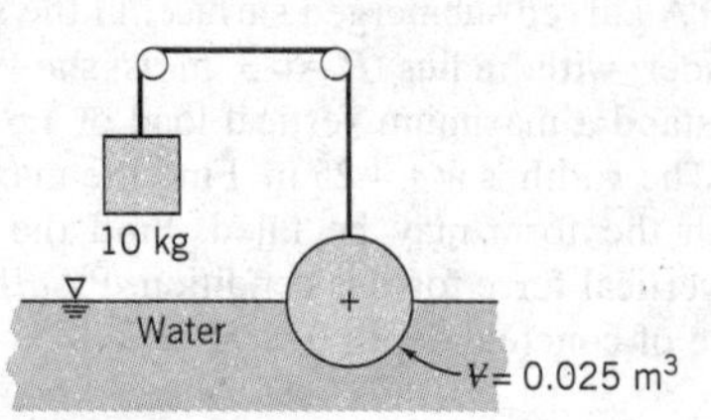

P3.89 P3.90

***3.90** Find the specific weight of the sphere shown if its volume is 0.025m³. State all assumptions. What is the equilibrium position of the sphere if the weight is removed?

***3.91** The fat-to-muscle ratio of a person may be determined from a specific gravity measurement. The measurement is made by immersing the body in a tank of water and measuring the net weight. Develop an expression for the specific gravity of a person in terms of their weight in air, net weight in water, and SG = $f(T)$ for water.

***3.92** Quantify the statement, "Only the tip of an iceberg shows (in seawater)."

***3.93** An open tank is filled to the top with water. A steel cylindrical container, wall thickness $\delta = 1$ mm, outside diameter $D = 100$ mm, and height $H = 1$ m, with an open top, is gently placed in the water. What is the volume of water that overflows from the tank? How many 1 kg weights must be placed in the container to make it sink? Neglect surface tension effects.

***3.94** Quantify the experiment performed by Archimedes to identify the material content of King Hiero's crown. Assume you can measure the weight of the king's crown in air, W_a, and the weight in water, W_w. Express the specific gravity of the crown as a function of these measured values.

***3.95** Gas bubbles are released from the regulator of a submerged scuba diver. What happens to the bubbles as they rise through the seawater? Explain.

***3.96** Hot-air ballooning is a popular sport. According to a recent article, "hot-air volumes must be large because air heated to 65°C over ambient lifts only 0.29 kg/m³ compared to 1.06 and 1.14 for helium and hydrogen, respectively." Check these statements for sea-level conditions. Calculate the effect of increasing the hot-air maximum temperature to 121°C above ambient.

***3.97** Hydrogen bubbles are used to visualize water flow streaklines in the video, *Flow Visualization*. A typical hydrogen bubble diameter is $d = 0.025$ mm. The bubbles tend to rise slowly in water because of buoyancy; eventually they reach terminal speed relative to the water. The drag force of the water on a bubble is given by $F_D = 3\pi\mu V d$, where μ is the viscosity of water and V is the bubble speed relative to the water. Find the buoyancy force that acts on a hydrogen bubble immersed in water. Estimate the terminal speed of a bubble rising in water.

***3.98** It is desired to use a hot air balloon with a volume of 9060 m³ for rides planned in summer morning hours when the air temperature is about 9°C. The torch will warm the air inside the balloon to a temperature of 70°C. Both inside and outside pressures will be "standard" (101.3 kPa). How much mass can be carried by the balloon (basket, fuel, passengers, personal items, and the component of the balloon itself) if neutral buoyancy is to be assured? What mass can be carried by the balloon to ensure vertical takeoff acceleration of 0.75 m/s²? For this, consider that both balloon and inside air have to be accelerated, as well as some of the surrounding air (to make way for the balloon). The rule of thumb is that the total mass subject to acceleration is the mass of the balloon, all its appurtenances, and twice its volume of air. Given that the volume of hot air is fixed during the flight, what can the balloonists do when they want to go down?

***3.99** Scientific balloons operating at pressure equilibrium with the surroundings have been used to lift instrument packages to extremely high altitudes. One such balloon,

*These problems require material from sections that may be omitted without loss of continuity in the text material.

filled with helium, constructed of polyester with a skin thickness of 0.013 mm and a diameter of 120 m, lifted a payload of 230 kg. The specific gravity of the skin material is 1.28. Determine the altitude to which the balloon would rise. Assume that the helium used in the balloon is in thermal equilibrium with the ambient air, and that the balloon is a perfect sphere.

***3.100** A helium balloon is to lift a payload to an altitude of 40 km, where the atmospheric pressure and temperature are 3.0 mbar and −25°C, respectively. The balloon skin is polyester with specific gravity of 1.28 and thickness of 0.015 mm. To maintain a spherical shape, the balloon is pressurized to a gage pressure of 0.45 mbar. Determine the maximum balloon diameter if the allowable tensile stress in the skin is limited to 62 MN/m^2. What payload can be carried?

***3.101** A block of volume 0.025 m^3 is allowed to sink in water as shown. A circular rod 5 m long and 20 cm^2 in cross-section is attached to the weight and also to the wall. If the rod mass is 1.25 kg and the rod makes an angle of 12 degrees with the horizontal at equilibrium, what is the mass of the block?

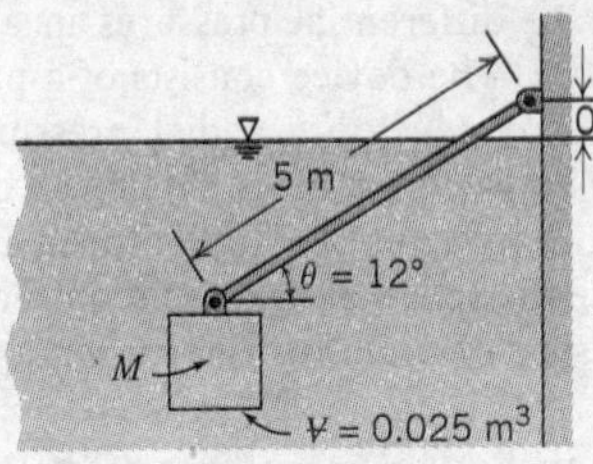

P3.101

***3.102** The stem of a glass hydrometer used to measure specific gravity is 5 mm in diameter. The distance between marks on the stem is 2 mm per 0.1 increment of specific gravity. Calculate the magnitude and direction of the error introduced by surface tension if the hydrometer floats in kerosene. (Assume the contact angle between kerosene and glass is 0°.)

***3.103** A sphere, of radius R, is partially immersed, to depth d, in a liquid of specific gravity SG. Obtain an algebraic expression for the buoyancy force acting on the sphere as a function of submersion depth d. Plot the results over the range of water depth $0 \le d \le 2R$.

***3.104** If the mass M in Problem 3.101 is released from the rod, at equilibrium how much of the rod will remain submerged? What will be the minimum required upward force at the tip of the rod to just lift it out of the water?

***3.105** In a logging operation, timber floats downstream to a lumber mill. It is a dry year, and the river is running low, as low as 60 cm in some locations. What is the largest diameter log that may be transported in this fashion (leaving a minimum 5 cm clearance between the log and the bottom of the river)? For the wood, SG = 0.8.

***3.106** A sphere of radius 25 mm, made from material of specific gravity of SG = 0.95, is submerged in a tank of water. The sphere is placed over a hole of radius 1.88 mm, in the tank bottom. When the sphere is released, will it stay on the bottom of the tank or float to the surface?

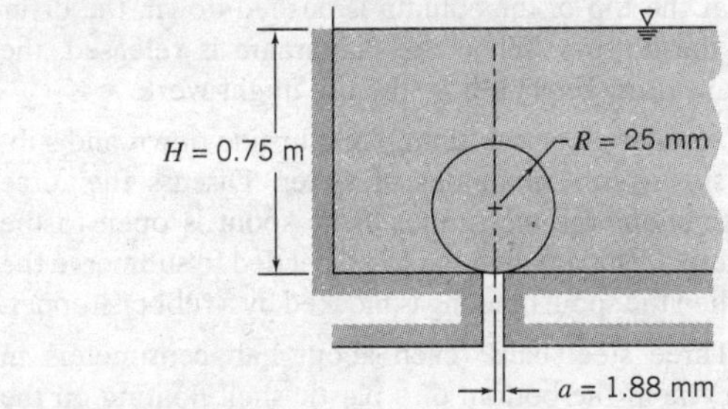

P3.106

***3.107** A cylindrical timber, with $D = 0.3$ m and $L = 4$ m, is weighted on its lower end so that it floats vertically with 3 m submerged in seawater. When displaced vertically from its equilibrium position, the timber oscillates or "heaves" in a vertical direction upon release. Estimate the frequency of oscillation in this heave mode. Neglect viscous effects and water motion.

***3.108** You are in the Bermuda Triangle when you see a bubble plume eruption (a large mass of air bubbles, similar to a foam) off to the side of the boat. Do you want to head toward it and be part of the action? What is the effective density of the water and air bubbles in the drawing on the right that will cause the boat to sink? Your boat is 3 m long, and weight is the same in both cases.

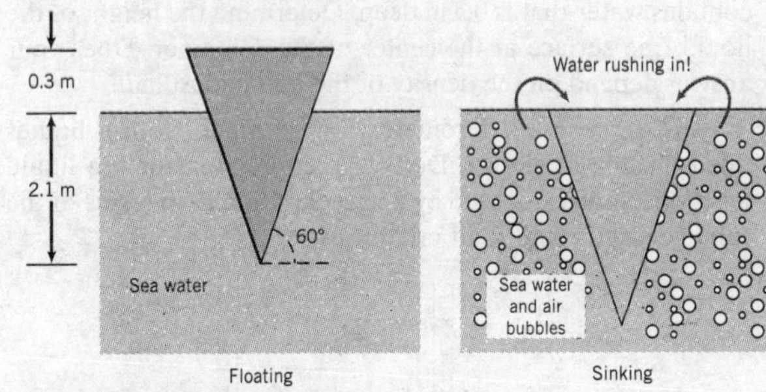

P3.108

***3.109** A bowl is inverted symmetrically and held in a dense fluid, SG = 15.6, to a depth of 200 mm measured along the centerline of the bowl from the bowl rim. The bowl height is 80 mm, and the fluid rises 20 mm inside the bowl. The bowl is 100 mm inside diameter, and it is made from an old clay recipe, SG = 6.1. The volume of the bowl itself is about 0.9 L. What is the force required to hold it in place?

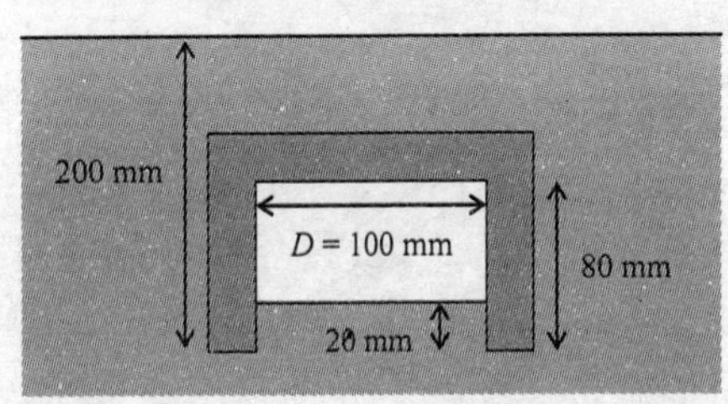

P3.109

*These problems require material from sections that may be omitted without loss of continuity in the text material.

*3.110 In the "Cartesian diver" child's toy, a miniature "diver" is immersed in a column of liquid. When a diaphragm at the top of the column is pushed down, the diver sinks to the bottom. When the diaphragm is released, the diver again rises. Explain how the toy might work.

*3.111 Consider a conical funnel held upside down and submerged slowly in a container of water. Discuss the force needed to submerge the funnel if the spout is open to the atmosphere. Compare with the force needed to submerge the funnel when the spout opening is blocked by a rubber stopper.

*3.112 Three steel balls (each about half centimeters in diameter) lie at the bottom of a plastic shell floating on the water surface in a partially filled bucket. Someone removes the steel balls from the shell and carefully lets them fall to the bottom of the bucket, leaving the plastic shell to float empty. What happens to the water level in the bucket? Does it rise, go down, or remain unchanged? Explain.

*3.113 A proposed ocean salvage scheme involves pumping air into "bags" placed within and around a wrecked vessel on the sea bottom. Comment on the practicality of this plan, supporting your conclusions with analyses.

Fluids in Rigid-Body Motion

*3.114 A cylindrical container, similar to that analyzed in Example 3.10 (on the Web), is rotated at a constant rate of 2 Hz about its axis. The cylinder is 0.5 m in diameter and initially contains water that is 0.3 m deep. Determine the height of the liquid free surface at the center of the container. Does your answer depend on the density of the liquid? Explain.

*3.115 A crude accelerometer can be made from a liquid-filled U-tube as shown. Derive an expression for the liquid level difference h caused by an acceleration $\vec{a}$, in terms of the tube geometry and fluid properties.

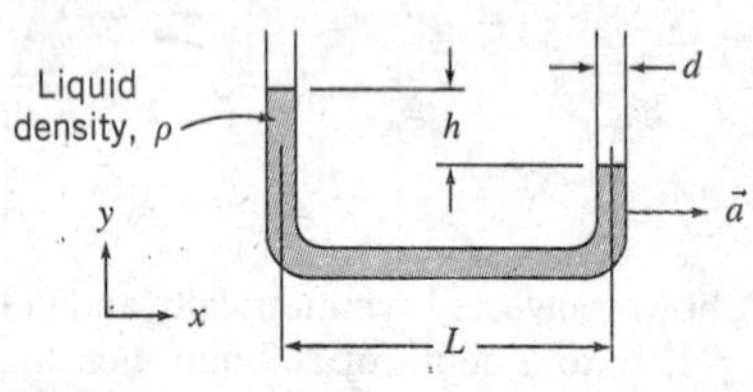

P3.115

*3.116 A rectangular container of water undergoes constant acceleration down an incline as shown. Determine the slope of the free surface using the coordinate system shown.

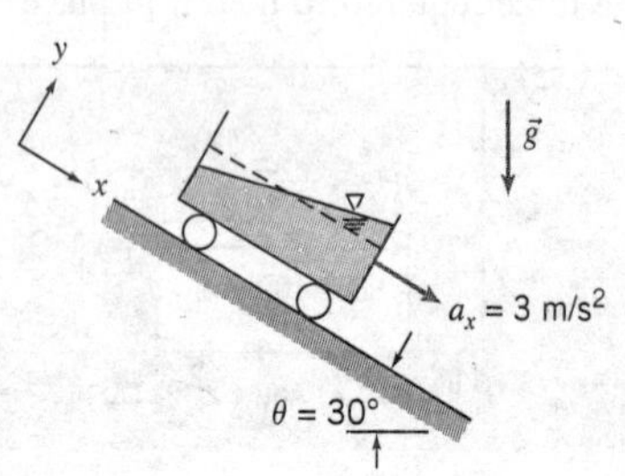

P3.116

*3.117 The U-tube shown is filled with water at $T = 20°C$. It is sealed at A and open to the atmosphere at D. The tube is rotated about vertical axis AB at 1600 rpm. For the dimensions shown, would cavitation occur in the tube?

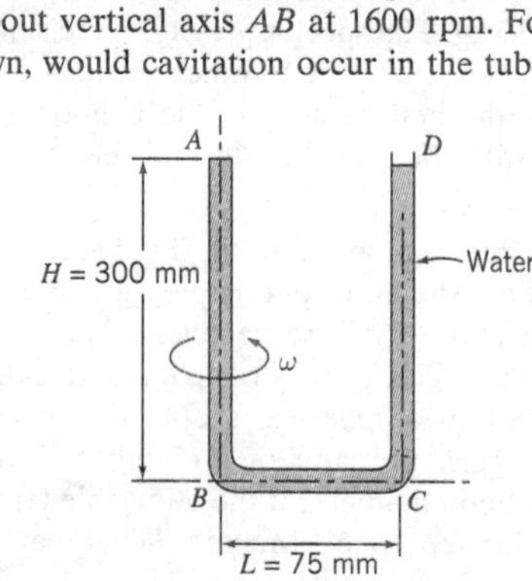

P3.117, P3.118

*3.118 If the U-tube of Problem 3.117 is spun at 300 rpm, what will the pressure be at A? If a small leak appears at A, how much water will be lost at D?

*3.119 A centrifugal micromanometer can be used to create small and accurate differential pressures in air for precise measurement work. The device consists of a pair of parallel disks that rotate to develop a radial pressure difference. There is no flow between the disks. Obtain an expression for pressure difference in terms of rotation speed, radius, and air density. Evaluate the speed of rotation required to develop a differential pressure of 8 μm of water using a device with a 50 mm radius.

*3.120 A test tube is spun in a centrifuge. The tube support is mounted on a pivot so that the tube swings outward as rotation speed increases. At high speeds, the tube is nearly horizontal. Find (a) an expression for the radial component of acceleration of a liquid element located at radius r, (b) the radial pressure gradient dp/dr, and (c) the required angular velocity to generate a pressure of 250 MPa in the bottom of a test tube containing water. (The free surface and bottom radii are 50 and 130 mm, respectively.)

*3.121 A rectangular container, of base dimensions 0.4 m × 0.2 m and height 0.4 m, is filled with water to a depth of 0.2 m; the mass of the empty container is 10 kg. The container is placed on a plane inclined at 30° to the horizontal. If the coefficient of sliding friction between the container and the plane is 0.3, determine the angle of the water surface relative to the horizontal.

*3.122 If the container of Problem 3.121 slides without friction, determine the angle of the water surface relative to the horizontal. What is the slope of the free surface for the same acceleration up the plane?

*3.123 A cubical box, 80 cm on a side, half-filled with oil (SG = 0.80), is given a constant horizontal acceleration of 0.25 g parallel to one edge. Determine the slope of the free surface and the pressure along the horizontal bottom of the box.

*3.124 Gas centrifuges are used in one process to produce enriched uranium for nuclear fuel rods. The maximum

*These problems require material from sections that may be omitted without loss of continuity in the text material.

peripheral speed of a gas centrifuge is limited by stress considerations to about 300 m/s. Assume a gas centrifuge containing uranium hexafluoride gas, with molecular gas M_m = 352, and ideal gas behavior. Develop an expression for the ratio of maximum pressure to pressure at the centrifuge axis. Evaluate the pressure ratio for a gas temperature of 325°C.

***3.125** A pail, 400 mm in diameter and 400 mm deep, weighs 15 N and contains 200 mm of water. The pail is swung in a vertical circle of 1-m radius at a speed of 5 m/s. Assume the water moves as a rigid body. At the instant when the pail is at the top of its trajectory, compute the tension in the string and the pressure on the bottom of the pail from the water.

***3.126** A partially full can of soda is placed at the outer edge of a child's merry-go-round, located $R = 1.5$ m from the axis of rotation. The can diameter and height are 65 mm and 120 mm, respectively. The can is half full, and the soda has specific gravity SG = 1.05. Evaluate the slope of the liquid surface in the can if the merry-go-round spins at 20 rpm. Calculate the spin rate at which the can would spill, assuming no slippage between the can bottom and the merry-go-round. Would the can most likely spill or slide off the merry-go-round?

***3.127** When a water polo ball is submerged below the surface in a swimming pool and released from rest, it is observed to pop out of the water. How would you expect the height to which it rises above the water to vary with depth of submersion below the surface? Would you expect the same results for a beach ball? For a table-tennis ball?

***3.128** Cast iron or steel molds are used in a horizontal-spindle machine to make tubular castings such as liners and tubes. A charge of molten metal is poured into the spinning mold. The radial acceleration permits nearly uniformly thick wall sections to form. A steel liner, of length $L = 2$ m, outer radius $r_o = 0.15$ m, and inner radius $r_i = 0.10$ m, is to be formed by this process. To attain nearly uniform thickness, the angular velocity should be at least 300 rpm. Determine (a) the resulting radial acceleration on the inside surface of the liner and (b) the maximum and minimum pressures on the surface of the mold.

***3.129** The analysis of Problem 3.121 suggests that it may be possible to determine the coefficient of sliding friction between two surfaces by measuring the slope of the free surface in a liquid-filled container sliding down an inclined surface. Investigate the feasibility of this idea.

*These problems require material from sections that may be omitted without loss of continuity in the text material.

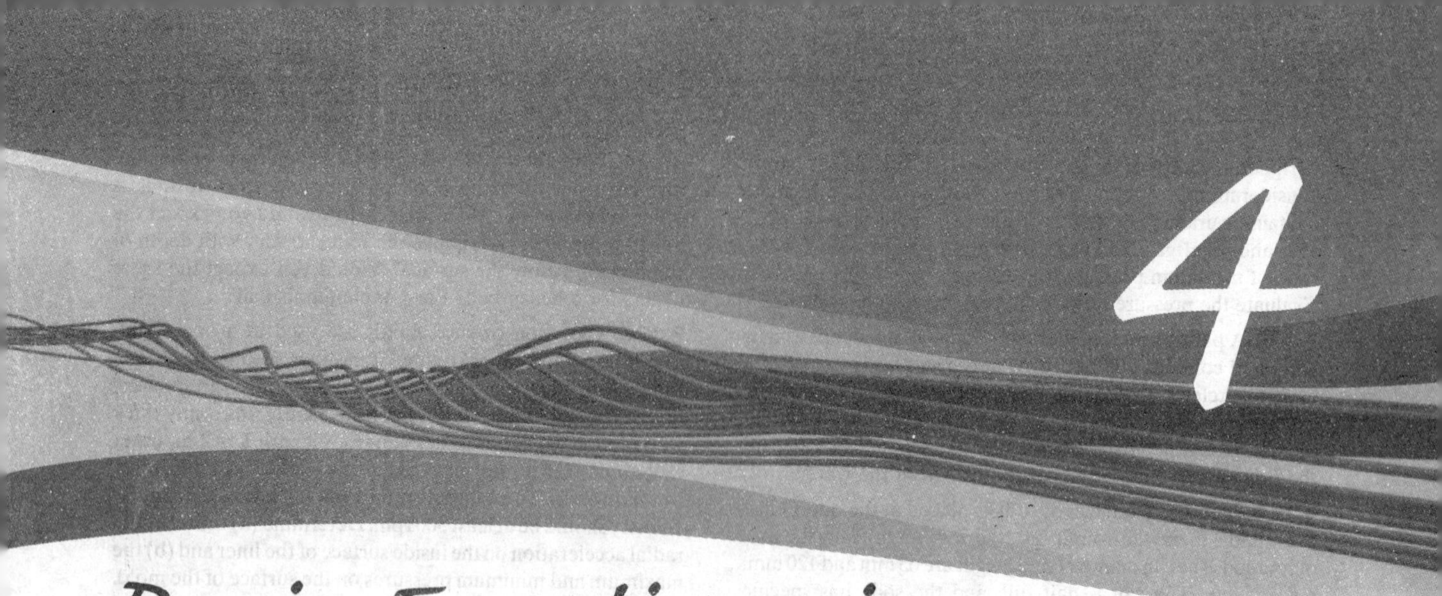

4

Basic Equations in Integral Form for a Control Volume

Case Study in Energy and the Environment

Wave Power: *Pelamis Wave Energy Converter*

As we have seen in earlier Case Studies in Energy and the Environment, there is a lot of renewable energy in ocean waves that could be exploited. A good example of a machine for doing this is the *Pelamis Wave Energy Converter* developed by *Pelamis Wave Power Ltd.* in Scotland. This machine was the world's first commercial-scale machine to generate power and supply it to the power grid from offshore wave energy, and the first to be used in a commercial wave farm project.

Schematic of possible *Pelamis* wave farm. (Picture courtesy of Pelamis Wave Power Ltd.)

The wave-powered electrical generating machine consists of a partially submerged, articulated structure made up of cylindrical sections connected by hinged joints. As waves pass over the structure, the flexing motion of the joints (generated by buoyancy forces, discussed in Chapter 3) is resisted by an arrangement of hydraulic "rams" inside the cylindrical sections; these rams are then used to pump high-pressure fluid through hydraulic motors, which ultimately drive electrical generators to produce electricity. The power that is generated in each joint is sent down a single cable to a junction device on the sea bed; several devices can be connected together (as suggested in the schematic) and linked to shore through a single seabed cable.

The latest generation of machines are 180 meters long (they have four sections, each 45 meters long) and 4 meters in diameter, with four power conversion modules. Each machine can generate up to 750 kilowatts, depending on the specific environmental conditions at the site; they will produce 25–40 percent of the full rated output, on average, over the course of a year. Hence each machine can provide sufficient power to meet the annual electricity demand of about 500 homes. This is not a future technology; three first-generation machines have already been installed off the coast of Portugal, and a single machine is being built and a four-unit machine (generating 3 megawatts of power) is planned for use off the northern coast of Scotland. *Pelamis Wave Power Ltd.* has also expressed interest in installing Pelamis machines off the coast of Cornwall in England, and in the Pacific Ocean off the coast of Tillamook, Oregon. The *Pelamis* machine has a number of advantages: It is durable and low maintenance, uses available technology, and generates electricity inexpensively.

We are now ready to study fluids in motion, so we have to decide how we are to examine a flowing fluid. There are two options available to us, discussed in Chapter 1:

1. We can study the motion of an *individual fluid particle or group of particles* as they move through space. This is the *system* approach, which has the advantage that the physical laws (e.g., Newton's second law, $\vec{F} = d\vec{P}/dt$, where $\vec{F}$ is the force and $d\vec{P}/dt$ is the rate of momentum change of the fluid) apply to matter and hence directly to the system. One disadvantage is that in practice the math associated with this approach can become somewhat complicated, usually leading to a set of partial differential equations. We will look at this approach in detail in Chapter 5. The system approach is needed if we are interested in studying the trajectory of particles over time, for example, in pollution studies.
2. We can study a *region of space* as fluid flows through it, which is the *control volume* approach. This is very often the method of choice, because it has widespread practical application; for example, in aerodynamics we are usually interested in the lift and drag on a wing (which we select as part of the control volume) rather than what happens to individual fluid particles. The disadvantage of this approach is that

the physical laws apply to matter and not directly to regions of space, so we have to perform some math to convert physical laws from their system formulation to a control volume formulation.

We will examine the control volume approach in this chapter. The alert reader will notice that this chapter has the word *integral* in its title, and Chapter 5 has the word *differential*. This is an important distinction: It indicates that we will study a finite region in this chapter and the motion of a particle (an infinitesimal) in Chapter 5 (although in Section 4.4 we will look at a differential control volume to derive the famous Bernoulli equation). The agenda for this chapter is to review the physical laws as they apply to a system (Section 4.1); develop some math to convert from a system to a control volume (Section 4.2) description; and obtain formulas for the physical laws for control volume analysis by combining the results of Sections 4.1 and 4.2.

4.1 Basic Laws for a System

The basic laws we will apply are conservation of mass, Newton's second law, the angular-momentum principle, and the first and second laws of thermodynamics. For converting these system equations to equivalent control volume formulas, it turns out we want to express each of the laws as a rate equation.

Conservation of Mass

For a system (by definition a fixed amount of matter, M, we have chosen) we have the simple result that $M = \text{constant}$. However, as discussed above, we wish to express each physical law as a rate equation, so we write

$$\left.\frac{dM}{dt}\right)_{\text{system}} = 0 \tag{4.1a}$$

where

$$M_{\text{system}} = \int_{M(\text{system})} dm = \int_{\forall(\text{system})} \rho\, d\forall \tag{4.1b}$$

Newton's Second Law

For a system moving relative to an inertial reference frame, Newton's second law states that the sum of all external forces acting on the system is equal to the time rate of change of linear momentum of the system,

$$\vec{F} = \left.\frac{d\vec{P}}{dt}\right)_{\text{system}} \tag{4.2a}$$

where the linear momentum of the system is given by

$$\vec{P}_{\text{system}} = \int_{M(\text{system})} \vec{V}\, dm = \int_{\forall(\text{system})} \vec{V}\rho\, d\forall \tag{4.2b}$$

The Angular-Momentum Principle

The angular-momentum principle for a system states that the rate of change of angular momentum is equal to the sum of all torques acting on the system,

$$\vec{T} = \left.\frac{d\vec{H}}{dt}\right)_{\text{system}} \tag{4.3a}$$

where the angular momentum of the system is given by

$$\vec{H}_{\text{system}} = \int_{M(\text{system})} \vec{r} \times \vec{V}\, dm = \int_{\mathcal{V}(\text{system})} \vec{r} \times \vec{V}\, \rho\, d\mathcal{V} \tag{4.3b}$$

Torque can be produced by surface and body forces (here gravity) and also by shafts that cross the system boundary,

$$\vec{T} = \vec{r} \times \vec{F}_s + \int_{M(\text{system})} \vec{r} \times \vec{g}\, dm + \vec{T}_{\text{shaft}} \tag{4.3c}$$

The First Law of Thermodynamics

The first law of thermodynamics is a statement of conservation of energy for a system,

$$\delta Q - \delta W = dE$$

The equation can be written in rate form as

$$\dot{Q} - \dot{W} = \left.\frac{dE}{dt}\right)_{\text{system}} \tag{4.4a}$$

where the total energy of the system is given by

$$E_{\text{system}} = \int_{M(\text{system})} e\, dm = \int_{\mathcal{V}(\text{system})} e\, \rho\, d\mathcal{V} \tag{4.4b}$$

and

$$e = u + \frac{V^2}{2} + gz \tag{4.4c}$$

In Eq. 4.4a, $\dot{Q}$ (the rate of heat transfer) is positive when heat is added to the system from the surroundings; $\dot{W}$ (the rate of work) is positive when work is done by the system on its surroundings. In Eq. 4.4c, u is the specific internal energy, V the speed, and z the height (relative to a convenient datum) of a particle of substance having mass dm.

The Second Law of Thermodynamics

If an amount of heat, δQ, is transferred to a system at temperature T, the second law of thermodynamics states that the change in entropy, dS, of the system satisfies

$$dS \geq \frac{\delta Q}{T}$$

On a rate basis we can write

$$\left.\frac{dS}{dt}\right)_{\text{system}} \geq \frac{1}{T}\dot{Q} \tag{4.5a}$$

where the total entropy of the system is given by

$$S_{\text{system}} = \int_{M(\text{system})} s\,dm = \int_{\text{V}(\text{system})} s\,\rho\,d\text{V} \tag{4.5b}$$

4.2 Relation of System Derivatives to the Control Volume Formulation

We now have the five basic laws expressed as system rate equations. Our task in this section is to develop a general expression for converting a system rate equation into an equivalent control volume equation. Instead of converting the equations for rates of change of M, $\vec{P}$, $\vec{H}$, E, and S (Eqs. 4.1a, 4.2a, 4.3a, 4.4a, and 4.5a) one by one, we let all of them be represented by the symbol N. Hence N represents the amount of mass, or momentum, or angular momentum, or energy, or entropy of the system. Corresponding to this extensive property, we will also need the intensive (i.e., per unit mass) property η. Thus

$$N_{\text{system}} = \int_{M(\text{system})} \eta\,dm = \int_{\text{V}(\text{system})} \eta\,\rho\,d\text{V} \tag{4.6}$$

Comparing Eq. 4.6 with Eqs. 4.1b, 4.2b, 4.3b, 4.4b, and 4.5b, we see that if:

$$\begin{aligned} N &= M, & \text{then } \eta &= 1 \\ N &= \vec{P}, & \text{then } \eta &= \vec{V} \\ N &= \vec{H}, & \text{then } \eta &= \vec{r}\times\vec{V} \\ N &= E, & \text{then } \eta &= e \\ N &= S, & \text{then } \eta &= s \end{aligned}$$

How can we derive a control volume description from a system description of a fluid flow? Before specifically answering this question, we can describe the derivation in general terms. We imagine selecting an arbitrary piece of the flowing fluid at some time t_0, as shown in Fig. 4.1*a*—we could imagine dyeing this piece of fluid, say, blue.

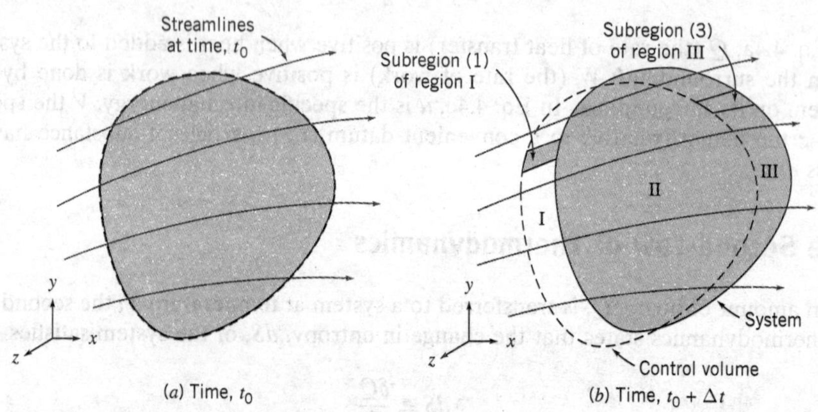

Fig. 4.1 System and control volume configuration.

This initial shape of the fluid system is chosen as our control volume, which is fixed in space relative to coordinates xyz. After an infinitesimal time Δt the system will have moved (probably changing shape as it does so) to a new location, as shown in Fig. 4.1*b*. The laws we discussed above apply to this piece of fluid—for example, its mass will be constant (Eq. 4.1a). By examining the geometry of the system/control volume pair at $t = t_0$ and at $t = t_0 + \Delta t$, we will be able to obtain control volume formulations of the basic laws.

Derivation

From Fig. 4.1 we see that the system, which was entirely within the control volume at time t_0, is partially out of the control volume at time $t_0 + \Delta t$. In fact, three regions can be identified. These are: regions I and II, which together make up the control volume, and region III, which, with region II, is the location of the system at time $t_0 + \Delta t$.

Recall that our objective is to relate the rate of change of any arbitrary extensive property, N, of the system to quantities associated with the control volume. From the definition of a derivative, the rate of change of N_{system} is given by

$$\left.\frac{dN}{dt}\right)_{\text{system}} \equiv \lim_{\Delta t \to 0} \frac{N_s)_{t_0+\Delta t} - N_s)_{t_0}}{\Delta t} \tag{4.7}$$

For convenience, subscript s has been used to denote the system in the definition of a derivative in Eq. 4.7.

From the geometry of Fig. 4.1,

$$N_s)_{t_0+\Delta t} = (N_{\text{II}} + N_{\text{III}})_{t_0+\Delta t} = (N_{\text{CV}} - N_{\text{I}} + N_{\text{III}})_{t_0+\Delta t}$$

and

$$N_s)_{t_0} = (N_{\text{CV}})_{t_0}$$

Substituting into the definition of the system derivative, Eq. 4.7, we obtain

$$\left.\frac{dN}{dt}\right)_s = \lim_{\Delta t \to 0} \frac{(N_{\text{CV}} - N_{\text{I}} + N_{\text{III}})_{t_0+\Delta t} - N_{\text{CV}})_{t_0}}{\Delta t}$$

Since the limit of a sum is equal to the sum of the limits, we can write

$$\left.\frac{dN}{dt}\right)_s = \underbrace{\lim_{\Delta t \to 0} \frac{N_{\text{CV}})_{t_0+\Delta t} - N_{\text{CV}})_{t_0}}{\Delta t}}_{①} + \underbrace{\lim_{\Delta t \to 0} \frac{N_{\text{III}})_{t_0+\Delta t}}{\Delta t}}_{②} - \underbrace{\lim_{\Delta t \to 0} \frac{N_{\text{I}})_{t_0+\Delta t}}{\Delta t}}_{③} \tag{4.8}$$

Our task now is to evaluate each of the three terms in Eq. 4.8.

Term ① in Eq. 4.8 simplifies to

$$\lim_{\Delta t \to 0} \frac{N_{\text{CV}})_{t_0+\Delta t} - N_{\text{CV}})_{t_0}}{\Delta t} = \frac{\partial N_{\text{CV}}}{\partial t} = \frac{\partial}{\partial t}\int_{\text{CV}} \eta\,\rho\, d\mathcal{V} \tag{4.9a}$$

To evaluate term ② we first develop an expression for $N_{\text{III}})_{t_0+\Delta t}$ by looking at the enlarged view of a typical subregion (subregion (3)) of region III shown in Fig. 4.2. The vector area element $d\vec{A}$ of the control surface has magnitude dA, and its direction is the *outward* normal of the area element. In general, the velocity vector $\vec{V}$ will be at some angle α with respect to $d\vec{A}$.

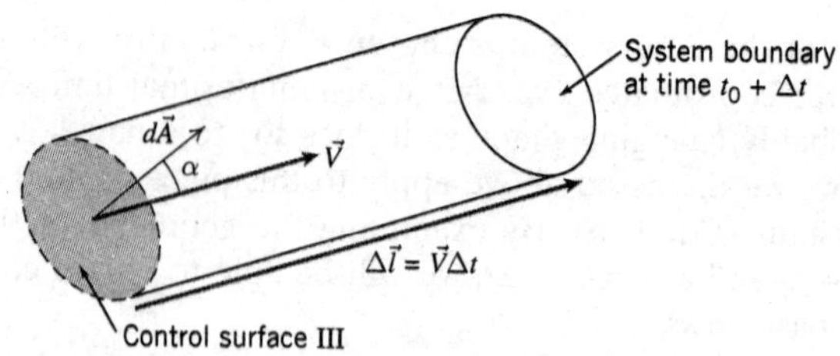

Fig. 4.2 Enlarged view of subregion (3) from Fig. 4.1.

For this subregion we have

$$dN_{\text{III}})_{t_0+\Delta t} = (\eta\,\rho\,d\mathcal{V})_{t_0+\Delta t}$$

We need to obtain an expression for the volume $d\mathcal{V}$ of this cylindrical element. The vector length of the cylinder is given by $\Delta\vec{l} = \vec{V}\Delta t$. The volume of a prismatic cylinder, whose area $d\vec{A}$ is at an angle α to its length $\Delta\vec{l}$, is given by $d\mathcal{V} = \Delta l\,dA\cos\alpha = \Delta\vec{l}\cdot d\vec{A} = \vec{V}\cdot d\vec{A}\Delta t$. Hence, for subregion (3) we can write

$$dN_{\text{III}})_{t_0+\Delta t} = \eta\,\rho\vec{V}\cdot d\vec{A}\,\Delta t$$

Then, for the entire region III we can integrate and for term ② in Eq. 4.8 obtain

$$\lim_{\Delta t\to 0}\frac{N_{\text{III}})_{t_0+\Delta t}}{\Delta t} = \lim_{\Delta t\to 0}\frac{\int_{\text{CS}_{\text{III}}} dN_{\text{III}})_{t_0+\Delta t}}{\Delta t} = \lim_{\Delta t\to 0}\frac{\int_{\text{CS}_{\text{III}}}\eta\rho\vec{V}\cdot d\vec{A}\,\Delta t}{\Delta t} = \int_{\text{CS}_{\text{III}}}\eta\rho\vec{V}\cdot d\vec{A} \tag{4.9b}$$

We can perform a similar analysis for subregion (1) of region I, and obtain for term in Eq. 4.8

$$\lim_{\Delta t\to 0}\frac{N_{\text{I}})_{t_0+\Delta t}}{\Delta t} = -\int_{\text{CS}_{\text{I}}}\eta\,\rho\vec{V}\cdot d\vec{A} \tag{4.9c}$$

For subregion (1), the velocity vector acts *into* the control volume, but the area normal *always* (by convention) points outward (angle $\alpha > \pi/2$), so the scalar product in Eq. 4.9c is negative. Hence the minus sign in Eq. 4.9c is needed to cancel the negative result of the scalar product to make sure we obtain a positive result for the amount of matter that was in region I (we can't have negative matter).

This concept of the sign of the scalar product is illustrated in Fig. 4.3 for (a) the general case of an inlet or exit, (b) an exit velocity parallel to the surface normal, and (c) an inlet velocity parallel to the surface normal. Cases (b) and (c) are obviously convenient special cases of (a); the value of the cosine in case (a) automatically generates the correct sign of either an inlet or an exit.

We can finally use Eqs. 4.9a, 4.9b, and 4.9c in Eq. 4.8 to obtain

$$\left.\frac{dN}{dt}\right)_{\text{system}} = \frac{\partial}{\partial t}\int_{\text{CV}}\eta\,\rho\,d\mathcal{V} + \int_{\text{CS}_{\text{I}}}\eta\,\rho\vec{V}\cdot d\vec{A} + \int_{\text{CS}_{\text{III}}}\eta\,\rho\vec{V}\cdot d\vec{A}$$

and the two last integrals can be combined because CS_{I} and CS_{III} constitute the entire control surface,

$$\left.\frac{dN}{dt}\right)_{\text{system}} = \frac{\partial}{\partial t}\int_{\text{CV}}\eta\,\rho\,d\mathcal{V} + \int_{\text{CS}}\eta\,\rho\vec{V}\cdot d\vec{A} \tag{4.10}$$

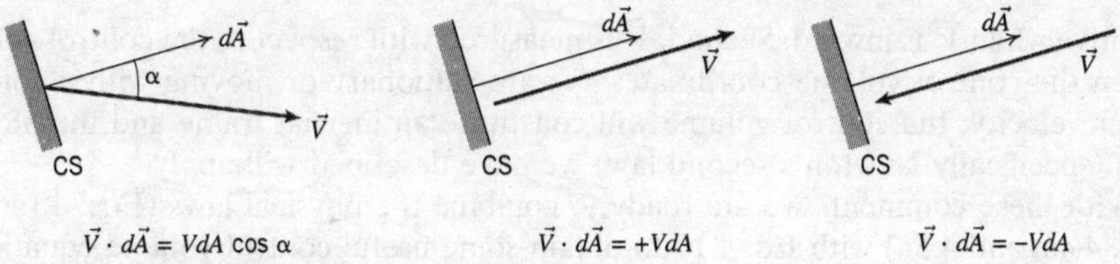

Fig. 4.3 Evaluating the scalar product.

Equation 4.10 is the relation we set out to obtain. It is the fundamental relation between the rate of change of any arbitrary extensive property, N, of a system and the variations of this property associated with a control volume. Some authors refer to Eq. 4.10 as the *Reynolds Transport Theorem*.

Physical Interpretation

It took several pages, but we have reached our goal: We now have a formula (Eq. 4.10) that we can use to convert the rate of change of any extensive property N of a system to an equivalent formulation for use with a control volume. We can now use Eq. 4.10 in the various basic physical law equations (Eqs. 4.1a, 4.2a, 4.3a, 4.4a, and 4.5a) one by one, with N replaced with each of the properties M, $\vec{P}$, $\vec{H}$, E, and S (with corresponding symbols for η), to replace system derivatives with control volume expressions. Because we consider the equation itself to be "basic" we repeat it to emphasize its importance:

$$\left.\frac{dN}{dt}\right)_{\text{system}} = \frac{\partial}{\partial t}\int_{\text{CV}} \eta\,\rho\,d\forall + \int_{\text{CS}} \eta\,\rho\vec{V}\cdot d\vec{A} \tag{4.10}$$

We need to be clear here: The system is the matter that happens to be passing through the chosen control volume, at the instant we chose. For example, if we chose as a control volume the region contained by an airplane wing and an imaginary rectangular boundary around it, the system would be the mass of the air that is instantaneously contained between the rectangle and the airfoil. Before applying Eq. 4.10 to the physical laws, let's discuss the meaning of each term of the equation:

$\left.\frac{dN}{dt}\right)_{\text{system}}$ is the rate of change of the system extensive property N. For example, if $N = \vec{P}$, we obtain the rate of change of momentum.

$\frac{\partial}{\partial t}\int_{\text{CV}} \eta\,\rho\,d\forall$ is the rate of change of the amount of property N in the control volume. The term $\int_{\text{CV}} \eta\,\rho\,d\forall$ computes the instantaneous value of N in the control volume ($\int_{\text{CV}} \rho\,d\forall$ is the instantaneous mass in the control volume). For example, if $N = \vec{P}$, then $\eta = \vec{V}$ and $\int_{\text{CV}} \vec{V}\rho d\forall$ computes the instantaneous amount of momentum in the control volume.

$\int_{\text{CS}} \eta\,\rho\vec{V}\cdot d\vec{A}$ is the rate at which property N is exiting the surface of the control volume. The term $\rho\vec{V}\cdot d\vec{A}$ computes the rate of mass transfer leaving across control surface area element $d\vec{A}$; multiplying by η computes the rate of flux of property N across the element; and integrating therefore computes the net flux of N out of the control volume. For example, if $N = \vec{P}$, then $\eta = \vec{V}$ and $\int_{\text{CS}} \vec{V}\rho\vec{V}\cdot d\vec{A}$ computes the net flux of momentum out of the control volume.

We make two comments about velocity $\vec{V}$ in Eq. 4.10. First, we reiterate the discussion for Fig. 4.3 that care should be taken in evaluating the dot product: Because $\vec{A}$ is always directed outwards, the dot product will be positive when $\vec{V}$ is outward and negative when $\vec{V}$ is inward. Second, $\vec{V}$ is measured with respect to the control volume: When the control volume coordinates xyz are stationary or moving with a constant linear velocity, the control volume will constitute an inertial frame and the physical laws (specifically Newton's second law) we have described will apply.[1]

With these comments we are ready to combine the physical laws (Eqs. 4.1a, 4.2a, 4.3a, 4.4a, and 4.5a) with Eq. 4.10 to obtain some useful control volume equations.

4.3 Conservation of Mass

The first physical principle to which we apply this conversion from a system to a control volume description is the mass conservation principle: The mass of the system remains constant,

$$\left.\frac{dM}{dt}\right)_{\text{system}} = 0 \tag{4.1a}$$

where

$$M_{\text{system}} = \int_{M(\text{system})} dm = \int_{V\!\!\!\!-(\text{system})} \rho\, dV\!\!\!\!- \tag{4.1b}$$

The system and control volume formulations are related by Eq. 4.10,

$$\left.\frac{dN}{dt}\right)_{\text{system}} = \frac{\partial}{\partial t}\int_{\text{CV}} \eta\,\rho\, dV\!\!\!\!- + \int_{\text{CS}} \eta\,\rho\vec{V}\cdot d\vec{A} \tag{4.10}$$

where

$$N_{\text{system}} = \int_{M(\text{system})} \eta\, dm = \int_{V\!\!\!\!-(\text{system})} \eta\,\rho\, dV\!\!\!\!- \tag{4.6}$$

To derive the control volume formulation of conservation of mass, we set

$$N = M \quad \text{and} \quad \eta = 1$$

With this substitution, we obtain

$$\left.\frac{dM}{dt}\right)_{\text{system}} = \frac{\partial}{\partial t}\int_{\text{CV}} \rho\, dV\!\!\!\!- + \int_{\text{CS}} \rho\vec{V}\cdot d\vec{A} \tag{4.11}$$

Comparing Eqs. 4.1a and 4.11, we arrive (after rearranging) at the control volume formulation of the conservation of mass:

$$\frac{\partial}{\partial t}\int_{\text{CV}} \rho\, dV\!\!\!\!- + \int_{\text{CS}} \rho\vec{V}\cdot d\vec{A} = 0 \tag{4.12}$$

In Eq. 4.12 the first term represents the rate of change of mass within the control volume; the second term represents the net rate of mass flux out through the control surface. Equation 4.12 indicates that the rate of change of mass in the control volume plus the net outflow is zero. The mass conservation equation is also called the

[1]For an accelerating control volume (one whose coordinates xyz are accelerating with respect to an "absolute" set of coordinates XYZ), we must modify the form of Newton's second law (Eq. 4.2a). We will do this in Sections 4.6 (linear acceleration) and 4.7 (arbitrary acceleration).

continuity equation. In common-sense terms, the rate of increase of mass in the control volume is due to the net inflow of mass:

$$\begin{array}{ccc}\text{Rate of increase} & = & \text{Net influx of} \\ \text{of mass in CV} & & \text{mass}\end{array}$$

$$\frac{\partial}{\partial t}\int_{CV}\rho\, dV\!\!\!\!/ \quad = \quad -\int_{CS}\rho\vec{V}\cdot d\vec{A}$$

VIDEO

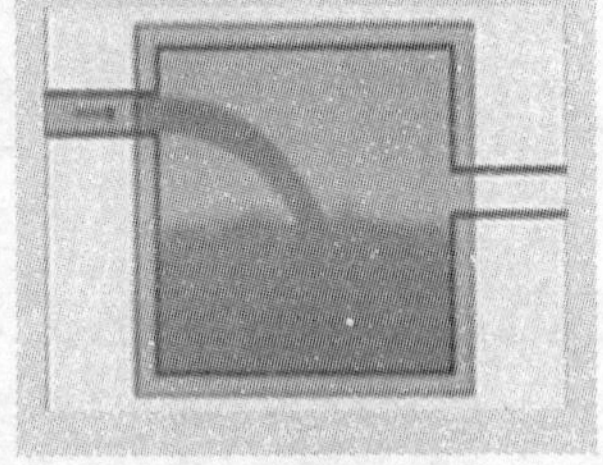

Mass Conservation: Filling a Tank.

Once again, we note that in using Eq. 4.12, care should be taken in evaluating the scalar product $\vec{V}\cdot d\vec{A} = VdA\cos\alpha$: It could be positive (outflow, $\alpha<\pi/2$), negative (inflow, $\alpha>\pi/2$), or even zero ($\alpha=\pi/2$). Recall that Fig. 4.3 illustrates the general case as well as the convenient cases $\alpha=0$ and $\alpha=\pi$.

Special Cases

In special cases it is possible to simplify Eq. 4.12. Consider first the case of an incompressible fluid, in which density remains constant. When ρ is constant, it is not a function of space or time. Consequently, for *incompressible fluids*, Eq. 4.12 may be written as

$$\rho\frac{\partial}{\partial t}\int_{CV} dV\!\!\!\!/ + \rho\int_{CS}\vec{V}\cdot d\vec{A} = 0$$

The integral of $dV\!\!\!\!/$ over the control volume is simply the volume of the control volume. Thus, on dividing through by ρ, we write

$$\frac{\partial V\!\!\!\!/}{\partial t} + \int_{CS}\vec{V}\cdot d\vec{A} = 0$$

For a nondeformable control volume of fixed size and shape, $V\!\!\!\!/$ = constant. The conservation of mass for incompressible flow through a fixed control volume becomes

$$\int_{CS}\vec{V}\cdot d\vec{A} = 0 \tag{4.13a}$$

A useful special case is when we have (or can approximate) uniform velocity at each inlet and exit. In this case Eq. 4.13a simplifies to

$$\sum_{CS}\vec{V}\cdot\vec{A} = 0 \tag{4.13b}$$

Note that we have not assumed the flow to be steady in reducing Eq. 4.12 to the forms 4.13a and 4.13b. We have only imposed the restriction of incompressible fluid. Thus Eqs. 4.13a and 4.13b are statements of conservation of mass for flow of an incompressible fluid that may be steady or unsteady.

The dimensions of the integrand in Eq. 4.13a are L^3/t. The integral of $\vec{V}\cdot d\vec{A}$ over a section of the control surface is commonly called the *volume flow rate* or *volume rate of flow*. Thus, for incompressible flow, the volume flow rate into a fixed control volume must be equal to the volume flow rate out of the control volume. The volume flow rate Q, through a section of a control surface of area A, is given by

$$Q = \int_{A}\vec{V}\cdot d\vec{A} \tag{4.14a}$$

The average velocity magnitude, $\vec{V}$, at a section is defined as

$$\vec{V} = \frac{Q}{A} = \frac{1}{A}\int_A \vec{V} \cdot d\vec{A} \tag{4.14b}$$

Consider now the general case of *steady, compressible flow* through a fixed control volume. Since the flow is steady, this means that at most $\rho = \rho(x, y, z)$. By definition, no fluid property varies with time in a steady flow. Consequently, the first term of Eq. 4.12 must be zero and, hence, for steady flow, the statement of conservation of mass reduces to

$$\int_{CS} \rho\vec{V} \cdot d\vec{A} = 0 \tag{4.15a}$$

A useful special case is when we have (or can approximate) uniform velocity at each inlet and exit. In this case, Eq. 4.15a simplifies to

$$\sum_{CS} \rho\vec{V} \cdot \vec{A} = 0 \tag{4.15b}$$

Thus, for steady flow, the mass flow rate into a control volume must be equal to the mass flow rate out of the control volume.

We will now look at three Examples to illustrate some features of the various forms of the conservation of mass equation for a control volume. Example 4.1 involves a problem in which we have uniform flow at each section, Example 4.2 involves a problem in which we do not have uniform flow at a location, and Example 4.3 involves a problem in which we have unsteady flow.

Example 4.1 MASS FLOW AT A PIPE JUNCTION

Consider the steady flow in a water pipe joint shown in the diagram. The areas are: $A_1 = 0.2\ \text{m}^2$, $A_2 = 0.2\ \text{m}^2$, and $A_3 = 0.15\ \text{m}^2$. In addition, fluid is lost out of a hole at ④, estimated at a rate of $0.1\ \text{m}^3/\text{s}$. The average speeds at sections ① and ③ are $V_1 = 5$ m/s and $V_3 = 12$ m/s, respectively. Find the velocity at section ②.

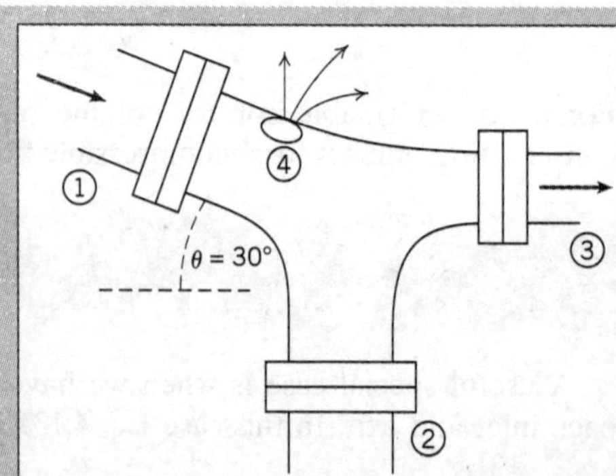

Given: Steady flow of water through the device.

$$A_1 = 0.2\,\text{m}^2 \quad A_2 = 0.2\,\text{m}^2 \quad A_3 = 0.15\,\text{m}^2$$
$$V_1 = 5\,\text{m/s} \quad V_3 = 12\,\text{m/s} \quad \rho = 999\,\text{kg/m}^3$$

Volume flow rate at ④ $= 0.1\ \text{m}^3/\text{s}$

Find: Velocity at section ②.

Solution: Choose a fixed control volume as shown. Make an assumption that the flow at section ② is outwards, and label the diagram accordingly (if this assumption is incorrect our final result will tell us).

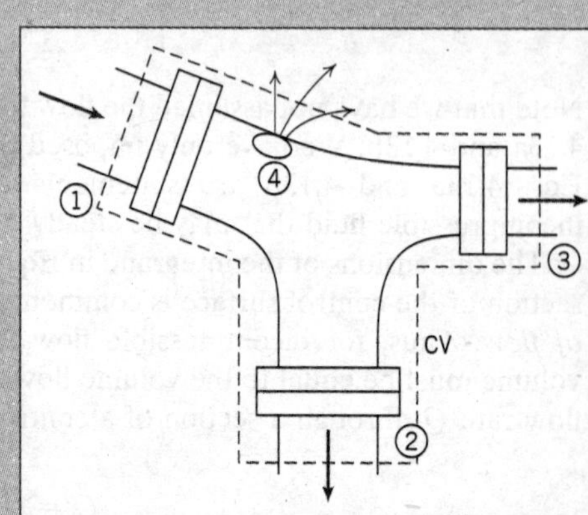

Governing equation: The general control volume equation is Eq. 4.12, but we can go immediately to Eq. 4.13b because of assumptions (2) and (3) below,

$$\sum_{CS} \vec{V} \cdot \vec{A} = 0$$

Assumptions: (1) Steady flow (given).
(2) Incompressible flow.
(3) Uniform properties at each section.

Hence (using Eq. 4.14a for the leak)

$$\vec{V}_1 \cdot \vec{A}_1 + \vec{V}_2 \cdot \vec{A}_2 + \vec{V}_3 \cdot \vec{A}_3 + Q_4 = 0 \qquad (1)$$

where Q_4 is the flow rate out of the leak.

Let us examine the first three terms in Eq. 1 in light of the discussion of Fig. 4.3 and the directions of the velocity vectors:

$$\vec{V}_1 \cdot \vec{A}_1 = -V_1 A_1$$

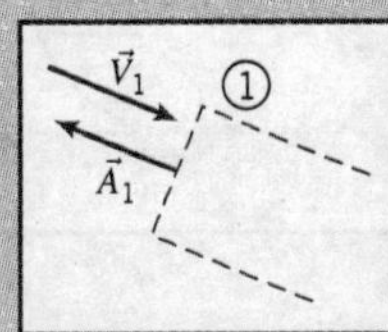

$\left\{ \begin{array}{l} \text{Sign of } \vec{V}_1 \cdot \vec{A}_1 \text{ is} \\ \text{negative at surface } ① \end{array} \right\}$

$$\vec{V}_2 \cdot \vec{A}_2 = +V_2 A_2$$

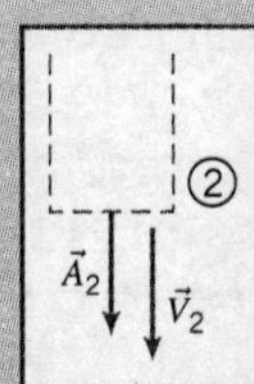

$\left\{ \begin{array}{l} \text{Sign of } \vec{V}_2 \cdot \vec{A}_2 \text{ is} \\ \text{positive at surface } ② \end{array} \right\}$

$$\vec{V}_3 \cdot \vec{A}_3 = +V_3 A_3$$

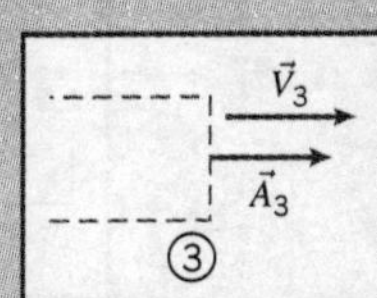

$\left\{ \begin{array}{l} \text{Sign of } \vec{V}_3 \cdot \vec{A}_3 \text{ is} \\ \text{positive at surface } ③ \end{array} \right\}$

Using these results in Eq. 1,

$$-V_1 A_1 + V_2 A_2 + V_3 A_3 + Q_4 = 0$$

or

$$V_2 = \frac{V_1 A_1 - V_3 A_3 - Q_4}{A_2}$$

$$= \frac{5\,\frac{\text{m}}{\text{s}} \times 0.2\,\text{m}^2 - 12\,\frac{\text{m}}{\text{s}} \times 0.15\,\text{m}^2 - \frac{0.1\,\text{m}^3}{\text{s}}}{0.2\,\text{m}^2}$$

$$= -4.5\,\text{m/s} \longleftarrow V_2$$

Recall that V_2 represents the magnitude of the velocity, which we assumed was outwards from the control volume. The fact that V_2 is negative means that in fact we have an *inflow* at location ②—our initial assumption was invalid.

This problem demonstrates use of the sign convention for evaluating $\int_A \vec{V} \cdot d\vec{A}$ or $\sum_{CS} \vec{V} \cdot \vec{A}$. In particular, the area normal is *always* drawn *outwards* from the control surface.

Example 4.2 MASS FLOW RATE IN BOUNDARY LAYER

The fluid in direct contact with a stationary solid boundary has zero velocity; there is no slip at the boundary. Thus the flow over a flat plate adheres to the plate surface and forms a boundary layer, as depicted below. The flow ahead of the plate is uniform with velocity $\vec{V} = U\hat{i}$; $U = 30$ m/s. The velocity distribution within the boundary layer $(0 \le y \le \delta)$ along cd is approximated as $u/U = 2(y/\delta) - (y/\delta)^2$.

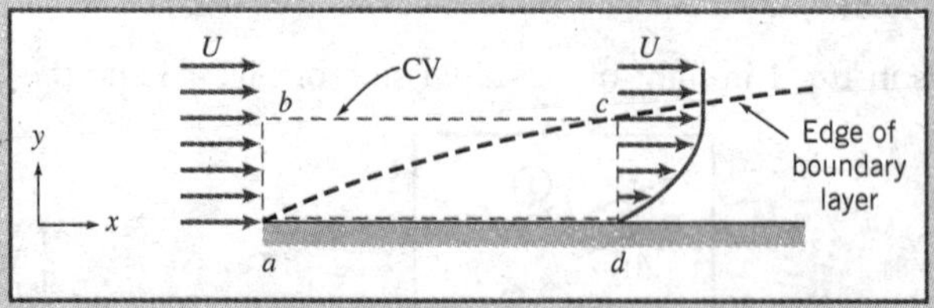

The boundary-layer thickness at location d is $\delta = 5$ mm. The fluid is air with density $\rho = 1.24$ kg/m^3. Assuming the plate width perpendicular to the paper to be $w = 0.6$ m, calculate the mass flow rate across surface bc of control volume $abcd$.

Given: Steady, incompressible flow over a flat plate, $\rho = 1.24$ kg/m^3. Width of plate, $w = 0.6$ m.
Velocity ahead of plate is uniform: $\vec{V} = U\hat{i}$, $U = 30$ m/s.

At $x = x_d$:

$$\delta = 5\text{ mm}$$

$$\frac{u}{U} = 2\left(\frac{y}{\delta}\right) - \left(\frac{y}{\delta}\right)^2$$

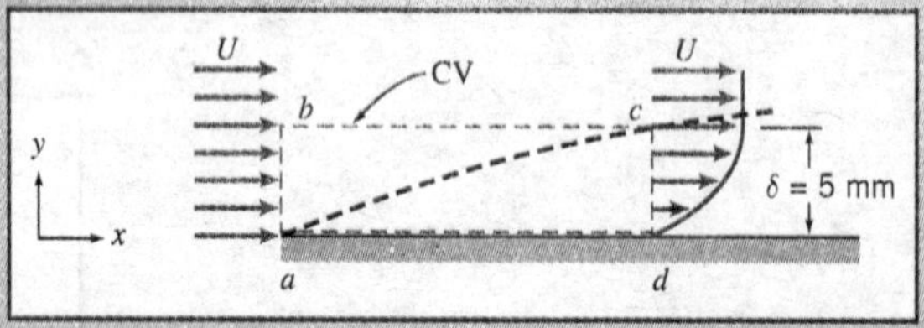

Find: Mass flow rate across surface bc.

Solution: The fixed control volume is shown by the dashed lines.

Governing equation: The general control volume equation is Eq. 4.12, but we can go immediately to Eq. 4.15a because of assumption (1) below,

$$\int_{CS} \rho \vec{V} \cdot d\vec{A} = 0$$

Assumptions: (1) Steady flow (given).
(2) Incompressible flow (given).
(3) Two-dimensional flow, given properties are independent of z.

Assuming that there is no flow in the z direction, then

$$\int_{A_{ab}} \rho \vec{V} \cdot d\vec{A} + \int_{A_{bc}} \rho \vec{V} \cdot d\vec{A} + \int_{A_{cd}} \rho \vec{V} \cdot d\vec{A} + \underbrace{\int_{A_{da}} \rho \vec{V} \cdot d\vec{A}}_{\text{no flow across } da} = 0$$

$$\therefore \dot{m}_{bc} = \int_{A_{bc}} \rho \vec{V} \cdot d\vec{A} = -\int_{A_{ab}} \rho \vec{V} \cdot d\vec{A} - \int_{A_{cd}} \rho \vec{V} \cdot d\vec{A} \qquad (1)$$

We need to evaluate the integrals on the right side of the equation.

For depth w in the z direction, we obtain

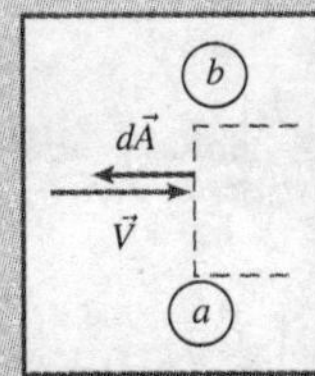

$$\int_{A_{ab}} \rho \vec{V} \cdot d\vec{A} = -\int_{A_{ab}} \rho u \, dA = -\int_{y_a}^{y_b} \rho u w \, dy \qquad \left\{ \begin{array}{c} \vec{V} \cdot d\vec{A} \text{ is negative} \\ dA = w dy \end{array} \right\}$$

$$= -\int_0^{\delta} \rho u w \, dy = -\int_0^{\delta} \rho U w \, dy \qquad \{u = U \text{ over area } ab\}$$

$$\int_{A_{ab}} \rho \vec{V} \cdot d\vec{A} = -\left[\rho U w y\right]_0^{\delta} = -\rho U w \delta$$

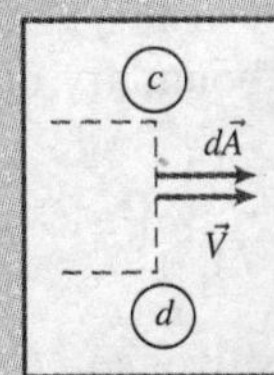

$$\int_{A_{cd}} \rho \vec{V} \cdot d\vec{A} = \int_{A_{cd}} \rho u \, dA = \int_{y_d}^{y_c} \rho u w \, dy \qquad \left\{ \begin{array}{c} \vec{V} \cdot d\vec{A} \text{ is positive} \\ dA = w dy \end{array} \right\}$$

$$= \int_0^{\delta} \rho u w \, dy = \int_0^{\delta} \rho w U \left[2\left(\frac{y}{\delta}\right) - \left(\frac{y}{\delta}\right)^2 \right] dy$$

$$\int_{A_{cd}} \rho \vec{V} \cdot d\vec{A} = \rho w U \left[\frac{y^2}{\delta} - \frac{y^3}{3\delta^2} \right]_0^{\delta} = \rho w U \delta \left[1 - \frac{1}{3} \right] = \frac{2\rho U w \delta}{3}$$

Substituting into Eq. 1, we obtain

$$\therefore \dot{m}_{bc} = \rho U w \delta - \frac{2\rho U w \delta}{3} = \frac{\rho U w \delta}{3}$$

$$= \frac{1}{3} \times 1.24 \frac{\text{kg}}{\text{m}^3} \times 30 \frac{\text{m}}{\text{s}} \times 0.6\,\text{m} \times 5\,\text{mm} \times \frac{\text{m}}{1000\,\text{mm}}$$

$$\left\{ \begin{array}{l} \text{Positive sign indicates flow} \\ \text{out across surface } bc. \end{array} \right\}$$

$$\dot{m}_{bc} = 0.0372\,\text{kg/s} \longleftarrow \dot{m}_{bc}$$

This problem demonstrates use of the conservation of mass equation when we have nonuniform flow at a section.

Example 4.3 DENSITY CHANGE IN VENTING TANK

A tank of 0.05 m^3 volume contains air at 800 kPa (absolute) and 15°C. At $t = 0$, air begins escaping from the tank through a valve with a flow area of 65 mm^2. The air passing through the valve has a speed of 300 m/s and a density of 6 kg/m^3. Determine the instantaneous rate of change of density in the tank at $t = 0$.

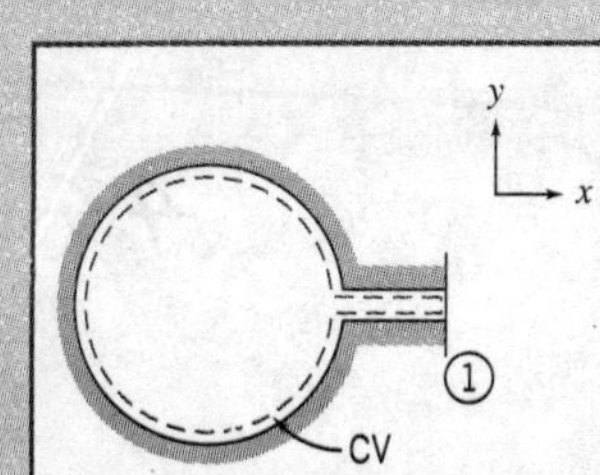

Given: Tank of volume $V\!\!\!\!-= 0.05\,\text{m}^3$ contains air at $p = 800$ kPa (absolute), $T =$ 15°C. At $t = 0$, air escapes through a valve. Air leaves with speed $V = 300$ m/s and density $\rho = 6$ kg/m^3 through area $A = 65$ mm^2.

Find: Rate of change of air density in the tank at $t = 0$.

Solution: Choose a fixed control volume as shown by the dashed line.

Governing equation: $\dfrac{\partial}{\partial t}\displaystyle\int_{\text{CV}} \rho \, dV\!\!\!\!- + \int_{\text{CS}} \rho \vec{V} \cdot d\vec{A} = 0$

Assumptions: (1) Properties in the tank are uniform, but time-dependent.
(2) Uniform flow at section ①.

Since properties are assumed uniform in the tank at any instant, we can take ρ out from within the volume integral of the first term,

$$\frac{\partial}{\partial t}\left[\rho_{\text{CV}}\int_{\text{CV}} d\mathcal{V}\right] + \int_{\text{CS}} \rho\vec{V}\cdot d\vec{A} = 0$$

Now, $\int_{\text{CV}} d\mathcal{V} = \mathcal{V}$, and hence

$$\frac{\partial}{\partial t}(\rho\mathcal{V})_{\text{CV}} + \int_{\text{CS}} \rho\vec{V}\cdot d\vec{A} = 0$$

The only place where mass crosses the boundary of the control volume is at surface ①. Hence

$$\int_{\text{CS}} \rho\vec{V}\cdot d\vec{A} = \int_{A_1} \rho\vec{V}\cdot d\vec{A} \qquad \text{and} \qquad \frac{\partial}{\partial t}(\rho\mathcal{V}) + \int_{A_1} \rho\vec{V}\cdot d\vec{A} = 0$$

At surface ① the sign of $\rho\vec{V}\cdot d\vec{A}$ is positive, so

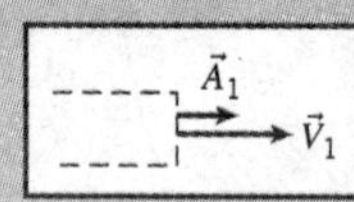

$$\frac{\partial}{\partial t}(\rho\mathcal{V}) + \int_{A_1} \rho V\, dA = 0$$

Since flow is assumed uniform over surface ①, then

$$\frac{\partial}{\partial t}(\rho\mathcal{V}) + \rho_1 V_1 A_1 = 0 \qquad \text{or} \qquad \frac{\partial}{\partial t}(\rho\mathcal{V}) = -\rho_1 V_1 A_1$$

Since the volume, $\mathcal{V}$, of the tank is not a function of time,

$$\mathcal{V}\frac{\partial\rho}{\partial t} = -\rho_1 V_1 A_1$$

and

$$\frac{\partial\rho}{\partial t} = -\frac{\rho_1 V_1 A_1}{\mathcal{V}}$$

At $t = 0$,

$$\frac{\partial\rho}{\partial t} = -6\frac{\text{kg}}{\text{m}^3} \times 300\frac{\text{m}}{\text{s}} \times 65\ \text{mm}^2 \times \frac{1}{0.05\ \text{m}^3} \times \frac{\text{m}^2}{10^6\ \text{mm}^2}$$

$$\frac{\partial\rho}{\partial t} = -2.34\ (\text{kg/m}^3)/\text{s} \longleftarrow \frac{\partial\rho}{\partial t}$$

{The density is decreasing.}

This problem demonstrates use of the conservation of mass equation for unsteady flow problems.

4.4 *Momentum Equation for Inertial Control Volume*

We now wish to obtain a control volume form of Newton's second law. We use the same procedure we just used for mass conservation, with one note of caution: the control volume coordinates (with respect to which we measure all velocities) are inertial; that is, the control volume coordinates xyz are either at rest or moving at constant speed with respect to an "absolute" set of coordinates XYZ. (Sections 4.6 and 4.7 will analyze noninertial control volumes.) We begin with the mathematical formulation for a system and then use Eq. 4.10 to go from the system to the control volume formulation.

Recall that Newton's second law for a system moving relative to an inertial coordinate system was given by Eq. 4.2a as

$$\vec{F} = \frac{d\vec{P}}{dt}\bigg)_{\text{system}} \tag{4.2a}$$

where the linear momentum of the system is given by

$$\vec{P}_{\text{system}} = \int_{M(\text{system})} \vec{V}\, dm = \int_{V\!\!\!-(\text{system})} \vec{V}\, \rho\, dV\!\!\!- \tag{4.2b}$$

and the resultant force, $\vec{F}$, includes all surface and body forces acting on the system,

$$\vec{F} = \vec{F}_S + \vec{F}_B$$

The system and control volume formulations are related using Eq. 4.10,

$$\frac{dN}{dt}\bigg)_{\text{system}} = \frac{\partial}{\partial t}\int_{\text{CV}} \eta\, \rho\, dV\!\!\!- + \int_{\text{CS}} \eta\, \rho \vec{V} \cdot d\vec{A} \tag{4.10}$$

To derive the control volume formulation of Newton's second law, we set

$$N = \vec{P} \qquad \text{and} \qquad \eta = \vec{V}$$

From Eq. 4.10, with this substitution, we obtain

$$\frac{d\vec{P}}{dt}\bigg)_{\text{system}} = \frac{\partial}{\partial t}\int_{\text{CV}} \vec{V}\, \rho\, dV\!\!\!- + \int_{\text{CS}} \vec{V} \rho \vec{V} \cdot d\vec{A} \tag{4.16}$$

From Eq. 4.2a

$$\frac{d\vec{P}}{dt}\bigg)_{\text{system}} = \vec{F}\Big)_{\text{on system}} \tag{4.2a}$$

Since, in deriving Eq. 4.10, the system and the control volume coincided at t_0, then

$$\vec{F}\Big)_{\text{on system}} = \vec{F}\Big)_{\text{on control volume}}$$

In light of this, Eqs. 4.2a and 4.16 may be combined to yield the control volume formulation of Newton's second law for a nonaccelerating control volume

$$\vec{F} = \vec{F}_S + \vec{F}_B = \frac{\partial}{\partial t}\int_{\text{CV}} \vec{V}\, \rho dV\!\!\!- + \int_{\text{CS}} \vec{V}\, \rho \vec{V} \cdot d\vec{A} \tag{4.17a}$$

For cases when we have uniform flow at each inlet and exit, we can use

$$\vec{F} = \vec{F}_S + \vec{F}_B = \frac{\partial}{\partial t}\int_{\text{CV}} \vec{V}\, \rho dV\!\!\!- + \sum\nolimits_{\text{CS}} \vec{V}\, \rho \vec{V} \cdot \vec{A} \tag{4.17b}$$

Equations 4.17a and 4.17b are our (nonaccelerating) control volume forms of Newton's second law. It states that the total force (due to surface and body forces) acting on the control volume leads to a rate of change of momentum within the control volume (the volume integral) and/or a net rate at which momentum is leaving the control volume through the control surface.

We must be a little careful in applying Eqs. 4.17. The first step will always be to carefully choose a control volume and its control surface so that we can evaluate the volume integral and the surface integral (or summation); each inlet and exit should be carefully labeled, as should the external forces acting. In fluid mechanics the body force is usually gravity, so

$$\vec{F}_B = \int_{\text{CV}} \rho \vec{g}\, d\forall = \vec{W}_{\text{CV}} = M\vec{g}$$

where $\vec{g}$ is the acceleration of gravity and $\vec{W}_{\text{CV}}$ is the instantaneous weight of the entire control volume. In many applications the surface force is due to pressure,

$$\vec{F}_S = \int_A -p d\vec{A}$$

Note that the minus sign is to ensure that we always compute pressure forces acting *onto* the control surface (recall $d\vec{A}$ was chosen to be a vector pointing *out* of the control volume). It is worth stressing that *even at points on the surface that have an outflow*, the pressure force acts *onto* the control volume.

In Eqs. 4.17 we must also be careful in evaluating $\int_{\text{CS}} \vec{V} \rho \vec{V} \cdot d\vec{A}$ or $\Sigma_{\text{CS}} \vec{V} \rho \vec{V} \cdot \vec{A}$ (this may be easier to do if we write them with the implied parentheses, $\int_{\text{CS}} \vec{V} \rho(\vec{V} \cdot d\vec{A})$ or $\Sigma_{\text{CS}} \vec{V} \rho(\vec{V} \cdot \vec{A})$). The velocity $\vec{V}$ must be measured with respect to the control volume coordinates xyz, with the appropriate signs for its vector components u, v, and w; recall also that the scalar product will be positive for outflow and negative for inflow (refer to Fig. 4.3).

The momentum equation (Eqs. 4.17) is a vector equation. We will usually write the three scalar components, as measured in the xyz coordinates of the control volume,

$$F_x = F_{S_x} + F_{B_x} = \frac{\partial}{\partial t} \int_{\text{CV}} u\, \rho\, d\forall + \int_{\text{CS}} u\, \rho \vec{V} \cdot d\vec{A} \tag{4.18a}$$

$$F_y = F_{S_y} + F_{B_y} = \frac{\partial}{\partial t} \int_{\text{CV}} v\, \rho\, d\forall + \int_{\text{CS}} v\, \rho \vec{V} \cdot d\vec{A} \tag{4.18b}$$

$$F_z = F_{S_z} + F_{B_z} = \frac{\partial}{\partial t} \int_{\text{CV}} w\, \rho\, d\forall + \int_{\text{CS}} w\, \rho \vec{V} \cdot d\vec{A} \tag{4.18c}$$

VIDEO

Momentum Effect: A Jet Impacting a Surface.

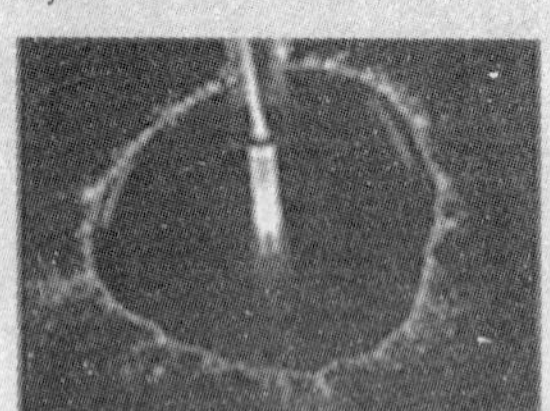

or, for uniform flow at each inlet and exit,

$$F_x = F_{S_x} + F_{B_x} = \frac{\partial}{\partial t} \int_{\text{CV}} u\, \rho\, d\forall + \sum\nolimits_{\text{CS}} u\, \rho \vec{V} \cdot \vec{A} \tag{4.18d}$$

$$F_y = F_{S_y} + F_{B_y} = \frac{\partial}{\partial t} \int_{\text{CV}} v\, \rho\, d\forall + \sum\nolimits_{\text{CS}} v\, \rho \vec{V} \cdot \vec{A} \tag{4.18e}$$

$$F_z = F_{S_z} + F_{B_z} = \frac{\partial}{\partial t} \int_{\text{CV}} w\, \rho\, d\forall + \sum\nolimits_{\text{CS}} w\, \rho \vec{V} \cdot \vec{A} \tag{4.18f}$$

Note that, as we found for the mass conservation equation (Eq. 4.12), for steady flow the first term on the right in Eqs. 4.17 and 4.18 is zero.

We will now look at five Examples to illustrate some features of the various forms of the momentum equation for a control volume. Example 4.4 demonstrates how intelligent choice of the control volume can simplify analysis of a problem, Example 4.5 involves a problem in which we have significant body forces, Example 4.6 explains how to simplify surface force evaluations by working in gage pressures, Example 4.7 involves nonuniform surface forces, and Example 4.8 involves a problem in which we have unsteady flow.

Example 4.4 CHOICE OF CONTROL VOLUME FOR MOMENTUM ANALYSIS

Water from a stationary nozzle strikes a flat plate as shown. The water leaves the nozzle at 15 m/s; the nozzle area is 0.01 m^2. Assuming the water is directed normal to the plate, and flows along the plate, determine the horizontal force you need to resist to hold it in place.

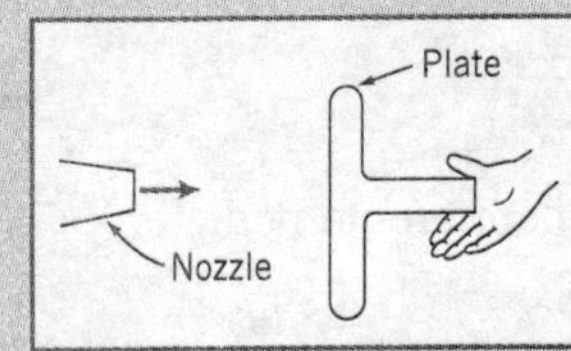

Given: Water from a stationary nozzle is directed normal to the plate; subsequent flow is parallel to plate.

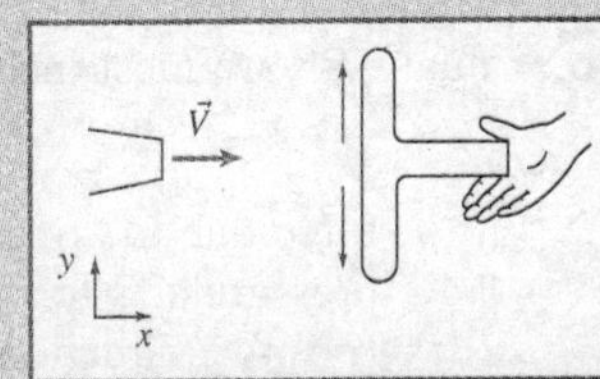

$$\text{Jet velocity, } \vec{V} = 15\hat{i}\ \text{m/s}$$

$$\text{Nozzle area, } A_n = 0.01\ \text{m}^2$$

Find: Horizontal force on your hand.

Solution: We chose a coordinate system in defining the problem above. We must now choose a suitable control volume. Two possible choices are shown by the dashed lines below.

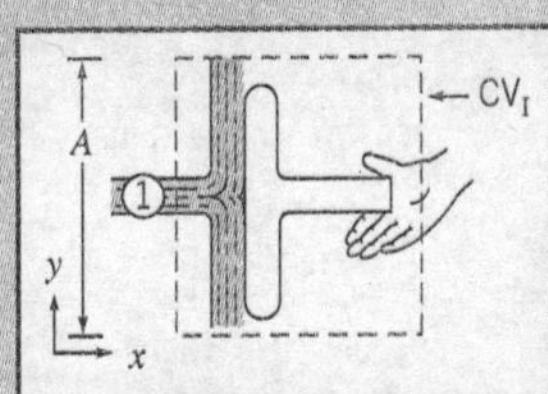

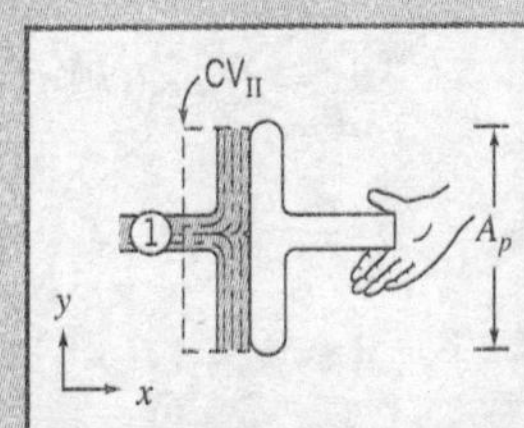

In both cases, water from the nozzle crosses the control surface through area A_1 (assumed equal to the nozzle area) and is assumed to leave the control volume tangent to the plate surface in the $+y$ or $-y$ direction. Before trying to decide which is the "best" control volume to use, let us write the governing equations.

$$\vec{F} = \vec{F}_S + \vec{F}_B = \frac{\partial}{\partial t}\int_{CV} \vec{V}\rho dV\!\!\!\!- + \int_{CS} \vec{V}\rho\vec{V}\cdot d\vec{A} \quad \text{and} \quad \frac{\partial}{\partial t}\int_{CV} \rho\, dV\!\!\!\!- + \int_{CS} \rho\vec{V}\cdot d\vec{A} = 0$$

Assumptions: (1) Steady flow.
(2) Incompressible flow.
(3) Uniform flow at each section where fluid crosses the CV boundaries.

Regardless of our choice of control volume, assumptions (1), (2), and (3) lead to

$$\vec{F} = \vec{F}_S + \vec{F}_B = \sum_{CS} \vec{V}\rho\vec{V}\cdot\vec{A} \quad \text{and} \quad \sum_{CS} \rho\vec{V}\cdot\vec{A} = 0$$

Evaluating the momentum flux term will lead to the same result for both control volumes. We should choose the control volume that allows the most straightforward evaluation of the forces.

Remember in applying the momentum equation that the force, $\vec{F}$, represents all forces acting *on* the control volume.

Let us solve the problem using each of the control volumes.

CV_I

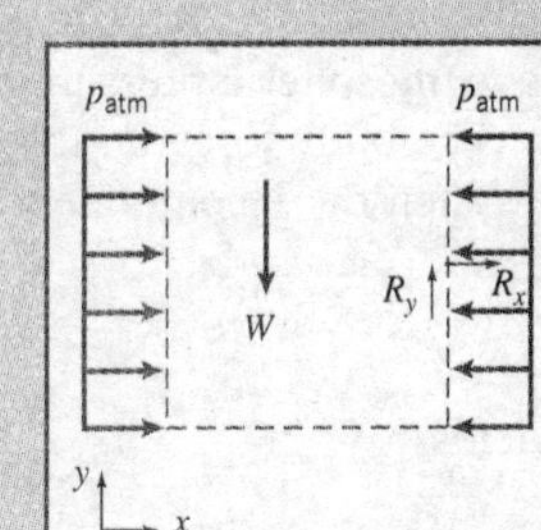

The control volume has been selected so that the area of the left surface is equal to the area of the right surface. Denote this area by A.

The control volume cuts through your hand. We denote the components of the reaction force of your hand on the control volume as R_x and R_y and assume both to be positive. (The force of the control volume on your hand is equal and opposite to R_x and R_y.)

Atmospheric pressure acts on all surfaces of the control volume. Note that *the pressure in a free jet is ambient*, i.e., in this case atmospheric. (The distributed force due to atmospheric pressure has been shown on the vertical faces only.)

The body force on the control volume is denoted as W.
Since we are looking for the horizontal force, we write the x component of the steady flow momentum equation

$$F_{S_x} + F_{B_x} = \sum_{CS} u\,\rho\vec{V}\cdot\vec{A}$$

There are no body forces in the x direction, so $F_{B_x} = 0$, and

$$F_{S_x} = \sum_{CS} u\,\rho\vec{V}\cdot\vec{A}$$

To evaluate F_{S_x}, we must include all surface forces acting on the control volume

$F_{S_x} =$	$p_{atm}A$	$-$	$p_{atm}A$	$+$	R_x
	force due to atmospheric pressure acts to right (positive direction) on left surface		force due to atmospheric pressure acts to left (negative direction) on right surface		force of your hand on control volume (assumed positive)

Consequently, $F_{S_x} = R_x$, and

$$R_x = \sum_{CS} u\,\rho\vec{V}\cdot\vec{A} = u\,\rho\vec{V}\cdot\vec{A}\Big|_1$$ {For top and bottom surfaces, $u = 0$}

$$R_x = -u_1\,\rho V_1 A_1$$ {At ①, $\rho\vec{V}_1\cdot\vec{A}_1 = \rho(-V_1A_1)$ since $\vec{V}_1$ and $\vec{A}_1$ are 180° apart. Note that $u_1 = V_1$}

$$R_x = -15\,\frac{\text{m}}{\text{s}} \times 999\,\frac{\text{kg}}{\text{m}^3} \times 15\,\frac{\text{m}}{\text{s}} \times 0.01\,\text{m}^2 \times \frac{\text{N}\cdot\text{s}^2}{\text{kg}\cdot\text{m}}$$ $\{u_1 = 15\,\text{m/s}\}$

$$R_x = -2.25\,\text{kN}$$ {R_x acts opposite to positive direction assumed.}

The horizontal force on your hand is

$$K_x = -R_x = 2.25\,\text{kN}$$ {force on your hand acts to the right} K_x

CV$_{II}$ with Horizontal Forces Shown

The control volume has been selected so the areas of the left surface and of the right surface are equal to the area of the plate. Denote this area by A_p.

The control volume is in contact with the plate over the entire plate surface. We denote the horizontal reaction force from the plate on the control volume as B_x (and assume it to be positive).

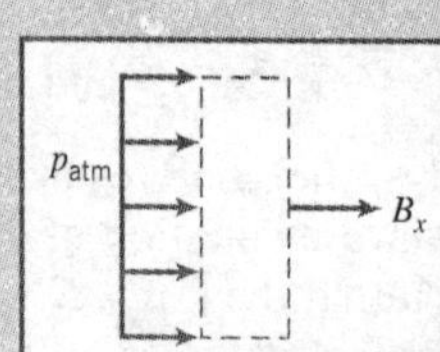

Atmospheric pressure acts on the left surface of the control volume (and on the two horizontal surfaces).
The body force on this control volume has no component in the x direction.
Then the x component of the momentum equation,

$$F_{S_x} = \sum_{CS} u\,\rho\vec{V}\cdot\vec{A}$$

yields

$$F_{S_x} = p_{atm}A_p + B_x = u\,\rho\vec{V}\cdot\vec{A}\Big|_1 = -u_1 V_1 A_1 = -2.25\,\text{kN}$$

Then

$$B_x = -p_{atm}\,A_p - 2.25\text{ kN}$$

To determine the net force on the plate, we need a free-body diagram of the plate:

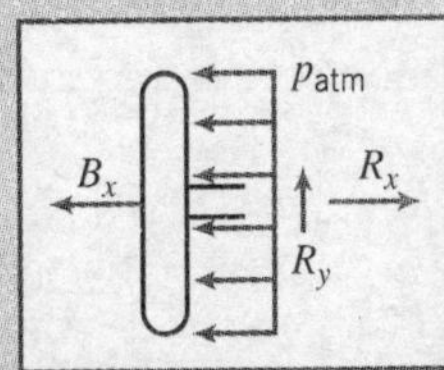

$$\sum F_x = 0 = -B_x - p_{atm}A_p + R_x$$
$$R_x = p_{atm}A_p + B_x$$
$$R_x = p_{atm}A_p + (-p_{atm}A_p - 2.25\text{ kN}) = -2.25\text{ kN}$$

Then the horizontal force on your hand is $K_x = -R_x = 2.25$ kN.

Note that the choice of CV_{II} meant we needed an additional free-body diagram. In general it is best to select the control volume so that the force sought acts explicitly on the control volume.

Notes:

- ✓ This problem demonstrates how thoughtful choice of the control volume can simplify use of the momentum equation.
- ✓ The analysis would have been greatly simplified if we had worked in gage pressures (see Example 4.6).
- ✓ For this problem the force generated was entirely due to the plate absorbing the jet's horizontal momentum.

Example 4.5 TANK ON SCALE: BODY FORCE

A metal container 0.61 m high, with an inside cross-sectional area of 0.09 m^2, weighs 22.2 N when empty. The container is placed on a scale and water flows in through an opening in the top and out through the two equal-area openings in the sides, as shown in the diagram. Under steady flow conditions, the height of the water in the tank is 0.58 m.

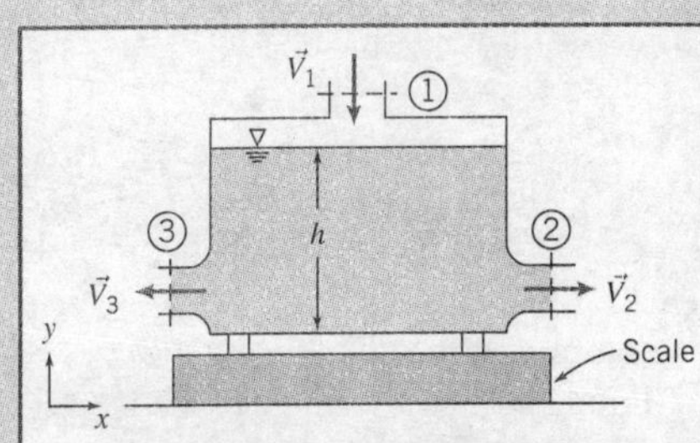

$$A_1 = 0.009\text{ m}^2$$
$$\vec{V}_1 = -3\hat{j}\text{ m/s}$$
$$A_2 = A_3 = 0.009\text{ m}^2$$

Your boss claims that the scale will read the weight of the volume of water in the tank plus the tank weight, i.e., that we can treat this as a simple statics problem. You disagree, claiming that a fluid flow analysis is required. Who is right, and what does the scale indicate?

Given: Metal container, of height 0.61 m and cross-sectional area $A =$ 0.09 m^2, weighs 22.2 N when empty. Container rests on scale. Under steady flow conditions water depth is $h = 0.58$ m. Water enters vertically at section ① and leaves horizontally through sections ② and ③

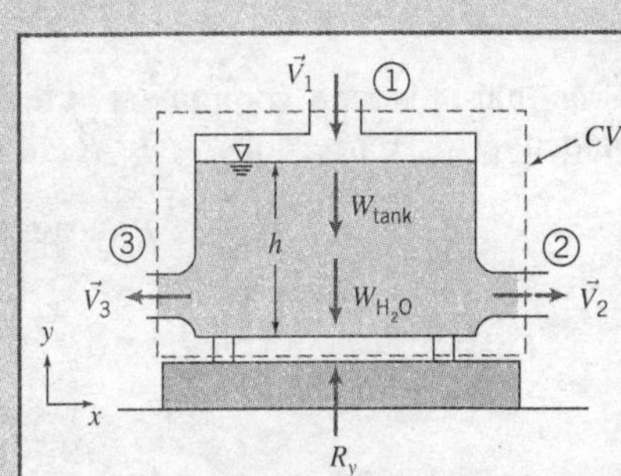

$$A_1 = 0.009\text{ m}^2$$
$$\vec{V}_1 = -3\hat{j}\text{ m/s}$$
$$A_2 = A_3 = 0.009\text{ m}^2$$

Find: Scale reading.

Solution:
Choose a control volume as shown; R_y is the force of the scale on the control volume (exerted on the control volume through the supports) and is assumed positive.

The weight of the tank is designated W_{tank}; the weight of the water in the tank is W_{H_2O}.

Atmospheric pressure acts uniformly on the entire control surface, and therefore has no net effect on the control volume. Because of this null effect we have not shown the pressure distribution in the diagram.

Governing equations: The general control volume momentum and mass conservation equations are Eqs. 4.17 and 4.12, respectively,

$$\vec{F}_S + \vec{F}_B = \underbrace{\frac{\partial}{\partial t}\int_{CV} \vec{V}\rho\, d\forall}_{=0(1)} + \int_{CS} \vec{V}\rho\vec{V}\cdot d\vec{A}$$

$$\underbrace{\frac{\partial}{\partial t}\int_{CV} \rho\, d\forall}_{=0(1)} + \int_{CS} \rho\vec{V}\cdot d\vec{A} = 0$$

Note that we usually start with the simplest forms (based on the problem assumptions, e.g., steady flow) of the mass conservation and momentum equations. However, in this problem, for illustration purposes, we start with the most general forms of the equations.

Assumptions:
(1) Steady flow (given).
(2) Incompressible flow.
(3) Uniform flow at each section where fluid crosses the CV boundaries.

We are only interested in the y component of the momentum equation

$$F_{S_y} + F_{B_y} = \int_{CS} v\rho\vec{V}\cdot d\vec{A} \qquad (1)$$

$$F_{S_y} = R_y \qquad \{\text{There is no net force due to atmosphere pressure.}\}$$

$$F_{B_y} = -W_{\text{tank}} - W_{H_2O} \qquad \{\text{Both body forces act in negative } y \text{ direction.}\}$$

$$W_{H_2O} = \rho g \forall = \gamma A h$$

$$\int_{CS} v\rho\vec{V}\cdot d\vec{A} = \int_{A_1} v\rho\vec{V}\cdot d\vec{A} = \int_{A_1} v(-\rho V_1 dA_1) \qquad \left\{\begin{array}{l}\vec{V}\cdot d\vec{A} \text{ is negative at } ①\\ v = 0 \text{ at sections } ② \text{ and } ③\end{array}\right\}$$

$$= v_1(-\rho V_1 A_1) \qquad \left\{\begin{array}{l}\text{We are assuming uniform}\\ \text{properties at } ①\end{array}\right\}$$

Using these results in Eq. 1 gives

$$R_y - W_{\text{tank}} - \gamma A h = v_1(-\rho V_1 A_1)$$

Note that v_1 is the y component of the velocity, so that $v_1 = -V_1$, where we recall that $V_1 = 3$ m/s is the magnitude of velocity $\vec{V}_1$. Hence, solving for R_y,

$$R_y = W_{\text{tank}} + \gamma A h + \rho V_1^2 A_1$$

$$= 22.2\,\text{N} + 9800\frac{\text{N}}{\text{m}^3} \times 0.09\,\text{m}^2 \times 0.58\,\text{m} + 1000\frac{\text{kg}}{\text{m}^3} \times 9\frac{\text{m}^2}{\text{s}^2} \times 0.009\,\text{m}^2 \times \frac{\text{N}\cdot\text{s}^2}{\text{kg}\cdot\text{m}}$$

$$= 22.2\,\text{N} + 511.6\,\text{N} + 81\,\text{N}$$

$$R_y = 614.8\,\text{N} \longleftarrow R_y$$

Note that this is the force of the scale on the control volume; it is also the reading on the scale. We can see that the scale reading is due to: the tank weight (22.2 N), the weight of water instantaneously in the tank (511.6 N), and the force involved in absorbing the downward momentum of the fluid at section ① (81 N). Hence your boss is wrong—neglecting the momentum results in an error of almost 13%.

This problem illustrates use of the momentum equation including significant body forces.

Example 4.6 FLOW THROUGH ELBOW: USE OF GAGE PRESSURES

Water flows steadily through the 90° reducing elbow shown in the diagram. At the inlet to the elbow, the absolute pressure is 220 kPa and the cross-sectional area is 0.01 m^2. At the outlet, the cross-sectional area is 0.0025 m^2 and the velocity is 16 m/s. The elbow discharges to the atmosphere. Determine the force required to hold the elbow in place.

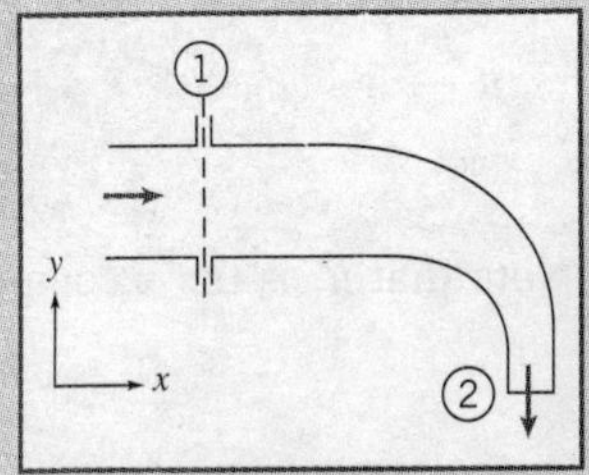

Given: Steady flow of water through 90° reducing elbow.

$$p_1 = 220\,\text{kPa (abs)} \qquad A_1 = 0.01\,\text{m}^2 \qquad \vec{V}_2 = -16\,\hat{j}\,\text{m/s} \qquad A_2 = 0.0025\,\text{m}^2$$

Find: Force required to hold elbow in place.

Solution:

Choose a fixed control volume as shown. Note that we have several surface force computations: p_1 on area A_1 and p_{atm} everywhere else. The exit at section ② is to a free jet, and so at ambient (i.e., atmospheric) pressure. We can use a simplification here: If we subtract p_{atm} from the entire surface (a null effect as far as forces are concerned) we can work in gage pressures, as shown.

Note that since the elbow is anchored to the supply line, in addition to the reaction forces R_x and R_y (shown), there would also be a reaction moment (not shown).

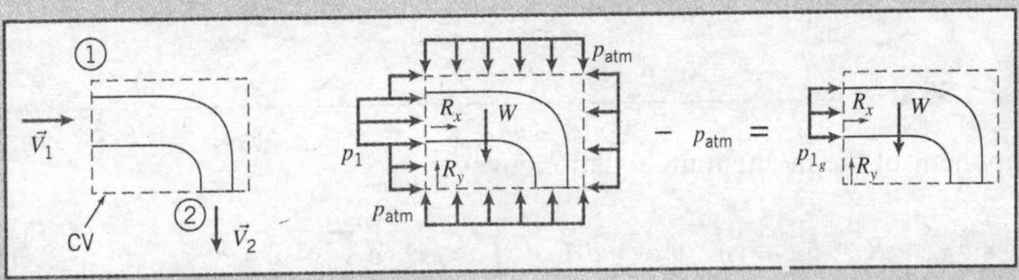

Governing equations:

$$\vec{F} = \vec{F}_S + \vec{F}_B = \overbrace{\frac{\partial}{\partial t}\int_{CV} \vec{V}\rho\, d\Psi}^{=0(4)} + \int_{CS} \vec{V}\rho\vec{V}\cdot d\vec{A}$$

$$\overbrace{\frac{\partial}{\partial t}\int_{CV} \rho\, d\Psi}^{=0(4)} + \int_{CS} \rho\vec{V}\cdot d\vec{A} = 0$$

Assumptions:
(1) Uniform flow at each section.
(2) Atmospheric pressure, $p_{atm} = 101$ kPa (abs).
(3) Incompressible flow.
(4) Steady flow (given).
(5) Neglect weight of elbow and water in elbow.

Once again (although we didn't need to) we started with the most general form of the governing equations. Writing the x component of the momentum equation results in

$$F_{S_x} = \int_{\text{CS}} u\rho\vec{V}\cdot d\vec{A} = \int_{A_1} u\rho\vec{V}\cdot d\vec{A} \qquad \{F_{B_x} = 0 \text{ and } u_2 = 0\}$$

$$p_{1_g}A_1 + R_x = \int_{A_1} u\,\rho\vec{V}\cdot d\vec{A}$$

so

$$R_x = -p_{1_g}A_1 + \int_{A_1} u\,\rho\vec{V}\cdot d\vec{A}$$
$$= -p_{1_g}A_1 + u_1(-\rho V_1 A_1)$$
$$R_x = -p_{1_g}A_1 - \rho V_1^2 A_1$$

Note that u_1 is the x component of the velocity, so that $u_1 = V_1$. To find V_1, use the mass conservation equation:

$$\int_{\text{CS}} \rho\vec{V}\cdot d\vec{A} = \int_{A_1} \rho\vec{V}\cdot d\vec{A} + \int_{A_2} \rho\vec{V}\cdot d\vec{A} = 0$$
$$\therefore (-\rho V_1 A_1) + (\rho V_2 A_2) = 0$$

and

$$V_1 = V_2\frac{A_2}{A_1} = 16\frac{\text{m}}{\text{s}} \times \frac{0.0025}{0.01} = 4\text{ m/s}$$

We can now compute R_x

$$R_x = -p_{1_g}A_1 - \rho V_1^2 A_1$$
$$= -1.19\times 10^5\frac{\text{N}}{\text{m}^2}\times 0.01\text{ m}^2 - 999\frac{\text{kg}}{\text{m}^3}\times 16\frac{\text{m}^2}{\text{s}^2}\times 0.01\text{ m}^2 \times \frac{\text{N}\cdot\text{s}^2}{\text{kg}\cdot\text{m}}$$

$R_x = -1.35\text{ kN}$ ← R_x

Writing the y component of the momentum equation gives

$$F_{S_y} + F_{B_y} = R_y + F_{B_y} = \int_{\text{CS}} v\,\rho\vec{V}\cdot d\vec{A} = \int_{A_2} v\,\rho\vec{V}\cdot d\vec{A} \qquad \{v_1 = 0\}$$

or

$$R_y = -F_{B_y} + \int_{A_2} v\,\rho\vec{V}\cdot d\vec{A}$$
$$= -F_{B_y} + v_2(\rho V_2 A_2)$$
$$R_y = -F_{B_y} - \rho V_2^2 A_2$$

Note that v_2 is the y component of the velocity, so that $v_2 = -V_2$, where V_2 is the magnitude of the exit velocity. Substituting known values

$$R_y = -F_{B_y} + -\rho V_2^2 A_2$$
$$= -F_{B_y} - 999\frac{\text{kg}}{\text{m}^3}\times(16)^2\frac{\text{m}^2}{\text{s}^2}\times 0.0025\text{ m}^2\times\frac{\text{N}\cdot\text{s}^2}{\text{kg}\cdot\text{m}}$$

$= -F_{B_y} - 639\text{ N}$ ← R_y

Neglecting F_{B_y} gives

$R_y = -639\text{ N}$ ← R_y

This problem illustrates how using gage pressures simplifies evaluation of the surface forces in the momentum equation.

Example 4.7 FLOW UNDER A SLUICE GATE: HYDROSTATIC PRESSURE FORCE

Water in an open channel is held in by a sluice gate. Compare the horizontal force of the water on the gate (a) when the gate is closed and (b) when it is open (assuming steady flow, as shown). Assume the flow at sections ① and ② is incompressible and uniform, and that (because the streamlines are straight there) the pressure distributions are hydrostatic.

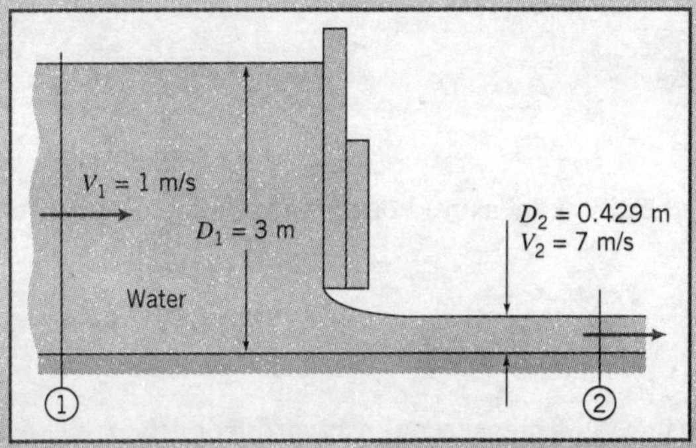

Given: Flow under sluice gate. Width $= w$.

Find: Horizontal force (per unit width) on the closed and open gate.

Solution:

Choose a control volume as shown for the open gate. Note that it is much simpler to work in gage pressures, as we learned in Example 4.6.

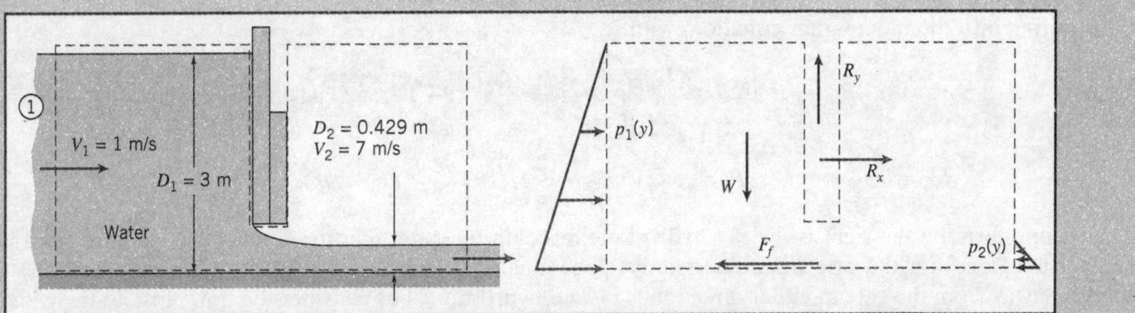

The forces acting on the control volume include:

- Force of gravity W.
- Friction force F_f.
- Components R_x and R_y of reaction force from gate.
- Hydrostatic pressure distribution on vertical surfaces, assumption (6).
- Pressure distribution $p_b(x)$ along bottom surface (not shown).

Apply the x component of the momentum equation.

Governing equation:

$$F_{S_x} + \underbrace{F_{B_x}}_{=0(2)} = \underbrace{\frac{\partial}{\partial t}\int_{CV} u\,\rho\, d\forall}_{=0(3)} + \int_{CS} u\,\rho\vec{V}\cdot d\vec{A}$$

Assumptions: (1) F_f negligible (neglect friction on channel bottom).
(2) $F_{B_x} = 0$.
(3) Steady flow.
(4) Incompressible flow (given).
(5) Uniform flow at each section (given).
(6) Hydrostatic pressure distributions at ① and ② (given).

Then

$$F_{S_x} = F_{R_1} + F_{R_2} + R_x = u_1(-\rho V_1 w D_1) + u_2(\rho V_2 w D_2)$$

The surface forces acting on the CV are due to the pressure distributions and the unknown force R_x. From assumption (6), we can integrate the gage pressure distributions on each side to compute the hydrostatic forces F_{R_1} and F_{R_2},

$$F_{R_1} = \int_0^{D_1} p_1\, dA = w \int_0^{D_1} \rho g y\, dy = \rho g w \frac{y^2}{2}\bigg|_0^{D_1} = \frac{1}{2}\rho g w D_1^2$$

where y is measured downward from the free surface of location ①, and

$$F_{R_2} = \int_0^{D_2} p_2\, dA = w \int_0^{D_2} \rho g y\, dy = \rho g w \frac{y^2}{2}\bigg|_0^{D_2} = \frac{1}{2}\rho g w D_2^2$$

where y is measured downward from the free surface of location ②. (Note that we could have used the hydrostatic force equation, Eq. 3.10b, directly to obtain these forces.)

Evaluating F_{S_x} gives

$$F_{S_x} = R_x + \frac{\rho g w}{2}(D_1^2 - D_2^2)$$

Substituting into the momentum equation, with $u_1 = V_1$ and $u_2 = V_2$, gives

$$R_x + \frac{\rho g w}{2}(D_1^2 - D_2^2) = -\rho V_1^2 w D_1 + \rho V_2^2 w D_2$$

or

$$R_x = \rho w(V_2^2 D_2 - V_1^2 D_1) - \frac{\rho g w}{2}(D_1^2 - D_2^2)$$

The second term on the right is the net hydrostatic force on the gate; the first term "corrects" this (and leads to a smaller net force) for the case when the gate is open. What is the nature of this "correction"? The pressure in the fluid far away from the gate in either direction is indeed hydrostatic, but consider the flow close to the gate: Because we have significant velocity variations here (in magnitude and direction), the pressure distributions deviate significantly from hydrostatic—for example, as the fluid accelerates under the gate there will be a significant pressure drop on the lower left side of the gate. Deriving this pressure field would be a difficult task, but by careful choice of our CV we have avoided having to do so!

We can now compute the horizontal force per unit width,

$$\frac{R_x}{w} = \rho(V_2^2 D_2 - V_1^2 D_1) - \frac{\rho g}{2}(D_1^2 - D_2^2)$$

$$= 999\frac{\text{kg}}{\text{m}^3} \times \left[(7)^2(0.429) - (1)^2(3)\right]\frac{\text{m}^2}{\text{s}^2}\text{m} \times \frac{\text{N}\cdot\text{s}^2}{\text{kg}\cdot\text{m}}$$

$$-\frac{1}{2} \times 999\frac{\text{kg}}{\text{m}^3} \times 9.81\frac{\text{m}}{\text{s}^2} \times [(3)^2 - (0.429)^2]\text{m}^2 \times \frac{\text{N}\cdot\text{s}^2}{\text{kg}\cdot\text{m}}$$

$$\frac{R_x}{w} = 18.0\,\text{kN/m} - 43.2\,\text{kN/m}$$

$$\frac{R_x}{w} = -25.2\,\text{kN/m}$$

R_x is the external force acting on the control volume, applied to the CV by the gate. Therefore, the force of the water on the gate is K_x, where $K_x = -R_x$. Thus,

$$\frac{K_x}{w} = -\frac{R_x}{w} = 25.2\,\text{kN/m} \longleftarrow \frac{K_x}{w}$$

This force can be compared to the force on the closed gate of 44.1 kN (obtained from the second term on the right in the equation above, evaluated with D_2 set to zero because for the closed gate there is no fluid on the right of the gate)—the force on the open gate is significantly less as the water accelerates out under the gate.

This problem illustrates the application of the momentum equation to a control volume for which the pressure is not uniform on the control surface.

Example 4.8 CONVEYOR BELT FILLING: RATE OF CHANGE OF MOMENTUM IN CONTROL VOLUME

A horizontal conveyor belt moving at 0.9 m/s receives sand from a hopper. The sand falls vertically from the hopper to the belt at a speed of 1.5 m/s and a flow rate of 225 kg/s (the density of sand is approximately 1580 kg/m^3). The conveyor belt is initially empty but begins to fill with sand. If friction in the drive system and rollers is negligible, find the tension required to pull the belt while the conveyor is filling.

Given: Conveyor and hopper shown in sketch.

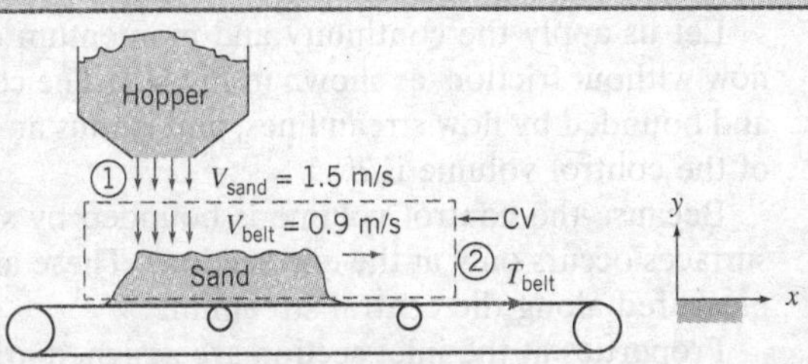

Find: T_{belt} at the instant shown.

Solution: Use the control volume and coordinates shown. Apply the x component of the momentum equation.

Governing equations:

$$F_{S_x} + \cancel{F_{B_x}}^{=0(2)} = \frac{\partial}{\partial t}\int_{CV} u\,\rho\, d\forall + \int_{CS} u\,\rho\vec{V}\cdot d\vec{A} \qquad \frac{\partial}{\partial t}\int_{CV}\rho\, d\forall + \int_{CS}\rho\vec{V}\cdot d\vec{A} = 0$$

Assumptions: (1) $F_{S_x} = T_{belt} = T$.
(2) $F_{B_x} = 0$.
(3) Uniform flow at section ①.
(4) All sand on belt moves with $V_{belt} = V_b$.

Then

$$T = \frac{\partial}{\partial t}\int_{CV} u\rho\, d\forall + u_1(-\rho V_1 A_1) + u_2(\rho V_2 A_2)$$

Since $u_1 = 0$, and there is no flow at section ②,

$$T = \frac{\partial}{\partial t}\int_{CV} u\rho\, d\forall$$

From assumption (4), inside the CV, $u = V_b =$ constant, and hence

$$T = V_b \frac{\partial}{\partial t}\int_{CV}\rho\, d\forall = V_b\frac{\partial M_s}{\partial t}$$

where M_s is the mass of sand on the belt (inside the control volume). This result is perhaps not surprising—the tension in the belt is the force required to increase the momentum inside the CV (which is increasing because even though the velocity of the mass in the CV is constant, the mass is not). From the continuity equation,

$$\frac{\partial}{\partial t}\int_{CV}\rho\, d\forall = \frac{\partial}{\partial t}M_s = -\int_{CS}\rho\vec{V}\cdot d\vec{A} = \dot{m}_s = 225\ \text{kg/s}$$

Then

$$T = V_b\dot{m}_s = 0.9\frac{\text{m}}{\text{s}} \times 225\frac{\text{kg}}{\text{s}} \times \frac{\text{N}\cdot\text{s}^2}{\text{kg}\cdot\text{m}}$$

$$T = 203.4\ \text{N} \longleftarrow T$$

This problem illustrates application of the momentum equation to a control volume in which the momentum is changing.

*Differential Control Volume Analysis

The control volume approach, as we have seen in the previous examples, provides useful results when applied to a finite region.

If we apply the approach to a differential control volume, we can obtain differential equations describing a flow field. In this section, we will apply the conservation of mass and momentum equations to such a control volume to obtain a simple differential equation describing flow in a steady, incompressible, frictionless flow, and integrate it along a streamline to obtain the famous Bernoulli equation.

Let us apply the continuity and momentum equations to a steady incompressible flow without friction, as shown in Fig. 4.4. The control volume chosen is fixed in space and bounded by flow streamlines, and is thus an element of a stream tube. The length of the control volume is ds.

Because the control volume is bounded by streamlines, flow across the bounding surfaces occurs only at the end sections. These are located at coordinates s and $s + ds$, measured along the central streamline.

Properties at the inlet section are assigned arbitrary symbolic values. Properties at the outlet section are assumed to increase by differential amounts. Thus at $s + ds$, the flow speed is assumed to be $V_s + dV_s$, and so on. The differential changes, dp, dV_s, and dA, all are assumed to be positive in setting up the problem. (As in a free-body analysis in statics or dynamics, the actual algebraic sign of each differential change will be determined from the results of the analysis.)

Now let us apply the continuity equation and the s component of the momentum equation to the control volume of Fig. 4.4.

a. Continuity Equation

Basic equation:

$$\overbrace{\frac{\partial}{\partial t}\int_{\text{CV}} \rho \, d\forall}^{=0(1)} + \int_{\text{CS}} \rho \vec{V} \cdot d\vec{A} = 0 \qquad (4.12)$$

Assumptions: (1) Steady flow.
(2) No flow across bounding streamlines.
(3) Incompressible flow, $\rho = \text{constant}$.

Then

$$(-\rho V_s A) + \{\rho(V_s + dV_s)(A + dA)\} = 0$$

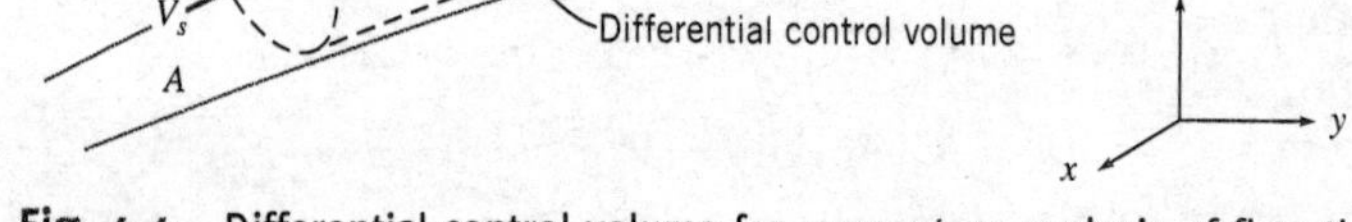

Fig. 4.4 Differential control volume for momentum analysis of flow through a stream tube.

*This section may be omitted without loss of continuity in the text material.

so

$$\rho(V_s + dV_s)(A + dA) = \rho V_s A \tag{4.19a}$$

On expanding the left side and simplifying, we obtain

$$V_s\,dA + A\,dV_s + dA\,dV_s = 0$$

But $dA\,dV_s$ is a product of differentials, which may be neglected compared with $V_s\,dA$ or $A\,dV_s$. Thus

$$V_s\,dA + A\,dV_s = 0 \tag{4.19b}$$

b. Streamwise Component of the Momentum Equation

Basic equation:

$$F_{S_s} + F_{B_s} = \overset{=0(1)}{\cancel{\frac{\partial}{\partial t}\int_{CV} u_s\,\rho\,d\forall}} + \int_{CS} u_s\,\rho\vec{V}\cdot d\vec{A} \tag{4.20}$$

Assumption: (4) No friction, so F_{S_b} is due to pressure forces only.

The surface force (due only to pressure) will have three terms:

$$F_{S_s} = pA - (p + dp)(A + dA) + \left(p + \frac{dp}{2}\right)dA \tag{4.21a}$$

The first and second terms in Eq. 4.21a are the pressure forces on the end faces of the control surface. The third term is F_{S_b}, the pressure force acting in the s direction on the bounding stream surface of the control volume. Its magnitude is the product of the average pressure acting on the stream surface, $p + \frac{1}{2}dp$, times the area component of the stream surface in the s direction, dA. Equation 4.21a simplifies to

$$F_{S_s} = -A\,dp - \frac{1}{2}dp\,dA \tag{4.21b}$$

The body force component in the s direction is

$$F_{B_s} = \rho g_s\,d\forall = \rho(-g\sin\theta)\left(A + \frac{dA}{2}\right)ds$$

But $\sin\theta\,ds = dz$, so that

$$F_{B_s} = -\rho g\left(A + \frac{dA}{2}\right)dz \tag{4.21c}$$

The momentum flux will be

$$\int_{CS} u_s\,\rho\vec{V}\cdot d\vec{A} = V_s(-\rho V_s A) + (V_s + dV_s)\{\rho(V_s + dV_s)(A + dA)\}$$

since there is no mass flux across the bounding stream surfaces. The mass flux factors in parentheses and braces are equal from continuity, Eq. 4.19a, so

$$\int_{CS} u_s\,\rho\vec{V}\cdot d\vec{A} = V_s(-\rho V_s A) + (V_s + dV_s)(\rho V_s A) = \rho V_s A\,dV_s \tag{4.22}$$

Substituting Eqs. 4.21b, 4.21c, and 4.22 into Eq. 4.20 (the momentum equation) gives

$$-A\,dp - \frac{1}{2}dp\,dA - \rho gA\,dz - \frac{1}{2}\rho g\,dA\,dz = \rho V_s A\,dV_s$$

Dividing by ρA and noting that products of differentials are negligible compared with the remaining terms, we obtain

$$-\frac{dp}{\rho} - g\,dz = V_s\,dV_s = d\left(\frac{V_s^2}{2}\right)$$

or

$$\frac{dp}{\rho} + d\left(\frac{V_s^2}{2}\right) + g\,dz = 0 \tag{4.23}$$

Because the flow is incompressible, this equation may be integrated to obtain

$$\frac{P}{\rho} + \frac{V_s^2}{2} + gz = \text{constant} \tag{4.24}$$

or, dropping subscript s,

$$\frac{P}{\rho} + \frac{V_s^2}{2} + gz = \text{constant} \tag{4.24}$$

This equation is subject to the restrictions:

1. Steady flow.
2. No friction.
3. Flow along a streamline.
4. Incompressible flow.

We have derived one form of perhaps the most famous (and misused) equation in fluid mechanics—the Bernoulli equation. It can be used *only* when the four restrictions listed above apply, at least to reasonable accuracy! Although no real flow satisfies all these restrictions (especially the second), we can approximate the behavior of many flows with Eq. 4.24.

For example, the equation is widely used in aerodynamics to relate the pressure and velocity in a flow (e.g., it explains the lift of a subsonic wing). It could also be used to find the pressure at the inlet of the reducing elbow analyzed in Example 4.6 or to determine the velocity of water leaving the sluice gate of Example 4.7 (both of these flows approximately satisfy the four restrictions). On the other hand, Eq. 4.24 does *not* correctly describe the variation of water pressure in pipe flow. According to it, for a horizontal pipe of constant diameter, the pressure will be constant, but in fact the pressure drops significantly along the pipe—we will need most of Chapter 8 to explain this.

The Bernoulli equation, and the limits on its use, is so important we will derive it again and discuss its limitations in detail in Chapter 6.

Example 4.9 NOZZLE FLOW: APPLICATION OF BERNOULLI EQUATION

Water flows steadily through a horizontal nozzle, discharging to the atmosphere. At the nozzle inlet the diameter is D_1; at the nozzle outlet the diameter is D_2. Derive an expression for the minimum gage pressure required at the nozzle inlet to produce a given volume flow rate, Q. Evaluate the inlet gage pressure if $D_1 = 75$ mm, $D_2 = 25$ mm, and the desired flow rate is 0.02 m³/s.

Given: Steady flow of water through a horizontal nozzle, discharging to the atmosphere.

$$D_1 = 75\text{ mm} \qquad D_2 = 25\text{ mm} \qquad p_2 = p_{\text{atm}}$$

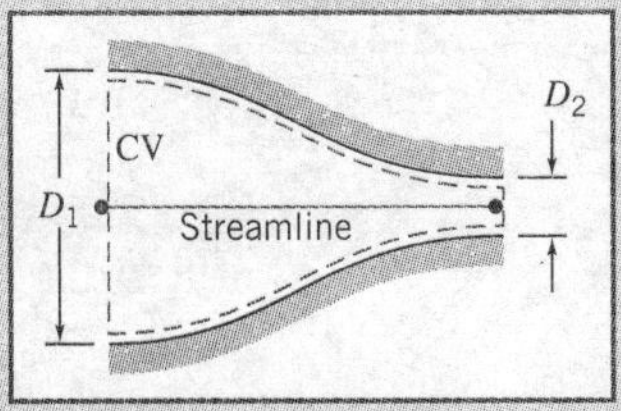

Find: (a) p_{1g} as a function of volume flow rate, Q.
(b) p_{1g} for $Q = 0.7\text{ ft}^3/\text{s}$.

Solution:

Governing equations:

$$\frac{p_1}{\rho} + \frac{V_1^2}{2} + gz_1 = \frac{p_2}{\rho} + \frac{V_2^2}{2} + gz_2$$

$$\overset{=0(1)}{\frac{\partial}{\partial t}\int_{\text{CV}} \rho\, d\forall} + \int_{\text{CS}} \rho \vec{V} \cdot d\vec{A} = 0$$

Assumptions:
(1) Steady flow (given).
(2) Incompressible flow.
(3) Frictionless flow.
(4) Flow along a streamline.
(5) $z_1 = z_2$.
(6) Uniform flow at sections ① and ②.

Apply the Bernoulli equation along a streamline between points ① and ② to evaluate p_1. Then

$$p_{1g} = p_1 - p_{\text{atm}} = p_1 - p_2 = \frac{\rho}{2}(V_2^2 - V_1^2) = \frac{\rho}{2}V_1^2\left[\left(\frac{V_2}{V_1}\right)^2 - 1\right]$$

Apply the continuity equation

$$(-\rho V_1 A_1) + (\rho V_2 A_2) = 0 \qquad \text{or} \qquad V_1 A_1 = V_2 A_2 = Q$$

so that

$$\frac{V_2}{V_1} = \frac{A_1}{A_2} \qquad \text{and} \qquad V_1 = \frac{Q}{A_1}$$

Then

$$p_{1g} = \frac{\rho Q^2}{2A_1^2}\left[\left(\frac{A_1}{A_2}\right)^2 - 1\right]$$

Since $A = \pi D^2/4$, then

$$p_{1g} = \frac{8\rho Q^2}{\pi^2 D_1^4}\left[\left(\frac{D_1}{D_2}\right)^4 - 1\right] \longleftarrow p_{1_g}$$

(Note that for a given nozzle the pressure required is proportional to the square of the flow rate—not surpising since we have used Eq. 4.24, which shows that $p \sim V^2 \sim Q^2$.) With $D_1 = 75$ mm $D_2 = 25$ mm, and $\rho = 1000\text{ kg/m}^3$,

$$p_{1g} = \frac{8}{\pi^2} \times 1000\frac{\text{kg}}{\text{m}^3} \times \frac{1}{(0.075)^4\text{m}^4} \times Q^2[(3.0)^4 - 1]\frac{\text{N}\cdot\text{s}^2}{\text{kg}\cdot\text{m}} \times \frac{\text{Pa}\cdot\text{m}^2}{\text{N}^2}$$

$$p_{1g} = 2049.44 \times 10^6\, Q^2 \frac{\text{N}\cdot\text{s}^2}{\text{m}^8} \times \frac{\text{Pa}\cdot\text{m}^2}{\text{N}}$$

With $Q = 0.02\text{ m}^3/\text{s}$, then $p_{1g} = 819{,}776$ kPa $\longleftarrow p_{1_g}$

This problem illustrates application of the Bernoulli equation to a flow where the restrictions of steady, incompressible, frictionless flow along a streamline are reasonable.

Control Volume Moving with Constant Velocity

In the preceding problems, which illustrate applications of the momentum equation to inertial control volumes, we have considered only stationary control volumes. Suppose we have a control volume moving at constant speed. We can set up two coordinate systems: XYZ, "absolute," or stationary (and therefore inertial), coordinates, and the xyz coordinates attached to the control volume (also inertial because the control volume is not accelerating with respect to XYZ).

Equation 4.10, which expresses system derivatives in terms of control volume variables, is valid for any motion of the control volume coordinate system xyz, provided that all velocities are measured *relative* to the control volume. To emphasize this point, we rewrite Eq. 4.10 as

$$\left.\frac{dN}{dt}\right)_{\text{system}} = \frac{\partial}{\partial t}\int_{\text{CV}} \eta\,\rho\, d\maltese + \int_{\text{CS}} \eta\,\rho\,\vec{V}_{xyz}\cdot d\vec{A} \tag{4.25}$$

Since all velocities must be measured relative to the control volume, in using this equation to obtain the momentum equation for an inertial control volume from the system formulation, we must set

$$N = \vec{P}_{xyz} \qquad \text{and} \qquad \eta = \vec{V}_{xyz}$$

The control volume equation is then written as

$$\vec{F} = \vec{F}_S + \vec{F}_B = \frac{\partial}{\partial t}\int_{\text{CV}} \vec{V}_{xyz}\,\rho\, d\maltese + \int_{\text{CS}} \vec{V}_{xyz}\,\rho\vec{V}_{xyz}\cdot d\vec{A} \tag{4.26}$$

Equation 4.26 is the formulation of Newton's second law applied to any inertial control volume (stationary or moving with a constant velocity). It is identical to Eq. 4.17a except that we have included subscript xyz to emphasize that velocities must be measured relative to the control volume. (It is helpful to imagine that the velocities are those that would be seen by an observer moving with the control volume.) Example 4.10 illustrates the use of Eq. 4.26 for a control volume moving at constant velocity.

Example 4.10 VANE MOVING WITH CONSTANT VELOCITY

The sketch shows a vane with a turning angle of 60°. The vane moves at constant speed, $U = 10$ m/s, and receives a jet of water that leaves a stationary nozzle with speed $V = 30$ m/s. The nozzle has an exit area of 0.003 m^2. Determine the force components that act on the vane.

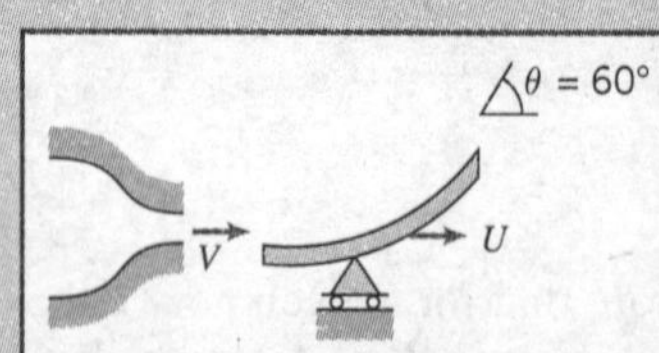

Given: Vane, with turning angle $\theta = 60°$, moves with constant velocity, $\vec{U} = 10\hat{i}$ m/s. Water from a constant area nozzle, $A = 0.003$ m^2, with velocity $\vec{V} = 30\hat{i}$ m/s, flows over the vane as shown.

Find: Force components acting on the vane.

Solution: Select a control volume moving with the vane at constant velocity, $\vec{U}$, as shown by the dashed lines. R_x and R_y are the components of force required to maintain the velocity of the control volume at $10\hat{i}$ m/s.

The control volume is inertial, since it is not accelerating (U = constant). Remember that all velocities must be measured relative to the control volume in applying the basic equations.

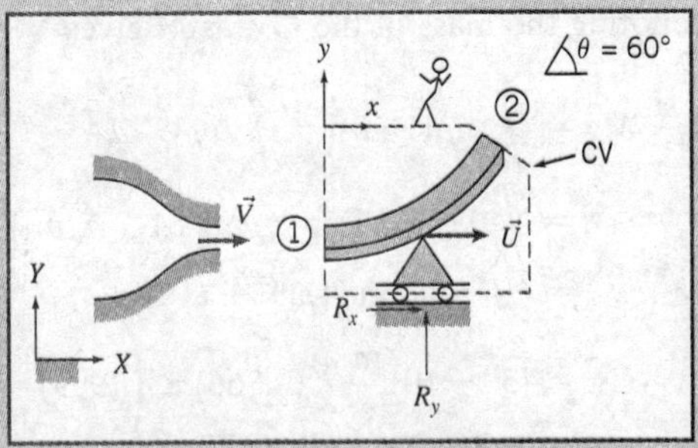

Governing equations:

$$\vec{F}_S + \vec{F}_B = \frac{\partial}{\partial t}\int_{CV} \vec{V}_{xyz}\rho\, d\forall + \int_{CS} \vec{V}_{xyz}\, \rho \vec{V}_{xyz} \cdot d\vec{A}$$

$$\frac{\partial}{\partial t}\int_{CV} \rho\, d\forall + \int_{CS} \rho \vec{V}_{xyz} \cdot d\vec{A} = 0$$

Assumptions:
(1) Flow is steady relative to the vane.
(2) Magnitude of relative velocity along the vane is constant: $|\vec{V}_1| = |\vec{V}_2| = V - U$.
(3) Properties are uniform at sections ① and ②.
(4) $F_{B_x} = 0$.
(5) Incompressible flow.

The x component of the momentum equation is

$$F_{S_x} + \underbrace{F_{B_x}}_{=0(4)} = \underbrace{\frac{\partial}{\partial t}\int_{CV} u_{xyz}\, \rho\, d\forall}_{=0(1)} + \int_{CS} u_{xyz}\, \rho \vec{V}_{xyz} \cdot d\vec{A}$$

There is no net pressure force, since p_{atm} acts on all sides of the CV. Thus

$$R_x = \int_{A_1} u(-\rho V dA) + \int_{A_2} u(\rho V dA) = +u_1(-\rho V_1 A_1) + u_2(\rho V_2 A_2)$$

(All velocities are measured relative to xyz.) From the continuity equation

$$\int_{A_1} (-\rho V dA) + \int_{A_2} (\rho V dA) = (-\rho V_1 A_1) + (\rho V_2 A_2) = 0$$

or

$$\rho V_1 A_1 = \rho V_2 A_2$$

Therefore,

$$R_x = (u_2 - u_1)(\rho V_1 A_1)$$

All velocities must be measured relative to the CV, so we note that

$$V_1 = V - U \qquad V_2 = V - U$$
$$u_1 = V - U \qquad u_2 = (V - U)\cos\theta$$

Substituting yields

$$R_x = [(V-U)\cos\theta - (V-U)](\rho(V-U)A_1) = (V-U)(\cos\theta - 1)\{\rho(V-U)A_1\}$$
$$= (30-10)\frac{\text{m}}{\text{s}} \times (0.50 - 1) \times \left(999\frac{\text{kg}}{\text{m}^3}(30-10)\frac{\text{m}}{\text{s}} \times 0.003\,\text{m}^2\right) \times \frac{\text{N}\cdot\text{s}^2}{\text{kg}\cdot\text{m}}$$

$$R_x = -599\text{N} \ \{\text{to the left}\}$$

Writing the y component of the momentum equation, we obtain

$$F_{S_y} + F_{B_y} = \underbrace{\frac{\partial}{\partial t}\int_{CV} v_{xyz}\rho\, d\forall}_{=0(1)} + \int_{CS} v_{xyz}\rho\, \vec{V}_{xyz} \cdot d\vec{A}$$

Denoting the mass of the CV as M gives

$$R_y - Mg = \int_{CS} v\rho\vec{V}\cdot d\vec{A} = \int_{A_2} v\rho\vec{V}\cdot d\vec{A} \qquad \{v_1 = 0\}$$

$$= \int_{A_2} v(\rho V dA) = v_2(\rho V_2 A_2) = v_2(\rho V_1 A_1)$$

$$= (V-U)\sin\theta\{\rho(V-U)A_1\}$$

$$= (30-10)\frac{\text{m}}{\text{s}} \times (0.866) \times \left((999)\frac{\text{kg}}{\text{m}^3}(30-10)\frac{\text{m}}{\text{s}} \times 0.003\text{m}^2\right) \times \frac{\text{N}\cdot\text{s}^2}{\text{kg}\cdot\text{m}}$$

$\left\{\begin{array}{l}\text{All velocities are}\\ \text{measured relative to}\\ xyz.\end{array}\right\}$

{Recall $\rho V_2 A_2 = \rho V_1 A_1$.}

$$R_y - Mg = 1.04\,\text{kN} \qquad \{\text{upward}\}$$

Thus the vertical force is

$$R_y = 1.04\,\text{kN} + Mg \qquad \{\text{upward}\}$$

Then the net force on the vane (neglecting the weight of the vane and water within the CV) is

$$\vec{R} = -0.599\hat{i} + 1.04\hat{j}\,\text{kN} \qquad \vec{R}$$

This problem illustrates how to apply the momentum equation for a control volume in constant velocity motion by evaluating all velocities relative to the control volume.

4.5 Momentum Equation for Control Volume with Rectilinear Acceleration

For an inertial control volume (having no acceleration relative to a stationary frame of reference), the appropriate formulation of Newton's second law is given by Eq. 4.26,

$$\vec{F} = \vec{F}_S + \vec{F}_B = \frac{\partial}{\partial t}\int_{CV} \vec{V}_{xyz}\,\rho\, d\forall + \int_{CS} \vec{V}_{xyz}\,\rho\vec{V}_{xyz}\cdot d\vec{A} \tag{4.26}$$

Not all control volumes are inertial; for example, a rocket must accelerate if it is to get off the ground. Since we are interested in analyzing control volumes that may accelerate relative to inertial coordinates, it is logical to ask whether Eq. 4.26 can be used for an accelerating control volume. To answer this question, let us briefly review the two major elements used in developing Eq. 4.26.

First, in relating the system derivatives to the control volume formulation (Eq. 4.25 or 4.10), the flow field, $\vec{V}(x, y, z, t)$, was specified relative to the control volume's coordinates x, y, and z. No restriction was placed on the motion of the xyz reference frame. Consequently, Eq. 4.25 (or Eq. 4.10) is valid at any instant for any arbitrary motion of the coordinates x, y, and z provided that all velocities in the equation are measured relative to the control volume.

Second, the system equation

$$\vec{F} = \left.\frac{d\vec{P}}{dt}\right)_{\text{system}} \tag{4.2a}$$

where the linear momentum of the system is given by

$$\vec{P}_{\text{system}} = \int_{M(\text{system})} \vec{V}\, dm = \int_{\forall(\text{system})} \vec{V}\, \rho\, d\forall \tag{4.2b}$$

is valid only for velocities measured relative to an inertial reference frame. Thus, if we denote the inertial reference frame by XYZ, then Newton's second law states that

$$\vec{F} = \left.\frac{d\vec{P}_{XYZ}}{dt}\right)_{\text{system}} \tag{4.27}$$

Since the time derivatives of $\vec{P}_{XYZ}$ and $\vec{P}_{xyz}$ are not equal when the control volume reference frame xyz is accelerating relative to the inertial reference frame, Eq. 4.26 is not valid for an accelerating control volume.

To develop the momentum equation for a linearly accelerating control volume, it is necessary to relate $\vec{P}_{XYZ}$ of the system to $\vec{P}_{xyz}$ of the system. The system derivative $d\vec{P}_{xyz}/dt$ can then be related to control volume variables through Eq. 4.25. We begin by writing Newton's second law for a system, remembering that the acceleration must be measured relative to an inertial reference frame that we have designated XYZ. We write

$$\vec{F} = \left.\frac{d\vec{P}_{XYZ}}{dt}\right)_{\text{system}} = \frac{d}{dt}\int_{M(\text{system})} \vec{V}_{XYZ}\, dm = \int_{M(\text{system})} \frac{d\vec{V}_{XYZ}}{dt}\, dm \tag{4.28}$$

The velocities with respect to the inertial (XYZ) and the control volume coordinates (xyz) are related by the relative-motion equation

$$\vec{V}_{XYZ} = \vec{V}_{xyz} + \vec{V}_{rf} \tag{4.29}$$

where $\vec{V}_{rf}$ is the velocity of the control volume coordinates xyz with respect to the "absolute" stationary coordinates XYZ.

Since we are assuming the motion of xyz is pure translation, without rotation, relative to inertial reference frame XYZ, then

$$\frac{d\vec{V}_{XYZ}}{dt} = \vec{a}_{XYZ} = \frac{d\vec{V}_{xyz}}{dt} + \frac{d\vec{V}_{rf}}{dt} = \vec{a}_{xyz} + \vec{a}_{rf} \tag{4.30}$$

where

$\vec{a}_{XYZ}$ is the rectilinear acceleration of the system relative to inertial reference frame XYZ,

$\vec{a}_{xyz}$ is the rectilinear acceleration of the system relative to noninertial reference frame xyz (i.e., relative to the control volume), and

$\vec{a}_{rf}$ is the rectilinear acceleration of noninertial reference frame xyz (i.e., of the control volume) relative to inertial frame XYZ.

Substituting from Eq. 4.30 into Eq. 4.28 gives

$$\vec{F} = \int_{M(\text{system})} \vec{a}_{rf}\, dm + \int_{M(\text{system})} \frac{d\vec{V}_{xyz}}{dt}\, dm$$

or

$$\vec{F} - \int_{M(\text{system})} \vec{a}_{rf}\, dm = \left.\frac{d\vec{P}_{xyz}}{dt}\right)_{\text{system}} \tag{4.31a}$$

where the linear momentum of the system is given by

$$\left.\vec{P}_{xyz}\right)_{\text{system}} = \int_{M(\text{system})} \vec{V}_{xyz}\, dm = \int_{\forall(\text{system})} \vec{V}_{xyz}\, \rho\, d\forall \tag{4.31b}$$

and the force, $\vec{F}$, includes all surface and body forces acting on the system.

To derive the control volume formulation of Newton's second law, we set

$$N = \vec{P}_{xyz} \qquad \text{and} \qquad \eta = \vec{V}_{xyz}$$

From Eq. 4.25, with this substitution, we obtain

$$\left.\frac{d\vec{P}_{xyz}}{dt}\right)_{\text{system}} = \frac{\partial}{\partial t}\int_{\text{CV}} \vec{V}_{xyz}\rho\, d\mathcal{V} + \int_{\text{CS}} \vec{V}_{xyz}\rho\vec{V}_{xyz}\cdot d\vec{A} \tag{4.32}$$

Combining Eq. 4.31a (the linear momentum equation for the system) and Eq. 4.32 (the system–control volume conversion), and recognizing that at time t_0 the system and control volume coincide, Newton's second law for a control volume accelerating, without rotation, relative to an inertial reference frame is

$$\vec{F} - \int_{\text{CV}} \vec{a}_{rf}\,\rho\, d\mathcal{V} = \frac{\partial}{\partial t}\int_{\text{CV}} \vec{V}_{xyz}\,\rho\, d\mathcal{V} + \int_{\text{CS}} \vec{V}_{xyz}\,\rho\vec{V}_{xyz}\cdot d\vec{A}$$

Since $\vec{F} = \vec{F}_S + \vec{F}_B$, this equation becomes

$$\vec{F}_S + \vec{F}_B - \int_{\text{CV}} \vec{a}_{rf}\,\rho\, d\mathcal{V} = \frac{\partial}{\partial t}\int_{\text{CV}} \vec{V}_{xyz}\,\rho\, d\mathcal{V} + \int_{\text{CS}} \vec{V}_{xyz}\,\rho\vec{V}_{xyz}\cdot d\vec{A} \tag{4.33}$$

Comparing this momentum equation for a control volume with rectilinear acceleration to that for a nonaccelerating control volume, Eq. 4.26, we see that the only difference is the presence of one additional term in Eq. 4.33. When the control volume is not accelerating relative to inertial reference frame *XYZ*, then $\vec{a}_{rf} = 0$, and Eq. 4.33 reduces to Eq. 4.26.

The precautions concerning the use of Eq. 4.26 also apply to the use of Eq. 4.33. Before attempting to apply either equation, one must draw the boundaries of the control volume and label appropriate coordinate directions. For an accelerating control volume, one must label two coordinate systems: one (*xyz*) on the control volume and the other (*XYZ*) an inertial reference frame.

In Eq. 4.33, $\vec{F}_S$ represents all surface forces acting on the control volume. Since the mass within the control volume may vary with time, both the remaining terms on the left side of the equation may be functions of time. Furthermore, the acceleration, $\vec{a}_{rf}$, of the reference frame *xyz* relative to an inertial frame will in general be a function of time.

All velocities in Eq. 4.33 are measured relative to the control volume. The momentum flux, $\vec{V}_{xyz}\rho\vec{V}_{xyz}\cdot d\vec{A}$, through an element of the control surface area, $d\vec{A}$, is a vector. As we saw for the nonaccelerating control volume, the sign of the scalar product, $\rho\vec{V}_{xyz}\cdot d\vec{A}$, depends on the direction of the velocity vector, $\vec{V}_{xyz}$, relative to the area vector, $d\vec{A}$.

The momentum equation is a vector equation. As with all vector equations, it may be written as three scalar component equations. The scalar components of Eq. 4.33 are

$$F_{S_x} + F_{B_x} - \int_{\text{CV}} a_{rf_x}\,\rho d\mathcal{V} = \frac{\partial}{\partial t}\int_{\text{CV}} u_{xyz}\,\rho d\mathcal{V} + \int_{\text{CS}} u_{xyz}\,\rho\vec{V}_{xyz}\cdot d\vec{A} \tag{4.34a}$$

$$F_{S_y} + F_{B_y} - \int_{\text{CV}} a_{rf_y}\,\rho d\mathcal{V} = \frac{\partial}{\partial t}\int_{\text{CV}} v_{xyz}\,\rho d\mathcal{V} + \int_{\text{CS}} v_{xyz}\,\rho\vec{V}_{xyz}\cdot d\vec{A} \tag{4.34b}$$

$$F_{S_z} + F_{B_z} - \int_{\text{CV}} a_{rf_z}\,\rho d\mathcal{V} = \frac{\partial}{\partial t}\int_{\text{CV}} w_{xyz}\,\rho d\mathcal{V} + \int_{\text{CS}} w_{xyz}\,\rho\vec{V}_{xyz}\cdot d\vec{A} \tag{4.34c}$$

We will consider two applications of the linearly accelerating control volume: Example 4.11 will analyze an accelerating control volume in which the mass contained in the control volume is constant; Example 4.12 will analyze an accelerating control volume in which the mass contained varies with time.

Example 4.11 VANE MOVING WITH RECTILINEAR ACCELERATION

A vane, with turning angle $\theta = 60°$, is attached to a cart. The cart and vane, of mass $M = 75$ kg, roll on a level track. Friction and air resistance may be neglected. The vane receives a jet of water, which leaves a stationary nozzle horizontally at $V = 35$ m/s. The nozzle exit area is $A = 0.003$ m^2. Determine the velocity of the cart as a function of time and plot the results.

Given: Vane and cart as sketched, with $M = 75$ kg.

Find: $U(t)$ and plot results.

Solution: Choose the control volume and coordinate systems shown for the analysis. Note that XY is a fixed frame, while frame xy moves with the cart. Apply the x component of the momentum equation.

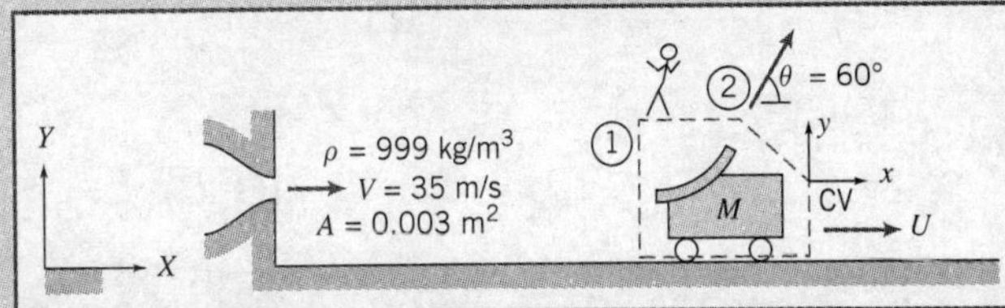

Governing equation:

$$\overset{=0(1)}{F_{S_x}} + \overset{=0(2)}{F_{B_x}} - \int_{CV} a_{rf_x}\,\rho\,d\forall = \overset{\simeq 0(4)}{\frac{\partial}{\partial t}\int_{CV} u_{xyz}\,\rho\,d\forall} + \int_{CS} u_{xyz}\,\rho\vec{V}_{xyz}\cdot d\vec{A}$$

Assumptions:
(1) $F_{S_x} = 0$, since no resistance is present.
(2) $F_{B_x} = 0$.
(3) Neglect the mass of water in contact with the vane compared to the cart mass.
(4) Neglect rate of change of momentum of liquid inside the CV.

$$\frac{\partial}{\partial t}\int_{CV} u_{xyz}\,\rho\,d\forall \simeq 0$$

(5) Uniform flow at sections ① and ②.
(6) Speed of water stream is not slowed by friction on the vane, so $|\vec{V}_{xyz_1}| = |\vec{V}_{xyz_2}|$.
(7) $A_2 = A_1 = A$.

Then, dropping subscripts rf and xyz for clarity (but remembering that all velocities are measured relative to the moving coordinates of the control volume),

$$\begin{aligned}-\int_{CV} a_x\,\rho\,d\forall &= u_1(-\rho V_1 A_1) + u_2(\rho V_2 A_2)\\ &= (V-U)\{-\rho(V-U)A\} + (V-U)\cos\theta\{\rho(V-U)A\}\\ &= -\rho(V-U)^2 A + \rho(V-U)^2 A\cos\theta\end{aligned}$$

For the left side of this equation we have

$$-\int_{CV} a_x\,\rho\,d\forall = -a_x M_{CV} = -a_x M = -\frac{dU}{dt}M$$

so that

$$-M\frac{dU}{dt} = -\rho(V-U)^2 A + \rho(V-U)^2 A\cos\theta$$

or

$$M\frac{dU}{dt} = (1-\cos\theta)\rho(V-U)^2 A$$

Separating variables, we obtain

$$\frac{dU}{(V-U)^2} = \frac{(1-\cos\theta)\rho A}{M}dt = b\,dt \quad \text{where } b = \frac{(1-\cos\theta)\rho A}{M}$$

Note that since V = constant, $dU = -d(V-U)$. Integrating between limits $U = 0$ at $t = 0$, and $U = U$ at $t = t$,

$$\int_0^U \frac{dU}{(V-U)^2} = \int_0^U \frac{-d(V-U)}{(V-U)^2} = \frac{1}{(V-U)}\bigg]_0^U = \int_0^t b\,dt = bt$$

or

$$\frac{1}{(V-U)} - \frac{1}{V} = \frac{U}{V(V-U)} = bt$$

Solving for U, we obtain

$$\frac{U}{V} = \frac{Vbt}{1+Vbt}$$

Evaluating Vb gives

$$Vb = V\frac{(1-\cos\theta)\rho A}{M}$$

$$Vb = 35\frac{\text{m}}{\text{s}} \times \frac{(1-0.5)}{75\text{ kg}} \times 999\frac{\text{kg}}{\text{m}^3} \times 0.003\text{ m}^2 = 0.699\text{ s}^{-1}$$

Thus

$$\frac{U}{V} = \frac{0.699t}{1+0.699t} \qquad (t \text{ in seconds}) \qquad U(t)$$

Plot:

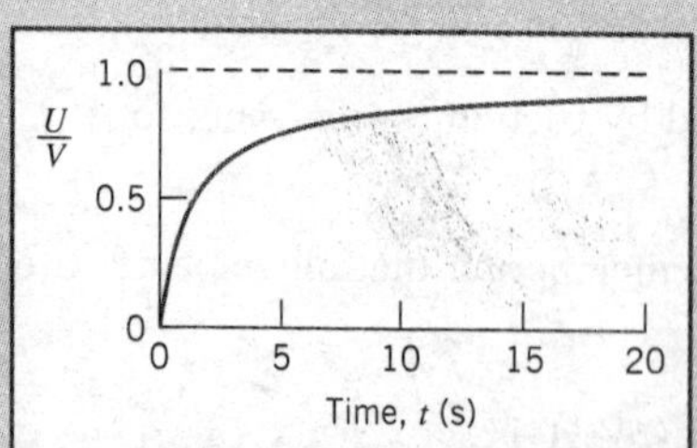

The graph was generated from an *Excel* workbook. This workbook is interactive: It allows one to see the effect of different values of ρ, A, M, and θ on U/V against time t, and also to determine the time taken for the cart to reach, for example, 95% of jet speed.

Example 4.12 ROCKET DIRECTED VERTICALLY

A small rocket, with an initial mass of 400 kg, is to be launched vertically. Upon ignition the rocket consumes fuel at the rate of 5 kg/s and ejects gas at atmospheric pressure with a speed of 3500 m/s relative to the rocket. Determine the initial acceleration of the rocket and the rocket speed after 10 s, if air resistance is neglected.

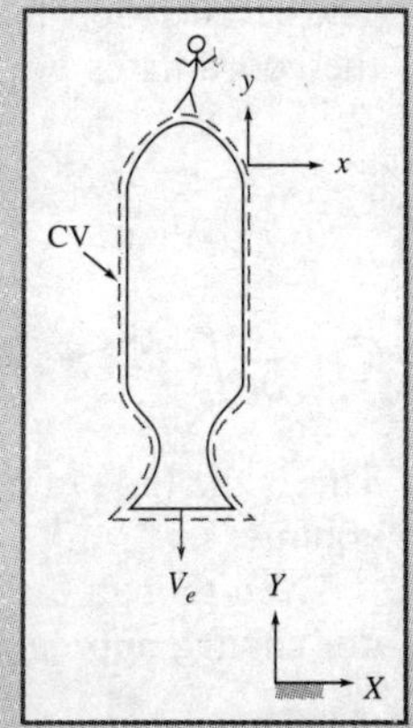

Given: Small rocket accelerates vertically from rest.
Initial mass, $M_0 = 400$ kg.
Air resistance may be neglected.
Rate of fuel consumption, $\dot{m}_e = 5$ kg/s.
Exhaust velocity, $V_e = 3500$ m/s, relative to rocket, leaving at atmospheric pressure.

Find: (a) Initial acceleration of the rocket.
(b) Rocket velocity after 10 s.

Solution:
Choose a control volume as shown by dashed lines. Because the control volume is accelerating, define inertial coordinate system XY and coordinate system xy attached to the CV. Apply the y component of the momentum equation.

Governing equation: $F_{S_y} + F_{B_y} - \int_{CV} a_{rf_y}\,\rho\, d\mathcal{V} = \frac{\partial}{\partial t}\int_{CV} v_{xyz}\,\rho\, d\mathcal{V} + \int_{CS} v_{xyz}\,\rho \vec{V}_{xyz}\cdot d\vec{A}$

Assumptions: (1) Atmospheric pressure acts on all surfaces of the CV; since air resistance is neglected, $F_{S_y} = 0$.
(2) Gravity is the only body force; g is constant.
(3) Flow leaving the rocket is uniform, and V_e is constant.

Under these assumptions the momentum equation reduces to

$$F_{B_y} - \int_{CV} a_{rf_y}\,\rho\, d\mathcal{V} = \frac{\partial}{\partial t}\int_{CV} v_{xyz}\,\rho\, d\mathcal{V} + \int_{CS} v_{xyz}\,\rho \vec{V}_{xyz}\cdot d\vec{A} \qquad (1)$$

Ⓐ Ⓑ Ⓒ Ⓓ

Let us look at the equation term by term:

Ⓐ $F_{B_y} = -\int_{CV} g\,\rho\, d\mathcal{V} = -g\int_{CV}\rho\, d\mathcal{V} = -gM_{CV}$ {since g is constant}

The mass of the CV will be a function of time because mass is leaving the CV at rate $\dot{m}_e$. To determine M_{CV} as a function of time, we use the conservation of mass equation

$$\frac{\partial}{\partial t}\int_{CV}\rho\, d\mathcal{V} + \int_{CS}\rho\vec{V}\cdot d\vec{A} = 0$$

Then

$$\frac{\partial}{\partial t}\int_{CV}\rho\, d\mathcal{V} = -\int_{CS}\rho\vec{V}\cdot d\vec{A} = -\int_{CS}(\rho V_{xyz} dA) = -\dot{m}_e$$

The minus sign indicates that the mass of the CV is decreasing with time. Since the mass of the CV is only a function of time, we can write

$$\frac{dM_{CV}}{dt} = -\dot{m}_e$$

To find the mass of the CV at any time, t, we integrate

$$\int_{M_0}^{M} dM_{CV} = -\int_0^t \dot{m}_e\, dt \qquad \text{where at } t = 0,\ M_{CV} = M_0, \text{ and at } t = t,\ M_{CV} = M$$

Then, $M - M_0 = -\dot{m}_e t$, or $M = M_0 - \dot{m}_e t$.

Substituting the expression for M into term Ⓐ, we obtain

$$F_{B_y} = -\int_{CV} g\,\rho\, d\mathcal{V} = -gM_{CV} = -g(M_0 - \dot{m}_e t)$$

Ⓑ $-\int_{CV} a_{rf_y}\,\rho\, d\mathcal{V}$

The acceleration, a_{rf_y}, of the CV is that seen by an observer in the XY coordinate system. Thus a_{rf_y} is not a function of the coordinates xyz, and

$$-\int_{CV} a_{rf_y}\,\rho\,d\mathcal{V} = -a_{rf_y}\int_{CV}\rho\,d\mathcal{V} = -a_{rf_y}M_{CV} = -a_{rf_y}(M_0 - \dot{m}_e t)$$

Ⓒ $\dfrac{\partial}{\partial t}\displaystyle\int_{CV} v_{xyz}\,\rho\,d\mathcal{V}$

This is the time rate of change of the y momentum of the fluid in the control volume measured relative to the control volume.

Even though the y momentum of the fluid inside the CV, measured relative to the CV, is a large number, it does not change appreciably with time. To see this, we must recognize that:

(1) The unburned fuel and the rocket structure have zero momentum relative to the rocket.
(2) The velocity of the gas at the nozzle exit remains constant with time as does the velocity at various points in the nozzle.

Consequently, it is reasonable to assume that

$$\frac{\partial}{\partial t}\int_{CV} v_{xyz}\,\rho\,d\mathcal{V} \approx 0$$

Ⓓ $$\int_{CS} v_{xyz}\,\rho\vec{V}_{xyz}\cdot d\vec{A} = \int_{CS} v_{xyz}(\rho V_{xyz}\,dA) = -V_e\int_{CS}(\rho V_{xyz}\,dA)$$

The velocity v_{xyz} (relative to the control volume) is $-V_e$ (it is in the negative y direction), and is a constant, so was taken outside the integral. The remaining integral is simply the mass flow rate at the exit (positive because flow is out of the control volume),

$$\int_{CS}(\rho V_{xyz}dA) = \dot{m}_e$$

and so

$$\int_{CS} v_{xyz}\,\rho\vec{V}_{xyz}\cdot d\vec{A} = -V_e\dot{m}_e$$

Substituting terms Ⓐ through Ⓓ into Eq. 1, we obtain

$$-g(M_0 - \dot{m}_e t) - a_{rf_y}(M_0 - \dot{m}_e t) = -V_e\dot{m}_e$$

or

$$a_{rf_y} = \frac{V_e\dot{m}_e}{M_0 - \dot{m}_e t} - g \tag{2}$$

At time $t = 0$,

$$a_{rf_y})_{t=0} = \frac{V_e\dot{m}_e}{M_0} - g = 3500\,\frac{\text{m}}{\text{s}} \times 5\,\frac{\text{kg}}{\text{s}} \times \frac{1}{400\,\text{kg}} - 9.81\,\frac{\text{m}}{\text{s}^2}$$

$$a_{rf_y})_{t=0} = 33.9\,\text{m/s}^2 \qquad a_{rf_y})_{t=0}$$

The acceleration of the CV is by definition

$$a_{rf_y} = \frac{dV_{CV}}{dt}$$

Substituting from Eq. 2,

$$\frac{dV_{CV}}{dt} = \frac{V_e \dot{m}_e}{M_0 - \dot{m}_e t} - g$$

Separating variables and integrating gives

$$V_{CV} = \int_0^{V_{CV}} dV_{CV} = \int_0^t \frac{V_e \dot{m}_e dt}{M_0 - \dot{m}_e t} - \int_0^t g dt = -V_e \ln\left[\frac{M_0 - \dot{m}_e t}{M_0}\right] - gt$$

At $t = 10$ s,

$$V_{CV} = -3500 \frac{\text{m}}{\text{s}} \times \ln\left[\frac{350\text{ kg}}{400\text{ kg}}\right] - 9.81 \frac{\text{m}}{\text{s}^2} \times 10\text{ s}$$

$$V_{CV} = 369\text{m/s} \qquad V_{CV})_{t=10\text{s}}$$

The velocity-time graph is shown in an *Excel* workbook. This workbook is interactive: It allows one to see the effect of different values of M_o, V_e, and $\dot{m}_e$ on V_{CV} versus time t. Also, the time at which the rocket attains a given speed, e.g., 2000 m/s, can be determined.

Momentum Equation for Control Volume with Arbitrary Acceleration (on the Web) 4.6

The Angular-Momentum Principle 4.7*

Our next task is to derive a control volume form of the angular-momentum principle. There are two obvious approaches we can use to express the angular-momentum principle: We can use an inertial (fixed) *XYZ* control volume; we can also use a rotating *xyz* control volume. For each approach we will: start with the principle in its system form (Eq. 4.3a), then write the system angular momentum in terms of *XYZ* or *xyz* coordinates, and finally use Eq. 4.10 (or its slightly different form, Eq. 4.25) to convert from a system to a control volume formulation. To verify that these two approaches are equivalent, we will use each approach to solve the same problem, in Examples 4.14 and 4.15 (on the Web), respectively.

There are two reasons for the material of this section: We wish to develop a control volume equation for each of the basic physical laws of Section 4.2; and we will need the results for use in Chapter 10, where we discuss rotating machinery.

Equation for Fixed Control Volume

The angular-momentum principle for a system in an inertial frame is

$$\vec{T} = \left.\frac{d\vec{H}}{dt}\right)_{\text{system}} \tag{4.3a}$$

where $\vec{T}$ = total torque exerted on the system by its surroundings, and
$\vec{H}$ = angular momentum of the system.

*This section may be omitted without loss of continuity in the text material.

$$\vec{H} = \int_{M(\text{system})} \vec{r} \times \vec{V}\, dm = \int_{V\!\!\!\!-(\text{system})} \vec{r} \times \vec{V}\, \rho\, dV\!\!\!\!- \tag{4.3b}$$

All quantities in the system equation must be formulated with respect to an inertial reference frame. Reference frames at rest, or translating with constant linear velocity, are inertial, and Eq. 4.3b can be used directly to develop the control volume form of the angular-momentum principle.

The position vector, $\vec{r}$, locates each mass or volume element of the system with respect to the coordinate system. The torque, $\vec{T}$, applied to a system may be written

$$\vec{T} = \vec{r} \times \vec{F}_s + \int_{M(\text{system})} \vec{r} \times \vec{g}\, dm + \vec{T}_{\text{shaft}} \tag{4.3c}$$

where $\vec{F}_s$ is the surface force exerted on the system.

The relation between the system and fixed control volume formulations is

$$\left.\frac{dN}{dt}\right)_{\text{system}} = \frac{\partial}{\partial t}\int_{\text{CV}} \eta\, \rho\, dV\!\!\!\!- + \int_{\text{CS}} \eta\, \rho \vec{V} \cdot d\vec{A} \tag{4.10}$$

where

$$N_{\text{system}} = \int_{M(\text{system})} \eta\, dm$$

If we set $N = \vec{H}$, then $\eta = \vec{r} \times \vec{V}$, and

$$\left.\frac{d\vec{H}}{dt}\right)_{\text{system}} = \frac{\partial}{\partial t}\int_{\text{CV}} \vec{r} \times \vec{V}\, \rho\, dV\!\!\!\!- + \int_{\text{CS}} \vec{r} \times \vec{V}\, \rho \vec{V} \cdot d\vec{A} \tag{4.45}$$

Combining Eqs. 4.3a, 4.3c, and 4.45, we obtain

$$\vec{r} \times \vec{F}_s + \int_{M(\text{system})} \vec{r} \times \vec{g}\, dm + \vec{T}_{\text{shaft}} = \frac{\partial}{\partial t}\int_{\text{CV}} \vec{r} \times \vec{V}\, \rho\, dV\!\!\!\!- + \int_{\text{CS}} \vec{r} \times \vec{V}\, \rho \vec{V} \cdot d\vec{A}$$

Since the system and control volume coincide at time t_0,

$$\vec{T} = \vec{T}_{\text{CV}}$$

and

$$\vec{r} \times \vec{F}_s + \int_{\text{CV}} \vec{r} \times \vec{g}\, \rho\, dV\!\!\!\!- + \vec{T}_{\text{shaft}} = \frac{\partial}{\partial t}\int_{\text{CV}} \vec{r} \times \vec{V}\, \rho\, dV\!\!\!\!- + \int_{\text{CS}} \vec{r} \times \vec{V}\, \rho \vec{V} \cdot d\vec{A} \tag{4.46}$$

Equation 4.46 is a general formulation of the angular-momentum principle for an inertial control volume. The left side of the equation is an expression for all the torques that act on the control volume. Terms on the right express the rate of change of angular momentum within the control volume and the net rate of flux of angular momentum from the control volume. All velocities in Eq. 4.46 are measured relative to the fixed control volume.

For analysis of rotating machinery, Eq. 4.46 is often used in scalar form by considering only the component directed along the axis of rotation. This application is illustrated in Chapter 10.

The application of Eq. 4.46 to the analysis of a simple lawn sprinkler is illustrated in Example 4.14. This same problem is considered in Example 4.15 (on the Web) using the angular-momentum principle expressed in terms of a *rotating* control volume.

Example 4.14 LAWN SPRINKLER: ANALYSIS USING FIXED CONTROL VOLUME

A small lawn sprinkler is shown in the sketch at right. At an inlet gage pressure of 20 kPa, the total volume flow rate of water through the sprinkler is 7.5 liters per minute and it rotates at 30 rpm. The diameter of each jet is 4 mm. Calculate the jet speed relative to each sprinkler nozzle. Evaluate the friction torque at the sprinkler pivot.

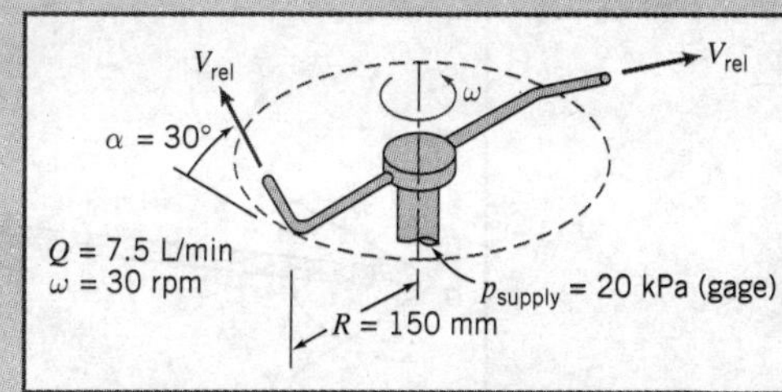

Given: Small lawn sprinkler as shown.

Find: (a) Jet speed relative to each nozzle.
(b) Friction torque at pivot.

Solution: Apply continuity and angular momentum equations using fixed control volume enclosing sprinkler arms.

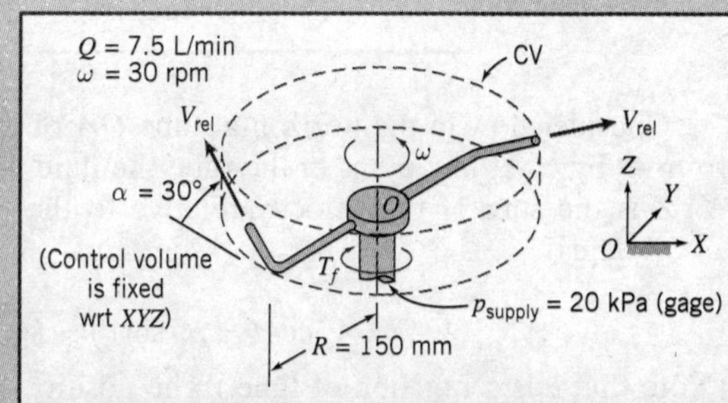

Governing equations:

$$\overbrace{\frac{\partial}{\partial t}\int_{CV} \rho\, d\forall}^{=0(1)} + \int_{CS} \rho \vec{V} \cdot d\vec{A} = 0$$

$$\vec{r} \times \vec{F}_s + \int_{CV} \vec{r} \times \vec{g}\, \rho\, d\forall + \vec{T}_{\text{shaft}} = \frac{\partial}{\partial t}\int_{CV} \vec{r} \times \vec{V} \rho\, d\forall + \int_{CS} \vec{r} \times \vec{V} \rho \vec{V} \cdot d\vec{A} \tag{1}$$

where all velocities are measured relative to the inertial coordinates XYZ.

Assumptions: (1) Incompressible flow.
(2) Uniform flow at each section.
(3) $\vec{\omega}$ = constant.

From continuity, the jet speed relative to the nozzle is given by

$$V_{\text{rel}} = \frac{Q}{2A_{\text{jet}}} = \frac{Q}{2}\frac{4}{\pi D_{\text{jet}}^2}$$

$$= \frac{1}{2} \times 7.5 \frac{\text{L}}{\text{min}} \times \frac{4}{\pi}\frac{1}{(4)^2\,\text{mm}^2} \times \frac{\text{m}^3}{1000\,\text{L}} \times 10^6 \frac{\text{mm}^2}{\text{m}^2} \times \frac{\text{min}}{60\,\text{s}}$$

$$V_{\text{rel}} = 4.97\,\text{m/s} \quad \longleftarrow \quad V_{\text{rel}}$$

Consider terms in the angular momentum equation separately. Since atmospheric pressure acts on the entire control surface, and the pressure force at the inlet causes no moment about O, $\vec{r} \times \vec{F}_s = 0$. The moments of the body (i.e., gravity) forces in the two arms are equal and opposite and hence the second term on the left side of the equation is zero. The only external torque acting on the CV is friction in the pivot. It opposes the motion, so

$$\vec{T}_{\text{shaft}} = -T_f \hat{K} \tag{2}$$

Our next task is to determine the two angular momentum terms on the right side of Eq. 1. Consider the unsteady term: This is the rate of change of angular momentum in the control volume. It is clear that although the position $\vec{r}$ and velocity $\vec{V}$ of fluid particles are functions of time in XYZ coordinates, because the sprinkler rotates at constant speed the control volume angular momentum *is* constant in XYZ coordinates, so this term is zero; however, as an exercise in manipulating vector quantities, let us derive this result. Before we can evaluate the control volume integral, we need to develop expressions for the instantaneous position vector, $\vec{r}$, and velocity vector, $\vec{V}$ (measured relative to the fixed coordinate system XYZ) of each element of fluid in the control volume. OA lies in the XY plane; AB is inclined at angle α to the XY plane; point B' is the projection of point B on the XY plane.

We assume that the length, L, of the tip AB is small compared with the length, R, of the horizontal arm OA. Consequently we neglect the angular momentum of the fluid in the tips compared with the angular momentum in the horizontal arms.

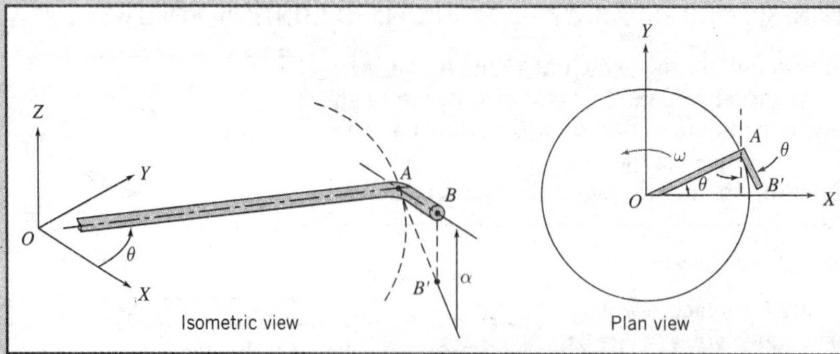

Consider flow in the horizontal tube OA of length R. Denote the radial distance from O by r. At any point in the tube the fluid velocity relative to fixed coordinates XYZ is the sum of the velocity relative to the tube $\vec{V}_t$ and the tangential velocity $\vec{\omega} \times \vec{r}$. Thus

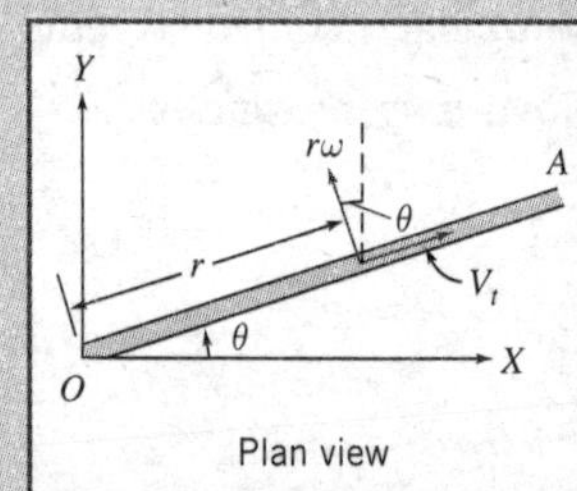

$$\vec{V} = \hat{I}(V_t \cos\theta - r\omega\sin\theta) + \hat{J}(V_t \sin\theta + r\omega\cos\theta)$$

(Note that θ is a function of time.) The position vector is

$$\vec{r} = \hat{I} r\cos\theta + \hat{J} r\sin\theta$$

and

$$\vec{r} \times \vec{V} = \hat{K}(r^2\omega\cos^2\theta + r^2\omega\sin^2\theta) = \hat{K} r^2\omega$$

Then

$$\int_{\mathcal{V}_{OA}} \vec{r} \times \vec{V} \rho \, d\mathcal{V} = \int_O^R \hat{K} r^2 \omega \rho A \, dr = \hat{K} \frac{R^3 \omega}{3} \rho A$$

and

$$\frac{\partial}{\partial t}\int_{\mathcal{V}_{OA}} \vec{r} \times \vec{V} \rho \, d\mathcal{V} = \frac{\partial}{\partial t}\left[\hat{K}\frac{R^3\omega}{3}\rho A\right] = 0 \tag{3}$$

where A is the cross-sectional area of the horizontal tube. Identical results are obtained for the other horizontal tube in the control volume. We have confirmed our insight that the angular momentum within the control volume does not change with time.

Now we need to evaluate the second term on the right, the flux of momentum across the control surface. There are three surfaces through which we have mass and therefore momentum flux: the supply line (for which $\vec{r} \times \vec{V} = 0$) because $\vec{r} = 0$ and the two nozzles. Consider the nozzle at the end of branch OAB. For $L \ll R$, we have

$$\vec{r}_{jet} = \vec{r}_B \approx \vec{r}|_{r=R} = (\hat{I} r\cos\theta + \hat{J} r\sin\theta)|_{r=R} = \hat{I} R\cos\theta + \hat{J} R\sin\theta$$

and for the instantaneous jet velocity $\vec{V}_j$ we have

$$\vec{V}_j = \vec{V}_{rel} + \vec{V}_{tip} = \hat{I} V_{rel}\cos\alpha\sin\theta - \hat{J} V_{rel}\cos\alpha\cos\theta + \hat{K} V_{rel}\sin\alpha - \hat{I}\omega R\sin\theta + \hat{J}\omega R\cos\theta$$

$$\vec{V}_j = \hat{I}(V_{rel}\cos\alpha - \omega R)\sin\theta - \hat{J}(V_{rel}\cos\alpha - \omega R)\cos\theta + \hat{K} V_{rel}\sin\alpha$$

$$\vec{r}_B \times \vec{V}_j = \hat{I} R V_{rel}\sin\alpha\sin\theta - \hat{J} R V_{rel}\sin\alpha\cos\theta - \hat{K} R(V_{rel}\cos\alpha - \omega R)(\sin^2\theta + \cos^2\theta)$$

$$\vec{r}_B \times \vec{V}_j = \hat{I} R V_{rel}\sin\alpha\sin\theta - \hat{J} R V_{rel}\sin\alpha\cos\theta - \hat{K} R(V_{rel}\cos\alpha - \omega R)$$

The flux integral evaluated for flow crossing the control surface at location B is then

$$\int_{CS} \vec{r} \times \vec{V}_j \rho \vec{V} \cdot d\vec{A} = \left[\hat{I} R V_{\text{rel}} \sin\alpha \sin\theta - \hat{J} R V_{\text{rel}} \sin\alpha \cos\theta - \hat{K} R (V_{\text{rel}} \cos\alpha - \omega R)\right] \rho \frac{Q}{2}$$

The velocity and radius vectors for flow in the left arm must be described in terms of the same unit vectors used for the right arm. In the left arm the $\hat{I}$ and $\hat{J}$ components of the cross product are of opposite sign, since $\sin(\theta + \pi) = -\sin(\theta)$ and $\cos(\theta + \pi) = -\cos(\theta)$. Thus for the complete CV,

$$\int_{CS} \vec{r} \times \vec{V}_j \rho \vec{V} \cdot d\vec{A} = -\hat{K} R (V_{\text{rel}} \cos\alpha - \omega R) \rho Q \tag{4}$$

Substituting terms (2), (3), and (4) into Eq. 1, we obtain

$$-T_f \hat{K} = -\hat{K} R (V_{\text{rel}} \cos\alpha - \omega R) \rho Q$$

or

$$T_f = R (V_{\text{rel}} \cos\alpha - \omega R) \rho Q$$

This expression indicates that when the sprinkler runs at constant speed the friction torque at the sprinkler pivot just balances the torque generated by the angular momentum of the two jets.

From the data given,

$$\omega R = 30 \frac{\text{rev}}{\text{min}} \times 150\,\text{mm} \times 2\pi \frac{\text{rad}}{\text{rev}} \times \frac{\text{min}}{60\,\text{s}} \times \frac{\text{m}}{1000\,\text{mm}} = 0.471\,\text{m/s}$$

Substituting gives

$$T_f = 150\,\text{mm} \times \left(4.97 \frac{\text{m}}{\text{s}} \times \cos 30^\circ - 0.471 \frac{\text{m}}{\text{s}}\right) 999 \frac{\text{kg}}{\text{m}^3} \times 7.5 \frac{\text{L}}{\text{min}}$$

$$\times \frac{\text{m}^3}{1000\,\text{L}} \times \frac{\text{min}}{60\,\text{s}} \times \frac{\text{N}\cdot\text{s}^2}{\text{kg}\cdot\text{m}} \times \frac{\text{m}}{1000\,\text{mm}}$$

$$T_f = 0.0718\,\text{N}\cdot\text{m} \longleftarrow T_f$$

This problem illustrates use of the angular momentum principle for an inertial control volume. Note that in this example the fluid particle position vector $\vec{r}$ and velocity vector $\vec{V}$ are time-dependent (through θ) in XYZ coordinates. This problem will be solved again using a noninertial (rotating) xyz coordinate system in Example 4.15 (on the Web).

Equation for Rotating Control Volume (on the Web)

The First Law of Thermodynamics 4.8

The first law of thermodynamics is a statement of conservation of energy. Recall that the system formulation of the first law was

$$\dot{Q} - \dot{W} = \left.\frac{dE}{dt}\right)_{\text{system}} \tag{4.4a}$$

where the total energy of the system is given by

$$E_{\text{system}} = \int_{M(\text{system})} e\, dm = \int_{\forall(\text{system})} e\, \rho\, d\forall \tag{4.4b}$$

and

$$e = u + \frac{V^2}{2} + gz$$

In Eq. 4.4a, the rate of heat transfer, $\dot{Q}$, is positive when heat is added *to* the system from the surroundings; the rate of work, $\dot{W}$, is positive when work is done *by* the system on its surroundings. (Note that some texts use the opposite notation for work.)

To derive the control volume formulation of the first law of thermodynamics, we set

$$N = E \quad \text{and} \quad \eta = e$$

in Eq. 4.10 and obtain

$$\left.\frac{dE}{dt}\right)_{\text{system}} = \frac{\partial}{\partial t}\int_{\text{CV}} e\,\rho\, d\forall + \int_{\text{CS}} e\,\rho \vec{V} \cdot d\vec{A} \tag{4.53}$$

Since the system and the control volume coincide at t_0,

$$[\dot{Q} - \dot{W}]_{\text{system}} = [\dot{Q} - \dot{W}]_{\text{control volume}}$$

In light of this, Eqs. 4.4a and 4.53 yield the control volume form of the first law of thermodynamics,

$$\dot{Q} - \dot{W} = \frac{\partial}{\partial t}\int_{\text{CV}} e\,\rho\, d\forall + \int_{\text{CS}} e\,\rho \vec{V} \cdot d\vec{A} \tag{4.54}$$

where

$$e = u + \frac{V^2}{2} + gz$$

Note that for steady flow the first term on the right side of Eq. 4.54 is zero.

Is Eq. 4.54 the form of the first law used in thermodynamics? Even for steady flow, Eq. 4.54 is not quite the same form used in applying the first law to control volume problems. To obtain a formulation suitable and convenient for problem solutions, let us take a closer look at the work term, $\dot{W}$.

Rate of Work Done by a Control Volume

The term $\dot{W}$ in Eq. 4.54 has a positive numerical value when work is done by the control volume on the surroundings. The rate of work done *on* the control volume is of opposite sign to the work done *by* the control volume.

The rate of work done by the control volume is conveniently subdivided into four classifications,

$$\dot{W} = \dot{W}_s + \dot{W}_{\text{normal}} + \dot{W}_{\text{shear}} + \dot{W}_{\text{other}}$$

Let us consider these separately:

1. *Shaft Work*

We shall designate shaft work W_s and hence the rate of work transferred out through the control surface by shaft work is designated $\dot{W}_s$. Examples of shaft work are the work produced by the steam turbine (positive shaft work) of a power plant, and the work input required to run the compressor of a refrigerator (negative shaft work).

2. Work Done by Normal Stresses at the Control Surface

Recall that work requires a force to act through a distance. Thus, when a force, $\vec{F}$, acts through an infinitesimal displacement, $d\vec{s}$, the work done is given by

$$\delta W = \vec{F} \cdot d\vec{s}$$

To obtain the rate at which work is done by the force, divide by the time increment, Δt, and take the limit as $\Delta t \to 0$. Thus the rate of work done by the force, $\vec{F}$, is

$$\dot{W} = \lim_{\Delta t \to 0} \frac{\delta W}{\Delta t} = \lim_{\Delta t \to 0} \frac{\vec{F} \cdot d\vec{s}}{\Delta t} \qquad \text{or} \qquad \dot{W} = \vec{F} \cdot \vec{V}$$

We can use this to compute the rate of work done by the normal and shear stresses. Consider the segment of control surface shown in Fig. 4.6. For an element of area $d\vec{A}$ we can write an expression for the normal stress force $d\vec{F}_{\text{normal}}$: It will be given by the normal stress σ_{nn} multiplied by the vector area element $d\vec{A}$ (normal to the control surface).

Hence the rate of work done on the area element is

$$d\vec{F}_{\text{normal}} \cdot \vec{V} = \sigma_{nn}\, d\vec{A} \cdot \vec{V}$$

Since the work out across the boundaries of the control volume is the negative of the work done on the control volume, the total rate of work out of the control volume due to normal stresses is

$$\dot{W}_{\text{normal}} = -\int_{\text{CS}} \sigma_{nn}\, d\vec{A} \cdot \vec{V} = -\int_{\text{CS}} \sigma_{nn} \vec{V} \cdot d\vec{A}$$

3. Work Done by Shear Stresses at the Control Surface

Just as work is done by the normal stresses at the boundaries of the control volume, so may work be done by the shear stresses.

As shown in Fig. 4.6, the shear force acting on an element of area of the control surface is given by

$$d\vec{F}_{\text{shear}} = \vec{\tau}\, dA$$

where the shear stress vector, $\vec{\tau}$, is the shear stress acting in some direction in the plane of dA.

The rate of work done on the entire control surface by shear stresses is given by

$$\int_{\text{CS}} \vec{\tau}\, dA \cdot \vec{V} = \int_{\text{CS}} \vec{\tau} \cdot \vec{V}\, dA$$

Since the work out across the boundaries of the control volume is the negative of the work done on the control volume, the rate of work out of the control volume due to shear stresses is given by

$$\dot{W}_{\text{shear}} = -\int_{\text{CS}} \vec{\tau} \cdot \vec{V}\, dA$$

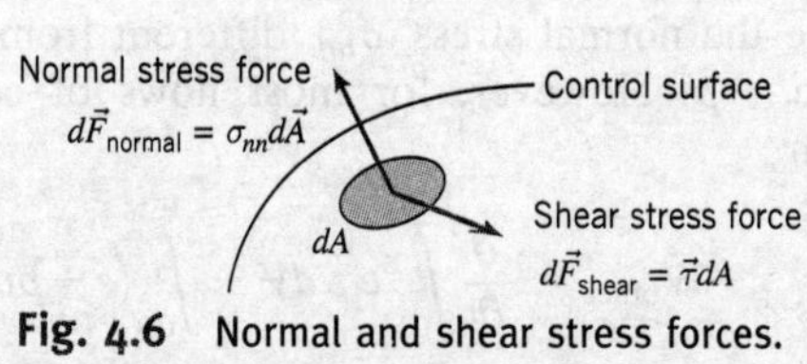

Fig. 4.6 Normal and shear stress forces.

This integral is better expressed as three terms

$$\begin{aligned}\dot{W}_{\text{shear}} &= -\int_{\text{CS}} \vec{\tau} \cdot \vec{V}\, dA \\ &= -\int_{A(\text{shafts})} \vec{\tau} \cdot \vec{V}\, dA - \int_{A(\text{solid surface})} \vec{\tau} \cdot \vec{V}\, dA - \int_{A(\text{ports})} \vec{\tau} \cdot \vec{V}\, dA\end{aligned}$$

We have already accounted for the first term, since we included $\dot{W}_s$ previously. At solid surfaces, $\vec{V} = 0$, so the second term is zero (for a fixed control volume). Thus,

$$\dot{W}_{\text{shear}} = -\int_{A(\text{ports})} \vec{\tau} \cdot \vec{V}\, dA$$

This last term can be made zero by proper choice of control surfaces. If we choose a control surface that cuts across each port perpendicular to the flow, then $d\vec{A}$ is parallel to $\vec{V}$. Since $\vec{\tau}$ is in the plane of dA, $\vec{\tau}$ is perpendicular to $\vec{V}$. Thus, for a control surface perpendicular to $\vec{V}$,

$$\vec{\tau} \cdot \vec{V} = 0 \quad \text{and} \quad \dot{W}_{\text{shear}} = 0$$

4. *Other Work*

Electrical energy could be added to the control volume. Also electromagnetic energy, e.g., in radar or laser beams, could be absorbed. In most problems, such contributions will be absent, but we should note them in our general formulation.

With all of the terms in $\dot{W}$ evaluated, we obtain

$$\dot{W} = \dot{W}_s - \int_{\text{CS}} \sigma_{nn}\, \vec{V} \cdot d\vec{A} + \dot{W}_{\text{shear}} + \dot{W}_{\text{other}} \tag{4.55}$$

Control Volume Equation

Substituting the expression for $\dot{W}$ from Eq. 4.55 into Eq. 4.54 gives

$$\dot{Q} - \dot{W}_s + \int_{\text{CS}} \sigma_{nn}\, \vec{V} \cdot d\vec{A} - \dot{W}_{\text{shear}} - \dot{W}_{\text{other}} = \frac{\partial}{\partial t}\int_{\text{CV}} e\,\rho\, d\forall + \int_{\text{CS}} e\,\rho\vec{V} \cdot d\vec{A}$$

Rearranging this equation, we obtain

$$\dot{Q} - \dot{W}_s - \dot{W}_{\text{shear}} - \dot{W}_{\text{other}} = \frac{\partial}{\partial t}\int_{\text{CV}} e\,\rho\, d\forall + \int_{\text{CS}} e\,\rho\vec{V} \cdot d\vec{A} - \int_{\text{CS}} \sigma_{nn}\, \vec{V} \cdot d\vec{A}$$

Since $\rho = 1/v$, where v is *specific volume*, then

$$\int_{\text{CS}} \sigma_{nn}\, \vec{V} \cdot d\vec{A} = \int_{\text{CS}} \sigma_{nn}\, v\, \rho\vec{V} \cdot d\vec{A}$$

Hence

$$\dot{Q} - \dot{W}_s - \dot{W}_{\text{shear}} - \dot{W}_{\text{other}} = \frac{\partial}{\partial t}\int_{\text{CV}} e\,\rho\, d\forall + \int_{\text{CS}} (e - \sigma_{nn} v)\, \rho\vec{V} \cdot d\vec{A}$$

Viscous effects can make the normal stress, σ_{nn}, different from the negative of the thermodynamic pressure, $-p$. However, for most flows of common engineering interest, $\sigma_{nn} \simeq -p$. Then

$$\dot{Q} - \dot{W}_s - \dot{W}_{\text{shear}} - \dot{W}_{\text{other}} = \frac{\partial}{\partial t}\int_{\text{CV}} e\,\rho\, d\forall + \int_{\text{CS}} (e + pv)\, \rho\vec{V} \cdot d\vec{A}$$

Finally, substituting $e = u + V^2/2 + gz$ into the last term, we obtain the familiar form of the first law for a control volume,

$$\dot{Q} - \dot{W}_s - \dot{W}_{\text{shear}} - \dot{W}_{\text{other}} = \frac{\partial}{\partial t}\int_{\text{CV}} e\,\rho\, d\forall + \int_{\text{CS}} \left(u + pv + \frac{V^2}{2} + gz\right)\rho \vec{V} \cdot d\vec{A} \quad (4.56)$$

Each work term in Eq. 4.56 represents the rate of work done by the control volume on the surroundings. Note that in thermodynamics, for convenience, the combination $u + pv$ (the fluid internal energy plus what is often called the "flow work") is usually replaced with enthalpy, $h \equiv u + pv$ (this is one of the reasons h was invented).

Example 4.16 COMPRESSOR: FIRST LAW ANALYSIS

Air at 101 kPa, 21°C, enters a compressor with negligible velocity and is discharged at 344 kPa, 38°C through a pipe with 0.09 m^2 area. The flow rate is 9 kg/s. The power input to the compressor is 447 kW. Determine the rate of heat transfer.

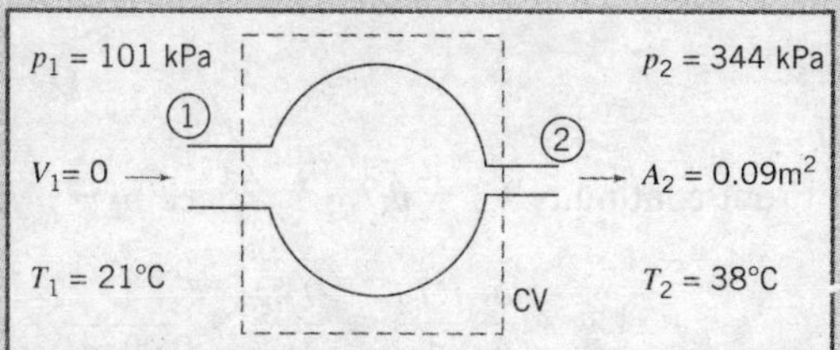

Given: Air enters a compressor at ① and leaves at ② with conditions as shown. The air flow rate is 9 kg/s and the power input to the compressor is 447 kW.

Find: Rate of heat transfer.

Solution:

Governing equations:

$$\overbrace{\frac{\partial}{\partial t}\int_{\text{CV}} \rho\, d\forall}^{=0(1)} + \int_{\text{CS}} \rho \vec{V} \cdot d\vec{A} = 0$$

$$\dot{Q} - \dot{W}_s - \overbrace{\dot{W}_{\text{shear}}}^{=0(4)} = \overbrace{\frac{\partial}{\partial t}\int_{\text{CV}} e\,\rho\, d\forall}^{=0(1)} + \int_{\text{CS}} \left(u + pv + \frac{V^2}{2} + gz\right)\rho \vec{V} \cdot d\vec{A}$$

Assumptions:
(1) Steady flow.
(2) Properties uniform over inlet and outlet sections.
(3) Treat air as an ideal gas, $p = \rho RT$.
(4) Area of CV at ① and ② perpendicular to velocity, thus $\dot{W}_{\text{shear}} = 0$.
(5) $z_1 = z_2$.
(6) Inlet kinetic energy is negligible.

Under the assumptions listed, the first law becomes

$$\dot{Q} - \dot{W}_s = \int_{\text{CV}} \left(u + pv + \frac{V^2}{2} + gz\right)\rho \vec{V} \cdot d\vec{A}$$

$$\dot{Q} - \dot{W}_s = \int_{\text{CS}} \left(h + \frac{V^2}{2} + gz\right)\rho \vec{V} \cdot d\vec{A}$$

or

$$\dot{Q} = \dot{W}_s + \int_{\text{CS}} \left(h + \frac{V^2}{2} + gz\right)\rho \vec{V} \cdot d\vec{A}$$

For uniform properties, assumption (2), we can write

$$\dot{Q} = \dot{W}_s + \left(h_1 + \overset{\approx 0(6)}{\cancel{\frac{V_1^2}{2}}} + gz_1\right)(-\rho_1 V_1 A_1) + \left(h_2 + \frac{V_2^2}{2} + gz_2\right)(\rho_2 V_2 A_2)$$

For steady flow, from conservation of mass,

$$\int_{\text{CS}} \rho \vec{V} \cdot d\vec{A} = 0$$

Therefore, $-(\rho_1 V_1 A_1) + (\rho_2 V_2 A_2) = 0$, or $\rho_1 V_1 A_1 = \rho_2 V_2 A_2 = \dot{m}$. Hence we can write

$$\dot{Q} = \dot{W}_s + \dot{m}\left[(h_2 - h_1) + \frac{V_2^2}{2} + g\overset{=0(5)}{\cancel{(z_2 - z_1)}}\right]$$

Assume that air behaves as an ideal gas with constant c_p. Then $h_2 - h_1 = c_p(T_2 - T_1)$, and

$$\dot{Q} = \dot{W}_s + \dot{m}\left[c_p(T_2 - T_1) + \frac{V_2^2}{2}\right]$$

From continuity $V_2 = \dot{m}/\rho_2 A_2$. Since $p_2 = \rho_2 R T_2$,

$$V_2 = \frac{\dot{m}\ RT_2}{A_2\ p_2} = \frac{9\,\text{kg}}{\text{s}} \times \frac{1}{0.09\,\text{m}^2} \times 287\frac{\text{j}}{\text{kg}\cdot{}^\circ\text{K}} \times (38 + 273)^\circ\text{K} \times \frac{1}{344,000\,\text{Pa}} \times \frac{\text{Pa}\cdot\text{m}^2}{\text{N}} \times \frac{\text{N}\cdot\text{m}}{\text{j}}$$

$$V_2 = 25.9\,\text{m/s}$$

Note that power input is *to* the CV, so $\dot{W}_s = -447\,\text{kW}$, and

$$\dot{Q} = \dot{W}_s + \dot{m}c_p\,(T_2 - T_1) + \dot{m}\,\frac{V_2^2}{2}$$

$$\dot{Q} = -447,000\,\text{W} + 9\frac{\text{kg}}{\text{s}} \times 1005\frac{\text{j}}{\text{kg}\cdot{}^\circ\text{K}} \times [(273 + 38) - (273 + 21)]^\circ\text{K} \times \frac{\text{W}\cdot\text{s}}{\text{j}}$$

$$+ 9\frac{\text{kg}}{\text{s}} \times \frac{(25.9)^2}{2}\frac{\text{m}^2}{\text{s}^2} \times \frac{\text{N}\cdot\text{s}^2}{\text{kg}\cdot\text{m}} \times \frac{\text{W}\cdot\text{s}^2}{\text{N}\cdot\text{m}}$$

$$\dot{Q} = -290.2\,\text{kW}$$ {heat rejection} $\dot{Q}$

This problem illustrates use of the first law of thermodynamics for a control volume. It is also an example of the care that must be taken with unit conversions for mass, energy, and power.

Example 4.17 TANK FILLING: FIRST LAW ANALYSIS

A tank of 0.1 m^3 volume is connected to a high-pressure air line; both line and tank are initially at a uniform temperature of 20°C. The initial tank gage pressure is 100 kPa. The absolute line pressure is 2.0 MPa; the line is large enough so that its temperature and pressure may be assumed constant. The tank temperature is monitored by a fast-response thermocouple. At the instant after the valve is opened, the tank temperature rises at the rate of 0.05°C/s. Determine the instantaneous flow rate of air into the tank if heat transfer is neglected.

Given: Air supply pipe and tank as shown. At $t = 0^+$, $\partial T/\partial t = 0.05°\text{C/s}$.

Find: $\dot{m}$ at $t = 0^+$.

Solution:
Choose CV shown, apply energy equation.

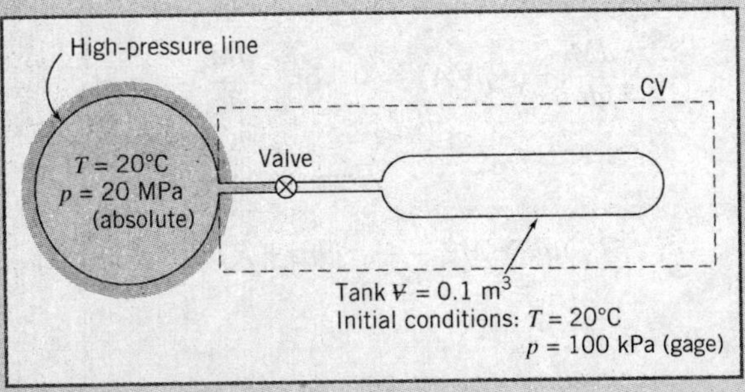

Governing equation:

$$\overset{=0(1)}{\cancel{\dot{Q}}} - \overset{=0(2)}{\cancel{\dot{W}_s}} - \overset{=0(3)}{\cancel{\dot{W}_{\text{shear}}}} - \overset{=0(4)}{\cancel{\dot{W}_{\text{other}}}} = \frac{\partial}{\partial t}\int_{\text{CV}} e\,\rho\, d\forall + \int_{\text{CS}} (e + pv)\rho \vec{V} \cdot d\vec{A}$$

$$e = u + \overset{\approx 0(5)}{\cancel{\frac{V^2}{2}}} + \overset{\approx 0(6)}{\cancel{gz}}$$

Assumptions:
(1) $\dot{Q} = 0$ (given).
(2) $\dot{W}_s = 0$.
(3) $\dot{W}_{\text{shear}} = 0$.
(4) $\dot{W}_{\text{other}} = 0$.
(5) Velocities in line and tank are small.
(6) Neglect potential energy.
(7) Uniform flow at tank inlet.
(8) Properties uniform in tank.
(9) Ideal gas, $p = \rho RT$, $du = c_v dT$.

Then

$$\frac{\partial}{\partial t}\int_{\text{CV}} u_{\text{tank}}\rho\, d\forall + (u + pv)|_{\text{line}}(-\rho VA) = 0$$

This expresses the fact that the gain in energy in the tank is due to influx of fluid energy (in the form of enthalpy $h = u + pv$) from the line. We are interested in the initial instant, when T is uniform at 20°C, so $u_{\text{tank}} = u_{\text{line}} = u$, the internal energy at T; also, $pv_{\text{line}} = RT_{\text{line}} = RT$, and

$$\frac{\partial}{\partial t}\int_{\text{CV}} u\,\rho\, d\forall + (u + RT)(-\rho VA) = 0$$

Since tank properties are uniform, $\partial/\partial t$ may be replaced by d/dt, and

$$\frac{d}{dt}(uM) = (u + RT)\dot{m}$$

(where M is the instantaneous mass in the tank and $\dot{m} = \rho VA$ is the mass flow rate), or

$$u\frac{dM}{dt} + M\frac{du}{dt} = u\dot{m} + RT\dot{m} \tag{1}$$

The term dM/dt may be evaluated from continuity:

Governing equation:

$$\frac{\partial}{\partial t}\int_{\text{CV}} \rho d\mathcal{V} + \int_{\text{CS}} \rho \vec{V} \cdot d\vec{A} = 0$$

$$\frac{dM}{dt} + (-\rho VA) = 0 \quad \text{or} \quad \frac{dM}{dt} = \dot{m}$$

Substituting in Eq. 1 gives

$$u\dot{m} + Mc_v \frac{dT}{dt} = u\dot{m} + RT\dot{m}$$

or

$$\dot{m} = \frac{Mc_v(dT/dt)}{RT} = \frac{\rho \mathcal{V} c_v(dT/dt)}{RT} \tag{2}$$

But at $t = 0$, $p_{\text{tank}} = 100$ kPa (gage), and

$$\rho = \rho_{\text{tank}} = \frac{p_{\text{tank}}}{RT} = (1.00 + 1.01)10^5 \frac{\text{N}}{\text{m}^2} \times \frac{\text{kg} \cdot \text{K}}{287\ \text{N} \cdot \text{m}} \times \frac{1}{293\ \text{K}}$$
$$= 2.39\ \text{kg/m}^3$$

Substituting into Eq. 2, we obtain

$$\dot{m} = 2.39 \frac{\text{kg}}{\text{m}^3} \times 0.1\ \text{m}^3 \times 717 \frac{\text{N} \cdot \text{m}}{\text{kg} \cdot \text{K}} \times 0.05 \frac{\text{K}}{\text{s}}$$
$$\times \frac{\text{kg} \cdot \text{K}}{287\ \text{N} \cdot \text{m}} \times \frac{1}{293\ \text{K}} \times 1000 \frac{\text{g}}{\text{kg}}$$

$$\dot{m} = 0.102\ \text{g/s} \longleftarrow \dot{m}$$

This problem illustrates use of the first law of thermodynamics for a control volume. It is also an example of the care that must be taken with unit conversions for mass, energy, and power.

4.9 The Second Law of Thermodynamics

Recall that the system formulation of the second law is

$$\left.\frac{dS}{dt}\right)_{\text{system}} \geq \frac{1}{T}\dot{Q} \tag{4.5a}$$

where the total entropy of the system is given by

$$S_{\text{system}} = \int_{M(\text{system})} s\, dm = \int_{\mathcal{V}(\text{system})} s\, \rho\, d\mathcal{V} \tag{4.5b}$$

To derive the control volume formulation of the second law of thermodynamics, we set

$$N = S \quad \text{and} \quad \eta = s$$

in Eq. 4.10 and obtain

$$\left.\frac{dS}{dt}\right)_{\text{system}} = \frac{\partial}{\partial t}\int_{\text{CV}} s\,\rho\, d\Psi + \int_{\text{CS}} s\,\rho\vec{V}\cdot d\vec{A} \tag{4.57}$$

The system and the control volume coincide at t_0; thus in Eq. 4.5a,

$$\left.\frac{1}{T}\dot{Q}\right)_{\text{system}} = \frac{1}{T}\dot{Q})_{\text{CV}} = \int_{\text{CS}} \frac{1}{T}\left(\frac{\dot{Q}}{A}\right) dA$$

In light of this, Eqs. 4.5a and 4.57 yield the control volume formulation of the second law of thermodynamics

$$\frac{\partial}{\partial t}\int_{\text{CV}} s\,\rho\, d\Psi + \int_{\text{CS}} s\,\rho\,\vec{V}\cdot d\vec{A} \geq \int_{\text{CS}} \frac{1}{T}\left(\frac{\dot{Q}}{A}\right) dA \tag{4.58}$$

In Eq. 4.58, the factor $(\dot{Q}/A)$ represents the heat flux per unit area into the control volume through the area element dA. To evaluate the term

$$\int_{\text{CS}} \frac{1}{T}\left(\frac{\dot{Q}}{A}\right) dA$$

both the local heat flux, $(\dot{Q}/A)$, and local temperature, T, must be known for each area element of the control surface.

4.10 Summary and Useful Equations

In this chapter we wrote the basic laws for a system: mass conservation (or continuity), Newton's second law, the angular-momentum equation, the first law of thermodynamics, and the second law of thermodynamics. We then developed an equation (sometimes called the Reynolds Transport Theorem) for relating system formulations to control volume formulations. Using this we derived control volume forms of:

- ✓ The mass conservation equation (sometimes called the continuity equation).
- ✓ Newton's second law (in other words, a momentum equation) for:
  - o An inertial control volume.
  - o A control volume with rectilinear acceleration.
  - o A control volume with arbitrary acceleration (on the Web).
- ✓ The angular-momentum equation for:*
  - o A fixed control volume.
  - o A rotating control volume (on the Web).
- ✓ The first law of thermodynamics (or energy equation).
- ✓ The second law of thermodynamics.

We discussed the physical meaning of each term appearing in these control volume equations, and used the equations for the solution of a variety of flow problems. In particular, we used a differential control volume* to derive a famous equation in fluid mechanics—the Bernoulli equation—and while doing so learned about the restrictions on its use in solving problems.

*These topics apply to a section that may be omitted without loss of continuity in the text material

Note: Most of the Useful Equations in the table below have a number of constraints or limitations—*be sure to refer to their page numbers for details*!

Useful Equations

Continuity (mass conservation), incompressible fluid:	$\int_{CS} \vec{V} \cdot d\vec{A} = 0$	**(4.13a)**	Page 105
Continuity (mass conservation), incompressible fluid, uniform flow:	$\sum_{CS} \vec{V} \cdot \vec{A} = 0$	**(4.13b)**	Page 105
Continuity (mass conservation), steady flow:	$\int_{CS} \rho\vec{V} \cdot d\vec{A} = 0$	**(4.15a)**	Page 106
Continuity (mass conservation), steady flow, uniform flow:	$\sum_{CS} \rho\vec{V} \cdot \vec{A} = 0$	**(4.15b)**	Page 106
Momentum (Newton's second law):	$\vec{F} = \vec{F}_S + \vec{F}_B = \frac{\partial}{\partial t}\int_{CV} \vec{V}\,\rho\, d\forall + \int_{CS} \vec{V}\,\rho\vec{V} \cdot d\vec{A}$	**(4.17a)**	Page 111
Momentum (Newton's second law), uniform flow:	$\vec{F} = \vec{F}_S + \vec{F}_B = \frac{\partial}{\partial t}\int_{CV} \vec{V}\,\rho\, d\forall + \sum_{CS} \vec{V}\rho\vec{V} \cdot \vec{A}$	**(4.17b)**	Page 111
Momentum (Newton's second law), scalar components:	$F_x = F_{S_x} + F_{B_x} = \frac{\partial}{\partial t}\int_{CV} u\,\rho\, d\forall + \int_{CS} u\,\rho\vec{V} \cdot d\vec{A}$	**(4.18a)**	Page 112
	$F_y = F_{S_y} + F_{B_y} = \frac{\partial}{\partial t}\int_{CV} v\,\rho\, d\forall + \int_{CS} v\,\rho\vec{V} \cdot d\vec{A}$	**(4.18b)**	
	$F_z = F_{S_z} + F_{B_z} = \frac{\partial}{\partial t}\int_{CV} w\,\rho\, d\forall + \int_{CS} w\,\rho\vec{V} \cdot d\vec{A}$	**(4.18c)**	
Momentum (Newton's second law), uniform flow, scalar components:	$F_x = F_{S_x} + F_{B_x} = \frac{\partial}{\partial t}\int_{CV} u\,\rho\, d\forall + \sum_{CS} u\,\rho\vec{V} \cdot \vec{A}$	**(4.18d)**	Page 112
	$F_y = F_{S_y} + F_{B_y} = \frac{\partial}{\partial t}\int_{CV} v\,\rho\, d\forall + \sum_{CS} v\,\rho\vec{V} \cdot \vec{A}$	**(4.18e)**	
	$F_z = F_{S_z} + F_{B_z} = \frac{\partial}{\partial t}\int_{CV} w\,\rho\, d\forall + \sum_{CS} w\,\rho\vec{V} \cdot \vec{A}$	**(4.18f)**	
Bernoulli equation (steady, incompressible, frictionless, flow along a streamline):	$\frac{p}{\rho} + \frac{V^2}{2} + gz = \text{constant}$	**(4.24)**	Page 124
Momentum (Newton's second law), inertial control volume (stationary or constant speed):	$\vec{F} = \vec{F}_S + \vec{F}_B = \frac{\partial}{\partial t}\int_{CV} \vec{V}_{xyz}\,\rho\, d\forall + \int_{CS} \vec{V}_{xyz}\,\rho\vec{V}_{xyz} \cdot d\vec{A}$	**(4.26)**	Page 126
Momentum (Newton's second law), rectilinear acceleration of control volume:	$\vec{F}_S + \vec{F}_B - \int_{CV} \vec{a}_{rf}\,\rho\, d\forall = \frac{\partial}{\partial t}\int_{CV} \vec{V}_{xyz}\,\rho\, d\forall \int_{CS} \vec{V}_{xyz}\,\rho\vec{V}_{xyz} \cdot d\vec{A}$	**(4.33)**	Page 130
Angular-momentum principle:	$\vec{r}\times\vec{F}_s + \int_{CV} \vec{r}\times\vec{g}\,\rho\, d\forall + \vec{T}_{\text{shaft}} = \frac{\partial}{\partial t}\int_{CV} \vec{r}\times\vec{V}\,\rho\, d\forall + \int_{CS} \vec{r}\times\vec{V}\,\rho\vec{V} \cdot d\vec{A}$	**(4.46)**	Page 136

First law of thermodynamics:	$\dot{Q} - \dot{W}_s - \dot{W}_{\text{shear}} - \dot{W}_{\text{other}} = \frac{\partial}{\partial t}\int_{\text{CV}} e\,\rho\, dV\!\!\!\!- + \int_{\text{CS}} \left(u + pv + \frac{V^2}{2} + gz\right)\rho \vec{V}\cdot d\vec{A}$	(4.56)	Page 143
Second law of thermodynamics:	$\frac{\partial}{\partial t}\int_{\text{CV}} s\,\rho\, dV\!\!\!\!- + \int_{\text{CS}} s\,\rho \vec{V}\cdot d\vec{A} \geq \int_{\text{CS}} \frac{1}{T}\left(\frac{\dot{Q}}{A}\right) dA$	(4.58)	Page 147

Case Study

"Lab-on-a-Chip"

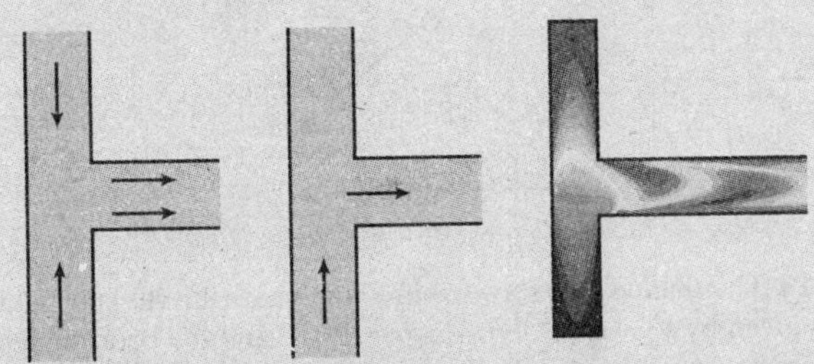

(a) (b) (c)

Mixing two fluids in a "lab-on-a-chip."

An exciting new area in fluid mechanics is microfluidics, applied to microelectromechanical systems (MEMS—the technology of very small devices, generally ranging in size from a micrometer to a millimeter). In particular, a lot of research is being done in "lab-on-a-chip" technology, which has many applications. An example of this is in medicine, with devices for use in the immediate point-of-care diagnosis of diseases, such as real-time detection of bacteria, viruses, and cancers in the human body. In the area of security, there are devices that continuously sample and test air or water samples for biochemical toxins and other dangerous pathogens such as those in always-on early warning systems.

Because of the extremely small geometry, flows in such devices will be very low Reynolds numbers and therefore laminar; surface tension effects will also be significant. In many common applications (for example, typical water pipes and air conditioning ducts), laminar flow would be desirable, but the flow is turbulent--it costs more to pump a turbulent as opposed to a laminar flow. In certain applications, turbulence is desirable instead because it acts as a mixing mechanism. If you couldn't generate turbulence in your coffee cup, it would take a lot of stirring before the cream and coffee were sufficiently blended; if your blood flow never became turbulent, you would not get sufficient oxygen to your organs and muscles! In the lab-on-a-chip, turbulent flow is usually desirable because the goal in these devices is often to mix minute amounts of two or more fluids.

How do we mix fluids in such devices that are inherently laminar? We could use complex geometries, or relatively long channels (relying on molecular diffusion), or some kind of MEM device with paddles. Research by professors Goullet, Glasgow, and Aubry at the New Jersey Institute of Technology instead suggests pulsing the two fluids. Part *a* of the figure shows a schematic of two fluids flowing at a constant rate (about 25 nL/s, average velocity less than 2 mm/s, in ducts about 200 μm wide) and meeting in a T junction. The two fluids do not mix because of the strongly laminar nature of the flow. Part *b* of the figure shows a schematic of an instant of a pulsed flow, and part *c* shows an instant computed using a computational fluid dynamics (CFD) model of the same flow. In this case, the interface between the two fluid samples is shown to stretch and fold, leading to good nonturbulent mixing within 2 mm downstream of the confluence (after about 1 s of contact). Such a compact mixing device would be ideal for many of the applications mentioned above.

Problems

Basic Laws for a System

4.1 A mass of 2.27 kg is released when it is just in contact with a spring of stiffness 365 kg/s^2 that is attached to the ground. What is the maximum spring compression? Compare this to the deflection if the mass was just resting on the compressed spring. What would be the maximum spring compression if the mass was released from a distance of 1.5 m above the top of the spring?

4.2 An ice-cube tray containing 250 mL of freshwater at 15°C is placed in a freezer at −5°C. Determine the change in internal energy (kJ) and entropy (kJ/K) of the water when it has frozen.

4.3 A small steel ball of radius r = 1 mm is placed on top of a horizontal pipe of outside radius R = 50 mm and begins to roll under the influence of gravity. Rolling resistance and air resistance are negligible. As the speed of the ball increases, it eventually leaves the surface of the pipe and becomes a projectile. Determine the speed and location at which the ball loses contact with the pipe.

4.4 A fully loaded Boeing 777-200 jet transport aircraft weighs 325,000 kg. The pilot brings the 2 engines to full takeoff thrust of 450 kN each before releasing the brakes. Neglecting aerodynamic and rolling resistance, estimate the minimum runway length and time needed to reach a takeoff speed of 225 km/hr. Assume engine thrust remains constant during ground roll.

4.5 A police investigation of tire marks showed that a car traveling along a straight and level street had skidded to a stop for a total distance of 50 m after the brakes were applied. The coefficient of friction between tires and pavement is estimated to be $\mu = .6$. What was the probable minimum speed (km/hr) of the car when the brakes were applied? How long did the car skid?

4.6 A high school experiment consists of a block of mass 2 kg sliding across a surface (coefficient of friction $\mu = 0.6$). If it is given an initial velocity of 5 m/s, how far will it slide, and how long will it take to come to rest? The surface is now roughened a little, so with the same initial speed it travels a distance of 2 m. What is the new coefficient of friction, and how long does it now slide?

4.7 A car traveling at 48 km/hr encounters a curve in the road. The radius of the road curve is 30 m. Find the maximum speeds (km/hr) before losing traction, if the coefficient of friction on a dry road is $\mu_{dry} = 0.7$ and on a wet road is $\mu_{wet} = 0.3$.

4.8 Air at 20°C and an absolute pressure of 101.3 kPa is compressed adiabatically in a piston-cylinder device, without friction, to an absolute pressure of 905.3 kPa in a piston-cylinder device. Find the work done (MJ).

4.9 In an experiment with a can of soda, it took 3 hr to cool from an initial temperature of 24°C to 10°C in a 4°C refrigerator. If the can is now taken from the refrigerator and placed in a room at 20°C, how long will the can take to reach 15°C? You may assume that for both processes the heat transfer is modeled by $\dot{Q} \approx k(T - T_{amb})$, where T is the can temperature, T_{amb} is the ambient temperature, and k is a heat transfer coefficient.

4.10 A block of copper of mass 5 kg is heated to 90°C and then plunged into an insulated container containing 4 L of water at 10°C. Find the final temperature of the system. For copper, the specific heat is 385 J/kg · K, and for water the specific heat is 4186 J/kg · K.

4.11 The average rate of heat loss from a person to the surroundings when not actively working is about 85 W. Suppose that in an auditorium with volume of approximately 3.5×10^5 m^3, containing 6000 people, the ventilation system fails. How much does the internal energy of the air in the auditorium increase during the first 15 min after the ventilation system fails? Considering the auditorium and people as a system, and assuming no heat transfer to the surroundings, how much does the internal energy of the system change? How do you account for the fact that the temperature of the air increases? Estimate the rate of temperature rise under these conditions.

Conservation of Mass

4.12 The velocity field in the region shown is given by $\vec{V} = (a\hat{j} + by\hat{k})$ where $a = 10$ m/s and $b = 5\,\text{s}^{-1}$. For the 1 m × 1 m triangular control volume (depth $w = 1$ m perpendicular to the diagram), an element of area ① may be represented by $d\vec{A}_1 = wdz\hat{j} - wdy\hat{k}$ and an element of area ② by $d\vec{A}_2 = -wdy\hat{k}$.
(a) Find an expression for $\vec{V} \cdot d\vec{A}_1$.
(b) Evaluate $\int_{A_1} \vec{V} \cdot d\vec{A}_1$.
(c) Find an expression for $\vec{V} \cdot d\vec{A}_2$.
(d) Find an expression for $\vec{V}(\vec{V} \cdot d\vec{A}_2)$.
(e) Evaluate $\int_{A_2} \vec{V}(\vec{V} \cdot d\vec{A}_2)$.

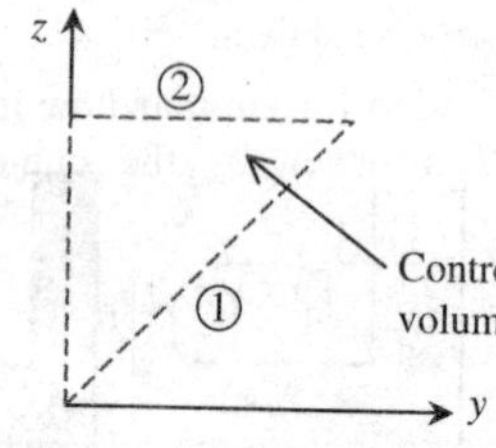

P4.12

4.13 The shaded area shown is in a flow where the velocity field is given by $\vec{V} = ax\hat{i} + by\hat{j}$; $a = b = 1\,\text{s}^{-1}$, and the coordinates are measured in meters. Evaluate the volume flow rate and the momentum flux through the shaded area ($\rho = 1$ kg/m^3).

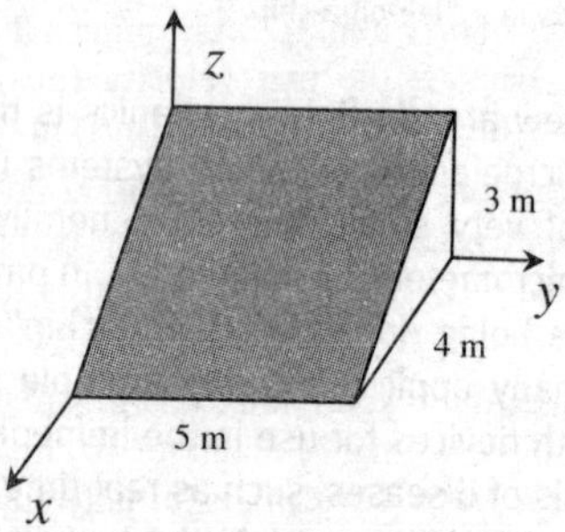

P4.13

4.14 The area shown shaded is in a flow where the velocity field is given by $\vec{V} = ax\hat{i} + by\hat{j} + c\hat{k}$; $a = b = 2\,\text{s}^{-1}$ and $c = 1$ m/s. Write a vector expression for an element of the shaded area. Evaluate the integrals $\int_A \vec{V} \cdot d\vec{A}$ and $\int_A \vec{V}(\vec{V} \cdot d\vec{A})$ over the shaded area.

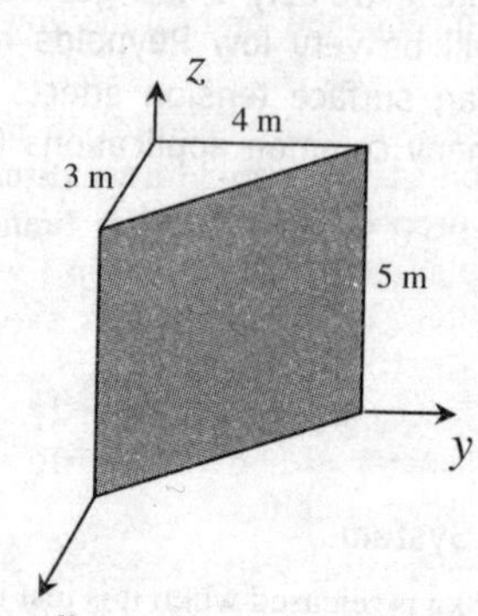

P4.14

4.15 Obtain expressions for the volume flow rate and the momentum flux through cross section ① of the control volume shown in the diagram.

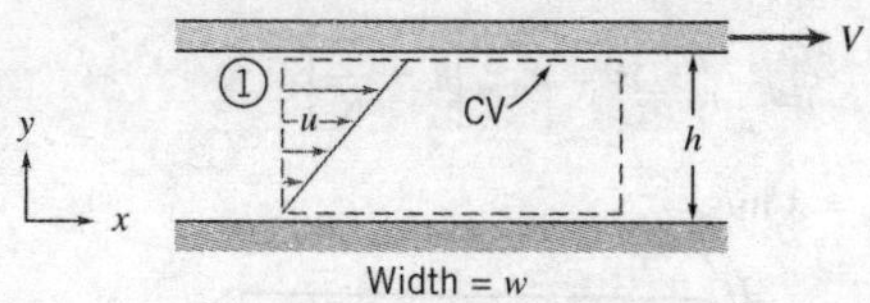

P4.15

4.16 For the flow of Problem 4.15, obtain an expression for the kinetic energy flux, $\int (V^2/2)\rho\vec{V}\cdot d\vec{A}$, through cross section ① of the control volume shown.

4.17 The velocity distribution for laminar flow in a long circular tube of radius R is given by the one-dimensional expression,

$$\vec{V} = u\hat{i} = u_{\max}\left[1 - \left(\frac{r}{R}\right)^2\right]\hat{i}$$

For this profile obtain expressions for the volume flow rate and the momentum flux through a section normal to the pipe axis.

4.18 For the flow of Problem 4.17, obtain an expression for the kinetic energy flux, $\int (V^2/2)\rho\vec{V}\cdot d\vec{A}$, through a section normal to the pipe axis.

4.19 A shower head fed by a 19.05 mm. ID water pipe consists of 50 nozzles of 0.79 mm. ID. Assuming a flow rate of 1.39×10^{-4} m^3/s, what is the exit velocity (m/s) of each jet of water? What is the average velocity (m/s) in the pipe?

4.20 A farmer is spraying a liquid through 10 nozzles, 3 mm. ID, at an average exit velocity of 3 m/s. What is the average velocity in the 25 mm. ID head feeder? What is the system flow rate, in L/m?

4.21 A cylindrical holding water tank has a 3 m ID and a height of 3 m. There is one inlet of diameter 10 cm, an exit of diameter 8 cm, and a drain. The tank is initially empty when the inlet pump is turned on, producing an average inlet velocity of 5 m/s. When the level in the tank reaches 0.7 m, the exit pump turns on, causing flow out of the exit; the exit average velocity is 3 m/s. When the water level reaches 2 m the drain is opened such that the level remains at 2 m. Find (a) the time at which the exit pump is switched on, (b) the time at which the drain is opened, and (c) the flow rate into the drain (m^3/min).

4.22 A university laboratory that generates 15 m^3/s of air flow at design condition wishes to build a wind tunnel with variable speeds. It is proposed to build the tunnel with a sequence of three circular test sections: section 1 will have a diameter of 1.5 m, section 2 a diameter of 1 m, and section 3 a diameter such that the average speed is 75 m/s.
(a) What will be the speeds in sections 1 and 2?
(b) What must the diameter of section 3 be to attain the desired speed at design condition?

4.23 A wet cooling tower cools warm water by spraying it into a forced dry-air flow. Some of the water evaporates in this air and is carried out of the tower into the atmosphere; the evaporation cools the remaining water droplets, which are collected at the exit pipe (150 mm ID) of the tower. Measurements indicate the warm water mass flow rate is 31.5 kg/s, and the cool water (21°C) flows at an average speed of 1.7 m/s in the exit pipe. The moist air density is 1.06 kg/m^3. Find (a) the volume flow rate (L/min) and mass flow rate (kg/s) of the cool water, (b) the mass flow rate (kg/s) of the moist air, and (c) the mass flow rate (kg/s) of the dry air. *Hint*: Google "density of moist air" for information on relating moist and dry air densities!

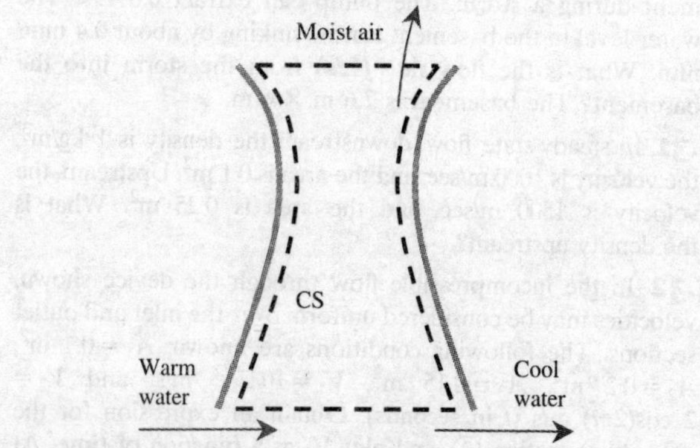

P4.23

4.24 Fluid with 1040 kg/m^3 density is flowing steadily through the rectangular box shown. Given $A_1 = 0.046$ m^2, $A_2 = 0.009$ m^2, $A_3 = 0.056$ m^2, $\vec{V}_1 = 3\hat{i}$ m/s, and $\vec{V}_2 = 6\hat{j}$ m/s, determine velocity $\vec{V}_3$.

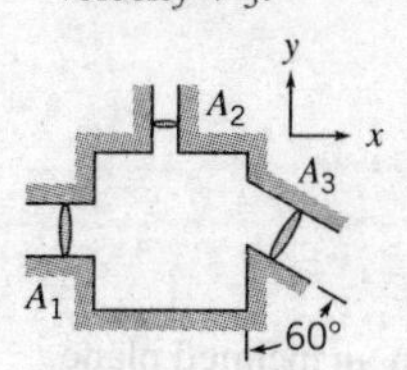

P4.24

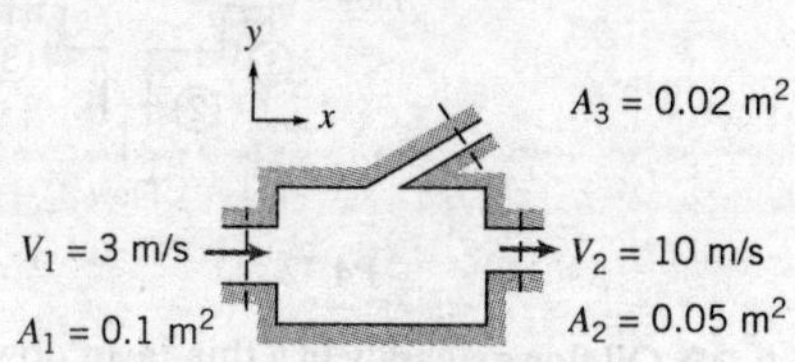

P4.25

4.25 Consider steady, incompressible flow through the device shown. Determine the magnitude and direction of the volume flow rate through port 3.

4.26 A rice farmer needs to fill her 150 m × 400 m field with water to a depth of 7.5 cm in 1 hr. How many 37.5 cm diameter supply pipes are needed if the average velocity in each must be less than 2.5 m/s?

4.27 You are making beer. The first step is filling the glass carboy with the liquid wort. The internal diameter of the carboy is 37.5 cm, and you wish to fill it up to a depth of 0.6 m. If your wort is drawn from the kettle using a siphon process that flows at 11.36 L/min, how long will it take to fill?

4.28 In your kitchen, the sink is .6 m by 45.7 cm by 30.5 cm deep. You are filling it with water at the rate of 252×10^{-6} m^3/s. How long will it take (in min) to half fill the sink? After this you turn off the faucet and open the drain slightly so that the tank starts to drain at 63×10^{-6} m^3/s. What is the rate (m/s) at which the water level drops?

4.29 Ventilation air specifications for classrooms require that at least 8.0 L/s of fresh air be supplied for each person in the room (students and instructor). A system needs to be designed that will supply ventilation air to 6 classrooms, each with a capacity of 20 students. Air enters through a central duct, with short branches successively leaving for each classroom. Branch registers are 200 mm high and 500 mm wide. Calculate the volume flow rate and air velocity entering each room. Ventilation noise increases with air velocity. Given a supply duct 500 mm high, find the narrowest supply duct that will limit air velocity to a maximum of 1.75 m/s.

4.30 You are trying to pump storm water out of your basement during a storm. The pump can extract 0.6 L/s. The water level in the basement is now sinking by about 0.4 mm/min. What is the flow rate (L/s) from the storm into the basement? The basement is 7.6 m × 6 m.

4.31 In steady-state flow, downstream the density is 1 kg/m^3, the velocity is 1000 m/sec, and the area is 0.1 m^2. Upstream, the velocity is 1500 m/sec, and the area is 0.25 m^2. What is the density upstream?

4.32 In the incompressible flow through the device shown, velocities may be considered uniform over the inlet and outlet sections. The following conditions are known: $A_1 = 0.1$ m^2, $A_2 = 0.2$ m^2, $A_3 = 0.15$ m^2, $V_1 = 10e^{-t/2}$ m/s, and $V_2 = 2\cos(2\pi t)$ m/s (t in seconds). Obtain an expression for the velocity at section ③, and plot V_3 as a function of time. At what instant does V_3 first become zero? What is the total mean volumetric flow at section ③?

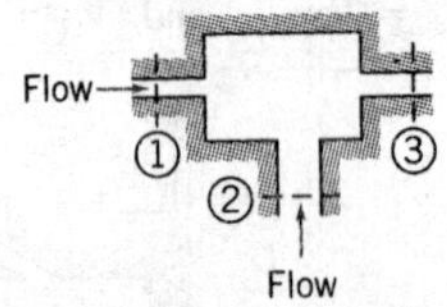

P4.32

4.33 Oil flows steadily in a thin layer down an inclined plane. The velocity profile is

$$u = \frac{\rho g \sin\theta}{\mu}\left[hy - \frac{y^2}{2}\right]$$

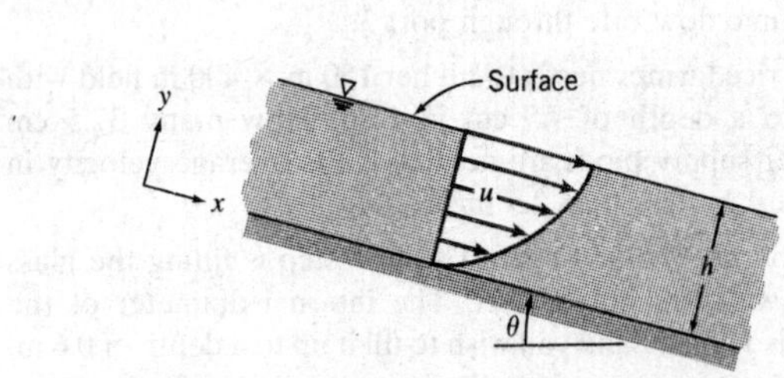

P4.33

Express the mass flow rate per unit width in terms of ρ, μ, g, θ, and h.

4.34 Water enters a wide, flat channel of height $2h$ with a uniform velocity of 2.5 m/s. At the channel outlet the velocity distribution is given by

$$\frac{u}{u_{\max}} = 1 - \left(\frac{y}{h}\right)^2$$

where y is measured from the centerline of the channel. Determine the exit centerline velocity, $u_{\max}$.

4.35 Water flows steadily through a pipe of length L and radius $R = 75$ mm. Calculate the uniform inlet velocity, U, if the velocity distribution across the outlet is given by

$$u = u_{\max}\left[1 - \frac{r^2}{R^2}\right]$$

and $u_{\max} = 3$ m/s.

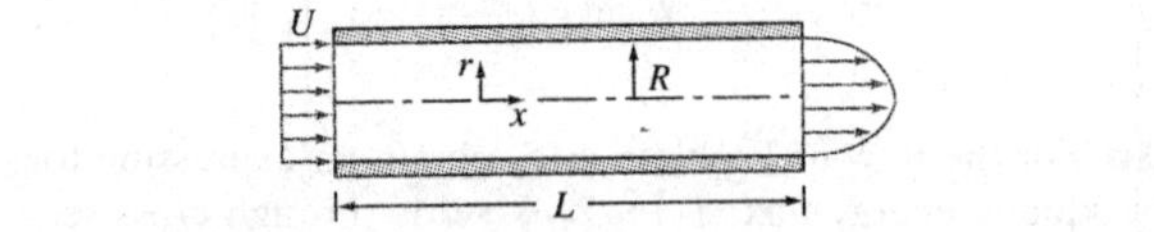

P4.35

4.36 Incompressible fluid flows steadily through a plane diverging channel. At the inlet, of height H, the flow is uniform with magnitude V_1. At the outlet, of height $2H$, the velocity profile is

$$V_2 = V_m \cos\left(\frac{\pi y}{2H}\right)$$

where y is measured from the channel centerline. Express V_m in terms of V_1.

4.37 The velocity profile for laminar flow in an annulus is given by

$$u(r) = -\frac{\Delta p}{4\mu L}\left[R_o^2 - r^2 + \frac{R_o^2 - R_i^2}{\ln(R_i/R_o)}\ln\frac{R_o}{r}\right]$$

where $\Delta p/L = -10$ kPa/m is the pressure gradient, μ is the viscosity (SAE 10 oil at 20°C), and $R_o = 5$ mm and $R_i = 1$ mm are the outer and inner radii. Find the volume flow rate, the average velocity, and the maximum velocity. Plot the velocity distribution.

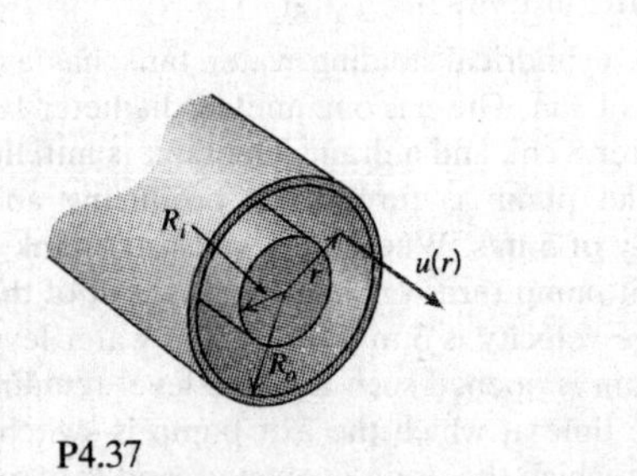

P4.37

4.38 A two-dimensional reducing bend has a linear velocity profile at section ①. The flow is uniform at sections ② and ③. The fluid is incompressible and the flow is steady. Find the maximum velocity, $V_{1,\max}$, at section ①.

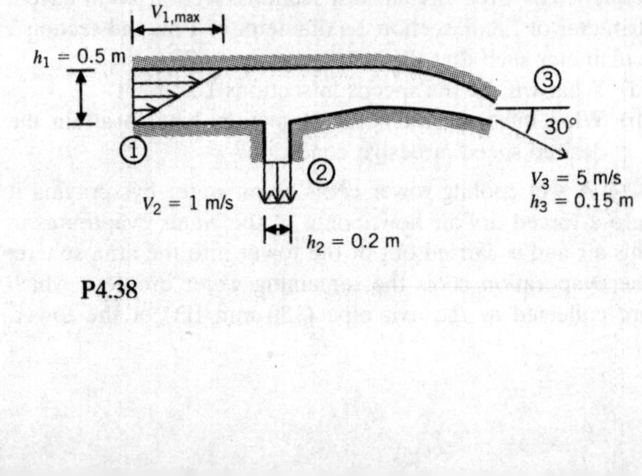

P4.38

4.39 Water enters a two-dimensional, square channel of constant width, $h = 75.5$ mm, with uniform velocity, U. The channel makes a 90° bend that distorts the flow to produce the linear velocity profile shown at the exit, with $V_{max} = 2\ V_{min}$. Evaluate V_{min}, if $U = 7.5$ m/s.

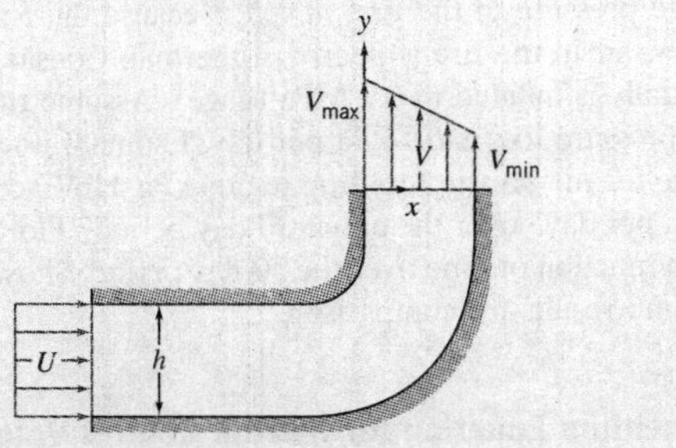

P4.39, 4.80, 4.98

4.40 Viscous liquid from a circular tank, $D = 300$ mm in diameter, drains through a long circular tube of radius $R = 50$ mm. The velocity profile at the tube discharge is

$$u = u_{max}\left[1 - \left(\frac{r}{R}\right)^2\right]$$

Show that the average speed of flow in the drain tube is $\overline{V} = \frac{1}{2}u_{max}$. Evaluate the rate of change of liquid level in the tank at the instant when $u_{max} = 0.155$ m/s.

4.41 A porous round tube with $D = 60$ mm carries water. The inlet velocity is uniform with $V_1 = 7.0$ m/s. Water flows radially and axisymmetrically outward through the porous walls with velocity distribution

$$v = V_0\left[1 - \left(\frac{x}{L}\right)^2\right]$$

where $V_0 = 0.03$ m/s and $L = 0.950$ m. Calculate the mass flow rate inside the tube at $x = L$.

4.42 A rectangular tank used to supply water for a Reynolds flow experiment is 230 mm deep. Its width and length are $W = 150$ mm and $L = 230$ mm. Water flows from the outlet tube (inside diameter $D = 6.35$ mm) at Reynolds number $Re = 2000$, when the tank is half full. The supply valve is closed. Find the rate of change of water level in the tank at this instant.

4.43 A *hydraulic accumulator* is designed to reduce pressure pulsations in a machine tool hydraulic system. For the instant shown, determine the rate at which the accumulator gains or loses hydraulic oil.

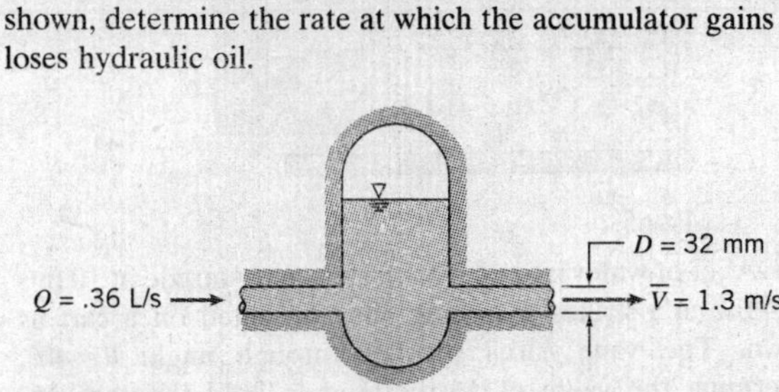

P4.43

4.44 A cylindrical tank, 0.3 m in diameter, drains through a hole in its bottom. At the instant when the water depth is 0.6 m, the flow rate from the tank is observed to be 4 kg/s. Determine the rate of change of water level at this instant.

4.45 A tank of 0.4 m^3 volume contains compressed air. A valve is opened and air escapes with a velocity of 250 m/s through an opening of 100 mm^2 area. Air temperature passing through the opening is −20°C and the absolute pressure is 300 kPa. Find the rate of change of density of the air in the tank at this moment.

4.46 Air enters a tank through an area of 0.018 m^2 with a velocity of 4.6 m/s and a density of 15.5 kg/m^3. Air leaves with a velocity of 1.5 m/s and a density equal to that in the tank. The initial density of the air in the tank is 10.3 kg/m^3. The total tank volume is 0.6 m^3 and the exit area is 0.04 m^2. Find the initial rate of change of density in the tank.

4.47 A recent TV news story about lowering Lake Shafer near Monticello, Indiana, by increasing the discharge through the dam that impounds the lake, gave the following information for flow through the dam:

Normal flow rate	8.2 m^3/s
Flow rate during draining of lake	57 m^3/s

(The flow rate during draining was stated to be equivalent to 60.5 m^3/s.) The announcer also said that during draining the lake level was expected to fall at the rate of 0.3 m every 8 hr. Calculate the actual flow rate during draining in m^3/s. Estimate the surface area of the lake.

4.48 A cylindrical tank, of diameter D = 50 mm, drains through an opening, d = 5 mm, in the bottom of the tank. The speed of the liquid leaving the tank is approximately $V = \sqrt{2gy}$ where y is the height from the tank bottom to the free surface. If the tank is initially filled with water to y_0 = 0.4 m, determine the water depths at t = 60 sec, t = 120 sec, and t = 180 sec. Plot y (m) versus t for the first 180 sec.

4.49 For the conditions of Problem 4.48, estimate the times required to drain the tank from initial depth to a depth $y = 0.3$ m (a change in depth of 0.1 m), and from $y = 0.3$ m to $y = 0.2$ m (also a change in depth of 0.1 m). Can you explain the discrepancy in these times? Plot the time to drain to a depth $y = 0.1$ m as a function of opening sizes ranging from $d = 2.5$ mm to 12.5 mm.

4.50 A conical flask contains water to height $H = 36.8$ mm, where the flask diameter is $D = 29.4$ mm. Water drains out through a smoothly rounded hole of diameter $d = 7.35$ mm at the apex of the cone. The flow speed at the exit is approximately $V = \sqrt{2gy}$, where y is the height of the liquid free surface above the hole. A stream of water flows into the top of the flask at constant volume flow rate, $Q = 3.75 \times 10^{-7}$ m^3/hr. Find the volume flow rate from the bottom of the flask. Evaluate the direction and rate of change of water surface level in the flask at this instant.

4.51 A conical funnel of half-angle $\theta = 15°$, with maximum diameter $D = 70$ mm and height H, drains through a hole (diameter $d = 3.12$ mm) in its bottom. The speed of the liquid leaving the funnel is approximately $V = \sqrt{2gy}$, where y is the height of the liquid free surface above the hole. Find the rate of change of surface level in the funnel at the instant when $y = H/2$.

4.52 Water flows steadily past a porous flat plate. Constant suction is applied along the porous section. The velocity profile at section *cd* is

$$\frac{u}{U_\infty} = 3\left[\frac{y}{\delta}\right] - 2\left[\frac{y}{\delta}\right]^{3/2}$$

Evaluate the mass flow rate across section *bc*.

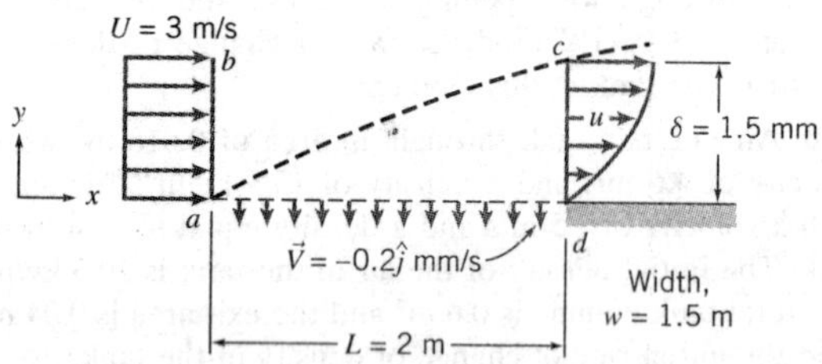

P4.52, P4.53

4.53 Consider incompressible steady flow of standard air in a boundary layer on the length of porous surface shown. Assume the boundary layer at the downstream end of the surface has an approximately parabolic velocity profile, $u/U_\infty = 2(y/\delta) - (y/\delta)^2$. Uniform suction is applied along the porous surface, as shown. Calculate the volume flow rate across surface *cd*, through the porous suction surface, and across surface *bc*.

4.54 A tank of fixed volume contains brine with initial density, ρ_i, greater than water. Pure water enters the tank steadily and mixes thoroughly with the brine in the tank. The liquid level in the tank remains constant. Derive expressions for (a) the rate of change of density of the liquid mixture in the tank and (b) the time required for the density to reach the value ρ_f, where $\rho_i > \rho_f > \rho_{H_2O}$.

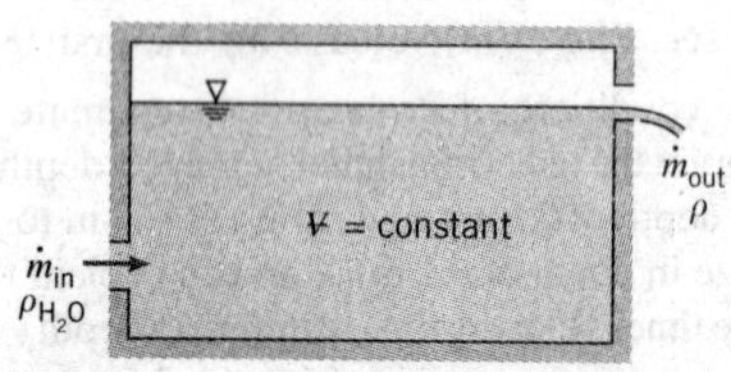

P4.54

4.55 A conical funnel of half-angle $\theta = 30°$ drains through a small hole of diameter $d = 6.25$ mm at the vertex. The speed of the liquid leaving the funnel is approximately $V = \sqrt{2gy}$, where *y* is the height of the liquid free surface above the hole. The funnel initially is filled to height $y_0 = 300$ mm. Obtain an expression for the time, *t*, for the funnel to completely drain, and evaluate. Find the time to drain from 300 mm to 150 mm (a change in depth of 150 mm), and from 150 mm to completely empty (also a change in depth of 150 mm). Can you explain the discrepancy in these times? Plot the drain time *t* as a function diameter *d* for *d* ranging from 6.25 mm to 12.5 mm.

4.56 For the funnel of Problem 4.55, find the diameter *d* required if the funnel is to drain in $t = 1$ min. from an initial depth $y_0 = 30$ cm. Plot the diameter *d* required to drain the funnel in 1 min as a function of initial depth y_0, for y_0 ranging from 2.5 cm to 60 cm.

4.57 Over time, air seeps through pores in the rubber of high-pressure bicycle tires. The saying is that a tire loses pressure at the rate of "6.9 kPa per day." The true rate of pressure loss is not constant; instead, the instantaneous leakage mass flow rate is proportional to the air density in the tire and to the gage pressure in the tire, $\dot{m} \propto \rho p$. Because the leakage rate is slow, air in the tire is nearly isothermal. Consider a tire that initially is inflated to 0.7 MPa (gage). Assume the initial rate of pressure loss is 6.9 kPa per day. Estimate how long it will take for the pressure to drop to 500 kPa. How accurate is "6.9 kPa per day" over the entire 30-day period? Plot the pressure as a function of time over the 30-day period. Show the rule-of-thumb results for comparison.

Momentum Equation for Inertial Control Volume

4.58 Evaluate the net rate of flux of momentum out through the control surface of Problem 4.24.

4.59 For the conditions of Problem 4.34, evaluate the ratio of the *x*-direction momentum flux at the channel outlet to that at the inlet.

4.60 For the conditions of Problem 4.35, evaluate the ratio of the *x*-direction momentum flux at the pipe outlet to that at the inlet.

4.61 Evaluate the net momentum flux through the bend of Problem 4.38, if the depth normal to the diagram is $w = 1$ m.

4.62 Evaluate the net momentum flux through the channel of Problem 4.39. Would you expect the outlet pressure to be higher, lower, or the same as the inlet pressure? Why?

4.63 Water jets are being used more and more for metal cutting operations. If a pump generates a flow of 63×10^{-6} m^3/s through an orifice of 0.254 mm diameter, what is the average jet speed? What force (N) will the jet produce at impact, assuming as an approximation that the water sprays sideways after impact?

4.64 Considering that in the fully developed region of a pipe, the integral of the axial momentum is the same at all cross sections, explain the reason for the pressure drop along the pipe.

4.65 Find the force required to hold the plug in place at the exit of the water pipe. The flow rate is 1.5 m^3/s, and the upstream pressure is 3.5 MPa.

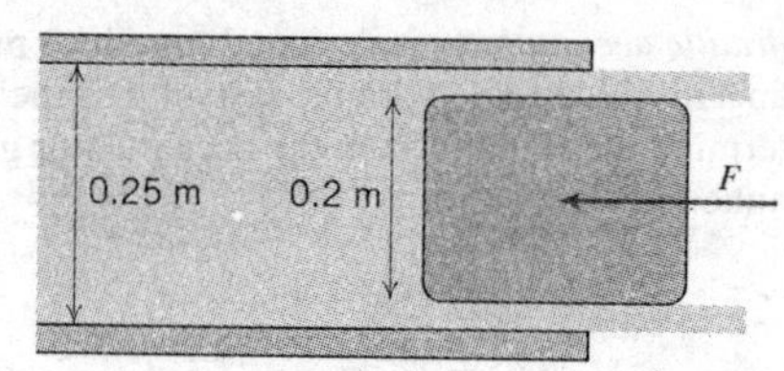

P4.65

4.66 A jet of water issuing from a stationary nozzle at 10 m/s ($A_j = 0.1$ m^2) strikes a turning vane mounted on a cart as shown. The vane turns the jet through angle $\theta = 40°$. Determine the value of *M* required to hold the cart stationary. If the vane angle θ is adjustable, plot the mass, *M*, needed to hold the cart stationary versus θ for $0 \le \theta \le 180°$.

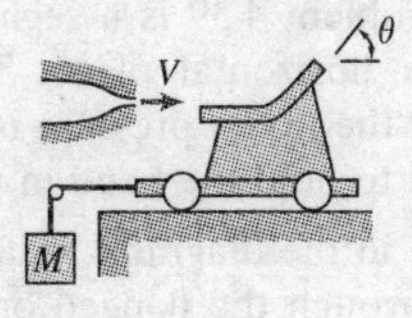

P4.66

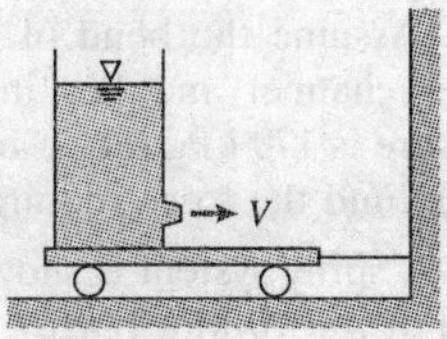

P4.67

4.67 A large tank of height $h = 1$ m and diameter $D = 0.75$ m is affixed to a cart as shown. Water issues from the tank through a nozzle of diameter $d = 15$ mm. The speed of the liquid leaving the tank is approximately $V = \sqrt{2gy}$, where y is the height from the nozzle to the free surface. Determine the tension in the wire when $y = 0.9$ m. Plot the tension in the wire as a function of water depth for $0 \leq y \leq 0.9$ m.

4.68 A circular cylinder inserted across a stream of flowing water deflects the stream through angle θ, as shown. (This is termed the "Coanda effect.") For $a = 12.5$ mm, $b = 2.5$ mm, $V = 3$ m/s, and $\theta = 20°$, determine the horizontal component of the force on the cylinder caused by the flowing water.

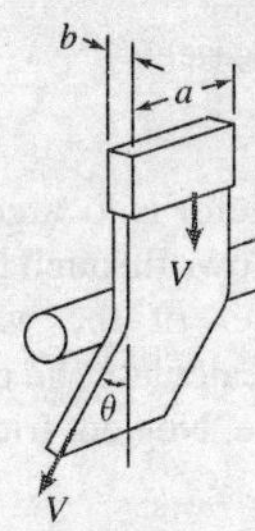

P4.68

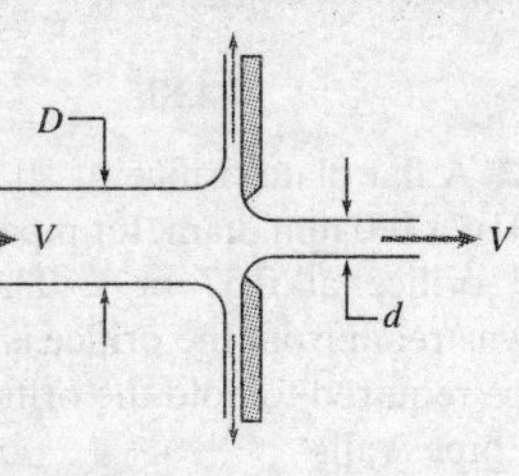

P4.69

4.69 A vertical plate has a sharp-edged orifice at its center. A water jet of speed V strikes the plate concentrically. Obtain an expression for the external force needed to hold the plate in place, if the jet leaving the orifice also has speed V. Evaluate the force for $V = 4.6$ m/s, $D = 100$ mm, and $d = 25$ mm. Plot the required force as a function of diameter ratio for a suitable range of diameter d.

4.70 In a laboratory experiment, the water flow rate is to be measured catching the water as it vertically exits a pipe into an empty open tank that is on a zeroed balance. The tank is 10 m directly below the pipe exit, and the pipe diameter is 50 mm. One student obtains a flow rate by noting that after 60 s the volume of water (at 4°C) in the tank was 3 m^3. Another student obtains a flow rate by reading the instantaneous weight accumulated of 3150 kg indicated at the 60-s point. Find the mass flow rate each student computes. Why do they disagree? Which one is more accurate? Show that the magnitude of the discrepancy can be explained by any concept you may have.

4.71 A tank of water sits on a cart with frictionless wheels as shown. The cart is attached using a cable to a mass $M = 10$ kg, and the coefficient of static friction of the mass with the ground is $\mu = 0.55$. If the gate blocking the tank exit is removed, will the resulting exit flow be sufficient to start the tank moving? (Assume the water flow is frictionless, and that the jet velocity is $V = \sqrt{2gh}$, where $h = 2$ m is the water depth.) Find the mass M that is just sufficient to hold the tank in place.

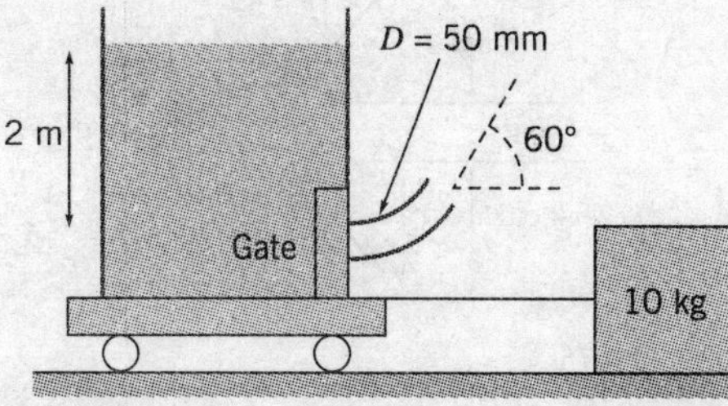

P4.71

4.72 A gate is 1 m wide and 1.2 m tall and hinged at the bottom. On one side the gate holds back a 1-m-deep body of water. On the other side, a 5-cm diameter water jet hits the gate at a height of 1 m. What jet speed V is required to hold the gate vertical? What will the required speed be if the body of water is lowered to 0.5 m? What will the required speed be if the water level is lowered to 0.25 m?

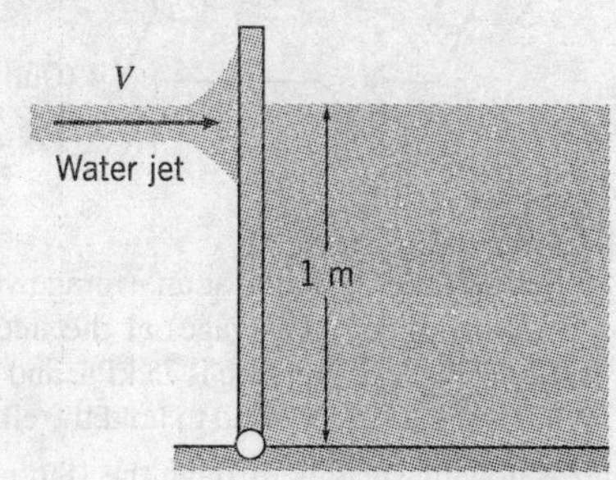

P4.72

4.73 A farmer purchases 675 kg of bulk grain from the local co-op. The grain is loaded into his pickup truck from a hopper with an outlet diameter of 0.3 m. The loading operator determines the payload by observing the indicated gross mass of the truck as a function of time. The grain flow from the hopper ($\dot{m} = 40$ kg/s) is terminated when the indicated scale reading reaches the desired gross mass. If the grain density is 600 kg/m^3, determine the true payload.

4.74 Water flows steadily through a fire hose and nozzle. The hose is 75 mm inside diameter, and the nozzle tip is 25 mm ID; water gage pressure in the hose is 510 kPa, and the stream leaving the nozzle is uniform. The exit speed and pressure are 32 m/s and atmospheric, respectively. Find the force transmitted by the coupling between the nozzle and hose. Indicate whether the coupling is in tension or compression.

4.75 A shallow circular dish has a sharp-edged orifice at its center. A water jet, of speed V, strikes the dish concentrically. Obtain an expression for the external force needed to hold the dish in place if the jet issuing from the orifice also has speed V. Evaluate the force for $V = 5$ m/s, $D = 100$ mm, and $d = 25$ mm. Plot the required force as a function of the angle θ ($0 \leq \theta \leq 90°$) with diameter ratio as a parameter for a suitable range of diameter d.

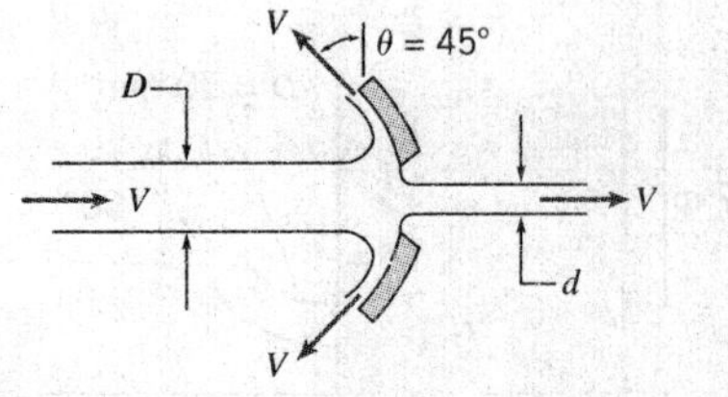

P4.75

4.76 Obtain expressions for the rate of change in mass of the control volume shown, as well as the horizontal and vertical forces required to hold it in place, in terms of p_1, A_1, V_1, p_2, A_2, V_2, p_3, A_3, V_3, p_4, A_4, V_4, and the constant density ρ.

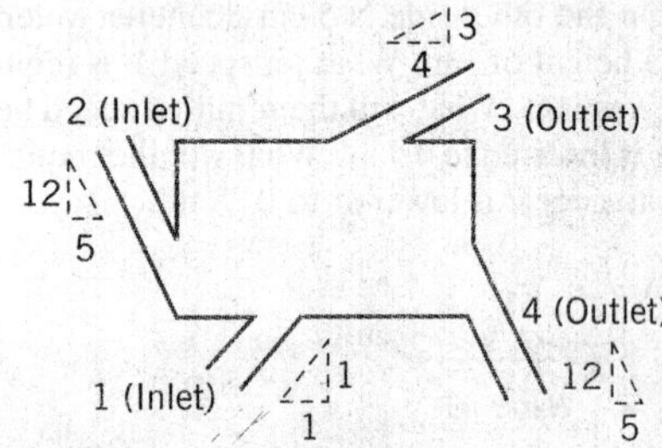

P4.76

4.77 A 180° elbow takes in water at an average velocity of 0.8 m/s and a pressure of 350 kPa (gage) at the inlet, where the diameter is 0.2 m. The exit pressure is 75 kPa, and the diameter is 0.04 m. What is the force required to hold the elbow in place?

4.78 Water is flowing steadily through the 180° elbow shown. At the inlet to the elbow the gage pressure is 103 kPa. The water discharges to atmospheric pressure. Assume properties are uniform over the inlet and outlet areas: $A_1 = 2500\ \text{mm}^2$, $A_2 = 650\ \text{mm}^2$, and $V_1 = 3$ m/s. Find the horizontal component of force required to hold the elbow in place.

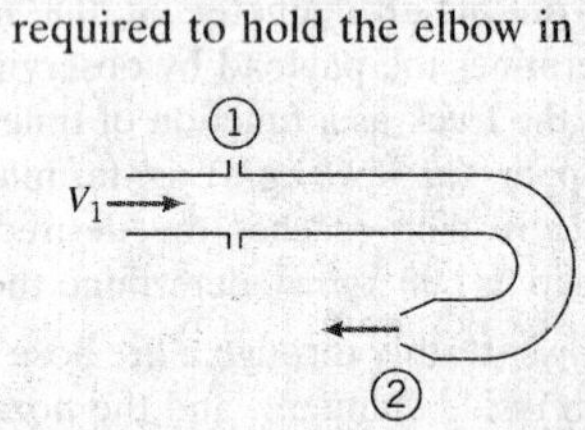

P4.78

4.79 Water flows steadily through the nozzle shown, discharging to atmosphere. Calculate the horizontal component of force in the flanged joint. Indicate whether the joint is in tension or compression.

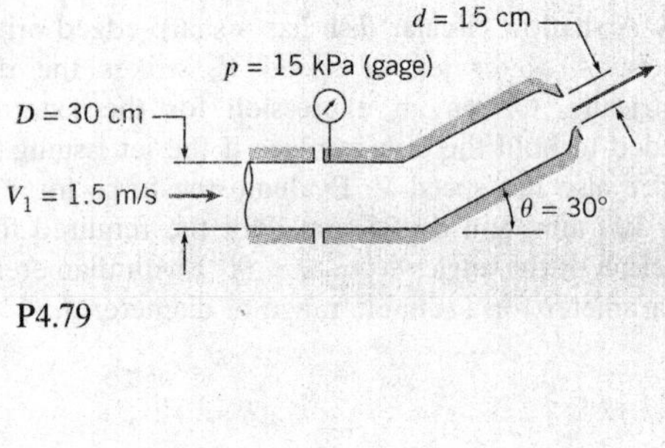

P4.79

4.80 Assume the bend of Problem 4.39 is a segment of a larger channel and lies in a horizontal plane. The inlet pressure is 170 kPa (abs), and the outlet pressure is 130 kPa (abs). Find the force required to hold the bend in place.

4.81 A spray system is shown in the diagram. Water is supplied at $p = 10$ kPa (gage), through the flanged opening of area $A = 1900\ \text{mm}^2$. The water leaves in a steady free jet at atmospheric pressure. The jet area and speed are $a = 650\ \text{mm}^2$ and $V = 4.6$ m/s. The mass of the spray system is 0.09 kg and it contains $\forall = 196\ \text{cm}^3$ of water. Find the force exerted on the supply pipe by the spray system.

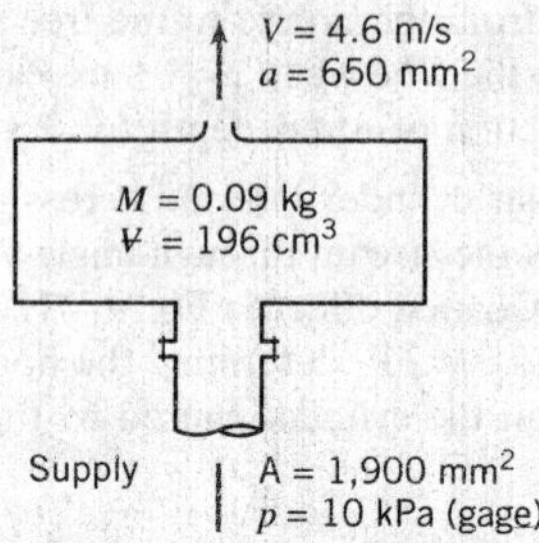

P4.81

4.82 A flat plate orifice of 50 mm diameter is located at the end of a 100 mm diameter pipe. Water flows through the pipe and orifice at $0.57\ \text{m}^3/\text{s}$. The diameter of the water jet downstream from the orifice is 38 mm. Calculate the external force required to hold the orifice in place. Neglect friction on the pipe wall.

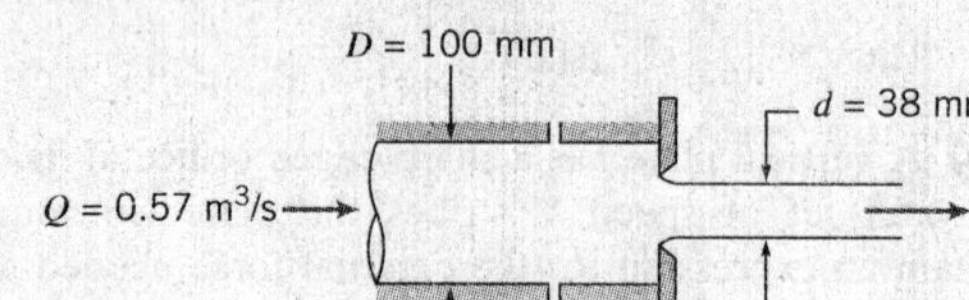

P4.82

4.83 The nozzle shown discharges a sheet of water through a 180° arc. The water speed is 15 m/s and the jet thickness is 30 mm at a radial distance of 0.3 m from the centerline of the supply pipe. Find (a) the volume flow rate of water in the jet sheet and (b) the y component of force required to hold the nozzle in place.

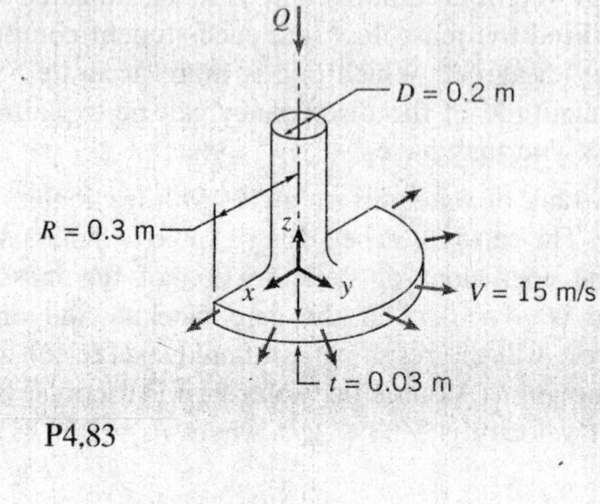

P4.83

4.84 At rated thrust, a liquid-fueled rocket motor consumes 80 kg/s of nitric acid as oxidizer and 32 kg/s of aniline as fuel. Flow leaves axially at 180 m/s relative to the nozzle and at 110 kPa. The nozzle exit diameter is $D = 0.6$ m. Calculate the thrust produced by the motor on a test stand at standard sea-level pressure.

4.85 A typical jet engine test stand installation is shown, together with some test data. Fuel enters the top of the engine vertically at a rate equal to 2 percent of the mass flow rate of the inlet air. For the given conditions, compute the air flow rate through the engine and estimate the thrust.

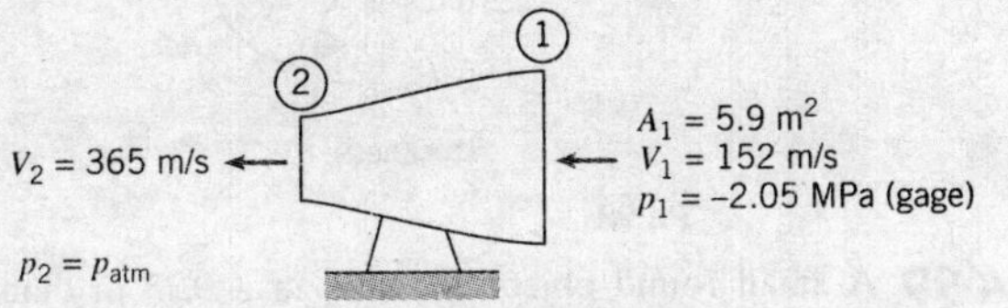

P4.85

4.86 Consider flow through the sudden expansion shown. If the flow is incompressible and friction is neglected, show that the pressure rise, $\Delta p = p_2 - p_1$, is given by

$$\frac{\Delta p}{\frac{1}{2}\rho \bar{V}_1^2} = 2\left(\frac{d}{D}\right)^2\left[1 - \left(\frac{d}{D}\right)^2\right]$$

Plot the nondimensional pressure rise versus diameter ratio to determine the optimum value of d/D and the corresponding value of the nondimensional pressure rise. *Hint:* Assume the pressure is uniform and equal to p_1 on the vertical surface of the expansion.

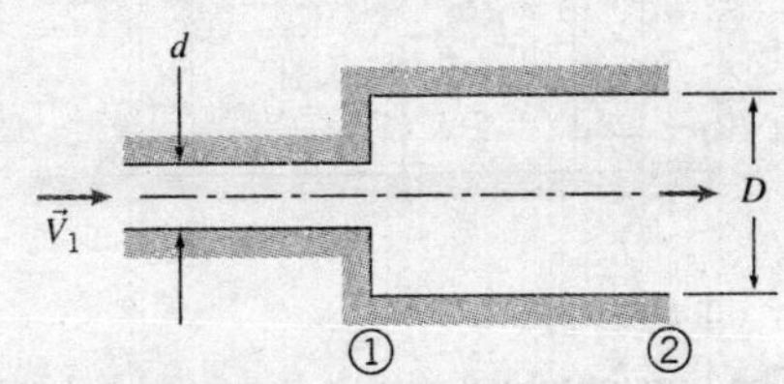

P4.86

4.87 A free jet of water with constant cross-section area 0.01 m^2 is deflected by a hinged plate of length 2 m supported by a spring with spring constant $k = 500$ N/m and uncompressed length $x_0 = 1$ m. Find and plot the deflection angle θ as a function of jet speed V. What jet speed has a deflection of $\theta = 5°$?

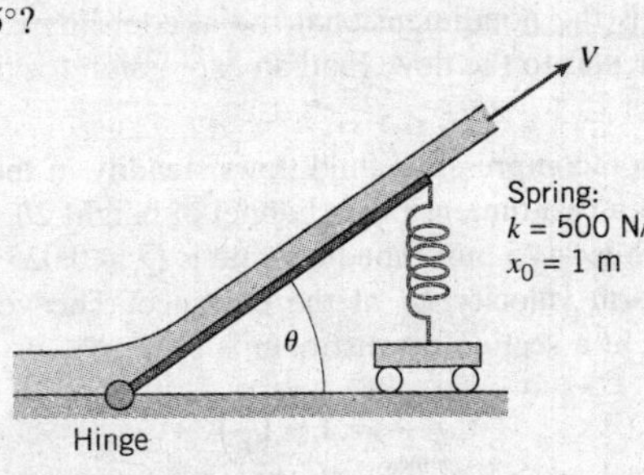

P4.87

4.88 A conical spray head is shown. The fluid is water and the exit stream is uniform. Evaluate (a) the thickness of the spray sheet at 400 mm radius and (b) the axial force exerted by the spray head on the supply pipe.

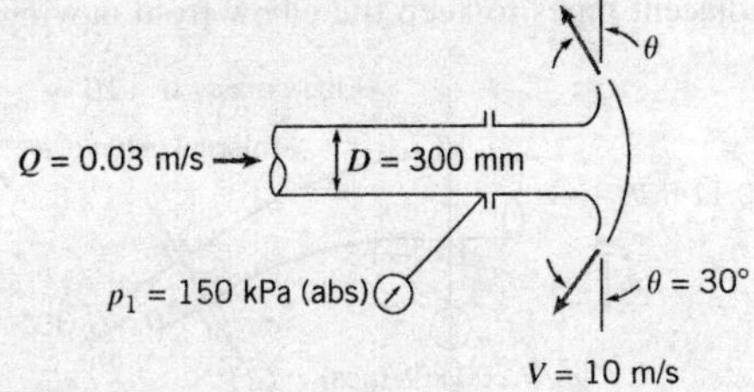

P4.88

4.89 A reducer in a piping system is shown. The internal volume of the reducer is 0.2 m^3 and its mass is 25 kg. Evaluate the total force that must be provided by the surrounding pipes to support the reducer. The fluid is gasoline.

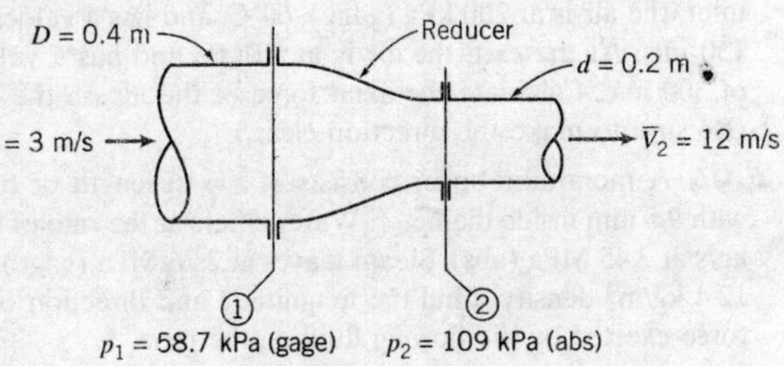

P4.89

4.90 A curved nozzle assembly that discharges to the atmosphere is shown. The nozzle mass is 4.5 kg and its internal volume is 0.002 m^3. The fluid is water. Determine the reaction force exerted by the nozzle on the coupling to the inlet pipe.

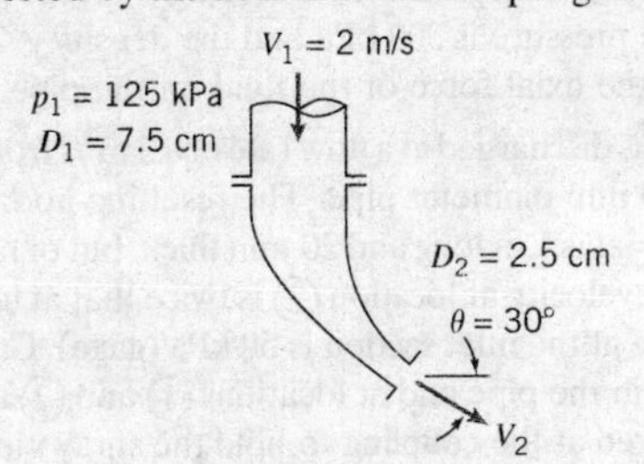

P4.90

4.91 A water jet pump has jet area 0.009 m^2 and jet speed 30.5 m/s. The jet is within a secondary stream of water having speed $V_s = 3$ m/s. The total area of the duct (the sum of the jet and secondary stream areas) is 0.07 m^2. The water is thoroughly mixed and leaves the jet pump in a uniform stream. The pressures of the jet and secondary stream are the same at the pump inlet. Determine the speed at the pump exit and the pressure rise, $p_2 - p_1$.

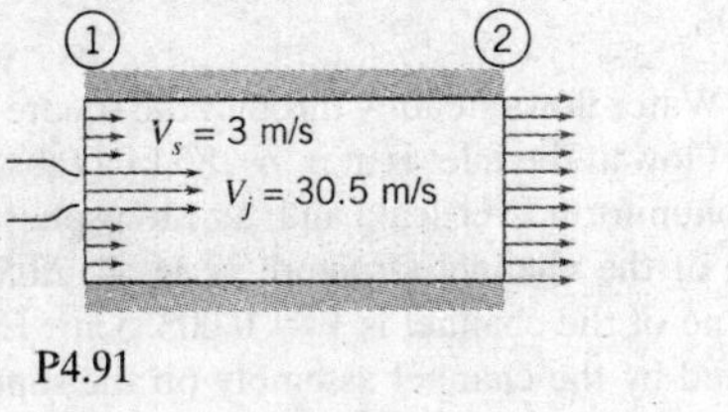

P4.91

4.92 A 30° reducing elbow is shown. The fluid is water. Evaluate the components of force that must be provided by the adjacent pipes to keep the elbow from moving.

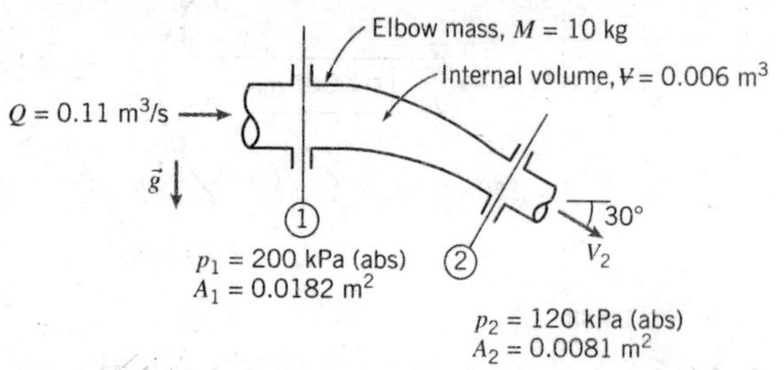

P4.92

4.93 Consider the steady adiabatic flow of air through a long straight pipe with 0.05 m^2 cross-sectional area. At the inlet, the air is at 200 kPa (gage), 60°C, and has a velocity of 150 m/s. At the exit, the air is at 80 kPa and has a velocity of 300 m/s. Calculate the axial force of the air on the pipe. (Be sure to make the direction clear.)

4.94 A monotube boiler consists of a 6 m length of tubing with 9.5 mm inside diameter. Water enters at the rate of 0.135 kg/s at 3.45 MPa (abs). Steam leaves at 2.76 MPa (gage) with 12.4 kg/m^3 density. Find the magnitude and direction of the force exerted by the flowing fluid on the tube.

4.95 A gas flows steadily through a heated porous pipe of constant 0.15 m^2 cross-sectional area. At the pipe inlet, the absolute pressure is 400 kPa, the density is 6 kg/m^3, and the mean velocity is 170 m/s. The fluid passing through the porous wall leaves in a direction normal to the pipe axis, and the total flow rate through the porous wall is 20 kg/s. At the pipe outlet, the absolute pressure is 300 kPa and the density is 2.75 kg/m^3. Determine the axial force of the fluid on the pipe.

4.96 Water is discharged at a flow rate of 0.3 m^3/s from a narrow slot in a 200 mm diameter pipe. The resulting horizontal two-dimensional jet is 1 m long and 20 mm thick, but of nonuniform velocity; the velocity at location ② is twice that at location ①. The pressure at the inlet section is 50 kPa (gage). Calculate (a) the velocity in the pipe and at locations ① and ② and (b) the forces required at the coupling to hold the spray pipe in place. Neglect the mass of the pipe and the water it contains.

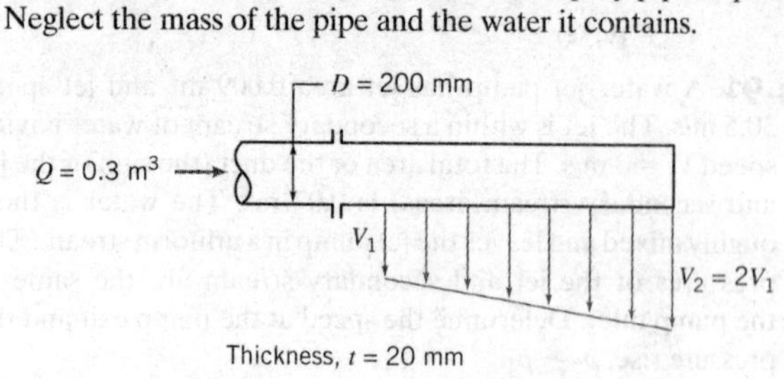

P4.96

4.97 Water flows steadily through the square bend of Problem 4.39. Flow at the inlet is at p_1 = 185 kPa (abs). Flow at the exit is nonuniform, vertical, and at atmospheric pressure. The mass of the channel structure is M_c = 2.05 kg; the internal volume of the channel is $\forall$ = 0.00355 m^3. Evaluate the force exerted by the channel assembly on the supply duct.

4.98 A nozzle for a spray system is designed to produce a flat radial sheet of water. The sheet leaves the nozzle at V_2 = 10 m/s, covers 180° of arc, and has thickness t = 1.5 mm. The nozzle discharge radius is R = 50 mm. The water supply pipe is 35 mm in diameter and the inlet pressure is p_1 = 150 kPa (abs). Evaluate the axial force exerted by the spray nozzle on the coupling.

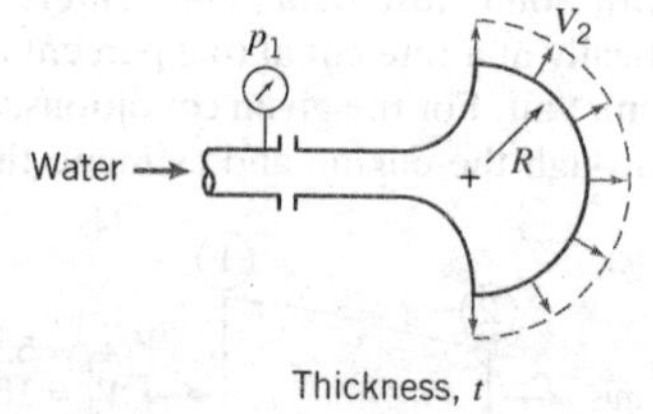

P4.98

4.99 A small round object is tested in a 0.75 m diameter wind tunnel. The pressure is uniform across sections ① and ②. The upstream pressure is 30 mm H_2O (gage), the downstream pressure is 15 mm H_2O (gage), and the mean air speed is 12.5 m/s. The velocity profile at section ② is linear; it varies from zero at the tunnel centerline to a maximum at the tunnel wall. Calculate (a) the mass flow rate in the wind tunnel, (b) the maximum velocity at section ②, and (c) the drag of the object and its supporting vane. Neglect viscous resistance at the tunnel wall.

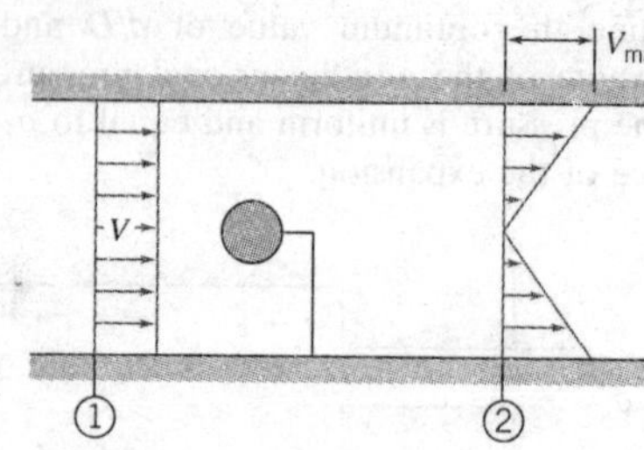

P4.99

4.100 The horizontal velocity in the wake behind an object in an air stream of velocity U is given by

$$u(r) = U\left[1 - \cos^2\left(\frac{\pi r}{2}\right)\right] \qquad |r| \le 1$$

$$u(r) = U \qquad |r| > 1$$

where r is the nondimensional radial coordinate, measured perpendicular to the flow. Find an expression for the drag on the object.

4.101 An incompressible fluid flows steadily in the entrance region of a two-dimensional channel of height $2h$ = 100 mm and width w = 25 mm. The flow rate is Q = 0.025 m^3/s. Find the uniform velocity U_1 at the entrance. The velocity distribution at a section downstream is

$$\frac{u}{u_{max}} = 1 - \left(\frac{y}{h}\right)^2$$

Evaluate the maximum velocity at the downstream section. Calculat the pressure drop that would exist in the channel if viscous friction at the walls could be neglected.

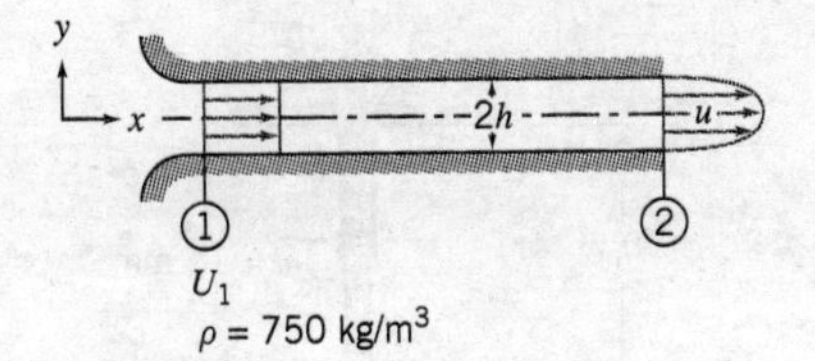

P4.101

4.102 An incompressible fluid flows steadily in the entrance region of a circular tube of radius R = 75 mm. The flow rate is Q = 0.1 m³/s. Find the uniform velocity U_1 at the entrance. The velocity distribution at a section downstream is

$$\frac{u}{u_{max}} = 1 - \left(\frac{r}{R}\right)^2$$

Evaluate the maximum velocity at the downstream section. Calculate the pressure drop that would exist in the channel if viscous friction at the walls could be neglected.

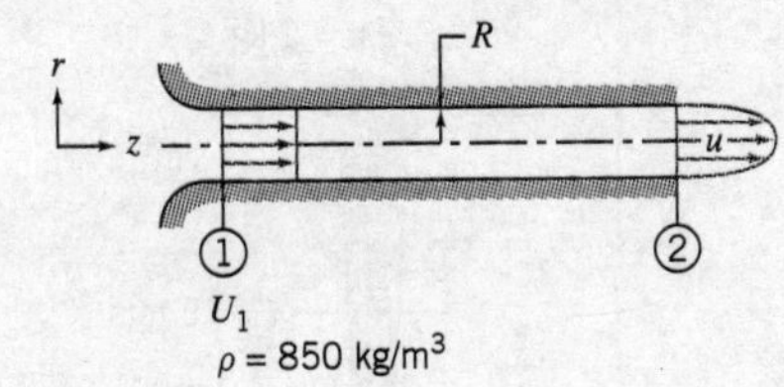

P4.102

4.103 Air enters a duct, of diameter D = 25.0 mm, through a well-rounded inlet with uniform speed, U_1 = 0.870 m/s. At a downstream section where L = 2.25 m, the fully developed velocity profile is

$$\frac{u(r)}{U_c} = 1 - \left(\frac{r}{R}\right)^2$$

The pressure drop between these sections is $p_1 - p_2 = 1.92\ \text{N/m}^2$. Find the total force of friction exerted by the tube on the air.

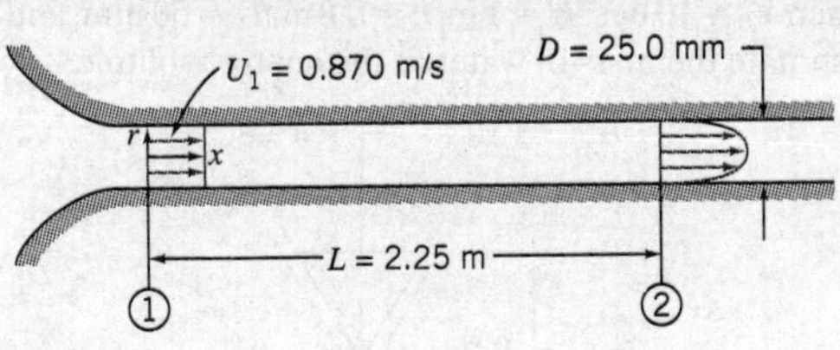

P4.103

4.104 Consider the incompressible flow of fluid in a boundary layer as depicted in Example 4.2. Show that the friction drag force of the fluid on the surface is given by

$$F_f = \int_0^\delta \rho u(U - u)w\, dy$$

Evaluate the drag force for the conditions of Example 4.2.

4.105 A fluid with density ρ = 750 kg/m³ flows along a flat plate of width 1 m. The undisturbed freestream speed is U_0 = 10 m/s. At L = 1 m downstream from the leading edge of the plate, the boundary-layer thickness is δ = 5 mm. The velocity profile at this location is

$$\frac{u}{U_0} = \frac{3}{2}\frac{y}{\delta} - \frac{1}{2}\left(\frac{y}{\delta}\right)^3$$

Plot the velocity profile. Calculate the horizontal component of force required to hold the plate stationary.

4.106 Air at standard conditions flows along a flat plate. The undisturbed freestream speed is U_0 = 20 m/s. At L = 0.4 m downstream from the leading edge of the plate, the boundary-layer thickness is δ = 2 mm. The velocity profile at this location is approximated as $u/U_0 = y/\delta$. Calculate the horizontal component of force per unit width required to hold the plate stationary.

4.107 A sharp-edged splitter plate inserted part way into a flat stream of flowing water produces the flow pattern shown. Analyze the situation to evaluate θ as a function of α, where $0 \le \alpha < 0.5$. Evaluate the force needed to hold the splitter plate in place. (Neglect any friction force between the water stream and the splitter plate.) Plot both θ and R_x as functions of α.

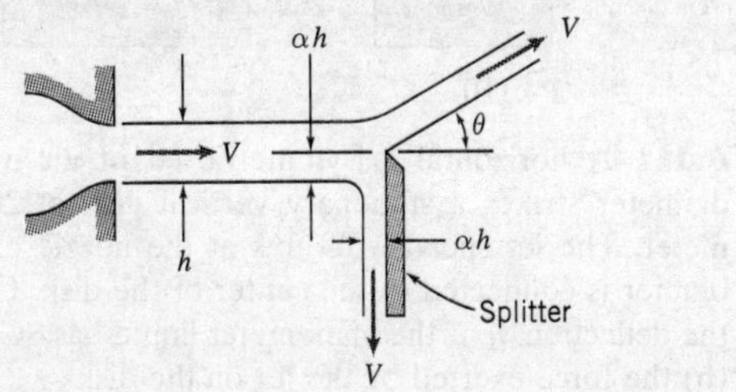

P4.107

4.108 Gases leaving the propulsion nozzle of a rocket are modeled as flowing radially outward from a point upstream from the nozzle throat. Assume the speed of the exit flow, V_e, has constant magnitude. Develop an expression for the axial thrust, T_a, developed by flow leaving the nozzle exit plane. Compare your result to the one-dimensional approximation, $T = \dot{m}V_e$. Evaluate the percent error for α = 15°. Plot the percent error versus α for $0 \le \alpha \le 22.5°$.

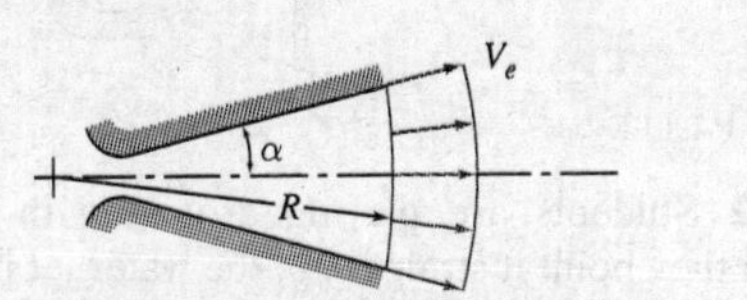

P4.108

4.109 When a plane liquid jet strikes an inclined flat plate, it splits into two streams of equal speed but unequal thickness. For frictionless flow there can be no tangential force on the plate surface. Use this assumption to develop an expression for h_2/h as a function of plate angle, θ. Plot your results and comment on the limiting cases, θ = 0 and θ = 90°.

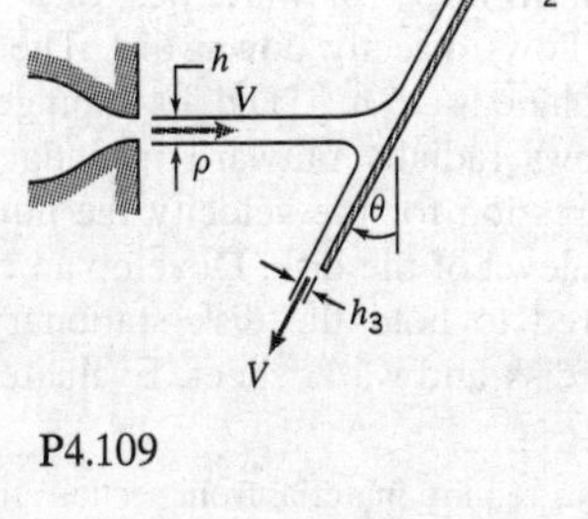

P4.109

***4.110** Two large tanks containing water have small smoothly contoured orifices of equal area. A jet of liquid issues from the left tank. Assume the flow is uniform and unaffected by friction. The jet impinges on a vertical flat plate covering the opening of the right tank. Determine the minimum value for the height, h, required to keep the plate in place over the opening of the right tank.

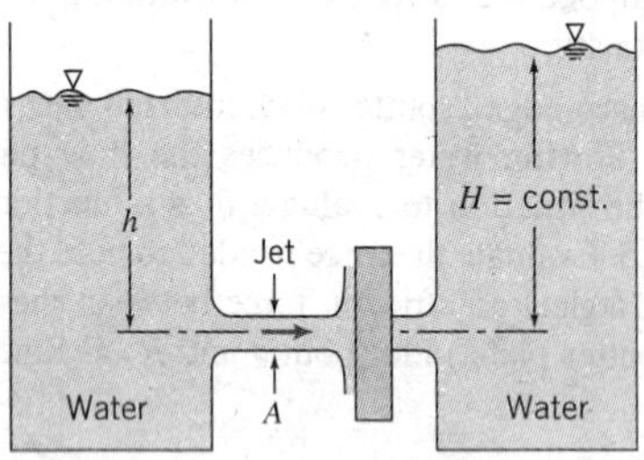

P4.110

***4.111** A horizontal axisymmetric jet of air with 13 mm. diameter strikes a stationary vertical disk of 203 mm diameter. The jet speed is 69 m/s at the nozzle exit. A manometer is connected to the center of the disk. Calculate (a) the deflection, h, if the manometer liquid has SG = 1.75 and (b) the force exerted by the jet on the disk.

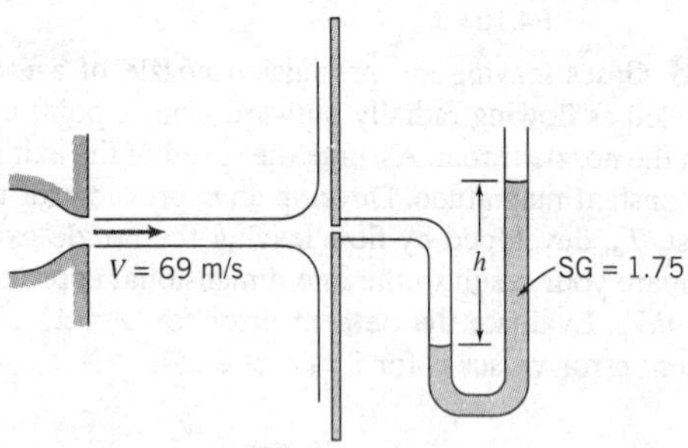

P4.111

***4.112** Students are playing around with a water hose. When they point it straight up, the water jet just reaches one of the windows of Professor Pritchard's office, 10 m above. If the hose diameter is 1 cm, estimate the water flow rate (L/min). Professor Pritchard happens to come along and places his hand just above the hose to make the jet spray sideways axisymmetrically. Estimate the maximum pressure, and the total force, he feels. The next day the students again are playing around, and this time aim at Professor Fox's window, 15 m above. Find the flow rate (L/min) and the total force and maximum pressure when he, of course, shows up and blocks the flow.

***4.113** A uniform jet of water leaves a 15 mm diameter nozzle and flows directly downward. The jet speed at the nozzle exit plane is 2.5 m/s. The jet impinges on a horizontal disk and flows radially outward in a flat sheet. Obtain a general expression for the velocity the liquid stream would reach at the level of the disk. Develop an expression for the force required to hold the disk stationary, neglecting the mass of the disk and water sheet. Evaluate for h = 3 m.

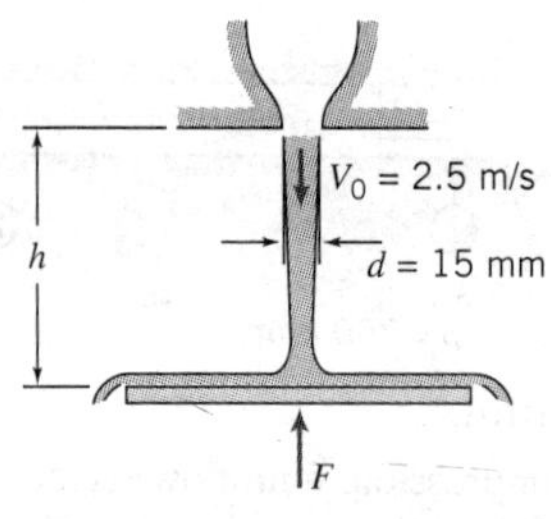

P4.113

***4.114** A 2-kg disk is constrained horizontally but is free to move vertically. The disk is struck from below by a vertical jet of water. The speed and diameter of the water jet are 10 m/s and 25 mm at the nozzle exit. Obtain a general expression for the speed of the water jet as a function of height, h. Find the height to which the disk will rise and remain stationary.

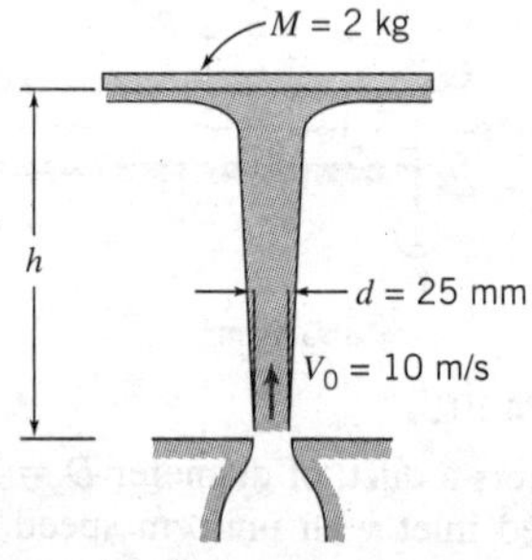

P4.114

***4.115** Water from a jet of diameter D is used to support the cone-shaped object shown. Derive an expression for the combined mass of the cone and water, M, that can be supported by the jet, in terms of parameters associated with a suitably chosen control volume. Use your expression to calculate M when $V_0 = 10\,\text{m/s}$, $H = 1\,\text{m}$, $h = 0.8\,\text{m}$, $D = 50\,\text{mm}$, and $\theta = 30°$. Estimate the mass of water in the control volume.

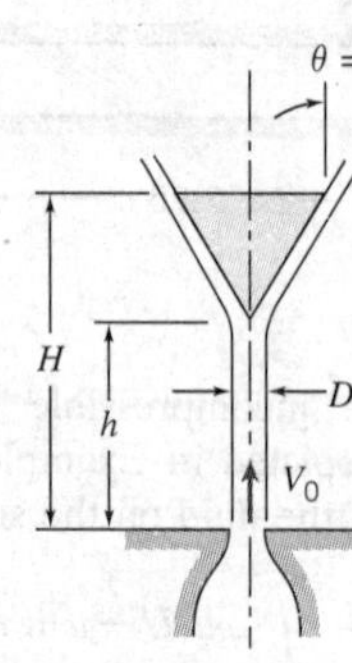

P4.115

***4.116** A stream of water from a 50 mm diameter nozzle strikes a curved vane, as shown. A stagnation tube connected to a mercury-filled U-tube manometer is located in the nozzle exit plane. Calculate the speed of the water leaving the nozzle. Estimate the horizontal component of force exerted on the vane by the jet. Comment on each assumption used to solve this problem.

*These problems require material from sections that may be omitted without loss of continuity in the text material.

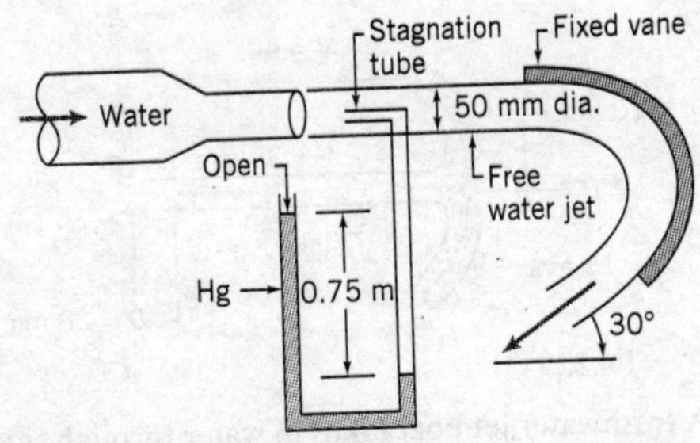

P4.116

*4.117 A Venturi meter installed along a water pipe consists of a convergent section, a constant-area throat, and a divergent section. The pipe diameter is $D = 100$ mm, and the throat diameter is $d = 50$ mm. Find the net fluid force acting on the convergent section if the water pressure in the pipe is 200 kPa (gage) and the flow rate is 1000 L/min. For this analysis, neglect viscous effects.

*4.118 A plane nozzle discharges vertically 1200 L/s per unit width downward to atmosphere. The nozzle is supplied with a steady flow of water. A stationary, inclined, flat plate, located beneath the nozzle, is struck by the water stream. The water stream divides and flows along the inclined plate; the two streams leaving the plate are of unequal thickness. Frictional effects are negligible in the nozzle and in the flow along the plate surface. Evaluate the minimum gage pressure required at the nozzle inlet and the force on the plate.

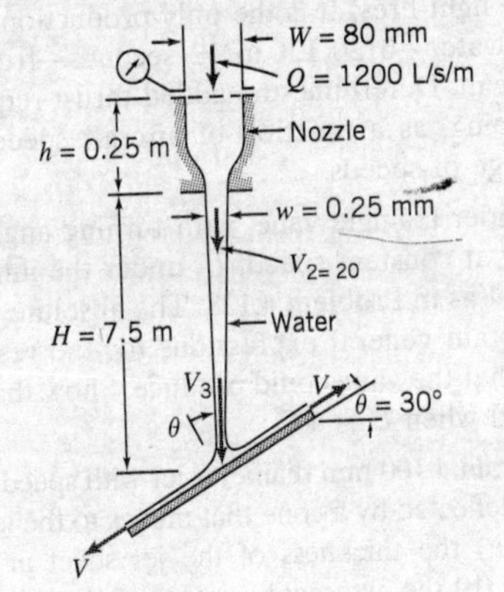

P4.118

*4.119 You turn on the kitchen faucet very slightly, so that a very narrow stream of water flows into the sink. You notice that it is "glassy" (laminar flow) and gets narrower and remains "glassy" for about the first 50 mm of descent. When you measure the flow, it takes three min to fill a 1-L bottle, and you estimate the stream of water is initially 5 mm in diameter. Assuming the speed at any cross section is uniform and neglecting viscous effects, derive expressions for and plot the variations of stream speed and diameter as functions of z (take the origin of coordinates at the faucet exit). What are the speed and diameter when it falls to the 50-mm point?

*4.120 In ancient Egypt, circular vessels filled with water sometimes were used as crude clocks. The vessels were shaped in such a way that, as water drained from the bottom, the surface level dropped at constant rate, s. Assume that water drains from a small hole of area A. Find an expression for the radius of the vessel, r, as a function of the water level, h. Obtain an expression for the volume of water needed so that the clock will operate for n hours.

*4.121 A stream of incompressible liquid moving at low speed leaves a nozzle pointed directly downward. Assume the speed at any cross section is uniform and neglect viscous effects. The speed and area of the jet at the nozzle exit are V_0 and A_0, respectively. Apply conservation of mass and the momentum equation to a differential control volume of length dz in the flow direction. Derive expressions for the variations of jet speed and area as functions of z. Evaluate the distance at which the jet area is half its original value. (Take the origin of coordinates at the nozzle exit.)

*4.122 Incompressible fluid of negligible viscosity is pumped, at total volume flow rate Q, through a porous surface into the small gap between closely spaced parallel plates as shown. The fluid has only horizontal motion in the gap. Assume uniform flow across any vertical section. Obtain an expression for the pressure variation as a function of x. *Hint:* Apply conservation of mass and the momentum equation to a differential control volume of thickness dx, located at position x.

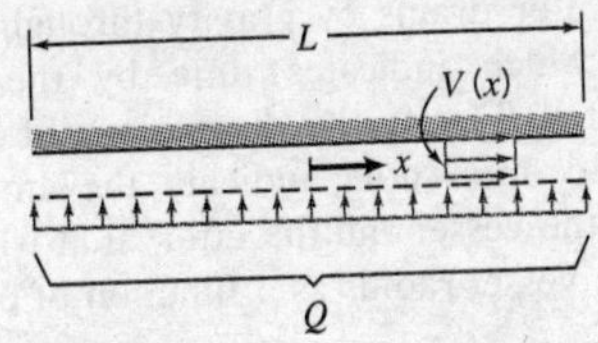

P4.122

*4.123 Incompressible liquid of negligible viscosity is pumped, at total volume flow rate Q, through two small holes into the narrow gap between closely spaced parallel disks as shown. The liquid flowing away from the holes has only radial motion. Assume uniform flow across any vertical section and discharge to atmospheric pressure at $r = R$. Obtain an expression for the pressure variation and plot as a function of radius. *Hint:* Apply conservation of mass and the momentum equation to a differential control volume of thickness dr located at radius r.

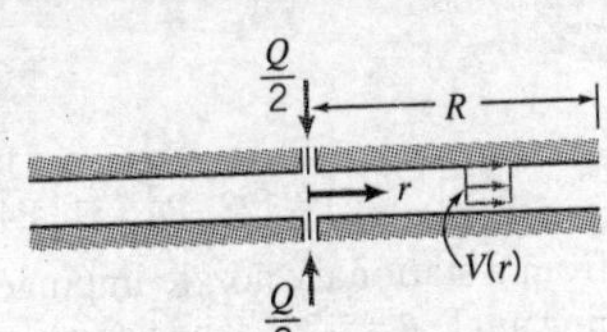

P4.123

*4.124 The narrow gap between two closely spaced circular plates initially is filled with incompressible liquid. At $t = 0$ the upper plate, initially h_0 above the lower plate, begins to move downward toward the lower plate with constant speed, V_0, causing the liquid to be squeezed from the narrow gap. Neglecting viscous effects and assuming uniform flow in the radial direction, develop an expression for the velocity field between the parallel plates. *Hint*: Apply conservation of mass to a control volume with the outer surface located at radius r. Note that even though the speed of the upper plate is

*These problems require material from sections that may be omitted without loss of continuity in the text material.

constant, the flow is unsteady. For $V_0 = 0.01$ m/s and $h_0 = 2$ mm, find the velocity at the exit radius $R = 100$ mm at $t = 0$ and $t = 0.1$ s. Plot the exit velocity as a function of time, and explain the trend.

***4.125** Liquid falls vertically into a short horizontal rectangular open channel of width b. The total volume flow rate, Q, is distributed uniformly over area bL. Neglect viscous effects. Obtain an expression for h_1 in terms of h_2, Q, and b. *Hint:* Choose a control volume with outer boundary located at $x = L$. Sketch the surface profile, $h(x)$. *Hint:* Use a differential control volume of width dx.

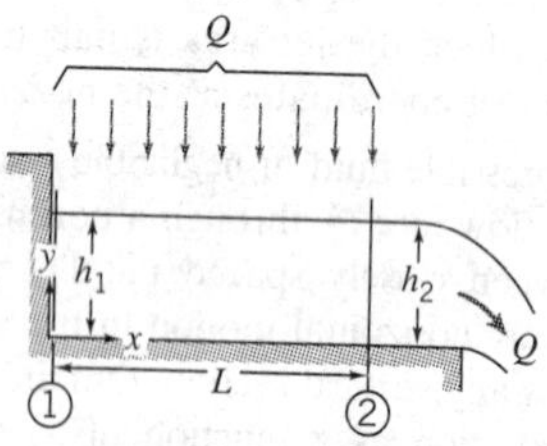

P4.125

***4.126** Design a clepsydra (Egyptian water clock)—a vessel from which water drains by gravity through a hole in the bottom and which indicates time by the level of the remaining water. Specify the dimensions of the vessel and the size of the drain hole; indicate the amount of water needed to fill the vessel and the interval at which it must be filled. Plot the vessel radius as a function of elevation.

4.127 A jet of water is directed against a vane, which could be a blade in a turbine or in any other piece of hydraulic machinery. The water leaves the stationary 40 mm diameter nozzle with a speed of 25 m/s and enters the vane tangent to the surface at A. The inside surface of the vane at B makes angle $\theta = 150°$ with the x direction. Compute the force that must be applied to maintain the vane speed constant at $U = 5$ m/s.

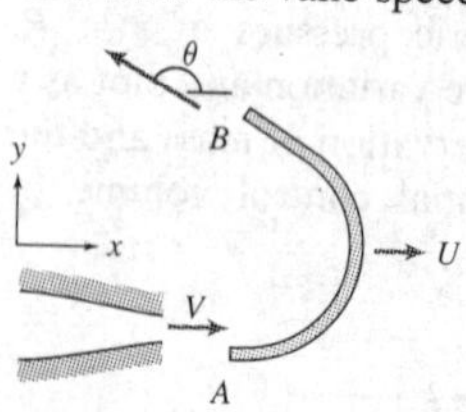

P4.127

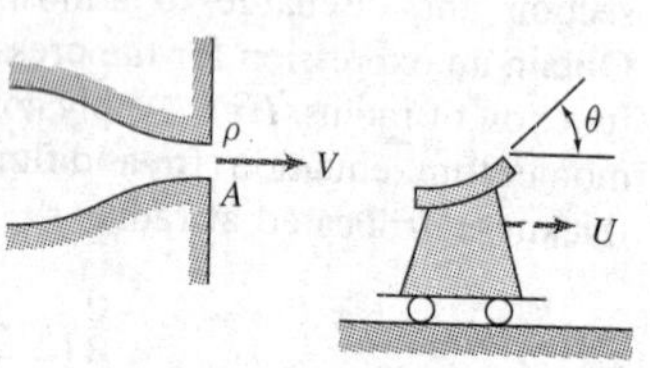

P4.128, P4.131, P4.133, P4.145

4.128 Water from a stationary nozzle impinges on a moving vane with turning angle $\theta = 120°$. The vane moves away from the nozzle with constant speed, $U = 10$ m/s, and receives a jet that leaves the nozzle with speed $V = 30$ m/s. The nozzle has an exit area of 0.004 m^2. Find the force that must be applied to maintain the vane speed constant.

4.129 The circular dish, whose cross section is shown, has an outside diameter of 0.20 m. A water jet with speed of 35 m/s strikes the dish concentrically. The dish moves to the left at 15 m/s. The jet diameter is 20 mm. The dish has a hole at its center that allows a stream of water 10 mm in diameter to pass through without resistance. The remainder of the jet is deflected and flows along the dish. Calculate the force required to maintain the dish motion.

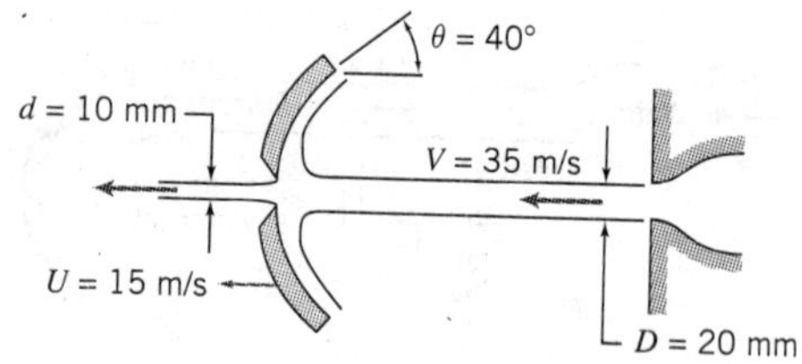

P4.129

4.130 A freshwater jet boat takes in water through side vents and ejects it through a nozzle of diameter $D = 75$ mm; the jet speed is V_j. The drag on the boat is given by $F_{drag} \propto kV^2$, where V is the boat speed. Find an expression for the steady speed, V, in terms of water density ρ, flow rate through the system of Q, constant k, and jet speed V_j. A jet speed $V_j = 15$ m/s produces a boat speed of $V = 10$ m/s.
(a) Under these conditions, what is the flow rate Q?
(b) Find the value of the constant k.
(c) What speed V will be produced if the jet speed is increased to $V_j = 25$ m/s?
(d) What will be the new flow rate?

4.131 A jet of oil (SG = 0.8) strikes a curved blade that turns the fluid through angle $\theta = 180°$. The jet area is 1200 mm^2 and its speed relative to the stationary nozzle is 20 m/s. The blade moves toward the nozzle at 10 m/s. Determine the force that must be applied to maintain the blade speed constant.

4.132 The Canadair CL-215T amphibious aircraft is specially designed to fight fires. It is the only production aircraft that can scoop water—6120 Lit in 12 seconds—from any lake, river, or ocean. Determine the added thrust required during water scooping, as a function of aircraft speed, for a reasonable range of speeds.

4.133 Consider a single vane, with turning angle θ, moving horizontally at constant speed, U, under the influence of an impinging jet as in Problem 4.128. The absolute speed of the jet is V. Obtain general expressions for the resultant force and power that the vane could produce. Show that the power is maximized when $U = V/3$.

4.134 Water, in a 100 mm diameter jet with speed of 30 m/s to the right, is deflected by a cone that moves to the left at 14 m/s. Determine (a) the thickness of the jet sheet at a radius of 230 mm. and (b) the external horizontal force needed to move the cone.

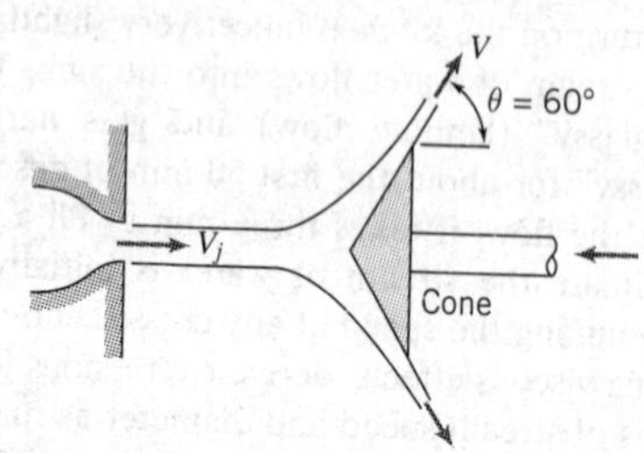

P4.134

4.135 The circular dish, whose cross section is shown, has an outside diameter of 0.15 m. A water jet strikes the dish concentrically and then flows outward along the surface of the dish. The jet speed is 45 m/s and the dish moves to the left at 10 m/s. Find the thickness of the jet sheet at a radius of

*These problems require material from sections that may be omitted without loss of continuity in the text material.

75 mm from the jet axis. What horizontal force on the dish is required to maintain this motion?

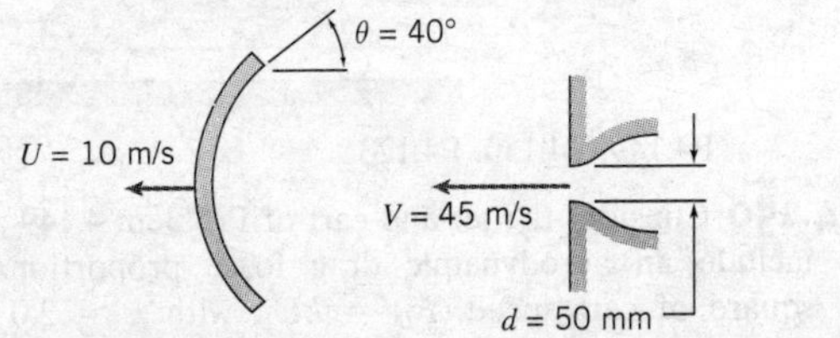

P4.135

4.136 Consider a series of turning vanes struck by a continuous jet of water that leaves a 50 mm diameter nozzle at constant speed, $V = 86.6$ m/s. The vanes move with constant speed, $U = 50$ m/s. Note that all the mass flow leaving the jet crosses the vanes. The curvature of the vanes is described by angles $\theta_1 = 30°$ and $\theta_2 = 45°$, as shown. Evaluate the nozzle angle, α, required to ensure that the jet enters tangent to the leading edge of each vane. Calculate the force that must be applied to maintain the vane speed constant.

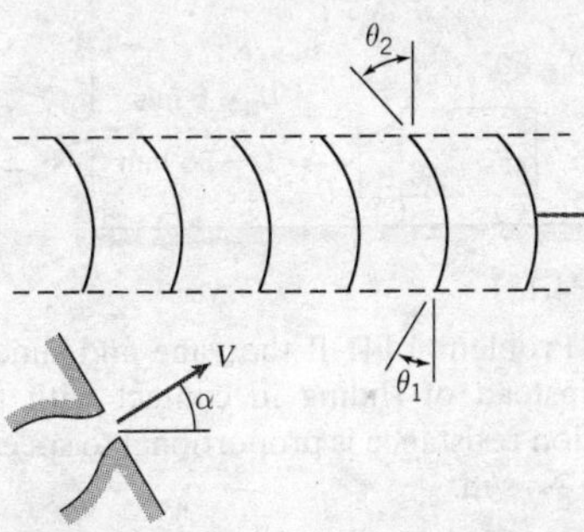

P4.136, P4.137

4.137 Consider again the moving multiple-vane system described in Problem 4.136. Assuming that a way could be found to make α nearly zero (and thus, θ_1 nearly 90°), evaluate the vane speed, U, that would result in maximum power output from the moving vane system.

4.138 A steady jet of water is used to propel a small cart along a horizontal track as shown. Total resistance to motion of the cart assembly is given by $F_D = kU^2$, where $k = 0.92 \text{ N} \cdot \text{s}^2/\text{m}^2$. Evaluate the acceleration of the cart at the instant when its speed is $U = 10$ m/s.

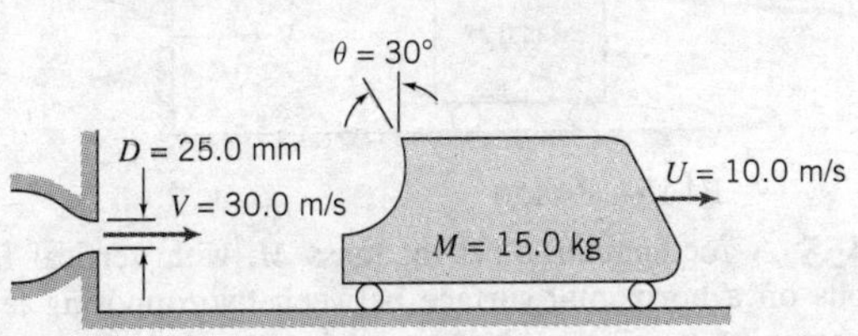

P4.138, P4.140, P4.144

4.139 A plane jet of water strikes a splitter vane and divides into two flat streams, as shown. Find the mass flow rate ratio, $\dot{m}_2/\dot{m}_3$, required to produce zero net vertical force on the splitter vane. If there is a resistive force of 16 N applied to the splitter vane, find the steady speed U of the vane.

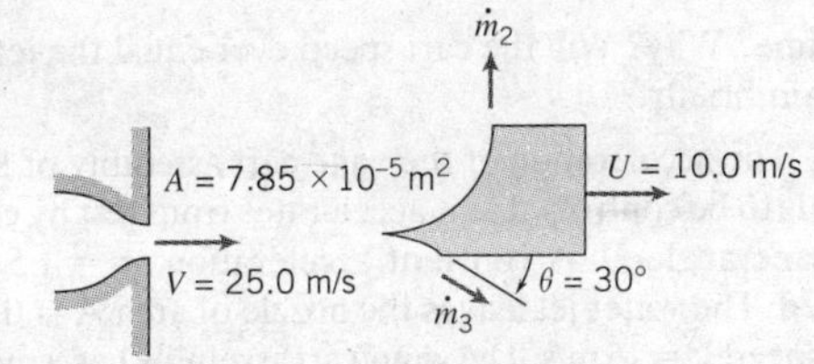

P4.139

Momentum Equation for Control Volume with Rectilinear Acceleration

4.140 The hydraulic catapult of Problem 4.138 is accelerated by a jet of water that strikes the curved vane. The cart moves along a level track with negligible resistance. At any time its speed is U. Calculate the time required to accelerate the cart from rest to $U = V/2$.

4.141 A vane/slider assembly moves under the influence of a liquid jet as shown. The coefficient of kinetic friction for motion of the slider along the surface is $\mu_k = 0.30$. Calculate the terminal speed of the slider.

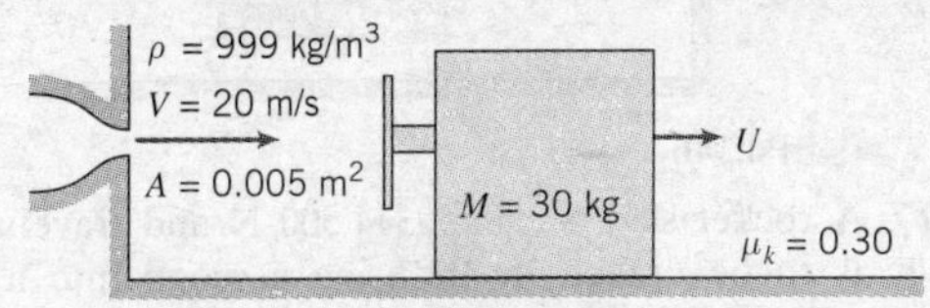

P4.141, P4.143, P4.152, P4.153

4.142 A cart is propelled by a liquid jet issuing horizontally from a tank as shown. The track is horizontal; resistance to motion may be neglected. The tank is pressurized so that the jet speed may be considered constant. Obtain a general expression for the speed of the cart as it accelerates from rest. If $M_0 = 100$ kg, $\rho = 999$ kg/m^3, and $A = 0.005$ m^2, find the jet speed V required for the cart to reach a speed of 1.5 m/s after 30 seconds. For this condition, plot the cart speed U as a function of time. Plot the cart speed after 30 seconds as a function of jet speed.

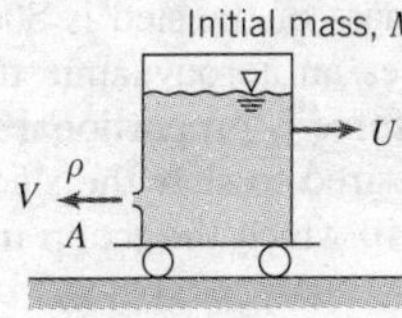

P4.142, P4.184

4.143 For the vane/slider problem of Problem 4.141, find and plot expressions for the acceleration and speed of the slider as a function of time.

4.144 If the cart of Problem 4.138 is released at $t = 0$, when would you expect the acceleration to be maximum? Sketch what you would expect for the curve of acceleration versus time. What value of θ would maximize the acceleration at any time? Why? Will the cart speed ever equal the jet speed? Explain briefly.

4.145 The acceleration of the vane/cart assembly of Problem 4.128 is to be controlled as it accelerates from rest by changing the vane angle, θ. A constant acceleration, a = 1.5 m/s^2, is desired. The water jet leaves the nozzle of area A = 0.025 m^2, with speed V = 15 m/s. The vane/cart assembly has a mass of 55 kg; neglect friction. Determine θ at t = 5 s. Plot $\theta(t)$ for the given constant acceleration over a suitable range of t.

4.146 The wheeled cart shown rolls with negligible resistance. The cart is to accelerate to the right at a constant rate of 2.5 m/s^2. This is to be accomplished by "programming" the water jet speed, $V(t)$, that hits the cart. The jet area remains constant at 50 mm^2. Find the initial jet speed, and the jet speed and cart speeds after 2.5 s and 5 s. Theoretically, what happens to the value of $(V - U)$ over time?

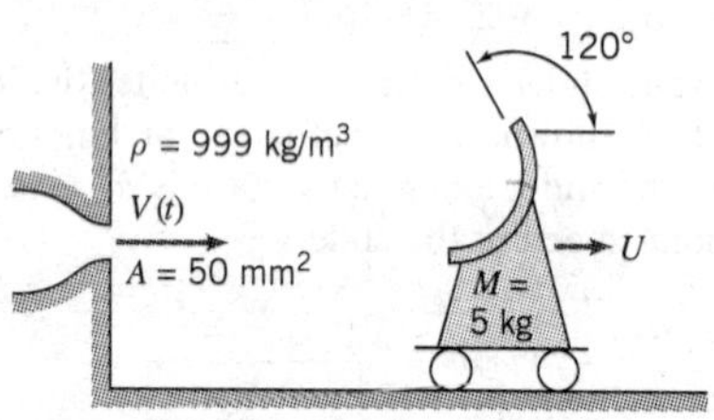

P4.146

4.147 A rocket sled, weighing 44,500 N and traveling 960 km/h, is to be braked by lowering a scoop into a water trough. The scoop is 150 mm wide. Determine the time required (after lowering the scoop to a depth of 75 mm into the water) to bring the sled to a speed of 32 km/h. Plot the sled speed as a function of time.

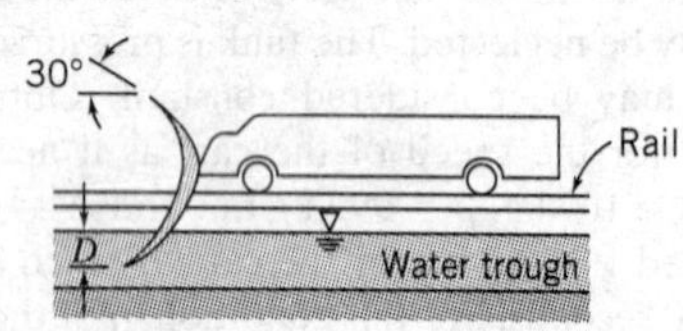

P4.147, P4.148

4.148 A rocket sled is to be slowed from an initial speed of 300 m/s by lowering a scoop into a water trough. The scoop is 0.3 m wide; it deflects the water through 150°. The trough is 800 m long. The mass of the sled is 8000 kg. At the initial speed it experiences an aerodynamic drag force of 90 kN. The aerodynamic force is proportional to the square of the sled speed. It is desired to slow the sled to 100 m/s. Determine the depth D to which the scoop must be lowered into the water.

4.149 Starting from rest, the cart shown is propelled by a hydraulic catapult (liquid jet). The jet strikes the curved surface and makes a 180° turn, leaving horizontally. Air and rolling resistance may be neglected. If the mass of the cart is 100 kg and the jet of water leaves the nozzle (of area 0.001 m^2) with a speed of 35 m/s, determine the speed of the cart 5 s after the jet is directed against the cart. Plot the cart speed as a function of time.

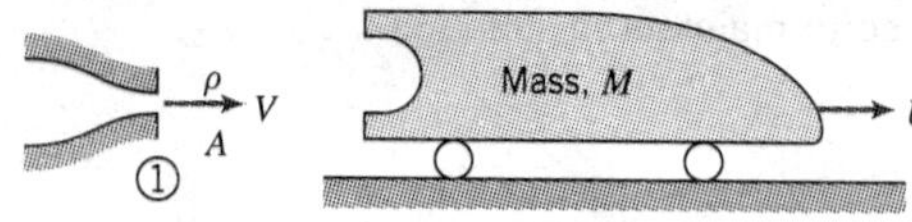

P4.149, P4.150, P4.173

4.150 Consider the jet and cart of Problem 4.149 again, but include an aerodynamic drag force proportional to the square of cart speed, $F_D = kU^2$, with k = 2.0 N·s^2/m^2. Derive an expression for the cart acceleration as a function of cart speed and other given parameters. Evaluate the acceleration of the cart at U = 10 m/s. What fraction is this speed of the terminal speed of the cart?

4.151 A small cart that carries a single turning vane rolls on a level track. The cart mass is M = 5 kg and its initial speed is U_0 = 5 m/s. At t = 0, the vane is struck by an opposing jet of water, as shown. Neglect any external forces due to air or rolling resistance. Determine the jet speed V required to bring the cart to rest in (a) 1 s and (b) 2 s. In each case find the total distance traveled.

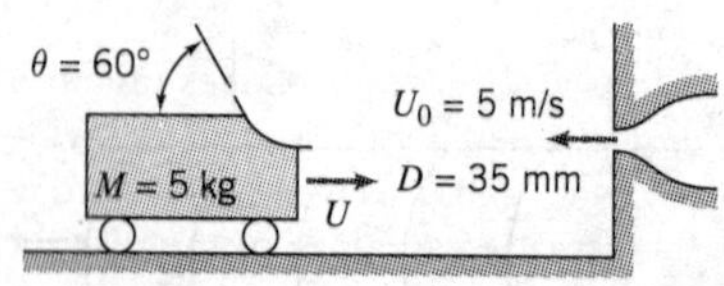

P4.151

4.152 Solve Problem 4.141 if the vane and slider ride on a film of oil instead of sliding in contact with the surface. Assume motion resistance is proportional to speed, $F_R = kU$, with k = 7.5 N·s/m.

4.153 For the vane/slider problem of Problem 4.152, plot the acceleration, speed, and position of the slider as functions of time. (Consider numerical integration.)

4.154 A rectangular block of mass M, with vertical faces, rolls without resistance along a smooth horizontal plane as shown. The block travels initially at speed U_0. At t = 0 the block is struck by a liquid jet and its speed begins to slow. Obtain an algebraic expression for the acceleration of the block for $t > 0$. Solve the equation to determine the time at which U = 0.

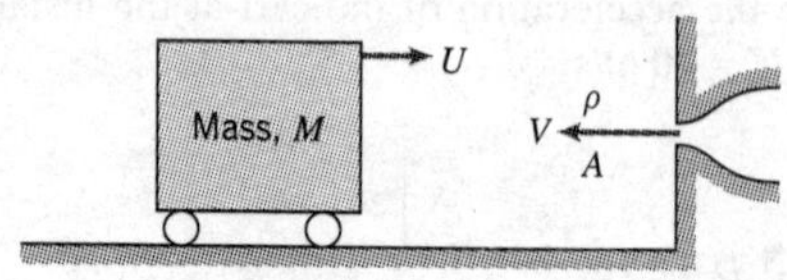

P4.154, P4.156

4.155 A rectangular block of mass M, with vertical faces, rolls on a horizontal surface between two opposing jets as shown. At t = 0 the block is set into motion at speed U_0. Subsequently, it moves without friction parallel to the jet axes with speed $U(t)$. Neglect the mass of any liquid adhering to the block compared with M. Obtain general expressions for the acceleration of the block, $a(t)$, and the block speed, $U(t)$.

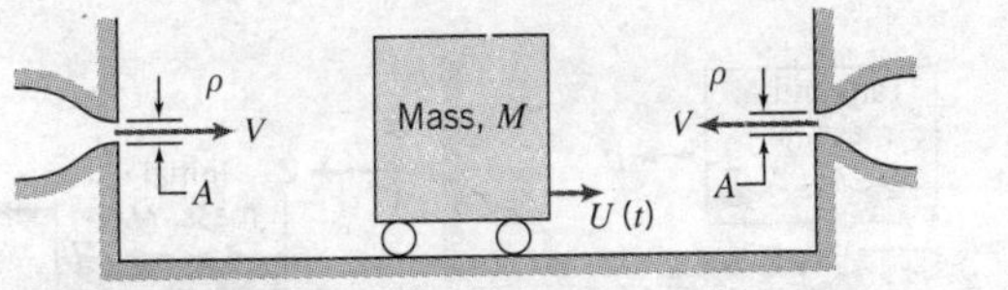

P4.155, P4.157

4.156 Consider the diagram of Problem 4.154. If $M = 100$ kg, $\rho = 999$ kg/m^3, and $A = 0.01$ m^2, find the jet speed V required for the cart to be brought to rest after one second if the initial speed of the cart is $U_0 = 5$ m/s. For this condition, plot the speed U and position x of the cart as functions of time. What is the maximum value of x, and how long does the cart take to return to its initial position?

4.157 Consider the statement and diagram of Problem 4.155. Assume that at $t = 0$, when the block of mass $M = 5$ kg is at $x = 0$, it is set into motion at speed $U_0 = 10$ m/s, to the right. The water jets have speed $V = 20$ m/s and area $A = 100$ mm^2. Calculate the time required to reduce the block speed to $U = 2.5$ m/s. Plot the block position x versus time. Compute the final rest position of the block. Explain why it comes to rest.

***4.158** A vertical jet of water impinges on a horizontal disk as shown. The disk assembly mass is 30 kg. When the disk is 3 m above the nozzle exit, it is moving upward at $U = 5$ m/s. Compute the vertical acceleration of the disk at this instant.

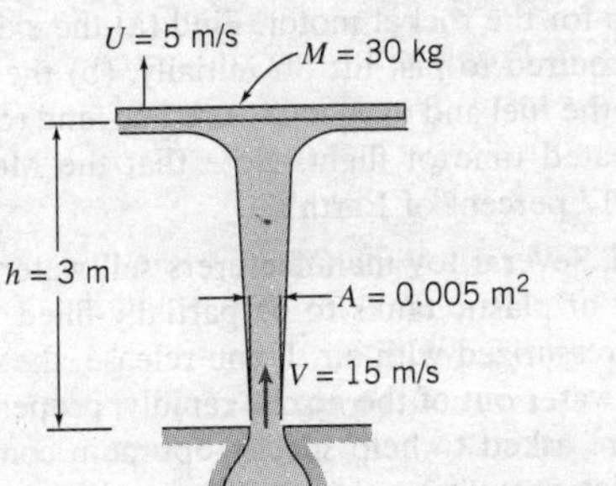

P4.158, P4.159, P4.180

4.159 A vertical jet of water leaves a 75 mm diameter nozzle. The jet impinges on a horizontal disk (see Problem 4.158). The disk is constrained horizontally but is free to move vertically. The mass of the disk is 35 kg. Plot disk mass versus flow rate to determine the water flow rate required to suspend the disk 3 m above the jet exit plane.

4.160 A rocket sled traveling on a horizontal track is slowed by a retro-rocket fired in the direction of travel. The initial speed of the sled is $U_0 = 500$ m/s. The initial mass of the sled is $M_0 = 1500$ kg. The retro-rocket consumes fuel at the rate of 7.75 kg/s, and the exhaust gases leave the nozzle at atmospheric pressure and a speed of 2500 m/s relative to the rocket. The retro-rocket fires for 20 s. Neglect aerodynamic drag and rolling resistance. Obtain and plot an algebraic expression for sled speed U as a function of firing time. Calculate the sled speed at the end of retro-rocket firing.

4.161 A manned space capsule travels in level flight above the Earth's atmosphere at initial speed $U_0 = 8.00$ km/s. The capsule is to be slowed by a retro-rocket to $U = 5.00$ km/s in preparation for a reentry maneuver. The initial mass of the capsule is $M_0 = 1600$ kg. The rocket consumes fuel at $\dot{m} = 8.0$ kg/s, and exhaust gases leave at $V_e = 3000$ m/s relative to the capsule and at negligible pressure. Evaluate the duration of the retro-rocket firing needed to accomplish this. Plot the final speed as a function of firing duration for a time range $\pm 10\%$ of this firing time.

4.162 A rocket sled accelerates from rest on a level track with negligible air and rolling resistances. The initial mass of the sled is $M_0 = 600$ kg. The rocket initially contains 150 kg of fuel. The rocket motor burns fuel at constant rate $\dot{m} = 15$ kg/s. Exhaust gases leave the rocket nozzle uniformly and axially at $V_e = 2900$ m/s relative to the nozzle, and the pressure is atmospheric. Find the maximum speed reached by the rocket sled. Calculate the maximum acceleration of the sled during the run.

4.163 A rocket sled has mass of 5000 kg, including 1000 kg of fuel. The motion resistance in the track on which the sled rides and that of the air total kU, where k is 50 N·s/m and U is the speed of the sled in m/s. The exit speed of the exhaust gas relative to the rocket is 1750 m/s, and the exit pressure is atmospheric. The rocket burns fuel at the rate of 50 kg/s.

(a) Plot the sled speed as a function of time.
(b) Find the maximum speed.
(c) What percentage increase in maximum speed would be obtained by reducing k by 10 percent?

4.164 A rocket sled with initial mass of 900 kg is to be accelerated on a level track. The rocket motor burns fuel at constant rate $\dot{m} = 13.5$ kg/s. The rocket exhaust flow is uniform and axial. Gases leave the nozzle at 2750 m/s relative to the nozzle, and the pressure is atmospheric. Determine the minimum mass of rocket fuel needed to propel the sled to a speed of 265 m/s before burnout occurs. As a first approximation, neglect resistance forces.

4.165 A rocket motor is used to accelerate a kinetic energy weapon to a speed of 5600 km/h in horizontal flight. The exit stream leaves the nozzle axially and at atmospheric pressure with a speed of 9600 km/h relative to the rocket. The rocket motor ignites upon release of the weapon from an aircraft flying horizontally at $U_0 = 960$ km/h. Neglecting air resistance, obtain an algebraic expression for the speed reached by the weapon in level flight. Determine the minimum fraction of the initial mass of the weapon that must be fuel to accomplish the desired acceleration.

4.166 A rocket sled with initial mass of 3 metric tons, including 1 ton of fuel, rests on a level section of track. At $t = 0$, the solid fuel of the rocket is ignited and the rocket burns fuel at the rate of 75 kg/s. The exit speed of the exhaust gas relative to the rocket is 2500 m/s, and the pressure is atmospheric. Neglecting friction and air resistance, calculate the acceleration and speed of the sled at $t = 10$ s.

4.167 A daredevil considering a record attempt—for the world's longest motorcycle jump—asks for your consulting

*These problems require material from sections that may be omitted without loss of continuity in the text material.

help: He must reach 875 km/hr (from a standing start on horizontal ground) to make the jump, so he needs rocket propulsion. The total mass of the motorcycle, the rocket motor without fuel, and the rider is 375 kg. Gases leave the rocket nozzle horizontally, at atmospheric pressure, with a speed of 2510 m/s. Evaluate the minimum amount of rocket fuel needed to accelerate the motorcycle and rider to the required speed.

4.168 A "home-made" solid propellant rocket has an initial mass of 9 kg; 6.8 kg of this is fuel. The rocket is directed vertically upward from rest, burns fuel at a constant rate of 0.225 kg/s, and ejects exhaust gas at a speed of 1980 m/s relative to the rocket. Assume that the pressure at the exit is atmospheric and that air resistance may be neglected. Calculate the rocket speed after 20 s and the distance traveled by the rocket in 20 s. Plot the rocket speed and the distance traveled as functions of time.

4.169 A large two-stage liquid rocket with mass of 30,000 kg is to be launched from a sea-level launch pad. The main engine burns liquid hydrogen and liquid oxygen in a stoichiometric mixture at 2450 kg/s. The thrust nozzle has an exit diameter of 2.6 m. The exhaust gases exit the nozzle at 2270 m/s and an exit plane pressure of 66 kPa absolute. Calculate the acceleration of the rocket at liftoff. Obtain an expression for speed as a function of time, neglecting air resistance.

4.170 Neglecting air resistance, what speed would a vertically directed rocket attain in 5 s if it starts from rest, has initial mass of 350 kg, burns 10 kg/s, and ejects gas at atmospheric pressure with a speed of 2500 m/s relative to the rocket? What would be the maximum velocity? Plot the rocket speed as a function of time for the first minute of flight.

4.171 Inflate a toy balloon with air and release it. Watch as the balloon darts about the room. Explain what causes the phenomenon you see.

4.172 The vane/cart assembly of mass $M = 30$ kg, shown in Problem 4.128, is driven by a water jet. The water leaves the stationary nozzle of area $A = 0.02\ \text{m}^2$, with a speed of 20 m/s. The coefficient of kinetic friction between the assembly and the surface is 0.10. Plot the terminal speed of the assembly as a function of vane turning angle, θ, for $0 \leq \theta \leq \pi/2$. At what angle does the assembly begin to move if the coefficient of static friction is 0.15?

4.173 Consider the vehicle shown in Problem 4.149. Starting from rest, it is propelled by a hydraulic catapult (liquid jet). The jet strikes the curved surface and makes a 180° turn, leaving horizontally. Air and rolling resistance may be neglected. Using the notation shown, obtain an equation for the acceleration of the vehicle at any time and determine the time required for the vehicle to reach $U = V/2$.

4.174 The moving tank shown is to be slowed by lowering a scoop to pick up water from a trough. The initial mass and speed of the tank and its contents are M_0 and U_0, respectively. Neglect external forces due to pressure or friction and assume that the track is horizontal. Apply the continuity and momentum equations to show that at any instant $U = U_0M_0/M$. Obtain a general expression for U/U_0 as a function of time.

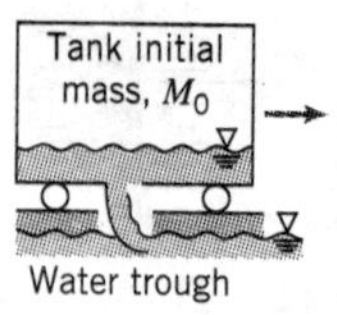

P4.174

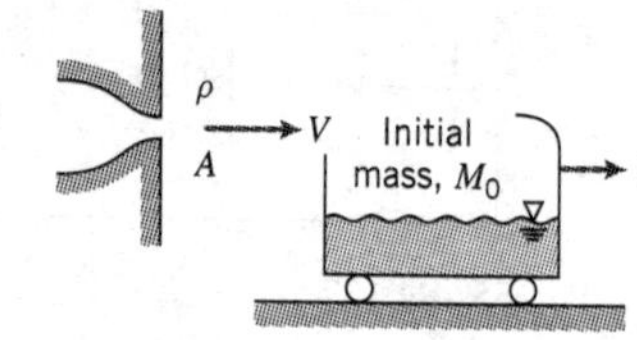

P4.175

4.175 The tank shown rolls with negligible resistance along a horizontal track. It is to be accelerated from rest by a liquid jet that strikes the vane and is deflected into the tank. The initial mass of the tank is M_0. Use the continuity and momentum equations to show that at any instant the mass of the vehicle and liquid contents is $M = M_0V/(V - U)$. Obtain a general expression for U/V as a function of time.

4.176 A model solid propellant rocket has a mass of 69.6 g, of which 12.5 g is fuel. The rocket produces 5.75 N of thrust for a duration of 1.7 s. For these conditions, calculate the maximum speed and height attainable in the absence of air resistance. Plot the rocket speed and the distance traveled as functions of time.

4.177 A small rocket motor is used to power a "jet pack" device to lift a single astronaut above the Moon's surface. The rocket motor produces a uniform exhaust jet with constant speed, $V_e = 3000$ m/s, and the thrust is varied by changing the jet size. The total initial mass of the astronaut and the jet pack is $M_0 = 200$ kg, 100 kg of which is fuel and oxygen for the rocket motor. Find (a) the exhaust mass flow rate required to just lift off initially, (b) the mass flow rate just as the fuel and oxygen are used up, and (c) the maximum anticipated time of flight. Note that the Moon's gravity is about 17 percent of Earth's.

***4.178** Several toy manufacturers sell water "rockets" that consist of plastic tanks to be partially filled with water and then pressurized with air. Upon release, the compressed air forces water out of the nozzle rapidly, propelling the rocket. You are asked to help specify optimum conditions for this water-jet propulsion system. To simplify the analysis, consider horizontal motion only. Perform the analysis and design needed to define the acceleration performance of the compressed air/water-propelled rocket. Identify the fraction of tank volume that initially should be filled with compressed air to achieve optimum performance (i.e., maximum speed from the water charge). Describe the effect of varying the initial air pressure in the tank.

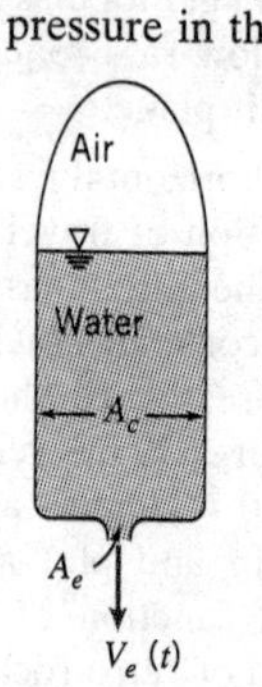

P4.178

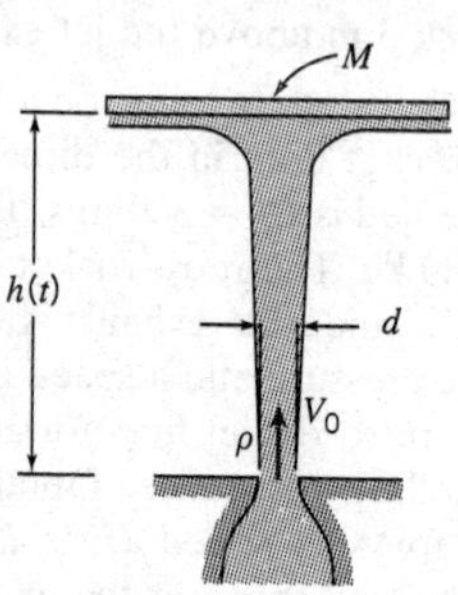

P4.179

*These problems require material from sections that may be omitted without loss of continuity in the text material.

*4.179 A disk, of mass M, is constrained horizontally but is free to move vertically. A jet of water strikes the disk from below. The jet leaves the nozzle at initial speed V_0. Obtain a differential equation for the disk height, $h(t)$, above the jet exit plane if the disk is released from large height, H. (You will not be able to solve this ODE, as it is highly nonlinear!) Assume that when the disk reaches equilibrium, its height above the jet exit plane is h_0.
(a) Sketch $h(t)$ for the disk released at $t = 0$ from $H > h_0$.
(b) Explain why the sketch is as you show it.

*4.180 Consider the configuration of the vertical jet impinging on a horizontal disk shown in Problem 4.158. Assume the disk is released from rest at an initial height of 2 m above the jet exit plane. Using a numerical method such as the Euler method (see Section 5.5), solve for the subsequent motion of this disk. Identify the steady-state height of the disk.

4.181 A small solid-fuel rocket motor is fired on a test stand. The combustion chamber is circular, with 100 mm diameter. Fuel, of density 1660 kg/m^3, burns uniformly at the rate of 12.7 mm/s. Measurements show that the exhaust gases leave the rocket at ambient pressure, at a speed of 2750 m/s. The absolute pressure and temperature in the combustion chamber are 7.0 MPa and 3610 K, respectively. Treat the combustion products as an ideal gas with molecular mass of 25.8. Evaluate the rate of change of mass and of linear momentum within the rocket motor. Express the rate of change of linear momentum within the motor as a percentage of the motor thrust.

*4.182 The capability of the Aircraft Landing Loads and Traction Facility at NASA's Langley Research Center is to be upgraded. The facility consists of a rail-mounted carriage propelled by a jet of water issuing from a pressurized tank. (The setup is identical in concept to the hydraulic catapult of Problem 4.138.) Specifications require accelerating the carriage with 49,000 kg mass to a speed of 407.09 km/hr in a distance of 122 m. (The vane turning angle is 170°.) Identify a range of water jet sizes and speeds needed to accomplish this performance. Specify the recommended operating pressure for the water-jet system and determine the shape and estimated size of tankage to contain the pressurized water.

*4.183 A classroom demonstration of linear momentum is planned, using a water-jet propulsion system for a cart traveling on a horizontal linear air track. The track is 5 m long, and the cart mass is 155 g. The objective of the design is to obtain the best performance for the cart, using 1 L of water contained in an open cylindrical tank made from plastic sheet with density of 0.0819 g/cm^2. For stability, the maximum height of the water tank cannot exceed 0.5 m. The diameter of the smoothly rounded water jet may not exceed 10 percent of the tank diameter. Determine the best dimensions for the tank and the water jet by modeling the system performance. Using a numerical method such as the Euler method (see Section 5.5), plot acceleration, velocity, and distance as functions of time. Find the optimum dimensions of the water tank and jet opening from the tank. Discuss the limitations on your analysis. Discuss how the assumptions affect the predicted performance of the cart. Would the actual performance of the cart be better or worse than predicted? Why? What factors account for the difference(s)?

*4.184 Analyze the design and optimize the performance of a cart propelled along a horizontal track by a water jet that issues under gravity from an open cylindrical tank carried on board the cart. (A water-jet-propelled cart is shown in the diagram for Problem 4.142.) Neglect any change in slope of the liquid free surface in the tank during acceleration. Analyze the motion of the cart along a horizontal track, assuming it starts from rest and begins to accelerate when water starts to flow from the jet. Derive algebraic equations or solve numerically for the acceleration and speed of the cart as functions of time. Present results as plots of acceleration and speed versus time, neglecting the mass of the tank. Determine the dimensions of a tank of minimum mass required to accelerate the cart from rest along a horizontal track to a specified speed in a specified time interval.

The Angular-Momentum Principle

*4.185 A large irrigation sprinkler unit, mounted on a cart, discharges water with a speed of 40 m/s at an angle of 30° to the horizontal. The 50 mm diameter nozzle is 3 m above the ground. The mass of the sprinkler and cart is $M = 350$ kg. Calculate the magnitude of the moment that tends to overturn the cart. What value of V will cause impending motion? What will be the nature of the impending motion? What is the effect of the angle of jet inclination on the results? For the case of impending motion, plot the jet velocity as a function of the angle of jet inclination over an appropriate range of the angles.

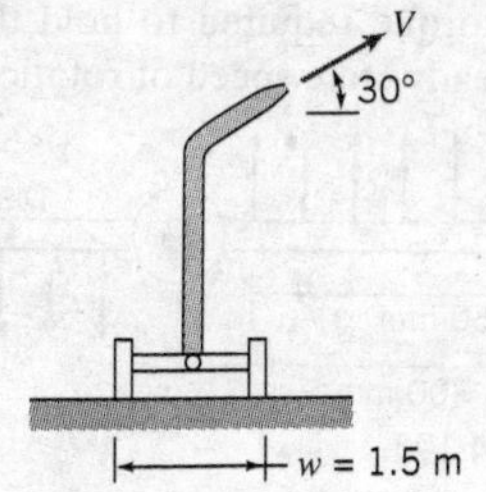

P4.185

*4.186 The 90° reducing elbow of Example 4.6 discharges to atmosphere. Section ② is located 0.3 m to the right of Section ①. Estimate the moment exerted by the flange on the elbow.

*4.187 Crude oil (SG = 0.95) from a tanker dock flows through a pipe of 0.25 m diameter in the configuration shown. The flow rate is 0.58 m^3/s, and the gage pressures are shown in the diagram. Determine the force and torque that are exerted by the pipe assembly on its supports.

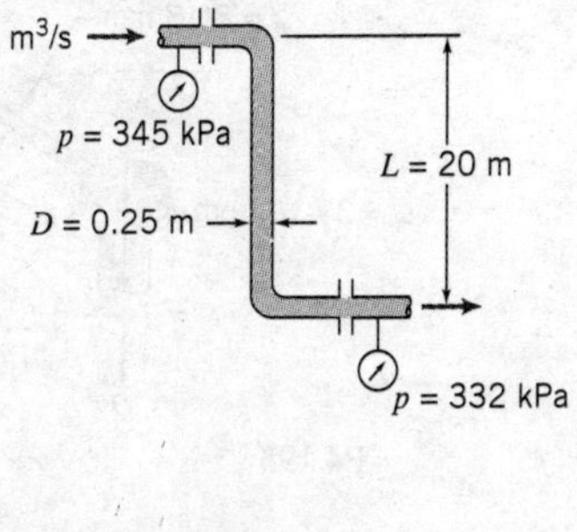

P4.187

*These problems require material from sections that may be omitted without loss of continuity in the text material.

***4.188** The simplified lawn sprinkler shown rotates in the horizontal plane. At the center pivot, $Q = 15$ L/min of water enters vertically. Water discharges in the horizontal plane from each jet. If the pivot is frictionless, calculate the torque needed to keep the sprinkler from rotating. Neglecting the inertia of the sprinkler itself, calculate the angular acceleration that results when the torque is removed.

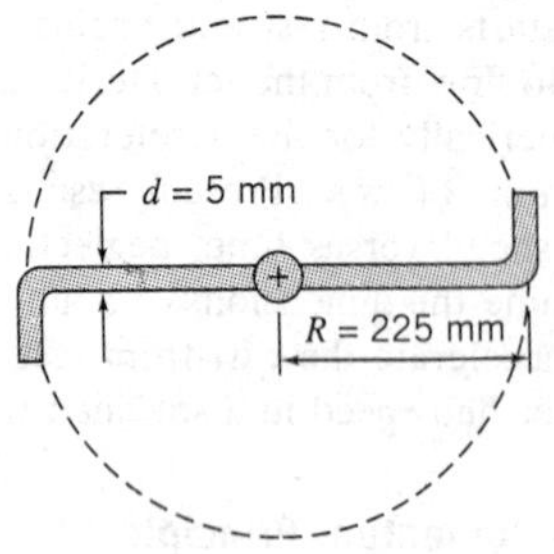

P4.188, P4.189, P4.190

***4.189** Consider the sprinkler of Problem 4.188 again. Derive a differential equation for the angular speed of the sprinkler as a function of time. Evaluate its steady-state speed of rotation if there is no friction in the pivot.

***4.190** Repeat Problem 4.189, but assume a constant retarding torque in the pivot of 0.05 N · m. At what retarding torque would the sprinkler not be able to rotate?

***4.191** Water flows in a uniform flow out of the 2.5-mm slots of the rotating spray system, as shown. The flow rate is 3 L/s. Find (a) the torque required to hold the system stationary and (b) the steady-state speed of rotation after it is released.

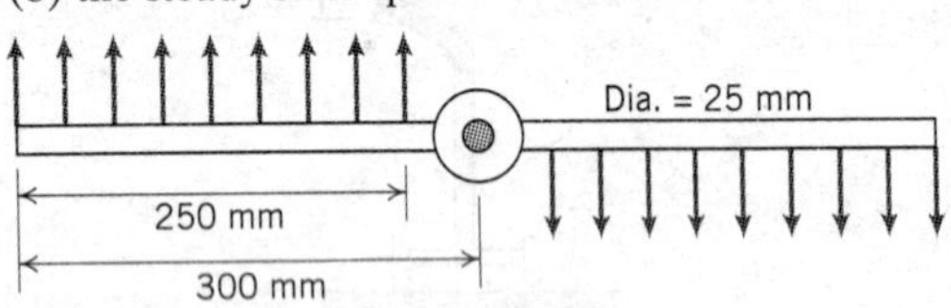

P4.191, P4.192

***4.192** If the same flow rate in the rotating spray system of Problem 4.191 is not uniform but instead varies linearly from a maximum at the outer radius to zero at the inner radius, find (a) the torque required to hold it stationary and (b) the steady-state speed of rotation.

***4.193** A single tube carrying water rotates at constant angular speed, as shown. Water is pumped through the tube at volume flow rate $Q = 13.8$ L/min. Find the torque that must be applied to maintain the steady rotation of the tube using two methods of analysis: (a) a rotating control volume and (b) a fixed control volume.

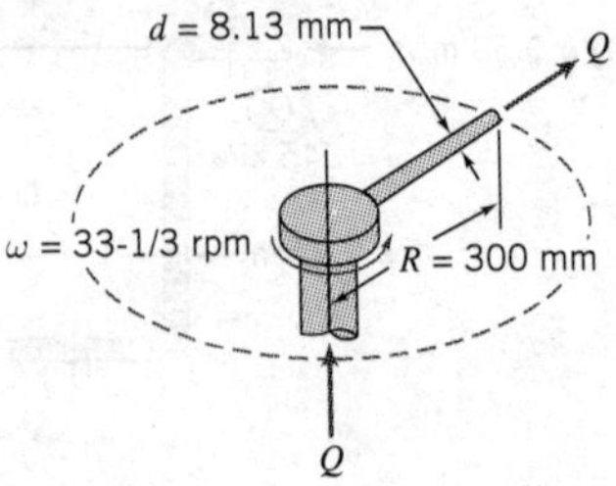

P4.193

***4.194** The lawn sprinkler shown is supplied with water at a rate of 68 L/min. Neglecting friction in the pivot, determine the steady-state angular speed for $\theta = 30°$. Plot the steady-state angular speed of the sprinkler for $0 \le \theta \le 90°$.

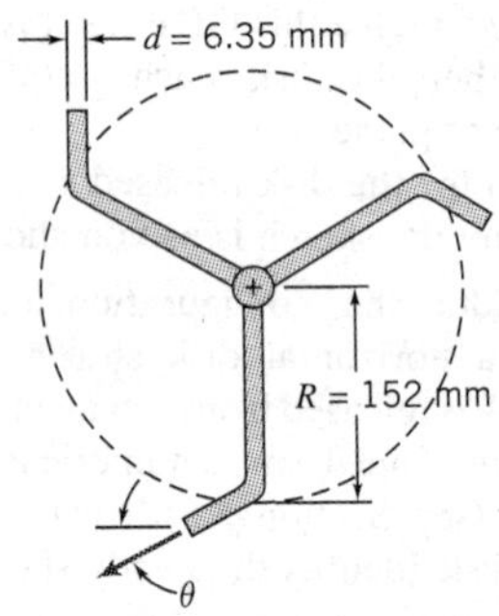

P4.194

***4.195** A small lawn sprinkler is shown. The sprinkler operates at a gage pressure of 140 kPa. The total flow rate of water through the sprinkler is 4 L/min. Each jet discharges at 17 m/s (relative to the sprinkler arm) in a direction inclined 30° above the horizontal. The sprinkler rotates about a vertical axis. Friction in the bearing causes a torque of 0.18 N · m opposing rotation. Evaluate the torque required to hold the sprinkler stationary.

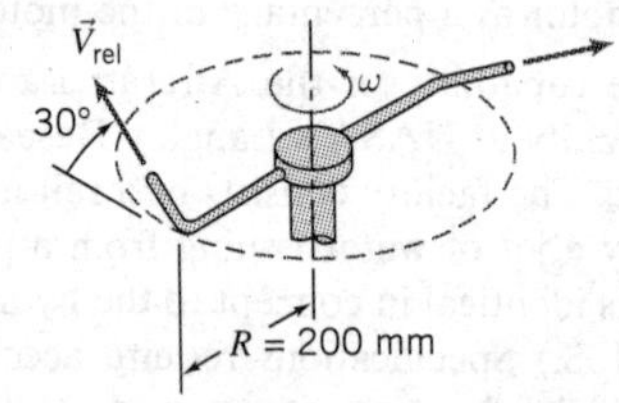

P4.195, P4.196, P4.197

***4.196** In Problem 4.195, calculate the initial acceleration of the sprinkler from rest if no external torque is applied and the moment of inertia of the sprinkler head is 0.1 kg · m^2 when filled with water.

***4.197** A small lawn sprinkler is shown (Problem 4.196). The sprinkler operates at an inlet gage pressure of 140 kPa. The total flow rate of water through the sprinkler is 4.0 L/min. Each jet discharges at 17 m/s (relative to the sprinkler arm) in a direction inclined 30° above the horizontal. The sprinkler rotates about a vertical axis. Friction in the bearing causes a torque of 0.18 N · m opposing rotation. Determine the steady speed of rotation of the sprinkler and the approximate area covered by the spray.

***4.198** When a garden hose is used to fill a bucket, water in the bucket may develop a swirling motion. Why does this happen? How could the amount of swirl be calculated approximately?

***4.199** Water flows at the rate of 0.15 m^3/s through a nozzle assembly that rotates steadily at 30 rpm. The arm and nozzle masses are negligible compared with the water inside. Determine the torque required to drive the device and the reaction torques at the flange.

*These problems require material from sections that may be omitted without loss of continuity in the text material.

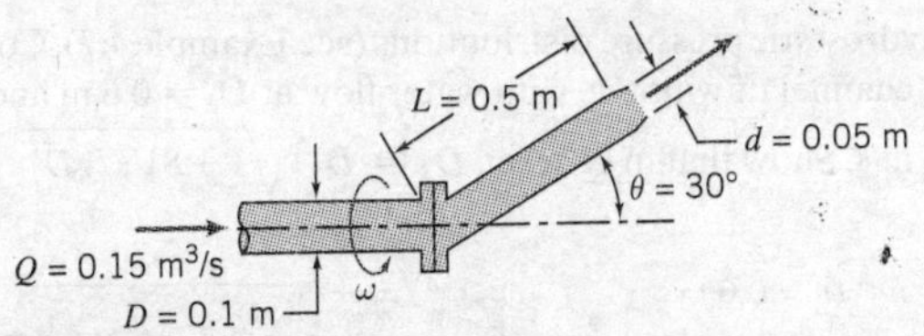

P4.199

***4.200** A pipe branches symmetrically into two legs of length L, and the whole system rotates with angular speed ω around its axis of symmetry. Each branch is inclined at angle α to the axis of rotation. Liquid enters the pipe steadily, with zero angular momentum, at volume flow rate Q. The pipe diameter, D, is much smaller than L. Obtain an expression for the external torque required to turn the pipe. What additional torque would be required to impart angular acceleration $\dot{\omega}$?

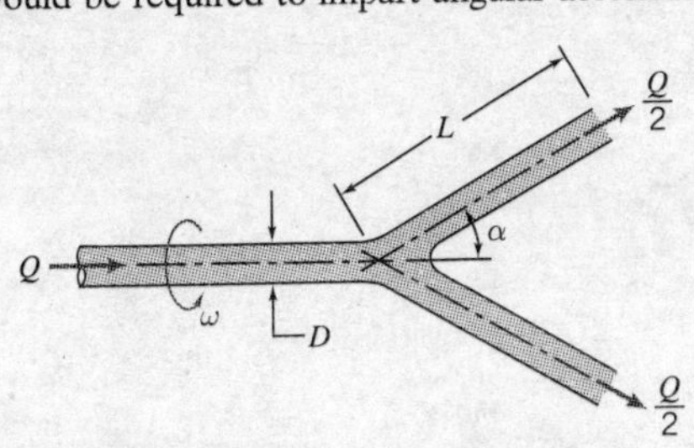

P4.200

***4.201** Liquid in a thin sheet, of width w and thickness h, flows from a slot and strikes a stationary inclined flat plate, as shown. Experiments show that the resultant force of the liquid jet on the plate does not act through point O, where the jet centerline intersects the plate. Determine the magnitude and line of application of the resultant force as functions of θ. Evaluate the equilibrium angle of the plate if the resultant force is applied at point O. Neglect any viscous effects.

***4.202** For the rotating sprinkler of Example 4.14, what value of α will produce the maximum rotational speed? What angle will provide the maximum area of coverage by the spray? Draw a velocity diagram (using an r, θ, z coordinate system) to indicate the absolute velocity of the water jet leaving the nozzle. What governs the steady rotational speed of the sprinkler? Does the rotational speed of the sprinkler affect the area covered by the spray? How would you estimate the area? For fixed α, what might be done to increase or decrease the area covered by the spray?

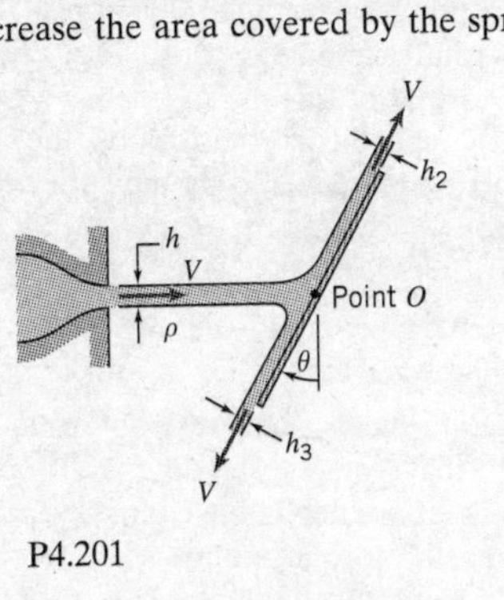

P4.201

The First Law of Thermodynamics

4.203 Air at standard conditions enters a compressor at 75 m/s and leaves at an absolute pressure and temperature of 200 kPa and 345 K, respectively, and speed $V = 125$ m/s. The flow rate is 1 kg/s. The cooling water circulating around the compressor casing removes 18 kJ/kg of air. Determine the power required by the compressor.

4.204 Compressed air is stored in a pressure bottle with a volume of 100 L, at 500 kPa and 20°C. At a certain instant, a valve is opened and mass flows from the bottle at $\dot{m} = 0.01$ kg/s. Find the rate of change of temperature in the bottle at this instant.

4.205 A centrifugal water pump with a 0.1 m diameter inlet and a 0.1 m diameter discharge pipe has a flow rate of 0.02 m^3/s. The inlet pressure is 0.2 m Hg vacuum and the exit pressure is 240 kPa. The inlet and outlet sections are located at the same elevation. The measured power input is 6.75 kW. Determine the pump efficiency.

4.206 A turbine is supplied with 0.6 m^3/s of water from a 0.3 m diameter pipe; the discharge pipe has a 0.4 m diameter. Determine the pressure drop across the turbine if it delivers 60 kW.

4.207 Air enters a compressor at 96 kPa, 27°C with negligible speed and is discharged at 480 kPa, 260°C with a speed of 152 m/s. If the power input is 2.39 MW and the flow rate is 9 kg/s, determine the rate of heat transfer.

4.208 Air is drawn from the atmosphere into a turbomachine. At the exit, conditions are 500 kPa (gage) and 130°C. The exit speed is 100 m/s and the mass flow rate is 0.8 kg/s. Flow is steady and there is no heat transfer. Compute the shaft work interaction with the surroundings.

4.209 All major harbors are equipped with fire boats for extinguishing ship fires. A 75 mm diameter hose is attached to the discharge of a 11 kW pump on such a boat. The nozzle attached to the end of the hose has a diameter of 25 mm. If the nozzle discharge is held 3 m above the surface of the water, determine the volume flow rate through the nozzle, the maximum height to which the water will rise, and the force on the boat if the water jet is directed horizontally over the stern.

4.210 A pump draws water from a reservoir through a 150 mm diameter suction pipe and delivers it to a 75 mm diameter discharge pipe. The end of the suction pipe is 2 m below the free surface of the reservoir. The pressure gage on the discharge pipe (2 m above the reservoir surface) reads 170 kPa. The average speed in the discharge pipe is 3 m/s. If the pump efficiency is 75 percent, determine the power required to drive it.

4.211 The total mass of the helicopter-type craft shown is 1000 kg. The pressure of the air is atmospheric at the outlet. Assume the flow is steady and one-dimensional. Treat the air as incompressible at standard conditions and calculate, for a hovering position, the speed of the air leaving the craft and the minimum power that must be delivered to the air by the propeller.

*These problems require material from sections that may be omitted without loss of continuity in the text material.

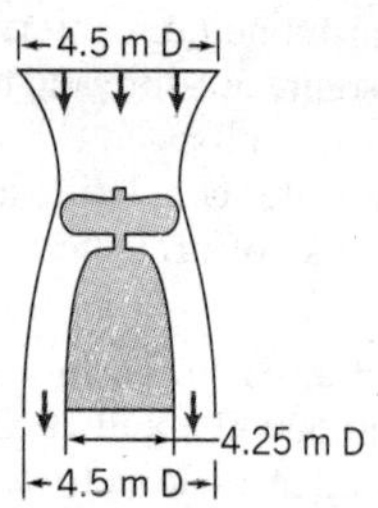

P4.211

4.212 Liquid flowing at high speed in a wide, horizontal open channel under some conditions can undergo a hydraulic jump, as shown. For a suitably chosen control volume, the flows entering and leaving the jump may be considered uniform with hydrostatic pressure distributions (see Example 4.7). Consider a channel of width w, with water flow at $D_1 = 0.6$ m and $V_1 = 5$ m/s. Show that in general, $D_2 = D_1\left[\sqrt{1+8V_1^2/gD_1}-1\right]/2$.

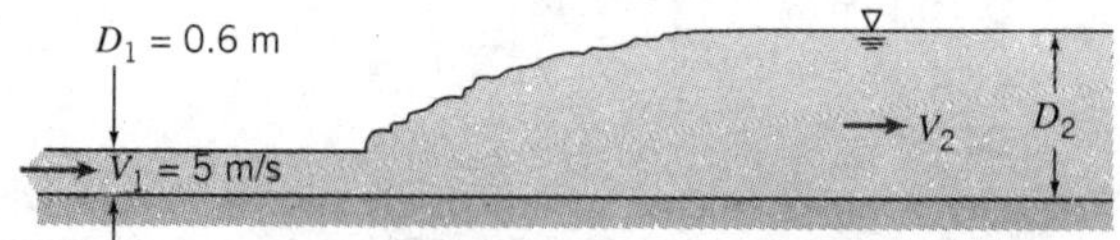

P4.212

Evaluate the change in mechanical energy through the hydraulic jump. If heat transfer to the surroundings is negligible, determine the change in water temperature through the jump.

5 Introduction to Differential Analysis of Fluid Motion

Case Study in Energy and the Environment

Wave Power: *Aquamarine Oyster* *Wave Energy Converter*

Aquamarine Power, a wave energy company located in Scotland, has developed an innovative hydroelectric wave energy converter, known as *Oyster*; a demonstration-scale model was installed in 2009 and began producing power for homes in some regions of Scotland. They eventually plan to have commercially viable *Oyster* wave farms around the world, the first planned for 2013. A farm of 20 *Oyster* devices would provide enough energy to power 9000 homes, offsetting carbon emissions of about 20,000 metric tons.

The *Oyster* device consists of a simple mechanical hinged flap, as shown in the figure, connected to the seabed at around a 10-m depth. As each wave passes by, it forces the flap to move; the flap in turn drives

A schematic of Aquamarine's Oyster device (Picture courtesy of Aquamarine Power)

hydraulic pistons to deliver high-pressure water, via a pipeline, to an onshore electrical turbine. *Oyster* farms using multiple devices are expected to be capable of generating 100 MW or more.

Oyster has a number of advantages: It has good efficiency and durability, and, with its low-cost operation, maintenance, and manufacture, it is hoped it will produce reliable *cost-competitive* electricity from the waves for the first time. The device uses simple and robust mechanical offshore component, combined with proven conventional onshore hydroelectric components. Designed with the notion that simple is best, less is more, it has a minimum of offshore submerged moving parts; there are no underwater generators, power electronics, or gearboxes. The *Oyster* is designed to take advantage of the more consistent waves found near the shore; for durability, any excess energy from exceptionally large waves simply spills over the top of *Oyster*'s flap. Its motion allows it to literally duck under such waves. *Aquamarine Power* believes its device is competitive with devices weighing up to five times as much, and, with multiple pumps feeding a single onshore generator, *Oyster* will offer good economies of scale. As a final bonus, *Oyster* uses water instead of oil as its hydraulic fluid for minimum environmental impact and generates essentially no noise pollution.

In Chapter 4, we developed the basic equations in integral form for a control volume. Integral equations are useful when we are interested in the gross behavior of a flow field and its effect on various devices. However, the integral approach does not enable us to obtain detailed point-by-point knowledge of the flow field. For example, the integral approach could provide information on the lift generated by a wing; it could not be used to determine the pressure distribution that produced the lift on the wing.

To see what is happening in a flow in detail, we need differential forms of the equations of motion. In this chapter we shall develop differential equations for the conservation of mass and Newton's second law of motion. Since we are interested in developing differential equations, we will need to analyze infinitesimal systems and control volumes.

5.1 Conservation of Mass

In Chapter 2, we developed the field representation of fluid properties. The property fields are defined by continuous functions of the space coordinates and time. The density and velocity fields were related through conservation of mass in integral form in Chapter 4 (Eq. 4.12). In this chapter we shall derive the differential equation for conservation of mass in rectangular and in cylindrical coordinates. In both cases the derivation is carried out by applying conservation of mass to a differential control volume.

Rectangular Coordinate System

In rectangular coordinates, the control volume chosen is an infinitesimal cube with sides of length dx, dy, dz as shown in Fig. 5.1. The density at the center, O, of the control volume is assumed to be ρ and the velocity there is assumed to be $\vec{V} = \hat{i}u + \hat{j}v + \hat{k}w$.

To evaluate the properties at each of the six faces of the control surface, we use a Taylor series expansion about point O. For example, at the right face,

$$\rho)_{x+dx/2} = \rho + \left(\frac{\partial \rho}{\partial x}\right)\frac{dx}{2} + \left(\frac{\partial^2 \rho}{\partial x^2}\right)\frac{1}{2!}\left(\frac{dx}{2}\right)^2 + \cdots$$

Neglecting higher-order terms, we can write

$$\rho)_{x+dx/2} = \rho + \left(\frac{\partial \rho}{\partial x}\right)\frac{dx}{2}$$

and

$$u)_{x+dx/2} = u + \left(\frac{\partial u}{\partial x}\right)\frac{dx}{2}$$

where ρ, u, $\partial\rho/\partial x$, and $\partial u/\partial x$ are all evaluated at point O. The corresponding terms at the left face are

$$\rho)_{x-dx/2} = \rho + \left(\frac{\partial \rho}{\partial x}\right)\left(-\frac{dx}{2}\right) = \rho - \left(\frac{\partial \rho}{\partial x}\right)\frac{dx}{2}$$

$$u)_{x-dx/2} = u + \left(\frac{\partial u}{\partial x}\right)\left(-\frac{dx}{2}\right) = u - \left(\frac{\partial u}{\partial x}\right)\frac{dx}{2}$$

We can write similar expressions involving ρ and v for the front and back faces and ρ and w for the top and bottom faces of the infinitesimal cube $dx\,dy\,dz$. These can then be used to evaluate the surface integral in Eq. 4.12 (recall that $\int_{CS}\rho\vec{V}\cdot d\vec{A}$ is the net flux of mass out of the control volume):

$$\frac{\partial}{\partial t}\int_{CV}\rho d\forall + \int_{CS}\rho\vec{V}\cdot d\vec{A} = 0 \tag{4.12}$$

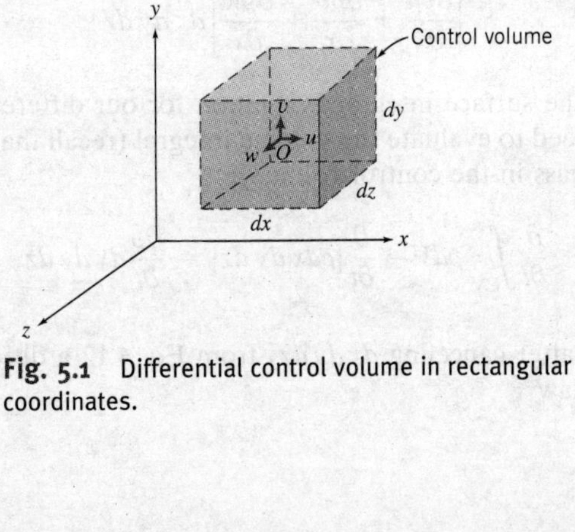

Fig. 5.1 Differential control volume in rectangular coordinates.

Table 5.1

Mass Flux Through the Control Surface of a Rectangular Differential Control Volume

Surface	Evaluation of $\int \rho \vec{V} \cdot d\vec{A}$
Left $(-x)$	$= -\left[\rho - \left(\frac{\partial \rho}{\partial x}\right)\frac{dx}{2}\right]\left[u - \left(\frac{\partial u}{\partial x}\right)\frac{dx}{2}\right] dy\,dz = -\rho u\,dy\,dz + \frac{1}{2}\left[u\left(\frac{\partial \rho}{\partial x}\right) + \rho\left(\frac{\partial u}{\partial x}\right)\right] dx\,dy\,dz$
Right $(+x)$	$= \left[\rho + \left(\frac{\partial \rho}{\partial x}\right)\frac{dx}{2}\right]\left[u + \left(\frac{\partial u}{\partial x}\right)\frac{dx}{2}\right] dy\,dz = \rho u\,dy\,dz + \frac{1}{2}\left[u\left(\frac{\partial \rho}{\partial x}\right) + \rho\left(\frac{\partial u}{\partial x}\right)\right] dx\,dy\,dz$
Bottom $(-y)$	$= -\left[\rho - \left(\frac{\partial \rho}{\partial y}\right)\frac{dy}{2}\right]\left[v - \left(\frac{\partial v}{\partial y}\right)\frac{dy}{2}\right] dx\,dz = -\rho v\,dx\,dz + \frac{1}{2}\left[v\left(\frac{\partial \rho}{\partial y}\right) + \rho\left(\frac{\partial v}{\partial y}\right)\right] dx\,dy\,dz$
Top $(+y)$	$= \left[\rho + \left(\frac{\partial \rho}{\partial y}\right)\frac{dy}{2}\right]\left[v + \left(\frac{\partial v}{\partial y}\right)\frac{dy}{2}\right] dx\,dz = \rho v\,dx\,dz + \frac{1}{2}\left[v\left(\frac{\partial \rho}{\partial y}\right) + \rho\left(\frac{\partial v}{\partial y}\right)\right] dx\,dy\,dz$
Back $(-z)$	$= -\left[\rho - \left(\frac{\partial \rho}{\partial z}\right)\frac{dz}{2}\right]\left[w - \left(\frac{\partial w}{\partial z}\right)\frac{dz}{2}\right] dx\,dy = -\rho w\,dx\,dy + \frac{1}{2}\left[w\left(\frac{\partial \rho}{\partial z}\right) + \rho\left(\frac{\partial w}{\partial z}\right)\right] dx\,dy\,dz$
Front $(+z)$	$= \left[\rho + \left(\frac{\partial \rho}{\partial z}\right)\frac{dz}{2}\right]\left[w + \left(\frac{\partial w}{\partial z}\right)\frac{dz}{2}\right] dx\,dy = \rho w\,dx\,dy + \frac{1}{2}\left[w\left(\frac{\partial \rho}{\partial z}\right) + \rho\left(\frac{\partial w}{\partial z}\right)\right] dx\,dy\,dz$

Adding the results for all six faces,

$$\int_{CS} \rho \vec{V} \cdot d\vec{A} = \left[\left\{u\left(\frac{\partial \rho}{\partial x}\right) + \rho\left(\frac{\partial u}{\partial x}\right)\right\} + \left\{v\left(\frac{\partial \rho}{\partial y}\right) + \rho\left(\frac{\partial v}{\partial y}\right)\right\} + \left\{w\left(\frac{\partial \rho}{\partial z}\right) + \rho\left(\frac{\partial w}{\partial z}\right)\right\}\right] dx\,dy\,dz$$

or

$$\int_{CS} \rho \vec{V} \cdot d\vec{A} = \left[\frac{\partial \rho u}{\partial x} + \frac{\partial \rho v}{\partial y} + \frac{\partial \rho w}{\partial z}\right] dx\,dy\,dz$$

Table 5.1 shows the details of this evaluation. Note: We assume that the velocity components u, v, and w are positive in the x, y, and z directions, respectively; the area normal is by convention positive out of the cube; and higher-order terms [e.g., $(dx)^2$] are neglected in the limit as dx, dy, and $dz \rightarrow 0$.

The result of all this work is

$$\left[\frac{\partial \rho u}{\partial x} + \frac{\partial \rho v}{\partial x} + \frac{\partial \rho w}{\partial x}\right] dx\,dy\,dz$$

This expression is the surface integral evaluation for our differential cube. To complete Eq. 4.12, we need to evaluate the volume integral (recall that $\partial/\partial t \int_{CV} \rho d\forall$ is the rate of change of mass in the control volume):

$$\frac{\partial}{\partial t}\int_{CV} \rho d\forall \rightarrow \frac{\partial}{\partial t}[\rho dx\,dy\,dz] = \frac{\partial \rho}{\partial t} dx\,dy\,dz$$

Hence, we obtain (after canceling $dx\,dy\,dz$) from Eq. 4.12 a differential form of the mass conservation law

$$\frac{\partial \rho u}{\partial x} + \frac{\partial \rho v}{\partial y} + \frac{\partial \rho w}{\partial z} + \frac{\partial \rho}{\partial t} = 0 \tag{5.1a}$$

Equation 5.1a is frequently called the *continuity equation.*

Since the vector operator, ∇, in rectangular coordinates, is given by

$$\nabla = \hat{i}\,\frac{\partial}{\partial x} + \hat{j}\,\frac{\partial}{\partial y} + \hat{k}\,\frac{\partial}{\partial z}$$

then

$$\frac{\partial \rho u}{\partial x} + \frac{\partial \rho v}{\partial y} + \frac{\partial \rho w}{\partial z} = \nabla \cdot \rho \vec{V}$$

Note that the del operator ∇ acts on ρ *and* $\vec{V}$. Think of it as $\nabla \cdot (\rho \vec{V})$. The conservation of mass may be written as

$$\nabla \cdot \rho \vec{V} + \frac{\partial \rho}{\partial t} = 0 \tag{5.1b}$$

Two flow cases for which the differential continuity equation may be simplified are worthy of note.

For an *incompressible* fluid, $\rho =$ constant; density is neither a function of space coordinates nor a function of time. For an incompressible fluid, the continuity equation simplifies to

$$\frac{\partial u}{\partial x} + \frac{\partial v}{\partial y} + \frac{\partial w}{\partial z} = \nabla \cdot \vec{V} = 0 \tag{5.1c}$$

Thus the velocity field, $\vec{V}(x, y, z, t)$, for incompressible flow must satisfy $\nabla \cdot \vec{V} = 0$.

For *steady* flow, all fluid properties are, by definition, independent of time. Thus $\partial\rho/\partial t = 0$ and at most $\rho = \rho(x, y, z)$. For steady flow, the continuity equation can be written as

$$\frac{\partial \rho u}{\partial x} + \frac{\partial \rho v}{\partial y} + \frac{\partial \rho w}{\partial z} = \nabla \cdot \rho \vec{V} = 0 \tag{5.1d}$$

(and remember that the del operator ∇ acts on ρ *and* $\vec{V}$).

Example 5.1 INTEGRATION OF TWO-DIMENSIONAL DIFFERENTIAL CONTINUITY EQUATION

For a two-dimensional flow in the xy plane, the x component of velocity is given by $u = Ax$. Determine a possible y component for incompressible flow. How many y components are possible?

Given: Two-dimensional flow in the xy plane for which $u = Ax$.

Find: (a) Possible y component for incompressible flow.
(b) Number of possible y components.

Solution:

Governing equation: $\nabla \cdot \rho \vec{V} + \dfrac{\partial \rho}{\partial t} = 0$

For incompressible flow this simplifies to $\nabla \cdot \vec{V} = 0$. In rectangular coordinates

$$\frac{\partial u}{\partial x} + \frac{\partial v}{\partial y} + \frac{\partial w}{\partial z} = 0$$

For two-dimensional flow in the xy plane, $\vec{V} = \vec{V}(x, y)$. Then partial derivatives with respect to z are zero, and

$$\frac{\partial u}{\partial x} + \frac{\partial v}{\partial y} = 0$$

Then

$$\frac{\partial v}{\partial y} = -\frac{\partial u}{\partial x} = -A$$

which gives an expression for the rate of change of v holding x constant. This equation can be integrated to obtain an expression for v. The result is

$$v = \int \frac{\partial v}{\partial y} dy + f(x, t) = -Ay + f(x, t) \qquad v$$

{The function of x and t appears because we had a partial derivative of v with respect to y.}

Any function $f(x, t)$ is allowable, since $\partial/\partial y \, f(x, t) = 0$. Thus any number of expressions for v could satisfy the differential continuity equation under the given conditions. The simplest expression for v would be obtained by setting $f(x, t) = 0$. Then $v = -Ay$, and

$$\vec{V} = Ax\hat{i} - Ay\hat{j} \qquad \vec{V}$$

This problem:

- ✓ Shows use of the differential continuity equation for obtaining information on a flow field.
- ✓ Demonstrates integration of a partial derivative.
- ✓ Proves that the flow originally discussed in Example 2.1 is indeed incompressible.

Example 5.2 UNSTEADY DIFFERENTIAL CONTINUITY EQUATION

A gas-filled pneumatic strut in an automobile suspension system behaves like a piston-cylinder apparatus. At one instant when the piston is $L = 0.15$ m away from the closed end of the cylinder, the gas density is uniform at $\rho = 18 \text{ kg/m}^3$ and the piston begins to move away from the closed end at $V = 12$ m/s. Assume as a simple model that the gas velocity is one-dimensional and proportional to distance from the closed end; it varies linearly from zero at the end to $u = V$ at the piston. Find the rate of change of gas density at this instant. Obtain an expression for the average density as a function of time.

Given: Piston-cylinder as shown.

Find: (a) Rate of change of density.
(b) $\rho(t)$.

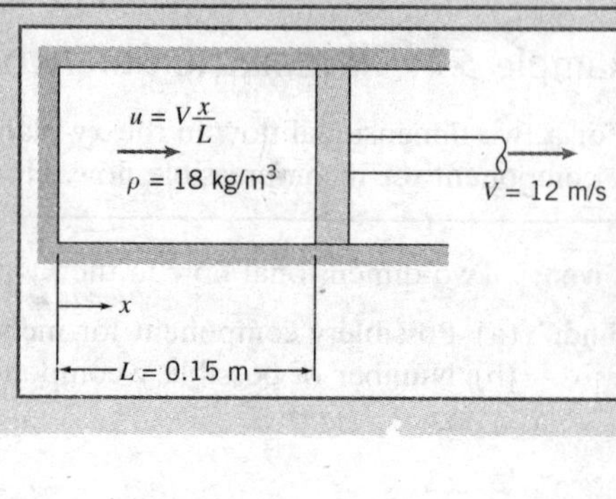

Solution:

Governing equation: $\nabla \cdot \rho \vec{V} + \dfrac{\partial \rho}{\partial t} = 0$

In rectangular coordinates, $\frac{\partial \rho u}{\partial x} + \frac{\partial \rho v}{\partial y} + \frac{\partial \rho w}{\partial z} + \frac{\partial \rho}{\partial t} = 0$

Since $u = u(x)$, partial derivatives with respect to y and z are zero, and

$$\frac{\partial \rho u}{\partial x} + \frac{\partial \rho}{\partial t} = 0$$

Then

$$\frac{\partial \rho}{\partial t} = -\frac{\partial \rho u}{\partial x} = -\rho\frac{\partial u}{\partial x} - u\frac{\partial \rho}{\partial x}$$

Since ρ is assumed uniform in the volume, $\frac{\partial \rho}{\partial x} = 0$, and $\frac{\partial \rho}{\partial t} = \frac{d\rho}{dt} = -\rho\frac{\partial u}{\partial x}$.

Since $u = V\frac{x}{L}$, $\frac{\partial u}{\partial x} = \frac{V}{L}$, then $\frac{d\rho}{dt} = -\rho\frac{V}{L}$. However, note that $L = L_0 + Vt$.

Separate variables and integrate,

$$\int_{\rho_0}^{\rho} \frac{d\rho}{\rho} = -\int_0^t \frac{V}{L}dt = -\int_0^t \frac{V\,dt}{L_0 + Vt}$$

$$\ln\frac{\rho}{\rho_0} = \ln\frac{L_0}{L_0 + Vt} \quad \text{and} \quad \rho(t) = \rho_0\left[\frac{1}{1 + Vt/L_0}\right] \longleftarrow \rho(t)$$

This problem demonstrates use of the differential continuity equation for obtaining the density variation with time for an unsteady flow.

The density-time graph is shown in an *Excel* workbook. This workbook is interactive: It allows one to see the effect of different values of ρ_o, L, and V on ρ versus t. Also, the time at which the density falls to any prescribed value can be determined.

At $t = 0$,

$$\frac{\partial \rho}{\partial t} = -\rho_0\frac{V}{L} = -18\frac{\text{kg}}{\text{m}^3} \times 12\frac{\text{m}}{\text{s}} \times \frac{1}{0.15\,\text{m}} = -1440\,\text{kg}/(\text{m}^3\cdot\text{s}) \longleftarrow \frac{\partial \rho}{\partial t}$$

Cylindrical Coordinate System

A suitable differential control volume for cylindrical coordinates is shown in Fig. 5.2. The density at the center, O, of the control volume is assumed to be ρ and the velocity there is assumed to be $\vec{V} = \hat{e}_r V_r + \hat{e}_\theta V_\theta + \hat{k} V_z$, where $\hat{e}_r$, $\hat{e}_\theta$, and $\hat{k}$ are unit vectors in the r, θ, and z directions, respectively, and V_r, V_θ, and V_z are the velocity components in the r, θ, and z directions, respectively. To evaluate $\int_{CS} \rho\vec{V}\cdot d\vec{A}$, we must consider the mass flux through each of the six faces of the control surface. The properties at each of the six faces of the control surface are obtained from a Taylor series expansion about point O. The details of the mass flux evaluation are shown in Table 5.2. Velocity components V_r, V_θ, and V_z are all assumed to be in the positive coordinate directions and we have again used the convention that the area normal is positive outwards on each face, and higher-order terms have been neglected.

We see that the net rate of mass flux out through the control surface (the term $\int_{CS} \rho\vec{V}\cdot d\vec{A}$ in Eq. 4.12) is given by

$$\left[\rho V_r + r\frac{\partial \rho V_r}{\partial r} + \frac{\partial \rho V_\theta}{\partial \theta} + r\frac{\partial \rho V_z}{\partial z}\right] dr\, d\theta\, dz$$

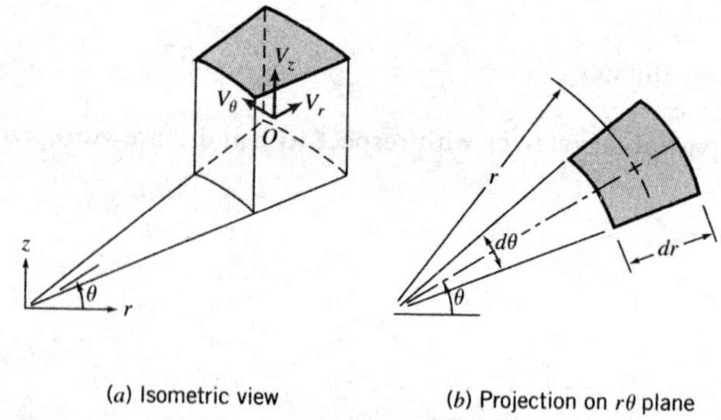

Fig. 5.2 Differential control volume in cylindrical coordinates.

The mass inside the control volume at any instant is the product of the mass per unit volume, ρ, and the volume, $rd\theta\,dr\,dz$. Thus the rate of change of mass inside the control volume (the term $\partial/\partial t\int_{CV}\rho d\forall$ in Eq. 4.12) is given by

$$\frac{\partial\rho}{\partial t}r\,d\theta\,dr\,dz$$

In cylindrical coordinates the differential equation for conservation of mass is then

$$\rho V_r + r\frac{\partial\rho V_r}{\partial r} + \frac{\partial\rho V_\theta}{\partial\theta} + r\frac{\partial\rho V_z}{\partial z} + r\frac{\partial\rho}{\partial t} = 0$$

or

$$\frac{\partial(r\rho V_r)}{\partial r} + \frac{\partial\rho V_\theta}{\partial\theta} + r\frac{\partial\rho V_z}{\partial z} + r\frac{\partial\rho}{\partial t} = 0$$

Dividing by r gives

$$\frac{1}{r}\frac{\partial(r\rho V_r)}{\partial r} + \frac{1}{r}\frac{\partial(\rho V_\theta)}{\partial\theta} + \frac{\partial(\rho V_z)}{\partial z} + \frac{\partial\rho}{\partial t} = 0 \tag{5.2a}$$

In cylindrical coordinates the vector operator ∇ is given by

$$\nabla = \hat{e}_r\frac{\partial}{\partial r} + \hat{e}_\theta\frac{1}{r}\frac{\partial}{\partial\theta} + \hat{k}\frac{\partial}{\partial z} \tag{3.19}$$

Equation 5.2a also may be written[1] in vector notation as

$$\nabla\cdot\rho\vec{V} + \frac{\partial\rho}{\partial t} = 0 \tag{5.1b}$$

For an *incompressible* fluid, ρ = constant, and Eq. 5.2a reduces to

$$\frac{1}{r}\frac{\partial(rV_r)}{\partial r} + \frac{1}{r}\frac{\partial V_\theta}{\partial\theta} + \frac{\partial V_z}{\partial z} = \nabla\cdot\vec{V} = 0 \tag{5.2b}$$

[1]To evaluate $\nabla\cdot\rho\vec{V}$ in cylindrical coordinates, we must remember that

$$\frac{\partial\hat{e}_r}{\partial\theta} = \hat{e}_\theta \quad\text{and}\quad \frac{\partial\hat{e}_\theta}{\partial\theta} = -\hat{e}_r$$

Table 5.2

Mass Flux Through the Control Surface of a Cylindrical Differential Control Volume

Surface	Evaluation of $\int \rho \vec{V} \cdot d\vec{A}$
Inside $(-r)$	$= -\left[\rho - \left(\frac{\partial \rho}{\partial r}\right)\frac{dr}{2}\right]\left[V_r - \left(\frac{\partial V_r}{\partial r}\right)\frac{dr}{2}\right]\left(r - \frac{dr}{2}\right)d\theta\, dz = -\rho V_r\, r d\theta\, dz + \rho V_r \frac{dr}{2} d\theta\, dz + \rho\left(\frac{\partial V_r}{\partial r}\right) r\frac{dr}{2} d\theta\, dz + V_r\left(\frac{\partial \rho}{\partial r}\right) r \frac{dr}{2} d\theta\, dz$
Outside $(+r)$	$= \left[\rho + \left(\frac{\partial \rho}{\partial r}\right)\frac{dr}{2}\right]\left[V_r + \left(\frac{\partial V_r}{\partial r}\right)\frac{dr}{2}\right]\left(r + \frac{dr}{2}\right)d\theta\, dz = \rho V_r\, r d\theta\, dz + \rho V_r \frac{dr}{2} d\theta\, dz + \rho\left(\frac{\partial V_r}{\partial r}\right) r\frac{dr}{2} d\theta\, dz + V_r\left(\frac{\partial \rho}{\partial r}\right) r \frac{dr}{2} d\theta\, dz$
Front $(-\theta)$	$= -\left[\rho - \left(\frac{\partial \rho}{\partial \theta}\right)\frac{d\theta}{2}\right]\left[V_\theta - \left(\frac{\partial V_\theta}{\partial \theta}\right)\frac{d\theta}{2}\right]dr\, dz = -\rho V_\theta\, dr\, dz + \rho\left(\frac{\partial V_\theta}{\partial \theta}\right)\frac{d\theta}{2} dr\, dz + V_\theta\left(\frac{\partial \rho}{\partial \theta}\right)\frac{d\theta}{2} dr\, dz$
Back $(+\theta)$	$= \left[\rho + \left(\frac{\partial \rho}{\partial \theta}\right)\frac{d\theta}{2}\right]\left[V_\theta + \left(\frac{\partial V_\theta}{\partial \theta}\right)\frac{d\theta}{2}\right]dr\, dz = \rho V_\theta\, dr\, dz + \rho\left(\frac{\partial V_\theta}{\partial \theta}\right)\frac{d\theta}{2} dr\, dz + V_\theta\left(\frac{\partial \rho}{\partial \theta}\right)\frac{d\theta}{2} dr\, dz$
Bottom $(-z)$	$= -\left[\rho - \left(\frac{\partial \rho}{\partial z}\right)\frac{dz}{2}\right]\left[V_z - \left(\frac{\partial V_z}{\partial z}\right)\frac{dz}{2}\right] r d\theta\, dr = -\rho V_z r d\theta\, dr + \rho\left(\frac{\partial V_z}{\partial z}\right)\frac{dz}{2} r d\theta\, dr + V_z\left(\frac{\partial \rho}{\partial z}\right)\frac{dz}{2} r d\theta\, dr$
Top $(+z)$	$= \left[\rho + \left(\frac{\partial \rho}{\partial z}\right)\frac{dz}{2}\right]\left[V_z + \left(\frac{\partial V_z}{\partial z}\right)\frac{dz}{2}\right] r d\theta\, dr = \rho V_z r d\theta\, dr + \rho\left(\frac{\partial V_z}{\partial z}\right)\frac{dz}{2} r d\theta\, dr + V_z\left(\frac{\partial \rho}{\partial z}\right)\frac{dz}{2} r d\theta\, dr$

Adding the results for all six faces,

$$\int_{CS} \rho \vec{V} \cdot d\vec{A} = \left[\rho V_r + r\left\{\rho\left(\frac{\partial V_r}{\partial r}\right) + V_r\left(\frac{\partial \rho}{\partial r}\right)\right\} + \left\{\rho\left(\frac{\partial V_\theta}{\partial \theta}\right) + V_\theta\left(\frac{\partial \rho}{\partial \theta}\right)\right\} + r\left\{\rho\left(\frac{\partial V_z}{\partial z}\right) + V_z\left(\frac{\partial \rho}{\partial z}\right)\right\}\right] dr\, d\theta\, dz$$

or

$$\int_{CS} \rho \vec{V} \cdot d\vec{A} = \left[\rho V_r + r\frac{\partial \rho V_r}{\partial r} + \frac{\partial \rho V_\theta}{\partial \theta} + r\frac{\partial \rho V_z}{\partial z}\right] dr\, d\theta\, dz$$

Thus the velocity field, $\vec{V}(x, y, z, t)$, for incompressible flow must satisfy $\nabla \cdot \vec{V} = 0$. For *steady* flow, Eq. 5.2a reduces to

$$\frac{1}{r}\frac{\partial(r\rho V_r)}{\partial r} + \frac{1}{r}\frac{\partial(\rho V_\theta)}{\partial \theta} + \frac{\partial(\rho V_z)}{\partial z} = \nabla \cdot \rho\vec{V} = 0 \tag{5.2c}$$

(and remember once again that the del operator ∇ acts on ρ *and* $\vec{V}$).

When written in vector form, the differential continuity equation (the mathematical statement of conservation of mass), Eq. 5.1b, may be applied in any coordinate system. We simply substitute the appropriate expression for the vector operator ∇. In retrospect, this result is not surprising since mass must be conserved regardless of our choice of coordinate system.

Example 5.3 DIFFERENTIAL CONTINUITY EQUATION IN CYLINDRICAL COORDINATES

Consider a one-dimensional radial flow in the $r\theta$ plane, given by $V_r = f(r)$ and $V_\theta = 0$. Determine the conditions on $f(r)$ required for the flow to be incompressible.

Given: One-dimensional radial flow in the $r\theta$ plane: $V_r = f(r)$ and $V_\theta = 0$.

Find: Requirements on $f(r)$ for incompressible flow.

Solution:

Governing equation: $\nabla \cdot \rho\vec{V} + \frac{\partial \rho}{\partial t} = 0$

For incompressible flow in cylindrical coordinates this reduces to Eq. 5.2b,

$$\frac{1}{r}\frac{\partial}{\partial r}(rV_r) + \frac{1}{r}\frac{\partial}{\partial \theta}V_\theta + \frac{\partial V_z}{\partial z} = 0$$

For the given velocity field, $\vec{V} = \vec{V}(r)$. $V_\theta = 0$ and partial derivatives with respect to z are zero, so

$$\frac{1}{r}\frac{\partial}{\partial r}(rV_r) = 0$$

Integrating with respect to r gives

$$rV_r = \text{constant}$$

Thus the continuity equation shows that the radial velocity must be $V_r = f(r) = C/r$ for one-dimensional radial flow of an incompressible fluid. This is not a surprising result: As the fluid moves outwards from the center, the volume flow rate (per unit depth in the z direction) $Q = 2\pi rV$ at any radius r is constant.

*5.2 Stream Function for Two-Dimensional Incompressible Flow

We already briefly discussed streamlines in Chapter 2, where we stated that they were lines tangent to the velocity vectors in a flow at an instant

*This section may be omitted without loss of continuity in the text material.

$$\left.\frac{dy}{dx}\right|_{\text{streamline}} = \frac{v}{u} \tag{2.8}$$

We can now develop a more formal definition of streamlines by introducing the *stream function*, ψ. This will allow us to represent two entities—the velocity components $u(x, y, t)$ and $v(x, y, t)$ of a two-dimensional incompressible flow—with a single function $\psi(x, y, t)$.

There are various ways to define the stream function. We start with the two-dimensional version of the continuity equation for incompressible flow (Eq. 5.1c):

$$\frac{\partial u}{\partial x} + \frac{\partial v}{\partial y} = 0 \tag{5.3}$$

We use what looks at first like a purely mathematical exercise (we will see a physical basis for it later) and define the stream function by

$$u \equiv \frac{\partial \psi}{\partial y} \quad \text{and} \quad v \equiv -\frac{\partial \psi}{\partial x} \tag{5.4}$$

VIDEO

An Example of Streamlines/Streaklines.

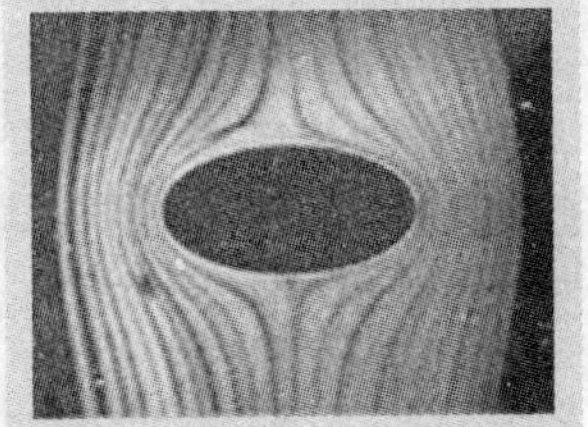

so that Eq. 5.3 is *automatically* satisfied for *any* $\psi(x, y, t)$ we choose! To see this, use Eq. 5.4 in Eq. 5.3:

$$\frac{\partial u}{\partial x} + \frac{\partial v}{\partial y} = \frac{\partial^2 \psi}{\partial x \partial y} - \frac{\partial^2 \psi}{\partial y \partial x} = 0$$

Using Eq. 2.8, we can obtain an equation valid only *along* a streamline

$$u dy - v dx = 0$$

or, using the definition of our stream function,

$$\frac{\partial \psi}{\partial x} dx + \frac{\partial \psi}{\partial y} dy = 0 \tag{5.5}$$

On the other hand, from a strictly mathematical point of view, at any instant in time t the variation in a function $\psi(x, y, t)$ in space (x, y) is given by

$$d\psi = \frac{\partial \psi}{\partial x} dx + \frac{\partial \psi}{\partial y} dy \tag{5.6}$$

Comparing Eqs. 5.5 and 5.6, we see that along an instantaneous streamline, $d\psi = 0$; in other words, *ψ is a constant along a streamline*. Hence we can specify individual streamlines by their stream function values: $\psi = 0, 1, 2$, etc. What is the significance of the ψ values? The answer is that they can be used to obtain the volume flow rate between any two streamlines. Consider the streamlines shown in Fig. 5.3. We can compute the volume flow rate between streamlines ψ_1 and ψ_2 by using line AB, BC, DE, or EF (recall that there is no flow *across* a streamline).

Let us compute the flow rate by using line AB, and also by using line BC—they should be the same!

For a unit depth (dimension perpendicular to the xy plane), the flow rate across AB is

$$Q = \int_{y_1}^{y_2} u\, dy = \int_{y_1}^{y_2} \frac{\partial \psi}{\partial y} dy$$

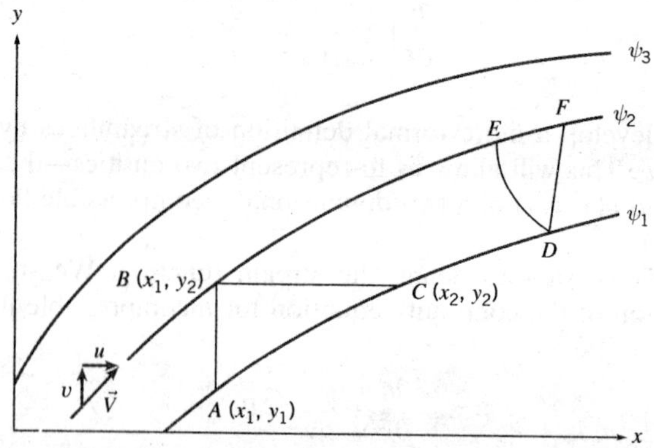

Fig. 5.3 Instantaneous streamlines in a two-dimensional flow.

But along AB, x = constant, and (from Eq. 5.6) $d\psi = \partial\psi/\partial y \; dy$. Therefore,

$$Q = \int_{y_1}^{y_2} \frac{\partial\psi}{\partial y} dy = \int_{\psi_1}^{\psi_2} d\psi = \psi_2 - \psi_1$$

For a unit depth, the flow rate across BC is

$$Q = \int_{x_1}^{x_2} v dx = -\int_{x_1}^{x_2} \frac{\partial\psi}{\partial x} dx$$

Along BC, y = constant, and (from Eq. 5.6) $d\psi = \partial\psi/\partial x \; dx$. Therefore,

$$Q = -\int_{x_1}^{x_2} \frac{\partial\psi}{\partial x} dx = -\int_{\psi_2}^{\psi_1} d\psi = \psi_2 - \psi_1$$

Hence, whether we use line AB or line BC (or for that matter lines DE or DF), we find that *the volume flow rate (per unit depth) between two streamlines is given by the difference between the two stream function values.*[2] (The derivations for lines AB and BC are the justification for using the stream function definition of Eq. 5.4.) If the streamline through the origin is designated $\psi = 0$, then the ψ value for any other streamline represents the flow between the origin and that streamline. [We are free to select any streamline as the zero streamline because the stream function is defined as a differential (Eq. 5.3); also, the flow rate will always be given by a *difference* of ψ values.] Note that because the volume flow between any two streamlines is constant, *the velocity will be relatively high wherever the streamlines are close together, and relatively low wherever the streamlines are far apart*—a very useful concept for "eyeballing" velocity fields to see where we have regions of high or low velocity.

For a two-dimensional, incompressible flow in the $r\theta$ plane, conservation of mass, Eq. 5.2b, can be written as

$$\frac{\partial(rV_r)}{\partial r} + \frac{\partial V_\theta}{\partial \theta} = 0 \tag{5.7}$$

[2]For two-dimensional steady compressible flow in the xy plane, the stream function, ψ, can be defined such that

$$\rho u \equiv \frac{\partial\psi}{\partial y} \quad \text{and} \quad \rho v \equiv -\frac{\partial\psi}{\partial x}$$

The difference between the constant values of ψ defining two streamlines is then the mass flow rate (per unit depth) between the two streamlines.

Using a logic similar to that used for Eq. 5.4, the stream function, $\psi(r, \theta, t)$, then is defined such that

$$V_r \equiv \frac{1}{r}\frac{\partial \psi}{\partial \theta} \quad \text{and} \quad V_\theta \equiv -\frac{\partial \psi}{\partial r} \tag{5.8}$$

With ψ defined according to Eq. 5.8, the continuity equation, Eq. 5.7, is satisfied exactly.

Example 5.4 STREAM FUNCTION FOR FLOW IN A CORNER

Given the velocity field for steady, incompressible flow in a corner (Example 2.1), $\vec{V} = Ax\hat{i} - Ay\hat{j}$, with $A = 0.3\,\text{s}^{-1}$, determine the stream function that will yield this velocity field. Plot and interpret the streamline pattern in the first and second quadrants of the xy plane.

Given: Velocity field, $\vec{V} = Ax\hat{i} - Ay\hat{j}$, with $A = 0.3\ \text{s}^{-1}$.

Find: Stream function ψ and plot in first and second quadrants; interpret the results.

Solution:

The flow is incompressible, so the stream function satisfies Eq. 5.4.

From Eq. 5.4, $u = \dfrac{\partial \psi}{\partial y}$ and $v = -\dfrac{\partial \psi}{\partial y}$. From the given velocity field,

$$u = Ax = \frac{\partial \psi}{\partial y}$$

Integrating with respect to y gives

$$\psi = \int \frac{\partial \psi}{\partial y} dy + f(x) = Axy + f(x) \tag{1}$$

where $f(x)$ is arbitrary. The function $f(x)$ may be evaluated using the equation for v. Thus, from Eq. 1,

$$v = -\frac{\partial \psi}{\partial x} = -Ay - \frac{df}{dx} \tag{2}$$

From the given velocity field, $v = -Ay$. Comparing this with Eq. 2 shows that $\dfrac{df}{dx} = 0$, or $f(x) = \text{constant}$. Therefore, Eq. 1 becomes

$$\psi = Axy + c \longleftarrow \psi$$

Lines of constant ψ represent streamlines in the flow field. The constant c may be chosen as any convenient value for plotting purposes. The constant is chosen as zero in order that the streamline through the origin be designated as $\psi = \psi_1 = 0$. Then the value for any other streamline represents the flow between the origin and that streamline. With $c = 0$ and $A = 0.3\ \text{s}^{-1}$, then

$$\psi = 0.3xy \quad (\text{m}^3/\text{s/m})$$

{This equation of a streamline is identical to the result ($xy = \text{constant}$) obtained in Example 2.1.}

Separate plots of the streamlines in the first and second quadrants are presented below. Note that in quadrant 1, $u > 0$, so ψ values are positive. In quadrant 2, $u < 0$, so ψ values are negative.

In the first quadrant, since $u > 0$ and $v < 0$, the flow is from left to right and down. The volume flow rate between the streamline $\psi = \psi_1$ through the origin and the streamline $\psi = \psi_2$ is

$$Q_{12} = \psi_2 - \psi_1 = 0.3\,\text{m}^3/\text{s/m}$$

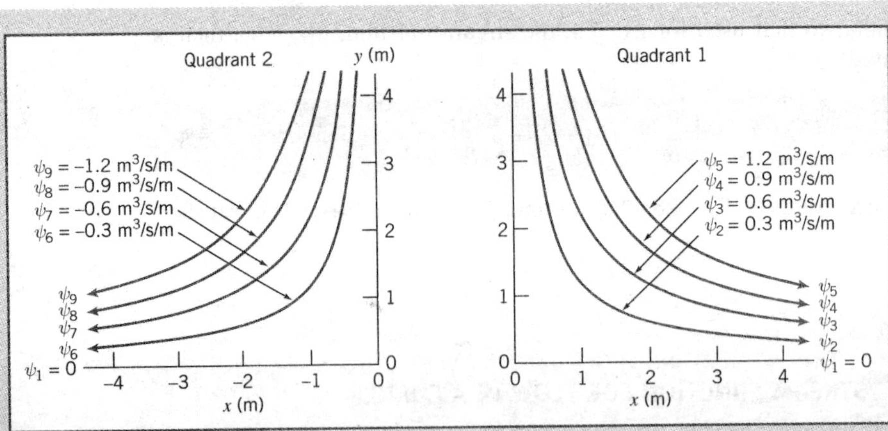

In the second quadrant, since $u<0$ and $v<0$, the flow is from right to left and down. The volume flow rate between streamlines ψ_7 and ψ_9 is

$$Q_{79} = \psi_9 - \psi_7 = [-1.2 - (-0.6)]\,\mathrm{m}^3/\mathrm{s/m} = -0.6\,\mathrm{m}^3/\mathrm{s/m}$$

The negative sign is consistent with flow having $u<0$.

As both the streamline spacing in the graphs and the equation for $\vec{V}$ indicate, the velocity is smallest near the origin (a "corner").

There is an *Excel* workbook for this problem that can be used to generate streamlines for this and many other stream functions.

5.3 Motion of a Fluid Particle (Kinematics)

Figure 5.4 shows a typical finite fluid element, within which we have selected an infinitesimal particle of mass dm and initial volume $dx\,dy\,dz$, at time t, and as it (and the infinitesimal particle) may appear after a time interval dt. The finite element has moved and changed its shape and orientation. Note that while the finite element has quite severe distortion, the infinitesimal particle has changes in shape limited to stretching/shrinking and rotation of the element's sides—this is because we are

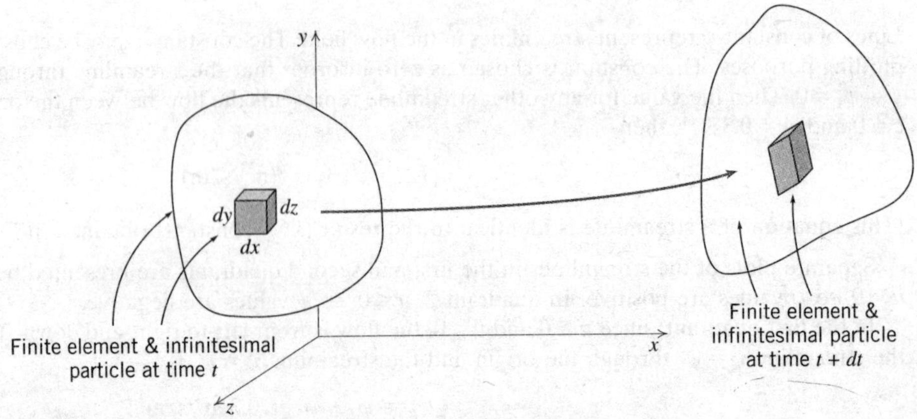

Fig. 5.4 Finite fluid element and infinitesimal particle at times t and $t + dt$.

considering both an infinitesimal time step and particle, so that the sides remain straight. We will examine the infinitesimal particle so that we will eventually obtain results applicable to a point. We can decompose this particle's motion into four components: *translation*, in which the particle moves from one point to another; *rotation* of the particle, which can occur about any or all of the x, y or z axes; *linear deformation*, in which the particle's sides stretch or contract; and *angular deformation*, in which the angles (which were initially 90° for our particle) between the sides change.

It may seem difficult by looking at Fig. 5.4 to distinguish between rotation and angular deformation of the infinitesimal fluid particle. It is important to do so, because pure rotation involves no deformation but angular deformation does and, as we learned in Chapter 2, fluid deformation generates shear stresses. Figure 5.5 shows a typical xy plane motion decomposed into the four components described above, and as we examine each of these four components in turn we will see that we *can* distinguish between rotation and angular deformation.

VIDEO

Particle Motion in a Channel.

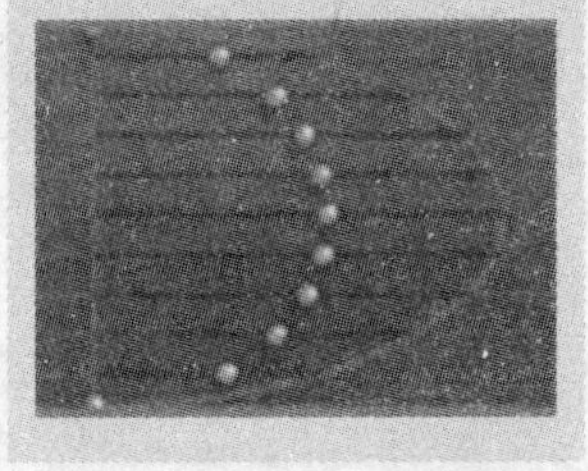

VIDEO

Particle Motion over a Cylinder.

Fluid Translation: Acceleration of a Fluid Particle in a Velocity Field

The translation of a fluid particle is obviously connected with the velocity field $\vec{V} = \vec{V}(x, y, z, t)$ that we previously discussed in Section 2.2. We will need the acceleration of a fluid particle for use in Newton's second law. It might seem that we could simply compute this as $\vec{a} = \partial\vec{V}/\partial t$. This is incorrect, because $\vec{V}$ is a *field*, i.e., it describes the whole flow and not just the motion of an individual particle. (We can see that this way of computing is incorrect by examining Example 5.4, in which particles are clearly accelerating and decelerating so $\vec{a} \neq 0$, but $\partial\vec{V}/\partial t = 0$.)

The problem, then, is to retain the field description for fluid properties and obtain an expression for the acceleration of a fluid particle as it moves in a flow field. Stated simply, the problem is:

> Given the velocity field, $\vec{V} = \vec{V}(x, y, z, t)$, find the acceleration of a fluid particle, $\vec{a}_p$.

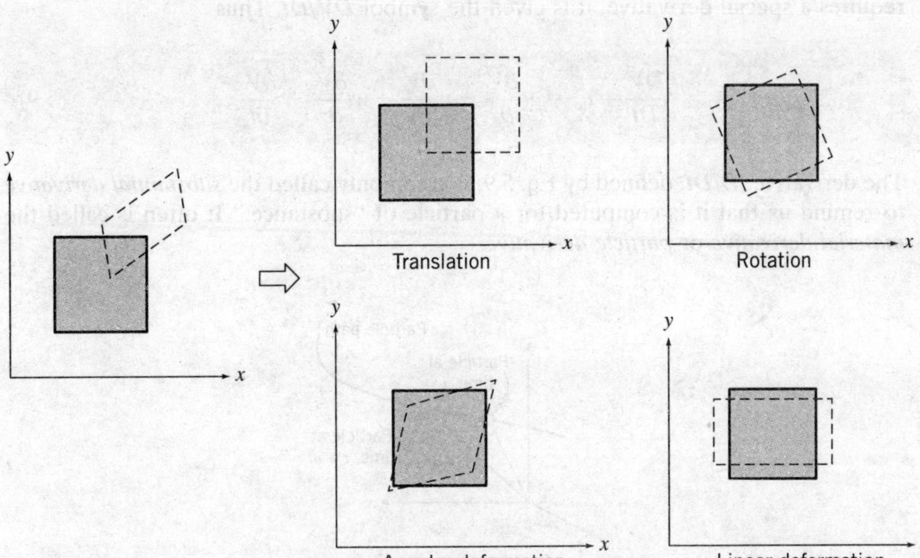

Fig. 5.5 Pictorial representation of the components of fluid motion.

Consider a particle moving in a velocity field. At time t, the particle is at the position x, y, z and has a velocity corresponding to the velocity at that point in space at time t,

$$\vec{V}_p\Big]_t = \vec{V}(x, y, z, t)$$

At $t + dt$, the particle has moved to a new position, with coordinates $x + dx$, $y + dy$, $z + dz$, and has a velocity given by

$$\vec{V}_p\Big]_{t+dt} = \vec{V}(x + dx, y + dy, z + dz, t + dt)$$

This is shown pictorially in Fig. 5.6.

The particle velocity at time t (position $\vec{r}$) is given by $\vec{V}_p = \vec{V}(x, y, z, t)$. Then $d\vec{V}_p$, the change in the velocity of the particle, in moving from location $\vec{r}$ to $\vec{r} + d\vec{r}$, in time dt, is given by the chain rule,

$$d\vec{V}_p = \frac{\partial \vec{V}}{\partial x} dx_p + \frac{\partial \vec{V}}{\partial y} dy_p + \frac{\partial \vec{V}}{\partial z} dz_p + \frac{\partial \vec{V}}{\partial t} dt$$

The total acceleration of the particle is given by

$$\vec{a}_p = \frac{d\vec{V}_p}{dt} = \frac{\partial \vec{V}}{\partial x}\frac{dx_p}{dt} + \frac{\partial \vec{V}}{\partial y}\frac{dy_p}{dt} + \frac{\partial \vec{V}}{\partial z}\frac{dz_p}{dt} + \frac{\partial \vec{V}}{\partial t}$$

Since

$$\frac{dx_p}{dt} = u, \quad \frac{dy_p}{dt} = v, \quad \text{and} \quad \frac{dz_p}{dt} = w,$$

we have

$$\vec{a}_p = \frac{d\vec{V}_p}{dt} = u\frac{\partial \vec{V}}{\partial x} + v\frac{\partial \vec{V}}{\partial y} + w\frac{\partial \vec{V}}{\partial z} + \frac{\partial \vec{V}}{\partial t}$$

To remind us that calculation of the acceleration of a fluid particle in a velocity field requires a special derivative, it is given the symbol $D\vec{V}/Dt$. Thus

$$\frac{D\vec{V}}{Dt} \equiv \vec{a}_p = u\frac{\partial \vec{V}}{\partial x} + v\frac{\partial \vec{V}}{\partial y} + w\frac{\partial \vec{V}}{\partial z} + \frac{\partial \vec{V}}{\partial t} \qquad (5.9)$$

The derivative, D/Dt, defined by Eq. 5.9, is commonly called the *substantial derivative* to remind us that it is computed for a particle of "substance." It often is called the *material derivative* or *particle derivative*.

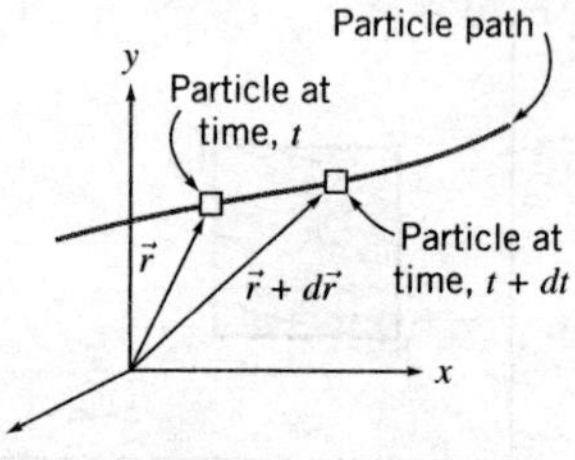

Fig. 5.6 Motion of a particle in a flow field.

The physical significance of the terms in Eq. 5.9 is

$$\vec{a}_p = \frac{D\vec{V}}{Dt} = \underbrace{u\frac{\partial \vec{V}}{\partial x} + v\frac{\partial \vec{V}}{\partial y} + w\frac{\partial \vec{V}}{\partial z}}_{\text{convective acceleration}} + \frac{\partial \vec{V}}{\partial t}$$

total acceleration of a particle — convective acceleration — local acceleration

From Eq. 5.9 we recognize that a fluid particle moving in a flow field may undergo acceleration for either of two reasons. As an illustration, refer to Example 5.4. This is a steady flow in which particles are *convected* toward the low-velocity region (near the "corner"), and then away to a high-velocity region. If a flow field is unsteady a fluid particle will undergo an additional *local* acceleration, because the velocity field is a function of time.

The convective acceleration may be written as a single vector expression using the gradient operator ∇. Thus

CLASSIC VIDEO

Eulerian and Lagrangian Descriptions in Fluid Mechanics.

$$u\frac{\partial \vec{V}}{\partial x} + v\frac{\partial \vec{V}}{\partial y} + w\frac{\partial \vec{V}}{\partial z} = (\vec{V}\cdot\nabla)\vec{V}$$

(We suggest that you check this equality by expanding the right side of the equation using the familiar dot product operation.) Thus Eq. 5.9 may be written as

$$\frac{D\vec{V}}{Dt} \equiv \vec{a}_p = (\vec{V}\cdot\nabla)\vec{V} + \frac{\partial \vec{V}}{\partial t} \tag{5.10}$$

For a *two-dimensional flow*, say $\vec{V} = \vec{V}(x, y, t)$, Eq. 5.9 reduces to

$$\frac{D\vec{V}}{Dt} = u\frac{\partial \vec{V}}{\partial x} + v\frac{\partial \vec{V}}{\partial y} + \frac{\partial \vec{V}}{\partial t}$$

For a *one-dimensional flow*, say $\vec{V} = \vec{V}(x, t)$, Eq. 5.9 becomes

$$\frac{D\vec{V}}{Dt} = u\frac{\partial \vec{V}}{\partial x} + \frac{\partial \vec{V}}{\partial t}$$

Finally, for a *steady flow in three dimensions*, Eq. 5.9 becomes

$$\frac{D\vec{V}}{Dt} = u\frac{\partial \vec{V}}{\partial x} + v\frac{\partial \vec{V}}{\partial y} + w\frac{\partial \vec{V}}{\partial z}$$

which, as we have seen, is not necessarily zero even though the flow is steady. Thus a fluid particle may undergo a convective acceleration due to its motion, even in a steady velocity field.

Equation 5.9 is a vector equation. As with all vector equations, it may be written in scalar component equations. Relative to an xyz coordinate system, the scalar components of Eq. 5.9 are written

$$a_{x_p} = \frac{Du}{Dt} = u\frac{\partial u}{\partial x} + v\frac{\partial u}{\partial y} + w\frac{\partial u}{\partial z} + \frac{\partial u}{\partial t} \tag{5.11a}$$

$$a_{y_p} = \frac{Dv}{Dt} = u\frac{\partial v}{\partial x} + v\frac{\partial v}{\partial y} + w\frac{\partial v}{\partial z} + \frac{\partial v}{\partial t} \tag{5.11b}$$

$$a_{z_p} = \frac{Dw}{Dt} = u\frac{\partial w}{\partial x} + v\frac{\partial w}{\partial y} + w\frac{\partial w}{\partial z} + \frac{\partial w}{\partial t} \tag{5.11c}$$

The components of acceleration in cylindrical coordinates may be obtained from Eq. 5.10 by expressing the velocity, $\vec{V}$, in cylindrical coordinates (Section 5.1) and utilizing the appropriate expression (Eq. 3.19, on the Web) for the vector operator ∇. Thus,[3]

$$a_{r_p} = V_r \frac{\partial V_r}{\partial r} + \frac{V_\theta}{r}\frac{\partial V_r}{\partial \theta} - \frac{V_\theta^2}{r} + V_z \frac{\partial V_r}{\partial z} + \frac{\partial V_r}{\partial t} \tag{5.12a}$$

$$a_{\theta_p} = V_r \frac{\partial V_\theta}{\partial r} + \frac{V_\theta}{r}\frac{\partial V_\theta}{\partial \theta} + \frac{V_r V_\theta}{r} + V_z \frac{\partial V_\theta}{\partial z} + \frac{\partial V_\theta}{\partial t} \tag{5.12b}$$

$$a_{z_p} = V_r \frac{\partial V_z}{\partial r} + \frac{V_\theta}{r}\frac{\partial V_z}{\partial \theta} + V_z \frac{\partial V_z}{\partial z} + \frac{\partial V_z}{\partial t} \tag{5.12c}$$

Equations 5.9, 5.11, and 5.12 are useful for computing the acceleration of a fluid particle anywhere in a flow from the velocity field (a function of x, y, z, and t); this is the *Eulerian* method of description, the most-used approach in fluid mechanics.

As an alternative (e.g., if we wish to track an individual particle's motion in, for example, pollution studies) we sometimes use the *Lagrangian* description of particle motion, in which the acceleration, position, and velocity of a particle are specified as a function of time only. Both descriptions are illustrated in Example 5.5.

Example 5.5 PARTICLE ACCELERATION IN EULERIAN AND LAGRANGIAN DESCRIPTIONS

Consider two-dimensional, steady, incompressible flow through the plane converging channel shown. The velocity on the horizontal centerline (x axis) is given by $\vec{V} = V_1[1 + (x/L)]\hat{i}$. Find an expression for the acceleration of a particle moving along the centerline using (a) the Eulerian approach and (b) the Lagrangian approach. Evaluate the acceleration when the particle is at the beginning and at the end of the channel.

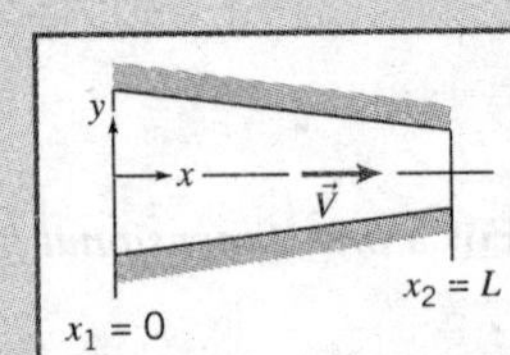

Given: Steady, two-dimensional, incompressible flow through the converging channel shown.

$$\vec{V} = V_1\left(1 + \frac{x}{L}\right)\hat{i} \quad \text{on } x \text{ axis}$$

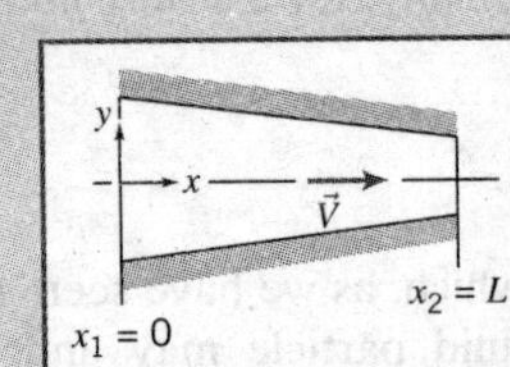

Find: (a) The acceleration of a particle moving along the centerline using the Eulerian approach.
(b) The acceleration of a particle moving along the centerline using the Lagrangian approach.
(c) Evaluate the acceleration when the particle is at the beginning and at the end of the channel.

Solution:

(a) The Eulerian approach

The governing equation for acceleration of a fluid particle is Eq. 5.9:

$$\vec{a}_p(x, y, z, t) = \frac{D\vec{V}}{Dt} = u\frac{\partial \vec{V}}{\partial x} + v\frac{\partial \vec{V}}{\partial y} + w\frac{\partial \vec{V}}{\partial z} + \frac{\partial \vec{V}}{\partial t} \tag{5.9}$$

[3]In evaluating $(\vec{V}\cdot\nabla)\vec{V}$, recall that $\hat{e}_r$ and $\hat{e}_\theta$ are functions of θ (see footnote 1 on p. 178).

In this case we are interested in the x component of acceleration (Eq. 5.11a):

$$a_{x_p}(x,y,z,t) = \frac{Du}{Dt} = u\frac{\partial u}{\partial x} + v\frac{\partial u}{\partial y} + w\frac{\partial u}{\partial z} + \frac{\partial u}{\partial t} \tag{5.11a}$$

On the x axis, $v = w = 0$ and $u = V_1\left(1 + \frac{x}{L}\right)$, so for steady flow we obtain

$$a_{x_p}(x) = \frac{Du}{Dt} = u\frac{\partial u}{\partial x} = V_1\left(1 + \frac{x}{L}\right)\frac{V_1}{L}$$

or

$$a_{x_p}(x) = \frac{V_1^2}{L}\left(1 + \frac{x}{L}\right) \longleftarrow \quad a_{x_p}(x)$$

This expression gives the acceleration of *any* particle that is at point x at an instant.

(b) The Lagrangian approach
In this approach we need to obtain the motion of a fluid particle as we would in particle mechanics; that is, we need the position $\vec{x}_p(t)$, and then we can obtain the velocity $\vec{V}_p(t) = d\vec{x}_p/dt$ and acceleration $\vec{a}_p(t) = d\vec{V}_p/dt$. Actually, we are considering motion along the x axis, so we want $x_p(t)$, $u_p(t) = dx_p/dt$, and $a_{x_p}(t) = du_p/dt$. We are not given $x_p(t)$, but we do have

$$u_p = \frac{dx_p}{dt} = V_1\left(1 + \frac{x_p}{L}\right)$$

Separating variables, and using limits $x_p(t = 0) = 0$ and $x_p(t = t) = x_p$,

$$\int_0^{x_p} \frac{dx_p}{\left(1 + \frac{x_p}{L}\right)} = \int_0^1 V_1 dt \quad \text{and} \quad L\ln\left(1 + \frac{x_p}{L}\right) = V_1 t \tag{1}$$

We can then solve for $x_p(t)$:

$$x_p(t) = L(e^{V_1 t/L} - 1)$$

The velocity and acceleration are then

$$u_p(t) = \frac{dx_p}{dt} = V_1 e^{V_1 t/L}$$

and

$$a_{x_p}(t) = \frac{du_p}{dt} = \frac{V_1^2}{L}e^{V_1 t/L} \longleftarrow \quad (2) \quad a_{x_p}(t)$$

This expression gives the acceleration at any time t of the particle that was initially at $x = 0$.

(c) We wish to evaluate the acceleration when the particle is at $x = 0$ and $x = L$. For the Eulerian approach this is straightforward:

$$a_{x_p}(x = 0) = \frac{V_1^2}{L}, \quad a_{x_p}(x = L) = 2\frac{V_1^2}{L} \longleftarrow \quad a_{x_p}$$

For the Lagrangian approach, we need to find the times at which $x = 0$ and $x = L$. Using Eq. 1, these are

$$t(x_p = 0) = \frac{L}{V_1} \qquad t(x_p = L) = \frac{L}{V_1}\ln(2)$$

Then, from Eq. 2,

$$a_{z_p}(t=0) = \frac{V_1^2}{L}, \quad \text{and}$$

$$a_{x_p}\left(t = \frac{L}{V_1}\ln(2)\right) = \frac{V_1^2}{L}e^{\ln(2)} = 2\frac{V_1^2}{L} \longleftarrow a_{x_p}$$

Note that both approaches yield the same results for particle acceleration, as they should.

This problem illustrates use of the Eulerian and Lagrangian descriptions of the motion of a fluid particle.

Fluid Rotation

A fluid particle moving in a general three-dimensional flow field may rotate about all three coordinate axes. Thus particle rotation is a vector quantity and, in general,

$$\vec{\omega} = \hat{i}\omega_x + \hat{j}\omega_y + \hat{k}\omega_z$$

where ω_x is the rotation about the x axis, ω_y is the rotation about the y axis, and ω_z is the rotation about the z axis. The positive sense of rotation is given by the right-hand rule.

We now see how we can extract the rotation component of the particle motion. Consider the xy plane view of the particle at time t. The left and lower sides of the particle are given by the two perpendicular line segments *oa* and *ob* of lengths Δx and Δy, respectively, shown in Fig. 5.7*a*. In general, after an interval Δt the particle will have translated to some new position, and also have rotated and deformed. A possible instantaneous orientation of the lines at time $t + \Delta t$ is shown in Fig. 5.7*b*.

We need to be careful here with our signs for angles. Following the right-hand rule, *counterclockwise rotation is positive*, and we have shown side *oa* rotating counterclockwise through angle $\Delta\alpha$, but be aware that we have shown edge *ob* rotating at a *clockwise* angle $\Delta\beta$. Both angles are obviously arbitrary, but it will help visualize the discussion if we assign values to these angles, e.g., let $\Delta\alpha = 6°$ and $\Delta\beta = 4°$.

How do we extract from $\Delta\alpha$ and $\Delta\beta$ a measure of the particle's rotation? The answer is that we take an average of the rotations $\Delta\alpha$ and $\Delta\beta$, so that the particle's rigid body counterclockwise rotation is $\frac{1}{2}(\Delta\alpha - \Delta\beta)$, as shown in Fig. 5.7*c*. The minus sign is needed because the *counterclockwise* rotation of *ob* is $-\Delta\beta$. Using the assigned values, the rotation of the particle is then $\frac{1}{2}(6° - 4°) = 1°$. (Given the two rotations, taking the average is the only way we can measure the particle's rotation, because any other approach would favor one side's rotation over the other, which doesn't make sense.)

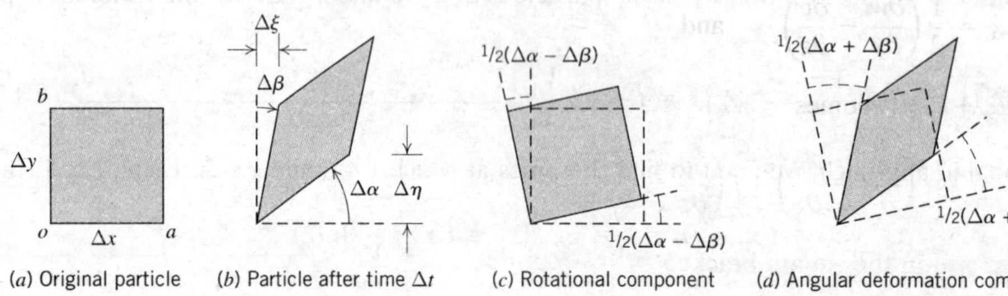

(*a*) Original particle (*b*) Particle after time Δt (*c*) Rotational component (*d*) Angular deformation component

Fig. 5.7 Rotation and angular deformation of perpendicular line segments in a two-dimensional flow.

Now we can determine from $\Delta\alpha$ and $\Delta\beta$ a measure of the particle's angular deformation, as shown in Fig. 5.7*d*. To obtain the deformation of side *oa* in Fig. 5.7*d*, we use Fig. 5.7*b* and 5.7*c*: If we subtract the particle rotation $\frac{1}{2}(\Delta\alpha - \Delta\beta)$, in Fig. 5.7*c*, from the actual rotation of *oa*, $\Delta\alpha$, in Fig. 5.7*b*, what remains must be pure deformation $[\Delta\alpha - \frac{1}{2}(\Delta\alpha - \Delta\beta) = \frac{1}{2}(\Delta\alpha + \Delta\beta)$, in Fig. 5.7*d*]. Using the assigned values, the deformation of side *oa* is $6° - \frac{1}{2}(6° - 4°) = 5°$. By a similar process, for side *ob* we end with $\Delta\beta - \frac{1}{2}(\Delta\alpha - \Delta\beta) = -\frac{1}{2}(\Delta\alpha + \Delta\beta)$, or a *clockwise* deformation $\frac{1}{2}(\Delta\alpha + \Delta\beta)$, as shown in Fig. 5.7*d*. The total deformation of the particle is the sum of the deformations of the sides, or $(\Delta\alpha + \Delta\beta)$ (with our example values, 10°). We verify that this leaves us with the correct value for the particle's deformation: Recall that in Section 2.4 we saw that deformation is measured by the change in a 90° angle. In Fig. 5.7*a* we see this is angle *aob*, and in Fig. 5.7*d* we see the total change of this angle is indeed $\frac{1}{2}(\Delta\alpha + \Delta\beta) + \frac{1}{2}(\Delta\alpha + \Delta\beta) = (\Delta\alpha + \Delta\beta)$.

We need to convert these angular measures to quantities obtainable from the flow field. To do this, we recognize that (for small angles) $\Delta\alpha = \Delta\eta/\Delta x$, and $\Delta\beta = \Delta\xi/\Delta y$. But $\Delta\xi$ arises because, if in interval Δt point *o* moves horizontally distance $u\Delta t$, then point *b* will have moved distance $(u + [\partial u/\partial y]\Delta y)\Delta t$ (using a Taylor series expansion). Likewise, $\Delta\eta$ arises because, if in interval Δt point *o* moves vertically distance $v\Delta t$, then point *a* will have moved distance $(v + [\partial v/\partial x]\Delta x)\Delta t$. Hence,

$$\Delta\xi = \left(u + \frac{\partial u}{\partial y}\Delta y\right)\Delta t - u\Delta t = \frac{\partial u}{\partial y}\Delta y\Delta t$$

and

$$\Delta\eta = \left(v + \frac{\partial v}{\partial x}\Delta x\right)\Delta t - v\Delta t = \frac{\partial v}{\partial x}\Delta x\Delta t$$

We can now compute the angular velocity of the particle about the *z* axis, ω_z, by combining all these results:

$$\omega_z = \lim_{\Delta t\to 0}\frac{\frac{1}{2}(\Delta\alpha - \Delta\beta)}{\Delta t} = \lim_{\Delta t\to 0}\frac{\frac{1}{2}\left(\frac{\Delta\eta}{\Delta x} - \frac{\Delta\xi}{\Delta y}\right)}{\Delta t} = \lim_{\Delta t\to 0}\frac{\frac{1}{2}\left(\frac{\partial v}{\partial x}\frac{\Delta x}{\Delta x}\Delta t - \frac{\partial u}{\partial y}\frac{\Delta y}{\Delta y}\Delta t\right)}{\Delta t}$$

$$\omega_z = \frac{1}{2}\left(\frac{\partial v}{\partial x} - \frac{\partial u}{\partial y}\right)$$

By considering the rotation of pairs of perpendicular line segments in the *yz* and *xz* planes, one can show similarly that

$$\omega_x = \frac{1}{2}\left(\frac{\partial w}{\partial y} - \frac{\partial v}{\partial z}\right) \quad \text{and} \quad \omega_y = \frac{1}{2}\left(\frac{\partial u}{\partial z} - \frac{\partial w}{\partial x}\right)$$

Then $\vec{\omega} = \hat{i}\omega_x + \hat{j}\omega_y + \hat{k}\omega_z$ becomes

$$\vec{\omega} = \frac{1}{2}\left[\hat{i}\left(\frac{\partial w}{\partial y} - \frac{\partial v}{\partial z}\right) + \hat{j}\left(\frac{\partial u}{\partial z} - \frac{\partial w}{\partial x}\right) + \hat{k}\left(\frac{\partial v}{\partial x} - \frac{\partial u}{\partial y}\right)\right] \tag{5.13}$$

We recognize the term in the square brackets as

$$\text{curl}\,\vec{V} = \nabla \times \vec{V}$$

Then, in vector notation, we can write

$$\vec{\omega} = \frac{1}{2}\nabla \times \vec{V} \tag{5.14}$$

It is worth noting here that we should not confuse rotation of a fluid particle with flow consisting of circular streamlines, or *vortex* flow. As we will see in Example 5.6, in such a flow the particles *could* rotate as they move in a circular motion, but they do not have to!

When might we expect to have a flow in which the particles rotate as they move ($\vec{\omega} \neq 0$)? One possibility is that we start out with a flow in which (for whatever reason) the particles already have rotation. On the other hand, if we assumed the particles are not initially rotating, particles will only begin to rotate if they experience a torque caused by surface shear stresses; the particle body forces and normal (pressure) forces may accelerate and deform the particle, but cannot generate a torque. We can conclude that rotation of fluid particles will *always* occur for flows in which we have shear stresses. We have already learned in Chapter 2 that shear stresses are present whenever we have a viscous fluid that is experiencing angular deformation (shearing). Hence we conclude that rotation of fluid particles only occurs in viscous flows[4] (unless the particles are initially rotating, as in Example 3.10).

CLASSIC VIDEO

Vorticity.

Flows for which no particle rotation occurs are called *irrotational* flows. Although no real flow is truly irrotational (all fluids have viscosity), it turns out that many flows can be successfully studied by assuming they are inviscid and irrotational, because viscous effects are often negligible. As we discussed in Chapter 1, and will again in Chapter 6, much of aerodynamics theory assumes inviscid flow. We just need to be aware that in any flow there will always be regions (e.g., the boundary layer for flow over a wing) in which viscous effects cannot be ignored.

The factor of $\frac{1}{2}$ can be eliminated from Eq. 5.14 by defining the *vorticity*, $\vec{\zeta}$, to be twice the rotation,

$$\vec{\zeta} \equiv 2\vec{\omega} = \nabla \times \vec{V} \tag{5.15}$$

The vorticity is a measure of the rotation of a fluid element as it moves in the flow field. In cylindrical coordinates the vorticity is[5]

$$\nabla \times \vec{V} = \hat{e}_r\left(\frac{1}{r}\frac{\partial V_z}{\partial \theta} - \frac{\partial V_\theta}{\partial z}\right) + \hat{e}_\theta\left(\frac{\partial V_r}{\partial z} - \frac{\partial V_z}{\partial r}\right) + \hat{k}\left(\frac{1}{r}\frac{\partial r V_\theta}{\partial r} - \frac{1}{r}\frac{\partial V_r}{\partial \theta}\right) \tag{5.16}$$

The *circulation*, Γ (which we will revisit in Example 6.12), is defined as the line integral of the tangential velocity component about any closed curve fixed in the flow,

$$\Gamma = \oint_c \vec{V} \cdot d\vec{s} \tag{5.17}$$

where $d\vec{s}$ is an elemental vector tangent to the curve and having length ds of the element of arc; a positive sense corresponds to a counterclockwise path of integration around the curve. We can develop a relationship between circulation and vorticity by considering the rectangular circuit shown in Fig. 5.8, where the velocity components at o are assumed to be (u, v), and the velocities along segments bc and ac can be derived using Taylor series approximations.

[4]A rigorous proof using the complete equations of motion for a fluid particle is given in Li and Lam, pp. 142–145.

[5]In carrying out the curl operation, recall that $\hat{e}_r$ and $\hat{e}_\theta$ are functions of θ (see footnote 1 on p. 178).

For the closed curve *oacb*,

$$\Delta\Gamma = u\Delta x + \left(v + \frac{\partial v}{\partial x}\Delta x\right)\Delta y - \left(u + \frac{\partial u}{\partial y}\Delta y\right)\Delta x - v\,\Delta y$$

$$\Delta\Gamma = \left(\frac{\partial v}{\partial x} - \frac{\partial u}{\partial y}\right)\Delta x\Delta y$$

$$\Delta\Gamma = 2\omega_z\Delta x\Delta y$$

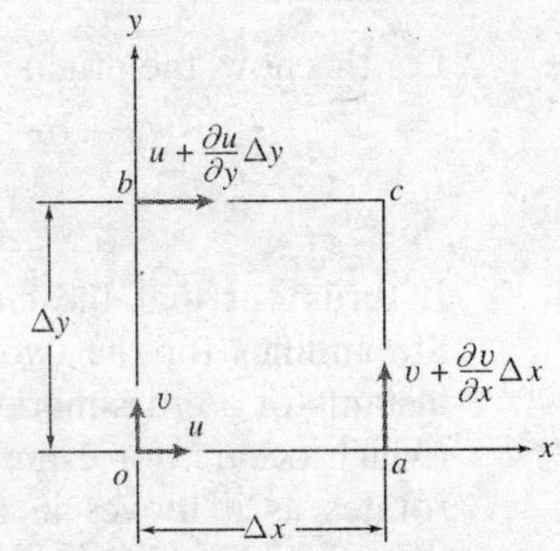

Fig. 5.8 Velocity components on the boundaries of a fluid element.

Then,

$$\Gamma = \oint_c \vec{V}\cdot d\vec{s} = \int_A 2\omega_z\,dA = \int_A (\nabla\times\vec{V})_z\,dA \qquad (5.18)$$

Equation 5.18 is a statement of the Stokes Theorem in two dimensions. Thus the circulation around a closed contour is equal to the total vorticity enclosed within it.

Example 5.6 FREE AND FORCED VORTEX FLOWS

Consider flow fields with purely tangential motion (circular streamlines): $V_r = 0$ and $V_\theta = f(r)$. Evaluate the rotation, vorticity, and circulation for rigid-body rotation, a *forced vortex*. Show that it is possible to choose $f(r)$ so that flow is irrotational, i.e., to produce a *free vortex*.

Given: Flow fields with tangential motion, $V_r = 0$ and $V_\theta = f(r)$.

Find: (a) Rotation, vorticity, and circulation for rigid-body motion (a *forced vortex*).
(b) $V_\theta = f(r)$ for irrotational motion (a *free vortex*).

Solution:

Governing equation: $\vec{\zeta} = 2\vec{\omega} = \nabla\times\vec{V}$ (5.15)

For motion in the $r\theta$ plane, the only components of rotation and vorticity are in the z direction,

$$\zeta_z = 2\omega_z = \frac{1}{r}\frac{\partial rV_\theta}{\partial r} - \frac{1}{r}\frac{\partial V_r}{\partial\theta}$$

Because $V_r = 0$ everywhere in these fields, this reduces to $\zeta_z = 2\omega_z = \frac{1}{r}\frac{\partial rV_\theta}{\partial r}$.

(a) For rigid-body rotation, $V_\theta = \omega r$.

Then $\omega_z = \frac{1}{2}\frac{1}{r}\frac{\partial rV_\theta}{\partial r} = \frac{1}{2}\frac{1}{r}\frac{\partial}{\partial r}\omega r^2 = \frac{1}{2r}(2\omega r) = \omega$ and $\zeta_z = 2\omega$.

The circulation is $\Gamma = \oint_c \vec{V}\cdot d\vec{s} = \int_A 2\omega_z dA.$ (5.18)

Since $\omega_z = \omega =$ constant, the circulation about any closed contour is given by $\Gamma = 2\omega A$, where A is the area enclosed by the contour. Thus for rigid-body motion (a forced vortex), the rotation and vorticity are constants; the circulation depends on the area enclosed by the contour.

(b) For irrotational flow, $\omega_z = \frac{1}{r}\frac{\partial}{\partial r}rV_\theta = 0$. Integrating, we find

$$rV_\theta = \text{constant} \quad \text{or} \quad V_\theta = f(r) = \frac{C}{r}$$

For this flow, the origin is a singular point where $V_\theta \to \infty$. The circulation for any contour enclosing the origin is

$$\Gamma = \oint_c \vec{V} \cdot d\vec{s} = \int_0^{2\pi} \frac{C}{r} r \, d\theta = 2\pi C$$

It turns out that the circulation around any contour *not* enclosing the singular point at the origin is zero. Streamlines for the two vortex flows are shown below, along with the location and orientation at different instants of a cross marked in the fluid that was initially at the 12 o'clock position. For the rigid-body motion (which occurs, for example, at the eye of a tornado, creating the "dead" region at the very center), the cross rotates as it moves in a circular motion; also, the streamlines are closer together as we move away from the origin. For the irrotational motion (which occurs, for example, outside the eye of a tornado—in such a large region viscous effects are negligible), the cross does not rotate as it moves in a circular motion; also, the streamlines are farther apart as we move away from the origin.

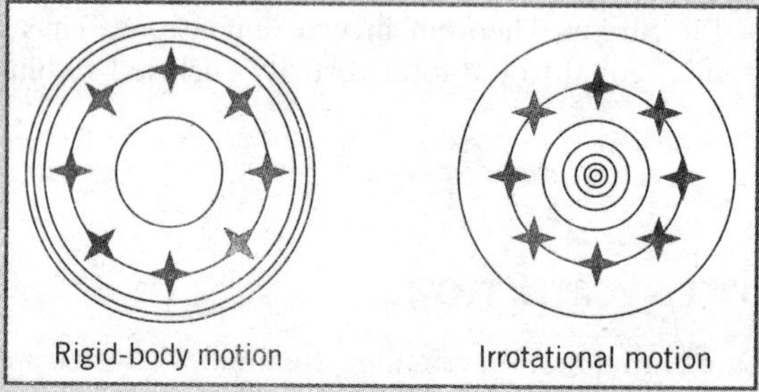

Fluid Deformation

a. Angular Deformation

As we discussed earlier (and as shown in Fig. 5.7*d*), the *angular deformation* of a particle is given by the sum of the two angular deformations, or in other words by $(\Delta\alpha + \Delta\beta)$.

We also recall that $\Delta\alpha = \Delta\eta/\Delta x, \Delta\beta = \Delta\xi/\Delta y$, and $\Delta\xi$ and $\Delta\eta$ are given by

$$\Delta\xi = \left(u + \frac{\partial u}{\partial y}\Delta y\right)\Delta t - u\Delta t = \frac{\partial u}{\partial y}\Delta y \Delta t$$

and

$$\Delta\eta = \left(v + \frac{\partial v}{\partial x}\Delta x\right)\Delta t - v\Delta t = \frac{\partial v}{\partial x}\Delta x \Delta t$$

We can now compute the rate of angular deformation of the particle in the *xy* plane by combining these results,

$$\begin{array}{l}\text{Rate of angular}\\ \text{deformation}\\ \text{in } xy \text{ plane}\end{array} = \lim_{\Delta t \to 0} \frac{(\Delta\alpha + \Delta\beta)}{\Delta t} = \lim_{\Delta t \to 0} \frac{\left(\dfrac{\Delta\eta}{\Delta x} + \dfrac{\Delta\xi}{\Delta y}\right)}{\Delta t}$$

$$\begin{array}{l}\text{Rate of angular}\\ \text{deformation}\\ \text{in } xy \text{ plane}\end{array} = \lim_{\Delta t \to 0} \frac{\left(\dfrac{\partial v}{\partial x}\dfrac{\Delta x}{\Delta x}\Delta t + \dfrac{\partial u}{\partial y}\dfrac{\Delta y}{\Delta y}\Delta t\right)}{\Delta t} = \left(\frac{\partial v}{\partial x} + \frac{\partial u}{\partial y}\right) \quad (5.19a)$$

Similar expressions can be written for the rate of angular deformation of the particle in the yz and zx planes,

CLASSIC VIDEO

Deformation of Continuous Media.

$$\text{Rate of angular deformation in } yz \text{ plane} = \left(\frac{\partial w}{\partial y} + \frac{\partial v}{\partial z}\right) \tag{5.19b}$$

$$\text{Rate of angular deformation in } zx \text{ plane} = \left(\frac{\partial w}{\partial x} + \frac{\partial u}{\partial z}\right) \tag{5.19c}$$

We saw in Chapter 2 that for one-dimensional laminar Newtonian flow the shear stress is given by the rate of deformation (du/dy) of the fluid particle,

$$\tau_{yx} = \mu\frac{du}{dy} \tag{2.15}$$

We will see shortly that we can generalize Eq. 2.15 to the case of three-dimensional laminar flow; this will lead to expressions for three-dimensional shear stresses involving the three rates of angular deformation given above. (Eq. 2.15 is a special case of Eq. 5.19a.)

Calculation of angular deformation is illustrated for a simple flow field in Example 5.7.

Example 5.7 ROTATION IN VISCOMETRIC FLOW

A viscometric flow in the narrow gap between large parallel plates is shown. The velocity field in the narrow gap is given by $\vec{V} = U(y/h)\hat{i}$, where $U = 4$ mm/s and $h = 4$ mm. At $t = 0$ line segments ac and bd are marked in the fluid to form a cross as shown. Evaluate the positions of the marked points at $t = 1.5$ s and sketch for comparison. Calculate the rate of angular deformation and the rate of rotation of a fluid particle in this velocity field. Comment on your results.

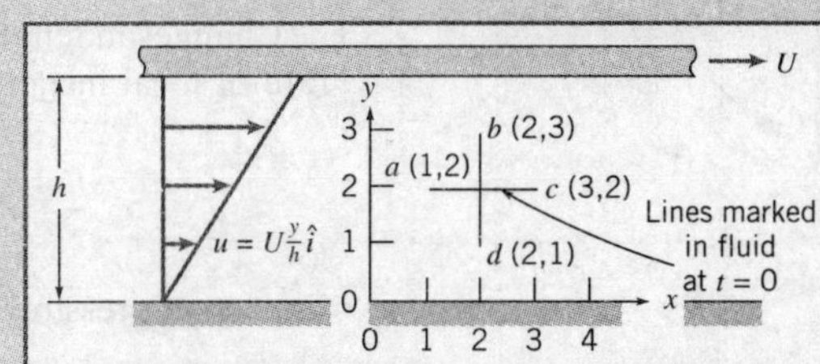

Given: Velocity field, $\vec{V} = U(y/h)\hat{i}$; $U = 4$ mm/s, and $h = 4$ mm. Fluid particles marked at $t = 0$ to form cross as shown.

Find: (a) Positions of points a', b', c', and d' at $t = 1.5$ s; plot.
(b) Rate of angular deformation.
(c) Rate of rotation of a fluid particle.
(d) Significance of these results.

Solution: For the given flow field $v = 0$, so there is no vertical motion. The velocity of each point stays constant, so $\Delta x = u\Delta t$ for each point. At point b, $u = 3$ mm/s, so

$$\Delta x_b = 3\frac{\text{mm}}{\text{s}} \times 1.5\,\text{s} = 4.5\,\text{mm}$$

Similarly, points a and c each move 3 mm, and point d moves 1.5 mm. Hence the plot at $t = 1.5$ s is

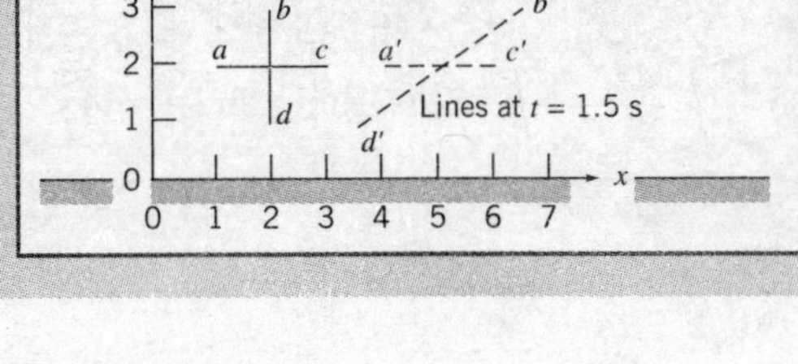

The rate of angular deformation is

$$\frac{\partial u}{\partial y} + \frac{\partial v}{\partial x} = U\frac{1}{h} + 0 = \frac{U}{h} = 4\frac{\text{mm}}{\text{s}} \times \frac{1}{4\text{ mm}} = 1\text{ s}^{-1} \longleftarrow$$

The rate of rotation is

$$\omega_z = \frac{1}{2}\left(\frac{\partial v}{\partial x} - \frac{\partial u}{\partial y}\right) = \frac{1}{2}\left(0 - \frac{U}{h}\right) = -\frac{1}{2} \times 4\frac{\text{mm}}{\text{s}} \times \frac{1}{4\text{ mm}} = -0.5\text{ s}^{-1} \longleftarrow \omega_z$$

In this problem we have a viscous flow, and hence should have expected both angular deformation and particle rotation.

b. Linear Deformation

VIDEO

Linear Deformation.

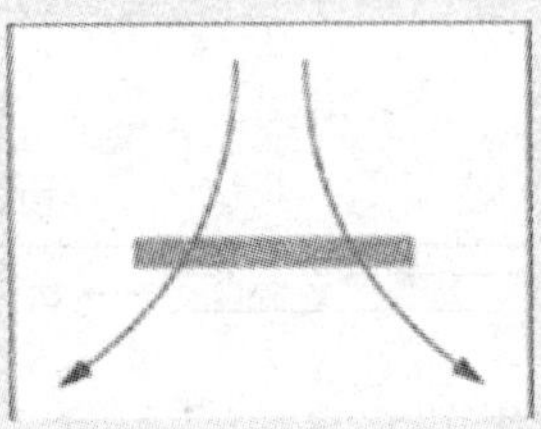

During linear deformation, the shape of the fluid element, described by the angles at its vertices, remains unchanged, since all right angles continue to be right angles (see Fig. 5.5). The element will change length in the x direction only if $\partial u/\partial x$ is other than zero. Similarly, a change in the y dimension requires a nonzero value of $\partial v/\partial y$ and a change in the z dimension requires a nonzero value of $\partial w/\partial z$. These quantities represent the components of longitudinal rates of strain in the x, y, and z directions, respectively.

Changes in length of the sides may produce changes in volume of the element. The rate of local instantaneous *volume dilation* is given by

$$\text{Volume dilation rate} = \frac{\partial u}{\partial x} + \frac{\partial v}{\partial y} + \frac{\partial w}{\partial z} = \nabla \cdot \vec{V} \tag{5.20}$$

For incompressible flow, the rate of volume dilation is zero (Eq. 5.1c).

Example 5.8 DEFORMATION RATES FOR FLOW IN A CORNER

The velocity field $\vec{V} = Ax\hat{i} - Ay\hat{j}$ represents flow in a "corner," as shown in Example 5.4, where $A = 0.3\text{ s}^{-1}$ and the coordinates are measured in meters. A square is marked in the fluid as shown at $t = 0$. Evaluate the new positions of the four corner points when point a has moved to $x = \frac{3}{2}$ m after τ seconds. Evaluate the rates of linear deformation in the x and y directions. Compare area $a'b'c'd'$ at $t = \tau$ with area $abcd$ at $t = 0$. Comment on the significance of this result.

Given: $\vec{V} = Ax\hat{i} - Ay\hat{j}$; $A = 0.3\text{ s}^{-1}$, x and y in meters.

Find: (a) Position of square at $t = \tau$ when a is at a' at $x = \frac{3}{2}$ m.

(b) Rates of linear deformation.

(c) Area $a'b'c'd'$ compared with area $abcd$.

(d) Significance of the results.

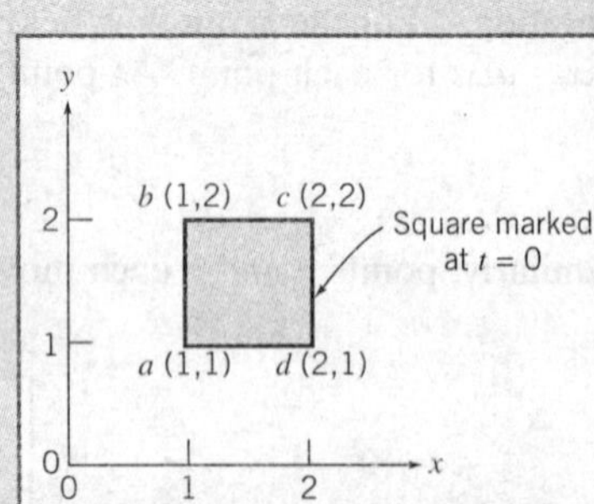

Solution:

First we must find τ, so we must follow a fluid particle using a Lagrangian description. Thus

$$u = \frac{dx_p}{dt} = Ax_p, \quad \frac{dx}{x} = A\,dt, \text{ so } \int_{x_0}^{x}\frac{dx}{x} = \int_0^{\tau} A\,dt \quad \text{and} \quad \ln\frac{x}{x_0} = A\tau$$

$$\tau = \frac{\ln x/x_0}{A} = \frac{\ln\left(\frac{3}{2}\right)}{0.3\text{ s}^{-1}} = 1.35\text{ s}$$

In the y direction

$$v = \frac{dy_p}{dt} = -Ay_p \qquad \frac{dy}{y} = -A\,dt \qquad \frac{y}{y_0} = e^{-A\tau}$$

The point coordinates at τ are:

Point	$t=0$	$t=\tau$
a	(1, 1)	$\left(\frac{3}{2}, \frac{2}{3}\right)$
b	(1, 2)	$\left(\frac{3}{2}, \frac{4}{3}\right)$
c	(2, 2)	$\left(3, \frac{4}{3}\right)$
d	(2, 1)	$\left(3, \frac{2}{3}\right)$

The plot is:

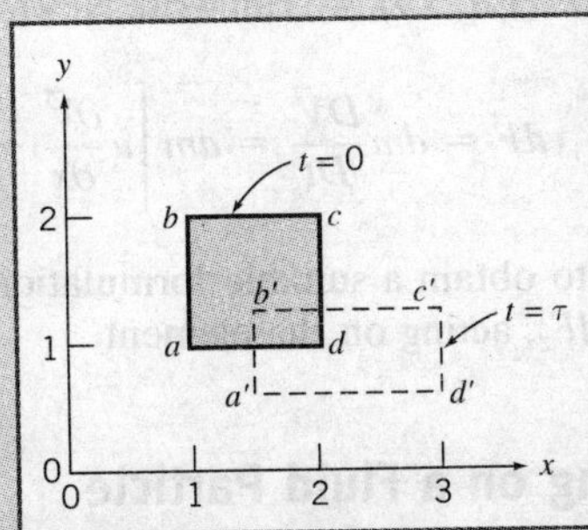

The rates of linear deformation are:

$$\frac{\partial u}{\partial x} = \frac{\partial}{\partial x}Ax = A = 0.3\text{ s}^{-1} \quad \text{in the } x \text{ direction}$$

$$\frac{\partial v}{\partial y} = \frac{\partial}{\partial y}(-Ay) = -A = -0.3\text{ s}^{-1} \quad \text{in the } y \text{ direction}$$

The rate of volume dilation is

$$\nabla \cdot \vec{V} = \frac{\partial u}{\partial x} + \frac{\partial v}{\partial y} = A - A = 0$$

Area $abcd = 1\text{ m}^2$ and area $a'b'c'd' = \left(3 - \frac{3}{2}\right)\left(\frac{4}{3} - \frac{2}{3}\right) = 1\text{ m}^2$.

Notes:

- ✓ Parallel planes remain parallel; there is linear deformation but no angular deformation.
- ✓ The flow is irrotational ($\partial v/\partial x - \partial u/\partial y = 0$).
- ✓ Volume is conserved because the two rates of linear deformation are equal and opposite.
- ✓ The NCFMF video *Flow Visualization* (see http://web.mit.edu/fluids/www/Shapiro/ncfmf.html for free online viewing of this film) uses hydrogen bubble time-streak markers to demonstrate experimentally that the area of a marked fluid square is conserved in two-dimensional incompressible flow.
- The *Excel* workbook for this problem shows an animation of this motion.

We have shown in this section that the velocity field can be used to find the acceleration, rotation, angular deformation, and linear deformation of a fluid particle in a flow field.

Momentum Equation 5.4

A dynamic equation describing fluid motion may be obtained by applying Newton's second law to a particle. To derive the differential form of the momentum equation, we shall apply Newton's second law to an infinitesimal fluid particle of mass dm.

Recall that Newton's second law for a finite system is given by

$$\vec{F} = \left.\frac{d\vec{P}}{dt}\right)_{\text{system}} \tag{4.2a}$$

where the linear momentum, $\vec{P}$, of the system is given by

$$\vec{P}_{\text{system}} = \int_{\text{mass (system)}} \vec{V}\,dm \tag{4.2b}$$

Then, for an infinitesimal system of mass dm, Newton's second law can be written

$$d\vec{F} = dm \left.\frac{d\vec{V}}{dt}\right)_{\text{system}} \tag{5.21}$$

Having obtained an expression for the acceleration of a fluid element of mass dm, moving in a velocity field (Eq. 5.9), we can write Newton's second law as the vector equation

$$d\vec{F} = dm\frac{D\vec{V}}{Dt} = dm\left[u\frac{\partial \vec{V}}{\partial x} + v\frac{\partial \vec{V}}{\partial y} + w\frac{\partial \vec{V}}{\partial z} + \frac{\partial \vec{V}}{\partial t}\right] \tag{5.22}$$

We now need to obtain a suitable formulation for the force, $d\vec{F}$, or its components, dF_x, dF_y, and dF_z, acting on the element.

Forces Acting on a Fluid Particle

Recall that the forces acting on a fluid element may be classified as body forces and surface forces; surface forces include both normal forces and tangential (shear) forces.

We shall consider the x component of the force acting on a differential element of mass dm and volume $d\mathcal{V} = dx\,dy\,dz$. Only those stresses that act in the x direction will give rise to surface forces in the x direction. If the stresses at the center of the differential element are taken to be σ_{xx}, τ_{yx}, and τ_{zx}, then the stresses acting in the x direction on all faces of the element (obtained by a Taylor series expansion about the center of the element) are as shown in Fig. 5.9.

To obtain the net surface force in the x direction, dF_{S_x}, we must sum the forces in the x direction. Thus,

$$\begin{aligned} dF_{S_x} = {} & \left(\sigma_{xx} + \frac{\partial \sigma_{xx}}{\partial x}\frac{dx}{2}\right)dy\,dz - \left(\sigma_{xx} - \frac{\partial \sigma_{xx}}{\partial x}\frac{dx}{2}\right)dy\,dz \\ & + \left(\tau_{yx} + \frac{\partial \tau_{yx}}{\partial y}\frac{dy}{2}\right)dx\,dz - \left(\tau_{yx} - \frac{\partial \tau_{yx}}{\partial y}\frac{dy}{2}\right)dx\,dz \\ & + \left(\tau_{zx} + \frac{\partial \tau_{zx}}{\partial z}\frac{dz}{2}\right)dx\,dy - \left(\tau_{zx} - \frac{\partial \tau_{zx}}{\partial z}\frac{dz}{2}\right)dx\,dy \end{aligned}$$

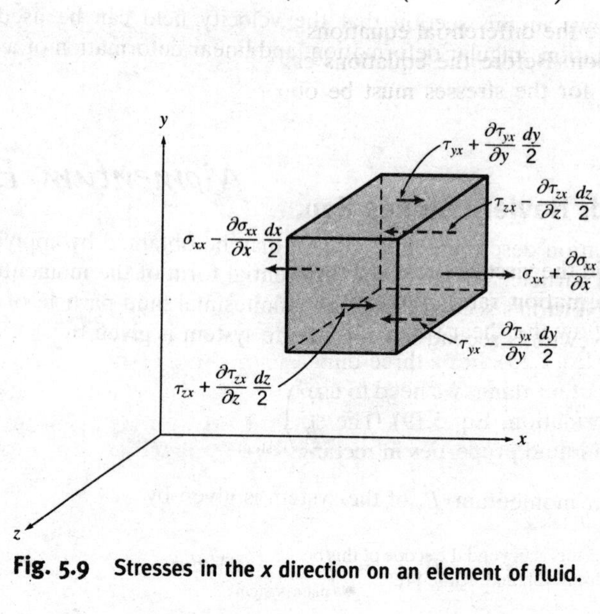

Fig. 5.9 Stresses in the x direction on an element of fluid.

On simplifying, we obtain

$$dF_{S_x} = \left(\frac{\partial \sigma_{xx}}{\partial x} + \frac{\partial \tau_{yx}}{\partial y} + \frac{\partial \tau_{zx}}{\partial z}\right) dx\,dy\,dz$$

When the force of gravity is the only body force acting, then the body force per unit mass is $\vec{g}$. The net force in the x direction, dF_x, is given by

$$dF_x = dF_{B_x} + dF_{S_x} = \left(\rho g_x + \frac{\partial \sigma_{xx}}{\partial x} + \frac{\partial \tau_{yx}}{\partial y} + \frac{\partial \tau_{zx}}{\partial z}\right) dx\,dy\,dz \tag{5.23a}$$

We can derive similar expressions for the force components in the y and z directions:

$$dF_y = dF_{B_y} + dF_{S_y} = \left(\rho g_y + \frac{\partial \tau_{xy}}{\partial x} + \frac{\partial \sigma_{yy}}{\partial y} + \frac{\partial \tau_{zy}}{\partial z}\right) dx\,dy\,dz \tag{5.23b}$$

$$dF_z = dF_{B_z} + dF_{S_z} = \left(\rho g_z + \frac{\partial \tau_{xz}}{\partial x} + \frac{\partial \tau_{yz}}{\partial y} + \frac{\partial \sigma_{zz}}{\partial z}\right) dx\,dy\,dz \tag{5.23c}$$

Differential Momentum Equation

We have now formulated expressions for the components, dF_x, dF_y, and dF_z, of the force, $d\vec{F}$, acting on the element of mass dm. If we substitute these expressions (Eqs. 5.23) for the force components into the x, y, and z components of Eq. 5.22, we obtain the differential equations of motion,

$$\rho g_x + \frac{\partial \sigma_{xx}}{\partial x} + \frac{\partial \tau_{yx}}{\partial y} + \frac{\partial \tau_{zx}}{\partial z} = \rho\left(\frac{\partial u}{\partial t} + u\frac{\partial u}{\partial x} + v\frac{\partial u}{\partial y} + w\frac{\partial u}{\partial z}\right) \tag{5.24a}$$

$$\rho g_y + \frac{\partial \tau_{xy}}{\partial x} + \frac{\partial \sigma_{yy}}{\partial y} + \frac{\partial \tau_{zy}}{\partial z} = \rho\left(\frac{\partial v}{\partial t} + u\frac{\partial v}{\partial x} + v\frac{\partial v}{\partial y} + w\frac{\partial v}{\partial z}\right) \tag{5.24b}$$

$$\rho g_z + \frac{\partial \tau_{xz}}{\partial x} + \frac{\partial \tau_{yz}}{\partial y} + \frac{\partial \sigma_{zz}}{\partial z} = \rho\left(\frac{\partial w}{\partial t} + u\frac{\partial w}{\partial x} + v\frac{\partial w}{\partial y} + w\frac{\partial w}{\partial z}\right) \tag{5.24c}$$

Equations 5.24 are the differential equations of motion for any fluid satisfying the continuum assumption. Before the equations can be used to solve for u, v, and w, suitable expressions for the stresses must be obtained in terms of the velocity and pressure fields.

Newtonian Fluid: Navier–Stokes Equations

For a Newtonian fluid the viscous stress is directly proportional to the rate of shearing strain (angular deformation rate). We saw in Chapter 2 that for one-dimensional laminar Newtonian flow the shear stress is proportional to the rate of angular deformation: $\tau_{yx} = du/dy$ (Eq. 2.15). For a three-dimensional flow the situation is a bit more complicated (among other things we need to use the more complicated expressions for rate of angular deformation, Eq. 5.19). The stresses may be expressed in terms of velocity gradients and fluid properties in rectangular coordinates as follows:[6]

[6]The derivation of these results is beyond the scope of this book. Detailed derivations may be found in Daily and Harleman [2], Schlichting [3], and White [4].

$$\tau_{xy} = \tau_{yx} = \mu\left(\frac{\partial v}{\partial x} + \frac{\partial u}{\partial y}\right) \tag{5.25a}$$

$$\tau_{yz} = \tau_{zy} = \mu\left(\frac{\partial w}{\partial y} + \frac{\partial v}{\partial z}\right) \tag{5.25b}$$

$$\tau_{zx} = \tau_{xz} = \mu\left(\frac{\partial u}{\partial z} + \frac{\partial w}{\partial x}\right) \tag{5.25c}$$

$$\sigma_{xx} = -p - \frac{2}{3}\mu\nabla\cdot\vec{V} + 2\mu\frac{\partial u}{\partial x} \tag{5.25d}$$

$$\sigma_{yy} = -p - \frac{2}{3}\mu\nabla\cdot\vec{V} + 2\mu\frac{\partial v}{\partial y} \tag{5.25e}$$

$$\sigma_{zz} = -p - \frac{2}{3}\mu\nabla\cdot\vec{V} + 2\mu\frac{\partial w}{\partial z} \tag{5.25f}$$

where p is the local thermodynamic pressure.[7] Thermodynamic pressure is related to the density and temperature by the thermodynamic relation usually called the equation of state.

If these expressions for the stresses are introduced into the differential equations of motion (Eqs. 5.24), we obtain

$$\rho\frac{Du}{Dt} = \rho g_x - \frac{\partial p}{\partial x} + \frac{\partial}{\partial x}\left[\mu\left(2\frac{\partial u}{\partial x} - \frac{2}{3}\nabla\cdot\vec{V}\right)\right] + \frac{\partial}{\partial y}\left[\mu\left(\frac{\partial u}{\partial y} + \frac{\partial v}{\partial x}\right)\right] + \frac{\partial}{\partial z}\left[\mu\left(\frac{\partial w}{\partial x} + \frac{\partial u}{\partial z}\right)\right] \tag{5.26a}$$

$$\rho\frac{Dv}{Dt} = \rho g_y - \frac{\partial p}{\partial y} + \frac{\partial}{\partial x}\left[\mu\left(\frac{\partial u}{\partial y} + \frac{\partial v}{\partial x}\right)\right] + \frac{\partial}{\partial y}\left[\mu\left(2\frac{\partial v}{\partial y} - \frac{2}{3}\nabla\cdot\vec{V}\right)\right] + \frac{\partial}{\partial z}\left[\mu\left(\frac{\partial v}{\partial z} + \frac{\partial w}{\partial y}\right)\right] \tag{5.26b}$$

$$\rho\frac{Dw}{Dt} = \rho g_z - \frac{\partial p}{\partial z} + \frac{\partial}{\partial x}\left[\mu\left(\frac{\partial w}{\partial x} + \frac{\partial u}{\partial z}\right)\right] + \frac{\partial}{\partial y}\left[\mu\left(\frac{\partial v}{\partial z} + \frac{\partial w}{\partial y}\right)\right] + \frac{\partial}{\partial z}\left[\mu\left(2\frac{\partial w}{\partial z} - \frac{2}{3}\nabla\cdot\vec{V}\right)\right] \tag{5.26c}$$

These equations of motion are called the *Navier–Stokes* equations. The equations are greatly simplified when applied to *incompressible flow* with *constant viscosity*. Under these conditions the equations reduce to

$$\rho\left(\frac{\partial u}{\partial t} + u\frac{\partial u}{\partial x} + v\frac{\partial u}{\partial y} + w\frac{\partial u}{\partial z}\right) = \rho g_x - \frac{\partial p}{\partial x} + \mu\left(\frac{\partial^2 u}{\partial x^2} + \frac{\partial^2 u}{\partial y^2} + \frac{\partial^2 u}{\partial z^2}\right) \tag{5.27a}$$

[7] Sabersky et al. [5] discuss the relation between the thermodynamic pressure and the average pressure defined as $p = -(\sigma_{xx} + \sigma_{yy} + \sigma_{zz})/3$.

$$\rho\left(\frac{\partial v}{\partial t}+u\frac{\partial v}{\partial x}+v\frac{\partial v}{\partial y}+w\frac{\partial v}{\partial z}\right)=\rho g_y-\frac{\partial p}{\partial y}+\mu\left(\frac{\partial^2 v}{\partial x^2}+\frac{\partial^2 v}{\partial y^2}+\frac{\partial^2 v}{\partial z^2}\right) \quad (5.27b)$$

$$\rho\left(\frac{\partial w}{\partial t}+u\frac{\partial w}{\partial x}+v\frac{\partial w}{\partial y}+w\frac{\partial w}{\partial z}\right)=\rho g_z-\frac{\partial p}{\partial z}+\mu\left(\frac{\partial^2 w}{\partial x^2}+\frac{\partial^2 w}{\partial y^2}+\frac{\partial^2 w}{\partial z^2}\right) \quad (5.27c)$$

This form of the Navier–Stokes equations is probably (next to the Bernoulli equation) the most famous set of equations in fluid mechanics, and has been widely studied. These equations, with the continuity equation (Eq. 5.1c), form a set of four coupled nonlinear partial differential equations for u, v, w, and p. In principle, these four equations describe many common flows; the only restrictions are that the fluid be Newtonian (with a constant viscosity) and incompressible. For example, lubrication theory (describing the behavior of machine bearings), pipe flows, and even the motion of your coffee as you stir it are explained by these equations. Unfortunately, they are impossible to solve analytically, except for the most basic cases [3], in which we have simple boundaries and initial or boundary conditions! We will solve the equations for such a simple problem in Example 5.9.

The Navier–Stokes equations for constant density and viscosity are given in cylindrical coordinates in Appendix B; they have also been derived for spherical coordinates [3]. We will apply the cylindrical coordinate form in solving Example 5.10.

In recent years computational fluid dynamics (CFD) computer applications (such as *Fluent* [6] and *STAR-CD* [7]) have been developed for analyzing the Navier–Stokes equations for more complicated, real-world problems. Although a detailed treatment of the topic is beyond the scope of this text, we shall have a brief introduction to CFD in the next section.

For the case of frictionless flow ($\mu = 0$) the equations of motion (Eqs. 5.26 or Eqs. 5.27) reduce to *Euler's equation*,

$$\rho\frac{D\vec{V}}{Dt}=\rho\vec{g}-\nabla p$$

We shall consider the case of frictionless flow in Chapter 6.

Example 5.9 ANALYSIS OF FULLY DEVELOPED LAMINAR FLOW DOWN AN INCLINED PLANE SURFACE

A liquid flows down an inclined plane surface in a steady, fully developed laminar film of thickness h. Simplify the continuity and Navier–Stokes equations to model this flow field. Obtain expressions for the liquid velocity profile, the shear stress distribution, the volume flow rate, and the average velocity. Relate the liquid film thickness to the volume flow rate per unit depth of surface normal to the flow. Calculate the volume flow rate in a film of water $h = 1$ mm thick, flowing on a surface $b = 1$ m wide, inclined at $\theta = 15°$ to the horizontal.

Given: Liquid flow down an inclined plane surface in a steady, fully developed laminar film of thickness h.

Find:
(a) Continuity and Navier–Stokes equations simplified to model this flow field.
(b) Velocity profile.
(c) Shear stress distribution.
(d) Volume flow rate per unit depth of surface normal to diagram.
(e) Average flow velocity.
(f) Film thickness in terms of volume flow rate per unit depth of surface normal to diagram.
(g) Volume flow rate in a film of water 1 mm thick on a surface 1 m wide, inclined at 15° to the horizontal.

Solution:

The geometry and coordinate system used to model the flow field are shown. (It is convenient to align one coordinate with the flow down the plane surface.)

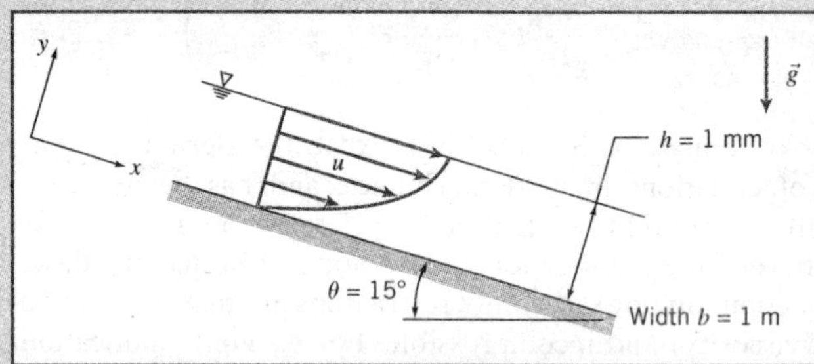

The governing equations written for incompressible flow with constant viscosity are

$$\frac{\partial u}{\partial x} + \frac{\partial v}{\partial y} + \frac{\partial w}{\partial z} = 0 \tag{5.1c}$$

$$\rho\left(\frac{\partial u}{\partial t} + u\frac{\partial u}{\partial x} + v\frac{\partial u}{\partial y} + w\frac{\partial u}{\partial z}\right) = \rho g_x - \frac{\partial p}{\partial x} + \mu\left(\frac{\partial^2 u}{\partial x^2} + \frac{\partial^2 u}{\partial y^2} + \frac{\partial^2 u}{\partial z^2}\right) \tag{5.27a}$$

$$\rho\left(\frac{\partial v}{\partial t} + u\frac{\partial v}{\partial x} + v\frac{\partial v}{\partial y} + w\frac{\partial v}{\partial z}\right) = \rho g_y - \frac{\partial p}{\partial y} + \mu\left(\frac{\partial^2 v}{\partial x^2} + \frac{\partial^2 v}{\partial y^2} + \frac{\partial^2 v}{\partial z^2}\right) \tag{5.27b}$$

$$\rho\left(\frac{\partial w}{\partial t} + u\frac{\partial w}{\partial x} + v\frac{\partial w}{\partial y} + w\frac{\partial w}{\partial z}\right) = \rho g_z - \frac{\partial p}{\partial z} + \mu\left(\frac{\partial^2 w}{\partial x^2} + \frac{\partial^2 w}{\partial y^2} + \frac{\partial^2 w}{\partial z^2}\right) \tag{5.27c}$$

The terms canceled to simplify the basic equations are keyed by number to the assumptions listed below. The assumptions are discussed in the order in which they are applied to simplify the equations.

Assumptions:
(1) Steady flow (given).
(2) Incompressible flow; ρ = constant.
(3) No flow or variation of properties in the z direction; $w = 0$ and $\partial/\partial z = 0$.
(4) Fully developed flow, so no properties vary in the x direction; $\partial/\partial x = 0$.

Assumption (1) eliminates time variations in any fluid property.

Assumption (2) eliminates space variations in density.

Assumption (3) states that there is no z component of velocity and no property variations in the z direction. All terms in the z component of the Navier–Stokes equation cancel.

After assumption (4) is applied, the continuity equation reduces to $\partial v/\partial y = 0$. Assumptions (3) and (4) also indicate that $\partial v/\partial z = 0$ and $\partial v/\partial x = 0$. Therefore v must be constant. Since v is zero at the solid surface, then v must be zero everywhere.

The fact that $v = 0$ reduces the Navier–Stokes equations further, as indicated by (5) in Eqs 5.27a and 5.27b. The final simplified equations are

$$0 = \rho g_x + \mu\frac{\partial^2 u}{\partial y^2} \tag{1}$$

$$0 = \rho g_y - \frac{\partial p}{\partial y} \tag{2}$$

Since $\partial u/\partial z = 0$ (assumption 3) and $\partial u/\partial x = 0$ (assumption 4), then u is at most a function of y, and $\partial^2 u/\partial y^2 = d^2u/dy^2$, and from Eq. 1, then

$$\frac{d^2u}{dy^2} = -\frac{\rho g_x}{\mu} = -\rho g\frac{\sin\theta}{\mu}$$

Integrating,

$$\frac{du}{dy} = -\rho g \frac{\sin\theta}{\mu} y + c_1 \tag{3}$$

and integrating again,

$$u = -\rho g \frac{\sin\theta}{\mu} \frac{y^2}{2} + c_1 y + c_2 \tag{4}$$

The boundary conditions needed to evaluate the constants are the no-slip condition at the solid surface ($u = 0$ at $y = 0$) and the zero-shear-stress condition at the liquid free surface ($du/dy = 0$ at $y = h$).

Evaluating Eq. 4 at $y = 0$ gives $c_2 = 0$. From Eq. 3 at $y = h$,

$$0 = -\rho g \frac{\sin\theta}{\mu} h + c_1$$

or

$$c_1 = \rho g \frac{\sin\theta}{\mu} h$$

Substituting into Eq. 4 we obtain the velocity profile

$$u = -\rho g \frac{\sin\theta}{\mu} \frac{y^2}{2} + \rho g \frac{\sin\theta}{\mu} hy$$

or

$$u = \rho g \frac{\sin\theta}{\mu} \left(hy - \frac{y^2}{2} \right) \longleftarrow \qquad u(y)$$

The shear stress distribution is (from Eq. 5.25a after setting $\partial v/\partial x$ to zero, or alternatively, for one-dimensional flow, from Eq. 2.15)

$$\tau_{yx} = \mu \frac{du}{dy} = \rho g \sin\theta\,(h - y) \longleftarrow \qquad \tau_{yx}(y)$$

The shear stress in the fluid reaches its maximum value at the wall ($y = 0$); as we expect, it is zero at the free surface ($y = h$). At the wall the shear stress τ_{yx} is positive but the surface normal *for the fluid* is in the negative y direction; hence the shear force acts in the negative x direction, and just balances the x component of the body force acting on the fluid. The volume flow rate is

$$Q = \int_A u\, dA = \int_0^h u\, b dy$$

where b is the surface width in the z direction. Substituting,

$$Q = \int_0^h \frac{\rho g \sin\theta}{\mu} \left(hy - \frac{y^2}{2} \right) b\, dy = \rho g \frac{\sin\theta\, b}{\mu} \left[\frac{hy^2}{2} - \frac{y^3}{6} \right]_0^h$$

$$Q = \frac{\rho g \sin\theta\, b}{\mu} \frac{h^3}{3} \longleftarrow \qquad (5)\,Q$$

The average flow velocity is $\overline{V} = Q/A = Q/bh$. Thus

$$\overline{V} = \frac{Q}{bh} = \frac{\rho g \sin\theta}{\mu} \frac{h^2}{3} \longleftarrow \qquad \overline{V}$$

Solving for film thickness gives

$$h = \left[\frac{3\mu Q}{\rho g \sin\theta\, b}\right]^{1/3} \qquad (6)\; h$$

A film of water $h = 1$ mm thick on a plane $b = 1$ m wide, inclined at $\theta = 15°$, would carry

$$Q = 999\frac{\text{kg}}{\text{m}^3} \times 9.81\frac{\text{m}}{\text{s}^2} \times \sin(15°) \times 1\,\text{m} \times \frac{\text{m}\cdot\text{s}}{1.00\times10^{-3}\,\text{kg}} \times \frac{(0.001)^3\,\text{m}^3}{3} \times 1000\frac{\text{L}}{\text{m}^3}$$

$$Q = 0.846\ \text{L/s} \qquad Q$$

Notes:

- ✓ This problem illustrates how the full Navier–Stokes equations (Eqs. 5.27) can sometimes be reduced to a set of solvable equations (Eqs. 1 and 2 in this problem).
- ✓ After integration of the simplified equations, boundary (or initial) conditions are used to complete the solution.
- ✓ Once the velocity field is obtained, other useful quantities (e.g., shear stress, volume flow rate) can be found.
- ✓ Equations (5) and (6) show that even for fairly simple problems the results can be quite complicated: The depth of the flow depends in a nonlinear way on flow rate ($h \propto Q^{1/3}$).

Example 5.10 ANALYSIS OF LAMINAR VISCOMETRIC FLOW BETWEEN COAXIAL CYLINDERS

A viscous liquid fills the annular gap between vertical concentric cylinders. The inner cylinder is stationary, and the outer cylinder rotates at constant speed. The flow is laminar. Simplify the continuity, Navier–Stokes, and tangential shear stress equations to model this flow field. Obtain expressions for the liquid velocity profile and the shear stress distribution. Compare the shear stress at the surface of the inner cylinder with that computed from a planar approximation obtained by "unwrapping" the annulus into a plane and assuming a linear velocity profile across the gap. Determine the ratio of cylinder radii for which the planar approximation predicts the correct shear stress at the surface of the inner cylinder within 1 percent.

Given: Laminar viscometric flow of liquid in annular gap between vertical concentric cylinders. The inner cylinder is stationary, and the outer cylinder rotates at constant speed.

Find:
(a) Continuity and Navier–Stokes equations simplified to model this flow field.
(b) Velocity profile in the annular gap.
(c) Shear stress distribution in the annular gap.
(d) Shear stress at the surface of the inner cylinder.
(e) Comparison with "planar" approximation for constant shear stress in the narrow gap between cylinders.
(f) Ratio of cylinder radii for which the planar approximation predicts shear stress within 1 percent of the correct value.

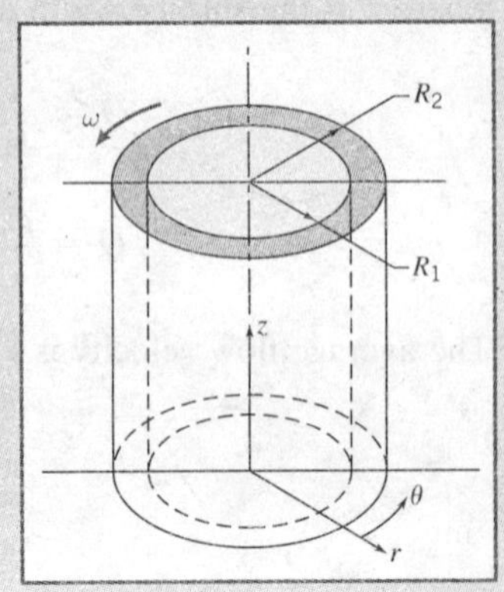

Solution:
The geometry and coordinate system used to model the flow field are shown. (The z coordinate is directed vertically upward; as a consequence, $g_r = g_\theta = 0$ and $g_z = -g$.)

The continuity, Navier–Stokes, and tangential shear stress equations (from Appendix B) written for incompressible flow with constant viscosity are

$$\frac{1}{r}\frac{\partial}{\partial r}(rv_r) + \frac{1}{r}\cancelto{4}{\frac{\partial}{\partial \theta}(v_\theta)} + \cancelto{3}{\frac{\partial}{\partial z}(v_z)} = 0 \tag{B.1}$$

r component:

$$\rho\left(\cancelto{1}{\frac{\partial v_r}{\partial t}} + v_r\cancelto{5}{\frac{\partial v_r}{\partial r}} + \frac{v_\theta}{r}\cancelto{4}{\frac{\partial v_r}{\partial \theta}} - \frac{v_\theta^2}{r} + v_z\cancelto{3}{\frac{\partial v_r}{\partial z}}\right)$$
$$= \cancelto{0}{\rho g_r} - \frac{\partial p}{\partial r} + \mu\left\{\cancelto{5}{\frac{\partial}{\partial r}\left(\frac{1}{r}\frac{\partial}{\partial r}[rv_r]\right)} + \frac{1}{r^2}\cancelto{4}{\frac{\partial^2 v_r}{\partial \theta^2}} - \frac{2}{r^2}\cancelto{4}{\frac{\partial v_\theta}{\partial \theta}} + \cancelto{3}{\frac{\partial^2 v_r}{\partial z^2}}\right\} \tag{B.3a}$$

θ component:

$$\rho\left(\cancelto{1}{\frac{\partial v_\theta}{\partial t}} + \cancelto{5}{v_r\frac{\partial v_\theta}{\partial r}} + \frac{v_\theta}{r}\cancelto{4}{\frac{\partial v_\theta}{\partial \theta}} + \cancelto{5}{\frac{v_r v_\theta}{r}} + v_z\cancelto{3}{\frac{\partial v_\theta}{\partial z}}\right)$$
$$= \cancelto{0}{\rho g_\theta} - \frac{1}{r}\cancelto{4}{\frac{\partial p}{\partial \theta}} + \mu\left\{\frac{\partial}{\partial r}\left(\frac{1}{r}\frac{\partial}{\partial r}[rv_\theta]\right) + \frac{1}{r^2}\cancelto{4}{\frac{\partial^2 v_\theta}{\partial \theta^2}} + \frac{2}{r^2}\cancelto{4}{\frac{\partial v_r}{\partial \theta}} + \cancelto{3}{\frac{\partial^2 v_\theta}{\partial z^2}}\right\} \tag{B.3b}$$

z component:

$$\rho\left(\cancelto{1}{\frac{\partial v_z}{\partial t}} + \cancelto{5}{v_r\frac{\partial v_z}{\partial r}} + \frac{v_\theta}{r}\cancelto{4}{\frac{\partial v_z}{\partial \theta}} + v_z\cancelto{3}{\frac{\partial v_z}{\partial z}}\right) = \rho g_z - \frac{\partial p}{\partial z} + \mu\left\{\frac{1}{r}\frac{\partial}{\partial r}\cancelto{3}{\left(r\frac{\partial v_z}{\partial r}\right)} + \frac{1}{r^2}\cancelto{3}{\frac{\partial^2 v_z}{\partial \theta^2}} + \cancelto{3}{\frac{\partial^2 v_z}{\partial z^2}}\right\} \tag{B.3c}$$

$$\tau_{r\theta} = \mu\left[r\frac{\partial}{\partial r}\left(\frac{v_\theta}{r}\right) + \frac{1}{r}\cancelto{4}{\frac{\partial v_r}{\partial \theta}}\right] \tag{B.2}$$

The terms canceled to simplify the basic equations are keyed by number to the assumptions listed below. The assumptions are discussed in the order in which they are applied to simplify the equations.

Assumptions:
(1) Steady flow; angular speed of outer cylinder is constant.
(2) Incompressible flow; $\rho = \text{constant}$.
(3) No flow or variation of properties in the z direction; $v_z = 0$ and $\partial/\partial z = 0$.
(4) Circumferentially symmetric flow, so properties do not vary with θ, so $\partial/\partial\theta = 0$.

Assumption (1) eliminates time variations in fluid properties.

Assumption (2) eliminates space variations in density.

Assumption (3) causes all terms in the z component of the Navier–Stokes equation (Eq. B.3c) to cancel, except for the hydrostatic pressure distribution.

After assumptions (3) and (4) are applied, the continuity equation (Eq. B.1) reduces to

$$\frac{1}{r}\frac{\partial}{\partial r}(rv_r) = 0$$

Because $\partial/\partial\theta = 0$ and $\partial/\partial z = 0$ by assumptions (3) and (4), then $\frac{\partial}{\partial r} \to \frac{d}{dr}$, so integrating gives

$$rv_r = \text{constant}$$

Since v_r is zero at the solid surface of each cylinder, then v_r must be zero everywhere.

The fact that $v_r = 0$ reduces the Navier–Stokes equations further, as indicated by cancellations (5). The final equations (Eqs. B.3a and B.3b) reduce to

$$-\rho\frac{v_\theta^2}{r} = -\frac{\partial p}{\partial r}$$

$$0 = \mu\left\{\frac{\partial}{\partial r}\left(\frac{1}{r}\frac{\partial}{\partial r}[rv_\theta]\right)\right\}$$

But since $\partial/\partial\theta = 0$ and $\partial/\partial z = 0$ by assumptions (3) and (4), then v_θ is a function of radius only, and

$$\frac{d}{dr}\left(\frac{1}{r}\frac{d}{dr}[rv_\theta]\right) = 0$$

Integrating once,

$$\frac{1}{r}\frac{d}{dr}[rv_\theta] = c_1$$

or

$$\frac{d}{dr}[rv_\theta] = c_1 r$$

Integrating again,

$$rv_\theta = c_1\frac{r^2}{2} + c_2 \qquad \text{or} \qquad v_\theta = c_1\frac{r}{2} + c_2\frac{1}{r}$$

Two boundary conditions are needed to evaluate constants c_1 and c_2. The boundary conditions are

$$v_\theta = \omega R_2 \quad \text{at} \quad r = R_2 \quad \text{and}$$
$$v_\theta = 0 \quad \text{at} \quad r = R_1$$

Substituting

$$\omega R_2 = c_1\frac{R_2}{2} + c_2\frac{1}{R_2}$$

$$0 = c_1\frac{R_1}{2} + c_2\frac{1}{R_1}$$

After considerable algebra

$$c_1 = \frac{2\omega}{1-\left(\frac{R_1}{R_2}\right)^2} \quad \text{and} \quad c_2 = \frac{-\omega R_1^2}{1-\left(\frac{R_1}{R_2}\right)^2}$$

Substituting into the expression for v_θ,

$$v_\theta = \frac{\omega r}{1-\left(\frac{R_1}{R_2}\right)^2} - \frac{\omega R_1^2/r}{1-\left(\frac{R_1}{R_2}\right)^2} = \frac{\omega R_1}{1-\left(\frac{R_1}{R_2}\right)^2}\left[\frac{r}{R_1} - \frac{R_1}{r}\right] \qquad v_\theta(r)$$

The shear stress distribution is obtained from Eq. B.2 after using assumption (4):

$$\tau_{r\theta} = \mu r \frac{d}{dr}\left(\frac{v_\theta}{r}\right) = \mu r \frac{d}{dr}\left\{\frac{\omega R_1}{1-\left(\frac{R_1}{R_2}\right)^2}\left[\frac{1}{R_1} - \frac{R_1}{r^2}\right]\right\} = \mu r \frac{\omega R_1}{1-\left(\frac{R_1}{R_2}\right)^2}(-2)\left(-\frac{R_1}{r^3}\right)$$

$$\tau_{r\theta} = \mu \frac{2\omega R_1^2}{1-\left(\frac{R_1}{R_2}\right)^2}\frac{1}{r^2}$$ $\tau_{r\theta}$

At the surface of the inner cylinder, $r = R_1$, so

$$\tau_{\text{surface}} = \mu \frac{2\omega}{1-\left(\frac{R_1}{R_2}\right)^2}$$ τ_{surface}

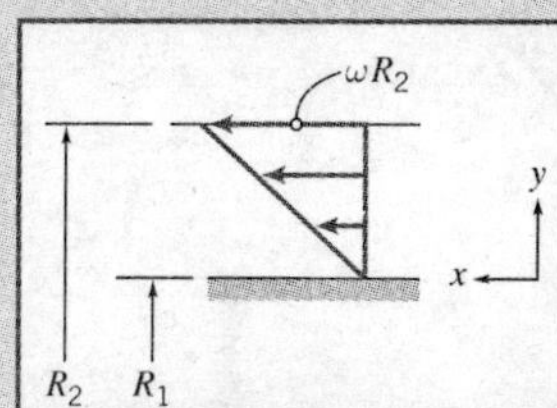

For a "planar" gap

$$\tau_{\text{planar}} = \mu \frac{\Delta v}{\Delta y} = \mu \frac{\omega R_2}{R_2 - R_1}$$

or

$$\tau_{\text{planar}} = \mu \frac{\omega}{1 - \frac{R_1}{R_2}}$$ τ_{planar}

Factoring the denominator of the exact expression for shear stress at the surface gives

$$\tau_{\text{surface}} = \mu \frac{2\omega}{\left(1-\frac{R_1}{R_2}\right)\left(1+\frac{R_1}{R_2}\right)} = \mu \frac{\omega}{1-\frac{R_1}{R_2}} \cdot \frac{2}{1+\frac{R_1}{R_2}}$$

Thus

$$\frac{\tau_{\text{surface}}}{\tau_{\text{planar}}} = \frac{2}{1+\frac{R_1}{R_2}}$$

For 1 percent accuracy,

$$1.01 = \frac{2}{1+\frac{R_1}{R_2}}$$

or

$$\frac{R_1}{R_2} = \frac{1}{1.01}(2 - 1.01) = 0.980$$ $\frac{R_1}{R_2}$

The accuracy criterion is met when the gap width is less than 2 percent of the cylinder radius.

Notes:

- ✓ This problem illustrates how the full Navier–Stokes equations in cylindrical coordinates (Eqs. B.1 to B.3) can sometimes be reduced to a set of solvable equations.
- ✓ As in Example 5.9, after integration of the simplified equations, boundary (or initial) conditions are used to complete the solution.
- ✓ Once the velocity field is obtained, other useful quantities (in this problem, shear stress) can be found.
- The *Excel* workbook for this problem compares the viscometer and linear velocity profiles. It also allows one to derive the appropriate value of the viscometer outer radius to meet a prescribed accuracy of the planar approximation. We will discuss the concentric cylinder–infinite parallel plates approximation again in Chapter 8.

*5.5 Introduction to Computational Fluid Dynamics

In this section we will discuss in a very basic manner the ideas behind *computational fluid dynamics* (CFD). We will first review some very basic ideas in numerically solving an ordinary and a partial differential equation using a spreadsheet such as *Excel*, with a couple of Examples. After studying these, the reader will be able to use the PC to numerically solve a range of simple CFD problems. Then, for those with further interest in CFD, we will review in more detail some concepts behind numerical methods, particularly CFD; this review will highlight some of the advantages and pitfalls of CFD. We will apply some of these concepts to a simple 1D model, but these concepts are so fundamental that they are applicable to almost any CFD calculation. As we apply the CFD solution procedure to the model, we'll comment on the extension to the general case. The goal is to enable the reader to apply the CFD solution procedure to simple nonlinear equations.

The Need for CFD

As discussed in Section 5.4, the equations describing fluid flow can be a bit intimidating. For example, even though we may limit ourselves to incompressible flows for which the viscosity is constant, we still end up with the following equations:

$$\frac{\partial u}{\partial x}+\frac{\partial v}{\partial y}+\frac{\partial w}{\partial z}=0 \tag{5.1c}$$

$$\rho\left(\frac{\partial u}{\partial t}+u\frac{\partial u}{\partial x}+v\frac{\partial u}{\partial y}+w\frac{\partial u}{\partial z}\right)=\rho g_x-\frac{\partial p}{\partial x}+\mu\left(\frac{\partial^2 u}{\partial x^2}+\frac{\partial^2 u}{\partial y^2}+\frac{\partial^2 u}{\partial z^2}\right) \tag{5.27a}$$

$$\rho\left(\frac{\partial v}{\partial t}+u\frac{\partial v}{\partial x}+v\frac{\partial v}{\partial y}+w\frac{\partial v}{\partial z}\right)=\rho g_y-\frac{\partial p}{\partial y}+\mu\left(\frac{\partial^2 v}{\partial x^2}+\frac{\partial^2 v}{\partial y^2}+\frac{\partial^2 v}{\partial z^2}\right) \tag{5.27b}$$

$$\rho\left(\frac{\partial w}{\partial t}+u\frac{\partial w}{\partial x}+v\frac{\partial w}{\partial y}+w\frac{\partial w}{\partial z}\right)=\rho g_z-\frac{\partial p}{\partial z}+\mu\left(\frac{\partial^2 w}{\partial x^2}+\frac{\partial^2 w}{\partial y^2}+\frac{\partial^2 w}{\partial z^2}\right) \tag{5.27c}$$

Equation 5.1c is the continuity equation (mass conservation) and Eqs. 5.27 are the Navier–Stokes equations (momentum), expressed in Cartesian coordinates. In principle, we can solve these equations for the velocity field $\vec{V}=\hat{i}u+\hat{j}v+\hat{k}w$ and pressure field p, given sufficient initial and boundary conditions. Note that in general, u, v, w, and p all depend on x, y, z, and t. In practice, there is no general analytic solution to these equations, for the combined effect of a number of reasons (none of which is insurmountable by itself):

1. They are coupled. The unknowns, u, v, w, and p, appear in all the equations (p is not in Eq. 5.1c) and we cannot manipulate the equations to end up with a single equation for any one of the unknowns. Hence we must solve for all unknowns simultaneously.
2. They are nonlinear. For example, in Eq. 5.27a, the convective acceleration term, $u\,\partial u/\partial x+v\,\partial u/\partial y+w\,\partial u/\partial z$, has products of u with itself as well as with v and w. The consequence of this is that we cannot take one solution to the equations and combine it with a second solution to obtain a third solution. We will see in Chapter 6 that if we can limit ourselves to frictionless flow, we *can* derive linear equations, which will allow us to do this combining procedure (you may wish to look at Table 6.3 for some beautiful examples of this).

*This section may be omitted without loss of continuity in the text material.

3. They are second-order partial differential equations. For example, in Eq. 5.27a, the viscous term, $\mu(\partial^2 u/\partial x^2 + \partial^2 u/\partial y^2 + \partial^2 u/\partial z^2)$, is second-order in u. These are obviously of a different order of complexity (no pun intended) than, say, a first-order ordinary differential equation.

These difficulties have led engineers, scientists, and mathematicians to adopt several approaches to the solution of fluid mechanics problems.

For relatively simple physical geometries and boundary or initial conditions, the equations can often be reduced to a solvable form. We saw two examples of this in Examples 5.9 and 5.10 (for cylindrical forms of the equations).

If we can neglect the viscous terms, the resulting incompressible, inviscid flow can often be successfully analyzed. This is the entire topic of Chapter 6.

Of course, most incompressible flows of interest do not have simple geometries and are not inviscid; for these, we are stuck with Eqs. 5.1c and 5.27. The only option remaining is to use numerical methods to analyze problems. It is possible to obtain approximate computer-based solutions to the equations for a variety of engineering problems. This is the main subject matter of CFD.

Applications of CFD

CFD is employed in a variety of applications and is now widely used in various industries. To illustrate the industrial applications of CFD, we present below some examples developed using FLUENT, a CFD software package from ANSYS, Inc. CFD is used to study the flow field around vehicles including cars, trucks, airplanes, helicopters, and ships. Figure 5.10 shows the paths taken by selected fluid particles around a Formula 1 car. By studying such pathlines and other flow attributes, engineers gain insights into how to design the car so as to reduce drag and enhance performance. The flow through a catalytic converter, a device used to clean automotive exhaust gases so that we can all breathe easier, is shown in Figure 5.11. This image shows path lines colored by velocity magnitude. CFD helps engineers develop more effective catalytic converters by allowing them to study how different chemical species mix and react in the device. Figure 5.12 presents contours of static pressure in a backward-inclined centrifugal fan used in ventilation applications.

VIDEO

Turbulent Flow in a Channel.

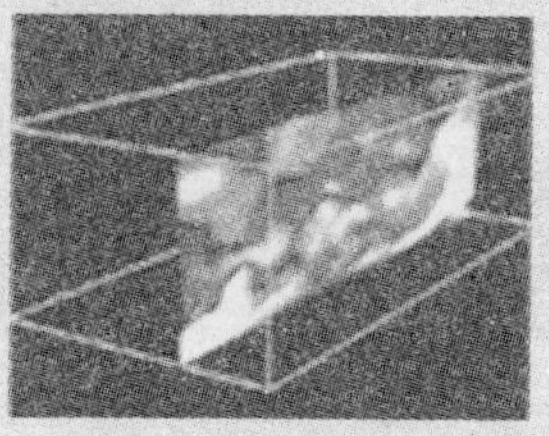

VIDEO

Flow Past a Cylinder.

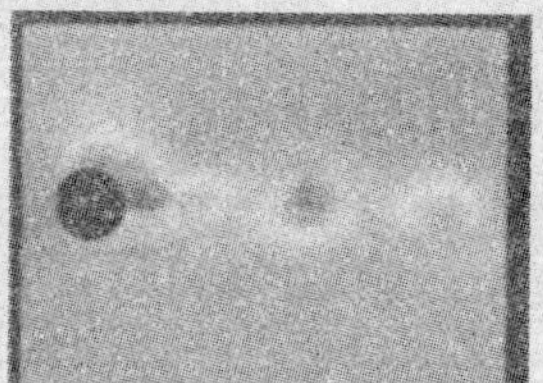

Fig. 5.10 Pathlines around a Formula 1 car. (image courtesy of ANSYS, Inc. © 2008.)

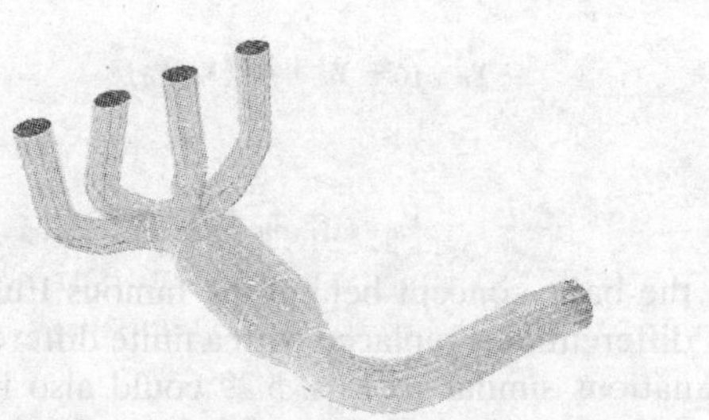

Fig. 5.11 Flow through a catalytic converter. (image courtesy of ANSYS, Inc. ©2008.)

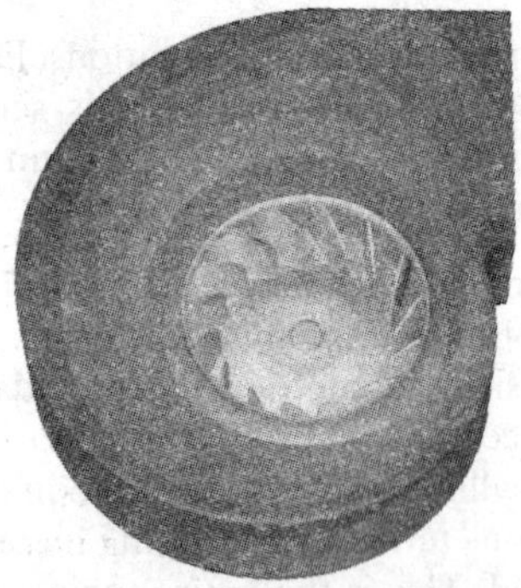

Fig. 5.12 Static pressure contours for flow through a centrifugal fan. (Image courtesy of ANSYS, Inc. © 2008.)

Fan performance characteristics obtained from the CFD simulations compared well with results from physical tests.

CFD is attractive to industry since it is more cost-effective than physical testing. However, we must note that complex flow simulations are challenging and error-prone, and it takes a lot of engineering expertise to obtain realistic solutions.

Some Basic CFD/Numerical Methods Using a Spreadsheet

Before discussing CFD in a little more detail, we can gain insight into numerical methods to solve some simple problems in fluid mechanics by using the spreadsheet. These methods will show how the student may perform elementary CFD using the PC. First, we consider solving the simplest form of a differential equation: a first-order ordinary differential equation:

$$\frac{dy}{dx} = f(x,y) \quad y(x_0) = y_0 \tag{5.28}$$

where $f(x,y)$ is a given function. We realize that graphically the derivative dy/dx is the slope of the (as yet unknown) solution curve $y(x)$. If we are at some point (x_n, y_n) on the curve, we can follow the tangent at that point, as an approximation to actually moving along the curve itself, to find a new value for y, y_{n+1}, corresponding to a new x, x_{n+1}, as shown in Fig. 5.13. We have

$$\frac{dy}{dx} = \frac{y_{n+1} - y_n}{x_{n+1} - x_n}$$

If we choose a *step size* $h = x_{n+1} - x_n$, then the above equation can be combined with the differential equation, Eq. 5.28, to give

$$\frac{dy}{dx} = \frac{y_{n+1} - y_n}{h} = f(x_n, y_n)$$

or

$$y_{n+1} = y_n + hf(x_n, y_n) \tag{5.29a}$$

with

$$x_{n+1} = x_n + h \tag{5.29b}$$

Equations 5.29 are the basic concept behind the famous Euler method for solving a first-order ODE: A differential is replaced with a finite difference. (As we'll see in the next subsection, equations similar to Eqs. 5.29 could also have been derived more formally as the result of a truncated Taylor series.) In these equations, y_{n+1} now represents our best effort to find the next point on the solution curve. From Fig. 5.13, we see that y_{n+1} is *not* on the solution curve but close to it; if we make the triangle

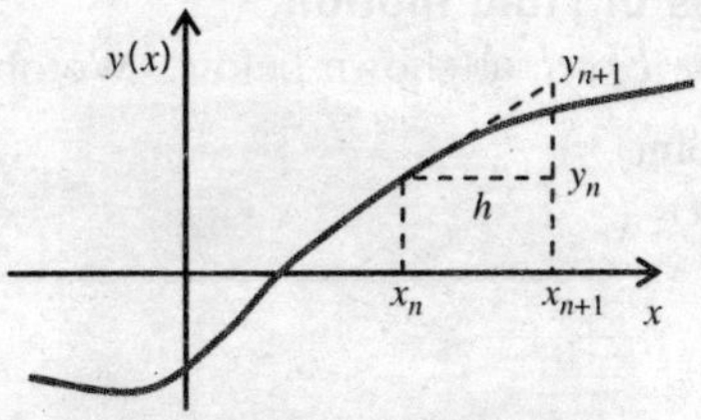

Fig. 5.13 The Euler method.

much smaller, by making the step size h smaller, then y_{n+1} will be even closer to the desired solution. We can repeatedly use the two Euler iteration equations to start at (x_0, y_0) and obtain (x_1, y_1), then (x_2, y_2), (x_3, y_3), and so on. We don't end up with an equation for the solution, but with a set of numbers; hence it is a numerical rather than an analytic method. This is the Euler method approach.

This method is very easy to set up, making it an attractive approach, but it is not very accurate: Following the tangent to a curve at each point, in an attempt to follow the curve, is pretty crude! If we make the step size h smaller, the accuracy of the method will generally increase, but obviously we then need more steps to achieve the solution. It turns out that, if we use too many steps (if h is extremely small), the accuracy of the results can actually *decrease* because, although each small step is very accurate, we will now need so many of them that round-off errors can build up. As with any numerical method, we are not guaranteed to get a solution or one that is very accurate! The Euler method is the simplest but least accurate numerical method for solving a first-order ODE; there are a number of more sophisticated ones available, as discussed in any good numerical methods text [8, 9].

Let's illustrate the method with an Example.

Example 5.11 THE EULER METHOD SOLUTION FOR DRAINING A TANK

A tank contains water at an initial depth $y_0 = 1$ m. The tank diameter is $D = 250$ mm. A hole of diameter $d = 2$ mm appears at the bottom of the tank. A reasonable model for the water level over time is

$$\frac{dy}{dt} = -\left(\frac{d}{D}\right)^2 \sqrt{2gy} \qquad y(0) = y_0$$

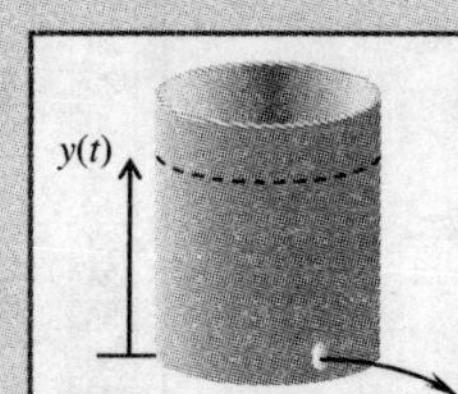

Using 11-point and 21-point Euler methods, estimate the water depth after $t = 100$ min, and compute the errors compared to the exact solution

$$y_{\text{exact}}(t) = \left[\sqrt{y_0} - \left(\frac{d}{D}\right)^2 \sqrt{\frac{g}{2}}\,t\right]^2$$

Plot the Euler and exact results.

Given: Water draining from a tank.

Find: Water depth after 100 min; plot of depth versus time; accuracy of results.

Solution: Use the Euler equations, Eq. (5.29).

Governing equations: $y_{n+1} = y_n + hf(t_n, y_n) \quad t_{n+1} = t_n + h$

with

$$f(t_n, y_n) = -\left(\frac{d}{D}\right)^2 \sqrt{2gy_n} \qquad y_0 = 1$$

(Note that in using Eqs. 5.29 we use t instead of x.)

This is convenient for solving using a spreadsheet such as *Excel*, as shown below. We obtain the following results:

Depth after 100 min = − 0.0021 m (Euler 11 point)
= 0.0102 m (Euler 21 point)
= 0.0224 m (Exact) ← $y(100\,\text{min})$

Error after 100 min = 110% (Euler 11 point)
= 54% (Euler 21 point) ← Error

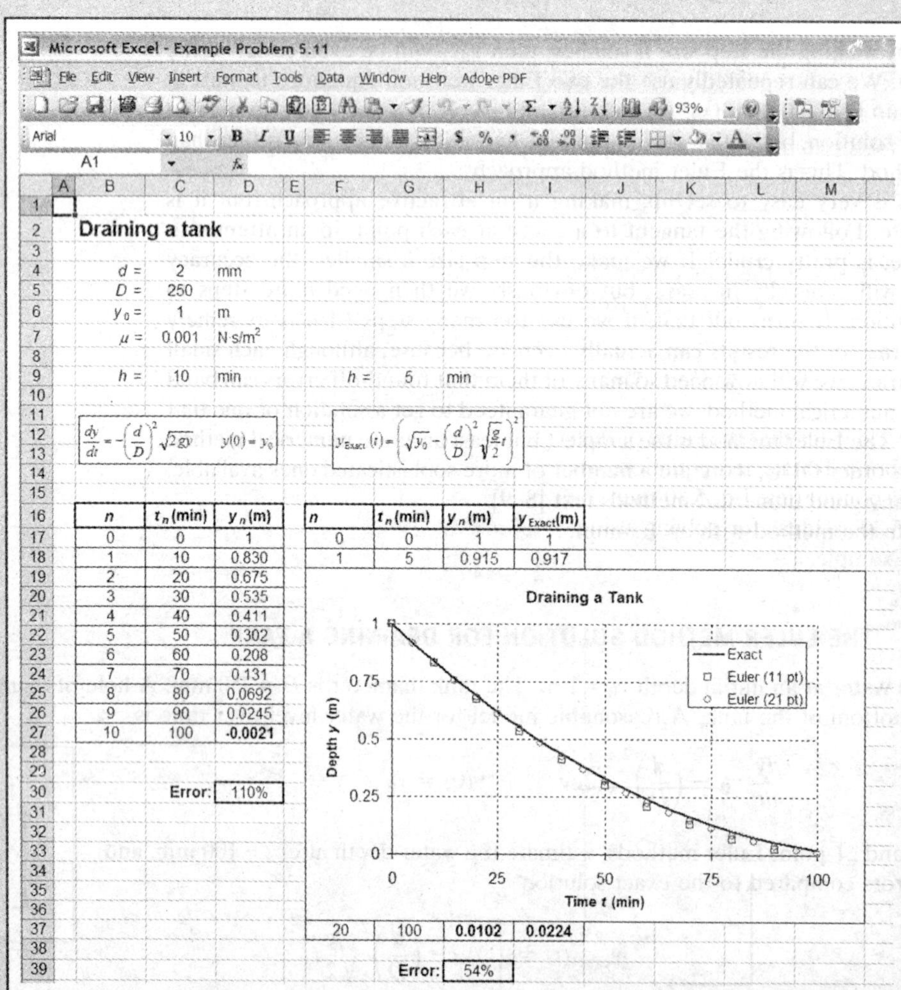

n	t_n (min)	y_n (m)
0	0	1
1	10	0.830
2	20	0.675
3	30	0.535
4	40	0.411
5	50	0.302
6	60	0.208
7	70	0.131
8	80	0.0692
9	90	0.0245
10	100	**-0.0021**

n	t_n (min)	y_n (m)	y_{Exact} (m)
0	0	1	1
1	5	0.915	0.917
20	100	**0.0102**	**0.0224**

This Example shows a simple application of the Euler method. Note that although the errors after 100 min are large for both Euler solutions, their plots are reasonably close to the exact solution.

The *Excel* workbook for this problem can be modified for solving a variety of fluids problems that involve first order ODEs.

Another basic application of a numerical method to a fluid mechanics problem is when we have two-dimensional, steady, incompressible, inviscid flow. These seem like a severe set of restrictions on the flow, but analysis of flows with these assumptions leads to very good predictions for real flows, for example, for the lift on a wing section. This is the topic of Chapter 6, but for now we simply state that under many circumstances such flows can be modeled with the Laplace equation,

$$\frac{\partial^2 \psi}{\partial x^2} + \frac{\partial^2 \psi}{\partial y^2} = 0$$

where ψ is the stream function. We leave out the steps here (they consist of approximating each differential with a Taylor series), but a numerical approximation of this equation is

$$\frac{\psi_{i+1,j} + \psi_{i-1,j}}{h^2} + \frac{\psi_{i,j+1} + \psi_{i,j-1}}{h^2} - 4\frac{\psi_{i,j}}{h^2} = 0$$

Here h is the step size in the x or y direction, and $\psi_{i,j}$ is the value of ψ at the ith value of x and jth value of y (see Fig. 5.14). Rearranging and simplifying,

$$\psi_{i,j} = \frac{1}{4}\left(\psi_{i+1,j} + \psi_{i-1,j} + \psi_{i,j+1} + \psi_{i,j-1}\right) \tag{5.30}$$

This equation indicates that the value of the stream function ψ is simply the average of its four neighbors! To use this equation, we need to specify the values of the stream function at all boundaries; Eq. 5.30 then allows computation of interior values.

Equation 5.30 is ideal for solving using a spreadsheet such as *Excel*. We again consider an Example.

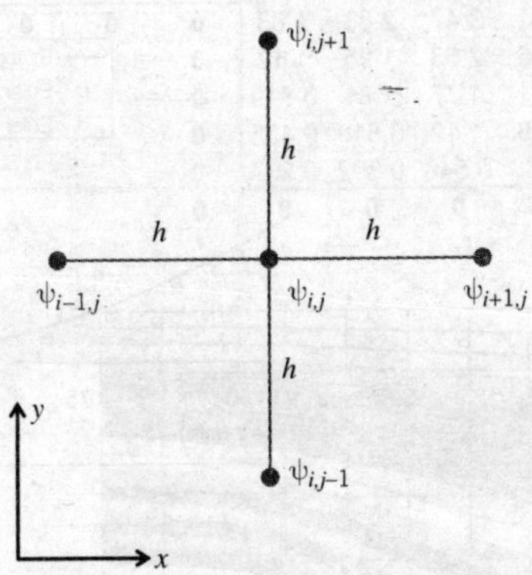

Fig. 5.14 Scheme for discretizing the Laplace equation.

Example 5.12 NUMERICAL MODELING OF FLOW OVER A CORNER

Consider a two-dimensional steady, incompressible, inviscid flow in a channel in which the area is reduced by half. Plot the streamlines.

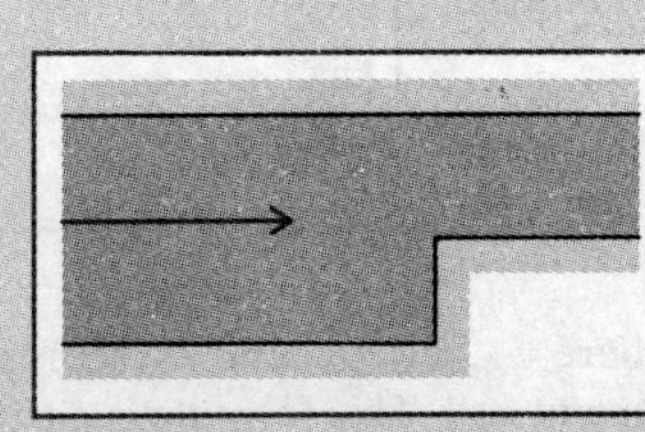

Given: Flow in a channel in which the area is reduced by half.

Find: Streamline plot.

Solution: Use the numerical approximation of the Laplace equation.

Governing equation: $\psi_{i,j} = \frac{1}{4}\left(\psi_{i+1,j} + \psi_{i-1,j} + \psi_{i,j+1} + \psi_{i,j-1}\right)$

This is again convenient for solving using a spreadsheet such as *Excel*. Each cell in the spreadsheet represents a location in physical space, and the value in the cell represents the value of the stream function ψ at that location. Referring to the figure, we assign values of zero to a range of cells that represent the bottom of the channel. We then assign a value of 10 to a second range of cells to represent the top of the channel. (The choice of 10 is arbitrary for plotting purposes; all it determines is the speed values, not the streamline shapes.) Next, we assign a uniform distribution of values at the left and right ends, to generate uniform flow at those locations. All inserted values are shown in bold in the figure.

We can now enter formulas in the "interior" cells to compute the stream function. Instead of the above governing equation, it is more intuitive to rephrase it as

$$\psi = \frac{1}{4}(\psi_A + \psi_R + \psi_B + \psi_L)$$

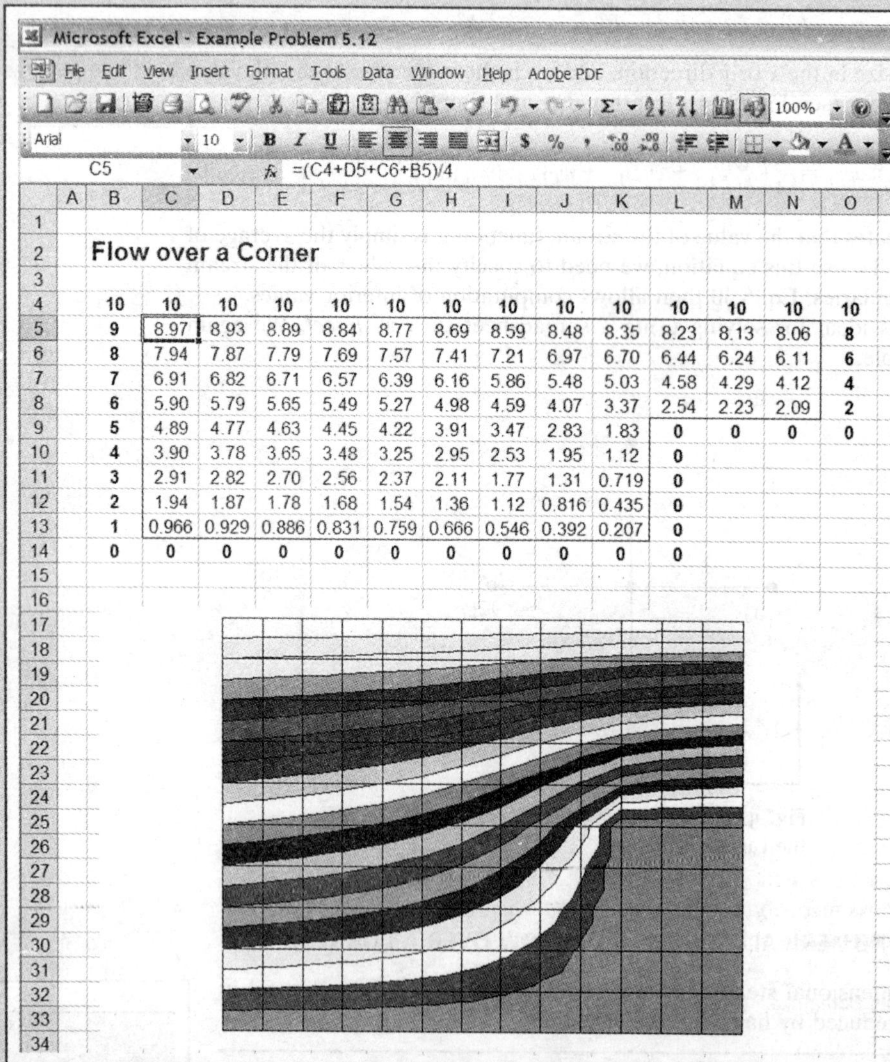

Microsoft Excel - Example Problem 5.12

C5 =(C4+D5+C6+B5)/4

	A	B	C	D	E	F	G	H	I	J	K	L	M	N	O
1															
2		Flow over a Corner													
3															
4		**10**	**10**	**10**	**10**	**10**	**10**	**10**	**10**	**10**	**10**	**10**	**10**	**10**	**10**
5		**9**	8.97	8.93	8.89	8.84	8.77	8.69	8.59	8.48	8.35	8.23	8.13	8.06	**8**
6		**8**	7.94	7.87	7.79	7.69	7.57	7.41	7.21	6.97	6.70	6.44	6.24	6.11	**6**
7		**7**	6.91	6.82	6.71	6.57	6.39	6.16	5.86	5.48	5.03	4.58	4.29	4.12	**4**
8		**6**	5.90	5.79	5.65	5.49	5.27	4.98	4.59	4.07	3.37	2.54	2.23	2.09	**2**
9		**5**	4.89	4.77	4.63	4.45	4.22	3.91	3.47	2.83	1.83	**0**	**0**	**0**	**0**
10		**4**	3.90	3.78	3.65	3.48	3.25	2.95	2.53	1.95	1.12	**0**			
11		**3**	2.91	2.82	2.70	2.56	2.37	2.11	1.77	1.31	0.719	**0**			
12		**2**	1.94	1.87	1.78	1.68	1.54	1.36	1.12	0.816	0.435	**0**			
13		**1**	0.966	0.929	0.886	0.831	0.759	0.666	0.546	0.392	0.207	**0**			
14		**0**	**0**	**0**	**0**	**0**	**0**	**0**	**0**	**0**	**0**	**0**			

where ψ_A, ψ_R, ψ_B, and ψ_L represent the values stored in the cells *A*bove, to the *R*ight, *B*elow, and to the *L*eft of the current cell. This formula is easy to enter—it is shown in cell C5 in the figure. Then it is copied into all interior cells, with one caveat: The spreadsheet will indicate an error of circular calculation. This is a warning that you appear to be making an error; for example, cell C5 needs cell C6 to compute, but cell C6 needs cell C5! Recall that each interior cell value is the average of its neighbors. Circular math is usually not what we want, but in this case we do wish it to occur. We need to switch on *iteration* in the spreadsheet. In the case of *Excel*, it's under menu item *Tools/Options/Calculation*. Finally, we need to repeatedly iterate (in *Excel*, press the F9 key several times) until we have convergence; the values in the interior cells will repeatedly update until the variations in values is zero or trivial. After all this, the results can be plotted (using a surface plot), as shown.

We can see that the streamlines look much as we would anticipate, although in reality there would probably be flow separation at the corner. Note also a mathematical artifact in that there is slight oscillations of streamlines as they flow up the vertical surface; using a finer grid (by using many more cells) would reduce this.

This Example shows a simple numerical modeling of the Laplace equation. The *Excel* workbook for this problem can be modified for solving a variety of fluids problems that involve the Laplace equation.

Examples 5.11 and 5.12 provide guidance in using the PC to solve some simple CFD problems. We now turn to a somewhat more detailed description of some of the concepts behind CFD.

The Strategy of CFD

Broadly, the strategy of CFD is to replace the continuous problem domain with a discrete domain using a "grid" or "mesh." In the continuous domain, each flow variable is defined at every point in the domain. For instance, the pressure p in the continuous 1D domain shown in Fig. 5.15 would be given as

$$p = p(x), \qquad 0 \le x \le 1$$

In the discrete domain, each flow variable is defined only at the grid points. So, in the discrete domain in Fig. 5.15, the pressure would be defined only at the N grid points,

$$p_i = p(x_i), \quad i = 1, 2, \ldots, N$$

We can extend this continuous-to-discrete conversion to two or three dimensions. Figure 5.16 shows a 2D grid used for solving the flow over an airfoil. The grid points are the locations where the grid lines cross. In a CFD solution, we would directly solve for the relevant flow variables only at the grid points. The values at other locations are determined by interpolating the values at the grid points. The governing partial differential equations and boundary conditions are defined in terms of the continuous variables p, $\vec{V}$, and so on. We can approximate these in the discrete domain in terms of the discrete variables p_i, $\vec{V}_i$, and so on. Using this procedure, we end up with a discrete system that consists of a large set of coupled, algebraic equations in the discrete variables. Setting up the discrete system and solving it (which is a matrix inversion problem) involves a very large number of repetitive calculations, a task made possible only with the advent of modern computers.

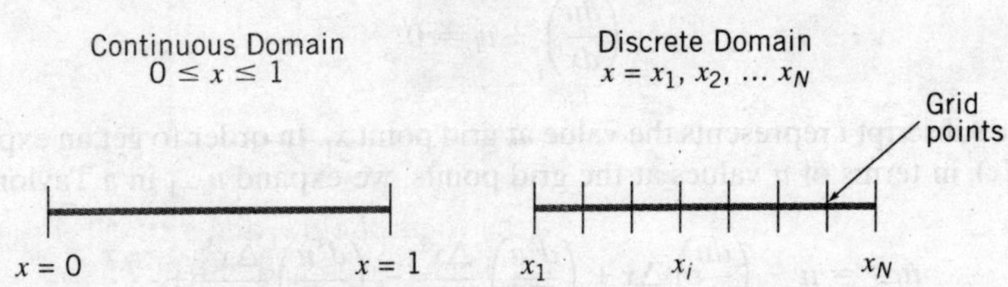

Fig. 5.15 Continuous and discrete domains for a one-dimensional problem.

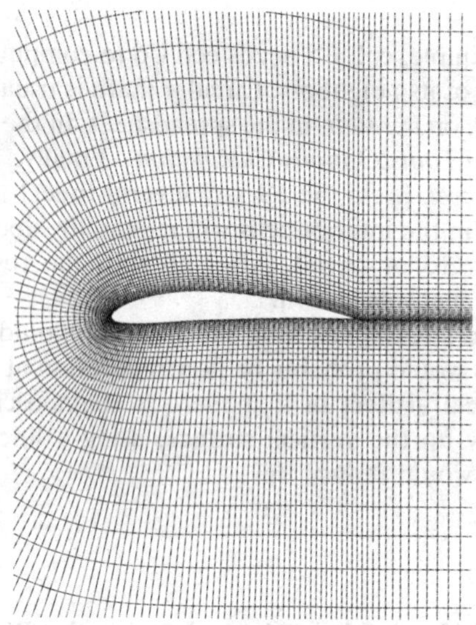

Fig. 5.16 Example of a grid used to solve for the flow around an airfoil.

Discretization Using the Finite-Difference Method

To keep the details simple, we will illustrate the process of going from the continuous domain to the discrete domain by applying it to the following simple 1D equation:

$$\frac{du}{dx} + u^m = 0; \qquad 0 \le x \le 1; \qquad u(0) = 1 \tag{5.31}$$

We'll first consider the case where $m = 1$, which is the case when the equation is linear. We'll later consider the nonlinear case $m = 2$. Keep in mind that the above problem is an initial-value problem, while the numerical solution procedure below is more suitable for boundary-value problems. Most CFD problems are boundary-value problems.

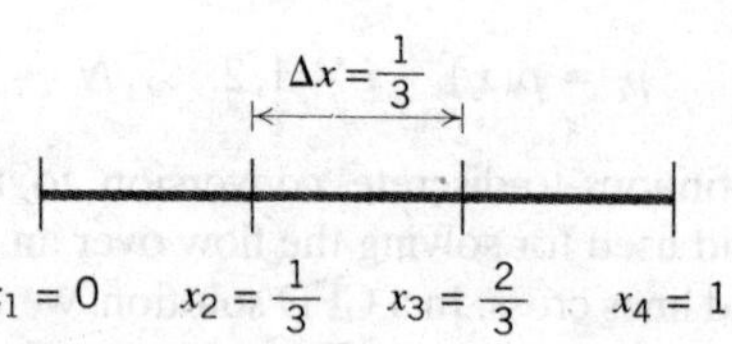

Fig. 5.17 A simple 1D grid with four grid points.

We'll derive a discrete representation of Eq. 5.31 with $m = 1$ on the rudimentary grid shown in Fig. 5.17. This grid has four equally spaced grid points, with $\Delta x = \frac{1}{3}$ being the spacing between successive points. Since the governing equation is valid at any grid point, we have

$$\left(\frac{du}{dx}\right)_i + u_i = 0 \tag{5.32}$$

where the subscript i represents the value at grid point x_i. In order to get an expression for $(du/dx)_i$ in terms of u values at the grid points, we expand u_{i-1} in a Taylor series:

$$u_{i-1} = u_i - \left(\frac{du}{dx}\right)_i \Delta x + \left(\frac{d^2u}{dx^2}\right)_i \frac{\Delta x^2}{2} - \left(\frac{d^3u}{dx^3}\right)_i \frac{\Delta x^3}{6} + \cdots$$

Rearranging this gives

$$\left(\frac{du}{dx}\right)_i = \frac{u_i - u_{i-1}}{\Delta x} + \left(\frac{d^2u}{dx^2}\right)_i \frac{\Delta x}{2} - \left(\frac{d^3u}{dx^3}\right)_i \frac{\Delta x^2}{6} + \cdots \tag{5.33}$$

We'll neglect the second-, third-, and higher-order terms on the right. Thus, the first term on the right is the finite-difference representation for $(du/dx)_i$ we are seeking. The error in $(du/dx)_i$ due to the neglected terms in the Taylor series is called the *truncation error.* In general, the truncation error is the difference between the differential equation and its finite-difference representation. The leading-order term in the truncation error in Eq. 5.33 is proportional to Δx. Equation 5.33 is rewritten as

$$\left(\frac{du}{dx}\right)_i = \frac{u_i - u_{i-1}}{\Delta x} + O(\Delta x) \tag{5.34}$$

where the last term is pronounced "order of delta x." The notation $O(\Delta x)$ has a precise mathematical meaning, which we will not go into here. Instead, in the interest of brevity, we'll return to it briefly later when we discuss the topic of grid convergence. Since the truncation error is proportional to the first power of Δx, this discrete representation is termed *first-order accurate.*

Using Eq. 5.34 in Eq. 5.32, we get the following discrete representation for our model equation:

$$\frac{u_i - u_{i-1}}{\Delta x} + u_i = 0 \tag{5.35}$$

Note that we have gone from a differential equation to an algebraic equation! Though we have not written it out explicitly, don't forget that the error in this representation is $O(\Delta x)$.

This method of deriving the discrete equation using Taylor's series expansions is called the *finite-difference method.* Keep in mind that most industrial CFD software packages use the *finite-volume* or *finite-element* discretization methods since they are better suited to modeling flow past complex geometries. We will stick with the finite-difference method in this text since it is the easiest to understand; the concepts discussed also apply to the other discretization methods.

Assembly of Discrete System and Application of Boundary Conditions

Rearranging the discrete equation, Eq. 5.35, we get

$$-u_{i-1} + (1 + \Delta x)u_i = 0$$

Applying this equation at grid points $i = 2, 3, 4$ for the 1D grid in Fig. 5.17 gives

$$-u_1 + (1 + \Delta x)u_2 = 0 \tag{5.36a}$$

$$-u_2 + (1 + \Delta x)u_3 = 0 \tag{5.36b}$$

$$-u_3 + (1 + \Delta x)u_4 = 0 \tag{5.36c}$$

The discrete equation cannot be applied at the left boundary ($i = 1$) since $u_{i-1} = u_0$ is not defined. Instead, we use the boundary condition to get

$$u_1 = 1 \tag{5.36d}$$

Equations 5.36 form a system of four simultaneous algebraic equations in the four unknowns u_1, u_2, u_3, and u_4. It's convenient to write this system in matrix form:

$$\begin{bmatrix} 1 & 0 & 0 & 0 \\ -1 & 1+\Delta x & 0 & 0 \\ 0 & -1 & 1+\Delta x & 0 \\ 0 & 0 & -1 & 1+\Delta x \end{bmatrix} \begin{bmatrix} u_1 \\ u_2 \\ u_3 \\ u_4 \end{bmatrix} = \begin{bmatrix} 1 \\ 0 \\ 0 \\ 0 \end{bmatrix} \tag{5.37}$$

In a general situation (e.g., 2D or 3D domains), we would apply the discrete equations to the grid points in the interior of the domain. For grid points at or near the boundary, we would apply a combination of the discrete equations and boundary conditions. In the end, one would obtain a system of simultaneous algebraic equations similar to Eqs. 5.36 and a matrix equation similar to Eq. 5.37, with the number of equations being equal to the number of independent discrete variables. The process is essentially the same as for the model equation above, with the details obviously being much more complex.

Solution of Discrete System

The discrete system (Eq. 5.37) for our own simple 1D example can be easily inverted, using any number of techniques of linear algebra, to obtain the unknowns at the grid points. For $\Delta x = \frac{1}{3}$, the solution is

$$u_1 = 1 \quad u_2 = \frac{3}{4} \quad u_3 = \frac{9}{16} \quad u_4 = \frac{27}{64}$$

The exact solution for Eq. 5.31 with $m = 1$ is easily shown to be

$$u_{\text{exact}} = e^{-x}$$

Figure 5.18 shows the comparison of the discrete solution obtained on the four-point grid with the exact solution, using *Excel*. The error is largest at the right boundary, where it is equal to 14.7 percent. [It also shows the results using eight points ($N = 8$, $\Delta x = \frac{1}{7}$) and sixteen points ($N = 16$, $\Delta x = \frac{1}{15}$), which we discuss below.]

In a practical CFD application, we would have thousands, even millions, of unknowns in the discrete system; if one were to use, say, a Gaussian elimination procedure to invert the calculations, it would be extremely time-consuming even with a fast computer. Hence a lot of work has gone into optimizing the matrix inversion in

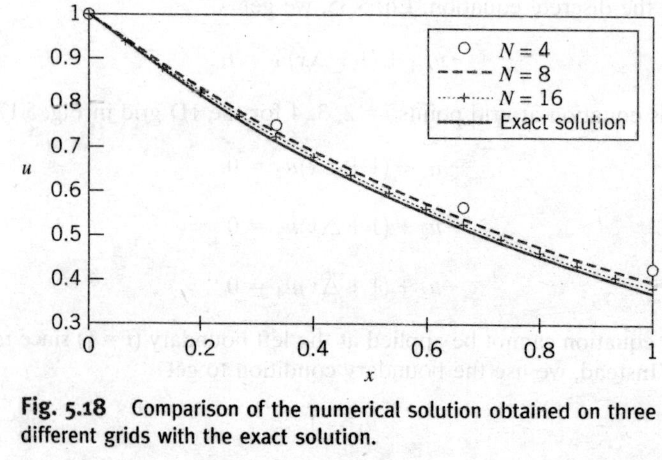

Fig. 5.18 Comparison of the numerical solution obtained on three different grids with the exact solution.

order to minimize the CPU time and memory required. The matrix to be inverted is sparse; that is, most of the entries in it are zeros. The nonzero entries are clustered around the diagonal since the discrete equation at a grid point contains only quantities at the neighboring grid points, as shown in Eq. 5.37. A CFD code would store only the nonzero values to minimize memory usage. It would also generally use an iterative procedure to invert the matrix; the longer one iterates, the closer one gets to the true solution for the matrix inversion. We'll return to this idea a little later.

Grid Convergence

While developing the finite-difference approximation for the 1D model problem (Eq. 5.37), we saw that the truncation error in our discrete system is $O(\Delta x)$. Hence we expect that as the number of grid points is increased and Δx is reduced, the error in the numerical solution would decrease and the agreement between the numerical and exact solutions would get better.

Let's consider the effect of increasing the number of grid points N on the numerical solution of the 1D problem. We'll consider $N = 8$ and $N = 16$ in addition to the $N = 4$ case solved previously. We repeat the above assembly and solution steps on each of these additional grids; instead of the 4×4 problem of Eq. 5.37, we end up with an 8×8 and a 16×16 problem, respectively. Figure 5.18 compares the results obtained (using *Excel*) on the three grids with the exact solution. As expected, the numerical error decreases as the number of grid points is increased (but this only goes so far—if we make Δx too small, we start to get round-off errors accumulating to make the results get worse!). When the numerical solutions obtained on different grids agree to within a level of tolerance specified by the user, they are referred to as "grid-converged" solutions. It is very important to investigate the effect of grid resolution on the solution in all CFD problems. We should never trust a CFD solution unless we are convinced that the solution is grid-converged to an acceptance level of tolerance (which will be problem dependent).

Let ε be some aggregate measure of the error in the numerical solution obtained on a specific grid. For the numerical solutions in Fig. 5.19, ε is, for instance, estimated as the RMS of the difference between the numerical and exact solutions:

$$\varepsilon = \sqrt{\frac{\sum_{i=1}^{N}\left(u_i - u_{i_{\text{exact}}}\right)^2}{N}}$$

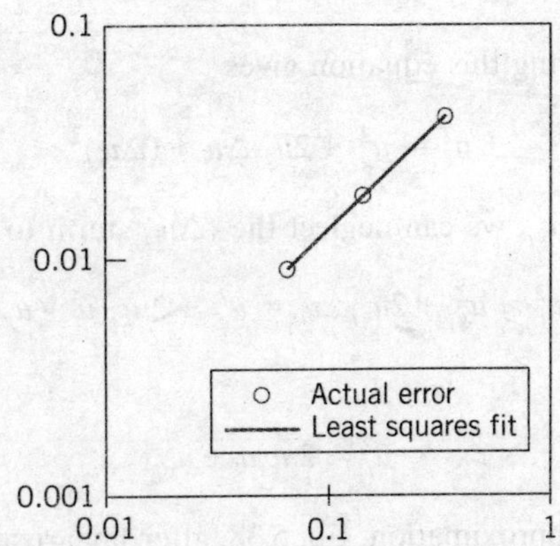

Fig. 5.19 The variation of the aggregate error ε with Δx.

It's reasonable to expect that

$$\varepsilon \propto \Delta x^n$$

Since the truncation error is $O(\Delta x)$ for our discretization scheme, we expect $n = 1$ (or more precisely, $n \to 1$ as $\Delta x \to 0$). The ε values for the three grids are plotted on a logarithmic scale in Fig. 5.19. The slope of the least squares fit gives the value of n. For Fig. 5.19, we get $n = 0.92$, which is quite close to 1. We expect that as the grid is refined further and Δx becomes progressively smaller, the value of n will approach 1. For a second-order scheme, we would expect $n \sim 2$; this means the discretization error will decrease twice as fast on refining the grid.

Dealing with Nonlinearity

The Navier–Stokes equations (Eqs. 5.27) contain nonlinear convection terms; for example, in Eq. 5.27a, the convective acceleration term, $u\partial u/\partial x + v\partial u/\partial y + w\partial u/\partial z$, has products of u with itself as well as with v and w. Phenomena such as turbulence and chemical reaction introduce additional nonlinearities. The highly nonlinear nature of the governing equations for a fluid makes it challenging to obtain accurate numerical solutions for complex flows of practical interest.

We will demonstrate the effect of nonlinearity by setting $m = 2$ in our simple 1D example, Eq. 5.31:

$$\frac{du}{dx} + u^2 = 0; \quad 0 \le x \le 1; \quad u(0) = 1$$

A first-order finite-difference approximation to this equation, analogous to that in Eq. 5.35 for $m = 1$, is

$$\frac{u_i - u_{i-1}}{\Delta x} + u_i^2 = 0 \tag{5.38}$$

This is a nonlinear algebraic equation with the u_i^2 term being the source of the nonlinearity.

The strategy that is adopted to deal with nonlinearity is to linearize the equations around a *guess value* of the solution and to iterate until the guess agrees with the solution to a specified tolerance level. We'll illustrate this on the above example. Let u_{g_i} be the guess for u_i. Define

$$\Delta u_i = u_i - u_{g_i}$$

Rearranging and squaring this equation gives

$$u_i^2 = u_{g_i}^2 + 2u_{g_i}\Delta u_i + (\Delta u_i)^2$$

Assuming that $\Delta u_i \ll u_{g_i}$, we can neglect the $(\Delta u_i)^2$ term to get

$$u_i^2 \approx u_{g_i}^2 + 2u_{g_i}\Delta u_i = u_{g_i}^2 + 2u_{g_i}(u_i - u_{g_i})$$

Thus

$$u_i^2 \approx 2u_{g_i}u_i - u_{g_i}^2 \tag{5.39}$$

The finite-difference approximation, Eq. 5.38, after linearization in u_i, becomes

$$\frac{u_i - u_{i-1}}{\Delta x} + 2u_{g_i}u_i - u_{g_i}^2 = 0 \tag{5.40}$$

Since the error due to linearization is $O(\Delta u^2)$, it tends to zero as $u_g \to u$.

In order to calculate the finite-difference approximation, Eq. 5.40, we need guess values u_g at the grid points. We start with an initial guess value in the first iteration. For each subsequent iteration, the u value obtained in the previous iteration is used as the guess value. We continue the iterations until they converge. We'll defer the discussion on how to evaluate convergence until a little later.

This is essentially the process used in CFD codes to linearize the nonlinear terms in the conservation equations, with the details varying depending on the code. The important points to remember are that the linearization is performed about a guess and that it is necessary to iterate through successive approximations until the iterations converge.

Direct and Iterative Solvers

We saw that we need to perform iterations to deal with the nonlinear terms in the governing equations. We next discuss another factor that makes it necessary to carry out iterations in practical CFD problems.

As an exercise, you can verify that the discrete equation system resulting from the finite-difference approximation of Eq. 5.40, on our four-point grid, is

$$\begin{bmatrix} 1 & 0 & 0 & 0 \\ -1 & 1+2\Delta x u_{g_2} & 0 & 0 \\ 0 & -1 & 1+2\Delta x u_{g_3} & 0 \\ 0 & 0 & -1 & 1+2\Delta x u_{g_4} \end{bmatrix} \begin{bmatrix} u_1 \\ u_2 \\ u_3 \\ u_4 \end{bmatrix} = \begin{bmatrix} 1 \\ \Delta x u_{g_2}^2 \\ \Delta x u_{g_3}^2 \\ \Delta x u_{g_4}^2 \end{bmatrix} \tag{5.41}$$

In a practical problem, one would usually have thousands to millions of grid points or cells so that each dimension of the above matrix would be of the order of a million (with most of the elements being zeros). Inverting such a matrix directly would take a prohibitively large amount of memory, so instead the matrix is inverted using an iterative scheme as discussed below.

Rearrange the finite-difference approximation, Eq. 5.40, at grid point i so that u_i is expressed in terms of the values at the neighboring grid points and the guess values:

$$u_i = \frac{u_{i-1} + \Delta x\, u_{g_i}^2}{1 + 2\Delta x\, u_{g_i}}$$

If a neighboring value at the current iteration level is not available, we use the guess value for it. Let's say that we sweep from right to left on our grid; that is, we update u_4, then u_3, and finally u_2 in each iteration. In any iteration, u_{i-1} is not available while updating u_i and so we use the guess value $u_{g_{i-1}}$ for it instead:

$$u_i = \frac{u_{g_{i-1}} + \Delta x\, u_{g_i}^2}{1 + 2\Delta x\, u_{g_i}} \tag{5.42}$$

Since we are using the guess values at neighboring points, we are effectively obtaining only an approximate solution for the matrix inversion in Eq. 5.41 during each iteration, but in the process we have greatly reduced the memory required for the inversion. This trade-off is a good strategy since it doesn't make sense to expend a great deal of resources to do an exact matrix inversion when the matrix elements depend on guess values that are continuously being refined. We have in effect combined the iteration to handle nonlinear terms with the iteration for matrix inversion into a single iteration process. Most importantly, as the iterations converge and $u_g \to u$, the approximate solution for the matrix inversion tends towards the exact solution for the inversion, since the error introduced by using u_g instead of u in Eq. 5.42 also tends to zero. We arrive at the solution without explicitly forming the matrix system (Eq. 5.41), which greatly simplifies the computer implementation.

Thus, iteration serves two purposes:

1. It allows for efficient matrix inversion with greatly reduced memory requirements.
2. It enables us to solve nonlinear equations.

In steady problems, a common and effective strategy used in CFD codes is to solve the unsteady form of the governing equations and "march" the solution in time until the solution converges to a steady value. In this case, each time step is effectively an iteration, with the guess value at any time level being given by the solution at the previous time level.

Iterative Convergence

Recall that as $u_g \to u$, the linearization and matrix inversion errors tend to zero. Hence we continue the iteration process until some selected measure of the difference between u_g and u, referred to as the residual, is "small enough." We could, for instance, define the residual R as the RMS value of the difference between u and u_g on the grid:

$$R \equiv \sqrt{\frac{\sum_{i=1}^{N} (u_i - u_{g_i})^2}{N}}$$

It's useful to scale this residual with the average value of u in the domain. Scaling ensures that the residual is a *relative* rather than an *absolute* measure. Scaling the above residual by dividing by the average value of u gives

$$R = \left(\sqrt{\frac{\sum_{i=1}^{N} (u_i - u_{g_i})^2}{N}}\right)\left(\frac{N}{\sum_{i=1}^{N} u_i}\right) = \frac{\sqrt{N \sum_{i=1}^{N} (u_i - u_{g_i})^2}}{\sum_{i=1}^{N} u_i} \tag{5.43}$$

In our nonlinear 1D example, we'll take the initial guess at all grid points to be equal to the value at the left boundary, that is, $u_g^{(1)} = 1$ (where $^{(1)}$ signifies the first iteration). In each iteration, we update u_g, sweep from right to left on the grid updating, in turn, u_4, u_3, and u_2 using Eq. 5.42, and calculate the residual using Eq. 5.43. We'll terminate the iterations when the residual falls below 10^{-9} (this is referred to as the *convergence criterion*). The variation of the residual with iterations is shown in Fig. 5.20. Note that a logarithmic scale is used for the ordinate. The iterative process converges to a level smaller than 10^{-9} in only six iterations. In more complex problems, many more iterations would be necessary for achieving convergence.

The solution after two, four, and six iterations and the exact solution are shown in Fig. 5.21. It can easily be verified that the exact solution is given by

$$u_{\text{exact}} = \frac{1}{x+1}$$

The solutions for four and six iterations are indistinguishable on the graph. This is another indication that the solution has converged. The converged solution doesn't agree well with the exact solution because we are using a coarse grid for which the truncation error is relatively large (we will repeat this problem with finer grids as problems at the end of the chapter). The iterative convergence error, which is of order 10^{-9}, is swamped by the truncation error, which is of order 10^{-1}. So driving the residual down to 10^{-9} when the truncation error is of order 10^{-1} is obviously a waste

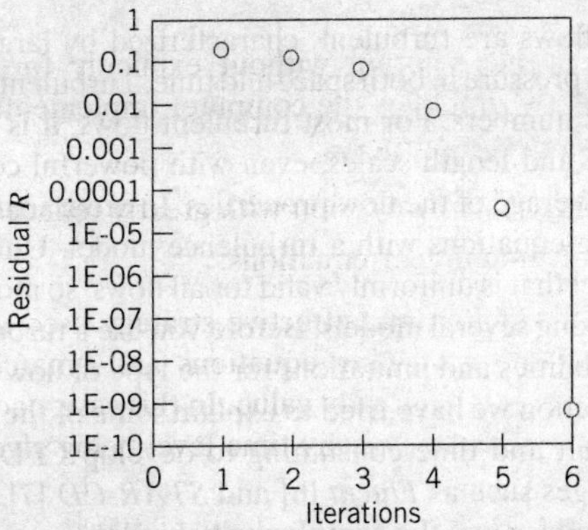

Fig. 5.20 Convergence history for the model nonlinear problem.

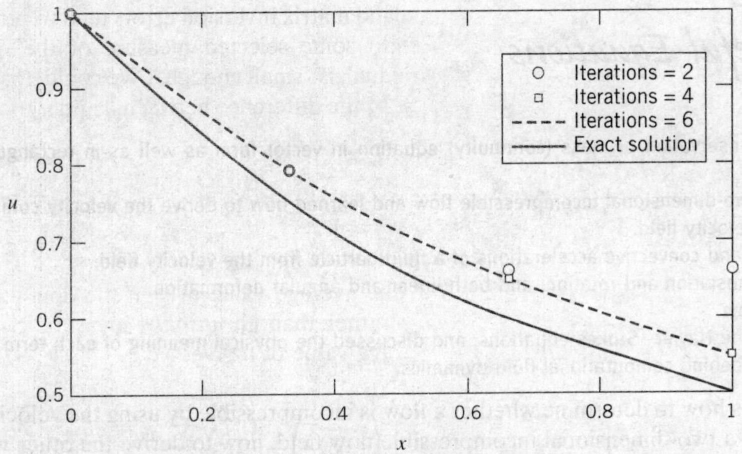

Fig. 5.21 Progression of the iterative solution.

of computing resources. In an efficient calculation, both errors would be set at comparable levels, and less than a tolerance level that was chosen by the user. The agreement between the numerical and exact solutions should get much better on refining the grid, as was the linear case (for $m = 1$). Different CFD codes use slightly different definitions for the residual. You should always read the documentation from the application to understand how the residual is calculated.

Concluding Remarks

In this section we have introduced some simple ways of using a spreadsheet for the numerical solution of two types of fluid mechanics problems. Examples 5.11 and 5.12 show how certain 1D and 2D flows may be computed. We then studied some concepts in more detail, such as convergence criteria, involved with numerical methods and CFD, by considering a first-order ODE. In our simple 1D example, the iterations converged very rapidly. In practice, one encounters many instances when the iterative process doesn't converge or converges lethargically. Hence, it's useful to know *a priori* the conditions under which a given numerical scheme converges. This is determined by performing a stability analysis of the numerical scheme. Stability analysis of numerical schemes and the various stabilization strategies used to overcome non-convergence are very important topics and necessary for you to explore if you decide to delve further into the topic of CFD.

VIDEO

Fully Turbulent Duct Flow.

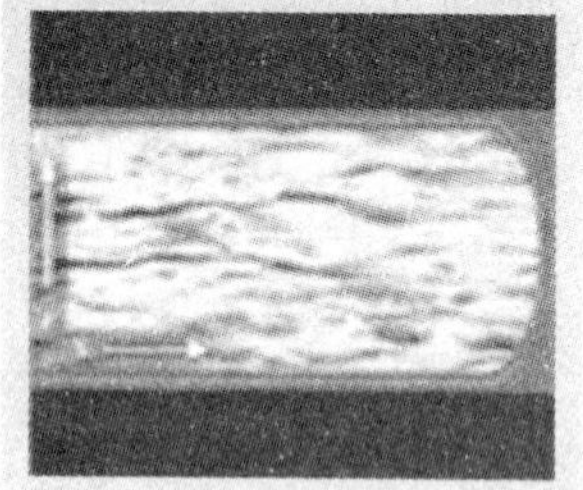

Many engineering flows are turbulent, characterized by large, nearly random fluctuations in velocity and pressure in both space and time. Turbulent flows often occur in the limit of high Reynolds numbers. For most turbulent flows, it is not possible to resolve the vast range of time and length scales, even with powerful computers. Instead, one solves for a statistical average of the flow properties. In order to do this, it is necessary to augment the governing equations with a turbulence model. Unfortunately, there is no single turbulence model that is uniformly valid for all flows, so most CFD packages allow you to choose from among several models. Before you use a turbulence model, you need to understand its possibilities and limitations for the type of flow being considered.

In this brief introduction we have tried to explain some of the concepts behind CFD. Because it is so difficult and time consuming to develop CFD code, most engineers use commercial packages such as *Fluent* [6] and *STAR-CD* [7]. This introduction will have hopefully indicated for you the complexity behind those applications, so that they are not completely a "black box" of magic tricks.

5.6 Summary and Useful Equations

In this chapter we have:

- ✓ Derived the differential form of the conservation of mass (continuity) equation in vector form as well as in rectangular and cylindrical coordinates.
- ✓ *Defined the stream function ψ for a two-dimensional incompressible flow and learned how to derive the velocity components from it, as well as to find ψ from the velocity field.
- ✓ Learned how to obtain the total, local, and convective accelerations of a fluid particle from the velocity field.
- ✓ Presented examples of fluid particle translation and rotation, and both linear and angular deformation.
- ✓ Defined vorticity and circulation of a flow.
- ✓ Derived, and solved for simple cases, the Navier–Stokes equations, and discussed the physical meaning of each term.
- ✓ *Been introduced to some basis ideas behind computational fluid dynamics.

We have also explored such ideas as how to determine whether a flow is incompressible by using the velocity field and, given one velocity component of a two-dimensional incompressible flow field, how to derive the other velocity component.

In this chapter we studied the effects of viscous stresses on fluid particle deformation and rotation; in the next chapter we examine flows for which viscous effects are negligible.

Note: Most of the Useful Equations in the table below have a number of constraints or limitations—*be sure to refer to their page numbers for details*!

Useful Equations

Continuity equation (general, rectangular coordinates):	$\dfrac{\partial \rho u}{\partial x} + \dfrac{\partial \rho v}{\partial y} + \dfrac{\partial \rho w}{\partial z} + \dfrac{\partial \rho}{\partial t} = 0$	**(5.1a)**	Page 175
	$\nabla \cdot \rho \vec{V} + \dfrac{\partial \rho}{\partial t} = 0$	**(5.1b)**	
Continuity equation (incompressible, rectangular coordinates):	$\dfrac{\partial u}{\partial x} + \dfrac{\partial v}{\partial y} + \dfrac{\partial w}{\partial z} = \nabla \cdot \vec{V} = 0$	**(5.1c)**	Page 175
Continuity equation (steady, rectangular coordinates):	$\dfrac{\partial \rho u}{\partial x} + \dfrac{\partial \rho v}{\partial y} + \dfrac{\partial \rho w}{\partial z} = \nabla \cdot \rho \vec{V} = 0$	**(5.1d)**	Page 175
Continuity equation (general, cylindrical coordinates):	$\dfrac{1}{r}\dfrac{\partial(r\rho V_r)}{\partial r} + \dfrac{1}{r}\dfrac{\partial(\rho V_\theta)}{\partial \theta} + \dfrac{\partial(\rho V_z)}{\partial z} + \dfrac{\partial \rho}{\partial t} = 0$	**(5.2a)**	Pages 178
	$\nabla \cdot \rho \vec{V} + \dfrac{\partial \rho}{\partial t} = 0$	**(5.1b)**	

*This section may be omitted without loss of continuity in the text material.

Continuity equation (incompressible, cylindrical coordinates):	$\frac{1}{r}\frac{\partial(rV_r)}{\partial r}+\frac{1}{r}\frac{\partial V_\theta}{\partial\theta}+\frac{\partial V_z}{\partial z}=\nabla\cdot\vec{V}=0$	**(5.2b)**	Page 178
Continuity equation (steady, cylindrical coordinates):	$\frac{1}{r}\frac{\partial(r\rho V_r)}{\partial r}+\frac{1}{r}\frac{\partial(\rho V_\theta)}{\partial\theta}+\frac{\partial(\rho V_z)}{\partial z}=\nabla\cdot\rho\vec{V}=0$	**(5.2c)**	Page 180
Continuity equation (2D, incompressible, rectangular coordinates):	$\frac{\partial u}{\partial x}+\frac{\partial v}{\partial y}=0$	**(5.3)**	Page 181
Stream function (2D, incompressible, rectangular coordinates):	$u\equiv\frac{\partial\psi}{\partial y}$ and $v\equiv-\frac{\partial\psi}{\partial x}$	**(5.4)**	Page 181
Continuity equation (2D, incompressible, cylindrical coordinates):	$\frac{\partial(rV_r)}{\partial r}+\frac{\partial V_\theta}{\partial\theta}=0$	**(5.7)**	Page 182
Stream function (2D, incompressible, cylindrical coordinates):	$V_r\equiv\frac{1}{r}\frac{\partial\psi}{\partial\theta}$ and $V_\theta\equiv-\frac{\partial\psi}{\partial r}$	**(5.8)**	Page 183
Particle acceleration (rectangular coordinates):	$\frac{D\vec{V}}{Dt}\equiv\vec{a}_p=u\frac{\partial\vec{V}}{\partial x}+v\frac{\partial\vec{V}}{\partial y}+w\frac{\partial\vec{V}}{\partial z}+\frac{\partial\vec{V}}{\partial t}$	**(5.9)**	Page 186
Particle acceleration components in rectangular coordinates:	$a_{x_p}=\frac{Du}{Dt}=u\frac{\partial u}{\partial x}+v\frac{\partial u}{\partial y}+w\frac{\partial u}{\partial z}+\frac{\partial u}{\partial t}$	**(5.11a)**	Page 187
	$a_{y_p}=\frac{Dv}{Dt}=u\frac{\partial v}{\partial x}+v\frac{\partial v}{\partial y}+w\frac{\partial v}{\partial z}+\frac{\partial v}{\partial t}$	**(5.11b)**	
	$a_{z_p}=\frac{Dw}{Dt}=u\frac{\partial w}{\partial x}+v\frac{\partial w}{\partial y}+w\frac{\partial w}{\partial z}+\frac{\partial w}{\partial t}$	**(5.11c)**	
Particle acceleration components in cylindrical coordinates:	$a_{r_p}=V_r\frac{\partial V_r}{\partial r}+\frac{V_\theta}{r}\frac{\partial V_r}{\partial\theta}-\frac{V_\theta^2}{r}+V_z\frac{\partial V_r}{\partial z}+\frac{\partial V_r}{\partial t}$	**(5.12a)**	Page 188
	$a_{\theta_p}=V_r\frac{\partial V_\theta}{\partial r}+\frac{V_\theta}{r}\frac{\partial V_\theta}{\partial\theta}+\frac{V_rV_\theta}{r}+V_z\frac{\partial V_\theta}{\partial z}+\frac{\partial V_\theta}{\partial t}$	**(5.12b)**	
	$a_{z_p}=V_r\frac{\partial V_z}{\partial r}+\frac{V_\theta}{r}\frac{\partial V_z}{\partial\theta}+V_z\frac{\partial V_z}{\partial z}+\frac{\partial V_z}{\partial t}$	**(5.12c)**	
Navier–Stokes equations (incompressible, constant viscosity):	$\rho\left(\frac{\partial u}{\partial t}+u\frac{\partial u}{\partial x}+v\frac{\partial u}{\partial y}+w\frac{\partial u}{\partial z}\right)=\rho g_x-\frac{\partial p}{\partial x}+\mu\left(\frac{\partial^2u}{\partial x^2}+\frac{\partial^2u}{\partial y^2}+\frac{\partial^2u}{\partial z^2}\right)$	**(5.27a)**	Page 200, 201
	$\rho\left(\frac{\partial v}{\partial t}+u\frac{\partial v}{\partial x}+v\frac{\partial v}{\partial y}+w\frac{\partial v}{\partial z}\right)=\rho g_y-\frac{\partial p}{\partial y}+\mu\left(\frac{\partial^2v}{\partial x^2}+\frac{\partial^2v}{\partial y^2}+\frac{\partial^2v}{\partial z^2}\right)$	**(5.27b)**	
	$\rho\left(\frac{\partial w}{\partial t}+u\frac{\partial w}{\partial x}+v\frac{\partial w}{\partial y}+w\frac{\partial w}{\partial z}\right)=\rho g_z-\frac{\partial p}{\partial z}+\mu\left(\frac{\partial^2w}{\partial x^2}+\frac{\partial^2w}{\partial y^2}+\frac{\partial^2w}{\partial z^2}\right)$	**(5.27c)**	

Case Study

Olympic Swimming and Bobsledding

Image is unavailable due to copyright restrictions

Athletes in many competitive sports are using technology to gain an advantage. In recent years, Fastskin® fabric has been developed by Speedo. This material allows the lowest-drag racing swimwear in the world to be developed. The fabric mimics the rough denticles of sharks' skin to reduce drag in key areas of the body. (Shark scales are tiny compared with those of most fishes and have a toothlike structure, called dermal denticles—literally, "tiny skin teeth." These denticles are nature's way of reducing drag on the shark.) Detailed design of swimsuits was based on tests in a water flume and on computational fluid dynamics (CFD) analyses. The figure shows an example of the results obtained. To optimize the suits, the results were used to guide the position of the seams; gripper panels on the underside of the forearms; and "vortex" riblets on the chest, shoulders, and back of the suit—as well as the positioning of different patches of fabric and fabric coatings.

The same technology is now being used to make outfits for athletes in the bobsled and luge events in the winter Olympics. The fabric has been modified, based on wind tunnel tests, to reduce drag based on the airflow direction unique to sledding sports. The new outfits also eliminate most of the fabric vibration (a major source of drag) found in other speed suits.

For both summer and winter sports, the ability to perform experimental and theoretical fluid dynamics analysis and make design changes based on these can make the difference in speed of several percent—the difference between silver and gold!

References

1. Li, W. H., and S. H. Lam, *Principles of Fluid Mechanics*. Reading, MA: Addison-Wesley, 1964.

2. Daily, J. W., and D. R. F. Harleman, *Fluid Dynamics*. Reading, MA: Addison-Wesley, 1966.

3. Schlichting, H., *Boundary-Layer Theory*, 7th ed. New York: McGraw-Hill, 1979.

4. White, F. M., *Viscous Fluid Flow*, 3rd ed. New York: McGraw-Hill, 2000.

5. Sabersky, R. H., A. J. Acosta, E. G. Hauptmann, and E. M. Gates, *Fluid Flow—A First Course in Fluid Mechanics*, 4th ed. New Jersey: Prentice Hall, 1999.

6. *Fluent*. Fluent Incorporated, Centerra Resources Park, 10 Cavendish Court, Lebanon, NH 03766 (www.fluent.com).

7. *STAR-CD*. Adapco, 60 Broadhollow Road, Melville, NY 11747 (www.cd-adapco.com).

8. Chapra, S. C., and R. P. Canale, *Numerical Methods for Engineers*, 5th ed. New York: McGraw-Hill, 2005.

9. Epperson, J. F., *An Introduction to Numerical Methods and Analysis*, rev. ed. New York: Wiley, 2007.

Problems

Conservation of Mass

5.1 Which of the following sets of equations represent possible two-dimensional incompressible flow cases?

(a) $u = 2x^2 + y^2 - x^2y;\ v = x^3 + x(y^2 - 4y)$

(b) $u = 2xy - x^2y;\ v = 2xy - y^2 + x^2$

(c) $u = x^2t + 2y;\ v = xt^2 - yt$

(d) $u = (2x + 4y)xt;\ v = -3(x + y)yt$

5.2 Which of the following sets of equations represent possible three-dimensional incompressible flow cases?

(a) $u = 2y^2 + 2xz;\ v = -2yz + 6x^2yz;\ w = 3x^2z^2 + x^3y^4$

(b) $u = xyzt;\ v = -xyzt^2;\ w = z^2(xt^2 - yt)$

(c) $u = x^2 + 2y + z^2;\ v = x - 2y + z;\ w = -2xz + y^2 + 2z$

5.3 For a flow in the xy plane, the x component of velocity is given by $u = Ax(y - B)$, where $A = 3.3\,\text{m}^{-1}\cdot\text{s}^{-1}$, $B = 1.8$ m, and x

and y are measured in meters. Find a possible y component for steady, incompressible flow. Is it also valid for unsteady, incompressible flow? Why? How many y components are possible?

5.4 The three components of velocity in a velocity field are given by $u = Ax + By + Cz$, $v = Dx + Ey + Fz$, and $w = Gx + Hy + Jz$. Determine the relationship among the coefficients A through J that is necessary if this is to be a possible incompressible flow field.

5.5 For a flow in the xy plane, the x component of velocity is given by $u = 3x^2y - y^3$. Determine a possible y component for steady, incompressible flow. Is it also valid for unsteady, incompressible flow? Why? How many possible y components are there?

5.6 The x component of velocity in a steady, incompressible flow field in the xy plane is $u = A/x$, where $A = 2\ \text{m}^2/\text{s}$, and x is measured in meters. Find the simplest y component of velocity for this flow field.

5.7 The y component of velocity in a steady, incompressible flow field in the xy plane is $v = Axy(x^2 - y^2)$, where $A = 3\ \text{m}^{-3}\cdot\text{s}^{-1}$ and x and y are measured in meters. Find the simplest x component of velocity for this flow field.

5.8 The y component of velocity in a steady incompressible flow field in the xy plane is

$$v = \frac{2xy}{(x^2 + y^2)^2}$$

Show that the simplest expression for the x component of velocity is

$$u = \frac{1}{(x^2 + y^2)} - \frac{2y^2}{(x^2 + y^2)^2}$$

5.9 The x component of velocity in a steady incompressible flow field in the xy plane is $u = Ae^{x/b}\cos(y/b)$, where $A = 10$ m/s, $b = 5$ m, and x and y are measured in meters. Find the simplest y component of velocity for this flow field.

5.10 A crude approximation for the x component of velocity in an incompressible laminar boundary layer is a linear variation from $u = 0$ at the surface ($y = 0$) to the freestream velocity, U, at the boundary-layer edge ($y = \delta$). The equation for the profile is $u = Uy/\delta$, where $\delta = cx^{1/2}$ and c is a constant. Show that the simplest expression for the y component of velocity is $v = uy/4x$. Evaluate the maximum value of the ratio v/U, at a location where $x = 0.5$ m and $\delta = 5$ mm.

5.11 A useful approximation for the x component of velocity in an incompressible laminar boundary layer is a parabolic variation from $u = 0$ at the surface ($y = 0$) to the freestream velocity, U, at the edge of the boundary layer ($y = \delta$). The equation for the profile is $u/U = 2(y/\delta) - (y/\delta)^2$, where $\delta = cx^{1/2}$ and c is a constant. Show that the simplest expression for the y component of velocity is

$$\frac{v}{U} = \frac{\delta}{x}\left[\frac{1}{2}\left(\frac{y}{\delta}\right)^2 - \frac{1}{3}\left(\frac{y}{\delta}\right)^3\right]$$

Plot v/U versus y/δ to find the location of the maximum value of the ratio v/U. Evaluate the ratio where $\delta = 5$ mm and $x = 0.5$ m.

5.12 A useful approximation for the x component of velocity in an incompressible laminar boundary layer is a sinusoidal variation from $u = 0$ at the surface ($y = 0$) to the freestream velocity, U, at the edge of the boundary layer ($y = \delta$). The equation for the profile is $u = U\sin(\pi y/2\delta)$, where $\delta = cx^{1/2}$ and c is a constant. Show that the simplest expression for the y component of velocity is

$$\frac{v}{U} = \frac{1}{\pi}\frac{\delta}{x}\left[\cos\left(\frac{\pi}{2}\frac{y}{\delta}\right) + \left(\frac{\pi}{2}\frac{y}{\delta}\right)\sin\left(\frac{\pi}{2}\frac{y}{\delta}\right) - 1\right]$$

Plot u/U and v/U versus y/δ, and find the location of the maximum value of the ratio v/U. Evaluate the ratio where $x = 0.5$ m and $\delta = 5$ mm.

5.13 A useful approximation for the x component of velocity in an incompressible laminar boundary layer is a cubic variation from $u = 0$ at the surface ($y = 0$) to the freestream velocity, U, at the edge of the boundary layer ($y = \delta$). The equation for the profile is $u/U = \frac{3}{2}(y/\delta) - \frac{1}{2}(y/\delta)^3$, where $\delta = cx^{1/2}$ and c is a constant. Derive the simplest expression for v/U, the y component of velocity ratio. Plot u/U and v/U versus y/δ, and find the location of the maximum value of the ratio v/U. Evaluate the ratio where $\delta = 5$ mm and $x = 0.5$ m.

5.14 For a flow in the xy plane, the x component of velocity is given by $u = Ax^2y^2$, where $A = 0.3\ \text{m}^{-3}\cdot\text{s}^{-1}$, and x and y are measured in meters. Find a possible y component for steady, incompressible flow. Is it also valid for unsteady, incompressible flow? Why? How many possible y components are there? Determine the equation of the streamline for the simplest y component of velocity. Plot the streamlines through points (1, 4) and (2, 4).

5.15 The y component of velocity in a steady, incompressible flow field in the xy plane is $v = -Bxy^3$, where $B = 0.2\ \text{m}^{-3}\cdot\text{s}^{-1}$, and x and y are measured in meters. Find the simplest x component of velocity for this flow field. Find the equation of the streamlines for this flow. Plot the streamlines through points (1, 4) and (2, 4).

5.16 Consider a water stream from a jet of an oscillating lawn sprinkler. Describe the corresponding pathline and streakline.

5.17 Derive the differential form of conservation of mass in rectangular coordinates by expanding the *products* of density and the velocity components, ρu, ρv, and ρw, in a Taylor series about a point O. Show that the result is identical to Eq. 5.1a.

5.18 Which of the following sets of equations represent possible incompressible flow cases?

(a) $V_r = U\cos\theta$; $V_\theta = -U\sin\theta$
(b) $V_r = -q/2\pi r$; $V_\theta = K/2\pi r$
(c) $V_r = U\cos\theta\,[1 - (a/r)^2]$; $V_\theta = -U\sin\theta[1 + (a/r)^2]$

5.19 Which of the following sets of equations represent(s) possible incompressible flow cases?

(a) $V_r = -K/r$; $V_\theta = 0$
(b) $V_r = 0$; $V_\theta = K/r$
(c) $V_r = -K\cos\theta/r^2$; $V_\theta = -K\sin\theta/r^2$

5.20 For an incompressible flow in the $r\theta$ plane, the r component of velocity is given as $V_r = U\cos\theta$.

(a) Determine a possible θ component of velocity.
(b) How many possible θ components are there?

5.21 For an incompressible flow in the $r\theta$ plane, the r component of velocity is given as $V_r = -\Lambda \cos\theta/r^2$. Determine a possible θ component of velocity. How many possible θ components are there?

5.22 A viscous liquid is sheared between two parallel disks of radius R, one of which rotates while the other is fixed. The velocity field is purely tangential, and the velocity varies linearly with z from $V_\theta = 0$ at $z = 0$ (the fixed disk) to the velocity of the rotating disk at its surface ($z = h$). Derive an expression for the velocity field between the disks.

5.23 Evaluate $\nabla \cdot \rho\vec{V}$ in cylindrical coordinates. Use the definition of ∇ in cylindrical coordinates. Substitute the velocity vector and perform the indicated operations, using the hint in footnote 1 on page 178. Collect terms and simplify; show that the result is identical to Eq. 5.2c.

Stream Function for Two-Dimensional Incompressible Flow

***5.24** A velocity field in cylindrical coordinates is given as $\vec{V} = \hat{e}_r A/r + \hat{e}_\theta B/r$, where A and B are constants with dimensions of m²/s. Does this represent a possible incompressible flow? Sketch the streamline that passes through the point $r_0 = 1$ m, $\theta = 90°$ if $A = B = 1$ m²/s, if $A = 1$ m²/s and $B = 0$, and if $B = 1$ m²/s and $A = 0$.

***5.25** The velocity field for the viscometric flow of Example 5.7 is $\vec{V} = U(y/h)\hat{i}$. Find the stream function for this flow. Locate the streamline that divides the total flow rate into two equal parts.

***5.26** Determine the family of stream functions ψ that will yield the velocity field $\vec{V} = 2y(2x+1)\hat{i} + [x(x+1) - 2y^2]\hat{j}$.

***5.27** Does the velocity field of Problem 5.24 represent a possible incompressible flow case? If so, evaluate and sketch the stream function for the flow. If not, evaluate the rate of change of density in the flow field.

***5.28** The stream function for a certain incompressible flow field is given by the expression $\psi = -Ur\sin\theta + q\theta/2\pi$. Obtain an expression for the velocity field. Find the stagnation point(s) where $|\vec{V}| = 0$, and show that $\psi = 0$ there.

***5.29** Consider a flow with velocity components $u = z(3x^2 - z^2)$, $v = 0$, and $w = x(x^2 - 3z^2)$.
(a) Is this a one-, two-, or three-dimensional flow?
(b) Demonstrate whether this is an incompressible flow.
(c) If possible, derive a stream function for this flow.

***5.30** An incompressible frictionless flow field is specified by the stream function $\psi = -5Ax - 2Ay$, where $A = 2$ m/s, and x and y are coordinates in meters.
(a) Sketch the streamlines $\psi = 0$ and $\psi = 5$, and indicate the direction of the velocity vector at the point (0, 0) on the sketch.
(b) Determine the magnitude of the flow rate between the streamlines passing through (2, 2) and (4, 1).

***5.31** A linear velocity profile was used to model flow in a laminar incompressible boundary layer in Problem 5.10. Derive the stream function for this flow field. Locate streamlines at one-quarter and one-half the total volume flow rate in the boundary layer.

***5.32** A parabolic velocity profile was used to model flow in a laminar incompressible boundary layer in Problem 5.11. Derive the stream function for this flow field. Locate streamlines at one-quarter and one-half the total volume flow rate in the boundary layer.

***5.33** Derive the stream function that represents the sinusoidal approximation used to model the x component of velocity for the boundary layer of Problem 5.12. Locate streamlines at one-quarter and one-half the total volume flow rate in the boundary layer.

***5.34** A cubic velocity profile was used to model flow in a laminar incompressible boundary layer in Problem 5.13. Derive the stream function for this flow field. Locate streamlines at one-quarter and one-half the total volume flow rate in the boundary layer.

***5.35** A rigid-body motion was modeled in Example 5.6 by the velocity field $\vec{V} = r\omega\hat{e}_\theta$. Find the stream function for this flow. Evaluate the volume flow rate per unit depth between $r_1 = 0.10$ m and $r_2 = 0.12$ m, if $\omega = 0.5$ rad/s. Sketch the velocity profile along a line of constant θ. Check the flow rate calculated from the stream function by integrating the velocity profile along this line.

***5.36** In a parallel one-dimensional flow in the positive x direction, the velocity varies linearly from zero at $y = 0$ to 30 m/s at $y = 1.5$ m. Determine an expression for the stream function, ψ. Also determine the y coordinate above which the volume flow rate is half the total between $y = 0$ and $y = 1.5$ m.

***5.37** Example 5.6 showed that the velocity field for a free vortex in the $r\theta$ plane is $\vec{V} = \hat{e}_\theta C/r$. Find the stream function for this flow. Evaluate the volume flow rate per unit depth between $r_1 = 0.20$ m and $r_2 = 0.24$ m, if $C = 0.3$ m²/s. Sketch the velocity profile along a line of constant θ. Check the flow rate calculated from the stream function by integrating the velocity profile along this line.

Motion of a Fluid Particle (Kinematics)

5.38 Consider the flow field given by $\vec{V} = xy^2\hat{i} - \frac{1}{3}y^3\hat{j} + xy\hat{k}$. Determine (a) the number of dimensions of the flow, (b) if it is a possible incompressible flow, and (c) the acceleration of a fluid particle at point $(x, y, z) = (1, 2, 3)$.

5.39 Consider the velocity field $\vec{V} = A(x^4 - 6x^2y^2 + y^4)\hat{i} + A(4xy^3 - 4x^3y)\hat{j}$ in the xy plane, where $A = 0.25\ \text{m}^{-3}\cdot\text{s}^{-1}$, and the coordinates are measured in meters. Is this a possible incompressible flow field? Calculate the acceleration of a fluid particle at point $(x, y) = (2, 1)$.

5.40 Consider the flow field given by $\vec{V} = ax^2y\hat{i} - by\hat{j} + cz^2\hat{k}$, where $a = 2\ \text{m}^{-2}\cdot\text{s}^{-1}$, $b = 2\ \text{s}^{-1}$, and $c = 1\ \text{m}^{-1}\cdot\text{s}^{-1}$. Determine (a) the number of dimensions of the flow, (b) if it is a possible incompressible flow, and (c) the acceleration of a fluid particle at point $(x, y, z) = (2, 1, 3)$.

5.41 The x component of velocity in a steady, incompressible flow field in the xy plane is $u = A(x^5 - 10x^3y^2 + 5xy^4)$, where

*These problems require material from sections that may be omitted without loss of continuity in the text material.

$A = 2\ m^{-4}\cdot s^{-1}$ and x is measured in meters. Find the simplest y component of velocity for this flow field. Evaluate the acceleration of a fluid particle at point $(x, y) = (1, 3)$.

5.42 The velocity field within a laminar boundary layer is approximated by the expression

$$\vec{V} = \frac{AUy}{x^{1/2}}\hat{i} + \frac{AUy^2}{4x^{3/2}}\hat{j}$$

In this expression, $A = 141\ m^{-1/2}$, and $U = 0.240$ m/s is the freestream velocity. Show that this velocity field represents a possible incompressible flow. Calculate the acceleration of a fluid particle at point $(x, y) = (0.5\ m, 5\ mm)$. Determine the slope of the streamline through the point.

5.43 Wave flow of an incompressible fluid into a solid surface follows a sinusoidal pattern. Flow is two-dimensional with the x axis normal to the surface and y axis along the wall. The x component of the flow follows the pattern

$$u = Ax \sin\left(\frac{2\pi t}{T}\right)$$

Determine the y component of flow (v) and the convective and local components of the acceleration vector.

5.44 The y component of velocity in a two-dimensional, incompressible flow field is given by $v = -Axy$, where v is in m/s, x and y are in meters, and A is a dimensional constant. There is no velocity component or variation in the z direction. Determine the dimensions of the constant, A. Find the simplest x component of velocity in this flow field. Calculate the acceleration of a fluid particle at point $(x, y) = (1, 2)$.

5.45 Consider the velocity field $\vec{V} = Ax/(x^2+y^2)\hat{i} + Ay/(x^2+y^2)\hat{j}$ in the xy plane, where $A = 10\ m^2/s$, and x and y are measured in meters. Is this an incompressible flow field? Derive an expression for the fluid acceleration. Evaluate the velocity and acceleration along the x axis, the y axis, and along a line defined by $y = x$. What can you conclude about this flow field?

5.46 An incompressible liquid with negligible viscosity flows steadily through a horizontal pipe of constant diameter. In a porous section of length $L = 0.3$ m, liquid is removed at a constant rate per unit length, so the uniform axial velocity in the pipe is $u(x) = U(1 - x/2L)$, where $U = 5$ m/s. Develop an expression for the acceleration of a fluid particle along the centerline of the porous section.

5.47 An incompressible liquid with negligible viscosity flows steadily through a horizontal pipe. The pipe diameter linearly varies from 10 cm to 2.5 cm over a length of 2 m. Develop an expression for the acceleration of a fluid particle along the pipe centerline. Plot the centerline velocity and acceleration versus position along the pipe, if the inlet centerline velocity is 1 m/s.

5.48 Consider the low-speed flow of air between parallel disks as shown. Assume that the flow is incompressible and inviscid, and that the velocity is purely radial and uniform at any section. The flow speed is $V = 15$ m/s at $R = 75$ mm. Simplify the continuity equation to a form applicable to this flow field. Show that a general expression for the velocity field is $\vec{V} = V(R/r)\hat{e}_r$ for $r_i \le r \le R$. Calculate the acceleration of a fluid particle at the locations $r = r_i$ and $r = R$.

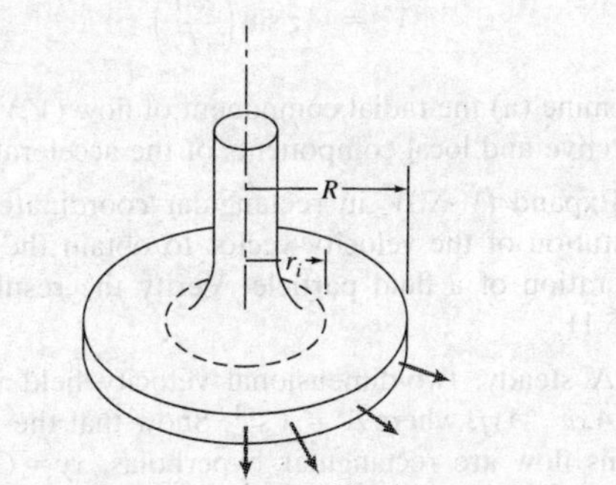

P5.48

5.49 Solve Problem 4.123 to show that the radial velocity in the narrow gap is $V_r = Q/2\pi rh$. Derive an expression for the acceleration of a fluid particle in the gap.

5.50 As part of a pollution study, a model concentration c as a function of position x has been developed,

$$c(x) = A(e^{-x/2a} - e^{-x/a})$$

where $A = 3 \times 10^{-5}$ ppm (parts per million) and $a = 1$ m. Plot this concentration from $x = 0$ to $x = 10$ m. If a vehicle with a pollution sensor travels through the area at $u = U = 20$ m/s, develop an expression for the measured concentration rate of change of c with time, and plot using the given data.

(a) At what location will the sensor indicate the most rapid rate of change?

(b) What is the value of this rate of change?

5.51 After a rainfall the sediment concentration at a certain point in a river increases at the rate of 100 parts per million (ppm) per hour. In addition, the sediment concentration increases with distance downstream as a result of influx from tributary streams; this rate of increase is 30 ppm per km. At this point the stream flows at 0.8 km/h. A boat is used to survey the sediment concentration. The operator is amazed to find three different apparent rates of change of sediment concentration when the boat travels upstream, drifts with the current, or travels downstream. Explain physically why the different rates are observed. If the speed of the boat is 4 km/h, compute the three rates of change.

5.52 As an aircraft flies through a cold front, an onboard instrument indicates that ambient temperature drops at the rate of 0.28°C per minute. Other instruments show an air speed of 154 m/s and a 18 m/s rate of climb. The front is stationary and vertically uniform. Compute the rate of change of temperature with respect to horizontal distance through the cold front.

5.53 An aircraft flies due north at 480 km/h ground speed. Its rate of climb is 15 m/s. The vertical temperature gradient is −5.6°C per km of altitude. The ground temperature varies with position through a cold front, falling at the rate of 0.345°C per km. Compute the rate of temperature change shown by a recorder on board the aircraft.

5.54 Wave flow of an incompressible fluid into a solid surface follows a sinusoidal pattern. Flow is axisymmetric about the z axis, which is normal to the surface. The z component of the flow follows the pattern

$$V_z = Az \sin\left(\frac{2\pi t}{T}\right)$$

Determine (a) the radial component of flow (V_r) and (b) the convective and local components of the acceleration vector.

5.55 Expand $(\vec{V}\cdot\nabla)\vec{V}$ in rectangular coordinates by direct substitution of the velocity vector to obtain the convective acceleration of a fluid particle. Verify the results given in Eqs. 5.11.

5.56 A steady, two-dimensional velocity field is given by $\vec{V} = Ax\hat{i} - Ay\hat{j}$, where $A = 1\ \text{s}^{-1}$. Show that the streamlines for this flow are rectangular hyperbolas, $xy = C$. Obtain a general expression for the acceleration of a fluid particle in this velocity field. Calculate the acceleration of fluid particles at the points $(x, y) = (\frac{1}{2}, 2)$, $(1, 1)$, and $(2, \frac{1}{2})$, where x and y are measured in meters. Plot streamlines that correspond to $C = 0$, 1, and 2 m^2 and show the acceleration vectors on the streamline plot.

5.57 A velocity field is represented by the expression $\vec{V} = (Ax - B)\hat{i} - Ay\hat{j}$, where $A = 0.2\ \text{s}^{-1}$, $B = 0.6\ \text{m}\cdot\text{s}^{-1}$, and the coordinates are expressed in meters. Obtain a general expression for the acceleration of a fluid particle in this velocity field. Calculate the acceleration of fluid particles at points $(x, y) = (0, \frac{4}{3})$, $(1, 2)$, and $(2, 4)$. Plot a few streamlines in the xy plane. Show the acceleration vectors on the streamline plot.

5.58 A velocity field is represented by the expression $\vec{V} = (Ax - B)\hat{i} + Cy\hat{j} + Dt\hat{k}$, where $A = 2\ \text{s}^{-1}$, $B = 4\ \text{m}\cdot\text{s}^{-1}$, $D = 5\ \text{m}\cdot\text{s}^{-2}$, and the coordinates are measured in meters. Determine the proper value for C if the flow field is to be incompressible. Calculate the acceleration of a fluid particle located at point $(x, y) = (3, 2)$. Plot a few flow streamlines in the xy plane.

5.59 A linear approximate velocity profile was used in Problem 5.10 to model a laminar incompressible boundary layer on a flat plate. For this profile, obtain expressions for the x and y components of acceleration of a fluid particle in the boundary layer. Locate the maximum magnitudes of the x and y accelerations. Compute the ratio of the maximum x magnitude to the maximum y magnitude for the flow conditions of Problem 5.10.

5.60 A parabolic approximate velocity profile was used in Problem 5.11 to model flow in a laminar incompressible boundary layer on a flat plate. For this profile, find the x component of acceleration, a_x, of a fluid particle within the boundary layer. Plot a_x at location $x = 0.8$ m, where $\delta = 1.2$ mm, for a flow with $U = 6$ m/s. Find the maximum value of a_x at this x location.

5.61 Show that the velocity field of Problem 2.18 represents a possible incompressible flow field. Determine and plot the streamline passing through point $(x, y) = (2, 4)$ at $t = 1.5$ s. For the particle at the same point and time, show on the plot the velocity vector and the vectors representing the local, convective, and total accelerations.

5.62 A sinusoidal approximate velocity profile was used in Problem 5.12 to model flow in a laminar incompressible boundary layer on a flat plate. For this profile, obtain an expression for the x and y components of acceleration of a fluid particle in the boundary layer. Plot a_x and a_y at location $x = 1$ m, where $\delta = 1$ mm, for a flow with $U = 5$ m/s. Find the maxima of a_x and a_y at this x location.

5.63 Air flows into the narrow gap, of height h, between closely spaced parallel disks through a porous surface as shown. Use a control volume, with outer surface located at position r, to show that the uniform velocity in the r direction is $V = v_0 r/2h$. Find an expression for the velocity component in the z direction ($v_0 \ll V$). Evaluate the components of acceleration for a fluid particle in the gap.

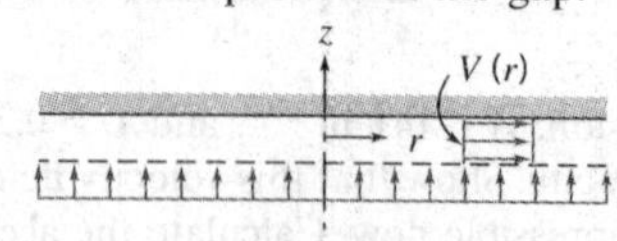

P5.63

5.64 The velocity field for steady inviscid flow from left to right over a circular cylinder, of radius R, is given by

$$\vec{V} = U\cos\theta\left[1 - \left(\frac{R}{r}\right)^2\right]\hat{e}_r - U\sin\theta\left[1 + \left(\frac{R}{r}\right)^2\right]\hat{e}_\theta$$

Obtain expressions for the acceleration of a fluid particle moving along the stagnation streamline ($\theta = \pi$) and for the acceleration along the cylinder surface ($r = R$). Plot a_r as a function of r/R for $\theta = \pi$, and as a function of θ for $r = R$; plot a_θ as a function of θ for $r = R$. Comment on the plots. Determine the locations at which these accelerations reach maximum and minimum values.

5.65 Air flows into the narrow gap, of height h, between closely spaced parallel plates through a porous surface as shown. Use a control volume, with outer surface located at position x, to show that the uniform velocity in the x direction is $u = v_0 x/h$. Find an expression for the velocity component in the y direction. Evaluate the acceleration of a fluid particle in the gap.

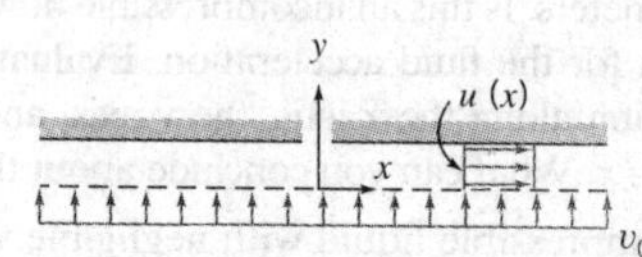

P5.65

5.66 Consider the incompressible flow of a fluid through a nozzle as shown. The area of the nozzle is given by $A = A_0(1 - bx)$ and the inlet velocity varies according to $U = U_0(0.5 + 0.5\cos\omega t)$ where $A_0 = 0.5\ \text{m}^2$, $L = 5$ m, $b = 0.1\ \text{m}^{-1}$, $\omega = 0.16$ rad/s, and $U_0 = 5$ m/s. Find and plot the acceleration on the centerline, with time as a parameter.

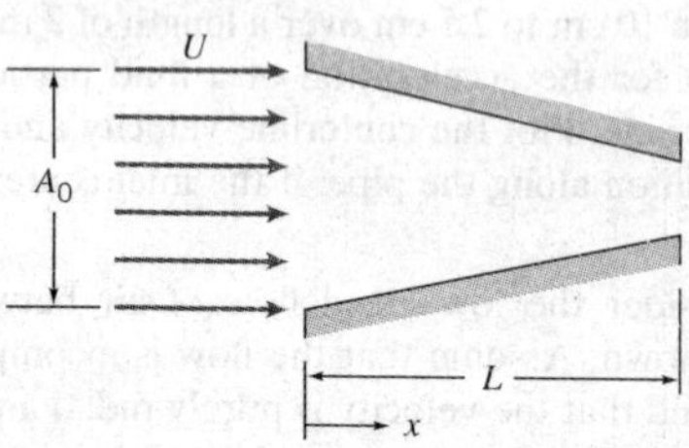

P5.66

5.67 Consider again the steady, two-dimensional velocity field of Problem 5.56. Obtain expressions for the particle coordinates, $x_p = f_1(t)$ and $y_p = f_2(t)$, as functions of time and the initial particle position, (x_0, y_0) at $t = 0$. Determine the time required for a particle to travel from initial position, $(x_0, y_0) = (\frac{1}{2}, 2)$ to positions $(x, y) = (1, 1)$ and $(2, \frac{1}{2})$. Compare the particle accelerations determined by differentiating $f_1(t)$ and $f_2(t)$ with those obtained in Problem 5.56.

5.68 Consider the one-dimensional, incompressible flow through the circular channel shown. The velocity at section ① is given by $U = U_0 + U_1 \sin \omega t$, where $U_0 = 20$ m/s, $U_1 = 2$ m/s, and $\omega = 0.3$ rad/s. The channel dimensions are $L = 1$ m, $R_1 = 0.2$ m, and $R_2 = 0.1$ m. Determine the particle acceleration at the channel exit. Plot the results as a function of time over a complete cycle. On the same plot, show the acceleration at the channel exit if the channel is constant area, rather than convergent, and explain the difference between the curves.

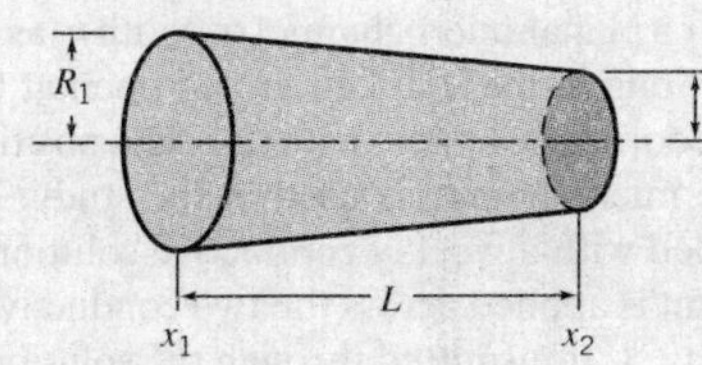

P5.68

5.69 Which, if any, of the following flow fields are irrotational?
(a) $u = 2x^2 + y^2 - x^2y$; $v = x^3 + x(y^2 - 2y)$
(b) $u = 2xy - x^2 + y$; $v = 2xy - y^2 + x^2$
(c) $u = xt + 2y$; $v = xt^2 - yt$
(d) $u = (x + 2y)xt$; $v = -(2x + y)yt$

5.70 Expand $(\vec{V} \cdot \nabla)\vec{V}$ in cylindrical coordinates by direct substitution of the velocity vector to obtain the convective acceleration of a fluid particle. (Recall the hint in footnote 1 on page 178.) Verify the results given in Eqs. 5.12.

5.71 Consider again the sinusoidal velocity profile used to model the x component of velocity for a boundary layer in Problem 5.12. Neglect the vertical component of velocity. Evaluate the circulation around the contour bounded by $x = 0.4$ m, $x = 0.6$ m, $y = 0$, and $y = 8$ mm. What would be the results of this evaluation if it were performed 0.2 m further downstream? Assume $U = 0.5$ m/s.

5.72 Consider the velocity field for flow in a rectangular "corner," $\vec{V} = Ax\hat{i} - Ay\hat{j}$, with $A = 0.3\ \text{s}^{-1}$, as in Example 5.8. Evaluate the circulation about the unit square of Example 5.8.

5.73 A flow is represented by the velocity field $\vec{V} = (x^7 - 21x^5y^2 + 35x^3y^4 - 7xy^6)\hat{i} + (7x^6y - 35x^4y^3 + 21x^2y^5 - y^7)\hat{j}$. Determine if the field is (a) a possible incompressible flow and (b) irrotational.

5.74 Consider the two-dimensional flow field in which $u = Ax^2$ and $v = Bxy$, where $A = 1.6\ \text{m}^{-1}\text{s}^{-1}$, $B = -3.3\ \text{m}^{-1}\text{s}^{-1}$, and the coordinates are measured in meters. Show that the velocity field represents a possible incompressible flow. Determine the rotation at point $(x, y) = (0.3, 0.3)$. Evaluate the circulation about the "curve" bounded by $y = 0$, $x = 0.3$, $y = 0.3$, and $x = 0$.

5.75 Consider the two-dimensional flow field in which $u = Axy$ and $v = By^2$, where $A = 1\text{m}^{-1} \cdot \text{s}^{-1}$, $B = -\frac{1}{2}\text{m}^{-1} \cdot \text{s}^{-1}$, and the coordinates are measured in meters. Show that the velocity field represents a possible incompressible flow. Determine the rotation at point $(x, y) = (1, 1)$. Evaluate the circulation about the "curve" bounded by $y = 0$, $x = 1$, $y = 1$, and $x = 0$.

***5.76** Consider a flow field represented by the stream function $\psi = 3x^5y - 10x^3y^3 + 3xy^5$. Is this a possible two-dimensional incompressible flow? Is the flow irrotational?

***5.77** Consider the flow field represented by the stream function $\psi = x^6 - 15x^4y^2 + 15x^2y^4 - y^6$. Is this a possible two-dimensional, incompressible flow? Is the flow irrotational?

***5.78** Consider a velocity field for motion parallel to the x axis with constant shear. The shear rate is $du/dy = A$, where $A = 0.1\ \text{s}^{-1}$. Obtain an expression for the velocity field, $\vec{V}$. Calculate the rate of rotation. Evaluate the stream function for this flow field.

***5.79** Consider a flow field represented by the stream function $\psi = -A/2(x^2 + y^2)$, where A = constant. Is this a possible two-dimensional incompressible flow? Is the flow irrotational?

***5.80** Consider the flow field represented by the stream function $\psi = Axy + Ay^2$, where $A = 1\ \text{s}^{-1}$. Show that this represents a possible incompressible flow field. Evaluate the rotation of the flow. Plot a few streamlines in the upper half plane.

***5.81** A flow field is represented by the stream function $\psi = x^2 - y^2$. Find the corresponding velocity field. Show that this flow field is irrotational. Plot several streamlines and illustrate the velocity field.

***5.82** Consider the velocity field given by $\vec{V} = Ax^2\hat{i} + Bxy\hat{j}$, where $A = 3.3\ \text{m}^{-1}\text{s}^{-1}$, $B = -6.6\ \text{m}^{-1}\text{s}^{-1}$, and the coordinates are measured in meters.
(a) Determine the fluid rotation.
(b) Evaluate the circulation about the "curve" bounded by $y = 0$, $x = 0.3$, $y = 0.3$, and $x = 0$.
(c) Obtain an expression for the stream function.
(d) Plot several streamlines in the first quadrant.

***5.83** Consider the flow represented by the velocity field $\vec{V} = (Ay + B)\hat{i} + Ax\hat{j}$, where $A = 10\ \text{s}^{-1}$, $B = 3$ m/s, and the coordinates are measured in meters.
(a) Obtain an expression for the stream function.
(b) Plot several streamlines (including the stagnation streamline) in the first quadrant.
(c) Evaluate the circulation about the "curve" bounded by $y = 0$, $x = 1$, $y = 1$, and $x = 0$.

5.84 Consider again the viscometric flow of Example 5.7. Evaluate the average rate of rotation of a pair of perpendicular line segments oriented at ±45° from the x axis. Show that this is the same as in the example.

5.85 Consider the pressure-driven flow between stationary parallel plates separated by distance b. Coordinate y is measured from the bottom plate. The velocity field is given by $u = U(y/b)[1 - (y/b)]$. Obtain an expression for the circulation about a closed contour of height h and length L. Evaluate when $h = b/2$ and when $h = b$. Show that the same result is obtained from the area integral of the Stokes Theorem (Eq. 5.18).

***5.86** The velocity field near the core of a tornado can be approximated as

*These problems require material from sections that may be omitted without loss of continuity in the text material.

$$\vec{V} = -\frac{q}{2\pi r}\hat{e}_r + \frac{K}{2\pi r}\hat{e}_\theta$$

Is this an irrotational flow field? Obtain the stream function for this flow.

5.87 The velocity profile for fully developed flow in a circular tube is $V_z = V_{max}[1 - (r/R)^2]$. Evaluate the rates of linear and angular deformation for this flow. Obtain an expression for the vorticity vector, $\vec{\zeta}$.

5.88 Consider the pressure-driven flow between stationary parallel plates separated by distance $2b$. Coordinate y is measured from the channel centerline. The velocity field is given by $u = u_{max}[1 - (y/b)^2]$. Evaluate the rates of linear and angular deformation. Obtain an expression for the vorticity vector, $\vec{\zeta}$. Find the location where the vorticity is a maximum.

Momentum Equation

5.89 Consider a steady, laminar, fully developed, incompressible flow between two infinite plates, as shown. The flow is due to the motion of the left plate as well a pressure gradient that is applied in the y direction. Given the conditions that $\vec{V} \neq \vec{V}(z)$, $w = 0$, and that gravity points in the negative y direction, prove that $u = 0$ and that the pressure gradient in the y direction must be constant.

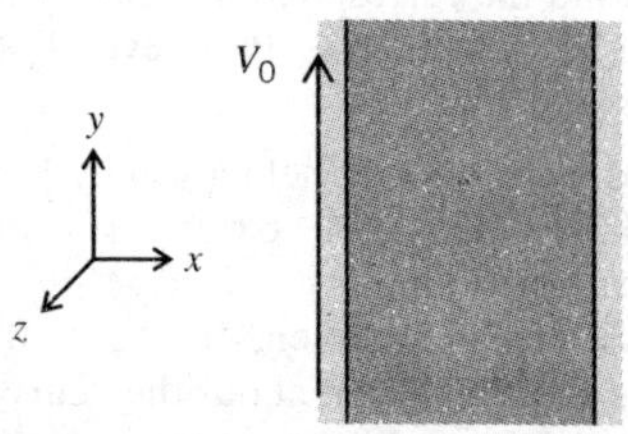

P5.89

5.90 Assume the liquid film in Example 5.9 is not isothermal, but instead has the following distribution:

$$T(y) = T_0 + (T_w - T_0)\left(1 - \frac{y}{h}\right)$$

where T_0 and T_w are, respectively, the ambient temperature and the wall temperature. The fluid viscosity decreases with increasing temperature and is assumed to be described by

$$\mu = \frac{\mu_0}{1 + a(T - T_0)}$$

with $a > 0$. In a manner similar to Example 5.9, derive an expression for the velocity profile.

5.91 The x component of velocity in a laminar boundary layer in water is approximated as $u = U \sin(\pi y/2\delta)$, where $U = 3$ m/s and $\delta = 2$ mm. The y component of velocity is much smaller than u. Obtain an expression for the net shear force per unit volume in the x direction on a fluid element. Calculate its maximum value for this flow.

5.92 A linear velocity profile was used to model flow in a laminar incompressible boundary layer in Problem 5.10. Express the rotation of a fluid particle. Locate the maximum rate of rotation. Express the rate of angular deformation for a fluid particle. Locate the maximum rate of angular deformation. Express the rates of linear deformation for a fluid particle. Locate the maximum rates of linear deformation. Express the shear force per unit volume in the x direction. Locate the maximum shear force per unit volume; interpret this result.

5.93 Problem 4.35 gave the velocity profile for fully developed laminar flow in a circular tube as $u = u_{max}[1 - (r/R)^2]$. Obtain an expression for the shear force per unit volume in the x direction for this flow. Evaluate its maximum value for the conditions of Problem 4.35.

5.94 Assume the liquid film in Example 5.9 is horizontal (i.e., $\theta = 0°$) and that the flow is driven by a constant shear stress on the top surface ($y = h$), $\tau_{yx} = C$. Assume that the liquid film is thin enough and flat and that the flow is fully developed with zero net flow rate (flow rate $Q = 0$). Determine the velocity profile $u(y)$ and the pressure gradient dp/dx.

5.95 Consider a planar microchannel of width h, as shown (it is actually very long in the x direction and open at both ends). A Cartesian coordinate system with its origin positioned at the center of the microchannel is used in the study. The microchannel is filled with a weakly conductive solution. When an electric current is applied across the two conductive walls, the current density, $\vec{J}$, transmitted through the solution is parallel to the y axis. The entire device is placed in a constant magnetic field, $\vec{B}$, which is pointed outward from the plane (the z direction), as shown. Interaction between the current density and the magnetic field induces a *Lorentz force* of density $\vec{J} \times \vec{B}$. Assume that the conductive solution is incompressible, and since the sample volume is very small in lab-on-a-chip applications, the gravitational body force is neglected. Under steady state, the flow driven by the Lorentz force is described by the continuity (Eq. 5.1a) and Navier–Stokes equations (Eqs. 5.27), except the x, y, and z components of the latter have extra Lorentz force components on the right. Assuming that the flow is fully developed and the velocity field $\vec{V}$ is a function of y only, find the three components of velocity.

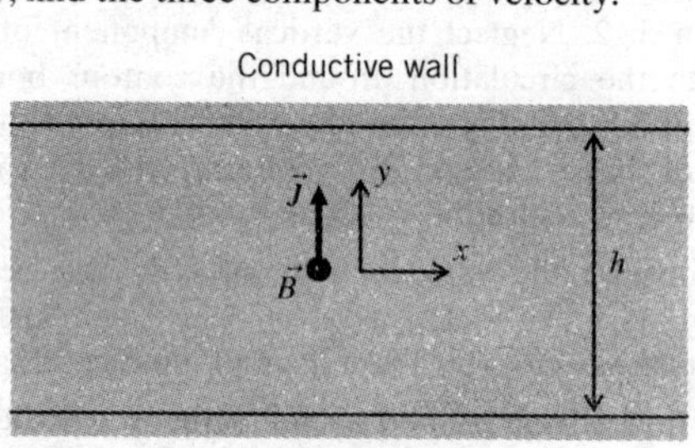

P5.95

5.96 The common thermal polymerase chain reaction (PCR) process requires the cycling of reagents through three distinct temperatures for denaturation (90–94°C), annealing (50–55°C), and extension (72°C). In continuous-flow PCR reactors, the temperatures of the three thermal zones are maintained as fixed while the reagents are cycled continuously through these zones. These temperature variations induce significant variations in the fluid density, which under appropriate conditions can be used to generate fluid motion. The figure depicts a thermosiphon-based PCR device (Chen et al., 2004, *Analytical Chemistry*, 76, 3707–3715).

The closed loop is filled with PCR reagents. The plan of the loop is inclined at an angle α with respect to the vertical. The loop is surrounded by three heaters and coolers that maintain different temperatures.

(a) Explain why the fluid automatically circulates in the closed loop along the counterclockwise direction.

(b) What is the effect of the angle α on the fluid velocity?

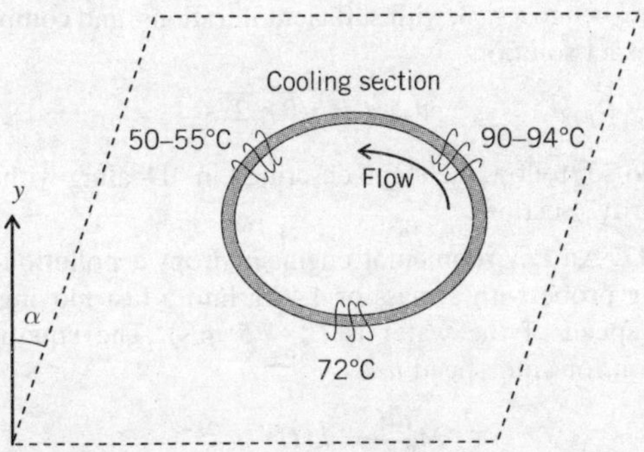

P5.96

5.97 Electro-osmotic flow (EOF) is the motion of liquid induced by an applied electric field across a charged capillary tube or microchannel. Assume the channel wall is negatively charged, a thin layer called the electric double layer (EDL) forms in the vicinity of the channel wall in which the number of positive ions is much larger than that of the negative ions. The net positively charged ions in the EDL then drag the electrolyte solution along with them and cause the fluid to flow toward the cathode. The thickness of the EDL is typically on the order of 10 nm. When the channel dimensions are much larger than the thickness of EDL, there is a slip velocity, $y - \frac{\varepsilon\zeta}{\mu}\vec{E}$, on the channel wall, where ε is the fluid permittivity, ζ is the negative surface electric potential, $\vec{E}$ is the electric field intensity, and μ is the fluid dynamic viscosity. Consider a microchannel formed by two parallel plates. The walls of the channel have a negative surface electric potential of ζ. The microchannel is filled with an electrolyte solution, and the microchannel ends are subjected to an electric potential difference that gives rise to a uniform electric field strength of E along the x direction. The pressure gradient in the channel is zero. Derive the velocity of the steady, fully developed electro-osmotic flow. Compare the velocity profile of the EOF to that of pressure-driven flow. Calculate the EOF velocity using $\varepsilon = 7.08 \times 10^{-10}$ C·V^{-1}m^{-1}, $\zeta = -0.1$ V, $\mu = 10^{-3}$ Pa·s, and $E = 1000$ V/m.

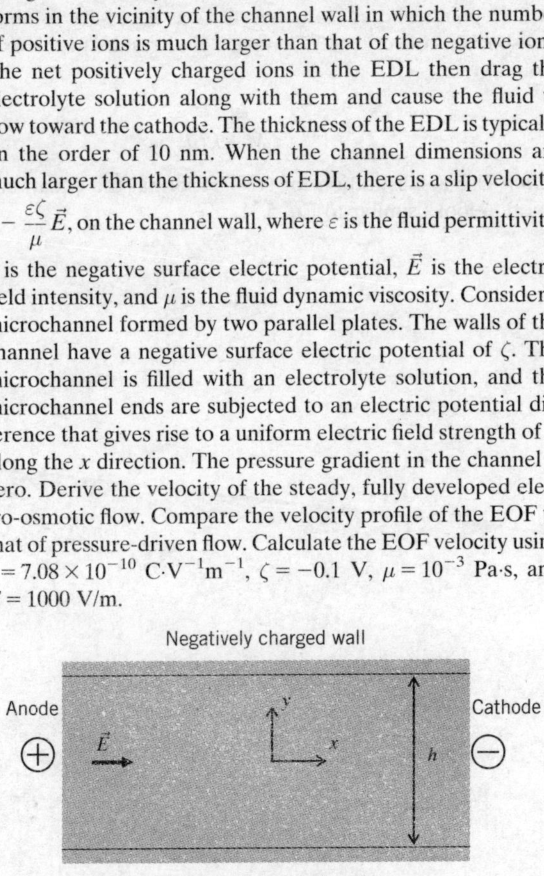

P5.97

Introduction to Computational Fluid Dynamics

***5.98** A tank contains water (20°C) at an initial depth $y_0 =$ 1 m. The tank diameter is $D = 250$ mm, and a tube of diameter $d = 3$ mm and length $L = 4$ m is attached to the bottom of the tank. For laminar flow a reasonable model for the water level over time is

$$\frac{dy}{dt} = -\frac{d^4 \rho g}{32 D^2 \mu L} y \qquad y(0) = y_0$$

Using Euler methods with time steps of 12 min and 6 min:

(a) Estimate the water depth after 120 min, and compute the errors compared to the exact solution

$$y_{\text{exact}}(t) = y_0 e^{-\frac{d^4 \rho g}{32 D^2 \mu L} t}$$

(b) Plot the Euler and exact results.

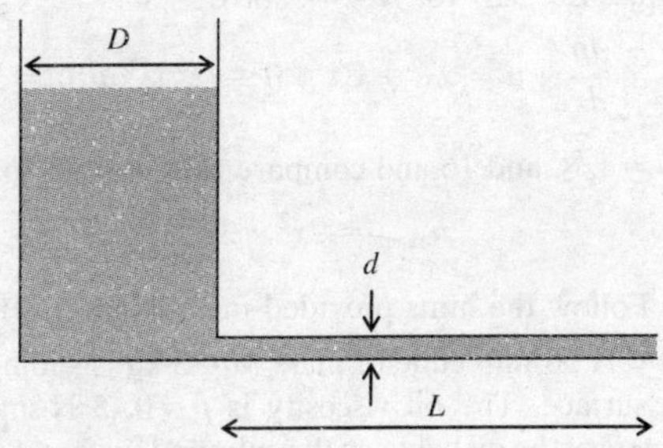

P5.98

***5.99** Use the Euler method to solve and plot

$$\frac{dy}{dx} = \cos(x) \qquad y(0) = 0$$

from $x = 0$ to $x = \pi/2$, using step sizes of $\pi/48$, $\pi/96$, and $\pi/144$. Also plot the exact solution,

$$y(x) = \sin(x)$$

and compute the errors at $x = \pi/2$ for the three Euler method solutions.

***5.100** Use *Excel* to generate the solution of Eq. 5.31 for $m = 1$ shown in Fig. 5.18. To do so, you need to learn how to perform linear algebra in *Excel*. For example, for $N = 4$ you will end up with the matrix equation of Eq. 5.37. To solve this equation for the u values, you will have to compute the inverse of the 4×4 matrix, and then multiply this inverse into the 4×1 matrix on the right of the equation. In *Excel*, to do *array operations*, you must use the following rules: Pre-select the cells that will contain the result; use the appropriate *Excel array function* (look at *Excel's* Help for details); press Ctrl+Shift+Enter, not just Enter. For example, to invert the 4×4 matrix you would: Pre-select a blank 4×4 array that will contain the inverse matrix; type = *minverse*([array containing matrix to be inverted]); press Ctrl+Shift+Enter. To multiply a 4×4 matrix into a 4×1 matrix you would: Pre-select a blank 4×1 array that will contain the result; type = *mmult*([array containing 4×4 matrix], [array containing 4×1 matrix]); press Ctrl+Shift+Enter.

*These problems require material from sections that may be omitted without loss of continuity in the text material.

***5.101** Following the steps to convert the differential equation Eq. 5.31 (for $m = 1$) into a difference equation (for example, Eq. 5.37 for $N = 4$), solve

$$\frac{du}{dx} + u = 2\cos(2x) \qquad 0 \le x \le 1 \quad u(0) = 0$$

for $N = 4$, 8, and 16 and compare to the exact solution

$$u_{\text{exact}} = \frac{2}{5}\cos(2x) + \frac{4}{5}\sin(2x) - \frac{2}{5}e^{-x}$$

Hints: Follow the rules for *Excel* array operations as described in Problem 5.100. Only the right side of the difference equations will change, compared to the solution method of Eq. 5.31 (for example, only the right side of Eq. 5.37 needs modifying).

***5.102** Following the steps to convert the differential equation Eq. 5.31 (for $m = 1$) into a difference equation (for example, Eq. 5.37 for $N = 4$), solve

$$\frac{du}{dx} + u = 2x^2 + x \qquad 0 \le x \le 1 \quad u(0) = 3$$

for $N = 4$, 8, and 16 and compare to the exact solution

$$u_{\text{exact}} = 2x^2 - 3x + 3$$

Hint: Follow the hints provided in Problem 5.101.

***5.103** A 50-mm cube of mass $M = 3$ kg is sliding across an oiled surface. The oil viscosity is $\mu = 0.45$ N·s/m^2, and the thickness of the oil between the cube and surface is $\delta = 0.2$ mm. If the initial speed of the block is $u_0 = 1$ m/s, use the numerical method that was applied to the linear form of Eq. 5.31 to predict the cube motion for the first second of motion. Use $N = 4$, 8, and 16 and compare to the exact solution

$$u_{\text{exact}} = u_0 e^{-(A\mu/M\delta)t}$$

where A is the area of contact. *Hint:* Follow the hints provided in Problem 5.101.

***5.104** Use *Excel* to generate the solutions of Eq. 5.31 for $m = 2$, as shown in Fig. 5.21.

***5.105** Use *Excel* to generate the solutions of Eq. 5.31 for $m = 2$, as shown in Fig. 5.21, except use 16 points and as many iterations as necessary to obtain reasonable convergence.

***5.106** Use *Excel* to generate the solutions of Eq. 5.31 for $m = -1$, with $u(0) = 3$, using 4 and 16 points over the interval from $x = 0$ to $x = 3$, with sufficient iterations, and compare to the exact solutior

$$u_{\text{exact}} = \sqrt{9 - 2x}$$

To do so, follow the steps described in "Dealing with Nonlinearity" section.

***5.107** An environmental engineer drops a pollution measuring probe with a mass of 4.4 kg into a fast moving river (the speed of the water is $U = 7.5$ m/s). The equation of motion for your speed u is

$$M\frac{du}{dt} = k(U - u)^2$$

where $k = 0.958$ kg/m is a constant indicating the drag of the water. Use *Excel* to generate and plot the probe speed versus time (for the first 10 s) using the same approach as the solutions of Eq. 5.31 for $m = 2$, as shown in Fig 5.21, except use 16 points and as many iterations as necessary to obtain reasonable convergence. Compare your results to the exact solution

$$u_{\text{exact}} = \frac{kU^2t}{M + kUt}$$

Hint: Use a substitution for $(U - u)$ so that the equation of motion looks similar to Eq. 5.31.

*These problems require material from sections that may be omitted without loss of continuity in the text material.

6 Incompressible Inviscid Flow

Case Study in Energy and the Environment

Wave Power: The *Limpet*

As we have discussed in previous *Case Studies in Energy and the Environment*, ocean waves contain a lot of energy; some regions of the world have an energy density (energy per width of water flow) of up to 75 kW/m in deep water, and up to 25 kW/m at the shoreline. Many ideas are being explored (some of which we have discussed earlier) for extracting this energy, from tethered buoys to articulated mechanisms. Technical issues are rapidly being resolved with many of these devices, but the Achilles heel of each of them is making the technologies work at a cost, for the power produced, that the consumer is willing to pay. Long-term, fossil fuels will become more expensive, and wave power will fall in cost, but we are not yet at the cross-over point. In the 1980s, wind power had the same kind of difficulty, but after several countries initially subsidized the industry, it is now becoming very

Two views of Wavegen's *Limpet* device (Pictures courtesy of Wavegen Ltd.)

cost-competitive. As with wind power, initial capital costs typically account for more than 90 percent of the cost of producing wave power; for fossil fuel plants the fuel supply itself is an ongoing part of the cost. To succeed, wave energy device developers have focused on driving down the initial capital costs.

The *Voith Hydro Wavegen Limited* company has been making big efforts in analyzing the costs and benefits of wave power with their *Limpet* (*L*and *I*nstalled *M*arine *P*owered *E*nergy *T*ransformer) device, shown in the photograph. This device was designed to be placed in onshore areas of high wave activity; in the long term, such devices will be designed for the higher-energy offshore regions. It is not a particularly impressive-looking device, but it has some interesting features. It looks like just a concrete block, but in fact is hollow and open to the sea on the underside, creating a trapped-air chamber; attached to it is an air turbine. It works pretty much like the swimming pool wave machine used at many amusement parks, except it runs in reverse. In these machines, air is blown in and out of a chamber beside the pool, which makes the water outside bob up and down, causing waves. For the Limpet, the arriving waves cause water in the chamber to rise and fall, which in turn compresses and expands the air trapped in the Limpet. If this is all we had, we would just have a device in which the water waves repeatedly compress and expand the trapped air. The clever innovation of the Limpet device is that a specially designed turbine is attached to the air chamber, so that the air flows through it first one way and then the other, extracting power. The *Wells* turbine (developed by Professor Alan Wells of Queen's University, Belfast) is a low-pressure air turbine that rotates continuously in one direction in spite of the direction of the air flow driving it. Its blades feature a symmetrical airfoil with its plane of symmetry in the plane of rotation and perpendicular to the air stream. Use of this bidirectional turbine allows power to be extracted as the air flows in *and* out of the chamber, avoiding the need for an expensive check valve system. The trade-off for the bidirectional turbine is that its efficiency is lower than that of a turbine with a constant air stream direction. However, the turbine is very simple and rugged: The blades are fixed onto the rotor and have no pitch-adjusting mechanism or gearbox and make no contact with the seawater. Turbines are discussed in some detail in Chapter 10, and some design concepts behind airfoil blade design in Chapter 9.

The whole device—concrete chamber, Wells turbine, and associated electronics—is rugged, inexpensive, and durable, so the goal of minimizing the initial capital cost is close to being realized. The technology used is called *OSW* (*o*scillating *w*ater column). A new project involving the installation of 16 turbines into a breakwater off the coast of Spain is being constructed and is intended to supply green electricity to around 250 households with a rated power of nearly 300 kW.

In Chapter 5 we devoted a great deal of effort to deriving the differential equations (Eqs. 5.24) that describe the behavior of any fluid satisfying the continuum assumption. We also saw how these equations reduced to various particular forms—the most well known being the Navier–Stokes equations for an incompressible, constant viscosity fluid (Eqs. 5.27). Although Eqs. 5.27 describe the behavior of common fluids (e.g., water, air, lubricating oil) for a wide range of problems, as we discussed in

Chapter 5, they are unsolvable analytically except for the simplest of geometries and flows. For example, even using the equations to predict the motion of your coffee as you slowly stir it would require the use of an advanced computational fluid dynamics computer application, and the prediction would take a lot longer to compute than the actual stirring! In this chapter, instead of the Navier–Stokes equations, we will study Euler's equation, which applies to an inviscid fluid. Although truly inviscid fluids do not exist, many flow problems (especially in aerodynamics) can be successfully analyzed with the approximation that $\mu = 0$.

6.1 Momentum Equation for Frictionless Flow: Euler's Equation

Euler's equation (obtained from Eqs. 5.27 after neglecting the viscous terms) is

$$\rho \frac{D\vec{V}}{Dt} = \rho \vec{g} - \nabla p \tag{6.1}$$

This equation states that for an inviscid fluid the change in momentum of a fluid particle is caused by the body force (assumed to be gravity only) and the net pressure force. For convenience we recall that the particle acceleration is

$$\frac{D\vec{V}}{Dt} = \frac{\partial \vec{V}}{\partial t} + (\vec{V} \cdot \nabla)\vec{V} \tag{5.10}$$

In this chapter we will apply Eq. 6.1 to the solution of incompressible, inviscid flow problems. In addition to Eq. 6.1 we have the incompressible form of the mass conservation equation,

$$\nabla \cdot \vec{V} = 0 \tag{5.1c}$$

Equation 6.1 expressed in rectangular coordinates is

$$\rho\left(\frac{\partial u}{\partial t} + u\frac{\partial u}{\partial x} + v\frac{\partial u}{\partial y} + w\frac{\partial u}{\partial z}\right) = \rho g_x - \frac{\partial p}{\partial x} \tag{6.2a}$$

$$\rho\left(\frac{\partial v}{\partial t} + u\frac{\partial v}{\partial x} + v\frac{\partial v}{\partial y} + w\frac{\partial v}{\partial z}\right) = \rho g_y - \frac{\partial p}{\partial y} \tag{6.2b}$$

$$\rho\left(\frac{\partial w}{\partial t} + u\frac{\partial w}{\partial x} + v\frac{\partial w}{\partial y} + w\frac{\partial w}{\partial z}\right) = \rho g_z - \frac{\partial p}{\partial z} \tag{6.2c}$$

If the z axis is assumed vertical, then $g_x = 0$, $g_y = 0$, and $g_z = -g$, so $\vec{g} = -g\hat{k}$.

In cylindrical coordinates, the equations in component form, with gravity the only body force, are

$$\rho a_r = \rho\left(\frac{\partial V_r}{\partial t} + V_r\frac{\partial V_r}{\partial r} + \frac{V_\theta}{r}\frac{\partial V_r}{\partial \theta} + V_z\frac{\partial V_r}{\partial z} - \frac{V_\theta^2}{r}\right) = \rho g_r - \frac{\partial p}{\partial r} \tag{6.3a}$$

$$\rho a_\theta = \rho\left(\frac{\partial V_\theta}{\partial t} + V_r\frac{\partial V_\theta}{\partial r} + \frac{V_\theta}{r}\frac{\partial V_\theta}{\partial \theta} + V_z\frac{\partial V_\theta}{\partial z} + \frac{V_r V_\theta}{r}\right) = \rho g_\theta - \frac{1}{r}\frac{\partial p}{\partial \theta} \tag{6.3b}$$

$$\rho a_z = \rho\left(\frac{\partial V_z}{\partial t} + V_r\frac{\partial V_z}{\partial r} + \frac{V_\theta}{r}\frac{\partial V_z}{\partial \theta} + V_z\frac{\partial V_z}{\partial z}\right) = \rho g_z - \frac{\partial p}{\partial z} \tag{6.3c}$$

If the z axis is directed vertically upward, then $g_r = g_\theta = 0$ and $g_z = -g$.

Equations 6.1, 6.2, and 6.3 apply to problems in which there are no viscous stresses. Before continuing with the main topic of this chapter (inviscid flow), let's consider for a moment when we have no viscous stresses, other than when $\mu = 0$. We recall from previous discussions that, in general, viscous stresses are present when we have fluid deformation (in fact this is how we initially defined a fluid); when we have no fluid deformation, i.e., when we have *rigid-body* motion, no viscous stresses will be present, even if $\mu \neq 0$. Hence Euler's equations apply to rigid-body motions as well as to inviscid flows. We discussed rigid-body motion in detail in Section 3.7 as a special case of fluid statics. As an exercise, you can show that Euler's equations can be used to solve Examples 3.9 and 3.10.

6.2 Euler's Equations in Streamline Coordinates

In Chapters 2 and 5 we pointed out that streamlines, drawn tangent to the velocity vectors at every point in the flow field, provide a convenient graphical representation. In steady flow a fluid particle will move along a streamline because, for steady flow, pathlines and streamlines coincide. Thus, in describing the motion of a fluid particle in a steady flow, in addition to using orthogonal coordinates x, y, z, the distance along a streamline is a logical coordinate to use in writing the equations of motion. "Streamline coordinates" also may be used to describe unsteady flow. Streamlines in unsteady flow give a graphical representation of the instantaneous velocity field.

For simplicity, consider the flow in the yz plane shown in Fig. 6.1. We wish to write the equations of motion in terms of the coordinate s, distance along a streamline, and the coordinate n, distance normal to the streamline. The pressure at the center of the fluid element is p. If we apply Newton's second law in the direction s of the streamline, to the fluid element of volume $ds\, dn\, dx$, then neglecting viscous forces we obtain

$$\left(p - \frac{\partial p}{\partial s}\frac{ds}{2}\right) dn\, dx - \left(p + \frac{\partial p}{\partial s}\frac{ds}{2}\right) dn\, dx - \rho g \sin\beta\, ds\, dn\, dx = \rho a_s\, ds\, dn\, dx$$

where β is the angle between the tangent to the streamline and the horizontal, and a_s is the acceleration of the fluid particle along the streamline. Simplifying the equation, we obtain

$$-\frac{\partial p}{\partial s} - \rho g \sin\beta = \rho a_s$$

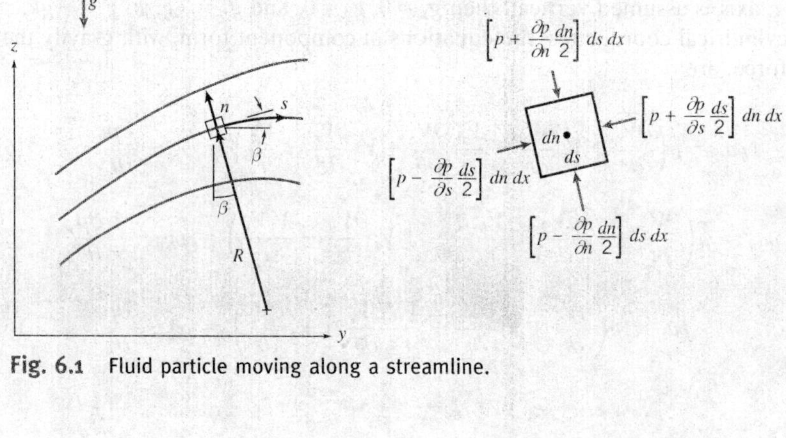

Fig. 6.1 Fluid particle moving along a streamline.

Since $\sin\beta = \partial z/\partial s$, we can write

$$-\frac{1}{\rho}\frac{\partial p}{\partial s} - g\frac{\partial z}{\partial s} = a_s$$

Along any streamline $V = V(s, t)$, and the material or total acceleration of a fluid particle in the streamwise direction is given by

$$a_s = \frac{DV}{Dt} = \frac{\partial V}{\partial t} + V\frac{\partial V}{\partial s}$$

Euler's equation in the streamwise direction with the z axis directed vertically upward is then

$$-\frac{1}{\rho}\frac{\partial p}{\partial s} - g\frac{\partial z}{\partial s} = \frac{\partial V}{\partial t} + V\frac{\partial V}{\partial s} \tag{6.4a}$$

For steady flow, and neglecting body forces, Euler's equation in the streamwise direction reduces to

$$\frac{1}{\rho}\frac{\partial p}{\partial s} = -V\frac{\partial V}{\partial s} \tag{6.4b}$$

which indicates that (for an incompressible, inviscid flow) *a decrease in velocity is accompanied by an increase in pressure* and conversely. This makes sense: The only force experienced by the particle is the net pressure force, so the particle accelerates toward low-pressure regions and decelerates when approaching high-pressure regions.

To obtain Euler's equation in a direction normal to the streamlines, we apply Newton's second law in the n direction to the fluid element. Again, neglecting viscous forces, we obtain

CLASSIC VIDEO

Pressure Fields and Fluid Acceleration.

$$\left(p - \frac{\partial p}{\partial n}\frac{dn}{2}\right)ds\,dx - \left(p + \frac{\partial p}{\partial n}\frac{dn}{2}\right)ds\,dx - \rho g\cos\beta\,dn\,dx\,ds = \rho a_n dn\,dx\,ds$$

where β is the angle between the n direction and the vertical, and a_n is the acceleration of the fluid particle in the n direction. Simplifying the equation, we obtain

$$-\frac{\partial p}{\partial n} - \rho g\cos\beta = \rho a_n$$

Since $\cos\beta = \partial z/\partial n$, we write

$$-\frac{1}{\rho}\frac{\partial p}{\partial n} - g\frac{\partial z}{\partial n} = a_n$$

The normal acceleration of the fluid element is toward the center of curvature of the streamline, in the minus n direction; thus in the coordinate system of Fig. 6.1, the familiar centripetal acceleration is written

$$a_n = -\frac{V^2}{R}$$

for steady flow, where R is the radius of curvature of the streamline at the point chosen. Then, Euler's equation normal to the streamline is written for steady flow as

$$\frac{1}{\rho}\frac{\partial p}{\partial n} + g\frac{\partial z}{\partial n} = \frac{V^2}{R} \tag{6.5a}$$

CLASSIC VIDEO

Pressure Fields and Fluid Acceleration.

For steady flow in a horizontal plane, Euler's equation normal to a streamline becomes

$$\frac{1}{\rho}\frac{\partial p}{\partial n} = \frac{V^2}{R} \qquad (6.5b)$$

Equation 6.5b indicates that *pressure increases in the direction outward from the center of curvature of the streamlines.* This also makes sense: Because the only force experienced by the particle is the net pressure force, the pressure field creates the centripetal acceleration. In regions where the streamlines are straight, the radius of curvature, R, is infinite so *there is no pressure variation normal to straight streamlines.*

Example 6.1 FLOW IN A BEND

The flow rate of air at standard conditions in a flat duct is to be determined by installing pressure taps across a bend. The duct is 0.3 m deep and 0.1 m wide. The inner radius of the bend is 0.25 m. If the measured pressure difference between the taps is 40 mm of water, compute the approximate flow rate.

Given: Flow through duct bend as shown.

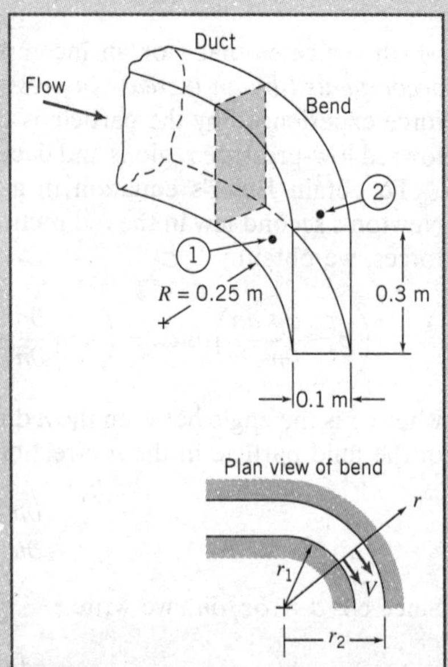

$$p_2 - p_1 = \rho_{H_2O}\, g\, \Delta h$$

where $\Delta h = 40$ mm H_2O. Air is at STP.

Find: Volume flow rate, Q.

Solution:

Apply Euler's n component equation across flow streamlines.

Governing equation: $\dfrac{\partial p}{\partial r} = \dfrac{\rho V^2}{r}$

Assumptions: (1) Frictionless flow.
(2) Incompressible flow.
(3) Uniform flow at measurement section.

For this flow, $p = p(r)$, so

$$\frac{\partial p}{\partial r} = \frac{dp}{dr} = \frac{\rho V^2}{r}$$

or

$$dp = \rho V^2 \frac{dr}{r}$$

Integrating gives

$$p_2 - p_1 = \rho V^2 \ln r\Big]_{r_1}^{r_2} = \rho V^2 \ln\frac{r_2}{r_1}$$

and hence

$$V = \left[\frac{p_2 - p_1}{\rho \ln(r_2/r_1)}\right]^{1/2}$$

But $\Delta p = p_2 - p_1 = \rho_{H_2O} g \Delta h$, so $V = \left[\dfrac{\rho_{H_2O} g \Delta h}{\rho \ln(r_2/r_1)}\right]^{1/2}$

Substituting numerical values,

$$V = \left[999 \frac{\text{kg}}{\text{m}^3} \times 9.81 \frac{\text{m}}{\text{s}^2} \times 0.04\,\text{m} \times \frac{\text{m}^3}{1.23\,\text{kg}} \times \frac{1}{\ln(0.35\,\text{m}/0.25\,\text{m})}\right]^{1/2}$$

$$= 30.8\ \text{m/s}$$

For uniform flow

$$Q = VA = 30.8 \frac{\text{m}}{\text{s}} \times 0.1\ \text{m} \times 0.3\,\text{m}$$

$$Q = 0.924\ \text{m}^3/\text{s} \qquad Q$$

In this problem we assumed that the velocity is uniform across the section. In fact, the velocity in the bend approximates a free vortex (irrotational) profile in which $V \propto 1/r$ (where r is the radius) instead of $V =$ const. Hence, this flow-measurement device could only be used to obtain approximate values of the flow rate (see Problem 6.32).

6.3 Bernoulli Equation: Integration of Euler's Equation Along a Streamline for Steady Flow

Compared to the viscous-flow equivalents, the momentum or Euler's equation for incompressible, inviscid flow (Eq. 6.1) is simpler mathematically, but solution (in conjunction with the mass conservation equation, Eq. 5.1c) still presents formidable difficulties in all but the most basic flow problems. One convenient approach for a steady flow is to integrate Euler's equation along a streamline. We will do this below using two different mathematical approaches, and each will result in the Bernoulli equation. Recall that in Section 4.4 we derived the Bernoulli equation by starting with a differential control volume; these two additional derivations will give us more insight into the restrictions inherent in use of the Bernoulli equation.

*Derivation Using Streamline Coordinates

Euler's equation for steady flow along a streamline (from Eq. 6.4a) is

$$-\frac{1}{\rho}\frac{\partial p}{\partial s} - g\frac{\partial z}{\partial s} = V\frac{\partial V}{\partial s} \tag{6.6}$$

If a fluid particle moves a distance, ds, along a streamline, then

$$\frac{\partial p}{\partial s} ds = dp \quad \text{(the change in pressure along } s\text{)}$$

$$\frac{\partial z}{\partial s} ds = dz \quad \text{(the change in elevation along } s\text{)}$$

$$\frac{\partial V}{\partial s} ds = dV \quad \text{(the change in speed along } s\text{)}$$

Thus, after multiplying Eq. 6.6 by ds, we can write

$$-\frac{dp}{\rho} - g\,dz = V\,dV \qquad \text{or} \qquad \frac{dp}{\rho} + V\,dV + g\,dz = 0 \qquad (\text{along } s)$$

Integration of this equation gives

$$\int \frac{dp}{\rho} + \frac{V^2}{2} + gz = \text{constant} \quad (\text{along } s) \tag{6.7}$$

Before Eq. 6.7 can be applied, we must specify the relation between pressure and density. For the special case of incompressible flow, ρ = constant, and Eq. 6.7 becomes the Bernoulli equation,

$$\frac{p}{\rho} + \frac{V^2}{2} + gz = \text{constant} \tag{6.8}$$

Restrictions: (1) Steady flow.
(2) Incompressible flow.
(3) Frictionless flow.
(4) Flow along a streamline.

The Bernoulli equation is probably the most famous, and abused, equation in all of fluid mechanics. It is always tempting to use because it is a simple algebraic equation for relating the pressure, velocity, and elevation in a fluid. For example, it is used to explain the lift of a wing: In aerodynamics the gravity term is usually negligible, so Eq. 6.8 indicates that wherever the velocity is relatively high (e.g., on the upper surface of a wing), the pressure must be relatively low, and wherever the velocity is relatively low (e.g., on the lower surface of a wing), the pressure must be relatively high, generating substantial lift. Equation 6.8 indicates that, in general (if the flow is not constrained in some way), if a particle increases its elevation ($z \uparrow$) or moves into a higher pressure region ($p \uparrow$), it will tend to decelerate ($V \downarrow$); this makes sense from a momentum point of view (recall that the equation was derived from momentum considerations). These comments *only* apply if the four restrictions listed are reasonable. For example, Eq. 6.8 cannot be used to explain the pressure drop in a horizontal constant diameter pipe flow: according to it, for z = constant and V = constant, p = constant! We cannot stress enough that you should *keep the restrictions firmly in mind whenever you consider using the Bernoulli equation*! (In general, the Bernoulli constant in Eq. 6.8 has different values along different streamlines.[1])

*Derivation Using Rectangular Coordinates

The vector form of Euler's equation, Eq. 6.1, also can be integrated along a streamline. We shall restrict the derivation to steady flow; thus, the end result of our effort should be Eq. 6.7.

For steady flow, Euler's equation in rectangular coordinates can be expressed as

$$\frac{D\vec{V}}{Dt} = u\frac{\partial \vec{V}}{\partial x} + v\frac{\partial \vec{V}}{\partial y} + w\frac{\partial \vec{V}}{\partial z} = (\vec{V}\cdot\nabla)\vec{V} = -\frac{1}{\rho}\nabla p - g\hat{k} \tag{6.9}$$

For steady flow the velocity field is given by $\vec{V} = \vec{V}(x, y, z)$. The streamlines are lines drawn in the flow field tangent to the velocity vector at every point. Recall again that for steady flow, streamlines, pathlines, and streaklines coincide. The motion of a

[1]For the case of irrotational flow, the constant has a single value throughout the entire flow field (Section 6.7).

*This section may be omitted without loss of continuity in the text material.

particle along a streamline is governed by Eq. 6.9. During time interval dt the particle has vector displacement $d\vec{s}$ along the streamline.

If we take the dot product of the terms in Eq. 6.9 with displacement $d\vec{s}$ along the streamline, we obtain a scalar equation relating pressure, speed, and elevation along the streamline. Taking the dot product of $d\vec{s}$ with Eq. 6.9 gives

$$(\vec{V} \cdot \nabla)\vec{V} \cdot d\vec{s} = -\frac{1}{\rho}\nabla p \cdot d\vec{s} - g\hat{k} \cdot d\vec{s} \tag{6.10}$$

where

$$d\vec{s} = dx\hat{i} + dy\hat{j} + dz\hat{k} \quad \text{(along } s\text{)}$$

Now we evaluate each of the three terms in Eq. 6.10, starting on the right,

$$\begin{aligned} -\frac{1}{\rho}\nabla p \cdot d\vec{s} &= -\frac{1}{\rho}\left[\hat{i}\frac{\partial p}{\partial x} + \hat{j}\frac{\partial p}{\partial y} + \hat{k}\frac{\partial p}{\partial z}\right] \cdot [dx\hat{i} + dy\hat{j} + dz\hat{k}] \\ &= -\frac{1}{\rho}\left[\frac{\partial p}{\partial x}dx + \frac{\partial p}{\partial y}dy + \frac{\partial p}{\partial z}dz\right] \quad \text{(along } s\text{)} \end{aligned}$$

$$-\frac{1}{\rho}\nabla p \cdot d\vec{s} = -\frac{1}{\rho}dp \quad \text{(along } s\text{)}$$

and

$$\begin{aligned} -g\hat{k} \cdot d\vec{s} &= -g\hat{k} \cdot [dx\hat{i} + dy\hat{j} + dz\hat{k}] \\ &= -g\,dz \qquad \text{(along } s\text{)} \end{aligned}$$

Using a vector identity,[2] we can write the third term as

$$\begin{aligned} (\vec{V} \cdot \nabla)\vec{V} \cdot d\vec{s} &= \left[\frac{1}{2}\nabla(\vec{V} \cdot \vec{V}) - \vec{V} \times (\nabla \times \vec{V})\right] \cdot d\vec{s} \\ &= \left\{\frac{1}{2}\nabla(\vec{V} \cdot \vec{V})\right\} \cdot d\vec{s} - \{\vec{V} \times (\nabla \times \vec{V})\} \cdot d\vec{s} \end{aligned}$$

The last term on the right side of this equation is zero, since $\vec{V}$ is parallel to $d\vec{s}$ [recall from vector math that $\vec{V} \times (\nabla \times \vec{V}) \cdot d\vec{s} = -(\nabla \times \vec{V}) \times \vec{V} \cdot d\vec{s} = -(\nabla \times \vec{V}) \cdot \vec{V} \times d\vec{s}$]. Consequently,

$$\begin{aligned} (\vec{V} \cdot \nabla)\vec{V} \cdot d\vec{s} &= \frac{1}{2}\nabla(\vec{V} \cdot \vec{V}) \cdot d\vec{s} = \frac{1}{2}\nabla(V^2) \cdot d\vec{s} \quad \text{(along } s\text{)} \\ &= \frac{1}{2}\left[\hat{i}\frac{\partial V^2}{\partial x} + \hat{j}\frac{\partial V^2}{\partial y} + \hat{k}\frac{\partial V^2}{\partial z}\right] \cdot [dx\hat{i} + dy\hat{j} + dz\hat{k}] \\ &= \frac{1}{2}\left[\frac{\partial V^2}{\partial x}dx + \frac{\partial V^2}{\partial y}dy + \frac{\partial V^2}{\partial z}dz\right] \end{aligned}$$

$$(\vec{V} \cdot \nabla)\vec{V} \cdot d\vec{s} = \frac{1}{2}d(V^2) \quad \text{(along } s\text{)}$$

[2]The vector identity

$$(\vec{V} \cdot \nabla)\vec{V} = \frac{1}{2}\nabla(\vec{V} \cdot \vec{V}) - \vec{V} \times (\nabla \times \vec{V})$$

may be verified by expanding each side into components.

Substituting these three terms into Eq. 6.10 yields

$$\frac{dp}{\rho} + \frac{1}{2}d(V^2) + g\,dz = 0 \quad (\text{along } s)$$

Integrating this equation, we obtain

$$\int \frac{dp}{\rho} + \frac{V^2}{2} + gz = \text{constant} \quad (\text{along } s)$$

If the density is constant, we obtain the Bernoulli equation

$$\frac{p}{\rho} + \frac{V^2}{2} + gz = \text{constant}$$

As expected, we see that the last two equations are identical to Eqs. 6.7 and 6.8 derived previously using streamline coordinates. The Bernoulli equation, derived using rectangular coordinates, is still subject to the restrictions: (1) steady flow, (2) incompressible flow, (3) frictionless flow, and (4) flow along a streamline.

Static, Stagnation, and Dynamic Pressures

The pressure, p, which we have used in deriving the Bernoulli equation, Eq. 6.8, is the thermodynamic pressure; it is commonly called the *static pressure*. The static pressure is the pressure experienced by the fluid particle as it moves (so it is something of a misnomer!)—we also have the stagnation and dynamic pressures, which we will define shortly. How do we measure the pressure in a fluid in motion?

In Section 6.2 we showed that there is no pressure variation normal to straight streamlines. This fact makes it possible to measure the static pressure in a flowing fluid using a wall pressure "tap," placed in a region where the flow streamlines are straight, as shown in Fig. 6.2*a*. The pressure tap is a small hole, drilled carefully in the wall, with its axis perpendicular to the surface. If the hole is perpendicular to the duct wall and free from burrs, accurate measurements of static pressure can be made by connecting the tap to a suitable pressure-measuring instrument [1].

In a fluid stream far from a wall, or where streamlines are curved, accurate static pressure measurements can be made by careful use of a static pressure probe, shown in Fig. 6.2*b*. Such probes must be designed so that the measuring holes are placed correctly with respect to the probe tip and stem to avoid erroneous results [2]. In use, the measuring section must be aligned with the local flow direction. (In these figures, it may appear that the pressure tap and small holes would allow flow to enter or leave or otherwise be entrained by the main flow, but each of these is ultimately attached to a pressure sensor or manometer and is therefore a dead-end, leading to no flow being possible—see Example 6.2.)

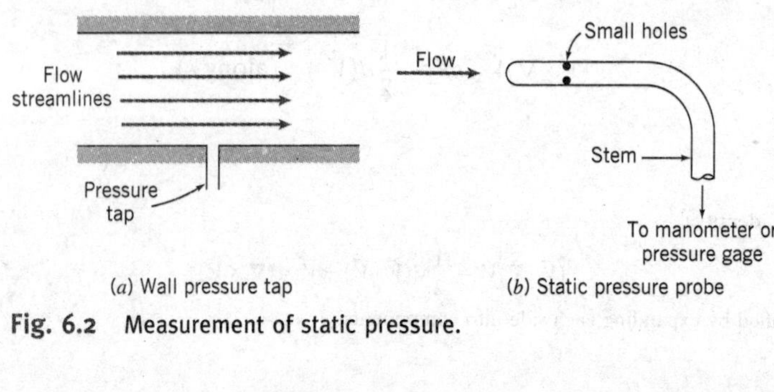

Fig. 6.2 Measurement of static pressure.

Static pressure probes, such as that shown in Fig 6.2*b*, and in a variety of other forms, are available commercially in sizes as small as 1.5 mm in diameter [3].

The *stagnation pressure* is obtained when a flowing fluid is decelerated to zero speed by a frictionless process. For incompressible flow, the Bernoulli equation can be used to relate changes in speed and pressure along a streamline for such a process. Neglecting elevation differences, Eq. 6.8 becomes

$$\frac{p}{\rho} + \frac{V^2}{2} = \text{constant}$$

If the static pressure is p at a point in the flow where the speed is V, then the stagnation pressure, p_0, where the stagnation speed, V_0, is zero, may be computed from

$$\frac{p_0}{\rho} + \overbrace{\frac{V_0^2}{2}}^{=0} = \frac{p}{\rho} + \frac{V^2}{2}$$

or

$$p_0 = p + \frac{1}{2}\rho V^2 \tag{6.11}$$

Equation 6.11 is a mathematical statement of the definition of stagnation pressure, valid for incompressible flow. The term $\frac{1}{2}\rho V^2$ generally is called the *dynamic pressure*. Equation 6.11 states that the stagnation (or *total*) pressure equals the static pressure plus the dynamic pressure. One way to picture the three pressures is to imagine you are standing in a steady wind holding up your hand: The static pressure will be atmospheric pressure; the larger pressure you feel at the center of your hand will be the stagnation pressure; and the buildup of pressure (the difference between the stagnation and static pressures) will be the dynamic pressure. Solving Eq. 6.11 for the speed,

$$V = \sqrt{\frac{2(p_0 - p)}{\rho}} \tag{6.12}$$

Thus, if the stagnation pressure and the static pressure could be measured at a point, Eq. 6.12 would give the local flow speed.

Stagnation pressure is measured in the laboratory using a probe with a hole that faces directly upstream as shown in Fig. 6.3. Such a probe is called a stagnation pressure probe, or pitot (pronounced *pea-toe*) tube. Again, the measuring section must be aligned with the local flow direction.

We have seen that static pressure at a point can be measured with a static pressure tap or probe (Fig. 6.2). If we knew the stagnation pressure at the same point, then the flow speed could be computed from Eq. 6.12. Two possible experimental setups are shown in Fig. 6.4.

In Fig. 6.4*a*, the static pressure corresponding to point *A* is read from the wall static pressure tap. The stagnation pressure is measured directly at *A* by the total head tube,

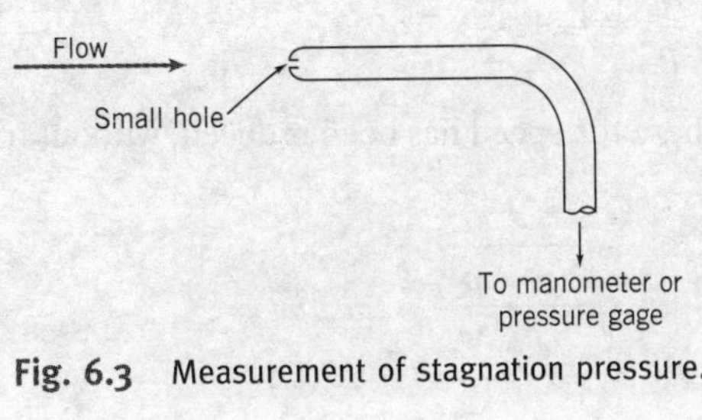

Fig. 6.3 Measurement of stagnation pressure.

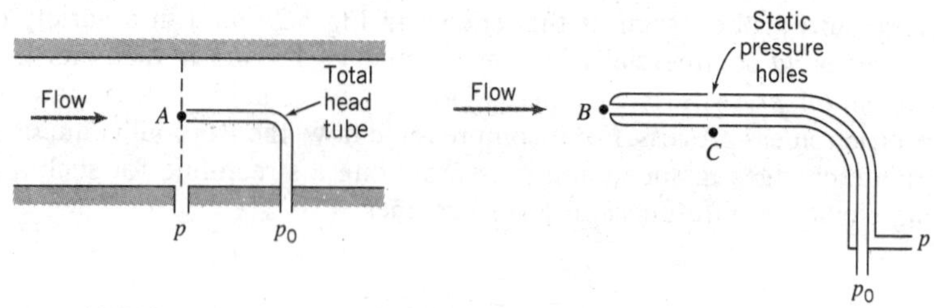

(*a*) Total head tube used with wall static tap

(*b*) Pitot-static tube

Fig. 6.4 Simultaneous measurement of stagnation and static pressures.

as shown. (The stem of the total head tube is placed downstream from the measurement location to minimize disturbance of the local flow.)

Two probes often are combined, as in the pitot-static tube shown in Fig. 6.4*b*. The inner tube is used to measure the stagnation pressure at point *B*, while the static pressure at *C* is sensed using the small holes in the outer tube. In flow fields where the static pressure variation in the streamwise direction is small, the pitot-static tube may be used to infer the speed at point *B* in the flow by assuming $p_B = p_C$ and using Eq. 6.12. (Note that when $p_B \neq p_C$, this procedure will give erroneous results.)

Remember that the Bernoulli equation applies only for incompressible flow (Mach number $M \leq 0.3$). The definition and calculation of the stagnation pressure for compressible flow will be discussed in Section 12.3.

Example 6.2 PITOT TUBE

A pitot tube is inserted in an air flow (at STP) to measure the flow speed. The tube is inserted so that it points upstream into the flow and the pressure sensed by the tube is the stagnation pressure. The static pressure is measured at the same location in the flow, using a wall pressure tap. If the pressure difference is 30 mm of mercury, determine the flow speed.

Given: A pitot tube inserted in a flow as shown. The flowing fluid is air and the manometer liquid is mercury.

Find: The flow speed.

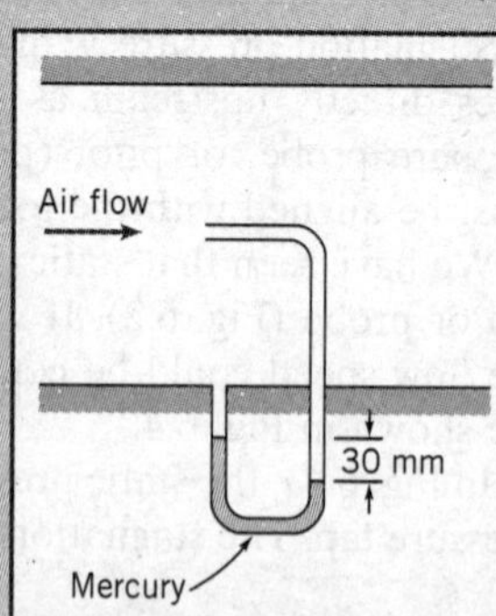

Solution:

Governing equation: $\frac{p}{\rho} + \frac{V^2}{2} + gz = \text{constant}$

Assumptions:
(1) Steady flow.
(2) Incompressible flow.
(3) Flow along a streamline.
(4) Frictionless deceleration along stagnation streamline.

Writing Bernoulli's equation along the stagnation streamline (with $\Delta z = 0$) yields

$$\frac{p_0}{\rho} = \frac{p}{\rho} + \frac{V^2}{2}$$

p_0 is the stagnation pressure at the tube opening where the speed has been reduced, without friction, to zero. Solving for V gives

$$V = \sqrt{\frac{2(p_0 - p)}{\rho_{\text{air}}}}$$

From the diagram,

$$p_0 - p = \rho_{Hg} g h = \rho_{H_2O} g h SG_{Hg}$$

and

$$V = \sqrt{\frac{2\rho_{H_2O} g h SG_{Hg}}{\rho_{air}}}$$

$$= \sqrt{2 \times 1000 \frac{kg}{m^3} \times 9.81 \frac{m}{s^2} \times 30\ mm \times 13.6 \times \frac{m^3}{1.23\ kg} \times \frac{1\ m}{1000\ mm}}$$

$V = 80.8\ m/s$ ← V

This problem illustrates use of a pitot tube to determine flow speed. Pitot (or pitot-static) tubes are often placed on the exterior of aircraft to indicate air speed relative to the aircraft, and hence aircraft speed relative to the air.

At $T = 20°C$, the speed of sound in air is 343 m/s. Hence, $M = 0.236$ and the assumption of incompressible flow is valid.

Applications

The Bernoulli equation can be applied between any two points on a streamline provided that the other three restrictions are satisfied. The result is

$$\frac{p_1}{\rho} + \frac{V_1^2}{2} + gz_1 = \frac{p_2}{\rho} + \frac{V_2^2}{2} + gz_2 \tag{6.13}$$

where subscripts 1 and 2 represent any two points on a streamline. Applications of Eqs. 6.8 and 6.13 to typical flow problems are illustrated in Examples 6.3 through 6.5.

In some situations, the flow appears unsteady from one reference frame, but steady from another, which translates with the flow. Since the Bernoulli equation was derived by integrating Newton's second law for a fluid particle, it can be applied in any inertial reference frame (see the discussion of translating frames in Section 4.4). The procedure is illustrated in Example 6.6.

Example 6.3 NOZZLE FLOW

Air flows steadily at low speed through a horizontal *nozzle* (by definition a device for accelerating a flow), discharging to atmosphere. The area at the nozzle inlet is 0.1 m². At the nozzle exit, the area is 0.02 m². Determine the gage pressure required at the nozzle inlet to produce an outlet speed of 50 m/s.

Given: Flow through a nozzle, as shown.

Find: $p_1 - p_{atm}$.

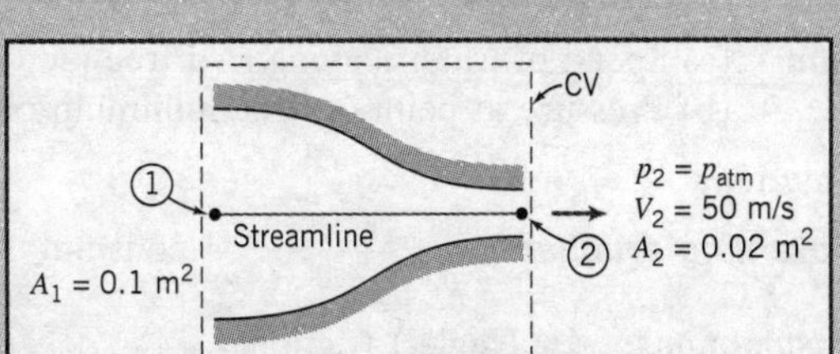

Solution:

Governing equations:

$$\frac{p_1}{\rho} + \frac{V_1^2}{2} + gz_1 = \frac{p_2}{\rho} + \frac{V_2^2}{2} + gz_2 \tag{6.13}$$

Continuity for incompressible and uniform flow:

$$\sum_{CS} \vec{V} \cdot \vec{A} = 0 \tag{4.13b}$$

Assumptions: (1) Steady flow.
(2) Incompressible flow.
(3) Frictionless flow.
(4) Flow along a streamline.
(5) $z_1 = z_2$.
(6) Uniform flow at sections ① and ②.

The maximum speed of 50 m/s is well below 100 m/s, which corresponds to Mach number $M \approx 0.3$ in standard air. Hence, the flow may be treated as incompressible.

Apply the Bernoulli equation along a streamline between points ① and ② to evaluate p_1. Then

$$p_1 - p_{atm} = p_1 - p_2 = \frac{\rho}{2}(V_2^2 - V_1^2)$$

Apply the continuity equation to determine V_1,

$$(-\rho V_1 A_1) + (\rho V_2 A_2) = 0 \qquad \text{or} \qquad V_1 A_1 = V_2 A_2$$

so that

$$V_1 = V_2 \frac{A_2}{A_1} = 50 \frac{\text{m}}{\text{s}} \times \frac{0.02 \text{ m}^2}{0.1 \text{ m}^2} = 10 \text{ m/s}$$

For air at standard conditions, $\rho = 1.23 \text{ kg/m}^3$. Then

$$p_1 - p_{atm} = \frac{\rho}{2}(V_2^2 - V_1^2)$$

$$= \frac{1}{2} \times 1.23 \frac{\text{kg}}{\text{m}^3}\left[(50)^2 \frac{\text{m}^2}{\text{s}^2} - (10)^2 \frac{\text{m}^2}{\text{s}^2}\right] \frac{\text{N} \cdot \text{s}^2}{\text{kg} \cdot \text{m}}$$

$$p_1 - p_{atm} = 1.48 \text{ kPa}$$

$p_1 - p_{atm}$

Notes:

✓ This problem illustrates a typical application of the Bernoulli equation.

✓ The streamlines must be straight at the inlet and exit in order to have uniform pressures at those locations.

Example 6.4 FLOW THROUGH A SIPHON

A U-tube acts as a water siphon. The bend in the tube is 1 m above the water surface; the tube outlet is 7 m below the water surface. The water issues from the bottom of the siphon as a free jet at atmospheric pressure. Determine (after listing the necessary assumptions) the speed of the free jet and the minimum absolute pressure of the water in the bend.

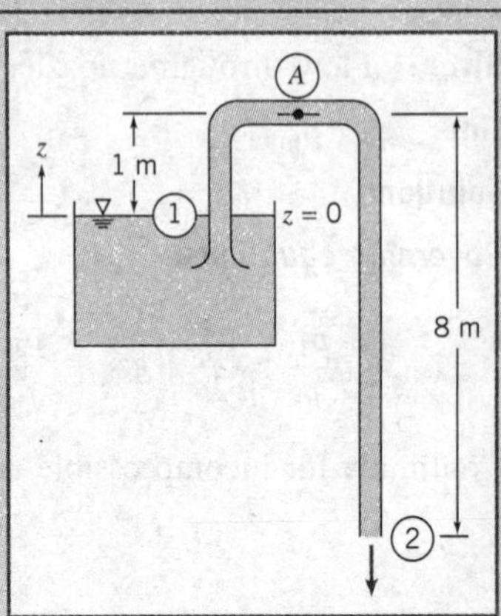

Given: Water flowing through a siphon as shown.

Find: (a) Speed of water leaving as a free jet.
(b) Pressure at point Ⓐ (the minimum pressure point) in the flow.

Solution:

Governing equation: $\dfrac{p}{\rho} + \dfrac{V^2}{2} + gz = \text{constant}$

Assumptions: (1) Neglect friction.
(2) Steady flow.
(3) Incompressible flow.
(4) Flow along a streamline.
(5) Reservoir is large compared with pipe.

Apply the Bernoulli equation between points ① and ②.

$$\frac{p_1}{\rho} + \frac{V_1^2}{2} + gz_1 = \frac{p_2}{\rho} + \frac{V_2^2}{2} + gz_2$$

Since $\text{area}_{\text{reservoir}} \gg \text{area}_{\text{pipe}}$, then $V_1 \approx 0$. Also $p_1 = p_2 = p_{\text{atm}}$, so

$$gz_1 = \frac{V_2^2}{2} + gz_2 \quad \text{and} \quad V_2^2 = 2g(z_1 - z_2)$$

$$V_2 = \sqrt{2g(z_1 - z_2)} = \sqrt{2 \times 9.81 \frac{\text{m}}{\text{s}^2} \times 7\ \text{m}}$$

$$= 11.7\ \text{m/s} \qquad V_2$$

To determine the pressure at location Ⓐ, we write the Bernoulli equation between ① and Ⓐ.

$$\frac{p_1}{\rho} + \frac{V_1^2}{2} + gz_1 = \frac{p_A}{\rho} + \frac{V_A^2}{2} + gz_A$$

Again $V_1 \approx 0$ and from conservation of mass $V_A = V_2$. Hence

$$\frac{p_A}{\rho} = \frac{p_1}{\rho} + gz_1 - \frac{V_2^2}{2} - gz_A = \frac{p_1}{\rho} + g(z_1 - z_A) - \frac{V_2^2}{2}$$

$$p_A = p_1 + \rho g(z_1 - z_A) - \rho\frac{V_2^2}{2}$$

$$= 1.01 \times 10^5 \frac{\text{N}}{\text{m}^2} + 999 \frac{\text{kg}}{\text{m}^3} \times 9.81 \frac{\text{m}}{\text{s}^2} \times (-1\ \text{m}) \frac{\text{N} \cdot \text{s}^2}{\text{kg} \cdot \text{m}}$$

$$- \frac{1}{2} \times 999 \frac{\text{kg}}{\text{m}^3} \times (11.7)^2 \frac{\text{m}^2}{\text{s}^2} \times \frac{\text{N} \cdot \text{s}^2}{\text{kg} \cdot \text{m}}$$

$p_A = 22.8$ kPa (abs) or -78.5 kPa (gage) $\qquad p_A$

Notes:

- ✓ This problem illustrates an application of the Bernoulli equation that includes elevation changes.
- ✓ It is interesting to note that when the Bernoulli equation applies between a reservoir and a free jet that it feeds at a location h below the reservoir surface, the jet speed will be $V = \sqrt{2gh}$; this is the same velocity a droplet (or stone) falling without friction from the reservoir level would attain if it fell a distance h. Can you explain why?
- ✓ Always take care when neglecting friction in any internal flow. In this problem, neglecting friction is reasonable if the pipe is smooth-surfaced and is relatively short. In Chapter 8 we will study frictional effects in internal flows.

Example 6.5 FLOW UNDER A SLUICE GATE

Water flows under a sluice gate on a horizontal bed at the inlet to a flume. Upstream from the gate, the water depth is 0.45 m and the speed is negligible. At the vena contracta downstream from the gate, the flow streamlines are straight and the depth is 50 mm. Determine the flow speed downstream from the gate and the discharge in cubic meter per second per meter of width.

Given: Flow of water under a sluice gate.

Find: (a) V_2.
(b) Q in $\text{m}^3/\text{s/m}$ of width.

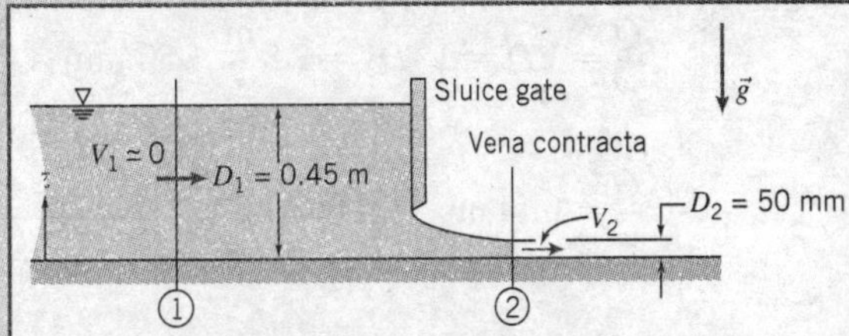

Solution:

Under the assumptions listed below, the flow satisfies all conditions necessary to apply the Bernoulli equation. The question is, what streamline do we use?

Governing equation: $\frac{p_1}{\rho} + \frac{V_1^2}{2} + gz_1 = \frac{p_2}{\rho} + \frac{V_2^2}{2} + gz_2$

Assumptions:
(1) Steady flow.
(2) Incompressible flow.
(3) Frictionless flow.
(4) Flow along a streamline.
(5) Uniform flow at each section.
(6) Hydrostatic pressure distribution (at each location, pressure increases linearly with depth).

If we consider the streamline that runs along the bottom of the channel ($z = 0$), because of assumption 6 the pressures at ① and ② are

$$p_1 = p_{atm} + \rho g D_1 \qquad \text{and} \qquad p_2 = p_{atm} + \rho g D_2$$

so that the Bernoulli equation for this streamline is

$$\frac{(p_{atm} + \rho g D_1)}{\rho} + \frac{V_1^2}{2} = \frac{(p_{atm} + \rho g D_2)}{\rho} + \frac{V_2^2}{2}$$

or

$$\frac{V_1^2}{2} + gD_1 = \frac{V_2^2}{2} + gD_2 \tag{1}$$

On the other hand, consider the streamline that runs along the free surface on both sides and down the inner surface of the gate. For this streamline

$$\frac{p_{atm}}{\rho} + \frac{V_1^2}{2} + gD_1 = \frac{p_{atm}}{\rho} + \frac{V_2^2}{2} + gD_2$$

or

$$\frac{V_1^2}{2} + gD_1 = \frac{V_2^2}{2} + gD_2 \tag{1}$$

We have arrived at the same equation (Eq. 1) for the streamline at the bottom and the streamline at the free surface, implying the Bernoulli constant is the same for both streamlines. We will see in Section 6.6 that this flow is one of a family of flows for which this is the case. Solving for V_2 yields

$$V_2 = \sqrt{2g(D_1 - D_2) + V_1^2}$$

But $V_1^2 \approx 0$, so

$$V_2 = \sqrt{2g(D_1 - D_2)} = \sqrt{2 \times 9.81\frac{\text{m}}{\text{s}^2} \times \left(0.45\ \text{m} - 50\ \text{mm} \times \frac{\text{m}}{1000\ \text{mm}}\right)}$$

$V_2 = 2.8\ \text{m/s}$ ← V_2

For uniform flow, $Q = VA = VDw$, or

$$\frac{Q}{w} = VD = V_2 D_2 = 2.8\frac{\text{m}}{\text{s}} + 50\ \text{mm} \times \frac{\text{m}}{1000\ \text{mm}} = 0.14\ \text{m}^2/\text{s}$$

$\frac{Q}{w} = 0.14\ \text{m}^3/\text{s/m of width}$ ← $\frac{Q}{w}$

Example 6.6 BERNOULLI EQUATION IN TRANSLATING REFERENCE FRAME

A light plane flies at 150 km/hr in standard air at an altitude of 1000 m. Determine the stagnation pressure at the leading edge of the wing. At a certain point close to the wing, the air speed *relative* to the wing is 60 m/s. Compute the pressure at this point.

Given: Aircraft in flight at 150 km/hr at 1000 m altitude in standard air.

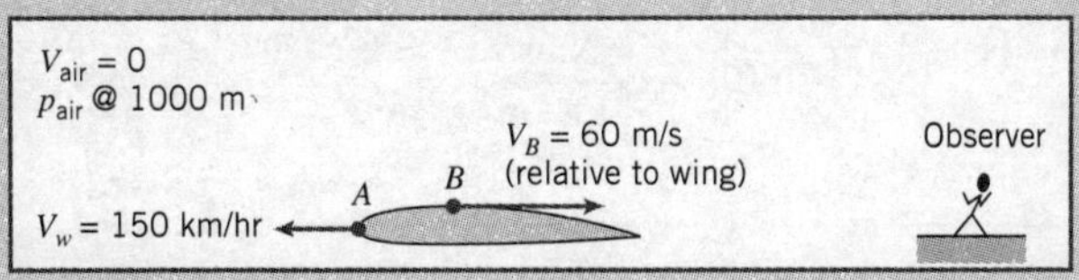

Find: Stagnation pressure, p_{0_A}, at point A and static pressure, p_B, at point B.

Solution:

Flow is unsteady when observed from a fixed frame, that is, by an observer on the ground. However, an observer *on* the wing sees the following steady flow:

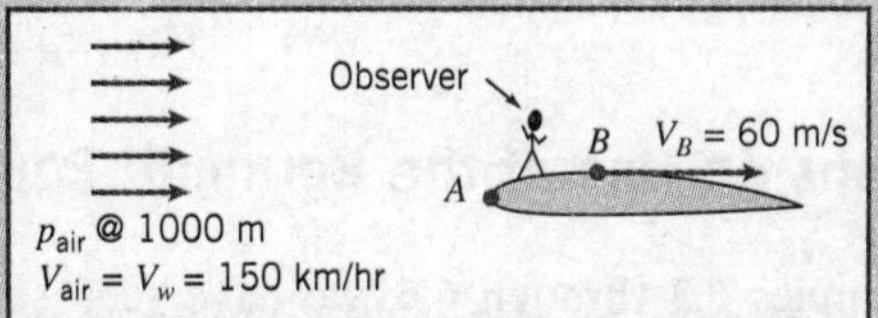

At $z = 1000$ m in standard air, the temperature is 281 K and the speed of sound is 336 m/s. Hence at point B, $M_B = V_B/c = 0.178$. This is less than 0.3, so the flow may be treated as incompressible. Thus the Bernoulli equation can be applied along a streamline in the moving observer's inertial reference frame.

Governing equation: $$\frac{p_{air}}{\rho} + \frac{V_{air}^2}{2} + gz_{air} = \frac{p_A}{\rho} + \frac{V_A^2}{2} + gz_A = \frac{p_B}{\rho} + \frac{V_B^2}{2} + gz_B$$

Assumptions:
(1) Steady flow.
(2) Incompressible flow ($V < 100$ m/s).
(3) Frictionless flow.
(4) Flow along a streamline.
(5) Neglect Δz.

Values for pressure and density may be found from Table A.3. Thus, at 1000 m, $p/p_{SL} = 0.8870$ and $\rho/\rho_{SL} = 0.9075$. Consequently,

$$p = 0.8870 p_{SL} = 0.8870 \times 1.01 \times 10^5 \frac{\text{N}}{\text{m}^2} = 8.96 \times 10^4 \text{ N/m}^2$$

and

$$\rho = 0.9075 \rho_{SL} = 0.9075 \times 1.23 \frac{\text{kg}}{\text{m}^3} = 1.12 \text{ kg/m}^3$$

Since the speed is $V_A = 0$ at the stagnation point,

$$p_{0_A} = p_{air} + \frac{1}{2}\rho V_{air}^2$$

$$= 8.96 \times 10^4 \frac{\text{N}}{\text{m}^2} + \frac{1}{2} \times 1.12 \frac{\text{kg}}{\text{m}^3}\left(150 \frac{\text{km}}{\text{hr}} \times 1000 \frac{\text{m}}{\text{km}} \times \frac{\text{hr}}{3600 \text{ s}}\right)^2 \times \frac{\text{N} \cdot \text{s}^2}{\text{kg} \cdot \text{m}}$$

$$p_{0_A} = 90.6 \text{ kPa(abs)}$$ ← p_{0_A}

Solving for the static pressure at B, we obtain

$$p_B = p_{\text{air}} + \frac{1}{2}\rho(V_{\text{air}}^2 - V_B^2)$$

$$p_B = 8.96 \times 10^4 \frac{\text{N}}{\text{m}^2} + \frac{1}{2} \times 1.12 \frac{\text{kg}}{\text{m}^3}\left[\left(150 \frac{\text{km}}{\text{hr}} \times 1000 \frac{\text{m}}{\text{km}} \times \frac{\text{hr}}{3600\ \text{s}}\right)^2 - (60)^2 \frac{\text{m}^2}{\text{s}^2}\right]\frac{\text{N}\cdot\text{s}^2}{\text{kg}\cdot\text{m}}$$

$$p_B = 88.6\ \text{kPa(abs)} \longleftarrow p_B$$

This problem gives a hint as to how a wing generates lift. The incoming air has a velocity $V_{\text{air}} = 150\ \text{km/hr} = 41.7\ \text{m/s}$ and *accelerates* to 60 m/s on the upper surface. This leads, through the Bernoulli equation, to a pressure *drop* of 1 kPa (from 89.6 kPa to 88.6 kPa). It turns out that the flow *decelerates* on the lower surface, leading to a pressure *rise* of about 1 kPa. Hence, the wing experiences a net upward pressure difference of about 2 kPa, a significant effect.

Cautions on Use of the Bernoulli Equation

In Examples 6.3 through 6.6, we have seen several situations where the Bernoulli equation may be applied because the restrictions on its use led to a reasonable flow model. However, in some situations you might be tempted to apply the Bernoulli equation where the restrictions are not satisfied. Some subtle cases that violate the restrictions are discussed briefly in this section.

Example 6.3 examined flow in a nozzle. In a *subsonic nozzle* (a converging section) the pressure drops, accelerating a flow. Because the pressure drops and the walls of the nozzle converge, there is no flow separation from the walls and the boundary layer remains thin. In addition, a nozzle is usually relatively short so frictional effects are not significant. All of this leads to the conclusion that the Bernoulli equation is suitable for use for subsonic nozzles.

CLASSIC VIDEO

Flow Visualization.

Sometimes we need to decelerate a flow. This can be accomplished using a *subsonic diffuser* (a diverging section), or by using a sudden expansion (e.g., from a pipe into a reservoir). In these devices the flow decelerates because of an adverse pressure gradient. As we discussed in Section 2.6, an adverse pressure gradient tends to lead to rapid growth of the boundary layer and its separation. Hence, we should be careful in applying the Bernoulli equation in such devices—at best, it will be an approximation. Because of area blockage caused by boundary-layer growth, pressure rise in actual diffusers always is less than that predicted for inviscid one-dimensional flow.

The Bernoulli equation was a reasonable model for the siphon of Example 6.4 because the entrance was well rounded, the bends were gentle, and the overall length was short. Flow separation, which can occur at inlets with sharp corners and in abrupt bends, causes the flow to depart from that predicted by a one-dimensional model and the Bernoulli equation. Frictional effects would not be negligible if the tube were long.

Example 6.5 presented an open-channel flow analogous to that in a nozzle, for which the Bernoulli equation is a good flow model. The hydraulic jump is an example of an open-channel flow with adverse pressure gradient. Flow through a hydraulic jump is mixed violently, making it impossible to identify streamlines. Thus the Bernoulli equation cannot be used to model flow through a hydraulic jump. We will see a more detailed presentation of open channel flows in Chapter 11.

The Bernoulli equation cannot be applied *through* a machine such as a propeller, pump, turbine, or windmill. The equation was derived by integrating along a stream tube (Section 4.4) or a streamline (Section 6.3) in the absence of moving surfaces such as blades or vanes. It is impossible to have locally steady flow or to identify streamlines during flow through a machine. Hence, while the Bernoulli equation may be applied between points *before* a machine, or between points *after* a machine (assuming its restrictions are satisfied), it cannot be applied *through* the machine. (In effect, a machine will change the value of the Bernoulli constant.)

CLASSIC VIDEO

Waves in Fluids and Stratified Flow.

Finally, compressibility must be considered for flow of gases. Density changes caused by dynamic compression due to motion may be neglected for engineering purposes if the local Mach number remains below about $M \approx 0.3$, as noted in Examples 6.3 and 6.6. Temperature changes can cause significant changes in density of a gas, even for low-speed flow. Thus the Bernoulli equation could not be applied to air flow through a heating element (e.g., of a hand-held hair dryer) where temperature changes are significant.

6.4 The Bernoulli Equation Interpreted as an Energy Equation

The Bernoulli equation, Eq. 6.8, was obtained by integrating Euler's equation along a streamline for steady, incompressible, frictionless flow. Thus Eq. 6.8 was derived from the momentum equation for a fluid particle.

An equation identical in form to Eq. 6.8 (although requiring very different restrictions) may be obtained from the first law of thermodynamics. Our objective in this section is to reduce the energy equation to the form of the Bernoulli equation given by Eq. 6.8. Having arrived at this form, we then compare the restrictions on the two equations to help us understand more clearly the restrictions on the use of Eq. 6.8.

Consider steady flow in the absence of shear forces. We choose a control volume bounded by streamlines along its periphery. Such a boundary, shown in Fig. 6.5, often is called a *stream tube*.

Basic equation:

$$\dot{Q} - \overset{=0(1)}{\cancel{\dot{W}_s}} - \overset{=0(2)}{\cancel{\dot{W}_{\text{shear}}}} - \overset{=0(3)}{\cancel{\dot{W}_{\text{other}}}} = \overset{=0(4)}{\cancel{\frac{\partial}{\partial t}}}\int_{\text{CV}} e\,\rho\, d\mathcal{V} + \int_{\text{CS}} (e + pv)\,\rho \vec{V} \cdot d\vec{A} \qquad (4.56)$$

$$e = u + \frac{V^2}{2} + gz$$

Restrictions: (1) $\dot{W}_s = 0$.
(2) $\dot{W}_{\text{shear}} = 0$.
(3) $\dot{W}_{\text{other}} = 0$.
(4) Steady flow.
(5) Uniform flow and properties at each section.

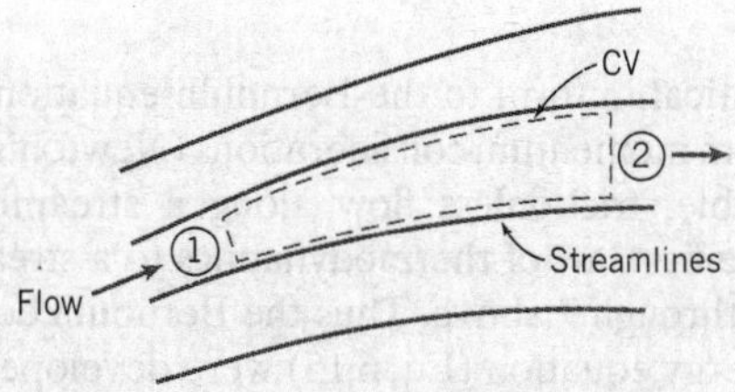

Fig. 6.5 Flow through a stream tube.

(Remember that here v represents the specific volume, and u represents the specific internal energy, not velocity!) Under these restrictions, Eq. 4.56 becomes

$$\left(u_1 + p_1 v_1 + \frac{V_1^2}{2} + gz_1\right)(-\rho_1 V_1 A_1) + \left(u_2 + p_2 v_2 + \frac{V_2^2}{2} + gz_2\right)(\rho_2 V_2 A_2) - \dot{Q} = 0$$

From continuity, with restrictions (4) and (5):

$$\sum_{\text{CS}} \rho \vec{V} \cdot \vec{A} = 0 \tag{4.15b}$$

or

$$(-\rho_1 V_1 A_1) + (\rho_2 V_2 A_2) = 0$$

That is,

$$\dot{m} = \rho_1 V_1 A_1 = \rho_2 V_2 A_2$$

Also

$$\dot{Q} = \frac{\delta Q}{dt} = \frac{\delta Q}{dm}\frac{dm}{dt} = \frac{\delta Q}{dm}\dot{m}$$

Thus, from the energy equation, after rearranging

$$\left[\left(p_2 v_2 + \frac{V_2^2}{2} + gz_2\right) - \left(p_1 v_1 + \frac{V_1^2}{2} + gz_1\right)\right]\dot{m} + \left(u_2 - u_1 - \frac{\delta Q}{dm}\right)\dot{m} = 0$$

or

$$p_1 v_1 + \frac{V_1^2}{2} + gz_1 = p_2 v_2 + \frac{V_2^2}{2} + gz_2 + \left(u_2 - u_1 - \frac{\delta Q}{dm}\right)$$

Under the additional assumption (6) of incompressible flow, $v_1 = v_2 = 1/\rho$ and hence

$$\frac{p_1}{\rho} + \frac{V_1^2}{2} + gz_1 = \frac{p_2}{\rho} + \frac{V_2^2}{2} + gz_2 + \left(u_2 - u_1 - \frac{\delta Q}{dm}\right) \tag{6.14}$$

Equation 6.14 would reduce to the Bernoulli equation if the term in parentheses were zero. Thus, under the further restriction,

$$(7) \quad \left(u_2 - u_1 - \frac{\delta Q}{dm}\right) = 0$$

the energy equation reduces to

$$\frac{p_1}{\rho} + \frac{V_1^2}{2} + gz_1 = \frac{p_2}{\rho} + \frac{V_2^2}{2} + gz_2$$

or

$$\frac{p}{\rho} + \frac{V^2}{2} + gz = \text{constant} \tag{6.15}$$

Equation 6.15 is identical in form to the Bernoulli equation, Eq. 6.8. The Bernoulli equation was derived from momentum considerations (Newton's second law), and is valid for steady, incompressible, frictionless flow along a streamline. Equation 6.15 was obtained by applying the first law of thermodynamics to a stream tube control volume, subject to restrictions 1 through 7 above. Thus the Bernoulli equation (Eq. 6.8) and the identical form of the energy equation (Eq. 6.15) were developed from entirely different models, coming from entirely different basic concepts, and involving different restrictions.

It looks like we needed restriction (7) to finally transform the energy equation into the Bernoulli equation. In fact, we didn't! It turns out that for an incompressible and frictionless flow [restriction (6), and the fact we are looking only at flows with no shear forces], restriction (7) is automatically satisfied, as we will demonstrate in Example 6.7.

Example 6.7 INTERNAL ENERGY AND HEAT TRANSFER IN FRICTIONLESS INCOMPRESSIBLE FLOW

Consider frictionless, incompressible flow with heat transfer. Show that

$$u_2 - u_1 = \frac{\delta Q}{dm}$$

Given: Frictionless, incompressible flow with heat transfer.

Show: $u_2 - u_1 = \dfrac{\delta Q}{dm}$.

Solution:
In general, internal energy can be expressed as $u = u(T, v)$. For incompressible flow, v = constant, and $u = u(T)$. Thus the thermodynamic state of the fluid is determined by the single thermodynamic property, T. For any process, the internal energy change, $u_2 - u_1$, depends only on the temperatures at the end states.

From the Gibbs equation, $Tds = du + \rho\, dv$, valid for a pure substance undergoing any process, we obtain

$$Tds = du$$

for incompressible flow, since $dv = 0$. Since the internal energy change, du, between specified end states, is independent of the process, we take a reversible process, for which $Tds = d(\delta Q/dm) = du$. Therefore,

$$u_2 - u_1 = \frac{\delta Q}{dm} \longleftarrow$$

For the steady, frictionless, and incompressible flow considered in this section, it is true that the first law of thermodynamics reduces to the Bernoulli equation. Each term in Eq. 6.15 has dimensions of energy per unit mass (we sometimes refer to the three terms in the equation as the "pressure" energy, kinetic energy, and potential energy per unit mass of the fluid). It is not surprising that Eq. 6.15 contains energy terms—after all, we used the first law of thermodynamics in deriving it. How did we end up with the same energy-like terms in the Bernoulli equation, which we derived from the momentum equation? The answer is because we integrated the momentum equation (which involves force terms) along a streamline (which involves distance), and by doing so ended up with work or energy terms (work being defined as force times distance): The work of gravity and pressure forces leads to a kinetic energy change (which came from integrating momentum over distance). In this context, we can think of the Bernoulli equation as a *mechanical energy balance*—the mechanical energy ("pressure" plus potential plus kinetic) will be constant. We must always bear in mind that for the Bernoulli equation to be valid along a streamline requires an incompressible inviscid flow, in addition to steady flow. It's interesting that these two properties of the flow—its compressibility and friction—are what "link" thermodynamic and mechanical energies. If a fluid is compressible, any flow-induced pressure changes will compress or expand the fluid, thereby doing work and changing the particle thermal energy; and friction, as we know from everyday experience, always converts mechanical to thermal energy. Their absence, therefore, breaks the link between the mechanical and thermal energies, and they are independent—it's as if they're in parallel universes!

In summary, when the conditions are satisfied for the Bernoulli equation to be valid, we can consider separately the mechanical energy and the internal thermal energy of a fluid particle (this is illustrated in Example 6.8); when they are not satisfied, there will be an interaction between these energies, the Bernoulli equation becomes invalid, and we must use the full first law of thermodynamics.

Example 6.8 FRICTIONLESS FLOW WITH HEAT TRANSFER

Water flows steadily from a large open reservoir through a short length of pipe and a nozzle with cross-sectional area $A = 560\ \text{mm}^2$. A well-insulated 10 kW heater surrounds the pipe. Find the temperature rise of the water.

Given: Water flows from a large reservoir through the system shown and discharges to atmospheric pressure. The heater is 10 kW; $A_4 = 560\ \text{mm}^2$.

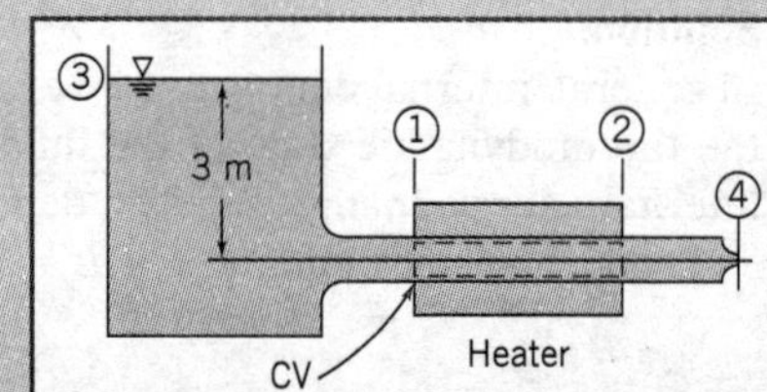

Find: The temperature rise of the water between points ① and ②.

Solution:

Governing equations:

$$\frac{p}{\rho} + \frac{V^2}{2} + gz = \text{constant} \tag{6.8}$$

$$\sum_{\text{CS}} \vec{V} \cdot \vec{A} = 0 \tag{4.13b}$$

$$\dot{Q} - \underset{=0(4)}{\dot{W}_s} - \underset{=0(4)}{\dot{W}_{\text{shear}}} = \underset{=0(1)}{\frac{\partial}{\partial t}} \int_{\text{CV}} e\,\rho\, d\forall + \int_{\text{CS}} \left(u + pv + \frac{V^2}{2} + gz\right) \rho \vec{V} \cdot d\vec{A} \tag{4.56}$$

Assumptions:
(1) Steady flow.
(2) Frictionless flow.
(3) Incompressible flow.
(4) No shaft work, no shear work.
(5) Flow along a streamline.
(6) Uniform flow at each section [a consequence of assumption (2)].

Under the assumptions listed, the first law of thermodynamics for the CV shown becomes

$$\dot{Q} = \int_{\text{CS}} \left(u + pv + \frac{V^2}{2} + gz\right) \rho \vec{V} \cdot d\vec{A}$$

$$= \int_{A_1} \left(u + pv + \frac{V^2}{2} + gz\right) \rho \vec{V} \cdot d\vec{A} + \int_{A_2} \left(u + pv + \frac{V^2}{2} + gz\right) \rho \vec{V} \cdot d\vec{A}$$

For uniform properties at ① and ②

$$\dot{Q} = -(\rho V_1 A_1)\left(u_1 + p_1 v + \frac{V_1^2}{2} + gz_1\right) + (\rho V_2 A_2)\left(u_2 + p_2 v + \frac{V_2^2}{2} + gz_2\right)$$

From conservation of mass, $\rho V_1 A_1 = \rho V_2 A_2 = \dot{m}$, so

$$\dot{Q} = \dot{m}\left[u_2 - u_1 + \left(\frac{p_2}{\rho} + \frac{V_2^2}{2} + gz_2\right) - \left(\frac{p_1}{\rho} + \frac{V_1^2}{2} + gz_1\right)\right]$$

For frictionless, incompressible, steady flow, along a streamline,

$$\frac{p}{\rho} + \frac{V^2}{2} + gz = \text{constant}$$

Therefore,

$$\dot{Q} = \dot{m}(u_2 - u_1)$$

Since, for an incompressible fluid, $u_2 - u_1 = c(T_2 - T_1)$, then

$$T_2 - T_1 = \frac{\dot{Q}}{\dot{m}c}$$

From continuity,

$$\dot{m} = \rho V_4 A_4$$

To find V_4, write the Bernoulli equation between the free surface at ③ and point ④.

$$\frac{p_3}{\rho} + \frac{V_3^2}{2} + gz_3 = \frac{p_4}{\rho} + \frac{V_4^2}{2} + gz_4$$

Since $p_3 = p_4$ and $V_3 \approx 0$, then

$$V_4 = \sqrt{2g(z_3 - z_4)} = \sqrt{2 \times 9.81 \frac{\text{m}}{\text{s}^2} \times 3\ \text{m}} = 7.7\ \text{m/s}$$

and

$$\dot{m} = \rho V_4 A_4 = 1000 \frac{\text{kg}}{\text{m}^3} \times 7.7 \frac{\text{m}}{\text{s}} \times 560\ \text{mm}^2 \times \frac{\text{m}^2}{10^6\ \text{mm}^2} = 4.31\ \text{kg/s}$$

Assuming no heat loss to the surroundings, we obtain

$$T_2 - T_1 = \frac{\dot{Q}}{\dot{m}c} = \text{W} \times \frac{1}{4.31\ \text{kg/s} \times 4179\ \text{J/kg} \cdot {}^\circ\text{K}} \times \frac{\text{J}}{\text{W} \cdot \text{s}}$$

$$T_2 - T_1 = 0.555\ {}^\circ\text{K} \qquad \longleftarrow \qquad T_2 - T_1$$

This problem illustrates that:
- ✓ In general, the first law of thermodynamics and the Bernoulli equation are independent equations.
- ✓ For an incompressible, inviscid flow the internal thermal energy is only changed by a heat transfer process, and is independent of the fluid mechanics.

Energy Grade Line and Hydraulic Grade Line 6.5

We have learned that for a steady, incompressible, frictionless flow, we may use the Bernoulli equation (Eq. 6.8), derived from the momentum equation, and also Eq. 6.15, derived from the energy equation:

$$\frac{p}{\rho} + \frac{V^2}{2} + gz = \text{constant} \tag{6.15}$$

We also interpreted the three terms comprised of "pressure," kinetic, and potential energies to make up the total mechanical energy, per unit mass, of the fluid. If we divide Eq. 6.15 by g, we obtain another form,

$$\frac{p}{\rho g} + \frac{V^2}{2g} + z = H \tag{6.16a}$$

Here H is the *total head* of the flow; it measures the total mechanical energy in units of meters or feet. We will learn in Chapter 8 that in a real fluid (one with friction) this head will *not* be constant but will continuously decrease in value as mechanical energy is converted to thermal; in this chapter H is constant. We can go one step further here and get a very useful graphical approach if we also define this to be the *energy grade line* (EGL),

$$EGL = \frac{p}{\rho g} + \frac{V^2}{2g} + z \tag{6.16b}$$

This can be measured using the pitot (total head) tube (shown in Fig. 6.3). Placing such a tube in a flow measures the total pressure, $p_0 = p + \frac{1}{2}\rho V^2$, so this will cause the height of a column *of the same fluid* to rise to a height $h = p_0/\rho g = p/\rho g + V^2/2g$. If the vertical location of the pitot tube is z, measured from some datum (e.g., the ground), the height of column of fluid *measured from the datum* will then be $h + z = p/\rho g + V^2/2g + z = EGL = H$. In summary, the height of the column, measured from the datum, attached to a pitot tube directly indicates the EGL.

We can also define the *hydraulic grade line* (HGL),

$$HGL = \frac{p}{\rho g} + z \tag{6.16c}$$

This can be measured using the static pressure tap (shown in Fig. 6.2*a*). Placing such a tube in a flow measures the static pressure, p, so this will cause the height of a column *of the same fluid* to rise to a height $h = p/\rho g$. If the vertical location of the tap is also at z, measured from some datum, the height of column of fluid *measured from the datum* will then be $h + z = p/\rho g + z = HGL$. The height of the column attached to a static pressure tap thus directly indicates the HGL.

From Eqs. 6.16b and 6.16c we obtain

$$EGL - HGL = \frac{V^2}{2g} \tag{6.16d}$$

which shows that the difference between the EGL and HGL is always the dynamic pressure term.

To see a graphical interpretation of the EGL and HGL, refer to the example shown in Fig. 6.6, which shows frictionless flow from a reservoir, through a pipe reducer.

At all locations the EGL is the same because there is no loss of mechanical energy. Station ① is at the reservoir, and here the EGL and HGL coincide with the free surface: in Eqs. 6.16b and 6.16c $p = 0$ (gage), $V = 0$, and $z = z_1$, so $EGL_1 = HGL_1 = H = z_1$; all of the mechanical energy is potential. (If we were to place a pitot tube in the fluid at station ①, the fluid would of course just rise to the free surface level.)

At station ② we have a pitot (total head) tube and a static head tap. The pitot tube's column indicates the correct value of the EGL ($EGL_1 = EGL_2 = H$), but *something* changed between the two stations: The fluid now has significant kinetic energy and has lost some potential energy (can you determine from the figure what happened to the pressure?). From Eq. 6.16d, we can see that the HGL is lower than the EGL by $V_2^2/2g$; the HGL at station ② shows this.

From station ② to station ③ there is a reduction in diameter, so continuity requires that $V_3 > V_2$; hence the gap between the EGL and HGL increases further, as shown.

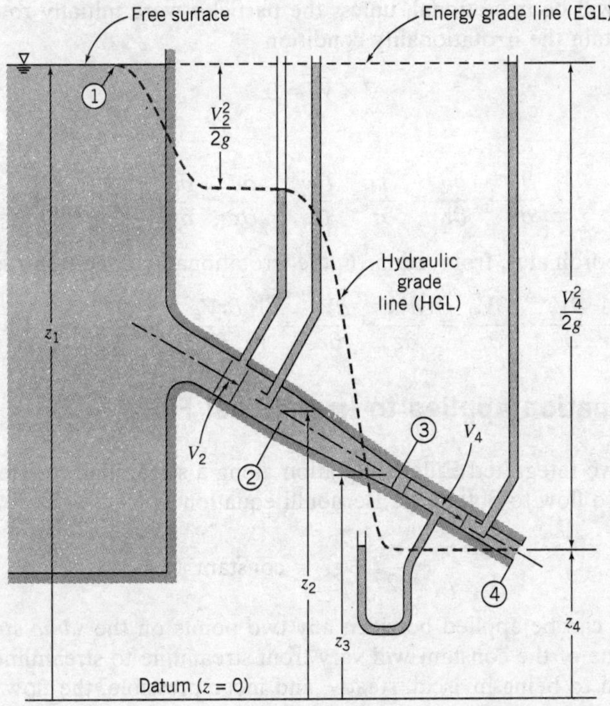

Fig. 6.6 Energy and hydraulic grade lines for frictionless flow.

Station ④ is at the exit (to the atmosphere). Here the pressure is zero (gage), so the EGL consists entirely of kinetic and potential energy terms, and $HGL_4 = HGL_3$. We can summarize two important ideas when sketching EGL and HGL curves:

1. The EGL is constant for incompressible, inviscid flow (in the absence of work devices). We will see in Chapter 8 that work devices may increase or decrease the EGL, and friction will always lead to a fall in the EGL.
2. The HGL is always lower than the EGL by distance $V^2/2g$. Note that the value of velocity V depends on the overall system (e.g., reservoir height, pipe diameter, etc.), but *changes* in velocity *only* occur when the diameter changes.

Unsteady Bernoulli Equation: Integration of Euler's Equation Along a Streamline (on the Web) 6.6*

Irrotational Flow 6.7*

We have already discussed irrotational flows in Section 5.3. These are flows in which the fluid particles do not rotate ($\vec{\omega} = 0$). We recall that the only stresses that can generate particle rotation are shear stresses; hence, inviscid flows (i.e., those with zero

*These sections may be omitted without loss of continuity in the text material. (Note that Section 5.2 contains background material needed for study of Section 6.7.)

shear stresses) will be irrotational, unless the particles were initially rotating. Using Eq. 5.14, we obtain the irrotationality condition

$$\nabla \times \vec{V} = 0 \tag{6.22}$$

leading to

$$\frac{\partial w}{\partial y} - \frac{\partial v}{\partial z} = \frac{\partial u}{\partial z} - \frac{\partial w}{\partial x} = \frac{\partial v}{\partial x} - \frac{\partial u}{\partial y} = 0 \tag{6.23}$$

In cylindrical coordinates, from Eq. 5.16, the irrotationality condition requires that

$$\frac{1}{r}\frac{\partial V_z}{\partial \theta} - \frac{\partial V_\theta}{\partial z} = \frac{\partial V_r}{\partial z} - \frac{\partial V_z}{\partial r} = \frac{1}{r}\frac{\partial r V_\theta}{\partial r} - \frac{1}{r}\frac{\partial V_r}{\partial \theta} = 0 \tag{6.24}$$

Bernoulli Equation Applied to Irrotational Flow

In Section 6.3, we integrated Euler's equation along a streamline for steady, incompressible, inviscid flow to obtain the Bernoulli equation

$$\frac{p}{\rho} + \frac{V^2}{2} + gz = \text{constant} \tag{6.8}$$

Equation 6.8 can be applied between any two points on the *same* streamline. In general, the value of the constant will vary from streamline to streamline.

If, in addition to being inviscid, steady, and incompressible, the flow field is also irrotational (i.e., the particles had no initial rotation), so that $\nabla \times \vec{V} = 0$ (Eq. 6.22), we can show that Bernoulli's equation can be applied between any and all points in the flow. Then the value of the constant in Eq. 6.8 is the same for all streamlines. To illustrate this, we start with Euler's equation in vector form,

$$(\vec{V} \cdot \nabla)\vec{V} = -\frac{1}{\rho}\nabla p - g\hat{k} \tag{6.9}$$

Using the vector identity

$$(\vec{V} \cdot \nabla)\vec{V} = \frac{1}{2}\nabla(\vec{V} \cdot \vec{V}) - \vec{V} \times (\nabla \times \vec{V})$$

we see for irrotational flow, where $\nabla \times \vec{V} = 0$, that

$$(\vec{V} \cdot \nabla)\vec{V} = \frac{1}{2}\nabla(\vec{V} \cdot \vec{V})$$

and Euler's equation for irrotational flow can be written as

$$\frac{1}{2}\nabla(\vec{V} \cdot \vec{V}) = \frac{1}{2}\nabla(V^2) = -\frac{1}{\rho}\nabla p - g\hat{k} \tag{6.25}$$

Consider a displacement in the flow field from position $\vec{r}$ to position $\vec{r} + d\vec{r}$; the displacement $d\vec{r}$ is an *arbitrary* infinitesimal displacement in *any* direction, not necessarily along a streamline. Taking the dot product of $d\vec{r} = dx\hat{i} + dy\hat{j} + dz\hat{k}$ with each of the terms in Eq. 6.25, we have

$$\frac{1}{2}\nabla(V^2) \cdot d\vec{r} = -\frac{1}{\rho}\nabla p \cdot d\vec{r} - g\hat{k} \cdot d\vec{r}$$

and hence

$$\frac{1}{2}d(V^2) = -\frac{dp}{\rho} - g dz$$

or

$$\frac{dp}{\rho} + \frac{1}{2}d(V^2) + gdz = 0$$

Integrating this equation for incompressible flow gives

$$\frac{p}{\rho} + \frac{V^2}{2} + gz = \text{constant} \tag{6.26}$$

Since $d\vec{r}$ was an arbitrary displacement, Eq. 6.26 is valid between *any* two points (i.e., not just along a streamline) in a steady, incompressible, inviscid flow that is also irrotational (see Example 6.5).

Velocity Potential

We were introduced in Section 5.2 to the notion of the stream function ψ for a two-dimensional incompressible flow.

For irrotational flow we can introduce a companion function, the *potential function* ϕ, defined by

$$\vec{V} = -\nabla\phi \tag{6.27}$$

VIDEO

An Example of Irrotational Flow.

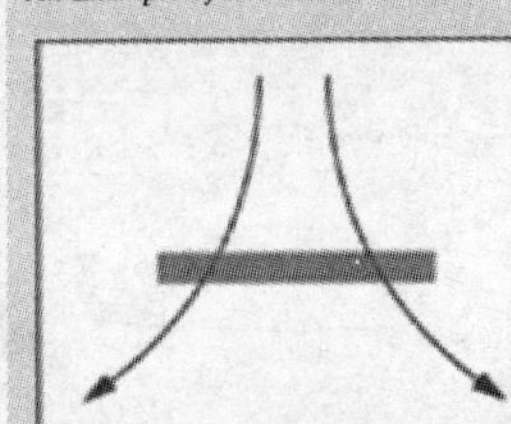

Why this definition? Because it guarantees that *any* continuous scalar function $\phi(x, y, z, t)$ *automatically* satisfies the irrotationality condition (Eq. 6.22) because of a fundamental identity:[3]

$$\nabla \times \vec{V} = -\nabla \times \nabla\phi = -\text{curl}(\text{grad}\,\phi) \equiv 0 \tag{6.28}$$

The minus sign (used in most textbooks) is inserted simply so that ϕ decreases in the flow direction (analogous to the temperature decreasing in the direction of heat flow in heat conduction). Thus,

$$u = -\frac{\partial\phi}{\partial x}, \quad v = -\frac{\partial\phi}{\partial y}, \quad \text{and} \quad w = -\frac{\partial\phi}{\partial z} \tag{6.29}$$

(You can check that the irrotationality condition, Eq. 6.22, is satisfied identically.)

In cylindrical coordinates,

$$\nabla = \hat{e}_r\frac{\partial}{\partial r} + \hat{e}_\theta\frac{1}{r}\frac{\partial}{\partial\theta} + \hat{k}\frac{\partial}{\partial z} \tag{3.19}$$

From Eq. 6.27, then, in cylindrical coordinates

$$V_r = -\frac{\partial\phi}{\partial r} \qquad V_\theta = -\frac{1}{r}\frac{\partial\phi}{\partial\theta} \qquad V_z = -\frac{\partial\phi}{\partial z} \tag{6.30}$$

CLASSIC VIDEO

Vorticity.

Because $\nabla \times \nabla\phi \equiv 0$ for all ϕ, the velocity potential exists only for irrotational flow.

Irrotationality may be a valid assumption for those regions of a flow in which viscous forces are negligible. (For example, such a region exists outside the boundary layer in the flow over a wing surface, and can be analyzed to find the lift produced by the wing.) The theory for irrotational flow is developed in terms of an imaginary ideal fluid whose viscosity is identically zero. Since, in an irrotational flow, the velocity field may be defined by the potential function ϕ, the theory is often referred to as potential flow theory.

[3] That $\nabla \times \nabla(\ ) \equiv 0$ can easily be demonstrated by expanding into components.

All real fluids possess viscosity, but there are many situations in which the assumption of inviscid flow considerably simplifies the analysis and, at the same time, gives meaningful results. Because of its relative simplicity and mathematical beauty, potential flow has been studied extensively.[4]

Stream Function and Velocity Potential for Two-Dimensional, Irrotational, Incompressible Flow: Laplace's Equation

For a two-dimensional, incompressible, irrotational flow we have expressions for the velocity components, u and v, in terms of both the stream function ψ, and the velocity potential ϕ,

$$u = \frac{\partial \psi}{\partial y} \qquad v = -\frac{\partial \psi}{\partial x} \tag{5.4}$$

$$u = -\frac{\partial \phi}{\partial x} \qquad v = -\frac{\partial \phi}{\partial y} \tag{6.29}$$

Substituting for u and v from Eq. 5.4 into the irrotationality condition,

$$\frac{\partial v}{\partial x} - \frac{\partial u}{\partial y} = 0 \tag{6.23}$$

we obtain

$$\frac{\partial^2 \psi}{\partial x^2} + \frac{\partial^2 \psi}{\partial y^2} = \nabla^2 \psi = 0 \tag{6.31}$$

Substituting for u and v from Eq. 6.29 into the continuity equation,

$$\frac{\partial u}{\partial x} + \frac{\partial v}{\partial y} = 0 \tag{5.3}$$

we obtain

$$\frac{\partial^2 \phi}{\partial x^2} + \frac{\partial^2 \phi}{\partial y^2} = \nabla^2 \phi = 0 \tag{6.32}$$

Equations 6.31 and 6.32 are forms of Laplace's equation—an equation that arises in many areas of the physical sciences and engineering. Any function ψ or ϕ that satisfies Laplace's equation represents a possible two-dimensional, incompressible, irrotational flow field.

Table 6.1 summarizes the results of our discussion of the stream function and velocity potential for two dimensional flows.

The same rules (of when incompressibility and irrotationality apply, and with the appropriate form of Laplace's equation) are valid for the stream function and velocity potential when expressed in cylindrical coordinates,

$$V_r = \frac{1}{r}\frac{\partial \psi}{\partial \theta} \quad \text{and} \quad V_\theta = -\frac{\partial \psi}{\partial r} \tag{5.8}$$

[4]Anyone interested in a detailed study of potential flow theory may find [4–6] of interest.

Table 6.1

Definitions of ψ and ϕ, and Conditions Necessary for Satisfying Laplace's Equation

Definition	Always satisfies ...	Satisfies Laplace equation ... $\frac{\partial^2()}{\partial x^2}+\frac{\partial^2()}{\partial y^2}=\nabla^2()=0$
Stream function ψ	... incompressibility:	... only if irrotational:
$u=\frac{\partial\psi}{\partial y}\quad v=-\frac{\partial\psi}{\partial x}$	$\frac{\partial u}{\partial x}+\frac{\partial v}{\partial y}=\frac{\partial^2\psi}{\partial x\partial y}-\frac{\partial^2\psi}{\partial y\partial x}\equiv 0$	$\frac{\partial v}{\partial x}-\frac{\partial u}{\partial y}=-\frac{\partial^2\psi}{\partial x\partial x}-\frac{\partial^2\psi}{\partial y\partial y}=0$
Velocity potential ϕ	... irrotationality:	... only if incompressible:
$u=-\frac{\partial\phi}{\partial x}\quad v=-\frac{\partial\phi}{\partial y}$	$\frac{\partial v}{\partial x}-\frac{\partial u}{\partial y}=-\frac{\partial^2\phi}{\partial x\partial y}-\frac{\partial^2\phi}{\partial y\partial x}\equiv 0$	$\frac{\partial u}{\partial x}+\frac{\partial v}{\partial y}=-\frac{\partial^2\phi}{\partial x\partial x}-\frac{\partial^2\phi}{\partial y\partial y}=0$

and

$$V_r=-\frac{\partial\phi}{\partial r}\quad\text{and}\quad V_\theta=-\frac{1}{r}\frac{\partial\phi}{\partial\theta}\tag{6.33}$$

In Section 5.2 we showed that the stream function ψ is constant along any streamline. For $\psi=\text{constant}$, $d\psi=0$ and

$$d\psi=\frac{\partial\psi}{\partial x}dx+\frac{\partial\psi}{\partial y}dy=0$$

The slope of a streamline—a line of constant ψ—is given by

$$\left.\frac{dy}{dx}\right)_\psi=-\frac{\partial\psi/dx}{\partial x/\partial y}=-\frac{-v}{u}=\frac{v}{u}\tag{6.34}$$

Along a line of constant ϕ, $d\phi=0$ and

$$d\phi=\frac{\partial\phi}{\partial x}dx+\frac{\partial\phi}{\partial y}dy=0$$

Consequently, the slope of a potential line — a line of constant ϕ — is given by

$$\left.\frac{dy}{dx}\right)_\phi=-\frac{\partial\phi/\partial x}{\partial\phi/\partial y}=-\frac{u}{v}\tag{6.35}$$

(The last equality of Eq. 6.35 follows from use of Eq. 6.29.)

Comparing Eqs. 6.34 and 6.35, we see that the slope of a constant ψ line at any point is the negative reciprocal of the slope of the constant ϕ line at that point; this means that *lines of constant ψ and constant ϕ are orthogonal*. This property of potential lines and streamlines is useful in graphical analyses of flow fields.

Example 6.10 VELOCITY POTENTIAL

Consider the flow field given by $\psi=ax^2-ay^2$, where $a=3\text{ s}^{-1}$. Show that the flow is irrotational. Determine the velocity potential for this flow.

Given: Incompressible flow field with $\psi=ax^2-ay^2$, where $a=3\text{ s}^{-1}$.

Find: (a) Whether or not the flow is irrotational.
(b) The velocity potential for this flow.

Solution: If the flow is irrotational, $\nabla^2\psi = 0$. Checking for the given flow,

$$\nabla^2\psi = \frac{\partial^2}{\partial x^2}(ax^2 - ay^2) + \frac{\partial^2}{\partial y^2}(ax^2 - ay^2) = 2a - 2a = 0$$

so that the flow *is* irrotational. As an alternative proof, we can compute the fluid particle rotation (in the xy plane, the only component of rotation is ω_z):

$$2\omega_z = \frac{\partial v}{\partial x} - \frac{\partial u}{\partial y} \quad \text{and} \quad u = \frac{\partial \psi}{\partial y} \quad v = -\frac{\partial \psi}{\partial x}$$

then

$$u = \frac{\partial}{\partial y}(ax^2 - ay^2) = -2ay \quad \text{and} \quad v = -\frac{\partial}{\partial x}(ax^2 - ay^2) = -2ax$$

so

$$2\omega_z = \frac{\partial v}{\partial x} - \frac{\partial u}{\partial y} = \frac{\partial}{\partial x}(-2ax) - \frac{\partial}{\partial y}(-2ay) = -2a + 2a = 0 \longleftarrow 2\omega_z$$

Once again, we conclude that the flow is irrotational. Because it is irrotational, ϕ must exist, and

$$u = -\frac{\partial \phi}{\partial x} \quad \text{and} \quad v = -\frac{\partial \phi}{\partial y}$$

Consequently, $u = -\frac{\partial \phi}{\partial x} = -2ay$ and $\frac{\partial \phi}{\partial x} = 2ay$. Integrating with respect to x gives $\phi = 2axy + f(y)$, where $f(y)$ is an arbitrary function of y. Then

$$v = -2ax = -\frac{\partial \phi}{\partial y} = -\frac{\partial}{\partial x}[2axy + f(y)]$$

Therefore, $-2ax = -2ax - \frac{\partial f(y)}{\partial y} = -2ax - \frac{df}{dy}$, so $\frac{df}{dy} = 0$ and $f =$ constant. Thus

$$\phi = 2axy + \text{constant} \longleftarrow \phi$$

We also can show that lines of constant ψ and constant ϕ are orthogonal.

$$\psi = ax^2 - ay^2 \quad \text{and} \quad \phi = 2axy$$

For $\psi =$ constant, $d\psi = 0 = 2axdx - 2aydy$; hence $\left.\frac{dy}{dx}\right)_{\psi = c} = \frac{x}{y}$

For $\phi =$ constant, $d\phi = 0 = 2aydx + 2axdy$; hence $\left.\frac{dy}{dx}\right)_{\phi = c} = -\frac{y}{x}$

The slopes of lines of constant ϕ and constant ψ are negative reciprocals. Therefore lines of constant ϕ are orthogonal to lines of constant ψ.

This problem illustrates the relations among the stream function, velocity potential, and velocity field.
The stream function ψ and velocity potential ϕ are shown in the *Excel* workbook. By entering the equations for ψ and ϕ, other fields can be plotted.

Elementary Plane Flows

The ψ and ϕ functions for five elementary two-dimensional flows—a uniform flow, a source, a sink, a vortex, and a doublet—are summarized in Table 6.2. The ψ and ϕ functions can be obtained from the velocity field for each elementary flow. (We saw in Example 6.10 that we can obtain ϕ from u and v.)

Table 6.2
Elementary Plane Flows

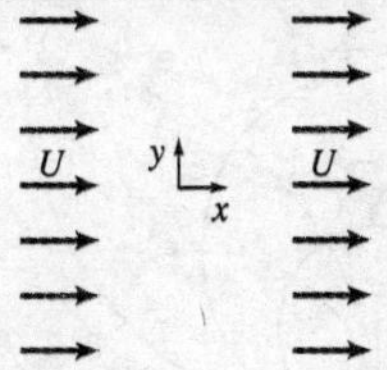

Uniform Flow (positive x direction)

$u = U \qquad \psi = Uy$

$v = 0 \qquad \phi = -Ux$

$\Gamma = 0$ around any closed curve

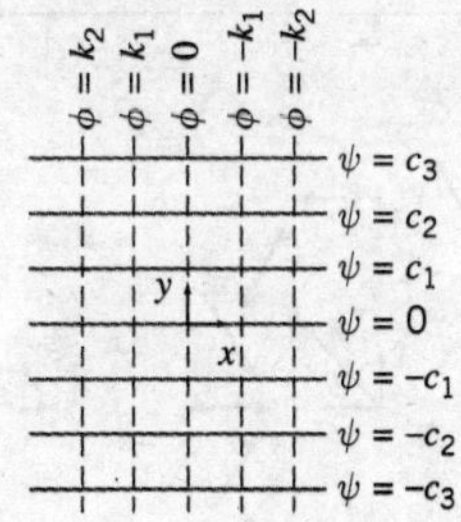

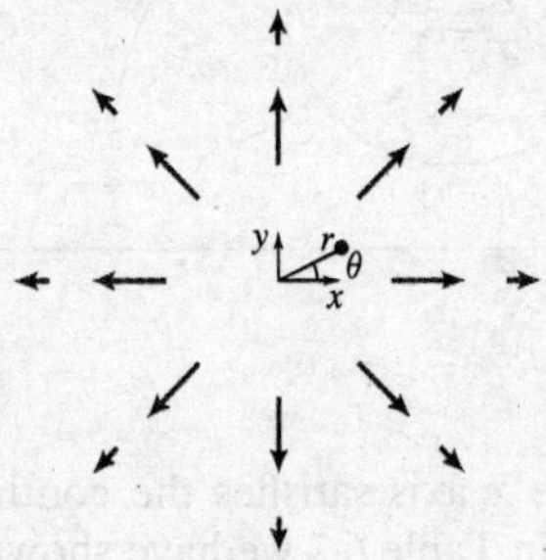

Source Flow (from origin)

$$V_r = \frac{q}{2\pi r} \qquad \psi = \frac{q}{2\pi}\theta$$

$$V_\theta = 0 \qquad \phi = -\frac{q}{2\pi}\ln r$$

Origin is singular point
q is volume flow rate per unit depth
$\Gamma = 0$ around any closed curve

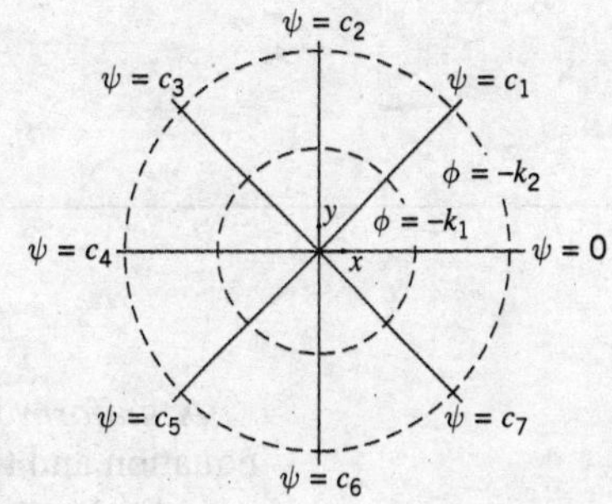

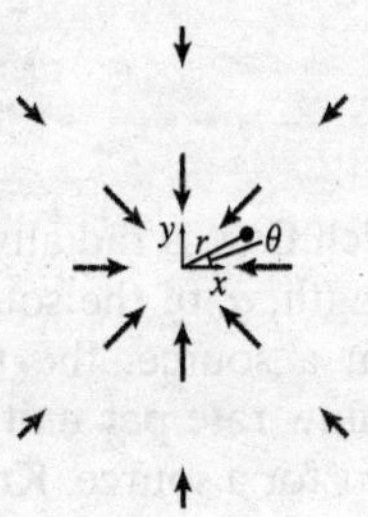

Sink Flow (toward origin)

$$V_r = -\frac{q}{2\pi r} \qquad \psi = -\frac{q}{2\pi}\theta$$

$$V_\theta = 0 \qquad \phi = \frac{q}{2\pi}\ln r$$

Origin is singular point
q is volume flow rate per unit depth
$\Gamma = 0$ around any closed curve

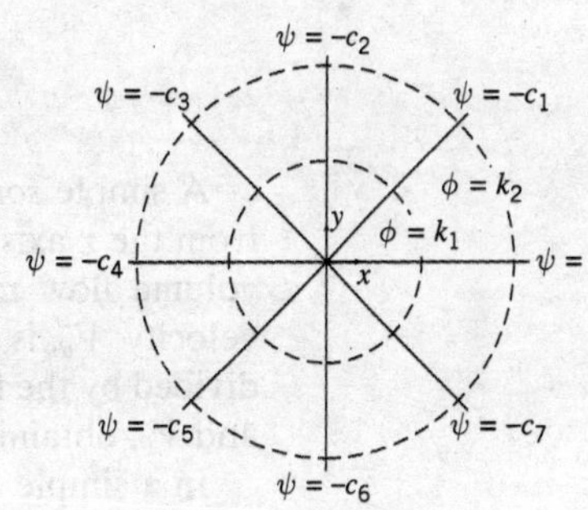

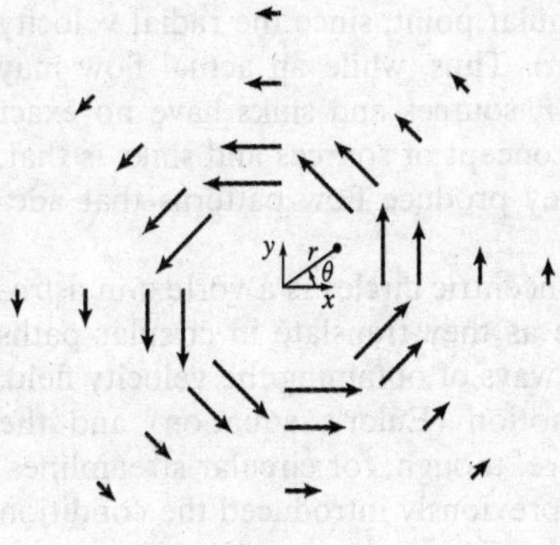

Irrotational Vortex (counterclockwise, center at origin)

$$V_r = 0 \qquad \psi = -\frac{K}{2\pi}\ln r$$

$$V_\theta = \frac{K}{2\pi r} \qquad \phi = \frac{K}{2\pi}\theta$$

Origin is singular point
K is strength of the vortex
$\Gamma = K$ around any closed curve enclosing origin
$\Gamma = 0$ around any closed curve not enclosing origin

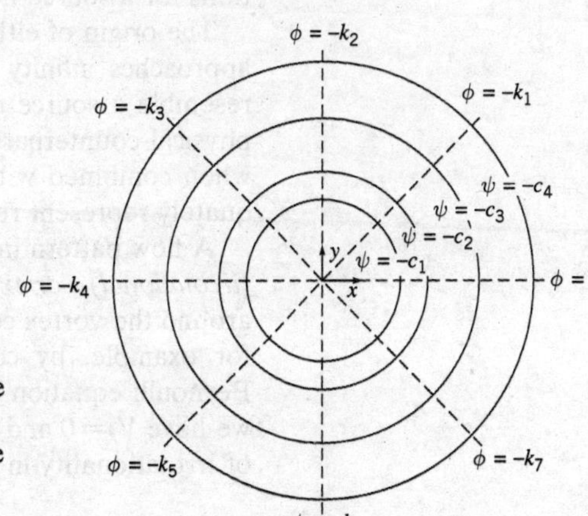

(Continued)

Table 6.2

Elementary Plane Flows (*Continued*)

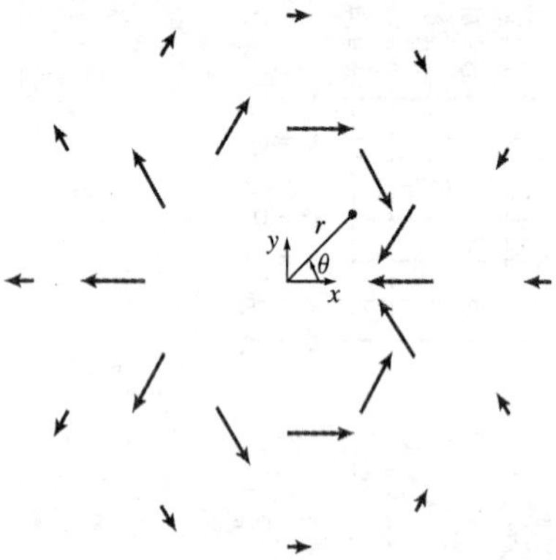

Doublet (center at origin)

$$V_r = -\frac{\Lambda}{r^2}\cos\theta \quad \psi = -\frac{\Lambda \sin\theta}{r}$$

$$V_\theta = -\frac{\Lambda}{r^2}\sin\theta \quad \phi = -\frac{\Lambda \cos\theta}{r}$$

Origin is singular point
Λ is strength of the doublet
Γ = around any closed curve

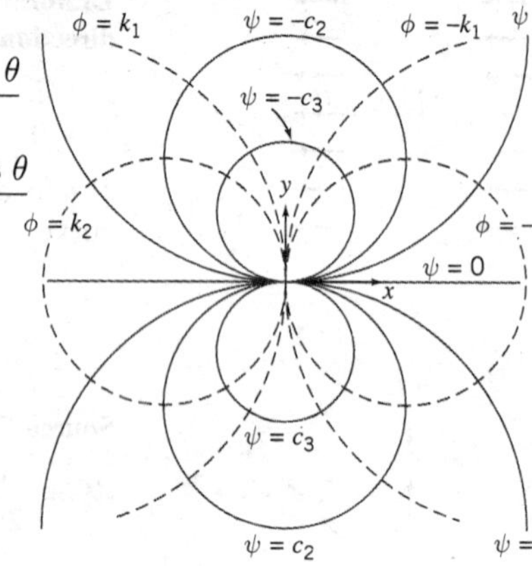

A *uniform flow* of constant velocity parallel to the x axis satisfies the continuity equation and the irrotationality condition identically. In Table 6.2 we have shown the ψ and ϕ functions for a uniform flow in the positive x direction.

For a uniform flow of constant magnitude V, inclined at angle α to the x axis,

$$\psi = (V\cos\alpha)y - (V\sin\alpha)x$$
$$\phi = -(V\sin\alpha)y - (V\cos\alpha)x$$

A simple *source* is a flow pattern in the xy plane in which flow is radially outward from the z axis and symmetrical in all directions. The strength, q, of the source is the volume flow rate per unit depth. At any radius, r, from a source, the tangential velocity, V_θ, is zero; the radial velocity, V_r, is the volume flow rate per unit depth, q, divided by the flow area per unit depth, $2\pi r$. Thus $V_r = q/2\pi r$ for a source. Knowing V_r and V_θ, obtaining ψ and ϕ from Eqs. 5.8 and 6.33, respectively, is straightforward.

In a simple *sink*, flow is radially inward; a sink is a negative source. The ψ and ϕ functions for a sink shown in Table 6.2 are the negatives of the corresponding functions for a source flow.

The origin of either a sink or a source is a singular point, since the radial velocity approaches infinity as the radius approaches zero. Thus, while an actual flow may resemble a source or a sink for some values of r, sources and sinks have no exact physical counterparts. The primary value of the concept of sources and sinks is that, when combined with other elementary flows, they produce flow patterns that adequately represent realistic flows.

A flow pattern in which the streamlines are concentric circles is a vortex; in a *free (irrotational) vortex*, fluid particles do not rotate as they translate in circular paths around the vortex center. There are a number of ways of obtaining the velocity field, for example, by combining the equation of motion (Euler's equation) and the Bernoulli equation to eliminate the pressure. Here, though, for circular streamlines, we have $V_r = 0$ and $V_\theta = f(\theta)$ only. We also have previously introduced the condition of irrotationality in cylindrical coordinates,

$$\frac{1}{r}\frac{\partial rV_\theta}{\partial r} - \frac{1}{r}\frac{\partial V_r}{\partial \theta} = 0 \tag{6.24}$$

Hence, using the known forms of V_r and V_θ, we obtain

$$\frac{1}{r}\frac{d(rV_\theta)}{dr} = 0$$

Integrating this equation gives

$$V_\theta r = \text{constant}$$

The strength, K, of the vortex is defined as $K = 2\pi r V_\theta$; the dimensions of K are L^2/t (volume flow rate per unit depth). Once again, knowing V_r and V_θ, obtaining ψ and ϕ from Eqs. 5.8 and 6.33, respectively, is straightforward. The irrotational vortex is a reasonable approximation to the flow field in a tornado (except in the region of the origin; the origin is a singular point).

The final "elementary" flow listed in Table 6.2 is the *doublet* of strength Λ. This flow is produced mathematically by allowing a source and a sink of numerically equal strengths to merge. In the limit, as the distance, δs, between them approaches zero, their strengths increase so the product $q\delta s/2\pi$ tends to a finite value, Λ, which is termed the strength of the doublet.

Superposition of Elementary Plane Flows

We saw earlier that both ϕ and ψ satisfy Laplace's equation for flow that is both incompressible and irrotational. Since Laplace's equation is a linear, homogeneous partial differential equation, solutions may be superposed (added together) to develop more complex and interesting patterns of flow. Thus if ψ_1 and ψ_2 satisfy Laplace's equation, then so does $\psi_3 = \psi_1 + \psi_2$. The elementary plane flows are the building blocks in this superposition process. There is one note of caution: While Laplace's equation for the stream function, and the stream function-velocity field equations (Eq. 5.3) are linear, the Bernoulli equation is not; hence, in the superposition process we will have $\psi_3 = \psi_1 + \psi_2$, $u_3 = u_1 + u_2$, and $v_3 = v_1 + v_2$, but $p_3 \neq p_1 + p_2$! We must use the Bernoulli equation, which is nonlinear in V, to find p_3.

We can add together elementary flows to try and generate recognizable flow patterns. The simplest superposition approach is called the *direct* method, in which we try different combinations of elementary flows and see what kinds of flow patterns are produced. This sounds like a random process, but with a little experience it becomes a quite logical process. For example, look at some of the classic examples listed in Table 6.3. The source and uniform flow combination makes sense—we would intuitively expect a source to partially push its way upstream, and to divert the flow around it. The source, sink, and uniform flow (generating what is called a Rankine body) is also not surprising—the entire flow out of the source makes its way into the sink, leading to a closed streamline. *Any streamline can be interpreted as a solid surface because there is no flow across it;* we can therefore pretend that this closed streamline represents a solid. We could easily generalize this source-sink approach to any number of sources and sinks distributed along the x axis, and as long as the sum of the source and sink strengths added up to zero, we would generate a closed streamline body shape. The doublet-uniform flow (with or without a vortex) generates a very interesting result: flow over a cylinder (with or without circulation)! We first saw the flow without circulation in Fig. 2.12*a*. The flow with a clockwise vortex produces a top-to-bottom asymmetry. This is because in the region above the cylinder the velocities due to the uniform flow and vortex are in the same overall direction and lead to a high velocity; below the cylinder they are in opposite directions and therefore lead to a low velocity. As we have learned, whenever velocities are high, streamlines will be close together, and vice versa—explaining the pattern shown. More importantly, from the Bernoulli equation we know that whenever the velocity is high the pressure will be low, and vice versa—hence, we can anticipate that the cylinder with circulation will experience a net upward force (lift) due to pressure. This approach, of looking at

Table 6.3

Superposition of Elementary Plane Flows

Source and Uniform Flow (flow past a half-body)

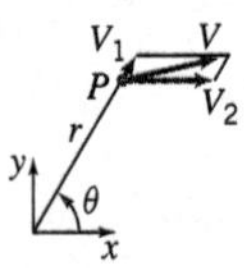

$$\psi = \psi_{so} + \psi_{uf} = \psi_1 + \psi_2 = \frac{q}{2\pi}\theta + Uy$$

$$\psi = \frac{q}{2\pi}\theta + Ur\sin\theta$$

$$\phi = \phi_{so} + \phi_{uf} = \phi_1 + \phi_2 = -\frac{q}{2\pi}\ln r - Ux$$

$$\phi = -\frac{q}{2\pi}\ln r - Ur\cos\theta$$

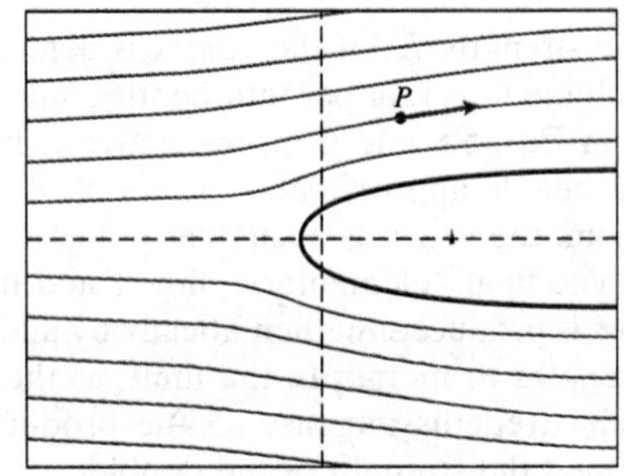

Source and Sink (equal strength, separation distance on x axis = 2a)

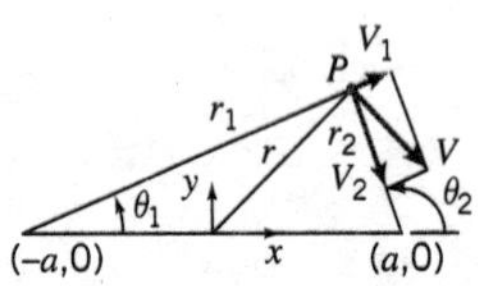

$$\psi = \psi_{so} + \psi_{si} = \psi_1 + \psi_2 = \frac{q}{2\pi}\theta_1 - \frac{q}{2\pi}\theta_2$$

$$\psi = \frac{q}{2\pi}(\theta_1 - \theta_2)$$

$$\phi = \phi_{so} + \phi_{si} = \phi_1 + \phi_2 = -\frac{q}{2\pi}\ln r_1 + \frac{q}{2\pi}\ln r_2$$

$$\phi = \frac{q}{2\pi}\ln\frac{r_2}{r_1}$$

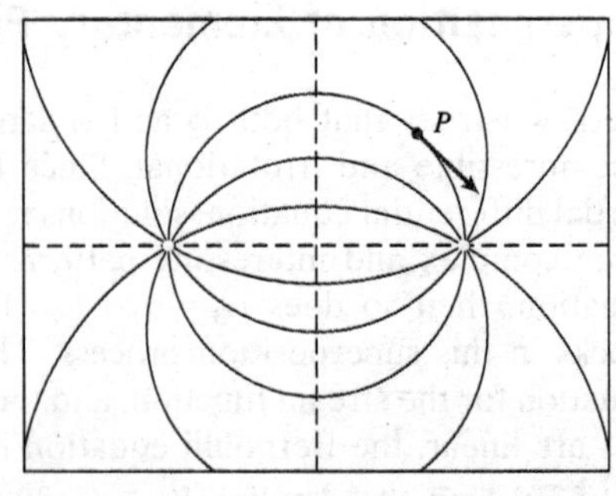

Source, Sink, and Uniform Flow (flow past a Rankine body)

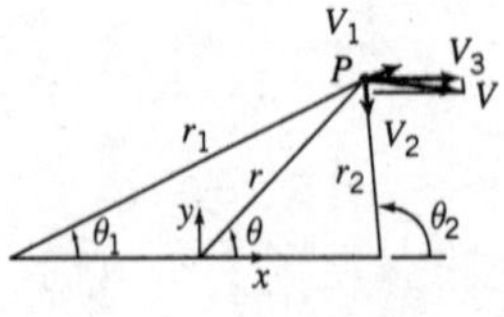

$$\psi = \psi_{so} + \psi_{si} + \psi_{uf} = \psi_1 + \psi_2 + \psi_3$$

$$= \frac{q}{2\pi}\theta_1 - \frac{q}{2\pi}\theta_2 + Uy$$

$$\psi = \frac{q}{2\pi}(\theta_1 - \theta_2) + Ur\sin\theta$$

$$\phi = \phi_{so} + \phi_{si} + \phi_{uf} = \phi_1 + \phi_2 + \phi_3$$

$$= -\frac{q}{2\pi}\ln r_1 + \frac{q}{2\pi}\ln r_2 - Ux$$

$$\phi = \frac{q}{2\pi}\ln\frac{r_2}{r_1} - Ur\cos\theta$$

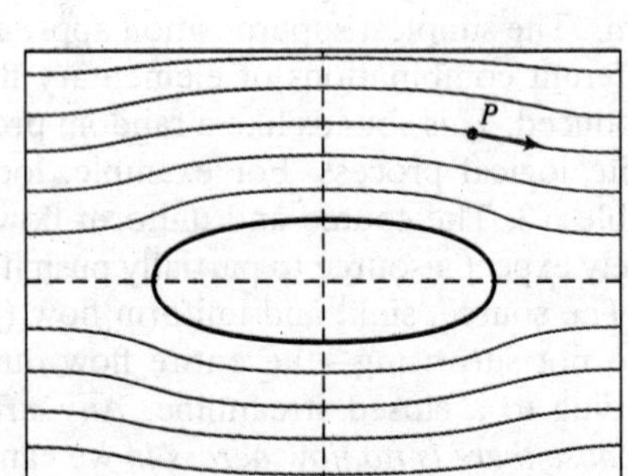

Vortex (clockwise) and Uniform Flow

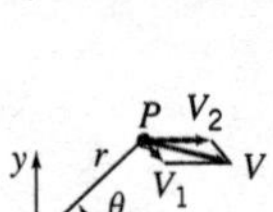

$$\psi = \psi_v + \psi_{uf} = \psi_1 + \psi_2 = \frac{K}{2\pi}\ln r + Uy$$

$$\psi = \frac{K}{2\pi}\ln r + Ur\sin\theta$$

$$\phi = \phi_v + \phi_{uf} = \phi_1 + \phi_2 = \frac{K}{2\pi}\theta - Ux$$

$$\phi = \frac{K}{2\pi}\theta - Ur\cos\theta$$

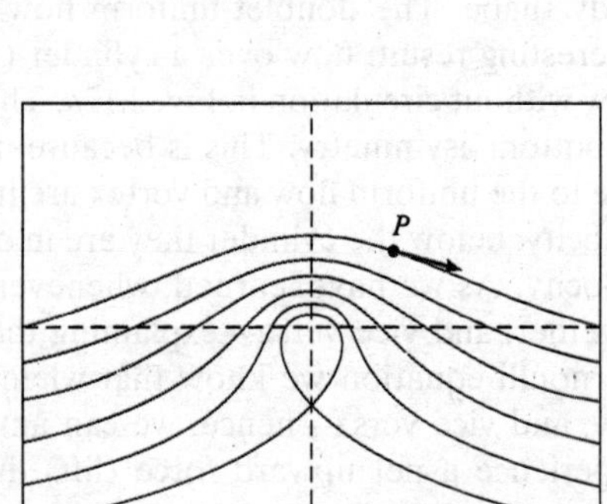

Table 6.3
Superposition of Elementary Plane Flows (*Continued*)

Doublet and Uniform Flow (flow past a cylinder)

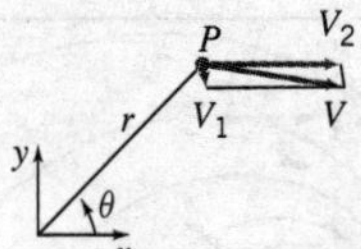

$$\psi = \psi_d + \psi_{uf} = \psi_1 + \psi_2 = -\frac{\Lambda \sin\theta}{r} + Uy$$

$$= -\frac{\Lambda \sin\theta}{r} + Ur\sin\theta$$

$$\psi = U\left(r - \frac{\Lambda}{Ur}\right)\sin\theta$$

$$\psi = Ur\left(1 - \frac{a^2}{r^2}\right)\sin\theta \quad a = \sqrt{\frac{\Lambda}{U}}$$

$$\phi = \phi_d + \phi_{uf} = \phi_1 + \phi_2 = -\frac{\Lambda \cos\theta}{r} - Ux$$

$$= -\frac{\Lambda \cos\theta}{r} - Ur\cos\theta$$

$$\phi = -U\left(r + \frac{\Lambda}{Ur}\right)\cos\theta = -Ur\left(1 + \frac{a^2}{r^2}\right)\cos\theta$$

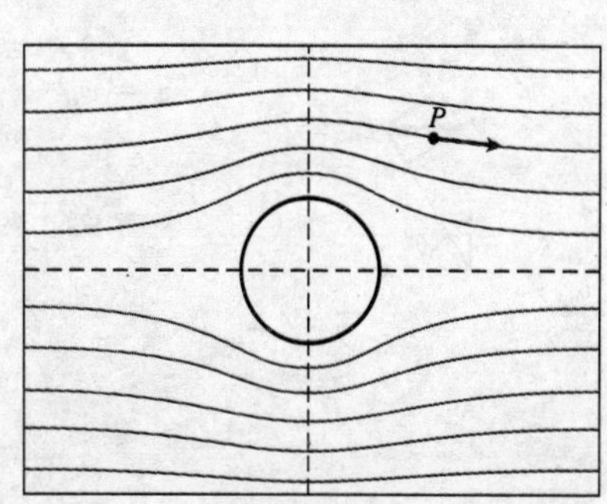

Doublet, Vortex (clockwise), and Uniform Flow (flow past a cylinder with circulation)

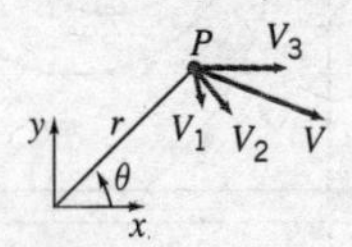

$$\psi = \psi_d + \psi_v + \psi_{uf} = \psi_1 + \psi_2 + \psi_3$$

$$= -\frac{\Lambda \sin\theta}{r} + \frac{K}{2\pi}\ln r + Uy$$

$$\psi = -\frac{\Lambda \sin\theta}{r} + \frac{K}{2\pi}\ln r + Ur\sin\theta$$

$$\psi = Ur\left(1 - \frac{a^2}{r^2}\right)\sin\theta + \frac{K}{2\pi}\ln r$$

$$\phi = \phi_d + \phi_v + \phi_{uf} = \phi_1 + \phi_2 + \phi_3$$

$$= -\frac{\Lambda \cos\theta}{r} + \frac{K}{2\pi}\theta - Ux$$

$$a = \sqrt{\frac{\Lambda}{U}};\ K < 4\pi a U \qquad \phi = -\frac{\Lambda \cos\theta}{r} + \frac{K}{2\pi}\theta - Ur\cos\theta$$

$$\phi = -Ur\left(1 + \frac{a^2}{r^2}\right)\cos\theta + \frac{K}{2\pi}\theta$$

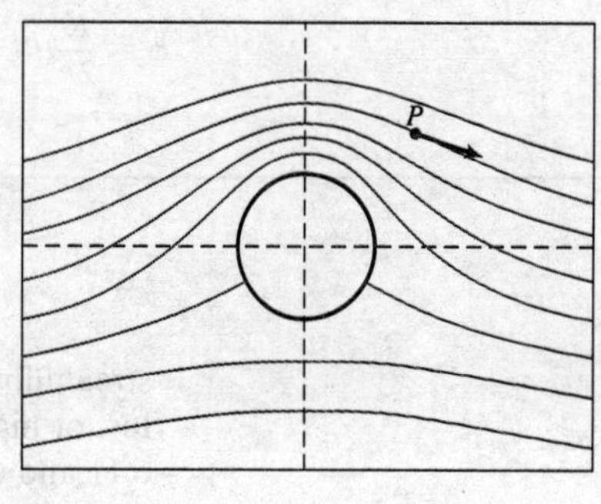

Source and Vortex (spiral vortex)

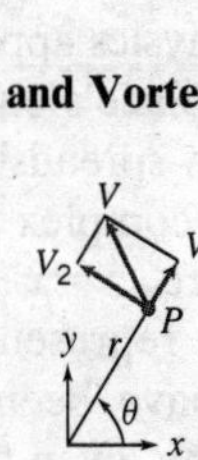

$$\psi = \psi_{so} + \psi_v = \psi_1 + \psi_2 = \frac{q}{2\pi}\theta - \frac{K}{2\pi}\ln r$$

$$\phi = \phi_{so} + \phi_v = \phi_1 + \phi_2 = -\frac{q}{2\pi}\ln r - \frac{K}{2\pi}\theta$$

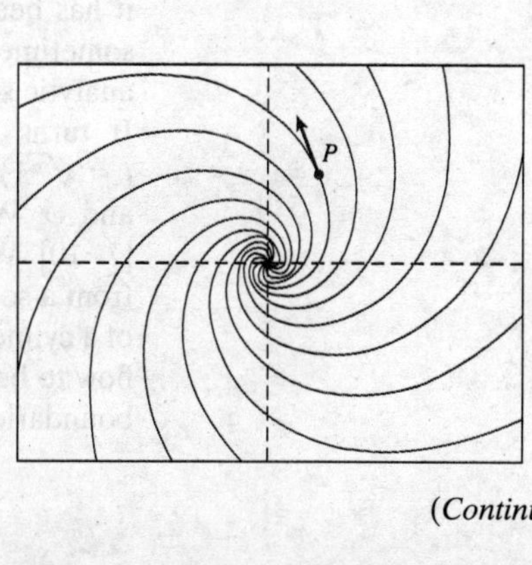

(*Continued*)

Table 6.3

Superposition of Elementary Plane Flows (*Continued*)

Sink and Vortex

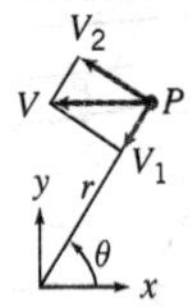

$$\psi = \psi_{si} + \psi_v = \psi_1 + \psi_2 = -\frac{q}{2\pi}\theta - \frac{K}{2\pi}\ln r$$

$$\phi = \phi_{si} + \phi_v = \phi_1 + \phi_2 = \frac{q}{2\pi}\ln r - \frac{K}{2\pi}\theta$$

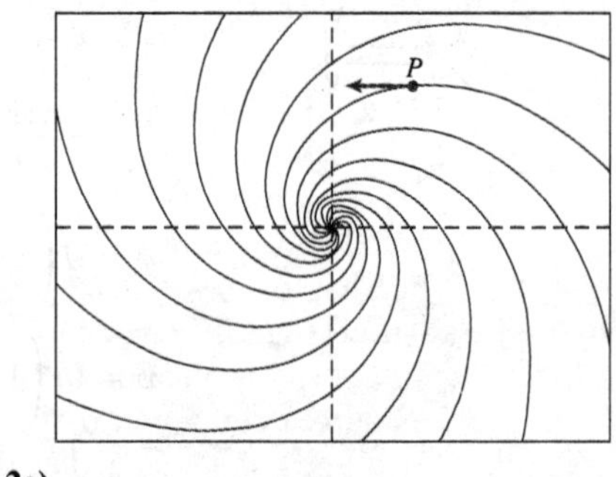

Vortex Pair (equal strength, opposite rotation, separation distance on x axis = 2a)

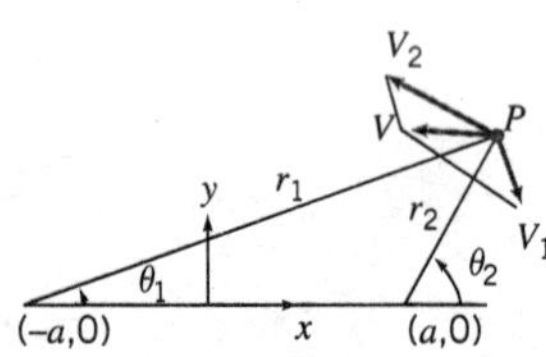

$$\psi = \psi_{v1} + \psi_{v2} = \psi_1 + \psi_2 = -\frac{K}{2\pi}\ln r_1 + \frac{K}{2\pi}\ln r_2$$

$$\psi = \frac{K}{2\pi}\ln\frac{r_2}{r_1}$$

$$\phi = \phi_{v1} + \phi_{v2} = \phi_1 + \phi_2 = -\frac{K}{2\pi}\theta_1 + \frac{K}{2\pi}\theta_2$$

$$\phi = \frac{K}{2\pi}(\theta_2 - \theta_1)$$

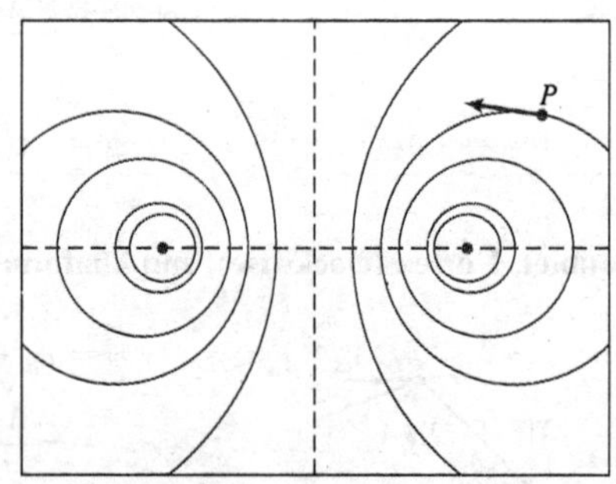

streamline patterns to see where we have regions of high or low velocity and hence low or high pressure, is very useful. We will examine these last two flows in Examples 6.11 and 6.12. The last example in Table 6.3, the vortex pair, hints at a way to create flows that simulate the presence of a wall or walls: for the y axis to be a streamline (and thus a wall), simply make sure that any objects (e.g., a source, a vortex) in the positive x quadrants have mirror-image objects in the negative x quadrants; the y axis will thus be a line of symmetry. For a flow pattern in a 90° corner, we need to place objects so that we have symmetry with respect to both the x and y axes. For flow in a corner whose angle is a fraction of 90° (e.g., 30°), we need to place objects in a radially symmetric fashion.

Because Laplace's equation appears in many engineering and physics applications, it has been extensively studied. We saw in Example 5.12 that Laplace's equation is sometimes amenable to a fairly simple numerical solution using a spreadsheet. For analytic solutions, one approach is to use conformal mapping with complex notation. It turns out that *any* continuous complex function $f(z)$ (where $z = x + iy$, and $i = \sqrt{-1}$) is a solution of Laplace's equation, and can therefore represent both ϕ and ψ. With this approach many elegant mathematical results have been derived [7–10]. We mention only two: the circle theorem, which enables any given flow [e.g., from a source at some point (a, b)] to be easily transformed to allow for the presence of a cylinder at the origin; and the Schwarz-Christoffel theorem, which enables a given flow to be transformed to allow for the presence of virtually unlimited stepwise linear boundaries (e.g., the presence on the x axis of the silhouette of a building).

Much of this analytical work was done centuries ago, when it was called "hydrodynamics" instead of potential theory. A list of famous contributors includes Bernoulli, Lagrange, d'Alembert, Cauchy, Rankine, and Euler [11]. As we discussed in Section 2.6, the theory immediately ran into difficulties: In an ideal fluid flow no body experiences drag—the d'Alembert paradox of 1752—a result completely counter to experience. Prandtl, in 1904, resolved this discrepancy by describing how real flows may be essentially inviscid almost everywhere, but there is always a "boundary layer" adjacent to the body. In this layer significant viscous effects occur, and the no-slip condition is satisfied (in potential flow theory the no-slip condition is not satisfied). Development of this concept, and the Wright brothers' historic first human flight, led to rapid developments in aeronautics starting in the 1900s. We will study boundary layers in detail in Chapter 9, where we will see that their existence leads to drag on bodies, and also affects the lift of bodies.

An alternative superposition approach is the *inverse* method in which distributions of objects such as sources, sinks, and vortices are used to model a body [12]. It is called inverse because the body shape is deduced based on a desired pressure distribution. Both the direct and inverse methods, including three-dimensional space, are today mostly analyzed using computer applications such as *Fluent* [13] and *STAR-CD* [14].

Example 6.11 FLOW OVER A CYLINDER: SUPERPOSITION OF DOUBLET AND UNIFORM FLOW

For two-dimensional, incompressible, irrotational flow, the superposition of a doublet and a uniform flow represents flow around a circular cylinder. Obtain the stream function and velocity potential for this flow pattern. Find the velocity field, locate the stagnation points and the cylinder surface, and obtain the surface pressure distribution. Integrate the pressure distribution to obtain the drag and lift forces on the circular cylinder.

Given: Two-dimensional, incompressible, irrotational flow formed from superposition of a doublet and a uniform flow.

Find: (a) Stream function and velocity potential.
(b) Velocity field.
(c) Stagnation points.
(d) Cylinder surface.
(e) Surface pressure distribution.
(f) Drag force on the circular cylinder.
(g) Lift force on the circular cylinder.

Solution: Stream functions may be added because the flow field is incompressible and irrotational. Thus from Table 6.2, the stream function for the combination is

$$\psi = \psi_d + \psi_{uf} = -\frac{\Lambda \sin\theta}{r} + Ur\sin\theta \qquad \psi$$

The velocity potential is

$$\phi = \phi_d + \phi_{uf} = -\frac{\Lambda \cos\theta}{r} - Ur\cos\theta \qquad \phi$$

The corresponding velocity components are obtained using Eqs. 6.30 as

$$V_r = -\frac{\partial \phi}{\partial r} = -\frac{\Lambda \cos\theta}{r^2} + U\cos\theta$$

$$V_\theta = -\frac{1}{r}\frac{\partial \phi}{\partial \theta} = -\frac{\Lambda \sin\theta}{r^2} - U\sin\theta$$

The velocity field is

$$\vec{V} = V_r\hat{e}_r + V_\theta\hat{e}_\theta = \left(-\frac{\Lambda\cos\theta}{r^2} + U\cos\theta\right)\hat{e}_r + \left(-\frac{\Lambda\sin\theta}{r^2} - U\sin\theta\right)\hat{e}_\theta \longleftarrow \vec{V}$$

Stagnation points are where $\vec{V} = V_r\hat{e}_r + V_\theta\hat{e}_\theta = 0$

$$V_r = -\frac{\Lambda\cos\theta}{r^2} + U\cos\theta = \cos\theta\left(U - \frac{\Lambda}{r^2}\right)$$

Thus $V_r = 0$ when $r = \sqrt{\frac{\Lambda}{U}} = a$. Also,

$$V_\theta = -\frac{\Lambda\sin\theta}{r^2} - U\sin\theta = -\sin\theta\left(U + \frac{\Lambda}{r^2}\right)$$

Thus $V_\theta = 0$ when $\theta = 0, \pi$.

Stagnation points are $(r, \theta) = (a, 0), (a, \pi)$. ⟵ Stagnation points

Note that $V_r = 0$ along $r = a$, so this represents flow around a circular cylinder, as shown in Table 6.3. Flow is irrotational, so the Bernoulli equation may be applied between any two points. Applying the equation between a point far upstream and a point on the surface of the cylinder (neglecting elevation differences), we obtain

$$\frac{p_\infty}{\rho} + \frac{U^2}{2} = \frac{p}{\rho} + \frac{V^2}{2}$$

Thus,

$$p - p_\infty = \frac{1}{2}\rho(U^2 - V^2)$$

Along the surface, $r = a$, and

$$V^2 = V_\theta^2 = \left(-\frac{\Lambda}{a^2} - U\right)^2 \sin^2\theta = 4U^2\sin^2\theta$$

since $\Lambda = Ua^2$. Substituting yields

$$p - p_\infty = \frac{1}{2}\rho(U^2 - 4U^2\sin^2\theta) = \frac{1}{2}\rho U^2(1 - 4\sin^2\theta)$$

or

$$\frac{p - p_\infty}{\frac{1}{2}\rho U^2} = 1 - 4\sin^2\theta \longleftarrow \text{Pressure distribution}$$

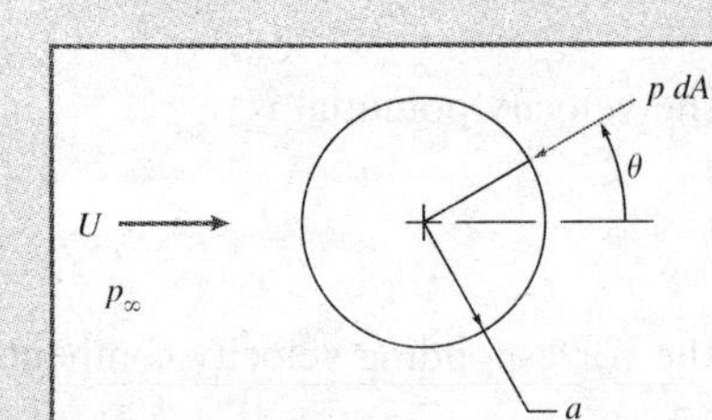

Drag is the force component parallel to the freestream flow direction. The drag force is given by

$$F_D = \int_A -p\,dA\cos\theta = \int_0^{2\pi} -pa\,d\theta\,b\cos\theta$$

since $dA = a\,d\theta\,b$, where b is the length of the cylinder normal to the diagram.

Substituting $p = p_\infty + \frac{1}{2}\rho U^2(1 - 4\sin^2\theta)$,

$$F_D = \int_0^{2\pi} -p_\infty ab \cos\theta \, d\theta + \int_0^{2\pi} -\frac{1}{2}\rho U^2(1 - 4\sin^2\theta) ab \cos\theta \, d\theta$$

$$= -p_\infty \, ab \sin\theta \Big]_0^{2\pi} - \frac{1}{2}\rho U^2 ab \sin\theta \Big]_0^{2\pi} + \frac{1}{2}\rho U^2 ab \frac{4}{3}\sin^3\theta \Big]_0^{2\pi}$$

$$F_D = 0 \qquad F_D$$

Lift is the force component normal to the freestream flow direction. (By convention, positive lift is an upward force.) The lift force is given by

$$F_L = \int_A p \, dA(-\sin\theta) = -\int_0^{2\pi} pa \, d\theta \, b \sin\theta$$

Substituting for p gives

$$F_L = -\int_0^{2\pi} p_\infty ab \sin\theta \, d\theta - \int_0^{2\pi} \frac{1}{2}\rho U^2(1 - 4\sin^2\theta) ab \sin\theta \, d\theta$$

$$= p_\infty a \, b \cos\theta \Big]_0^{2\pi} + \frac{1}{2}\rho U^2 ab \cos\theta \Big]_0^{2\pi} + \frac{1}{2}\rho U^2 ab \left[\frac{4\cos^3\theta}{3} - 4\cos\theta\right]_0^{2\pi}$$

$$F_L = 0 \qquad F_L$$

This problem illustrates:

- ✓ How elementary plane flows can be combined to generate interesting and useful flow patterns.
- ✓ d'Alembert's paradox, that potential flows over a body do not generate drag.

The stream function and pressure distribution are plotted in the *Excel* workbook.

Example 6.12 FLOW OVER A CYLINDER WITH CIRCULATION: SUPERPOSITION OF DOUBLET, UNIFORM FLOW, AND CLOCKWISE FREE VORTEX

For two-dimensional, incompressible, irrotational flow, the superposition of a doublet, a uniform flow, and a free vortex represents the flow around a circular cylinder with circulation. Obtain the stream function and velocity potential for this flow pattern, using a clockwise free vortex. Find the velocity field, locate the stagnation points and the cylinder surface, and obtain the surface pressure distribution. Integrate the pressure distribution to obtain the drag and lift forces on the circular cylinder. Relate the lift force on the cylinder to the circulation of the free vortex.

Given: Two-dimensional, incompressible, irrotational flow formed from superposition of a doublet, a uniform flow, and a clockwise free vortex.

Find: (a) Stream function and velocity potential.
(b) Velocity field.
(c) Stagnation points.
(d) Cylinder surface.
(e) Surface pressure distribution.
(f) Drag force on the circular cylinder.
(g) Lift force on the circular cylinder.
(h) Lift force in terms of circulation of the free vortex.

Solution:
Stream functions may be added because the flow field is incompressible and irrotational. From Table 6.2, the stream function and velocity potential for a clockwise free vortex are

$$\psi_{fv} = \frac{K}{2\pi}\ln r \qquad \phi_{fv} = \frac{K}{2\pi}\theta$$

Using the results of Example 6.11, the stream function for the combination is

$$\psi = \psi_d + \psi_{uf} + \psi_{fv}$$

$$\psi = -\frac{\Lambda \sin\theta}{r} + Ur\sin\theta + \frac{K}{2\pi}\ln r \quad \longleftarrow \quad \psi$$

The velocity potential for the combination is

$$\phi = \phi_d + \phi_{uf} + \phi_{fv}$$

$$\phi = -\frac{\Lambda \cos\theta}{r} - Ur\cos\theta + \frac{K}{2\pi}\theta \quad \longleftarrow \quad \phi$$

The corresponding velocity components are obtained using Eqs. 6.30 as

$$V_r = -\frac{\partial \phi}{\partial r} = -\frac{\Lambda\cos\theta}{r^2} + U\cos\theta \tag{1}$$

$$V_\theta = -\frac{1}{r}\frac{\partial\phi}{\partial\theta} = -\frac{\Lambda\sin\theta}{r^2} - U\sin\theta - \frac{K}{2\pi r} \tag{2}$$

The velocity field is

$$\vec{V} = V_r\,\hat{e}_r + V_\theta\,\hat{e}_\theta$$

$$\vec{V} = \left(-\frac{\Lambda\cos\theta}{r^2} + U\cos\theta\right)\hat{e}_r + \left(-\frac{\Lambda\sin\theta}{r} - U\sin\theta - \frac{K}{2\pi r}\right)\hat{e}_\theta \quad \longleftarrow \quad \vec{V}$$

Stagnation points are located where $\vec{V} = V_r\,\hat{e}_r + V_\theta\,\hat{e}_\theta = 0$. From Eq. 1,

$$V_r = -\frac{\Lambda\cos\theta}{r^2} + U\cos\theta = \cos\theta\left(U - \frac{\Lambda}{r^2}\right)$$

Thus $V_r = 0$ when $r = \sqrt{\Lambda/U} = a$ $\longleftarrow$ Cylinder surface

The stagnation points are located on $r = a$. Substituting into Eq. 2 with $r = a$,

$$V_\theta = -\frac{\Lambda\sin\theta}{a^2} - U\sin\theta - \frac{K}{2\pi a}$$

$$= -\frac{\Lambda\sin\theta}{\Lambda/U} - U\sin\theta - \frac{K}{2\pi a}$$

$$V_\theta = -2U\sin\theta - \frac{K}{2\pi a}$$

Thus $V_\theta = 0$ along $r = a$ when

$$\sin\theta = -\frac{K}{4\pi Ua} \qquad \text{or} \qquad \theta = \sin^{-1}\left[\frac{-K}{4\pi Ua}\right]$$

Stagnation points: $r = a \quad \theta = \sin^{-1}\left[\frac{-K}{4\pi Ua}\right]$ $\longleftarrow$ Stagnation points

As in Example 6.11, $V_r = 0$ along $r = a$, so this flow field once again represents flow around a circular cylinder, as shown in Table 6.3. For $K = 0$ the solution is identical to that of Example 6.11.

The presence of the free vortex ($K > 0$) moves the stagnation points below the center of the cylinder. Thus the free vortex alters the vertical symmetry of the flow field. The flow field has two stagnation points for a range of vortex strengths between $K = 0$ and $K = 4\pi Ua$.

A single stagnation point is located at $\theta = -\pi/2$ when $K = 4\pi Ua$.

Even with the free vortex present, the flow field is irrotational, so the Bernoulli equation may be applied between any two points. Applying the equation between a point far upstream and a point on the surface of the cylinder we obtain

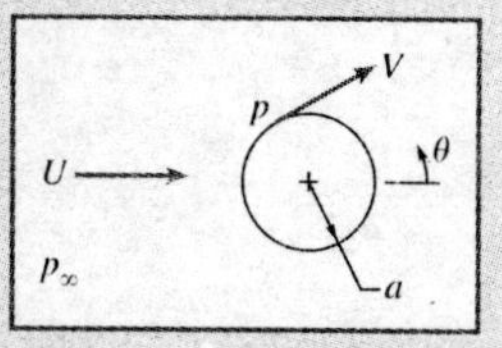

$$\frac{p_\infty}{\rho} + \frac{U^2}{2} + gz = \frac{p}{\rho} + \frac{V^2}{2} + gz$$

Thus, neglecting elevation differences,

$$p - p_\infty = \frac{1}{2}\rho\,(U^2 - V^2) = \frac{1}{2}\rho U^2\left[1 - \left(\frac{U}{V}\right)^2\right]$$

Along the surface $r = a$ and $V_r = 0$, so

$$V^2 = V_\theta^2 = \left(-2U\sin\theta - \frac{K}{2\pi a}\right)^2$$

and

$$\left(\frac{V}{U}\right)^2 = 4\sin^2\theta + \frac{2K}{\pi Ua}\sin\theta + \frac{K^2}{4\pi^2U^2a^2}$$

Thus

$$p = p_\infty + \frac{1}{2}\rho U^2\left(1 - 4\sin^2\theta - \frac{2K}{\pi Ua}\sin\theta - \frac{K^2}{4\pi^2U^2a^2}\right)$$ $p(\theta)$

Drag is the force component parallel to the freestream flow direction. As in Example 6.11, the drag force is given by

$$F_D = \int_A -p\,dA\cos\theta = \int_0^{2\pi} -pa\,d\theta b\cos\theta$$

since $dA = a\,d\theta\,b$, where b is the length of the cylinder normal to the diagram.

Comparing pressure distributions, the free vortex contributes only to the terms containing the factor K. The contribution of these terms to the drag force is

$$\frac{F_{D_{fv}}}{\frac{1}{2}\rho U^2} = \int_0^{2\pi}\left(\frac{2K}{\pi Ua}\sin\theta + \frac{K^2}{4\pi^2U^2a^2}\right)ab\cos\theta\,d\theta \tag{3}$$

$$\frac{F_{D_{fv}}}{\frac{1}{2}\rho U^2} = \frac{2K}{\pi Ua}ab\left.\frac{\sin^2\theta}{2}\right]_0^{2\pi} + \frac{K^2}{4\pi^2U^2a^2}ab\left.\sin\theta\right]_0^{2\pi} = 0$$ F_D

Lift is the force component normal to the freestream flow direction. (Upward force is defined as positive lift.) The lift force is given by

$$F_L = \int_A -p\,dA\sin\theta = \int_0^{2\pi} -pa\,d\theta\,b\sin\theta$$

Comparing pressure distributions, the free vortex contributes only to the terms containing the factor K. The contribution of these terms to the lift force is

$$\frac{F_{L_{fv}}}{\frac{1}{2}\rho U^2} = \int_0^{2\pi}\left(\frac{2K}{\pi Ua}\sin\theta + \frac{K^2}{4\pi^2U^2a^2}\right)ab\sin\theta\,d\theta$$

$$= \frac{2K}{\pi Ua}\int_0^{2\pi} ab\sin^2\theta d\theta + \frac{K^2}{4\pi^2U^2a^2}\int_0^{2\pi} ab\sin\theta\,d\theta$$

$$= \frac{2Kb}{\pi U}\left[\frac{\theta}{2} - \frac{\sin^2\theta}{4}\right]_0^{2\pi} - \frac{K^2b}{4\pi^2U^2a}\cos\theta\Big]_0^{2\pi}$$

$$\frac{F_{L_{fv}}}{\frac{1}{2}\rho U^2} = \frac{2Kb}{\pi U}\left[\frac{2\pi}{2}\right] = \frac{2Kb}{U}$$

Then $F_{L_{fv}} = \rho UKb$ ← F_L

The *circulation* is defined by Eq. 5.18 as

$$\Gamma \equiv \oint \vec{V}\cdot d\vec{s}$$

On the cylinder surface, $r = a$, and $\vec{V} = V_\theta\,\hat{e}_\theta$, so

$$\Gamma = \int_0^{2\pi}\left(-2U\sin\theta - \frac{K}{2\pi a}\right)\hat{e}_\theta\cdot a\,d\theta\,\hat{e}_\theta$$

$$= -\int_0^{2\pi} 2Ua\sin\theta\,d\theta - \int_0^{2\pi}\frac{K}{2\pi}\,d\theta$$

$\Gamma = -K$ ← Circulation

Substituting into the expression for lift,

$$F_L = \rho UKb = \rho U(-\Gamma)b = -\rho U\,\Gamma b$$

or the lift force per unit length of cylinder is

$\frac{F_L}{b} = -\rho U\Gamma$ ← $\frac{F_L}{b}$

This problem illustrates:
- ✓ Once again d'Alembert's paradox, that potential flows do not generate drag on a body.
- ✓ That the lift per unit length is $-\rho U\Gamma$. It turns out that this expression for lift is the same for *all* bodies in an ideal fluid flow, regardless of shape!

The stream function and pressure distribution are plotted in the *Excel* workbook.

6.8 Summary and Useful Equations

In this chapter we have:

- ✓ Derived Euler's equations in vector form and in rectangular, cylindrical, and streamline coordinates.
- ✓ Obtained Bernoulli's equation by integrating Euler's equation along a steady-flow streamline, and discussed its restrictions. We have also seen how for a steady, incompressible flow through a stream tube the first law of thermodynamics reduces to the Bernoulli equation if certain restrictions apply.

✓ Defined the static, dynamic, and stagnation (or total) pressures.
✓ Defined the energy and hydraulic grade lines.
✓ *Derived an unsteady flow Bernoulli equation, and discussed its restrictions.
✓ *Observed that for an irrotational flow that is steady and incompressible, the Bernoulli equation applies between *any* two points in the flow.
✓ *Defined the velocity potential ϕ and discussed its restrictions.

We have also explored in detail two-dimensional, incompressible, and irrotational flows, and learned that for these flows: the stream function ψ and the velocity potential ϕ satisfy Laplace's equation; ψ and ϕ can be derived from the velocity components, and vice versa, and the iso-lines of the stream function ψ and the velocity potential ϕ are orthogonal. We explored for such flows how to combine potential flows to generate various flow patterns, and how to determine the pressure distribution and lift and drag on, for example, a cylindrical shape.

Note: Most of the Useful Equations in the table below have a number of constraints or limitations—*be sure to refer to their page numbers for details*!

Useful Equations

The Euler equation for incompressible, inviscid flow:	$\rho \dfrac{D\vec{V}}{Dt} = \rho\vec{g} - \nabla p$	(6.1)	Page 237
The Euler equation (rectangular coordinates):	$\rho\left(\dfrac{\partial u}{\partial t} + u\dfrac{\partial u}{\partial x} + v\dfrac{\partial u}{\partial y} + w\dfrac{\partial u}{\partial z}\right) = \rho g_x - \dfrac{\partial p}{\partial x}$	(6.2a)	Page 237
	$\rho\left(\dfrac{\partial v}{\partial t} + u\dfrac{\partial v}{\partial x} + v\dfrac{\partial v}{\partial y} + w\dfrac{\partial v}{\partial z}\right) = \rho g_y - \dfrac{\partial p}{\partial y}$	(6.2b)	
	$\rho\left(\dfrac{\partial w}{\partial t} + u\dfrac{\partial w}{\partial x} + v\dfrac{\partial w}{\partial y} + w\dfrac{\partial w}{\partial z}\right) = \rho g_z - \dfrac{\partial p}{\partial z}$	(6.2c)	
The Euler equation (cylindrical coordinates):	$\rho a_r = \rho\left(\dfrac{\partial V_r}{\partial t} + V_r\dfrac{\partial V_r}{\partial r} + \dfrac{V_\theta}{r}\dfrac{\partial V_r}{\partial \theta} + V_z\dfrac{\partial V_r}{\partial z} - \dfrac{V_\theta^2}{r}\right) = \rho g_r - \dfrac{\partial p}{\partial r}$	(6.3a)	Page 237
	$\rho a_\theta = \rho\left(\dfrac{\partial V_\theta}{\partial t} + V_r\dfrac{\partial V_\theta}{\partial r} + \dfrac{V_\theta}{r}\dfrac{\partial V_\theta}{\partial \theta} + V_z\dfrac{\partial V_\theta}{\partial z} + \dfrac{V_r V_\theta}{r}\right) = \rho g_\theta - \dfrac{1}{r}\dfrac{\partial p}{\partial \theta}$	(6.3b)	
	$\rho a_z = \rho\left(\dfrac{\partial V_z}{\partial t} + V_r\dfrac{\partial V_z}{\partial r} + \dfrac{V_\theta}{r}\dfrac{\partial V_z}{\partial \theta} + V_z\dfrac{\partial V_z}{\partial z}\right) = \rho g_z - \dfrac{\partial p}{\partial z}$	(6.3c)	
The Bernoulli equation (steady, incompressible, inviscid, along a streamline):	$\dfrac{p}{\rho} + \dfrac{V^2}{2} + gz = \text{constant}$	(6.8)	Page 242
Definition of total head of a flow:	$\dfrac{p}{\rho g} + \dfrac{V^2}{2g} + z = H$	(6.16a)	Page 258
Definition of energy grade line (EGL):	$EGL = \dfrac{p}{\rho g} + \dfrac{V^2}{2g} + z$	(6.16b)	Page 258
Definition of hydraulic grade line (HGL):	$HGL = \dfrac{p}{\rho g} + z$	(6.16c)	Page 258

*These topics apply to sections that may be omitted without loss of continuity in the text material.

Relation between EGL, HGL, and dynamic pressure:	$EGL - HGL = \dfrac{V^2}{2g}$	**(6.16d)**	Page 258
The unsteady Bernoulli equation (incompressible, inviscid, along a streamline):	$\dfrac{p_1}{\rho} + \dfrac{V_1^2}{2} + gz_1 = \dfrac{p_2}{\rho} + \dfrac{V_2^2}{2} + gz_2 + \int_1^2 \dfrac{\partial V}{\partial t}\, ds$	**(6.21)**	Page W-16
Definition of stream function (2D, incompressible flow):	$u = \dfrac{\partial \psi}{\partial y} \qquad v = -\dfrac{\partial \psi}{\partial x}$	**(5.4)**	Page 262
Definition of velocity potential (2D irrotational flow):	$u = -\dfrac{\partial \phi}{\partial x} \qquad v = -\dfrac{\partial \phi}{\partial y}$	**(6.29)**	Page 262
Definition of stream function (2D, incompressible flow, cylindrical coordinates):	$V_r = \dfrac{1}{r}\dfrac{\partial \psi}{\partial \theta}$ and $V_\theta = -\dfrac{\partial \psi}{\partial r}$	**(5.8)**	Page 262
Definition of velocity potential (2D irrotational flow, cylindrical coordinates):	$V_r = -\dfrac{\partial \phi}{\partial r}$ and $V_\theta = -\dfrac{1}{r}\dfrac{\partial \phi}{\partial \theta}$	**(6.33)**	Page 263

Case Study

The Blended Wing-Body Aircraft

The X-48B prototype in the full-scale NASA tunnel. (Credit: Boeing/Bob Ferguson)

Boeing Phantom Works has partnered with NASA and the U.S. Air Force Research Laboratory to study an advanced-concept, fuel-efficient, blended wing-body. It is called a blended wing-body (BWB) because it looks more like a modified triangular-shaped wing than traditional aircraft, which essentially consist of a tube and wing with a tail. The concept of a BWB actually goes back to the 1940s, but developments in composite materials and fly-by-wire controls are making it more feasible. Researchers have tested a 6.3-m wingspan (8.5 percent scale) prototype of the X-48B, a BWB aircraft that could have military and commercial applications. The next step is for NASA to flight-test a scale-model variant called X-48C. The X-48C will be used to examine how engines mounted to the rear and above the body help to shield the ground from engine noise on takeoff and approach. It also features tail fins for additional noise shielding and for flight control.

The big difference between BWB aircraft and the traditional tube-and-wing aircraft, apart from the fact that the tube is absorbed into the wing shape, is that it does not have a tail. Traditional aircraft need a tail for stability and control; the BWB uses a number of different multiple-control surfaces and possibly tail fins to control the vehicle. There will be a number of advantages to the BWB if it proves feasible. Because the entire structure generates lift, less power is needed for takeoff. Studies have also shown that BWB designs can fit into the 80-m envelope that is the current standard for airplane maneuver at airports. A BWB could carry up to 1000 people, making such a future U.S. product a challenge to Airbus's A380 and future stretched versions.

Apart from possible fuel savings of up to 30 percent due to improved streamlining, the interior of a commercial BWB airplane would be radically different from that of current airplanes. Passengers would enter a room like a movie theater rather than a cramped half-cylinder, there would be no windows (video screens would be connected to external cameras instead), and passengers would be seated in the large movie theater—like room (because seating is not only in the central core but also well out into the blended wings).

References

1. Shaw, R., "The Influence of Hole Dimensions on Static Pressure Measurements," *J. Fluid Mech., 7*, Part 4, April 1960, pp. 550–564.
2. Chue, S. H., "Pressure Probes for Fluid Measurement," *Progress in Aerospace Science, 16*, 2, 1975, pp. 147–223.
3. United Sensor Corporation, 3 Northern Blvd., Amherst, NH 03031.
4. Robertson, J. M., *Hydrodynamics in Theory and Application*. Englewood Cliffs, NJ: Prentice-Hall, 1965.
5. Streeter, V. L., *Fluid Dynamics*. New York: McGraw-Hill, 1948.
6. Vallentine, H. R., *Applied Hydrodynamics*. London: Butterworths, 1959.
7. Lamb, H., *Hydrodynamics*. New York: Dover, 1945.
8. Milne-Thomson, L. M., *Theoretical Hydrodynamics*, 4th ed. New York: Macmillan, 1960.
9. Karamcheti, K., *Principles of Ideal-Fluid Aerodynamics*. New York: Wiley, 1966.
10. Kirchhoff, R. H., *Potential Flows: Computer Graphic Solutions*. New York: Marcel Dekker, 1985.
11. Rouse, H., and S. Ince, *History of Hydraulics*. New York: Dover, 1957.
12. Kuethe, A. M., and C.-Y. Chow, *Foundations of Aerodynamics: Bases of Aerodynamic Design*, 4th ed. New York: Wiley, 1986.
13. *Fluent*. Fluent Incorporated, Centerra Resources Park, 10 Cavendish Court, Lebanon, NH 03766 (www.fluent.com).
14. *STAR-CD*. Adapco, 60 Broadhollow Road, Melville, NY 11747 (www.cd-adapco.com).

Problems

Euler's Equation

6.1 Consider the flow field with velocity given by $\vec{V} = [A(y^2 - x^2) - Bx]\hat{i} + [2Axy + By]\hat{j}$; $A = 3.28\ \text{m}^{-1}\cdot\text{s}^{-1}$, $B = 3.28\ \text{m}^{-1}\cdot\text{s}^{-1}$; the coordinates are measured in meters. The density is 1,030 kg/m^3, and gravity acts in the negative y direction. Calculate the acceleration of a fluid particle and the pressure gradient at point $(x, y) = (0.3, 0.3)$.

6.2 An incompressible frictionless flow field is given by $\vec{V} = (Ax + By)\hat{i} + (Bx - Ay)\hat{j}$, where $A = 2\ \text{s}^{-1}$ and $B = 2\ \text{s}^{-1}$, and the coordinates are measured in meters. Find the magnitude and direction of the acceleration of a fluid particle at point $(x, y) = (2, 2)$. Find the pressure gradient at the same point, if $\vec{g} = -g\hat{j}$ and the fluid is water.

6.3 A horizontal flow of water is described by the velocity field $\vec{V} = (-Ax + Bt)\hat{i} + (Ay + Bt)\hat{j}$, where $A = 1\ \text{s}^{-1}$ and $B = 2\ \text{m/s}^2$, x and y are in meters, and t is in seconds. Find expressions for the local acceleration, the convective acceleration, and the total acceleration. Evaluate these at point (1, 2) at $t = 5$ seconds. Evaluate ∇p at the same point and time.

6.4 A velocity field in a fluid with density of 1000 kg/m^3 is given by $\vec{V} = (-Ax + By)t\hat{i} + (Ay + Bx)t\hat{j}$, where $A = 2\ \text{s}^{-2}$ and $B = 1\ \text{s}^{-2}$, x and y are in meters, and t is in seconds. Body forces are negligible. Evaluate ∇p at point $(x, y) = (1, 1)$ at $t = 1$ s.

6.5 Consider the flow field with velocity given by $\vec{V} = [A(x^2 - y^2) - 3Bx]\hat{i} - [2Axy - 3By]\hat{j}$, where $A = 3.28\ \text{m}^{-1}\cdot\text{s}^{-1}$, $B = 1\ \text{s}^{-1}$, and the coordinates are measured in meters. The density is 1,030 kg/m^3 and gravity acts in the negative y direction. Determine the acceleration of a fluid particle and the pressure gradient at point $(x, y) = (1, 1)$.

6.6 The x component of velocity in an incompressible flow field is given by $u = Ax$, where $A = 2\ \text{s}^{-1}$ and the coordinates are measured in meters. The pressure at point $(x, y) = (0, 0)$ is $p_0 = 190$ kPa (gage). The density is $\rho = 1.50$ kg/m^3 and the z axis is vertical. Evaluate the simplest possible y component of velocity. Calculate the fluid acceleration and determine the pressure gradient at point $(x, y) = (2, 1)$. Find the pressure distribution along the positive x axis.

6.7 Consider the flow field with velocity given by $\vec{V} = Ax\sin(2\pi\omega t)\hat{i} - Ay\sin(2\pi\omega t)\hat{j}$, where $A = 2\ \text{s}^{-1}$ and $\omega = 1\ \text{s}^{-1}$. The fluid density is 2 kg/m^3. Find expressions for the local acceleration, the convective acceleration, and the total acceleration. Evaluate these at point (1, 1) at $t = 0$, 0.5, and 1 seconds. Evaluate ∇p at the same point and times.

6.8 The velocity field for a plane source located distance $h = 1$ m above an infinite wall aligned along the x axis is given by

$$\vec{V} = \frac{q}{2\pi[x^2 + (y-h)^2]}[x\hat{i} + (y-h)\hat{j}] + \frac{q}{2\pi[x^2 + (y+h)^2]}[x\hat{i} + (y+h)\hat{j}]$$

where $q = 2$ m^3/s/m. The fluid density is 1000 kg/m^3 and body forces are negligible. Derive expressions for the velocity and acceleration of a fluid particle that moves along the wall, and plot from $x = 0$ to $x = +10h$. Verify that the velocity and acceleration normal to the wall are zero. Plot the pressure gradient $\partial p/\partial x$ along the wall. Is the pressure gradient along the wall adverse (does it oppose fluid motion) or not?

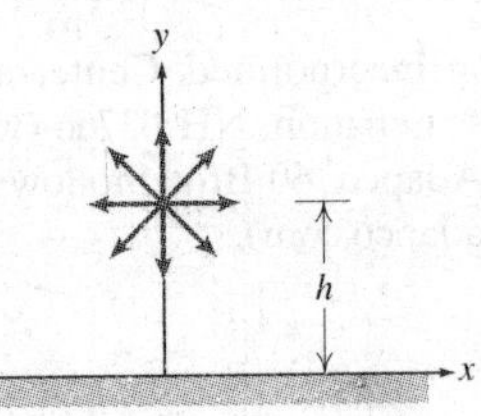

P6.8

6.9 The velocity distribution in a two-dimensional steady flow field in the xy plane is $\vec{V} = (Ax - B)\hat{i} + (C - Ay)\hat{j}$, where $A = 2\ \text{s}^{-1}$, $B = 5\ \text{m}\cdot\text{s}^{-1}$, and $C = 3\ \text{m}\cdot\text{s}^{-1}$; the coordinates are measured in meters, and the body force distribution is $\vec{g} = -g\hat{k}$. Does the velocity field represent the flow of an incompressible fluid? Find the stagnation point of the flow field. Obtain an expression for the pressure gradient in the flow field. Evaluate the difference in pressure between point $(x, y) = (1, 3)$ and the origin, if the density is 1.2 kg/m^3.

6.10 In a two-dimensional frictionless, incompressible ($\rho = 1500$ kg/m^3) flow, the velocity field in meters per second is given by $\vec{V} = (Ax + By)\hat{i} + (Bx - Ay)\hat{j}$; the coordinates are measured in meters, and $A = 4\ \text{s}^{-1}$ and $B = 2\ \text{s}^{-1}$. The pressure is $p_0 = 200$ kPa at point $(x, y) = (0, 0)$. Obtain an expression for the pressure field, $p(x, y)$ in terms of p_0, A, and B, and evaluate at point $(x, y) = (2, 2)$.

6.11 An incompressible liquid with a density of 1250 kg/m^3 and negligible viscosity flows steadily through a horizontal pipe of constant diameter. In a porous section of length $L = 5$ m, liquid is removed at a constant rate per unit length so that the uniform axial velocity in the pipe is $u(x) = U(1 - x/L)$, where $U = 15$ m/s. Develop expressions for and plot the pressure gradient along the centerline. Evaluate the outlet pressure if the pressure at the inlet to the porous section is 100 kPa (gage).

6.12 An incompressible liquid with a density of 900 kg/m^3 and negligible viscosity flows steadily through a horizontal pipe of constant diameter. In a porous section of length $L = 2$ m, liquid is removed at a variable rate along the length so that the uniform axial velocity in the pipe is $u(x) = Ue^{-x/L}$, where $U = 20$ m/s. Develop expressions for and plot the acceleration of a fluid particle along the centerline of the porous section and the pressure gradient along the centerline. Evaluate the outlet pressure if the pressure at the inlet to the porous section is 50 kPa (gage).

6.13 For the flow of Problem 4.123 show that the uniform radial velocity is $V_r = Q/2\pi rh$. Obtain expressions for the r component of acceleration of a fluid particle in the gap and for the pressure variation as a function of radial distance from the central holes.

6.14 The velocity field for a plane vortex sink is given by $\vec{V} = (-q/2\pi r)\hat{e}_r + (K/2\pi r)\hat{e}_\theta$, where $q = 2$ m^3/s/m and $K = 1$ m^3/s/m. The fluid density is 1000 kg/m^3. Find the acceleration at (1, 0), (1, $\pi/2$), and (2, 0). Evaluate ∇p under the same conditions.

6.15 An incompressible, inviscid fluid flows into a horizontal round tube through its porous wall. The tube is closed at the left end and the flow discharges from the tube to the atmosphere at the right end. For simplicity, consider the x component of velocity in the tube uniform across any cross section. The density of the fluid is ρ, the tube diameter and length are D and L, respectively, and the uniform inflow velocity is v_0. The flow is steady. Obtain an algebraic expression for the x component of acceleration of a fluid particle located at position x, in terms of v_0, x, and D. Find an expression for the pressure gradient, $\partial p/\partial x$, at position x. Integrate to obtain an expression for the gage pressure at $x = 0$.

6.16 An incompressible liquid with negligible viscosity and density $\rho = 850$ kg/m^3 flows steadily through a horizontal pipe. The pipe cross-section area linearly varies from 100 cm^2 to 25 cm^2 over a length of 2 m. Develop an expression for and plot the pressure gradient and pressure versus position along the pipe, if the inlet centerline velocity is 1 m/s and inlet pressure is 250 kPa. What is the exit pressure? *Hint:* Use relation

$$u\frac{\partial u}{\partial x} = \frac{1}{2}\frac{\partial}{\partial x}(u^2)$$

6.17 An incompressible liquid with negligible viscosity and density $\rho = 1250$ kg/m^3 flows steadily through a 5 m long convergent-divergent section of pipe for which the area varies as

$$A(x) = A_0(1 + e^{-x/a} - e^{-x/2a})$$

where $A_0 = 0.25$ m^2 and $a = 1.5$ m. Plot the area for the first 5 m. Develop an expression for and plot the pressure gradient and pressure versus position along the pipe, for the first 5 m, if the inlet centerline velocity is 10 m/s and inlet pressure is 300 kPa. *Hint:* Use relation

$$u\frac{\partial u}{\partial x} = \frac{1}{2}\frac{\partial}{\partial x}(u^2)$$

6.18 A nozzle for an incompressible, inviscid fluid of density $\rho = 1000 \text{ kg/m}^3$ consists of a converging section of pipe. At the inlet the diameter is $D_i = 100$ mm, and at the outlet the diameter is $D_o = 20$ mm. The nozzle length is $L = 500$ mm, and the diameter decreases linearly with distance x along the nozzle. Derive and plot the acceleration of a fluid particle, assuming uniform flow at each section, if the speed at the inlet is $V_i = 1$ m/s. Plot the pressure gradient through the nozzle, and find its maximum absolute value. If the pressure gradient must be no greater than 5 MPa/m in absolute value, how long would the nozzle have to be?

6.19 A diffuser for an incompressible, inviscid fluid of density $\rho = 1000 \text{ kg/m}^3$ consists of a diverging section of pipe. At the inlet the diameter is $D_i = 0.25$ m, and at the outlet the diameter is $D_o = 0.75$ m. The diffuser length is $L = 1$ m, and the diameter increases linearly with distance x along the diffuser. Derive and plot the acceleration of a fluid particle, assuming uniform flow at each section, if the speed at the inlet is $V_i = 5$ m/s. Plot the pressure gradient through the diffuser, and find its maximum value. If the pressure gradient must be no greater than 25 kPa/m, how long would the diffuser have to be?

6.20 Consider the flow of Problem 5.48. Evaluate the magnitude and direction of the net pressure force that acts on the upper plate between r_i and R, if $r_i = R/2$.

6.21 Consider again the flow field of Problem 5.65. Assume the flow is incompressible with $\rho = 1.23 \text{ kg/m}^3$ and friction is negligible. Further assume the vertical air flow velocity is $v_0 = 15$ mm/s, the half-width of the cavity is $L = 22$ mm, and its height is $h = 1.2$ mm. Calculate the pressure gradient at $(x, y) = (L, h)$. Obtain an equation for the flow streamlines in the cavity.

6.22 A liquid layer separates two plane surfaces as shown. The lower surface is stationary; the upper surface moves downward at constant speed V. The moving surface has width w, perpendicular to the plane of the diagram, and $w \gg L$. The incompressible liquid layer, of density ρ, is squeezed from between the surfaces. Assume the flow is uniform at any cross section and neglect viscosity as a first approximation. Use a suitably chosen control volume to show that $u = Vx/b$ within the gap, where $b = b_0 - Vt$. Obtain an algebraic expression for the acceleration of a fluid particle located at x. Determine the pressure gradient, $\partial p/\partial x$, in the liquid layer. Find the pressure distribution, $p(x)$. Obtain an expression for the net pressure force that acts on the upper (moving) flat surface.

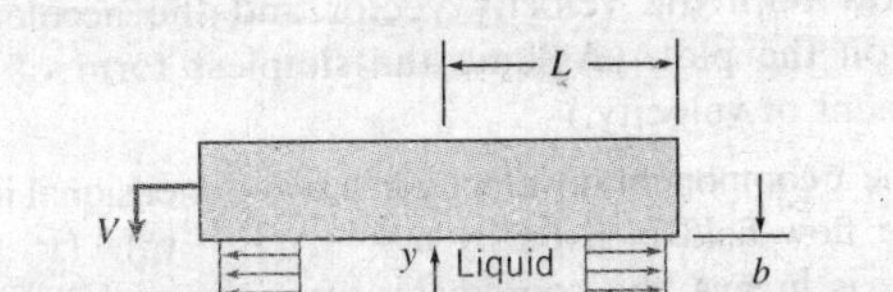

P6.22

6.23 A rectangular computer chip floats on a thin layer of air, $h = 0.5$ mm thick, above a porous surface. The chip width is $b = 40$ mm, as shown. Its length, L, is very long in the direction perpendicular to the diagram. There is no flow in the z direction. Assume flow in the x direction in the gap under the chip is uniform. Flow is incompressible, and frictional effects may be neglected. Use a suitably chosen control volume to show that $U(x) = qx/h$ in the gap. Find a general expression for the (2D) acceleration of a fluid particle in the gap in terms of $q, h, x,$ and y. Obtain an expression for the pressure gradient $\partial p/\partial x$. Assuming atmospheric pressure on the chip upper surface, find an expression for the net pressure force on the chip; is it directed upward or downward? Explain. Find the required flow rate q ($\text{m}^3/\text{s}/\text{m}^2$) and the maximum velocity, if the mass per unit length of the chip is 0.005 kg/m. Plot the pressure distribution as part of your explanation of the direction of the net force.

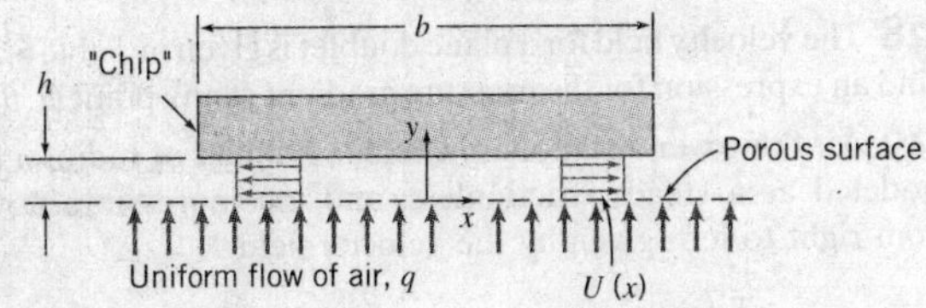

P6.23

6.24 Heavy weights can be moved with relative ease on air cushions by using a load pallet as shown. Air is supplied from the plenum through porous surface AB. It enters the gap vertically at uniform speed, q. Once in the gap, all air flows in the positive x direction (there is no flow across the plane at $x = 0$). Assume air flow in the gap is incompressible and uniform at each cross section, with speed $u(x)$, as shown in the enlarged view. Although the gap is narrow ($h \ll L$), neglect frictional effects as a first approximation. Use a suitably chosen control volume to show that $u(x) = qx/h$ in the gap. Calculate the acceleration of a fluid particle in the gap. Evaluate the pressure gradient, $\partial p/\partial x$, and sketch the pressure distribution within the gap. Be sure to indicate the pressure at $x = L$.

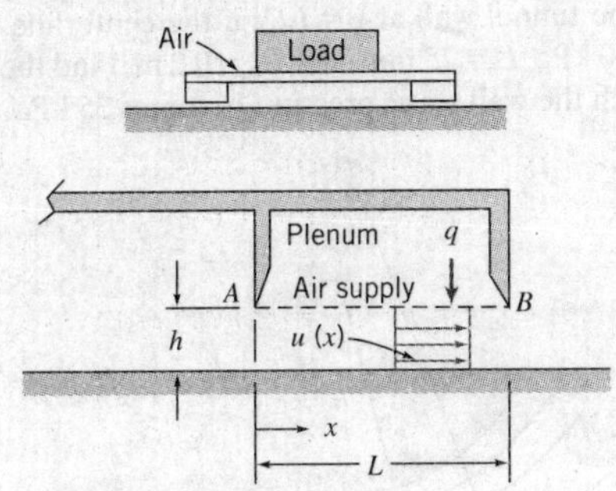

P6.24

6.25 A velocity field is given by $\vec{V} = [Ax^3 + Bxy^2]\hat{i} + [Ay^3 + Bx^2y]\hat{j}$; $A = 0.2 \text{ m}^{-2}\cdot\text{s}^{-1}$, B is a constant, and the coordinates are measured in meters. Determine the value and units for B if this velocity field is to represent an incompressible flow. Calculate the acceleration of a fluid particle at point $(x, y) = (2, 1)$. Evaluate the component of particle acceleration normal to the velocity vector at this point.

6.26 The y component of velocity in a two-dimensional incompressible flow field is given by $v = -Axy$, where v is in m/s, the coordinates are measured in meters, and $A = 1 \text{ m}^{-1}\cdot\text{s}^{-1}$. There is no velocity component or variation in the z direction. Calculate the acceleration of a fluid particle at point $(x, y) = (1, 2)$. Estimate the radius of curvature of the streamline passing through this point. Plot the streamline and show both the velocity vector and the acceleration vector on the plot. (Assume the simplest form of the x component of velocity.)

6.27 Consider the velocity field $\vec{V} = A[x^4 - 6x^2y^2 + y^4]\hat{i} + B[x^3y - xy^3]\hat{j}$; $A = 2\ \text{m}^{-3}\cdot\text{s}^{-1}$, B is a constant, and the coordinates are measured in meters. Find B for this to be an incompressible flow. Obtain the equation of the streamline through point $(x, y) = (1, 2)$. Derive an algebraic expression for the acceleration of a fluid particle. Estimate the radius of curvature of the streamline at $(x, y) = (1, 2)$.

6.28 The velocity field for a plane doublet is given in Table 6.2. Find an expression for the pressure gradient at any point (r, θ).

6.29 Air flow over a stationary circular cylinder of radius a is modeled as a steady, frictionless, and incompressible flow from right to left, given by the velocity field

$$\vec{V} = U\left[\left(\frac{a}{r}\right)^2 - 1\right]\cos\theta\,\hat{e}_r + U\left[\left(\frac{a}{r}\right)^2 + 1\right]\sin\theta\,\hat{e}_\theta$$

Consider flow along the streamline forming the cylinder surface, $r = a$. Express the components of the pressure gradient in terms of angle θ. Obtain an expression for the variation of pressure (gage) on the surface of the cylinder. For $U = 75$ m/s and $a = 150$ mm, plot the pressure distribution (gage) and explain, and find the minimum pressure. Plot the speed V as a function of r along the radial line $\theta = \pi/2$ for $r > a$ (that is, directly above the cylinder), and explain.

6.30 To model the velocity distribution in the curved inlet section of a water channel, the radius of curvature of the streamlines is expressed as $R = LR_0/2y$. As an approximation, assume the water speed along each streamline is $V = 10$ m/s. Find an expression for and plot the pressure distribution from $y = 0$ to the tunnel wall at $y = L/2$, if the centerline pressure (gage) is 50 kPa, $L = 75$ mm, and $R_0 = 0.2$ m. Find the value of V for which the wall static pressure becomes 35 kPa.

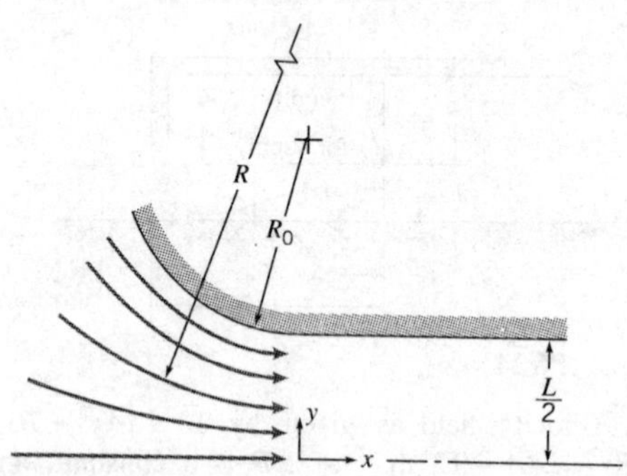

P6.30

6.31 Air at 138 kPa (abs) and 38°C flows around a smooth corner at the inlet to a diffuser. The air speed is 46 m/s, and the radius of curvature of the streamlines is 75 mm. Determine the magnitude of the centripetal acceleration experienced by a fluid particle rounding the corner. Express your answer in gs. Evaluate the pressure gradient, $\partial p/\partial r$.

6.32 Repeat Example 6.1, but with the somewhat more realistic assumption that the flow is similar to a free vortex (irrotational) profile, $V_\theta = c/r$ (where c is a constant), as shown in Fig. P6.32. In doing so, prove that the flow rate is given by $Q = k\sqrt{\Delta p}$, where k is

$$k = w \ln\left(\frac{r_2}{r_1}\right)\sqrt{\frac{2r_2^2r_1^2}{\rho(r_2^2 - r_1^2)}}$$

and w is the depth of the bend.

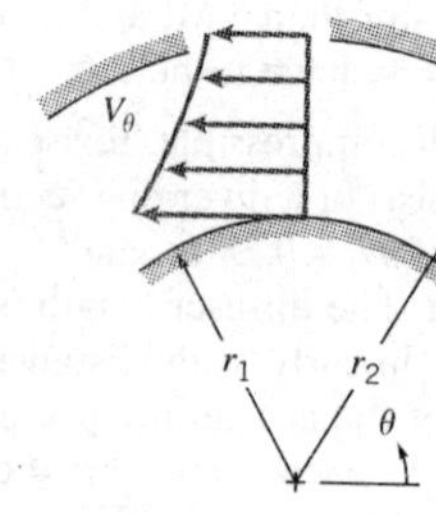

P6.32

6.33 The velocity field in a two-dimensional, steady, inviscid flow field in the horizontal xy plane is given by $\vec{V} = (Ax + B)\hat{i} - Ay\hat{j}$, where $A = 1\ \text{s}^{-1}$ and $B = 2$ m/s; x and y are measured in meters. Show that streamlines for this flow are given by $(x + B/A)y = \text{constant}$. Plot streamlines passing through points $(x, y) = (1, 1)$, $(1, 2)$, and $(2, 2)$. At point $(x, y) = (1, 2)$, evaluate and plot the acceleration vector and the velocity vector. Find the component of acceleration along the streamline at the same point; express it as a vector. Evaluate the pressure gradient along the streamline at the same point if the fluid is air. What statement, if any, can you make about the relative value of the pressure at points (1, 1) and (2, 2)?

6.34 Using the analyses of Example 6.1 and Problem 6.32, plot the discrepancy (percent) between the flow rates obtained from assuming uniform flow and the free vortex (irrotational) profile as a function of r_2/r_1.

6.35 The x component of velocity in a two-dimensional, incompressible flow field is given by $u = Ax^2$; the coordinates are measured in meters and $A = 3.28\ \text{m}^{-1}\cdot\text{s}^{-1}$. There is no velocity component or variation in the z direction. Calculate the acceleration of a fluid particle at point $(x, y) = (0.3, 0.6)$. Estimate the radius of curvature of the streamline passing through this point. Plot the streamline and show both the velocity vector and the acceleration vector on the plot. (Assume the simplest form of the y component of velocity.)

6.36 The x component of velocity in a two-dimensional incompressible flow field is given by $u = -\Lambda(x^2 - y^2)/(x^2 + y^2)^2$, where u is in m/s, the coordinates are measured in meters, and $\Lambda = 2\ \text{m}^3\cdot\text{s}^{-1}$. Show that the simplest form of the y component of velocity is given by $v = -2\Lambda xy/(x^2 + y^2)^2$. There is no velocity component or variation in the z direction. Calculate the acceleration of fluid particles at points $(x, y) = (0, 1)$, $(0, 2)$, and $(0, 3)$. Estimate the radius of curvature of the streamlines passing through these points. What does the relation among the three points and their radii of curvature suggest to you about the flow field? Verify this by plotting these streamlines. [*Hint*: You will need to use an integrating factor.]

6.37 The x component of velocity in a two-dimensional, incompressible flow field is given by $u = Axy$; the coordinates are measured in meters and $A = 2\ \text{m}^{-1} \cdot \text{s}^{-1}$. There is no velocity component or variation in the z direction. Calculate the acceleration of a fluid particle at point $(x, y) = (2, 1)$. Estimate the radius of curvature of the streamline passing through this point. Plot the streamline and show both the velocity vector and the acceleration vector on the plot. (Assume the simplest form of the y component of velocity.)

The Bernoulli Equation

6.38 Water flows at a speed of 3 m/s. Calculate the dynamic pressure of this flow. Express your answer in inches of mercury.

6.39 Calculate the dynamic pressure that corresponds to a speed of 100 km/hr in standard air. Express your answer in millimeters of water.

6.40 Plot the speed of air versus the dynamic pressure (in millimeters of mercury), up to a dynamic pressure of 250 mm Hg.

6.41 You present your open hand out of the window of an automobile perpendicular to the airflow. Assuming for simplicity that the air pressure on the entire front surface is stagnation pressure (with respect to automobile coordinates), with atmospheric pressure on the rear surface, estimate the net force on your hand when driving at (a) 48 km/h and (b) 96 km/h. Do these results roughly correspond with your experience? Do the simplifications tend to make the calculated force an over- or underestimate?

6.42 A jet of air from a nozzle is blown at right angles against a wall in which two pressure taps are located. A manometer connected to the tap directly in front of the jet shows a head of 25 mm of mercury above atmospheric. Determine the approximate speed of the air leaving the nozzle if it is at $-10°$C and 200 kPa. At the second tap a manometer indicates a head of 5 mm of mercury above atmospheric; what is the approximate speed of the air there?

6.43 A pitot-static tube is used to measure the speed of air at standard conditions at a point in a flow. To ensure that the flow may be assumed incompressible for calculations of engineering accuracy, the speed is to be maintained at 100 m/s or less. Determine the manometer deflection, in millimeters of water, that corresponds to the maximum desirable speed.

6.44 The inlet contraction and test section of a laboratory wind tunnel are shown. The air speed in the test section is $U = 50$ m/s. A total-head tube pointed upstream indicates that the stagnation pressure on the test section centerline is 10 mm of water below atmospheric. The laboratory is maintained at atmospheric pressure and a temperature of $-5°$C. Evaluate the dynamic pressure on the centerline of the wind tunnel test section. Compute the static pressure at the same point. Qualitatively compare the static pressure at the tunnel wall with that at the centerline. Explain why the two may not be identical.

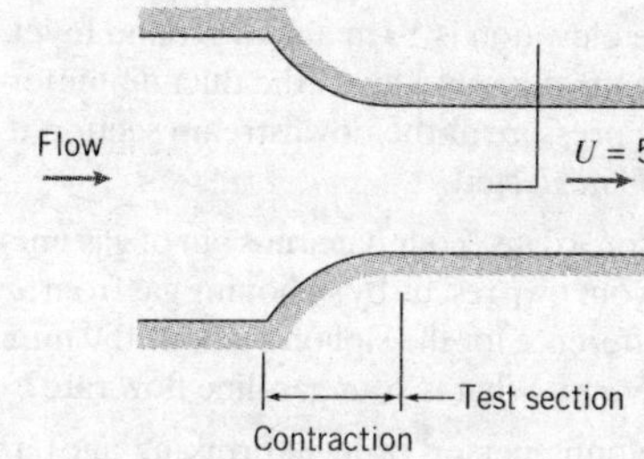

P6.44

6.45 Maintenance work on high-pressure hydraulic systems requires special precautions. A small leak can result in a high-speed jet of hydraulic fluid that can penetrate the skin and cause serious injury (therefore troubleshooters are cautioned to use a piece of paper or cardboard, *not a finger*, to search for leaks). Calculate and plot the jet speed of a leak versus system pressure, for pressures up to 40 MPa (gage). Explain how a high-speed jet of hydraulic fluid can cause injury.

6.46 An open-circuit wind tunnel draws in air from the atmosphere through a well-contoured nozzle. In the test section, where the flow is straight and nearly uniform, a static pressure tap is drilled into the tunnel wall. A manometer connected to the tap shows that static pressure within the tunnel is 45 mm of water below atmospheric. Assume that the air is incompressible, and at 25°C, 100 kPa (abs). Calculate the air speed in the wind-tunnel test section.

6.47 The wheeled cart shown in Problem 4.128 rolls with negligible resistance. The cart is to accelerate to the right. The jet speed is $V = 40$ m/s. The jet area remains constant at $A = 25\ \text{mm}^2$. Neglect viscous forces between the water and vane. When the cart attains speed $U = 15$ m/s, calculate the stagnation pressure of the water leaving the nozzle with respect to a fixed observer, the stagnation pressure of the water jet leaving the nozzle with respect to an observer on the vane, the absolute velocity of the jet leaving the vane with respect to a fixed observer, and the stagnation pressure of the jet leaving the vane with respect to a fixed observer. How would viscous forces affect the latter stagnation pressure, i.e., would viscous forces increase, decrease, or leave unchanged this stagnation pressure? Justify your answer.

6.48 Water flows steadily up the vertical 0.1m diameter pipe and out the nozzle, which is 0.05 m in diameter, discharging to atmospheric pressure. The stream velocity at the nozzle exit must be 20 m/s. Calculate the minimum gage pressure required at section ①. If the device were inverted, what would be the required minimum pressure at section ① to maintain the nozzle exit velocity at 20 m/s?

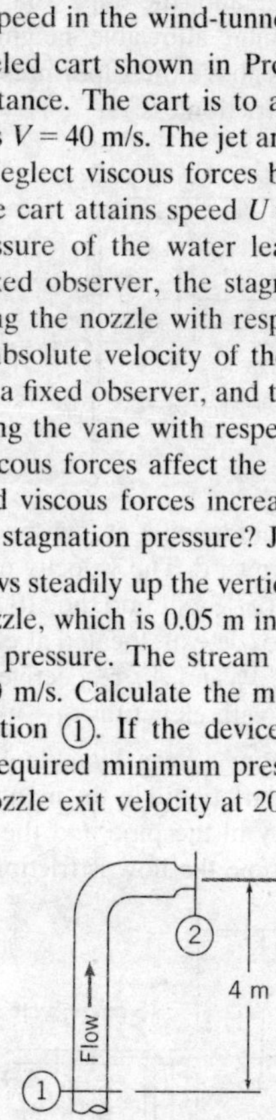

P6.48

6.49 Water flows in a circular duct. At one section the diameter is 0.3 m, the static pressure is 260 kPa (gage), the velocity is

3 m/s, and the elevation is 10 m above ground level. At a section downstream at ground level, the duct diameter is 0.15 m. Find the gage pressure at the downstream section if frictional effects may be neglected.

6.50 You are on a date. Your date runs out of gas unexpectedly. You come to your own rescue by siphoning gas from another car. The height difference for the siphon is about 150 mm. The hose diameter is 25 mm. What is your gasoline flow rate?

6.51 You (a young person of legal drinking age) are making homemade beer. As part of the process you have to siphon the wort (the fermenting beer with sediment at the bottom) into a clean tank using a 5-mm ID tubing. Being a young engineer, you're curious about the flow you can produce. Find an expression for and plot the flow rate Q (liters per minute) versus the differential in height h (millimeters) between the wort free surface and the location of the hose exit. Find the value of h for which $Q = 2$ L/min.

6.52 A tank at a pressure of 50 kPa (gage) gets a pinhole rupture and benzene shoots into the air. Ignoring losses, to what height will the benzene rise?

6.53 A can of Coke (you're not sure if it's diet or regular) has a small pinhole leak in it. The Coke sprays vertically into the air to a height of 0.5 m. What is the pressure inside the can of Coke? (Estimate for both kinds of Coke.)

6.54 The water flow rate through the siphon is 5 L/s, its temperature is 20°C, and the pipe diameter is 25 mm. Compute the maximum allowable height, h, so that the pressure at point A is above the vapor pressure of the water. (Assume the flow is frictionless.)

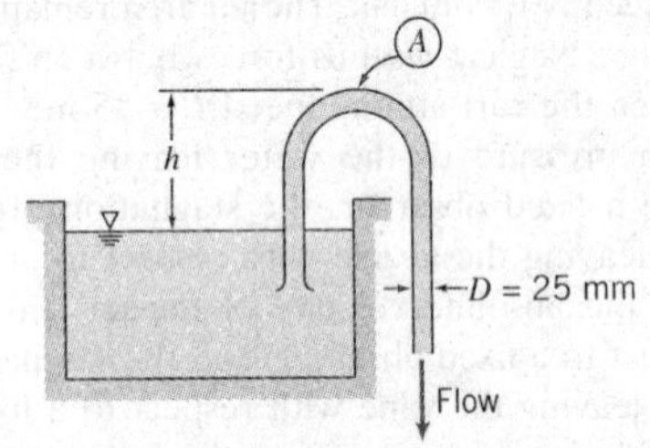

P6.54

6.55 A stream of liquid moving at low speed leaves a nozzle pointed directly downward. The velocity may be considered uniform across the nozzle exit and the effects of friction may be ignored. At the nozzle exit, located at elevation z_0, the jet velocity and area are V_0 and A_0, respectively. Determine the variation of jet area with elevation.

6.56 Water flows from a very large tank through a 5 cm diameter tube. The dark liquid in the manometer is mercury. Estimate the velocity in the pipe and the rate of discharge from the tank. (Assume the flow is frictionless.)

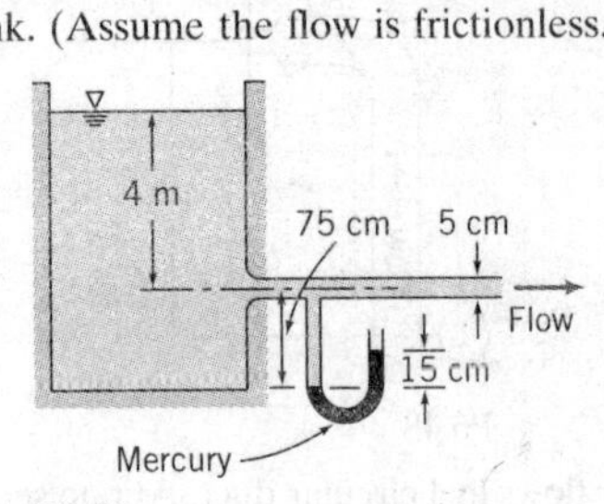

P6.56

6.57 In a laboratory experiment, water flows radially outward at moderate speed through the space between circular plane parallel disks. The perimeter of the disks is open to the atmosphere. The disks have diameter $D = 150$ mm and the spacing between the disks is $h = 0.8$ mm. The measured mass flow rate of water is $\dot{m} = 305$ g/s. Assuming frictionless flow in the space between the disks, estimate the theoretical static pressure between the disks at radius $r = 50$ mm. In the laboratory situation, where *some* friction is present, would the pressure measured at the same location be above or below the theoretical value? Why?

6.58 Consider frictionless, incompressible flow of air over the wing of an airplane flying at 200 km/hr. The air approaching the wing is at 65 kPa and −10°C. At a certain point in the flow, the pressure is 60 kPa. Calculate the speed of the air relative to the wing at this point and the absolute air speed.

6.59 A speedboat on hydrofoils is moving at 20 m/s in a freshwater lake. Each hydrofoil is 3 m below the surface. Assuming, as an approximation, frictionless, incompressible flow, find the stagnation pressure (gage) at the front of each hydrofoil. At one point on a hydrofoil, the pressure is −75 kPa (gage). Calculate the speed of the water relative to the hydrofoil at this point and the absolute water speed.

6.60 A mercury barometer is carried in a car on a day when there is no wind. The temperature is 20°C and the corrected barometer height is 761 mm of mercury. One window is open slightly as the car travels at 105 km/hr. The barometer reading in the moving car is 5 mm lower than when the car is stationary. Explain what is happening. Calculate the local speed of the air flowing past the window, *relative to* the automobile.

6.61 A fire nozzle is coupled to the end of a hose with inside diameter $D = 75$ mm. The nozzle is contoured smoothly and has outlet diameter $d = 25$ mm. The design inlet pressure for the nozzle is $p_1 = 690$ kPa (gage). Evaluate the maximum flow rate the nozzle could deliver.

6.62 A racing car travels at 98.3 m/s along a straightaway. The team engineer wishes to locate an air inlet on the body of the car to obtain cooling air for the driver's suit. The plan is to place the inlet at a location where the air speed is 25.5 m/s along the surface of the car. Calculate the static pressure at the proposed inlet location. Express the pressure rise above ambient as a fraction of the freestream dynamic pressure.

6.63 Steady, frictionless, and incompressible flow from left to right over a stationary circular cylinder, of radius a, is represented by the velocity field

$$\vec{V} = U\left[1 - \left(\frac{a}{r}\right)^2\right]\cos\theta\,\hat{e}_r - U\left[1 + \left(\frac{a}{r}\right)^2\right]\sin\theta\,\hat{e}_\theta$$

Obtain an expression for the pressure distribution along the streamline forming the cylinder surface, $r = a$. Determine the locations where the static pressure on the cylinder is equal to the freestream static pressure.

6.64 The velocity field for a plane source at a distance h above an infinite wall aligned along the x axis was given in Problem 6.8. Using the data from that problem, plot the pressure distribution along the wall from $x = -10h$ to $x = +10h$ (assume the pressure at infinity is atmospheric). Find the net force on the wall if the pressure on the lower surface is atmospheric. Does the force tend to pull the wall towards the source, or push it away?

6.65 The velocity field for a plane doublet is given in Table 6.2. If $\Lambda = 3\ m^3 \cdot s^{-1}$, the fluid density is $\rho = 1.5\ kg/m^3$, and the pressure at infinity is 100 kPa, plot the pressure along the x axis from $x = -2.0$ m to -0.5 m and $x = 0.5$ m to 2.0 m.

6.66 A smoothly contoured nozzle, with outlet diameter $d = 20$ mm, is coupled to a straight pipe by means of flanges. Water flows in the pipe, of diameter $D = 50$ mm, and the nozzle discharges to the atmosphere. For steady flow and neglecting the effects of viscosity, find the volume flow rate in the pipe corresponding to a calculated axial force of 45.5 N needed to keep the nozzle attached to the pipe.

6.67 A fire nozzle is coupled to the end of a hose with inside diameter $D = 75$ mm. The nozzle is smoothly contoured and its outlet diameter is $d = 25$ mm. The nozzle is designed to operate at an inlet water pressure of 700 kPa (gage). Determine the design flow rate of the nozzle. (Express your answer in L/s.) Evaluate the axial force required to hold the nozzle in place. Indicate whether the hose coupling is in tension or compression.

6.68 Water flows steadily through a 82 mm diameter pipe and discharges through a 32 mm diameter nozzle to atmospheric pressure. The flow rate is 93 L/min. Calculate the minimum static pressure required in the pipe to produce this flow rate. Evaluate the axial force of the nozzle assembly on the pipe flange.

6.69 Water flows steadily through the reducing elbow shown. The elbow is smooth and short, and the flow accelerates; so the effect of friction is small. The volume flow rate is $Q = 2.5$ L/s. The elbow is in a horizontal plane. Estimate the gage pressure at section ①. Calculate the x component of the force exerted by the reducing elbow on the supply pipe.

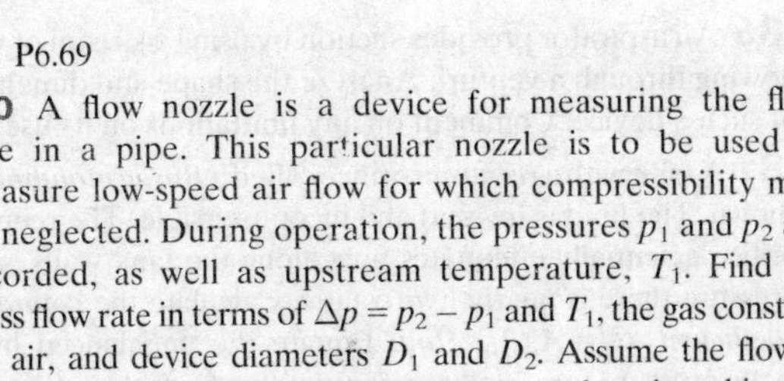

P6.69

6.70 A flow nozzle is a device for measuring the flow rate in a pipe. This particular nozzle is to be used to measure low-speed air flow for which compressibility may be neglected. During operation, the pressures p_1 and p_2 are recorded, as well as upstream temperature, T_1. Find the mass flow rate in terms of $\Delta p = p_2 - p_1$ and T_1, the gas constant for air, and device diameters D_1 and D_2. Assume the flow is frictionless. Will the actual flow be more or less than this predicted flow? Why?

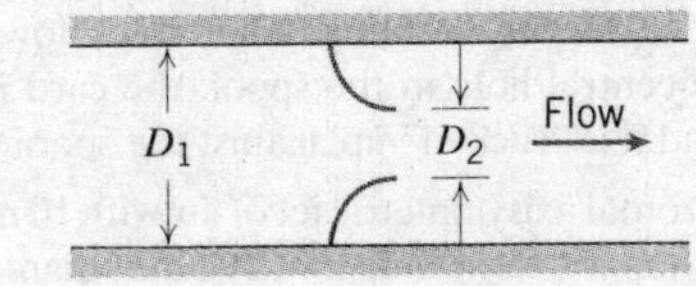

P6.70

6.71 The branching of a blood vessel is shown. Blood at a pressure of 140 mm Hg flows in the main vessel at 4.5 L/min. Estimate the blood pressure in each branch, assuming that blood vessels behave as rigid tubes, that we have frictionless flow, and that the vessel lies in the horizontal plane. What is the force generated at the branch by the blood? You may approximate blood to have a density of 1060 kg/m³.

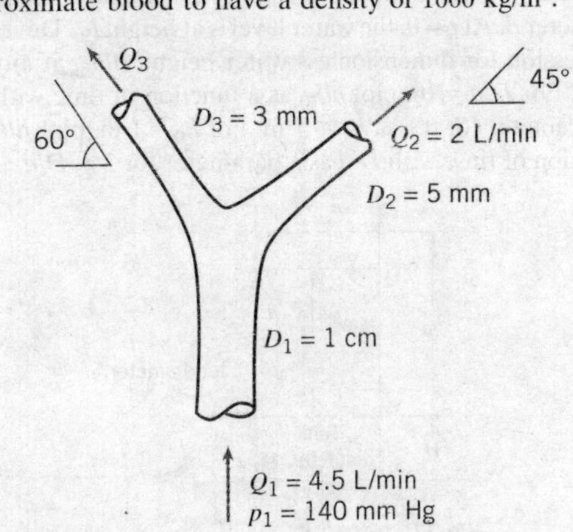

P6.71

6.72 An object, with a flat horizontal lower surface, moves downward into the jet of the spray system of Problem 4.81 with speed $U = 1.5$ m/s. Determine the minimum supply pressure needed to produce the jet leaving the spray system at $V = 4.6$ m/s. Calculate the maximum pressure exerted by the liquid jet on the flat object at the instant when the object is $h = 0.46$ m above the jet exit. Estimate the force of the water jet on the flat object.

6.73 A water jet is directed upward from a well-designed nozzle of area $A_1 = 600\ mm^2$; the exit jet speed is $V_1 = 6.3$ m/s. The flow is steady and the liquid stream does not break up. Point ② is located $H = 1.55$ m above the nozzle exit plane. Determine the velocity in the undisturbed jet at point ②. Calculate the pressure that would be sensed by a stagnation tube located there. Evaluate the force that would be exerted on a flat plate placed normal to the stream at point ②. Sketch the pressure distribution on the plate.

6.74 Water flows out of a kitchen faucet of 1.25 cm diameter at the rate of 0.1 L/s. The bottom of the sink is 45 cm below the faucet outlet. Will the cross-sectional area of the fluid stream increase, decrease, or remain constant between the faucet outlet and the bottom of the sink? Explain briefly. Obtain an expression for the stream cross section as a function of distance y above the sink bottom. If a plate is held directly under the faucet, how will the force required to hold the plate in a horizontal position vary with height above the sink? Explain briefly.

6.75 An old magic trick uses an empty thread spool and a playing card. The playing card is placed against the bottom of the spool. Contrary to intuition, when one blows downward through the central hole in the spool, the card is not blown away. Instead it is "sucked" up against the spool. Explain.

6.76 A horizontal axisymmetric jet of air with 10 mm diameter strikes a stationary vertical disk of 190 mm diameter. The jet speed is 69 m/s at the nozzle exit. A manometer is connected to the center of the disk. Calculate (a) the deflection, if the manometer liquid has SG = 1.75, (b) the force exerted by the jet on the disk, and (c) the force exerted on the disk if it is assumed that the stagnation pressure acts on the entire forward surface of the disk. Sketch the streamline pattern and plot the distribution of pressure on the face of the disk.

6.77 The tank, of diameter D, has a well-rounded nozzle with diameter d. At $t = 0$, the water level is at height h_0. Develop an expression for dimensionless water height, h/h_0, at any later time. For $D/d = 10$, plot h/h_0 as a function of time with h_0 as a parameter for $0.1 \leq h_0 \leq 1$ m. For $h_0 = 1$ m, plot h/h_0 as a function of time with D/d as a parameter for $2 \leq D/d \leq 10$.

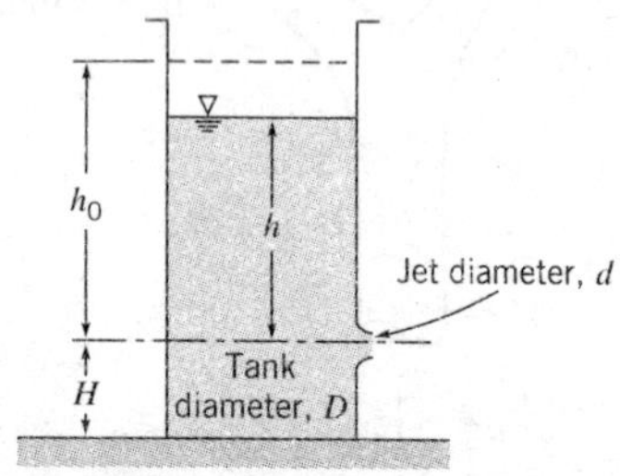

P6.77

6.78 The water level in a large tank is maintained at height H above the surrounding level terrain. A rounded nozzle placed in the side of the tank discharges a horizontal jet. Neglecting friction, determine the height h at which the orifice should be placed so the water strikes the ground at the maximum horizontal distance X from the tank. Plot jet speed V and distance X as functions of h $(0 < h < H)$.

6.79 The flow over a Quonset hut may be approximated by the velocity distribution of Problem 6.63 with $0 \leq \theta \leq \pi$. During a storm the wind speed reaches 100 km/hr; the outside temperature is 5°C. A barometer inside the hut reads 720 mm of mercury; pressure p_∞ is also 720 mm Hg. The hut has a diameter of 6 m and a length of 18 m. Determine the net force tending to lift the hut off its foundation.

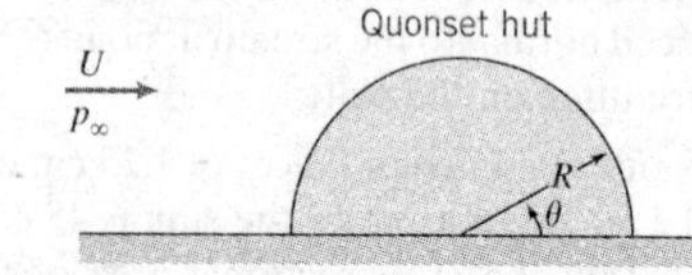

P6.79

6.80 Many recreation facilities use inflatable "bubble" structures. A tennis bubble to enclose four courts is shaped roughly as a circular semicylinder with a diameter of 30 m and a length of 70 m. The blowers used to inflate the structure can maintain the air pressure inside the bubble at 10 mm of water above ambient pressure. The bubble is subjected to a wind that blows at 60 km/hr in a direction perpendicular to the axis of the semicylindrical shape. Using polar coordinates, with angle θ measured from the ground on the upwind side of the structure, the resulting pressure distribution may be expressed as

$$\frac{p - p_\infty}{\frac{1}{2}\rho V_w^2} = 1 - 4\sin^2\theta$$

where p is the pressure at the surface, p_∞ the atmospheric pressure, and V_w the wind speed. Determine the net vertical force exerted on the structure.

6.81 High-pressure air forces a stream of water from a tiny, rounded orifice, of area A, in a tank. The pressure is high enough that gravity may be neglected. The air expands slowly, so that the expansion may be considered isothermal. The initial volume of air in the tank is $\mathcal{V}_0$. At later instants the volume of air is $\mathcal{V}(t)$; the total volume of the tank is $\mathcal{V}_t$. Obtain an algebraic expression for the mass flow rate of water leaving the tank. Find an algebraic expression for the rate of change in mass of the water inside the tank. Develop an ordinary differential equation and solve for the water mass in the tank at any instant. If $\mathcal{V}_0 = 5\ \text{m}^3$, $\mathcal{V}_t = 10\ \text{m}^3$, $A = 25\ \text{mm}^2$, and $p_0 = 1$ MPa, plot the water mass in the tank versus time for the first forty minutes.

6.82 Water flows at low speed through a circular tube with inside diameter of 50 mm. A smoothly contoured body of 38 mm. diameter is held in the end of the tube where the water discharges to atmosphere. Neglect frictional effects and assume uniform velocity profiles at each section. Determine the pressure measured by the gage and the force required to hold the body.

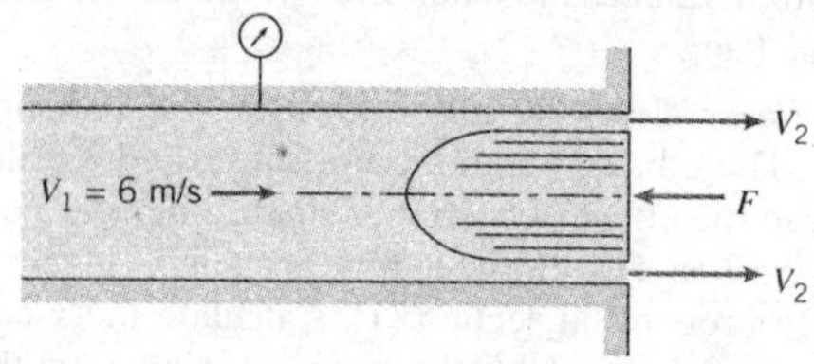

P6.82

6.83 Repeat Problem 6.81 assuming the air expands so rapidly that the expansion may be treated as adiabatic.

6.84 Describe the pressure distribution on the exterior of a multistory building in a steady wind. Identify the locations of the maximum and minimum pressures on the outside of the building. Discuss the effect of these pressures on infiltration of outside air into the building.

6.85 Imagine a garden hose with a stream of water flowing out through a nozzle. Explain why the end of the hose may be unstable when held a half meter or so from the nozzle end.

6.86 An aspirator provides suction by using a stream of water flowing through a venturi. Analyze the shape and dimensions of such a device. Comment on any limitations on its use.

6.87 A tank with a *reentrant* orifice called a *Borda mouthpiece* is shown. The fluid is inviscid and incompressible. The reentrant orifice essentially eliminates flow along the tank walls, so the pressure there is nearly hydrostatic. Calculate the *contraction coefficient*, $C_c = A_j/A_0$. *Hint:* Equate the unbalanced hydrostatic pressure force and momentum flux from the jet.

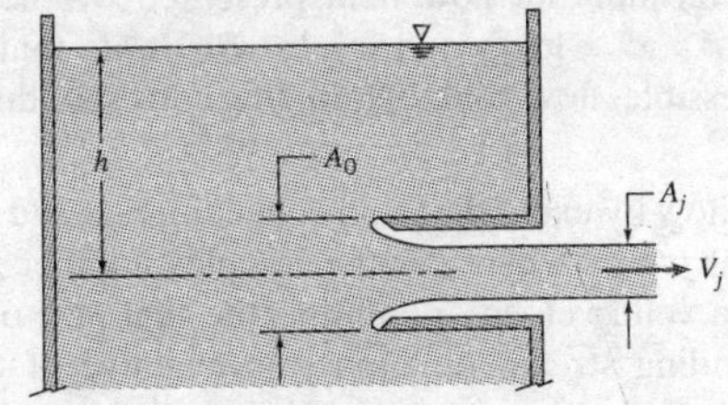

P6.87

Unsteady Bernoulli Equation

***6.88** Apply the unsteady Bernoulli equation to the U-tube manometer of constant diameter shown. Assume that the manometer is initially deflected and then released. Obtain a differential equation for l as a function of time.

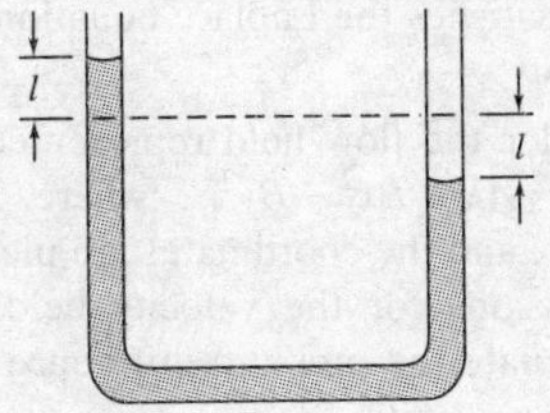

P6.88

***6.89** Compressed air is used to accelerate water from a tube. Neglect the velocity in the reservoir and assume the flow in the tube is uniform at any section. At a particular instant, it is known that $V = 1.8$ m/s and $dV/dt = 2.3$ m/s^2. The cross-sectional area of the tube is $A = 20{,}645$ mm^2. Determine the pressure in the tank at this instant.

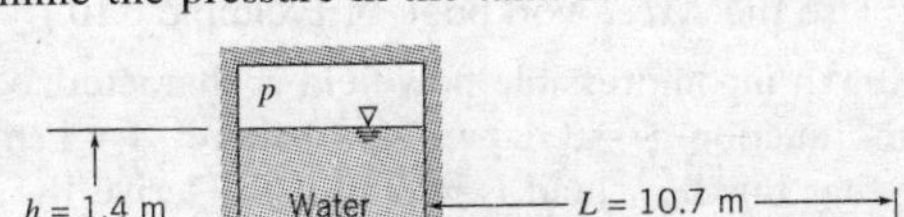

P6.89, P6.90, P6.92

***6.90** If the water in the pipe in Problem 6.89 is initially at rest and the air pressure is 21 kPa (gage), what will be the initial acceleration of the water in the pipe?

***6.91** Consider the reservoir and disk flow system with the reservoir level maintained constant. Flow between the disks is started from rest at $t = 0$. Evaluate the rate of change of volume flow rate at $t = 0$, if $r_1 = 50$ mm.

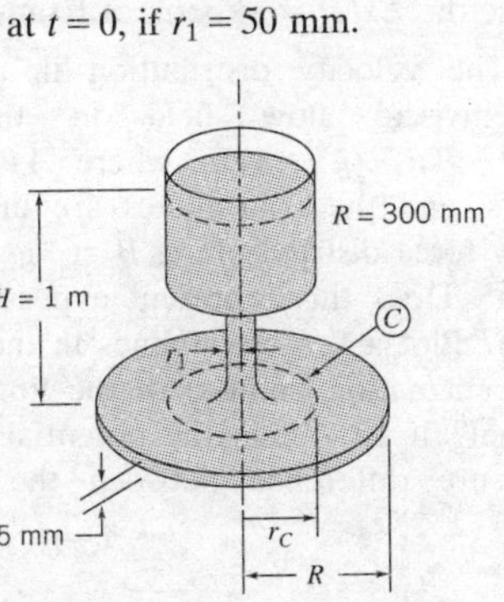

P6.91

***6.92** If the water in the pipe of Problem 6.89 is initially at rest, and the air pressure is maintained at 10.3 kPa (gage), derive a differential equation for the velocity V in the pipe as a function of time, integrate, and plot V versus t for $t = 0$ to 5 s.

***6.93** Consider the tank of Problem 4.48. Using the Bernoulli equation for unsteady flow along a streamline, evaluate the minimum diameter ratio, D/d, required to justify the assumption that flow from the tank is quasi-steady.

***6.94** Two circular disks, of radius R, are separated by distance b. The upper disk moves toward the lower one at constant speed V. The space between the disks is filled with a frictionless, incompressible fluid, which is squeezed out as the disks come together. Assume that, at any radial section, the velocity is uniform across the gap width b. However, note that b is a function of time. The pressure surrounding the disks is atmospheric. Determine the gage pressure at $r = 0$.

Energy Grade Line And Hydraulic Grade Line

6.95 Carefully sketch the energy grade lines (EGL) and hydraulic grade lines (HGL) for the system shown in Fig. 6.6 if the pipe is horizontal (i.e., the outlet is at the base of the reservoir), and a water turbine (extracting energy) is located at point ②, or at point ③. In Chapter 8 we will investigate the effects of friction on internal flows. Can you anticipate and sketch the effect of friction on the EGL and HGL for the two cases?

6.96 Carefully sketch the energy grade lines (EGL) and hydraulic grade lines (HGL) for the system shown in Fig. 6.6 if a pump (adding energy to the fluid) is located at point ②, or at point ③, such that flow is *into* the reservoir. In Chapter 8 we will investigate the effects of friction on internal flows. Can you anticipate and sketch the effect of friction on the EGL and HGL for the two cases?

Irrotational Flow

***6.97** Determine whether the Bernoulli equation can be applied between different radii for the vortex flow fields (a) $\vec{V} = \omega r\, \hat{e}_\theta$ and (b) $\vec{V} = \hat{e}_\theta K/2\pi r$.

***6.98** Consider a two-dimensional fluid flow: $u = ax + by$ and $v = cx + dy$, where a, b, c and d are constant. If the flow is incompressible and irrotational, find the relationships among a, b, c, and d. Find the stream function and velocity potential function of this flow.

***6.99** Consider the flow represented by the stream function $\psi = Ax^2y$, where A is a dimensional constant equal to 2.5 m$^{-1}\cdot$s^{-1}. The density is 1200 kg/m^3. Is the flow rotational? Can the pressure difference between points $(x, y) = (1, 4)$ and $(2, 1)$ be evaluated? If so, calculate it, and if not, explain why.

***6.100** The velocity field for a two-dimensional flow is $\vec{V} = (Ax - By)t\hat{i} - (Bx + Ay)t\hat{j}$, where $A = 1$ s^{-2} $B = 2$ s^{-2}, t is in seconds, and the coordinates are measured in meters. Is this a possible incompressible flow? Is the flow steady or unsteady? Show that the flow is irrotational and derive an expression for the velocity potential.

***6.101** Using Table 6.2, find the stream function and velocity potential for a plane source, of strength q, near a 90° corner.

*These problems require material from sections that may be omitted without loss of continuity in the text material.

The source is equidistant h from each of the two infinite planes that make up the corner. Find the velocity distribution along one of the planes, assuming $p = p_0$ at infinity. By choosing suitable values for q and h, plot the streamlines and lines of constant velocity potential. (*Hint*: Use the *Excel* workbook of Example 6.10.)

***6.102** The flow field for a plane source at a distance h above an infinite wall aligned along the x axis is given by

$$\vec{V} = \frac{q}{2\pi[x^2 + (y-h)^2]}[x\hat{i} + (y-h)\hat{j}] + \frac{q}{2\pi[x^2 + (y+h)^2]}[x\hat{i} + (y+h)\hat{j}]$$

where q is the strength of the source. The flow is irrotational and incompressible. Derive the stream function and velocity potential. By choosing suitable values for q and h, plot the streamlines and lines of constant velocity potential. (*Hint*: Use the *Excel* workbook of Example 6.10.)

***6.103** Using Table 6.2, find the stream function and velocity potential for a plane vortex, of strength K, near a 90° corner. The vortex is equidistant h from each of the two infinite planes that make up the corner. Find the velocity distribution along one of the planes, assuming $p = p_0$ at infinity. By choosing suitable values for K and h, plot the streamlines and lines of constant velocity potential. (*Hint*: Use the *Excel* workbook of Example Problem 6.10.)

***6.104** The stream function of a flow field is $\psi = Ax^2y - By^3$, where $A = 1\ \text{m}^{-1}\cdot\text{s}^{-1}$, $B = \frac{1}{3}\ \text{m}^{-1}\cdot\text{s}^{-1}$, and the coordinates are measured in meters. Find an expression for the velocity potential.

***6.105** A flow field is represented by the stream function $\psi = x^5 - 10x^3y^2 + 5xy^4$. Find the corresponding velocity field. Show that this flow field is irrotational and obtain the potential function.

***6.106** The stream function of a flow field is $\psi = Ax^3 - Bxy^2$, where $A = 1\ \text{m}^{-1}\cdot\text{s}^{-1}$ and $B = 3\ \text{m}^{-1}\cdot\text{s}^{-1}$, and coordinates are measured in meters. Find an expression for the velocity potential.

***6.107** The stream function of a flow field is $\psi = Ax^3 + B(xy^2 + x^2 - y^2)$, where ψ, x, y, A, and B are all dimensionless. Find the relation between A and B for this to be an irrotational flow. Find the velocity potential.

***6.108** A flow field is represented by the stream function $\psi = x^5 - 15x^4y^2 + 15x^2y^4 - y^6$. Find the corresponding velocity field. Show that this flow field is irrotational and obtain the potential function.

***6.109** Consider the flow field represented by the potential function $\phi = Ax^2 + Bxy - Ay^2$. Verify that this is an incompressible flow and determine the corresponding stream function.

***6.110** Consider the flow field presented by the potential function $\phi = x^5 - 10x^3y^2 + 5xy^4 - x^2 + y^2$. Verify that this is an incompressible flow, and obtain the corresponding stream function.

***6.111** Consider the flow field presented by the potential function $\phi = x^6 - 15x^4y^2 + 15x^2y^4 - y^6$. Verify that this is an incompressible flow and obtain the corresponding stream function.

***6.112** Show by expanding and collecting real and imaginary terms that $f = z^6$ (where z is the complex number $z = x + iy$) leads to a valid velocity potential (the real part of f) and a corresponding stream function (the negative of the imaginary part of f) of an irrotational and incompressible flow. Then show that the real and imaginary parts of df/dz yield $-u$ and v, respectively.

***6.113** Show that *any* differentiable function $f(z)$ of the complex number $z = x + iy$ leads to a valid potential (the real part of f) and a corresponding stream function (the negative of the imaginary part of f) of an incompressible, irrotational flow. To do so, prove using the chain rule that $f(z)$ automatically satisfies the Laplace equation. Then show that $df/dz = -u + iv$.

***6.114** Consider the flow field represented by the velocity potential $\phi = Ax + Bx^2 - By^2$, where $A = 1\ \text{m}\cdot\text{s}^{-1}$, $B = 1\ \text{m}^{-1}\cdot\text{s}^{-1}$, and the coordinates are measured in meters. Obtain expressions for the velocity field and the stream function. Calculate the pressure difference between the origin and point $(x, y) = (1, 2)$.

***6.115** A flow field is represented by the potential function $\phi = Ay^3 - Bx^2y$, where $A = 1/3\ \text{m}^{-1}\cdot\text{s}^{-1}$, $B = 1\ \text{m}^{-1}\cdot\text{s}^{-1}$, and the coordinates are measured in meters. Obtain an expression for the magnitude of the velocity vector. Find the stream function for the flow. Plot the streamlines and potential lines, and visually verify that they are orthogonal. (*Hint*: Use the *Excel* workbook of Example 6.10.)

***6.116** An incompressible flow field is characterized by the stream function $\psi = 3Ax^2y - Ay^3$, where $A = 1\ \text{m}^{-1}\cdot\text{s}^{-1}$. Show that this flow field is irrotational. Derive the velocity potential for the flow. Plot the streamlines and potential lines, and visually verify that they are orthogonal. (*Hint*: Use the *Excel* workbook of Example 6.10.)

***6.117** A certain irrotational flow field in the xy plane has the stream function $\psi = Bxy$, where $B = 0.25\ \text{s}^{-1}$, and the coordinates are measured in meters. Determine the rate of flow between points $(x, y) = (2, 2)$ and $(3, 3)$. Find the velocity potential for this flow. Plot the streamlines and potential lines, and visually verify that they are orthogonal. (*Hint*: Use the *Excel* workbook of Example 6.10.)

***6.118** The velocity distribution in a two-dimensional, steady, inviscid flow field in the xy plane is $\vec{V} = (Ax + B)\hat{i} + (C - Ay)\hat{j}$, where $A = 3\ \text{s}^{-1}$, $B = 6$ m/s, $C = 4$ m/s, and the coordinates are measured in meters. The body force distribution is $\vec{B} = -g\hat{k}$ and the density is 825 kg/m^3. Does this represent a possible incompressible flow field? Plot a few streamlines in the upper half plane. Find the stagnation point(s) of the flow field. Is the flow irrotational? If so, obtain the potential function. Evaluate the pressure difference between the origin and point $(x, y, z) = (2, 2, 2)$.

*These problems require material from sections that may be omitted without loss of continuity in the text material.

***6.119** Consider flow around a circular cylinder with freestream velocity from right to left and a counterclockwise free vortex. Show that the lift force on the cylinder can be expressed as $F_L = -\rho U\Gamma$, as illustrated in Example 6.12.

***6.120** Consider the flow past a circular cylinder, of radius a, used in Example 6.11. Show that $V_r = 0$ along the lines $(r, \theta) = (r, \pm\pi/2)$. Plot V_θ/U versus radius for $r \geq a$, along the line $(r, \theta) = (r, \pi/2)$. Find the distance beyond which the influence of the cylinder is less than 1 percent of U.

***6.121** A crude model of a tornado is formed by combining a sink, of strength $q = 2800\ \text{m}^2/\text{s}$, and a free vortex, of strength $K = 5600\ \text{m}^2/\text{s}$. Obtain the stream function and velocity potential for this flow field. Estimate the radius beyond which the flow may be treated as incompressible. Find the gage pressure at that radius.

***6.122** A source and a sink with strengths of equal magnitude, $q = 3\pi\ \text{m}^2/\text{s}$, are placed on the x axis at $x = -a$ and $x = a$, respectively. A uniform flow, with speed $U = 20$ m/s, in the positive x direction, is added to obtain the flow past a Rankine body. Obtain the stream function, velocity potential, and velocity field for the combined flow. Find the value of $\psi =$ constant on the stagnation streamline. Locate the stagnation points if $a = 0.3$ m.

***6.123** Consider again the flow past a Rankine body of Problem 6.122. The half-width, h, of the body in the y direction is given by the transcendental equation

$$\frac{h}{a} = \cot\left(\frac{\pi Uh}{q}\right)$$

Evaluate the half-width, h. Find the local velocity and the pressure at points $(x, y) = (0, \pm h)$. Assume the fluid density is that of standard air.

***6.124** A flow field is formed by combining a uniform flow in the positive x direction, with $U = 10$ m/s, and a counterclockwise vortex, with strength $K = 16\pi\ \text{m}^2/\text{s}$, located at the origin. Obtain the stream function, velocity potential, and velocity field for the combined flow. Locate the stagnation point(s) for the flow. Plot the streamlines and potential lines. (*Hint*: Use the *Excel* workbook of Example 6.10.)

***6.125** Consider the flow field formed by combining a uniform flow in the positive x direction with a sink located at the origin. Let $U = 50$ m/s and $q = 90\ \text{m}^2/\text{s}$. Use a suitably chosen control volume to evaluate the net force per unit depth needed to hold in place (in standard air) the surface shape formed by the stagnation streamline.

***6.126** Consider the flow field formed by combining a uniform flow in the positive x direction and a source located at the origin. Obtain expressions for the stream function, velocity potential, and velocity field for the combined flow. If $U = 25$ m/s, determine the source strength if the stagnation point is located at $x = -1$ m. Plot the streamlines and potential lines. (*Hint*: Use the *Excel* workbook of Example 6.10.)

***6.127** Consider the flow field formed by combining a uniform flow in the positive x direction and a source located at the origin. Let $U = 30$ m/s and $q = 150\ \text{m}^2/\text{s}$. Plot the ratio of the local velocity to the freestream velocity as a function of θ along the stagnation streamline. Locate the points on the stagnation streamline where the velocity reaches its maximum value. Find the gage pressure there if the fluid density is 1.2 kg/m³.

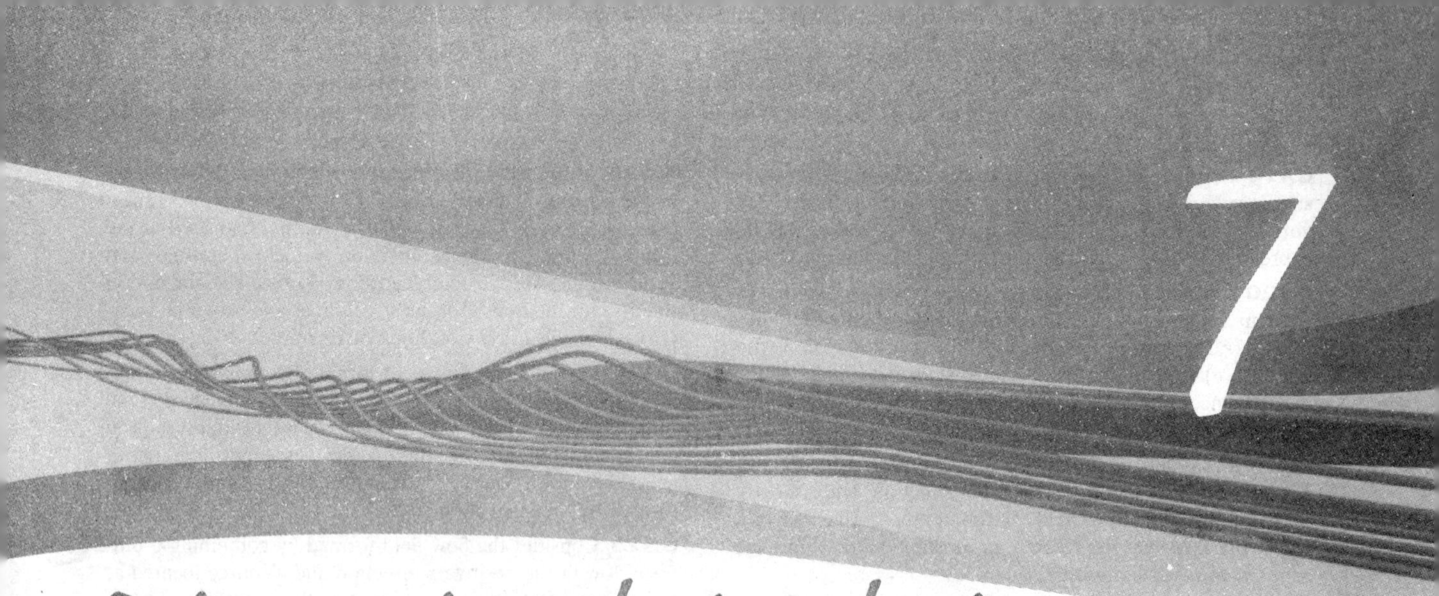

7 Dimensional Analysis and Similitude

Case Study in Energy and the Environment

Ocean Current Power: *The Vivace*

We have so far presented *Case Studies in Energy and the Environment* mostly on wave power, but many developments are taking place in ocean current power—actually, in the power available wherever there is a current, such as in estuaries and rivers, not just in the ocean. Plenty of power is available. Although ocean and river currents move slowly compared to typical wind speeds, they carry a great deal of energy because water is about 1000 times as dense as air, and the energy flux in a current is directly proportional to density. Hence water moving at 16 km/h exerts about the same amount of force as a 160 km/h wind. Ocean and river currents thus contain an enormous amount of energy that can be captured and converted to a usable form. For example, near the surface of the Florida Straits Current, the relatively constant extractable energy density is about 1 kW/m^2 of flow area. It has been estimated that capturing just 1/1000th of the available energy from the Gulf Stream could supply Florida with 35 percent of its electrical needs.

A horizontal- and a vertical-axis turbine, and an oscillating foil device (Courtesy the University of Strathclyde)

Ocean current energy is at an early stage of development, and only a small number of prototypes and demonstration units have so far been tested. A team of young engineers at the University of Strathclyde in Scotland recently did a survey of current developments. They found that perhaps the most obvious approach is to use submerged turbines. The first figure shows a horizontal-axis turbine (which is similar to a wind turbine) and a vertical-axis turbine. In each case, columns, cables, or anchors are required to keep the turbines stationary relative to the currents with which they interact. For example, they may be tethered with cables, in such a way that the current interacts with the turbine to maintain its location and stability; this is analogous to underwater kite flying in which the turbine plays the role of kite and the ocean-bottom anchor, the role of kite flyer. Turbines can include venturi-shaped shrouds around the blades to increase the flow speed and power output from the turbine. In regions with powerful currents over a large area, turbines could be assembled in clusters, similar to wind turbine farms. Space would be needed between the water turbines to eliminate wake-interaction effects and to allow access by maintenance vessels. The engineers at Strathclyde also discuss the third device shown in the figure, an oscillating foil design, in which a hydrofoil's angle of attack would be repeatedly adjusted to generate a lift force that is upward, then downward. The mechanism and controls would use this oscillating force to generate power. The advantage of this design is that there are no rotating parts that could become fouled, but the disadvantage is that the control systems involved would be quite complex.

For ocean current energy to be commercially successful, a number of technical challenges need to be addressed, including cavitation problems, prevention of marine growth buildup on turbine blades, and corrosion resistance. Environmental concerns include the protection of wildlife (fish and marine mammals) from turning turbine blades.

As the research in these types of turbines and foils continues, engineers are also looking at alternative devices. A good example is the work of Professor Michael Bernitsas, of the Department of Naval Architecture and Marine Engineering at the University of Michigan. He has developed a novel device, called the *Vivace Converter*, which uses the well-known phenomenon of vortex-induced vibrations to extract power from a flowing current. We are all familiar with vortex-induced vibrations, in which an object in a flow is made to vibrate due to vortices shedding first from one side and then the other side of the object's rear. For example, cables or wires often vibrate in the wind, sometimes sufficiently to make noise (*Aeolian tones*); many factory chimneys and car antennas have a spiral surface built into them specifically to suppress this vibration. Another famous example is the collapse of the Tacoma Narrows Bridge in Washington State in 1940, which many engineers believe was due to vortex-shedding of cross winds (a quite scary, but

The *Vivace Converter* (Courtesy Professor Michael Bernitsas)

fascinating, video of this can easily be found on the Internet). Professor Bernitas has made a source of energy from a phenomenon that is usually a nuisance or a danger!

The figure shows a conceptualization of his device, which consists of an assemblage of horizontal submerged cylinders. As the current flows across these, vortex shedding occurs, generating an oscillating up-and-down force on each cylinder. Instead of the cylinders being rigidly mounted, they are attached to a hydraulic system designed in such a way that, as the cylinders are forced up and down, they generate power. Whereas existing turbine systems need a current of about 2.5 m/s to operate efficiently, the Vivace can generate energy using currents that are as slow as 0.5 m/s (most of the earth's currents are slower than 1.5 m/s). The device also does not obstruct views or access on the water's surface because it can be installed on the river or ocean floor. It's probable that this new technology is gentler on aquatic life because it is slow moving and mimics the natural vortex patterns created by the movement of swimming fish. An installation of 1 × 1.5 km in a current of 1.5 m/s could generate enough power for 100,000 homes. A prototype, funded by the U.S. Department of Energy and the Office Naval Research, is currently operating in the Marine Hydrodynamics Laboratory at the University of Michigan. The phenomenon of vortex shedding is discussed in Chapter 9; the vortex flow meter, which exploits the phenomenon to measure flow rate, is discussed in Chapter 8. We will discuss airfoil design in Chapter 9 and concepts behind the operation of turbines and propellers in Chapter 10.

In previous chapters we have mentioned several instances in which we claim a simplified flow exists. For example, we have stated that a flow with typical speed V will be essentially incompressible if the Mach number, $M \equiv V/c$ (where c is the speed of sound), is less than about 0.3 and that we can neglect viscous effects in most of a flow if the Reynolds number, $Re = \rho VL/\mu$ (L is a typical or "characteristic" size scale of the flow), is "large." We will also make extensive use of the Reynolds number based on pipe diameter, D ($Re = \rho VD/\mu$), to predict with a high degree of accuracy whether the pipe flow is laminar or turbulent. It turns out that there are many such interesting dimensionless groupings in engineering science—for example, in heat transfer, the value of the Biot number, $Bi = hL/k$, of a hot body, size L and conductivity k, indicates whether that body will tend to cool on the outside surface first or will basically cool uniformly when it's plunged into a cool fluid with convection coefficient h. (Can you figure out what a high Bi number predicts?) How do we obtain these groupings, and why do their values have such powerful predictive power?

The answers to these questions will be provided in this chapter when we introduce the method of dimensional analysis. This is a technique for gaining insight into fluid flows (in fact, into many engineering and scientific phenomena) before we do either extensive theoretical analysis or experimentation; it also enables us to extract trends from data that would otherwise remain disorganized and incoherent.

We will also discuss modeling. For example, how do we correctly perform tests on the drag on a 3/8-scale model of an automobile in a wind tunnel to predict what the drag would be on the full-size automobile at the same speed? *Must* we use the same speed for model and full-size automobile? How do we scale up the measured model drag to find the automobile drag?

7.1 Nondimensionalizing the Basic Differential Equations

Before describing dimensional analysis let us see what we can learn from our previous analytical descriptions of fluid flow. Consider, for example, a steady incompressible two-dimensional flow of a Newtonian fluid with constant viscosity (already quite a list of assumptions!). The mass conservation equation (Eq. 5.1c) becomes

$$\frac{\partial u}{\partial x} + \frac{\partial v}{\partial y} = 0 \tag{7.1}$$

and the Navier–Stokes equations (Eqs. 5.27) reduce to

$$\rho\left(u\frac{\partial u}{\partial x} + v\frac{\partial u}{\partial y}\right) = -\frac{\partial p}{\partial x} + \mu\left(\frac{\partial^2 u}{\partial x^2} + \frac{\partial^2 u}{\partial y^2}\right) \tag{7.2}$$

and

$$\rho\left(u\frac{\partial v}{\partial x} + v\frac{\partial v}{\partial y}\right) = -\rho g - \frac{\partial p}{\partial y} + \mu\left(\frac{\partial^2 v}{\partial x^2} + \frac{\partial^2 v}{\partial y^2}\right) \tag{7.3}$$

As we discussed in Section 5.4, these equations form a set of coupled nonlinear partial differential equations for u, v, and p, and are difficult to solve for most flows. Equation 7.1 has dimensions of 1/time, and Eqs. 7.2 and 7.3 have dimensions of force/volume. Let us see what happens when we convert them into dimensionless equations. (Even if you did not study Section 5.4 you will be able to understand the following material.)

To nondimensionalize these equations, divide all lengths by a reference length, L, and all velocities by a reference speed, V_∞, which usually is taken as the freestream velocity. Make the pressure nondimensional by dividing by ρV_∞^2 (twice the freestream dynamic pressure). Denoting nondimensional quantities with asterisks, we obtain

$$x^* = \frac{x}{L}, \quad y^* = \frac{y}{L}, \quad u^* = \frac{u}{V_\infty}, \quad v^* = \frac{v}{V_\infty}, \quad \text{and} \quad p^* = \frac{p}{\rho V_\infty^2} \tag{7.4}$$

so that $x = x^*L$, $y = y^*L$, $u = u^*V_\infty$, and so on. We can then substitute into Eqs. 7.1 through 7.3; below we show two representative substitutions:

$$u\frac{\partial u}{\partial x} = u^*V_\infty\frac{\partial(u^*V_\infty)}{\partial(x^*L)} = \frac{V_\infty^2}{L}u^*\frac{\partial u^*}{\partial x^*}$$

and

$$\frac{\partial^2 u}{\partial x^2} = \frac{\partial(u^*V_\infty)}{\partial(x^*L)^2} = \frac{V_\infty}{L^2}\frac{\partial^2 u^*}{\partial x^{*2}}$$

Using this procedure, the equations become

$$\frac{V_\infty}{L}\frac{\partial u^*}{\partial x^*} + \frac{V_\infty}{L}\frac{\partial v^*}{\partial y^*} = 0 \tag{7.5}$$

$$\frac{\rho V_\infty^2}{L}\left(u^*\frac{\partial u^*}{\partial x^*} + v^*\frac{\partial u^*}{\partial y^*}\right) = -\frac{\rho V_\infty^2}{L}\frac{\partial p^*}{\partial x^*} + \frac{\mu V_\infty}{L^2}\left(\frac{\partial^2 u^*}{\partial x^{*2}} + \frac{\partial^2 u^*}{\partial y^{*2}}\right) \tag{7.6}$$

$$\frac{\rho V_\infty^2}{L}\left(u^*\frac{\partial v^*}{\partial x^*} + v^*\frac{\partial v^*}{\partial y^*}\right) = -\rho g - \frac{\rho V_\infty^2}{L}\frac{\partial p^*}{\partial y^*} + \frac{\mu V_\infty}{L^2}\left(\frac{\partial^2 v^*}{\partial x^{*2}} + \frac{\partial^2 v^*}{\partial y^{*2}}\right) \tag{7.7}$$

Dividing Eq. 7.5 by V_∞/L and Eqs. 7.6 and 7.7 by $\rho V_\infty^2/L$ gives

$$\frac{\partial u^*}{\partial x^*} + \frac{\partial v^*}{\partial y^*} = 0 \tag{7.8}$$

$$u^* \frac{\partial u^*}{\partial x^*} + v^* \frac{\partial u^*}{\partial y^*} = -\frac{\partial p^*}{\partial x^*} + \frac{\mu}{\rho V_\infty L}\left(\frac{\partial^2 u^*}{\partial x^{*2}} + \frac{\partial^2 u^*}{\partial y^{*2}}\right) \tag{7.9}$$

$$u^* \frac{\partial v^*}{\partial x^*} + v^* \frac{\partial v^*}{\partial y^*} = -\frac{gL}{V_\infty^2} - \frac{\partial p^*}{\partial y^*} + \frac{\mu}{\rho V_\infty L}\left(\frac{\partial^2 v^*}{\partial x^{*2}} + \frac{\partial^2 v^*}{\partial y^{*2}}\right) \tag{7.10}$$

Equations 7.8, 7.9, and 7.10 are the nondimensional forms of our original equations (Eqs. 7.1, 7.2, 7.3). As such, we can think about their solution (with appropriate boundary conditions) as an exercise in applied mathematics. Equation 7.9 contains a dimensionless coefficient $\mu/\rho V_\infty L$ (which we recognize as the inverse of the Reynolds number) in front of the second-order (viscous) terms; Eq. 7.10 contains this and another dimensionless coefficient, gL/V_∞^2 (which we will discuss shortly) for the gravity force term. We recall from the theory of differential equations that the mathematical form of the solution of such equations is very sensitive to the values of the coefficients in the equations (e.g., certain second-order partial differential equations can be elliptical, parabolic, or hyperbolic depending on coefficient values).

These equations tell us that the solution, and hence the actual flow pattern they describe, depends on the values of the two coefficients. For example, if $\mu/\rho V_\infty L$ is very small (i.e., we have a high Reynolds number), the second-order differentials, representing viscous forces, can be neglected, at least in most of the flow, and we end up with a form of Euler's equations (Eqs. 6.2). We say "in most of the flow" because we have already learned that in reality for this case we will have a boundary layer in which there *is* significant effect of viscosity; in addition, from a mathematical point of view, it is always dangerous to neglect higher-order derivatives, even if their coefficients are small, because reduction to a lower-order equation means we lose a boundary condition (specifically the no-slip condition). We can predict that if $\mu/\rho V_\infty L$ is large or small, then viscous forces will be significant or not, respectively; if gLV_∞^2 is large or small, we can predict that gravity forces will be significant or not, respectively. We can thus gain insight even before attempting a solution to the differential equations. Note that for completeness, we would have to apply the same nondimensionalizing approach to the boundary conditions of the problem, which often introduce further dimensionless coefficients.

Writing nondimensional forms of the governing equations, then, can yield insight into the underlying physical phenomena, and indicate which forces are dominant. If we had two geometrically similar but different scale flows satisfying Eqs. 7.8, 7.9, and 7.10 (for example, a model and a prototype), the equations would only yield the same mathematical results if the two flows had the same values for the two coefficients (i.e., had the same relative importance of gravity, viscous, and inertia forces). This nondimensional form of the equations is also the starting point in numerical methods, which is very often the only way of obtaining their solution. Additional derivations and examples of establishing similitude from the governing equations of a problem are presented in Kline [1] and Hansen [2].

We will now see how the method of dimensional analysis can be used instead of the above procedure to find appropriate dimensionless groupings of physical parameters. As we have mentioned, using dimensionless groupings is very useful for experimental measurements, and we will see in the next two sections that we can obtain them even when we do not have the governing equations such as Eqs. 7.1, 7.2, and 7.3 to work from.

7.2 Nature of Dimensional Analysis

Most phenomena in fluid mechanics depend in a complex way on geometric and flow parameters. For example, consider the drag force on a stationary smooth sphere immersed in a uniform stream. What experiments must be conducted to determine the drag force on the sphere? To answer this question, we must specify what we believe are the parameters that are important in determining the drag force. Clearly, we would expect the drag force to depend on the size of the sphere (characterized by the diameter, D), the fluid speed, V, and the fluid viscosity, μ. In addition, the density of the fluid, ρ, also might be important. Representing the drag force by F, we can write the symbolic equation

$$F = f(D, V, \rho, \mu)$$

Although we may have neglected parameters on which the drag force depends, such as surface roughness (or may have included parameters on which it does not depend), we have set up the problem of determining the drag force for a stationary sphere in terms of quantities that are both controllable and measurable in the laboratory.

We could set up an experimental procedure for finding the dependence of F on V, D, ρ, and μ. To see how the drag, F, is affected by fluid speed, V, we could place a sphere in a wind tunnel and measure F for a range of V values. We could then run more tests in which we explore the effect on F of sphere diameter, D, by using different diameter spheres. We are already generating a lot of data: If we ran the wind tunnel at, say, 10 different speeds, for 10 different sphere sizes, we'd have 100 data points. We could present these results on one graph (e.g., we could plot 10 curves of F vs. V, one for each sphere size), but acquiring the data would already be time consuming: If we assume each run takes $\frac{1}{2}$ hour, we have already accumulated 50 hours of work! We still wouldn't be finished—we would have to book time using, say, a water tank, where we could repeat all these runs for a different value of ρ and of μ. In principle, we would next have to search out a way to use other fluids to be able to do experiments for a range of ρ and μ values (say, 10 of each). At the end of the day (actually, at the end of about $2\frac{1}{2}$ years of 40-hour weeks!) we would have performed about 10^4 tests. Then we would have to try and make sense of the data: How do we plot, say, curves of F vs. V, with D, ρ, and μ all being parameters? This is a daunting task, even for such a seemingly simple phenomenon as the drag on a sphere!

Fortunately we do not have to do all this work. As we will see in Example 7.1, using dimensional analysis, all the data for drag on a smooth sphere can be plotted as a single relationship between two nondimensional parameters in the form

$$\frac{F}{\rho V^2 D^2} = f\left(\frac{\rho V D}{\mu}\right)$$

VIDEO

Flow Around a Sphere 1.

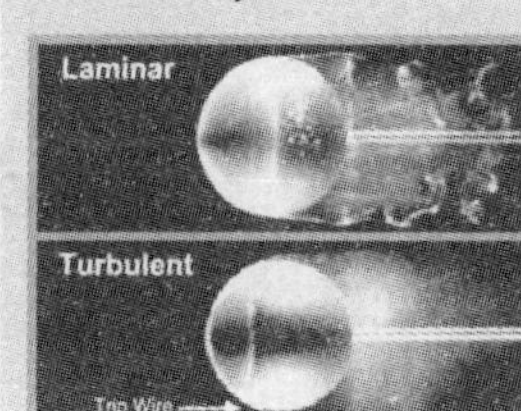

The form of the function f still must be determined experimentally, but the point is that all spheres, in all fluids, for most velocities will fall on the same curve. Rather than needing to conduct 10^4 experiments, we could establish the nature of the function as accurately with only about 10 tests. The time saved in performing only 10 rather than 10^4 tests is obvious. Even more important is the greater experimental convenience. No longer must we find fluids with 10 different values of density and viscosity. Nor must we make 10 spheres of different diameters. Instead, only the parameter $\rho VD/\mu$ must be varied. This can be accomplished simply by using *one* sphere (e.g., 25 mm diameter), in *one* fluid (e.g., air), and only changing the speed, for example.

VIDEO

Flow Around a Sphere 2.

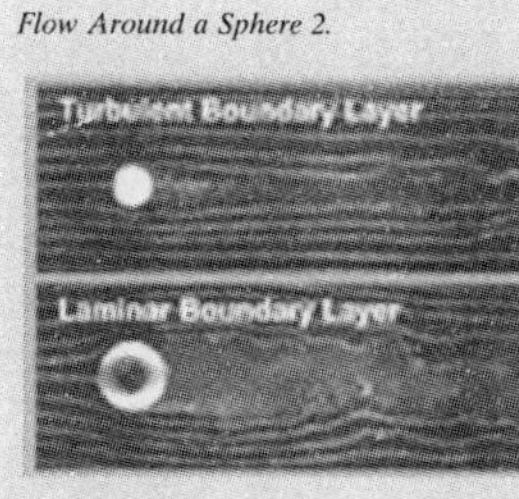

Figure 7.1 shows some classic data for flow over a sphere (the factors $\frac{1}{2}$ and $\pi/4$ have been added to the denominator of the parameter on the left to make it take the form of a commonly used nondimensional group, the drag coefficient, C_D, that we will discuss in detail in Chapter 9). If we performed the experiments as outlined above, our results would fall on the same curve, within experimental error. The data points represent results obtained by various workers for several different fluids and spheres. Note that we end up with a curve that can be used to obtain the drag force on a very wide range of sphere/fluid combinations. For example, it could be used to obtain the drag on a hot-air balloon due to a crosswind, or on a red blood cell (assuming it could be modeled as a

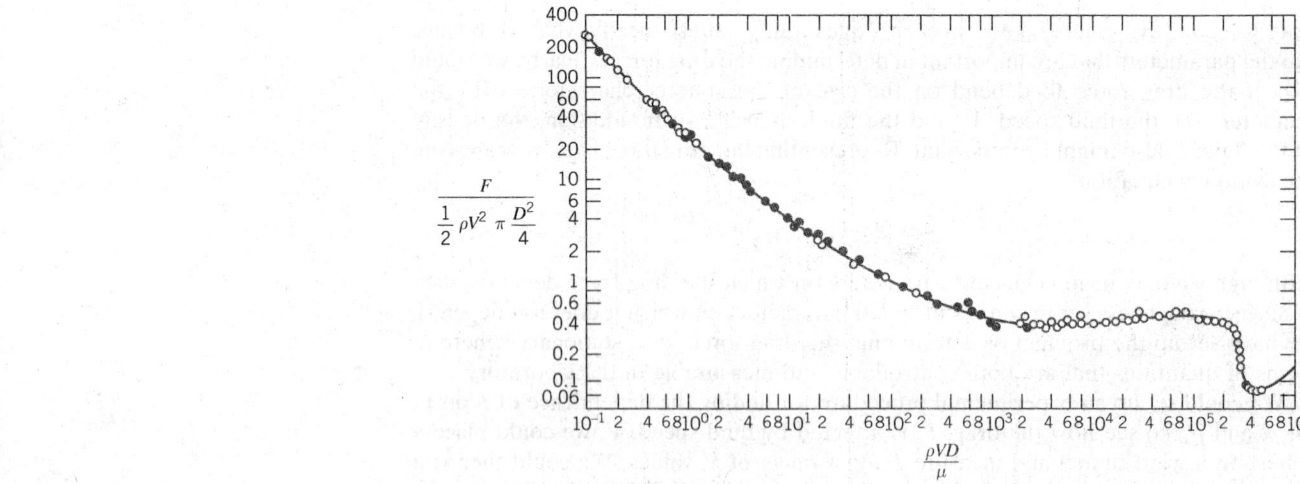

Fig. 7.1 Experimentally derived relation between the nondimensional parameters [3].

sphere) as it moves through the aorta—in either case, given the fluid (ρ and μ), the flow speed V, and the sphere diameter D, we could compute a value for $\rho VD/\mu$, then read the corresponding value for C_D, and finally compute the drag force F.

In Section 7.3 we introduce the *Buckingham Pi* theorem, a formalized procedure for deducing the dimensionless groups appropriate for a given fluid mechanics or other engineering problem. This section, and Section 7.4, may seem a bit difficult to follow; we suggest you read them once, then study Examples 7.1, 7.2, and 7.3 to see how practical and useful the method in fact is, before returning to the two sections for a reread.

The Buckingham Pi theorem is a statement of the relation between a function expressed in terms of dimensional parameters and a related function expressed in terms of nondimensional parameters. The Buckingham Pi theorem allows us to develop the important nondimensional parameters quickly and easily.

7.3 Buckingham Pi Theorem

In the previous section we discussed how the drag F on a sphere depends on the sphere diameter D, fluid density ρ and viscosity μ, and fluid speed V, or

$$F = F(D, \rho, \mu, V)$$

with theory or experiment being needed to determine the nature of function f. More formally, we write

$$g(F, D, \rho, \mu, V) = 0$$

where g is an unspecified function, different from f. The Buckingham Pi theorem [4] states that we can transform a relationship between n parameters of the form

$$g(q_1, q_2, \ldots, q_n) = 0$$

into a corresponding relationship between $n - m$ independent dimensionless Π parameters in the form

$$G(\Pi_1, \Pi_2, \ldots, \Pi_{n-m}) = 0$$

or

$$\Pi_1 = G_1(\Pi_2, \ldots, \Pi_{n-m})$$

where m is *usually* the minimum number, r, of independent dimensions (e.g., mass, length, time) required to define the dimensions of all the parameters $q_1, q_2, \ldots, q_n$. (Sometimes $m \neq r$; we will see this in Example 7.3.) For example, for the sphere problem, we will see (in Example 7.1) that

$$g(F, D, \rho, \mu, V) = 0 \quad \text{or} \quad F = F(D, \rho, \mu, V)$$

leads to

$$G\left(\frac{F}{\rho V^2 D^2}, \frac{\mu}{\rho V D}\right) = 0 \quad \text{or} \quad \frac{F}{\rho V^2 D^2} = G_1\left(\frac{\mu}{\rho V D}\right)$$

The theorem does not predict the functional form of G or G_1. The functional relation among the independent dimensionless Π parameters must be determined experimentally.

The $n - m$ dimensionless Π parameters obtained from the procedure are independent. A Π parameter is not independent if it can be formed from any combination of one or more of the other Π parameters. For example, if

$$\Pi_5 = \frac{2\Pi_1}{\Pi_2 \Pi_3} \quad \text{or} \quad \Pi_6 = \frac{\Pi_1^{3/4}}{\Pi_3^2}$$

then neither Π_5 nor Π_6 is independent of the other dimensionless parameters.

Several methods for determining the dimensionless parameters are available. A detailed procedure is presented in the next section.

Determining the Π Groups 7.4

Regardless of the method to be used to determine the dimensionless parameters, one begins by listing all dimensional parameters that are known (or believed) to affect the given flow phenomenon. Some experience admittedly is helpful in compiling the list. Students, who do not have this experience, often are troubled by the need to apply engineering judgment in an apparent massive dose. However, it is difficult to go wrong if a generous selection of parameters is made.

If you suspect that a phenomenon depends on a given parameter, include it. If your suspicion is correct, experiments will show that the parameter must be included to get consistent results. If the parameter is extraneous, an extra Π parameter may result, but experiments will later show that it may be eliminated. Therefore, do not be afraid to include *all* the parameters that you feel are important.

The six steps listed below (which may seem a bit abstract but are actually easy to do) outline a recommended procedure for determining the Π parameters:

Step 1. *List all the dimensional parameters involved.* (Let n be the number of parameters.) If all of the pertinent parameters are not included, a relation may be obtained, but it will not give the complete story. If parameters that actually have no effect on the physical phenomenon are included, either the process of dimensional analysis will show that these do not enter the relation sought, or one or more dimensionless groups will be obtained that experiments will show to be extraneous.

Step 2. *Select a set of fundamental (primary) dimensions, e.g., MLt or FLt.* (Note that for heat transfer problems you may also need T for temperature, and in electrical systems, q for charge.)

Step 3. *List the dimensions of all parameters in terms of primary dimensions.* (Let r be the number of primary dimensions.) Either force or mass may be selected as a primary dimension.

Step 4. *Select a set of r dimensional parameters that includes all the primary dimensions.* These parameters will all be combined with each of the remaining parameters, one of those at a time, and so will be called repeating parameters. No repeating parameter should have dimensions that are a power of the dimensions of another repeating parameter; for example, do not include both an area (L^2) and a second moment of area (L^4) as repeating parameters. The repeating parameters chosen may appear in all the dimensionless groups obtained; consequently, do *not* include the dependent parameter among those selected in this step.

Step 5. *Set up dimensional equations, combining the parameters selected in Step 4 with each of the other parameters in turn, to form dimensionless groups.* (There will be $n - m$ equations.) Solve the dimensional equations to obtain the $n - m$ dimensionless groups.

Step 6. *Check to see that each group obtained is dimensionless.* If mass was initially selected as a primary dimension, it is wise to check the groups using force as a primary dimension, or vice versa.

The functional relationship among the Π parameters must be determined experimentally. The detailed procedure for determining the dimensionless Π parameters is illustrated in Examples 7.1 and 7.2.

Example 7.1 DRAG FORCE ON A SMOOTH SPHERE

As noted in Section 7.2, the drag force, F, on a smooth sphere depends on the relative speed, V, the sphere diameter, D, the fluid density, ρ, and the fluid viscosity, μ. Obtain a set of dimensionless groups that can be used to correlate experimental data.

Given: $F = f(\rho, V, D, \mu)$ for a smooth sphere.

Find: An appropriate set of dimensionless groups.

Solution:

(Circled numbers refer to steps in the procedure for determining dimensionless Π parameters.)

① $F \quad V \quad D \quad \rho \quad \mu$ $\qquad n = 5$ dimensional parameters

② Select primary dimensions M, L, and t.

③ $F \quad V \quad D \quad \rho \quad \mu$

$$\frac{ML}{t^2} \quad \frac{L}{t} \quad L \quad \frac{M}{L^3} \quad \frac{M}{Lt} \qquad r = 3 \text{ primary dimensions}$$

④ Select repeating parameters ρ, V, D. $\qquad m = r = 3$ repeating parameters

⑤ Then $n - m = 2$ dimensionless groups will result. Setting up dimensional equations, we obtain

$$\Pi_1 = \rho^a V^b D^c F \quad \text{and} \quad \left(\frac{M}{L^3}\right)^a \left(\frac{L}{t}\right)^b (L)^c \left(\frac{ML}{t^2}\right) = M^0 L^0 t^0$$

Equating the exponents of M, L, and t results in

$$\left.\begin{array}{lrr} M: & a + 1 = 0 & a = -1 \\ L: & -3a + b + c + 1 = 0 & c = -2 \\ t: & -b - 2 = 0 & b = -2 \end{array}\right\} \quad \text{Therefore,} \quad \Pi_1 = \frac{F}{\rho V^2 D^2}$$

Similarly,

$$\Pi_2 = \rho^d V^e D^f \mu \quad \text{and} \quad \left(\frac{M}{L^3}\right)^d \left(\frac{L}{t}\right)^e (L)^f \left(\frac{M}{Lt}\right) = M^0 L^0 t^0$$

$$\left.\begin{array}{lll} M: & d+1=0 & d=-1 \\ L: & -3d+e+f-1=0 & f=-1 \\ t: & -e-1=0 & e=-1 \end{array}\right\} \quad \text{Therefore, } \Pi_2 = \frac{\mu}{\rho V D}$$

⑥ Check using F, L, t dimensions

$$[\Pi_1] = \left[\frac{F}{\rho V^2 D^2}\right] \quad \text{and} \quad F\frac{L^4}{Ft^2}\left(\frac{t}{L}\right)^2 \frac{1}{L^2} = 1$$

where [ ] means "has dimensions of," and

$$[\Pi_2] = \left[\frac{\mu}{\rho V D}\right] \quad \text{and} \quad \frac{Ft}{L^2}\,\frac{L^4}{Ft^2}\,\frac{t}{L}\,\frac{1}{L} = 1$$

The functional relationship is $\Pi_1 = f(\Pi_2)$, or

$$\frac{F}{\rho V^2 D^2} = f\left(\frac{\mu}{\rho V D}\right)$$

as noted before. The form of the function, f, must be determined experimentally (see Fig. 7.1).

The *Excel* workbook for this Example is convenient for computing the values of a, b, and c for this and other problems.

Example 7.2 PRESSURE DROP IN PIPE FLOW

The pressure drop, Δp, for steady, incompressible viscous flow through a straight horizontal pipe depends on the pipe length, l, the average velocity, $\overline{V}$, the fluid viscosity, μ, the pipe diameter, D, the fluid density, ρ, and the average "roughness" height, e. Determine a set of dimensionless groups that can be used to correlate data.

Given: $\Delta p = f(\rho, \overline{V}, D, l, \mu, e)$ for flow in a circular pipe.

Find: A suitable set of dimensionless groups.

Solution:

(Circled numbers refer to steps in the procedure for determining dimensionless Π parameters.)

① $\Delta p \quad \rho \quad \mu \quad \overline{V} \quad l \quad D \quad e \qquad n = 7$ dimensional parameters

② Choose primary dimensions M, L, and t.

③ $\Delta p \quad \rho \quad \mu \quad \overline{V} \quad l \quad D \quad e$

$$\frac{M}{Lt^2} \quad \frac{M}{L^3} \quad \frac{M}{Lt} \quad \frac{L}{t} \quad L \quad L \quad L \qquad r = 3 \text{ primary dimensions}$$

④ Select repeating parameters ρ, $\overline{V}$, D. $\qquad m = r = 3$ repeating parameters

⑤ Then $n - m = 4$ dimensionless groups will result. Setting up dimensional equations we have:

$$\Pi_1 = \rho^a \overline{V}^b D^c \Delta p \quad \text{and}$$

$$\left(\frac{M}{L^3}\right)^a \left(\frac{L}{t}\right)^b (L)^c \left(\frac{M}{Lt^2}\right) = M^0 L^0 t^0$$

$$\left.\begin{array}{ll} M: & 0 = a + 1 \\ L: & 0 = -3a + b + c - 1 \\ t: & 0 = -b - 2 \end{array}\right\} \quad \begin{array}{l} a = -1 \\ b = -2 \\ c = 0 \end{array}$$

Therefore, $\Pi_1 = \rho^{-1} \overline{V}^{-2} D^0 \Delta p = \dfrac{\Delta p}{\rho \overline{V}^2}$

$$\Pi_2 = \rho^d \overline{V}^e D^f \mu \quad \text{and}$$

$$\left(\frac{M}{L^3}\right)^d \left(\frac{L}{t}\right)^e (L)^f \frac{M}{Lt} = M^0 L^0 t^0$$

$$\left.\begin{array}{ll} M: & 0 = d + 1 \\ L: & 0 = -3d + e + f - 1 \\ t: & 0 = -e - 1 \end{array}\right\} \quad \begin{array}{l} d = -1 \\ e = -1 \\ f = -1 \end{array}$$

Therefore, $\Pi_2 = \dfrac{\mu}{\rho \overline{V} D}$

$$\Pi_3 = \rho^g \overline{V}^h D^i l \quad \text{and}$$

$$\left(\frac{M}{L^3}\right)^g \left(\frac{L}{t}\right)^h (L)^i L = M^0 L^0 t^0$$

$$\left.\begin{array}{ll} M: & 0 = g \\ L: & 0 = -3g + h + i + 1 \\ t: & 0 = -h \end{array}\right\} \quad \begin{array}{l} g = 0 \\ h = 0 \\ i = -1 \end{array}$$

Therefore, $\Pi_3 = \dfrac{l}{D}$

$$\Pi_4 = \rho^j \overline{V}^k D^l e \quad \text{and}$$

$$\left(\frac{M}{L^3}\right)^j \left(\frac{L}{t}\right)^k (L)^l L = M^0 L^0 t^0$$

$$\left.\begin{array}{ll} M: & 0 = j \\ L: & 0 = -3j + k + l + 1 \\ t: & 0 = -k \end{array}\right\} \quad \begin{array}{l} j = 0 \\ k = 0 \\ l = -1 \end{array}$$

Therefore, $\Pi_4 = \dfrac{e}{D}$

⑥ Check, using F, L, t dimensions

$$[\Pi_1] = \left[\frac{\Delta p}{\rho \overline{V}^2}\right] \quad \text{and} \quad \frac{F}{L^2}\frac{L^4}{Ft^2}\frac{t^2}{L^2} = 1 \qquad [\Pi_3] = \left[\frac{l}{D}\right] \quad \text{and} \quad \frac{L}{L} = 1$$

$$[\Pi_2] = \left[\frac{\mu}{\rho \overline{V} D}\right] \quad \text{and} \quad \frac{Ft}{L^2}\frac{L^4}{Ft^2}\frac{t}{L}\frac{1}{L} = 1 \qquad [\Pi_4] = \left[\frac{e}{D}\right] \quad \text{and} \quad \frac{L}{L} = 1$$

Finally, the functional relationship is

$$\Pi_1 = f(\Pi_2,\ \Pi_3,\ \Pi_4)$$

or

$$\frac{\Delta p}{\rho \overline{V}^2} = f\left(\frac{\mu}{\rho \overline{V} D}, \frac{l}{D}, \frac{e}{D}\right)$$

Notes:

- ✓ As we shall see when we study pipe flow in detail in Chapter 8, this relationship correlates the data well.
- ✓ Each Π group is unique (e.g., there is only *one* possible dimensionless grouping of μ, ρ, $\overline{V}$, and D).
- ✓ We can often deduce Π groups by inspection, e.g., l/D is the obvious unique grouping of l with ρ, $\overline{V}$, and D.
- The *Excel* workbook for Example 7.1 is convenient for computing the values of a, b, and c for this problem.

The procedure outlined above, where m is taken equal to r (the fewest independent dimensions required to specify the dimensions of all parameters involved), almost always produces the correct number of dimensionless Π parameters. In a few cases, trouble arises because the number of primary dimensions differs when variables are expressed in terms of different systems of dimensions (e.g., MLt or FLt). The value of m can be established with certainty by determining the rank of the dimensional matrix; that rank is m. Although not needed in most applications, for completeness, this procedure is illustrated in Example 7.3.

The $n - m$ dimensionless groups obtained from the procedure are independent but not unique. If a different set of repeating parameters is chosen, different groups result. The repeating parameters are so named because they may appear in all the dimensionless groups obtained. Based on experience, viscosity should appear in only one dimensionless parameter. Therefore μ should *not* be chosen as a repeating parameter.

When we have a choice, it usually works out best to choose density ρ (dimensions M/L^3 in the MLt system), speed V (dimensions L/t), and characteristic length L (dimension L) as repeating parameters because experience shows this generally leads to a set of dimensionless parameters that are suitable for correlating a wide range of experimental data; in addition, ρ, V, and L are usually fairly easy to measure or otherwise obtain. The values of the dimensionless parameters obtained using these repeating parameters almost always have a very tangible meaning, telling you the relative strength of various fluid forces (e.g., viscous) to inertia forces—we will discuss several "classic" ones shortly.

It's also worth stressing that, given the parameters you're combining, *we can often determine the unique dimensional parameters by inspection*. For example, if we had repeating parameters ρ, V, and L and were combining them with a parameter A_f, representing the frontal area of an object, it's fairly obvious that only the combination A_f/L^2 is dimensionless; experienced fluid mechanicians also know that ρV^2 produces dimensions of stress, so any time a stress or force parameter arises, dividing by ρV^2 or $\rho V^2 L^2$ will produce a dimensionless quantity.

We will find useful a measure of the magnitude of fluid inertia forces, obtained from Newton's second law, $F = ma$; the dimensions of inertia force are thus MLt^{-2}. Using ρ, V, and L to build the dimensions of ma leads to the unique combination $\rho V^2 L^2$ (only ρ has dimension M, and only V^2 will produce dimension t^{-2}; L^2 is then required to leave us with MLt^{-2}).

If $n - m = 1$, then a single dimensionless Π parameter is obtained. In this case, the Buckingham Pi theorem indicates that the single Π parameter must be a constant.

Example 7.3 CAPILLARY EFFECT: USE OF DIMENSIONAL MATRIX

When a small tube is dipped into a pool of liquid, surface tension causes a meniscus to form at the free surface, which is elevated or depressed depending on the contact angle at the liquid-solid-gas interface. Experiments indicate that the magnitude of this capillary effect, Δh, is a function of the tube diameter, D, liquid specific weight, γ, and surface tension, σ. Determine the number of independent Π parameters that can be formed and obtain a set.

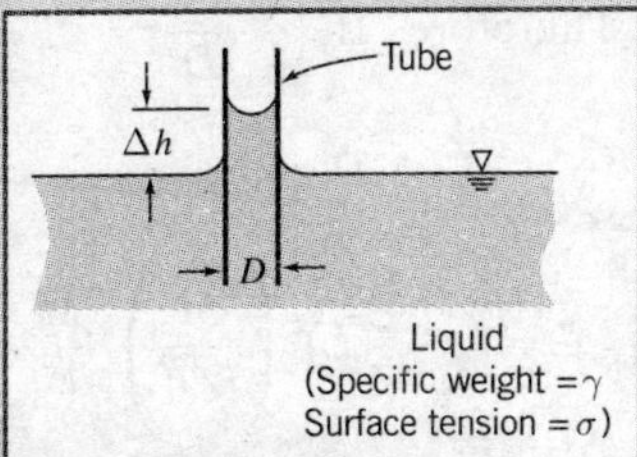

Given: $\Delta h = f(D, \gamma, \sigma)$

Find: (a) Number of independent Π parameters.
(b) One set of Π parameters.

Solution:

(Circled numbers refer to steps in the procedure for determining dimensionless Π parameters.)

① $\Delta h \quad D \quad \gamma \quad \sigma \qquad n = 4$ dimensional parameters

② Choose primary dimensions (use both M, L, t and F, L, t dimensions to illustrate the problem in determining m).

③ (a) M, L, t

Δh	D	γ	σ
L	L	$\frac{M}{L^2t^2}$	$\frac{M}{t^2}$

$r = 3$ primary dimensions

(b) F, L, t

Δh	D	γ	σ
L	L	$\frac{F}{L^3}$	$\frac{F}{L}$

$r = 2$ primary dimensions

Thus for each set of primary dimensions we ask, "Is m equal to r?" Let us check each dimensional matrix to find out. The dimensional matrices are

	Δh	D	γ	σ
M	0	0	1	1
L	1	1	−2	0
t	0	0	−2	−2

	Δh	D	γ	σ
F	0	0	1	1
L	1	1	−3	−1

The rank of a matrix is equal to the order of its largest nonzero determinant.

(a)

$$\begin{vmatrix} 0 & 1 & 1 \\ 1 & -2 & 0 \\ 0 & -2 & -2 \end{vmatrix} = 0 - (1)(-2) + (1)(-2) = 0$$

$$\begin{vmatrix} -2 & 0 \\ -2 & -2 \end{vmatrix} = 4 \neq 0 \qquad \therefore\ m = 2 \quad m \neq r$$

④ $m = 2$. Choose D, γ as repeating parameters.

⑤ $n - m = 2$ dimensionless groups will result.

$$\Pi_1 = D^a \gamma^b \Delta h \quad \text{and}$$

$$(L)^a \left(\frac{M}{L^2t^2}\right)^b (L) = M^0L^0t^0$$

$$\left.\begin{aligned} M&: \quad b + 0 = 0 \\ L&: \quad a - 2b + 1 = 0 \\ t&: \quad -2b + 0 = 0 \end{aligned}\right\} \qquad \begin{aligned} b &= 0 \\ a &= -1 \end{aligned}$$

Therefore, $\Pi_1 = \dfrac{\Delta h}{D}$

$$\Pi_2 = D^c \gamma^d \sigma \quad \text{and}$$

$$(L)^c \left(\frac{M}{L^2t^2}\right)^d \frac{M}{t^2} = M^0L^0t^0$$

$$\left.\begin{aligned} M&: \quad d + 1 = 0 \\ L&: \quad c - 2d = 0 \\ t&: \quad -2d - 2 = 0 \end{aligned}\right\} \qquad \begin{aligned} d &= -1 \\ c &= -2 \end{aligned}$$

Therefore, $\Pi_2 = \dfrac{\sigma}{D^2\gamma}$

(b)

$$\begin{vmatrix} 1 & 1 \\ -3 & -1 \end{vmatrix} = -1 + 3 = 2 \neq 0$$

$$\therefore\ m = 2 \quad m = r$$

$m = 2$. Choose D, γ as repeating parameters.

$n - m = 2$ dimensionless groups will result.

$$\Pi_1 = D^e \gamma^f \Delta h \quad \text{and}$$

$$(L)^e \left(\frac{F}{L^3}\right)^f L = F^0L^0t^0$$

$$\left.\begin{aligned} F&: \quad f = 0 \\ L&: \quad e - 3f + 1 = 0 \end{aligned}\right\} \qquad e = -1$$

Therefore, $\Pi_1 = \dfrac{\Delta h}{D}$

$$\Pi_2 = D^g \gamma^h \sigma \quad \text{and}$$

$$(L)^g \left(\frac{F}{L^3}\right)^h \frac{F}{L} = F^0L^0t^0$$

$$\left.\begin{aligned} F&: \quad h + 1 = 0 \\ L&: \quad g - 3h - 1 = 0 \end{aligned}\right\} \qquad \begin{aligned} h &= -1 \\ g &= -2 \end{aligned}$$

Therefore, $\Pi_2 = \dfrac{\sigma}{D^2\gamma}$

⑥ Check, using F, L, t dimensions

$$[\Pi_1] = \left[\frac{\Delta h}{D}\right] \quad \text{and} \quad \frac{L}{L} = 1$$

$$[\Pi_2] = \left[\frac{\sigma}{D^2\gamma}\right] \quad \text{and} \quad \frac{F}{L}\frac{1}{L^2}\frac{L^3}{F} = 1$$

Check, using M, L, t dimensions

$$[\Pi_1] = \left[\frac{\Delta h}{D}\right] \quad \text{and} \quad \frac{L}{L} = 1$$

$$[\Pi_2] = \left[\frac{\sigma}{D^2\gamma}\right] \quad \text{and} \quad \frac{M}{t^2}\frac{1}{L^2}\frac{L^2t^2}{M} = 1$$

Therefore, both systems of dimensions yield the same dimensionless Π parameters. The predicted functional relationship is

$$\Pi_1 = f(\Pi_2) \quad \text{or} \quad \frac{\Delta h}{D} = f\left(\frac{\sigma}{D^2\gamma}\right)$$

Notes:

- ✓ This result is reasonable on physical grounds. The fluid is static; we would not expect time to be an important dimension.
- ✓ We analyzed this problem in Example 2.3, where we found that $\Delta h = 4\sigma\cos(\theta)/\rho g D$ (θ is the contact angle). Hence $\Delta h/D$ is directly proportional to $\sigma/D^2\gamma$.
- ✓ The purpose of this problem is to illustrate use of the dimensional matrix to determine the required number of repeating parameters.

Significant Dimensionless Groups in Fluid Mechanics 7.5

Over the years, several hundred different dimensionless groups that are important in engineering have been identified. Following tradition, each such group has been given the name of a prominent scientist or engineer, usually the one who pioneered its use. Several are so fundamental and occur so frequently in fluid mechanics that we should take time to learn their definitions. Understanding their physical significance also gives insight into the phenomena we study.

Forces encountered in flowing fluids include those due to inertia, viscosity, pressure, gravity, surface tension, and compressibility. The ratio of any two forces will be dimensionless. We have previously shown that the inertia force is proportional to $\rho V^2 L^2$.

We can now compare the relative magnitudes of various fluid forces to the inertia force, using the following scheme:

Viscous force $\sim \quad \tau A = \mu \dfrac{du}{dy} A \propto \mu \dfrac{V}{L} L^2 = \mu V L \quad$ so $\quad \dfrac{\text{viscous}}{\text{inertia}} \sim \dfrac{\mu V L}{\rho V^2 L^2} = \dfrac{\mu}{\rho V L}$

Pressure force $\sim \quad \Delta p A \propto \Delta p L^2 \quad$ so $\quad \dfrac{\text{pressure}}{\text{inertia}} \sim \dfrac{\Delta p L^2}{\rho V^2 L^2} = \dfrac{\Delta p}{\rho V^2}$

Gravity force $\sim \quad mg \propto g \rho L^3 \quad$ so $\quad \dfrac{\text{gravity}}{\text{inertia}} \sim \dfrac{g \rho L^3}{\rho V^2 L^2} = \dfrac{gL}{V^2}$

Surface tension ~	σL	so	$\dfrac{\text{surface tension}}{\text{inertia}}$ ~	$\dfrac{\sigma L}{\rho V^2 L^2}$	$= \dfrac{\sigma}{\rho V^2 L}$
Compressibility force ~	$E_v A \propto E_v L^2$	so	$\dfrac{\text{compressibility force}}{\text{inertia}}$ ~	$\dfrac{E_v L^2}{\rho V^2 L^2}$	$= \dfrac{E_v}{\rho V^2}$

All of the dimensionless parameters listed above occur so frequently, and are so powerful in predicting the relative strengths of various fluid forces, that they (slightly modified—usually by taking the inverse) have been given identifying names.

The first parameter, $\mu/\rho VL$, is by tradition inverted to the form $\rho VL/\mu$, and was actually explored independently of dimensional analysis in the 1880s by Osborne Reynolds, the British engineer, who studied the transition between laminar and turbulent flow regimes in a tube. He discovered that the parameter (later named after him)

$$Re = \frac{\rho \overline{V} D}{\mu} = \frac{\overline{V} D}{\nu}$$

is a criterion by which the flow regime may be determined. Later experiments have shown that the *Reynolds number* is a key parameter for other flow cases as well. Thus, in general,

$$Re = \frac{\rho VL}{\mu} = \frac{VL}{\nu} \tag{7.11}$$

where L is a characteristic length descriptive of the flow field geometry. The Reynolds number is the ratio of inertia forces to viscous forces. Flows with "large" Reynolds number generally are turbulent. Flows in which the inertia forces are "small" compared with the viscous forces are characteristically laminar flows.

In aerodynamic and other model testing, it is convenient to modify the second parameter, $\Delta p/\rho V^2$, by inserting a factor $\frac{1}{2}$ to make the denominator represent the dynamic pressure (the factor, of course, does not affect the dimensions). The ratio

$$Eu = \frac{\Delta p}{\frac{1}{2}\rho V^2} \tag{7.12}$$

is formed, where Δp is the local pressure minus the freestream pressure, and ρ and V are properties of the freestream flow. This ratio has been named after Leonhard Euler, the Swiss mathematician who did much early analytical work in fluid mechanics. Euler is credited with being the first to recognize the role of pressure in fluid motion; the Euler equations of Chapter 6 demonstrate this role. The *Euler number* is the ratio of pressure forces to inertia forces. The Euler number is often called the *pressure coefficient*, C_p.

In the study of cavitation phenomena, the pressure difference, Δp, is taken as $\Delta p = p - p_v$, where p is the pressure in the liquid stream, and p_v is the liquid vapor pressure at the test temperature. Combining these with ρ and V in the stream yields the dimensionless parameter called the *cavitation number*,

$$Ca = \frac{p - p_v}{\frac{1}{2}\rho V^2} \tag{7.13}$$

The smaller the cavitation number, the more likely cavitation is to occur. This is usually an unwanted phenomenon.

William Froude was a British naval architect. Together with his son, Robert Edmund Froude, he discovered that the parameter

$$Fr = \frac{V}{\sqrt{gL}} \tag{7.14}$$

was significant for flows with free surface effects. Squaring the *Froude number* gives

$$Fr^2 = \frac{V^2}{gL}$$

which may be interpreted as the ratio of inertia forces to gravity forces (it is the inverse of the third force ratio, V^2/gL, that we discussed above). The length, L, is a characteristic length descriptive of the flow field. In the case of open-channel flow, the characteristic length is the water depth; Froude numbers less than unity indicate subcritical flow and values greater than unity indicate supercritical flow. We will have much more to say on this in Chapter 11.

By convention, the inverse of the fourth force ratio, $\sigma/\rho V^2 L$, discussed above, is called the *Weber* number; it indicates the ratio of inertia to surface tension forces

$$We = \frac{\rho V^2 L}{\sigma} \tag{7.15}$$

The value of the Weber number is indicative of the existence of, and frequency of, capillary waves at a free surface.

In the 1870s, the Austrian physicist Ernst Mach introduced the parameter

$$M = \frac{V}{c} \tag{7.16}$$

where V is the flow speed and c is the local sonic speed. Analysis and experiments have shown that the *Mach number* is a key parameter that characterizes compressibility effects in a flow. The Mach number may be written

$$M = \frac{V}{c} = \frac{V}{\sqrt{\dfrac{dp}{d\rho}}} = \frac{V}{\sqrt{\dfrac{E_v}{\rho}}} \quad \text{or} \quad M^2 = \frac{\rho V^2 L^2}{E_v L^2} = \frac{\rho V^2}{E_v}$$

which is the inverse of the final force ratio, $E_v/\rho V^2$, discussed above, and can be interpreted as a ratio of inertia forces to forces due to compressibility. For truly incompressible flow (and note that under some conditions even liquids are quite compressible), $c = \infty$ so that $M = 0$.

Equations 7.11 through 7.16 are some of the most commonly used dimensionless groupings in fluid mechanics because for any flow pattern they immediately (even before performing any experiments or analysis) indicate the relative importance of inertia, viscosity, pressure, gravity, surface tension, and compressibility.

7.6 Flow Similarity and Model Studies

To be useful, a model test must yield data that can be scaled to obtain the forces, moments, and dynamic loads that would exist on the full-scale prototype. What conditions must be met to ensure the similarity of model and prototype flows?

VIDEO

Geometric, Not Dynamic, Similarity: Flow Past a Block 1.

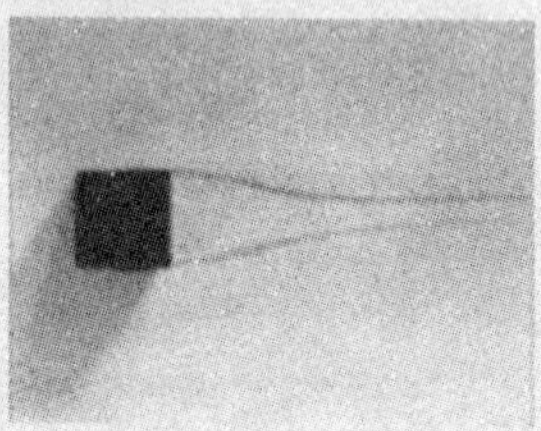

VIDEO

Geometric, Not Dynamic, Similarity: Flow Past a Block 2.

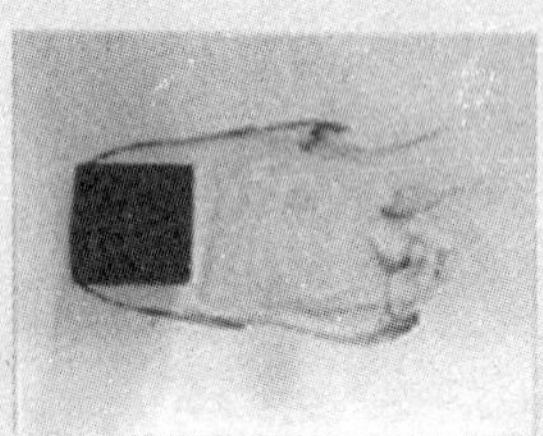

Perhaps the most obvious requirement is that the model and prototype must be geometrically similar. *Geometric similarity* requires that the model and prototype be the same shape, and that all linear dimensions of the model be related to corresponding dimensions of the prototype by a constant scale factor.

A second requirement is that the model and prototype flows must be *kinematically similar*. Two flows are kinematically similar when the velocities at corresponding points are in the same direction and differ only by a constant scale factor. Thus two flows that are kinematically similar also have streamline patterns related by a constant scale factor. Since the boundaries form the bounding streamlines, flows that are kinematically similar must be geometrically similar.

In principle, in order to model the performance in an infinite flow field correctly, kinematic similarity would require that a wind tunnel of infinite cross section be used to obtain data for drag on an object. In practice, this restriction may be relaxed considerably, permitting use of equipment of reasonable size.

Kinematic similarity requires that the regimes of flow be the same for model and prototype. If compressibility or cavitation effects, which may change even the qualitative patterns of flow, are not present in the prototype flow, they must be avoided in the model flow.

When two flows have force distributions such that identical types of forces are parallel and are related in magnitude by a constant scale factor at all corresponding points, the flows are *dynamically similar*.

The requirements for dynamic similarity are the most restrictive. Kinematic similarity requires geometric similarity; kinematic similarity is a necessary, but not sufficient, requirement for dynamic similarity.

To establish the conditions required for complete dynamic similarity, all forces that are important in the flow situation must be considered. Thus the effects of viscous forces, of pressure forces, of surface tension forces, and so on, must be considered. Test conditions must be established such that all important forces are related by the same scale factor between model and prototype flows. When dynamic similarity exists, data measured in a model flow may be related quantitatively to conditions in the prototype flow. What, then, are the conditions that ensure dynamic similarity between model and prototype flows?

The Buckingham Pi theorem may be used to obtain the governing dimensionless groups for a flow phenomenon; to achieve dynamic similarity between geometrically similar flows, we must make sure that each independent dimensionless group has the same value in the model and in the prototype. Then not only will the forces have the same relative importance, but also the dependent dimensionless group will have the same value in the model and prototype.

For example, in considering the drag force on a sphere in Example 7.1, we began with

$$F = f(D,\ V,\ \rho,\ \mu)$$

The Buckingham Pi theorem predicted the functional relation

$$\frac{F}{\rho V^2 D^2} = f_1\left(\frac{\rho V D}{\mu}\right)$$

In Section 7.5 we showed that the dimensionless parameters can be viewed as ratios of forces. Thus, in considering a model flow and a prototype flow about a sphere (the flows are geometrically similar), the flows also will be dynamically similar if the value of the independent parameter, $\rho V D/\mu$, is duplicated between model and prototype, i.e., if

$$\left(\frac{\rho V D}{\mu}\right)_{\text{model}} = \left(\frac{\rho V D}{\mu}\right)_{\text{prototype}}$$

Furthermore, if

$$Re_{\text{model}} = Re_{\text{prototype}}$$

then the value of the dependent parameter, $F/\rho V^2 D^2$, in the functional relationship, will be duplicated between model and prototype, i.e.,

$$\left(\frac{F}{\rho V^2 D^2}\right)_{\text{model}} = \left(\frac{F}{\rho V^2 D^2}\right)_{\text{prototype}}$$

and the results determined from the model study can be used to predict the drag on the full-scale prototype.

The actual force on the object caused by the fluid is not the same for the model and prototype, but the value of its dimensionless group is. The two tests can be run using different fluids, if desired, as long as the Reynolds numbers are matched. For experimental convenience, test data can be measured in a wind tunnel in air and the results used to predict drag in water, as illustrated in Example 7.4.

Example 7.4 SIMILARITY: DRAG OF A SONAR TRANSDUCER

The drag of a sonar transducer is to be predicted, based on wind tunnel test data. The prototype, a 0.3 m-diameter sphere, is to be towed at 2.57 m/s in seawater at 4.5°C. The model is 152 mm. in diameter. Determine the required test speed in air. If the drag of the model at these test conditions is 2.7 N estimate the drag of the prototype.

Given: Sonar transducer to be tested in a wind tunnel.

Find: (a) V_m.
(b) F_p.

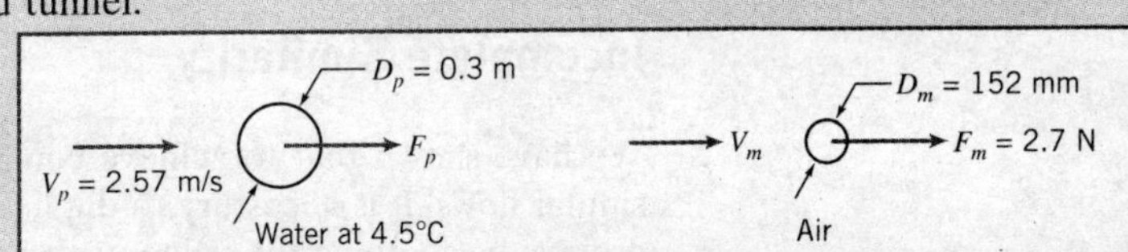

Solution:
Since the prototype operates in water and the model test is to be performed in air, useful results can be expected only if cavitation effects are absent in the prototype flow and compressibility effects are absent from the model test. Under these conditions,

$$\frac{F}{\rho V^2 D^2} = f\left(\frac{\rho V D}{\mu}\right)$$

and the test should be run at

$$Re_{\text{model}} = Re_{\text{prototype}}$$

to ensure dynamic similarity. For seawater at 4.5°C, $\rho = 1000\ \text{kg/m}^3$ and $\nu \approx 1.57 \times 10^{-6}\ \text{m}^2/\text{s}$. At prototype conditions,

$$V_p = 2.57\ \text{m/s}$$

$$Re_p = \frac{V_p D_p}{\nu_p} = 2.57\frac{\text{m}}{\text{s}} \times 0.3\ \text{m} \times \frac{1}{1.57 \times 10^{-6}\ \text{m}^2/\text{s}} = 4.91 \times 10^5$$

The model test conditions must duplicate this Reynolds number. Thus

$$Re_m = \frac{V_m D_m}{\nu_m} = 4.91 \times 10^5$$

For air at STP, $\rho = 1.227\ \text{kg/m}^3$ and $\nu = 1.46 \times 10^{-5}\ \text{m}^2/\text{s}$. The wind tunnel must be operated at

$$V_m = Re_m \frac{\nu_m}{D_m} = 4.91 \times 10^5 \times 1.46 \times 10^{-5} \frac{\text{m}^2}{\text{s}} \times \frac{1}{152\ \text{mm}} \times \frac{1000\ \text{mm}}{\text{m}}$$

$$V_m = 47.16\ \text{m/s} \longleftarrow V_m$$

This speed is low enough to neglect compressibility effects.

At these test conditions, the model and prototype flows are dynamically similar. Hence

$$\left.\frac{F}{\rho V^2 D^2}\right)_m = \left.\frac{F}{\rho V^2 D^2}\right)_p$$

and

$$F_p = F_m \frac{\rho_p}{\rho_m} \frac{V_p^2}{V_m^2} \frac{D_p^2}{D_m^2} = 2.7\ \text{N} \times \frac{1000}{1.227} \times \frac{(2.57)^2}{(47.16)^2} \times \frac{(0.3)^2}{(0.152)^2}$$

$$F_p = 25.5\ \text{N} \longleftarrow F_p$$

If cavitation were expected—if the sonar probe were operated at high speed near the free surface of the seawater—then useful results could not be obtained from a model test in air.

This problem:

✓ Demonstrates the calculation of prototype values from model test data.

✓ "Reinvented the wheel": the results for drag on a smooth sphere are very well known, so we did not need to do a model experiment but instead could have simply read from the graph of Fig. 7.1 the value of $C_D = F_p / \left(\frac{1}{2}\rho V_p^2 \frac{\pi}{4} D_p^2\right) \approx 0.1$, corresponding to a Reynolds number of 4.91×10^5. Then $F_p \approx 23.3$ N can easily be computed. We will have more to say on drag coefficients in Chapter 9.

Incomplete Similarity

We have shown that to achieve complete dynamic similarity between geometrically similar flows, it is necessary to duplicate the values of the independent dimensionless groups; by so doing the value of the dependent parameter is then duplicated.

In the simplified situation of Example 7.4, duplicating the Reynolds number value between model and prototype ensured dynamically similar flows. Testing in air allowed the Reynolds number to be duplicated exactly (this also could have been accomplished in a water tunnel for this situation). The drag force on a sphere actually depends on the nature of the boundary-layer flow. Therefore, geometric similarity requires that the relative surface roughness of the model and prototype be the same. This means that relative roughness also is a parameter that must be duplicated between model and prototype situations. If we assume that the model was constructed carefully, measured values of drag from model tests could be scaled to predict drag for the operating conditions of the prototype.

In many model studies, to achieve dynamic similarity requires duplication of several dimensionless groups. In some cases, complete dynamic similarity between model and prototype may not be attainable. Determining the drag force (resistance) of a surface ship is an example of such a situation. Resistance on a surface ship arises from skin friction on the hull (viscous forces) and surface wave resistance (gravity forces). Complete dynamic similarity requires that both Reynolds and Froude numbers be duplicated between model and prototype.

In general it is not possible to predict wave resistance analytically, so it must be modeled. This requires that

$$Fr_m = \frac{V_m}{(gL_m)^{1/2}} = Fr_p = \frac{V_p}{(gL_p)^{1/2}}$$

To match Froude numbers between model and prototype therefore requires a velocity ratio of

$$\frac{V_m}{V_p} = \left(\frac{L_m}{L_p}\right)^{1/2}$$

to ensure dynamically similar surface wave patterns.

Hence for any model length scale, matching the Froude numbers determines the velocity ratio. Only the kinematic viscosity can then be varied to match Reynolds numbers. Thus

$$Re_m = \frac{V_m L_m}{\nu_m} = Re_p = \frac{V_p L_p}{\nu_p}$$

leads to the condition that

$$\frac{\nu_m}{\nu_p} = \frac{V_m}{V_p}\frac{L_m}{L_P}$$

If we use the velocity ratio obtained from matching the Froude numbers, equality of Reynolds numbers leads to a kinematic viscosity ratio requirement of

$$\frac{\nu_m}{\nu_p} = \left(\frac{L_m}{L_p}\right)^{1/2}\frac{L_m}{L_p} = \left(\frac{L_m}{L_p}\right)^{3/2}$$

If $L_m/L_p = \frac{1}{100}$ (a typical length scale for ship model tests), then ν_m/ν_p must be $\frac{1}{1000}$. Figure A.3 shows that mercury is the only liquid with kinematic viscosity less than that of water. However, it is only about an order of magnitude less, so the kinematic viscosity ratio required to duplicate Reynolds numbers cannot be attained.

We conclude that we have a problem: it is impossible in practice for this model/prototype scale of $\frac{1}{100}$ to satisfy both the Reynolds number and Froude number criteria; at best we will be able to satisfy only one of them. In addition, water is the only practical fluid for most model tests of free-surface flows. To obtain complete dynamic similarity then would require a full-scale test. However, all is not lost: Model studies do provide useful information even though complete similarity cannot be obtained. As an example, Fig. 7.2 shows data from a test of a 1:80 scale model of a ship conducted at the U.S. Naval Academy Hydromechanics Laboratory. The plot displays "resistance coefficient" data versus Froude number. The square points are calculated from values of total resistance measured in the test. We would like to obtain the corresponding total resistance curve for the full-scale ship.

If you think about it, we can *only* measure the total drag (the square data points). The total drag is due to both wave resistance (dependent on the Froude number) and friction resistance (dependent on the Reynolds number), and it's not possible to determine experimentally how much each contributes. We *cannot* use the total drag curve of Fig. 7.2 for the full-scale ship because, as we have discussed above, we can never set up the model conditions so that its Reynolds number *and* Froude number match those of the full-scale ship. Nevertheless, we would like to extract from Fig. 7.2 the corresponding total drag curve for the full-scale ship. In many experimental situations we need to use a creative "trick" to come up with a solution. In this case, the experimenters used boundary-layer theory (which we discuss in Chapter 9) to *predict* the viscous resistance component of the model (shown as diamonds in Fig. 7.2); then they estimated the wave resistance (not obtainable from theory) by simply subtracting this theoretical viscous resistance from the experimental total resistance, point by point (shown as circles in Fig. 7.2).

Using this clever idea (typical of the kind of experimental and analytical approaches experimentalists need to employ), Fig. 7.2 therefore gives the wave

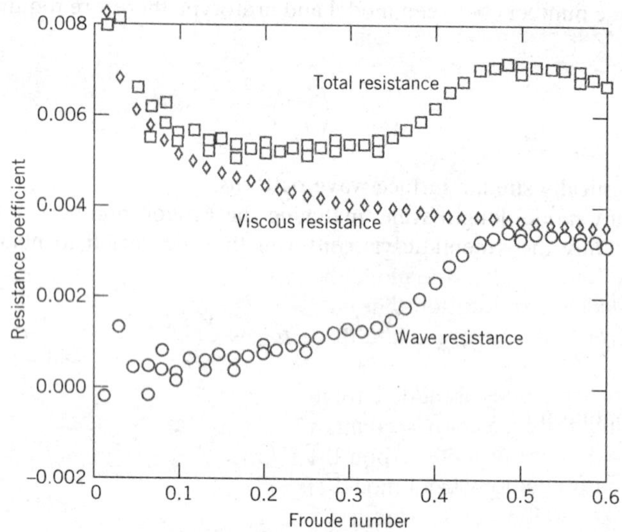

Fig. 7.2 Data from test of 1:80 scale model of U.S. Navy guided missile frigate *Oliver Hazard Perry* (FFG-7). (Data from U.S. Naval Academy Hydromechanics Laboratory, courtesy of Professor Bruce Johnson.)

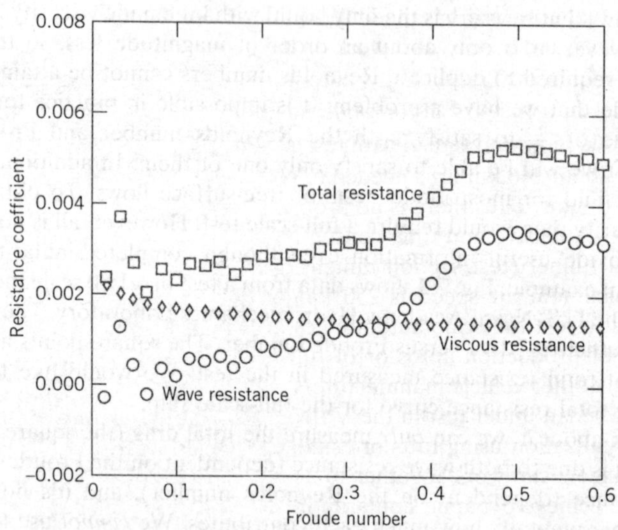

Fig. 7.3 Resistance of full-scale ship predicted from model test results. (Data from U.S. Naval Academy Hydromechanics Laboratory, courtesy of Professor Bruce Johnson.)

resistance of the model as a function of Froude number. It is *also* valid for the full-scale ship, because wave resistance depends only on the Froude number! We can now build a graph similar to Fig. 7.2 valid for the full-scale ship: Simply compute from boundary-layer theory the viscous resistance of the full-scale ship and add this to the wave resistance values, point by point. The result is shown in Fig. 7.3. The wave resistance points are identical to those in Fig. 7.2; the viscous resistance points are computed from theory (and are different from those of Fig. 7.2); and the predicted total resistance curve for the full-scale ship is finally obtained.

In this example, incomplete modeling was overcome by using analytical computations; the model experiments modeled the Froude number, but not the Reynolds number.

Because the Reynolds number cannot be matched for model tests of surface ships, the boundary-layer behavior is not the same for model and prototype. The model Reynolds number is only $(L_m/L_p)^{3/2}$ as large as the prototype value, so the extent of laminar flow in the boundary layer on the model is too large by a corresponding factor. The method just described assumes that boundary-layer behavior can be scaled. To make this possible, the model boundary layer is "tripped" or "stimulated" to become turbulent at a location that corresponds to the behavior on the full-scale vessel. "Studs" were used to stimulate the boundary layer for the model test results shown in Fig. 7.2.

A correction sometimes is added to the full-scale coefficients calculated from model test data. This correction accounts for roughness, waviness, and unevenness that inevitably are more pronounced on the full-scale ship than on the model. Comparisons between predictions from model tests and measurements made in full-scale trials suggest an overall accuracy within ± 5 percent [5].

As we will see in Chapter 11, the Froude number is an important parameter in the modeling of rivers and harbors. In these situations it is not practical to obtain complete similarity. Use of a reasonable model scale would lead to extremely small water depths, so that viscous forces and surface tension forces would have much larger relative effects in the model flow than in the prototype. Consequently, different length scales are used for the vertical and horizontal directions. Viscous forces in the deeper model flow are increased using artificial roughness elements.

Emphasis on fuel economy has made reduction of aerodynamic drag important for automobiles, trucks, and buses. Most work on development of low-drag configurations is done using model tests. Traditionally, automobile models have been built to $\frac{3}{8}$ scale, at which a model of a full-size automobile has a frontal area of about 0.3 m^2. Thus testing can be done in a wind tunnel with test section area of 6 m^2 or larger. At $\frac{3}{8}$ scale, a wind speed of about 240 km/hr is needed to model a prototype automobile traveling at the legal speed limit. Thus there is no problem with compressibility effects, but the scale models are expensive and time-consuming to build.

A large wind tunnel (test section dimensions are 5.4 m high, 10.4 m wide, and 21.3 m long; maximum air speed is 250 km/hr with the tunnel empty) is used by General Motors to test full-scale automobiles at highway speeds. The large test section allows use of production autos or of full-scale clay mockups of proposed auto body styles. Many other vehicle manufacturers are using comparable facilities; Fig. 7.4 shows a full-size sedan under test in the Volvo wind tunnel. The relatively low speed permits flow visualization using tufts or "smoke" streams.[1] Using full-size "models," stylists and engineers can work together to achieve optimum results.

It is harder to achieve dynamic similarity in tests of trucks and buses; models must be made to smaller scale than those for automobiles.[2] A large scale for truck and bus testing is 1:8. To achieve complete dynamic similarity by matching Reynolds numbers at this scale would require a test speed of 700 km/hr. This would introduce unwanted compressibility effects, and model and prototype flows would not be kinematically similar. Fortunately, trucks and buses are "bluff" objects. Experiments show that above a certain Reynolds number, their nondimensional drag becomes independent

[1] A mixture of liquid nitrogen and steam may be used to produce "smoke" streaklines that evaporate and do not clog the fine mesh screens used to reduce the turbulence level in a wind tunnel. Streaklines may be made to appear "colored" in photos by placing a filter over the camera lens. This and other techniques for flow visualization are detailed in Reference [6] and Merzkirch [7].

[2] The vehicle length is particularly important in tests at large yaw angles to simulate crosswind behavior. Tunnel blockage considerations limit the acceptable model size. See Reference [8] for recommended practices.

Fig. 7.4 Full-scale automobile under test in Volvo wind tunnel, using smoke streaklines for flow visualization. (Photograph courtesy of Volvo Cars of North America, Inc.)

of Reynolds number [8]. (Figure 7.1 actually shows an example of this—for a sphere, the dimensionless drag is approximately constant for $2000 < Re < 2 \times 10^5$.) Although similarity is not complete, measured test data can be scaled to predict prototype drag forces. The procedure is illustrated in Example 7.5.

Example 7.5 INCOMPLETE SIMILARITY: AERODYNAMIC DRAG ON A BUS

The following wind tunnel test data from a 1:16 scale model of a bus are available:

Air Speed (m/s)	18.0	21.8	26.0	30.1	35.0	38.5	40.9	44.1	46.7
Drag Force (N)	3.10	4.41	6.09	7.97	10.7	12.9	14.7	16.9	18.9

Using the properties of standard air, calculate and plot the dimensionless aerodynamic drag coefficient,

$$C_D = \frac{F_D}{\frac{1}{2}\rho V^2 A}$$

versus Reynolds number $Re = \rho V w/\mu$, where w is model width. Find the minimum test speed above which C_D remains constant. Estimate the aerodynamic drag force and power requirement for the prototype vehicle at 100 km/hr. (The width and frontal area of the prototype are 2.44 m and 7.8 m^2, respectively.)

Given: Data from a wind tunnel test of a model bus. Prototype dimensions are width of 2.44 m and frontal area of 7.8 m^2. Model scale is 1:16. Standard air is the test fluid.

Find: (a) Aerodynamic drag coefficient, $C_D = F_D/\frac{1}{2}\rho V^2 A$, versus Reynolds number, $Re = \rho V w/\mu$; plot.
(b) Speed above which C_D is constant.
(c) Estimated aerodynamic drag force and power required for the full-scale vehicle at 100 km/hr.

Solution:
The model width is

$$w_m = \frac{1}{16} w_p = \frac{1}{16} \times 2.44 \text{ m} = 0.152 \text{ m}$$

The model area is

$$A_m = \left(\frac{1}{16}\right)^2 A_p = \left(\frac{1}{16}\right)^2 \times 7.8 \text{ m}^2 = 0.0305 \text{ m}^2$$

The aerodynamic drag coefficient may be calculated as

$$C_D = \frac{F_D}{\frac{1}{2}\rho V^2 A}$$

$$= 2 \times F_D(\text{N}) \times \frac{\text{m}^3}{1.23 \text{ kg}} \times \frac{\text{s}^2}{(V)^2 \text{ m}^2} \times \frac{1}{0.0305 \text{ m}^2} \times \frac{\text{kg} \cdot \text{m}}{\text{N} \cdot \text{s}^2}$$

$$C_D = \frac{53.3\, F_D(\text{N})}{[V(\text{m/s})]^2}$$

The Reynolds number may be calculated as

$$Re = \frac{\rho V w}{\mu} = \frac{V w}{\nu} = V \frac{\text{m}}{\text{s}} \times 0.152 \text{ m} \times \frac{\text{s}}{1.46 \times 10^{-5} \text{ m}^2}$$

$$Re = 1.04 \times 10^4\ V(\text{m/s})$$

The calculated values are plotted in the following figure:

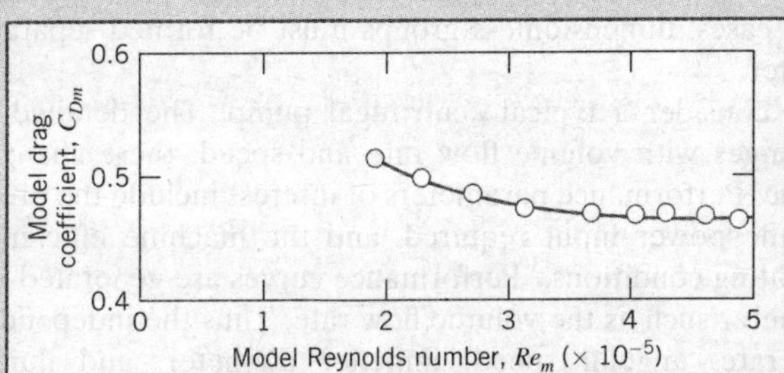

C_{D_m} versus Re_m

The plot shows that the model drag coefficient becomes constant at $C_{Dm} \approx 0.46$ above $Re_m = 4 \times 10^5$, which corresponds to an air speed of approximately 40 m/s. Since the drag coefficient is independent of Reynolds number above $Re \approx 4 \times 10^5$, then for the prototype vehicle ($Re \approx 4.5 \times 10^6$), $C_D \approx 0.46$. The drag force on the full-scale vehicle is

$$F_{D_p} = C_D \frac{1}{2} \rho V_p^2 A_p$$

$$= \frac{0.46}{2} \times 1.23 \frac{\text{kg}}{\text{m}^3} \left(100 \frac{\text{km}}{\text{hr}} \times 1000 \frac{\text{m}}{\text{km}} \times \frac{\text{hr}}{3600 \text{ s}}\right)^2 \times 7.8 \text{ m}^2 \times \frac{\text{N} \cdot \text{s}^2}{\text{kg} \cdot \text{m}}$$

$$F_{D_p} = 1.71 \text{ kN}$$

F_{Dp}

The corresponding power required to overcome aerodynamic drag is

$$\mathcal{P}_p = F_{D_p} V_p$$

$$= 1.71 \times 10^3 \text{ N} \times 100 \frac{\text{km}}{\text{hr}} \times 1000 \frac{\text{m}}{\text{km}} \times \frac{\text{hr}}{3600 \text{ s}} \times \frac{\text{W} \cdot \text{s}}{\text{N} \cdot \text{m}}$$

$$\mathcal{P}_p = 47.5 \text{ kW} \qquad \mathcal{P}_p$$

This problem illustrates a common phenomenon in aerodynamics: Above a certain minimum Reynolds number the drag coefficient of an object usually approaches a constant—that is, becomes independent of the Reynolds number. Hence, in these situations we do not have to match the Reynolds numbers of the model and prototype in order for them to have the same drag coefficient—a considerable advantage. However, the SAE Recommended Practices [8] suggests $Re \geq 2 \times 10^6$ for truck and bus testing.

For additional details on techniques and applications of dimensional analysis consult [9–12].

Scaling with Multiple Dependent Parameters

In some situations of practical importance there may be more than one dependent parameter. In such cases, dimensionless groups must be formed separately for each dependent parameter.

As an example, consider a typical centrifugal pump. The detailed flow pattern within a pump changes with volume flow rate and speed; these changes affect the pump's performance. Performance parameters of interest include the pressure rise (or head) developed, the power input required, and the machine efficiency measured under specific operating conditions.[3] Performance curves are generated by varying an independent parameter such as the volume flow rate. Thus the independent variables are volume flow rate, angular speed, impeller diameter, and fluid properties. Dependent variables are the several performance quantities of interest.

Finding dimensionless parameters begins from the symbolic equations for the dependence of head, h (energy per unit mass, L^2/t^2), and power, $\mathcal{P}$, on the independent parameters, given by

$$h = g_1(Q, \ \rho, \ \omega, \ D, \ \mu)$$

and

$$\mathcal{P} = g_2(Q, \ \rho, \ \omega, \ D, \ \mu)$$

Straightforward use of the Pi theorem gives the dimensionless *head coefficient* and *power coefficient* as

$$\frac{h}{\omega^2 D^2} = f_1\left(\frac{Q}{\omega D^3}, \ \frac{\rho \omega D^2}{\mu}\right) \tag{7.17}$$

[3]Efficiency is defined as the ratio of power delivered to the fluid divided by input power, $\eta = \mathcal{P}/\mathcal{P}_{in}$. For incompressible flow, we will see in Chapter 8 that the energy equation reduces to $\mathcal{P} = \rho Q h$ (when "head" h is expressed as energy per unit mass) or to $\mathcal{P} = \rho g Q H$ (when head H is expressed as energy per unit weight).

and

$$\frac{\mathscr{P}}{\rho\omega^3 D^5} = f_2\left(\frac{Q}{\omega D^3}, \frac{\rho\omega D^2}{\mu}\right) \tag{7.18}$$

The dimensionless parameter $Q/\omega D^3$ in these equations is called the *flow coefficient.* The dimensionless parameter $\rho\omega D^2/\mu$ ($\propto \rho VD/\mu$) is a form of Reynolds number.

Head and power in a pump are developed by inertia forces. Both the flow pattern within a pump and the pump performance change with volume flow rate and speed of rotation. Performance is difficult to predict analytically except at the design point of the pump, so it is measured experimentally. Typical characteristic curves plotted from experimental data for a centrifugal pump tested at constant speed are shown in Fig. 7.5 as functions of volume flow rate. The head, power, and efficiency curves in Fig. 7.5 are smoothed through points calculated from measured data. Maximum efficiency usually occurs at the design point.

Complete similarity in pump performance tests would require identical flow coefficients and Reynolds numbers. In practice, it has been found that viscous effects are relatively unimportant when two geometrically similar machines operate under "similar" flow conditions. Thus, from Eqs. 7.17 and 7.18, when

$$\frac{Q_1}{\omega_1 D_1^3} = \frac{Q_2}{\omega_2 D_2^3} \tag{7.19}$$

it follows that

$$\frac{h_1}{\omega_1^2 D_1^2} = \frac{h_2}{\omega_2^2 D_2^2} \tag{7.20}$$

and

$$\frac{\mathscr{P}_1}{\rho_1 \omega_1^3 D_1^5} = \frac{\mathscr{P}_2}{\rho_2 \omega_2^3 D_2^5} \tag{7.21}$$

The empirical observation that viscous effects are unimportant under similar flow conditions allows use of Eqs. 7.19 through 7.21 to scale the performance characteristics of machines to different operating conditions, as either the speed or diameter is changed. These useful scaling relationships are known as pump or fan "laws." If operating conditions for one machine are known, operating conditions for any geometrically similar machine can be found by changing D and ω according to Eqs. 7.19 through 7.21. (More details on dimensional analysis, design, and performance curves for fluid machinery are presented in Chapter 10.)

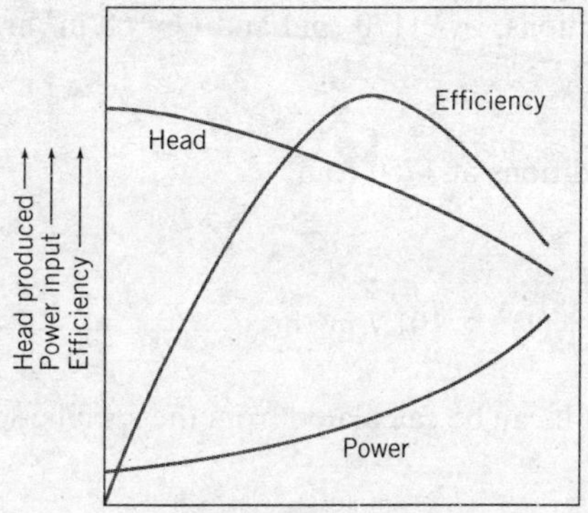

Fig. 7.5 Typical characteristic curves for centrifugal pump tested at constant speed.

Another useful pump parameter can be obtained by eliminating the machine diameter from Eqs. 7.19 and 7.20. If we designate $\Pi_1 = Q/\omega D^3$ and $\Pi_2 = h/\omega^2 D^2$, then the ratio $\Pi_1^{1/2}/\Pi_2^{3/4}$ is another dimensionless parameter; this parameter is the *specific speed*, N_s,

$$N_s = \frac{\omega Q^{1/2}}{h^{3/4}} \tag{7.22a}$$

The specific speed, as defined in Eq. 7.22a, is a dimensionless parameter (provided that the head, h, is expressed as energy per unit mass). You may think of specific speed as the speed required for a machine to produce unit head at unit volume flow rate. A constant specific speed describes all operating conditions of geometrically similar machines with similar flow conditions.

Although specific speed is a dimensionless parameter, it is common practice to use a convenient but inconsistent set of units in specifying the variables ω and Q, and to use the energy per unit weight H in place of energy per unit mass h in Eq. 7.22a. When this is done the specific speed,

$$N_{s_{cu}} = \frac{\omega Q^{1/2}}{H^{3/4}} \tag{7.22b}$$

is not a unitless parameter and its magnitude depends on the units used to calculate it. Metric units used in engineering practice for pumps are rpm for ω, m^3/h for Q, and meters (energy per unit weight) for H. In these metric units, "low" specific speed means $580 < N_{s_{cu}} < 4645$ and "high" means $11{,}615 < N_{s_{cu}} < 17{,}420$. Example 7.6 illustrates use of the pump scaling laws and specific speed parameter. More details of specific speed calculations and additional examples of applications to fluid machinery are presented in Chapter 10.

Example 7.6 PUMP "LAWS"

A centrifugal pump has an efficiency of 80 percent at its design-point specific speed of 2323 (units of rpm, m^3/hr, and meters). The impeller diameter is 200 mm. At design-point flow conditions, the volume flow rate is 68 m^3/hr of water at 1170 rpm. To obtain a higher flow rate, the pump is to be fitted with a 1750 rpm motor. Use the pump "laws" to find the design-point performance characteristics of the pump at the higher speed. Show that the specific speed remains constant for the higher operating speed. Determine the motor size required.

Given: Centrifugal pump with design specific speed of 2323 (in rpm, m^3/hr, and meter units). Impeller diameter is $D = 200$ mm. At the pump's design-point flow conditions, $\omega = 1170$ rpm and $Q = 68\ m^3/hr$, with water.

Find: (a) Performance characteristics,
(b) specific speed, and
(c) motor size required, for similar flow conditions at 1750 rpm.

Solution: From pump "laws," $Q/\omega D^3$ = constant, so

$$Q_2 = Q_1 \frac{\omega_2}{\omega_1}\left(\frac{D_2}{D_1}\right)^3 = 68\ m^3/hr \left(\frac{1750}{1170}\right)(1)^3 = 101.7\ m^3/hr \qquad Q_2$$

The pump head is not specified at $\omega_1 = 1170$ rpm, but it can be calculated from the specific speed, $N_{s_{cu}} = 2323$. Using the given units and the definition of $N_{s_{cu}}$,

$$N_{s_{cu}} = \frac{\omega Q^{1/2}}{H^{3/4}} \quad \text{so} \quad H_1 = \left(\frac{\omega_1 Q_1^{1/2}}{N_{s_{cu}}}\right)^{4/3} = 6.68\ m$$

Then $H/\omega^2 D^2 =$ constant, so

$$H_2 = H_1\left(\frac{\omega_2}{\omega_1}\right)^2\left(\frac{D_2}{D_1}\right)^2 = 6.68\text{ m}\left(\frac{1750}{1170}\right)^2(1)^2 = 14.94\text{ m} \qquad H_2$$

The pump output power is $\mathcal{P}_1 = \rho g Q_1 H_1$, so at $\omega_1 = 1170$ rpm,

$$\mathcal{P}_1 = 1000\frac{\text{kg}}{\text{m}^3}\times 9.81\frac{\text{m}}{\text{s}^2}\times 68\frac{\text{m}^3}{\text{hr}}\times 6.68\text{ m}\times\frac{\text{hr}}{60\text{ min}}\times\frac{\text{min}}{60\text{ s}}$$

$$\mathcal{P}_1 = 1.24\text{ kW}$$

But $\mathcal{P}/\rho\omega^3 D^5 =$ constant, so

$$\mathcal{P}_2 = \mathcal{P}_1\left(\frac{\rho_2}{\rho_1}\right)\left(\frac{\omega_2}{\omega_1}\right)^3\left(\frac{D_2}{D_1}\right)^5 = 1.24\text{ kW}(1)\left(\frac{1750}{1170}\right)^3(1)^5 = 4.15\text{ kW} \qquad \mathcal{P}_2$$

The required input power may be calculated as

$$\mathcal{P}_{in} = \frac{\mathcal{P}_2}{\eta} = \frac{4.15\text{ kW}}{0.80} = 5.19\text{ kW} \qquad \mathcal{P}_{in}$$

Thus a 5.6 kW motor (the next larger standard size) probably would be specified.

The specific speed at $\omega_2 = 1750$ rpm is

$$N_{S_{cu}} = \frac{\omega Q^{1/2}}{H^{3/4}} = \frac{1750(101.7)^{1/2}}{(14.94)^{3/4}} = 2323 \qquad N_{S_{cu}}$$

This problem illustrates application of the pump "laws" and specific speed to scaling of performance data. Pump and fan "laws" are used widely in industry to scale performance curves for families of machines from a single performance curve, and to specify drive speed and power in machine applications.

Comments on Model Testing

While outlining the procedures involved in model testing, we have tried not to imply that testing is a simple task that automatically gives results that are easily interpreted, accurate, and complete. As in all experimental work, careful planning and execution are needed to obtain valid results. Models must be constructed carefully and accurately, and they must include sufficient detail in areas critical to the phenomenon being measured. Aerodynamic balances or other force measuring systems must be aligned carefully and calibrated correctly. Mounting methods must be devised that offer adequate rigidity and model motion, yet do not interfere with the phenomenon being measured. References [13–15] are considered the standard sources for details of wind tunnel test techniques. More specialized techniques for water impact testing are described in Waugh and Stubstad [16].

Experimental facilities must be designed and constructed carefully. The quality of flow in a wind tunnel must be documented. Flow in the test section should be as nearly uniform as possible (unless the desire is to simulate a special profile such as an atmospheric boundary layer), free from angularity, and with little swirl. If they interfere with measurements, boundary layers on tunnel walls must be removed by suction or energized by blowing. Pressure gradients in a wind tunnel test section may cause erroneous drag-force readings due to pressure variations in the flow direction.

Special facilities are needed for unusual conditions or for special test requirements, especially to achieve large Reynolds numbers. Many facilities are so large or specialized that they cannot be supported by university laboratories or private industry. A few examples include [17–19]:

- National Full-Scale Aerodynamics Complex, NASA, Ames Research Center, Moffett Field, California.
  Two wind tunnel test sections, powered by a 93255 kW electric drive system:
  - 12 m high and 24 m wide test section, maximum wind speed of 154 m/s.
  - 24 m high and 36 m wide test section, maximum wind speed of 70.5 m/s.
- U.S. Navy, David Taylor Research Center, Carderock, Maryland.
  - High-Speed Towing Basin 905 m long, 6.4 m wide, and 4.9 m deep. Towing carriage can travel at up to 51 m/s while measuring drag loads to 35,600 N and side loads to 8,900 N.
  - 0.91 m variable-pressure water tunnel with 25.7 m/s maximum test speed at pressures between 13.8 and 413.4 kPa (abs).
  - Anechoic Flow Facility with quiet, low-turbulence air flow in 0.75 m^2 by 6.4 m-long open-jet test section. Flow noise at maximum speed of 61 m/s is less than that of conversational speech.
- U.S. Army Corps of Engineers, Sausalito, California.
  - San Francisco Bay and Delta Model with slightly more than 4,047 m^2 in area, 1:1000 horizontal scale and 1:100 vertical scale, 0.85 m^3/s of pumping capacity, use of fresh and salt water, and tide simulation.
- NASA, Langley Research Center, Hampton, Virginia.
  - National Transonic Facility (NTF) with cryogenic technology (temperatures as low as −184°C) to reduce gas viscosity, raising Reynolds number by a factor of 6, while halving drive power.

7.7 Summary and Useful Equations

In this chapter we have:

✓ Obtained dimensionless coefficients by nondimensionalizing the governing differential equations of a problem.
✓ Stated the Buckingham Pi theorem and used it to determine the independent and dependent dimensionless parameters from the physical parameters of a problem.
✓ Defined a number of important dimensionless groups: the Reynolds number, Euler number, cavitation number, Froude number, Weber number, and Mach number, and discussed their physical significance.

We have also explored some ideas behind modeling: geometric, kinematic, and dynamic similarity, incomplete modeling, and predicting prototype results from model tests.

Note: Most of the Useful Equations in the table below have a number of constraints or limitations—*be sure to refer to their page numbers for details*!

Useful Equations

Reynolds number (inertia to viscous):	$Re = \dfrac{\rho VL}{\mu} = \dfrac{VL}{\nu}$	**(7.11)**	Page 304
Euler number (pressure to inertia):	$Eu = \dfrac{\Delta p}{\frac{1}{2}\rho V^2}$	**(7.12)**	Page 304
Cavitation number:	$Ca = \dfrac{p - p_v}{\frac{1}{2}\rho V^2}$	**(7.13)**	Page 304

Froude number (inertia to gravity):	$Fr = \dfrac{V}{\sqrt{gL}}$	**(7.14)**	Page 305
Weber number (inertia to surface tension):	$We = \dfrac{\rho V^2 L}{\sigma}$	**(7.15)**	Page 305
Mach number (inertia to compressibility):	$M = \dfrac{V}{c}$	**(7.16)**	Page 305
Centrifugal pump specific speed (in terms of head h):	$N_s = \dfrac{\omega Q^{1/2}}{h^{3/4}}$	**(7.22a)**	Page 316
Centrifugal pump specific speed (in terms of head H):	$N_{S_{cu}} = \dfrac{\omega Q^{1/2}}{H^{3/4}}$	**(7.22b)**	Page 316

Case Study

T. Rex

Tyrannosaurus rex. (California Academy of Sciences)

Dimensional analysis, the main topic of this chapter, is used in many scientific pursuits. It has even been used by Professor Alexander McNeil, now at Heriot-Watt University in Scotland, to try to determine the speed at which dinosaurs such as *Tyrannosaurus rex* may have been able to run. The only data available on these creatures are in the fossil record—the most pertinent data being the dinosaurs' average leg length l and stride s. Could these data be used to extract the dinosaurs' speed? Comparing data on l and s and the speed V of quadrupeds (e.g., horses, dogs) and bipeds (e.g., humans) does not indicate a pattern, unless dimensional analysis is used to learn that all of the data should be plotted in the following way: Plot the dimensionless quantity V^2/gl (where V is the measured speed of the animal and g is the acceleration of gravity) against the dimensionless ratio s/l. When this is done, "magically" the data for most animals fall approximately on one curve! Hence, the running behavior of most animals can be obtained from the graph: In this case, the dinosaurs' value of s/l allows a corresponding value of V^2/gl to be interpolated from the curve, leading to an estimate for V of dinosaurs (because l and g are known). Based on this, in contrast to the *Jurassic Park* movies, it seems likely that humans could easily outrun *T. rex*!

References

1. Kline, S. J., *Similitude and Approximation Theory*. New York: McGraw-Hill, 1965.
2. Hansen, A. G., *Similarity Analysis of Boundary-Value Problems in Engineering*. Englewood Cliffs, NJ: Prentice-Hall, 1964.
3. Schlichting, H., *Boundary Layer Theory*, 7th ed. New York: McGraw-Hill, 1979.
4. Buckingham, E., "On Physically Similar Systems: Illustrations of the Use of Dimensional Equations," *Physical Review*, *4*, 4, 1914, pp. 345–376.
5. Todd, L. H., "Resistance and Propulsion," in *Principles of Naval Architecture*, J. P. Comstock, ed. New York: Society of Naval Architects and Marine Engineers, 1967.

6. "Aerodynamic Flow Visualization Techniques and Procedures." Warrendale, PA: Society of Automotive Engineers, SAE Information Report HS J1566, January 1986.
7. Merzkirch, W., *Flow Visualization*, 2nd ed. New York: Academic Press, 1987.
8. "SAE Wind Tunnel Test Procedure for Trucks and Buses," *Recommended Practice* SAE J1252, Warrendale, PA: Society of Automotive Engineers, 1981.
9. Sedov, L. I., *Similarity and Dimensional Methods in Mechanics*. New York: Academic Press, 1959.
10. Birkhoff, G., *Hydrodynamics—A Study in Logic, Fact, and Similitude*, 2nd ed. Princeton, NJ: Princeton University Press, 1960.
11. Ipsen, D. C., *Units, Dimensions, and Dimensionless Numbers*. New York: McGraw-Hill, 1960.
12. Yalin, M. S., *Theory of Hydraulic Models*. New York: Macmillan, 1971.
13. Pankhurst, R. C., and D. W. Holder, *Wind-Tunnel Technique*. London: Pitman, 1965.
14. Rae, W. H., and A. Pope, *Low-Speed Wind Tunnel Testing*, 2nd ed. New York: Wiley-Interscience, 1984.
15. Pope, A., and K. L. Goin, *High-Speed Wind Tunnel Testing*. New York: Krieger, 1978.
16. Waugh, J. G., and G. W. Stubstad, *Hydroballistics Modeling*. San Diego, CA: U.S. Naval Undersea Center, *ca.* 1965.
17. Baals, D. W., and W. R. Corliss, *Wind Tunnels of NASA*. Washington, D.C.: National Aeronautics and Space Administration, SP-440, 1981.
18. Vincent, M., "The Naval Ship Research and Development Center." Carderock, MD: Naval Ship Research and Development Center, Report 3039 (Revised), November 1971.
19. Smith, B. E., P. T. Zell, and P. M. Shinoda, "Comparison of Model- and Full-Scale Wind-Tunnel Performance," *Journal of Aircraft*, *27*, 3, March 1990, pp. 232–238.

Problems

Nondimensionalizing the Basic Differential Equations

Many of the Problems in this chapter involve obtaining the Π groups that characterize a problem. The *Excel* workbook used in Example 7.1 is useful for performing the computations involved. To avoid needless duplication, the computer symbol will only be used next to Problems when they have an *additional* benefit (e.g., for graphing).

7.1 The propagation speed of small-amplitude surface waves in a region of uniform depth is given by

$$c^2 = \left(\frac{\sigma}{\rho}\frac{2\pi}{\lambda} + \frac{g\lambda}{2\pi}\right)\tanh\frac{2\pi h}{\lambda}$$

where h is depth of the undisturbed liquid and λ is wavelength. Using L as a characteristic length and V_0 as a characteristic velocity, obtain the dimensionless groups that characterize the equation.

7.2 The equation describing small-amplitude vibration of a beam is

$$\rho A\frac{\partial^2 y}{\partial t^2} + EI\frac{\partial^4 y}{\partial x^4} = 0$$

where y is the beam deflection at location x and time t, ρ and E are the density and modulus of elasticity of the beam material, respectively, and A and I are the beam cross-section area and second moment of area, respectively. Use the beam length L, and frequency of vibration ω, to nondimensionalize this equation. Obtain the dimensionless groups that characterize the equation.

7.3 The slope of the free surface of a steady wave in one-dimensional flow in a shallow liquid layer is described by the equation

$$\frac{\partial h}{\partial x} = -\frac{u}{g}\frac{\partial u}{\partial x}$$

Use a length scale, L, and a velocity scale, V_0, to nondimensionalize this equation. Obtain the dimensionless groups that characterize this flow.

7.4 One-dimensional unsteady flow in a thin liquid layer is described by the equation

$$\frac{\partial u}{\partial t} + u\frac{\partial u}{\partial x} = -g\frac{\partial h}{\partial x}$$

Use a length scale, L, and a velocity scale, V_0, to nondimensionalize this equation. Obtain the dimensionless groups that characterize this flow.

7.5 A two-dimensional steady flow in a viscous liquid is described by the equation:

$$u\frac{\partial u}{\partial x} = -g\frac{\partial h}{\partial x} + \frac{\mu}{\rho}\left(\frac{\partial^2 u}{\partial x^2} + \frac{\partial^2 u}{\partial y^2}\right)$$

Use a length scale, L, and a velocity scale, V_0, to nondimensionalize this equation. Obtain the dimensionless groups that characterize this flow.

7.6 In atmospheric studies the motion of the earth's atmosphere can sometimes be modeled with the equation

$$\frac{D\vec{V}}{Dt} + 2\vec{\Omega}\times\vec{V} = -\frac{1}{\rho}\nabla p$$

where $\vec{V}$ is the large-scale velocity of the atmosphere across the Earth's surface, ∇p is the climatic pressure gradient, and $\vec{\Omega}$ is the Earth's angular velocity. What is the meaning of the term $\vec{\Omega}\times\vec{V}$? Use the pressure difference, Δp, and typical length scale, L (which could, for example, be the magnitude of, and distance between, an atmospheric high and low, respectively), to nondimensionalize this equation. Obtain the dimensionless groups that characterize this flow.

7.7 By using order of magnitude analysis, the continuity and Navier–Stokes equations can be simplified to the Prandtl boundary-layer equations. For steady, incompressible, and two-dimensional flow, neglecting gravity, the result is

$$\frac{\partial u}{\partial x} + \frac{\partial v}{\partial y} = 0$$

$$u\frac{\partial u}{\partial x} + v\frac{\partial u}{\partial y} = -\frac{1}{\rho}\frac{\partial p}{\partial x} + \nu\frac{\partial^2 u}{\partial y^2}$$

Use L and V_0 as characteristic length and velocity, respectively. Nondimensionalize these equations and identify the similarity parameters that result.

7.8 An unsteady, two-dimensional, compressible, inviscid flow can be described by the equation

$$\frac{\partial^2 \psi}{\partial t^2} + \frac{\partial}{\partial t}(u^2 + v^2) + (u^2 - c^2)\frac{\partial^2 \psi}{\partial x^2} + (v^2 - c^2)\frac{\partial^2 \psi}{\partial y^2} + 2uv\frac{\partial^2 \psi}{\partial x \partial y} = 0$$

where ψ is the stream function, u and v are the x and y components of velocity, respectively, c is the local speed of sound, and t is the time. Using L as a characteristic length and c_0 (the speed of sound at the stagnation point) to nondimensionalize this equation, obtain the dimensionless groups that characterize the equation.

7.9 The equation describing motion of fluid in a pipe due to an applied pressure gradient, when the flow starts from rest, is

$$\frac{\partial u}{\partial t} = -\frac{1}{\rho}\frac{\partial p}{\partial x} + \nu\left(\frac{\partial^2 u}{\partial r^2} + \frac{1}{r}\frac{\partial u}{\partial r}\right)$$

Use the average velocity $\overline{V}$, pressure drop Δp, pipe length L, and diameter D to nondimensionalize this equation. Obtain the dimensionless groups that characterize this flow.

Determining the Π Groups

7.10 Experiments show that the pressure drop for flow through an orifice plate of diameter d mounted in a length of pipe of diameter D may be expressed as $\Delta p = p_1 - p_2 = f(\rho, \mu, \overline{V}, d, D)$. You are asked to organize some experimental data. Obtain the resulting dimensionless parameters.

7.11 At relatively high speeds the drag on an object is independent of fluid viscosity. Thus the aerodynamic drag force, F, on an automobile, is a function only of speed, V, air density ρ, and vehicle size, characterized by its frontal area A. Use dimensional analysis to determine how the drag force F depends on the speed V.

7.12 At very low speeds, the drag on an object is independent of fluid density. Thus the drag force, F, on a small sphere is a function only of speed, V, fluid viscosity, μ, and sphere diameter, D. Use dimensional analysis to determine how the drag force F depends on the speed V.

7.13 The drag force on the International Space Station depends on the mean free path of the molecules λ (a length), the density ρ, a characteristic length L, and the mean speed of the air molecules c. Find a nondimensional form of this functional relationship.

7.14 We saw in Chapter 3 that the buoyant force, F_B, on a body submerged in a fluid is directly proportional to the specific weight of the fluid, γ. Demonstrate this using dimensional analysis, by starting with the buoyant force as a function of the volume of the body and the specific weight of the fluid.

7.15 When an object travels at supersonic speeds, the aerodynamic drag force F acting on the object is a function of the velocity V, air density ρ, object size (characterized by some reference area A), and the speed of sound c (note that all of the variables except c were considered when traveling at subsonic speeds as in Problem 7.11). Develop a functional relationship between a set of dimensionless variables to describe this problem.

7.16 The speed, V, of a free-surface wave in shallow liquid is a function of depth, D, density, ρ, gravity, g, and surface tension, σ. Use dimensional analysis to find the functional dependence of V on the other variables. Express V in the simplest form possible.

7.17 The wall shear stress, τ_w, in a boundary layer depends on distance from the leading edge of the body, x, the density, ρ, and viscosity, μ, of the fluid, and the freestream speed of the flow, U. Obtain the dimensionless groups and express the functional relationship among them.

7.18 The boundary-layer thickness, δ, on a smooth flat plate in an incompressible flow without pressure gradients depends on the freestream speed, U, the fluid density, ρ, the fluid viscosity, μ, and the distance from the leading edge of the plate, x. Express these variables in dimensionless form.

7.19 If an object is light enough it can be supported on the surface of a fluid by surface tension. Tests are to be done to investigate this phenomenon. The weight, W, supportable in this way depends on the object's perimeter, p, and the fluid's density, ρ, surface tension σ, and gravity, g. Determine the dimensionless parameters that characterize this problem.

7.20 The speed, V, of a free-surface gravity wave in deep water is a function of wavelength, λ, depth, D, density, ρ, and acceleration of gravity, g. Use dimensional analysis to find the functional dependence of V on the other variables. Express V in the simplest form possible.

7.21 The mean velocity, $\bar{u}$, for turbulent flow in a pipe or a boundary layer may be correlated using the wall shear stress, τ_w, distance from the wall, y, and the fluid properties, ρ and μ. Use dimensional analysis to find one dimensionless parameter containing $\bar{u}$ and one containing y that are suitable for organizing experimental data. Show that the result may be written

$$\frac{\bar{u}}{u_*} = f\left(\frac{yu_*}{\nu}\right)$$

where $u_* = (\tau_w/\rho)^{1/2}$ is the *friction velocity*.

7.22 The energy released during an explosion, E, is a function of the time after detonation t, the blast radius R at time t, and the ambient air pressure p, and density ρ. Determine, by dimensional analysis, the general form of the expression for E in terms of the other variables.

7.23 Capillary waves are formed on a liquid free surface as a result of surface tension. They have short wavelengths. The speed of a capillary wave depends on surface tension, σ, wavelength, λ, and liquid density, ρ. Use dimensional analysis to express wave speed as a function of these variables.

7.24 Measurements of the liquid height upstream from an obstruction placed in an open-channel flow can be used to determine volume flow rate. (Such obstructions, designed and calibrated to measure rate of open-channel flow, are called *weirs*.) Assume the volume flow rate, Q, over a weir is a function of upstream height, h, gravity, g, and channel width, b. Use dimensional analysis to find the functional dependence of Q on the other variables.

7.25 The torque, T, of a handheld automobile buffer is a function of rotational speed, ω, applied normal force, F, automobile surface roughness, e, buffing paste viscosity, μ, and surface tension, σ. Determine the dimensionless parameters that characterize this problem.

7.26 The power, $\mathcal{P}$, used by a vacuum cleaner is to be correlated with the amount of suction provided (indicated by the pressure drop, Δp, below the ambient room pressure). It also depends on impeller diameter, D, and width, d, motor speed, ω, air density, ρ, and cleaner inlet and exit widths, d_i and d_o, respectively. Determine the dimensionless parameters that characterize this problem.

7.27 The load-carrying capacity, W, of a journal bearing is known to depend on its diameter, D, length, l, and clearance, c, in addition to its angular speed, ω, and lubricant viscosity, μ. Determine the dimensionless parameters that characterize this problem.

7.28 The time, t, for oil to drain out of a viscosity calibration container depends on the fluid viscosity, μ, and density, ρ, the orifice diameter, d, and gravity, g. Use dimensional analysis to find the functional dependence of t on the other variables. Express t in the simplest possible form.

7.29 The power per unit cross-sectional area, E, transmitted by a sound wave is a function of wave speed, V, medium density, ρ, wave amplitude, r, and wave frequency, n. Determine, by dimensional analysis, the general form of the expression for E in terms of the other variables.

7.30 You are asked to find a set of dimensionless parameters to organize data from a laboratory experiment, in which a tank is drained through an orifice from initial liquid level h_0. The time, τ, to drain the tank depends on tank diameter, D, orifice diameter, d, acceleration of gravity, g, liquid density, ρ, and liquid viscosity, μ. How many dimensionless parameters will result? How many repeating variables must be selected to determine the dimensionless parameters? Obtain the Π parameter that contains the viscosity.

7.31 A continuous belt moving vertically through a bath of viscous liquid drags a layer of liquid, of thickness h, along with it. The volume flow rate of liquid, Q, is assumed to depend on μ, ρ, g, h, and V, where V is the belt speed. Apply dimensional analysis to predict the form of dependence of Q on the other variables.

7.32 The power, $\mathcal{P}$, required to drive a fan is believed to depend on fluid density, ρ, volume flow rate, Q, impeller diameter, D, and angular velocity, ω. Use dimensional analysis to determine the dependence of $\mathcal{P}$ on the other variables.

7.33 In a fluid mechanics laboratory experiment a tank of water, with diameter D, is drained from initial level h_0. The smoothly rounded drain hole has diameter d. Assume the mass flow rate from the tank is a function of h, D, d, g, ρ, and μ, where g is the acceleration of gravity and ρ and μ are fluid properties. Measured data are to be correlated in dimensionless form. Determine the number of dimensionless parameters that will result. Specify the number of repeating parameters that must be selected to determine the dimensionless parameters. Obtain the Π parameter that contains the viscosity.

7.34 Cylindrical water tanks are frequently found on the tops of tall buildings. When a tank is filled with water, the bottom of the tank typically deflects under the weight of the water inside. The deflection δ is a function of the tank diameter D, the height of water h, the thickness of the tank bottom d, the specific weight of the water γ, and the modulus of elasticity of the tank material E. Determine the functional relationship among these parameters using dimensionless groups.

7.35 Small droplets of liquid are formed when a liquid jet breaks up in spray and fuel injection processes. The resulting droplet diameter, d, is thought to depend on liquid density, viscosity, and surface tension, as well as jet speed, V, and diameter, D. How many dimensionless ratios are required to characterize this process? Determine these ratios.

7.36 The sketch shows an air jet discharging vertically. Experiments show that a ball placed in the jet is suspended in a stable position. The equilibrium height of the ball in the jet is found to depend on D, d, V, ρ, μ, and W, where W is the weight of the ball. Dimensional analysis is suggested to correlate experimental data. Find the Π parameters that characterize this phenomenon.

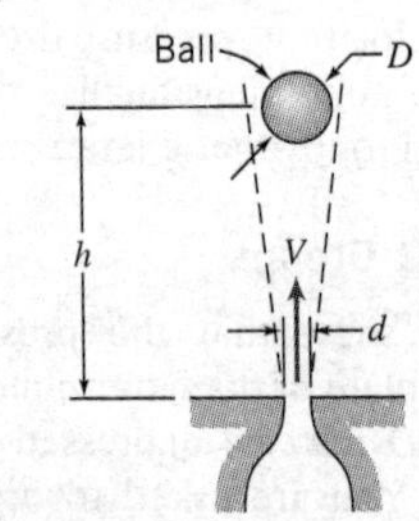

P7.36

7.37 The diameter, d, of the dots made by an ink jet printer depends on the ink viscosity, μ, density, ρ, and surface tension, σ, the nozzle diameter, D, the distance, L, of the nozzle from the paper surface, and the ink jet velocity, V. Use dimensional analysis to find the Π parameters that characterize the ink jet's behavior.

7.38 The diameter, d, of bubbles produced by a bubble-making toy depends on the soapy water viscosity, μ, density, ρ, and surface tension, σ, the ring diameter, D, and the pressure differential, Δp, generating the bubbles. Use dimensional analysis to find the Π parameters that characterize this phenomenon.

7.39 The terminal speed V of shipping boxes sliding down an incline on a layer of air (injected through numerous pinholes in the incline surface) depends on the box mass, m, and base area, A, gravity, g, the incline angle, θ, the air viscosity, μ, and the air layer thickness, δ. Use dimensional analysis to find the Π parameters that characterize this phenomenon.

7.40 The length of the wake w behind an airfoil is a function of the flow speed V, chord length L, thickness t, and fluid density ρ and viscosity μ. Find the dimensionless parameters that characterize this phenomenon.

7.41 A washing machine agitator is to be designed. The power, $\mathcal{P}$, required for the agitator is to be correlated with the amount of water used (indicated by the depth, H, of the water). It also depends on the agitator diameter, D, height, h, maximum angular velocity, ω_{max}, and frequency of oscillations, f, and water density, ρ, and viscosity, μ. Determine the dimensionless parameters that characterize this problem.

7.42 Choked-flow nozzles are often used to meter the flow of gases through piping systems. The mass flow rate of gas is thought to depend on nozzle area A, pressure p, and temperature T upstream of the meter, and the gas constant R. Determine how many independent Π parameters can be formed for this problem. State the functional relationship for the mass flow rate in terms of the dimensionless parameters.

7.43 The time, t, for a flywheel, with moment of inertia, I, to reach angular velocity, ω, from rest, depends on the applied torque, T, and the following flywheel bearing properties: the oil viscosity, μ, gap, δ, diameter, D, and length, L. Use dimensional analysis to find the Π parameters that characterize this phenomenon.

7.44 A large tank of liquid under pressure is drained through a smoothly contoured nozzle of area A. The mass flow rate is thought to depend on nozzle area, A, liquid density, ρ, difference in height between the liquid surface and nozzle, h, tank gage pressure, Δp, and gravitational acceleration, g. Determine how many independent Π parameters can be formed for this problem. Find the dimensionless parameters. State the functional relationship for the mass flow rate in terms of the dimensionless parameters.

7.45 Spin plays an important role in the flight trajectory of golf, Ping-Pong, and tennis balls. Therefore, it is important to know the rate at which spin decreases for a ball in flight. The aerodynamic torque, T, acting on a ball in flight, is thought to depend on flight speed, V, air density, ρ, air viscosity, μ, ball diameter, D, spin rate (angular speed), ω, and diameter of the dimples on the ball, d. Determine the dimensionless parameters that result.

7.46 The ventilation in the clubhouse on a cruise ship is insufficient to clear cigarette smoke (the ship is not yet completely smoke-free). Tests are to be done to see if a larger extractor fan will work. The concentration of smoke, c (particles per cubic meter) depends on the number of smokers, N, the pressure drop produced by the fan, Δp, the fan diameter, D, motor speed, ω, the particle and air densities, ρ_p, and ρ, respectively, gravity, g, and air viscosity, μ. Determine the dimensionless parameters that characterize this problem.

7.47 The mass burning rate of flammable gas $\dot{m}$ is a function of the thickness of the flame δ, the gas density ρ, the thermal diffusivity α, and the mass diffusivity D. Using dimensional analysis, determine the functional form of this dependence in terms of dimensionless parameters. Note that α and D have the dimensions L^2/t.

7.48 The power loss, $\mathcal{P}$, in a journal bearing depends on length, l, diameter, D, and clearance, c, of the bearing, in addition to its angular speed, ω. The lubricant viscosity and mean pressure are also important. Obtain the dimensionless parameters that characterize this problem. Determine the functional form of the dependence of $\mathcal{P}$ on these parameters.

7.49 In a fan-assisted convection oven, the heat transfer rate to a roast, $\dot{Q}$ (energy per unit time), is thought to depend on the specific heat of air, c_p, temperature difference, Θ, a length scale, L, air density, ρ, air viscosity, μ, and air speed, V. How many basic dimensions are included in these variables? Determine the number of Π parameters needed to characterize the oven. Evaluate the Π parameters.

7.50 The thrust of a marine propeller is to be measured during "open-water" tests at a variety of angular speeds and forward speeds ("speeds of advance"). The thrust, F_T, is thought to depend on water density, ρ, propeller diameter, D, speed of advance, V, acceleration of gravity, g, angular speed, ω, pressure in the liquid, p, and liquid viscosity, μ. Develop a set of dimensionless parameters to characterize the performance of the propeller. (One of the resulting parameters, gD/V^2, is known as the *Froude speed of advance*.)

7.51 The rate dT/dt at which the temperature T at the center of a rice kernel falls during a food technology process is critical—too high a value leads to cracking of the kernel, and too low a value makes the process slow and costly. The rate depends on the rice specific heat, c, thermal conductivity, k, and size, L, as well as the cooling air specific heat, c_p, density, ρ, viscosity, μ, and speed, V. How many basic dimensions are included in these variables? Determine the Π parameters for this problem.

7.52 The power, $\mathcal{P}$, required to drive a propeller is known to depend on the following variables: freestream speed, V, propeller diameter, D, angular speed, ω, fluid viscosity, μ, fluid density, ρ, and speed of sound in the fluid, c. How many dimensionless groups are required to characterize this situation? Obtain these dimensionless groups.

7.53 The fluid velocity u at any point in a boundary layer depends on the distance y of the point above the surface, the free-stream velocity U and free-stream velocity gradient dU/dx, the fluid kinematic viscosity ν, and the boundary layer thickness δ. How many dimensionless groups are required to describe this problem? Find: (a) two Π groups by inspection, (b) one Π that is a standard fluid mechanics group, and (c) any remaining Π groups using the Buckingham Pi theorem.

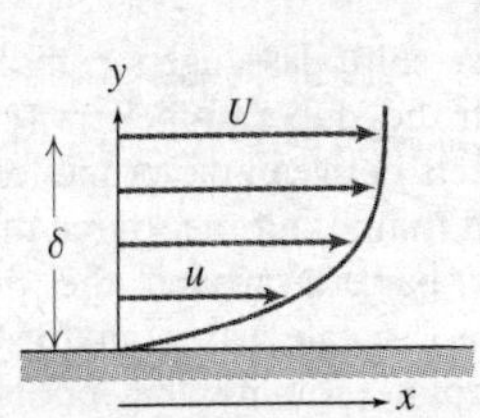

P7.53

7.54 When a valve is closed suddenly in a pipe with flowing water, a water hammer pressure wave is set up. The very high pressures generated by such waves can damage the pipe. The maximum pressure, p_{max}, generated by water hammer is a function of liquid density, ρ, initial flow speed, U_0, and liquid bulk modulus, E_v. How many dimensionless groups are needed to characterize water hammer? Determine the functional relationship among the variables in terms of the necessary Π groups.

Flow Similarity and Model Studies

7.55 The designers of a large tethered pollution-sampling balloon wish to know what the drag will be on the balloon for the maximum anticipated wind speed of 5 m/s (the air is assumed to be at 20°C). A $\frac{1}{20}$-scale model is built for testing in water at 20°C. What water speed is required to model the prototype? At this speed the model drag is measured to be 2 kN. What will be the corresponding drag on the prototype?

7.56 An airship is to operate at 20 m/s in air at standard conditions. A model is constructed to $\frac{1}{20}$-scale and tested in a wind tunnel at the same air temperature to determine drag. What criterion should be considered to obtain dynamic similarity? If the model is tested at 75 m/s, what pressure should be used in the wind tunnel? If the model drag force is 250 N, what will be the drag of the prototype?

7.57 To match the Reynolds number in an air flow and a water flow using the same size model, which flow will require the higher flow speed? How much higher must it be?

7.58 An ocean-going vessel is to be powered by a rotating circular cylinder. Model tests are planned to estimate the power required to rotate the prototype cylinder. A dimensional analysis is needed to scale the power requirements from model test results to the prototype. List the parameters that should be included in the dimensional analysis. Perform a dimensional analysis to identify the important dimensionless groups.

7.59 Measurements of drag force are made on a model automobile in a towing tank filled with fresh water. The model length scale is $\frac{1}{5}$ that of the prototype. State the conditions required to ensure dynamic similarity between the model and prototype. Determine the fraction of the prototype speed in air at which the model test should be made in water to ensure dynamically similar conditions. Measurements made at various speeds show that the dimensionless force ratio becomes constant at model test speeds above $V_m = 4$ m/s. The drag force measured during a test at this speed is $F_{D_m} = 182$ N. Calculate the drag force expected on the prototype vehicle operating at 90 km/hr in air.

7.60 On a cruise ship, passengers complain about the noise emanating from the ship's propellers (probably due to turbulent flow effects between the propeller and the ship). You have been hired to find out the source of this noise. You will study the flow pattern around the propellers and have decided to use a 1:9-scale water tank. If the ship's propellers rotate at 100 rpm, estimate the model propeller rotation speed if (a) the Froude number or (b) the Reynolds number is the governing dimensionless group. Which is most likely to lead to the best modeling?

7.61 A $\frac{1}{5}$-scale model of a torpedo is tested in a wind tunnel to determine the drag force. The prototype operates in water, has 533 mm diameter, and is 6.7 m long. The desired operating speed of the prototype is 28 m/s. To avoid compressibility effects in the wind tunnel, the maximum speed is limited to 110 m/s. However, the pressure in the wind tunnel can be varied while holding the temperature constant at 20°C. At what minimum pressure should the wind tunnel be operated to achieve a dynamically similar test? At dynamically similar test conditions, the drag force on the model is measured as 618 N. Evaluate the drag force expected on the full-scale torpedo.

7.62 The drag of an airfoil at zero angle of attack is a function of density, viscosity, and velocity, in addition to a length parameter. A 1:5-scale model of an airfoil was tested in a wind tunnel at a speed of 40 m/s, temperature of 15°C, and 3800 mm Hg pressure. The prototype airfoil has a chord length of 1.8 m and is to be flown in air at standard conditions. Determine the Reynolds number at which the wind tunnel model was tested and the corresponding prototype speed at the same Reynolds number.

7.63 Consider a smooth sphere, of diameter D, immersed in a fluid moving with speed V. The drag force on a 3 m diameter weather balloon in air moving at 1.5 m/s is to be calculated from test data. The test is to be performed in water using a 50 mm diameter model. Under dynamically similar conditions, the model drag force is measured as 3.8 N. Evaluate the model test speed and the drag force expected on the full-scale balloon.

7.64 An airplane wing, with chord length of 1.5 m and span of 9 m, is designed to move through standard air at a speed of 7.5 m/s. A $\frac{1}{10}$-scale model of this wing is to be tested in a water tunnel. What speed is necessary in the water tunnel to achieve dynamic similarity? What will be the ratio of forces measured in the model flow to those on the prototype wing?

7.65 The fluid dynamic characteristics of a golf ball are to be tested using a model in a wind tunnel. Dependent parameters are the drag force, F_D, and lift force, F_L, on the ball. The independent parameters should include angular speed, ω, and dimple depth, d. Determine suitable dimensionless parameters and express the functional dependence among them. A golf pro can hit a ball at $V = 75$ m/s and $\omega = 8100$ rpm. To model these conditions in a wind tunnel with a maximum speed of 25 m/s, what diameter model should be used? How fast must the model rotate? (The diameter of a U.S. golf ball is 4.27 cm.)

7.66 A water pump with impeller diameter 60 cm is to be designed to move 0.4 m^3/s when running at 750 rpm. Testing is performed on a 1:4 scale model running at 2400 rpm using air (20°C) as the fluid. For similar conditions (neglecting Reynolds number effects), what will be the model flow rate? If the model draws 75 W, what will be the power requirement of the prototype?

7.67 A model test is performed to determine the flight characteristics of a Frisbee. Dependent parameters are drag force, F_D, and lift force, F_L. The independent parameters should include angular speed, ω, and roughness height, h. Determine suitable dimensionless parameters, and express the functional dependence among them. The test (using air) on a 1:7-scale model Frisbee is to be geometrically, kinematically, and dynamically similar to the prototype. The wind tunnel test conditions are $V_m = 42$ m/s and $\omega_m = 5000$ rpm. What are the corresponding values of V_p and ω_p?

7.68 A model hydrofoil is to be tested at 1:20 scale. The test speed is chosen to duplicate the Froude number corresponding to the 30 m/s prototype speed. To model cavitation correctly, the cavitation number also must be duplicated. At what ambient pressure must the test be run? Water in the model test basin can be heated to 54°C, compared to 7°C for the prototype.

7.69 SAE 10W oil at 25°C flowing in a 25 mm diameter horizontal pipe, at an average speed of 1 m/s, produces a pressure drop of 450 kPa (gage) over a 150 m length. Water at 15°C flows through the same pipe under dynamically similar conditions. Using the results of Example 7.2, calculate the average speed of the water flow and the corresponding pressure drop.

7.70 In some speed ranges, vortices are shed from the rear of bluff cylinders placed across a flow. The vortices alternately leave the top and bottom of the cylinder, as shown, causing an alternating force normal to the freestream velocity. The vortex shedding frequency, f, is thought to depend on ρ, d, V, and μ. Use dimensional analysis to develop a functional relationship for f. Vortex shedding occurs in standard air on two cylinders with a diameter ratio of 2. Determine the velocity ratio for dynamic similarity, and the ratio of vortex shedding frequencies.

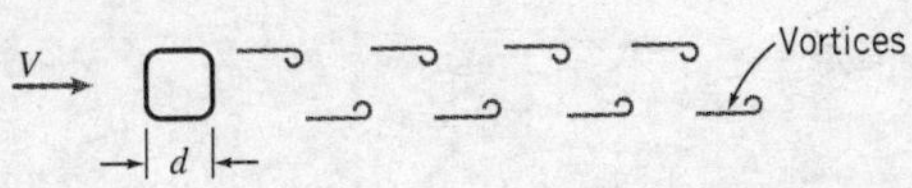

P7.70

7.71 A $\frac{1}{8}$-scale model of a tractor-trailer rig is tested in a pressurized wind tunnel. The rig width, height, and length are $W = 0.305$ m, $H = 0.476$ m, and $L = 2.48$ m, respectively. At wind speed $V = 75.0$ m/s, the model drag force is $F_D = 128$ N. (Air density in the tunnel is $\rho = 3.23$ kg/m^3.) Calculate the aerodynamic drag coefficient for the model. Compare the Reynolds numbers for the model test and for the prototype vehicle at 88 km/hr. Calculate the aerodynamic drag force on the prototype vehicle at a road speed of 88 km/hr into a headwind of 16 km/hr.

7.72 On a cruise ship, passengers complain about the amount of smoke that becomes entrained behind the cylindrical smoke stack. You have been hired to study the flow pattern around the stack, and have decided to use a 1:15 scale model of the 4.75 m smoke stack. What range of wind tunnel speeds could you use if the ship speed for which the problem occurs is 6 to 12 m/s?

7.73 The aerodynamic behavior of a flying insect is to be investigated in a wind tunnel using a 1:8-scale model. If the insect flaps its wings 60 times per second when flying at 1.5 m/s, determine the wind tunnel air speed and wing oscillation required for dynamic similarity. Do you expect that this would be a successful or practical model for generating an easily measurable wing lift? If not, can you suggest a different fluid (e.g., water, or air at a different pressure or temperature) that would produce a better modeling?

7.74 A model test of a tractor-trailer rig is performed in a wind tunnel. The drag force, F_D, is found to depend on frontal area A, wind speed V, air density ρ, and air viscosity μ. The model scale is 1:4; frontal area of the model is 0.625 m^2. Obtain a set of dimensionless parameters suitable to characterize the model test results. State the conditions required to obtain dynamic similarity between model and prototype flows. When tested at wind speed $V = 89.6$ m/s in standard air, the measured drag force on the model was $F_D = 2.46$ kN. Assuming dynamic similarity, estimate the aerodynamic drag force on the full-scale vehicle at $V = 22.4$ m/s. Calculate the power needed to overcome this drag force if there is no wind.

7.75 Tests are performed on a 1:10-scale boat model. What must be the kinematic viscosity of the model fluid if friction and wave drag phenomena are to be correctly modeled? The full-size boat will be used in a freshwater lake where the average water temperature is 10°C.

7.76 Your favorite professor likes mountain climbing, so there is always a possibility that the professor may fall into a crevasse in some glacier. If that happened today, and the professor was trapped in a slowly moving glacier, you are curious to know whether the professor would reappear at the downstream drop-off of the glacier during this academic year. Assuming ice is a Newtonian fluid with the density of glycerine but a million times as viscous, you decide to build a glycerin model and use dimensional analysis and similarity to estimate when the professor would reappear. Assume the real glacier is 15 m deep and is on a slope that falls 1.5 m in a horizontal distance of 1850 m. Develop the dimensionless parameters and conditions expected to govern dynamic similarity in this problem. If the model professor reappears in the laboratory after 9.6 hours, when should you return to the end of the real glacier to provide help to your favorite professor?

7.77 An automobile is to travel through standard air at 96 km/hr. To determine the pressure distribution, a $\frac{1}{5}$-scale model is to be tested in water. What factors must be considered to ensure kinematic similarity in the tests? Determine the water speed that should be used. What is the corresponding ratio of drag force between prototype and model flows? The lowest pressure coefficient is $C_p = -1.4$ at the location of the minimum static pressure on the surface. Estimate the minimum tunnel pressure required to avoid cavitation, if the onset of cavitation occurs at a cavitation number of 0.5.

7.78 A 1:50-scale model of a submarine is to be tested in a towing tank under two conditions: motion at the free surface and motion far below the surface. The tests are performed in freshwater. On the surface, the submarine cruises at 12 m/s. At what speed should the model be towed to ensure dynamic similarity? Far below the surface, the sub cruises at 0.18 m/s. At what speed should the model be towed to ensure dynamic similarity? What must the drag of the model be multiplied by under each condition to give the drag of the full-scale submarine?

7.79 A wind tunnel is being used to study the aerodynamics of a full-scale model rocket that is 30 cm long. Scaling for drag calculations are based on the Reynolds number. The rocket has an expected maximum velocity of 190 km/hr. What is the Reynolds number at this speed? Assume ambient air is at 20°C. The wind tunnel is capable of speeds up to 160 km/hr; so an attempt is made to improve this top speed by varying the air temperature. Calculate the equivalent speed for the wind tunnel using air at 5°C and 65°C. Would replacing air with carbon dioxide provide higher equivalent speeds?

7.80 Consider water flow around a circular cylinder, of diameter D and length l. In addition to geometry, the drag force is known to depend on liquid speed, V, density, ρ, and viscosity, μ. Express drag force, F_D, in dimensionless form as a function of all relevant variables. The static pressure distribution on a circular cylinder, measured in the laboratory, can be expressed in terms of the dimensionless pressure coefficient; the lowest pressure coefficient is $C_p = -2.4$ at the location of the minimum static pressure on the cylinder surface. Estimate the maximum speed at which a cylinder could be towed in water at atmospheric pressure, without causing cavitation, if the onset of cavitation occurs at a cavitation number of 0.5.

7.81 A circular container, partially filled with water, is rotated about its axis at constant angular speed, ω. At any time, τ, from the start of rotation, the speed, V_θ, at distance r from the axis of rotation, was found to be a function of τ, ω, and the properties of the liquid. Write the dimensionless parameters that characterize this problem. If, in another experiment, honey is rotated in the same cylinder at the same angular speed, determine from your dimensionless parameters whether honey will attain steady motion as quickly as water. Explain why the Reynolds number would not be an important dimensionless parameter in scaling the steady-state motion of liquid in the container.

7.82 A $\frac{1}{10}$-scale model of a tractor-trailer rig is tested in a wind tunnel. The model frontal area is $A_m = 0.1\ \text{m}^2$. When tested at $V_m = 75$ m/s in standard air, the measured drag force is $F_D = 350$ N. Evaluate the drag coefficient for the model conditions given. Assuming that the drag coefficient is the same for model and prototype, calculate the drag force on a prototype rig at a highway speed of 90 km/hr. Determine the air speed at which a model should be tested to ensure dynamically similar results if the prototype speed is 90 km/hr. Is this air speed practical? Why or why not?

7.83 It is recommended in [8] that the frontal area of a model be less than 5 percent of the wind tunnel test section area and $Re = Vw/\nu > 2 \times 10^6$, where w is the model width. Further, the model height must be less than 30 percent of the test section height, and the maximum projected width of the model at maximum yaw (20°) must be less than 30 percent of the test section width. The maximum air speed should be less than 91 m/s to avoid compressibility effects. A model of a tractor-trailer rig is to be tested in a wind tunnel that has a test section 0.46 m high and 0.61 m wide. The height, width, and length of the full-scale rig are 4.1 m, 2.4 m and 19.8 m, respectively. Evaluate the scale ratio of the largest model that meets the recommended criteria. Assess whether an adequate Reynolds number can be achieved in this test facility.

7.84 The power, $\mathcal{P}$, required to drive a fan is assumed to depend on fluid density ρ, volume flow rate Q, impeller diameter D, and angular speed ω. If a fan with $D_1 = 200$ mm delivers $Q_1 = 0.4\ \text{m}^3/\text{s}$ of air at $\omega_1 = 2500$ rpm, what size diameter fan could be expected to deliver $Q_2 = 2.38\ \text{m}^3/\text{s}$ of air at $\omega_2 = 1800$ rpm, provided they were geometrically and dynamically similar?

7.85 Over a certain range of air speeds, V, the lift, F_L, produced by a model of a complete aircraft in a wind tunnel depends on the air speed, air density, ρ, and a characteristic length (the wing base chord length, $c = 150$ mm). The following experimental data is obtained for air at standard atmospheric conditions:

V (m/s)	10	15	20	25	30	35	40	45	50
F_L (N)	2.2	4.8	8.7	13.3	19.6	26.5	34.5	43.8	54

Plot the lift versus speed curve. By using *Excel* to perform a trendline analysis on this curve, generate and plot data for the lift produced by the prototype, which has a wing base chord length of 5 m, over a speed range of 75 m/s to 250 m/s.

7.86 The pressure rise, Δp, of a liquid flowing steadily through a centrifugal pump depends on pump diameter D, angular speed of the rotor ω, volume flow rate Q, and density ρ. The table gives data for the prototype and for a geometrically similar model pump. For conditions corresponding to dynamic similarity between the model and prototype pumps, calculate the missing values in the table.

Variable	Prototype	Model
Δp	52.5 kPa	
Q		0.0928 m³/min
ρ	800 kg/m³	999 kg/m³
ω	183 rad/s	367 rad/s
D	150 mm	50 mm

7.87 Tests are performed on a 1 m long ship model in a water tank. Results obtained (after doing some data analysis) are as follows:

V (m/s)	3	6	9	12	15	18	20
D_{Wave} (N)	0	0.125	0.5	1.5	3	4	5.5
D_{Friction} (N)	0.1	0.35	0.75	1.25	2	2.75	3.25

The assumption is that wave drag is done using the Froude number and friction drag by the Reynolds number. The full-size ship will be 50 m long when built. Estimate the total drag when it is cruising at 7.7 m/s and at 10.3 m/s in a freshwater lake.

7.88 A centrifugal water pump running at speed $\omega = 800$ rpm has the following data for flow rate, Q, and pressure head, Δp.

Q (m³/hr)	0	100	150	200	250	300	325	350
Δp (kpa)	361	349	328	293	230	145	114	59

The pressure head is a function of the flow rate, speed, impeller diameter D, and water density ρ. Plot the pressure head versus flow rate curve. Find the two Π parameters for this problem, and, from the above data, plot one against the other. By using *Excel* to perform a trendline analysis on this latter curve, generate and plot data for pressure head versus flow rate for impeller speeds of 600 rpm and 1200 rpm.

7.89 An axial-flow pump is required to deliver 0.75 m³/s of water at a head of 15 J/kg. The diameter of the rotor is 0.25 m, and it is to be driven at 500 rpm. The prototype is to be modeled on a small test apparatus having a 2.25 kW, 1000 rpm power supply. For similar performance between the prototype and the model, calculate the head, volume flow rate, and diameter of the model.

7.90 A model propeller 1 m in diameter is tested in a wind tunnel. Air approaches the propeller at 50 m/s when it rotates at 1800 rpm. The thrust and torque measured under these conditions are 100 N and 10 N·m, respectively. A prototype 8 times as large as the model is to be built. At a dynamically similar operating point, the approach air speed is to be 130 m/s. Calculate the speed, thrust, and torque of the prototype propeller under these conditions, neglecting the effect of viscosity but including density.

7.91 Consider again Problem 7.51. Experience shows that for ship-size propellers, viscous effects on scaling are small. Also, when cavitation is not present, the nondimensional parameter containing pressure can be ignored. Assume that torque, T, and power, $\mathcal{P}$, depend on the same parameters as thrust. For conditions under which effects of μ and p can be neglected, derive scaling "laws" for propellers, similar to the pump "laws" of Section 7.6, that relate thrust, torque, and power to the angular speed and diameter of the propeller.

7.92 Water drops are produced by a mechanism that it is believed follows the pattern $d_p = D(\text{We})^{-3/5}$. In this formula, d_p is the drop size, D is proportional to a length scale, and We is the Weber number. In scaling up, if the large-scale characteristic length scale was increased by a factor of 20 and the large-scale velocity decreased by a factor of 5, how would the small- and large-scale drops differ from each other for the same material, for example, water?

7.93 Closed-circuit wind tunnels can produce higher speeds than open-circuit tunnels with the same power input because energy is recovered in the diffuser downstream from the test section. The *kinetic energy ratio* is a figure of merit defined as the ratio of the kinetic energy flux in the test section to the drive power. Estimate the kinetic energy ratio for the 12.2 m × 24.4 m wind tunnel at NASA-Ames described on page 318.

7.94 A 1:16 model of a 20 m long truck is tested in a wind tunnel at a speed of 80 m/s, where the axial static pressure gradient is -11.17 N/m^2 per meter. The frontal area of the prototype is 9.9 m^2. Estimate the horizontal buoyancy correction for this situation. Express the correction as a fraction of the measured C_D, of $C_D = 0.85$.

7.95 Frequently one observes a flag on a pole flapping in the wind. Explain why this occurs.

7.96 A 1:16 model of a bus is tested in a wind tunnel in standard air. The model is 152 mm wide, 200 mm high, and 762 mm long. The measured drag force at 26.5 m/s wind speed is 6.09 N. The longitudinal pressure gradient in the wind tunnel test section is -11.8 N/m^2/m. Estimate the correction that should be made to the measured drag force to correct for horizontal buoyancy caused by the pressure gradient in the test section. Calculate the drag coefficient for the model. Evaluate the aerodynamic drag force on the prototype at 100 km/hr on a calm day.

7.97 Explore the variation in wave propagation speed given by the equation of Problem 7.1 for a free-surface flow of water. Find the operating depth to minimize the speed of capillary waves (waves with small wavelength, also called ripples). First assume wavelength is much smaller than water depth. Then explore the effect of depth. What depth do you recommend for a water table used to visualize compressible-flow wave phenomena? What is the effect of reducing surface tension by adding a surfactant?

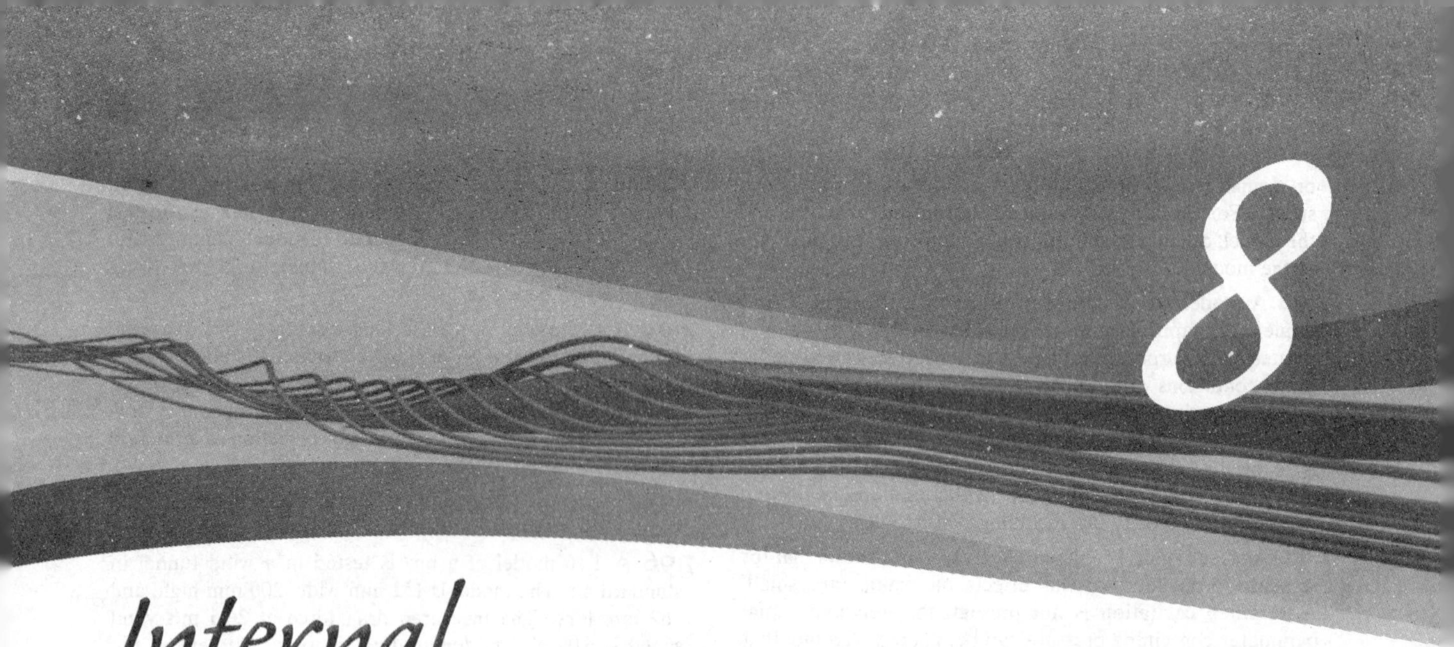

8 Internal Incompressible Viscous Flow

Case Study in Energy and the Environment

Wind Power: *The FloDesign Wind Turbine*

We are all now familiar with the ubiquitous three-bladed wind turbines that are being used to generate increasing amounts of power. The technology is already quite mature, so new developments will be incremental: improved blade designs, better controls, and composite materials to allow larger turbines. The largest in the world, being built by a Norwegian team, will be 162 m tall with a rotor diameter of 145 m, and it will generate about 10 MW, sufficient for more than 2000 homes. Bearing in mind that the Empire State Building is 381 m tall, this wind turbine will be huge—so big it must be installed offshore.

Engineers are still investigating alternatives to these designs. *FloDesign Wind Turbine*, a spin-off from the aerospace company *FloDesign* based in Wilbraham, Massachusetts, is developing a prototype that, according to CEO Stanley Kowalski III, will be up to three times more efficient than conventional wind turbines. From the front, the wind turbine looks something like the air intake of a jet engine (not surprisingly, considering *FloDesign*'s history). The shaped cowlings shown in the figure guide the air into spinning vortices as it exits the device, accelerating the flow and causing a significant pressure drop. The incoming wind first meets a set of fixed stator blades, which direct it onto the rotor blades to extract power from the flow. The exiting air hence has lower energy and velocity than the air flowing around the turbine, but the device's shroud is so shaped that the relatively fast-moving outside air is blended with the exiting air in the area just behind the rotors, creating a low-pressure region behind the turbine blades. This is where the device has an advantage over conventional turbines; the induced low-pressure region actually draws air into the device at an increased rate, generating more power. This idea is not new, but past attempts to build similar turbines were limited by the fact that such a turbine had to be very precisely aligned with the wind's direction (within

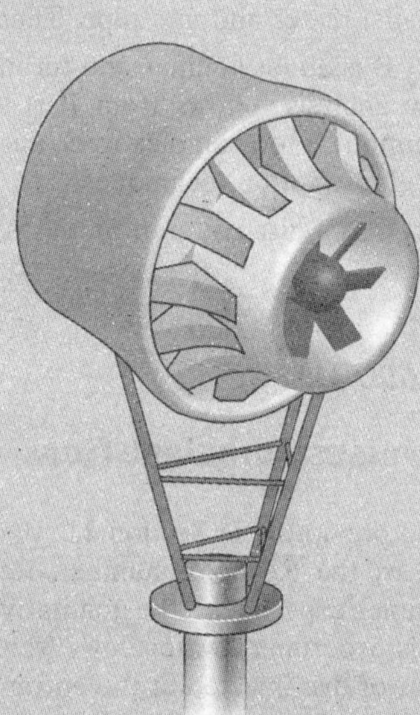

FloDesign Wind Turbine

about 4°); this device will work at angles of up to 20° off the wind.

Theoretically (as we'll learn in Chapter 10), conventional wind turbines capture a maximum of 59.3 percent of the wind energy. The new design generates as much power as a conventional wind turbine with blades twice as big. The smaller blade size of the new design means the *FloDesign Wind Turbine* could be packed closer together than conventional turbines, increasing the amount of power that can be generated per acre of land. Because its blades are lighter and smaller, the design starts spinning and generating power at lower wind speeds, and it is more tolerant of unstable wind patterns, making it excellent for windy regions where large turbines cannot be used, such as in cities. Smaller blades can also be allowed to spin faster, reducing the need for expensive gearboxes that conventional wind turbines must use to connect slow-moving rotors to a high-speed generator. With fewer gears and other moving parts, the company claims it can reduce the number of components by up to 75 percent, reducing costs and making maintenance easier.

FloDesign has already built a small prototype for wind-tunnel tests. Their next step is to build a 3.6 m diameter, 10-kW system for field tests. The prototype will be finished in 2010, with commercial wind turbines to follow.

VIDEO

The Reynolds Transition Experiment.

VIDEO

Variable Viscosity Experiment (Animation).

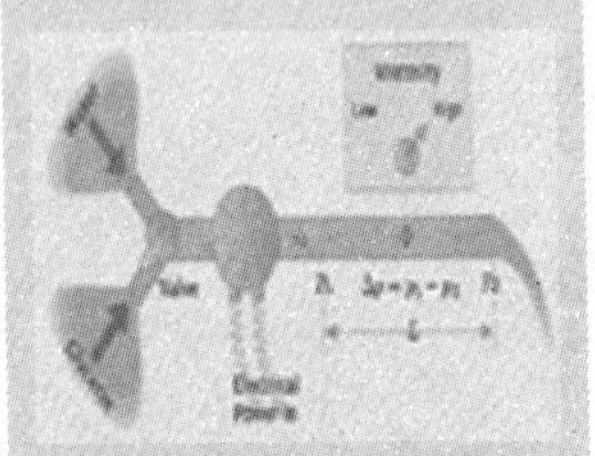

Flows completely bounded by solid surfaces are called internal flows. Thus internal flows include many important and practical flows such as those through pipes, ducts, nozzles, diffusers, sudden contractions and expansions, valves, and fittings.

Internal flows may be laminar or turbulent. Some laminar flow cases may be solved analytically. In the case of turbulent flow, analytical solutions are not possible, and we must rely heavily on semi-empirical theories and on experimental data. The nature of laminar and turbulent flows was discussed in Section 2.6. For internal flows, the flow regime (laminar or turbulent) is primarily a function of the Reynolds number.

In this chapter we will only consider incompressible flows; hence we will study the flow of liquids as well as gases that have negligible heat transfer and for which the Mach number $M < 0.3$; a value of $M = 0.3$ in air corresponds to a speed of approximately 100 m/s. Following a brief introduction, this chapter is divided into the following parts:

Part A: Part A discusses fully developed laminar flow of a Newtonian fluid between parallel plates and in a pipe. These two cases can be studied analytically.

Part B: Part B is about laminar and turbulent flows in pipes and ducts. The laminar flow analysis follows from Part A; the turbulent flow (which is the most common) is too complex to be analyzed, so experimental data will be used to develop solution techniques.

Part C: Part C is a discussion of methods of flow measurement.

8.1 Introduction

VIDEO

Variable Viscosity Experiment: Pressure Drop.

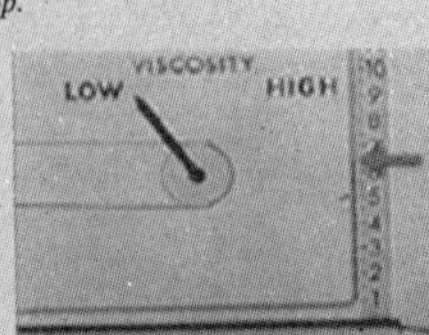

Laminar versus Turbulent Flow

As discussed previously in Section 2.6, the pipe flow regime (laminar or turbulent) is determined by the Reynolds number, $Re = \rho\bar{V}D/\mu$. One can demonstrate, by the classic Reynolds experiment, the qualitative difference between laminar and turbulent flows. In this experiment water flows from a large reservoir through a clear tube. A thin filament of dye injected at the entrance to the tube allows visual observation of the flow. At low flow rates (low Reynolds numbers) the dye injected into the flow remains in a single filament along the tube; there is little dispersion of dye because the flow is laminar. A laminar flow is one in which the fluid flows in laminae, or layers; there is no macroscopic mixing of adjacent fluid layers.

As the flow rate through the tube is increased, the dye filament eventually becomes unstable and breaks up into a random motion throughout the tube; the line of dye is stretched and twisted into myriad entangled threads, and it quickly disperses throughout the entire flow field. This behavior of turbulent flow is caused by small, high-frequency velocity fluctuations superimposed on the mean motion of a turbulent flow, as illustrated earlier in Fig. 2.15; the mixing of fluid particles from adjacent layers of fluid results in rapid dispersion of the dye. We mentioned in Chapter 2 an everyday example of the difference between laminar and turbulent flow—when you gently turn on the kitchen faucet (not aerated). For very low flow rates, the water exits smoothly (indicating laminar flow in the pipe); for higher flow rates, the flow is churned up (turbulent flow).

Under normal conditions, transition to turbulence occurs at $Re \approx 2300$ for flow in pipes: For water flow in a 25 mm diameter pipe, this corresponds to an average speed of 0.091 m/s. With great care to maintain the flow free from disturbances, and with smooth surfaces, experiments have been able to maintain laminar flow in a pipe to a Reynolds number of about 100,000! However, most engineering flow situations are not so carefully controlled, so we will take $Re \approx 2300$ as our benchmark for transition to turbulence. Transition Reynolds numbers for some other flow situations are given in the Examples. Turbulence occurs when the viscous forces in the fluid are unable to damp out random fluctuations in the fluid motion (generated, for example, by roughness of a pipe wall), and the flow becomes chaotic. For example, a high-viscosity fluid such as motor oil is able to damp out fluctuations more effectively than a low viscosity fluid such as water and therefore remains laminar even at relatively high flow rates. On the other hand, a high-density fluid will generate significant inertia forces due to the random fluctuations in the motion, and this fluid will transition to turbulence at a relatively low flow rate.

CLASSIC VIDEO

Turbulence.

VIDEO

Laminar Pipe Flow: Velocity Profile.

VIDEO

Pipe Flow: Laminar.

The Entrance Region

Figure 8.1 illustrates laminar flow in the entrance region of a circular pipe. The flow has uniform velocity U_0 at the pipe entrance. Because of the no-slip condition at the wall, we know that the velocity at the wall must be zero along the entire length of the pipe. A boundary layer (Section 2.6) develops along the walls of the channel. The solid surface exerts a retarding shear force on the flow; thus the speed of the fluid in the neighborhood of the surface is reduced. At successive sections along the pipe in this entry region, the effect of the solid surface is felt farther out into the flow.

For incompressible flow, mass conservation requires that, as the speed close to the wall is reduced, the speed in the central frictionless region of the pipe must increase slightly to compensate; for this inviscid central region, then, the pressure (as indicated by the Bernoulli equation) must also drop somewhat.

Sufficiently far from the pipe entrance, the boundary layer developing on the pipe wall reaches the pipe centerline and the flow becomes entirely viscous. The velocity profile shape then changes slightly after the inviscid core disappears. When the profile shape no longer changes with increasing distance x, the flow is called *fully developed*. The distance

VIDEO

Pipe Flow: Transitional.

VIDEO

Pipe Flow: Turbulent.

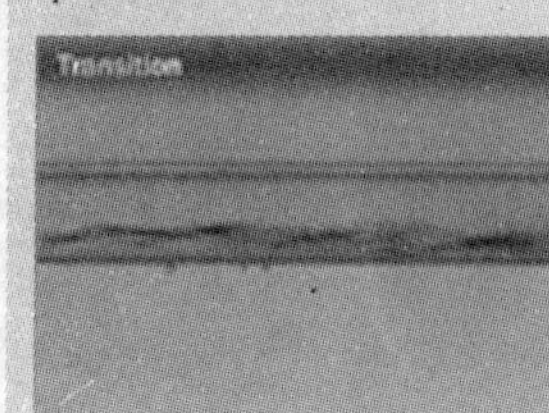

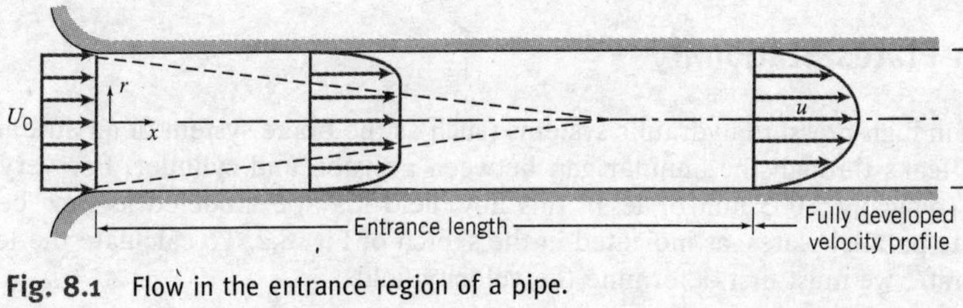

Fig. 8.1 Flow in the entrance region of a pipe.

downstream from the entrance to the location at which fully developed flow begins is called the *entrance length*. The actual shape of the fully developed velocity profile depends on whether the flow is laminar or turbulent. In Fig. 8.1 the profile is shown qualitatively for a laminar flow. Although the velocity profiles for some fully developed laminar flows can be obtained by simplifying the complete equations of motion from Chapter 5, turbulent flows cannot be so treated.

CLASSIC VIDEO

Turbulence.

For laminar flow, it turns out that entrance length, L, is a function of Reynolds number,

$$\frac{L}{D} \simeq 0.06 \frac{\rho \overline{V} D}{\mu} \tag{8.1}$$

where $\overline{V} \equiv Q/A$ is the average velocity (because flow rate $Q = A\overline{V} = AU_0$, we have $\overline{V} = U_0$). Laminar flow in a pipe may be expected only for Reynolds numbers less than 2300. Thus the entrance length for laminar pipe flow may be as long as

VIDEO

Fully-Developed Pipe Flow.

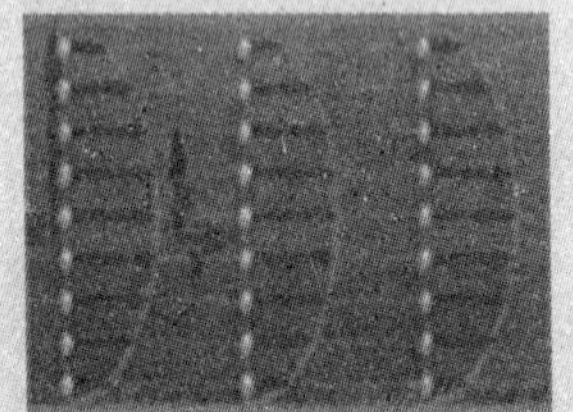

$$L \simeq 0.06\, ReD \leq (0.06)(2300)\, D = 138D$$

or nearly 140 pipe diameters. If the flow is turbulent, enhanced mixing among fluid layers causes more rapid growth of the boundary layer. Experiments show that the mean velocity profile becomes fully developed within 25 to 40 pipe diameters from the entrance. However, the details of the turbulent motion may not be fully developed for 80 or more pipe diameters. We are now ready to study laminar internal flows (Part A), as well as laminar and turbulent flows in pipes and ducts (Part B). For these we will be focusing on what happens after the entrance region, i.e., fully developed flows.

Part A Fully Developed Laminar Flow

In this section we consider a few classic examples of fully developed laminar flows. Our intent is to obtain detailed information about the velocity field because knowledge of the velocity field permits calculation of shear stress, pressure drop, and flow rate.

8.2 Fully Developed Laminar Flow Between Infinite Parallel Plates

The flow between parallel plates is appealing because the geometry is the simplest possible, but why *would* there be a flow at all? The answer is that flow could be generated by applying a pressure gradient parallel to the plates, or by moving one plate parallel with respect to the other, or by having a body force (e.g., gravity) parallel to the plates, or by a combination of these driving mechanisms. We will consider all of these possibilities.

Both Plates Stationary

Fluid in high-pressure hydraulic systems (such as the brake system of an automobile) often leaks through the annular gap between a piston and cylinder. For very small gaps (typically 0.005 mm or less), this flow field may be modeled as flow between infinite parallel plates, as indicated in the sketch of Fig. 8.2. To calculate the leakage flow rate, we must first determine the velocity field.

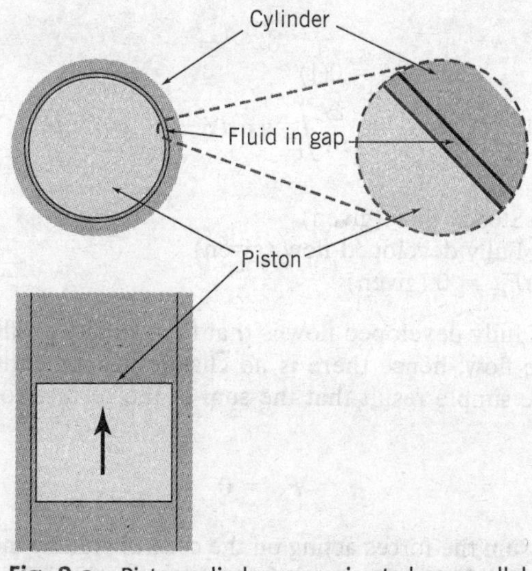

Fig. 8.2 Piston-cylinder approximated as parallel plates.

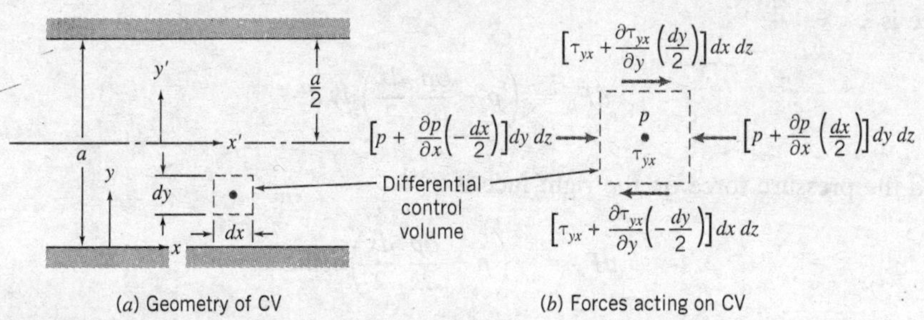

Fig. 8.3 Control volume for analysis of laminar flow between stationary infinite parallel plates.

Let us consider the fully developed laminar flow between horizontal infinite parallel plates. The plates are separated by distance a, as shown in Fig. 8.3. The plates are considered infinite in the z direction, with no variation of any fluid property in this direction. The flow is also assumed to be steady and incompressible. Before starting our analysis, what do we know about the flow field? For one thing we know that the x component of velocity must be zero at both the upper and lower plates as a result of the no-slip condition at the wall. The boundary conditions are then

$$\text{at} \quad y = 0 \qquad u = 0$$
$$\text{at} \quad y = a \qquad u = 0$$

Since the flow is fully developed, the velocity cannot vary with x and, hence, depends on y only, so that $u = u(y)$. Furthermore, there is no component of velocity in either the y or z direction ($v = w = 0$). In fact, for fully developed flow only the pressure can and will change (in a manner to be determined from the analysis) in the x direction.

This is an obvious case for using the Navier–Stokes equations in rectangular coordinates (Eqs. 5.27). Using the above assumptions, these equations can be greatly simplified and then solved using the boundary conditions (see Problem 8.17). In this section we will instead take a longer route—using a differential control volume—to bring out some important features of the fluid mechanics.

For our analysis we select a differential control volume of size $d\forall = dx\,dy\,dz$, and apply the x component of the momentum equation.

Basic equation:

$$F_{S_x} + \cancel{F_{B_x}} = \cancel{\frac{\partial}{\partial t}\int_{CV} u\,\rho\,dV} + \int_{CS} u\,\rho\vec{V}\cdot d\vec{A} \tag{4.18a}$$

(with $F_{B_x} = 0(3)$ and $\frac{\partial}{\partial t}\int_{CV} = 0(1)$)

Assumptions: (1) Steady flow (given)
(2) Fully developed flow (given)
(3) $F_{B_x} = 0$ (given)

The very nature of fully developed flow is that the velocity profile is the same at all locations along the flow; hence there is no change in momentum. Equation 4.18a then reduces to the simple result that the sum of the surface forces on the control volume is zero,

$$F_{S_x} = 0 \tag{8.2}$$

The next step is to sum the forces acting on the control volume in the x direction. We recognize that normal forces (pressure forces) act on the left and right faces and tangential forces (shear forces) act on the top and bottom faces.

If the pressure at the center of the element is p, then the pressure force on the left face is

$$dF_L = \left(p - \frac{\partial p}{\partial x}\frac{dx}{2}\right)dy\,dz$$

and the pressure force on the right face is

$$dF_R = -\left(p + \frac{\partial p}{\partial x}\frac{dx}{2}\right)dy\,dz$$

If the shear stress at the center of the element is τ_{yx}, then the shear force on the bottom face is

$$dF_B = -\left(\tau_{yx} - \frac{d\tau_{yx}}{dy}\frac{dy}{2}\right)dx\,dz$$

and the shear force on the top face is

$$dF_T = \left(\tau_{yx} + \frac{d\tau_{yx}}{dy}\frac{dy}{2}\right)dx\,dz$$

Note that in expanding the shear stress, τ_{yx}, in a Taylor series about the center of the element, we have used the total derivative rather than a partial derivative. We did this because we recognized that τ_{yx} is only a function of y, since $u = u(y)$.

Using the four surface forces dF_L, dF_R, dF_B, and dF_T in Eq. 8.2, this equation simplifies to

$$\frac{\partial p}{\partial x} = \frac{d\tau_{yx}}{dy} \tag{8.3}$$

This equation states that because there is no change in momentum, the net pressure force (which is actually $-\partial p/\partial x$) balances the net friction force (which is actually $-d\tau_{yx}/dy$). Equation 8.3 has an interesting feature: The left side is at most a function of x only (this follows immediately from writing the y component of the momentum equation); the right side is at most a function of y only (the flow is fully developed, so it does not change with x). Hence, the only way the equation can be valid for all x and y is for each side to in fact be constant:

$$\frac{d\tau_{yx}}{dy} = \frac{\partial p}{\partial x} = \text{constant}$$

Integrating this equation, we obtain

$$\tau_{yx} = \left(\frac{\partial p}{\partial x}\right) y + c_1$$

which indicates that the shear stress varies linearly with y. We wish to find the velocity distribution. To do so, we need to relate the shear stress to the velocity field. For a Newtonian fluid we can use Eq. 2.15 because we have a one-dimensional flow [or we could have started with the full stress equation (Eq. 5.25a) and simplified],

$$\tau_{yx} = \mu \frac{du}{dy} \tag{2.15}$$

so we get

$$\mu \frac{du}{dy} = \left(\frac{\partial p}{\partial x}\right) y + c_1$$

Integrating again

$$u = \frac{1}{2\mu}\left(\frac{\partial p}{\partial x}\right) y^2 + \frac{c_1}{\mu} y + c_2 \tag{8.4}$$

It is interesting to note that if we had started with the Navier–Stokes equations (Eqs. 5.27) instead of using a differential control volume, after only a few steps (i.e., simplifying and integrating twice) we would have obtained Eq. 8.4 (see Problem 8.17). To evaluate the constants, c_1 and c_2, we must apply the boundary conditions. At $y = 0$, $u = 0$. Consequently, $c_2 = 0$. At $y = a$, $u = 0$. Hence

$$0 = \frac{1}{2\mu}\left(\frac{\partial p}{\partial x}\right) a^2 + \frac{c_1}{\mu} a$$

This gives

$$c_1 = -\frac{1}{2}\left(\frac{\partial p}{\partial x}\right) a$$

and hence,

$$u = \frac{1}{2\mu}\left(\frac{\partial p}{\partial x}\right) y^2 - \frac{1}{2\mu}\left(\frac{\partial p}{\partial x}\right) ay = \frac{a^2}{2\mu}\left(\frac{\partial p}{\partial x}\right)\left[\left(\frac{y}{a}\right)^2 - \left(\frac{y}{a}\right)\right] \tag{8.5}$$

At this point we have the velocity profile. This is key to finding other flow properties, as we next discuss.

Shear Stress Distribution

The shear stress distribution is given by

$$\tau_{yx} = \left(\frac{\partial p}{\partial x}\right) y + c_1 = \left(\frac{\partial p}{\partial x}\right) y - \frac{1}{2}\left(\frac{\partial p}{\partial x}\right) a = a\left(\frac{\partial p}{\partial x}\right)\left[\frac{y}{a} - \frac{1}{2}\right] \tag{8.6a}$$

Volume Flow Rate

The volume flow rate is given by

$$Q = \int_A \vec{V} \cdot d\vec{A}$$

For a depth l in the z direction,

$$Q = \int_0^a ul\,dy \qquad \text{or} \qquad \frac{Q}{l} = \int_0^a \frac{1}{2\mu}\left(\frac{\partial p}{\partial x}\right)(y^2 - ay)\,dy$$

Thus the volume flow rate per unit depth is given by

$$\frac{Q}{l} = -\frac{1}{12\mu}\left(\frac{\partial p}{\partial x}\right)a^3 \tag{8.6b}$$

Flow Rate as a Function of Pressure Drop

Since $\partial p/\partial x$ is constant, the pressure varies linearly with x and

$$\frac{\partial p}{\partial x} = \frac{p_2 - p_1}{L} = \frac{-\Delta p}{L}$$

Substituting into the expression for volume flow rate gives

$$\frac{Q}{l} = -\frac{1}{12\mu}\left[\frac{-\Delta p}{L}\right]a^3 = \frac{a^3 \Delta p}{12\mu L} \tag{8.6c}$$

Average Velocity

The average velocity magnitude, $\overline{V}$, is given by

$$\overline{V} = \frac{Q}{A} = -\frac{1}{12\mu}\left(\frac{\partial p}{\partial x}\right)\frac{a^3 l}{la} = -\frac{1}{12\mu}\left(\frac{\partial p}{\partial x}\right)a^2 \tag{8.6d}$$

Point of Maximum Velocity

To find the point of maximum velocity, we set du/dy equal to zero and solve for the corresponding y. From Eq. 8.5

$$\frac{du}{dy} = \frac{a^2}{2\mu}\left(\frac{\partial p}{\partial x}\right)\left[\frac{2y}{a^2} - \frac{1}{a}\right]$$

Thus,

$$\frac{du}{dy} = 0 \quad \text{at} \quad y = \frac{a}{2}$$

At

$$y = \frac{a}{2}, \qquad u = u_{\max} = -\frac{1}{8\mu}\left(\frac{\partial p}{\partial x}\right)a^2 = \frac{3}{2}\overline{V} \tag{8.6c}$$

Transformation of Coordinates

In deriving the above relations, the origin of coordinates, $y = 0$, was taken at the bottom plate. We could just as easily have taken the origin at the centerline of the channel. If we

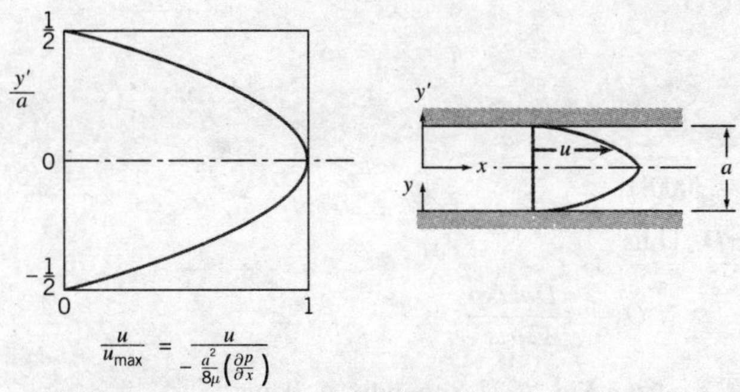

Fig. 8.4 Dimensionless velocity profile for fully developed laminar flow between infinite parallel plates.

denote the coordinates with origin at the channel centerline as x, y', the boundary conditions are $u = 0$ at $y' = \pm a/2$.

To obtain the velocity profile in terms of x, y', we substitute $y = y' + a/2$ into Eq. 8.5. The result is

$$u = \frac{a^2}{2\mu}\left(\frac{\partial p}{\partial x}\right)\left[\left(\frac{y'}{a}\right)^2 - \frac{1}{4}\right] \tag{8.7}$$

Equation 8.7 shows that the velocity profile for laminar flow between stationary parallel plates is parabolic, as shown in Fig. 8.4.

Since all stresses were related to velocity gradients through Newton's law of viscosity, and the additional stresses that arise as a result of turbulent fluctuations have not been accounted for, *all of the results in this section are valid for laminar flow only*. Experiments show that laminar flow between stationary parallel plates becomes turbulent for Reynolds numbers (defined as $Re = \rho \vec{V} a/\mu$) greater than approximately 1400. Consequently, the Reynolds number should be checked after using Eqs. 8.6 to ensure a valid solution.

Example 8.1 LEAKAGE FLOW PAST A PISTON

A hydraulic system operates at a gage pressure of 20 MPa and 55°C. The hydraulic fluid is SAE 10W oil. A control valve consists of a piston 25 mm in diameter, fitted to a cylinder with a mean radial clearance of 0.005 mm. Determine the leakage flow rate if the gage pressure on the low-pressure side of the piston is 1.0 MPa. (The piston is 15 mm long.)

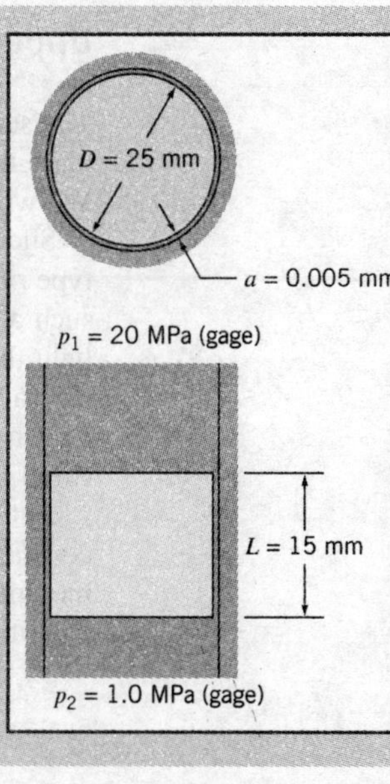

Given: Flow of hydraulic oil between piston and cylinder, as shown. Fluid is SAE 10W oil at 55°C.

Find: Leakage flow rate, Q.

Solution:

The gap width is very small, so the flow may be modeled as flow between parallel plates. Equation 8.6c may be applied.

Governing equation:

$$\frac{Q}{l} = \frac{a^3 \Delta p}{12\mu L} \tag{8.6c}$$

Assumptions:
(1) Laminar flow.
(2) Steady flow.
(3) Incompressible flow.
(4) Fully developed flow.
(Note $L/a = 15/0.005 = 3000!$)

The plate width, l, is approximated as $l = \pi D$. Thus

$$Q = \frac{\pi D a^3 \Delta p}{12 \mu L}$$

For SAE 10W oil at 55°C, $\mu = 0.018\ \text{kg}/(\text{m}\cdot\text{s})$, from Fig. A.2, Appendix A. Thus

$$Q = \frac{\pi}{12} \times 25\ \text{mm} \times (0.005)^3\ \text{mm}^3 \times (20-1)10^6\ \frac{\text{N}}{\text{m}^2} \times \frac{\text{m}\cdot\text{s}}{0.018\ \text{kg}} \times \frac{1}{15\ \text{mm}} \times \frac{\text{kg}\cdot\text{m}}{\text{N}\cdot\text{s}^2}$$

$$Q = 57.6\ \text{mm}^3/\text{s} \qquad Q$$

To ensure that flow is laminar, we also should check the Reynolds number.

$$\overline{V} = \frac{Q}{A} = \frac{Q}{\pi D a} = 57.6\ \frac{\text{mm}^3}{\text{s}} \times \frac{1}{\pi} \times \frac{1}{25\ \text{mm}} \times \frac{1}{0.005\ \text{mm}} \times \frac{\text{m}}{10^3\ \text{mm}} = 0.147\ \text{m/s}$$

and

$$Re = \frac{\rho \overline{V} a}{\mu} = \frac{\text{SG}\rho_{\text{H}_2\text{O}} \overline{V} a}{\mu}$$

For SAE 10W oil, SG = 0.92, from Table A.2, Appendix A. Thus

$$Re = 0.92 \times 1000 \frac{\text{kg}}{\text{m}^3} \times 0.147 \frac{\text{m}}{\text{s}} \times 0.005\ \text{mm} \times \frac{\text{m}\cdot\text{s}}{0.018\ \text{kg}} \times \frac{\text{m}}{10^3\ \text{mm}} = 0.0375$$

Thus flow is surely laminar, since $Re \ll 1400$.

Upper Plate Moving with Constant Speed, *U*

The second basic way to generate flow between infinite parallel plates is to have one plate move parallel to the other, either with or without an applied pressure gradient. We will next analyze this problem for the case of laminar flow.

Such a flow commonly occurs, for example, in a journal bearing (a commonly used type of bearing, e.g., the main crankshaft bearings in the engine of an automobile). In such a bearing, an inner cylinder, the journal, rotates inside a stationary member. At light loads, the centers of the two members essentially coincide, and the small clearance gap is symmetric. Since the gap is small, it is reasonable to "unfold" the bearing and to model the flow field as flow between infinite parallel plates, as indicated in the sketch of Fig. 8.5.

Let us now consider a case where the upper plate is moving to the right with constant speed, U. All we have done in going from a stationary upper plate to a moving upper plate is to change one of the boundary conditions. The boundary conditions for the moving plate case are

$$u = 0 \quad \text{at} \quad y = 0$$
$$u = U \quad \text{at} \quad y = a$$

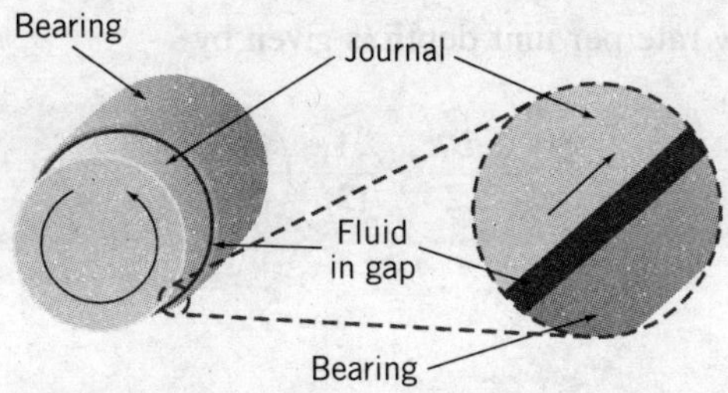

Fig. 8.5 Journal bearing approximated as parallel plates.

Since only the boundary conditions have changed, there is no need to repeat the entire analysis of the previous section. The analysis leading to Eq. 8.4 is equally valid for the moving plate case. Thus the velocity distribution is given by

$$u = \frac{1}{2\mu}\left(\frac{\partial p}{\partial x}\right)y^2 + \frac{c_1}{\mu}y + c_2 \tag{8.4}$$

and our only task is to evaluate constants c_1 and c_2 by using the appropriate boundary conditions. [Note once again that using the full Navier–Stokes equations (Eqs. 5.27) would have led very quickly to Eq. 8.4.]

At $y = 0$, $u = 0$. Consequently, $c_2 = 0$.
At $y = a$, $u = U$. Consequently,

$$U = \frac{1}{2\mu}\left(\frac{\partial p}{\partial x}\right)a^2 + \frac{c_1}{\mu}a \quad \text{and thus} \quad c_1 = \frac{U\mu}{a} - \frac{1}{2}\left(\frac{\partial p}{\partial x}\right)a$$

Hence,

$$u = \frac{1}{2\mu}\left(\frac{\partial p}{\partial x}\right)y^2 + \frac{Uy}{a} - \frac{1}{2\mu}\left(\frac{\partial p}{\partial x}\right)ay = \frac{Uy}{a} + \frac{1}{2\mu}\left(\frac{\partial p}{\partial x}\right)(y^2 - ay)$$

$$u = \frac{Uy}{a} + \frac{a^2}{2\mu}\left(\frac{\partial p}{\partial x}\right)\left[\left(\frac{y}{a}\right)^2 - \left(\frac{y}{a}\right)\right] \tag{8.8}$$

It is reassuring to note that Eq. 8.8 reduces to Eq. 8.5 for a stationary upper plate (set $U = 0$). From Eq. 8.8, for zero pressure gradient (for $\partial p/\partial x = 0$) the velocity varies linearly with y. This was the case treated earlier in Chapter 2; this linear profile is called a *Couette* flow, after a 19th-century physicist.

We can obtain additional information about the flow from the velocity distribution of Eq. 8.8.

Shear Stress Distribution

The shear stress distribution is given by $\tau_{yx} = \mu(du/dy)$,

$$\tau_{yx} = \mu\frac{U}{a} + \frac{a^2}{2}\left(\frac{\partial p}{\partial x}\right)\left[\frac{2y}{a^2} - \frac{1}{a}\right] = \mu\frac{U}{a} + a\left(\frac{\partial p}{\partial x}\right)\left[\frac{y}{a} - \frac{1}{2}\right] \tag{8.9a}$$

Volume Flow Rate

The volume flow rate is given by $Q = \int_A \vec{V}\cdot d\vec{A}$. For depth l in the z direction

$$Q = \int_0^a ul\,dy \quad \text{or} \quad \frac{Q}{l} = \int_0^a \left[\frac{Uy}{a} + \frac{1}{2\mu}\left(\frac{\partial p}{\partial x}\right)(y^2 - ay)\right]dy$$

Thus the volume flow rate per unit depth is given by

$$\frac{Q}{l} = \frac{Ua}{2} - \frac{1}{12\mu}\left(\frac{\partial p}{\partial x}\right)a^3 \tag{8.9b}$$

Average Velocity

The average velocity magnitude, $\overline{V}$, is given by

$$\overline{V} = \frac{Q}{A} = l\left[\frac{Ua}{2} - \frac{1}{12\mu}\left(\frac{\partial p}{\partial x}\right)a^3\right] \Big/ la = \frac{U}{2} - \frac{1}{12\mu}\left(\frac{\partial p}{\partial x}\right)a^2 \tag{8.9c}$$

Point of Maximum Velocity

To find the point of maximum velocity, we set du/dy equal to zero and solve for the corresponding y. From Eq. 8.8

$$\frac{du}{dy} = \frac{U}{a} + \frac{a^2}{2\mu}\left(\frac{\partial p}{\partial x}\right)\left[\frac{2y}{a^2} - \frac{1}{a}\right] = \frac{U}{a} + \frac{a}{2\mu}\left(\frac{\partial p}{\partial x}\right)\left[2\left(\frac{y}{a}\right) - 1\right]$$

Thus,

$$\frac{du}{dy} = 0 \quad \text{at} \quad y = \frac{a}{2} - \frac{U/a}{(1/\mu)(\partial p/\partial x)}$$

There is no simple relation between the maximum velocity, u_{max}, and the mean velocity, $\overline{V}$, for this flow case.

Equation 8.8 suggests that the velocity profile may be treated as a combination of a linear and a parabolic velocity profile; the last term in Eq. 8.8 is identical to that in Eq. 8.5. The result is a family of velocity profiles, depending on U and $(1/\mu)(\partial p/\partial x)$; three profiles are sketched in Fig. 8.6. (As shown in Fig. 8.6, some reverse flow—flow in the negative x direction—can occur when $\partial p/\partial x > 0$.)

Again, all of the results developed in this section are valid for laminar flow only. Experiments show that this flow becomes turbulent (for $\partial p/\partial x = 0$) at a Reynolds number of approximately 1500, where $Re = \rho U a/\mu$ for this flow case. Not much information is available for the case where the pressure gradient is not zero.

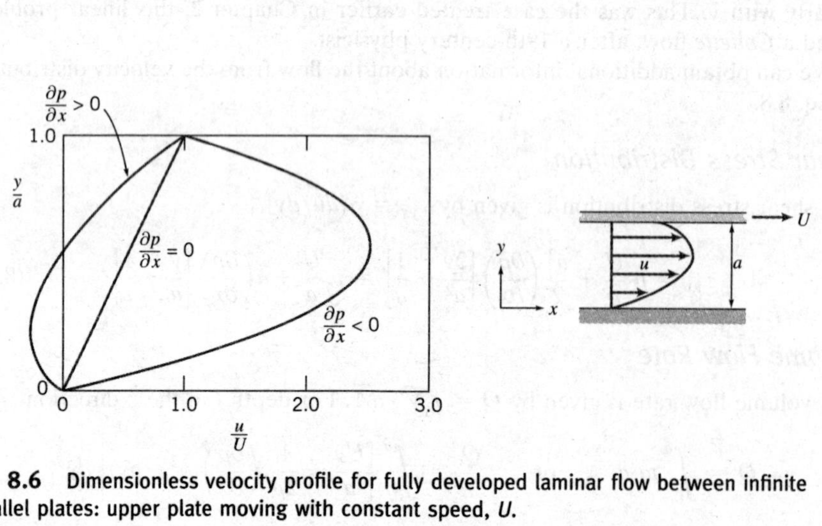

Fig. 8.6 Dimensionless velocity profile for fully developed laminar flow between infinite parallel plates: upper plate moving with constant speed, U.

Example 8.2 TORQUE AND POWER IN A JOURNAL BEARING

A crankshaft journal bearing in an automobile engine is lubricated by SAE 30 oil at 99°C. The bearing diameter is 76 mm, the diametral clearance is 0.0635 mm, and the shaft rotates at 3600 rpm; it is 31.8 mm long. The bearing is under no load, so the clearance is symmetric. Determine the torque required to turn the journal and the power dissipated.

Given: Journal bearing, as shown. Note that the gap width, a, is *half* the diametral clearance. Lubricant is SAE 30 oil at 99°C. Speed is 3600 rpm.

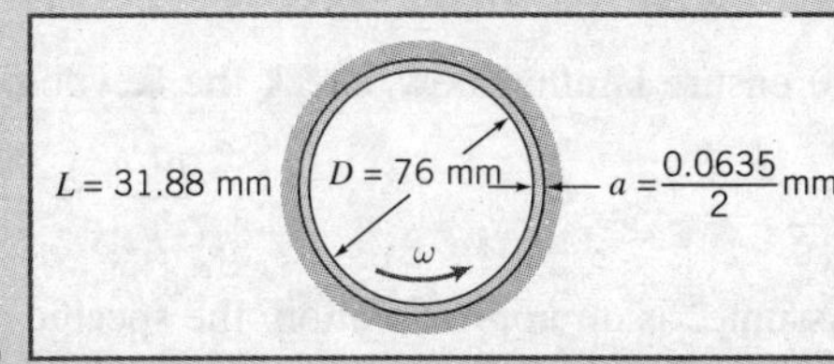

Find: (a) Torque, T.
(b) Power dissipated.

Solution:
Torque on the journal is caused by viscous shear in the oil film. The gap width is small, so the flow may be modeled as flow between infinite parallel plates:

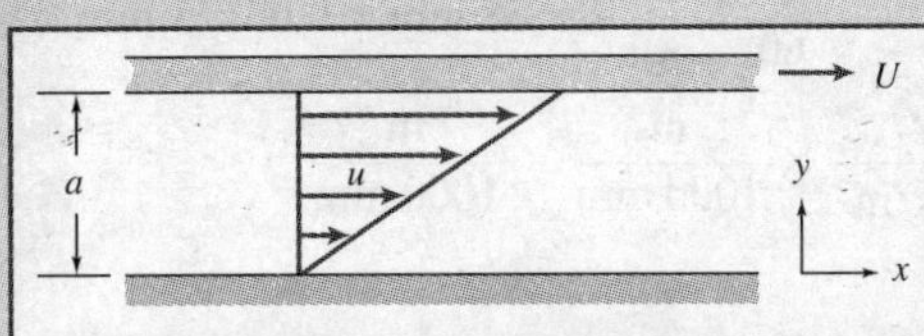

Governing equation:

$$\tau_{yx} = \mu \frac{U}{a} + a\overbrace{\left(\frac{\partial p}{\partial x}\right)}^{=0(6)}\left[\frac{y}{a} - \frac{1}{2}\right] \qquad (8.9a)$$

Assumptions:
(1) Laminar flow.
(2) Steady flow.
(3) Incompressible flow.
(4) Fully developed flow.
(5) Infinite width ($L/a = 31.8/0.03175 = 1000$, so this is a reasonable assumption).
(6) $\partial p/\partial x = 0$ (flow is symmetric in the actual bearing at no load).

Then

$$\tau_{yx} = \mu \frac{U}{a} = \mu \frac{\omega R}{a} = \mu \frac{\omega D}{2a}$$

For SAE 30 oil at 99°C, $\mu = 9.6 \times 10^{-3}\,\mathrm{N \cdot s/m^2}$ from Fig. A.2, Appendix A. Thus,

$$\tau_{yx} = 9.6 \times 10^{-3} \frac{\mathrm{N \cdot S}}{\mathrm{m^2}} \times 3600 \frac{\mathrm{rev}}{\mathrm{min}} \times 2\pi \frac{\mathrm{rad}}{\mathrm{rev}} \times \frac{\mathrm{min}}{60\,\mathrm{s}} \times 76\,\mathrm{mm} \times \frac{\mathrm{m}}{1000\,\mathrm{mm}} \times \frac{1}{2} \times \frac{1}{0.03175\,\mathrm{mm}} \times \frac{1000\,\mathrm{mm}}{\mathrm{m}} \times \frac{\mathrm{Pa \cdot m^2}}{\mathrm{N}}$$

$$\tau_{yx} = 4331.6\,\mathrm{Pa}$$

The total shear force is given by the shear stress times the area. It is applied to the journal surface. Therefore, for the torque

$$T = FR = \tau_{yx}\pi DLR = \frac{\pi}{2}\tau_{yx}D^2L$$

$$= \frac{\pi}{2} \times 4331.6\,\mathrm{Pa} \times (76)^2\mathrm{mm^2} \times 31.8\,\mathrm{min} \times \frac{\mathrm{N}}{\mathrm{m^2}} \times \frac{\mathrm{m^2}}{10^6\mathrm{mm^2}} \times \frac{\mathrm{m}}{1000\,\mathrm{mm}}$$

$$T = 1.25\,\mathrm{N \cdot m} \longleftarrow T$$

The power dissipated in the bearing is

$$\dot{W} = FU = FR\omega = T\omega$$
$$= 1.25\ \text{N}\cdot\text{m} \times 3600\frac{\text{rev}}{\text{min}} \times \frac{\text{min}}{60\ \text{s}} \times 2\pi\frac{\text{rad}}{\text{rev}} \times \frac{\text{j}}{\text{N}\cdot\text{M}} \times \frac{\text{W}}{\text{j}\cdot\text{s}}$$

$$\dot{W} = 471\ \text{W} \qquad \dot{W}$$

To ensure laminar flow, check the Reynolds number.

$$Re = \frac{\rho U a}{\mu} = \frac{SG\rho_{H_2O}Ua}{\mu} = \frac{SG\rho_{H_2O}\omega Ra}{\mu}$$

Assume, as an approximation, the specific gravity of SAE 30 oil is the same as that of SAE 10W oil. From Table A.2, Appendix A, SG = 0.92. Thus

$$Re = 0.92 \times 1000\frac{\text{kg}}{\text{m}^3} \times \frac{(3600)2\pi}{60}\frac{\text{rad}}{\text{s}} \times 38\ \text{mm} \times 0.03175\ \text{mm}$$
$$\times \frac{1}{9.6\times 10^{-3}\text{N}\cdot\text{s/m}^2} \times \frac{\text{m}}{1000\ \text{mm}} \times \frac{\text{m}}{1000\ \text{mm}}$$
$$Re = 43.6$$

Therefore, the flow is laminar, since $Re \ll 1500$.

In this problem we approximated the circular-streamline flow in a small annular gap as a linear flow between infinite parallel plates. As we saw in Example 5.10, for the small value of the gap width a to radius R ratio a/R (in this problem $<1\%$), the error in shear stress is about $\frac{1}{2}$ of this ratio. Hence, the error introduced is insignificant—much less than the uncertainty associated with obtaining a viscosity for the oil.

We have seen how steady, one-dimensional laminar flows between two plates can be generated by applying a pressure gradient, by moving one plate with respect to the other, or by having both driving mechanisms present. To finish our discussion of this type of flow, Example 8.3 examines a *gravity-driven* steady, one-dimensional laminar flow down a vertical wall. Once again, the direct approach would be to start with the two-dimensional rectangular coordinate form of the Navier–Stokes equations (Eqs. 5.27; see Problem 8.44); instead we will use a differential control volume.

Example 8.3 LAMINAR FILM ON A VERTICAL WALL

A viscous, incompressible, Newtonian liquid flows in steady, laminar flow down a vertical wall. The thickness, δ, of the liquid film is constant. Since the liquid free surface is exposed to atmospheric pressure, there is no pressure gradient. For this gravity-driven flow, apply the momentum equation to differential control volume $dx\,dy\,dz$ to derive the velocity distribution in the liquid film.

Given: Fully developed laminar flow of incompressible, Newtonian liquid down a vertical wall; thickness, δ, of the liquid film is constant and $\partial p/\partial x = 0$.

Find: Expression for the velocity distribution in the film.

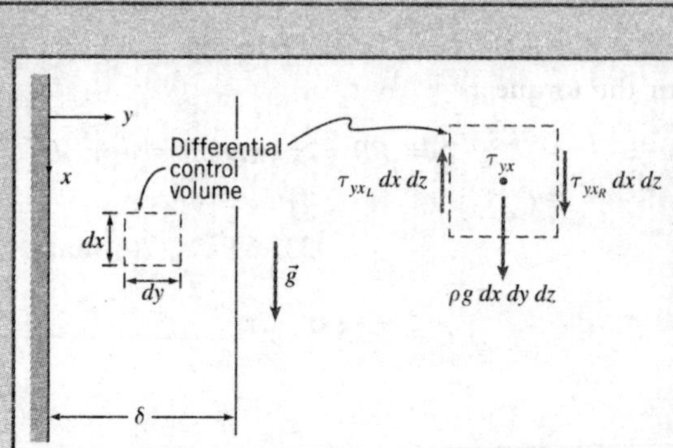

Solution:

The x component of the momentum equation for a control volume is

$$F_{S_x} + F_{B_x} = \frac{\partial}{\partial t}\int_{CV} u\,\rho\,d\forall + \int_{CS} u\,\rho\vec{V}\cdot d\vec{A} \qquad (4.18a)$$

Under the conditions given we are dealing with a steady, incompressible, fully developed laminar flow.

For steady flow, $\frac{\partial}{\partial t}\int_{CV} u\,\rho\,d\mathcal{V} = 0$

For fully developed flow, $\int_{CS} u\,\rho\vec{V}\cdot d\vec{A} = 0$

Thus the momentum equation for the present case reduces to

$$F_{S_x} + F_{B_x} = 0$$

The body force, F_{B_x}, is given by $F_{B_x} = \rho g\,d\mathcal{V} = \rho g\,dx\,dy\,dz$. The only surface forces acting on the differential control volume are shear forces on the vertical surfaces. (Since we have a free-surface flow, with straight streamlines, the pressure is atmospheric throughout; no net pressure forces act on the control volume.)

If the shear stress at the center of the differential control volume is τ_{yx}, then,

$$\text{shear stress on left face is } \tau_{yx_L} = \left(\tau_{yx} - \frac{d\tau_{yx}}{dy}\frac{dy}{2}\right)$$

and

$$\text{shear stress on right face is } \tau_{yx_R} = \left(\tau_{yx} + \frac{d\tau_{yx}}{dy}\frac{dy}{2}\right)$$

The direction of the shear stress vectors is taken consistent with the sign convention of Section 2.3. Thus on the left face, a minus y surface, τ_{yx_L} acts upward, and on the right face, a plus y surface, τ_{yx_R} acts downward.

The surface forces are obtained by multiplying each shear stress by the area over which it acts. Substituting into $F_{S_x} + F_{B_x} = 0$, we obtain

$$-\tau_{yx_L}\,dx\,dz + \tau_{yx_R}\,dx\,dz + \rho g\,dx\,dy\,dz = 0$$

or

$$-\left(\tau_{yx} - \frac{d\tau_{yx}}{dy}\frac{dy}{2}\right)dx\,dz + \left(\tau_{yx} + \frac{d\tau_{yx}}{dy}\frac{dy}{2}\right)dx\,dz + \rho g\,dx\,dy\,dz = 0$$

Simplifying gives

$$\frac{d\tau_{yx}}{dy} + \rho g = 0 \quad \text{or} \quad \frac{d\tau_{yx}}{dy} = -\rho g$$

Since

$$\tau_{yx} = \mu\frac{du}{dy} \quad \text{then} \quad \mu\frac{d^2u}{dy^2} = -\rho g \quad \text{and} \quad \frac{d^2u}{dy^2} = -\frac{\rho g}{\mu}$$

Integrating with respect to y gives

$$\frac{du}{dy} = -\frac{\rho g}{\mu}y + c_1$$

Integrating again, we obtain

$$u = -\frac{\rho g}{\mu}\frac{y^2}{2} + c_1 y + c_2$$

To evaluate constants c_1 and c_2, we apply appropriate boundary conditions:

(i) $y = 0, \quad u = 0$ (no-slip)

(ii) $y = \delta, \quad \frac{du}{dy} = 0$ (neglect air resistance, i.e., assume zero shear stress at free surface)

From boundary condition (i), $c_2 = 0$

From boundary condition (ii), $0 = -\frac{\rho g}{\mu}\delta + c_1$ or $c_1 = \frac{\rho g}{\mu}\delta$

Hence,

$$u = -\frac{\rho g}{\mu}\frac{y^2}{2} + \frac{\rho g}{\mu}\delta y \quad \text{or} \quad u = \frac{\rho g}{\mu}\delta^2\left[\left(\frac{y}{\delta}\right) - \frac{1}{2}\left(\frac{y}{\delta}\right)^2\right] \longleftarrow u(y)$$

Using the velocity profile it can be shown that:

the volume flow rate is $Q/l = \frac{\rho g}{3\mu}\delta^3$

the maximum velocity is $U_{\max} = \frac{\rho g}{2\mu}\delta^2$

the average velocity is $\overline{V} = \frac{\rho g}{3\mu}\delta^2$

Flow in the liquid film is laminar for $Re = \overline{V}\delta/\nu \leq 1000$ [1].

Notes:

✓ This problem is a special case ($\theta = 90°$) of the inclined plate flow analyzed in Example 5.9 that we solved using the Navier–Stokes equations.

✓ This problem and Example 5.9 demonstrate that use of the differential control volume approach or the Navier–Stokes equations leads to the same result.

8.3 Fully Developed Laminar Flow in a Pipe

VIDEO

Laminar Flow Exiting from a Tube.

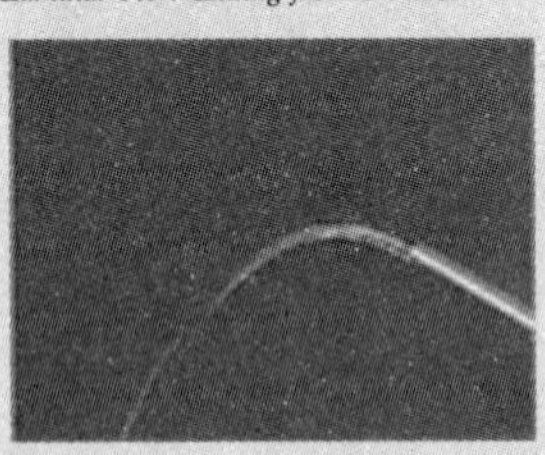

As a final example of fully developed laminar flow, let us consider fully developed laminar flow in a pipe. Here the flow is axisymmetric. Consequently it is most convenient to work in cylindrical coordinates. This is yet another case where we could use the Navier–Stokes equations, this time in cylindrical coordinates (Eqs. B.3). Instead we will again take the longer route—using a differential control volume—to bring out some important features of the fluid mechanics. The development will be very similar to that for parallel plates in the previous section; cylindrical coordinates just make the analysis a little trickier mathematically. Since the flow is axisymmetric, the control volume will be a differential annulus, as shown in Fig. 8.7. The control volume length is dx and its thickness is dr.

For a fully developed steady flow, the x component of the momentum equation (Eq. 4.18a), when applied to the differential control volume, once again reduces to

$$F_{S_x} = 0$$

The next step is to sum the forces acting on the control volume in the x direction. We know that normal forces (pressure forces) act on the left and right ends of the control volume, and that tangential forces (shear forces) act on the inner and outer cylindrical surfaces.

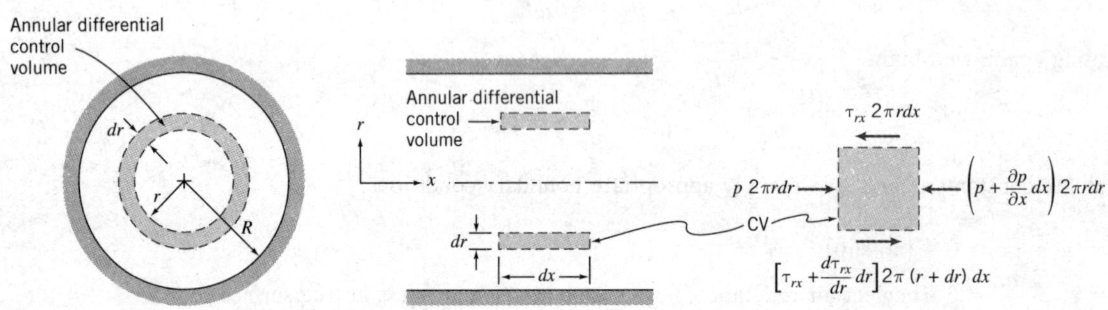

(*a*) End view of CV (*b*) Side view of CV (*c*) Forces on CV

Fig. 8.7 Differential control volume for analysis of fully developed laminar flow in a pipe.

If the pressure at the left face of the control volume is p, then the pressure force on the left end is

$$dF_L = p2\pi r\,dr$$

The pressure force on the right end is

$$dF_R = -\left(p + \frac{\partial p}{\partial x}\,dx\right) 2\pi r\,dr$$

If the shear stress at the inner surface of the annular control volume is τ_{rx}, then the shear force on the inner cylindrical surface is

$$dF_I = -\tau_{rx} 2\pi r\,dx$$

The shear force on the outer cylindrical surface is

$$dF_O = \left(\tau_{rx} + \frac{d\tau_{rx}}{dr}\,dr\right) 2\pi\,(r + dr)dx$$

The sum of the x components of force, dF_L, dF_R, dF_I, and dF_O, acting on the control volume must be zero. This leads to the condition that

$$-\frac{\partial p}{\partial x}\,2\pi r\,dr\,dx + \tau_{rx}2\pi\,dr\,dx + \frac{d\tau_{rx}}{dr}\,2\pi r\,dr\,dx = 0$$

Dividing this equation by $2\pi r\,dr\,dx$ and solving for $\partial p/\partial x$ gives

$$\frac{\partial p}{\partial x} = \frac{\tau_{rx}}{r} + \frac{d\tau_{rx}}{dr} = \frac{1}{r}\,\frac{d(r\tau_{rx})}{dr}$$

Comparing this to the corresponding equation for parallel plates (Eq. 8.3) shows the mathematical complexity introduced because we have cylindrical coordinates. The left side of the equation is at most a function of x only (the pressure is uniform at each section); the right side is at most a function of r only (because the flow is fully developed). Hence, the only way the equation can be valid for all x and r is for both sides to in fact be constant:

$$\frac{1}{r}\,\frac{d(r\tau_{rx})}{dr} = \frac{\partial p}{\partial x} = \text{constant} \quad \text{or} \quad \frac{d(r\tau_{rx})}{dr} = r\frac{\partial p}{\partial x}$$

We are not quite finished, but already we have an important result: *In a constant diameter pipe, the pressure drops uniformly along the pipe length* (except for the entrance region).

Integrating this equation, we obtain

$$r\tau_{rx} = \frac{r^2}{2}\left(\frac{\partial p}{\partial x}\right) + c_1$$

or

$$\tau_{rx} = \frac{r}{2}\left(\frac{\partial p}{\partial x}\right) + \frac{c_1}{r} \qquad (8.10)$$

Since $\tau_{rx} = \mu du/dr$, we have

$$\mu\frac{du}{dr} = \frac{r}{2}\left(\frac{\partial p}{\partial x}\right) + \frac{c_1}{r}$$

and

$$u = \frac{r^2}{4\mu}\left(\frac{\partial p}{\partial x}\right) + \frac{c_1}{\mu}\ln r + c_2 \tag{8.11}$$

We need to evaluate constants c_1 and c_2. However, we have only the one boundary condition that $u = 0$ at $r = R$. What do we do? Before throwing in the towel, let us look at the solution for the velocity profile given by Eq. 8.11. Although we do not know the velocity at the pipe centerline, we do know from physical considerations that the velocity must be finite at $r = 0$. The only way that this can be true is for c_1 to be zero. (We could have also concluded that $c_1 = 0$ from Eq. 8.10—which would otherwise yield an infinite stress at $r = 0$.) Thus, from physical considerations, we conclude that $c_1 = 0$, and hence

$$u = \frac{r^2}{4\mu}\left(\frac{\partial p}{\partial x}\right) + c_2$$

The constant, c_2, is evaluated by using the available boundary condition at the pipe wall: at $r = R$, $u = 0$. Consequently,

$$0 = \frac{R^2}{4\mu}\left(\frac{\partial p}{\partial x}\right) + c_2$$

This gives

$$c_2 = -\frac{R^2}{4\mu}\left(\frac{\partial p}{\partial x}\right)$$

and hence

$$u = \frac{r^2}{4\mu}\left(\frac{\partial p}{\partial x}\right) - \frac{R^2}{4\mu}\left(\frac{\partial p}{\partial x}\right) = \frac{1}{4\mu}\left(\frac{\partial p}{\partial x}\right)(r^2 - R^2)$$

or

$$u = -\frac{R^2}{4\mu}\left(\frac{\partial p}{\partial x}\right)\left[1 - \left(\frac{r}{R}\right)^2\right] \tag{8.12}$$

Since we have the velocity profile, we can obtain a number of additional features of the flow.

Shear Stress Distribution

The shear stress is

$$\tau_{rx} = \mu\frac{du}{dr} = \frac{r}{2}\left(\frac{\partial p}{\partial x}\right) \tag{8.13a}$$

Volume Flow Rate

The volume flow rate is

$$Q = \int_A \vec{V} \cdot d\vec{A} = \int_0^R u 2\pi r\, dr = \int_0^R \frac{1}{4\mu}\left(\frac{\partial p}{\partial x}\right)(r^2 - R^2)2\pi r\, dr$$

$$Q = -\frac{\pi R^4}{8\mu}\left(\frac{\partial p}{\partial x}\right) \tag{8.13b}$$

Flow Rate as a Function of Pressure Drop

We proved that in fully developed flow the pressure gradient, $\partial p/\partial x$, is constant. Therefore, $\partial p/\partial x = (p_2 - p_1)/L = -\Delta p/L$. Substituting into Eq. 8.13b for the volume flow rate gives

$$Q = -\frac{\pi R^4}{8\mu}\left[\frac{-\Delta p}{L}\right] = \frac{\pi \Delta p R^4}{8\mu L} = \frac{\pi \Delta p D^4}{128\mu L} \tag{8.13c}$$

for laminar flow in a horizontal pipe. Note that Q is a sensitive function of D; $Q \sim D^4$, so, for example, doubling the diameter D increases the flow rate Q by a factor of 16.

Average Velocity

The average velocity magnitude, $\overline{V}$, is given by

$$\overline{V} = \frac{Q}{A} = \frac{Q}{\pi R^2} = -\frac{R^2}{8\mu}\left(\frac{\partial p}{\partial x}\right) \tag{8.13d}$$

Point of Maximum Velocity

To find the point of maximum velocity, we set du/dr equal to zero and solve for the corresponding r. From Eq. 8.12

$$\frac{du}{dr} = \frac{1}{2\mu}\left(\frac{\partial p}{\partial x}\right)r$$

Thus,

$$\frac{du}{dr} = 0 \quad \text{at} \quad r = 0$$

At $r = 0$,

$$u = u_{\max} = U = -\frac{R^2}{4\mu}\left(\frac{\partial p}{\partial x}\right) = 2\overline{V} \tag{8.13e}$$

The velocity profile (Eq. 8.12) may be written in terms of the maximum (centerline) velocity as

$$\frac{u}{U} = 1 - \left(\frac{r}{R}\right)^2 \tag{8.14}$$

The parabolic velocity profile, given by Eq. 8.14 for fully developed laminar pipe flow, was sketched in Fig. 8.1.

Example 8.4 CAPILLARY VISCOMETER

A simple and accurate viscometer can be made from a length of capillary tubing. If the flow rate and pressure drop are measured, and the tube geometry is known, the viscosity of a Newtonian liquid can be computed from Eq. 8.13c. A test of a certain liquid in a capillary viscometer gave the following data:

Flow rate:	880 mm^3/s	Tube length:	1 m
Tube diameter:	0.50 mm	Pressure drop:	1.0 MPa

Determine the viscosity of the liquid.

Given: Flow in a capillary viscometer.
The flow rate is $Q = 880\ \text{mm}^3/\text{s}$.

Find: The fluid viscosity.

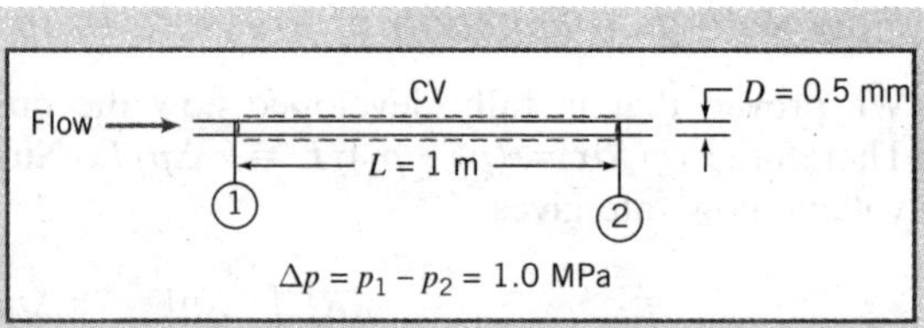

Solution:

Equation 8.13c may be applied.

Governing equation:
$$Q = \frac{\pi \Delta p D^4}{128 \mu L} \tag{8.13c}$$

Assumptions:
(1) Laminar flow.
(2) Steady flow.
(3) Incompressible flow.
(4) Fully developed flow.
(5) Horizontal tube.

Then

$$\mu = \frac{\pi \Delta p D^4}{128\, LQ} = \frac{\pi}{128} \times 1.0 \times 10^6 \frac{\text{N}}{\text{m}^2} \times (0.50)^4\ \text{mm}^4 \times \frac{\text{s}}{880\ \text{mm}^3} \times \frac{1}{1\ \text{m}} \times \frac{\text{m}}{10^3\ \text{mm}}$$

$$\mu = 1.74 \times 10^{-3}\ \text{N}\cdot\text{s/m}^2 \longleftarrow \mu$$

Check the Reynolds number. Assume the fluid density is similar to that of water, 999 kg/m^3. Then

$$\overline{V} = \frac{Q}{A} = \frac{4Q}{\pi D^2} = \frac{4}{\pi} \times 880 \frac{\text{mm}^3}{\text{s}} \times \frac{1}{(0.50)^2\ \text{mm}^2} \times \frac{\text{m}}{10^3\ \text{mm}} = 4.48\ \text{m/s}$$

and

$$Re = \frac{\rho \overline{V} D}{\mu} = 999 \frac{\text{kg}}{\text{m}^3} \times 4.48 \frac{\text{m}}{\text{s}} \times 0.50\ \text{mm} \times \frac{\text{m}^2}{1.74 \times 10^{-3}\ \text{N}\cdot\text{s}} \times \frac{\text{m}}{10^3\ \text{mm}} \times \frac{\text{N}\cdot\text{s}^2}{\text{kg}\cdot\text{m}}$$

$$Re = 1290$$

Consequently, since $Re < 2300$, the flow is laminar.

This problem is a little oversimplified. To design a capillary viscometer the entrance length, liquid temperature, and kinetic energy of the flowing liquid would all need to be considered.

Part B Flow in Pipes and Ducts

In this section we will be interested in determining the factors that affect the pressure in an incompressible fluid as it flows in a pipe or duct (we will refer to "pipe" but imply "duct," too). If we ignore friction for a moment (and assume steady flow and consider a streamline in the flow), the Bernoulli equation from Chapter 6 applies,

$$\frac{p}{\rho} + \frac{V^2}{2} + gz = \text{constant} \tag{6.8}$$

From this equation we can see what *tends* to lead to a *pressure decrease* along the streamline in this frictionless flow: a *reduction of area* at some point in the pipe (causing an increase in the velocity V), or the pipe having a *positive incline* (so z increases). Conversely, the pressure will tend to increase if the flow area is increased or the pipe slopes downward. We say "tends to" because one factor may counteract another; for example, we may have a downward sloping pipe (tending to increase pressure) with a reduction in diameter (tending to decrease pressure).

In reality, flows in pipes and ducts experience significant friction and are often turbulent, so the Bernoulli equation does not apply (it doesn't even make sense to use V; instead we will use $\overline{V}$, to represent the average velocity at a section along the pipe). We will learn that, in effect, friction effects lead to a continual reduction in the value of the Bernoulli constant of Eq. 6.8 (this represents a "loss" of mechanical energy). We have already seen that, in contrast to the Bernoulli equation, for laminar flow there is a pressure drop even for a horizontal, constant diameter pipe; in this section we will see that turbulent flows experience an even larger pressure drop. We will need to replace the Bernoulli equation with an energy equation that incorporates the effects of friction.

In summary, we can state that *three* factors tend to reduce the pressure in a pipe flow: a decrease in pipe area, an upward slope, and friction. For now we will focus on pressure loss due to friction and so will analyze pipes that are of constant area and that are horizontal.

We have already seen in the previous section that for laminar flow we can theoretically deduce the pressure drop. Rearranging Eq. 8.13c to solve for the pressure drop Δp,

$$\Delta p = \frac{128\mu L Q}{\pi D^4}$$

We would like to develop a similar expression that applies for turbulent flows, but we will see that this is not possible analytically; instead, we will develop expressions based on a combination of theoretical and experimental approaches. Before proceeding, we note that it is conventional to break losses due to friction into two categories: *major losses*, which are losses due to friction in the constant-area sections of the pipe; and *minor losses* (sometimes larger than "major" losses), which are losses due to valves, elbows, and so on (and we will treat the pressure drop at the entrance region as a minor loss term).

Since circular pipes are most common in engineering applications, the basic analysis will be performed for circular geometries. The results can be extended to other geometries by introducing the hydraulic diameter, which is treated in Section 8.7. (Open channel flows will be treated in Chapter 11, and compressible flow in ducts will be treated in Chapter 13.)

8.4 Shear Stress Distribution in Fully Developed Pipe Flow

We consider again fully developed flow in a horizontal circular pipe, except now we may have laminar or turbulent flow. In Section 8.3 we showed that a force balance between friction and pressure forces leads to Eq. 8.10:

$$\tau_{rx} = \frac{r}{2}\left(\frac{\partial p}{\partial x}\right) + \frac{c_1}{r} \tag{8.10}$$

Because we cannot have infinite stress at the centerline, the constant of integration c_1 must be zero, so

$$\tau_{rx} = \frac{r}{2}\frac{\partial p}{\partial x} \tag{8.15}$$

Equation 8.15 indicates that *for both laminar and turbulent fully developed flows the shear stress varies linearly across the pipe,* from zero at the centerline to a maximum at the pipe wall. The stress on the wall, τ_w (equal and opposite to the stress in the fluid at the wall), is given by

$$\tau_w = -\left[\tau_{rx}\right]_{r=R} = -\frac{R}{2}\frac{\partial p}{\partial x} \tag{8.16}$$

For *laminar* flow we used our familiar stress equation $\tau_{rx} = \mu\, du/dr$ in Eq. 8.15 to eventually obtain the laminar velocity distribution. This led to a set of usable equations, Eqs. 8.13, for obtaining various flow characteristics; e.g., Eq. 8.13c gave a relationship for the flow rate Q, a result first obtained experimentally by Jean Louis Poiseuille, a French physician, and independently by Gotthilf H. L. Hagen, a German engineer, in the 1850s [2].

Unfortunately there is no equivalent stress equation for *turbulent* flow, so we cannot replicate the laminar flow analysis to derive turbulent equivalents of Eqs. 8.13. All we can do in this section is indicate some classic semi-empirical results [3].

As we discussed in Section 2.6, and illustrated in Fig. 2.17, turbulent flow is represented at each point by the time-mean velocity $\bar{u}$ plus (for a two-dimensional flow) randomly fluctuating velocity components u' and v' in the x and y directions (in this context y is the distance inwards from the pipe wall). These components continuously transfer momentum between adjacent fluid layers, tending to reduce any velocity gradient present. This effect shows up as an apparent stress, first introduced by Osborne Reynolds, and called the *Reynolds stress*.[1] This stress is given by $-\rho\overline{u'v'}$, where the overbar indicates a time average. Hence, we find

$$\tau = \tau_{\text{lam}} + \tau_{\text{turb}} = \mu\frac{d\bar{u}}{dy} - \rho\overline{u'v'} \tag{8.17}$$

Do not misunderstand the minus sign in Eq. 8.17—it turns out that u' and v' are negatively correlated, so that $\tau_{\text{turb}} = -\rho\overline{u'v'}$ is positive. In Fig. 8.8, experimental measurements of the Reynolds stress for fully developed turbulent pipe flow at two Reynolds numbers are presented; $Re_U = UD/\nu$, where U is the centerline velocity. The turbulent shear stress has been nondimensionalized with the wall shear stress. Recall that Eq. 8.15 showed that the shear stress in the fluid varies linearly from τ_w at the pipe wall ($y/R \to 0$) to zero at the centerline ($y/R = 1$); from Fig. 8.8 we see that the Reynolds stress has almost the same trend, so that the friction is almost all due to Reynolds stress. What Fig. 8.8 *doesn't* show is that close to the wall ($y/R \to 0$) the Reynolds stress drops to zero. This is because the no-slip condition holds, so not

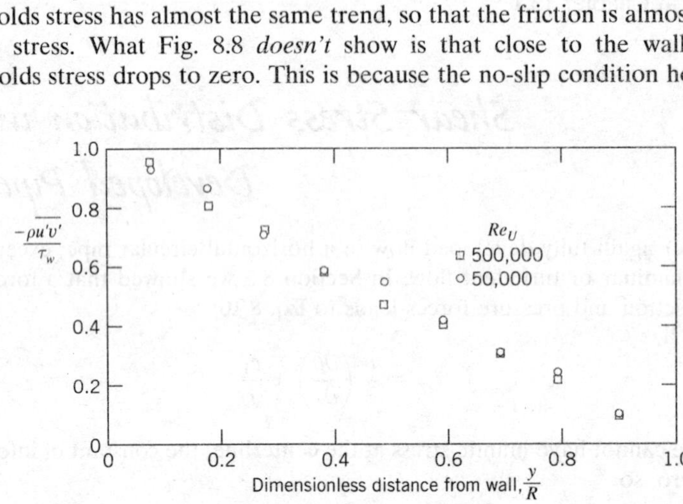

Fig. 8.8 Turbulent shear stress (Reynolds stress) for fully developed turbulent flow in a pipe. (Data from Laufer [5].)

[1]The Reynolds stress terms arise from consideration of the complete equations of motion for turbulent flow [4].

only does the mean velocity $\overline{u} \to 0$, but also the fluctuating velocity components u' and $v' \to 0$ (the wall tends to suppress the fluctuations). Hence, the turbulent stress, $\tau_{\text{turb}} = -\rho\overline{u'v'} \to 0$, as we approach the wall, and is zero at the wall. Since the Reynolds stress is zero at the wall, Eq. 8.17 shows that the wall shear is given by $\tau_w = \mu(d\overline{u}/dy)_{y=0}$. In the region very close to the wall, called the *wall layer*, viscous shear is dominant. In the region between the wall layer and the central portion of the pipe both viscous and turbulent shear are important.

Turbulent Velocity Profiles in Fully Developed Pipe Flow 8.5

Except for flows of very viscous fluids in small diameter ducts, internal flows generally are turbulent. As noted in the discussion of shear stress distribution in fully developed pipe flow (Section 8.4), in turbulent flow there is no universal relationship between the stress field and the mean velocity field. Thus, for turbulent flows we are forced to rely on experimental data.

VIDEO

The Glen Canyon Dam: A Turbulent Pipe Flow.

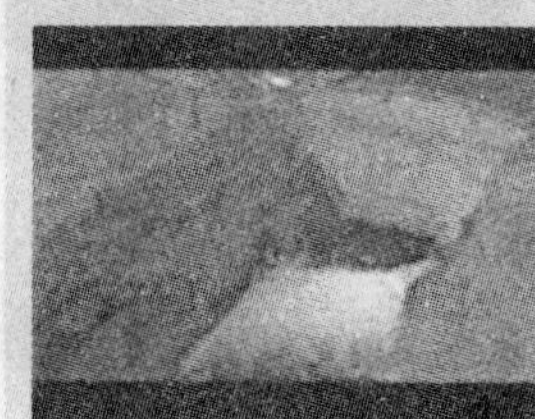

Dividing Eq. 8.17 by ρ gives

$$\frac{\tau}{\rho} = \nu \frac{d\overline{u}}{dy} - \overline{u'v'} \tag{8.18}$$

The term τ/ρ arises frequently in the consideration of turbulent flows; it has dimensions of velocity squared. In particular, the quantity $(\tau_w/\rho)^{1/2}$ is called the *friction velocity* and is denoted by the symbol u_*. It is a constant for a given flow.

The velocity profile for fully developed turbulent flow through a smooth pipe is shown in Fig. 8.9. The plot is semilogarithmic; $\overline{u}/u_*$ is plotted against $\log(yu_*/\nu)$. The nondimensional parameters $\overline{u}/u_*$ and yu_*/ν arise from dimensional analysis if one reasons that the velocity in the neighborhood of the wall is determined by the conditions at the wall, the fluid properties, and the distance from the wall. It is simply fortuitous that the dimensionless plot of Fig. 8.9 gives a fairly accurate representation of the velocity profile in a pipe away from the wall; note the small deviations in the region of the pipe centerline.

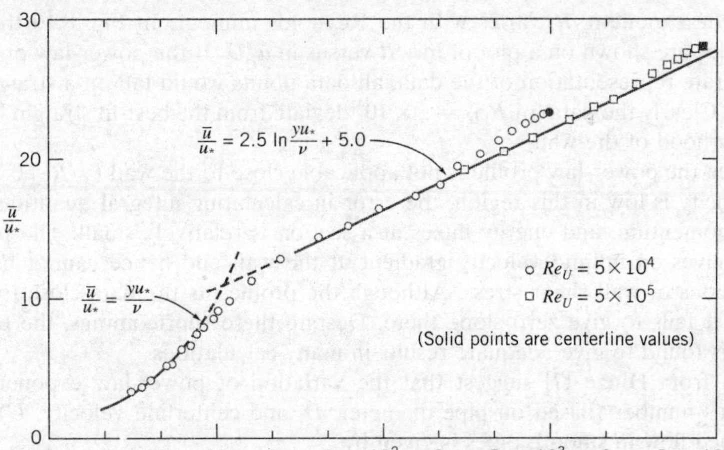

Fig. 8.9 Turbulent velocity profile for fully developed flow in a smooth pipe. (Data from Laufer [5].)

In the region very close to the wall where viscous shear is dominant, the mean velocity profile follows the linear viscous relation

$$u^+ = \frac{\bar{u}}{u_*} = \frac{yu_*}{\nu} = y^+ \tag{8.19}$$

where y is distance measured from the wall ($y = R - r$; R is the pipe radius), and $\bar{u}$ is mean velocity. Equation 8.19 is valid for $0 \le y^+ \le 5\text{–}7$; this region is called the *viscous sublayer*.

For values of $yu_*/\nu > 30$, the data are quite well represented by the semilogarithmic curve-fit equation

$$\frac{\bar{u}}{u_*} = 2.5 \ln \frac{yu_*}{\nu} + 5.0 \tag{8.20}$$

VIDEO

Computer Simulation: Turbulent Channel Flow 1.

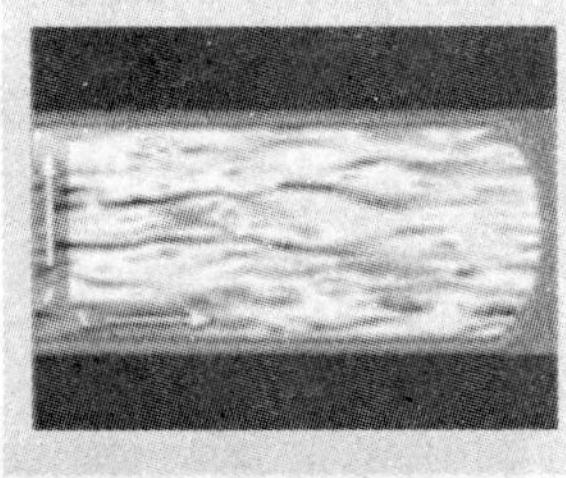

In this region both viscous and turbulent shear are important (although turbulent shear is expected to be significantly larger). There is considerable scatter in the numerical constants of Eq. 8.20; the values given represent averages over many experiments [6]. The region between $y^+ = 5\text{–}7$ and $y^+ = 30$ is referred to as the *transition region*, or *buffer layer*.

If Eq. 8.20 is evaluated at the centerline ($y = R$ and $u = U$) and the general expression of Eq. 8.20 is subtracted from the equation evaluated at the centerline, we obtain

$$\frac{U - \bar{u}}{u_*} = 2.5 \ln \frac{R}{y} \tag{8.21}$$

where U is the centerline velocity. Equation 8.21, referred to as the *defect law*, shows that the velocity defect (and hence the general shape of the velocity profile in the neighborhood of the centerline) is a function of the distance ratio only and does not depend on the viscosity of the fluid.

VIDEO

Computer Simulation: Turbulent Channel Flow 2.

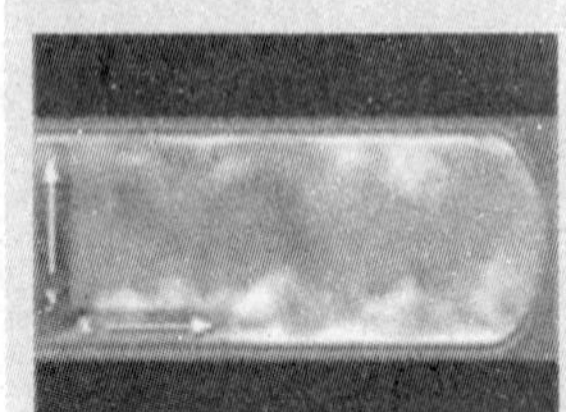

The velocity profile for turbulent flow through a smooth pipe may also be approximated by the empirical *power-law* equation

$$\frac{\bar{u}}{U} = \left(\frac{y}{R}\right)^{1/n} = \left(1 - \frac{r}{R}\right)^{1/n} \tag{8.22}$$

where the exponent, n, varies with the Reynolds number. In Fig. 8.10 the data of Laufer [5] are shown on a plot of $\ln y/R$ versus $\ln \bar{u}/U$. If the power-law profile were an accurate representation of the data, all data points would fall on a straight line of slope n. Clearly the data for $Re_U = 5 \times 10^4$ deviate from the best-fit straight line in the neighborhood of the wall.

VIDEO

Computer Simulation: Turbulent Channel Flow 3.

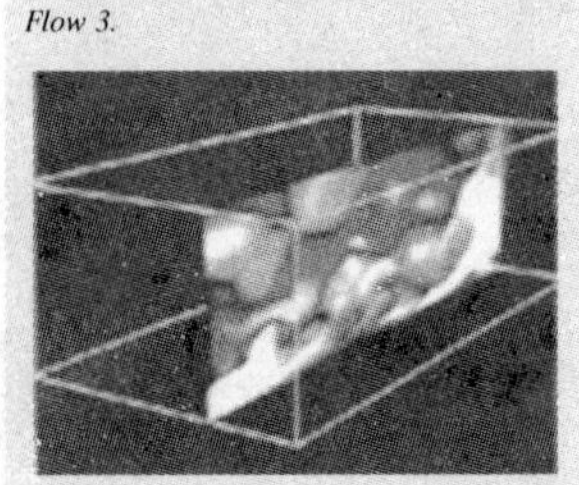

Hence the power-law profile is not applicable close to the wall ($y/R < 0.04$). Since the velocity is low in this region, the error in calculating integral quantities such as mass, momentum, and energy fluxes at a section is relatively small. The power-law profile gives an infinite velocity gradient at the wall and hence cannot be used in calculations of wall shear stress. Although the profile fits the data close to the centerline, it fails to give zero slope there. Despite these shortcomings, the power-law profile is found to give adequate results in many calculations.

Data from Hinze [7] suggest that the variation of power-law exponent n with Reynolds number (based on pipe diameter, D, and centerline velocity, U) for fully developed flow in smooth pipes is given by

$$n = -1.7 + 1.8 \log Re_U \tag{8.23}$$

for $Re_U > 2 \times 10^4$.

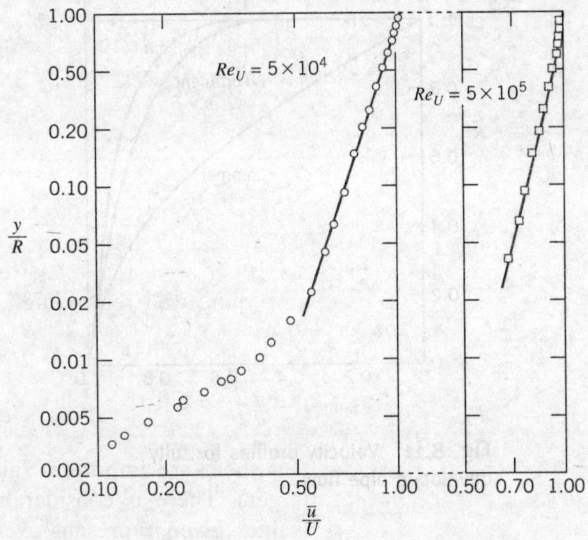

Fig. 8.10 Power-law velocity profiles for fully developed turbulent flow in a smooth pipe. (Data from Laufer [5].)

Since the average velocity is $\overline{V} = Q/A$, and

$$Q = \int_A \vec{V} \cdot d\vec{A}$$

the ratio of the average velocity to the centerline velocity may be calculated for the power-law profiles of Eq. 8.22 assuming the profiles to be valid from wall to centerline. The result is

$$\frac{\overline{V}}{U} = \frac{2n^2}{(n+1)(2n+1)} \tag{8.24}$$

From Eq. 8.24, we see that as n increases (with increasing Reynolds number) the ratio of the average velocity to the centerline velocity increases; with increasing Reynolds number the velocity profile becomes more blunt or "fuller" (for $n = 6$, $\overline{V}/U = 0.79$ and for $n = 10$, $\overline{V}/U = 0.87$). As a representative value, 7 often is used for the exponent; this gives rise to the term "a one-seventh power profile" for fully developed turbulent flow:

$$\frac{\bar{u}}{U} = \left(\frac{y}{R}\right)^{1/7} = \left(1 - \frac{r}{R}\right)^{1/7}$$

Velocity profiles for $n = 6$ and $n = 10$ are shown in Fig. 8.11. The parabolic profile for fully developed laminar flow is included for comparison. It is clear that the turbulent profile has a much steeper slope near the wall. This is consistent with our discussion leading to Eq. 8.17—the fluctuating velocity components u' and v' continuously transfer momentum between adjacent fluid layers, tending to reduce the velocity gradient.

Energy Considerations in Pipe Flow 8.6

We have so far used the momentum and conservation of mass equations, in control volume form, to discuss viscous flow. It is obvious that viscous effects will have an

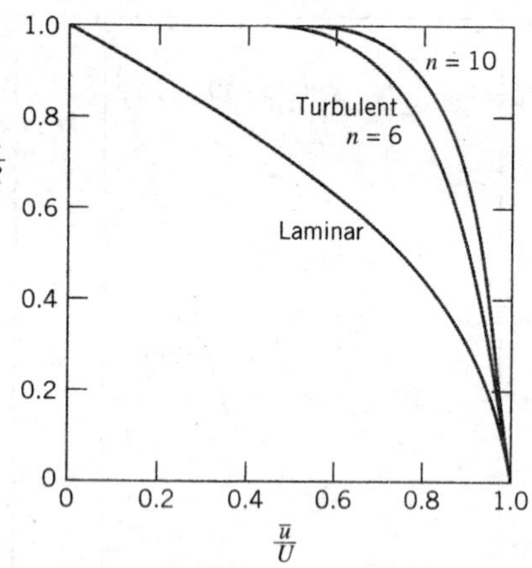

Fig. 8.11 Velocity profiles for fully developed pipe flow.

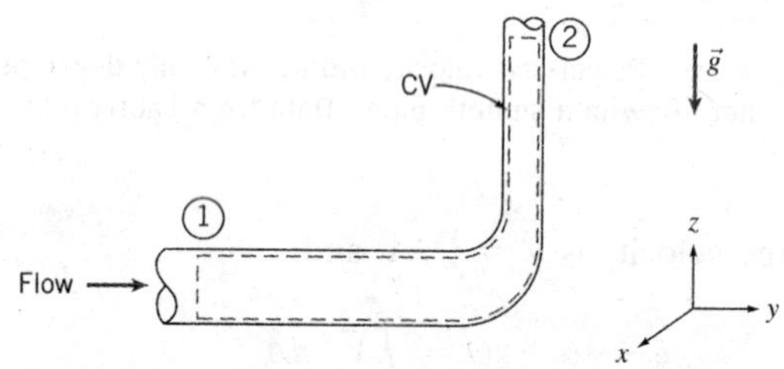

Fig. 8.12 Control volume and coordinates for energy analysis of flow through a 90° reducing elbow.

important effect on energy considerations. In Section 6.5 we discussed the Energy Grade Line (EGL),

$$EGL = \frac{p}{\rho g} + \frac{V^2}{2g} + z \tag{6.16b}$$

and saw that this is a measure of the total mechanical energy ("pressure," kinetic and potential, per unit mass) in a flow. We can expect that instead of being constant (which it was for inviscid flow), the EGL will continuously decrease in the direction of flow as friction "eats" the mechanical energy (Examples 8.9 and 8.10 have sketches of such EGL—and HGL—curves; you may wish to preview them). We can now consider the energy equation (the first law of thermodynamics) to obtain information on the effects of friction.

Consider, for example, steady flow through the piping system, including a reducing elbow, shown in Fig. 8.12. The control volume boundaries are shown as dashed lines. They are normal to the flow at sections ① and ② and coincide with the inside surface of the pipe wall elsewhere.

Basic equation:

$$= 0(1) = 0(2) = 0(1) = 0(3)$$

$$\dot{Q} - \dot{W}_s - \dot{W}_{\text{shear}} - \dot{W}_{\text{other}} = \frac{\partial}{\partial t}\int_{\text{CV}} e\,\rho\, dV + \int_{\text{CS}} (e + pv)\rho \vec{V} \cdot d\vec{A} \tag{4.56}$$

$$e = u + \frac{V^2}{2} + gz$$

Assumptions: (1) $\dot{W}_s = 0,\ \dot{W}_{\text{other}} = 0$.
(2) $\dot{W}_{\text{shear}} = 0$ (although shear stresses are present at the walls of the elbow, the velocities are zero there, so there is no possibility of work).
(3) Steady flow.
(4) Incompressible flow.
(5) Internal energy and pressure uniform across sections ① and ②.

Under these assumptions the energy equation reduces to

$$\dot{Q} = \dot{m}(u_2 - u_1) + \dot{m}\left(\frac{p_2}{\rho} - \frac{p_1}{\rho}\right) + \dot{m}g(z_2 - z_1) + \int_{A_2} \frac{V_2^2}{2}\rho V_2\,dA_2 - \int_{A_1} \frac{V_1^2}{2}\rho V_1\,dA_1 \tag{8.25}$$

Note that we have *not* assumed the velocity to be uniform at sections ① and ②, since we know that for viscous flows the velocity at a cross-section cannot be uniform. However, it is convenient to introduce the average velocity into Eq. 8.25 so that we can eliminate the integrals. To do this, we define a kinetic energy coefficient.

Kinetic Energy Coefficient

The *kinetic energy coefficient*, α, is defined such that

$$\int_A \frac{V^2}{2}\rho V\,dA = \alpha\int_A \frac{\overline{V}^2}{2}\rho V dA = \alpha\dot{m}\frac{\overline{V}^2}{2} \tag{8.26a}$$

or

$$\alpha = \frac{\int_A \rho V^3 dA}{\dot{m}\overline{V}^2} \tag{8.26b}$$

We can think of α as a correction factor that allows us to use the average velocity $\overline{V}$ in the energy equation to compute the kinetic energy at a cross section.

For laminar flow in a pipe (velocity profile given by Eq. 8.12), $\alpha = 2.0$.

In turbulent pipe flow, the velocity profile is quite flat, as shown in Fig. 8.11. We can use Eq. 8.26b together with Eqs. 8.22 and 8.24 to determine α. Substituting the power-law velocity profile of Eq. 8.22 into Eq. 8.26b, we obtain

$$\alpha = \left(\frac{U}{\overline{V}}\right)^3 \frac{2n^2}{(3+n)(3+2n)} \tag{8.27}$$

Equation 8.24 gives $\overline{V}/U$ as a function of the power-law exponent n; combining this with Eq. 8.27 leads to a fairly complicated expression in n. The overall result is that in the realistic range of n, from $n = 6$ to $n = 10$ for high Reynolds numbers, α varies from 1.08 to 1.03; for the one-seventh power profile ($n = 7$), $\alpha = 1.06$. Because α is reasonably close to unity for high Reynolds numbers, and because the change in kinetic energy is usually small compared with the dominant terms in the energy equation, *we shall almost always use the approximation* $\alpha = 1$ *in our pipe flow calculations.*

Head Loss

Using the definition of α, the energy equation (Eq. 8.25) can be written

$$\dot{Q} = \dot{m}(u_2 - u_1) + \dot{m}\left(\frac{p_2}{\rho} - \frac{p_1}{\rho}\right) + \dot{m}g(z_2 - z_1) + \dot{m}\left(\frac{\alpha_2\overline{V}_2^2}{2} - \frac{\alpha_1\overline{V}_1^2}{2}\right)$$

Dividing by the mass flow rate gives

$$\frac{\delta Q}{dm} = u_2 - u_1 + \frac{p_2}{\rho} - \frac{p_1}{\rho} + gz_2 - gz_1 + \frac{\alpha_2 \overline{V}_2^2}{2} - \frac{\alpha_1 \overline{V}_1^2}{2}$$

Rearranging this equation, we write

$$\left(\frac{p_1}{\rho} + \alpha_1 \frac{\overline{V}_1^2}{2} + gz_1\right) - \left(\frac{p_2}{\rho} + \alpha_2 \frac{\overline{V}_2^2}{2} + gz_2\right) = (u_2 - u_1) - \frac{\delta Q}{dm} \tag{8.28}$$

In Eq. 8.28, the term

$$\left(\frac{p}{\rho} + \alpha \frac{\overline{V}^2}{2} + gz\right)$$

represents the mechanical energy per unit mass at a cross section. (Compare it to the EGL expression, Eq. 6.16b, for computing "mechanical" energy, which we discussed at the beginning of this section. The differences are that in the EGL we divide by g to obtain the EGL in units of meters, and here $\alpha \overline{V}^2$ allows for the fact that in a pipe flow we have a velocity profile, not a uniform flow.) The term $u_2 - u_1 - \delta Q/dm$ is equal to the difference in mechanical energy per unit mass between sections ① and ②. It represents the (irreversible) conversion of mechanical energy at section ① to unwanted thermal energy $(u_2 - u_1)$ and loss of energy via heat transfer $(-\delta Q/dm)$. We identify this group of terms as the total energy loss per unit mass and designate it by the symbol h_{l_T}. Then

$$\left(\frac{p_1}{\rho} + \alpha_1 \frac{\overline{V}_1^2}{2} + gz_1\right) - \left(\frac{p_2}{\rho} + \alpha_2 \frac{\overline{V}_2^2}{2} + gz_2\right) = h_{l_T} \tag{8.29}$$

The dimensions of energy per unit mass FL/M are equivalent to dimensions of L^2/t^2. Equation 8.29 is one of the most important and useful equations in fluid mechanics. It enables us to compute the loss of mechanical energy caused by friction between two sections of a pipe. We recall our discussion at the beginning of Part B, where we discussed what would cause the pressure to change. We hypothesized a frictionless flow (i.e., described by the Bernoulli equation, or Eq. 8.29 with $\alpha = 1$ and $h_{l_T} = 0$) so that the pressure could only change if the velocity changed (if the pipe had a change in diameter), or if the potential changed (if the pipe was not horizontal). Now, with friction, Eq. 8.29 indicates that the pressure will change even for a constant-area horizontal pipe—mechanical energy will be continuously changed into thermal energy.

As the empirical science of hydraulics developed during the 19th century, it was common practice to express the energy balance in terms of energy per unit *weight* of flowing liquid (e.g., water) rather than energy per unit *mass*, as in Eq. 8.29. When Eq. 8.29 is divided by the acceleration of gravity, g, we obtain

$$\left(\frac{p_1}{\rho g} + \alpha_1 \frac{\overline{V}_1^2}{2g} + z_1\right) - \left(\frac{p_2}{\rho g} + \alpha_2 \frac{\overline{V}_2^2}{2g} + z_2\right) = \frac{h_{l_T}}{g} = H_{l_T} \tag{8.30}$$

Each term in Eq. 8.30 has dimensions of energy per unit weight of flowing fluid. Then the net dimensions of $H_{l_T} = h_{l_T}/g$ are $(L^2/t^2)(t^2/L) = L$, or meters of flowing liquid. Since the term head loss is in common use, we shall use it when referring to either H_{l_T} (with dimensions of energy per unit weight or length) or $h_{l_T} = gH_{l_T}$ (with dimensions of energy per unit mass).

Equation 8.29 (or Eq. 8.30) can be used to calculate the pressure difference between any two points in a piping system, provided the head loss, h_{l_T} (or H_{l_T}), can be determined. We shall consider calculation of head loss in the next section.

Calculation of Head Loss 8.7

Total head loss, h_{l_T}, is regarded as the sum of major losses, h_l, due to frictional effects in fully developed flow in constant-area tubes, and minor losses, h_{l_m}, resulting from entrances, fittings, area changes, and so on. Consequently, we consider the major and minor losses separately.

Major Losses: Friction Factor

The energy balance, expressed by Eq. 8.29, can be used to evaluate the major head loss. For fully developed flow through a constant-area pipe, $h_{l_m} = 0$, and $\alpha_1 (\overline{V}_1^2/2) = \alpha_2 (\overline{V}_2^2/2)$; Eq. 8.29 reduces to

$$\frac{p_1 - p_2}{\rho} = g(z_2 - z_1) + h_l \tag{8.31}$$

If the pipe is horizontal, then $z_2 = z_1$ and

$$\frac{p_1 - p_2}{\rho} = \frac{\Delta p}{\rho} = h_l \tag{8.32}$$

Thus the major head loss can be expressed as the pressure loss for fully developed flow through a horizontal pipe of constant area.

Since head loss represents the energy converted by frictional effects from mechanical to thermal energy, head loss for fully developed flow in a constant-area duct depends only on the details of the flow through the duct. Head loss is independent of pipe orientation.

a. Laminar Flow

In laminar flow, we saw in Section 8.3 that the pressure drop may be computed analytically for fully developed flow in a horizontal pipe. Thus, from Eq. 8.13c,

$$\Delta p = \frac{128\mu L Q}{\pi D^4} = \frac{128\mu L \overline{V}(\pi D^2/4)}{\pi D^4} = 32\frac{L}{D}\frac{\mu \overline{V}}{D}$$

Substituting in Eq. 8.32 gives

$$h_l = 32\frac{L}{D}\frac{\mu \overline{V}}{\rho D} = \frac{L}{D}\frac{\overline{V}^2}{2}\left(64\frac{\mu}{\rho \overline{V} D}\right) = \left(\frac{64}{Re}\right)\frac{L}{D}\frac{\overline{V}^2}{2} \tag{8.33}$$

(We shall see the reason for writing h_l in this form shortly.)

b. Turbulent Flow

In turbulent flow we cannot evaluate the pressure drop analytically; we must resort to experimental results and use dimensional analysis to correlate the experimental data. In fully developed turbulent flow, the pressure drop, Δp, caused by friction in a horizontal constant-area pipe is known to depend on pipe diameter, D, pipe length, L, pipe roughness, e, average flow velocity, $\overline{V}$, fluid density, ρ, and fluid viscosity, μ. In functional form

$$\Delta p = \Delta p(D, L, e, \overline{V}, \rho, \mu)$$

We applied dimensional analysis to this problem in Example 7.2. The results were a correlation of the form

$$\frac{\Delta p}{\rho \overline{V}^2} = f\left(\frac{\mu}{\rho \overline{V} D}, \frac{L}{D}, \frac{e}{D}\right)$$

We recognize that $\mu/\rho\overline{V}D = 1/Re$, so we could just as well write

$$\frac{\Delta p}{\rho \overline{V}^2} = \phi\left(Re, \frac{L}{D}, \frac{e}{D}\right)$$

Substituting from Eq. 8.32, we see that

$$\frac{h_l}{\overline{V}^2} = \phi\left(Re, \frac{L}{D}, \frac{e}{D}\right)$$

Although dimensional analysis predicts the functional relationship, we must obtain actual values experimentally.

Experiments show that the nondimensional head loss is directly proportional to L/D. Hence we can write

$$\frac{h_l}{\overline{V}^2} = \frac{L}{D}\,\phi_1\left(Re, \frac{e}{D}\right)$$

Since the function, ϕ_1, is still undetermined, it is permissible to introduce a constant into the left side of the above equation. By convention the number $\frac{1}{2}$ is introduced into the denominator so that the left side of the equation is the ratio of the head loss to the kinetic energy per unit mass of flow. Then

$$\frac{h_l}{\frac{1}{2}\overline{V}^2} = \frac{L}{D}\phi_2\left(Re, \frac{e}{D}\right)$$

The unknown function, $\phi_2(Re, e/D)$, is defined as the *friction factor*, *f*,

$$f \equiv \phi_2\left(Re, \frac{e}{D}\right)$$

and

$$h_l = f\,\frac{L}{D}\,\frac{\overline{V}^2}{2} \tag{8.34}$$

or

$$H_l = f\,\frac{L}{D}\,\frac{\overline{V}^2}{2g} \tag{8.35}$$

The friction factor[2] is determined experimentally. The results, published by L. F. Moody [8], are shown in Fig. 8.13.

To determine head loss for fully developed flow with known conditions, the Reynolds number is evaluated first. Roughness, *e*, is obtained from Table 8.1. Then the friction factor, *f*, can be read from the appropriate curve in Fig. 8.13, at the known values of *Re* and *e*/*D*. Finally, head loss can be found using Eq. 8.34 or Eq. 8.35.

[2]The friction factor defined by Eq. 8.34 is the *Darcy friction factor*. The *Fanning friction factor*, less frequently used, is defined in Problem 8.95.

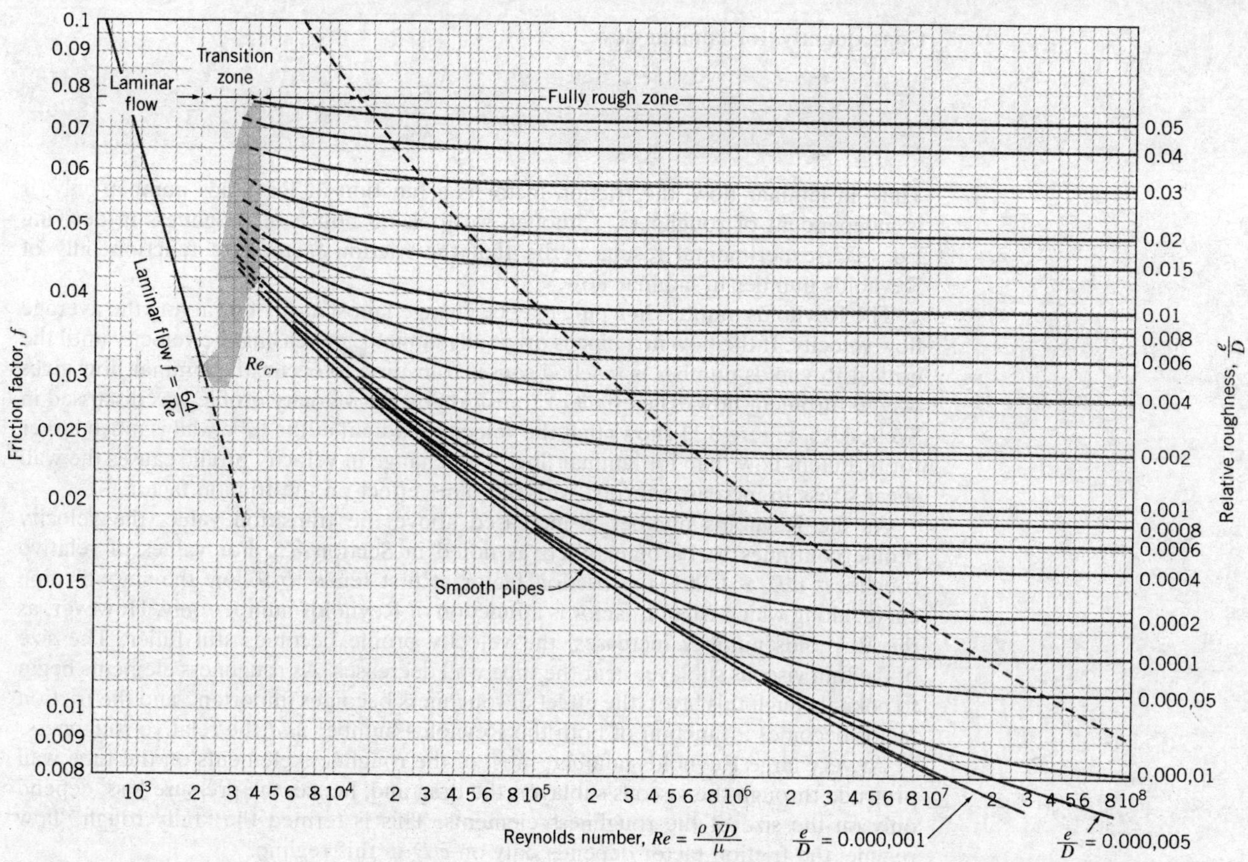

Fig. 8.13 Friction factor for fully developed flow in circular pipes. (Data from Moody [8], used by permission.)

Table 8.1

Roughness for Pipes of Common Engineering Materials

Pipe	Roughness, e Millimeters
Riveted steel	0.9–9
Concrete	0.3–3
Wood stave	0.2–0.9
Cast iron	0.26
Galvanized iron	0.15
Asphalted cast iron	0.12
Commercial steel or wrought iron	0.046
Drawn tubing	0.0015

Source: Data from Moody [8].

Several features of Fig. 8.13 require some discussion. The friction factor for laminar flow may be obtained by comparing Eqs. 8.33 and 8.34:

$$h_l = \left(\frac{64}{Re}\right)\frac{L}{D}\frac{\overline{V}^2}{2} = f\,\frac{L}{D}\frac{\overline{V}^2}{2}$$

Consequently, for laminar flow

$$f_{\text{laminar}} = \frac{64}{Re} \tag{8.36}$$

Thus, in laminar flow, the friction factor is a function of Reynolds number only; it is independent of roughness. Although we took no notice of roughness in deriving Eq. 8.33, experimental results verify that the friction factor is a function only of Reynolds number in laminar flow.

The Reynolds number in a pipe may be changed most easily by varying the average flow velocity. If the flow in a pipe is originally laminar, increasing the velocity until the critical Reynolds number is reached causes transition to occur; the laminar flow gives way to turbulent flow. The effect of transition on the velocity profile was discussed in Section 8.5. Figure 8.11 shows that the velocity gradient at the tube wall is much larger for turbulent flow than for laminar flow. This change in velocity profile causes the wall shear stress to increase sharply, with the same effect on the friction factor.

As the Reynolds number is increased above the transition value, the velocity profile continues to become fuller, as noted in Section 8.5. For values of relative roughness $e/D \leq 0.001$, the friction factor at first tends to follow the smooth pipe curve, along which friction factor is a function of Reynolds number only. However, as the Reynolds number increases, the velocity profile becomes still fuller. The size of the thin viscous sublayer near the tube wall decreases. As roughness elements begin to poke through this layer, the effect of roughness becomes important, and the friction factor becomes a function of both the Reynolds number *and* the relative roughness.

At very large Reynolds number, most of the roughness elements on the tube wall protrude through the viscous sublayer; the drag and, hence, the pressure loss, depend only on the size of the roughness elements. This is termed the "fully rough" flow regime; the friction factor depends only on e/D in this regime.

For values of relative roughness $e/D \geq 0.001$, as the Reynolds number is increased above the transition value, the friction factor is greater than the smooth pipe value. As was the case for lower values of e/D, the value of Reynolds number at which the flow regime becomes fully rough decreases with increasing relative roughness.

To summarize the preceding discussion, we see that as Reynolds number is increased, the friction factor decreases as long as the flow remains laminar. At transition, f increases sharply. In the turbulent flow regime, the friction factor decreases gradually and finally levels out at a constant value for large Reynolds number.

Bear in mind that the actual loss of energy is h_l (Eq. 8.34), which is proportional to f and $\overline{V}^2$. Hence, for laminar flow $h_l \propto \overline{V}$ (because $f = 64/Re$, and $Re \propto \overline{V}$); for the transition region there is a sudden increase in h_l; for the fully rough zone $h_l \propto \overline{V}^2$ (because $f \approx$ const.), and for the rest of the turbulent region h_l increases at a rate somewhere between $\overline{V}$ and $\overline{V}^2$. We conclude that the head loss *always* increases with flow rate, and more rapidly when the flow is turbulent.

To avoid having to use a graphical method for obtaining f for turbulent flows, various mathematical expressions have been fitted to the data. The most widely used formula for friction factor is from Colebrook [9],

$$\frac{1}{\sqrt{f}} = -2.0 \log\left(\frac{e/D}{3.7} + \frac{2.51}{Re\sqrt{f}}\right) \tag{8.37}$$

Equation 8.37 is implicit in f, but these days most scientific calculators have an equation-solving feature that can be easily used to find f for a given roughness ratio e/D and Reynolds number Re (and some calculators have the Colebrook equation itself built in!). Certainly a spreadsheet such as *Excel*, or other mathematical computer applications, can also be used (there is an *Excel* add-in for computing f for laminar and turbulent flows available on the Web site). Even without using these automated approaches, Eq. 8.37 is not difficult to solve for f—all we need to do is iterate. Equation 8.37 is quite stable—almost any initial guess value for f in the right side will, after very few iterations, lead to a converged value for f to three significant figures. From Fig. 8.13, we can see that for turbulent flows $f < 0.1$; hence $f = 0.1$ would make a good initial value. Another strategy is to use Fig. 8.13 to obtain a good first estimate; then usually one iteration using Eq. 8.37 yields a good value for f. As an alternative, Haaland [10] developed the following equation,

$$\frac{1}{\sqrt{f}} = -1.8 \log\left[\left(\frac{e/D}{3.7}\right)^{1.11} + \frac{6.9}{Re}\right]$$

as an approximation to the Colebrook equation; for $Re > 3000$, it gives results within about 2 percent of the Colebrook equation, without the need to iterate.

For turbulent flow in smooth pipes, the Blasius correlation, valid for $Re \leq 10^5$, is

$$f = \frac{0.316}{Re^{0.25}} \tag{8.38}$$

When this relation is combined with the expression for wall shear stress (Eq. 8.16), the expression for head loss (Eq. 8.32), and the definition of friction factor (Eq. 8.34), a useful expression for the wall shear stress is obtained as

$$\tau_w = 0.0332\rho\bar{V}^2\left(\frac{\nu}{R\bar{V}}\right)^{0.25} \tag{8.39}$$

This equation will be used later in our study of turbulent boundary-layer flow over a flat plate (Chapter 9).

All of the e values given in Table 8.1 are for new pipes, in relatively good condition. Over long periods of service, corrosion takes place and, particularly in hard water areas, lime deposits and rust scale form on pipe walls. Corrosion can weaken pipes, eventually leading to failure. Deposit formation increases wall roughness appreciably, and also decreases the effective diameter. These factors combine to cause e/D to increase by factors of 5 to 10 for old pipes (see Problem 10.63). An example is shown in Fig. 8.14.

Curves presented in Fig. 8.13 represent average values for data obtained from numerous experiments. The curves should be considered accurate within approximately ±10 percent, which is sufficient for many engineering analyses. If more accuracy is needed, actual test data should be used.

Minor Losses

The flow in a piping system may be required to pass through a variety of fittings, bends, or abrupt changes in area. Additional head losses are encountered, primarily as a result of flow separation. (Energy eventually is dissipated by violent mixing in the separated zones.) These losses will be minor (hence the term *minor losses*) if the piping system includes long lengths of constant-area pipe. Depending on the device, minor losses traditionally are computed in one of two ways, either

$$h_{l_m} = K\frac{\bar{V}^2}{2} \tag{8.40a}$$

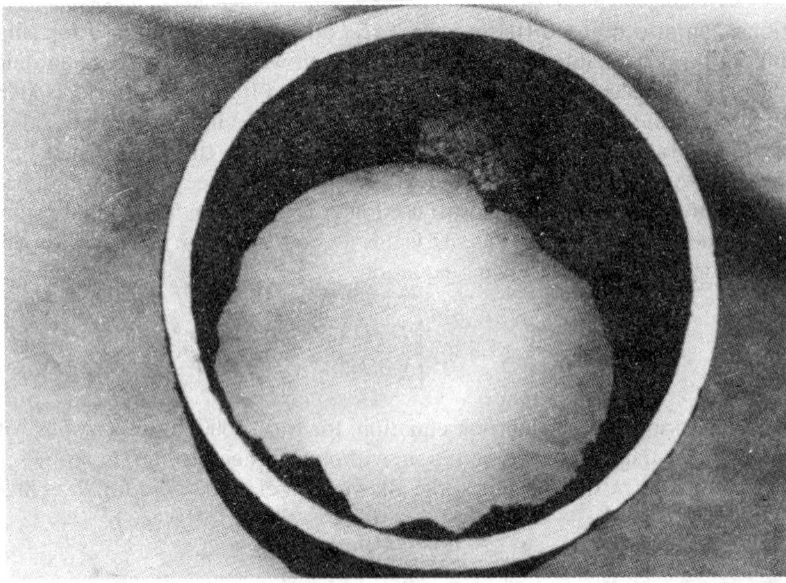

Fig. 8.14 Pipe section removed after 40 years of service as a water line, showing formation of scale. (Photo courtesy of Alan T. McDonald.)

where the *loss coefficient*, K, must be determined experimentally for each situation, or

$$h_{l_m} = f\frac{L_e}{D}\frac{\bar{V}^2}{2} \tag{8.40b}$$

where L_e is an *equivalent length* of straight pipe.

For flow through pipe bends and fittings, the loss coefficient, K, is found to vary with pipe size (diameter) in much the same manner as the friction factor, f, for flow through a straight pipe. Consequently, the equivalent length, L_e/D, tends toward a constant for different sizes of a given type of fitting.

Experimental data for minor losses are plentiful, but they are scattered among a variety of sources. Different sources may give different values for the same flow configuration. The data presented here should be considered as representative for some commonly encountered situations; in each case the source of the data is identified.

a. Inlets and Exits

A poorly designed inlet to a pipe can cause appreciable head loss. If the inlet has sharp corners, flow separation occurs at the corners, and a *vena contracta* is formed. The fluid must accelerate locally to pass through the reduced flow area at the vena contracta. Losses in mechanical energy result from the unconfined mixing as the flow stream decelerates again to fill the pipe. Three basic inlet geometries are shown in Table 8.2. From the table it is clear that the loss coefficient is reduced significantly when the inlet is rounded even slightly. For a well-rounded inlet ($r/D \geq 0.15$) the entrance loss coefficient is almost negligible. Example 8.9 illustrates a procedure for experimentally determining the loss coefficient for a pipe inlet.

The kinetic energy per unit mass, $\alpha\bar{V}^2/2$, is completely dissipated by mixing when flow discharges from a duct into a large reservoir or plenum chamber. The situation corresponds to flow through an abrupt expansion with $AR = 0$ (Fig. 8.15). The minor loss coefficient thus equals α, which as we saw in the previous section we usually set

Table 8.2

Minor Loss Coefficients for Pipe Entrances

Entrance Type		Minor Loss Coefficient, K^a
Reentrant	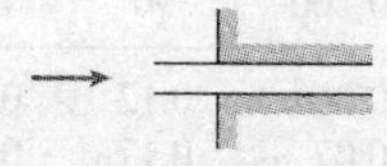	0.78
Square-edged		0.5
Rounded	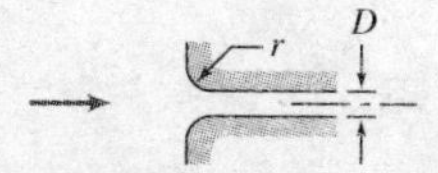	r/D: 0.02, 0.06, ≥ 0.15; K: 0.28, 0.15, 0.04

[a]Based on $h_{l_m} = K(\overline{V}^2/2)$, where $\overline{V}$ is the mean velocity in the pipe.
Source: Data from Reference [11].

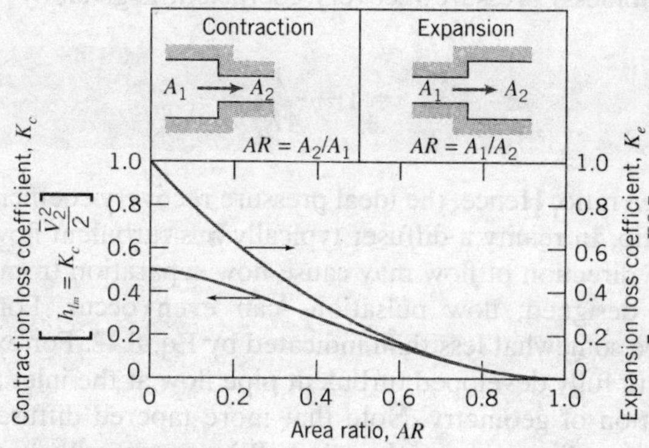

Fig. 8.15 **Loss coefficients for flow through sudden area changes. (Data from Streeter [1].)**

to 1 for turbulent flow. No improvement in minor loss coefficient for an exit is possible; however, addition of a diffuser can reduce $\overline{V}^2/2$ and therefore h_{l_m} considerably (see Example 8.10).

b. Enlargements and Contractions

Minor loss coefficients for sudden expansions and contractions in circular ducts are given in Fig. 8.15. Note that both loss coefficients are based on the *larger* $\overline{V}^2/2$. Thus losses for a sudden expansion are based on $\overline{V}_1^2/2$, and those for a contraction are based on $\overline{V}_2^2/2$.

Losses caused by area change can be reduced somewhat by installing a nozzle or diffuser between the two sections of straight pipe. Data for nozzles are given in Table 8.3. Note that the final column (data for the included angle $\theta = 180°$) agrees with the data of Fig. 8.15.

Losses in diffusers depend on a number of geometric and flow variables. Diffuser data most commonly are presented in terms of a pressure recovery coefficient, C_p, defined as the ratio of static pressure rise to inlet dynamic pressure,

$$C_p \equiv \frac{p_2 - p_1}{\frac{1}{2}\rho\overline{V}_1^2} \tag{8.41}$$

Table 8.3

Loss Coefficients (*K*) for Gradual Contractions: Round and Rectangular Ducts

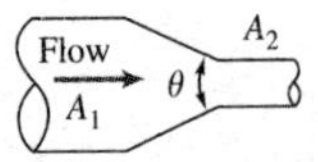

	Included Angle, θ, Degrees							
A_2/A_1	10	15–40	50–60	90	120	150	180
0.50	0.05	0.05	0.06	0.12	0.18	0.24	0.26
0.25	0.05	0.04	0.07	0.17	0.27	0.35	0.41
0.10	0.05	0.05	0.08	0.19	0.29	0.37	0.43

Note: Coefficients are based on $h_{l_m} = K(\bar{V}_2^2/2)$.
Source: Data from ASHRAE [12].

This shows what fraction of the inlet kinetic energy shows up as a pressure rise. It is not difficult to show (using the Bernoulli and continuity equations; see Problem 8.201) that the ideal (frictionless) pressure recovery coefficient is given by

$$C_{p_i} = 1 - \frac{1}{AR^2} \tag{8.42}$$

where AR is the area ratio. Hence, the ideal pressure recovery coefficient is a function only of the area ratio. In reality a diffuser typically has turbulent flow, and the static pressure rise in the direction of flow may cause flow separation from the walls if the diffuser is poorly designed; flow pulsations can even occur. For these reasons the actual C_p will be somewhat less than indicated by Eq. 8.42. For example, data for conical diffusers with fully developed turbulent pipe flow at the inlet are presented in Fig. 8.16 as a function of geometry. Note that more tapered diffusers (small divergence angle ϕ or large dimensionless length N/R_1) are more likely to approach the ideal constant value for C_p. As we make the cone shorter, for a given fixed area ratio we start to see a drop in C_p—we can consider the cone length at which this starts to happen the optimum length (it is the shortest length for which we obtain the maximum coefficient for a given area ratio—closest to that predicted by Eq. 8.42). We can relate C_p to the head loss. If gravity is neglected, and $\alpha_1 = \alpha_2 = 1.0$, the head loss equation, Eq. 8.29, reduces to

$$\left[\frac{p_1}{\rho} + \frac{\bar{V}_1^2}{2}\right] - \left[\frac{p_2}{\rho} + \frac{\bar{V}_2^2}{2}\right] = h_{l_T} = h_{l_m}$$

Thus,

$$h_{l_m} = \frac{\bar{V}_1^2}{2} - \frac{\bar{V}_2^2}{2} - \frac{p_2 - p_1}{\rho}$$

$$h_{l_m} = \frac{\bar{V}_1^2}{2}\left[\left(1 - \frac{\bar{V}_2^2}{\bar{V}_1^2}\right) - \frac{p_2 - p_1}{\frac{1}{2}\rho\bar{V}_1^2}\right] = \frac{\bar{V}_1^2}{2}\left[\left(1 - \frac{\bar{V}_2^2}{\bar{V}_1^2}\right) - C_p\right]$$

From continuity, $A_1\bar{V}_1 = A_2\bar{V}_2$, so

$$h_{l_m} = \frac{\bar{V}_1^2}{2}\left[1 - \left(\frac{A_1}{A_2}\right)^2 - C_p\right]$$

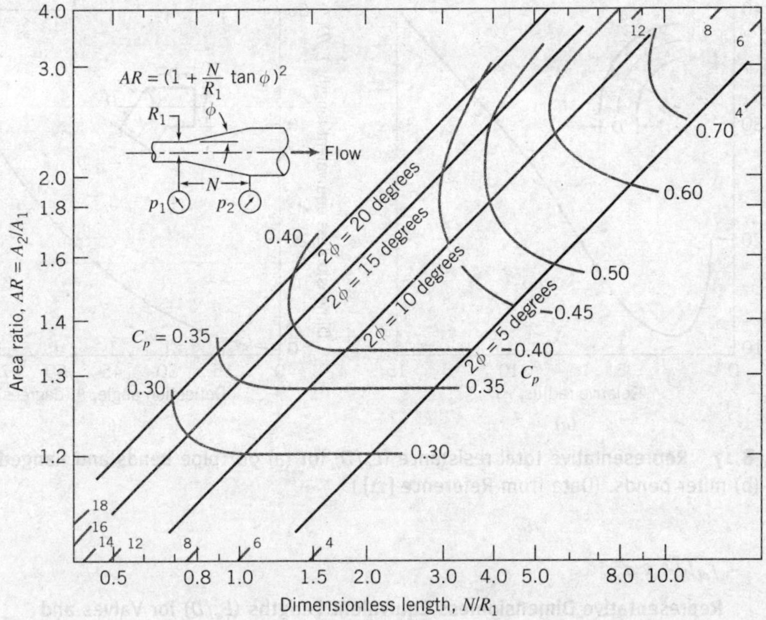

Fig. 8.16 Pressure recovery for conical diffusers with fully developed turbulent pipe flow at inlet. (Data from Cockrell and Bradley [13].)

or

$$h_{l_m} = \frac{\overline{V}_1^2}{2}\left[\left(1 - \frac{1}{(AR)^2}\right) - C_p\right] \tag{8.43}$$

The frictionless result (Eq. 8.42) is obtained from Eq. 8.43 if $h_{l_m} = 0$. We can combine Eqs. 8.42 and 8.43 to obtain an expression for the head loss in terms of the actual and ideal C_p values:

$$h_{l_m} = (C_{p_i} - C_p)\frac{\overline{V}_1^2}{2} \tag{8.44}$$

Performance maps for plane wall and annular diffusers [14] and for radial diffusers [15] are available in the literature.

Diffuser pressure recovery is essentially independent of Reynolds number for inlet Reynolds numbers greater than 7.5×10^4 [16]. Diffuser pressure recovery with uniform inlet flow is somewhat better than that for fully developed inlet flow. Performance maps for plane wall, conical, and annular diffusers for a variety of inlet flow conditions are presented in [17].

CLASSIC VIDEO

Flow Visualization.

Since static pressure rises in the direction of flow in a diffuser, flow may separate from the walls. For some geometries, the outlet flow is distorted. For wide angle diffusers, vanes or splitters can be used to suppress stall and improve pressure recovery [18].

c. Pipe Bends

The head loss of a bend is larger than for fully developed flow through a straight section of equal length. The additional loss is primarily the result of secondary flow, and is represented most conveniently by an equivalent length of straight pipe. The equivalent length depends on the relative radius of curvature of the bend, as shown in

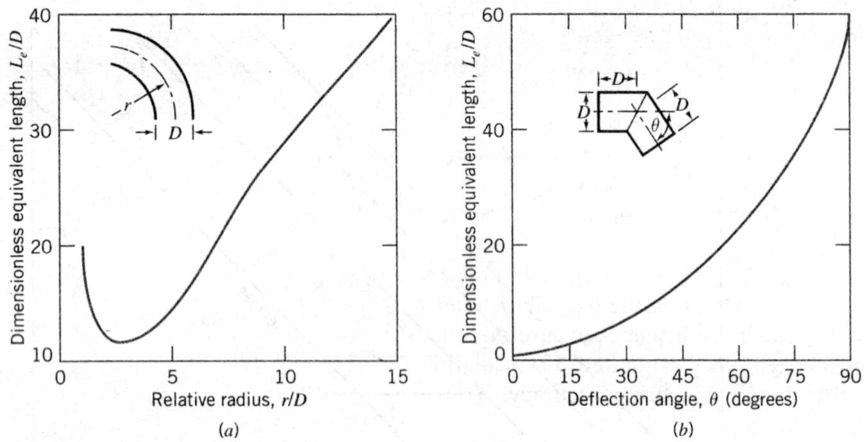

Fig. 8.17 Representative total resistance (L_e/D) for (a) 90° pipe bends and flanged elbows, and (b) miter bends. (Data from Reference [11].)

Table 8.4

Representative Dimensionless Equivalent Lengths (L_e/D) for Valves and Fittings

Fitting Type	Equivalent Length,[a] L_e/D
Valves (fully open)	
Gate valve	8
Globe valve	340
Angle valve	150
Ball valve	3
Lift check valve: globe lift	600
angle lift	55
Foot valve with strainer: poppet disk	420
hinged disk	75
Standard elbow: 90°	30
45°	16
Return bend, close pattern	50
Standard tee: flow through run	20
flow through branch	60

[a]Based on $h_{l_m} = f(L_e/D)(\bar{V}^2/2)$.

Source: Data from Reference [11].

Fig. 8.17*a* for 90° bends. An approximate procedure for computing the resistance of bends with other turning angles is given in [11].

Because they are simple and inexpensive to construct in the field, miter bends often are used in large pipe systems. Design data for miter bends are given in Fig. 8.17*b*. Note that you get what you pay for: From Fig. 8.17a the equivalent length for pipe bends varies from about 10 to about 40-diameters; for the cheaper 90° miter bend of Fig. 8.17b we get a much larger equivalent length of 60 diameters.

d. Valves and Fittings

Losses for flow through valves and fittings also may be expressed in terms of an equivalent length of straight pipe. Some representative data are given in Table 8.4.

All resistances are given for fully open valves; losses increase markedly when valves are partially open. Valve design varies significantly among manufacturers.

Whenever possible, resistances furnished by the valve supplier should be used if accurate results are needed.

Fittings in a piping system may have threaded, flanged, or welded connections. For small diameters, threaded joints are most common; large pipe systems frequently have flanged or welded joints.

In practice, insertion losses for fittings and valves vary considerably, depending on the care used in fabricating the pipe system. If burrs from cutting pipe sections are allowed to remain, they cause local flow obstructions, which increase losses appreciably.

Although the losses discussed in this section were termed "minor losses," they can be a large fraction of the overall system loss. Thus a system for which calculations are to be made must be checked carefully to make sure all losses have been identified and their magnitudes estimated. If calculations are made carefully, the results will be of satisfactory engineering accuracy. You may expect to predict actual losses within ±10 percent.

We include here one more device that changes the energy of the fluid—except this time the energy of the fluid will be increased, so it creates a "negative energy loss."

Pumps, Fans, and Blowers in Fluid Systems

In many practical flow situations (e.g., the cooling system of an automobile engine, the HVAC system of a building), the driving force for maintaining the flow against friction is a pump (for liquids) or a fan or blower (for gases). Here we will consider pumps, although all the results apply equally to fans and blowers. We generally neglect heat transfer and internal energy changes of the fluid (we will incorporate them later into the definition of the pump efficiency), so the first law of thermodynamics applied across the pump is

$$\dot{W}_{\text{pump}} = \dot{m}\left[\left(\frac{p}{\rho} + \frac{\overline{V}^2}{2} + gz\right)_{\text{discharge}} - \left(\frac{p}{\rho} + \frac{\overline{V}^2}{2} + gz\right)_{\text{suction}}\right]$$

We can also compute the head Δh_{pump} (energy/mass) produced by the pump,

$$\Delta h_{\text{pump}} = \frac{\dot{W}_{\text{pump}}}{\dot{m}} = \left(\frac{p}{\rho} + \frac{\overline{V}^2}{2} + gz\right)_{\text{discharge}} - \left(\frac{p}{\rho} + \frac{\overline{V}^2}{2} + gz\right)_{\text{suction}} \tag{8.45}$$

In many cases the inlet and outlet diameters (and therefore velocities) and elevations are the same or negligibly different, so Eq. 8.45 simplifies to

$$\Delta h_{\text{pump}} = \frac{\Delta p_{\text{pump}}}{\rho} \tag{8.46}$$

It is interesting to note that a pump adds energy to the fluid in the form of a gain in pressure—the everyday, invalid perception is that pumps add kinetic energy to the fluid. (It *is* true that when a pump-pipe system is first started up, the pump does work to accelerate the fluid to its steady speed; this is when a pump driven by an electric motor is most in danger of burning out the motor.)

The idea is that in a pump-pipe system the head produced by the pump (Eq. 8.45 or 8.46) is needed to overcome the head loss for the pipe system. Hence, the flow rate in such a system depends on the pump characteristics and the major and minor losses of the pipe system. We will learn in Chapter 10 that the head produced by a given pump is not constant, but varies with flow rate through the pump, leading to the notion of "matching" a pump to a given system to achieve the desired flow rate.

A useful relation is obtained from Eq. 8.46 if we multiply by $\dot{m} = \rho Q$ (Q is the flow rate) and recall that $\dot{m}\Delta h_{\text{pump}}$ is the power supplied to the fluid,

$$\dot{W}_{\text{pump}} = Q\Delta p_{\text{pump}} \tag{8.47}$$

We can also define the pump efficiency:

$$\eta = \frac{\dot{W}_{\text{pump}}}{\dot{W}_{\text{in}}} \tag{8.48}$$

where $\dot{W}_{\text{pump}}$ is the power reaching the fluid, and $\dot{W}_{\text{in}}$ is the power input (usually electrical) to the pump.

We note that, when applying the energy equation (Eq. 8.29) to a pipe system, we may sometimes choose points 1 and 2 so that a pump is included in the system. For these cases we can simply include the head of the pump as a "negative loss":

$$\left(\frac{p_1}{\rho} + \alpha_1\frac{\bar{V}_1^2}{2} + gz_1\right) - \left(\frac{p_2}{\rho} + \alpha_2\frac{\bar{V}_2^2}{2} + gz_2\right) = h_{l_T} - \Delta h_{\text{pump}} \tag{8.49}$$

Noncircular Ducts

The empirical correlations for pipe flow also may be used for computations involving noncircular ducts, provided their cross sections are not too exaggerated. Thus ducts of square or rectangular cross section may be treated if the ratio of height to width is less than about 3 or 4.

The correlations for turbulent pipe flow are extended for use with noncircular geometries by introducing the *hydraulic diameter*, defined as

$$D_h \equiv \frac{4A}{P} \tag{8.50}$$

in place of the diameter, D. In Eq. 8.50, A is cross-sectional area, and P is *wetted perimeter*, the length of wall in contact with the flowing fluid at any cross-section. The factor 4 is introduced so that the hydraulic diameter will equal the duct diameter for a circular cross section. For a circular duct, $A = \pi D^2/4$ and $P = \pi D$, so that

$$D_h = \frac{4A}{P} = \frac{4\left(\frac{\pi}{4}\right)D^2}{\pi D} = D$$

For a rectangular duct of width b and height h, $A = bh$ and $P = 2(b+h)$, so

$$D_h = \frac{4bh}{2(b+h)}$$

If the *aspect ratio*, *ar*, is defined as $ar = h/b$, then

$$D_h = \frac{2h}{1+ar}$$

for rectangular ducts. For a square duct, $ar = 1$ and $D_h = h$.

As noted, the hydraulic diameter concept can be applied in the approximate range $\frac{1}{4} < ar < 4$. Under these conditions, the correlations for pipe flow give acceptably accurate results for rectangular ducts. Since such ducts are easy and cheap to fabricate from sheet metal, they are commonly used in air conditioning, heating, and ventilating applications. Extensive data on losses for air flow are available (e.g., see [12, 19]).

Losses caused by secondary flows increase rapidly for more extreme geometries, so the correlations are not applicable to wide, flat ducts, or to ducts of triangular or other irregular shapes. Experimental data must be used when precise design information is required for specific situations.

Solution of Pipe Flow Problems 8.8

Section 8.7 provides us with a complete scheme for solving many different pipe flow problems. For convenience we collect together the relevant computing equations.

The *energy equation*, relating the conditions at any two points 1 and 2 for a single-path pipe system, is

$$\left(\frac{p_1}{\rho} + \alpha_1 \frac{\bar{V}_1^2}{2} + gz_1\right) - \left(\frac{p_2}{\rho} + \alpha_2 \frac{\bar{V}_2^2}{2} + gz_2\right) = h_{l_T} = \sum h_l + \sum h_{l_m} \tag{8.29}$$

This equation expresses the fact that there will be a loss of mechanical energy ("pressure," kinetic and/or potential) in the pipe. Recall that for turbulent flows $\alpha \approx 1$. Note that by judicious choice of points 1 and 2 we can analyze not only the entire pipe system, but also just a certain section of it that we may be interested in. The *total head loss* is given by the sum of the major and minor losses. (Remember that we can also include "negative losses" for any pumps present between points 1 and 2. The relevant form of the energy equation is then Eq. 8.49.)

Each *major loss* is given by

$$h_l = f \frac{L}{D} \frac{\bar{V}^2}{2} \tag{8.34}$$

where the *friction factor* is obtained from

$$f = \frac{64}{Re} \quad \text{for laminar flow } (Re < 2300) \tag{8.36}$$

or

$$\frac{1}{\sqrt{f}} = -2.0 \log\left(\frac{e/D}{3.7} + \frac{2.51}{Re\sqrt{f}}\right) \quad \text{for turbulent flow } (Re \geq 2300) \tag{8.37}$$

and Eqs. 8.36 and 8.37 are presented graphically in the Moody chart (Fig. 8.13).

Each *minor loss* is given either by

$$h_{l_m} = K \frac{\bar{V}^2}{2} \tag{8.40a}$$

where K is the device *loss coefficient*, or

$$h_{l_m} = f \frac{L_e}{D} \frac{\bar{V}^2}{2} \tag{8.40b}$$

where L_e is the additional *equivalent length* of pipe.

We also note that the flow rate Q is related to the average velocity $\bar{V}$ at each pipe cross section by

$$Q = \pi \frac{D^2}{4} \bar{V}$$

We will apply these equations first to single-path systems.

Single-Path Systems

In single-path pipe problems we generally know the system configuration (type of pipe material and hence pipe roughness, the number and type of elbows, valves, and other fittings, etc., and changes of elevation), as well as the fluid (ρ and μ) we will be working with. Although not the only possibilities, usually the goal is to determine one of the following:

(a) The pressure drop Δp, for a given pipe (L and D), and flow rate Q.

(b) The pipe length L, for a given pressure drop Δp, pipe diameter D, and flow rate Q.

(c) The flow rate Q, for a given pipe (L and D), and pressure drop Δp.

(d) The pipe diameter D, for a given pipe length L, pressure drop Δp, and flow rate Q.

Each of these cases often arises in real-world situations. For example, case (a) is a necessary step in selecting the correct size pump to maintain the desired flow rate in a system—the pump must be able to produce the system Δp at the specified flow rate Q. (We will discuss this in more detail in Chapter 10.) Cases (a) and (b) are computationally straightforward; we will see that cases (c) and (d) can be a little tricky to evaluate. We will discuss each case, and present an Example for each. The Examples present solutions as you might do them using a calculator, but there is also an *Excel* workbook for each. (Remember that the Web site has an *Excel* add-in that once installed will automatically compute f from Re and e/D!) The advantage of using a computer application such as a spreadsheet is that we do not have to use either the Moody chart (Fig. 8.13) or solve the implicit Colebrook equation (Eq. 8.37) to obtain turbulent friction factors—the application can find them for us! In addition, as we'll see, cases (c) and (d) involve significant iterative calculations that can be avoided by use of a computer application. Finally, once we have a solution using a computer application, engineering "what-ifs" become easy, e.g., if we double the head produced by a pump, how much will the flow rate in a given system increase?

a. Find Δp for a Given L, D, and Q

These types of problems are quite straightforward—the energy equation (Eq. 8.29) can be solved directly for $\Delta p = (p_1 - p_2)$ in terms of known or computable quantities. The flow rate leads to the Reynolds number (or numbers if there is a diameter change) and hence the friction factor (or factors) for the flow; tabulated data can be used for minor loss coefficients and equivalent lengths. The energy equation can then be used to directly obtain the pressure drop. Example 8.5 illustrates this type of problem.

b. Find L for a Given Δp, D, and Q

These types of problems are also straightforward—the energy equation (Eq. 8.29) can be solved directly for L in terms of known or computable quantities. The flow rate again leads to the Reynolds number and hence the friction factor for the flow. Tabulated data can be used for minor loss coefficients and equivalent lengths. The energy equation can then be rearranged and solved directly for the pipe length. Example 8.6 illustrates this type of problem.

c. Find Q for a Given Δp, L, and D

These types of problems require either manual iteration or use of a computer application such as *Excel*. The unknown flow rate or velocity is needed before the Reynolds number and hence the friction factor can be found. To manually iterate we first solve the energy equation directly for $\overline{V}$ in terms of known quantities and the unknown friction factor f. To start the iterative process we make a guess for f (a good choice is to take a value from the fully turbulent region of the Moody chart because many practical flows are in this region) and obtain a value for $\overline{V}$. Then we can compute a Reynolds number and hence obtain a new value for f. We repeat the iteration process $f \rightarrow \overline{V} \rightarrow Re \rightarrow f$ until convergence (usually only two or three iterations are necessary). A much quicker procedure is to use a computer application. For example, spreadsheets (such as *Excel*) have built-in solving features for solving one or more algebraic equations for one or more unknowns. Example 8.7 illustrates this type of problem.

d. Find D for a Given Δp, L, and Q

These types of problems arise, for example, when we have designed a pump-pipe system and wish to choose the best pipe diameter—the best being the minimum diameter (for minimum pipe cost) that will deliver the design flow rate. We need to manually iterate, or use a computer application such as *Excel*. The unknown diameter is needed before the Reynolds number and relative roughness, and hence the friction factor, can be found. To manually iterate we could first solve the energy equation directly for D in terms of known quantities and the unknown friction factor f, and then iterate from a starting guess for f in a way similar to case (c) above: $f \rightarrow D \rightarrow Re$ and $e/D \rightarrow f$. In practice this is a little unwieldy, so instead to manually find a solution we make successive guesses for D until the corresponding pressure drop Δp (for the given flow rate Q) computed from the energy equation matches the design Δp. As in case (c) a much quicker procedure is to use a computer application. For example, spreadsheets (such as *Excel*) have built-in solving features for solving one or more algebraic equations for one or more unknowns. Example 8.8 illustrates this type of problem.

In choosing a pipe size, it is logical to work with diameters that are available commercially. Pipe is manufactured in a limited number of standard sizes. Some data for standard pipe sizes are given in Table 8.5. For data on extra strong or double extra strong pipes, consult a handbook, e.g., [11]. Pipe larger than 300 mm nominal diameter is produced in multiples of 50 mm up to a nominal diameter of 900 mm and in multiples of 150 mm for still larger sizes.

Table 8.5

Standard Sizes for Carbon Steel, Alloy Steel, and Stainless Steel Pipe

Nominal Pipe Size (mm)	Inside Diameter (mm)	Nominal Pipe Size (mm)	Inside Diameter (mm)
3.175	6.832	63.500	62.712
6.350	9.245	76.200	77.927
3.525	12.522	101.600	102.260
12.700	15.798	127.000	128.193
19.050	20.929	152.400	154.051
25.400	26.644	203.200	202.717
38.100	40.894	254.000	254.508
50.800	52.501	304.800	304.800

Source: Data from Reference [11].

Example 8.5 PIPE FLOW INTO A RESERVOIR: PRESSURE DROP UNKNOWN

A 100-m length of smooth horizontal pipe is attached to a large reservoir. A pump is attached to the end of the pipe to pump water into the reservoir at a volume flow rate of 0.01 m^3/s. What pressure (gage) must the pump produce at the pipe to generate this flow rate? The inside diameter of the smooth pipe is 75 mm.

Given: Water is pumped at 0.01 m^3/s through a 75 mm-diameter smooth pipe, with $L = 100$ m, into a constant-level reservoir of depth $d = 10$ m.

Find: Pump pressure, p_1, required to maintain the flow.

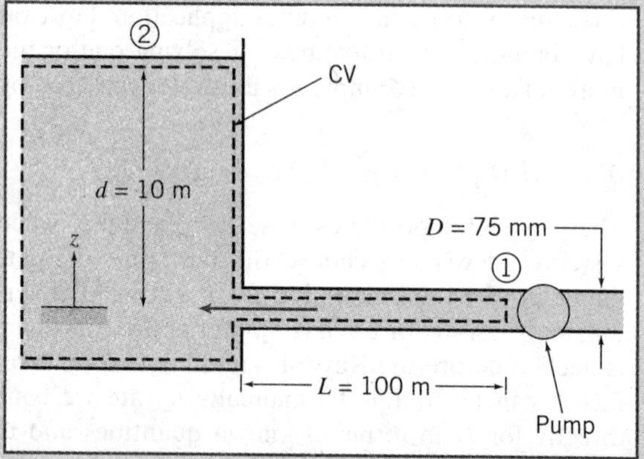

Solution:

Governing equations:

$$\left(\frac{p_1}{\rho} + \alpha_1\frac{\bar{V}_1^2}{2} + gz_1\right) - \left(\frac{p_2}{\rho} + \alpha_2\frac{\bar{V}_2^2}{2} + gz_2\right) = h_{l_T} = h_l + h_{l_m} \quad (8.29)$$

where

$$h_l = f\frac{L}{D}\frac{\bar{V}^2}{2} \quad (8.34) \qquad \text{and} \qquad h_{l_m} = K\frac{\bar{V}^2}{2} \quad (8.40a)$$

For the given problem, $p_1 = p_{\text{pump}}$ and $p_2 = 0$ (gage), so $\Delta p = p_1 - p_2 = p_{\text{pump}}$, $\bar{V}_1 = \bar{V}$, $\bar{V}_2 \approx 0$, K (exit loss) $= 1.0$, and $\alpha_1 \approx 1.0$. If we set $z_1 = 0$, then $z_2 = d$. Simplifying Eq. 8.29 gives

$$\frac{\Delta p}{\rho} + \frac{\bar{V}^2}{2} - gd = f\frac{L}{D}\frac{\bar{V}^2}{2} + \frac{\bar{V}^2}{2} \quad (1)$$

The left side of the equation is the loss of mechanical energy between points ① and ②; the right side is the major and minor losses that contributed to the loss. Solving for the pressure drop, $\Delta p = p_{\text{pump}}$,

$$p_{\text{pump}} = \Delta p = \rho\left(gd + f\frac{L}{D}\frac{\bar{V}^2}{2}\right)$$

Everything on the right side of the equation is known or can be readily computed. The flow rate Q leads to $\bar{V}$,

$$\bar{V} = \frac{Q}{A} = \frac{4Q}{\pi D^2} = \frac{4}{\pi} \times 0.01\frac{\text{m}^3}{\text{s}} \times \frac{1}{(0.075)^2\ \text{m}^2} = 2.26\ \text{m/s}$$

This in turn [assuming water at 20°C, $\rho = 999\ \text{kg/m}^3$, and $\mu = 1.0 \times 10^{-3}\ \text{kg/(m}\cdot\text{s)}$] leads to the Reynolds number

$$Re = \frac{\rho\bar{V}D}{\mu} = 999\frac{\text{kg}}{\text{m}^3} \times 2.26\frac{\text{m}}{\text{s}} \times 0.075\ \text{m} \times \frac{\text{m}\cdot\text{s}}{1.0 \times 10^{-3}\text{kg}} = 1.70 \times 10^5$$

For turbulent flow in a smooth pipe ($e = 0$), from Eq. 8.37, $f = 0.0162$. Then

$$p_{\text{pump}} = \Delta p = \rho\left(gd + f\frac{L}{D}\frac{\bar{V}^2}{2}\right)$$

$$= 999\frac{\text{kg}}{\text{m}^3}\left(9.81\frac{\text{m}}{\text{s}^2} \times 10\ \text{m} + (0.0162) \times \frac{100\ \text{m}}{0.075\ \text{m}} \times \frac{(2.26)^2\text{m}^2}{2\ \text{s}^2}\right) \times \frac{\text{N}\cdot\text{s}^2}{\text{kg}\cdot\text{m}}$$

$$p_{\text{pump}} = 1.53 \times 10^5\ \text{N/m}^2\ \text{(gage)}$$

Hence,

$$p_{\text{pump}} = 153\ \text{kPa (gage)} \longleftarrow p_{\text{pump}}$$

This problem illustrates the method for manually calculating pressure drop. The *Excel* workbook for this problem automatically computes *Re* and *f* from the given data. It then solves Eq. 1 directly for pressure p_{pump} without having to explicitly solve for it first. The workbook can be easily used to see, for example, how the pump pressure p_{pump} required to maintain flow *Q* is affected by changing the diameter *D*; it is easily editable for other case (a) type problems.

Example 8.6 FLOW IN A PIPELINE: LENGTH UNKNOWN

Crude oil flows through a level section of the Alaskan pipeline at a rate of 2.944 m^3/s. The pipe inside diameter is 1.22 m; its roughness is equivalent to galvanized iron. The maximum allowable pressure is 8.27 MPa; the minimum pressure required to keep dissolved gases in solution in the crude oil is 344.5 kPa. The crude oil has SG = 0.93; its viscosity at the pumping temperature of 60°C is $\mu = 10.0168\ \text{N}\cdot\text{s/m}^2$. For these conditions, determine the maximum possible spacing between pumping stations. If the pump efficiency is 85 percent, determine the power that must be supplied at each pumping station.

Given: Flow of crude oil through horizontal section of Alaskan pipeline.

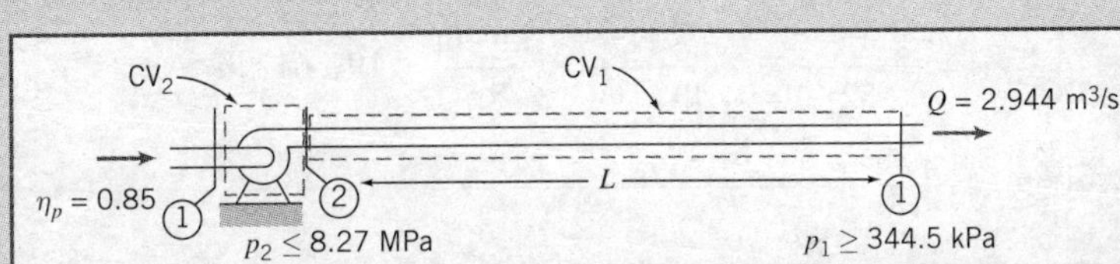

$D = 1.22$ m. (roughness of galvanized iron), SG = 0.93, $\mu = 0.0168\ \text{N}\cdot\text{s/m}^2$

Find: (a) Maximum spacing, *L*.
(b) Power needed at each pump station.

Solution:
As shown in the figure, we assume that the Alaskan pipeline is made up of repeating pump-pipe sections. We can draw two control volumes: CV_1, for the pipe flow (state ② to state ①); CV_2, for the pump (state ① to state ②).

First we apply the energy equation for steady, incompressible pipe flow to CV_1.

Governing equations:

$$\left(\frac{p_2}{\rho} + \alpha_2\frac{\bar{V}_2^2}{2} + gz_2\right) - \left(\frac{p_1}{\rho} + \alpha_1\frac{\bar{V}_1^2}{2} + gz_1\right) = h_{l_T} = h_l + h_{l_m} \qquad (8.29)$$

where

$$h_l = f\frac{L}{D}\frac{\bar{V}^2}{2} \quad (8.34) \qquad \text{and} \qquad h_{l_m} = K\frac{\bar{V}^2}{2} \quad (8.40a)$$

Assumptions: (1) $\alpha_1 \overline{V}_1^2 = \alpha_2 \overline{V}_2^2$.
(2) Horizontal pipe, $z_1 = z_2$.
(3) Neglect minor losses.
(4) Constant viscosity.

Then, using CV_1

$$\Delta p = p_2 - p_1 = f \frac{L}{D} \rho \frac{\overline{V}^2}{2} \tag{1}$$

or

$$L = \frac{2D}{f} \frac{\Delta p}{\rho \overline{V}^2} \quad \text{where } f = f(Re, e/D)$$

so

$$\overline{V} = \frac{Q}{A} = 2.944 \frac{\text{m}^3}{\text{s}} \times \frac{4}{\pi (1.22)^2 \text{m}^2} = 2.52 \text{ m/s}$$

$$Re = \frac{\rho \overline{V} D}{\mu} = 0.93 \times 1000 \frac{\text{kg}}{\text{m}^3} \times 2.52 \frac{\text{m}}{\text{s}} \times 1.22 \text{ m} \times \frac{1}{0.0168 \text{ N·s/m}^2} \times \frac{\text{N.s}^2}{\text{kg} \cdot \text{m}}$$

$$Re = 1.71 \times 10^5$$

From Table 8.1, $e = 0.00015$ m and hence $e/D = 0.00012$. Then from Eq. 8.37, $f = 0.017$ and thus

$$L = \frac{2}{0.017} \times 1.22 \text{ m} \times (8.27 \times 10^6 - 3.445 \times 10^5) \text{Pa} \times \frac{1}{0.93 \times 1000 \times \text{kg/m}^3}$$

$$\times \frac{1}{(2.52)^2} \frac{\text{s}^2}{\text{m}^2} \times \frac{\text{N}}{\text{m}^2 \cdot \text{Pa}} \times \frac{\text{kg} \cdot \text{m}}{\text{N·s}^2} = 192,612 \text{ m}$$

$$L = 192,612 \text{ m} \qquad L$$

To find the pumping power we can apply the first law of thermodynamics to CV_2. This control volume consists only of the pump, and we saw in Section 8.7 that this law simplifies to

$$\dot{W}_{\text{pump}} = Q \Delta p_{\text{pump}} \tag{8.47}$$

and the pump efficiency is

$$\eta = \frac{\dot{W}_{\text{pump}}}{\dot{W}_{\text{in}}} \tag{8.48}$$

We recall that $\dot{W}_{\text{pump}}$ is the power reaching the fluid, and $\dot{W}_{\text{in}}$ is the power input. Because we have a repeating system the pressure rise through the pump (i.e., from state ① to state ②) equals the pressure drop in the pipe (i.e., from state ② to state ①),

$$\Delta p_{\text{pump}} = \Delta p$$

so that

$$\dot{W}_{\text{pump}} = Q\Delta p_{\text{pump}} = 2.944\frac{\text{m}^3}{\text{s}} \times (8.27\times 10^6 - 3.445\times 10^5)\text{Pa}$$

$$\times \frac{\text{N}}{\text{m}^2\cdot\text{Pa}} \times \frac{\text{j}}{\text{N}\cdot\text{m}} \times \frac{\text{W}\cdot\text{S}}{\text{j}} \approx 23.13\ \text{MW}$$

and the required power input is

$$\dot{W}_{\text{in.}} = \frac{\dot{W}_{\text{pump}}}{\eta} = \frac{23.13}{0.85} = 27.21\ \text{MW} \longleftarrow \dot{W}_{\text{needed}}$$

This problem illustrates the method for manually calculating pipe length L. The *Excel* workbook for this problem automatically computes Re and f from the given data. It then solves Eq. 1 directly for L without having to explicitly solve for it first. The workbook can be easily used to see, for example, how the flow rate Q depends on L; it may be edited for other case (b) type problems.

Example 8.7 FLOW FROM A WATER TOWER: FLOW RATE UNKNOWN

A fire protection system is supplied from a water tower and standpipe 24.4 m tall. The longest pipe in the system is 182.9 m and is made of cast iron about 20 years old. The pipe contains one gate valve; other minor losses may be neglected. The pipe diameter is 101.6 mm. Determine the maximum rate of flow (gpm) through this pipe.

Given: Fire protection system, as shown.

Find: Q, gpm.

Solution:

Governing equations:

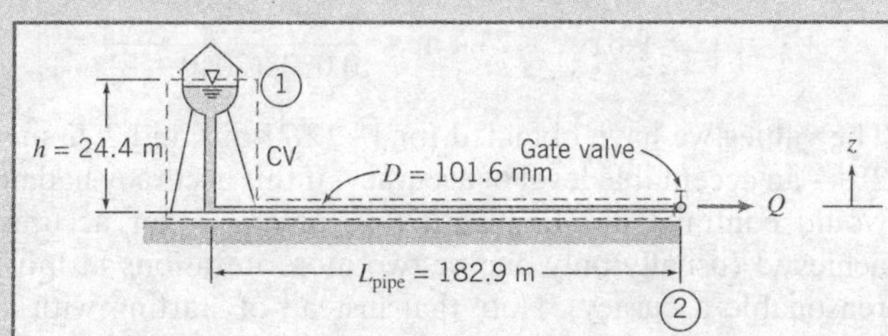

$$\left(\frac{p_1}{\rho} + \alpha_1\frac{\overline{V}_1^2}{2} + gz_1\right) - \left(\frac{p_2}{\rho} + \alpha_2\frac{\overline{V}_2^2}{2} + gz_2\right) = h_{l_T} = h_l + h_{l_m} \quad (8.29)$$

($\approx 0(2)$)

where

$$h_l = f\frac{L}{D}\frac{\overline{V}^2}{2} \quad (8.34) \qquad \text{and} \qquad h_{l_m} = f\frac{L_e}{D}\frac{\overline{V}^2}{2} \quad (8.40b)$$

Assumptions: (1) $p_1 = p_2 = p_{\text{atm}}$
(2) $\overline{V}_1 = 0$, and $\alpha_2 \simeq 1.0$.

Then Eq. 8.29 can be written as

$$g(z_1 - z_2) - \frac{\overline{V}_2^2}{2} = h_{l_T} = f\left(\frac{L}{D} + \frac{L_e}{D}\right)\frac{\overline{V}_2^2}{2} \quad (1)$$

For a fully open gate valve, from Table 8.4, $L_e/D = 8$. Thus

$$g(z_1 - z_2) = \frac{\overline{V}_2^2}{2}\left[f\left(\frac{L}{D} + 8\right) + 1\right]$$

To manually iterate, we solve for $\overline{V}_2$ and obtain

$$\overline{V}_2 = \left[\frac{2g(z_1 - z_2)}{f(L/D + 8) + 1}\right]^{1/2} \qquad (2)$$

To be conservative, assume the standpipe is the same diameter as the horizontal pipe. Then

$$\frac{L}{D} = \frac{182.9 \text{ m} + 24.4 \text{ m}}{101.6 \text{ mm}} \times \frac{1000 \text{ mm}}{\text{m}} = 2040$$

Also

$$z_1 - z_2 = h = 24.4 \text{ m}$$

To solve Eq. 2 manually we need to iterate. To start, we make an estimate for f by assuming the flow is fully turbulent (where f is constant). This value can be obtained from solving Eq. 8.37 using a calculator or from Fig. 8.13. For a large value of Re (e.g., 10^8), and a roughness ratio $e/D \approx 0.005$ ($e = 0.26$ mm for cast iron is obtained from Table 8.1, and doubled to allow for the fact that the pipe is old), we find that $f \approx 0.03$. Thus a first iteration for $\overline{V}_2$ from Eq. 2 is

$$\overline{V}_2 = \left[2 \times 9.81 \frac{\text{m}}{\text{s}^2} \times 24.4 \text{ m} \times \frac{1}{0.03(2040 + 8) + 1}\right]^{\frac{1}{2}} = 2.77 \text{ m/s}$$

Now obtain a new value for f:

$$Re = \frac{\rho \overline{V} D}{\mu} = \frac{\overline{V} D}{\nu} = 2.77 \frac{\text{m}}{\text{s}} \times 101.6 \text{ mm} \times \frac{\text{s}}{1.124 \times 10^{-6} \text{ m}^2} \times \frac{\text{m}}{1000 \text{ mm}} = 2.50 \times 10^5$$

For $e/D = 0.005$, $f = 0.0308$ from Eq. 8.37. Thus we obtain

$$\overline{V}_2 = \left[2 \times 9.81 \frac{\text{m}}{\text{s}^2} \times 24.4 \text{ m} \times \frac{1}{0.0308(2040 + 8) + 1}\right]^{\frac{1}{2}} = 2.73 \text{ m/s}$$

The values we have obtained for $\overline{V}_2$ (2.77 m/s and 2.73 m/s) differ by less than 2%—an acceptable level of accuracy. If this accuracy had not been achieved we would continue iterating until this, or any other accuracy we desired, was achieved (usually only one or two more iterations at most are necessary for reasonable accuracy). Note that instead of starting with a fully rough value for f, we could have started with a guess value for $\overline{V}_2$ of, say, 0.3 m/s or 3 m/s. The volume flow rate is

$$Q = \overline{V}_2 A = \overline{V}_2 \frac{\pi D^2}{4} = 2.73 \frac{\text{m}}{\text{s}} \times \frac{\pi}{4}(100.6 \text{ mm})^2 \times \frac{\text{m}^2}{10^6 \text{ mm}^2}$$

$Q = 0.022 \text{ m}^3\text{s}$ ← Q

This problem illustrates the method for manually iterating to calculate flow rate.

The *Excel* workbook for this problem automatically iterates to solve for the flow rate Q. It solves Eq. 1 without having to obtain the explicit equation (Eq. 2) for $\overline{V}_2$ (or Q) first. The workbook can be easily used to perform numerous "what-ifs" that would be extremely time-consuming to do manually, e.g., to see how Q is affected by changing the roughness e/D. For example, it shows that replacing the old cast-iron pipe with a new pipe ($e/D \approx 0.0025$) would increase the flow rate from 0.0221 m³/s to about 0.0244 m³/s, a 10% increase! The workbook can be modified to solve other case (c) type problems.

Example 8.8 FLOW IN AN IRRIGATION SYSTEM: DIAMETER UNKNOWN

Spray heads in an agricultural spraying system are to be supplied with water through 152.4 m of drawn aluminum tubing from an engine-driven pump. In its most efficient operating range, the pump output is $0.0946 \text{ m}^3/\text{s}$ at a discharge pressure not exceeding 448.2 kPa(gage). For satisfactory operation, the sprinklers must operate at 206.8 kPa(gage) or higher pressure. Minor losses and elevation changes may be neglected. Determine the smallest standard pipe size that can be used.

Given: Water supply system, as shown.

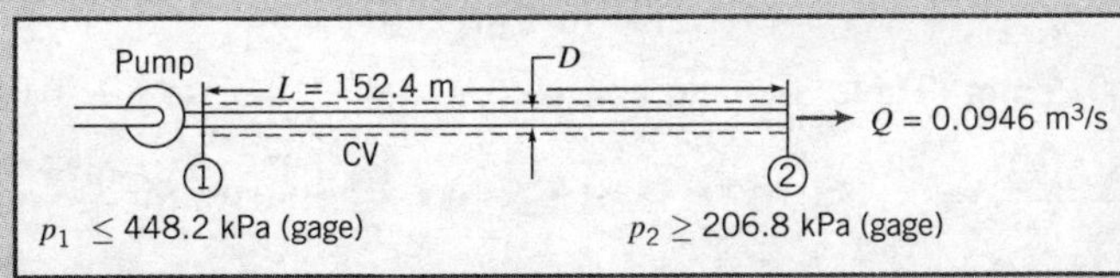

Find: Smallest standard D.

Solution:

Δp, L, and Q are known. D is unknown, so iteration is needed to determine the minimum standard diameter that satisfies the pressure drop constraint at the given flow rate. The maximum allowable pressure drop over the length, L, is

$$\Delta p_{\max} = p_{1_{\max}} - p_{2_{\min}} = (448.2 - 206.8)\ \text{kPa} = 241.4\ \text{kPa}$$

Governing equations:

$$\left(\frac{p_1}{\rho} + \alpha_1 \cancel{\frac{\bar{V}_1^2}{2}} + \cancel{gz_1}\right) - \left(\frac{p_2}{\rho} + \alpha_2 \cancel{\frac{\bar{V}_2^2}{2}} + \cancel{gz_2}\right) = h_{l_T} \tag{8.29}$$

$$= 0(3)$$

$$h_{l_T} = h_l + \cancel{h_{l_m}} = f\frac{L}{D}\frac{\bar{V}_2^2}{2}$$

Assumptions:
(1) Steady flow.
(2) Incompressible flow.
(3) $h_{l_T} = h_l$, i.e., $h_{l_m} = 0$.
(4) $z_1 = z_2$.
(5) $\bar{V}_1 = \bar{V}_2 = \bar{V}; \alpha_1 \simeq \alpha_2$.

Then

$$\Delta p = p_1 - p_2 = f\frac{L}{D}\frac{\rho \bar{V}^2}{2} \tag{1}$$

Equation 1 is difficult to solve for D because both $\bar{V}$ and f depend on D! The best approach is to use a computer application such as *Excel* to automatically solve for D. For completeness here we show the manual iteration procedure. The first step is to express Eq. 1 and the Reynolds number in terms of Q instead of $\bar{V}$ (Q is constant but $\bar{V}$ varies with D). We have $\bar{V} = Q/A = 4Q/\pi D^2$ so that

$$\Delta p = f\frac{L}{D}\frac{\rho}{2}\left(\frac{4Q}{\pi D^2}\right)^2 = \frac{8fL\rho Q^2}{\pi^2 D^5} \tag{2}$$

The Reynolds number in terms of Q is

$$Re = \frac{\rho \bar{V} D}{\mu} = \frac{\bar{V} D}{\nu} = \frac{4Q}{\pi D^2}\frac{D}{\nu} = \frac{4Q}{\pi \nu D}$$

For an initial guess, take nominal 100 mm (102.3 mm i.d.) pipe:

$$Re = \frac{4Q}{\pi \nu D} = \frac{4}{\pi} \times 0.094 \frac{m^3}{s} \times \frac{s}{1.21 \times 10^{-6} m^2} \times \frac{1}{102.3\ mm} \times \frac{1000\ mm}{m} = 1.06 \times 10^6$$

For drawn tubing, $e = 0.0015$ mm (Table 8.1) and hence $e/D = 1.47 \times 10^{-5}$, so $f \simeq 0.012$ (Eq. 8.37), and

$$\Delta p = \frac{8fL\rho Q^2}{\pi^2 D^5} = \frac{8}{\pi^2} \times 0.012 \times 152.4\ m \times 1000 \frac{kg}{m^3} \times (0.0946)^2 \frac{m^6}{s^2}$$

$$\times \frac{1}{(102.3)^5 mm^5} \times \frac{10^{15} mm^5}{m^5} \times \frac{N \cdot s^2}{kg \cdot m} \times \frac{Pa \cdot m^2}{N} = 1184\ kPa$$

$$\Delta p = 1184\ kPa > \Delta p_{max}$$

Since this pressure drop is too large, try $D = 150$ mm (actually 154 mm i.d.):

$$Re = \frac{4}{\pi} \times (0.0946) \frac{m^3}{s} \times \frac{s}{1.12 \times 10^{-6}\ m^2} \times \frac{1}{150\ mm} \times \frac{1000\ mm}{1\ m} = 7.17 \times 10^5$$

For drawn tubing with $D = 150$ mm, $e/D = 9.7 \times 10^{-6}$, so $f \simeq 0.0125$ (Eq. 8.37), and

$$\Delta p = \frac{8}{\pi^2} \times 0.0125 \times 152.4\ m \times 1000\ kg/m^3 \times (0.0946)^2 \frac{m^6}{s^2}$$

$$\times \frac{1}{(154)^5 mm^5} \times \frac{10^{15} mm^5}{m^5} \times \frac{N \cdot s^2}{kg \cdot m} \times \frac{Pa \cdot m^2}{N}$$

$$\Delta p = 159.5\ kPa < \Delta p_{max}$$

Since this is less than the allowable pressure drop, we should check a 125 mm (nominal) pipe. With an actual i.d. of 128 mm,

$$Re = \frac{4}{\pi} \times 0.0946 \times \frac{s}{1.12 \times 10^{-6} m^2} \times \frac{1}{128\ mm} \times \frac{1000\ mm}{1m} = 8.4 \times 10^5$$

For drawn tubing with $D = 125$ mm, $e/D = 1.17 \times 10^{-5}$, so $f \simeq 0.0125$ (Eq. 8.37), and

$$\Delta p = \frac{8}{\pi^2} \times 0.0125 \times 152.4m \times 1000 \frac{kg}{m^3} \times (0.0946)^2 \frac{m^6}{s^2}$$

$$\times \frac{1}{(128)^5 mm^5} \times \frac{10^{15} mm^5}{m^5} \times \frac{N \cdot s^2}{kg \cdot m} \times \frac{Pa \cdot m^2}{N}$$

$$\Delta p = 402.2\ kPa > \Delta p_{max}$$

Thus the criterion for pressure drop is satisfied for a minimum nominal diameter of 150 mm pipe. ← D

This problem illustrates the method for manually iterating to calculate pipe diameter.

The *Excel* workbook for this problem automatically iterates to solve for the exact pipe diameter D that satisfies Eq. 1, without having to obtain the explicit equation (Eq. 2) for D first. Then all that needs to be done is to select the smallest standard pipe size that is equal to or greater than this value. For the given data, $D = 142$ mm, so the appropriate pipe size is 150 mm. The workbook can be used to perform numerous "what-ifs" that would be extremely time-consuming to do manually, e.g., to see how the required D is affected by changing the pipe length L. For example, it shows that reducing L to 76 m would allow 125 mm. (nominal) pipe to be used. The workbook can be modified for solving other case (d) type problems.

We have solved Examples 8.7 and 8.8 by iteration (manual, or using *Excel*). Several specialized forms of friction factor versus Reynolds number diagrams have been introduced to solve problems of this type without the need for iteration. For examples of these specialized diagrams, see Daily and Harleman [20] and White [21].

Examples 8.9 and 8.10 illustrate the evaluation of minor loss coefficients and the application of a diffuser to reduce exit kinetic energy from a flow system.

Example 8.9 CALCULATION OF ENTRANCE LOSS COEFFICIENT

Hamilton [22] reports results of measurements made to determine entrance losses for flow from a reservoir to a pipe with various degrees of entrance rounding. A copper pipe 3 m long, with 38 mm i.d., was used for the tests. The pipe discharged to atmosphere. For a square-edged entrance, a discharge of $0.016\ \mathrm{m^3/s}$ was measured when the reservoir level was 25.9 m above the pipe centerline. From these data, evaluate the loss coefficient for a square-edged entrance.

Given: Pipe with square-edged entrance discharging from reservoir as shown.

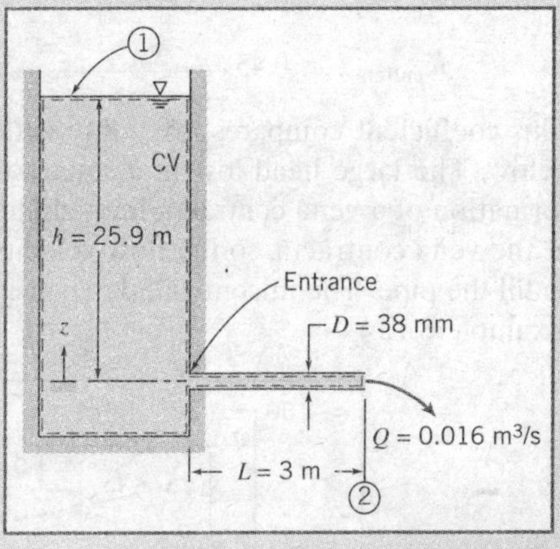

Find: K_{entrance}.

Solution:
Apply the energy equation for steady, incompressible pipe flow.

Governing equations:

$$\cancel{\frac{p_1}{\rho}}+\alpha_1\overset{\approx 0(2)}{\cancel{\frac{\bar{V}_1^2}{2}}}+gz_1=\cancel{\frac{p_2}{\rho}}+\alpha_2\frac{\bar{V}_2^2}{2}+\overset{=0}{\cancel{gz_2}}+h_{l_T}$$

$$h_{l_T}=f\frac{L}{D}\frac{\bar{V}_2^2}{2}+K_{\text{entrance}}\frac{\bar{V}_2^2}{2}$$

Assumptions: (1) $p_1=p_2=p_{\text{atm}}$.
(2) $\bar{V}_1\approx 0$.

Substituting for h_{l_T} and dividing by g gives $z_1=h=\alpha_2\dfrac{\bar{V}_2^2}{2g}+f\dfrac{L}{D}\dfrac{\bar{V}_2^2}{2g}+K_{\text{entrance}}\dfrac{\bar{V}_2^2}{2g}$

or

$$K_{\text{entrance}}=\frac{2gh}{\bar{V}_2^2}-f\frac{L}{D}-\alpha_2 \tag{1}$$

The average velocity is

$$\bar{V}_2=\frac{Q}{A}=\frac{4Q}{\pi D^2}$$

$$\bar{V}_2=\frac{4}{\pi}\times 0.016\frac{\mathrm{m^3}}{\mathrm{s}}\times\frac{1}{(38)^2\ \mathrm{mm^2}}\times 10^6\frac{\mathrm{mm^2}}{\mathrm{m^2}}=14.1\ \mathrm{m/s}$$

Assume $T=21°\mathrm{C}$, so $\nu=9.75\times10^{-7}\ \mathrm{m^2/s}$ (Table A.7). Then

$$Re=\frac{\bar{V}D}{\nu}=14.1\frac{\mathrm{m}}{\mathrm{s}}\times 38\ \mathrm{mm}\times\frac{\mathrm{m}}{1000\ \mathrm{mm}}\times\frac{\mathrm{s}}{9.75\times10^{-7}\mathrm{m^2}}$$

For drawn tubing, $e=0.0015$ mm (Table 8.1), so $e/D=0.000{,}04$ and $f=0.0135$ (Eq. 8.37).

In this problem we need to be careful in evaluating the kinetic energy correction factor α_2, as it is a significant factor in computing K_{entrance} from Eq. 1. We recall from Section 8.6 and previous Examples that we have usually assumed $\alpha\approx 1$, but here we will compute a value from Eq. 8.27:

$$\alpha=\left(\frac{U}{\bar{V}}\right)^3\frac{2n^2}{(3+n)(3+2n)} \tag{8.27}$$

To use this equation we need values for the turbulent power-law coefficient n and the ratio of centerline to mean velocity $U/\bar{V}$. For n, from Section 8.5

$$n=-1.7+1.8\log(Re_U)\approx 8.63 \tag{8.23}$$

where we have used the approximation $Re_U \approx Re_{\overline{V}}$. For $\overline{V}/U$, we have

$$\frac{\overline{V}}{U} = \frac{2n^2}{(n+1)(2n+1)} = 0.847 \tag{8.24}$$

Using these results in Eq. 8.27 we find $\alpha = 1.04$. Substituting into Eq. 1, we obtain

$$K_{\text{entrance}} = 2 \times 9.81 \frac{\text{m}}{\text{s}^2} \times 25.9 \text{ m} \times \frac{\text{s}^2}{(14.1)^2 \text{m}^2} - 0.0135 \frac{3\text{m}}{38 \text{ mm}} \times 1000 \frac{\text{mm}}{\text{m}} - 1.04$$

$$K_{\text{entrance}} = 0.45$$

K_{entrance}

This coefficient compares favorably with that shown in Table 8.2. The hydraulic and energy grade lines are shown below. The large head loss in a square-edged entrance is due primarily to separation at the sharp inlet corner and formation of a vena contracta immediately downstream from the corner. The effective flow area reaches a minimum at the vena contracta, so the flow velocity is a maximum there. The flow expands again following the vena contracta to fill the pipe. The uncontrolled expansion following the vena contracta is responsible for most of the head loss. (See Example 8.12.)

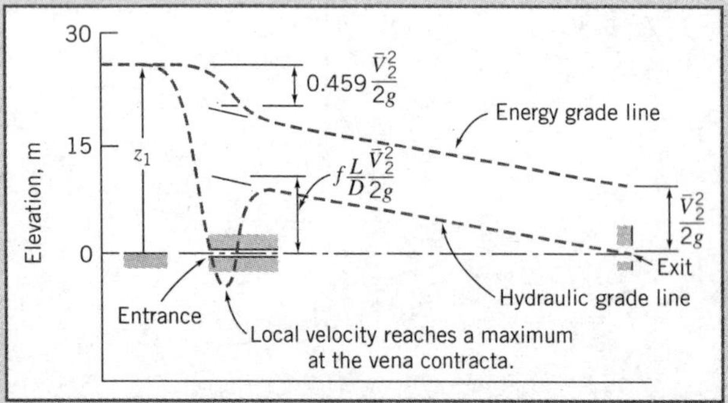

This problem:

- ✓ Illustrates a method for obtaining the value of a minor loss coefficient from experimental data.
- ✓ Shows how the EGL and HGL lines first introduced in Section 6.5 for inviscid flow are modified by the presence of major and minor losses. The EGL line continuously drops as mechanical energy is consumed—quite sharply when, for example, we have a square-edged entrance loss; the HGL at each location is lower than the EGL by an amount equal to the local dynamic head $\overline{V}^2/2g$—at the vena contracta, for example, the HGL experiences a large drop, then recovers.

Rounding the inlet corner reduces the extent of separation significantly. This reduces the velocity increase through the vena contracta and consequently reduces the head loss caused by the entrance. A "well-rounded" inlet almost eliminates flow separation; the flow pattern approaches that shown in Fig. 8.1. The added head loss in a well-rounded inlet compared with fully developed flow is the result of higher wall shear stresses in the entrance length.

Example 8.10 USE OF DIFFUSER TO INCREASE FLOW RATE

Water rights granted to each citizen by the Emperor of Rome gave permission to attach to the public water main a calibrated, circular, tubular bronze nozzle [23]. Some citizens were clever enough to take unfair advantage of a law that regulated flow rate by such an indirect method. They installed diffusers on the outlets of the nozzles to increase their discharge. Assume the static head available from the main is $z_0 = 1.5$ m and the nozzle exit diameter is $D = 25$ mm. (The discharge is to atmospheric pressure.) Determine the increase in flow rate when a diffuser with $N/R_1 = 3.0$ and $AR = 2.0$ is attached to the end of the nozzle.

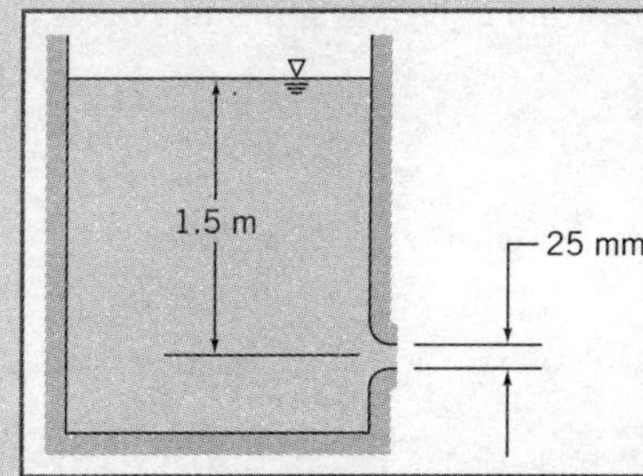

Given: Nozzle attached to water main as shown.

Find: Increase in discharge when diffuser with $N/R_1 = 3.0$ and $AR = 2.0$ is installed.

Solution: Apply the energy equation for steady, incompressible pipe flow.

Governing equation:
$$\frac{p_0}{\rho} + \alpha_0 \frac{\bar{V}_0^2}{2} + gz_0 = \frac{p_1}{\rho} + \alpha_1 \frac{\bar{V}_1^2}{2} + gz_1 + h_{l_T} \qquad (8.29)$$

Assumptions: (1) $\bar{V}_0 \approx 0$.
(2) $\alpha_1 \approx 1$.

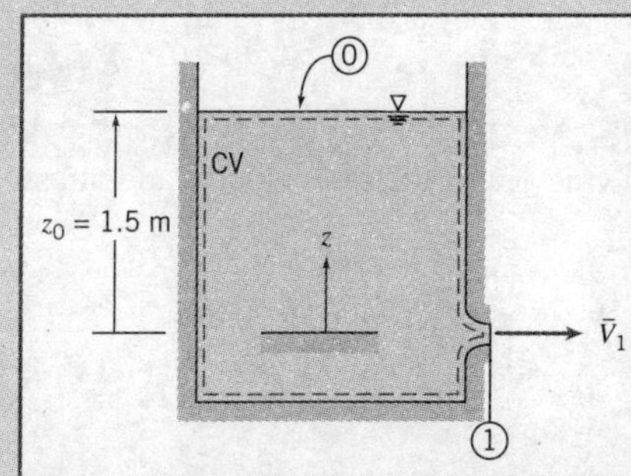

For the nozzle alone,

$$\frac{p_0}{\rho} + \alpha_0 \frac{\bar{V}_0^2}{2} + gz_0 = \frac{p_1}{\rho} + \alpha_1 \frac{\bar{V}_1^2}{2} + gz_1 + h_{l_T} \qquad \text{with } \bar{V}_0 \approx 0(1),\ \alpha_1 \approx 1(2),\ z_1 = 0$$

$$h_{l_T} = K_{\text{entrance}} \frac{\bar{V}_1^2}{2}$$

Thus

$$gz_0 = \frac{\bar{V}_1^2}{2} + K_{\text{entrance}} \frac{\bar{V}_1^2}{2} = (1 + K_{\text{entrance}}) \frac{\bar{V}_1^2}{2} \qquad (1)$$

Solving for the velocity and substituting the value of $K_{\text{entrance}} \approx 0.04$ (from Table 8.2),

$$\bar{V}_1 = \sqrt{\frac{2gz_0}{1.04}} = \sqrt{\frac{2}{1.04} \times 9.81 \frac{\text{m}}{\text{s}^2} \times 1.5 \text{ m}} = 5.32 \text{ m/s}$$

$$Q = \bar{V}_1 A_1 = \bar{V}_1 \frac{\pi D_1^2}{4} = 5.32 \frac{\text{m}}{\text{s}} \times \frac{\pi}{4} \times (0.025)^2 \text{ m}^2 = 0.00261 \text{ m}^3/\text{s} \longleftarrow Q$$

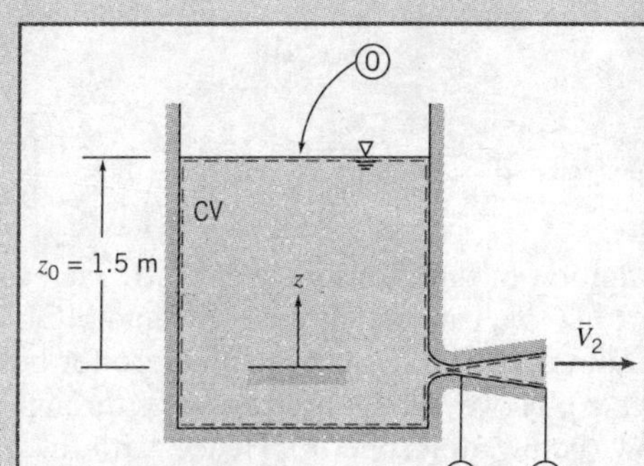

For the nozzle with diffuser attached,

$$\frac{p_0}{\rho} + \alpha_0 \frac{\bar{V}_0^2}{2} + gz_0 = \frac{p_2}{\rho} + \alpha_2 \frac{\bar{V}_2^2}{2} + gz_2 = h_{l_T} \qquad \text{with } \bar{V}_0 \approx 0(1),\ \alpha_2 \approx 1(2),\ z_2 = 0$$

$$h_{l_T} = K_{\text{entrance}} \frac{\bar{V}_1^2}{2} + K_{\text{diffuser}} \frac{\bar{V}_1^2}{2}$$

or

$$gz_0 = \frac{\bar{V}_2^2}{2} + (K_{\text{entrance}} + K_{\text{diffuser}}) \frac{\bar{V}_1^2}{2} \qquad (2)$$

From continuity $\bar{V}_1 A_1 = \bar{V}_2 A_2$, so

$$\bar{V}_2 = \bar{V}_1 \frac{A_1}{A_2} = \bar{V}_1 \frac{1}{AR}$$

and Eq. 2 becomes

$$gz_0 = \left[\frac{1}{(AR)^2} + K_{\text{entrance}} + K_{\text{diffuser}} \right] \frac{\bar{V}_1^2}{2} \qquad (3)$$

Figure 8.16 gives data for $C_p = \dfrac{p_2 - p_1}{\frac{1}{2}\rho \bar{V}_1^2}$ for diffusers.

To obtain K_{diffuser}, apply the energy equation from ① to ②.

$$\frac{p_1}{\rho} + \alpha_1\frac{\overline{V}_1^2}{2} + g\not{z}_1 = \frac{p_2}{\rho} + \alpha_2\frac{\overline{V}_2^2}{2} + g\not{z}_2 + K_{\text{diffuser}}\frac{\overline{V}_1^2}{2}$$

Solving, with $\alpha_2 \approx 1$, we obtain

$$K_{\text{diffuser}} = 1 - \frac{\overline{V}_2^2}{\overline{V}_1^2} - \frac{p_2 - p_1}{\frac{1}{2}\rho\overline{V}_1^2} = 1 - \left(\frac{A_1}{A_2}\right)^2 - C_p = 1 - \frac{1}{(AR)^2} - C_p$$

From Fig. 8.16, $C_p = 0.45$, so

$$K_{\text{diffuser}} = 1 - \frac{1}{(2.0)^2} - 0.45 = 0.75 - 0.45 = 0.3$$

Solving Eq. 3 for the velocity and substituting the values of K_{entrance} and K_{diffuser}, we obtain

$$\overline{V}_1^2 = \frac{2gz_0}{0.25 + 0.04 + 0.3}$$

so

$$\overline{V}_1 = \sqrt{\frac{2gz_0}{0.59}} = \sqrt{\frac{2}{0.59} \times 9.81\frac{\text{m}}{\text{s}^2} \times 1.5\ \text{m}} = 7.06\ \text{m/s}$$

and

$$Q_d = \overline{V}_1 A_1 = \overline{V}_1\frac{\pi D_1^2}{4} = 7.06\frac{\text{m}}{\text{s}} \times \frac{\pi}{4} \times (0.025)^2\ \text{m}^2 = 0.00347\ \text{m}^3/\text{s} \qquad Q_d$$

The flow rate increase that results from adding the diffuser is

$$\frac{\Delta Q}{Q} = \frac{Q_d - Q}{Q} = \frac{Q_d}{Q} - 1 = \frac{0.00347}{0.00261} - 1 = 0.330 \quad \text{or} \quad 33\ \text{percent} \qquad \frac{\Delta Q}{Q}$$

Addition of the diffuser significantly increases the flow rate! There are two ways to explain this.

First, we can sketch the EGL and HGL curves—approximately to scale—as shown below. We can see that, as required, the HGL at the exit is zero for both flows (recall that the HGL is the sum of static pressure and potential heads). However, the pressure rises through the diffuser, so the pressure at the diffuser inlet will be, as shown, quite low (below atmospheric). Hence, with the diffuser, the Δp driving force for the nozzle is much larger than that for the bare nozzle, leading to a much greater velocity, and flow rate, at the nozzle exit plane—it is as if the diffuser acted as a suction device on the nozzle.

Second, we can examine the energy equations for the two flows (for the bare nozzle Eq. 1, and for the nozzle with diffuser Eq. 3). These equations can be rearranged to yield equations for the velocity at the nozzle exit,

$$\overline{V}_1 = \sqrt{\frac{2gz_0}{1 + K_{\text{entrance}}}} \quad \text{(bare nozzle)} \qquad \overline{V}_1 = \sqrt{\frac{2gz_0}{\frac{1}{(AR)^2} + K_{\text{diffuser}} + K_{\text{entrance}}}} \quad \text{(nozzle + diffuser)}$$

Comparing these two expressions, we see that the diffuser introduces an extra term (its loss coefficient $K_{\text{diffuser}} = 0.3$) to the denominator, tending to reduce the nozzle velocity, but on the other hand we replace the term 1 (representing loss of the bare nozzle exit plane kinetic energy) with $1/(AR)^2 = 0.25$ (representing a smaller loss, of the diffuser exit plane kinetic energy). The net effect is that we replace 1 in the denominator with $0.25 + 0.3 = 0.55$, leading to a net increase in the nozzle velocity. The resistance to flow introduced by adding the diffuser is more than made up by the fact that we "throw away" much less kinetic energy at the exit of the device (the exit velocity for the bare nozzle is 5.32 m/s, whereas for the diffuser it is 1.77 m/s).

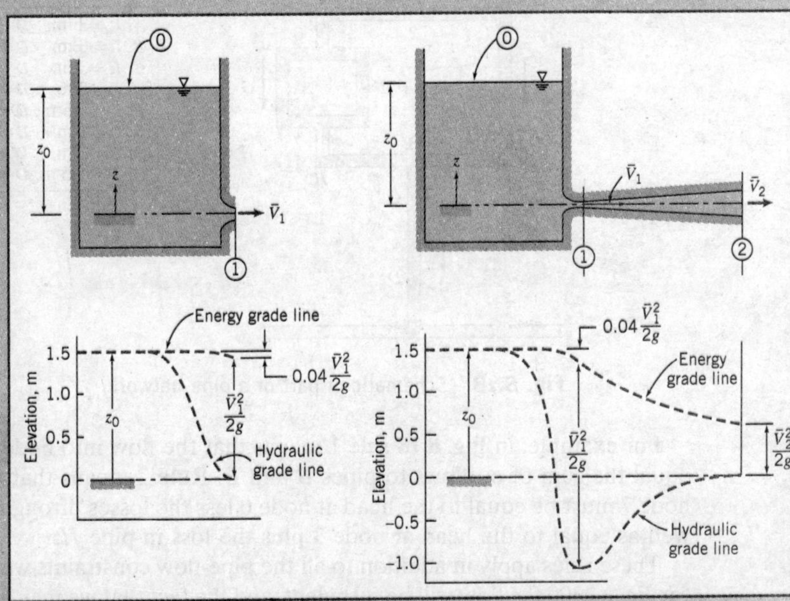

Water Commissioner Frontinus standardized conditions for all Romans in 97 A.D. He required that the tube attached to the nozzle of each customer's pipe be the same diameter for at least 15 lineal meters from the public water main (see Problem 8.157).

*Multiple-Path Systems

Many real-world pipe systems (e.g., the pipe network that supplies water to the apartments in a large building) consist of a network of pipes of various diameters assembled in a complicated configuration that may contain parallel and series connections. As an example, consider part of a system as shown in Fig. 8.18. Water is supplied at some pressure from a manifold at point 1, and flows through the components shown to the drain at point 5. Some water flows through pipes *A*, *B*, *C*, and *D*, constituting a *series* of pipes (and pipe *B* has a lower flow rate than the others); some flows through *A*, *E*, *F* or *G*, *H*, *C*, and *D* (*F* and *G* are *parallel*), and these two main branches are in *parallel*. We analyze this type of problem in a similar way to how we analyze DC resistor circuits in electrical theory: by applying a few basic rules to the system. The electrical potential at each point in the circuit is analogous to the HGL (or static pressure head if we neglect gravity) at corresponding points in the system. The current in each resistor is analogous to the flow rate in each pipe section. We have the additional difficulty in pipe systems that the resistance to flow in each pipe is a function of the flow rate (electrical resistors are usually considered constant).

The simple rules for analyzing networks can be expressed in various ways. We will express them as follows:

1. The net flow out of any node (junction) is zero.
2. Each node has a unique pressure head (HGL).

*This section may be omitted without loss of continuity in the text material.

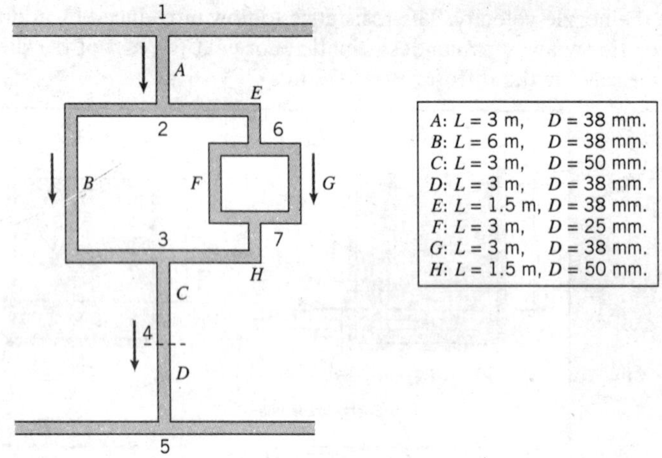

Fig. 8.18 Schematic of part of a pipe network.

For example, in Fig. 8.18 rule 1 means that the flow into node 2 from pipe A must equal the sum of outflows to pipes B and E. Rule 2 means that the pressure head at node 7 must be equal to the head at node 6 less the losses through pipe F or pipe G, as well as equal to the head at node 3 plus the loss in pipe H.

These rules apply in addition to all the pipe-flow constraints we have discussed (e.g., for $Re \geq 2300$ the flow will be turbulent, and the fact that we may have significant minor losses from features such as sudden expansions). We can anticipate that the flow in pipe F (diameter 25 mm) will be a good deal less than the flow in pipe G (diameter 38 mm), and the flow through branch E will be larger than that through branch B (why?).

The problems that arise with pipe networks can be as varied as those we discussed when studying single-path systems, but the most common involve finding the flow delivered to each pipe, given an applied pressure difference. We examine this case in Example 8.11. Obviously, pipe networks are much more difficult and time-consuming to analyze than single-path problems, almost always requiring iterative solution methods, and in practice are usually only solved using the computer. A number of computer schemes for analyzing networks have been developed [24], and many engineering consulting companies use proprietary software applications for such analysis. A spreadsheet such as *Excel* is also very useful for setting up and solving such problems.

Example 8.11 FLOW RATES IN A PIPE NETWORK

In the section of a cast-iron water pipe network shown in Fig. 8.18, the static pressure head (gage) available at point 1 is 30 m of water, and point 5 is a drain (atmospheric pressure). Find the flow rates (L/min) in each pipe.

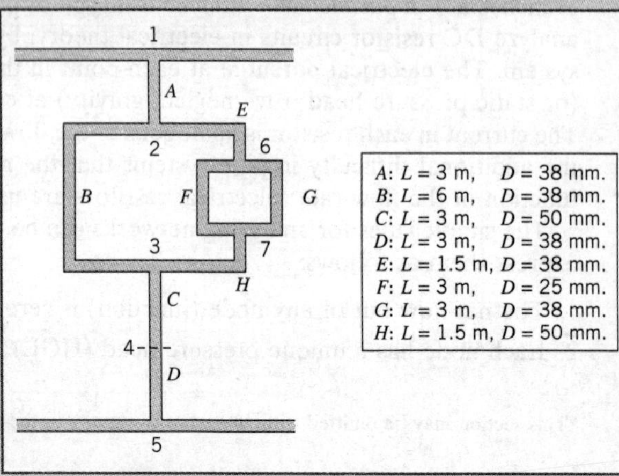

Given: Pressure head h_{1-5} of 30 m across pipe network.

Find: The flow rate in each pipe.

Solution:

Governing equations:

For each pipe section,

$$\left(\frac{p_1}{\rho} + \alpha_1 \frac{\bar{V}_1^2}{2} + \cancelto{0(1)}{gz_1}\right) - \left(\frac{p_2}{\rho} + \alpha_2 \frac{\bar{V}_2^2}{2} + \cancelto{0(1)}{gz_2}\right) = h_{l_T} = h_l + \cancelto{0(2)}{\Sigma h_{l_m}} \qquad (8.29)$$

where

$$h_l = f\frac{L}{D}\frac{\overline{V}^2}{2} \tag{8.34}$$

and f is obtained from either Eq. 8.36 (laminar) or Eq. 8.37 (turbulent). For the cast-iron pipe, Table 8.1 gives a roughness for cast iron of $e = 0.26$ mm.

Assumptions: (1) Ignore gravity effects.
(2) Ignore minor losses.

(Assumption 2 is applied to make the analysis clearer—minor losses can be incorporated easily later.)

In addition we have mathematical expressions for the basic rules

1. The net flow out of any node (junction) is zero.
2. Each node has a unique pressure head (HGL).

We can apply basic rule 1 to nodes 2 and 6:

Node 2: $Q_A = Q_B + Q_E$ (1) Node 6: $Q_E = Q_F + Q_G$ (2)

and we also have the obvious constraints

$Q_A = Q_C$ (3) $Q_A = Q_D$ (4) $Q_E = Q_H$ (5)

We can apply basic rule 2 to obtain the following pressure drop constraints:

$h_{1-5}: h = h_A + h_B + h_C + h_D$ (6) $h_{2-3}: h_B = h_E + h_F + h_H$ (7) $h_{6-7}: h_F = h_G$ (8)

This set of eight equations (as well as Eqs. 8.29 and 8.34 for each pipe section!) must be solved iteratively. If we were to manually iterate, we would use Eqs. 3, 4, and 5 to immediately reduce the number of unknowns and equations to five (Q_A, Q_B, Q_E, Q_F, Q_G). There are several approaches to the iteration, one of which is:

1. Make a guess for Q_A, Q_B, and Q_F.
2. Eqs. 1 and 2 then lead to values for Q_E and Q_G.
3. Eqs. 6, 7, and 8 are finally used as a check to see if rule 2 (for unique pressure heads at the nodes) is satisfied.
4. If any of Eqs. 6, 7, or 8 are not satisfied, use knowledge of pipe flow to adjust the values of Q_A, Q_B, or Q_F.
5. Repeat steps 2 through 5 until convergence occurs.

An example of applying step 4 would be if Eq. 8 were not satisfied. Suppose $h_F > h_G$; then we would have selected too large a value for Q_F, and would reduce this slightly, and recompute all flow rates and heads.

This iterative process is obviously quite unrealistic for manual calculation (remember that obtaining each head loss h from each Q involves a good amount of calculation). Fortunately, we can use a spreadsheet such as *Excel* to automate all these calculations—it will simultaneously solve for all eight unknowns automatically! The first step is to set up one worksheet for each pipe section for computing the pipe head h given the flow rate Q. A typical such worksheet is shown below:

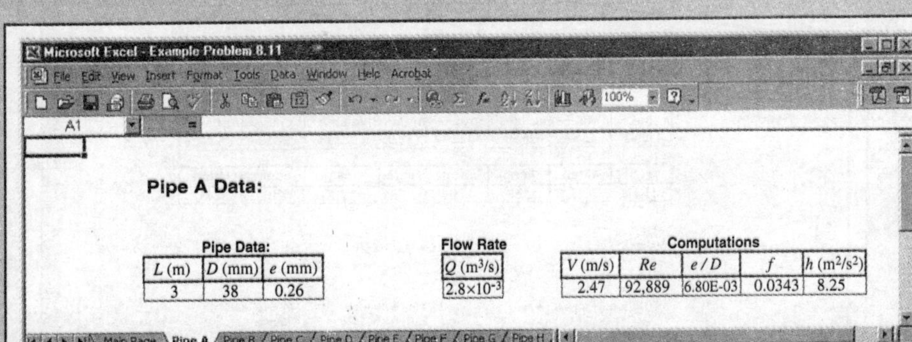

In these worksheets, knowing L, D, and e, a given flow rate Q is used to compute $\overline{V}$, Re, f, and finally h from L, D, and e.

The next step is to set up a calculation page that collects together the flow rates and corresponding head losses for all of the pipe sections, and then use these to check whether Eqs. 1 through 8 are satisfied. Shown below is this page with initial guess values of $2.8 \times 10^{-3}\,\mathrm{m^3/s}$ for each of the flow rates. The logic of the workbook is that the eight values entered for Q_A through Q_H determine all the other values—that is, h_A through h_H, and the values of the constraint equations. The absolute errors for each of the constraint equations are shown, as well as their sum. We can then use *Excel*'s *Solver* feature (repeatedly as necessary) to minimize the total error (currently 768.1%) by varying Q_A through Q_H.

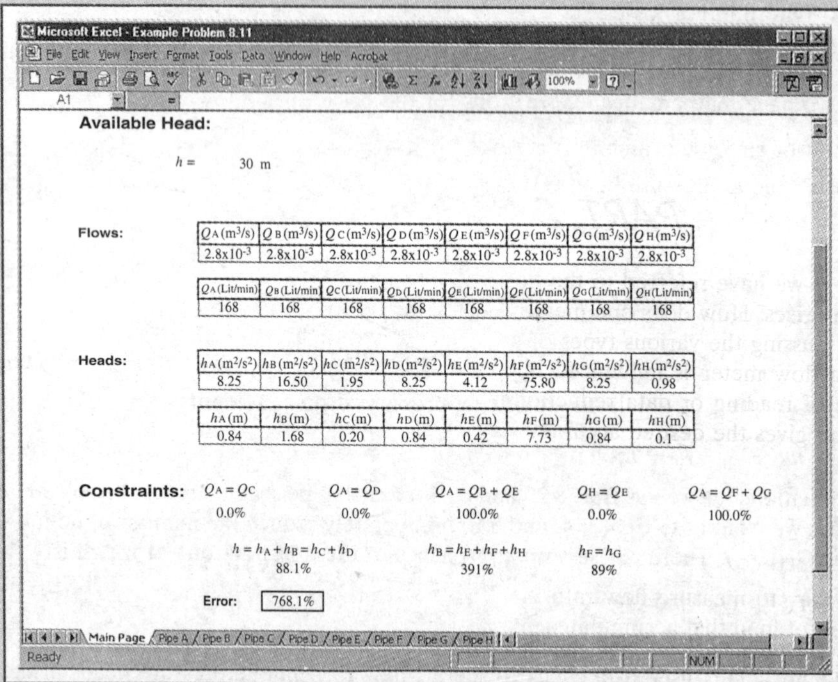

Microsoft Excel - Example Problem 8.11

Available Head:

$h =$ 30 m

Flows:

Q_A (m³/s)	Q_B (m³/s)	Q_C (m³/s)	Q_D (m³/s)	Q_E (m³/s)	Q_F (m³/s)	Q_G (m³/s)	Q_H (m³/s)
2.8x10⁻³	2.8x10⁻³	2.8x10⁻³	2.8x10⁻³	2.8x10⁻³	2.8x10⁻³	2.8x10⁻³	2.8x10⁻³

Q_A (Lit/min)	Q_B (Lit/min)	Q_C (Lit/min)	Q_D (Lit/min)	Q_E (Lit/min)	Q_F (Lit/min)	Q_G (Lit/min)	Q_H (Lit/min)
168	168	168	168	168	168	168	168

Heads:

h_A (m²/s²)	h_B (m²/s²)	h_C (m²/s²)	h_D (m²/s²)	h_E (m²/s²)	h_F (m²/s²)	h_G (m²/s²)	h_H (m²/s²)
8.25	16.50	1.95	8.25	4.12	75.80	8.25	0.98

h_A (m)	h_B (m)	h_C (m)	h_D (m)	h_E (m)	h_F (m)	h_G (m)	h_H (m)
0.84	1.68	0.20	0.84	0.42	7.73	0.84	0.1

Constraints:

$Q_A = Q_C$	$Q_A = Q_D$	$Q_A = Q_B + Q_E$	$Q_H = Q_E$	$Q_A = Q_F + Q_G$
0.0%	0.0%	100.0%	0.0%	100.0%

$h = h_A + h_B = h_C + h_D$	$h_B = h_E + h_F + h_H$	$h_F = h_G$
88.1%	391%	89%

Error: 768.1%

Main Page / Pipe A / Pipe B / Pipe C / Pipe D / Pipe E / Pipe F / Pipe G / Pipe H

Ready NUM

The final results obtained by *Excel* are:

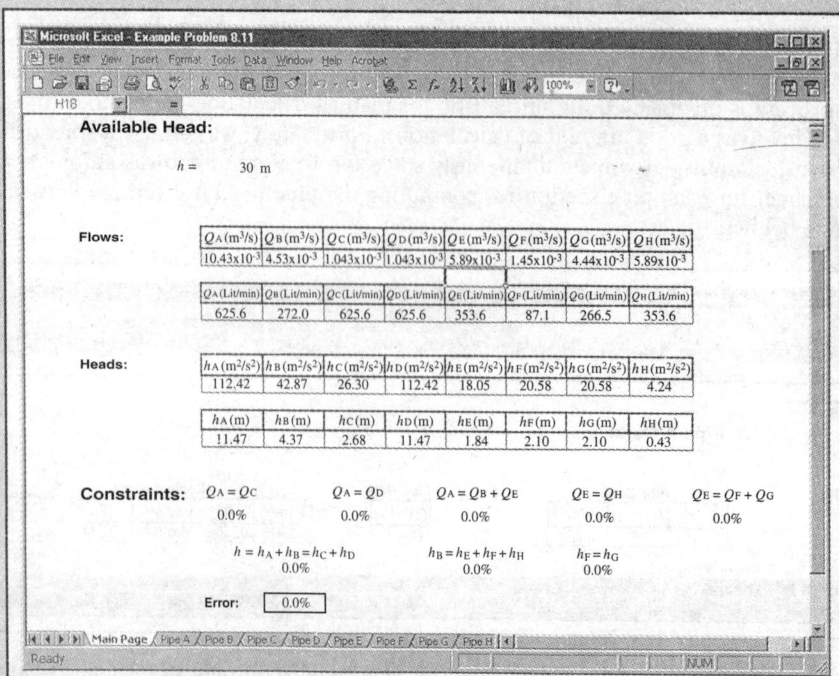

Microsoft Excel - Example Problem 8.11

Available Head:

$h =$ 30 m

Flows:

Q_A (m³/s)	Q_B (m³/s)	Q_C (m³/s)	Q_D (m³/s)	Q_E (m³/s)	Q_F (m³/s)	Q_G (m³/s)	Q_H (m³/s)
10.43x10⁻³	4.53x10⁻³	1.043x10⁻³	1.043x10⁻³	5.89x10⁻³	1.45x10⁻³	4.44x10⁻³	5.89x10⁻³

Q_A (Lit/min)	Q_B (Lit/min)	Q_C (Lit/min)	Q_D (Lit/min)	Q_E (Lit/min)	Q_F (Lit/min)	Q_G (Lit/min)	Q_H (Lit/min)
625.6	272.0	625.6	625.6	353.6	87.1	266.5	353.6

Heads:

h_A (m²/s²)	h_B (m²/s²)	h_C (m²/s²)	h_D (m²/s²)	h_E (m²/s²)	h_F (m²/s²)	h_G (m²/s²)	h_H (m²/s²)
112.42	42.87	26.30	112.42	18.05	20.58	20.58	4.24

h_A (m)	h_B (m)	h_C (m)	h_D (m)	h_E (m)	h_F (m)	h_G (m)	h_H (m)
11.47	4.37	2.68	11.47	1.84	2.10	2.10	0.43

Constraints:

$Q_A = Q_C$	$Q_A = Q_D$	$Q_A = Q_B + Q_E$	$Q_E = Q_H$	$Q_E = Q_F + Q_G$
0.0%	0.0%	0.0%	0.0%	0.0%

$h = h_A + h_B = h_C + h_D$	$h_B = h_E + h_F + h_H$	$h_F = h_G$
0.0%	0.0%	0.0%

Error: 0.0%

Main Page / Pipe A / Pipe B / Pipe C / Pipe D / Pipe E / Pipe F / Pipe G / Pipe H

Ready NUM

The flow rates are:

$$Q_A = Q_C = Q_D = 625.6\ \text{L/min}$$
$$Q_B(\text{L/min}) = 272.0\ \text{L/min}$$
$$Q_E(\text{L/min}) = Q_H(\text{L/min}) = 353.6\ \text{L/min}$$
$$Q_F(\text{L/min}) = 87.1\ \text{L/min}$$
$$Q_G(\text{L/min}) = 266.5\ \text{L/min}$$

This problem illustrates use of *Excel* to solve a set of coupled, nonlinear equations for unknown flow rates. The *Excel* workbook for this problem can be modified for solving a variety of other multiple-path systems.

PART C Flow Measurement

Throughout this text we have referred to the flow rate Q or average velocity $\overline{V}$ in a pipe. The question arises: How does one measure these quantities? We will address this question by discussing the various types of flow meters available.

The choice of a flow meter is influenced by the accuracy required, range, cost, complication, ease of reading or data reduction, and service life. The simplest and cheapest device that gives the desired accuracy should be chosen.

Direct Methods 8.9

The most obvious way to measure flow rate in a pipe is the *direct method*—simply measure the amount of fluid that accumulates in a container over a fixed time period! Tanks can be used to determine flow rate for steady liquid flows by measuring the volume or mass of liquid collected during a known time interval. If the time interval is long enough to be measured accurately, flow rates may be determined precisely in this way.

Compressibility must be considered in volume measurements for gas flows. The densities of gases generally are too small to permit accurate direct measurement of mass flow rate. However, a volume sample often can be collected by displacing a "bell," or inverted jar over water (if the pressure is held constant by counterweights). If volume or mass measurements are set up carefully, no calibration is required; this is a great advantage of direct methods.

In specialized applications, particularly for remote or recording uses, *positive displacement* flow meters may be specified, in which the fluid moves a component such as a reciprocating piston or oscillating disk as it passes through the device. Common examples include household water and natural gas meters, which are calibrated to read directly in units of product, or gasoline metering pumps, which measure total flow and automatically compute the cost. Many positive-displacement meters are available commercially. Consult manufacturers' literature or References (e.g., [25]) for design and installation details.

Restriction Flow Meters for Internal Flows 8.10

Most restriction flow meters for internal flow (except the laminar flow element, discussed shortly) are based on acceleration of a fluid stream through some form of nozzle, as shown schematically in Fig. 8.19. The idea is that the change in velocity

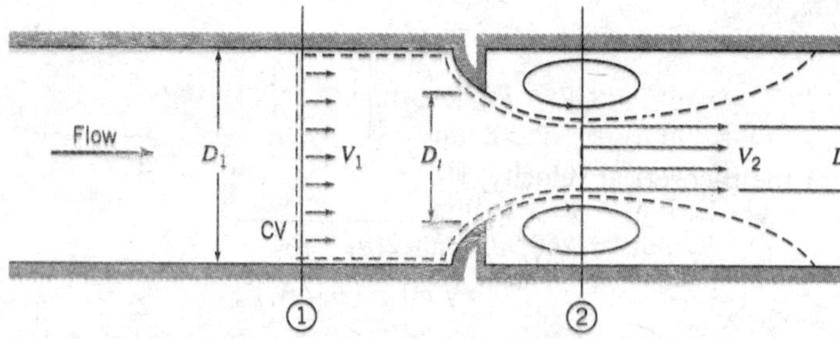

Fig. 8.19 Internal flow through a generalized nozzle, showing control volume used for analysis.

leads to a change in pressure. This Δp can be measured using a pressure gage (electronic or mechanical) or a manometer, and the flow rate inferred using either a theoretical analysis or an experimental correlation for the device. Flow separation at the sharp edge of the nozzle throat causes a recirculation zone to form, as shown by the dashed lines downstream from the nozzle. The mainstream flow continues to accelerate from the nozzle throat to form a vena contracta at section ② and then decelerates again to fill the duct. At the vena contracta, the flow area is a minimum, the flow streamlines are essentially straight, and the pressure is uniform across the channel section.

The theoretical flow rate may be related to the pressure differential between sections ① and ② by applying the continuity and Bernoulli equations. Then empirical correction factors may be applied to obtain the actual flow rate.

Basic equations:

We will need mass-conservation,

$$\sum_{CS} \vec{V} \cdot \vec{A} = 0 \tag{4.13b}$$

[we can use this instead of Eq. 4.12, because of assumption (5) below] and the Bernoulli equation,

$$\frac{p_1}{\rho} + \frac{V_1^2}{2} + g\not{z}_1 = \frac{p_2}{\rho} + \frac{V_2^2}{2} + g\not{z}_2 \tag{6.8}$$

which we can use if assumption (4) is valid. For the short section of pipe considered, this is reasonable.

Assumptions:
(1) Steady flow.
(2) Incompressible flow.
(3) Flow along a streamline.
(4) No friction.
(5) Uniform velocity at sections ① and ②.
(6) No streamline curvature at sections ① or ②, so pressure is uniform across those sections.
(7) $z_1 = z_2$.

Then, from the Bernoulli equation,

$$p_1 - p_2 = \frac{\rho}{2}(V_2^2 - V_1^2) = \frac{\rho V_2^2}{2}\left[1 - \left(\frac{V_1}{V_2}\right)^2\right]$$

and from continuity

$$(-\rho V_1 A_1) + (\rho V_2 A_2) = 0$$

or

$$V_1 A_1 = V_2 A_2 \qquad \text{so} \qquad \left(\frac{V_1}{V_2}\right)^2 = \left(\frac{A_2}{A_1}\right)^2$$

Substituting gives

$$p_1 - p_2 = \frac{\rho V_2^2}{2}\left[1 - \left(\frac{A_2}{A_1}\right)^2\right]$$

Solving for the theoretical velocity, V_2,

$$V_2 = \sqrt{\frac{2(p_1 - p_2)}{\rho[1 - (A_2/A_1)^2]}} \tag{8.51}$$

The theoretical mass flow rate is then given by

$$\begin{aligned} \dot{m}_{\text{theoretical}} &= \rho V_2 A_2 \\ &= \rho \sqrt{\frac{2(p_1 - p_2)}{\rho[1 - (A_2/A_1)^2]}} A_2 \end{aligned}$$

or

$$\dot{m}_{\text{theoretical}} = \frac{A_2}{\sqrt{1 - (A_2/A_1)^2}} \sqrt{2\rho(p_1 - p_2)} \tag{8.52}$$

Equation 8.52 shows that, under our set of assumptions, for a given fluid (ρ) and flow meter geometry (A_1 and A_2), the flow rate is directly proportional to the square root of the pressure drop across the meter taps,

$$\dot{m}_{\text{theoretical}} \propto \sqrt{\Delta p}$$

which is the basic idea of these devices. This relationship limits the flow rates that can be measured accurately to approximately a 4:1 range.

Several factors limit the utility of Eq. 8.52 for calculating the actual mass flow rate through a meter. The actual flow area at section ② is unknown when the vena contracta is pronounced (e.g., for orifice plates when D_t is a small fraction of D_1). The velocity profiles approach uniform flow only at large Reynolds numbers. Frictional effects can become important (especially downstream from the meter) when the meter contours are abrupt. Finally, the location of pressure taps influences the differential pressure reading.

The theoretical equation is adjusted for Reynolds number and diameter ratio D_t/D_1 by defining an empirical *discharge coefficient* C such that, replacing Eq. 8.52, we have

$$\dot{m}_{\text{actual}} = \frac{CA_t}{\sqrt{1 - (A_t/A_1)^2}} \sqrt{2\rho(p_1 - p_2)} \tag{8.53}$$

Letting $\beta = D_t/D_1$, then $(A_t/A_1)^2 = (D_t/D_1)^4 = \beta^4$, so

$$\dot{m}_{\text{actual}} = \frac{CA_t}{\sqrt{1 - \beta^4}} \sqrt{2\rho(p_1 - p_2)} \tag{8.54}$$

In Eq. 8.54, $1/\sqrt{1 - \beta^4}$ is the *velocity-of-approach factor*. The discharge coefficient and velocity-of-approach factor frequently are combined into a single *flow coefficient*,

$$K \equiv \frac{C}{\sqrt{1 - \beta^4}} \tag{8.55}$$

In terms of this flow coefficient, the actual mass flow rate is expressed as

$$\dot{m}_{\text{actual}} = KA_t\sqrt{2\rho(p_1 - p_2)} \tag{8.56}$$

For standardized metering elements, test data [25, 26] have been used to develop empirical equations that predict discharge and flow coefficients from meter bore, pipe diameter, and Reynolds number. The accuracy of the equations (within specified ranges) usually is adequate so that the meter can be used without calibration. If the Reynolds number, pipe size, or bore diameter fall outside the specified range of the equation, the coefficients must be measured experimentally.

For the turbulent flow regime (pipe Reynolds number greater than 4000) the discharge coefficient may be expressed by an equation of the form [25]

$$C = C_\infty + \frac{b}{Re_{D_1}^n} \tag{8.57}$$

The corresponding form for the flow-coefficient equation is

$$K = K_\infty + \frac{1}{\sqrt{1-\beta^4}}\frac{b}{Re_{D_1}^n} \tag{8.58}$$

In Eqs. 8.57 and 8.58, subscript ∞ denotes the coefficient at infinite Reynolds number; constants b and n allow for scaling to finite Reynolds numbers. Correlating equations and curves of coefficients versus Reynolds number are given in the next three subsections, following a general comparison of the characteristics of specific metering elements.

As we have noted, selection of a flow meter depends on factors such as cost, accuracy, need for calibration, and ease of installation and maintenance. Some of these factors are compared for *orifice plate*, *flow nozzle*, and *venturi* meters in Table 8.6. Note that a high head loss means that the running cost of the device is high—it will consume a lot of the fluid energy. A high initial cost must be amortized over the life of the device. This is an example of a common cost calculation for a company (and an individual!)—between a high initial but low running cost, or low initial but high running cost.

Flow meter coefficients reported in the literature have been measured with fully developed turbulent velocity distributions at the meter inlet (Section ①). If a flow meter is to be installed downstream from a valve, elbow, or other disturbance, a straight section of pipe must be placed in front of the meter. Approximately 10 diameters of straight pipe are required for venturi meters, and up to 40 diameters for orifice plate or flow nozzle meters. When a meter has been properly installed, the flow rate may be computed from Eq. 8.54 or 8.56, after choosing an appropriate value for the empirical discharge coefficient, C, or flow coefficient, K, defined in Eqs. 8.53 and 8.55, respectively. Some design data for incompressible flow are given in the next few sections. The same basic methods can be extended to compressible flows, but these will not be treated here. For complete details, see Miller [25] or Bean [26].

Table 8.6

Characteristics of Orifice, Flow Nozzle, and Venturi Flow Meters

Flow Meter Type	Diagram	Head Loss	Initial Cost
Orifice	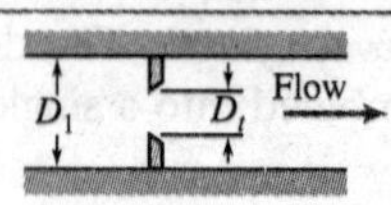	High	Low
Flow Nozzle	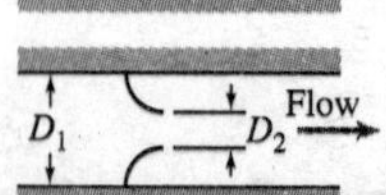	Intermediate	Intermediate
Venturi	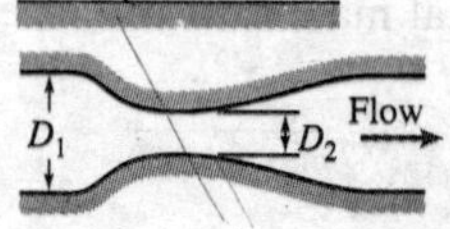	Low	High

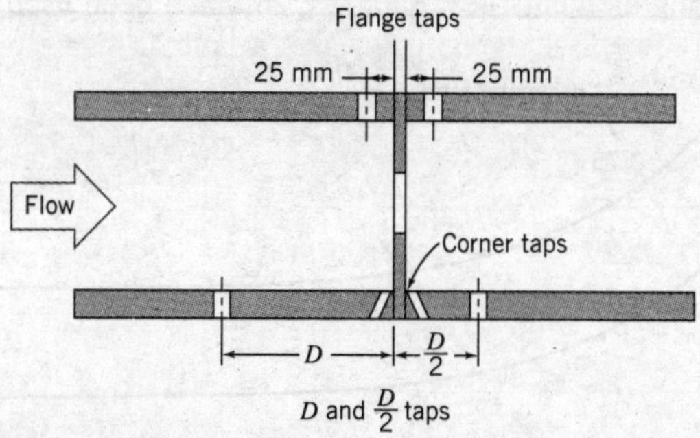

Fig. 8.20 Orifice geometry and pressure tap locations [25].

The Orifice Plate

The orifice plate (Fig. 8.20) is a thin plate that may be clamped between pipe flanges. Since its geometry is simple, it is low in cost and easy to install or replace. The sharp edge of the orifice will not foul with scale or suspended matter. However, suspended matter can build up at the inlet side of a concentric orifice in a horizontal pipe; an eccentric orifice may be placed flush with the bottom of the pipe to avoid this difficulty. The primary disadvantages of the orifice are its limited capacity and the high permanent head loss caused by the uncontrolled expansion downstream from the metering element.

Pressure taps for orifices may be placed in several locations, as shown in Fig. 8.20 (see [25] or [26] for additional details). Since the location of the pressure taps influences the empirically determined flow coefficient, one must select handbook values of C or K consistent with the location of pressure taps.

The correlating equation recommended for a concentric orifice with corner taps [25] is

$$C = 0.5959 + 0.0312\beta^{2.1} - 0.184\beta^{8} + \frac{91.71\beta^{2.5}}{Re_{D_1}^{0.75}} \tag{8.59}$$

Equation 8.59 is the form of Eq. 8.57 for the discharge coefficient C for the orifice plate; it predicts orifice discharge coefficients within ± 0.6 percent for $0.2 < \beta < 0.75$ and for $10^4 < Re_{D_1} < 10^7$. Some flow coefficients calculated from Eq. 8.59 and 8.55 are presented in Fig. 8.21.

A similar correlating equation is available for orifice plates with D and $D/2$ taps. Flange taps require a different correlation for every line size. Pipe taps, located at $2\frac{1}{2}$ and 8 D, no longer are recommended for accurate work.

Example 8.12, which appears later in this section, illustrates the application of flow coefficient data to orifice sizing.

The Flow Nozzle

Flow nozzles may be used as metering elements in either plenums or ducts, as shown in Fig. 8.22; the nozzle section is approximately a quarter ellipse. Design details and recommended locations for pressure taps are given in [26].

The correlating equation recommended for an ASME long-radius flow nozzle [25] is

$$C = 0.9975 - \frac{6.53\beta^{0.5}}{Re_{D_1}^{0.5}} \tag{8.60}$$

Equation 8.60 is the form of Eq. 8.57 for the discharge coefficient C for the flow nozzle; it predicts discharge coefficients for flow nozzles within ± 2.0 percent for

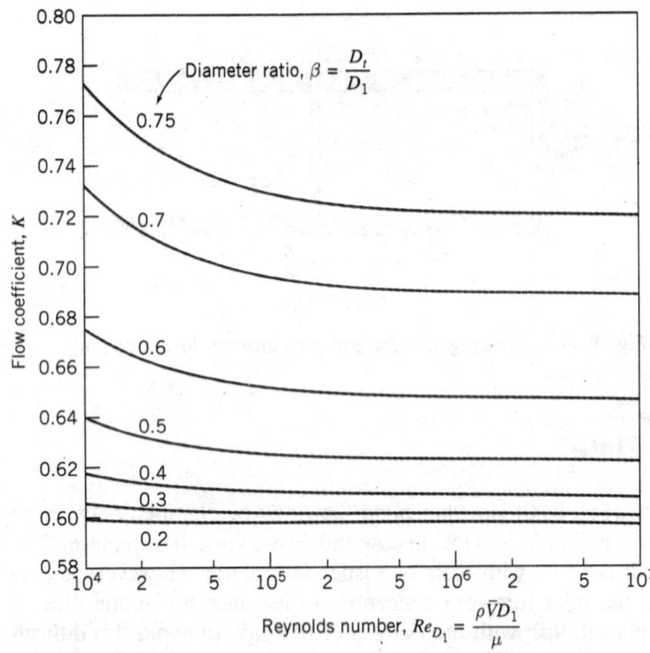

Fig. 8.21 Flow coefficients for concentric orifices with corner taps.

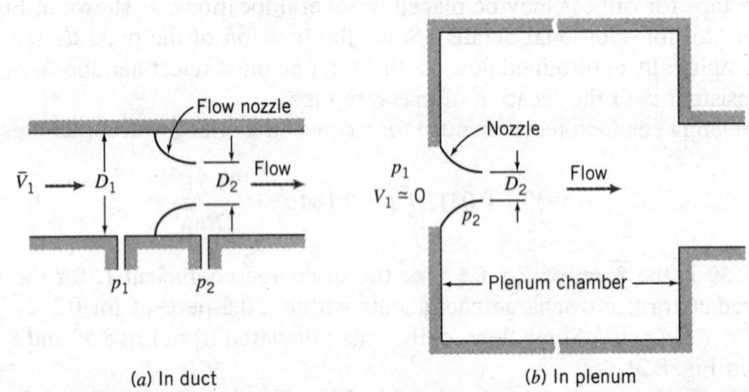

Fig. 8.22 Typical installations of nozzle flow meters.

$0.25 < \beta < 0.75$ for $10^4 < Re_{D_1} < 10^7$. Some flow coefficients calculated from Eq. 8.60 and Eq. 8.55 are presented in Fig. 8.23. (K can be greater than one when the velocity-of-approach factor exceeds one.)

a. Pipe Installation

For pipe installation, K is a function of β and Re_{D_1}. Figure 8.23 shows that K is essentially independent of Reynolds number for $Re_{D_1} > 10^6$. Thus at high flow rates, the flow rate may be computed directly using Eq. 8.56. At lower flow rates, where K is a weak function of Reynolds number, iteration may be required.

b. Plenum Installation

For plenum installation, nozzles may be fabricated from spun aluminum, molded fiberglass, or other inexpensive materials. Thus they are simple and cheap to make

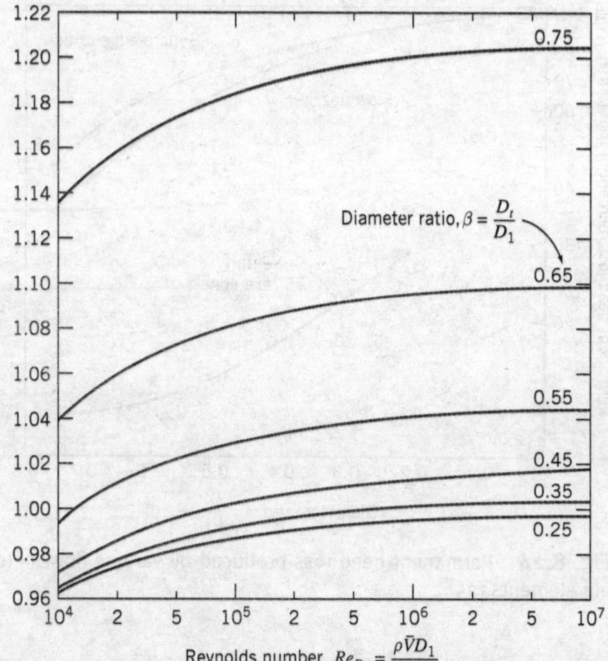

Fig. 8.23 Flow coefficients for ASME long-radius flow nozzles.

and install. Since the plenum pressure is equal to p_2, the location of the downstream pressure tap is not critical. Meters suitable for a wide range of flow rates may be made by installing several nozzles in a plenum. At low flow rates, most of them may be plugged. For higher flow rates, more nozzles may be used.

For plenum nozzles $\beta = 0$, which is outside the range of applicability of Eq. 8.58. Typical flow coefficients are in the range, $0.95 < K < 0.99$; the larger values apply at high Reynolds numbers. Thus the mass rate of flow can be computed within approximately ± 2 percent using Eq. 8.56 with $K = 0.97$.

The Venturi

Venturi meters, as sketched in Table 8.6, are generally made from castings and machined to close tolerances to duplicate the performance of the standard design. As a result, venturi meters are heavy, bulky, and expensive. The conical diffuser section downstream from the throat gives excellent pressure recovery; therefore, overall head loss is low. Venturi meters are also self-cleaning because of their smooth internal contours.

Experimental data show that discharge coefficients for venturi meters range from 0.980 to 0.995 at high Reynolds numbers ($Re_{D_1} > 2 \times 10^5$). Thus $C = 0.99$ can be used to measure mass flow rate within about ± 1 percent at high Reynolds number [25]. Consult manufacturers' literature for specific information at Reynolds numbers below 10^5.

The orifice plate, flow nozzle, and venturi all produce pressure differentials proportional to the square of the flow rate, according to Eq. 8.56. In practice, a meter size must be chosen to accommodate the highest flow rate expected. Because the relationship of pressure drop to flow rate is nonlinear, the range of flow rate that can be measured accurately is limited. Flow meters with single throats usually are considered only for flow rates over a 4:1 range [25].

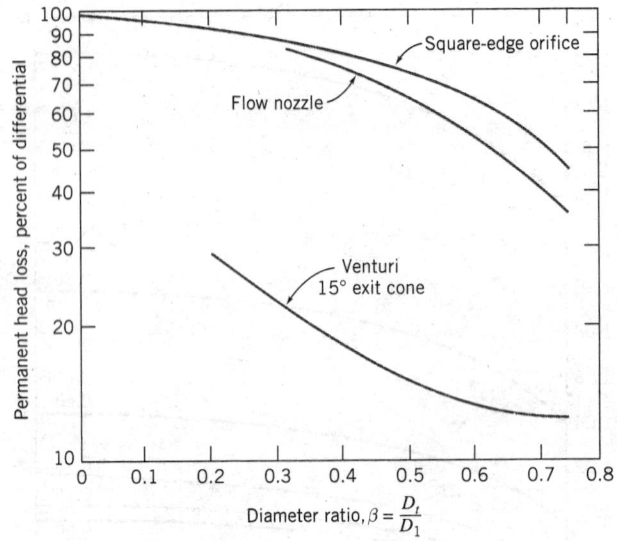

Fig. 8.24 Permanent head loss produced by various flow metering elements [25].

The unrecoverable loss in head across a metering element may be expressed as a fraction of the differential pressure, Δp, across the element. Pressure losses are displayed as functions of diameter ratio in Fig. 8.24 [25]. Note that the venturi meter has a much lower permanent head loss than the orifice (which has the highest loss) or nozzle, confirming the trends we summarized in Table 8.6.

The Laminar Flow Element

The *laminar flow element*[3] is designed to produce a pressure differential directly proportional to flow rate. The idea is that the laminar flow element (LFE) contains a metering section in which the flow passes through a large number of tubes or passages (these often look like a bunch of straws) that are each narrow enough that the flow through them is laminar, regardless of the flow conditions in the main pipe (recall that $Re_{\text{tube}} = \rho V_{\text{tube}} D_{\text{tube}}/\mu$, so if D_{tube} is made small enough we can ensure that $Re_{\text{tube}} < Re_{\text{crit}} \approx 2300$). For each laminar flow tube we can apply the results of Section 8.3, specifically

$$Q_{\text{tube}} = \frac{\pi D_{\text{tube}}^4}{128\mu L_{\text{tube}}}\Delta p \propto \Delta p \tag{8.13c}$$

so the flow rate in each tube is a linear function of the pressure drop across the device. The flow rate in the whole pipe will be the sum of each of these tube flows, and so will also be a linear function of pressure drop. Usually this linear relation is provided in a calibration from the manufacturer, and the LFE can be used over a 10:1 range of flow rates. The relationship between pressure drop and flow rate for laminar flow also depends on viscosity, which is a strong function of temperature. Therefore, the fluid temperature must be known to obtain accurate metering with an LFE.

A laminar flow element costs approximately as much as a venturi, but it is much lighter and smaller. Thus the LFE is becoming widely used in applications where compactness and extended range are important.

[3]Patented and manufactured by Meriam Instrument Co., 10920 Madison Ave., Cleveland, Ohio 44102.

Example 8.12 FLOW THROUGH AN ORIFICE METER

An air flow rate of 1 m^3/s at standard conditions is expected in a 0.25 m diameter duct. An orifice meter is used to measure the rate of flow. The manometer available to make the measurement has a maximum range of 300 mm of water. What diameter orifice plate should be used with corner taps? Analyze the head loss if the flow area at the vena contracta is $A_2 = 0.65\,A_t$. Compare with data from Fig. 8.24.

Given: Flow through duct and orifice as shown.

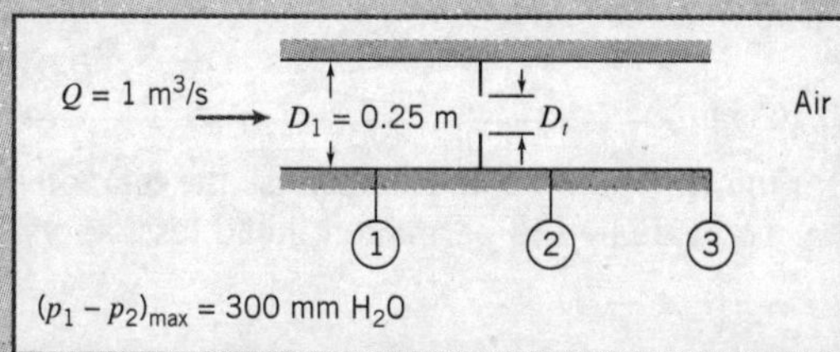

Find: (a) D_t.
(b) Head loss between sections ① and ②.
(c) Degree of agreement with data from Fig. 8.24.

Solution:

The orifice plate may be designed using Eq. 8.56 and data from Fig. 8.21.

Governing equation:

$$\dot{m}_{\text{actual}} = KA_t\sqrt{2\rho(p_1 - p_2)} \tag{8.56}$$

Assumptions: (1) Steady flow.
(2) Incompressible flow.

Since $A_t/A_1 = (D_t/D_1)^2 = \beta^2$,

$$\dot{m}_{\text{actual}} = K\beta^2 A_1\sqrt{2\rho(p_1 - p_2)}$$

or

$$K\beta^2 = \frac{\dot{m}_{\text{actual}}}{A_1\sqrt{2\rho(p_1 - p_2)}} = \frac{\rho Q}{A_1\sqrt{2\rho(p_1 - p_2)}} = \frac{Q}{A_1}\sqrt{\frac{\rho}{2(p_1 - p_2)}}$$

$$= \frac{Q}{A_1}\sqrt{\frac{\rho}{2g\rho_{H_2O}\Delta h}}$$

$$= 1\,\frac{m^3}{s} \times \frac{4}{\pi}\,\frac{1}{(0.25)^2\,m^2}\left[\frac{1}{2} \times 1.23\,\frac{kg}{m^3} \times \frac{s^2}{9.81\,m} \times \frac{m^3}{999\,kg} \times \frac{1}{0.30\,m}\right]^{1/2}$$

$$K\beta^2 = 0.295 \quad \text{or} \quad K = \frac{0.295}{\beta^2} \tag{1}$$

Since K is a function of both β (Eq. 1) and Re_{D_1} (Fig. 8.21), we must iterate to find β. The duct Reynolds number is

$$Re_{D_1} = \frac{\rho \bar{V}_1 D_1}{\mu} = \frac{\rho\,(Q/A_1)\,D_1}{\mu} = \frac{4Q}{\pi \nu D_1}$$

$$Re_{D_1} = \frac{4}{\pi} \times 1\,\frac{m^3}{s} \times \frac{s}{1.46 \times 10^{-5} m^2} \times \frac{1}{0.25 m} = 3.49 \times 10^5$$

Guess $\beta = 0.75$. From Fig. 8.21, K should be 0.72. From Eq. 1,

$$K = \frac{0.295}{(0.75)^2} = 0.524$$

Thus our guess for β is too large. Guess $\beta = 0.70$. From Fig. 8.21, K should be 0.69. From Eq. 1,

$$K = \frac{0.295}{(0.70)^2} = 0.602$$

Thus our guess for β is still too large. Guess $\beta = 0.65$. From Fig. 8.21, K should be 0.67. From Eq. 1,

$$K = \frac{0.295}{(0.65)^2} = 0.698$$

There is satisfactory agreement with $\beta \simeq 0.66$ and

$$D_t = \beta D_1 = 0.66(0.25\,\text{m}) = 0.165\,\text{m} \longleftarrow D_t$$

To find the permanent head loss for this device, we could simply use the diameter ratio $\beta \approx 0.66$ in Fig. 8.24; but instead we will find it from the given data. To evaluate the permanent head loss, apply Eq. 8.29 between sections ① and ③.

Governing equation:

$$\left(\frac{p_1}{\rho} + \alpha_1 \frac{\bar{V}_1^2}{2} + gz_1\right) - \left(\frac{p_3}{\rho} + \alpha_3 \frac{\bar{V}_3^2}{2} + gz_3\right) = h_{l_T} \tag{8.29}$$

Assumptions: (3) $\alpha_1 \bar{V}_1^2 = \alpha_3 \bar{V}_3^2$.
(4) Neglect Δz.

Then

$$h_{l_T} = \frac{p_1 - p_3}{\rho} = \frac{p_1 - p_2 - (p_3 - p_2)}{\rho} \tag{2}$$

Equation 2 indicates our approach: We will find $p_1 - p_3$ by using $p_1 - p_2 = 300\,\text{mm}\,H_2O$, and obtain a value for $p_3 - p_2$ by applying the x component of the momentum equation to a control volume between sections ② and ③.

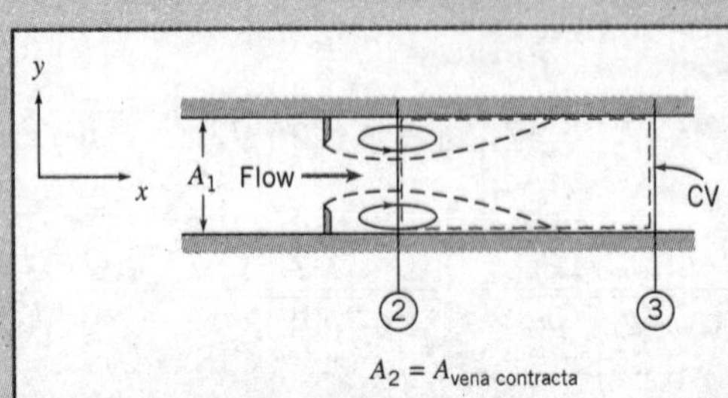

Governing equation:

$$F_{S_x} + F_{B_x} = \frac{\partial}{\partial t}\int_{CV} u\,\rho\,dV + \int_{CS} u\,\rho \vec{V} \cdot d\vec{A} \tag{4.18a}$$

(with $F_{B_x} = 0(5)$ and $\frac{\partial}{\partial t}\int_{CV} = 0(1)$)

Assumptions: (5) $F_{B_x} = 0$
(6) Uniform flow at sections ② and ③.
(7) Pressure uniform across duct at sections ② and ③.
(8) Neglect friction force on CV.

Then, simplifying and rearranging,

$$(p_2 - p_3)A_1 = u_2(-\rho \bar{V}_2 A_2) + u_3(\rho \bar{V}_3 A_3) = (u_3 - u_2)\rho Q = (\bar{V}_3 - \bar{V}_2)\rho Q$$

or

$$p_3 - p_2 = (\bar{V}_2 - \bar{V}_3)\frac{\rho Q}{A_1}$$

Now $\overline{V}_3 = Q/A_1$, and

$$\overline{V}_2 = \frac{Q}{A_2} = \frac{Q}{0.65\,A_t} = \frac{Q}{0.65\beta^2\,A_1}$$

Thus,

$$p_3 - p_2 = \frac{\rho Q^2}{A_1^2}\left[\frac{1}{0.65\,\beta^2} - 1\right]$$

$$p_3 - p_2 = 1.23\frac{\text{kg}}{\text{m}^3} \times (1)^2\frac{\text{m}^6}{\text{s}^2} \times \frac{4^2}{\pi^2}\frac{1}{(0.25)^4\,\text{m}^4}\left[\frac{1}{0.65(0.66)^2} - 1\right]\frac{\text{N}\cdot\text{s}^2}{\text{kg}\cdot\text{m}}$$

$$p_3 - p_2 = 1290\,\text{N/m}^2$$

The diameter ratio, β, was selected to give maximum manometer deflection at maximum flow rate. Thus

$$p_1 - p_2 = \rho_{\text{H}_2\text{O}}g\Delta h = 999\frac{\text{kg}}{\text{m}^3} \times 9.81\frac{\text{m}}{\text{s}^2} \times 0.30\,\text{m} \times \frac{\text{N}\cdot\text{s}^2}{\text{kg}\cdot\text{m}} = 2940\,\text{N/m}^2$$

Substituting into Eq. 2 gives

$$h_{l_T} = \frac{p_1 - p_3}{\rho} = \frac{p_1 - p_2 - (p_3 - p_2)}{\rho}$$

$$h_{l_T} = (2940 - 1290)\frac{\text{N}}{\text{m}^2} \times \frac{\text{m}^3}{1.23\,\text{kg}} = 1340\,\text{N}\cdot\text{m/kg}$$ h_{l_T}

To compare with Fig. 8.24, express the permanent pressure loss as a fraction of the meter differential

$$\frac{p_1 - p_3}{p_1 - p_2} = \frac{(2940 - 1290)\,\text{N/m}^2}{2940\,\text{N/m}^2} = 0.561$$

The fraction from Fig. 8.24 is about 0.57. This is satisfactory agreement!

This problem illustrates flow meter calculations and shows use of the momentum equation to compute the pressure rise in a sudden expansion.

Linear Flow Meters 8.11

The disadvantage of restriction flow meters (except the LFE) is that the measured output (Δp) is not linear with the flow rate Q. Several flow meter types produce outputs that are directly proportional to flow rate. These meters produce signals without the need to measure differential pressure. The most common linear flow meters are discussed briefly in the following paragraphs.

Float meters may be used to indicate flow rate directly for liquids or gases. An example is shown in Fig. 8.25 In operation, the ball or float is carried upward in the tapered clear tube by the flowing fluid until the drag force and float weight are in equilibrium. Such meters (often called *rotameters*) are available with factory calibration for a number of common fluids and flow rate ranges.

A free-running vaned impeller may be mounted in a cylindrical section of tube (Fig. 8.26) to make a *turbine flow meter*. With proper design, the rate of rotation of the impeller may be made closely proportional to volume flow rate over a wide range.

Rotational speed of the turbine element can be sensed using a magnetic or modulated carrier pickup external to the meter. This sensing method therefore requires no penetrations or seals in the duct. Thus turbine flow meters can be used

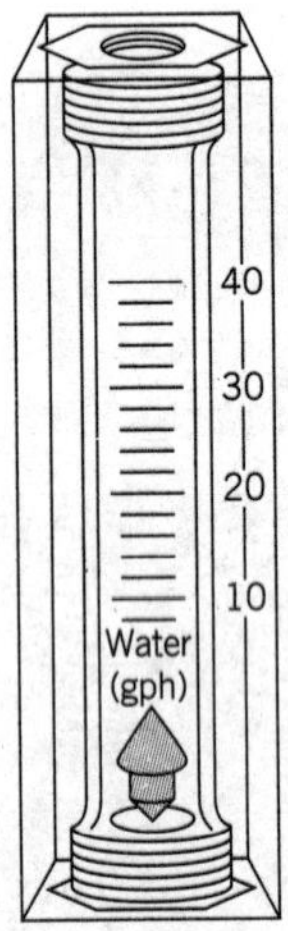

Fig. 8.25 Float-type variable-area flow meter. (Courtesy of Dwyer Instrument Co., Michigan City, Indiana.)

safely to measure flow rates in corrosive or toxic fluids. The electrical signal can be displayed, recorded, or integrated to provide total flow information.

An interesting device is the *vortex flow meter*. This device takes advantage of the fact that a uniform flow will generate a vortex street when it encounters a bluff body such as a cylinder perpendicular to the flow. A vortex street is a series of alternating vortices shed from the rear of the body; the alternation generates an oscillating sideways force on, and therefore oscillation of, the cylinder (the classic example of this being the "singing" of telephone wires in a high wind). It turns out that the dimensionless group characterizing this phenomenon is the Strouhal number, $St = fL/V$ (f is the vortex shedding frequency, L is the cylinder diameter, and V is the freestream velocity), and it is approximately constant ($St \approx 0.21$). Hence we have a device for which $V \propto f$. Measurement of f thus directly indicates the velocity $\vec{V}$ (however, the velocity profile does affect the shedding frequency so calibration is required). The cylinder used in a flow meter is usually quite short in length—10 mm or less—and placed perpendicular to the flow (and for some devices is not a cylinder at all but some other small bluff object). The oscillation can be measured using a strain gage or other sensor. Vortex flow meters can be used over a 20:1 range of flow rates [25].

The *electromagnetic flow meter* uses the principle of magnetic induction. A magnetic field is created across a pipe. When a conductive fluid passes through the field, a voltage is generated at right angles to the field and velocity vectors. Electrodes placed on a pipe diameter are used to detect the resulting signal voltage. The signal voltage is proportional to the average axial velocity when the profile is axisymmetric.

Magnetic flow meters may be used with liquids that have electrical conductivities above 100 microsiemens per meter (1 siemen = 1 ampere per volt). The minimum flow speed should be above about 0.3 m/s, but there are no restrictions on Reynolds number. The flow rate range normally quoted is 10:1 [25].

Ultrasonic flow meters also respond to average velocity at a pipe cross section. Two principal types of ultrasonic meters are common: Propagation time is measured for clean liquids, and reflection frequency shift (Doppler effect) is measured for flows carrying particulates. The speed of an acoustic wave increases in the flow direction and decreases when transmitted against the flow. For clean liquids, an acoustic path inclined to the pipe axis is used to infer flow velocity. Multiple paths are used to estimate the volume flow rate accurately.

Doppler effect ultrasonic flow meters depend on reflection of sonic waves (in the MHz range) from scattering particles in the fluid. When the particles move at flow speed, the frequency shift is proportional to flow speed; for a suitably chosen path, output is proportional to volume flow rate. One or two transducers may be used; the meter may be clamped to the outside of the pipe. Ultrasonic meters may require calibration in place. Flow rate range is 10:1 [25].

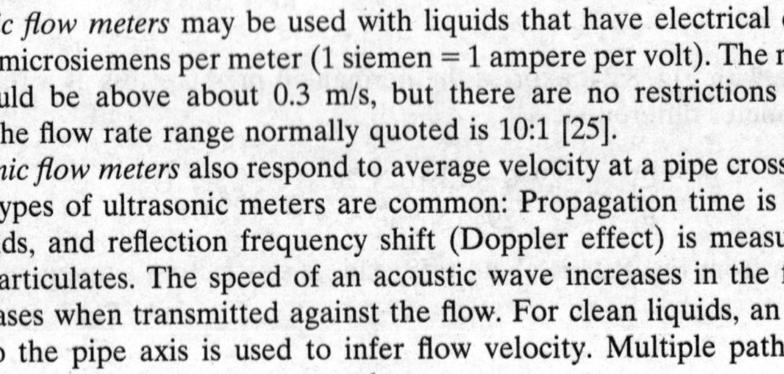

Fig. 8.26 Turbine flow meter. (Courtesy of Potter Aeronautical Corp., Union, New Jersey.)

Traversing Methods 8.12

In situations such as in air handling or refrigeration equipment, it may be impractical or impossible to install fixed flow meters. In such cases it may be possible to obtain flow rate data using traversing techniques.

To make a flow rate measurement by traverse, the duct cross section is conceptually subdivided into segments of equal area. The velocity is measured at the center of each area segment using a pitot tube, a total head tube, or a suitable anemometer. The volume flow rate for each segment is approximated by the product of the measured velocity and the segment area. The flow rate through the entire duct is the sum of these segmental flow rates. Details of recommended procedures for flow rate measurements by the traverse method are given in [27].

Use of *pitot* or *pitot-static tubes* for traverse measurements requires direct access to the flow field. Pitot tubes give uncertain results when pressure gradients or streamline curvature are present, and their response times are slow. Two types of anemometers—*thermal anemometers* and *laser Doppler anemometers* (LDAs)—overcome these difficulties partially, although they introduce new complications.

Thermal anemometers use tiny elements (either hot-wire or hot-film elements) that are heated electrically. Sophisticated electronic feedback circuits are used to maintain the temperature of the element constant and to sense the input heating rate needed to do this. The heating rate is related to the local flow velocity by calibration (a higher velocity leads to more heat transfer). The primary advantage of thermal anemometers is the small size of the sensing element. Sensors as small as 0.002 mm in diameter and 0.1 mm long are available commercially. Because the thermal mass of such tiny elements is extremely small, their response to fluctuations in flow velocity is rapid. Frequency responses to the 50 kHz range have been quoted [28]. Thus thermal anemometers are ideal for measuring turbulence quantities. Insulating coatings may be applied to permit their use in conductive or corrosive gases or liquids.

Because of their fast response and small size, thermal anemometers are used extensively for research. Numerous schemes have been published for treating the resulting data [29]. Digital processing techniques, including fast Fourier transforms, can be applied to the signals to obtain mean values and moments, and to analyze frequency content and correlations.

VIDEO

Flow Visualization: Laser Induced Fluorescence.

VIDEO

Laser Doppler Anemometry (Animation).

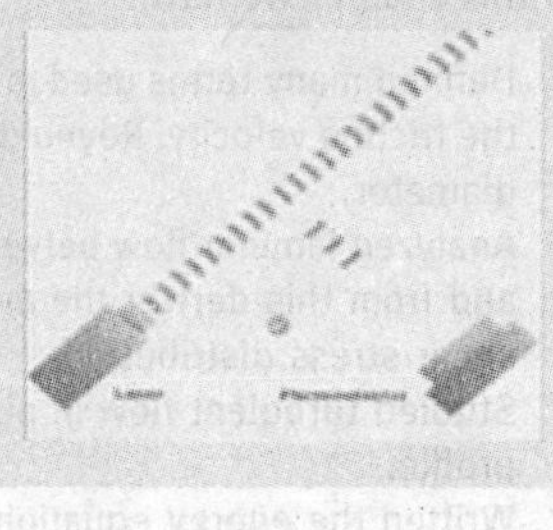

Fig. 8.27 A 2-component Laser Doppler Anemometer Probe Volume. (Courtesy Dr. Frank W. Chambers, Oklahoma State University)

Laser Doppler anemometers are becoming widely used for specialized applications where direct physical access to the flow field is difficult or impossible. One or more laser beams are focused to a small volume in the flow at the location of interest (as shown in Fig 8.27). Laser light is scattered from particles that are present in the flow (dust or particulates) or introduced for this purpose. A frequency shift is caused by the local flow speed (Doppler effect). Scattered light and a reference beam are collected by receiving optics. The frequency shift is proportional to the flow speed; this relationship may be calculated, so there is no need for calibration. Since velocity is measured directly, the signal is unaffected by changes in temperature, density, or composition in the flow field. The primary disadvantages of LDAs are that the optical equipment is expensive and fragile, and that extremely careful alignment is needed (as the authors can attest).

8.13 Summary and Useful Equations

In this chapter we have:

- ✓ Defined many terms used in the study of internal incompressible viscous flow, such as: the entrance length, fully developed flow, the friction velocity, Reynolds stress, the kinetic energy coefficient, the friction factor, major and minor head losses, and hydraulic diameter.
- ✓ Analyzed laminar flow between parallel plates and in pipes and observed that we can obtain the velocity distribution analytically, and from this derive: the average velocity, the maximum velocity and its location, the flow rate, the wall shear stress, and the shear stress distribution.
- ✓ Studied turbulent flow in pipes and ducts and learned that semi-empirical approaches are needed, e.g., the power-law velocity profile.
- ✓ Written the energy equation in a form useful for analyzing pipe flows.
- ✓ Discussed how to incorporate pumps, fans, and blowers into a pipe flow analysis.
- ✓ Described various flow measurement devices: direct measurement, restriction devices (orifice plate, nozzle, and venturi), linear flow meters (rotameters, various electromagnetic or acoustic devices, and the vortex flow meter), and traversing devices (pitot tubes and laser-Doppler anemometers).

We have learned that pipe and duct flow problems often need iterative solution—the flow rate Q is not a linear function of the driving force (usually Δp), except for laminar flows (which are not common in practice). *We have also seen that pipe networks can be analyzed using the same techniques as a single-pipe system, with the addition of a few basic rules, and that in practice a computer application such as *Excel* is needed to solve all but the simplest networks.

Note: Most of the Useful Equations in the table below have a number of constraints or limitations—*be sure to refer to their page numbers for details*!

Useful Equations

Velocity profile for pressure-driven laminar flow between stationary parallel plates:	$u = \frac{a^2}{2\mu}\left(\frac{\partial p}{\partial x}\right)\left[\left(\frac{y}{a}\right)^2 - \left(\frac{y}{a}\right)\right]$	**(8.5)**	Page 335
Flow rate for pressure-driven laminar flow between stationary parallel plates:	$\frac{Q}{l} = -\frac{1}{12\mu}\left[\frac{-\Delta p}{L}\right]a^3 = \frac{a^3\,\Delta p}{12\mu\, L}$	**(8.6c)**	Page 336

*This topic applies to a section that may be omitted without loss of continuity in the text material.

Velocity profile for pressure-driven laminar flow between stationary parallel plates (centered coordinates):	$u = \frac{a^2}{2\mu}\left(\frac{\partial p}{\partial x}\right)\left[\left(\frac{y'}{a}\right)^2 - \frac{1}{4}\right]$	**(8.7)**	Page 337
Velocity profile for pressure-driven laminar flow between parallel plates (upper plate moving):	$u = \frac{Uy}{a} + \frac{a^2}{2\mu}\left(\frac{\partial p}{\partial x}\right)\left[\left(\frac{y}{a}\right)^2 - \left(\frac{y}{a}\right)\right]$	**(8.8)**	Page 339
Flow rate for pressure-driven laminar flow between parallel plates (upper plate moving):	$\frac{Q}{l} = \frac{Ua}{2} - \frac{1}{12\mu}\left(\frac{\partial p}{\partial x}\right)a^3$	**(8.9b)**	Page 340
Velocity profile for laminar flow in a pipe:	$u = -\frac{R^2}{4\mu}\left(\frac{\partial p}{\partial x}\right)\left[1 - \left(\frac{r}{R}\right)^2\right]$	**(8.12)**	Page 346
Flow rate for laminar flow in a pipe:	$Q = -\frac{\pi R^4}{8\mu}\left[\frac{-\Delta p}{L}\right] = \frac{\pi \Delta p R^4}{8\mu L} = \frac{\pi \Delta p D^4}{128\mu L}$	**(8.13c)**	Page 347
Velocity profile for laminar flow in a pipe (normalized form):	$\frac{u}{U} = 1 - \left(\frac{r}{R}\right)^2$	**(8.14)**	Page 347
Velocity profile for turbulent flow in a smooth pipe (power-law equation):	$\frac{\bar{u}}{U} = \left(\frac{y}{R}\right)^{1/n} = \left(1 - \frac{r}{R}\right)^{1/n}$	**(8.22)**	Page 352
Head loss equation:	$\left(\frac{p_1}{\rho} + \alpha_1 \frac{\overline{V}_1^2}{2} + gz_1\right) - \left(\frac{p_2}{\rho} + \alpha_2 \frac{\overline{V}_2^2}{2} + gz_2\right) = h_{l_T}$	**(8.29)**	Page 356
Major head loss equation:	$h_l = f\frac{L}{D}\frac{\overline{V}^2}{2}$	**(8.34)**	Page 358
Friction factor (laminar flow):	$f_{\text{laminar}} = \frac{64}{Re}$	**(8.36)**	Page 360
Friction factor (turbulent flow—Colebrook equation):	$\frac{1}{\sqrt{f}} = -2.0\log\left(\frac{e/D}{3.7} + \frac{2.51}{Re\sqrt{f}}\right)$	**(8.37)**	Page 360
Minor loss using loss coefficient K:	$h_{l_m} = K\frac{\overline{V}^2}{2}$	**(8.40a)**	Page 361
Minor loss using equivalent length L_e:	$h_{l_m} = f\frac{L_e}{D}\frac{\overline{V}^2}{2}$	**(8.40b)**	Page 362
Diffuser pressure recovery coefficient:	$C_p \equiv \frac{p_2 - p_1}{\frac{1}{2}\rho\overline{V}_1^2}$	**(8.41)**	Page 363
Ideal diffuser pressure recovery coefficient:	$C_{p_i} = 1 - \frac{1}{AR^2}$	**(8.42)**	Page 364
Head loss in diffuser in terms of pressure recovery coefficients:	$h_{l_m} = (C_{p_i} - C_p)\frac{\overline{V}_1^2}{2}$	**(8.44)**	Page 365
Pump work:	$\dot{W}_{\text{pump}} = Q\Delta p_{\text{pump}}$	**(8.47)**	Page 368

Pump efficiency:	$$\eta = \frac{\dot{W}_{\text{pump}}}{\dot{W}_{\text{in}}}$$	**(8.48)**	Page 368
Hydraulic diameter:	$$D_h \equiv \frac{4A}{P}$$	**(8.50)**	Page 368
Mass flow rate equation for a flow meter (in terms of discharge coefficient C):	$$\dot{m}_{\text{actual}} = \frac{CA_t}{\sqrt{1-\beta^4}}\sqrt{2\rho(p_1 - p_2)}$$	**(8.54)**	Page 389
Mass flow rate equation for a flow meter (in terms of flow coefficient K):	$$\dot{m}_{\text{actual}} = KA_t\sqrt{2\rho(p_1 - p_2)}$$	**(8.56)**	Page 389
Discharge coefficient (as a function of Re):	$$C = C_\infty + \frac{b}{Re^n_{D_1}}$$	**(8.57)**	Page 390
Flow coefficient (as a function of Re):	$$K = K_\infty + \frac{1}{\sqrt{1-\beta^4}}\frac{b}{Re^n_{D_1}}$$	**(8.58)**	Page 390

References

1. Streeter, V. L., ed., *Handbook of Fluid Dynamics*. New York: McGraw-Hill, 1961.
2. Rouse, H., and S. Ince, *History of Hydraulics*. New York: Dover, 1957.
3. Moin, P., and J. Kim, "Tackling Turbulence with Supercomputers," *Scientific American, 276*, 1, January 1997, pp. 62–68.
4. Panton, R. L., *Incompressible Flow*, 2nd ed. New York: Wiley, 1996.
5. Laufer, J., "The Structure of Turbulence in Fully Developed Pipe Flow," U.S. National Advisory Committee for Aeronautics (NACA), Technical Report 1174, 1954.
6. Tennekes, H., and J. L. Lumley, *A First Course in Turbulence*. Cambridge, MA: The MIT Press, 1972.
7. Hinze, J. O., *Turbulence*, 2nd ed. New York: McGraw-Hill, 1975.
8. Moody, L. F., "Friction Factors for Pipe Flow," *Transactions of the ASME, 66*, 8, November 1944. pp. 671–684.
9. Colebrook, C. F., "Turbulent Flow in Pipes, with Particular Reference to the Transition Region between the Smooth and Rough Pipe Laws," *Journal of the Institution of Civil Engineers, London, 11*, 1938–39, pp. 133–156.
10. Haaland, S. E., "Simple and Explicit Formulas for the Friction Factor in Turbulent Flow," *Transactions of ASME, Journal of Fluids Engineering, 103*, 1983, pp. 89–90.
11. "Flow of Fluids Through Valves, Fittings, and Pipe," New York: Crane Company, Technical Paper No. 410, 1982.
12. *ASHRAE Handbook—Fundamentals*. Atlanta, GA: American Society of Heating, Refrigerating, and Air Conditioning Engineers, Inc., 1981.
13. Cockrell, D. J., and C. I. Bradley, "The Response of Diffusers to Flow Conditions at Their Inlet," Paper No. 5, *Symposium on Internal Flows*, University of Salford, Salford, England, April 1971. pp. A32–A41.
14. Sovran, G., and E. D. Klomp, "Experimentally Determined Optimum Geometries for Rectilinear Diffusers with Rectangular, Conical, or Annular Cross-Sections," in *Fluid Mechanics of Internal Flows*, G. Sovran, ed. Amsterdam: Elsevier, 1967.
15. Feiereisen, W. J., R. W. Fox, and A. T. McDonald, "An Experimental Investigation of Incompressible Flow Without Swirl in R-Radial Diffusers," *Proceedings, Second International Japan Society of Mechanical Engineers Symposium on Fluid Machinery and Fluidics*, Tokyo, Japan, September 4–9, 1972. pp. 81–90.
16. McDonald, A. T., and R. W. Fox, "An Experimental Investigation of Incompressible Flow in Conical Diffusers," *International Journal of Mechanical Sciences, 8*, 2, February 1966. pp. 125–139.
17. Runstadler, P. W., Jr., "Diffuser Data Book," Hanover, NH: Creare, Inc., Technical Note 186, 1975.
18. Reneau, L. R., J. P. Johnston, and S. J. Kline, "Performance and Design of Straight, Two-Dimensional Diffusers," *Transactions of the ASME, Journal of Basic Engineering, 89D*, 1, March 1967. pp. 141–150.
19. *Aerospace Applied Thermodynamics Manual*. New York: Society of Automotive Engineers, 1969.
20. Daily, J. W., and D. R. F. Harleman, *Fluid Dynamics*. Reading, MA: Addison-Wesley, 1966.
21. White, F. M., *Fluid Mechanics*, 6th ed. New York: McGraw-Hill, 2007.
22. Hamilton, J. B., "The Suppression of Intake Losses by Various Degrees of Rounding," University of Washington, Seattle, WA, Experiment Station Bulletin 51, 1929.
23. Herschel, C., *The Two Books on the Water Supply of the City of Rome, from Sextus Julius Frontinus (ca. 40–103 A.D.)*. Boston, 1899.

24. Lam, C. F., and M. L. Wolla, "Computer Analysis of Water Distribution Systems: Part 1, Formulation of Equations," *Proceedings of the ASCE, Journal of the Hydraulics Division, 98*, HY2, February 1972. pp. 335–344.
25. Miller, R. W., *Flow Measurement Engineering Handbook*. 3rd ed. New York: McGraw Hill, 1996.
26. Bean, H. S., ed., *Fluid Meters, Their Theory and Application*. New York: American Society of Mechanical Engineers, 1971.
27. ISO 7145, *Determination of Flowrate of Fluids in Closed Conduits or Circular Cross Sections—Method of Velocity Determination at One Point in the Cross Section*, ISO UDC 532.57.082.25:532.542, 1st ed. Geneva: International Standards Organization, 1982.
28. Goldstein, R. J., ed., *Fluid Mechanics Measurements*, 2nd ed. Washington, D.C.: Taylor & Francis, 1996.
29. Bruun, H. H., *Hot-Wire Anemometry—Principles and Signal Analysis*. New York: Oxford University Press, 1995.
30. Bruus, H., *Theoretical Microfluidics* (Oxford University Press, 2007).
31. Swamee, P. K., and A. K. Jain, "Explicit Equations for Pipe-Flow Problems," *Proceedings of the ASCE, Journal of the Hydraulics Division, 102*, HY5, May 1976. pp. 657–664.
32. Potter, M. C., and J. F. Foss, *Fluid Mechanics*. New York: Ronald, 1975.

Case Study

The Fountains at the Bellagio in Las Vegas

The fountains at the Bellagio in Las Vegas.

Any visitor to Las Vegas will be familiar with the water fountains at the Bellagio hotel. These are a set of high-powered water jets designed and built by the WET Design Company that are choreographed to vary in their strength and direction to selected pieces of music.

WET developed many innovations to make the fountains. Traditional fountains use pumps and pipes, which must be matched for optimum flow (one of the topics we discussed in this chapter). Many of WET's designs use compressed air instead of water pumps, which allows energy to be continuously generated and accumulated, ready for instant output. This innovative use of compressed air allowed the fountains to become a reality—with the traditional systems of pipes or pumps, a fountain such as the Bellagio's would be impractical and expensive. For example, it would be difficult to obtain the 73 m heights the fountains achieve without expensive, large, and noisy water pumps. The "Shooter" that WET developed works on the principle of introducing a large bubble of compressed air into the piping, which forces trapped water through a nozzle at high pressure. The ones installed at the Bellagio are able to shoot about 284 liters per second of water over 73 m in the air. In addition to providing a spectacular effect, they require only about 1/10th the energy of traditional water pumps to produce the same effect. Other air-powered devices produce pulsing water jets, achieving a maximum height of 38 m. In addition to their power, these innovations lead to a saving of 80 percent or more in energy costs and have project construction costs that are about 50 percent less than traditional pipe-pump fountains.

Problems

Laminar versus Turbulent Flow

8.1 Air at 100°C enters a 125 mm diameter duct. Find the volume flow rate at which the flow becomes turbulent. At this flow rate, estimate the entrance length required to establish fully developed flow.

8.2 Consider incompressible flow in a circular channel. Derive general expressions for Reynolds number in terms of (a) volume flow rate and tube diameter and (b) mass flow rate and tube diameter. The Reynolds number is 1800 in a section where the tube diameter is 10 mm. Find the Reynolds number for the same flow rate in a section where the tube diameter is 6 mm.

8.3 Air at 40°C flows in a pipe system in which diameter is decreased in two stages from 25 mm to 15 mm to 10 mm. Each section is 2 m long. As the flow rate is increased, which section will become turbulent first? Determine the flow rates at which one, two, and then all three sections first become turbulent. At each of these flow rates, determine which sections, if any, attain fully developed flow.

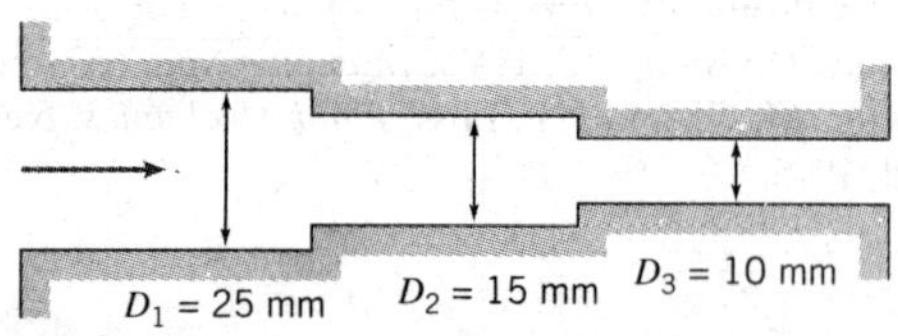

P8.3

8.4 For flow in circular tubes, transition to turbulence usually occurs around $Re \approx 2300$. Investigate the circumstances under which the flows of (a) standard air and (b) water at 15°C become turbulent. On log-log graphs, plot: the average velocity, the volume flow rate, and the mass flow rate, at which turbulence first occurs, as functions of tube diameter.

Laminar Flow between Parallel Plates

8.5 For the laminar flow in the section of pipe shown in Fig. 8.1, sketch the expected wall shear stress, pressure, and centerline velocity as functions of distance along the pipe. Explain significant features of the plots, comparing them with fully developed flow. Can the Bernoulli equation be applied anywhere in the flow field? If so, where? Explain briefly.

8.6 An incompressible fluid flows between two infinite stationary parallel plates. The velocity profile is given by $u = u_{\text{max}}(Ay^2 + By + C)$, where A, B, and C are constants and y is measured upward from the lower plate. The total gap width is h units. Use appropriate boundary conditions to express the magnitude and units of the constants in terms of h. Develop an expression for volume flow rate per unit depth and evaluate the ratio $\overline{V}/u_{\text{max}}$.

8.7 The velocity profile for fully developed flow between stationary parallel plates is given by $u = a(h^2/4 - y^2)$, where a is a constant, h is the total gap width between plates, and y is the distance measured from the center of the gap. Determine the ratio $\overline{V}/u_{\text{max}}$.

8.8 A fluid flows steadily between two parallel plates. The flow is fully developed and laminar. The distance between the plates is h.

(a) Derive an equation for the shear stress as a function of y. Sketch this function.

(b) For μ=1.15 N · s/m², $\partial p/\partial x$=−58 Pa/m, and h=1.3 mm, calculate the maximum shear stress, in Pa.

8.9 Oil is confined in a 100 mm diameter cylinder by a piston having a radial clearance of 0.025 mm and a length of 50 mm. A steady force of 20,000 N is applied to the piston. Assume the properties of SAE 30 oil at 49°C. Estimate the rate at which oil leaks past the piston.

8.10 A viscous oil flows steadily between stationary parallel plates. The flow is laminar and fully developed. The total gap width between the plates is h = 5 mm. The oil viscosity is 0.5 N · s/m² and the pressure gradient is −1000 N/m²/m. Find the magnitude and direction of the shear stress on the upper plate and the volume flow rate through the channel, per meter of width.

8.11 Viscous oil flows steadily between parallel plates. The flow is fully developed and laminar. The pressure gradient is 1.25 kPa/m and the channel half-width is h = 1.5 mm. Calculate the magnitude and direction of the wall shear stress at the upper plate surface. Find the volume flow rate through the channel (μ = 0.50 N · s/m²).

8.12 A large mass is supported by a piston of diameter D = 100 mm and length L = 100 mm. The piston sits in a cylinder closed at the bottom, and the gap a = 0.025 mm between the cylinder wall and piston is filled with SAE 10 oil at 20°C. The piston slowly sinks due to the mass, and oil is forced out at a rate of 6×10^{-6} m³/s What is the mass (kg)?

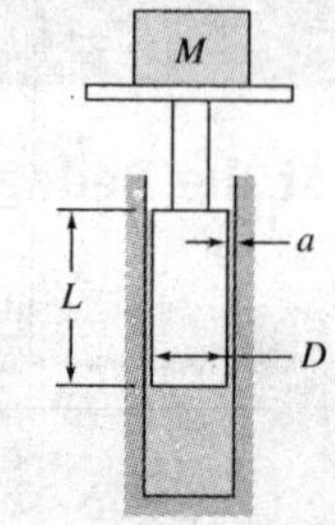

P8.12, P8.16

8.13 A high pressure in a system is created by a small piston-cylinder assembly. The piston diameter is 6 mm and it extends 50 mm into the cylinder. The radial clearance between the piston and cylinder is 0.002 mm. Neglect elastic deformations of the piston and cylinder caused by pressure. Assume the fluid properties are those of SAE 10W oil at 35°C. When the pressure in the cylinder is 600 MPa, estimate the leakage rate.

8.14 A hydraulic jack supports a load of 9000 kg. The following data are given:

Diameter of piston	100 mm
Radial clearance between piston and cylinder	0.05 mm
Length of piston	120 mm

Estimate the rate of leakage of hydraulic fluid past the piston, assuming the fluid is SAE 30 oil at 30°C.

8.15 A hydrostatic bearing is to support a load of 50,000 N/m of length perpendicular to the diagram. The bearing is supplied with SAE 10W-30 oil at 35°C and 700 kPa through the central slit. Since the oil is viscous and the gap is small, the flow may be considered fully developed. Calculate (a) the required width of the bearing pad, (b) the resulting pressure gradient, dp/dx, and (c) the gap height, if the flow rate is Q = 1 mL/min/m.

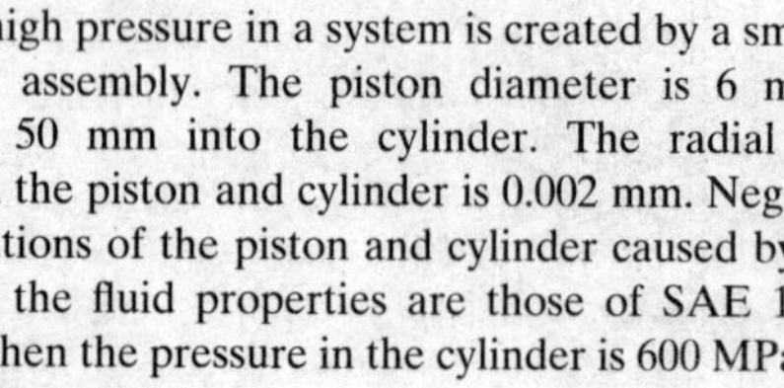

P8.15

8.16 The basic component of a pressure gage tester consists of a piston-cylinder apparatus as shown. The piston, 6 mm in diameter, is loaded to develop a pressure of known magnitude. (The piston length is 25 mm.) Calculate the mass, M, required to produce 1.5 MPa (gage) in the cylinder. Determine the leakage flow rate as a function of radial clearance, a, for this load if the liquid is SAE 30 oil at 20°C. Specify the maximum allowable radial clearance so the vertical movement of the piston due to leakage will be less than 1 mm/min.

8.17 In Section 8.2 we derived the velocity profile between parallel plates (Eq. 8.5) by using a differential control volume. Instead, following the procedure we used in Example 5.9, derive Eq. 8.5 by starting with the Navier–Stokes equations (Eqs. 5.27). Be sure to state all assumptions.

8.18 Consider the simple power-law model for a non-Newtonian fluid given by Eq. 2.16. Extend the analysis of Section 8.2 to show that the velocity profile for fully developed laminar flow of a power-law fluid between stationary parallel plates separated by distance $2h$ may be written

$$u = \left(\frac{h}{k}\frac{\Delta p}{L}\right)^{1/n} \frac{nh}{n+1}\left[1 - \left(\frac{y}{h}\right)^{(n+1)/n}\right]$$

where y is the coordinate measured from the channel centerline. Plot the profiles u/U_{max} versus y/h for n = 0.7, 1.0, and 1.3.

8.19 Viscous liquid, at volume flow rate Q, is pumped through the central opening into the narrow gap between the parallel disks shown. The flow rate is low, so the flow is laminar, and the pressure gradient due to convective acceleration in the gap is negligible compared with the gradient caused by viscous forces (this is termed *creeping flow*). Obtain a general expression for the variation of average velocity in the gap between the disks. For creeping flow, the velocity profile at any cross section in the gap is the same as for fully developed flow between stationary parallel plates. Evaluate the pressure gradient, dp/dr, as a function of radius. Obtain an expression for $p(r)$. Show that the net force required to hold the upper plate in the position shown is

$$F = \frac{3\mu Q R^2}{h^3}\left[1 - \left(\frac{R_0}{R}\right)^2\right]$$

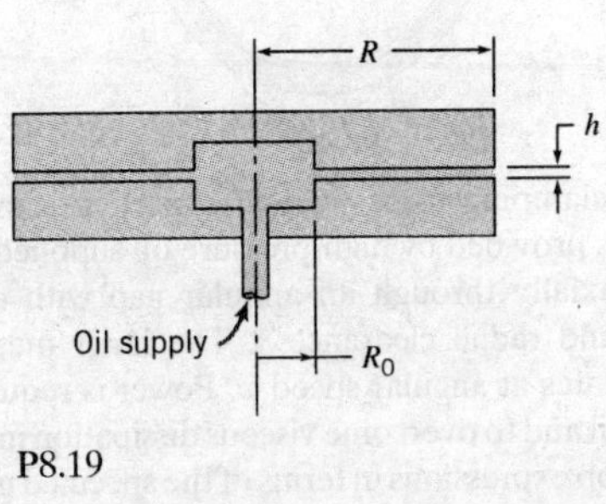

P8.19

8.20 A sealed journal bearing is formed from concentric cylinders. The inner and outer radii are 25 and 26 mm, the journal length is 100 mm, and it turns at 2800 rpm. The gap is filled with oil in laminar motion. The velocity profile is linear across the gap. The torque needed to turn the journal is 0.2 N·m. Calculate the viscosity of the oil. Will the torque increase or decrease with time? Why?

8.21 Using the profile of Problem 8.18, show that the flow rate for fully developed laminar flow of a power-law fluid between stationary parallel plates may be written as

$$Q = \left(\frac{h}{k}\frac{\Delta p}{L}\right)^{1/n} \frac{2nwh^2}{2n+1}$$

Here w is the plate width. In such an experimental setup the following data on applied pressure difference Δp and flow rate Q were obtained:

Δp (kPa)	10	20	30	40	50	60	70	80	90	100
Q (L/min)	0.451	0.759	1.01	1.15	1.41	1.57	1.66	1.85	2.05	2.25

Determine if the fluid is pseudoplastic or dilatant, and obtain an experimental value for n.

8.22 Consider fully developed laminar flow between infinite parallel plates separated by gap width d = 10 mm. The upper plate moves to the right with speed U_2 = 0.5 m/s; the lower plate moves to the left with speed U_1 = 0.25 m/s. The pressure gradient in the direction of flow is zero. Develop an expression for the velocity distribution in the gap. Find the volume flow rate per unit depth (m³/sec/m) passing a given cross section.

8.23 Water at 60°C flows between two large flat plates. The lower plate moves to the left at a speed of 0.3 m/s; the upper plate is stationary. The plate spacing is 3 mm, and the flow is laminar. Determine the pressure gradient required to produce zero net flow at a cross section.

8.24 Two immiscible fluids are contained between infinite parallel plates. The plates are separated by distance $2h$, and the two fluid layers are of equal thickness h = 5 mm. The dynamic viscosity of the upper fluid is four times that of the lower fluid, which is μ_{lower} = 0.1 N·s/m². If the plates are stationary and the applied pressure gradient is −50 kPa/m, find the velocity at the interface. What is the maximum velocity of the flow? Plot the velocity distribution.

8.25 Two immiscible fluids are contained between infinite parallel plates. The plates are separated by distance $2h$, and the two fluid layers are of equal thickness h; the dynamic viscosity of the upper fluid is three times that of the lower fluid. If the lower plate is stationary and the upper plate moves at constant speed U = 6.1 m/s, what is the velocity at the interface? Assume laminar flows, and that the pressure gradient in the direction of flow is zero.

8.26 The record-read head for a computer disk-drive memory storage system rides above the spinning disk on a very thin film of air (the film thickness is 0.25 μm). The head location is 25 mm from the disk centerline; the disk spins at 8500 rpm. The record-read head is 5 mm square. For standard air in the gap between the head and disk, determine (a) the Reynolds number of the flow, (b) the viscous shear stress, and (c) the power required to overcome viscous shear.

8.27 The dimensionless velocity profile for fully developed laminar flow between infinite parallel plates with the upper plate moving at constant speed U is shown in Fig. 8.6. Find the pressure gradient $\partial p/\partial x$ at which (a) the upper plate and

(b) the lower plate experience zero shear stress, in terms of U, a, and μ. Plot the dimensionless velocity profiles for these cases.

8.28 Consider steady, fully developed laminar flow of a viscous liquid down an inclined surface. The liquid layer is of constant thickness, h. Use a suitably chosen differential control volume to obtain the velocity profile. Develop an expression for the volume flow rate.

8.29 Consider steady, incompressible, and fully developed laminar flow of a viscous liquid down an incline with no pressure gradient. The velocity profile was derived in Example 5.9. Plot the velocity profile. Calculate the kinematic viscosity of the liquid if the film thickness on a 30° slope is 0.8 mm and the maximum velocity is 15.7 mm/s.

8.30 Two immiscible fluids of equal density are flowing down a surface inclined at a 60° angle. The two fluid layers are of equal thickness $h = 10$ mm; the kinematic viscosity of the upper fluid is 1/5th that of the lower fluid, which is $\nu_{\text{lower}} = 0.01$ m²/s. Find the velocity at the interface and the velocity at the free surface. Plot the velocity distribution.

8.31 The velocity distribution for flow of a thin viscous film down an inclined plane surface was developed in Example 5.9. Consider a film 7 mm thick, of liquid with SG = 1.2 and dynamic viscosity of 1.60 N · s/m². Derive an expression for the shear stress distribution within the film. Calculate the maximum shear stress within the film and indicate its direction. Evaluate the volume flow rate in the film, in mm³/s per millimeter of surface width. Calculate the film Reynolds number based on average velocity.

8.32 Consider fully developed flow between parallel plates with the upper plate moving at $U = 1.5$ m/s; the spacing between the plates is $a = 2.5$ mm. Determine the flow rate per unit depth for the case of zero pressure gradient. If the fluid is air, evaluate the shear stress on the lower plate and plot the shear stress distribution across the channel for the zero pressure gradient case. Will the flow rate increase or decrease if the pressure gradient is adverse? Determine the pressure gradient that will give zero shear stress at $y = 0.25a$. Plot the shear stress distribution across the channel for the latter case.

8.33 Glycerin at 15°C flows between parallel plates with gap width $b = 2.5$ mm. The upper plate moves with speed $U = 0.6$ m/s in the positive x direction. The pressure gradient is $\partial p/\partial x = -1150$ kPa/m. Locate the point of maximum velocity and determine its magnitude (let $y = 0$ at the bottom plate). Determine the volume of flow (m²) that passes a given cross section (x = constant) in 10 s. Plot the velocity and shear stress distributions.

8.34 The velocity profile for fully developed flow of castor oil at 20°C between parallel plates with the upper plate moving is given by Eq. 8.8. Assume $U = 1.5$ m/s and $a = 5$ mm. Find the pressure gradient for which there is no net flow in the x direction. Plot the expected velocity distribution and the expected shear stress distribution across the channel for this flow. For the case where $u = U$ at $y/a = 0.5$, plot the expected velocity distribution and shear stress distribution across the channel. Comment on features of the plots.

8.35 The velocity profile for fully developed flow of carbon tetrachloride at 15°C between parallel plates (gap $a = 1.25$ mm), with the upper plate moving, is given by Eq. 8.8. Assuming a volume flow rate per unit depth is 3.15×10^{-4} m³/sec/m for zero pressure gradient, find U. Evaluate the shear stress on the lower plate. Would the volume flow rate increase or decrease with a mild adverse pressure gradient? Calculate the pressure gradient that will give zero shear stress at $y/a = 0.25$. Plot the velocity distribution and the shear stress distribution for this case.

8.36 Free-surface waves begin to form on a laminar liquid film flowing down an inclined surface whenever the Reynolds number, based on mass flow per unit width of film, is larger than about 33. Estimate the maximum thickness of a laminar film of water that remains free from waves while flowing down a vertical surface.

8.37 Microchips are supported on a thin air film on a smooth horizontal surface during one stage of the manufacturing process. The chips are 11.7 mm long and 9.35 mm wide and have a mass of 0.325 g. The air film is 0.125 mm thick. The initial speed of a chip is $V_0 = 1.75$ mm/s; the chip slows as the result of viscous shear in the air film. Analyze the chip motion during deceleration to develop a differential equation for chip speed V versus time t. Calculate the time required for a chip to lose 5 percent of its initial speed. Plot the variation of chip speed versus time during deceleration. Explain why it looks as you have plotted it.

8.38 A viscous-shear pump is made from a stationary housing with a close-fitting rotating drum inside. The clearance is small compared with the diameter of the drum, so flow in the annular space may be treated as flow between parallel plates. Fluid is dragged around the annulus by viscous forces. Evaluate the performance characteristics of the shear pump (pressure differential, input power, and efficiency) as functions of volume flow rate. Assume that the depth normal to the diagram is b.

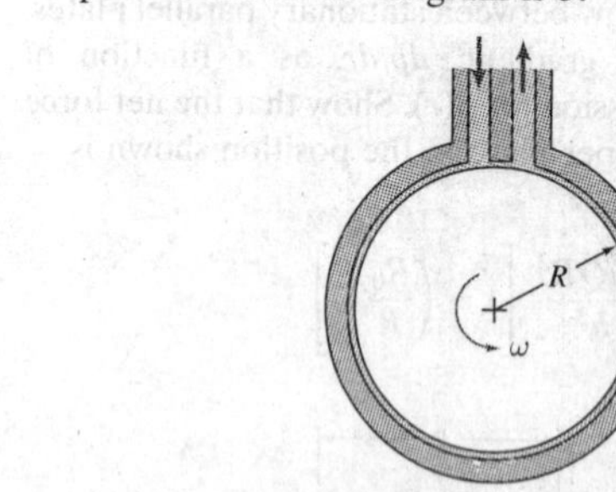

P8.38, P8.40

8.39 The clamping force to hold a part in a metal-turning operation is provided by high-pressure oil supplied by a pump. Oil leaks axially through an annular gap with diameter D, length L, and radial clearance a. The inner member of the annulus rotates at angular speed ω. Power is required both to pump the oil and to overcome viscous dissipation in the annular gap. Develop expressions in terms of the specified geometry for the pump power, $\mathcal{P}_p$, and the viscous dissipation power, $\mathcal{P}_v$. Show that the total power requirement is minimized when the radial clearance, a, is chosen such that $\mathcal{P}_v = 3\mathcal{P}_p$.

8.40 The efficiency of the viscous-shear pump of Fig. P8.39 is given by

$$\eta = 6q\frac{(1-2q)}{(4-6q)}$$

where $q = Q/abR\omega$ is a dimensionless flow rate (Q is the flow rate at pressure differential Δp, and b is the depth normal to the diagram). Plot the efficiency versus dimensionless flow rate, and find the flow rate for maximum efficiency. Explain why the efficiency peaks, and why it is zero at certain values of q.

8.41 Automotive design is tending toward all-wheel drive to improve vehicle performance and safety when traction is poor. An all-wheel drive vehicle must have an interaxle differential to allow operation on dry roads. Numerous vehicles are being built using multiplate viscous drives for interaxle differentials. Perform the analysis and design needed to define the torque transmitted by the differential for a given speed difference, in terms of the design parameters. Identify suitable dimensions for a viscous differential to transmit a torque of 150 N · m at a speed loss of 125 rpm, using lubricant with the properties of SAE 30 oil. Discuss how to find the minimum material cost for the viscous differential, if the plate cost per square meter is constant.

8.42 An inventor proposes to make a "viscous timer" by placing a weighted cylinder inside a slightly larger cylinder containing viscous liquid, creating a narrow annular gap close to the wall. Analyze the flow field created when the apparatus is inverted and the mass begins to fall under gravity. Would this system make a satisfactory timer? If so, for what range of time intervals? What would be the effect of a temperature change on measured time?

8.43 A journal bearing consists of a shaft of diameter $D = 35$ mm and length $L = 50$ mm (moment of inertia $I = 0.125$ kg · m^2) installed symmetrically in a stationary housing such that the annular gap is $\delta = 1$ mm. The fluid in the gap has viscosity $\mu = 0.1$ N · s/m^2. If the shaft is given an initial angular velocity of $\omega = 500$ rpm, determine the time for the shaft to slow to 100 rpm. On another day, an unknown fluid is tested in the same way, but takes 10 minutes to slow from 500 to 100 rpm. What is its viscosity?

8.44 In Example 8.3 we derived the velocity profile for laminar flow on a vertical wall by using a differential control volume. Instead, following the procedure we used in Example 5.9, derive the velocity profile by starting with the Navier–Stokes equations (Eqs. 5.27). Be sure to state all assumptions.

8.45 A continuous belt, passing upward through a chemical bath at speed U_0, picks up a liquid film of thickness h, density ρ, and viscosity μ. Gravity tends to make the liquid drain down, but the movement of the belt keeps the liquid from running off completely. Assume that the flow is fully developed and laminar with zero pressure gradient, and that the atmosphere produces no shear stress at the outer surface of the film. State clearly the boundary conditions to be satisfied by the velocity at $y = 0$ and $y = h$. Obtain an expression for the velocity profile.

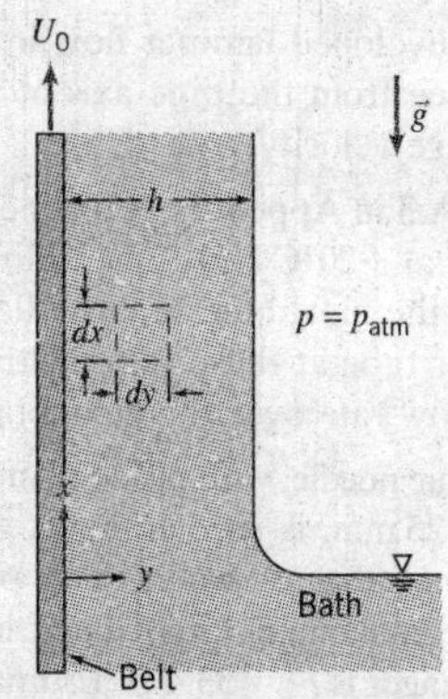

P8.45

8.46 A wet paint film of uniform thickness, δ, is painted on a vertical wall. The wet paint can be approximated as a Bingham fluid with a yield stress, τ_y, and density, ρ. Derive an expression for the maximum value of δ that can be sustained without having the paint flow down the wall. Calculate the maximum thickness for lithographic ink whose yield stress $\tau_y = 40$ Pa and density is approximately 1000 kg/m^3.

8.47 When dealing with the lubrication of bearings, the governing equation describing pressure is the Reynolds equation, generally written in 1D as

$$\frac{d}{dx}\left(\frac{h^3}{\mu}\frac{dp}{dx}\right) + 6U\frac{dh}{dx} = 0$$

where h is the step height and U is the velocity of the lower surface. Step bearings have a relatively simple design and are used with low-viscosity fluids such as water, gasoline, and solvents. The minimum film thickness in these applications is quite small. The step height must be small enough for good load capacity, yet large enough for the bearing to accommodate some wear without losing its load capacity by becoming smooth and flat. Beginning with the 1D equation for fluid motion in the x direction, show that the pressure distribution in the step bearing is as shown, where

$$p_s = \frac{6\mu(h_2 - h_1)}{\dfrac{h_1^3}{L_1} + \dfrac{h_2^3}{L_2}}$$

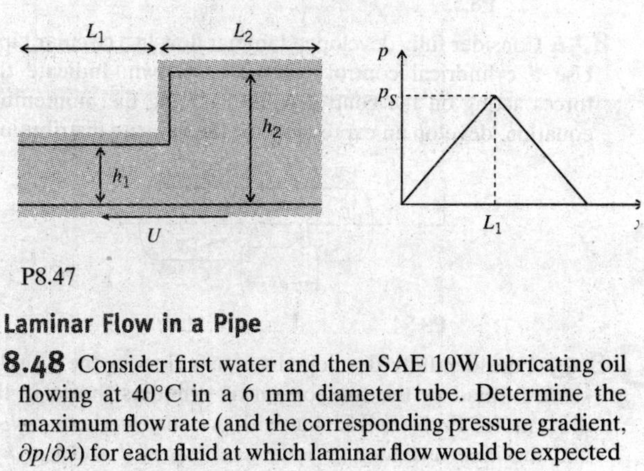

P8.47

Laminar Flow in a Pipe

8.48 Consider first water and then SAE 10W lubricating oil flowing at 40°C in a 6 mm diameter tube. Determine the maximum flow rate (and the corresponding pressure gradient, $\partial p/\partial x$) for each fluid at which laminar flow would be expected

8.49 For fully developed laminar flow in a pipe, determine the radial distance from the pipe axis at which the velocity equals the average velocity.

8.50 Using Eq. A.3 in Appendix A for the viscosity of water, find the viscosity at −20°C and 120°C. Plot the viscosity over this range. Find the maximum laminar flow rate (L/hr) in a 7.5 mm diameter tube at these temperatures. Plot the maximum laminar flow rate over this temperature range.

8.51 A hypodermic needle, with inside diameter $d = 0.127$ mm and length $L = 25$ mm, is used to inject saline solution with viscosity five times that of water. The plunger diameter is $D = 10$ mm; the maximum force that can be exerted by a thumb on the plunger is $F = 33.4$ N. Estimate the volume flow rate of saline that can be produced.

8.52 In engineering science, there are often analogies to be made between disparate phenomena. For example, the applied pressure difference, Δp, and corresponding volume flow rate, Q, in a tube can be compared to the applied DC voltage, V, across and current, I, through an electrical resistor, respectively. By analogy, find a formula for the "resistance" of laminar flow of fluid of viscosity, μ, in a tube length of L and diameter D, corresponding to electrical resistance, R. For a tube 250 mm long with inside diameter 7.5 mm, find the maximum flow rate and pressure difference for which this analogy will hold for (a) kerosene and (b) castor oil (both at 40°C). When the flow exceeds this maximum, why does the analogy fail?

8.53 Consider fully developed laminar flow in the annulus between two concentric pipes. The outer pipe is stationary, and the inner pipe moves in the x direction with speed V. Assume the axial pressure gradient is zero ($\partial p/\partial x = 0$). Obtain a general expression for the shear stress, τ, as a function of the radius, r, in terms of a constant, C_1. Obtain a general expression for the velocity profile, $u(r)$, in terms of two constants, C_1 and C_2. Obtain expressions for C_1 and C_2.

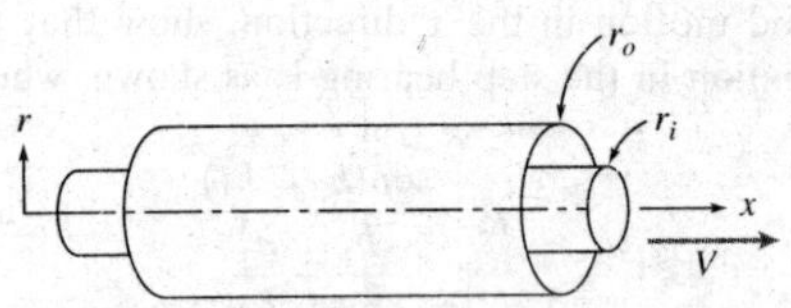

P8.53

8.54 Consider fully developed laminar flow in a circular pipe. Use a cylindrical control volume as shown. Indicate the forces acting on the control volume. Using the momentum equation, develop an expression for the velocity distribution.

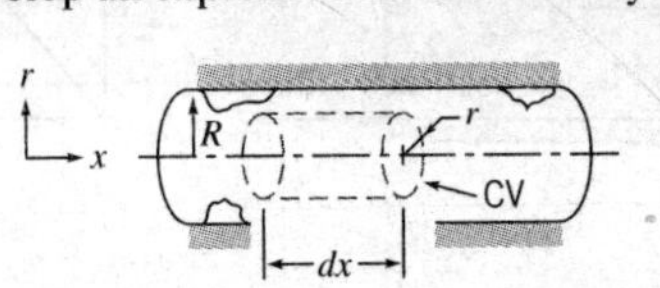

P8.54

8.55 Consider fully developed laminar flow in the annular space formed by the two concentric cylinders shown in the diagram for Problem 8.53, but with pressure gradient, $\partial p/\partial x$, and the inner cylinder stationary. Let $r_0 = R$ and $r_i = kR$. Show that the velocity profile is given by

$$u = -\frac{R^2}{4\mu}\frac{\partial p}{\partial x}\left[1-\left(\frac{r}{R}\right)^2+\left(\frac{1-k^2}{\ln(1/k)}\right)\ln\frac{r}{R}\right]$$

Obtain an expression for the location of the maximum velocity as a function of k. Plot the location of maximum velocity ($\alpha = r/R$) as a function of radius ratio k. Compare the limiting case, $k \to 0$, with the corresponding expression for flow in a circular pipe.

8.56 For the flow of Problem 8.55 show that the volume flow rate is given by

$$Q = -\frac{\pi R^4}{8\mu}\frac{\partial p}{\partial x}\left[(1-k^4)-\frac{(1-k^2)^2}{\ln(1/k)}\right]$$

Find an expression for the average velocity. Compare the limiting case, $k \to 0$, with the corresponding expression for flow in a circular pipe.

8.57 It has been suggested in the design of an agricultural sprinkler that a structural member be held in place by a wire placed along the centerline of a pipe; it is surmised that a relatively small wire would have little effect on the pressure drop for a given flow rate. Using the result of Problem 8.56, derive an expression giving the percentage change in pressure drop as a function of the ratio of wire diameter to pipe diameter for laminar flow. Plot the percentage change in pressure drop as a function of radius ratio k for $0.001 \le k \le 0.10$.

8.58 Consider fully developed pressure-driven flow in a cylindrical tube of radius, R, and length, $L = 10$ mm, with flow generated by an applied pressure gradient, Δp. Tests are performed with room temperature water for various values of R, with a fixed flow rate of $Q = 10$ μL/min. The hydraulic resistance is defined as $R_{\text{hyd}} = \Delta p/Q$ (by analogy with the electrical resistance $R_{\text{elec}} = \Delta V/I$, where ΔV is the electrical potential drop and I is the electric current). Calculate the required pressure gradient and hydraulic resistance for the range of tube radii listed in the table. Based on the results, is it appropriate to use a pressure gradient to pump fluids in microchannels, or should some other driving mechanism be used?

R (mm)	Δp (Pa)	R_{hyd} (Pa · s/m^3)
1		
10^{-1}		
10^{-2}		
10^{-3}		
10^{-4}		

8.59 The figure schematically depicts a conical diffuser, which is designed to increase pressure and reduce kinetic energy. We assume the angle α is small ($\alpha < 10°$) so that $\tan\alpha \approx \alpha$ and $r_e = r_i + \alpha l$, where r_i is the radius at the diffuser inlet, r_e is the radius at the exit, and l is the length of the diffuser. The flow in a diffuser is complex, but here we assume that each layer of fluid in the diffuser flow is laminar, as in a cylindrical tube with constant cross-sectional area. Based on reasoning similar to that in Section 8.3, the pressure difference Δp between the ends of a cylindrical pipe is

$$\Delta p = \frac{8\mu}{\pi} Q \int_0^x \frac{1}{r^4} dx$$

where x is the location in the diffuser, μ is the fluid dynamic viscosity, and Q is the flow rate. The equation above is applicable to flows in a diffuser assuming that the inertial force and exit effects are negligible. Derive the hydraulic resistance, $R_{hyd} = \Delta p/Q$, of the diffuser.

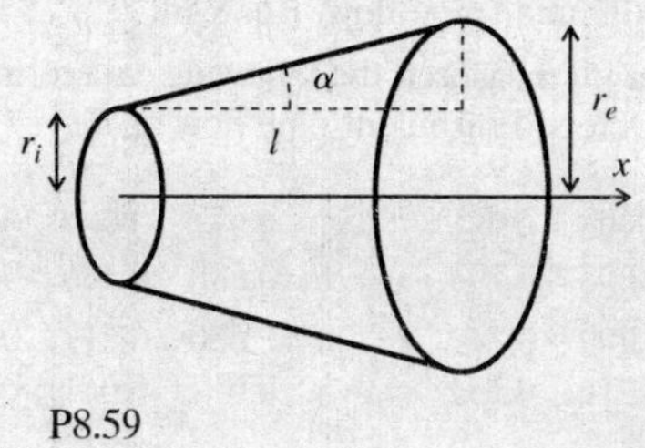

P8.59

8.60 Consider blood flow in an artery. Blood is non-Newtonian; the shear stress versus shear rate is described by the Casson relationship:

$$\begin{cases} \sqrt{\tau} = \sqrt{\tau_c} + \sqrt{\mu \dfrac{du}{dr}}, & \text{for } \tau \geq \tau_c \\ \tau = 0 & \text{for } \tau < \tau_c \end{cases}$$

where τ_c is the critical shear stress, and μ is a constant having the same dimensions as dynamic viscosity. The Casson relationship shows a linear relationship between $\sqrt{\tau}$ and $\sqrt{du/dr}$, with intercept $\sqrt{\tau_c}$ and slope $\sqrt{\mu}$. The Casson relationship approaches Newtonian behavior at high values of deformation rate. Derive the velocity profile of steady fully developed blood flow in an artery of radius R. Determine the flow rate in the blood vessel. Calculate the flow rate due to a pressure gradient $dp/dx = -100$ Pa/m, in an artery of radius $R = 1$ mm, using the following blood data: $\mu = 3.5$ cP, $\tau_c = 0.05$ dynes/cm^2.

8.61 Using Eq. 2.16, derive the velocity profile, flow rate, and average velocity of a non-Newtonian fluid in a circular tube. For a flow rate of $Q = 1\ \mu$L/min and $R = 1$ mm, with k having a value of unity in standard SI units, compare the required pressure gradients for $n = 0.5$, 1.0, and 1.5. Which fluid requires the smallest pump for the same pipe length?

8.62 The classic Poiseuille flow (Eq. 8.12), is for no-slip conditions at the walls. If the fluid is a gas, and when the mean free path, l (the average distance a molecule travels before collision with another molecule), is comparable to the length-scale L of the flow, slip will occur at the walls, and the flow rate and velocity will be increased for a given pressure gradient. In Eq. 8.11, c_1 will still be zero, but c_2 must satisfy the slip condition $u = l\,\partial u/\partial r$ at $r = R$. Derive the velocity profile and flow rate of gas flow in a micro- or nanotube which has such a slip velocity on the wall. Calculate the flow rate when $R = 10$ m, $\mu = 1.84 \times 10^{-5}$ N·s/m^2, the mean free path $l = 68$ nm, and $-\partial p/\partial x = 1.0 \times 10^6$ Pa/m.

8.63 The following solution:

$$u = u_0\left(1 - \frac{y^2}{a^2} - \frac{z^2}{b^2}\right)$$

can be used as a model for the velocity profile of fully developed pressure-driven flow in a channel with an elliptic cross section. The center of the ellipse is $(y, z) = (0, 0)$, and the major axis of length a and the minor axis of length b are parallel to the y axis and z axes, respectively. The axial pressure gradient, $\partial p/\partial x$, is constant. Based on the Navier–Stokes equations, determine the maximum velocity u_0 in terms of a, b, viscosity μ, and $\partial p/\partial x$. Letting (ρ, ϕ) be the radial and azimuthal polar coordinates, respectively, of a unit disk ($0 \leq \rho \leq 1$ and $0 \leq \phi \leq 2\pi$), the coordinates (y, z) and the velocity $u(y, z)$ can be expressed as functions of (ρ, ϕ):

$$y(\rho, \phi) = a\rho\cos\phi \qquad z(\rho, \phi) = b\rho\sin\phi \qquad u(\rho, \phi) = u_0(1 - \rho^2)$$

The flow rate is $Q = \int u(y, z)dydz = ab\int_0^{2\pi}\int_0^1 \rho u(\rho, \phi)d\rho d\phi$. Derive the flow rate of fully developed pressure-driven flow in an elliptic pipe. Compare the flow rates in a channel with an elliptic cross section with $a = 1.5R$ and $b = R$ and in a pipe of radius R with the same pressure gradient.

8.64 For pressure-driven, steady, fully developed laminar flow of an incompressible fluid through a straight channel of length L, we can define the hydraulic resistance as $R_{hyd} = \Delta p/Q$, where Δp is the pressure drop and Q is the flow rate (analogous to the electrical resistance $R_{elec} = \Delta V/I$, where ΔV is the electrical potential drop and I is the electric current). The following table summarizes the hydraulic resistance of channels with different cross sectional shapes [30]:

Shape	Formula for R_{hyd}	Computed R_{hyd}
Circle 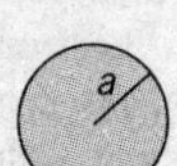	$\dfrac{8\mu L}{\pi a^4}$	
Ellipse 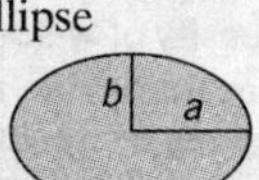	$\dfrac{4\mu L[1 + (b/a)^2]}{\pi a b^3}$	
Triangle 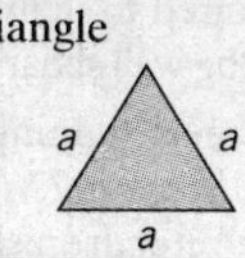	$\dfrac{320\mu L}{\sqrt{3}a^4}$	
Two plates 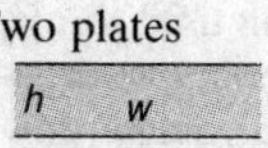	$\dfrac{12\mu L}{h^3 w}$	
Rectangle 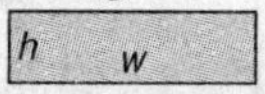	$\dfrac{12\mu L}{h^3 w[1 - 0.63(h/w)]}$	
Square 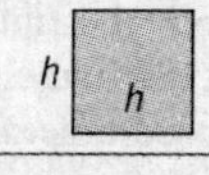	$\dfrac{12\mu L}{0.37h^4}$	

Calculate the hydraulic resistance of a straight channel with the listed cross-sectional shapes using the following parameters: $\mu = 1$ mPa·s (water), $L = 10$ mm, $a = 100\ \mu$m, $b = 33\ \mu$m,

$h = 100\ \mu m$, and $w = 300\ \mu m$. Based on the calculated hydraulic resistance, which shape is the most energy efficient to pump water?

8.65 In a food industry plant, two immiscible fluids are pumped through a tube such that fluid 1 ($\mu_1 = 0.5\ N \cdot s/m^2$) forms an inner core ($r = D/4$) and fluid 2 ($\mu_2 = 5\ N \cdot s/m^2$) forms an outer annulus. The tube has $D = 5$ mm diameter and length $L = 5$ m. Derive and plot the velocity distribution if the applied pressure difference, Δp, is 5 MPa.

8.66 A horizontal pipe carries fluid in fully developed turbulent flow. The static pressure difference measured between two sections is 35 kPa. The distance between the sections is 10 m, and the pipe diameter is 150 mm. Calculate the shear stress, τ_w, that acts on the walls.

8.67 One end of a horizontal pipe is attached using glue to a pressurized tank containing liquid, and the other has a cap attached. The inside diameter of the pipe is 2.5 cm, and the tank pressure is 250 kPa(gage). Find the force the glue must withstand with the cap on, and the force it must withstand when the cap is off and the liquid is discharging to atmosphere.

8.68 Kerosene is pumped through a smooth tube with inside diameter $D = 30$ mm at close to the critical Reynolds number. The flow is unstable and fluctuates between laminar and turbulent states, causing the pressure gradient to intermittently change from approximately -4.5 kPa/m to -11 kPa/m. Which pressure gradient corresponds to laminar, and which to turbulent, flow? For each flow, compute the shear stress at the tube wall, and sketch the shear stress distributions.

8.69 The pressure drop between two taps separated in the streamwise direction by 9 m in a horizontal, fully developed channel flow of water is 6.9 kPa. The cross section of the channel is a 25 mm × 240 mm rectangle. Calculate the average wall shear stress.

8.70 A liquid drug, with the viscosity and density of water, is to be administered through a hypodermic needle. The inside diameter of the needle is 0.25 mm and its length is 50 mm. Determine (a) the maximum volume flow rate for which the flow will be laminar, (b) the pressure drop required to deliver the maximum flow rate, and (c) the corresponding wall shear stress.

8.71 The "pitch-drop" experiment has been running continuously at the University of Queensland since 1927 (http://www.physics.uq.edu.au/physics_museum/pitchdrop.shtml). In this experiment, a funnel pitch is being used to measure the viscosity of pitch. Flow averages at about one drop—*per decade!* Viscosity is calculated using the volume flow rate equation

$$Q = \frac{\forall}{t} = \frac{\pi D^4 \rho g}{128\mu}\left(1 + \frac{h}{L}\right)$$

where D is the diameter of the flow from the funnel, h is the depth to the pitch in the main body of the funnel, L is the length of the funnel stem, and t is the elapsed time. Compare this equation with Eq. 8.13c using hydrostatic force instead of a pressure gradient. After the 6th drop in 1979, they measured that it took 17,708 days for $4.7 \times 10^{-5}\ m^3$ of pitch to fall. Given the measurements $D = 9.4$ mm, $h = 75$ mm, $L = 29$ mm, and $\rho_{pitch} = 1.1 \times 10^3\ kg/m^3$, what is the viscosity of the pitch?

Turbulent Velocity Profiles in Fully Developed Pipe Flow

8.72 Consider the empirical "power-law" profile for turbulent pipe flow, Eq. 8.22. For $n = 7$ determine the value of r/R at which u is equal to the average velocity, $\overline{V}$. Plot the results over the range $6 \le n \le 10$ and compare with the case of fully developed laminar pipe flow, Eq. 8.14.

8.73 Laufer [5] measured the following data for mean velocity in fully developed turbulent pipe flow at $Re_U = 50{,}000$:

$\bar{u}/U$	0.996	0.981	0.963	0.937	0.907	0.866	0.831
y/r	0.898	0.794	0.691	0.588	0.486	0.383	0.280
$\bar{u}/U$	0.792	0.742	0.700	0.650	0.619	0.551	
y/R	0.216	0.154	0.093	0.062	0.041	0.024	

In addition, Laufer measured the following data for mean velocity in fully developed turbulent pipe flow at $Re_U = 500{,}000$:

$\bar{u}/U$	0.997	0.988	0.975	0.959	0.934	0.908
y/R	0.898	0.794	0.691	0.588	0.486	0.383
$\bar{u}/U$	0.874	0.847	0.818	0.771	0.736	0.690
y/R	0.280	0.216	0.154	0.093	0.062	0.037

Using *Excel*'s trendline analysis, fit each set of data to the "power-law" profile for turbulent flow, Eq. 8.22, and obtain a value of n for each set. Do the data tend to confirm the validity of Eq. 8.22? Plot the data and their corresponding trendlines on the same graph.

8.74 Equation 8.23 gives the power-law velocity profile exponent, n, as a function of centerline Reynolds number, Re_U, for fully developed turbulent flow in smooth pipes. Equation 8.24 relates mean velocity, $\overline{V}$, to centerline velocity, U, for various values of n. Prepare a plot of $\overline{V}/U$ as a function of Reynolds number, $Re_{\overline{V}}$.

8.75 A momentum coefficient, β, is defined by

$$\int_A u\,\rho u\,dA = \beta \int_A \overline{V}\,\rho u\,dA = \beta \dot{m} \overline{V}$$

Evaluate β for a laminar velocity profile, Eq. 8.14, and for a "power-law" turbulent velocity profile, Eq. 8.22. Plot β as a function of n for turbulent power-law profiles over the range $6 \le n \le 10$ and compare with the case of fully developed laminar pipe flow.

Energy Considerations in Pipe Flow

8.76 Consider fully developed laminar flow of water between stationary parallel plates. The maximum flow speed, plate spacing, and width are 6.1 m/s, 1.9 mm and 38 mm, respectively. Find the kinetic energy coefficient, α.

8.77 Consider fully developed laminar flow in a circular tube. Evaluate the kinetic energy coefficient for this flow.

8.78 Show that the kinetic energy coefficient, α, for the "power law" turbulent velocity profile of Eq. 8.22 is given by Eq. 8.27.

Plot α as a function of $Re_{\bar{V}}$, for $Re_{\bar{V}} = 1 \times 10^4$ to 1×10^7. When analyzing pipe flow problems it is common practice to assume $\alpha \approx 1$. Plot the error associated with this assumption as a function of $Re_{\bar{V}}$, for $Re_{\bar{V}} = 1 \times 10^4$ to 1×10^7.

8.79 Measurements are made for the flow configuration shown in Fig. 8.12. At the inlet, section ①, the pressure is 70 kPa (gage), the average velocity is 1.75 m/s, and the elevation is 2.25 m. At the outlet, section ②, the pressure, average velocity, and elevation are 45 kPa (gage), 3.5 m/s, and 3 m, respectively. Calculate the head loss in meters. Convert to units of energy per unit mass.

8.80 Water flows in a horizontal constant-area pipe; the pipe diameter is 75 mm and the average flow speed is 5 m/s. At the pipe inlet, the gage pressure is 275 kPa, and the outlet is at atmospheric pressure. Determine the head loss in the pipe. If the pipe is now aligned so that the outlet is 15 m above the inlet, what will the inlet pressure need to be to maintain the same flow rate? If the pipe is now aligned so that the outlet is 15 m below the inlet, what will the inlet pressure need to be to maintain the same flow rate? Finally, how much lower than the inlet must the outlet be so that the same flow rate is maintained if both ends of the pipe are at atmospheric pressure (i.e., gravity feed)?

8.81 For the flow configuration of Fig. 8.12, it is known that the head loss is 1 m. The pressure drop from inlet to outlet is 50 kPa, the velocity doubles from inlet to outlet, and the elevation increase is 2 m. Compute the inlet water velocity.

Calculation of Head Loss

8.82 For a given volume flow rate and piping system, will the pressure loss be greater for hot water or cold water? Explain.

8.83 Consider the pipe flow from the water tower of Example 8.7. After another 5 years the pipe roughness has increased such that the flow is fully turbulent and $f = 0.04$. Find by how much the flow rate is decreased.

8.84 Consider the pipe flow from the water tower of Problem 8.83. To increase delivery, the pipe length is reduced from 183 m to 91 m (the flow is still fully turbulent and $f = 0.04$). What is the flow rate?

8.85 Water flows from a horizontal tube into a large tank. The tube is located 2.5 m below the free surface of water in the tank. The head loss is 2 J/kg. Compute the average flow speed in the tube.

8.86 The average flow speed in a constant-diameter section of the Alaskan pipeline is 2.5 m/s. At the inlet, the pressure is 8.25 MPa (gage) and the elevation is 45 m; at the outlet, the pressure is 350 kPa (gage) and the elevation is 115 m. Calculate the head loss in this section of pipeline.

8.87 At the inlet to a constant-diameter section of the Alaskan pipeline, the pressure is 8.5 MPa and the elevation is 45 m; at the outlet the elevation is 115 m. The head loss in this section of pipeline is 6.9 kJ/kg. Calculate the outlet pressure.

8.88 Water flows at 10 L/min through a horizontal 15-mm-diameter tube. The pressure drop along a 20-m length of tube is 85 kPa. Calculate the head loss.

8.89 Laufer [5] measured the following data for mean velocity near the wall in fully developed turbulent pipe flow at $Re_U = 50{,}000$ ($U = 3$ m/s and $R = 123$ mm) in air:

$\bar{u}/U$	0.343	0.318	0.300	0.264	0.228	0.221	0.179	0.152	0.140
y/R	0.0082	0.0075	0.0071	0.0061	0.0055	0.0051	0.0041	0.0034	0.0030

Plot the data and obtain the best-fit slope, $d\bar{u}/dy$. Use this to estimate the wall shear stress from $\tau_w = \mu\, d\bar{u}/dy$. Compare this value to that obtained using the friction factor f computed using (a) the Colebrook formula (Eq. 8.37), and (b) the Blasius correlation (Eq. 8.38).

8.90 Water is pumped at the rate of 0.075 m³/s from a reservoir 20 m above a pump to a free discharge 35 m above the pump. The pressure on the intake side of the pump is 150 kPa and the pressure on the discharge side is 450 kPa. All pipes are commercial steel of 15 cm diameter. Determine (a) the head supplied by the pump and (b) the total head loss between the pump and point of free discharge.

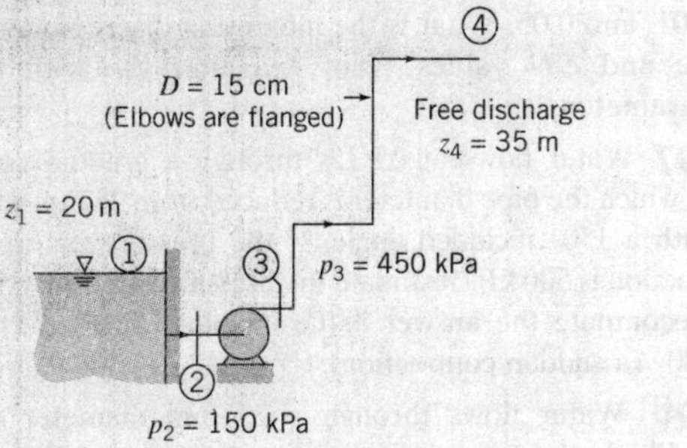

P8.90

8.91 A smooth, 75 mm diameter pipe carries water (65°C) horizontally. When the mass flow rate is 0.075 kg/s, the pressure drop is measured to be 7.5 Pa per 100 m of pipe. Based on these measurements, what is the friction factor? What is the Reynolds number? Does this Reynolds number generally indicate laminar or turbulent flow? Is the flow actually laminar or turbulent?

8.92 A small-diameter capillary tube made from drawn aluminum is used in place of an expansion valve in a home refrigerator. The inside diameter is 0.5 mm. Calculate the corresponding relative roughness. Comment on whether this tube may be considered "smooth" with regard to fluid flow.

8.93 The Colebrook equation (Eq. 8.37) for computing the turbulent friction factor is implicit in f. An explicit expression [31] that gives reasonable accuracy is

$$f_0 = 0.25\left[\log\left(\frac{e/D}{3.7} + \frac{5.74}{Re^{0.9}}\right)\right]^{-2}$$

Compare the accuracy of this expression for f with Eq. 8.37 by computing the percentage discrepancy as a function of Re and e/D, for $Re = 10^4$ to 10^8, and $e/D = 0$, 0.0001, 0.001, 0.01, and 0.05. What is the maximum discrepancy for these Re and e/D values? Plot f against Re with e/D as a parameter.

8.94 Using Eqs. 8.36 and 8.37, generate the Moody chart of Fig. 8.13.

8.95 The Moody diagram gives the Darcy friction factor, f, in terms of Reynolds number and relative roughness. The *Fanning friction factor* for pipe flow is defined as

$$f_F = \frac{\tau_w}{\frac{1}{2}\rho \bar{V}^2}$$

where τ_w is the wall shear stress in the pipe. Show that the relation between the Darcy and Fanning friction factors for fully developed pipe flow is given by $f = 4f_F$.

8.96 We saw in Section 8.7 that instead of the implicit Colebrook equation (Eq. 8.37) for computing the turbulent friction factor f, an explicit expression that gives reasonable accuracy is

$$\frac{1}{\sqrt{f}} = -1.8 \log\left[\left(\frac{e/D}{3.7}\right)^{1.11} + \frac{6.9}{Re}\right]$$

Compare the accuracy of this expression for f with Eq. 8.37 by computing the percentage discrepancy as a function of Re and e/D, for $Re = 10^4$ to 10^8, and $e/D = 0, 0.0001, 0.001, 0.01$, and 0.05. What is the maximum discrepancy for these Re and e/D values? Plot f against Re with e/D as a parameter.

8.97 Water flows at 25 L/s through a gradual contraction, in which the pipe diameter is reduced from 75 mm to 37.5 mm, with a 150° included angle. If the pressure before the contraction is 500 kPa, estimate the pressure after the contraction. Recompute the answer if the included angle is changed to 180° (a sudden contraction).

8.98 Water flows through a 25 mm diameter tube that suddenly enlarges to a diameter of 50 mm. The flow rate through the enlargement is 1.25 L/s. Calculate the pressure rise across the enlargement. Compare with the value for frictionless flow.

8.99 Water flows through a 50 mm diameter tube that suddenly contracts to 25 mm diameter. The pressure drop across the contraction is 3.45 kPa. Determine the volume flow rate.

8.100 Air at standard conditions flows through a sudden expansion in a circular duct. The upstream and downstream duct diameters are 75 mm and 225 mm, respectively. The pressure downstream is 5 mm of water higher than that upstream. Determine the average speed of the air approaching the expansion and the volume flow rate.

8.101 In an undergraduate laboratory, you have been assigned the task of developing a crude flow meter for measuring the flow in a 45 mm diameter water pipe system. You are to install a 22.5 mm diameter section of pipe and a water manometer to measure the pressure drop at the sudden contraction. Derive an expression for the theoretical calibration constant k in $Q = k\sqrt{\Delta h}$, where Q is the volume flow rate in L/min, and Δh is the manometer deflection in mm. Plot the theoretical calibration curve for a flow rate range of 10 to 50 L/min. Would you expect this to be a practical device for measuring flow rate?

8.102 Water flows from a larger pipe, diameter $D_1 = 100$ mm, into a smaller one, diameter $D_2 = 50$ mm, by way of a reentrant device. Find the head loss between points ① and ②.

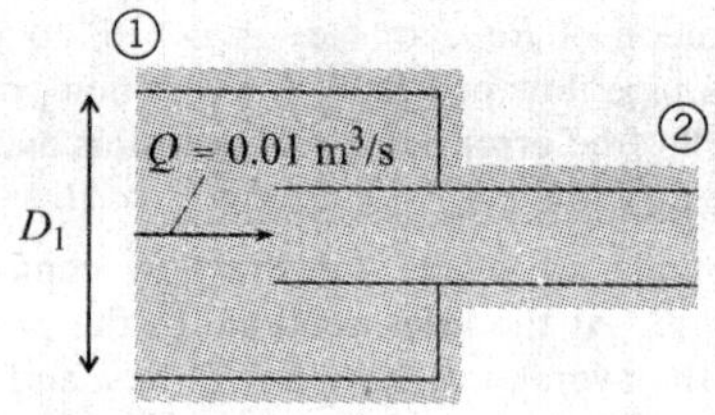

P8.102

8.103 Flow through a sudden contraction is shown. The minimum flow area at the vena contracta is given in terms of the area ratio by the contraction coefficient [32],

$$C_c = \frac{A_c}{A_2} = 0.62 + 0.38\left(\frac{A_2}{A_1}\right)^3$$

The loss in a sudden contraction is mostly a result of the vena contracta: The fluid accelerates into the contraction, there is flow separation (as shown by the dashed lines), and the vena contracta acts as a miniature sudden expansion with significant secondary flow losses. Use these assumptions to obtain and plot estimates of the minor loss coefficient for a sudden contraction, and compare with the data presented in Fig. 8.15.

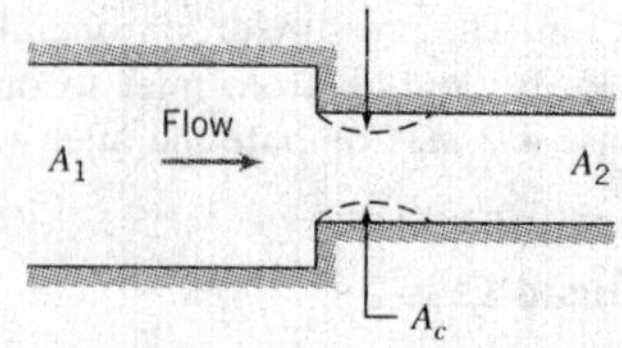

P8.103

8.104 Water flows from the tank shown through a very short pipe. Assume the flow is quasi-steady. Estimate the flow rate at the instant shown. How could you improve the flow system if a larger flow rate were desired?

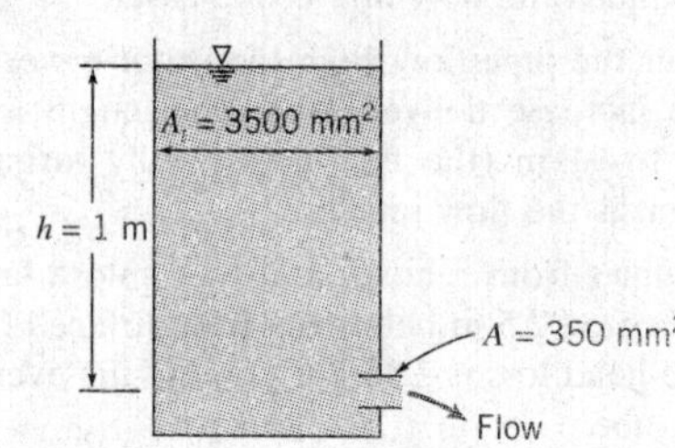

P8.104

8.105 Consider again flow through the elbow analyzed in Example 4.6. Using the given conditions, calculate the minor head loss coefficient for the elbow.

8.106 Air flows out of a clean room test chamber through a 150 mm diameter duct of length L. The original duct had a square edged entrance, but this has been replaced with a well-rounded one. The pressure in the chamber is 2.5 mm of water above ambient. Losses from friction are negligible compared with the entrance and exit losses. Estimate the increase in volume flow rate that results from the change in entrance contour.

8.107 A water tank (open to the atmosphere) contains water to a depth of 5 m. A 25 mm diameter hole is punched in the bottom. Modeling the hole as square-edged, estimate the flow rate (L/s) exiting the tank. If you were to stick a short section of pipe into the hole, by how much would the flow rate change? If instead you were to machine the inside of the hole to give it a rounded edge ($r = 5$ mm), by how much would the flow rate change?

8.108 A conical diffuser is used to expand a pipe flow from a diameter of 100 mm to a diameter of 150 mm. Find the minimum length of the diffuser if we want a loss coefficient (a) $K_{diffuser} \leq 0.2$, (b) $K_{diffuser} \leq 0.35$.

8.109 A conical diffuser of length 150 mm is used to expand a pipe flow from a diameter of 50 mm to a diameter of 89 mm. For a water flow rate of 47 L/s, estimate the static pressure rise. What is the approximate value of the loss coefficient?

8.110 Space has been found for a conical diffuser 0.45 m long in the clean room ventilation system described in Problem 8.106. The best diffuser of this size is to be used. Assume that data from Fig. 8.16 may be used. Determine the appropriate diffuser angle and area ratio for this installation and estimate the volume flow rate that will be delivered after it is installed.

8.111 By applying the basic equations to a control volume starting at the expansion and ending downstream, analyze flow through a sudden expansion (assume the inlet pressure p_1 acts on the area A_2 at the expansion). Develop an expression for and plot the minor head loss across the expansion as a function of area ratio, and compare with the data of Fig. 8.15.

8.112 Water at 45°C enters a shower head through a circular tube with 15.8 mm inside diameter. The water leaves in 24 streams, each of 1.05 mm diameter. The volume flow rate is 5.67 L/min. Estimate the minimum water pressure needed at the inlet to the shower head. Evaluate the force needed to hold the shower head onto the end of the circular tube. Indicate clearly whether this is a compression or a tension force.

8.113 Analyze flow through a sudden expansion to obtain an expression for the upstream average velocity $\overline{V}_1$ in terms of the pressure change $\Delta p = p_2 - p_1$, area ratio AR, fluid density ρ, and loss coefficient K. If the flow were frictionless, would the flow rate indicated by a measured pressure change be higher or lower than a real flow, and why? Conversely, if the flow were frictionless, would a given flow generate a larger or smaller pressure change, and why?

8.114 Water discharges to atmosphere from a large reservoir through a moderately rounded horizontal nozzle of 25 mm diameter. The free surface is 2.5 m above the nozzle exit plane. Calculate the change in flow rate when a short section of 50 mm diameter pipe is attached to the end of the nozzle to form a sudden expansion. Determine the location and estimate the magnitude of the minimum pressure with the sudden expansion in place. If the flow were frictionless (with the sudden expansion in place), would the minimum pressure be higher, lower, or the same? Would the flow rate be higher, lower, or the same?

8.115 Water flows steadily from a large tank through a length of smooth plastic tubing, then discharges to atmosphere. The tubing inside diameter is 3.18 mm, and its length is 15.3 m. Calculate the maximum volume flow rate for which flow in the tubing will remain laminar. Estimate the water level in the tank below which flow will be laminar (for laminar flow, $\alpha = 2$ and $K_{ent} = 1.4$).

8.116 You are asked to compare the behavior of fully developed laminar flow and fully developed turbulent flow in a horizontal pipe under different conditions. For the same flow rate, which will have the larger centerline velocity? Why? If the pipe discharges to atmosphere, what would you expect the trajectory of the discharge stream to look like (for the same flow rate)? Sketch your expectations for each case. For the same flow rate, which flow would give the larger wall shear stress? Why? Sketch the shear stress distribution τ/τ_w as a function of radius for each flow. For the same Reynolds number, which flow would have the larger pressure drop per unit length? Why? For a given imposed pressure differential, which flow would have the larger flow rate? Why?

Most of the remaining problems in this chapter involve determination of the turbulent friction factor f from the Reynolds number Re and dimensionless roughness e/D. For approximate calculations, f can be read from Fig. 8.13; a more accurate approach is to use this value (or some other value, even $f = 1$) as the first value for iterating in Eq. 8.37. The most convenient approach is to use solution of Eq. 8.37 programmed into (or built-into) your calculator, or programmed into an *Excel* workbook. Hence, most of the remaining problems benefit from use of *Excel*. To avoid needless duplication, the computer symbol will only be used next to remaining problems in this chapter when it has an *additional* benefit (e.g., for iterating to a solution, or for graphing).

8.117 Estimate the minimum level in the water tank of Problem 8.115 such that the flow will be turbulent.

8.118 A laboratory experiment is set up to measure pressure drop for flow of water through a smooth tube. The tube diameter is 15.9 mm, and its length is 3.56 m. Flow enters the tube from a reservoir through a square-edged entrance. Calculate the volume flow rate needed to obtain turbulent flow in the tube. Evaluate the reservoir height differential required to obtain turbulent flow in the tube.

8.119 A benchtop experiment consists of a reservoir with a 500 mm long horizontal tube of diameter 7.5 mm attached to its base. The tube exits to a sink. A flow of water at 10°C is to be generated such that the Reynolds number is 10,000. What is the flow rate? If the entrance to the tube is square-edged, how deep should the reservoir be? If the entrance to the tube is well-rounded, how deep should the reservoir be?

8.120 As discussed in Problem 8.52, the applied pressure difference, Δp, and corresponding volume flow rate, Q, for laminar flow in a tube can be compared to the applied DC voltage V across, and current I through, an electrical resistor, respectively. Investigate whether or not this analogy is valid for turbulent flow by plotting the "resistance" $\Delta p/Q$ as a function of Q for turbulent flow of kerosene (at 40°C) in a tube 250 mm long with inside diameter 7.5 mm.

8.121 Plot the required reservoir depth of water to create flow in a smooth tube of diameter 10 mm and length 100 m, for a flow rate range of 1 L/min through 10 L/min.

8.122 Oil with kinematic viscosity $\nu = 0.00005\ \text{m}^2/\text{s}$ flows at $0.003\ \text{m}^3/\text{s}$ in a 25 m long horizontal drawn-tubing pipe of 4 cm diameter. By what percentage ratio will the energy loss increase if the same flow rate is maintained while the pipe diameter is reduced to 1 cm?

8.123 A water system is used in a laboratory to study flow in a smooth pipe. The water is at 10°C. To obtain a reasonable range, the maximum Reynolds number in the pipe must be 100,000. The system is supplied from an overhead constant-head tank. The pipe system consists of a square-edged entrance, two 45° standard elbows, two 90° standard elbows, and a fully open gate valve. The pipe diameter is 7.5 mm, and the total length of pipe is 1 m. Calculate the minimum height of the supply tank above the pipe system discharge to reach the desired Reynolds number. If a pressurized chamber is used instead of the reservoir, what will be the required pressure?

8.124 Water from a pump flows through a 230 mm diameter commercial steel pipe for a distance of 6400 m from the pump discharge to a reservoir open to the atmosphere. The level of the water in the reservoir is 15 m above the pump discharge, and the average speed of the water in the pipe is 3 m/s. Calculate the pressure at the pump discharge.

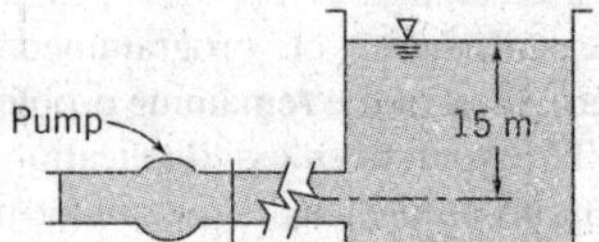

P8.124

8.125 Water is to flow by gravity from one reservoir to a lower one through a straight, inclined galvanized iron pipe. The pipe diameter is 50 mm, and the total length is 250 m. Each reservoir is open to the atmosphere. Plot the required elevation difference Δz as a function of flow rate Q, for Q ranging from 0 to $0.01\ \text{m}^3/\text{s}$. Estimate the fraction of Δz due to minor losses.

8.126 A 5 cm diameter potable water line is to be run through a maintenance room in a commercial building. Three possible layouts for the water line are proposed, as shown. Which is the best option, based on minimizing losses? Assume galvanized iron, and a flow rate of 350 L/min.

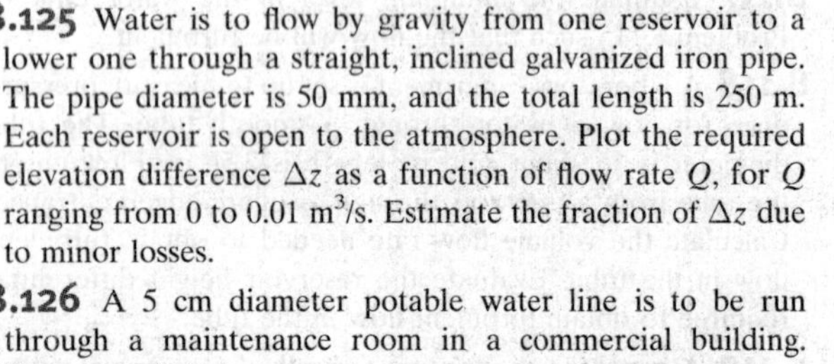

P8.126

8.127 In an air-conditioning installation, a flow rate of $35\ \text{m}^3/\text{min}$ of air at 10°C is required. A smooth sheet metal duct of rectangular section (0.23 m by 0.75 m) is to be used. Determine the pressure drop (mm of water) for a 30 m horizontal duct section.

8.128 A system for testing variable-output pumps consists of the pump, four standard elbows, and an open gate valve forming a closed circuit as shown. The circuit is to absorb the energy added by the pump. The tubing is 75 mm diameter cast iron, and the total length of the circuit is 20 m. Plot the pressure difference required from the pump for water flow rates Q ranging from $0.01\ \text{m}^3/\text{s}$ to $0.06\ \text{m}^3/\text{s}$.

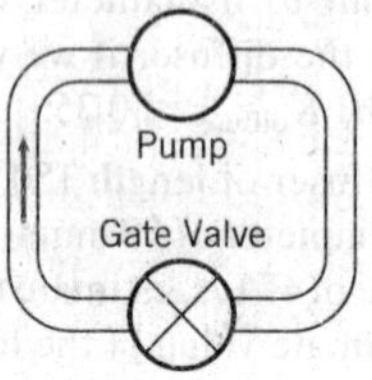

P8.128

8.129 A pipe friction experiment is to be designed, using water, to reach a Reynolds number of 100,000. The system will use 5 cm smooth PVC pipe from a constant-head tank to the flow bench and 20 m of smooth 2.5 cm PVC line mounted horizontally for the test section. The water level in the constant-head tank is 0.5 m above the entrance to the 5-cm PVC line. Determine the required average speed of water in the 2.5 cm pipe. Estimate the feasibility of using a constant-head tank. Calculate the pressure difference expected between taps 5 m apart in the horizontal test section.

8.130 Two reservoirs are connected by three clean cast-iron pipes in series, $L_1 = 600$ m, $D_1 = 0.3$ m, $L_2 = 900$ m, $D_2 = 0.4$ m, $L_3 = 1500$ m, and $D_3 = 0.45$ m. When the discharge is $0.11\ \text{m}^3/\text{s}$ of water at 15°C, determine the difference in elevation between the reservoirs.

8.131 Consider flow of standard air at $0.6\ \text{m}^3/\text{s}$. Compare the pressure drop per unit length of a round duct with that for rectangular ducts of aspect ratio 1, 2, and 3. Assume that all ducts are smooth, with cross-sectional areas of $0.09\ \text{m}^2$.

8.132 Data were obtained from measurements on a vertical section of old, corroded, galvanized iron pipe of 50 mm inside diameter. At one section the pressure was $p_1 = 750$ kPa (gage); at a second section, 40 m lower, the pressure was $p_2 = 250$ kPa (gage). The volume flow rate of water was $0.015\ \text{m}^3/\text{s}$. Estimate the relative roughness of the pipe. What percentage savings in pumping power would result if the pipe were restored to its new, clean relative roughness?

8.133 Water, at volume flow rate $Q = 21$ L/sec, is delivered by a fire hose and nozzle assembly. The hose ($L = 76$ m, $D = 75$ mm, and $e/D = 0.004$) is made up of four 18 m sections joined by couplings. The entrance is square-edged; the minor loss coefficient for each coupling is $K_c = 0.5$, based on mean velocity through the hose. The nozzle loss coefficient is $K_n = 0.02$, based on velocity in the exit jet, of $D_2 = 25$ mm diameter. Estimate the supply pressure required at this flow rate.

8.134 Flow in a tube may alternate between laminar and turbulent states for Reynolds numbers in the transition zone. Design a bench-top experiment consisting of a constant-head cylindrical transparent plastic tank with depth graduations, and a length of plastic tubing (assumed smooth) attached at the base of the tank through which the water flows to a measuring container. Select tank and tubing dimensions so that the system is compact, but will operate in the transition zone range. Design the experiment so that you can easily increase the tank head from a low range (laminar flow) through transition to turbulent flow, and vice versa. (Write instructions for students on recognizing when the flow is laminar or turbulent.) Generate plots (on the same graph) of tank depth against Reynolds number, assuming laminar or turbulent flow.

8.135 A small swimming pool is drained using a garden hose. The hose has 20 mm inside diameter, a roughness height of 0.2 mm, and is 30 m long. The free end of the hose is located 3 m below the elevation of the bottom of the pool. The average velocity at the hose discharge is 1.2 m/s. Estimate the depth of the water in the swimming pool. If the flow were inviscid, what would be the velocity?

8.136 When you drink you beverage with a straw, you need to overcome both gravity and friction in the straw. Estimate the fraction of the total effort you put into quenching your thirst of each factor, making suitable assumptions about the liquid and straw properties, and your drinking rate (for example, how long it would take you to drink a 350 ml drink if you drank it all in one go (quite a feat with a straw). Is the flow laminar or turbulent? (Ignore minor losses.)

Solution of Pipe Flow Problems

8.137 The hose in Problem 8.135 is replaced with a larger-diameter hose, diameter 25 mm (same length and roughness). Assuming a pool depth of 1.5 m, what will be the new average velocity and flow rate?

8.138 What flow rate (gpm) will be produced in a 75 mm diameter water pipe for which there is a pressure drop of 425 kPa over a 200 m length? The pipe roughness is 2.5 mm. The water is at 0°C.

8.139 A compressed air drill requires 0.25 kg/s of air at 650 kPa (gage) at the drill. The hose from the air compressor to the drill is 40 mm inside diameter. The maximum compressor discharge gage pressure is 670 kPa; air leaves the compressor at 40°C. Neglect changes in density and any effects of hose curvature. Calculate the longest hose that may be used.

8.140 You recently bought a house and want to improve the flow rate of water on your top floor. The poor flow rate is due to three reasons: The city water pressure at the water meter is poor ($p = 200$ kPa gage); the piping has a small diameter ($D = 1.27$ cm) and has been crudded up, increasing its roughness ($e/D = 0.05$); and the top floor of the house is 15 m higher than the water meter. You are considering two options to improve the flow rate: Option 1 is replacing all the piping after the water meter with new smooth piping with a diameter of 1.9 cm; and option 2 is installing a booster pump while keeping the original pipes. The booster pump has an outlet pressure of 300 kPa. Which option would be more effective? Neglect minor losses.

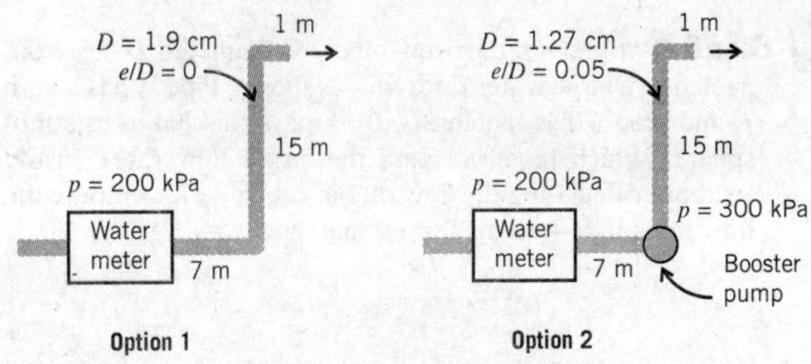

P8.140

8.141 The students of Phi Gamma Delta are putting a kiddy pool on a porch attached to the second story of their house and plan to fill it with water from a garden hose. The kiddy pool has a diameter of 1.5 m, and is 0.76 m deep. The porch is 5.5 m above the faucet. The garden hose is very smooth on the inside, has a length of 15 m, and a diameter of 1.6 cm. If the water pressure at the faucet is 414 kPa, how long will it take to fill the pool? Neglect minor losses.

8.142 Gasoline flows in a long, underground pipeline at a constant temperature of 15°C. Two pumping stations at the same elevation are located 13 km apart. The pressure drop between the stations is 1.4 MPa. The pipeline is made from 0.6 m diameter pipe. Although the pipe is made from commercial steel, age and corrosion have raised the pipe roughness to approximately that for galvanized iron. Compute the volume flow rate.

8.143 Water flows steadily in a horizontal 125 mm diameter cast-iron pipe. The pipe is 150 m long and the pressure drop between sections ① and ② is 150 kPa. Find the volume flow rate through the pipe.

8.144 Water flows steadily in a 125 mm diameter cast-iron pipe 150 m long. The pressure drop between sections ① and ② is 150 kPa, and section ② is located 15 m above section ①. Find the volume flow rate.

8.145 Two open standpipes of equal diameter are connected by a straight tube, as shown. Water flows by gravity from one standpipe to the other. For the instant shown, estimate the rate of change of water level in the left standpipe.

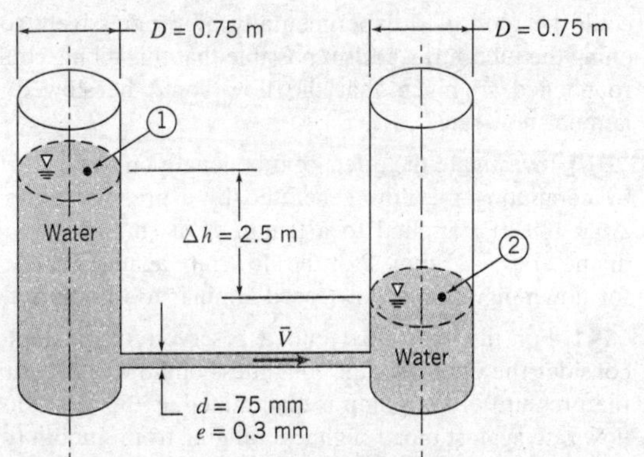

P8.145

8.146 Two galvanized iron pipes of diameter D are connected to a large water reservoir, as shown. Pipe A has length L and pipe B has length $2L$. Both pipes discharge to atmosphere. Which pipe will pass the larger flow rate? Justify (without calculating the flow rate in each pipe). Compute the flow rates if $H = 10$ m, $D = 50$ mm, and $L = 10$ m.

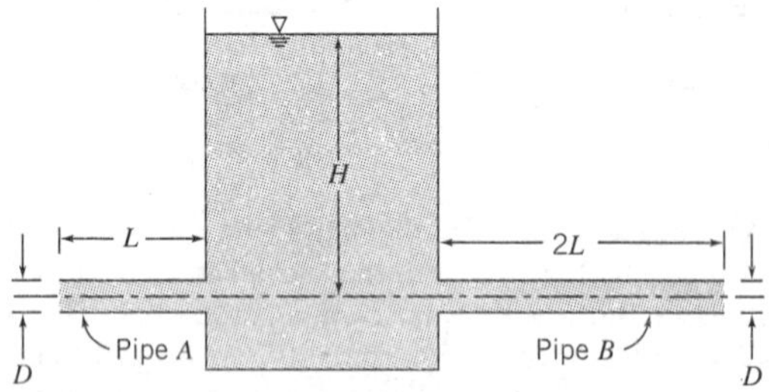

P8.146, P8.161

8.147 Galvanized iron drainpipes of diameter 50 mm are located at the four corners of a building, but three of them become clogged with debris. Find the rate of downpour (cm/min) at which the single functioning drainpipe can no longer drain the roof. The building roof area is 500 m^2, and the height is 5 m. Assume the drainpipes are the same height as the building, and that both ends are open to atmosphere. Ignore minor losses.

8.148 A mining engineer plans to do hydraulic mining with a high-speed jet of water. A lake is located $H = 300$ m above the mine site. Water will be delivered through $L = 900$ m of fire hose; the hose has inside diameter $D = 75$ mm and relative roughness $e/D = 0.01$. Couplings, with equivalent length $L_e = 20\ D$, are located every 10 m along the hose. The nozzle outlet diameter is $d = 25$ mm. Its minor loss coefficient is $K = 0.02$ based on outlet velocity. Estimate the maximum outlet velocity that this system could deliver. Determine the maximum force exerted on a rock face by this water jet.

8.149 Investigate the effect of tube roughness on water flow rate by computing the flow generated by a pressure difference $\Delta p = 100$ kPa applied to a length $L = 100$ m of tubing, with diameter $D = 25$ mm. Plot the flow rate against tube relative roughness e/D for e/D ranging from 0 to 0.05 (this could be replicated experimentally by progressively roughening the tube surface). Is it possible that this tubing could be roughened so much that the flow could be slowed to a laminar flow rate?

8.150 Investigate the effect of tube length on water flow rate by computing the flow generated by a pressure difference $\Delta p = 100$ kPa applied to a length L of smooth tubing, of diameter $D = 25$ mm. Plot the flow rate against tube length for flow ranging from low speed laminar to fully turbulent.

8.151 For the pipe flow into a reservoir of Example 8.5 consider the effect of pipe roughness on flow rate, assuming the pressure of the pump is maintained at 153 kPa. Plot the flow rate against pipe roughness ranging from smooth ($e = 0$) to very rough ($e = 3.75$ mm). Also consider the effect of pipe length (again assuming the pump always produces 153 kPa) for smooth pipe. Plot the flow rate against pipe length for $L = 100$ m through $L = 1000$ m.

8.152 Water for a fire protection system is supplied from a water tower through a 150-mm cast-iron pipe. A pressure gage at a fire hydrant indicates 600 kPa when no water is flowing. The total pipe length between the elevated tank and the hydrant is 200 m. Determine the height of the water tower above the hydrant. Calculate the maximum volume flow rate that can be achieved when the system is flushed by opening the hydrant wide (assume minor losses are 10 percent of major losses at this condition). When a fire hose is attached to the hydrant, the volume flow rate is 0.75 m^3/min. Determine the reading of the pressure gage at this flow condition.

8.153 The siphon shown is fabricated from 50-mm-i.d. drawn aluminum tubing. The liquid is water at 15°C. Compute the volume flow rate through the siphon. Estimate the minimum pressure inside the tube.

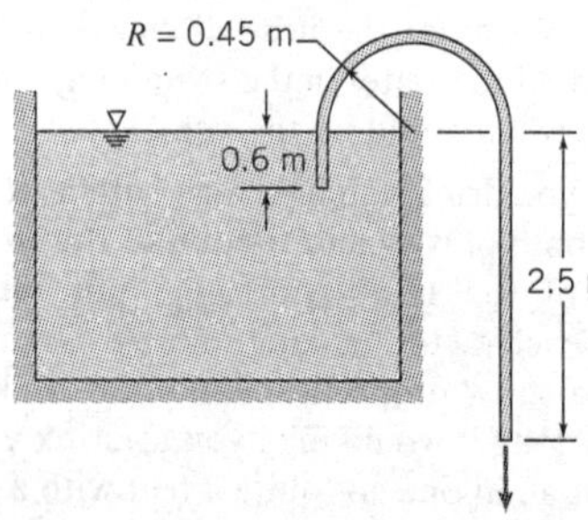

P8.153

8.154 A large open water tank has a horizontal cast iron drainpipe of diameter $D = 2.5$ cm and length $L = 1.5$ m attached at its base. If the depth of water is $h = 3.5$ m, find the flow rate (m^3/hr) if the pipe entrance is (a) reentrant, (b) square-edged, and (c) rounded ($r = 3.75$ mm).

8.155 Repeat Problem 8.154, except now the pipe is vertical, as shown.

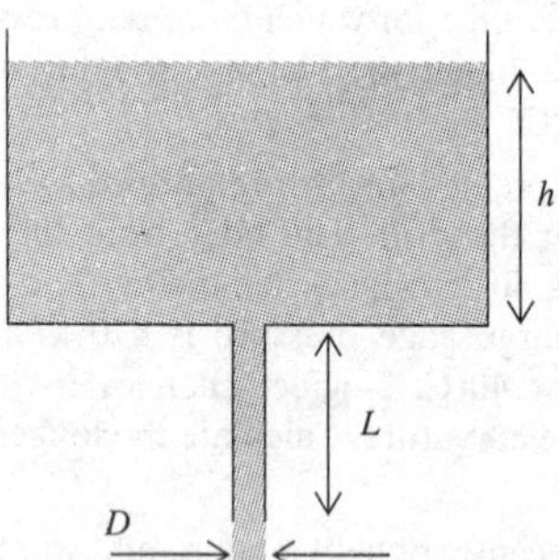

P8.155

8.156 A tank containing 30 m^3 of kerosene is to be emptied by a gravity feed using a drain hose of diameter 15 mm, roughness 0.2 mm, and length 1 m. The top of the tank is open to the atmosphere and the hose exits to an open chamber. If the kerosene level is initially 10 m above the drain exit, estimate (by assuming steady flow) the initial drainage rate. Estimate the flow rate when the kerosene level is down to 5 m, and then down to 1 m. Based on these three estimates, make a rough estimate of the time it took to drain to the 1-m level.

8.157 Consider again the Roman water supply discussed in Example 8.10. Assume that the 15 m length of horizontal constant-diameter pipe required by law has been installed. The relative roughness of the pipe is 0.01. Estimate the flow rate of water delivered by the pipe under the inlet conditions of the example. What would be the effect of adding the same diffuser to the end of the 15 m pipe?

8.158 You are watering your lawn with an *old* hose. Because lime deposits have built up over the years, the 19 mm-i.d. hose now has an average roughness height of 0.56 mm. One 15 m length of the hose, attached to your spigot, delivers 57 L/min of water (15°C). Compute the pressure at the spigot, in kPa. Estimate the delivery if two 15 m lengths of the hose are connected. Assume that the pressure at the spigot varies with flow rate and the water main pressure remains constant at 345 kPa.

8.159 In Example 8.10 we found that the flow rate from a water main could be increased (by as much as 33 percent) by attaching a diffuser to the outlet of the nozzle installed into the water main. We read that the Roman water commissioner required that the tube attached to the nozzle of each customer's pipe be the same diameter for at least 15 m from the public water main. Was the commissioner overly conservative? Using the data of the problem, estimate the length of pipe (with $e/D = 0.01$) at which the system of pipe and diffuser would give a flow rate equal to that with the nozzle alone. Plot the volume flow ratio Q/Q_i as a function of L/D, where L is the length of pipe between the nozzle and the diffuser, Q_i is the volume flow rate for the nozzle alone, and Q is the actual volume flow rate with the pipe inserted between nozzle and diffuser.

8.160 Your boss, from the "old school," claims that for pipe flow the flow rate, $Q \propto \sqrt{\Delta p}$, where Δp is the pressure difference driving the flow. You dispute this, so perform some calculations. You take a 25 mm diameter commercial steel pipe and assume an initial flow rate of 4.7 L/min of water. You then increase the applied pressure in equal increments and compute the new flow rates so you can plot Q versus Δp, as computed by you and your boss. Plot the two curves on the same graph. Was your boss right?

8.161 For Problem 8.146, what would the diameter of the pipe of length $2L$ need to be to generate the same flow as the pipe of length L?

8.162 A hydraulic press is powered by a remote high-pressure pump. The gage pressure at the pump outlet is 20.7 Mpa, whereas the pressure required for the press is 18.9 MPa (gage), at a flow rate of 0.00057 m^3/s. The press and pump are connected by 50.3 m of smooth, drawn steel tubing. The fluid is SAE 10W oil at 38°C. Determine the minimum tubing diameter that may be used.

8.163 A pump is located 4.5 m to one side of, and 3.5 m above a reservoir. The pump is designed for a flow rate of 6 L/s. For satisfactory operation, the static pressure at the pump inlet must not be lower than −6 m of water gage. Determine the smallest standard commercial steel pipe that will give the required performance.

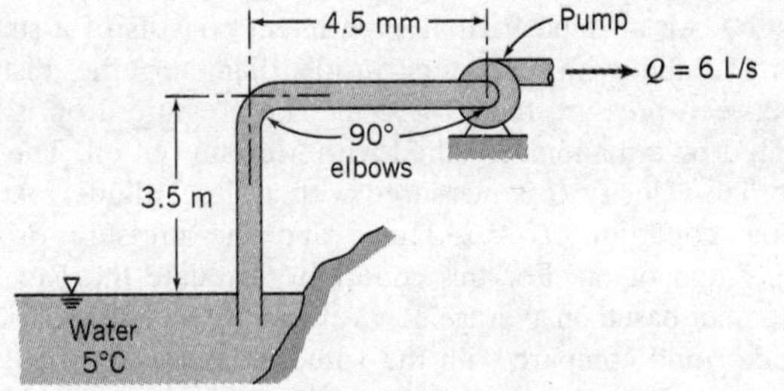

P8.163

8.164 Determine the minimum size smooth rectangular duct with an aspect ratio of 3 that will pass 1 m^3/s of 10°C air with a head loss of 25 mm of water per 100 m of duct.

8.165 A new industrial plant requires a water flow rate of 5.7 m^3/min. The gage pressure in the water main, located in the street 50 m from the plant, is 800 kPa. The supply line will require installation of 4 elbows in a total length of 65 m. The gage pressure required in the plant is 500 kPa. What size galvanized iron line should be installed?

8.166 Air at 20°C flows in a horizontal square cross-section duct made from commercial steel. The duct is 25 m long. What size (length of a side) duct is required to convey 2 m^3/s of air with a pressure drop of 1.5 cm H_2O?

8.167 Investigate the effect of tube diameter on water flow rate by computing the flow generated by a pressure difference, $\Delta p = 100$ kPa, applied to a length $L = 100$ m of smooth tubing. Plot the flow rate against tube diameter for a range that includes laminar and turbulent flow.

8.168 What diameter water pipe is required to handle 0.075 m^3/s and a 500 kPa pressure drop? The pipe length is 175 m, and roughness is 2.5 mm.

8.169 A large reservoir supplies water for a community. A portion of the water supply system is shown. Water is pumped from the reservoir to a large storage tank before being sent on to the water treatment facility. The system is designed to provide 1310 L/s of water at 20°C. From B to C the system consists of a square-edged entrance, 760 m of pipe, three gate valves, four 45° elbows, and two 90° elbows. Gage pressure at C is 197 kPa. The system between F and G contains 760 m of pipe, two gate valves, and four 90° elbows. All pipe is 508 mm diameter, cast iron. Calculate the average velocity of water in the pipe, the gage pressure at section F, the power input to the pump (its efficiency is 80 percent), and the wali shear stress in section FG.

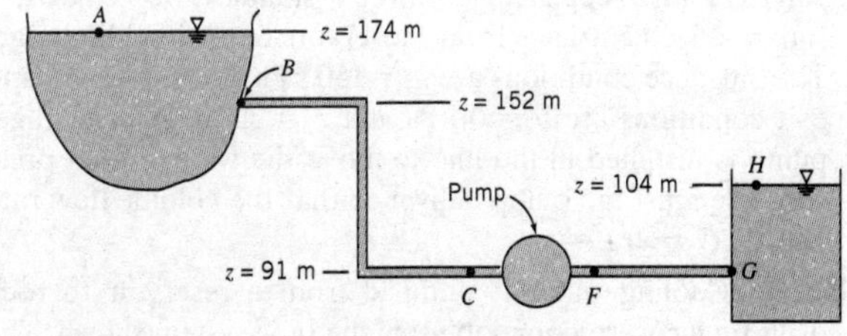

P8.169

8.170 An air-pipe friction experiment consists of a smooth brass tube with 63.5 mm inside diameter; the distance between pressure taps is 1.52 m. The pressure drop is indicated by a manometer filled with Meriam red oil. The centerline velocity U is measured with a pitot cylinder. At one flow condition, $U = 23.1$ m/s and the pressure drop is 12.3 mm of oil. For this condition, evaluate the Reynolds number based on average flow velocity. Calculate the friction factor and compare with the value obtained from Eq. 8.37 (use $n = 7$ in the power-law velocity profile).

***8.171** Oil has been flowing from a large tank on a hill to a tanker at the wharf. The compartment in the tanker is nearly full and an operator is in the process of stopping the flow. A valve on the wharf is closed at a rate such that 1 MPa is maintained in the line immediately upstream of the valve. Assume:

Length of line from tank to valve	3 km
Inside diameter of line	200 mm
Elevation of oil surface in tank	60 m
Elevation of valve on wharf	6 m
Instantaneous flow rate	2.5 m^3/min
Head loss in line (exclusive of valve being closed) at this rate of flow	23 m of oil
Specific gravity of oil	0.88

Calculate the initial instantaneous rate of change of volume flow rate.

***8.172** Problem 8.171 describes a situation in which flow in a long pipeline from a hilltop tank is slowed gradually to avoid a large pressure rise. Expand this analysis to predict and plot the closing schedule (valve loss coefficient versus time) needed to maintain the maximum pressure at the valve at or below a given value throughout the process of stopping the flow from the tank.

8.173 A pump draws water at a steady flow rate of 11.3 kg/s through a piping system. The pressure on the suction side of the pump is −17.2 kPa(gage). The pump outlet pressure is 345 kPa(gage). The inlet pipe diameter is 75 mm; the outlet pipe diameter is 50 mm. The pump efficiency is 70 percent. Calculate the power required to drive the pump.

8.174 The pressure rise across a water pump is 75 kPa when the volume flow rate is 25 L/s. If the pump efficiency is 80 percent, determine the power input to the pump.

8.175 A 125 mm diameter pipeline conveying water at 10°C contains 50 m of straight galvanized pipe, 5 fully open gate valves, 1 fully open angle valve, 7 standard 90° elbows, 1 square-edged entrance from a reservoir, and 1 free discharge. The entrance conditions are $p_1 = 150$ kPa and $z_1 = 15$ m, and exit conditions are $p_2 = 0$ kPa and $z_2 = 30$ m. A centrifugal pump is installed in the line to move the water. What pressure rise must the pump deliver so that the volume flow rate will be $Q = 50$ L/s?

8.176 Cooling water is pumped from a reservoir to rock drills on a construction job using the pipe system shown. The flow rate must be 38 L/s and water must leave the spray nozzle at 37 m/s. Calculate the minimum pressure needed at the pump outlet. Estimate the required power input if the pump efficiency is 70 percent.

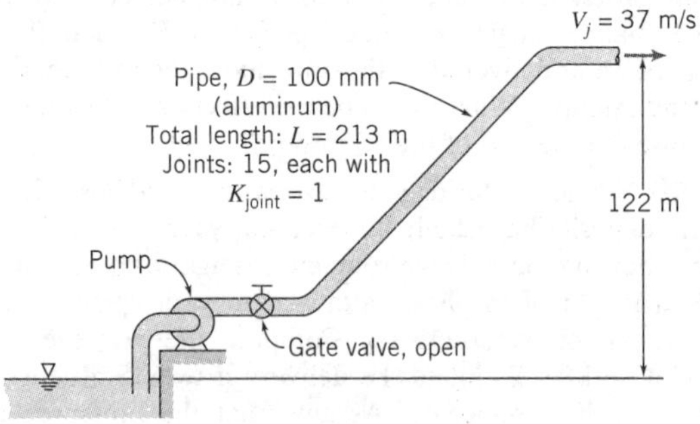

P8.176

8.177 You are asked to size a pump for installation in the water supply system of the Willis Tower (formerly the Sears Tower) in Chicago. The system requires 100 gpm of water pumped to a reservoir at the top of the tower 340 m above the street. City water pressure at the street-level pump inlet is 400 kPa (gage). Piping is to be commercial steel. Determine the minimum diameter required to keep the average water velocity below 3.5 m/s in the pipe. Calculate the pressure rise required across the pump. Estimate the minimum power needed to drive the pump.

8.178 Air conditioning on a university campus is provided by chilled water (10°C) pumped through a main supply pipe. The pipe makes a loop 5 km in length. The pipe diameter is 0.75 m and the material is steel. The maximum design volume flow rate is 0.65 m^3/s. The circulating pump is driven by an electric motor. The efficiencies of pump and motor are $\eta_p = 85$ percent and $\eta_m = 85$ percent, respectively. Electricity cost is 14¢/(kW·hr). Determine (a) the pressure drop, (b) the rate of energy addition to the water, and (c) the daily cost of electrical energy for pumping.

8.179 A fire nozzle is supplied through 100 m of 3.5 cm diameter, smooth, rubber-lined hose. Water from a hydrant is supplied to a booster pump on board the pumper truck at 350 kPa (gage). At design conditions, the pressure at the nozzle inlet is 700 kPa (gage), and the pressure drop along the hose is 750 kPa per 100 m of length. Determine (a) the design flow rate, (b) the nozzle exit velocity, assuming no losses in the nozzle, and (c) the power required to drive the booster pump, if its efficiency is 70 percent.

8.180 Heavy crude oil (SG = 0.925 and $\nu = 1.0 \times 10^{-4}$ m^2/s) is pumped through a pipeline laid on flat ground. The line is made from steel pipe with 600 mm i.d. and has a wall thickness of 12 mm. The allowable tensile stress in the pipe wall is limited to 275 MPa by corrosion considerations. It is important to keep the oil under pressure to ensure that gases remain in solution. The minimum recommended pressure is 500 kPa. The pipeline carries a flow of 400,000 barrels (in the petroleum industry, a "barrel" is 42 gal) per day. Determine the maximum spacing between pumping stations. Compute the power added to the oil at each pumping station.

8.181 The volume flow rate through a water fountain on a college campus is 0.075 m^3/s. Each water stream can rise to a height of 10 m. Estimate the daily cost to operate the

*These problems require material from sections that may be omitted without loss of continuity in the text material.

fountain. Assume that the pump motor efficiency is 85 percent, the pump efficiency is 85 percent, and the cost of electricity is 14¢/(kW · hr).

8.182 Petroleum products are transported over long distances by pipeline, e.g., the Alaskan pipeline (see Example 8.6). Estimate the energy needed to pump a typical petroleum product, expressed as a fraction of the throughput energy carried by the pipeline. State and critique your assumptions clearly.

8.183 The pump testing system of Problem 8.128 is run with a pump that generates a pressure difference given by $\Delta p = 750 - 15 \times 10^4 Q^2$ where Δp is in kPa, and the generated flow rate is Q m³/s. Find the water flow rate, pressure difference, and power supplied to the pump if it is 70 percent efficient.

8.184 A water pump can generate a pressure difference Δp (kPa) given by $\Delta p = 999 - 859\, Q^2$, where the flow rate is Q m³/s. It supplies a pipe of diameter 0.5 m, roughness 13 mm, and length 760 m. Find the flow rate, pressure difference, and the power supplied to the pump if it is 70 percent efficient. If the pipe were replaced with one of roughness 6 mm, how much would the flow increase, and what would the required power be?

8.185 A square cross-section duct (0.35 m × 0.35 m × 175 m) is used to convey air ($\rho = 1.1$ kg/m³) into a clean room in an electronics manufacturing facility. The air is supplied by a fan and passes through a filter installed in the duct. The duct friction factor is $f = 0.003$, the filter has a loss coefficient of $K = 3$. The fan performance is given by $\Delta p = 2250 - 250Q - 150Q^2$, where Δp (Pa) is the pressure generated by the fan at flow rate Q (m³/s). Determine the flow rate delivered to the room.

8.186 The head versus capacity curve for a certain fan may be approximated by the equation $H = 762 - 11.4\, Q^2$, where H is the output static head in mm of water and Q is the air flow rate in m³/s. The fan outlet dimensions are 200 × 400 mm. Determine the air flow rate delivered by the fan into a 61 m straight length of 200 × 400 mm rectangular duct.

***8.187** The water pipe system shown is constructed from galvanized iron pipe. Minor losses may be neglected. The inlet is at 400 kPa (gage), and all exits are at atmospheric pressure. Find the flow rates Q_0, Q_1, Q_2, Q_3, and Q_4.

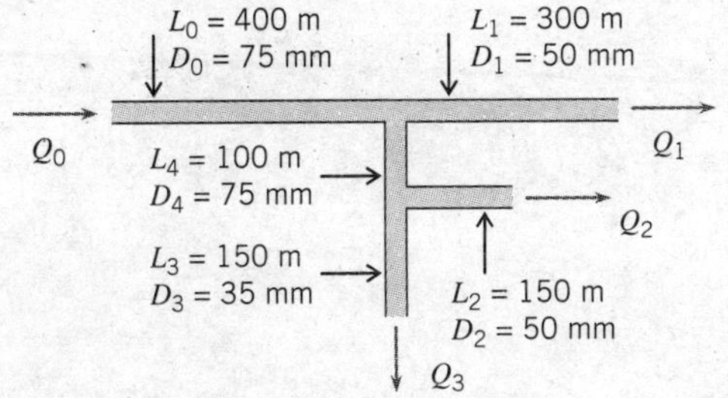

P8.187, P8.188

***8.188** Find flow rates Q_0, Q_1, Q_2, and Q_4 if pipe 3 becomes blocked.

***8.189** A cast-iron pipe system consists of a 46 m section of water pipe, after which the flow branches into two 46 m sections. The two branches then meet in a final 46 m section. Minor losses may be neglected. All sections are 38 mm diameter, except one of the two branches, which is 25 mm diameter. If the applied pressure across the system is 345 kPa, find the overall flow rate and the flow rates in each of the two branches.

***8.190** A swimming pool has a partial-flow filtration system. Water at 24°C is pumped from the pool through the system shown. The pump delivers 1.9 L/s. The pipe is nominal 20 mm PVC (i.d. = 20.93 mm). The pressure loss through the filter is approximately $\Delta p = 1039\, Q^2$, where Δp is in kPa and Q is in L/s. Determine the pump pressure and the flow rate through each branch of the system.

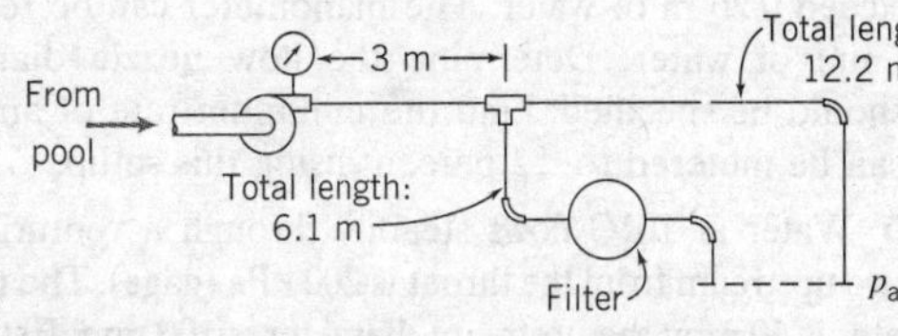

P8.190

8.191 Why does the shower temperature change when a toilet is flushed? Sketch pressure curves for the hot and cold water supply systems to explain what happens.

Flow Meters

8.192 A square-edged orifice with corner taps and a water manometer are used to meter compressed air. The following data are given:

Inside diameter of air line	150 mm
Orifice plate diameter	100 mm
Upstream pressure	600 kPa
Temperature of air	25°C
Manometer deflection	750 mm H_2O

Calculate the volume flow rate in the line, expressed in cubic meters per hour.

8.193 Water at 65°C flows through a 75 mm diameter orifice installed in a 150-mm-i.d. pipe. The flow rate is 20 L/s. Determine the pressure difference between the corner taps.

8.194 A smooth 200-m pipe, 100 mm diameter connects two reservoirs (the entrance and exit of the pipe are sharp-edged). At the midpoint of the pipe is an orifice plate with diameter 40 mm. If the water levels in the reservoirs differ by 30 m, estimate the pressure differential indicated by the orifice plate and the flow rate.

8.195 A venturi meter with a 76.2 mm diameter throat is placed in a 152 mm diameter line carrying water at 24°C. The pressure drop between the upstream tap and the venturi throat is 305 mm of mercury. Compute the rate of flow.

8.196 Consider a horizontal 50 mm × 25 mm venturi with water flow. For a differential pressure of 150 kPa calculate the volume flow rate.

8.197 Gasoline flows through a 50 mm × 25 mm venturi meter. The differential pressure is 380 mm of mercury. Find the volume flow rate.

*These problems require material from sections that may be omitted without loss of continuity in the text material.

8.198 Air flows through the venturi meter described in Problem 8.195. Assume that the upstream pressure is 413 kPa, and that the temperature is everywhere constant at 20°C. Determine the maximum possible mass flow rate of air for which the assumption of incompressible flow is a valid engineering approximation. Compute the corresponding differential pressure reading on a mercury manometer.

8.199 Air flow rate in a test of an internal combustion engine is to be measured using a flow nozzle installed in a plenum. The engine displacement is 1.6 liters, and its maximum operating speed is 6000 rpm. To avoid loading the engine, the maximum pressure drop across the nozzle should not exceed 0.25 m of water. The manometer can be read to ± 0.5 mm of water. Determine the flow nozzle diameter that should be specified. Find the minimum rate of air flow that can be metered to ± 2 percent using this setup.

8.200 Water at 10°C flows steadily through a venturi. The pressure upstream from the throat is 200 kPa (gage). The throat diameter is 50 mm; the upstream diameter is 100 mm. Estimate the maximum flow rate this device can handle without cavitation.

8.201 Derive Eq. 8.42, the pressure loss coefficient for a diffuser assuming ideal (frictionless) flow.

8.202 Consider a flow nozzle installation in a pipe. Apply the basic equations to the control volume indicated, to show that the permanent head loss across the meter can be expressed, in dimensionless form, as the head loss coefficient,

$$C_l = \frac{p_1 - p_3}{p_1 - p_2} = \frac{1 - A_2/A_1}{1 + A_2/A_1}$$

Plot C_l as a function of diameter ratio, D_2/D_1.

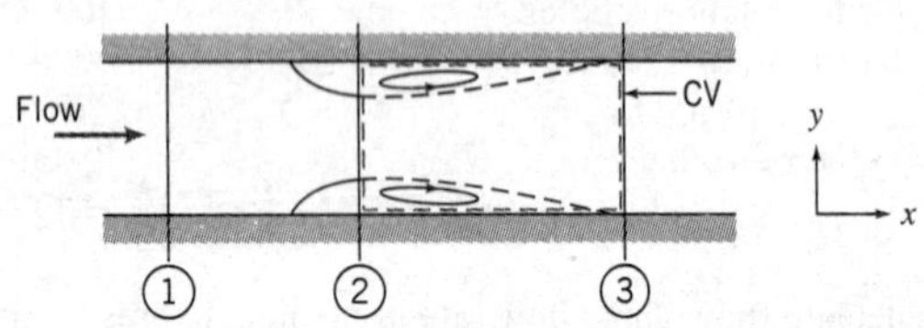

P8.202

8.203 Drinking straws are to be used to improve the air flow in a pipe-flow experiment. Packing a section of the air pipe with drinking straws to form a "laminar flow element" might allow the air flow rate to be measured directly, and simultaneously would act as a flow straightener. To evaluate this idea, determine (a) the Reynolds number for flow in each drinking straw, (b) the friction factor for flow in each straw, and (c) the gage pressure at the exit from the drinking straws. (For laminar flow in a tube, the entrance loss coefficient is $K_{\text{ent}} = 1.4$ and $\alpha = 2.0$.) Comment on the utility of this idea.

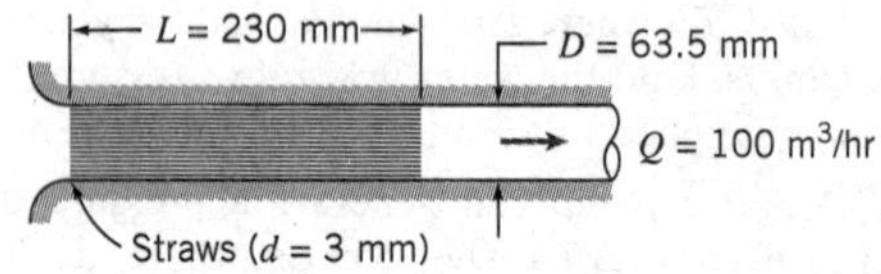

P8.203

8.204 In some western states, water for mining and irrigation was sold by the "miner's inch," the rate at which water flows through an opening in a vertical plank of 645 mm^2 area, up to 102 mm tall, under a head of 152 to 229 mm. Develop an equation to predict the flow rate through such an orifice. Specify clearly the aspect ratio of the opening, thickness of the plank, and datum level for measurement of head (top, bottom, or middle of the opening). Show that the unit of measure varies from 38.4 (in Colorado) to 50 (in Arizona, Idaho, Nevada, and Utah) miner's inches equal to 0.0283 m^3/s.

8.205 The volume flow rate in a circular duct may be measured by "pitot traverse," i.e., by measuring the velocity in each of several area segments across the duct, then summing. Comment on the way such a traverse should be set up. Quantify and plot the expected error in measurement of flow rate as a function of the number of radial locations used in the traverse.

9

External Incompressible Viscous Flow

Case Study in Energy and the Environment

Wind Power: *The Magenn Air Rotor System (MARS)*

In Chapter 8's *Case Study in Energy and the Environment*, we looked at an alternative to the three-bladed wind turbine farms that are cropping up all over the world. In this *Case Study* we look at a second unique idea for wind power, the *Magenn Air Rotor System* (MARS). *Magenn Power* is a company that's been in business several decades and that began designing the *Magnus Spherical Airship* in 1978. The founder of Magenn Power, Fred Ferguson, patented the Magnus Airship in the 1980s. This unique airship generates extra lift due to the Magnus effect. We will discuss the Magnus effect in this chapter, but for now we simply state that the effect is one in which a lift force is generated whenever a sphere or cylinder rotates in a cross flow; the classic example is in golf, where a ball hit with backspin will travel much further due to the lift generated by the spin. The airship was a large spherical envelope filled with helium to achieve static, buoyant lift. It was designed to rotate as it moved so that Magnus lift was generated. With Magnus lift, as we'll learn in this chapter, the faster the spin or the vehicle's forward motion, the larger the Magnus lift.

The *MARS* shown in the rendering is a high-altitude lighter-than-air, tethered device that rotates about a horizontal axis in response to wind; it is essentially a horizontal-axis *Savonius* type of wind turbine (see Figure P9.97 in the problem set). Buoyancy is provided by the helium contained in the device, but its rotation also generates lift via the Magnus effect. The extra lift allows the device to be stabilized in flight, keeping it precisely located, and prevents downwind drift on its tether; the stronger the wind is, the greater the Magnus lift. Rotation of the *MARS* drives a generator inside the device, and the electrical power travels down the tether to a transformer at a ground station.

The *MARS* has a number of advantages over conventional three-blade wind turbines. Magenn Power believes it will develop power less expensively and will have time-averaged output much closer to its rated capacity than the capacity factor typical with conventional designs; it is also operable over a wide range of wind speeds (2 m/s to greater than about 30 m/s). It's thought the efficiency will be 25 to 60 percent. *MARS* wind farms could be placed closer to demand centers than conventional wind turbines, reducing power transmission line initial costs and operation losses. The devices can be raised to a high altitude, where the winds are stronger; altitudes from about 100 m to about 300 m above ground level are possible, without having to build expensive infrastructure. *MARS* are mobile, and could be very useful in emergency situations and in disaster relief. The 10 to 25 kW prototype shown in the figure is now undergoing tests.

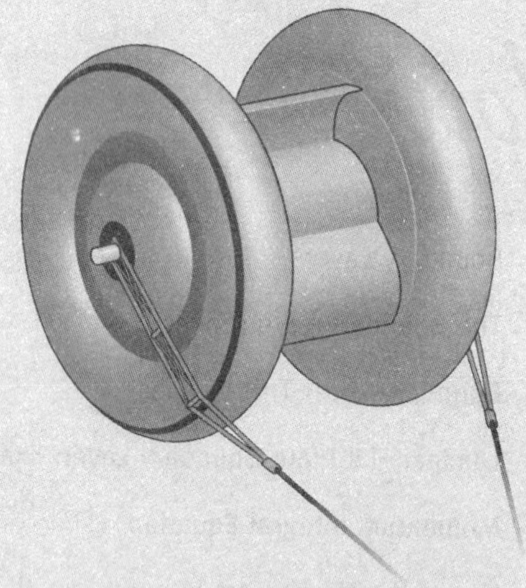

An example of the MARS device

External flows are flows over bodies immersed in an unbounded fluid. The flow over a sphere (Fig. 2.14*b*) and the flow over a streamlined body (Fig. 2.16) are examples of external flows, which were discussed qualitatively in Chapter 2. More interesting examples are the flow fields around such objects as airfoils (Fig. 9.1), automobiles, and airplanes. Our objective in this chapter is to quantify the behavior of viscous, incompressible fluids in external flow.

A number of phenomena that occur in external flow over a body are illustrated in the sketch of viscous flow at high Reynolds number over an airfoil (Fig. 9.1). The freestream flow divides at the stagnation point and flows around the body. Fluid at the surface takes on the velocity of the body as a result of the no-slip condition. Boundary layers form on both the upper and lower surfaces of the body. (The boundary-layer thickness on both surfaces in Fig. 9.1 is exaggerated greatly for clarity.) The flow in the boundary layers initially is

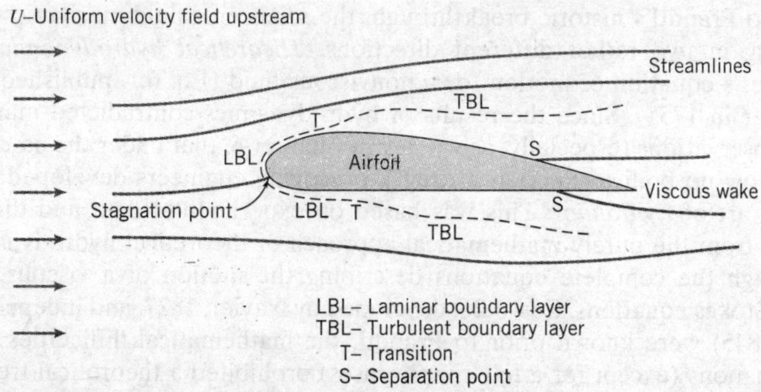

Fig. 9.1 Details of viscous flow around an airfoil.

laminar. Transition to turbulent flow occurs at some distance from the stagnation point, depending on freestream conditions, surface roughness, and pressure gradient. The transition points are indicated by "T" in the figure. The turbulent boundary layer following transition grows more rapidly than the laminar layer. A slight displacement of the streamlines of the external flow is caused by the thickening boundary layers on the surface. In a region of increasing pressure (an *adverse pressure gradient*—so called because it opposes the fluid motion, tending to decelerate the fluid particles) flow separation may occur. Separation points are indicated by "S" in the figure. Fluid that was in the boundary layers on the body surface forms the viscous wake behind the separation points.

VIDEO

Flow around an Airfoil.

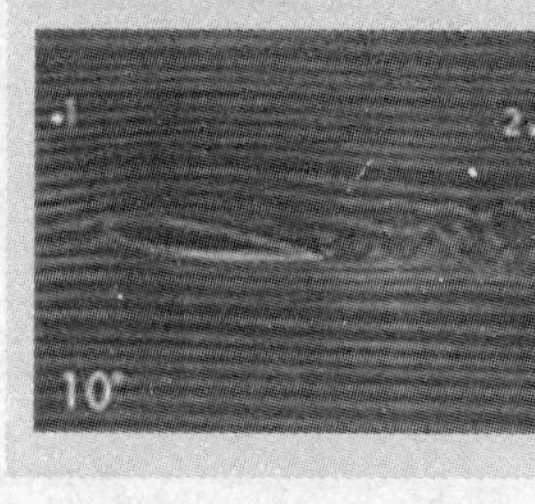

This chapter has two parts. Part A is a review of boundary-layer flows. Here we discuss in a little more detail the ideas introduced in Chapter 2, and then apply the fluid mechanics concepts we have learned to analyze the boundary layer for flow along a flat plate—the simplest possible boundary layer, because the pressure field is constant. We will be interested in seeing how the boundary-layer thickness grows, what the surface friction will be, and so on. We will explore a classic analytical solution for a laminar boundary layer, and see that we need to resort to approximate methods when the boundary layer is turbulent (and we will also be able to use these approximate methods for laminar boundary layers, to avoid using the somewhat difficult analytical method). This will conclude our introduction to boundary layers, except we will briefly discuss the effect of pressure gradients (present for *all* body shapes except flat plates) on boundary-layer behavior.

VIDEO

Flow Separation on an Airfoil.

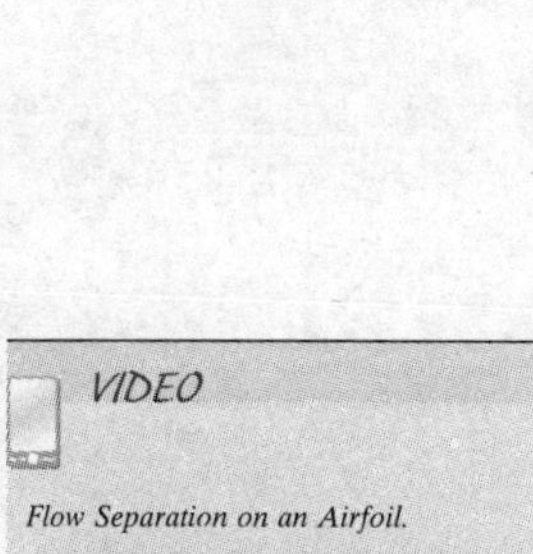

In Part B we will discuss the force on a submerged body, such as the airfoil of Fig. 9.1. We will see that this force results from both shear and pressure forces acting on the body surface, and that both of these are profoundly affected by the fact that we have a boundary layer, especially when this causes flow separation and a wake. Traditionally the force a body experiences is decomposed into the component parallel to the flow, the *drag*, and the component perpendicular to the flow, the *lift*. Because most bodies do have a point of separation and a wake, it is difficult to use analysis to determine the force components, so we will present approximate analyses and experimental data for various interesting body shapes.

Part A Boundary Layers

9.1 The Boundary-Layer Concept

The concept of a boundary layer was first introduced by Ludwig Prandtl [1], a German aerodynamicist, in 1904.

Prior to Prandtl's historic breakthrough, the science of fluid mechanics had been developing in two rather different directions. *Theoretical hydrodynamics* evolved from Euler's equation of motion for a nonviscous fluid (Eq. 6.1, published by Leonhard Euler in 1755). Since the results of hydrodynamics contradicted many experimental observations (especially, as we saw in Chapter 6, that under the assumption of inviscid flow no bodies experience drag!), practicing engineers developed their own empirical art of *hydraulics*. This was based on experimental data and differed significantly from the purely mathematical approach of theoretical hydrodynamics.

Although the complete equations describing the motion of a viscous fluid (the Navier–Stokes equations, Eqs. 5.26, developed by Navier, 1827, and independently by Stokes, 1845) were known prior to Prandtl, the mathematical difficulties in solving these equations (except for a few simple cases) prohibited a theoretical treatment of viscous flows. Prandtl showed [1] that many viscous flows can be analyzed by dividing the flow into two regions, one close to solid boundaries, the other covering the rest of the flow. Only in the thin region adjacent to a solid boundary (the boundary layer) is the effect of viscosity important. In the region outside of the boundary layer, the effect of viscosity is negligible and the fluid may be treated as inviscid.

The boundary-layer concept provided the link that had been missing between theory and practice (for one thing, it introduced the theoretical possibility of drag!). Furthermore, the boundary-layer concept permitted the solution of viscous flow problems that would have been impossible through application of the Navier–Stokes equations to the complete flow field.[1] Thus the introduction of the boundary-layer concept marked the beginning of the modern era of fluid mechanics.

The development of a boundary layer on a solid surface was discussed in Section 2.6. In the boundary layer both viscous and inertia forces are important. Consequently, it is not surprising that the Reynolds number (which represents the ratio of inertia to viscous forces) is significant in characterizing boundary-layer flows. The characteristic length used in the Reynolds number is either the length in the flow direction over which the boundary layer has developed or some measure of the boundary-layer thickness.

As is true for flow in a duct, flow in a boundary layer may be laminar or turbulent. There is no unique value of Reynolds number at which transition from laminar to turbulent flow occurs in a boundary layer. Among the factors that affect boundary-layer transition are pressure gradient, surface roughness, heat transfer, body forces, and freestream disturbances. Detailed consideration of these effects is beyond the scope of this book.

In many real flow situations, a boundary layer develops over a long, essentially flat surface. Examples include flow over ship and submarine hulls, aircraft wings, and atmospheric motions over flat terrain. Since the basic features of all these flows are illustrated in the simpler case of flow over a flat plate, we consider this first. The simplicity of the flow over an infinite flat plate is that the velocity U outside the boundary layer is constant, and therefore, because this region is steady, inviscid, and incompressible, the pressure will also be constant. This constant pressure is the pressure felt by the boundary layer—obviously the simplest pressure field possible. This is a *zero pressure gradient flow*.

A qualitative picture of the boundary-layer growth over a flat plate is shown in Fig. 9.2. The boundary layer is laminar for a short distance downstream from the leading edge; transition occurs over a region of the plate rather than at a single line across the plate. The transition region extends downstream to the location where the boundary-layer flow becomes completely turbulent.

For incompressible flow over a smooth flat plate (zero pressure gradient), in the absence of heat transfer, transition from laminar to turbulent flow in the boundary layer can be delayed to a Reynolds number, $Re_x = \rho Ux/\mu$, greater than one million if

[1]Today, computer solutions of the Navier–Stokes equations are common.

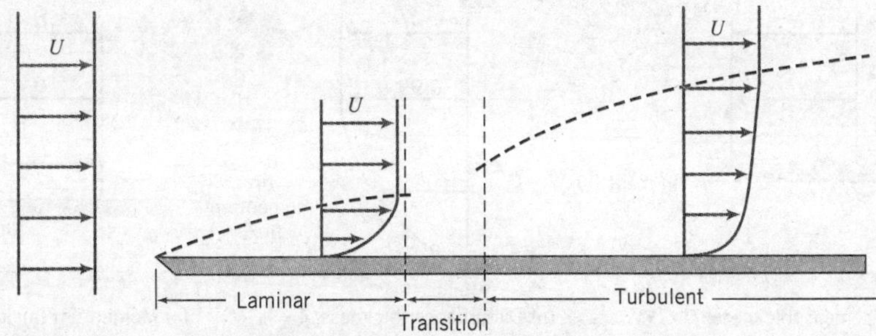

Fig. 9.2 Boundary layer on a flat plate (vertical thickness exaggerated greatly).

VIDEO

Laminar and Turbulent Boundary Layers.

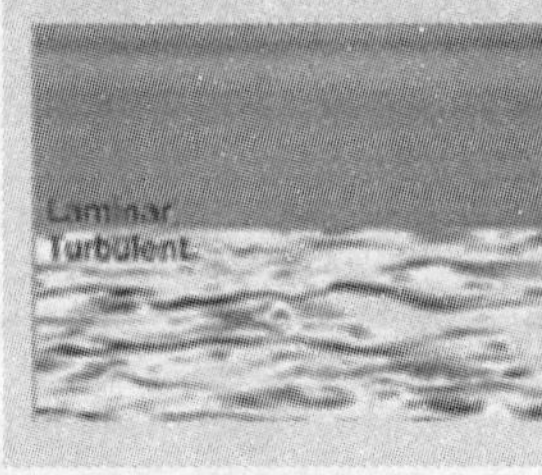

external disturbances are minimized. (The length x is measured from the leading edge.) For calculation purposes, under typical flow conditions, transition usually is considered to occur at a length Reynolds number of 500,000. For air at standard conditions, with freestream velocity $U = 30$ m/s, this corresponds to $x \approx 0.24$ m. In the qualitative picture of Fig. 9.2, we have shown the turbulent boundary layer growing faster than the laminar layer. In later sections of this chapter we shall show that this is indeed true.

In the next section we discuss various ways to quantify the thickness of a boundary layer.

Boundary-Layer Thicknesses 9.2

VIDEO

Growth of the Boundary Layer.

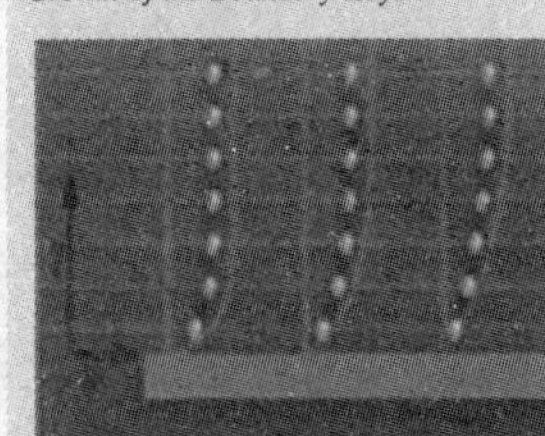

The boundary layer is the region adjacent to a solid surface in which viscous stresses are present, as opposed to the free stream where viscous stresses are negligible. These stresses are present because we have shearing of the fluid layers, i.e., a velocity gradient, in the boundary layer. As indicated in Fig. 9.2, both laminar and turbulent layers have such gradients, but the difficulty is that the gradients only asymptotically approach zero as we reach the edge of the boundary layer. Hence, the location of the edge, i.e., of the boundary-layer thickness, is not very obvious—we cannot simply define it as where the boundary-layer velocity u equals the freestream velocity U. Because of this, several boundary-layer definitions have been developed: the disturbance thickness δ, the displacement thickness δ^*, and the momentum thickness θ. (Each of these increases as we move down the plate, in a manner we have yet to determine.)

VIDEO

Effect of Viscosity on Boundary Layer Growth.

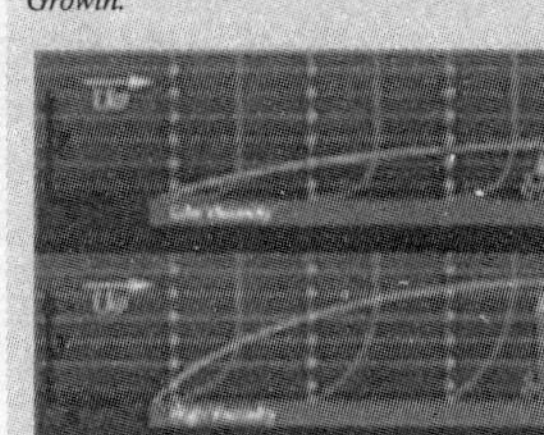

The most straightforward definition is the disturbance thickness, δ. This is usually defined as the distance from the surface at which the velocity is within 1 percent of the free stream, $u \approx 0.99U$ (as shown in Fig. 9.3*b*). The other two definitions are based on the notion that the boundary layer retards the fluid, so that the mass flux and momentum flux are both less than they would be in the absence of the boundary layer. We imagine that the flow remains at uniform velocity U, but the surface of the plate is moved upwards to reduce either the mass or momentum flux by the same amount that the boundary layer actually does. The *displacement thickness*, δ^*, is the distance the plate would be moved so that the loss of mass flux (due to reduction in uniform flow area) is equivalent to the loss the boundary layer causes. The mass flux if we had no boundary layer would be $\int_0^\infty \rho U \, dy \, w$, where w is the width of the plate perpendicular to the flow. The actual flow mass flux is $\int_0^\infty \rho u \, dy \, w$. Hence, the loss due to the boundary layer is $\int_0^\infty \rho(U-u)\, dy \, w$. If we imagine keeping the velocity at a constant U, and instead move the plate up a distance δ^* (as shown in Fig. 9.3*a*), the loss of mass flux would be $\rho U \delta^* w$. Setting these losses equal to one another gives

$$\rho U \delta^* w = \int_0^\infty \rho (U-u)\, dy \, w$$

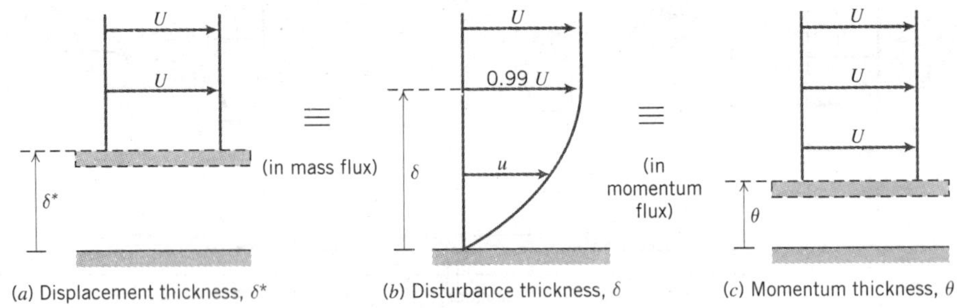

Fig. 9.3 Boundary-layer thickness definitions.

For incompressible flow, ρ = constant, and

$$\delta^* = \int_0^\infty \left(1 - \frac{u}{U}\right) dy \approx \int_0^\delta \left(1 - \frac{u}{U}\right) dy \tag{9.1}$$

Since $u \approx U$ at $y = \delta$, the integrand is essentially zero for $y \geq \delta$. Application of the displacement-thickness concept is illustrated in Example 9.1.

The *momentum thickness*, θ, is the distance the plate would be moved so that the loss of momentum flux is equivalent to the loss the boundary layer actually causes. The momentum flux if we had no boundary layer would be $\int_0^\infty \rho u U \, dy \, w$ (the actual mass flux is $\int_0^\infty \rho u \, dy \, w$, and the momentum per unit mass flux of the uniform flow is U itself). The actual momentum flux of the boundary layer is $\int_0^\infty \rho u^2 \, dy \, w$. Hence, the loss of momentum in the boundary layer is $\int_0^\infty \rho u(U - u) \, dy \, w$. If we imagine keeping the velocity at a constant U, and instead move the plate up a distance θ (as shown in Fig. 9.3*c*), the loss of momentum flux would be $\int_0^\theta \rho UU \, dy \, w = \rho U^2 \theta w$. Setting these losses equal to one another gives

$$\rho U^2 \theta = \int_0^\infty \rho u(U - u) \, dy$$

and

$$\theta = \int_0^\infty \frac{u}{U}\left(1 - \frac{u}{U}\right) dy \approx \int_0^\delta \frac{u}{U}\left(1 - \frac{u}{U}\right) dy \tag{9.2}$$

Again, the integrand is essentially zero for $y \geq \delta$.

The displacement and momentum thicknesses, δ^* and θ, are *integral thicknesses*, because their definitions, Eqs. 9.1 and 9.2, are in terms of integrals across the boundary layer. Because they are defined in terms of integrals for which the integrand vanishes in the freestream, they are appreciably easier to evaluate accurately from experimental data than the boundary-layer disturbance thickness, δ. This fact, coupled with their physical significance, accounts for their common use in specifying boundary-layer thickness.

We have seen that the velocity profile in the boundary layer merges into the local freestream velocity asymptotically. Little error is introduced if the slight difference between velocities at the edge of the boundary layer is ignored for an approximate analysis. Simplifying assumptions usually made for engineering analyses of boundary-layer development are:

1. $u \to U$ at $y = \delta$
2. $\partial u/\partial y \to 0$ at $y = \delta$
3. $v \ll U$ within the boundary layer

Results of the analyses developed in the next two sections show that the boundary layer is very thin compared with its development length along the surface. Therefore it is also reasonable to assume:

4. Pressure variation across the thin boundary layer is negligible. The freestream pressure distribution is *impressed* on the boundary layer.

Example 9.1 BOUNDARY LAYER IN CHANNEL FLOW

A laboratory wind tunnel has a test section that is 305 mm square. Boundary-layer velocity profiles are measured at two cross-sections and displacement thicknesses are evaluated from the measured profiles. At section ①, where the freestream speed is $U_1 = 26$ m/s, the displacement thickness is $\delta_1^* = 1.5$ mm. At section ②, located downstream from section ①, $\delta_2^* = 2.1$ mm. Calculate the change in static pressure between sections ① and ②. Express the result as a fraction of the freestream dynamic pressure at section ①. Assume standard atmosphere conditions.

Given: Flow of standard air in laboratory wind tunnel. Test section is $L = 305$ mm square. Displacement thicknesses are $\delta_1^* = 1.5$ mm and $\delta_2^* = 2.1$ mm. Freestream speed is $U_1 = 26$ m/s.

Find: Change in static pressure between sections ① and ②. (Express as a fraction of freestream dynamic pressure at section ①.)

Solution:
The idea here is that at each location the boundary-layer displacement thickness effectively reduces the area of uniform flow, as indicated in the following figures: Location ② has a smaller effective flow area than location ① (because $\delta_2^* > \delta_1^*$). Hence, from mass conservation the uniform velocity at location ② will be higher. Finally, from the Bernoulli equation the pressure at location ② will be lower than that at location ①.

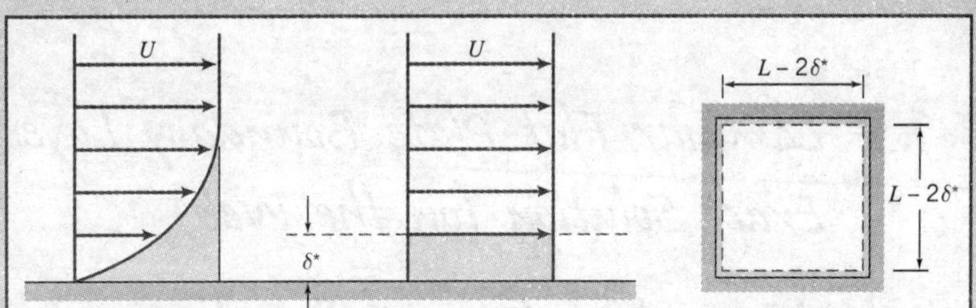

(*a*) Actual velocity profile (*b*) Hypothetical velocity profile (*c*) Cross section of wind tunnel

Apply the continuity and Bernoulli equations to freestream flow outside the boundary-layer displacement thickness, where viscous effects are negligible.

Governing equations:

$$\overset{=0(1)}{\cancel{\frac{\partial}{\partial t}\int_{CV} \rho \, d\forall}} + \int_{CS} \rho \vec{V} \cdot d\vec{A} = 0 \tag{4.12}$$

$$\frac{p_1}{\rho} + \frac{V_1^2}{2} + \cancel{gz_1} = \frac{p_2}{\rho} + \frac{V_2^2}{2} + \cancel{gz_2} \tag{4.24}$$

Assumptions: (1) Steady flow.
(2) Incompressible flow.
(3) Flow uniform at each section outside δ^*.

(4) Flow along a streamline between sections ① and ②.
(5) No frictional effects in freestream.
(6) Negligible elevation changes.

From the Bernoulli equation we obtain

$$p_1 - p_2 = \frac{1}{2}\rho(V_2^2 - V_1^2) = \frac{1}{2}\rho(U_2^2 - U_1^2) = \frac{1}{2}\rho U_1^2\left[\left(\frac{U_2}{U_1}\right)^2 - 1\right]$$

or

$$\frac{p_1 - p_2}{\frac{1}{2}\rho U_1^2} = \left(\frac{U_2}{U_1}\right)^2 - 1$$

From continuity, $V_1 A_1 = U_1 A_1 = V_2 A_2 = U_2 A_2$, so $U_2/U_1 = A_1/A_2$, where $A = (L - 2\delta^*)^2$ is the effective flow area. Substituting gives

$$\frac{p_1 - p_2}{\frac{1}{2}\rho U_1^2} = \left(\frac{A_1}{A_2}\right)^2 - 1 = \left[\frac{(L - 2\delta_1^*)^2}{(L - 2\delta_2^*)^2}\right]^2 - 1$$

$$\frac{p_1 - p_2}{\frac{1}{2}\rho U_1^2} = \left[\frac{305 - 2(1.5)}{305 - 2(2.1)}\right]^4 - 1 = 0.0161 \quad \text{or}$$

$$\frac{p_1 - p_2}{\frac{1}{2}\rho U_1^2} = 1.61 \text{ percent} \longleftarrow \frac{p_1 - p_2}{\frac{1}{2}\rho U_1^2}$$

Notes:

✓ This problem illustrates a basic application of the displacement-thickness concept. It is somewhat unusual in that, because the flow is confined, the reduction in flow area caused by the boundary layer leads to the result that the pressure in the inviscid flow region drops (if only slightly). In most applications the pressure distribution is determined from the inviscid flow and *then* applied to the boundary layer.

✓ We saw a similar phenomenon in Section 8.1, where we discovered that the centerline velocity at the entrance of a pipe increases due to the boundary layer "squeezing" the effective flow area.

9.3 Laminar Flat-Plate Boundary Layer: Exact Solution (on the Web)

9.4 Momentum Integral Equation

Blasius' exact solution, discussed in Section 9.3 (on the Web), analyzed a laminar boundary layer on a flat plate. Even this simplest case (i.e., constant freestream velocity U and pressure p, laminar flow) involved a rather subtle mathematical transformation of two differential equations. The solution was based on the insight that the laminar boundary-layer velocity profile is self-similar—only its scale changes as we move along the plate. Numerical integration was necessary to obtain results for the boundary-layer thickness $\delta(x)$, velocity profile u/U versus y/δ, and wall shear stress $\tau_w(x)$.

We would like to obtain a method for analyzing the general case—that is, for laminar *and* turbulent boundary layers, for which the freestream velocity $U(x)$ and pressure $p(x)$ are known functions of position along the surface x (such as on the curved surface of an airfoil or on the flat but divergent surfaces of a flow diffuser). The approach is one in which we will again apply the basic equations to a control volume. The derivation, from the mass conservation (or continuity) equation and the momentum equation, will take several pages.

Consider incompressible, steady, two-dimensional flow over a solid surface. The boundary-layer thickness, δ, grows in some manner with increasing distance, x.

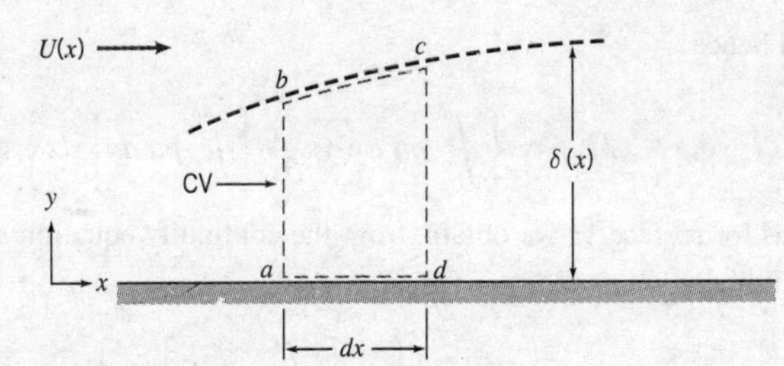

Fig. 9.4 Differential control volume in a boundary layer.

For our analysis we choose a differential control volume, of length dx, width w, and height $\delta(x)$, as shown in Fig. 9.4. The freestream velocity is $U(x)$.

We wish to determine the boundary-layer thickness, δ, as a function of x. There will be mass flow across surfaces ab and cd of differential control volume $abcd$. What about surface bc? Surface bc is *not* a streamline (we showed this in Example 9.2, on the Web); it is the imaginary boundary that separates the viscous boundary layer and the inviscid freestream flow. Thus there will be mass flow across surface bc. Since control surface ad is adjacent to a solid boundary, there will not be flow across ad. Before considering the forces acting on the control volume and the momentum fluxes through the control surface, let us apply the continuity equation to determine the mass flux through each portion of the control surface.

a. Continuity Equation

Basic equation:

$$\overbrace{\frac{\partial}{\partial t}\int_{\text{CV}} \rho \, d\forall}^{=0(1)} + \int_{\text{CS}} \rho \vec{V} \cdot d\vec{A} = 0 \tag{4.12}$$

Assumptions: (1) Steady flow.
(2) Two-dimensional flow.

Then

$$\int_{\text{CS}} \rho \vec{V} \cdot d\vec{A} = 0$$

Hence

$$\dot{m}_{ab} + \dot{m}_{bc} + \dot{m}_{cd} = 0$$

or

$$\dot{m}_{bc} = -\dot{m}_{ab} - \dot{m}_{cd}$$

Now let us evaluate these terms for the differential control volume of width w:

Surface	Mass Flux
ab	Surface ab is located at x. Since the flow is two-dimensional (no variation with z), the mass flux through ab is $$\dot{m}_{ab} = -\left\{\int_0^\delta \rho u \, dy\right\} w$$
cd	Surface cd is located at $x + dx$. Expanding $\dot{m}$ in a Taylor series about location x, we obtain $$\dot{m}_{x+dx} = \dot{m}_x + \left.\frac{\partial \dot{m}}{\partial x}\right]_x dx$$

and hence

$$\dot{m}_{cd} = \left\{ \int_0^\delta \rho u \, dy + \frac{\partial}{\partial x}\left[\int_0^\delta \rho u \, dy\right] dx \right\} w$$

bc Thus for surface *bc* we obtain, from the continuity equation and the above results,

$$\dot{m}_{bc} = -\left\{ \frac{\partial}{\partial x}\left[\int_0^\delta \rho u \, dy\right] dx \right\} w$$

(Note that the velocity u and boundary-layer thickness δ, the upper limit on the integral, both depend on x.)

Now let us consider the momentum fluxes and forces associated with control volume *abcd*. These are related by the momentum equation.

b. Momentum Equation

Apply the x component of the momentum equation to control volume *abcd*:

Basic equation:

$$\overset{=0(3)}{} \quad \overset{=0(1)}{}$$
$$F_{S_x} + \cancel{F_{B_x}} = \cancel{\frac{\partial}{\partial t}} \int_{\mathrm{CV}} u \, \rho \, d\forall + \int_{\mathrm{CS}} u \, \rho \vec{V} \cdot d\vec{A} \qquad (4.18a)$$

Assumption: (3) $F_{B_x} = 0$.

Then

$$F_{S_x} = \mathrm{mf}_{ab} + \mathrm{mf}_{bc} + \mathrm{mf}_{cd}$$

where mf represents the x component of momentum flux.

To apply this equation to differential control volume *abcd*, we must obtain expressions for the x momentum flux through the control surface and also the surface forces acting on the control volume in the x direction. Let us consider the momentum flux first and again consider each segment of the control surface.

Surface	Momentum Flux (mf)
ab	Surface *ab* is located at x. Since the flow is two-dimensional, the x momentum flux through *ab* is $$\mathrm{mf}_{ab} = -\left\{ \int_0^\delta u \, \rho u \, dy \right\} w$$
cd	Surface *cd* is located at $x + dx$. Expanding the x momentum flux (mf) in a Taylor series about location x, we obtain $$\mathrm{mf}_{x+dx} = \mathrm{mf}_x + \left.\frac{\partial \mathrm{mf}}{\partial x}\right]_x dx$$ or $$\mathrm{mf}_{cd} = \left\{ \int_0^\delta u \, \rho u \, dy + \frac{\partial}{\partial x}\left[\int_0^\delta u \, \rho u \, dy\right] dx \right\} w$$

bc Since the mass crossing surface *bc* has velocity component U in the x direction, the x momentum flux across *bc* is given by

$$\text{mf}_{bc} = U\dot{m}_{bc}$$

$$\text{mf}_{bc} = -U\left\{\frac{\partial}{\partial x}\left[\int_0^\delta \rho u\,dy\right]dx\right\}w$$

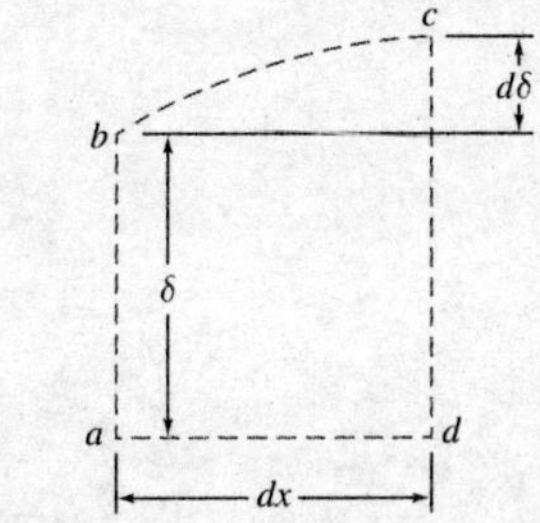

Fig. 9.5 Differential control volume.

From the above we can evaluate the net x momentum flux through the control surface as

$$\int_{\text{CS}} u\,\rho\vec{V}\cdot d\vec{A} = -\left\{\int_0^\delta u\,\rho u\,dy\right\}w + \left\{\int_0^\delta u\,\rho u\,dy\right\}w$$
$$+\left\{\frac{\partial}{\partial x}\left[\int_0^\delta u\,\rho u\,dy\right]dx\right\}w - U\left\{\frac{\partial}{\partial x}\left[\int_0^\delta \rho u\,dy\right]dx\right\}w$$

Collecting terms, we find that

$$\int_{\text{CS}} u\,\rho\vec{V}\cdot d\vec{A} = \left\{\frac{\partial}{\partial x}\left[\int_0^\delta u\,\rho u\,dy\right]dx - U\frac{\partial}{\partial x}\left[\int_0^\delta \rho u\,dy\right]dx\right\}w$$

Now that we have a suitable expression for the x momentum flux through the control surface, let us consider the surface forces acting on the control volume in the x direction. (For convenience the differential control volume has been redrawn in Fig. 9.5.) Note that surfaces *ab*, *bc*, and *cd* all experience normal forces (i.e., pressure) that generate force in the x direction. In addition, a shear force acts on surface *ad*. Since, by definition of the boundary layer, the velocity gradient goes to zero at the edge of the boundary layer, the shear force acting along surface *bc* is negligible.

Surface	Force

ab If the pressure at x is p, then the force acting on surface *ab* is given by

$$F_{ab} = pw\delta$$

[The boundary layer is very thin; its thickness has been greatly exaggerated in all the sketches we have made. Because it is thin, pressure variations in the y direction may be neglected, and we assume that within the boundary layer, $p = p(x)$ only.]

cd Expanding in a Taylor series, the pressure at $x + dx$ is given by

$$p_{x+dx} = p + \left.\frac{dp}{dx}\right]_x dx$$

The force on surface *cd* is then given by

$$F_{cd} = -\left(p + \left.\frac{dp}{dx}\right]_x dx\right)w(\delta + d\delta)$$

bc The average pressure acting over surface *bc* is

$$p + \frac{1}{2}\left.\frac{dp}{dx}\right]_x dx$$

Then the x component of the normal force acting over *bc* is given by

$$F_{bc} = \left(p + \frac{1}{2}\left.\frac{dp}{dx}\right]_x dx\right)w\,d\delta$$

ad The average shear force acting on *ad* is given by

$$F_{ad} = -\left(\tau_w + \frac{1}{2}\,d\tau_w\right)w\,dx$$

Summing these x components, we obtain the total force acting in the x direction on the control volume,

$$F_{S_x} = \left\{ -\frac{dp}{dx}\,\delta\, dx - \frac{1}{2}\frac{dp}{dx}\,\overset{\approx 0}{dx\, d\delta} - \tau_w\, dx - \frac{1}{2}\,\overset{\approx 0}{d\tau_w\, dx} \right\} w$$

where we note that $dx\, d\delta \ll \delta\, dx$ and $d\tau_w \ll \tau_w$, and so neglect the second and fourth terms.

Substituting the expressions, for $\int_{CS} u\, \rho \vec{V} \cdot d\vec{A}$ and F_{S_x} into the x momentum equation (Eq. 4.18*a*), we obtain

$$\left\{ -\frac{dp}{dx}\,\delta\, dx - \tau_w\, dx \right\} w = \left\{ \frac{\partial}{\partial x}\left[\int_0^\delta u\, \rho u\, dy \right] dx - U \frac{\partial}{\partial x}\left[\int_0^\delta \rho u\, dy \right] dx \right\} w$$

Dividing this equation by $w\, dx$ gives

$$-\delta \frac{dp}{dx} - \tau_w = \frac{\partial}{\partial x}\int_0^\delta u\, \rho u\, dy - U \frac{\partial}{\partial x}\int_0^\delta \rho u\, dy \tag{9.16}$$

Equation 9.16 is a "momentum integral" equation that gives a relation between the x components of the forces acting in a boundary layer and the x momentum flux.

The pressure gradient, dp/dx, can be determined by applying the Bernoulli equation to the inviscid flow outside the boundary layer: $dp/dx = -\rho U\, dU/dx$. If we recognize that $\delta = \int_0^\delta dy$, then Eq. 9.16 can be written as

$$\tau_w = -\frac{\partial}{\partial x}\int_0^\delta u\, \rho u\, dy + U \frac{\partial}{\partial x}\int_0^\delta \rho u\, dy + \frac{dU}{dx}\int_0^\delta \rho U\, dy$$

Since

$$U \frac{\partial}{\partial x}\int_0^\delta \rho u\, dy = \frac{\partial}{\partial x}\int_0^\delta \rho u U\, dy - \frac{dU}{dx}\int_0^\delta \rho u\, dy$$

we have

$$\tau_w = \frac{\partial}{\partial x}\int_0^\delta \rho u (U - u)\, dy + \frac{dU}{dx}\int_0^\delta \rho (U - u)\, dy$$

and

$$\tau_w = \frac{\partial}{\partial x} U^2 \int_0^\delta \rho \frac{u}{U}\left(1 - \frac{u}{U}\right) dy + U \frac{dU}{dx}\int_0^\delta \rho \left(1 - \frac{u}{U}\right) dy$$

Using the definitions of displacement thickness, δ^* (Eq. 9.1), and momentum thickness, θ (Eq. 9.2), we obtain

$$\frac{\tau_w}{\rho} = \frac{d}{dx}(U^2 \theta) + \delta^* U \frac{dU}{dx} \tag{9.17}$$

Equation 9.17 is the *momentum integral equation*. This equation will yield an ordinary differential equation for boundary-layer thickness δ as a function of x. Where does δ appear in Eq. 9.17? It appears in the upper limits of the integrals that define δ^* and θ! All we need to do is provide a suitable expression for the velocity profile u/U and somehow relate the wall stress τ_w to other variables—not necessarily easy tasks! Once the boundary-layer thickness is determined, expressions for the momentum thickness, displacement thickness, and wall shear stress can then be obtained.

Equation 9.17 was obtained by applying the basic equations (continuity and x momentum) to a differential control volume. Reviewing the assumptions we made in the derivation, we see that the equation is restricted to steady, incompressible, two-dimensional flow with no body forces parallel to the surface.

We have not made any specific assumption relating the wall shear stress, τ_w, to the velocity field. Thus Eq. 9.17 is valid for either a laminar or turbulent boundary-layer flow. In order to use this equation to estimate the boundary-layer thickness as a function of x, we must first:

1. Obtain a first approximation to the freestream velocity distribution, $U(x)$. This is determined from inviscid flow theory (the velocity that would exist in the absence of a boundary layer) and depends on body shape.
2. Assume a reasonable velocity-profile shape inside the boundary layer.
3. Derive an expression for τ_w using the results obtained from item **2**.

To illustrate the application of Eq. 9.17 to boundary-layer flows, we consider first the case of flow with zero pressure gradient over a flat plate (Section 9.5)—the results we obtain for a laminar boundary layer can then be compared to the exact Blasius results. The effects of pressure gradients in boundary-layer flow are then discussed in Section 9.6.

9.5 Use of the Momentum Integral Equation for Flow with Zero Pressure Gradient

For the special case of a flat plate (zero pressure gradient) the freestream pressure p and velocity U are both constant, so for item **1** we have $U(x) = U =$ constant.

The momentum integral equation then reduces to

$$\tau_w = \rho U^2 \frac{d\theta}{dx} = \rho U^2 \frac{d}{dx} \int_0^\delta \frac{u}{U}\left(1 - \frac{u}{U}\right) dy \tag{9.18}$$

The velocity distribution, u/U, in the boundary layer is assumed to be similar for all values of x and normally is specified as a function of y/δ. (Note that u/U is dimensionless and δ is a function of x only.) Consequently, it is convenient to change the variable of integration from y to y/δ. Defining

$$\eta = \frac{y}{\delta}$$

we get

$$dy = \delta \, d\eta$$

and the momentum integral equation for zero pressure gradient is written

$$\tau_w = \rho U^2 \frac{d\theta}{dx} = \rho U^2 \frac{d\delta}{dx} \int_0^1 \frac{u}{U}\left(1 - \frac{u}{U}\right) d\eta \tag{9.19}$$

We wish to solve this equation for the boundary-layer thickness as a function of x. To do this, we must satisfy the remaining items:

2. Assume a velocity distribution in the boundary layer—a functional relationship of the form

$$\frac{u}{U} = f\left(\frac{y}{\delta}\right)$$

a. The assumed velocity distribution should satisfy the following approximate physical boundary conditions:

$$\text{at } y = 0, \qquad u = 0$$

$$\text{at } y = \delta, \qquad u = U$$

$$\text{at } y = \delta, \qquad \frac{\partial u}{\partial y} = 0$$

b. Note that once we have assumed a velocity distribution, from the definition of the momentum thickness (Eq. 9.2), the numerical value of the integral in Eq. 9.19 is simply

$$\int_0^1 \frac{u}{U}\left(1 - \frac{u}{U}\right)d\eta = \frac{\theta}{\delta} = \text{constant} = \beta$$

and the momentum integral equation becomes

$$\tau_w = \rho U^2 \frac{d\delta}{dx}\beta$$

3. Obtain an expression for τ_w in terms of δ. This will then permit us to solve for $\delta(x)$, as illustrated below.

Laminar Flow

For laminar flow over a flat plate, a reasonable assumption for the velocity profile is a polynomial in y:

$$u = a + by + cy^2$$

The physical boundary conditions are:

$$\text{at } y = 0, \qquad u = 0$$

$$\text{at } y = \delta, \qquad u = U$$

$$\text{at } y = \delta, \qquad \frac{\partial u}{\partial y} = 0$$

Evaluating constants a, b, and c gives

$$\frac{u}{U} = 2\left(\frac{y}{\delta}\right) - \left(\frac{y}{\delta}\right)^2 = 2\eta - \eta^2 \tag{9.20}$$

Equation 9.20 satisfies item 2. For item 3, we recall that the wall shear stress is given by

$$\tau_w = \mu \left.\frac{\partial u}{\partial y}\right)_{y=0}$$

Substituting the assumed velocity profile, Eq. 9.20, into this expression for τ_w gives

$$\tau_w = \mu \left.\frac{\partial u}{\partial y}\right]_{y=0} = \mu \left.\frac{U\,\partial(u/U)}{\delta\,\partial(y/\delta)}\right]_{y/\delta=0} = \left.\frac{\mu U}{\delta}\frac{d(u/U)}{d\eta}\right]_{\eta=0}$$

or

$$\tau_w = \left.\frac{\mu U}{\delta}\frac{d}{d\eta}(2\eta - \eta^2)\right]_{\eta=0} = \left.\frac{\mu U}{\delta}(2 - 2\eta)\right]_{\eta=0} = \frac{2\mu U}{\delta}$$

Note that this shows that the wall stress τ_w is a function of x, since the boundary-layer thickness $\delta = \delta(x)$. Now that we have completed items **1**, **2**, and **3**, we can return to the momentum integral equation

$$\tau_w = \rho U^2 \frac{d\delta}{dx} \int_0^1 \frac{u}{U}\left(1 - \frac{u}{U}\right) d\eta \tag{9.19}$$

Substituting for τ_w and u/U, we obtain

$$\frac{2\mu U}{\delta} = \rho U^2 \frac{d\delta}{dx} \int_0^1 (2\eta - \eta^2)(1 - 2\eta + \eta^2)\, d\eta$$

or

$$\frac{2\mu U}{\delta \rho U^2} = \frac{d\delta}{dx} \int_0^1 (2\eta - 5\eta^2 + 4\eta^3 - \eta^4)\, d\eta$$

Integrating and substituting limits yields

$$\frac{2\mu}{\delta \rho U} = \frac{2}{15}\frac{d\delta}{dx} \quad \text{or} \quad \delta\, d\delta = \frac{15\mu}{\rho U}\, dx$$

which is a differential equation for δ. Integrating again gives

$$\frac{\delta^2}{2} = \frac{15\mu}{\rho U} x + c$$

If we assume that $\delta = 0$ at $x = 0$, then $c = 0$, and thus

$$\delta = \sqrt{\frac{30\mu x}{\rho U}}$$

Note that this shows that the laminar boundary-layer thickness δ grows as $\sqrt{x}$; it has a parabolic shape. Traditionally this is expressed in dimensionless form:

$$\frac{\delta}{x} = \sqrt{\frac{30\mu}{\rho U x}} = \frac{5.48}{\sqrt{Re_x}} \tag{9.21}$$

VIDEO

Examples of Boundary Layer Growth.

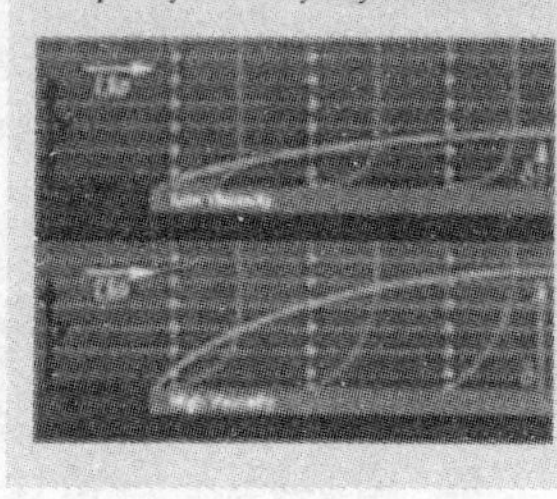

Equation 9.21 shows that the ratio of laminar boundary-layer thickness to distance along a flat plate varies inversely with the square root of length Reynolds number. It has the same form as the exact solution derived from the complete differential equations of motion by H. Blasius in 1908. Remarkably, Eq. 9.21 is only in error (the constant is too large) by about 10 percent compared with the exact solution (Section 9.3 on the Web). Table 9.2 summarizes corresponding results calculated using other approximate velocity profiles and lists results obtained from the exact solution. (The only thing that changes in the analysis when we choose a different velocity profile is the value of β in $\tau_w = \rho U^2 d\delta/dx\beta$ in item **2b** on page 434.) The shapes of the approximate profiles may be compared readily by plotting u/U versus y/δ.

Once we know the boundary-layer thickness, all details of the flow may be determined. The wall shear stress, or "skin friction," coefficient is defined as

Table 9.2

Results of the Calculation of Laminar Boundary-Layer Flow over a Flat Plate at Zero Incidence Based on Approximate Velocity Profiles

Velocity Distribution $\frac{u}{U} = f\left(\frac{y}{\delta}\right) = f(\eta)$	$\beta \equiv \frac{\theta}{\delta}$	$\frac{\delta^*}{\delta}$	$H \equiv \frac{\delta^*}{\theta}$	Constant a in $\frac{\delta}{x} = \frac{a}{\sqrt{Re_x}}$	Constant b in $C_f = \frac{b}{\sqrt{Re_x}}$
$f(\eta) = \eta$	$\frac{1}{6}$	$\frac{1}{2}$	3.00	3.46	0.577
$f(\eta) = 2\eta - \eta^2$	$\frac{2}{15}$	$\frac{1}{3}$	2.50	5.48	0.730
$f(\eta) = \frac{3}{2}\eta - \frac{1}{2}\eta^3$	$\frac{39}{280}$	$\frac{3}{8}$	2.69	4.64	0.647
$f(\eta) = 2\eta - 2\eta^3 + \eta^4$	$\frac{37}{315}$	$\frac{3}{10}$	2.55	5.84	0.685
$f(\eta) = \sin\left(\frac{\pi}{2}\eta\right)$	$\frac{4-\pi}{2\pi}$	$\frac{\pi-2}{\pi}$	2.66	4.80	0.654
Exact	0.133	0.344	2.59	5.00	0.664

$$C_f \equiv \frac{\tau_w}{\frac{1}{2}\rho U^2} \tag{9.22}$$

Substituting from the velocity profile and Eq. 9.21 gives

$$C_f = \frac{\tau_w}{\frac{1}{2}\rho U^2} = \frac{2\mu(U/\delta)}{\frac{1}{2}\rho U^2} = \frac{4\mu}{\rho U\delta} = 4\frac{\mu}{\rho Ux}\frac{x}{\delta} = 4\frac{1}{Re_x}\frac{\sqrt{Re_x}}{5.48}$$

Finally,

$$C_f = \frac{0.730}{\sqrt{Re_x}} \tag{9.23}$$

Once the variation of τ_w is known, the viscous drag on the surface can be evaluated by integrating over the area of the flat plate, as illustrated in Example 9.3.

Equation 9.21 can be used to calculate the thickness of the laminar boundary layer at transition. At $Re_x = 5 \times 10^5$, with $U = 30$ m/s, for example, $x = 0.24$ m for air at standard conditions. Thus

$$\frac{\delta}{x} = \frac{5.48}{\sqrt{Re_x}} = \frac{5.48}{\sqrt{5 \times 10^5}} = 0.00775$$

and the boundary-layer thickness is

$$\delta = 0.00775x = 0.00775(0.24\text{ m}) = 1.86\text{ mm}$$

The boundary-layer thickness at transition is less than 1 percent of the development length, x. These calculations confirm that viscous effects are confined to a very thin layer near the surface of a body.

The results in Table 9.2 indicate that reasonable results may be obtained with a variety of approximate velocity profiles.

Example 9.3 LAMINAR BOUNDARY LAYER ON A FLAT PLATE: APPROXIMATE SOLUTION USING SINUSOIDAL VELOCITY PROFILE

Consider two-dimensional laminar boundary-layer flow along a flat plate. Assume the velocity profile in the boundary layer is sinusoidal,

$$\frac{u}{U} = \sin\left(\frac{\pi}{2}\frac{y}{\delta}\right)$$

Find expressions for:

(a) The rate of growth of δ as a function of x.
(b) The displacement thickness, δ^*, as a function of x.
(c) The total friction force on a plate of length L and width b.

Given: Two-dimensional, laminar boundary-layer flow along a flat plate. The boundary-layer velocity profile is

$$\frac{u}{U} = \sin\left(\frac{\pi}{2}\frac{y}{\delta}\right) \qquad \text{for } 0 \le y \le \delta$$

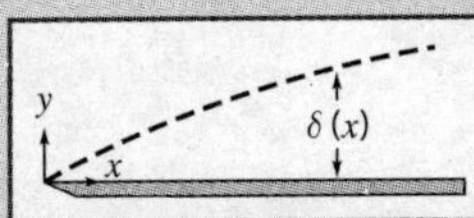

and

$$\frac{u}{U} = 1 \qquad \text{for } y > \delta$$

Find: (a) $\delta(x)$. (b) $\delta^*(x)$.
(c) Total friction force on a plate of length L and width b.

Solution:
For flat plate flow, U = constant, $dp/dx = 0$, and

$$\tau_w = \rho U^2 \frac{d\theta}{dx} = \rho U^2 \frac{d\delta}{dx} \int_0^1 \frac{u}{U}\left(1 - \frac{u}{U}\right) d\eta \tag{9.19}$$

Assumptions: (1) Steady flow.
(2) Incompressible flow.

Substituting $\dfrac{u}{U} = \sin\dfrac{\pi}{2}\eta$ into Eq. 9.19, we obtain

$$\tau_w = \rho U^2 \frac{d\delta}{dx}\int_0^1 \sin\frac{\pi}{2}\eta\left(1 - \sin\frac{\pi}{2}\eta\right)d\eta = \rho U^2 \frac{d\delta}{dx}\int_0^1\left(\sin\frac{\pi}{2}\eta - \sin^2\frac{\pi}{2}\eta\right)d\eta$$

$$= \rho U^2 \frac{d\delta}{dx}\frac{2}{\pi}\left[-\cos\frac{\pi}{2}\eta - \frac{1}{2}\frac{\pi}{2}\eta + \frac{1}{4}\sin\pi\eta\right]_0^1 = \rho U^2 \frac{d\delta}{dx}\frac{2}{\pi}\left[0 + 1 - \frac{\pi}{4} + 0 + 0 - 0\right]$$

$$\tau_w = 0.137\rho U^2 \frac{d\delta}{dx} = \beta\rho U^2 \frac{d\delta}{dx}; \quad \beta = 0.137$$

Now

$$\tau_w = \mu\frac{\partial u}{\partial y}\bigg]_{y=0} = \mu\frac{U}{\delta}\frac{\partial(u/U)}{\partial(y/\delta)}\bigg]_{y=0} = \mu\frac{U}{\delta}\frac{\pi}{2}\cos\frac{\pi}{2}\eta\bigg]_{\eta=0} = \frac{\pi\mu U}{2\delta}$$

Therefore,

$$\tau_w = \frac{\pi\mu U}{2\delta} = 0.137\rho U^2 \frac{d\delta}{dx}$$

Separating variables gives

$$\delta\, d\delta = 11.5\frac{\mu}{\rho U}dx$$

Integrating, we obtain

$$\frac{\delta^2}{2} = 11.5 \frac{\mu}{\rho U} x + c$$

But $c = 0$, since $\delta = 0$ at $x = 0$, so

$$\delta = \sqrt{23.0 \frac{x\mu}{\rho U}}$$

or

$$\frac{\delta}{x} = 4.80\sqrt{\frac{\mu}{\rho Ux}} = \frac{4.80}{\sqrt{Re_x}} \longleftarrow \qquad \delta(x)$$

The displacement thickness, δ^*, is given by

$$\delta^* = \delta \int_0^1 \left(1 - \frac{u}{U}\right) d\eta$$

$$= \delta \int_0^1 \left(1 - \sin \frac{\pi}{2} \eta\right) d\eta = \delta \left[\eta + \frac{2}{\pi} \cos \frac{\pi}{2} \eta\right]_0^1$$

$$\delta^* = \delta \left[1 - 0 + 0 - \frac{2}{\pi}\right] = \delta \left[1 - \frac{2}{\pi}\right]$$

Since, from part (a),

$$\frac{\delta}{x} = \frac{4.80}{\sqrt{Re_x}}$$

then

$$\frac{\delta^*}{x} = \left(1 - \frac{2}{\pi}\right) \frac{4.80}{\sqrt{Re_x}} = \frac{1.74}{\sqrt{Re_x}} \longleftarrow \qquad \delta^*(x)$$

The total friction force on one side of the plate is given by

$$F = \int_{A_p} \tau_w \, dA$$

Since $dA = b\,dx$ and $0 \le x \le L$, then

$$F = \int_0^L \tau_w b\,dx = \int_0^L \rho U^2 \frac{d\theta}{dx} b\,dx = \rho U^2 b \int_0^{\theta_L} d\theta = \rho U^2 b \theta_L$$

$$\theta_L = \int_0^{\delta_L} \frac{u}{U}\left(1 - \frac{u}{U}\right) dy = \delta_L \int_0^1 \frac{u}{U}\left(1 - \frac{u}{U}\right) d\eta = \beta \delta_L$$

From part (a), $\beta = 0.137$ and $\delta_L = \dfrac{4.80L}{\sqrt{Re_L}}$, so

$$F = \frac{0.658 \rho U^2 bL}{\sqrt{Re_L}} \longleftarrow \qquad F$$

This problem illustrates application of the momentum integral equation to the laminar boundary layer on a flat plate.
The *Excel* workbook for this Example plots the growth of δ and δ^* in the boundary layer, and the exact solution (Eq. 9.13 on the Web). It also shows wall shear stress distributions for the sinusoidal velocity profile and the exact solution.

Turbulent Flow

For the flat plate, we still have for item **1** that U = constant. As for the laminar boundary layer, we need to satisfy item **2** (an approximation for the turbulent velocity profile) and item **3** (an expression for τ_w) in order to solve Eq. 9.19 for $\delta(x)$:

$$\tau_w = \rho U^2 \frac{d\delta}{dx} \int_0^1 \frac{u}{U}\left(1 - \frac{u}{U}\right) d\eta \tag{9.19}$$

Details of the turbulent velocity profile for boundary layers at zero pressure gradient are very similar to those for turbulent flow in pipes and channels. Data for turbulent boundary layers plot on the universal velocity profile using coordinates of $\bar{u}/u_*$ versus yu_*/ν, as shown in Fig. 8.9. However, this profile is rather complex mathematically for easy use with the momentum integral equation. The momentum integral equation is approximate; hence, an acceptable velocity profile for turbulent boundary layers on smooth flat plates is the empirical power-law profile. An exponent of $\frac{1}{7}$ is typically used to model the turbulent velocity profile. Thus

$$\frac{u}{U} = \left(\frac{y}{\delta}\right)^{1/7} = \eta^{1/7} \tag{9.24}$$

However, this profile does not hold in the immediate vicinity of the wall, since at the wall it predicts $du/dy = \infty$. Consequently, we cannot use this profile in the definition of τ_w to obtain an expression for τ_w in terms of δ as we did for laminar boundary-layer flow. For turbulent boundary-layer flow we adapt the expression developed for pipe flow,

$$\tau_w = 0.0332\rho\bar{V}^2 \left[\frac{\nu}{R\bar{V}}\right]^{0.25} \tag{8.39}$$

For a $\frac{1}{7}$-power profile in a pipe, Eq. 8.24 gives $\bar{V}/U = 0.817$. Substituting $\bar{V} = 0.817U$ and $R = \delta$ into Eq. 8.39, we obtain

$$\tau_w = 0.0233\rho U^2 \left(\frac{\nu}{U\delta}\right)^{1/4} \tag{9.25}$$

Substituting for τ_w and u/U into Eq. 9.19 and integrating, we obtain

$$0.0233\left(\frac{\nu}{U\delta}\right)^{1/4} = \frac{d\delta}{dx} \int_0^1 \eta^{1/7}\left(1 - \eta^{1/7}\right) d\eta = \frac{7}{72}\frac{d\delta}{dx}$$

Thus we obtain a differential equation for δ:

$$\delta^{1/4}\, d\delta = 0.240\left(\frac{\nu}{U}\right)^{1/4} dx$$

Integrating gives

$$\frac{4}{5}\delta^{5/4} = 0.240\left(\frac{\nu}{U}\right)^{1/4} x + c$$

If we assume that $\delta \simeq 0$ at $x = 0$ (this is equivalent to assuming turbulent flow from the leading edge), then $c = 0$ and

$$\delta = 0.382\left(\frac{\nu}{U}\right)^{1/5} x^{4/5}$$

Note that this shows that the turbulent boundary-layer thickness δ grows as $x^{4/5}$; it grows almost linearly (recall that δ grows more slowly, as $\sqrt{x}$, for the laminar boundary layer). Traditionally this is expressed in dimensionless form:

$$\frac{\delta}{x} = 0.382\left(\frac{\nu}{Ux}\right)^{1/5} = \frac{0.382}{Re_x^{1/5}} \tag{9.26}$$

Using Eq. 9.25, we obtain the skin friction coefficient in terms of δ:

$$C_f = \frac{\tau_w}{\frac{1}{2}\rho U^2} = 0.0466\left(\frac{\nu}{U\delta}\right)^{1/4}$$

Substituting for δ, we obtain

$$C_f = \frac{\tau_w}{\frac{1}{2}\rho U^2} = \frac{0.0594}{Re_x^{1/5}} \tag{9.27}$$

Experiments show that Eq. 9.27 predicts turbulent skin friction on a flat plate very well for $5 \times 10^5 < Re_x < 10^7$. This agreement is remarkable in view of the approximate nature of our analysis.

Application of the momentum integral equation for turbulent boundary-layer flow is illustrated in Example 9.4.

Example 9.4 TURBULENT BOUNDARY LAYER ON A FLAT PLATE: APPROXIMATE SOLUTION USING $\frac{1}{7}$-POWER VELOCITY PROFILE

Water flows at $U = 1$ m/s past a flat plate with $L = 1$ m in the flow direction. The boundary layer is tripped so it becomes turbulent at the leading edge. Evaluate the disturbance thickness, δ, displacement thickness, δ^*, and wall shear stress, τ_w, at $x = L$. Compare with laminar flow maintained to the same position. Assume a $\frac{1}{7}$-power turbulent velocity profile.

Given: Flat-plate boundary-layer flow; turbulent flow from the leading edge. Assume $\frac{1}{7}$-power velocity profile.

Find: (a) Disturbance thickness, δ_L.
(b) Displacement thickness, δ_L^*.
(c) Wall shear stress, $\tau_w(L)$.
(d) Comparison with results for laminar flow from the leading edge.

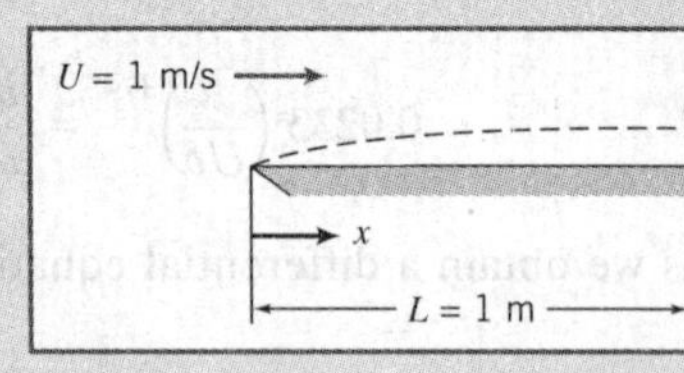

Solution:
Apply results from the momentum integral equation.

Governing equations:

$$\frac{\delta}{x} = \frac{0.382}{Re_x^{1/5}} \tag{9.26}$$

$$\delta^* = \int_0^\infty \left(1 - \frac{u}{U}\right)dy \tag{9.1}$$

$$C_f = \frac{\tau_w}{\frac{1}{2}\rho U^2} = \frac{0.0594}{Re_x^{1/5}} \tag{9.27}$$

At $x = L$, with $\nu = 1.00 \times 10^{-6}$ m^2/s for water ($T = 20°$C),

$$Re_L = \frac{UL}{\nu} = 1\frac{\text{m}}{\text{s}} \times 1\,\text{m} \times \frac{\text{s}}{10^{-6}\,\text{m}^2} = 10^6$$

From Eq. 9.26,

$$\delta_L = \frac{0.382}{Re_L^{1/5}} L = \frac{0.382}{(10^6)^{1/5}} \times 1\,\text{m} = 0.0241\,\text{m} \quad \text{or} \quad \delta_L = 24.1\,\text{mm} \longleftarrow \qquad \delta_L$$

Using Eq. 9.1, with $u/U = (y/\delta)^{1/7} = \eta^{1/7}$, we obtain

$$\delta_L^* = \int_0^\infty \left(1 - \frac{u}{U}\right) dy = \delta_L \int_0^1 \left(\frac{u}{U}\right) d\left(\frac{y}{\delta}\right) = \delta_L \int_0^1 \left(1 - \eta^{1/7}\right) d\eta = \delta_L \left[\eta - \frac{7}{8}\eta^{8/7}\right]_0^1$$

$$\delta_L^* = \frac{\delta_L}{8} = \frac{24.1\,\text{mm}}{8} = 3.01\,\text{mm} \longleftarrow \qquad \delta_L^*$$

From Eq. 9.27,

$$C_f = \frac{0.0594}{(10^6)^{1/5}} = 0.00375$$

$$\tau_w = C_f \frac{1}{2}\rho U^2 = 0.00375 \times \frac{1}{2} \times 999 \frac{\text{kg}}{\text{m}^3} \times (1)^2 \frac{\text{m}^2}{\text{s}^2} \times \frac{\text{N} \cdot \text{s}^2}{\text{kg} \cdot \text{m}}$$

$$\tau_w = 1.87\,\text{N/m}^2 \longleftarrow \qquad \tau_w(L)$$

For laminar flow, use Blasius solution values. From Eq. 9.13 (on the Web),

$$\delta_L = \frac{5.0}{\sqrt{Re_L}} L = \frac{5.0}{(10^6)^{1/2}} \times 1\,\text{m} = 0.005\,\text{m} \quad \text{or} \quad 5.00\,\text{mm}$$

From Example 9.2, $\delta^*/\delta = 0.344$, so

$$\delta^* = 0.344 \quad \delta = 0.344 \times 5.0\,\text{mm} = 1.72\,\text{mm}$$

From Eq. 9.15, $C_f = \dfrac{0.664}{\sqrt{Re_x}}$, so

$$\tau_w = C_f \frac{1}{2}\rho U^2 = \frac{0.664}{\sqrt{10^6}} \times \frac{1}{2} \times 999 \frac{\text{kg}}{\text{m}^3} \times (1)^2 \frac{\text{m}^2}{\text{s}^2} \times \frac{\text{N} \cdot \text{s}^2}{\text{kg} \cdot \text{m}} = 0.332\,\text{N/m}^2$$

Comparing values at $x = L$, we obtain

$$\text{Disturbance thickness,} \quad \frac{\delta_{\text{turbulent}}}{\delta_{\text{laminar}}} = \frac{24.1\,\text{mm}}{5.00\,\text{mm}} = 4.82$$

$$\text{Displacement thickness,} \quad \frac{\delta^*_{\text{turbulent}}}{\delta^*_{\text{laminar}}} = \frac{3.01\,\text{mm}}{1.72\,\text{mm}} = 1.75$$

$$\text{Wall shear stress,} \quad \frac{\tau_{w,\,\text{turbulent}}}{\tau_{w,\,\text{laminar}}} = \frac{1.87\,\text{N/m}^2}{0.332\,\text{N/m}^2} = 5.63$$

This problem illustrates application of the momentum integral equation to the turbulent boundary layer on a flat plate. Compared to a laminar boundary layer, it is clear that the turbulent boundary layer grows much more rapidly—because the turbulent wall stress is significantly greater than the laminar wall stress.

The *Excel* workbook for this Example plots the $\frac{1}{7}$-power-law turbulent boundary layer (Eq. 9.26) and the laminar boundary layer (Eq. 9.13 on the Web). It also shows the wall stress distributions for both cases.

Summary of Results for Boundary-Layer Flow with Zero Pressure Gradient

Use of the momentum integral equation is an approximate technique to predict boundary-layer development; the equation predicts trends correctly. Parameters of the laminar boundary layer vary as $Re_x^{-1/2}$; those for the turbulent boundary layer vary as $Re_x^{-1/5}$. Thus the turbulent boundary layer develops more rapidly than the laminar boundary layer.

Laminar and turbulent boundary layers were compared in Example 9.4. Wall shear stress is much higher in the turbulent boundary layer than in the laminar layer. This is the primary reason for the more rapid development of turbulent boundary layers.

The agreement we have obtained with experimental results shows that use of the momentum integral equation is an effective approximate method that gives us considerable insight into the general behavior of boundary layers.

9.6 Pressure Gradients in Boundary-Layer Flow

The boundary layer (laminar or turbulent) with a uniform flow along an infinite flat plate is the easiest one to study because the pressure gradient is zero—the fluid particles in the boundary layer are slowed only by shear stresses, leading to boundary-layer growth. We now consider the effects caused by a pressure gradient, which will be present for all bodies except, as we have seen, a flat plate.

A *favorable pressure gradient* is one in which the pressure decreases in the flow direction (i.e., $\partial p/\partial x < 0$); it is called favorable because it tends to overcome the slowing of fluid particles caused by friction in the boundary layer. This pressure gradient arises when the freestream velocity U is increasing with x, for example, in the converging flow field in a nozzle. On the other hand, an *adverse pressure gradient* is one in which pressure increases in the flow direction (i.e., $\partial p/\partial x > 0$); it is called adverse because it will cause fluid particles in the boundary-layer to slow down at a greater rate than that due to boundary-layer friction alone. If the adverse pressure gradient is severe enough, the fluid particles in the boundary layer will actually be brought to rest. When this occurs, the particles will be forced away from the body surface (a phenomenon called *flow separation*) as they make room for following particles, ultimately leading to a *wake* in which flow is turbulent. Examples of this are when the walls of a diffuser diverge too rapidly and when an airfoil has too large an angle of attack; both of these are generally very undesirable!

VIDEO

Flow Separation: Sudden Expansion.

This description, of the adverse pressure gradient and friction in the boundary layer together forcing flow separation, certainly makes intuitive sense; the question arises whether we can more formally see when this occurs. For example, can we have flow separation and a wake for uniform flow over a flat plate, for which $\partial p/\partial x = 0$? We can gain insight into this question by considering when the velocity in the boundary layer will become zero. Consider the velocity u in the boundary layer at an infinitesimal distance Δy above the plate. This will be

$$u_{y=\Delta y} = u_0 + \left.\frac{\partial u}{\partial y}\right)_{y=0} \Delta y = \left.\frac{\partial u}{\partial y}\right)_{y=0} \Delta y$$

where $u_0 = 0$ is the velocity at the surface of the plate. It is clear that $u_{y=\Delta y}$ will be zero (i.e., separation will occur) only when $\partial u/\partial y)_{y=0} = 0$. Hence, we can use this as our litmus test for flow separation. We recall that the velocity gradient near the surface in a laminar boundary layer, and in the viscous sublayer of a turbulent boundary layer, was related to the wall shear stress by

VIDEO

Flow Separation: Airfoil.

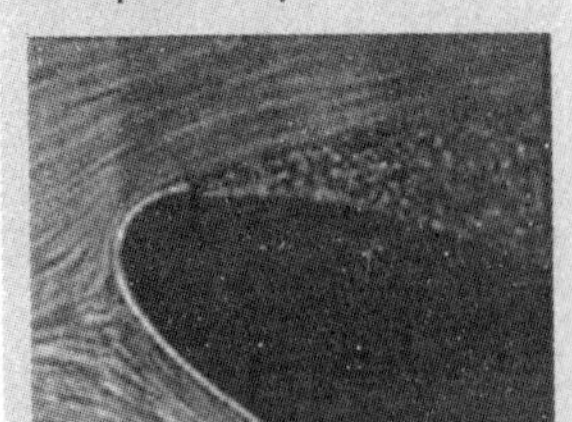

$$\tau_w = \mu \left.\frac{\partial u}{\partial y}\right)_{y=0}$$

Further, we learned in the previous section that the wall shear stress for the flat plate is given by

$$\frac{\tau_w(x)}{\rho U^2} = \frac{\text{constant}}{\sqrt{Re_x}}$$

for a laminar boundary layer and

$$\frac{\tau_w(x)}{\rho U^2} = \frac{\text{constant}}{Re_x^{1/5}}$$

for a turbulent boundary layer. We see that for the flow over a flat plate, the wall stress is always $\tau_w > 0$. Hence, $\partial u/\partial y)_{y=0} > 0$ always; and therefore, finally, $u_{y=\Delta y} > 0$ always. We conclude that for uniform flow over a flat plate the flow *never* separates, and we never develop a wake region, whether the boundary layer is laminar or turbulent, regardless of plate length.

We conclude that flow will not separate for flow over a flat plate, when $\partial p/\partial x = 0$. Clearly, for flows in which $\partial p/\partial x < 0$ (whenever the freestream velocity is increasing), we can be sure that there will be no flow separation; for flows in which $\partial p/\partial x > 0$ (i.e., adverse pressure gradients) we *could* have flow separation. We should not conclude that an adverse pressure gradient *always* leads to flow separation and a wake; we have only concluded that it is a necessary condition for flow separation to occur.

To illustrate these results consider the variable cross-sectional flow shown in Fig. 9.6. Outside the boundary layer the velocity field is one in which the flow accelerates (Region 1), has a constant velocity region (Region 2), and then a deceleration region (Region 3). Corresponding to these, the pressure gradient is favorable, zero, and adverse, respectively, as shown. (Note that the straight wall is not a simple flat plate—it has these various pressure gradients because the flow above the wall is not a uniform flow.) From our discussions above, we conclude that separation cannot occur in Region 1 or 2, but can (as shown) occur in Region 3. Could we avoid flow separation in a device like this? Intuitively, we can see that if we make the divergent section less severe, we may be able to eliminate flow separation. In other words, we may eliminate flow separation if we make the adverse pressure gradient $\partial p/\partial x$ small enough. The final question remaining is how small the adverse pressure gradient needs to be to accomplish this. This, and a more rigorous proof that we must have $\partial p/\partial x > 0$ for a chance of flow separation, is beyond the scope of this text [3]. We conclude that flow separation is possible, but not guaranteed, when we have an adverse pressure gradient.

The nondimensional velocity profiles for laminar and turbulent boundary-layer flow over a flat plate are shown in Fig. 9.7*a*. The turbulent profile is much fuller (more blunt) than the laminar profile. At the same freestream speed, the momentum flux within the turbulent boundary layer is greater than within the laminar layer (Fig. 9.7*b*). Separation occurs when the momentum of fluid layers near the surface is reduced to zero by the combined action of pressure and viscous forces. As shown in Fig. 9.7*b*, the momentum of the fluid near the surface is significantly greater for the turbulent profile.

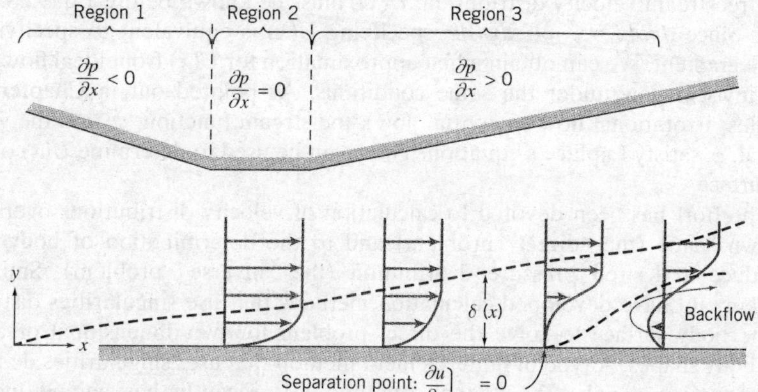

Fig. 9.6 Boundary-layer flow with pressure gradient (boundary-layer thickness exaggerated for clarity).

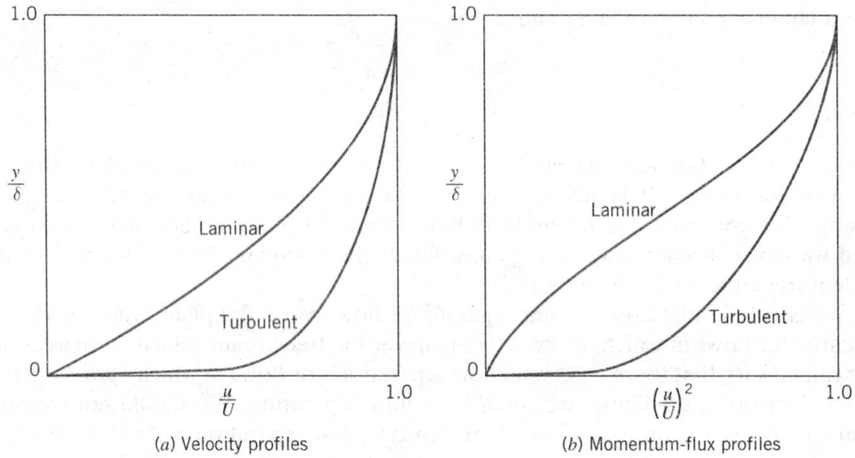

Fig. 9.7 Nondimensional profiles for flat plate boundary-layer flow.

Consequently, the turbulent layer is better able to resist separation in an adverse pressure gradient. We shall discuss some consequences of this behavior in Section 9.7.

Adverse pressure gradients cause significant changes in velocity profiles for both laminar and turbulent boundary-layer flows. Approximate solutions for nonzero pressure gradient flow may be obtained from the momentum integral equation

$$\frac{\tau_w}{\rho} = \frac{d}{dx}(U^2\theta) + \delta^* U \frac{dU}{dx} \tag{9.17}$$

Expanding the first term, we can write

$$\frac{\tau_w}{\rho} = U^2 \frac{d\theta}{dx} + (\delta^* + 2\theta)\, U \frac{dU}{dx}$$

or

$$\frac{\tau_w}{\rho U^2} = \frac{C_f}{2} = \frac{d\theta}{dx} + (H+2)\, \frac{\theta}{U} \frac{dU}{dx} \tag{9.28}$$

where $H = \delta^*/\theta$ is a velocity-profile "shape factor." The shape factor increases in an adverse pressure gradient. For turbulent boundary-layer flow, H increases from 1.3 for a zero pressure gradient to approximately 2.5 at separation. For laminar flow with zero pressure gradient, $H = 2.6$; at separation $H = 3.5$.

The freestream velocity distribution, $U(x)$, must be known before Eq. 9.28 can be applied. Since $dp/dx = -\rho U\, dU/dx$, specifying $U(x)$ is equivalent to specifying the pressure gradient. We can obtain a first approximation for $U(x)$ from ideal flow theory for an inviscid flow under the same conditions. As pointed out in Chapter 6, for frictionless irrotational flow (potential flow), the stream function, ψ, and the velocity potential, ϕ, satisfy Laplace's equation. These can be used to determine $U(x)$ over the body surface.

Much effort has been devoted to calculation of velocity distributions over bodies of known shape (the "direct" problem) and to the determination of body shapes to produce a desired pressure distribution (the "inverse" problem). Smith and co-workers [6] have developed calculation methods that use singularities distributed over the body surface to solve the direct problem for two-dimensional or axisymmetric body shapes. A type of finite-element method that uses singularities defined on discrete surface panels (the "panel" method [7]) recently has gained increased popularity for application to three-dimensional flows. Recall also that in Section 5.5 we briefly reviewed some basic ideas of CFD (Computational Fluid Dynamics).

Once the velocity distribution, $U(x)$, is known, Eq. 9.28 can be integrated to determine $\theta(x)$, if H and C_f can be correlated with θ. A detailed discussion of various calculation methods for flows with nonzero pressure gradient is beyond the scope of this book. Numerous solutions for laminar flows are given in Kraus [8]. Calculation methods for turbulent boundary-layer flow based on the momentum integral equation are reviewed in Rotta [9].

Because of the importance of turbulent boundary layers in engineering flow situations, the state of the art of calculation schemes is advancing rapidly. Numerous calculation schemes have been proposed [10, 11]; most such schemes for turbulent flow use models to predict turbulent shear stress and then solve the boundary-layer equations numerically [12, 13]. Continuing improvement in size and speed of computers is beginning to make possible the solution of the full Navier–Stokes equations using numerical methods [14, 15].

Part B Fluid Flow About Immersed Bodies

Whenever there is relative motion between a solid body and the viscous fluid surrounding it, the body will experience a net force $\vec{F}$. The magnitude of this force depends on many factors—certainly the relative velocity $\vec{V}$, but also the body shape and size, and the fluid properties (ρ, μ, etc.). As the fluid flows around the body, it will generate surface stresses on each element of the surface, and it is these that lead to the net force. The surface stresses are composed of tangential stresses due to viscous action and normal stresses due to the local pressure. We might be tempted to think that we can analytically derive the net force by integrating these over the body surface. The first step might be: Given the shape of the body (and assuming that the Reynolds number is high enough that we can use inviscid flow theory), compute the pressure distribution. Then integrate the pressure over the body surface to obtain the contribution of pressure forces to the net force $\vec{F}$. (As we discussed in Chapter 6, this step was developed very early in the history of fluid mechanics; it led to the result that no bodies experience drag!) The second step might be: Use this pressure distribution to find the surface viscous stress τ_w (at least in principle, using, for example, Eq. 9.17). Then integrate the viscous stress over the body surface to obtain its contribution to the net force $\vec{F}$. This procedure sounds conceptually straightforward, but in practice is quite difficult except for the simplest body shapes. In addition, even if possible, it leads to erroneous results in most cases because it takes no account of a very important consequence of the existence of boundary layers—flow separation. This causes a wake, which not only creates a low-pressure region usually leading to large drag on the body, but also radically changes the overall flow field and hence the inviscid flow region and pressure distribution on the body.

VIDEO

Flow about a Sports Car.

For these reasons we must usually resort to experimental methods to determine the net force for most body shapes (although CFD approaches are improving rapidly). Traditionally the net force $\vec{F}$ is resolved into the drag force, F_D, defined as the component of the force parallel to the direction of motion, and the lift force, F_L (if it exists for a body), defined as the component of the force perpendicular to the direction of motion. In Sections 9.7 and 9.8 we will examine these forces for a number of different body shapes.

Drag 9.7

Drag is the component of force on a body acting parallel to the direction of relative motion. In discussing the need for experimental results in fluid mechanics (Chapter 7), we considered the problem of determining the drag force, F_D, on a smooth sphere of diameter d, moving through a viscous, incompressible fluid with speed V; the fluid density and viscosity were ρ and μ, respectively. The drag force, F_D, was written in the functional form

$$F_D = f_1(d, V, \mu, \rho)$$

Application of the Buckingham Pi theorem resulted in two dimensionless Π parameters that were written in functional form as

$$\frac{F_D}{\rho V^2 d^2} = f_2\left(\frac{\rho V d}{\mu}\right)$$

Note that d^2 is proportional to the cross-sectional area ($A = \pi d^2/4$) and therefore we could write

$$\frac{F_D}{\rho V^2 A} = f_3\left(\frac{\rho V d}{\mu}\right) = f_3(Re) \tag{9.29}$$

Although Eq. 9.29 was obtained for a sphere, the form of the equation is valid for incompressible flow over *any* body; the characteristic length used in the Reynolds number depends on the body shape.

The *drag coefficient*, C_D, is defined as

$$C_D \equiv \frac{F_D}{\frac{1}{2}\rho V^2 A} \tag{9.30}$$

The number $\frac{1}{2}$ has been inserted (as was done in the defining equation for the friction factor) to form the familiar dynamic pressure. Then Eq. 9.29 can be written as

$$C_D = f(Re) \tag{9.31}$$

We have not considered compressibility or free-surface effects in this discussion of the drag force. Had these been included, we would have obtained the functional form

$$C_D = f(Re, Fr, M)$$

At this point we shall consider the drag force on several bodies for which Eq. 9.31 is valid. The total drag force is the sum of friction drag and pressure drag. However, the drag coefficient is a function only of the Reynolds number.

We now consider the drag force and drag coefficient for a number of bodies, starting with the simplest: a flat plate parallel to the flow (which has only friction drag); a flat plate normal to the flow (which has only pressure drag); and cylinders and spheres (the simplest 2D and 3D bodies, which have both friction and pressure drag). We will also briefly discuss streamlining.

Pure Friction Drag: Flow over a Flat Plate Parallel to the Flow

This flow situation was considered in detail in Section 9.5. Since the pressure gradient is zero (and in any event the pressure forces are perpendicular to the plate and therefore do not contribute to drag), the total drag is equal to the friction drag. Thus

$$F_D = \int_{\text{plate surface}} \tau_w \, dA$$

and

$$C_D = \frac{F_D}{\frac{1}{2}\rho V^2 A} = \frac{\int_{PS} \tau_w \, dA}{\frac{1}{2}\rho V^2 A} \tag{9.32}$$

where A is the total surface area in contact with the fluid (i.e., the *wetted area*). The drag coefficient for a flat plate parallel to the flow depends on the shear stress distribution along the plate.

For laminar flow over a flat plate, the shear stress coefficient was given by

$$C_f = \frac{\tau_w}{\frac{1}{2}\rho U^2} = \frac{0.664}{\sqrt{Re_x}} \tag{9.15}$$

The drag coefficient for flow with freestream velocity V, over a flat plate of length L and width b, is obtained by substituting for τ_w from Eq. 9.15 into Eq. 9.32. Thus

$$C_D = \frac{1}{A}\int_A 0.664\, Re_x^{-0.5}\, dA = \frac{1}{bL}\int_0^L 0.664 \left(\frac{V}{\nu}\right)^{-0.5} x^{-0.5} b\, dx$$

$$= \frac{0.664}{L}\left(\frac{\nu}{V}\right)^{0.5}\left[\frac{x^{0.5}}{0.5}\right]_0^L = 1.33\left(\frac{\nu}{VL}\right)^{0.5}$$

$$C_D = \frac{1.33}{\sqrt{Re_L}} \tag{9.33}$$

Assuming the boundary layer is turbulent from the leading edge, the shear stress coefficient, based on the approximate analysis of Section 9.5, is given by

$$C_f = \frac{\tau_w}{\frac{1}{2}\rho U^2} = \frac{0.0594}{Re_x^{1/5}} \tag{9.27}$$

Substituting for τ_w from Eq. 9.27 into Eq. 9.32, we obtain

$$C_D = \frac{1}{A}\int_A 0.0594\, Re_x^{-0.2}\, dA = \frac{1}{bL}\int_0^L 0.0594 \left(\frac{V}{\nu}\right)^{-0.2} x^{-0.2} b\, dx$$

$$= \frac{0.0594}{L}\left(\frac{\nu}{V}\right)^{0.2}\left[\frac{x^{0.8}}{0.8}\right]_0^L = 0.0742\left(\frac{\nu}{VL}\right)^{0.2}$$

$$C_D = \frac{0.0742}{Re_L^{1/5}} \tag{9.34}$$

Equation 9.34 is valid for $5 \times 10^5 < Re_L < 10^7$.

For $Re_L < 10^9$ the empirical equation given by Schlichting [3]

$$C_D = \frac{0.455}{(\log Re_L)^{2.58}} \tag{9.35}$$

fits experimental data very well.

For a boundary layer that is initially laminar and undergoes transition at some location on the plate, the turbulent drag coefficient must be adjusted to account for the laminar flow over the initial length. The adjustment is made by subtracting the quantity B/Re_L from the C_D determined for completely turbulent flow. The value of B depends on the Reynolds number at transition; B is given by

$$B = Re_{tr}(C_{D_{\text{turbulent}}} - C_{D_{\text{laminar}}}) \tag{9.36}$$

For a transition Reynolds number of 5×10^5, the drag coefficient may be calculated by making the adjustment to Eq. 9.34, in which case

$$C_D = \frac{0.0742}{Re_L^{1/5}} - \frac{1740}{Re_L} \qquad (5 \times 10^5 < Re_L < 10^7) \tag{9.37a}$$

or to Eq. 9.35, in which case

$$C_D = \frac{0.455}{(\log Re_L)^{2.58}} - \frac{1610}{Re_L} \qquad (5 \times 10^5 < Re_L < 10^9) \tag{9.37b}$$

The variation in drag coefficient for a flat plate parallel to the flow is shown in Fig. 9.8.

In the plot of Fig. 9.8, transition was assumed to occur at $Re_x = 5 \times 10^5$ for flows in which the boundary layer was initially laminar. The actual Reynolds number at which transition occurs depends on a combination of factors, such as surface roughness and freestream disturbances. Transition tends to occur earlier (at lower Reynolds number) as surface roughness or freestream turbulence is increased. For transition at other than $Re_x = 5 \times 10^5$, the constant in the second term of Eqs. 9.37 is modified using Eq. 9.36. Figure 9.8 shows that the drag coefficient is less, for a given length of plate, when laminar flow is maintained over the longest possible distance. However, at large Re_L ($> 10^7$) the contribution of the laminar drag is negligible.

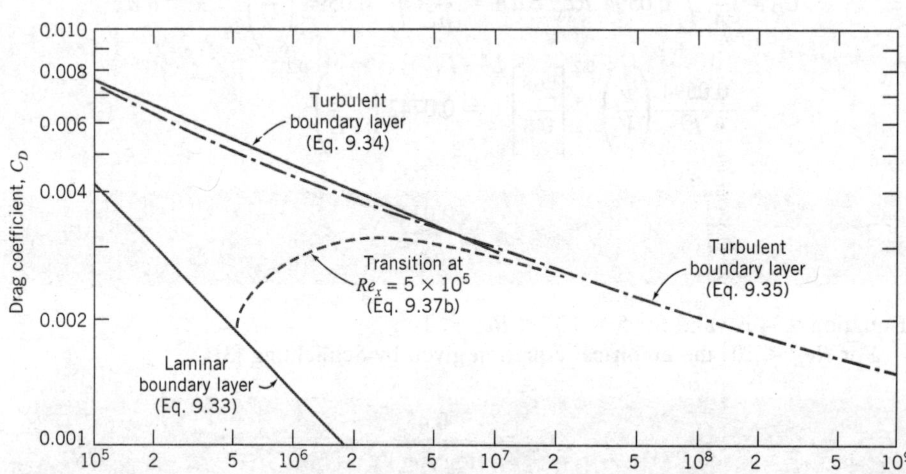

Fig. 9.8 Variation of drag coefficient with Reynolds number for a smooth flat plate parallel to the flow.

Example 9.5 SKIN FRICTION DRAG ON A SUPERTANKER

A supertanker is 360 m long and has a beam width of 70 m and a draft of 25 m. Estimate the force and power required to overcome skin friction drag at a cruising speed of 6.69 m/s in seawater at 10°C.

Given: Supertanker cruising at $U = 6.69$ m/s.

Find: (a) Force.
(b) Power required to overcome skin friction drag.

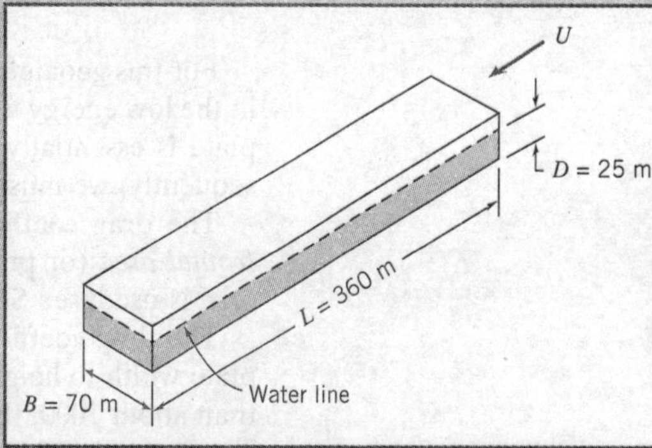

Solution:
Model the tanker hull as a flat plate, of length L and width $b = B + 2D$, in contact with water. Estimate skin friction drag from the drag coefficient.

Governing equations:

$$C_D = \frac{F_D}{\frac{1}{2}\rho U^2 A} \quad (9.32)$$

$$C_D = \frac{0.455}{(\log Re_L)^{2.58}} - \frac{1610}{Re_L} \quad (9.37b)$$

The ship speed is 6.69 m/s, so

$$U = 6.69\,\text{m/s}$$

From Appendix A, at 10°C, $\nu = 1.37 \times 10^{-6}$ m²/s for seawater. Then

$$Re_L = \frac{UL}{\nu} = 6.69\frac{\text{m}}{\text{s}} \times 360\,\text{m} \times \frac{\text{s}}{1.37 \times 10^{-6}\,\text{m}^2} = 1.76 \times 10^9$$

Assuming Eq. 9.37b is valid,

$$C_D = \frac{0.455}{(\log 1.76 \times 10^9)^{2.58}} - \frac{1610}{1.76 \times 10^9} = 0.00147$$

and from Eq. 9.32,

$$F_D = C_D A \frac{1}{2}\rho U^2$$

$$= 0.00147 \times (360\,\text{m})(70 + 50)\text{m} \times \frac{1}{2} \times 1020\frac{\text{kg}}{\text{m}^3} \times (6.69)^2\frac{\text{m}^2}{\text{s}^2} \times \frac{\text{N}\cdot\text{s}^2}{\text{kg}\cdot\text{m}}$$

$$F_D = 1.45\,\text{MN} \qquad \longleftarrow F_D$$

The corresponding power is

$$\mathcal{P} = F_D U = 1.45 \times 10^6\,\text{N} \times 6.69\frac{\text{m}}{\text{s}} \times \frac{\text{W}\cdot\text{s}}{\text{N}\cdot\text{m}}$$

$$\mathcal{P} = 9.70\,\text{MW} \qquad \longleftarrow \mathcal{P}$$

> This problem illustrates application of drag coefficient equations for a flat plate parallel to the flow.
> ✓ The power required (about 9.70 MW) is very large because although the friction stress is small, it acts over a substantial area.
> ✓ The boundary layer is turbulent for almost the entire length of the ship (transition occurs at $x \approx 0.1$ m).

Pure Pressure Drag: Flow over a Flat Plate Normal to the Flow

In flow over a flat plate normal to the flow (Fig. 9.9), the wall shear stress is perpendicular to the flow direction and therefore does not contribute to the drag force. The drag is given by

$$F_D = \int_{\text{surface}} p\,dA$$

For this geometry the flow separates from the edges of the plate; there is back-flow in the low energy wake of the plate. Although the pressure over the rear surface of the plate is essentially constant, its magnitude cannot be determined analytically. Consequently, we must resort to experiments to determine the drag force.

The drag coefficient for flow over an immersed object usually is based on the *frontal area* (or projected area) of the object. (For airfoils and wings, the *planform area* is used; see Section 9.8.)

The drag coefficient for a finite plate normal to the flow depends on the ratio of plate width to height and on the Reynolds number. For Re (based on height) greater than about 1000, the drag coefficient is essentially independent of Reynolds number. The variation of C_D with the ratio of plate width to height (b/h) is shown in Fig. 9.10. (The ratio b/h is defined as the *aspect ratio* of the plate.) For $b/h = 1.0$, the drag coefficient is a minimum at $C_D = 1.18$; this is just slightly higher than for a circular disk ($C_D = 1.17$) at large Reynolds number.

The drag coefficient for all objects with sharp edges is essentially independent of Reynolds number (for $Re \gtrsim 1000$) because the separation points and therefore the size of the wake are fixed by the geometry of the object. Drag coefficients for selected objects are given in Table 9.3.

Friction and Pressure Drag: Flow over a Sphere and Cylinder

We have looked at two special flow cases in which either friction or pressure drag was the sole form of drag present. In the former case, the drag coefficient was a strong function of Reynolds number, while in the latter case, C_D was essentially independent of Reynolds number for $Re \gtrsim 1000$.

VIDEO

Plate Normal to the Flow.

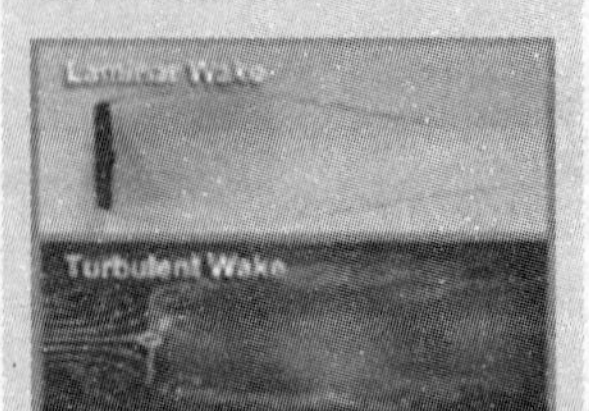

In the case of flow over a sphere, both friction drag and pressure drag contribute to total drag. The drag coefficient for flow over a smooth sphere is shown in Fig. 9.11 as a function of Reynolds number.[2]

At very low Reynolds number,[3] $Re \le 1$, there is no flow separation from a sphere; the wake is laminar and the drag is predominantly friction drag. Stokes has shown analytically, for very low Reynolds number flows where inertia forces may be

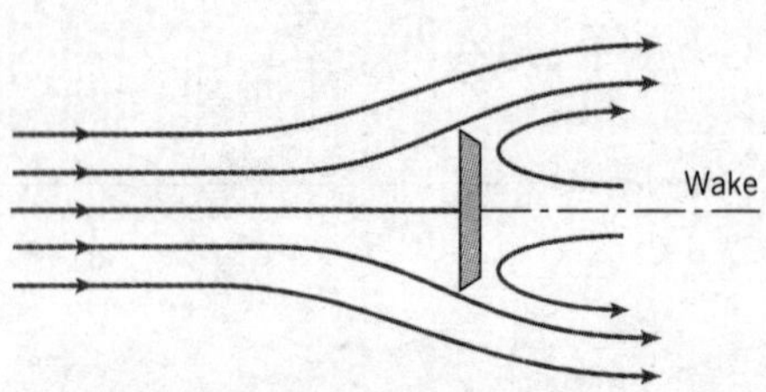

Fig. 9.9 Flow over a flat plate normal to the flow.

[2]An approximate curve fit to the data of Fig. 9.11 is presented in Problem 9.132.

[3]See Shapiro [17] for a good discussion of drag on spheres and other shapes. See also Fage [18].

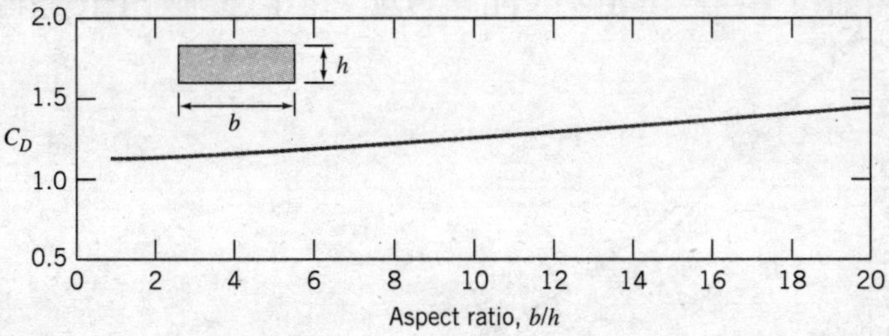

Fig. 9.10 Variation of drag coefficient with aspect ratio for a flat plate of finite width normal to the flow with $Re_h > 1000$ [16].

Table 9.3

Drag Coefficient Data for Selected Objects ($Re \gtrsim 10^3$)[a]

Object	Diagram		$C_D(Re \gtrsim 10^3)$
Square prism	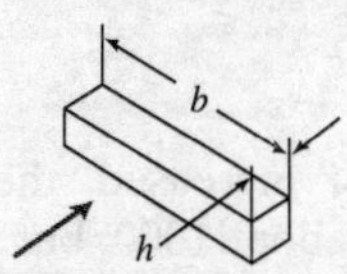	$b/h = \infty$ $b/h = 1$	2.05 1.05
Disk	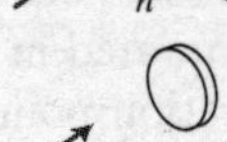		1.17
Ring			1.20[b]
Hemisphere (open end facing flow)	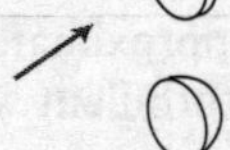		1.42
Hemisphere (open end facing downstream)			0.38
C-section (open side facing flow)			2.30
C-section (open side facing downstream)	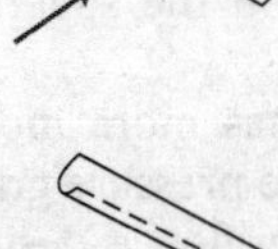		1.20

[a]Data from Hoerner [16].
[b]Based on ring area.

VIDEO

An Object with a High Drag Coefficient.

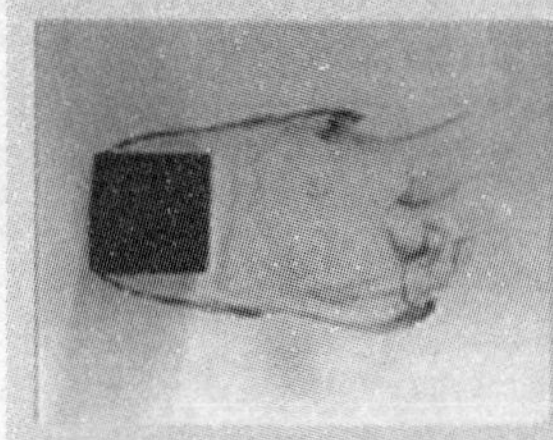

neglected, that the drag force on a sphere of diameter d, moving at speed V, through a fluid of viscosity μ, is given by

$$F_D = 3\pi\mu Vd$$

The drag coefficient, C_D, defined by Eq. 9.30, is then

$$C_D = \frac{24}{Re}$$

As shown in Fig. 9.11, this expression agrees with experimental values at low Reynolds number but begins to deviate significantly from the experimental data for $Re > 1.0$.

CLASSIC VIDEO

The Fluid Dynamics of Drag.

CLASSIC VIDEO

Low Reynolds Number Flows.

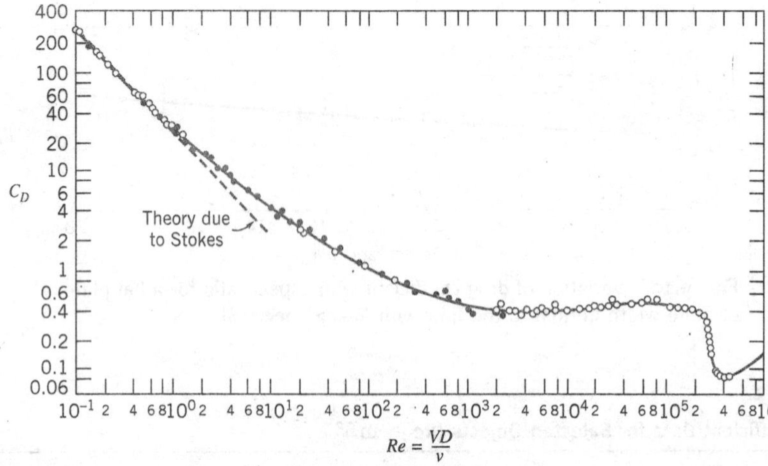

Fig. 9.11 Drag coefficient of a smooth sphere as a function of Reynolds number [3].

VIDEO

Examples of Flow around a Sphere.

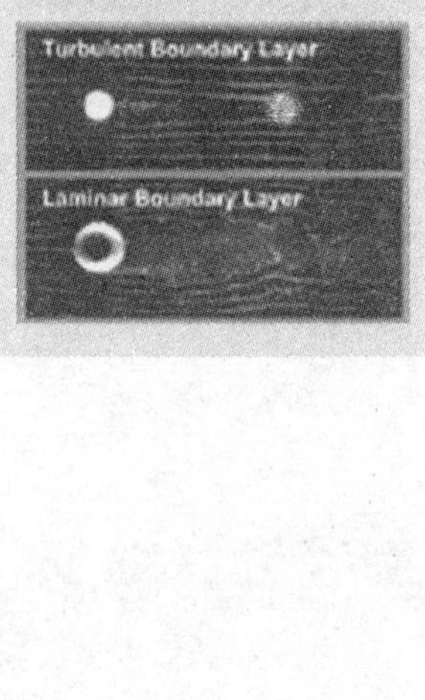

As the Reynolds number is further increased, the drag coefficient drops continuously up to a Reynolds number of about 1000, but not as rapidly as predicted by Stokes' theory. A turbulent wake (not incorporated in Stokes' theory) develops and grows at the rear of the sphere as the separation point moves from the rear of the sphere toward the front; this wake is at a relatively low pressure, leading to a large pressure drag. By the time $Re \approx 1000$, about 95% of total drag is due to pressure. For $10^3 < Re < 3 \times 10^5$ the drag coefficient is approximately constant. In this range the entire rear of the sphere has a low-pressure turbulent wake, as indicated in Fig. 9.12, and most of the drag is caused by the front-rear pressure asymmetry. Note that $C_D \propto 1/Re$ corresponds to $F_D \propto V$, and that $C_D \sim$ const. corresponds to $F_D \propto V^2$, indicating a quite rapid increase in drag.

For Reynolds numbers larger than about 3×10^5, transition occurs and the boundary layer on the forward portion of the sphere becomes turbulent. The point of separation then moves downstream from the sphere midsection, and the size of the wake decreases. The net pressure force on the sphere is reduced (Fig. 9.12), and the drag coefficient decreases abruptly.

A turbulent boundary layer, since it has more momentum flux than a laminar boundary layer, can better resist an adverse pressure gradient, as discussed in Section 9.6. Consequently, turbulent boundary-layer flow is desirable on a blunt body because it delays separation and thus reduces the pressure drag.

Transition in the boundary layer is affected by roughness of the sphere surface and turbulence in the flow stream. Therefore, the reduction in drag associated with a turbulent boundary layer does not occur at a unique value of Reynolds number. Experiments with smooth spheres in a flow with low turbulence level show that transition may be delayed to a critical Reynolds number, Re_D, of about 4×10^5. For rough surfaces and/or highly turbulent freestream flow, transition can occur at a critical Reynolds number as low as 50,000.

The drag coefficient of a sphere with turbulent boundary-layer flow is about one-fifth that for laminar flow near the critical Reynolds number. The corresponding reduction in drag force can affect the range of a sphere (e.g., a golf ball) appreciably. The "dimples" on a golf ball are designed to "trip" the boundary layer and, thus, to guarantee turbulent boundary-layer flow and minimum drag. To illustrate this effect graphically, we obtained samples of golf balls without dimples some years ago. One of our students volunteered to hit drives with the smooth balls. In 50 tries with each type of ball, the average distance with the standard balls was 197 m; the average with the smooth balls was only 114 m!

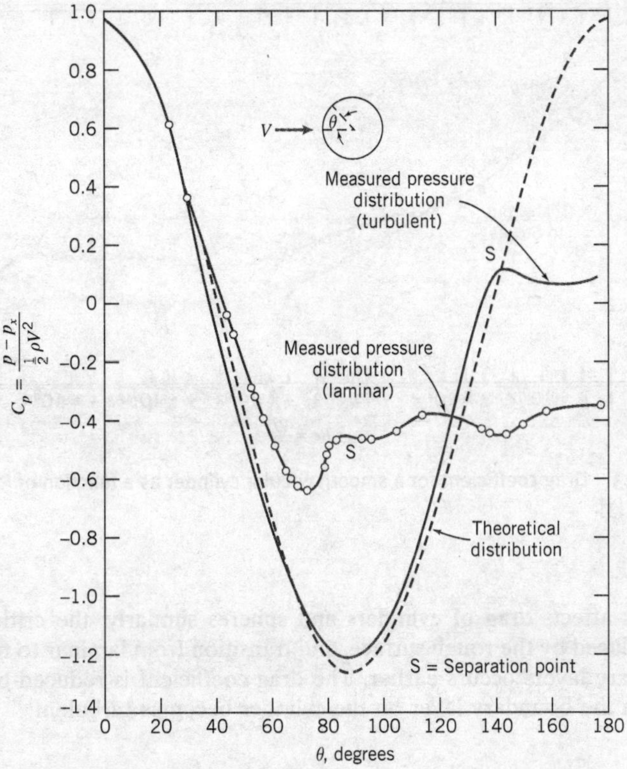

Fig. 9.12 Pressure distribution around a smooth sphere for laminar and turbulent boundary-layer flow, compared with inviscid flow [18].

VIDEO

Laminar and Turbulent Flow Past a Sphere.

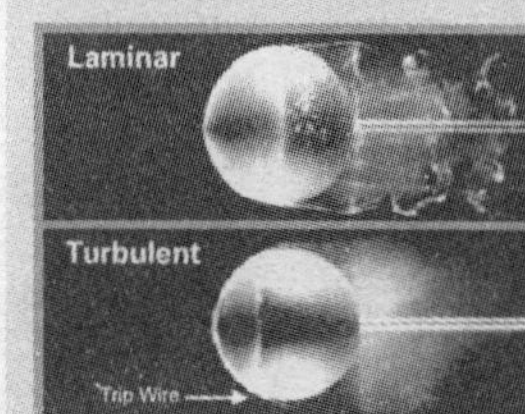

VIDEO

Flow Separation on a Cylinder.

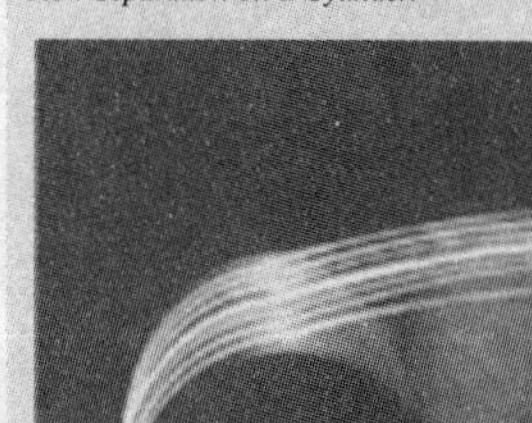

VIDEO

Vortex Trail behind a Cylinder.

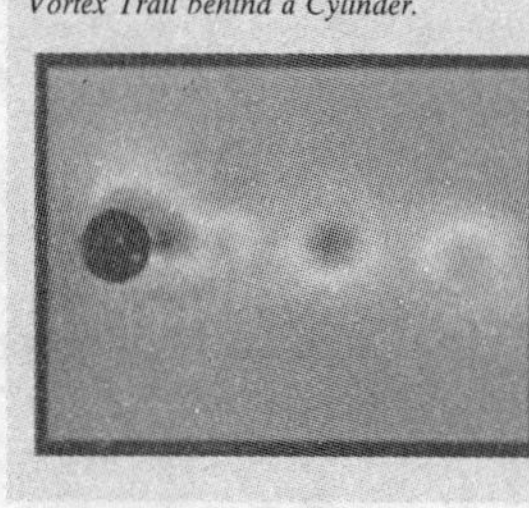

Adding roughness elements to a sphere also can suppress local oscillations in location of the transition between laminar and turbulent flow in the boundary layer. These oscillations can lead to variations in drag and to random fluctuations in lift (see Section 9.8). In baseball, the "knuckle ball" pitch is intended to behave erratically to confuse the batter. By throwing the ball with almost no spin, the pitcher relies on the seams to cause transition in an unpredictable fashion as the ball moves on its way to the batter. This causes the desired variation in the flight path of the ball.

Figure 9.13 shows the drag coefficient for flow over a smooth cylinder. The variation of C_D with Reynolds number shows the same characteristics as observed in the flow over a smooth sphere, but the values of C_D are about twice as high.

Flow about a smooth circular cylinder may develop a regular pattern of alternating vortices downstream. The *vortex trail*[4] causes an oscillatory lift force on the cylinder perpendicular to the stream motion. Vortex shedding excites oscillations that cause telegraph wires to "sing" and ropes on flag poles to "slap" annoyingly. Sometimes structural oscillations can reach dangerous magnitudes and cause high stresses; they can be reduced or eliminated by applying roughness elements or fins—either axial or helical (sometimes seen on chimneys or automobile antennas)—that destroy the symmetry of the cylinder and stabilize the flow.

Experimental data show that regular vortex shedding occurs most strongly in the range of Reynolds number from about 60 to 5000. For $Re > 1000$ the dimensionless frequency of vortex shedding, expressed as a Strouhal number, $St = fD/V$, is approximately equal to 0.21 [3].

[4]The regular pattern of vortices in the wake of a cylinder sometimes is called a *Karman vortex street* in honor of the prominent fluid mechanician, Theodore von Kármán, who was first to predict the stable spacing of the vortex trail on theoretical grounds in 1911; see Goldstein [19].

VIDEO

Low Reynolds Number Flow over a Cylinder.

VIDEO

Flow Separation behind a Cylinder.

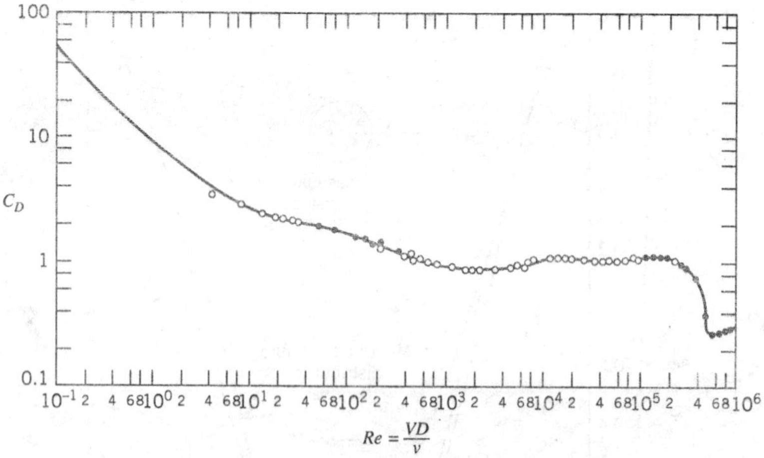

Fig. 9.13 Drag coefficient for a smooth circular cylinder as a function of Reynolds number [3].

Roughness affects drag of cylinders and spheres similarly: the critical Reynolds number is reduced by the rough surface, and transition from laminar to turbulent flow in the boundary layers occurs earlier. The drag coefficient is reduced by a factor of about 4 when the boundary layer on the cylinder becomes turbulent.

Example 9.6 AERODYNAMIC DRAG AND MOMENT ON A CHIMNEY

A cylindrical chimney 1 m in diameter and 25 m tall is exposed to a uniform 50 km/hr wind at standard atmospheric conditions. End effects and gusts may be neglected. Estimate the bending moment at the base of the chimney due to wind forces.

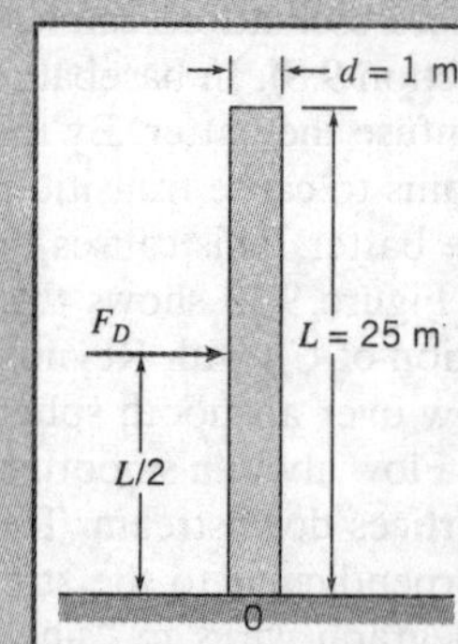

Given: Cylindrical chimney, $D = 1$ m, $L = 25$ m, in uniform flow with

$$V = 50 \text{ km/hr} \qquad p = 101 \text{ kPa (abs)} \qquad T = 15°\text{C}$$

Neglect end effects.

Find: Bending moment at bottom of chimney.

Solution:
The drag coefficient is given by $C_D = F_D / \frac{1}{2}\rho V^2 A$, and thus $F_D = C_D A \frac{1}{2} \rho V^2$. Since the force per unit length is uniform over the entire length, the resultant force, F_D, will act at the midpoint of the chimney. Hence the moment about the chimney base is

$$M_0 = F_D \frac{L}{2} = C_D A \frac{1}{2} \rho V^2 \frac{L}{2} = C_D A \frac{L}{4} \rho V^2$$

$$V = 50 \frac{\text{km}}{\text{hr}} \times 10^3 \frac{\text{m}}{\text{km}} \times \frac{\text{hr}}{3600 \text{ s}} = 13.9 \text{ m/s}$$

For air at standard conditions, $\rho = 1.23$ kg/m^3, and $\mu = 1.79 \times 10^{-5}$ kg/(m · s). Thus

$$Re = \frac{\rho V D}{\mu} = 1.23 \frac{\text{kg}}{\text{m}^3} \times 13.9 \frac{\text{m}}{\text{s}} \times 1 \text{ m} \times \frac{\text{m} \cdot \text{s}}{1.79 \times 10^{-5} \text{ kg}} = 9.55 \times 10^5$$

From Fig. 9.13, $C_D \approx 0.35$. For a cylinder, $A = DL$, so

$$M_0 = C_D A \frac{L}{4} \rho V^2 = C_D DL \frac{L}{4} \rho V^2 = C_D D \frac{L^2}{4} \rho V^2$$

$$= \frac{1}{4} \times 0.35 \times 1 \text{ m} \times (25)^2 \text{ m}^2 \times 1.23 \frac{\text{kg}}{\text{m}^3} \times (13.9)^2 \frac{\text{m}^2}{\text{s}^2} \times \frac{\text{N} \cdot \text{s}^2}{\text{kg} \cdot \text{m}}$$

$$M_0 = 13.0 \text{ kN} \cdot \text{m} \longleftarrow M_0$$

This problem illustrates application of drag-coefficient data to calculate the force and moment on a structure. We modeled the wind as a uniform flow; more realistically, the lower atmosphere is often modeled as a huge turbulent boundary layer, with a power-law velocity profile, $u \sim y^{1/n}$ (y is the elevation). See Problem 9.135, where this is analyzed for the case $n = 7$.

Example 9.7 DECELERATION OF AN AUTOMOBILE BY A DRAG PARACHUTE

A dragster weighing 7120 N attains a speed of 430 km/hr in the quarter mile. Immediately after passing through the timing lights, the driver opens the drag chute, of area $A = 2.3 \text{ m}^2$. Air and rolling resistance of the car may be neglected. Find the time required for the machine to decelerate to 160 km/hr in standard air.

Given: Dragster weighing 7120 N, moving with initial speed $V_0 = 430$ km/hr, is slowed by the drag force on a chute of area $A = 2.3 \text{ m}^2$. Neglect air and rolling resistance of the car. Assume standard air.

Find: Time required for the machine to decelerate to 160 km/hr.

Solution:
Taking the car as a system and writing Newton's second law in the direction of motion gives

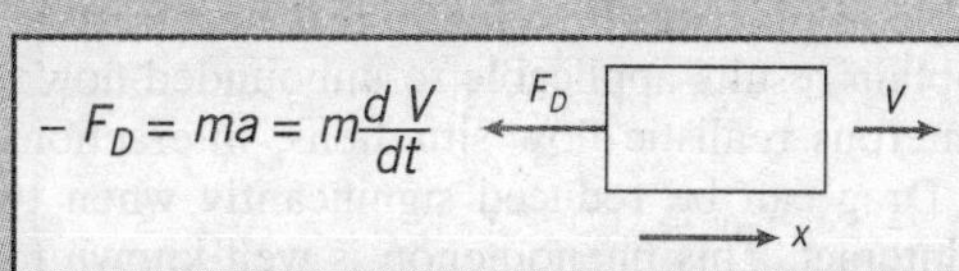

$V_0 = 430 \text{ km/hr}$
$V_f = 160 \text{ km/hr}$
$\rho = 1.227 \text{ kg/m}^3$

Since $C_D = \dfrac{F_D}{\frac{1}{2}\rho V^2 A}$, then $F_D = \frac{1}{2} C_D \rho V^2 A$.

Substituting into Newton's second law gives

$$-\frac{1}{2} C_D \rho V^2 A = m \frac{dV}{dt}$$

Separating variables and integrating, we obtain

$$-\frac{1}{2} C_D \rho \frac{A}{m} \int_0^t dt = \int_{V_0}^{V_f} \frac{dV}{V^2}$$

$$-\frac{1}{2} C_D \rho \frac{A}{m} t = -\frac{1}{V}\Bigg]_{V_0}^{V_f} = -\frac{1}{V_f} + \frac{1}{V_0} = -\frac{(V_0 - V_f)}{V_f V_0}$$

Finally,

$$t = \frac{(V_0 - V_f)}{V_f V_0} \frac{2m}{C_D \rho A} = \frac{(V_0 - V_f)}{V_f V_0} \frac{2W}{C_D \rho A g}$$

Model the drag chute as a hemisphere (with open end facing flow). From Table 9.3, $C_D = 1.42$ (assuming $Re > 10^3$). Then, substituting numerical values,

$$t = \frac{(430 - 160)\ \text{km/hr}}{160\ \text{km/hr} \times 430\ \text{km/hr}} \times \frac{2 \times 7120\ \text{N}}{1.42 \times 1.227\ \text{kg/m}^3 \times 2.3\ \text{m}^2 \times 9.81\ \text{m/s}^2}$$

$$\times \frac{\text{km}}{1000\ \text{m}} \times \frac{3600\ \text{s}}{\text{hr}} \times \frac{\text{kg} \cdot \text{m}}{\text{N} \cdot \text{s}^2}$$

$$t = 5.12\ \text{s} \qquad t$$

Check the assumption on Re:

$$Re = \frac{DV}{\nu} = \left[\frac{4A}{\pi}\right]^{1/2} \frac{V}{\nu}$$

$$= \left[\frac{4}{\pi} \times 2.3\ \text{m}^2\right]^{1/2} \times 160\ \frac{\text{km}}{\text{hr}} \times \frac{1000\ \text{m}}{\text{km}} \times \frac{\text{hr}}{3600\ \text{s}} \times \frac{\text{s}}{1.46 \times 10^{-5}\ \text{m}^2}$$

$$Re = 5.21 \times 10^6$$

Hence the assumption is valid.

This problem illustrates application of drag-coefficient data to calculate the drag on a vehicle parachute.

The *Excel* workbook for this Example plots the dragster velocity (and distance traveled) as a function of time; it also allows "what-ifs," e.g., we can find the parachute area A required to slow the dragster to 96 km/hr in 5 sec.

All experimental data presented in this section are for single objects immersed in an unbounded fluid stream. The objective of wind tunnel tests is to simulate the conditions of an unbounded flow. Limitations on equipment size make this goal unreachable in practice. Frequently it is necessary to apply corrections to measured data to obtain results applicable to unbounded flow conditions.

In numerous realistic flow situations, interactions occur with nearby objects or surfaces. Drag can be reduced significantly when two or more objects, moving in tandem, interact. This phenomenon is well known to bicycle riders and those interested in automobile racing, where "drafting" is a common practice. Drag reductions of 80 percent may be achieved with optimum spacing [20]. Drag also can be increased significantly when spacing is not optimum.

Drag can be affected by neighbors alongside as well. Small particles falling under gravity travel more slowly when they have neighbors than when they are isolated. This phenomenon has important applications to mixing and sedimentation processes.

Experimental data for drag coefficients on objects must be selected and applied carefully. Due regard must be given to the differences between the actual conditions and the more controlled conditions under which measurements were made.

CLASSIC VIDEO

Low Reynolds Number Flows.

Streamlining

The extent of the separated flow region behind many of the objects discussed in the previous section can be reduced or eliminated by streamlining, or fairing, the body shape. We have seen that due to the convergent body shape at the rear of any object (after all, every object is of finite length!), the streamlines will diverge, so that the velocity will decrease, and therefore, more importantly (as shown by the Bernoulli equation, applicable in the freestream region) the pressure will increase. Hence, we

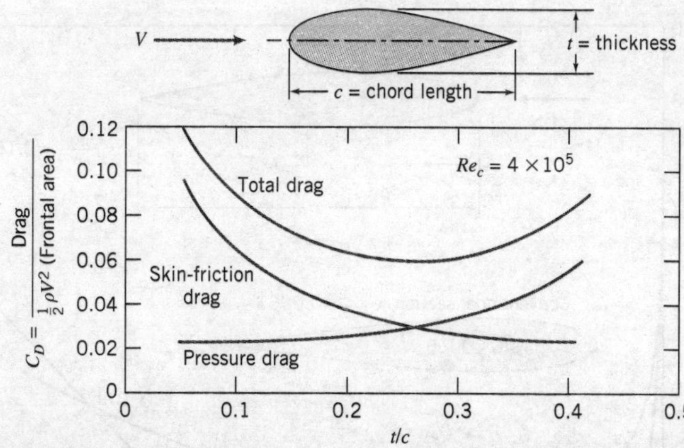

Fig. 9.14 Drag coefficient on a streamlined strut as a function of thickness ratio, showing contributions of skin friction and pressure to total drag [19].

initially have an adverse pressure gradient at the rear of the body, leading to boundary-layer separation and ultimately to a low-pressure wake leading to large pressure drag. Streamlining is the attempt to reduce the drag on a body. We can reduce the drag on a body by making the rear of the body more tapered (e.g., we can reduce the drag on a sphere by making it "teardrop" shaped), which will reduce the adverse pressure gradient and hence make the turbulent wake smaller. However, as we do so, we are in danger of increasing the skin friction drag simply because we have increased the surface area. In practice, there is an optimum amount of fairing or tapering at which the total drag (the sum of pressure and skin friction drag) is minimized.

CLASSIC VIDEO

Low Reynolds Number Flows.

The pressure gradient around a "teardrop" shape (a "streamlined" cylinder) is less severe than that around a cylinder of circular section. The trade-off between pressure and friction drag for this case is illustrated by the results presented in Fig. 9.14, for tests at $Re_c = 4 \times 10^5$. (This Reynolds number is typical of that for a strut on an early aircraft.) From the figure, the minimum drag coefficient is $C_D \approx 0.06$, which occurs when the ratio of thickness to chord is $t/c \approx 0.25$. This value is approximately 20 percent of the minimum drag coefficient for a circular cylinder of the same thickness! Hence, even a small aircraft will typically have fairings on many structural members, e.g., the struts that make up the landing wheel assembly, leading to significant fuel savings.

CLASSIC VIDEO

The Fluid Dynamics of Drag.

The maximum thickness for the shapes shown in Fig. 9.14 is located approximately 25 percent of the chord distance from the leading edge. Most of the drag on the thinner sections is due to skin friction in the turbulent boundary layers on the tapered rear sections. Interest in low-drag airfoils increased during the 1930s. The National Advisory Committee for Aeronautics (NACA) developed several series of "laminar-flow" airfoils for which transition was postponed to 60 or 65 percent chord length aft from the airfoil nose.

Pressure distribution and drag coefficients[5] for two symmetric airfoils of infinite span and 15 percent thickness at zero angle of attack are presented in Fig. 9.15. Transition on the conventional (NACA 0015) airfoil takes place where the pressure gradient becomes adverse, at $x/c = 0.13$, near the point of maximum thickness. Thus most of the airfoil surface is covered with a turbulent boundary layer; the drag coefficient is $C_D \approx 0.0061$. The point of maximum thickness has been moved aft on the

[5]Note that drag coefficients for airfoils are based on the *planform* area, i.e., $C_D = F_D / \frac{1}{2}\rho V^2 A_p$, where A_p is the maximum projected wing area.

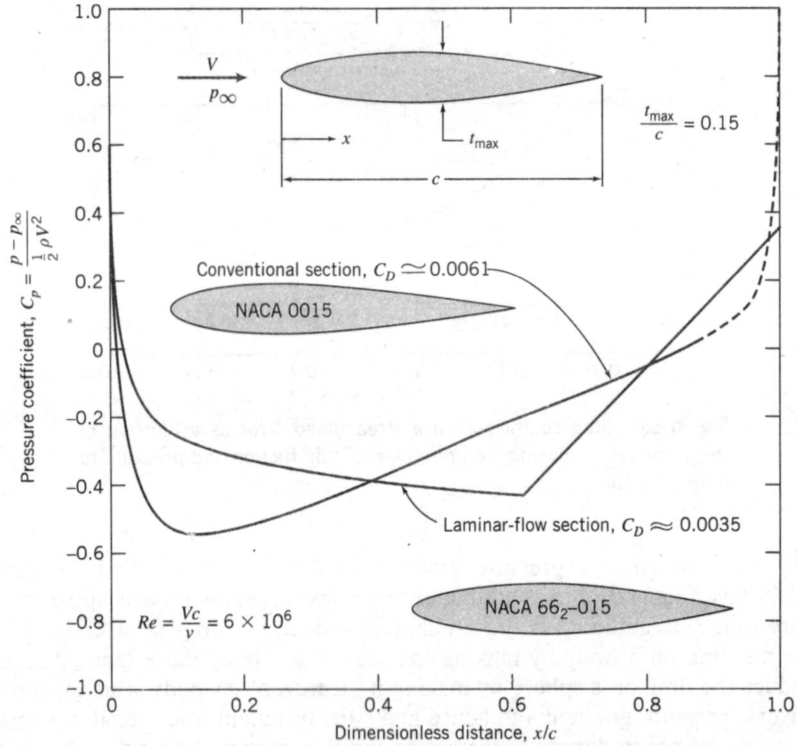

Fig. 9.15 Theoretical pressure distributions at zero angle of attack for two symmetric airfoil sections of 15 percent thickness ratio. (Data from Abbott and von Doenhoff [21].)

VIDEO

Flow past a Model A and a Sports Car.

airfoil (NACA 66_2–015) designed for laminar flow. The boundary layer is maintained in the laminar regime by the favorable pressure gradient to $x/c = 0.63$. Thus the bulk of the flow is laminar; $C_D \approx 0.0035$ for this section, based on planform area. The drag coefficient based on frontal area is $C_{D_f} = C_D/0.15 = 0.0233$, or about 40 percent of the optimum for the shapes shown in Fig. 9.14.

Tests in special wind tunnels have shown that laminar flow can be maintained up to length Reynolds numbers as high as 30 million by appropriate profile shaping. Because they have favorable drag characteristics, laminar-flow airfoils are used in the design of most modern subsonic aircraft.

Recent advances have made possible development of low-drag shapes even better than the NACA 60-series shapes. Experiments [21, 22] led to the development of a pressure distribution that prevented separation while maintaining the turbulent boundary layer in a condition that produces negligible skin friction. Improved methods for calculating body shapes that produced a desired pressure distribution [23, 24] led to development of nearly optimum shapes for thick struts with low drag. Figure 9.16 shows an example of the results.

Reduction of aerodynamic drag also is important for road vehicle applications. Interest in fuel economy has provided significant incentive to balance efficient aerodynamic performance with attractive design for automobiles. Drag reduction also has become important for buses and trucks.

Practical considerations limit the overall length of road vehicles. Fully streamlined tails are impractical for all but land-speed-record cars. Consequently, it is not possible to achieve results comparable to those for optimum airfoil shapes. However, it is possible to optimize both front and rear contours within given constraints on overall length [25–27].

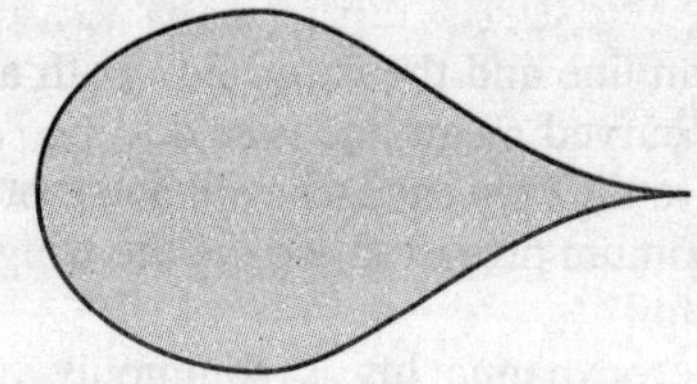

Fig. 9.16 Nearly optimum shape for low-drag strut [24].

Much attention has been focused on front contours. Studies on buses have shown that drag reductions up to 25 percent are possible with careful attention to front contour [27]. Thus it is possible to reduce the drag coefficient of a bus from about 0.65 to less than 0.5 with practical designs. Highway tractor-trailer rigs have higher drag coefficients—C_D values from 0.90 to 1.1 have been reported. Commercially available add-on devices offer improvements in drag of up to 15 percent, particularly for windy conditions where yaw angles are nonzero. The typical fuel saving is half the percentage by which aerodynamic drag is reduced.

Front contours and details are important for automobiles. A low nose and smoothly rounded contours are the primary features that promote low drag. Radii of "A-pillar" and windshield header, and blending of accessories to reduce parasite and interference drag have received increased attention. As a result, drag coefficients have been reduced from about 0.55 to 0.30 or less for recent production vehicles. Recent advances in computational methods have led to development of computer-generated optimum shapes. A number of designs have been proposed, with claims of C_D values below 0.2 for vehicles complete with running gear.

Lift 9.8

For most objects in relative motion in a fluid, the most significant fluid force is the drag. However, there are some objects, such as airfoils, for which the lift is significant. Lift is defined as the component of fluid force perpendicular to the fluid motion. For an airfoil, the *lift coefficient*, C_L, is defined as

$$C_L \equiv \frac{F_L}{\frac{1}{2}\rho V^2 A_p} \tag{9.38}$$

It is worth noting that the lift coefficient defined above and the drag coefficient (Eq. 9.30) are each defined as the ratio of an actual force (lift or drag) divided by the product of dynamic pressure and area. This denominator can be viewed as the force that would be generated if we imagined bringing to rest the fluid directly approaching the area (recall that the dynamic pressure is the difference between total and static pressures). This gives us a "feel" for the meaning of the coefficients: They indicate the ratio of the actual force to this (unrealistic but nevertheless intuitively meaningful) force. We note also that the coefficient definitions include V^2 in the denominator, so that F_L (or F_D) being proportional to V^2 corresponds to a constant C_L (or C_D), and that F_L (or F_D) increasing with V at a lower rate than quadratic corresponds to a decrease in C_L (or C_D) with V.

CLASSIC VIDEO

Dimensionless Lift and Drag Coefficients.

The lift and drag coefficients for an airfoil are functions of both Reynolds number and angle of attack; the angle of attack, α, is the angle between the airfoil chord and the freestream velocity vector. The *chord* of an airfoil is the straight line joining the leading edge and the trailing edge. The wing section shape is obtained by combining a *mean line* and a thickness distribution (see [21] for details). When the airfoil has a

CLASSIC VIDEO

Dimensionless Lift and Drag Coefficients.

CLASSIC VIDEO

Boundary Layer Control.

VIDEO

Flow Past an Airfoil ($\alpha = 0°$).

VIDEO

Flow Past an Airfoil ($\alpha = 10°$).

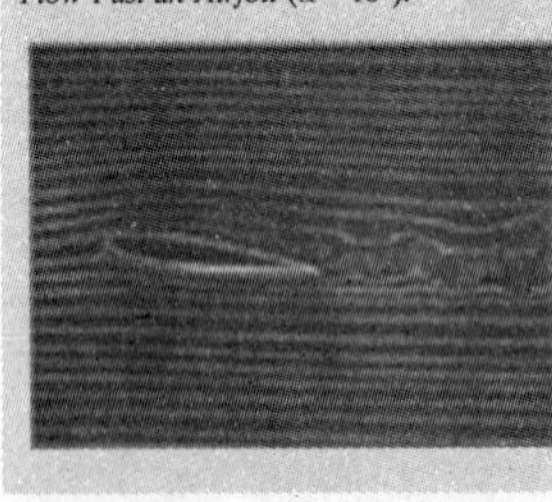

VIDEO

Flow Past an Airfoil ($\alpha = 20°$).

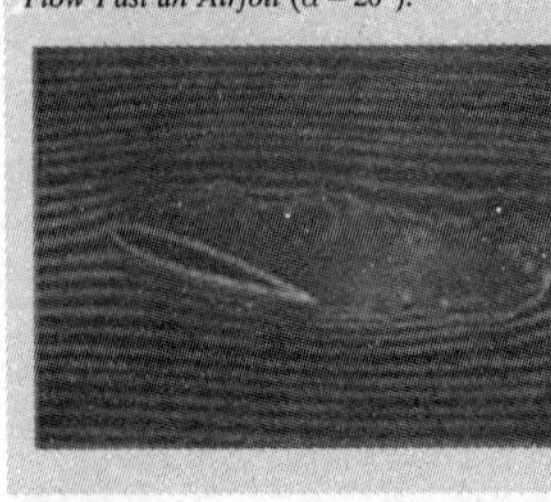

symmetric section, the mean line and the chord line both are straight lines, and they coincide. An airfoil with a curved mean line is said to be *cambered*.

The area at right angles to the flow changes with angle of attack. Consequently, the planform area, A_p (the maximum projected area of the wing), is used to define lift and drag coefficients for an airfoil.

The phenomenon of aerodynamic lift is commonly explained by the velocity increase causing pressure to decrease (the Bernoulli effect) over the top surface of the airfoil and the velocity decrease (causing pressure to increase) along the bottom surface of the airfoil. The resulting pressure distributions are shown clearly in the video *Boundary Layer Control.* Because of the pressure differences relative to atmosphere, the upper surface of the airfoil may be called the *suction surface* and the lower surface the *pressure surface*.

As shown in Example 6.12, lift on a body can also be related to the circulation around the profile: In order for lift to be generated, there must be a net circulation around the profile. One may imagine the circulation to be caused by a vortex "bound" within the profile.

Advances continue in computational methods and computer hardware. However, most airfoil data available in the literature were obtained from wind tunnel tests. Reference [21] contains results from a large number of tests conducted by NACA (the National Advisory Committee for Aeronautics—the predecessor to NASA). Data for some representative NACA profile shapes are described in the next few paragraphs.

Lift and drag coefficient data for typical conventional and laminar-flow profiles are plotted in Fig. 9.17 for a Reynolds number of 9×10^6 based on chord length. The section shapes in Fig. 9.17 are designated as follows:

Conventional—23015

2 30 15

- 15 — section thickness (15 percent chord)
- 30 — maximum camber location ($\frac{1}{2} \times 30 = 15$ percent chord)
- 2 — design lift coefficient ($\frac{3}{2} \times 0.2 = 0.3$)

Laminar Flow—$66_2 - 215$

6 6 $_2$ − 2 15

- 15 — section thickness (15 percent)
- 2 — design lift coefficient (0.2)
- $_2$ — maximum lift coefficient for favorable pressure gradient (0.2)
- 6 (second) — location of minimum pressure ($x/c \approx 0.6$)
- 6 (first) — series designation (laminar flow)

Both sections are cambered to give lift at zero angle of attack. As the angle of attack is increased, the Δp between the upper and lower surfaces increases, causing the lift coefficient to increase smoothly until a maximum is reached. Further increases in angle of attack produce a sudden decrease in C_L. The airfoil is said to have *stalled* when C_L drops in this fashion.

Airfoil stall results when flow separation occurs over a major portion of the upper surface of the airfoil. As the angle of attack is increased, the stagnation point moves

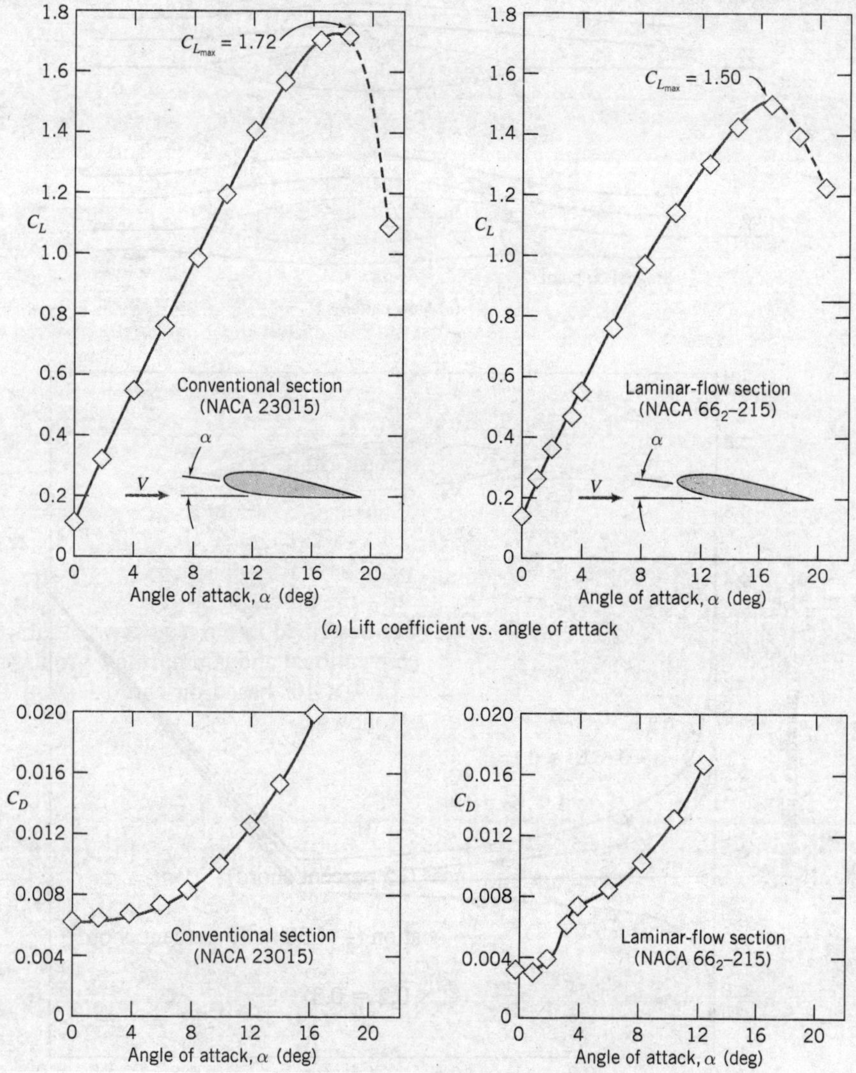

Fig. 9.17 Lift and drag coefficients versus angle of attack for two airfoil sections of 15 percent thickness ratio at $Re_c = 9 \times 10^6$. (Data from Abbott and von Doenhoff [21].)

back along the lower surface of the airfoil, as shown schematically for the symmetric laminar-flow section in Fig. 9.18*a*. Flow on the upper surface then must accelerate sharply to round the nose of the airfoil. The effect of angle of attack on the theoretical upper-surface pressure distribution is shown in Fig. 9.18*b*. The minimum pressure becomes lower, and its location moves forward on the upper surface. A severe adverse pressure gradient appears following the point of minimum pressure; finally, the adverse pressure gradient causes the flow to separate completely from the upper surface and the airfoil stalls (the uniform pressure in the turbulent wake will be approximately equal to the pressure just before separation, i.e., low).

Movement of the minimum pressure point and accentuation of the adverse pressure gradient are responsible for the sudden increase in C_D for the laminar-flow section, which is apparent in Fig. 9.17. The sudden rise in C_D is caused by early transition from laminar to turbulent boundary-layer flow on the upper surface. Aircraft with laminar-flow sections are designed to cruise in the low-drag region.

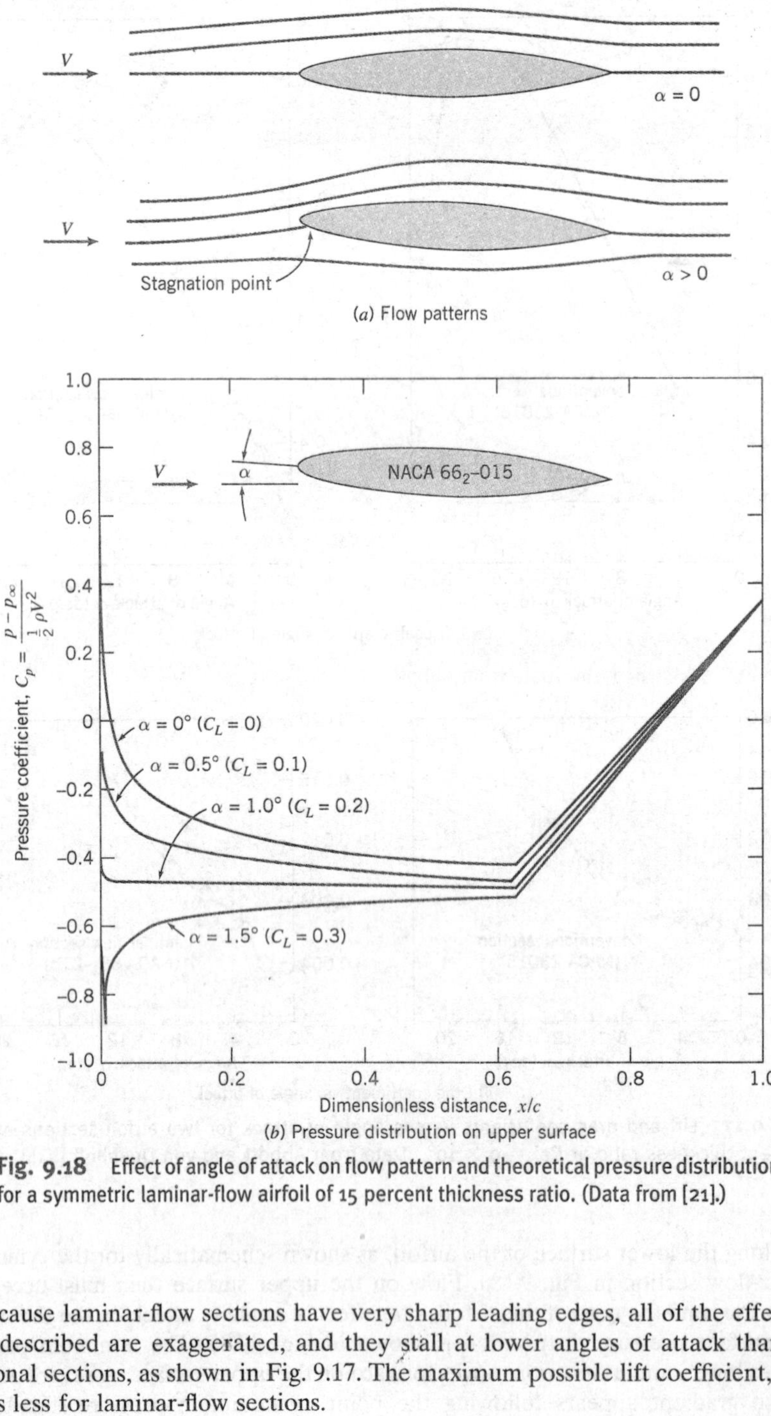

Fig. 9.18 **Effect of angle of attack on flow pattern and theoretical pressure distribution for a symmetric laminar-flow airfoil of 15 percent thickness ratio. (Data from [21].)**

Because laminar-flow sections have very sharp leading edges, all of the effects we have described are exaggerated, and they stall at lower angles of attack than conventional sections, as shown in Fig. 9.17. The maximum possible lift coefficient, $C_{L_{max}}$, also is less for laminar-flow sections.

Plots of C_L versus C_D (called lift-drag polars) often are used to present airfoil data in compact form. A polar plot is given in Fig. 9.19 for the two sections we have discussed. The lift/drag ratio, C_L/C_D, is shown at the design lift coefficient for both sections. This ratio is very important in the design of aircraft: The lift coefficient determines the lift of the wing and hence the load that can be carried, and the drag coefficient indicates a large part (in addition to that caused by the fuselage, etc.) of the drag the airplane engines have to work against in order to generate the needed lift; hence, in general, a high C_L/C_D is the goal, at which the laminar airfoil clearly excels.

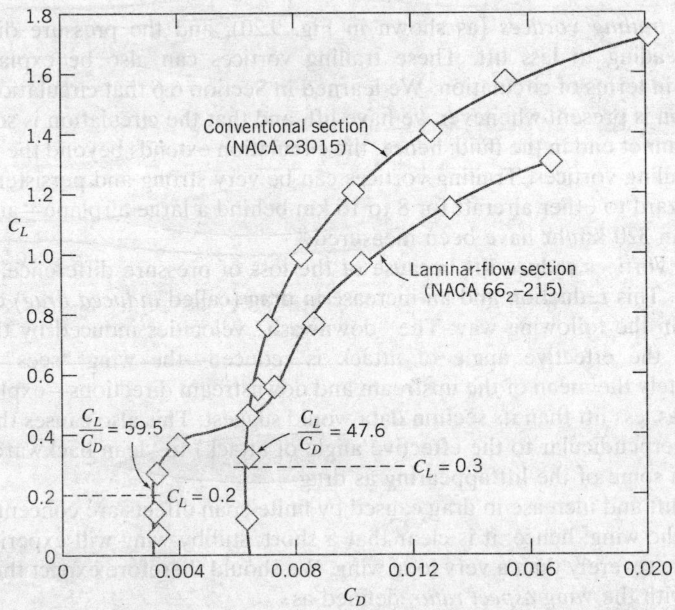

Fig. 9.19 Lift-drag polars for two airfoil sections of 15 percent thickness ratio. (Data from Abbott and von Doenhoff [21].)

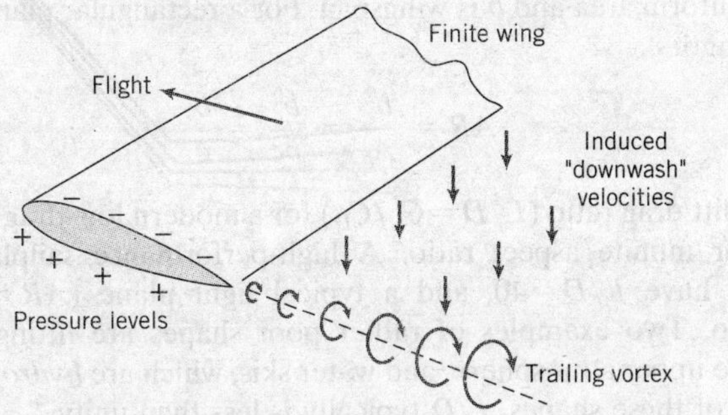

Fig. 9.20 Schematic representation of the trailing vortex system of a finite wing.

Recent improvements in modeling and computational capabilities have made it possible to design airfoil sections that develop high lift while maintaining very low drag [23, 24]. Boundary-layer calculation codes are used with inverse methods for calculating potential flow to develop pressure distributions and the resulting body shapes that postpone transition to the most rearward location possible. The turbulent boundary layer following transition is maintained in a state of incipient separation with nearly zero skin friction by appropriate shaping of the pressure distribution.

Such computer-designed airfoils have been used on racing cars to develop very high negative lift (downforce) to improve high-speed stability and cornering performance [23]. Airfoil sections especially designed for operation at low Reynolds number were used for the wings and propeller on the Kremer prize-winning man-powered "Gossamer Condor" [28], which now hangs in the National Air and Space Museum in Washington, D.C.

All real airfoils—*wings*—are of finite span and have less lift and more drag than their airfoil section data would indicate. There are several ways to explain this. If we consider the pressure distribution near the end of the wing, the low pressure on the upper and high pressure on the lower surface cause flow to occur around the wing tip,

VIDEO

Wing Tip Vortices.

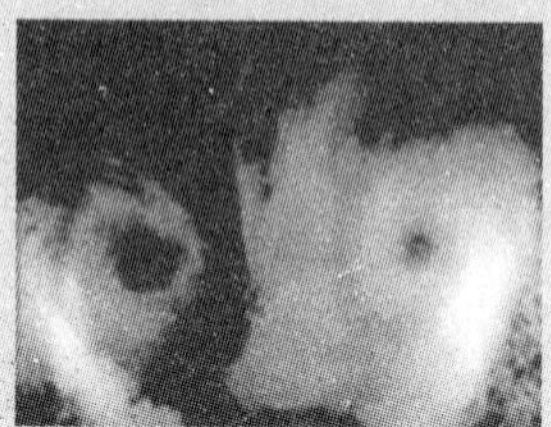

leading to *trailing vortices* (as shown in Fig. 9.20), and the pressure difference is reduced, leading to less lift. These trailing vortices can also be explained more abstractly, in terms of circulation: We learned in Section 6.6 that circulation around a wing section is present whenever we have lift, and that the circulation is solenoidal—that is, it cannot end in the fluid; hence, the circulation extends beyond the wing in the form of trailing vortices. Trailing vortices can be very strong and persistent, possibly being a hazard to other aircraft for 8 to 16 km behind a large airplane—air speeds of greater than 320 km/hr have been measured.[6]

Trailing vortices reduce lift because of the loss of pressure difference, as we just mentioned. This reduction and an increase in drag (called *induced drag*) can also be explained in the following way: The "downwash" velocities induced by the vortices mean that the effective angle of attack is reduced—the wing "sees" a flow at approximately the mean of the upstream and downstream directions—explaining why the wing has less lift than its section data would suggest. This also causes the lift force (which is perpendicular to the effective angle of attack) to "lean backwards" a little, resulting in some of the lift appearing as drag.

Loss of lift and increase in drag caused by finite-span effects are concentrated near the tip of the wing; hence, it is clear that a short, stubby wing will experience these effects more severely than a very long wing. We should therefore expect the effects to correlate with the wing *aspect ratio*, defined as

$$AR \equiv \frac{b^2}{A_p} \tag{9.39}$$

where A_p is planform area and b is wingspan. For a rectangular planform of wingspan b and chord length c,

$$AR = \frac{b^2}{A_p} = \frac{b^2}{bc} = \frac{b}{c}$$

The maximum lift/drag ratio ($L/D = C_L/C_D$) for a modern low-drag section may be as high as 400 for infinite aspect ratio. A high-performance sailplane (glider) with $AR = 40$ might have $L/D = 40$, and a typical light plane ($AR \approx 12$) might have $L/D \approx 20$ or so. Two examples of rather poor shapes are lifting bodies used for reentry from the upper atmosphere, and water skis, which are *hydrofoils* of low aspect ratio. For both of these shapes, L/D typically is less than unity.

Variations in aspect ratio are seen in nature. Soaring birds, such as the albatross or California condor, have thin wings of long span. Birds that must maneuver quickly to catch their prey, such as owls, have wings of relatively short span, but large area, which gives low *wing loading* (ratio of weight to planform area) and thus high maneuverability.

It makes sense that as we try to generate more lift from a finite wing (by, for example, increasing the angle of attack), the trailing vortices and therefore the downwash increase; we also learned that the downwash causes the effective angle of attack to be less than that of the corresponding airfoil section (i.e., when $AR = \infty$), ultimately leading to loss of lift and to induced drag. Hence, we conclude that the effects of the finite aspect ratio can be characterized as a reduction $\Delta\alpha$ in the effective angle of attack, and that this (which is usually undesirable) becomes worse as we generate more lift (i.e., as the lift coefficient C_L increases) and as the aspect ratio AR is made smaller. Theory and experiment indicate that

$$\Delta\alpha \approx \frac{C_L}{\pi AR} \tag{9.40}$$

[6]Sforza, P. M., "Aircraft Vortices: Benign or Baleful?" *Space/Aeronautics, 53*, 4, April 1970, pp. 42–49. See also the University of Iowa video *Form Drag, Lift, and Propulsion.*

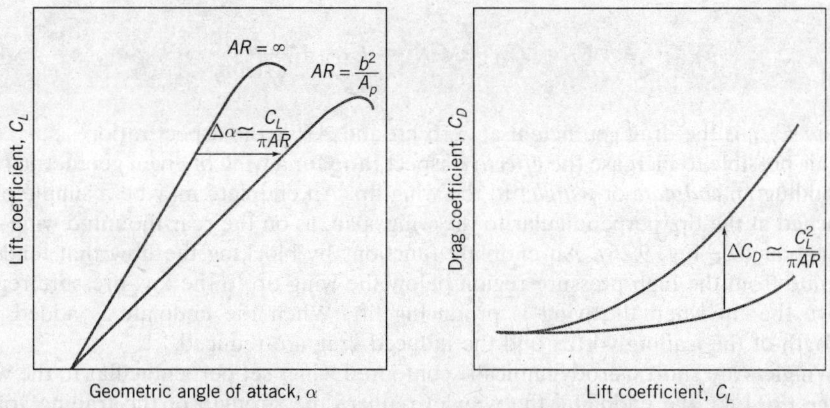

Fig. 9.21 Effect of finite aspect ratio on lift and drag coefficients for a wing.

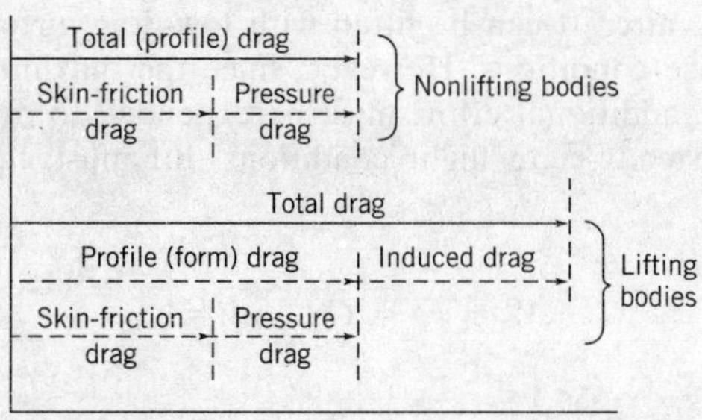

Fig. 9.22 Drag breakdown on nonlifting and lifting bodies.

Compared with an airfoil section ($AR = \infty$), the geometric angle of attack of a wing (finite AR) must be increased by this amount to get the same lift, as shown in Fig. 9.21. It also means that instead of being perpendicular to the motion, the lift force leans angle $\Delta\alpha$ backwards from the perpendicular—we have an induced drag component of the drag coefficient. From simple geometry

$$\Delta C_D \approx C_L \Delta\alpha \approx \frac{C_L^2}{\pi AR} \tag{9.41}$$

This also is shown in Fig. 9.21.

When written in terms of aspect ratio, the drag of a wing of finite span becomes [21]

$$C_D = C_{D,\infty} + C_{D,i} = C_{D,\infty} + \frac{C_L^2}{\pi AR} \tag{9.42}$$

where $C_{D,\infty}$ is the section drag coefficient at C_L, $C_{D,i}$ is the induced drag coefficient at C_L, and AR is the aspect ratio of the finite-span wing.

Drag on airfoils arises from viscous and pressure forces. Viscous drag changes with Reynolds number but only slightly with angle of attack. These relationships and some commonly used terminology are illustrated in Fig. 9.22.

A useful approximation to the drag polar for a complete aircraft may be obtained by adding the induced drag to the drag at zero lift. The drag at any lift coefficient is obtained from

$$C_D = C_{D,0} + C_{D,i} = C_{D,0} + \frac{C_L^2}{\pi AR} \tag{9.43}$$

where $C_{D,0}$ is the drag coefficient at zero lift and AR is the aspect ratio.

It is possible to increase the *effective* aspect ratio for a wing of given geometric ratio by adding an *endplate* or *winglet* to the wing tip. An endplate may be a simple plate attached at the tip, perpendicular to the wing span, as on the rear-mounted wing of a racing car (see Fig. 9.26). An endplate functions by blocking the flow that tends to migrate from the high-pressure region below the wing tip to the low-pressure region above the tip when the wing is producing lift. When the endplate is added, the strength of the trailing vortex and the induced drag are reduced.

Winglets are short, aerodynamically contoured wings set perpendicular to the wing at the tip. Like the endplate, the winglet reduces the strength of the trailing vortex system and the induced drag. The winglet also produces a small component of force in the flight direction, which has the effect of further reducing the overall drag of the aircraft. The contour and angle of attack of the winglet are adjusted based on wind tunnel tests to provide optimum results.

As we have seen, aircraft can be fitted with low-drag airfoils to give excellent performance at cruise conditions. However, since the maximum lift coefficient is low for thin airfoils, additional effort must be expended to obtain acceptably low landing speeds. In steady-state flight conditions, lift must equal aircraft weight. Thus,

$$W = F_L = C_L \frac{1}{2} \rho V^2 A$$

Minimum flight speed is therefore obtained when $C_L = C_{L_{max}}$. Solving for V_{min},

$$V_{min} = \sqrt{\frac{2W}{\rho C_{L_{max}} A}} \tag{9.44}$$

According to Eq. 9.44, the minimum landing speed can be reduced by increasing either $C_{L_{max}}$ or wing area. Two basic techniques are available to control these variables: variable-geometry wing sections (e.g., obtained through the use of flaps) or boundary-layer control techniques.

Flaps are movable portions of a wing trailing edge that may be extended during landing and takeoff to increase effective wing area. The effects on lift and drag of two typical flap configurations are shown in Fig. 9.23, as applied to the NACA 23012 airfoil section. The maximum lift coefficient for this section is increased from 1.52 in the "clean" condition to 3.48 with double-slotted flaps. From Eq. 9.44, the corresponding reduction in landing speed would be 34 percent.

Figure 9.23 shows that section drag is increased substantially by high-lift devices. From Fig. 9.23*b*, section drag at $C_{L_{max}}$ ($C_D \approx 0.28$) with double-slotted flaps is about 5 times as high as section drag at $C_{L_{max}}$ ($C_D \approx 0.055$) for the clean airfoil. Induced drag due to lift must be added to section drag to obtain total drag. Because induced drag is proportional to C_L^2 (Eq. 9.41), total drag rises sharply at low aircraft speeds. At speeds near stall, drag may increase enough to exceed the thrust available from the engines. To avoid this dangerous region of unstable operation, the Federal Aviation Administration (FAA) limits operation of commercial aircraft to speeds above 1.2 times stall speed.

Although details are beyond the scope of this book, the basic purpose of all boundary-layer control techniques is to delay separation or reduce drag, by adding momentum to the boundary layer through blowing, or by removing low-momentum boundary-layer fluid by suction. Many examples of practical boundary-layer control systems may be seen on commercial transport aircraft at your local airport. Two typical systems are shown in Fig. 9.24.

VIDEO

Leading Edge Slats.

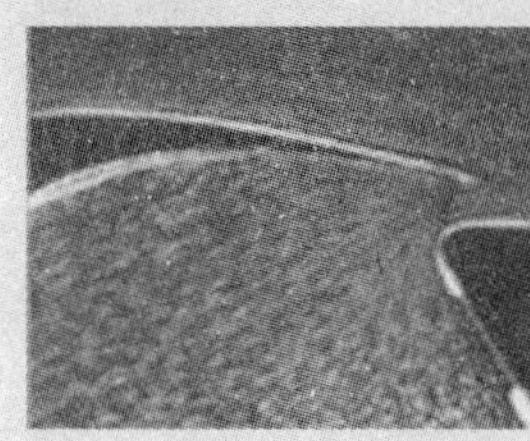

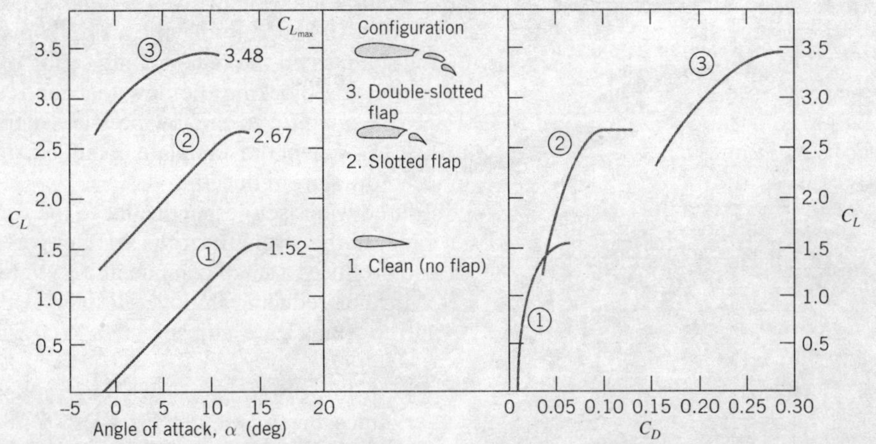

(*a*) Lift coefficient versus angle of attack (*b*) Lift-drag polar

Fig. 9.23 Effect of flaps on aerodynamic characteristics of NACA 23012 airfoil section. (Data from Abbott and von Doenhoff [21].)

Fig. 9.24 (a) Application of high-lift boundary-layer control devices to reduce landing speed of a jet transport aircraft. The wing of the Boeing 777 is highly mechanized. In the landing configuration, large slotted trailing-edge flaps roll out from under the wing and deflect downward to increase wing area and camber, thus increasing the lift coefficient. Slats at the leading edge of the wing move forward and down, to increase the effective radius of the leading edge and prevent flow separation, and to open a slot that helps keep air flow attached to the wing's upper surface. After touchdown, spoilers (not shown in use) are raised in front of each flap to decrease lift and ensure that the plane remains on the ground, despite use of the lift-augmenting devices. (This photograph was taken during a flight test. Flow cones are attached to the flaps and ailerons to identify regions of separated flow on these surfaces.) (Photograph courtesy of Chad Slattery/ Getty Images, Inc.)

Fig. 9.24 (b) Application of high-lift boundary-layer control devices to reduce takeoff speed of a jet transport aircraft. This is another view of the Boeing 777 wing. In the takeoff configuration, large slotted trailing-edge flaps deflect to increase the lift coefficient. The low-speed aileron near the wingtip also deflects to improve span loading during takeoff. This view also shows the single-slotted outboard flap, the high-speed aileron, and nearest the fuselage, the double-slotted inboard flap. (Photograph courtesy of Jeremy Simms/Getty Images, Inc.)

Example 9.8 OPTIMUM CRUISE PERFORMANCE OF A JET TRANSPORT

Jet engines burn fuel at a rate proportional to thrust delivered. The optimum cruise condition for a jet aircraft is at maximum speed for a given thrust. In steady level flight, thrust and drag are equal. Hence, optimum cruise occurs at the speed when the ratio of drag force to air speed is minimized.

A Boeing 727-200 jet transport has wing planform area $A_p = 149\ \text{m}^2$, and aspect ratio $AR = 6.5$. Stall speed at sea level for this aircraft with flaps up and a gross weight of 667,500 N is 280 km/hr. Below $M = 0.6$, drag due to compressibility effects is negligible, so Eq. 9.43 may be used to estimate total drag on the aircraft. $C_{D,0}$ for the aircraft is constant at 0.0182. Assume sonic speed at sea level is $c = 1,214$ km/hr.

Evaluate the performance envelope for this aircraft at sea level by plotting drag force versus speed, between stall and $M = 0.6$. Use this graph to estimate optimum cruise speed for the aircraft at sea-level conditions. Comment on stall speed and optimum cruise speed for the aircraft at 9,140 m altitude on a standard day.

Given: Boeing 727-200 jet transport at sea-level conditions.

$$W = 667,500\ \text{N},\ A = 149\ \text{m}^2,\ AR = 6.5,\ \text{and}\ C_{D,0} = 0.0182$$

Stall speed is $V_{\text{stall}} = 280$ km/hr, and compressibility effects on drag are negligible for $M \leq 0.6$ (sonic speed at sea level is $c = 1,124$ km/hr).

Find: (a) Drag force as a function of speed from V_{stall} to $M = 0.6$; plot results.
(b) Estimate of optimum cruise speed at sea level.
(c) Stall speed and optimum cruise speed at 9,140 m altitude.

Solution:
For steady, level flight, weight equals lift and thrust equals drag.

Governing equations:

$$F_L = C_L\, A\, \frac{1}{2}\, \rho V^2 = W \qquad C_D = C_{D,0} + \frac{C_L^2}{\pi AR}$$

$$F_D = C_D A\, \frac{1}{2}\, \rho V^2 = T \qquad M = \frac{V}{c}$$

At sea level, $\rho = 1.227\ \text{kg/m}^3$, and $c = 1,214\ \text{km/hr}$.

Since $F_L = W$ for level flight at any speed, then

$$C_L = \frac{W}{\frac{1}{2}\rho V^2 A} = \frac{2W}{\rho V^2 A}$$

At stall speed, $V = 280\ \text{km/hr}$, so

$$C_L = \times \frac{2 \times 667,500\ \text{N}}{1.227\ \text{kg/m}^3 \times (280\ )^2 (\text{km/hr})^2 \times 149\ \text{m}^2} \times \frac{\text{kg}\cdot\text{m}}{\text{N}\cdot\text{s}^2} \times \left(\frac{3.600\ \text{s}}{\text{hr}}\right)^2 \times \left(\frac{\text{km}}{1,000\ \text{m}}\right)^2$$

$$C_L = \frac{9.46\times 10^4}{[V(\text{km/hr})]^2} = \frac{9.46\times 10^4}{(280)^2} = 1.207, \text{ and}$$

$$C_D = C_{D,0} + \frac{C_L^2}{\pi AR} = 0.0182 + \frac{(1.207)^2}{\pi(6.5)} = 0.0895$$

Then

$$F_D = W\frac{C_D}{C_L} = 667,500\left(\frac{0.0895}{1.207}\right) = 49,496\ \text{N}$$

At $M = 0.6$, $V = Mc = (0.6)1,214\ \text{km/hr} = 728\ \text{km/hr}$, so $C_L = 0.178$ and

$$C_D = 0.0182 + \frac{(0.178)^2}{\pi(6.5)} = 0.0198$$

so

$$F_D = 667,500\ \text{N}\left(\frac{0.0198}{0.178}\right) = 74,250\ \text{N}$$

Similar calculations lead to the following table (computed using *Excel*):

V (km/hr)	280	320	480	640	730
C_L	1.207	0.924	0.411	0.231	0.178
C_D	0.0895	0.0600	0.0265	0.0208	0.0197
F_D (N)	49,510	43,348	43,009	60,150	74,237

These data may be plotted as:

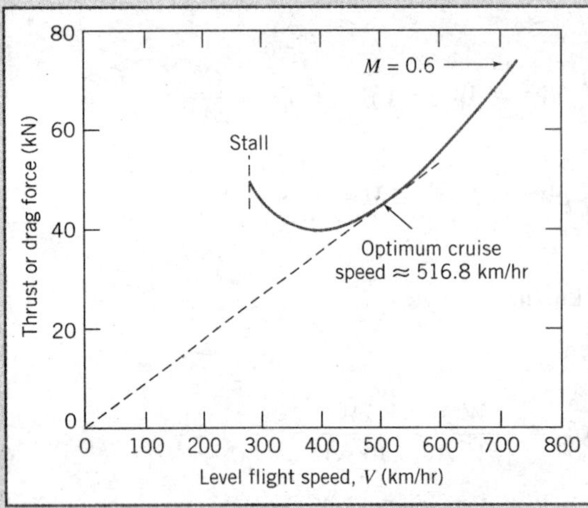

From the plot, the optimum cruise speed at sea level is estimated as 516.8 km/hr (and using *Excel* we obtain 518.4 km/hr).

At 9140 m altitude, the density is only about 0.375 times sea level density, from Table A.3. The speeds for corresponding forces are calculated from

$$F_L = C_L A \frac{1}{2} \rho V^2 \quad \text{or} \quad V = \sqrt{\frac{2F_L}{C_L \rho A}} \quad \text{or} \quad \frac{V_{30}}{V_{SL}} = \sqrt{\frac{\rho_{SL}}{\rho_{30}}} = \sqrt{\frac{1}{0.375}} = 1.63$$

Thus, speeds increase 63 percent at 9140 m altitude: $V_{\text{stall}} \approx 456$ km/hr
$V_{\text{cruise}} \approx 845$ km/hr

This problem illustrates that high-altitude flight increases the optimum cruising speed—in general this speed depends on aircraft configuration, gross weight, segment length, and winds aloft.

The *Excel* workbook for this Example plots the drag or thrust and power as functions of speed. It also allows "what-ifs," e.g., what happens to the optimum speed if altitude is increased, or if the aspect ratio is increased, and so on.

CLASSIC VIDEO

Boundary Layer Control.

Aerodynamic lift is an important consideration in the design of high-speed land vehicles such as racing cars and land-speed-record machines. A road vehicle generates lift by virtue of its shape [29]. A representative centerline pressure distribution measured in the wind tunnel for an automobile is shown in Fig. 9.25 [30].

The pressure is low around the nose because of streamline curvature as the flow rounds the nose. The pressure reaches a maximum at the base of the windshield, again as a result of streamline curvature. Low-pressure regions also occur at the windshield header and over the top of the automobile. The air speed across the top is approximately 30 percent higher than the freestream air speed. The same effect occurs around

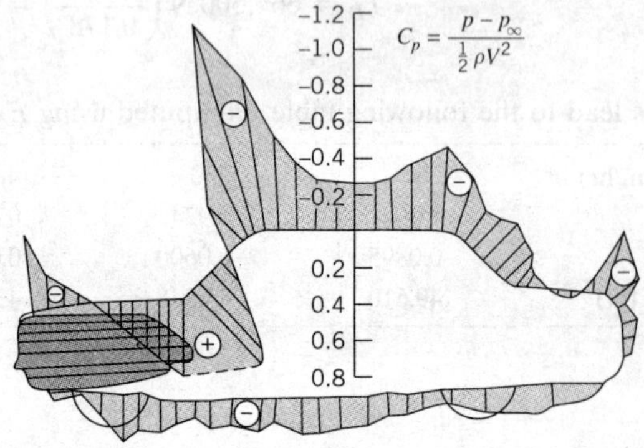

Fig. 9.25 Pressure distribution along the centerline of an automobile [30].

Fig. 9.26 Contemporary racing car showing aerodynamic features. The car's front and rear wings are designed to provide significant downforce at speed to improve traction. Also visible are fairings to direct hot air from the radiators around the rear tires, and at the front of the car, cool air toward the brakes. Not shown are other aerodynamic features such as the fuselage bottom, which is designed to route the airflow carefully, using diffusers, to develop the most negative pressure, and to cause this negative pressure to act over the largest possible area under the car, to develop additional downforce. (Photograph courtesy of Mark Evans/iStock photo.)

the "A-pillars" at the windshield edges. The drag increase caused by an added object, such as an antenna, spotlight, or mirror at that location, thus would be $(1.3)^2 \approx 1.7$ times the drag the object would experience in an undisturbed flow field. Thus the *parasite drag* of an added component can be much higher than would be predicted from its drag calculated for free flow.

At high speeds, aerodynamic lift forces can unload tires, causing serious reductions in steering control and reducing stability to a dangerous extent. Lift forces on early racing cars were counteracted somewhat by "spoilers," at considerable penalty in drag. In 1965 Jim Hall introduced the use of movable inverted airfoils on his Chaparral sports cars to develop aerodynamic downforce and provide aerodynamic braking [31]. Since then the developments in application of aerodynamic devices have been rapid. Aerodynamic design is used to reduce lift on all modern racing cars, as exemplified in Fig. 9.26. Liebeck airfoils [23] are used frequently for high-speed automobiles. Their high lift coefficients and relatively low drag allow downforce equal to or greater than the car weight to be developed at racing speeds. "Ground effect" cars use venturi-shaped ducts under the car and side skirts to seal leakage flows. The net result of these aerodynamic effects is that the downward force (which increases with speed) generates excellent traction without adding significant weight to the vehicle, allowing faster speeds through curves and leading to lower lap times.

Another method of boundary-layer control is use of moving surfaces to reduce skin friction effects on the boundary layer [32]. This method is hard to apply to practical devices, because of geometric and weight complications, but it is very important in recreation. Most golfers, tennis players, soccer players, and baseball pitchers can attest to this! Tennis and soccer players use spin to control the trajectory and bounce of a shot. In golf, a drive can leave the tee at 84 m/s or more, with backspin of 9000 rpm! Spin provides significant aerodynamic lift that substantially increases the carry of a drive. Spin is also largely responsible for hooking and slicing when shots are not hit squarely. The baseball pitcher uses spin to throw a curve ball.

Flow about a spinning sphere is shown in Fig. 9.27*a*. Spin alters the pressure distribution and also affects the location of boundary-layer separation. Separation is delayed on the upper surface of the sphere in Fig. 9.27*a*, and it occurs earlier on the lower surface. Thus pressure (because of the Bernoulli effect) is reduced on the upper surface and increased on the lower surface; the wake is deflected downward as shown. Pressure forces cause a lift in the direction shown; spin in the opposite direction would produce negative lift—a downward force. The force is directed perpendicular to both V and the spin axis.

Lift and drag data for spinning smooth spheres are presented in Fig. 9.27*b*. The most important parameter is the *spin ratio*, $\omega D/2V$, the ratio of surface speed to

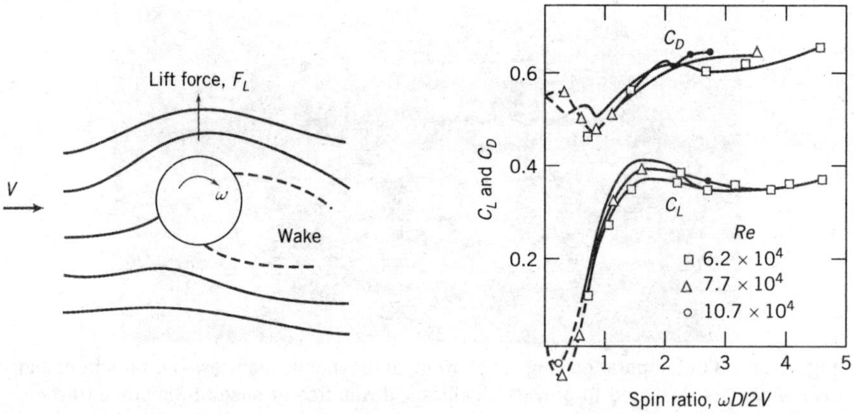

(a) Flow pattern (b) Lift and drag coefficients

Fig. 9.27 Flow pattern, lift, and drag coefficients for a smooth spinning sphere in uniform flow. (Data from [19].)

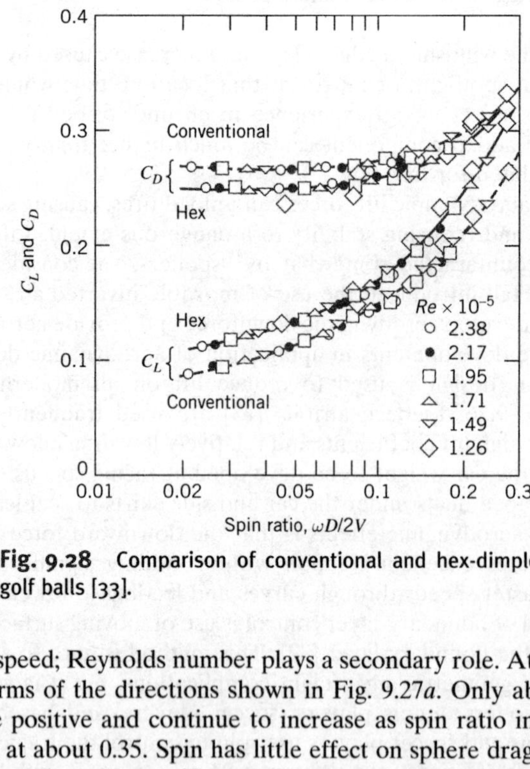

Fig. 9.28 Comparison of conventional and hex-dimpled golf balls [33].

freestream flow speed; Reynolds number plays a secondary role. At low spin ratio, lift is negative in terms of the directions shown in Fig. 9.27*a*. Only above $\omega D/2V \approx 0.5$ does lift become positive and continue to increase as spin ratio increases. Lift coefficient levels out at about 0.35. Spin has little effect on sphere drag coefficient, which varies from about 0.5 to about 0.65 over the range of spin ratio shown.

Earlier we mentioned the effect of dimples on the drag of a golf ball. Experimental data for lift and drag coefficients for spinning golf balls are presented in Fig. 9.28 for subcritical Reynolds numbers between 126,000 and 238,000. Again the independent variable is spin ratio; a much smaller range of spin ratio, typical of golf balls, is presented in Fig. 9.28.

A clear trend is evident: The lift coefficient increases consistently with spin ratio for both hexagonal and "conventional" (round) dimples. The lift coefficient on a golf ball with hexagonal dimples is significantly—as much as 15 percent—higher than on a ball with round dimples. The advantage for hexagonal dimples continues to the largest spin ratios that were measured. The drag coefficient for a ball with hexagonal dimples is consistently 5 to 7 percent lower than the drag coefficient for a ball with round dimples at low spin ratios, but the difference becomes less pronounced as spin ratio increases.

The combination of higher lift and lower drag increases the carry of a golf shot. A recent design—the Callaway HX—has improved performance further by using a "tubular lattice network" using ridges of hexagons and pentagons (at a precise height of 0.21 mm) instead of dimples, so that there are no flat spots at all on the surface [34]. Callaway claims the HX flies farther than any ball they ever tested.

Example 9.9 LIFT OF A SPINNING BALL

A smooth tennis ball, with 57 g mass and 64 mm diameter, is hit at 25 m/s with topspin of 7500 rpm. Calculate the aerodynamic lift acting on the ball. Evaluate the radius of curvature of its path at maximum elevation in a vertical plane. Compare with the radius for no spin.

Given: Tennis ball in flight, with $m = 57$ g and $D = 64$ mm, hit with $V = 25$ m/s and topspin of 7500 rpm.

Find: (a) Aerodynamic lift acting on ball.
(b) Radius of curvature of path in vertical plane.
(c) Comparison with radius for no spin.

Solution:

Assume ball is smooth.

Use data from Fig. 9.27 to find lift: $C_L = f\left(\frac{\omega D}{2V}, Re_D\right)$.

From given data (for standard air, $\nu = 1.46 \times 10^{-5}\,\text{m}^2/\text{s}$),

$$\frac{\omega D}{2V} = \frac{1}{2} \times 7500 \frac{\text{rev}}{\text{min}} \times 0.064\,\text{m} \times \frac{\text{s}}{25\,\text{m}} \times 2\pi \frac{\text{rad}}{\text{rev}} \times \frac{\text{min}}{60\,\text{s}} = 1.01$$

$$Re_D = \frac{VD}{\nu} = 25\frac{\text{m}}{\text{s}} \times 0.064\,\text{m} \times \frac{\text{s}}{1.46 \times 10^{-5}\,\text{m}^2} = 1.10 \times 10^5$$

From Fig. 9.27, $C_L \approx 0.3$, so

$$F_L = C_L A \frac{1}{2}\rho V^2$$

$$= C_L \frac{\pi D^2}{4} \frac{1}{2}\rho V^2 = \frac{\pi}{8} C_L D^2 \rho V^2$$

$$F_L = \frac{\pi}{8} \times 0.3 \times (0.064)^2\,\text{m}^2 \times 1.23\frac{\text{kg}}{\text{m}^3} \times (25)^2 \frac{\text{m}^2}{\text{s}^2} \times \frac{\text{N}\cdot\text{s}^2}{\text{kg}\cdot\text{m}} = 0.371\,\text{N} \qquad F_L$$

Because the ball is hit with topspin, this force acts downward.

Use Newton's second law to evaluate the curvature of the path. In the vertical plane,

$$\sum F_z = -F_L - mg = ma_z = -m\frac{V^2}{R} \quad \text{or} \quad R = \frac{V^2}{g + F_L/m}$$

$$R = (25)^2 \frac{\text{m}^2}{\text{s}^2}\left[\frac{1}{9.81\frac{\text{m}}{\text{s}^2} + 0.371\,\text{N} \times \frac{1}{0.057\,\text{kg}} \times \frac{\text{kg}\cdot\text{m}}{\text{N}\cdot\text{s}^2}}\right]$$

$$R = 38.3\,\text{m (with spin)} \qquad R$$

$$R = (25)^2 \frac{\text{m}^2}{\text{s}^2} \times \frac{\text{s}^2}{9.81\,\text{m}} = 63.7\,\text{m (without spin)} \qquad R$$

Thus topspin has a significant effect on trajectory of the shot!

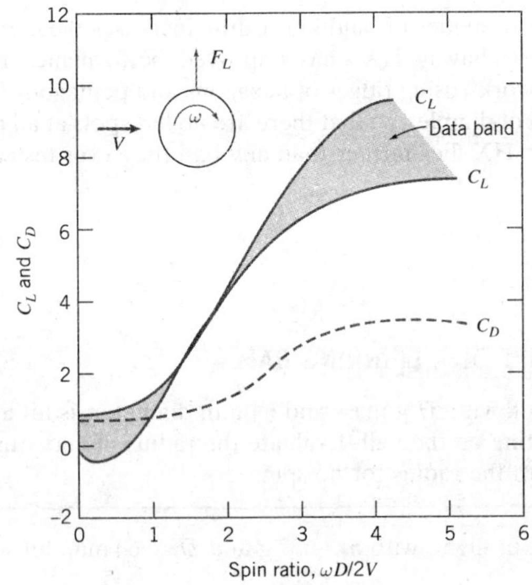

Fig. 9.29 Lift and drag of a rotating cylinder as a function of relative rotational speed; Magnus force. (Data from [35].)

It has long been known that a spinning projectile in flight is affected by a force perpendicular to the direction of motion and to the spin axis. This effect, known as the *Magnus effect*, is responsible for the systematic drift of artillery shells.

Cross flow about a rotating circular cylinder is qualitatively similar to flow about the spinning sphere shown schematically in Fig. 9.27*a*. If the velocity of the upper surface of a cylinder is in the same direction as the freestream velocity, separation is delayed on the upper surface; it occurs earlier on the lower surface. Thus the wake is deflected and the pressure distribution on the cylinder surface is altered when rotation is present. Pressure is reduced on the upper surface and increased on the lower surface, causing a net lift force acting upward. Spin in the opposite direction reverses these effects and causes a downward lift force.

Lift and drag coefficients for the rotating cylinder are based on projected area, LD. Experimentally measured lift and drag coefficients for subcritical Reynolds numbers between 40,000 and 660,000 are shown as functions of spin ratio in Fig. 9.29. When surface speed exceeds flow speed, the lift coefficient increases to surprisingly high values, while in two-dimensional flow, drag is affected only moderately. Induced drag, which must be considered for finite cylinders, can be reduced by using end disks larger in diameter than the body of the cylinder.

The power required to rotate a cylinder may be estimated from the skin friction drag of the cylinder surface. Hoerner [35] suggests basing the skin friction drag estimate on the tangential surface speed and surface area. Goldstein [19] suggests that the power required to spin the cylinder, when expressed as an equivalent drag coefficient, may represent 20 percent or more of the aerodynamic C_D of a stationary cylinder.

9.9 Summary and Useful Equations

In this chapter we have:

- ✓ Defined and discussed various terms commonly used in aerodynamics, such as: boundary-layer disturbance, displacement and momentum thicknesses; flow separation; streamlining; skin friction and pressure drag and drag coefficient; lift and lift coefficient; wing chord, span and aspect ratio; and induced drag.
- ✓ Derived expressions for the boundary-layer thickness on a flat plate (zero pressure gradient) using exact* and approximate methods (using the momentum integral equation).
- ✓ Learned how to estimate the lift and drag from published data for a variety of objects.

*This topic applies to a section that may be omitted without loss of continuity in the material.

While investigating the above phenomena, we developed insight into some of the basic concepts of aerodynamic design, such as how to minimize drag, how to determine the optimum cruising speed of an airplane, and so on.

Note: Most of the Useful Equations in the table below have a number of constraints or limitations—*be sure to refer to their page numbers for details*!

Useful Equations

Definition of displacement thickness:	$\delta^* = \int_0^\infty \left(1 - \frac{u}{U}\right) dy \approx \int_0^\delta \left(1 - \frac{u}{U}\right) dy$	**(9.1)**	Page 426
Definition of momentum thickness:	$\theta = \int_0^\infty \frac{u}{U}\left(1 - \frac{u}{U}\right) dy \approx \int_0^\delta \frac{u}{U}\left(1 - \frac{u}{U}\right) dy$	**(9.2)**	Page 426
Boundary-layer thickness (laminar, exact—Blasius):	$\delta \approx \frac{5.0}{\sqrt{U/\nu x}} = \frac{5.0x}{\sqrt{Re_x}}$	**(9.13)**	Page W-20
Wall stress (laminar, exact—Blasius):	$\tau_w = 0.332U\sqrt{\rho\mu U/x} = \frac{0.332\rho U^2}{\sqrt{Re_x}}$	**(9.14)**	Page W-21
Skin friction coefficient (laminar, exact—Blasius):	$C_f = \frac{\tau_w}{\frac{1}{2}\rho U^2} = \frac{0.664}{\sqrt{Re_x}}$	**(9.15)**	Page W-21
Momentum integral equation:	$\frac{\tau_w}{\rho} = \frac{d}{dx}(U^2\theta) + \delta^* U \frac{dU}{dx}$	**(9.17)**	Page 432
Boundary-layer thickness for flat plate (laminar, approximate—polynomial velocity profile):	$\frac{\delta}{x} = \sqrt{\frac{30\mu}{\rho U x}} = \frac{5.48}{\sqrt{Re_x}}$	**(9.21)**	Page 435
Definition of skin friction coefficient:	$C_f \equiv \frac{\tau_w}{\frac{1}{2}\rho U^2}$	**(9.22)**	Page 436
Skin friction coefficient for flat plate (laminar, approximate—polynomial velocity profile):	$C_f = \frac{0.730}{\sqrt{Re_x}}$	**(9.23)**	Page 436
Boundary-layer thickness for flat plate (turbulent, approximate—$\frac{1}{7}$-power-law velocity profile):	$\frac{\delta}{x} = 0.382\left(\frac{\nu}{Ux}\right)^{1/5} = \frac{0.382}{Re_x^{1/5}}$	**(9.26)**	Page 440
Skin friction coefficient for flat plate (turbulent, approximate—$\frac{1}{7}$-power-law velocity profile):	$C_f = \frac{\tau_w}{\frac{1}{2}\rho U^2} = \frac{0.0594}{Re_x^{1/5}}$	**(9.27)**	Page 440
Definition of drag coefficient:	$C_D \equiv \frac{F_D}{\frac{1}{2}\rho V^2 A}$	**(9.30)**	Page 446
Drag coefficient for flat plate (entirely laminar, based on Blasius solution):	$C_D = \frac{1.33}{\sqrt{Re_L}}$	**(9.33)**	Page 447
Drag coefficient for flat plate (entirely turbulent, based on $\frac{1}{7}$-power-law velocity profile):	$C_D = \frac{0.0742}{Re_L^{1/5}}$	**(9.34)**	Page 447
Drag coefficient for flat plate (empirical, $Re_L < 10^9$):	$C_D = \frac{0.455}{(\log Re_L)^{2.58}}$	**(9.35)**	Page 447

Drag coefficient for flat plate (based on $\frac{1}{7}$th power-law velocity profile, $5 \times 10^5 \leq Re_L \leq 10^7$):	$C_D = \dfrac{0.0742}{Re_L^{1/5}} - \dfrac{1740}{Re_L}$	**(9.37a)**	Page 448
Drag coefficient for flat plate (empirical, $5 \times 10^5 < Re_L < 10^9$):	$C_D = \dfrac{0.455}{(\log Re_L)^{2.58}} - \dfrac{1610}{Re_L}$	**(9.37b)**	Page 448
Definition of lift coefficient:	$C_L \equiv \dfrac{F_L}{\frac{1}{2}\rho V^2 A_p}$	**(9.38)**	Page 459
Definition of aspect ratio:	$AR \equiv \dfrac{b^2}{A_p}$	**(9.39)**	Page 464
Drag coefficient of a wing (finite span airfoil, using $C_{D,\infty}$):	$C_D = C_{D,\infty} + C_{D,i} = C_{D,\infty} + \dfrac{C_L^2}{\pi AR}$	**(9.42)**	Page 465
Drag coefficient of a wing (finite span airfoil, using $C_{D,0}$):	$C_D = C_{D,0} + C_{D,i} = C_{D,0} + \dfrac{C_L^2}{\pi AR}$	**(9.43)**	Page 465

Case Study

The Humpback Whale Flipper

Image is unavailable due to copyright restrictions

In Chapter 5 we developed the Navier–Stokes equations for describing many of the flow fields that we are likely to study, and in Chapter 6 we developed Euler's equation and the Bernoulli equation, which are useful when analyzing flows we can model as inviscid, such as much of aerodynamics. In Chapter 9 we have expanded on this material, considering many real-world phenomena such as boundary layers, flow separation, and so on. However, we still have a lot to learn about lots of different flow problems. For example, it is conventional knowledge that airfoils and hydrofoils should have leading edges that are smooth and streamlined—even insect debris stuck to the leading edge of a wind turbine rotor, for example, can reduce performance. However, Dr. Frank E. Fish at West Chester University, Pennsylvania, and research colleagues at Duke University and the U.S. Naval Academy have studied the fluid mechanics of the flippers of the humpback whale, which are anything but smooth, as shown in the figure. The researchers were curious that even allowing for the natural variability that any animal's profile would have, the whale seems to have evolved flippers that have a unique row of bumps, or tubercles, along their leading edges that produce a serrated look.

What are the bumps doing on those flippers? Testing and analysis (using many of the ideas discussed in Chapters 5, 6, and 9) have been done comparing airfoils with tubercles to similar airfoils with a traditional smooth leading edge. This research showed that the stall angle (recall that this is the angle when flow separation occurs, leading to a sudden loss in lift) was greatly increased and that when stall occurs, it does so gradually, not suddenly as in most traditional airfoils. In addition, the tubercles airfoil was more efficient: It had significantly less drag and more lift.

It's believed that this occurs because the tubercles channel the wind as it hits the leading edge of the airfoil, causing vortices to develop as the flow moves along the airfoil surface, stabilizing the flow and, among other things, preventing secondary flow from moving across the span of the airfoil to its tip, which causes noise and loss of lift.

Possible future uses of tubercles airfoil sections include virtually any application that needs efficient, quiet performance, with excellent stall behavior: wind turbines, airplane wings, ship propellers and rudders, household fans, and so on. It is even possible that existing devices such as large-scale wind turbines may be retrofitted with tubercles to improve their performance and reduce their noise. We will have more to say about new developments in this area in the next *Case Study in Energy and the Environment*, at the beginning of Chapter 10.

References

1. Prandtl, L., "Fluid Motion with Very Small Friction (in German)," *Proceedings of the Third International Congress on Mathematics*, Heidelberg, 1904; English translation available as NACA TM 452, March 1928.
2. Blasius, H., "The Boundary Layers in Fluids with Little Friction (in German)," *Zeitschrift für Mathematik und Physik, 56*, 1, 1908, pp. 1–37; English translation available as NACA TM 1256, February 1950.
3. Schlichting, H., *Boundary-Layer Theory*, 7th ed. New York: McGraw-Hill, 1979.
4. Stokes, G. G., "On the Effect of the Internal Friction of Fluids on the Motion of Pendulums," *Cambridge Philosophical Transactions, IX*, 8, 1851.
5. Howarth, L., "On the Solution of the Laminar Boundary-Layer Equations," *Proceedings of the Royal Society of London, A164*, 1938, pp. 547–579.
6. Hess, J. L., and A. M. O. Smith, "Calculation of Potential Flow About Arbitrary Bodies," in *Progress in Aeronautical Sciences*, Vol. 8, D. Kuchemann et al., eds. Elmsford, NY: Pergamon Press, 1966.
7. Kraus, W., "Panel Methods in Aerodynamics," in *Numerical Methods in Fluid Dynamics*, H. J. Wirz and J. J. Smolderen, eds. Washington, DC: Hemisphere, 1978.
8. Rosenhead, L., ed., *Laminar Boundary Layers*. London: Oxford University Press, 1963.
9. Rotta, J. C., "Turbulent Boundary Layers in Incompressible Flow," in *Progress in Aeronautical Sciences*, A. Ferri, et al., eds. New York: Pergamon Press, 1960, pp. 1–220.
10. Kline, S. J., et al., eds., *Proceedings, Computation of Turbulent Boundary Layers—1968 AFOSR-IFP-Stanford Conference*, Vol. I: Methods, Predictions, Evaluation, and Flow Structure, and Vol. II: Compiled Data. Stanford, CA: Thermosciences Division, Department of Mechanical Engineering, Stanford University, 1969.
11. Kline, S. J., et al., eds., *Proceedings, 1980–81 AFOSR-HTTM-Stanford Conference on Complex Turbulent Flows: Comparison of Computation and Experiment*, three volumes. Stanford, CA: Thermosciences Division, Department of Mechanical Engineering, Stanford University, 1982.
12. Cebeci, T., and P. Bradshaw, *Momentum Transfer in Boundary Layers*. Washington, D.C.: Hemisphere, 1977.
13. Bradshaw, P., T. Cebeci, and J. H. Whitelaw, *Engineering Calculation Methods for Turbulent Flow*. New York: Academic Press, 1981.
14. *Fluent*. Fluent Incorporated, Centerra Resources Park, 10 Cavendish Court, Lebanon, NH 03766 (www.fluent.com).
15. *STAR-CD*. Adapco, 60 Broadhollow Road, Melville, NY 11747 (www.cd-adapco.com).
16. Hoerner, S. F., *Fluid-Dynamic Drag*, 2nd ed. Midland Park, NJ: Published by the author, 1965.
17. Shapiro, A. H., *Shape and Flow, the Fluid Dynamics of Drag*. New York: Anchor, 1961 (paperback).
18. Fage, A., "Experiments on a Sphere at Critical Reynolds Numbers," Great Britain, *Aeronautical Research Council, Reports and Memoranda*, No. 1766, 1937.
19. Goldstein, S., ed., *Modern Developments in Fluid Dynamics*, Vols. I and II. Oxford: Clarendon Press, 1938. (Reprinted in paperback by Dover, New York, 1967.)
20. Morel, T., and M. Bohn, "Flow over Two Circular Disks in Tandem," *Transactions of the ASME, Journal of Fluids Engineering, 102*, 1, March 1980, pp. 104–111.
21. Abbott, I. H., and A. E. von Doenhoff, *Theory of Wing Sections, Including a Summary of Airfoil Data*. New York: Dover, 1959 (paperback).
22. Stratford, B. S., "An Experimental Flow with Zero Skin Friction," *Journal of Fluid Mechanics, 5*, Pt. 1, January 1959, pp. 17–35.
23. Liebeck, R. H., "Design of Subsonic Airfoils for High Lift," *AIAA Journal of Aircraft, 15*, 9, September 1978, pp. 547–561.
24. Smith, A. M. O., "Aerodynamics of High-Lift Airfoil Systems," in *Fluid Dynamics of Aircraft Stalling*, AGARD CP-102, 1973, pp. 10–1 through 10–26.
25. Morel, T., "Effect of Base Slant on Flow in the Near Wake of an Axisymmetric Cylinder," *Aeronautical Quarterly, XXXI*, Pt. 2, May 1980, pp. 132–147.
26. Hucho, W. H., "The Aerodynamic Drag of Cars—Current Understanding, Unresolved Problems, and Future Prospects," in *Aerodynamic Drag Mechanisms of Bluff Bodies and Road Vehicles*, G. Sovran, T. Morel, and W. T. Mason, eds. New York: Plenum, 1978.
27. McDonald, A. T., and G. M. Palmer, "Aerodynamic Drag Reduction of Intercity Buses," *Transactions, Society of Automotive Engineers, 89*, Section 4, 1980, pp. 4469–4484 (SAE Paper No. 801404).
28. Grosser, M., *Gossamer Odyssey*. Boston: Houghton Mifflin, 1981.
29. Carr, G. W., "The Aerodynamics of Basic Shapes for Road Vehicles. Part 3: Streamlined Bodies," The Motor

Industry Research Association, Warwickshire, England, Report No. 107/4, 1969.
30. Goetz, H., "The Influence of Wind Tunnel Tests on Body Design, Ventilation, and Surface Deposits of Sedans and Sports Cars," SAE Paper No. 710212, 1971.
31. Hall, J., "What's Jim Hall Really Like?" *Automobile Quarterly, VIII*, 3, Spring 1970, pp. 282–293.
32. Moktarian, F., and V. J. Modi, "Fluid Dynamics of Airfoils with Moving Surface Boundary-Layer Control," *AIAA Journal of Aircraft, 25*, 2, February 1988, pp. 163–169.
33. Mehta, R. D., "Aerodynamics of Sports Balls," in *Annual Review of Fluid Mechanics*, ed. by M. van Dyke, et al. Palo Alto, CA: Annual Reviews, 1985, *17*, pp. 151–189.
34. "The Year in Ideas," *New York Times Magazine*, December 9, 2001, pp. 58–60.
35. Hoerner, S. F., and H. V. Borst, *Fluid-Dynamic Lift*. Bricktown, NJ: Hoerner Fluid Dynamics, 1975.
36. Chow, C.-Y., *An Introduction to Computational Fluid Mechanics*. New York: Wiley, 1980.
37. Carr, G. W., "The Aerodynamics of Basic Shapes for Road Vehicles, Part 1: Simple Rectangular Bodies," The Motor Industry Research Association, Warwickshire, England, Report No. 1968/2, 1967.

Problems

The Boundary-Layer Concept

9.1 The roof of a minivan is approximated as a horizontal flat plate. Plot the length of the laminar boundary layer as a function of minivan speed, V, as the minivan accelerates from 16 km/hr to 144 km/hr.

9.2 A model of a river towboat is to be tested at 1:18 scale. The boat is designed to travel at 3.5 m/s in fresh water at 10°C. Estimate the distance from the bow where transition occurs. Where should transition be stimulated on the model towboat?

9.3 The takeoff speed of a Boeing 757 is 260 km/hr. At approximately what distance will the boundary layer on the wings become turbulent? If it cruises at 850 km/hr at 10,000 m, at approximately what distance will the boundary layer on the wings become turbulent?

9.4 A student is to design an experiment involving dragging a sphere through a tank of fluid to illustrate (a) "creeping flow" ($Re_D < 1$) and (b) flow for which the boundary layer becomes turbulent ($Re_D \approx 2.5 \times 10^5$). She proposes to use a smooth sphere of diameter 1 cm in SAE 10 oil at room temperature. Is this realistic for both cases? If either case is unrealistic, select an alternative reasonable sphere diameter and common fluid for that case.

9.5 For flow around a sphere the boundary layer becomes turbulent around $Re_D \approx 2.5 \times 10^5$. Find the speeds at which (a) an American golf ball (D = 43 mm), (b) a British golf ball (D = 41.1 mm), and (c) a soccer ball (D = 222 mm) develop turbulent boundary layers. Assume standard atmospheric conditions.

9.6 1 m × 2 m sheet of plywood is attached to the roof of your vehicle after being purchased at the hardware store. At what speed (in kilometers per hour, in 20°C air) will the boundary layer first start becoming turbulent? At what speed is about 90 percent of the boundary layer turbulent?

9.7 Plot on one graph the length of the laminar boundary layer on a flat plate, as a function of freestream velocity, for (a) water and standard air at (b) sea level and (c) 10 km altitude. Use log-log axes, and compute data for the boundary-layer length ranging from 0.01 m to 10 m.

9.8 The extent of the laminar boundary layer on the surface of an aircraft or missile varies with altitude. For a given speed, will the laminar boundary-layer length increase or decrease with altitude? Why? Plot the ratio of laminar boundary-layer length at altitude z, to boundary-layer length at sea level, as a function of z, up to altitude z = 30 km, for a standard atmosphere.

Boundary-Layer Thickness

9.9 The most general sinusoidal velocity profile for laminar boundary-layer flow on a flat plate is $u = A\sin(By) + C$. State three boundary conditions applicable to the laminar boundary-layer velocity profile. Evaluate constants A, B, and C.

9.10 Velocity profiles in laminar boundary layers often are approximated by the equations

$$\text{Linear}: \quad \frac{u}{U} = \frac{y}{\delta}$$

$$\text{Sinusoidal}: \quad \frac{u}{U} = \sin\left(\frac{\pi y}{2\delta}\right)$$

$$\text{Parabolic}: \quad \frac{u}{U} = 2\left(\frac{y}{\delta}\right) - \left(\frac{y}{\delta}\right)^2$$

Compare the shapes of these velocity profiles by plotting y/δ (on the ordinate) versus u/U (on the abscissa).

9.11 An approximation for the velocity profile in a laminar boundary layer is

$$\frac{u}{U} = \frac{3}{2}\frac{y}{\delta} - \frac{1}{2}\left(\frac{y}{\delta}\right)^3$$

Does this expression satisfy boundary conditions applicable to the laminar boundary-layer velocity profile? Evaluate δ^*/δ and θ/δ.

9.12 An approximation for the velocity profile in a laminar boundary layer is

$$\frac{u}{U} = 2\frac{y}{\delta} - 2\left(\frac{y}{\delta}\right)^3 + \left(\frac{y}{\delta}\right)^4$$

Does this expression satisfy boundary conditions applicable to the laminar boundary-layer velocity profile? Evaluate δ^*/δ and θ/δ.

9.13 A simplistic laminar boundary-layer model is

$$\frac{u}{U} = \sqrt{2}\frac{y}{\delta} \qquad 0 < y \le \frac{\delta}{2}$$

$$\frac{u}{U} = (2-\sqrt{2})\frac{y}{\delta} + (\sqrt{2}-1) \qquad \frac{\delta}{2} < y \le \delta$$

Does this expression satisfy boundary conditions applicable to the laminar boundary-layer velocity profile? Evaluate δ^*/δ and θ/δ.

9.14 The velocity profile in a turbulent boundary layer often is approximated by the $\frac{1}{7}$-power-law equation

$$\frac{u}{U} = \left(\frac{y}{\delta}\right)^{1/7}$$

Compare the shape of this profile with the parabolic laminar boundary-layer velocity profile (Problem 9.10) by plotting y/δ (on the ordinate) versus u/U (on the abscissa) for both profiles.

9.15 Evaluate θ/δ for each of the laminar boundary-layer velocity profiles given in Problem 9.10.

9.16 Evaluate δ^*/δ for each of the laminar boundary-layer velocity profiles given in Problem 9.10.

9.17 Evaluate δ^*/δ and θ/δ for the turbulent $\frac{1}{7}$-power-law velocity profile given in Problem 9.14. Compare with ratios for the parabolic laminar boundary-layer velocity profile given in Problem 9.10.

9.18 A fluid, with density $\rho = 800$ kg/m^3, flows at $U = 3$ m/s over a flat plate 3 m long and 1 m wide. At the trailing edge, the boundary-layer thickness is $\delta = 25$ mm. Assume the velocity profile is linear, as shown, and that the flow is two-dimensional (flow conditions are independent of z). Using control volume $abcd$, shown by the dashed lines, compute the mass flow rate across surface ab. Determine the drag force on the upper surface of the plate. Explain how this (viscous) drag can be computed from the given data even though we do not know the fluid viscosity (see Problem 9.41).

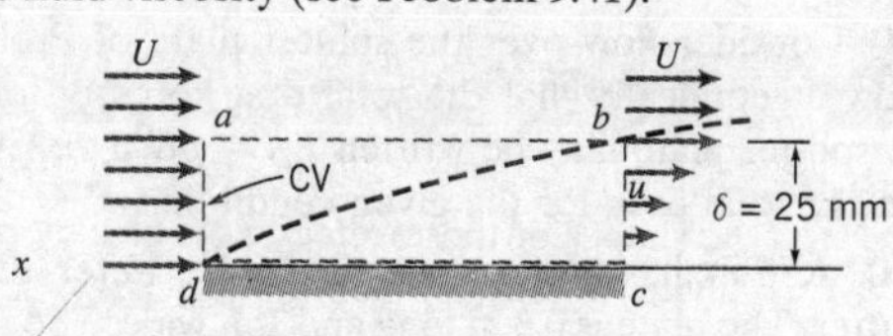

P9.18

9.19 The flat plate of Problem 9.18 is turned so that the 1 m side is parallel to the flow (the width becomes 3 m). Should we expect that the drag increases or decreases? Why? The trailing edge boundary-layer thickness is now $\delta = 14$ mm. Assume again that the velocity profile is linear and that the flow is two-dimensional (flow conditions are independent of z). Repeat the analysis of Problem 9.18.

9.20 Solve Problem 9.18 again with the velocity profile at section bc given by the parabolic expression from Problem 9.10.

9.21 The test section of a low-speed wind tunnel is 1.5 m long, preceded by a nozzle with a diffuser at the outlet. The tunnel cross section is 20 cm × 20 cm. The wind tunnel is to operate with 40°C air and have a design velocity of 50 m/s in the test section. A potential problem with such a wind tunnel is boundary-layer blockage. The boundary-layer displacement thickness reduces the effective cross-sectional area (the test area, in which we have uniform flow); in addition, the uniform flow will be accelerated. If these effects are pronounced, we end up with a smaller useful test cross section with a velocity somewhat higher than anticipated. If the boundary layer thickness is 10 mm at the entrance and 25 mm at the exit, and the boundary layer velocity profile is given by $u/U = (y/\delta)^{1/7}$, estimate the displacement thickness at the end of the test section and the percentage change in the uniform velocity between the inlet and outlet.

9.22 Air flows in a horizontal cylindrical duct of diameter $D = 100$ mm. At a section a few meters from the entrance, the turbulent boundary layer is of thickness $\delta_1 = 5.25$ mm, and the velocity in the inviscid central core is $U_1 = 12.5$ m/s. Farther downstream the boundary layer is of thickness $\delta_2 = 24$ mm. The velocity profile in the boundary layer is approximated well by the $\frac{1}{7}$-power expression. Find the velocity, U_2, in the inviscid central core at the second section, and the pressure drop between the two sections.

9.23 Laboratory wind tunnels have test sections 25 cm square and 50 cm long. With nominal air speed $U_1 = 25$ m/s at the test section inlet, turbulent boundary layers form on the top, bottom, and side walls of the tunnel. The boundary-layer thickness is $\delta_1 = 20$ mm at the inlet and $\delta_2 = 30$ mm at the outlet from the test section. The boundary-layer velocity profiles are of power-law form, with $u/U = (y/\delta)^{1/7}$. Evaluate the freestream velocity, U_2, at the exit from the wind-tunnel test section. Determine the change in static pressure along the test section.

9.24 The square test section of a small laboratory wind tunnel has sides of width $W = 40$ cm. At one measurement location, the turbulent boundary layers on the tunnel walls are $\delta_1 = 1$ cm thick. The velocity profile is approximated well by the $\frac{1}{7}$-power expression. At this location, the freestream air speed is $U_1 = 20$ m/s, and the static pressure is $p_1 = -250$ Pa (gage). At a second measurement location downstream, the boundary layer thickness is $\delta_2 = 1.3$ cm. Evaluate the air speed in the freestream in the second section. Calculate the difference in static pressure from section ① to section ②.

9.25 Air flows in the entrance region of a square duct, as shown. The velocity is uniform, $U_0 = 30$ m/s, and the duct is 76 mm square. At a section 0.3 m downstream from the entrance, the displacement thickness, δ^*, on each wall measures 0.9 mm. Determine the pressure change between sections ① and ②.

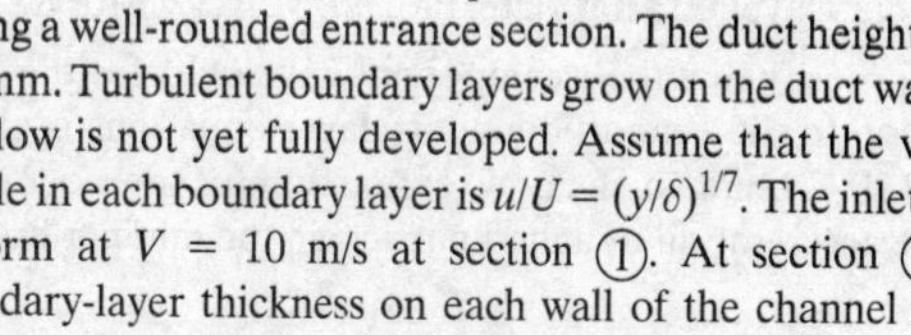

P9.25

9.26 Flow of 20°C air develops in a flat horizontal duct following a well-rounded entrance section. The duct height is $H = 300$ mm. Turbulent boundary layers grow on the duct walls, but the flow is not yet fully developed. Assume that the velocity profile in each boundary layer is $u/U = (y/\delta)^{1/7}$. The inlet flow is uniform at $V = 10$ m/s at section ①. At section ②, the boundary-layer thickness on each wall of the channel is $\delta_2 = 100$ mm. Show that, for this flow, $\delta^* = \delta/8$. Evaluate the static gage pressure at section ①. Find the average wall shear stress between the entrance and section ②, located at $L = 5$ m.

9.27 A laboratory wind tunnel has a square test section with sides of width $W = 305$ mm and length $L = 610$ mm. When the freestream air speed at the test section entrance is $U_1 = 24.4$ m/s, the head loss from the atmosphere is 6.5 mm. H_2O. Turbulent boundary layers form on the top, bottom, and side walls of the test section. Measurements show the boundary-layer thicknesses are $\delta_1 = 20.3$ mm at the entrance and $\delta_2 = 25.4$ mm at the outlet of the test section. The velocity profiles are of $\frac{1}{7}$-power form. Evaluate the freestream air speed at the outlet from the test section. Determine the static pressures at the test section inlet and outlet.

9.28 Flow of air develops in a horizontal cylindrical duct, of diameter $D = 400$ mm, following a well-rounded entrance. A turbulent boundary grows on the duct wall, but the flow is not yet fully developed. Assume that the velocity profile in the boundary layer is $u/U = (y/\delta)^{1/7}$. The inlet flow is $U = 15$ m/s at section ①. At section ②, the boundary-layer thickness is $\delta_2 = 100$ mm. Evaluate the static gage pressure at section ②, located at $L = 6$ m. Find the average wall shear stress.

9.29 Air flows into the inlet contraction section of a wind tunnel in an undergraduate laboratory. From the inlet the air enters the test section, which is square in cross-section with side dimensions of 305 mm. The test section is 609 mm long. At one operating condition air leaves the contraction at 50.2 m/s with negligible boundary-layer thickness. Measurements show that boundary layers at the downstream end of the test section are 20.3 mm thick. Evaluate the displacement thickness of the boundary layers at the downstream end of the wind tunnel test section. Calculate the change in static pressure along the wind tunnel test section. Estimate the approximate total drag force caused by skin friction on each wall of the wind tunnel.

Laminar Flat-Plate Boundary Layer: Exact Solution

***9.30** Using numerical results for the Blasius exact solution for laminar boundary-layer flow on a flat plate, plot the dimensionless velocity profile, u/U (on the abscissa), versus dimensionless distance from the surface, y/δ (on the ordinate). Compare with the approximate parabolic velocity profile of Problem 9.10.

***9.31** Using numerical results obtained by Blasius (Table 9.1), evaluate the distribution of shear stress in a laminar boundary layer on a flat plate. Plot τ/τ_w versus y/δ. Compare with results derived from the approximate sinusoidal velocity profile given in Problem 9.10.

***9.32** Using numerical results obtained by Blasius (Table 9.1), evaluate the distribution of shear stress in a laminar boundary layer on a flat plate. Plot τ/τ_w versus y/δ. Compare with results derived from the approximate parabolic velocity profile given in Problem 9.10.

***9.33** Using numerical results obtained by Blasius (Table 9.1), evaluate the vertical component of velocity in a laminar boundary layer on a flat plate. Plot v/U versus y/δ for $Re_x = 10^5$.

***9.34** Verify that the y component of velocity for the Blasius solution to the Prandtl boundary-layer equations is given by Eq. 9.10. Obtain an algebraic expression for the x component of the acceleration of a fluid particle in the laminar boundary layer. Plot a_x versus η to determine the maximum x component of acceleration at a given x.

***9.35** Numerical results of the Blasius solution to the Prandtl boundary-layer equations are presented in Table 9.1. Consider steady, incompressible flow of standard air over a flat plate at freestream speed $U = 5$ m/s. At $x = 20$ cm, estimate the distance from the surface at which $u = 0.95\ U$. Evaluate the slope of the streamline through this point. Obtain an algebraic expression for the local skin friction, $\tau_w(x)$. Obtain an algebraic expression for the total skin friction drag force on the plate. Evaluate the momentum thickness at $L = 1$ m.

***9.36** Consider flow of air over a flat plate. On one graph, plot the laminar boundary-layer thickness as a function of distance along the plate (up to transition) for freestream speeds $U = 1$ m/s, 2 m/s, 3 m/s, 4 m/s, 5 m/s, and 10 m/s.

***9.37** The Blasius exact solution involves solving a nonlinear equation, Eq. 9.11, with initial and boundary conditions given by Eq. 9.12. Set up an *Excel* workbook to obtain a numerical solution of this system. The workbook should consist of columns for η, f, f', and f''. The rows should consist of values of these, with a suitable step size for η (e.g., for 1000 rows the step size for η would be 0.01 to generate data through $\eta = 10$, to go a little beyond the data in Table 9.1). The values of f and f' for the first row are zero (from the initial conditions, Eq. 9.12); a guess value is needed for f'' (try 0.5). Subsequent row values for f, f', and f'' can be obtained from previous row values using the Euler method of Section 5.5 for approximating first derivatives (and Eq. 9.11). Finally, a solution can be found by using *Excel*'s *Goal Seek* or *Solver* functions to vary the initial value of f'' until $f' = 1$ for large η (e.g., $\eta = 10$, boundary condition of Eq. 9.12). Plot the results. Note: Because the Euler method is relatively crude, the results will agree with Blasius' only to within about 1%.

***9.38** A thin flat plate, $L = 0.25$ m. long and $b = 1$ m wide, is installed in a water tunnel as a splitter. The freestream speed is $U = 1.75$ m/s, and the velocity profile in the boundary layer is approximated as parabolic. Plot δ, δ^*, and τ_w versus x/L for the plate.

9.39 Consider flow over the splitter plate of Problem 9.38. Show algebraically that the total drag force on one side of the splitter plate may be written $F_D = \rho U^2 \theta_L b$. Evaluate θ_L and the total drag for the given conditions.

9.40 A thin flat plate is installed in a water tunnel as a splitter. The plate is 0.3 m long and 1 m wide. The freestream speed is 1.6 m/s. Laminar boundary layers form on both sides of the plate. The boundary-layer velocity profile is approximated as parabolic. Determine the total viscous drag force on the plate assuming that pressure drag is negligible.

9.41 In Problems 9.18 and 9.19 the drag on the upper surface of a flat plate, with flow (fluid density $\rho = 800$ kg/m^3) at freestream speed $U = 3$ m/s, was determined from momentum flux calculations. The drag was determined for the plate with its long edge (3 m) and its short edge (1 m) parallel to the flow. If the fluid viscosity $\mu = 0.02$ N·s/m^2, compute the drag using boundary-layer calculations.

9.42 Assume laminar boundary-layer flow to estimate the drag on the plate shown when it is placed parallel to a 5 m/s air flow. The air is 20°C and 101.3 kPa.

*These problems require material from sections that may be omitted without loss of continuity in the text material.

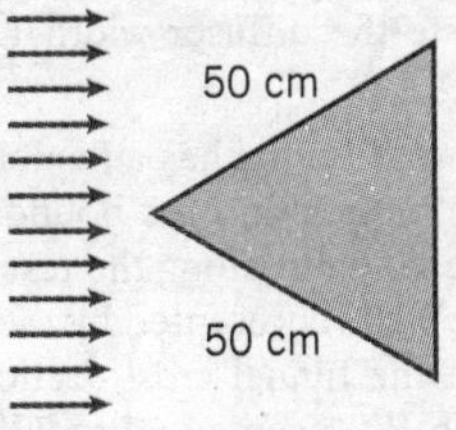

P9.42, P9.43

9.43 Assume laminar boundary-layer flow to estimate the drag on the plate shown when it is placed parallel to a 5 m/s air flow, except that the base rather than the tip faces the flow. Would you expect this to be larger than, the same as, or lower than the drag for Problem 9.42?

9.44 Assume laminar boundary-layer flow to estimate the drag on the plate shown when it is placed parallel to a 7.5 m/s air flow. The air is at 20°C and 101.3 kPa. (Note that the shape is given by $x = y^2/25$, where x and y are in cm.)

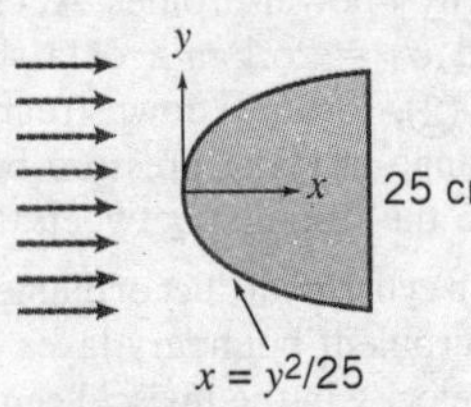

P9.44, P9.45

9.45 Assume laminar boundary-layer flow to estimate the drag on the plate shown when it is placed parallel to a 7.5 m/s flow, except that the base rather than the tip faces the flow. Would you expect this to be large than, the same as, or lower than the drag for Problem 9.44?

9.46 Assume laminar boundary-layer flow to estimate the drag on four square plates (each 7.5 cm × 7.5 cm) placed parallel to a 1 m/s water flow, for the two configurations shown. Before calculating, which configuration do you expect to experience the lower drag? Assume that the plates attached with string are far enough apart for wake effects to be negligible and that the water is at 20°C.

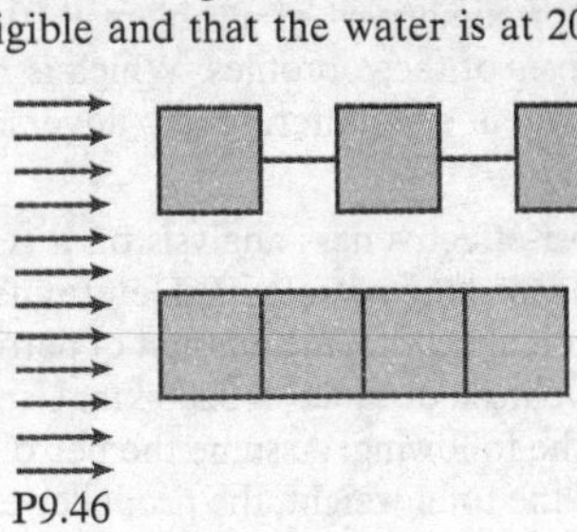

P9.46

Momentum Integral Equation

9.47 The velocity profile in a laminar boundary-layer flow at zero pressure gradient is approximated by the linear expression given in Problem 9.10. Use the momentum integral equation with this profile to obtain expressions for δ/x and C_f.

9.48 A horizontal surface, with length $L = 1.8$ m and width $b = 0.9$ m, is immersed in a stream of standard air flowing at $U = 3.2$ m/s. Assume a laminar boundary layer forms and approximate the velocity profile as sinusoidal. Plot δ, δ^*, and τ_w versus x/L for the plate.

9.49 Water at 10°C flows over a flat plate at a speed of 0.8 m/s. The plate is 0.35 m long and 1 m wide. The boundary layer on each surface of the plate is laminar. Assume that the velocity profile may be approximated as linear. Determine the drag force on the plate.

9.50 A horizontal surface, with length $L = 0.8$ m and width $b = 1.9$ m, is immersed in a stream of standard air flowing at $U = 5.3$ m/s. Assume a laminar boundary layer forms and approximate the velocity profile as linear. Plot δ, δ^*, and τ_w versus x/L for the plate.

9.51 For the flow conditions of Problem 9.50, develop an algebraic expression for the variation of wall shear stress with distance along the surface. Integrate to obtain an algebraic expression for the total skin-friction drag on the surface. Evaluate the drag for the given conditions.

9.52 Standard air flows from the atmosphere into the wide, flat channel shown. Laminar boundary layers form on the top and bottom walls of the channel (ignore boundary-layer effects on the side walls). Assume the boundary layers behave as on a flat plate, with linear velocity profiles. At any axial distance from the inlet, the static pressure is uniform across the channel. Assume uniform flow at section ①. Indicate where the Bernoulli equation can be applied in this flow field. Find the static pressure (gage) and the displacement thickness at section ②. Plot the stagnation pressure (gage) across the channel at section ②, and explain the result. Find the static pressure (gage) at section ①, and compare to the static pressure (gage) at section ②.

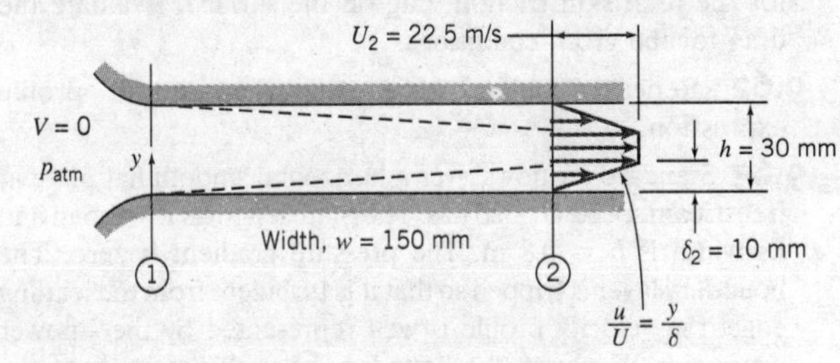

P9.52

9.53 For the flow conditions of Example 9.4, develop an algebraic expression for the variation of wall shear stress with distance along the surface. Integrate to obtain an algebraic expression for the total skin friction drag on the surface. Evaluate the drag for the given conditions.

***9.54** A developing boundary layer of standard air on a flat plate is shown in Fig. P9.18. The freestream flow outside the boundary layer is undisturbed with $U = 50$ m/s. The plate is 3 m wide perpendicular to the diagram. Assume flow in the boundary layer is turbulent, with a $\frac{1}{7}$-power velocity profile, and that $\delta = 19$ mm at surface bc. Calculate the mass flow rate across surface ad and the mass flux across surface ab. Evaluate the x momentum flux across surface ab. Determine the drag force exerted on the flat plate between d and c. Estimate the distance from the leading edge at which transition from laminar to turbulent flow may be expected.

*This problem requires material from sections that may be omitted without loss of continuity in the text material.

9.55 Consider flow of air over a flat plate of length 5 m. On one graph, plot the boundary-layer thickness as a function of distance along the plate for freestream speed $U = 10$ m/s assuming (a) a completely laminar boundary layer, (b) a completely turbulent boundary layer, and (c) a laminar boundary layer that becomes turbulent at $Re_x = 5 \times 10^5$. Use *Excel's Goal Seek* or *Solver* to find the speeds U for which transition occurs at the trailing edge, and at $x = 4$ m, 3 m, 2 m, and 1 m.

9.56 Assume the flow conditions given in Example 9.4. Plot δ, δ^*, and τ_w versus x/L for the plate.

9.57 Repeat Problem 9.42, except that the air flow is now at 25 m/s (assume turbulent boundary-layer flow).

9.58 Repeat Problem 9.44, except that the air flow is now at 25 m/s (assume turbulent boundary-layer flow).

9.59 Repeat Problem 9.46, except that the air flow is now at 10 m/s (assume turbulent boundary-layer flow).

9.60 The velocity profile in a turbulent boundary-layer flow at zero pressure gradient is approximated by the $\frac{1}{6}$-power profile expression,

$$\frac{u}{U} = \eta^{1/6}, \quad \text{where} \quad \eta = \frac{y}{\delta}$$

Use the momentum integral equation with this profile to obtain expressions for δ/x and C_f. Compare with results obtained in Section 9.5 for the $\frac{1}{7}$-power profile.

9.61 For the flow conditions of Example 9.4, but using the $\frac{1}{6}$-power velocity profile of Problem 9.60, develop an algebraic expression for the variation of wall shear stress with distance along the surface. Integrate to obtain an algebraic expression for the total skin friction drag on the surface. Evaluate the drag for the given conditions.

9.62 Repeat Problem 9.60, using the $\frac{1}{8}$-power profile expression.

9.63 Standard air flows over a horizontal smooth flat plate at freestream speed $U = 20$ m/s. The plate length is $L = 1.5$ m and its width is $b = 0.8$ m. The pressure gradient is zero. The boundary layer is tripped so that it is turbulent from the leading edge; the velocity profile is well represented by the $\frac{1}{7}$-power expression. Evaluate the boundary-layer thickness, δ, at the trailing edge of the plate. Calculate the wall shear stress at the trailing edge of the plate. Estimate the skin friction drag on the portion of the plate between $x = 0.5$ m and the trailing edge.

9.64 Air at standard conditions flows over a flat plate. The freestream speed is 10 m/s. Find δ and τ_w at $x = 1$ m from the leading edge assuming (a) completely laminar flow (assume a parabolic velocity profile) and (b) completely turbulent flow (assume a $\frac{1}{7}$-power velocity profile).

Use of the Momentum Integral Equation for Flow with Zero Pressure Gradient

9.65 A uniform flow of standard air at 60 m/s enters a plane-wall diffuser with negligible boundary-layer thickness. The inlet width is 75 mm. The diffuser walls diverge slightly to accommodate the boundary-layer growth so that the pressure gradient is negligible. Assume flat-plate boundary-layer behavior. Explain why the Bernoulli equation is applicable to this flow. Estimate the diffuser width 1.2 m downstream from the entrance.

9.66 A laboratory wind tunnel has a flexible upper wall that can be adjusted to compensate for boundary-layer growth, giving zero pressure gradient along the test section. The wall boundary layers are well represented by the $\frac{1}{7}$-power-velocity profile. At the inlet the tunnel cross section is square, with height H_1 and width W_1, each equal to 305 mm. With freestream speed $U_1 = 26.5$ m/s, measurements show that $\delta_1 = 12.2$ mm and downstream $\delta_6 = 16.6$ mm. Calculate the height of the tunnel walls at ⑥. Determine the equivalent length of a flat plate that would produce the inlet boundary layer thickness. Estimate the streamwise distance between sections ① and ⑥ in the tunnel. Assume standard air.

Pressure Gradients in Boundary-Layer Flow

9.67 Small wind tunnels in an undergraduate laboratory have 305-mm square test sections. Measurements show the boundary layers on the tunnel walls are fully turbulent and well represented by $\frac{1}{7}$-power profiles. At cross section ① with freestream speed $U_1 = 26.1$ m/s, data show that $\delta_1 = 12.2$ mm; at section ②, located downstream, $\delta_2 = 16.6$ mm. Evaluate the change in static pressure between sections ① and ②. Estimate the distance between the two sections.

9.68 Air flows in a cylindrical duct of diameter $D = 150$ mm. At section ①, the turbulent boundary layer is of thickness $\delta_1 = 10$ mm and the velocity in the inviscid central core is $U_1 = 25$ m/s. Further downstream, at section ②, the boundary layer is of thickness $\delta_2 = 30$ mm. The velocity profile in the boundary layer is approximated well by the $\frac{1}{7}$-power expression. Find the velocity, U_2, in the inviscid central core at the second section, and the pressure drop between the two sections. Does the magnitude of the pressure drop indicate that we are justified in approximating the flow between sections ① and ② as one with zero pressure gradient? Estimate the length of duct between sections ① and ②. Estimate the distance downstream from section ① at which the boundary layer thickness is $\delta = 20$ mm. Assume standard air.

9.69 Consider the linear, sinusoidal, and parabolic laminar boundary-layer approximations of Problem 9.10. Compare the momentum fluxes of these profiles. Which is most likely to separate first when encountering an adverse pressure gradient?

9.70 Perform a cost-effectiveness analysis on a typical large tanker used for transporting petroleum. Determine, as a percentage of the petroleum cargo, the amount of petroleum that is consumed in traveling a distance of 3200 km. Use data from Example 9.5, and the following: Assume the petroleum cargo constitutes 75% of the total weight, the propeller efficiency is 70%, the wave drag and power to run auxiliary equipment constitute losses equivalent to an additional 20%, the engines have a thermal efficiency of 40%, and the petroleum energy is 46,520 kJ/kg. Also compare the performance of this tanker to that of the Alaskan Pipeline, which requires about 79 kJ of energy for each ton-mile of petroleum delivery.

9.71 Consider the plane-wall diffuser shown in Fig. P9.71. First, assume the fluid is inviscid. Describe the flow pattern, including the pressure distribution, as the diffuser angle ϕ is increased from zero degrees (parallel walls). Second, modify your description to allow for boundary layer effects. Which fluid (inviscid or viscous) will generally have the highest exit pressure?

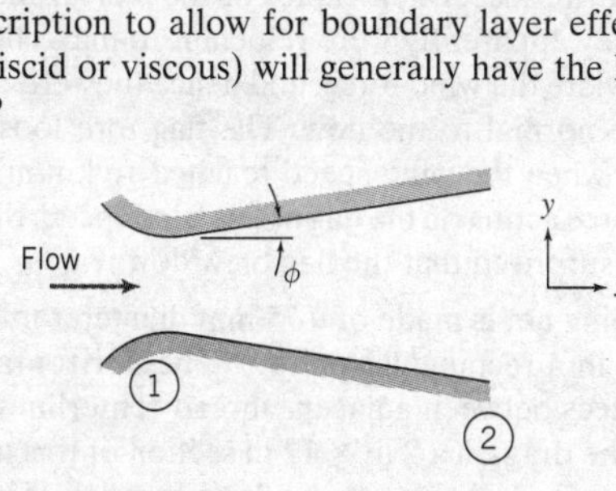

P9.71

***9.72** Table 9.1 shows the numerical results obtained from Blasius exact solution of the laminar boundary-layer equations. Plot the velocity distribution (note that from Eq. 9.13 we see that $\eta \approx 5.0\frac{y}{\delta}$). On the same graph, plot the turbulent velocity distribution given by the $\frac{1}{7}$-power expression of Eq. 9.24. Which is most likely to separate first when encountering an adverse pressure gradient? To justify your answer, compare the momentum fluxes of these profiles (the laminar data can be integrated using a numerical method such as Simpson's rule).

9.73 Cooling air is supplied through the wide, flat channel shown. For minimum noise and disturbance of the outlet flow, laminar boundary layers must be maintained on the channel walls. Estimate the maximum inlet flow speed at which the outlet flow will be laminar. Assuming parabolic velocity profiles in the laminar boundary layers, evaluate the pressure drop, $p_1 - p_2$. Express your answer in inches of water.

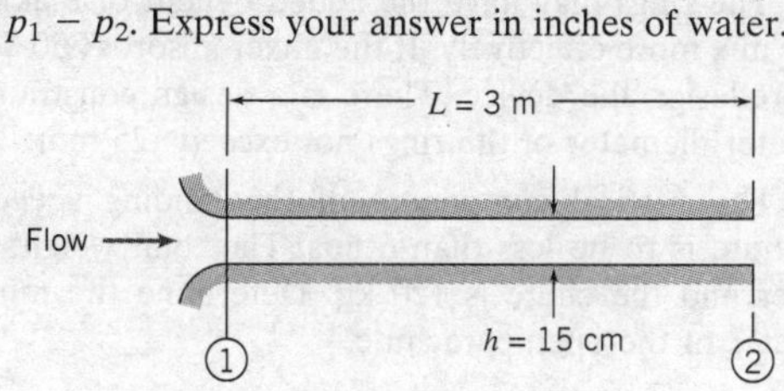

P9.73

9.74 Boundary-layer separation occurs when the shear stress at the surface becomes zero. Assume a polynomial representation for the laminar boundary layer of the form, $u/U = a + b\lambda + c\lambda^2 + d\lambda^3$, where $\lambda = y/\delta$. Specify boundary conditions on the velocity profile at separation. Find appropriate constants, a, b, c, and d, for the separation profile. Calculate the shape factor H at separation. Plot the profile and compare with the parabolic approximate profile.

9.75 For flow over a flat plate with zero pressure gradient, will the shear stress increase, decrease, or remain constant along the plate? Justify your answer. Does the momentum flux increase, decrease, or remain constant as the flow proceeds along the plate? Justify your answer. Compare the behavior of laminar flow and turbulent flow (both from the leading edge) over a flat plate. At a given distance from the leading edge, which flow will have the larger boundary-layer thickness? Does your answer depend on the distance along the plate? How would you justify your answer?

9.76 A laboratory wind tunnel has a test section that is square in cross section, with inlet width W_1 and height H_1, each equal to 305 mm. At freestream speed $U_1 = 24.5$ m/s, measurements show the boundary-layer thickness is $\delta_1 = 9.75$ mm with a $\frac{1}{7}$-power turbulent velocity profile. The pressure gradient in this region is given approximately by $dp/dx = -0.035$ mm H_2O/mm. Evaluate the reduction in effective flow area caused by the boundary layers on the tunnel bottom, top, and walls at section ①. Calculate the rate of change of boundary-layer momentum thickness, $d\theta/dx$, at section ①. Estimate the momentum thickness at the end of the test section, located at $L = 254$ mm downstream.

9.77 The variable-wall concept is proposed to maintain constant boundary-layer thickness in the wind tunnel of Problem 9.76. Beginning with the initial conditions of Problem 9.76, evaluate the freestream velocity distribution needed to maintain constant boundary-layer thickness. Assume constant width, W_1. Estimate and plot the top-height settings along the test section from $x = 0$ at section ① to $x = 254$ mm at section ② downstream.

Drag

9.78 A flat-bottomed barge, 24 m long and 10.7 m wide, submerged to a depth of 1.5 m, is to be pushed up a river (the river water is at 15.5°C). Estimate and plot the power required to overcome skin friction for speeds ranging up to 24 km/hr.

9.79 Repeat Problem 9.46, except that the water flow is now at 10 m/s. (Use formulas for C_D from Section 9.7.)

9.80 A towboat for river barges is tested in a towing tank. The towboat model is built at a scale ratio of 1:13.5. Dimensions of the model are overall length 3.5 m, beam 1 m, and draft 0.2 m. (The model displacement in fresh water is 5500 N.) Estimate the average length of wetted surface on the hull. Calculate the skin friction drag force of the prototype at a speed of 3.601 m/s relative to the water.

9.81 A jet transport aircraft cruises at 12 km in steady level flight at 800 km/h. Model the aircraft fuselage as a circular cylinder with diameter $D = 4$ m and length $L = 38$ m. Neglecting compressibility effects, estimate the skin friction drag force on the fuselage. Evaluate the power needed to overcome this force.

9.82 Resistance of a barge is to be determined from model test data. The model is constructed to a scale ratio of 1:13.5 and has length, beam, and draft of 7.00 m, 1.4 m, and 0.2 m, respectively. The test is to simulate performance of the prototype at 18.5 km/hr. What must the model speed be for the model and prototype to exhibit similar wave drag behavior? Is the boundary layer on the prototype predominantly laminar or turbulent? Does the model boundary layer become turbulent at the comparable point? If not, the model boundary layer could be artificially triggered to turbulent by placing a tripwire across the hull. Where could this be placed? Estimate the skin-friction drag on model and prototype.

9.83 A vertical stabilizing fin on a land-speed-record car is $L = 1.65$ m long and $H = 0.785$ m tall. The automobile is to

*This problem requires material from sections that may be omitted without loss of continuity in the text material.

be driven at the Bonneville Salt Flats in Utah, where the elevation is 1340 m and the summer temperature reaches 50°C. The car speed is 560 km/hr. Evaluate the length Reynolds number of the fin. Estimate the location of transition from laminar to turbulent flow in the boundary layers. Calculate the power required to overcome skin friction drag on the fin.

9.84 A nuclear submarine cruises fully submerged at 13.9 m/s. The hull is approximately a circular cylinder with diameter D = 11.0 m and length L = 107 m. Estimate the percentage of the hull length for which the boundary layer is laminar. Calculate the skin friction drag on the hull and the power consumed.

9.85 You are asked by your college crew to estimate the skin friction drag on their eight-seat racing shell. The hull of the shell may be approximated as half a circular cylinder with 457 mm diameter and 7.32 m length. The speed of the shell through the water is 6.71 m/s. Estimate the location of transition from laminar to turbulent flow in the boundary layer on the hull of the shell. Calculate the thickness of the turbulent boundary layer at the rear of the hull. Determine the total skin friction drag on the hull under the given conditions.

9.86 A sheet of plastic material 10 mm thick, with specific gravity SG = 1.5, is dropped into a large tank containing water. The sheet is 0.5 m × 1 m. Estimate the terminal speed of the sheet as it falls with (a) the short side vertical and (b) the long side vertical. Assume that the drag is due only to skin friction, and that the boundary layers are turbulent from the leading edge.

9.87 The 600-seat jet transport aircraft proposed by Airbus Industrie has a fuselage that is 70 m long and 7.5 m in diameter. The aircraft is to operate 14 hours per day, 6 days per week; it will cruise at 257 m/s (M = 0.87) at 12-km altitude. The engines consume fuel at the rate of 0.06 kg/hr for each N of thrust produced. Estimate the skin friction drag force on the aircraft fuselage at cruise. Calculate the annual fuel savings for the aircraft if friction drag on the fuselage could be reduced by 1 percent by modifying the surface coating.

9.88 A supertanker displacement is approximately 600,000 metric tons. The ship has length L = 300 m, beam (width) b = 80 m, and draft (depth) D = 25 m. The ship steams at 7.20 m/s through seawater at 4°C. For these conditions, estimate (a) the thickness of the boundary layer at the stern of the ship, (b) the total skin friction drag acting on the ship, and (c) the power required to overcome the drag force.

9.89 In Section 7.6 the wave resistance and viscous resistance on a model and prototype ship were discussed. For the prototype, L = 130 m and A = 1800 m^2. From the data of Figs 7.2 and 7.3, plot on one graph the wave, viscous, and total resistance (N) experienced by the prototype, as a function of speed. Plot a similar graph for the model. Discuss your results. Finally, plot the power (kW) required for the prototype ship to overcome the total resistance.

9.90 As part of the 1976 bicentennial celebration, an enterprising group hung a giant American flag (59 m high and 112 m wide) from the suspension cables of the Verrazano Narrows Bridge. They apparently were reluctant to make holes in the flag to alleviate the wind force, and hence they effectively had a flat plate normal to the flow. The flag tore loose from its mountings when the wind speed reached 16 km/hr. Estimate the wind force acting on the flag at this wind speed. Should they have been surprised that the flag blew down?

9.91 A fishing net is made of 0.75 mm diameter nylon thread assembled in a rectangular pattern. The horizontal and vertical distances between adjacent thread centerlines are 1 cm. Estimate the drag on a 2 m × 12 m section of this net when it is dragged (perpendicular to the flow) through 15°C water at 6 knots. What is the power required to maintain this motion?

9.92 A rotary mixer is constructed from two circular disks as shown. The mixer is rotated at 60 rpm in a large vessel containing a brine solution (SG = 1.1). Neglect the drag on the rods and the motion induced in the liquid. Estimate the minimum torque and power required to drive the mixer.

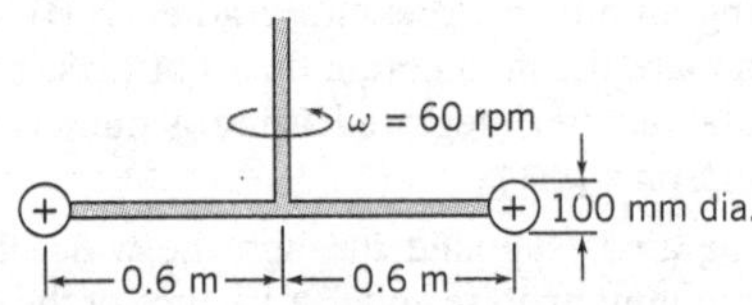

P9.92, P9.93

9.93 As a young design engineer you decide to make the rotary mixer look more "cool" by replacing the disks with rings. The rings may have the added benefit of making the mixer mix more effectively. If the mixer absorbs 350 W at 60 rpm, redesign the device. There is a design constraint that the outer diameter of the rings not exceed 125 mm.

9.94 The vertical component of the landing speed of a parachute is to be less than 6 m/s. The total weight of the jumper and the chute is 120 kg. Determine the minimum diameter of the open parachute.

9.95 As a young design engineer you are asked to design an emergency braking parachute system for use with a military aircraft of mass 9500 kg. The plane lands at 350 km/hr, and the parachute system alone must slow the airplane to 100 km/hr in less than 1200 m. Find the minimum diameter required for a single parachute, and for three non-interfering parachutes. Plot the airplane speed versus distance and versus time. What is the maximum "g-force" experienced?

9.96 An emergency braking parachute system on a military aircraft consists of a large parachute of diameter 6 m. If the airplane mass is 8500 kg, and it lands at 400 km/hr, find the time and distance at which the airplane is slowed to 100 km/hr by the parachute alone. Plot the aircraft speed versus distance and versus time. What is the maximum "g-force" experienced? An engineer proposes that less space would be taken up by replacing the large parachute with three non-interfering parachutes each of diameter 3.75 m. What effect would this have on the time and distance to slow to 100 km/hr?

9.97 It has been proposed to use surplus 208 liters oil drums to make simple windmills for underdeveloped countries.

(It is a simple Savonius turbine.) Two possible configurations are shown. Estimate which would be better, why, and by how much. The diameter and length of a 208 L drum are D = 610 mm and H = 737 mm.

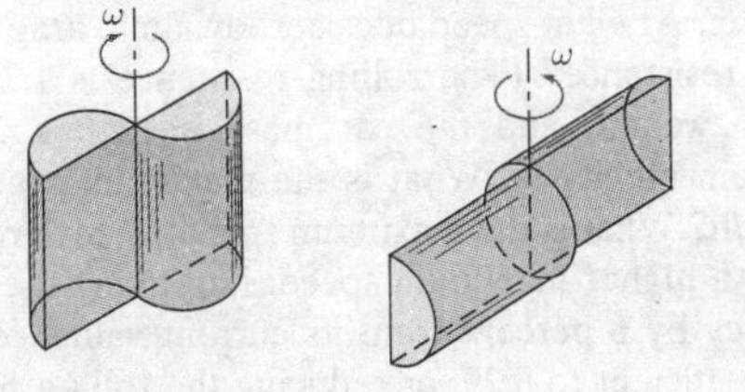

P9.97

9.98 The resistance to motion of a good bicycle on smooth pavement is nearly all due to aerodynamic drag. Assume that the total weight of rider and bike is M = 100 kg. The frontal area measured from a photograph is A = 0.46 m^2. Experiments on a hill, where the road grade is 8 percent, show that terminal speed is V_t = 15 m/s. From these data, the drag coefficient is estimated as C_D = 1.2. Verify this calculation of drag coefficient. Estimate the distance needed for the bike and rider to decelerate from 15 m/s to 10 m/s while coasting after reaching level road.

9.99 A cyclist is able to attain a maximum speed of 30 km/hr on a calm day. The total mass of rider and bike is 65 kg. The rolling resistance of the tires is F_R = 7.5 N, and the drag coefficient and frontal area are C_D = 1.2 and A = 0.25 m^2. The cyclist bets that today, even though there is a headwind of 10 km/hr, she can maintain a speed of 24 km/hr. She also bets that, cycling with wind support, she can attain a top speed of 40 km/hr. Which, if any, bets does she win?

9.100 Ballistic data obtained on a firing range show that aerodynamic drag reduces the speed of a .44 magnum revolver bullet from 250 m/s to 210 m/s as it travels over a horizontal distance of 150 m. The diameter and mass of the bullet are 11.2 mm and 15.6 g, respectively. Evaluate the average drag coefficient for the bullet.

9.101 Consider the cyclist in Problem 9.99. She is having a bad day, because she has to climb a hill with a 5° slope. What is the speed she is able to attain? What is the maximum speed if there is also a headwind of 10 km/hr? She reaches the top of the hill, and turns around and heads down the hill. If she still pedals as hard as possible, what will be her top speed (when it is calm, and when the wind is present)? What will be her maximum speed if she decides to coast down the hill (with and without the aid of the wind)?

9.102 Consider the cyclist in Problem 9.99. Determine the maximum speeds she is actually able to attain today (with the 10 km/hr wind) cycling into the wind, and cycling with the wind. If she were to replace the tires with high-tech ones that had a rolling resistance of only 3.5 N, determine her maximum speed on a calm day, cycling into the wind, and cycling with the wind. If she in addition attaches an aerodynamic fairing that reduces the drag coefficient to C_D = 0.9, what will be her new maximum speeds?

9.103 At a surprise party for a friend you've tied a series of 20 cm diameter helium balloons to a flagpole, each tied with a short string. The first one is tied 1 m above the ground, and the other eight are tied at 1 m spacings, so that the last is tied at a height of 9 m. Being quite a nerdy engineer, you notice that in the steady wind, each balloon is blown by the wind so it looks like the angles that the strings make with the vertical are about 10°, 20°, 30°, 35°, 40°, 45°, 50°, 60°, and 65°. Estimate and plot the wind velocity profile for the 9-m range. Assume the helium is at 20°C and 10 kPa gage and that each balloon is made of 3 g of latex.

9.104 A 0.5 m diameter hollow plastic sphere containing pollution test equipment is being dragged through the Hudson River in New York by a diver riding an underwater jet device. The sphere (with an effective specific gravity of SG = 0.30) is fully submerged, and it is tethered to the diver by a thin 1.5-m-long wire. What is the angle the wire makes with the horizontal if the velocity of the diver and sphere relative to the water is 5 m/s? The water is at 10°C.

9.105 A simple but effective anemometer to measure wind speed can be made from a thin plate hinged to deflect in the wind. Consider a thin plate made from brass that is 20 mm high and 10 mm wide. Derive a relationship for wind speed as a function of deflection angle, θ. What thickness of brass should be used to give θ = 30° at 10 m/s?

9.106 An anemometer to measure wind speed is made from four hemispherical cups of 50 mm diameter, as shown. The center of each cup is placed at R = 80 mm from the pivot. Find the theoretical calibration constant, k, in the calibration equation $V = k\omega$, where V (km/hr) is the wind speed and ω (rpm) is the rotation speed. In your analysis, base the torque calculations on the drag generated at the instant when two of the cups are orthogonal and the other two cups are parallel, and ignore friction in the bearings. Explain why, in the absence of friction, at any given wind speed, the anemometer runs at constant speed rather than accelerating without limit. If the actual anemometer bearing has (constant) friction such that the anemometer needs a minimum wind speed of 1 km/hr to begin rotating, compare the rotation speeds with and without friction for V = 10 km/hr.

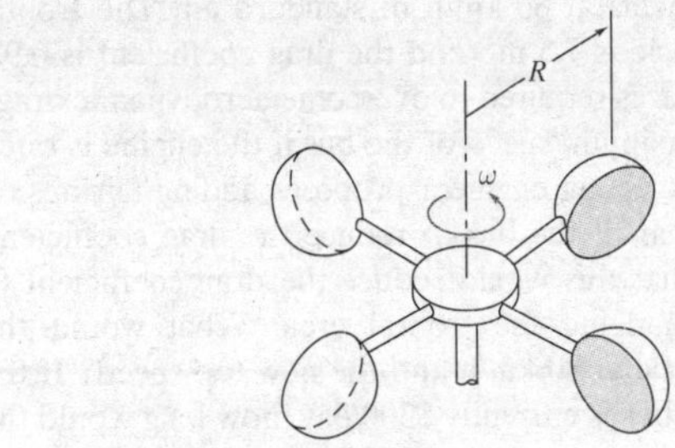

P9.106

9.107 A circular disk is hung in an air stream from a pivoted strut as shown. In a wind-tunnel experiment, performed in air at 15 m/s with a 25 mm diameter disk, α was measured at 10°. For these conditions determine the mass of the disk. Assume the drag coefficient for the disk applies when the component of wind speed normal to the disk is used. Assume drag on the strut and friction in the pivot are negligible. Plot a theoretical curve of α as a function of air speed.

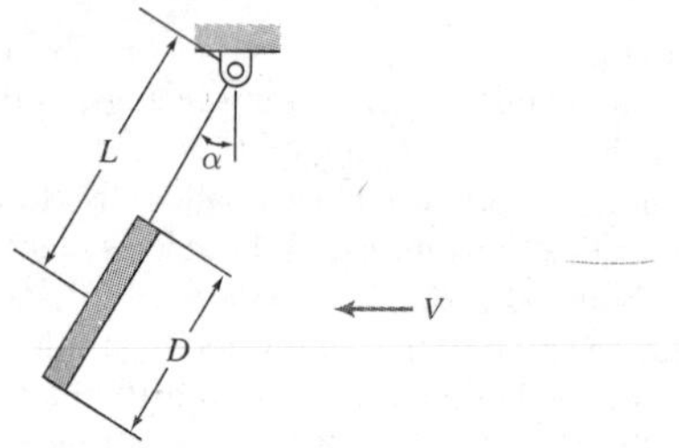

P9.107

9.108 Experimental data [16] suggest that the maximum and minimum drag area (C_DA) for a skydiver varies from about 0.85 m^2 for a prone, spread-eagle position to about 0.11 m^2 for vertical fall. Estimate the terminal speeds for a 75 kg skydiver in each position. Calculate the time and distance needed for the skydiver to reach 90 percent of terminal speed at an altitude of 3000 m on a standard day.

9.109 A vehicle is built to try for the land-speed record at the Bonneville Salt Flats, elevation 1340 m. The engine delivers 373 kW to the rear wheels, and careful streamlining has resulted in a drag coefficient of 0.15, based on a 1.4 m^2 frontal area. Compute the theoretical maximum ground speed of the car (a) in still air and (b) with a 32 km/hr headwind.

9.110 An F-4 aircraft is slowed after landing by dual parachutes deployed from the rear. Each parachute is 3.7 m in diameter. The F-4 weighs 142,400 N and lands at 160 m/s. Estimate the time and distance required to decelerate the aircraft to 100 m/s, assuming that the brakes are not used and the drag of the aircraft is negligible.

9.111 A tractor-trailer rig has frontal area $A = 9.5$ m^2 and drag coefficient $C_D = 0.9$. Rolling resistance is 6 N per 1000 N of vehicle weight. The specific fuel consumption of the diesel engine is 0.206 kg of fuel per kW hour, and drivetrain efficiency is 92 percent. The density of diesel fuel is 812 kg/m^3. Estimate the fuel economy of the rig at 88 km/hr if its gross weight is 320,400 N. An air fairing system reduces aerodynamic drag 15 percent. The truck travels 192,000 km per year. Calculate the fuel saved per year by the roof fairing.

9.112 A bus travels at 80 km/h in standard air. The frontal area of the vehicle is 7.5 m^2, and the drag coefficient is 0.92. How much power is required to overcome aerodynamic drag? Estimate the maximum speed of the bus if the engine is rated at 346.75 kW. A young engineer proposes adding fairings on the front and rear of the bus to reduce the drag coefficient. Tests indicate that this would reduce the drag coefficient to 0.86 without changing the frontal area. What would the required power be at 80 km/h and the new top speed? If the fuel cost for the bus is currently $300/day, how long would the modification take to pay for itself if it costs $4800 to install?

9.113 Compare and plot the power (kW) required by a typical large American sedan of the 1970s and a current midsize sedan to overcome aerodynamic drag versus road speed in standard air, for a speed range of 32 km/hr to 160 km/hr. Use the following as representative values:

	Weight (N)	Drag Coefficient	Frontal Area (m^2)
1970s Sedan	20,025	0.5	2.23
Current Sedan	15,575	0.3	1.86

If rolling resistance is 1.5 percent of curb weight, determine for each vehicle the speed at which the aerodynamic force exceeds frictional resistance.

9.114 A 134.23 kW sports car of frontal area 1.72 m^2, with a drag coefficient of 0.31, requires 12.677 kW to cruise at 100 km/h. At what speed does aerodynamic drag first exceed rolling resistance? (The rolling resistance is 1.2 percent of the car weight, and the car mass is 1250 kg.) Find the drivetrain efficiency. What is the maximum acceleration at 100 km/h? What is the maximum speed? Which redesign will lead to a higher maximum speed: improving the drive train efficiency by 6 percent from its current value, reducing the drag coefficient to 0.29, or reducing the rolling resistance to 0.91 percent of the car weight?

9.115 Consider a negatively charged spherical particle of radius a bearing a charge, Q_s, suspended in a pure dielectric fluid (containing no ions). When subject to a uniform electric field, $\vec{E}_\infty$, the particle will translate under the influence of the electric force acting on it. The induced particle motion refers to electrophoresis, which has been widely used to characterize and purify molecules and colloidal particles. The net electrical force on the charged particle will simply be $\vec{F}_E = Q_S\vec{E}_\infty$. As soon as the particle starts to move under the influence of this electric force, it encounters an oppositely directed fluid drag force.

(a) Under the Stokes flow regime and neglecting the gravitational force and the buoyancy force acting on the microparticle, derive an expression to calculate the particle's steady-state translational velocity.

(b) Based on the above results, explain why electrophoresis can be used to separate biological samples.

(c) Calculate the translational velocities of two particles of radius a = 1 μm and 10 μm using $Q_s = -10^{-12}$ C, E_∞ = 1000 V/m, and $\mu = 10^{-3}$ Pa · s.

9.116 Repeat the analysis for the frictionless anemometer of Problem 9.106, except this time base the torque calculations on the more realistic model that the average torque is obtained by integrating, over one revolution, the instantaneous torque generated by each cup (i.e., as the cup's orientation to the wind varies).

9.117 A round thin disk of radius R is oriented perpendicular to a fluid stream. The pressure distributions on the front and back surfaces are measured and presented in the form of pressure coefficients. The data are modeled with the following expressions for the front and back surfaces, respectively:

Front Surface $\quad C_p = 1 - \left(\frac{r}{R}\right)^6$

Rear Surface $\quad C_p = -0.42$

Calculate the drag coefficient for the disk.

9.118 An object falls in air down a long vertical chute. The speed of the object is constant at 3 m/s. The flow pattern around the object is shown. The static pressure is uniform across sections ① and ②; pressure is atmospheric at section ①. The effective flow area at section ② is 20 percent of the chute area. Frictional effects between sections ① and ② are negligible. Evaluate the flow speed relative to the object at section ②. Calculate the static pressure at section ②. Determine the mass of the object.

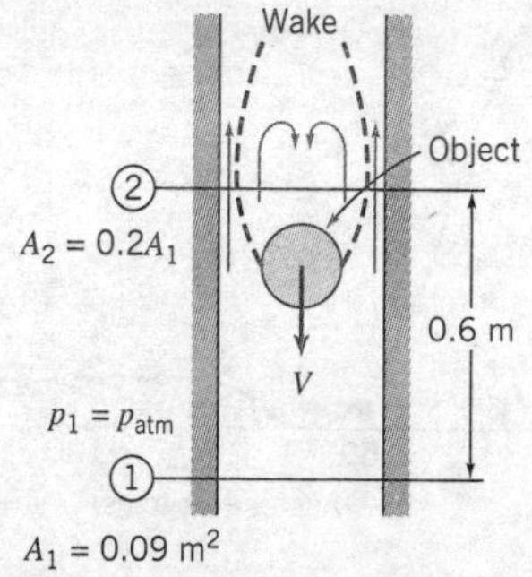

P9.118

9.119 An object of mass m, with cross-sectional area equal to half the size of the chute, falls down a mail chute. The motion is steady. The wake area is $\frac{3}{4}$ the size of the chute at its maximum area. Use the assumption of constant pressure in the wake. Apply the continuity, Bernoulli, and momentum equations to develop an expression for terminal speed of the object in terms of its mass and other quantities.

9.120 A light plane tows an advertising banner over a football stadium on a Saturday afternoon. The banner is 1.2 m tall and 13.7 m long. According to Hoerner [16], the drag coefficient based on area (Lh) for such a banner is approximated by $C_D = 0.05\ L/h$, where L is the banner length and h is the banner height. Estimate the power required to tow the banner at V = 88 km/hr. Compare with the drag of a rigid flat plate. Why is the drag larger for the banner?

9.121 A large paddle wheel is immersed in the current of a river to generate power. Each paddle has area A and drag coefficient C_D; the center of each paddle is located at radius R from the centerline of the paddle wheel. Assume the equivalent of one paddle is submerged continuously in the flowing stream. Obtain an expression for the drag force on a single paddle in terms of geometric variables, current speed, V, and linear speed of the paddle center, $U = R\omega$. Develop expressions for the torque and power produced by the paddle wheel. Find the speed at which the paddle wheel should rotate to obtain maximum power output from the wheel in a given current.

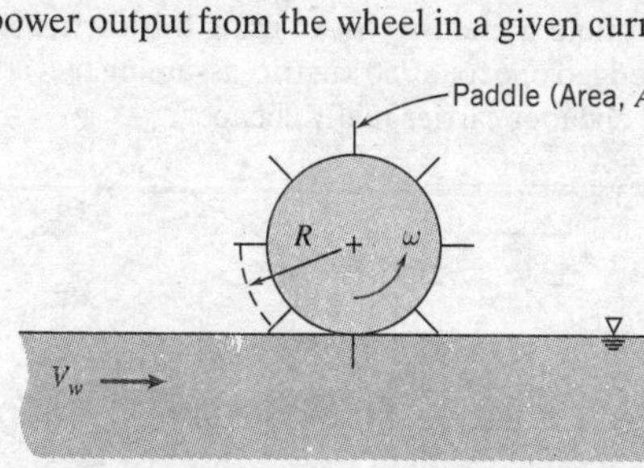

P9.121

9.122 The antenna on a car is 10 mm in diameter and 1.8 m long. Estimate the bending moment that tends to snap it off if the car is driven at 120 km/hr on a standard day.

9.123 A large three-blade horizontal axis wind turbine (HAWT) can be damaged if the wind speed is too high. To avoid this, the blades of the turbine can be oriented so that they are parallel to the flow. Find the bending moment at the base of each blade when the wind speed is 45 m/s. Model each blade as a flat plate 35 m wide × 0.45 m long.

9.124 The HAWT of Problem 9.123 is not self-starting. The generator is used as an electric motor to get the turbine up to the operating speed of 20 units. To make this easier, the blades are aligned so that they lie in the plane of rotation. Assuming an overall efficiency of motor and drive train of 65 percent, find the power required to maintain the turbine at the operating speed. As an approximation, model each blade as a series of flat plates (the outer region of each blade moves at a significantly higher speed than the inner region).

9.125 A runner maintains a speed of 12 km/hr during a 6.4 km run. The runner's route consists of running straight down a road for 3.22 km, then turning around and returning the 3.22 km straight home. The C_DA for the runner is 0.84 m^2. On a windless day, how many calories (kcal) will the runner burn overcoming drag? On a day in which the wind is blowing 8 km/hr directly along the runner's route how many calories (kcal) will the runner burn overcoming drag?

9.126 Consider small oil droplets (SG = 0.85) rising in water. Develop a relation for calculating terminal speed of a droplet (in m/s) as a function of droplet diameter (in mm) assuming Stokes flow. For what range of droplet diameter is Stokes flow a reasonable assumption?

9.127 Standard air is drawn into a low-speed wind tunnel. A 30-mm diameter sphere is mounted on a force balance to measure lift and drag. An oil-filled manometer is used to measure static pressure inside the tunnel; the reading is −40 mm of oil (SG = 0.85). Calculate the freestream air speed in the tunnel, the Reynolds number of flow over the sphere, and the drag force on the sphere. Are the boundary layers on the sphere laminar or turbulent? Explain.

9.128 A spherical helium-filled balloon, 0.6 m in diameter, exerts an upward force of 1.3 N on a restraining string when held stationary in standard air with no wind. With a wind speed of 3 m/s, the string holding the balloon makes an angle of 60° with the horizontal. Calculate the drag coefficient of the balloon under these conditions, neglecting the weight of the string.

9.129 A field hockey ball has diameter D = 73 mm and mass m = 160 g. When struck well, it leaves the stick with initial speed U_0 = 50 m/s. The ball is essentially smooth. Estimate the distance traveled in horizontal flight before the speed of the ball is reduced 10 percent by aerodynamic drag.

9.130 Compute the terminal speed of a 3 mm-diameter raindrop (assume spherical) in standard air.

9.131 A small sphere (D = 6 mm) is observed to fall through castor oil at a terminal speed of 60 mm/s. The temperature is 20°C. Compute the drag coefficient for the sphere. Determine the density of the sphere. If dropped in water, would the sphere fall slower or faster? Why?

9.132 The following curve-fit for the drag coefficient of a smooth sphere as a function of Reynolds number has been proposed by Chow [36]:

$C_D = 24/Re$	$Re \le 1$
$C_D = 24/Re^{0.646}$	$1 < Re \le 400$
$C_D = 0.5$	$400 < Re \le 3 \times 10^5$
$C_D = 0.000366\, Re^{0.4275}$	$3 \times 10^5 < Re \le 2 \times 10^6$
$C_D = 0.18$	$Re > 2 \times 10^6$

Use data from Fig. 9.11 to estimate the magnitude and location of the maximum error between the curve fit and data.

9.133 Problem 9.107 showed a circular disk hung in an air stream from a cylindrical strut. Assume the strut is L = 40 mm long and d = 3 mm diameter. Solve Problem 9.107 including the effect of drag on the support.

9.134 A tennis ball with a mass of 57 g and diameter of 64 mm is dropped in standard sea level air. Calculate the terminal velocity of the ball. Assuming as an approximation that the drag coefficient remains constant at its terminal-velocity value, estimate the time and distance required for the ball to reach 95% of its terminal speed.

9.135 Consider a cylindrical flag pole of height H. For constant drag coefficient, evaluate the drag force and bending moment on the pole if wind speed varies as $u/U = (y/H)^{1/7}$, where y is distance measured from the ground. Compare with drag and moment for a uniform wind profile with constant speed U.

9.136 A water tower consists of a 12 m diameter sphere on top of a vertical tower 30 m tall and 2 m in diameter. Estimate the bending moment exerted on the base of the tower due to the aerodynamic force imposed by a 100 km/hr wind on a standard day. Neglect interference at the joint between the sphere and tower.

9.137 A model airfoil of chord 15 cm and span 60 cm is placed in a wind tunnel with an air flow of 30 m/s (the air is at 20°C). It is mounted on a cylindrical support rod 2 cm in diameter and 25 cm tall. Instruments at the base of the rod indicate a vertical force of 50 N and a horizontal force of 6 N. Calculate the lift and drag coefficients of the airfoil.

9.138 A cast-iron "12-pounder" cannonball rolls off the deck of a ship and falls into the ocean at a location where the depth is 1000 m. Estimate the time that elapses before the cannonball hits the sea bottom.

9.139 The Stokes drag law for smooth spheres is to be verified experimentally by dropping steel ball bearings in glycerin. Evaluate the largest diameter steel ball for which $Re < 1$ at terminal speed. Calculate the height of glycerin column needed for a bearing to reach 95 percent of terminal speed.

9.140 The plot shows pressure difference versus angle, measured for air flow around a circular cylinder at Re = 80,000. Use these data to estimate C_D for this flow. Compare with data from Fig. 9.13. How can you explain the difference?

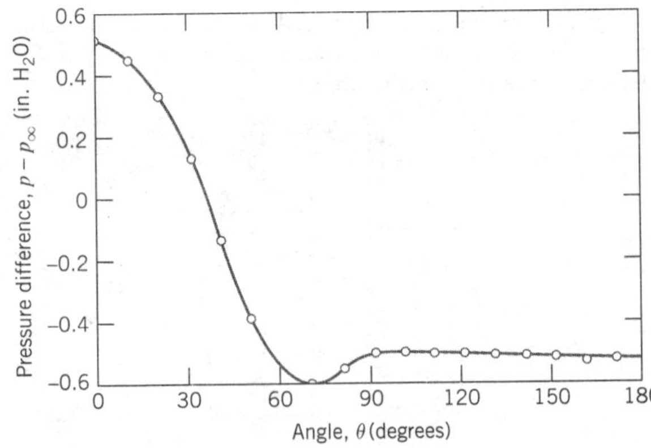

P9.140

9.141 Consider the tennis ball of Problem 9.134. Use the equations for drag coefficient given in Problem 9.132, and a numerical integration scheme (e.g., Simpson's rule) to compute the time and distance required for the ball to reach 95% of its terminal speed.

9.142 The air bubble of Problem 3.10 expands as it rises in water. Find the time it takes for the bubble to reach the surface. Repeat for bubbles of diameter 5 mm and 15 mm. Compute and plot the depth of the bubbles as a function of time.

9.143 Consider the tennis ball of Problem 9.134. Suppose it is hit so that it has an initial upward speed of 50 m/s. Estimate the maximum height of the ball, assuming (a) a constant drag coefficient and (b) using the equations for drag coefficient given in Problem 9.132, and a numerical integration scheme (e.g., a Simpson's rule).

9.144 Why is it possible to kick a football farther in a spiral motion than in an end-over-end tumbling motion?

9.145 Approximate dimensions of a rented rooftop carrier are shown. Estimate the drag force on the carrier (r = 10 cm) at 100 km/hr. If the drivetrain efficiency of the vehicle is 0.85 and the brake specific fuel consumption of its engine is 0.3 kg/(kW · hr), estimate the additional rate of fuel consumption due to the carrier. Compute the effect on fuel economy if the auto achieves 12.75 km/L without the carrier. The rental company offers you a cheaper, square-edged carrier at a price $5 less than the current carrier. Estimate the extra cost of using this carrier instead of the round-edged one for a 750 km trip, assuming fuel is $0.92 per liter. Is the cheaper carrier really cheaper?

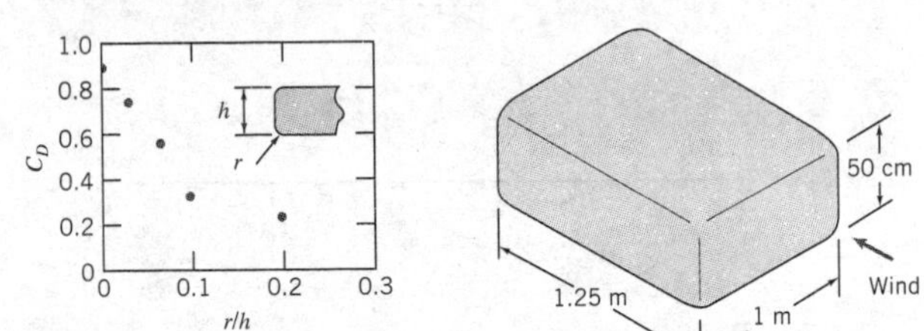

Drag coefficient v. radius ratio [37]

P9.145

9.146 A barge weighing 8820 kN that is 10 m wide, 30 m long, and 7 m tall has come free from its tug boat in the Mississippi River. It is in a section of river which has a current of 1 m/s, and there is a wind blowing straight upriver at 10 m/s. Assume that the drag coefficient is 1.3 for both the part of the barge in the wind as well as the part below water. Determine the speed at which the barge will be steadily moving. Is it moving upriver or downriver?

9.147 Coastdown tests, performed on a level road on a calm day, can be used to measure aerodynamic drag and rolling resistance coefficients for a full-scale vehicle. Rolling resistance is estimated from dV/dt measured at low speed, where aerodynamic drag is small. Rolling resistance then is deducted from dV/dt measured at high speed to determine the aerodynamic drag. The following data were obtained during a test with a vehicle, of weight $W = 111{,}250$ N and frontal area $A = 7.34\ \text{m}^2$:

V(km/hr)	8	88
$\dfrac{dV}{dt}\left(\dfrac{\text{km/hr}}{\text{s}}\right)$	0.24	0.76

Estimate the aerodynamic drag coefficient for this vehicle. At what speed does the aerodynamic drag first exceed rolling resistance?

9.148 A spherical sonar transducer with 0.375 m diameter is to be towed in seawater. The transducer must be fully submerged at 16 m/s. To avoid cavitation, the minimum pressure on the surface of the transducer must be greater than 30 kPa (abs). Calculate the hydrodynamic drag force acting on the transducer at the required towing speed. Estimate the minimum depth to which the transducer must be submerged to avoid cavitation.

9.149 While walking across campus one windy day, Floyd Fluids speculates about using an umbrella as a "sail" to propel a bicycle along the sidewalk. Develop an algebraic expression for the speed a bike could reach on level ground with the umbrella "propulsion system." The frontal area of bike and rider is estimated as $0.3\ \text{m}^2$, and the drag coefficient is about 1.2. Assume the rolling resistance is 0.75 percent of the bike and rider weight; the combined mass is 75 kg. Evaluate the bike speed that could be achieved with an umbrella 1.22 m in diameter in a wind that blows at 24 km/hr. Discuss the practicality of this propulsion system.

9.150 Motion of a small rocket was analyzed in Example 4.12 assuming negligible aerodynamic drag. This was not realistic at the final calculated speed of 369 m/s. Use the Euler method of Section 5.5 for approximating the first derivatives, in an *Excel* workbook, to solve the equation of motion for the rocket. Plot the rocket speed as a function of time, assuming $C_D = 0.3$ and a rocket diameter of 700 mm. Compare with the results for $C_D = 0$.

9.151 A baseball is popped straight up with an initial velocity of 25 m/s. The baseball has a diameter of 0.073 m and a mass of 0.143 kg. The drag coefficient for the baseball can be estimated as 0.47 for $Re < 10^4$ and 0.10 for $Re > 10^4$. Determine how long the ball will be in the air and how high it will go.

9.152 Wiffle™ balls made from light plastic with numerous holes are used to practice baseball and golf. Explain the purpose of the holes and why they work. Explain how you could test your hypothesis experimentally.

9.153 Towers for television transmitters may be up to 500 m in height. In the winter, ice forms on structural members. When the ice thaws, chunks break off and fall to the ground. How far from the base of a tower would you recommend placing a fence to limit danger to pedestrians from falling ice chunks?

9.154 The "shot tower," used to produce spherical lead shot, has been recognized as a mechanical engineering landmark. In a shot tower, molten lead is dropped from a high tower; as the lead solidifies, surface tension pulls each shot into a spherical shape. Discuss the possibility of increasing the "hang time," or of using a shorter tower, by dropping molten lead into an air stream that is moving upward. Support your discussion with appropriate calculations.

9.155 Design a wind anemometer that uses aerodynamic drag to move or deflect a member or linkage, producing an output that can be related to wind speed, for the range from 1 to 10 m/s in standard air. Consider three alternative design concepts. Select the best concept and prepare a detailed design. Specify the shape, size, and material for each component. Quantify the relation between wind speed and anemometer output. Present results as a theoretical "calibration curve" of anemometer output versus wind speed. Discuss reasons why you rejected the alternative designs and chose your final design concept.

9.156 A model airfoil of chord 150 mm and span 750 mm is placed in a wind tunnel with an air flow of 30 m/s (the air is at 20°C). It is mounted on a cylindrical support rod 25 mm in diameter and 250 mm tall. Instruments at the base of the rod indicate a vertical force of 44.5 N and a horizontal force of 6.7 N. Calculate the lift and drag coefficients of the airfoil.

9.157 An antique airplane carries 50 m of external guy wires stretched normal to the direction of motion. The wire diameter is 5 mm. Estimate the maximum power saving that results from an optimum streamlining of the wires at a plane speed of 175 km/hr in standard air at sea level.

9.158 Why do modern guns have rifled barrels?

9.159 How do cab-mounted wind deflectors for tractor-trailer trucks work? Explain using diagrams of the flow pattern around the truck and pressure distribution on the surface of the truck.

9.160 An airplane with an effective lift area of $25\ \text{m}^2$ is fitted with airfoils of NACA 23012 section (Fig. 9.23). The maximum flap setting that can be used at takeoff corresponds to configuration ② in Fig. 9.23. Determine the maximum gross mass possible for the airplane if its takeoff speed is 150 km/hr at sea level (neglect added lift due to ground effect). Find the minimum takeoff speed required for this gross mass if the airplane is instead taking off from Denver (elevation approximately 1.6 km).

9.161 An aircraft is in level flight at 225 km/h through air at standard conditions. The lift coefficient at this speed is 0.45 and the drag coefficient is 0.065. The mass of the aircraft is 900 kg. Calculate the effective lift area for the craft, and the required engine thrust and power.

9.162 The foils of a surface-piercing hydrofoil watercraft have a total effective area of $0.7\ \text{m}^2$. Their coefficients of lift and drag are 1.6 and 0.5, respectively. The total mass of the craft in running trim is 1800 kg. Determine the minimum speed at which the craft is supported by the hydrofoils. At this speed, find the power required to overcome water resistance. If the craft is fitted with a 110 kW engine, estimate its top speed.

9.163 A high school project involves building a model ultralight airplane. Some of the students propose making an airfoil from a sheet of plastic 1.5 m long × 2 m wide at an angle of attack of 12°. At this airfoil's aspect ratio and angle of attack the lift and drag coefficients are C_L = 0.72 and C_D = 0.17. If the airplane is designed to fly at 12 m/s, what is the maximum total payload? What will be the required power to maintain flight? Does this proposal seem feasible?

9.164 The U.S. Air Force F-16 fighter aircraft has wing planform area $A = 27.9\text{ m}^2$; it can achieve a maximum lift coefficient of $C_L = 1.6$. When fully loaded, its mass is M = 11,600 kg. The airframe is capable of maneuvers that produce $9g$ vertical accelerations. However, student pilots are restricted to $5g$ maneuvers during training. Consider a turn flown in level flight with the aircraft banked. Find the minimum speed in standard air at which the pilot can produce a $5g$ total acceleration. Calculate the corresponding flight radius. Discuss the effect of altitude on these results.

9.165 The teacher of the students designing the airplane of Problem 9.163 is not happy with the idea of using a sheet of plastic for the airfoil. He asks the students to evaluate the expected maximum total payload, and required power to maintain flight, if the sheet of plastic is replaced with a conventional section (NACA 23015) airfoil with the same aspect ratio and angle of attack. What are the results of the analysis?

9.166 A light airplane, with mass M = 1000 kg, has a conventional-section (NACA 23015) wing of planform area $A = 10\text{ m}^2$. Find the angle of attack of the wing for a cruising speed of V = 63 m/s. What is the required power? Find the maximum instantaneous vertical "g force" experienced at cruising speed if the angle of attack is suddenly increased.

9.167 A light airplane has 10 m effective wingspan and 1.8 m chord. It was originally designed to use a conventional (NACA 23015) airfoil section. With this airfoil, its cruising speed on a standard day near sea level is 225 km/hr. A conversion to a laminar-flow (NACA 66_2–215) section airfoil is proposed. Determine the cruising speed that could be achieved with the new airfoil section for the same power.

9.168 Instead of a new laminar-flow airfoil, a redesign of the light airplane of Problem 9.167 is proposed in which the current conventional airfoil section is replaced with another conventional airfoil section of the same area, but with aspect ratio AR = 8. Determine the cruising speed that could be achieved with this new airfoil for the same power.

9.169 Assume the Boeing 727 aircraft has wings with NACA 23012 section, planform area of 150 m^2, double-slotted flaps, and effective aspect ratio of 6.5. If the aircraft flies at 77.2 m/s in standard air at 778,750 N gross weight, estimate the thrust required to maintain level flight.

9.170 An airplane with mass of 4500 kg is flown at constant elevation and speed on a circular path at 240 km/hr. The flight circle has a radius of 990 m. The plane has lifting area of 23 m^2 and is fitted with NACA 23015 section airfoils with effective aspect ratio of 7. Estimate the drag on the aircraft and the power required.

9.171 Find the minimum and maximum speeds at which the airplane of Problem 9.170 can fly on a 990 m radius circular flight path, and estimate the drag on the aircraft and power required at these extremes.

9.172 Jim Hall's Chaparral 2F sports-racing cars in the 1960s pioneered use of airfoils mounted above the rear suspension to enhance stability and improve braking performance. The airfoil was effectively 1.8 m wide (span) and had a 0.3 m chord. Its angle of attack was variable between 0 and minus 12 degrees. Assume lift and drag coefficient data are given by curves (for conventional section) in Fig. 9.17. Consider a car speed of 192 km/hr on a calm day. For an airfoil deflection of 12° down, calculate (a) the maximum downward force and (b) the maximum increase in deceleration force produced by the airfoil.

9.173 The glide angle for unpowered flight is such that lift, drag, and weight are in equilibrium. Show that the glide slope angle, θ, is such that $\tan\theta = C_D/C_L$. The minimum glide slope occurs at the speed where C_L/C_D is a maximum. For the conditions of Example 9.8, evaluate the minimum glide slope angle for a Boeing 727-200. How far could this aircraft glide from an initial altitude of 10 km on a standard day?

9.174 The wing loading of the Gossamer Condor is 19 N/m^2 of wing area. Crude measurements showed drag was approximately 27 N at 19.2 km/hr. The total weight of the Condor was 890 N. The effective aspect ratio of the Condor is 17. Estimate the minimum power required to fly the aircraft. Compare to the 290 W that pilot Brian Allen could sustain for 2 hours.

9.175 Some cars come with a "spoiler," a wing section mounted on the rear of the vehicle that salespeople sometimes claim significantly increases traction of the tires at highway speeds. Investigate the validity of this claim. Are these devices really just cosmetic?

9.176 Roadside signs tend to oscillate in a twisting motion when a strong wind blows. Discuss the phenomena that must occur to cause this behavior.

9.177 How does a Frisbee™ fly? What causes it to curve left or right? What is the effect of spin on its flight?

9.178 Air moving over an automobile is accelerated to speeds higher than the travel speed, as shown in Fig. 9.25. This causes changes in interior pressure when windows are opened or closed. Use the data of Fig. 9.25 to estimate the pressure reduction when a window is opened slightly at a speed of 100 km/hr. What is the air speed in the freestream near the window opening?

9.179 An automobile travels down the road with a bicycle attached to a carrier across the rear of the trunk. The bicycle wheels rotate slowly. Explain why and in what direction the rotation occurs.

9.180 A class demonstration showed that lift is present when a cylinder rotates in an air stream. A string wrapped around a paper cylinder and pulled causes the cylinder to spin and move forward simultaneously. Assume a cylinder of 5 cm diameter and 30 cm length is given a rotational speed of 240 rpm and a forward speed of 1.5 m/s. Estimate the approximate lift force that acts on the cylinder.

9.181 A golf ball (diameter D = 43 mm) with circular dimples is hit from a sand trap at 20 m/s with backspin of 2000 rpm. The mass of the ball is 48 g. Evaluate the lift and drag forces acting on the ball. Express your results as fractions of the weight of the ball.

9.182 Rotating cylinders were proposed as a means of ship propulsion in 1924 by the German engineer, Flettner. The original Flettner rotor ship had two rotors, each about 3 m in diameter and 15 m high, rotating at up to 750 rpm. Calculate the maximum lift and drag forces that act on each rotor in a 50 km/hr wind. Compare the total force to that produced at the optimum L/D at the same wind speed. Estimate the power needed to spin the rotor at 750 rpm.

9.183 A baseball pitcher throws a ball at 128 km/hr. Home plate is 18.3 m away from the pitcher's mound. What spin should be placed on the ball for maximum horizontal deviation from a straight path? (A baseball has a mass of 142 gm and a circumference of 230 mm.) How far will the ball deviate from a straight line?

9.184 American and British golf balls have slightly different diameters but the same mass (see Problems 1.39 and 1.42). Assume a professional golfer hits each type of ball from a tee at 85 m/s with backspin of 9000 rpm. Evaluate the lift and drag forces on each ball. Express your answers as fractions of the weight of each ball. Estimate the radius of curvature of the trajectory of each ball. Which ball should have the longer range for these conditions?

9.185 A soccer player takes a free kick. Over a distance of 10 m, the ball veers to the right by about 1 m. Estimate the spin the player's kick put on the ball if its speed is 30 m/s. The ball has a mass of 420 gm and has a circumference of 70 cm.

10

Fluid Machinery

10.1 Introduction and Classification of Fluid Machines

10.2 Turbomachinery Analysis

10.3 Pumps, Fans, and Blowers

10.4 Positive Displacement Pumps

10.5 Hydraulic Turbines

10.6 Propellers and Wind-Power Machines

10.7 Compressible Flow Turbomachines

10.8 Summary and Useful Equations

Case Study in Energy and the Environment

Wind Power: Wind Turbine and Fan Design Using Tubercles

In the *Case Study* of Chapter 9, we learned that humpback whales' incredible agility to a large degree comes from the bumps on the leading edges of their flippers, known as *tubercles*. Ernst van Nierop, a PhD candidate at the School of Engineering and Applied Sciences at Harvard University, coauthored a study to explain this phenomenon with mathematics professor Michael Brenner and researcher Silas Alben.

As with the airfoils discussed in Chapter 9, when the angle of attack of a whale flipper becomes too steep, the result is stall. Previous experiments have shown, however, that the angle of attack before stall occurs for a humpback-whale flipper is much steeper than that of a smooth flipper. The Harvard team showed that the tubercles change the distribution of pressure on the flipper so that some parts of it stall before others, thereby avoiding abrupt stalling and giving the whale more freedom to attack at higher angles and the ability to better predict its hydrodynamic limitations.

Studying living things in order to come up with ideas to improve technology is a practice known as *biomimicry*, and it is becoming more widely used to increase efficiency in machines. In this instance we have a practical application of tubercle technology, specifically as it applies to turbomachinery, the topic of this chapter. A company in Toronto, Ontario, called

The tubercles on *WhalePower*'s turbine blade (Photo courtesy of J. Subirana, *WhalePower*)

WhalePower, has demonstrated the advantages of tubercles when they are integrated into the leading edges of wind-turbine and fan blades. The photograph shows a prototype of a wind turbine blade incorporating tubercles on its leading edge. Tests of these prototypes have shown that the delayed stall doubles the performance of the turbines and allows the turbine to capture more energy out of lower-speed winds. As we alluded to in the *Case Study* of Chapter 9, this increase in performance can be explained by looking at how stall affects the flow over rotating blades. In particular, it is well-known that the stall experienced by conventional blades causes air to travel from the hub to the tips of the blades, rather than parallel to the axis of rotation. The result of this effect, sometimes referred to as *span-wise pumping*, is that additional energy is needed to move the air in the desired direction, decreasing efficiency of the fan. Similarly, a wind turbine that experiences stall will generate less lift, and therefore less power can extracted from the air. In addition, the radial component of the air flow increases vibration of the blades, causing noise and increased wear and tear.

Ongoing tests at the Wind Energy Institute of Canada have shown that because the tubercle-lined blades delay stall, they are more stable, quiet, and durable than conventional blades. Recent studies show that the addition of tubercles to the leading edges of turbine blades makes them generate more stable lift, with lower drag even at high pitch angles, and when there is stall it is a gradual, not catastrophic, change. *WhalePower* claims that there is improved power generation at low wind speeds, the blades are quieter than conventional blades, and there is decreased tip chatter (vibrations at the blade tip due to flow instabilities); in other words, in real-world conditions the tubercle-enhanced blades appear to be more stable and responsive than any previous turbine. *WhalePower* has also shown that tubercle-lined blades on industrial ceiling fans can operate 20 percent more efficiently than conventional blades and that they do a better job at circulating air in a building. The results were dramatic enough to convince Envira-North, Canada's largest maker of ventilation fans, to license the design.

Humans have sought to control nature since antiquity. Early humans carried water by the bucket; as larger groups formed, this process was mechanized. The first fluid machines developed as bucket wheels and screw pumps to lift water. The Romans introduced paddle wheels around 70 B.C. to obtain energy from streams [1]. Later, windmills were developed to harness wind power, but the low power density of the wind limited output to a few hundred kW. Development of waterwheels made it possible to extract thousands of kW at a single site.

Today we take many fluid machines for granted. On a typical day we draw pressurized water from the tap, use a blower to dry our hair, drive a car (in which fluid machines operate the lubrication, cooling, and power steering systems), and work in a comfortable environment provided by air circulation. The list could be extended indefinitely.

A fluid machine is a device that either performs work on or extracts work (or power) from a fluid. As you can imagine, this is a very large field of study, so we will limit ourselves mostly to incompressible flows. First the terminology of the field is introduced and machines are classified by operating principle and physical characteristics. Rather than attempt a treatment of the entire field, we focus on machines in which energy transfer to or from the fluid is through a rotating element. Basic equations are reviewed and then simplified to forms useful for analysis of fluid machines. Performance characteristics of typical machines are considered. Examples are given of pump and turbine applications in typical systems. Next, we will discuss propellers and wind turbines, which are unique in that they achieve energy transfer with a fluid without the benefit of an external housing. A discussion of compressible flow machines concludes the chapter.

10.1 Introduction and Classification of Fluid Machines

Fluid machines may be broadly classified as either *positive displacement* or *dynamic*. In positive-displacement machines, energy transfer is accomplished by volume changes that occur due to movement of the boundary in which the fluid is confined. This includes piston-cylinder arrangements, gear pumps (for example, the oil pump for a car engine), and lobe pumps (for example, those used in medicine for circulating blood through a machine). We will not analyze these devices in this chapter; we will review them briefly in Section 10.4. Dynamic fluid-handling devices that direct the flow with blades or vanes attached to a rotating member are termed *turbomachines*. In contrast to positive displacement machinery, there is no closed volume in a turbomachine. These devices are very widely used in industry for power generation (for example, water and steam turbines) and in numerous other applications (for example, the turbocharger of a high-performance car). The emphasis in this chapter is on dynamic machines.

A further distinction among types of turbomachines is based on the geometry of the flow path. In *radial-flow* machines, the flow path is essentially radial, with significant changes in radius from inlet to outlet. (Such machines sometimes are called *centrifugal* machines.) In *axial-flow* machines, the flow path is nearly parallel to the machine centerline, and the radius of the flow path does not vary significantly. In *mixed-flow* machines the flow-path radius changes only moderately. Diagrams and photographs of typical turbomachines are shown in Figs. 10.1 through 10.5.

All work interactions in a turbomachine result from dynamic effects of the rotor on the fluid stream; that is, the transfer of work between the fluid and the rotating machine either increases or decreases the speed of the flow. However, in conjunction with this kinetic energy transfer, machines that include external housings (e.g., compressors, pumps, and turbines) also involve either the conversion of pressure energy to kinetic energy, or vice versa. This acceleration or deceleration of the flow allows for maximum pressure rise in pumps and compressors and for maximum power output from turbines.

Machines for Doing Work on a Fluid

Machines that add energy to a fluid by performing work on it are called *pumps* when the flow is liquid or slurry, and *fans*, *blowers*, or *compressors* for gas- or vapor-handling units, depending on pressure rise. Fans usually have small pressure rise (less than 25 mm of water) and blowers have moderate pressure rise (perhaps 25 mm of mercury); pumps and compressors may have very high pressure rises. Current industrial systems operate at pressures up to 1 GPa (10^4 atmospheres).

Pumps and compressors consist of a rotating wheel (called an *impeller* or *rotor*, depending on the type of machine) driven by an external power source (e.g., a motor or another fluid machine) to increase the flow kinetic energy, followed by an element to decelerate the flow, thereby increasing its pressure. This combination is known as a *stage*. These elements are contained within a *housing* or *casing*. A single pump or compressor might consist of several stages within a single housing, depending on the amount of pressure rise required of the machine. The shaft must penetrate the

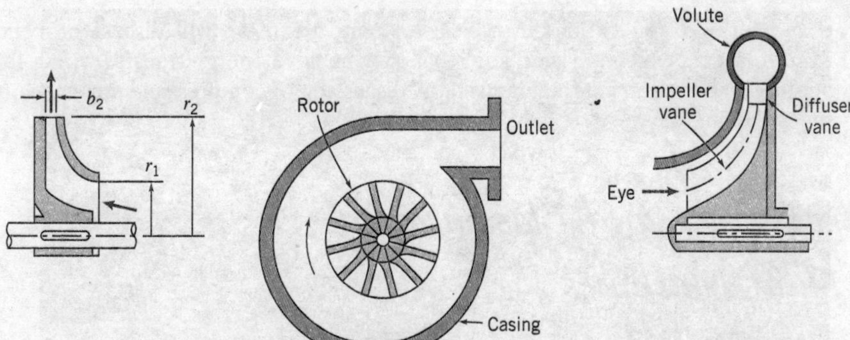

(*a*) Centrifugal pump (*b*) Centrifugal blower (*c*) Centrifugal compressor

Fig. 10.1 Schematic diagrams of typical centrifugal-flow turbomachines, adapted from [2].

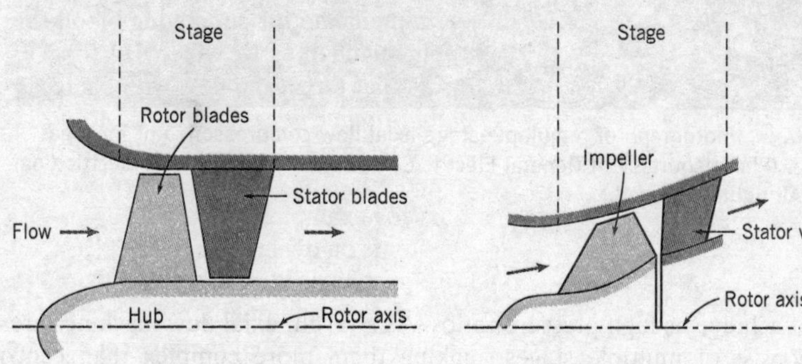

(*a*) Axial-flow compressor stage (*b*) Mixed-flow pump stage

Fig. 10.2 Schematic diagrams of typical axial-flow and mixed-flow turbomachines, adapted from [2].

housing in order to receive mechanical work from the external power source. Bearings and seals are needed to minimize frictional (mechanical) losses and prevent leakage of the working fluid.

VIDEO

Flow in an Axial Flow Compressor (Animation).

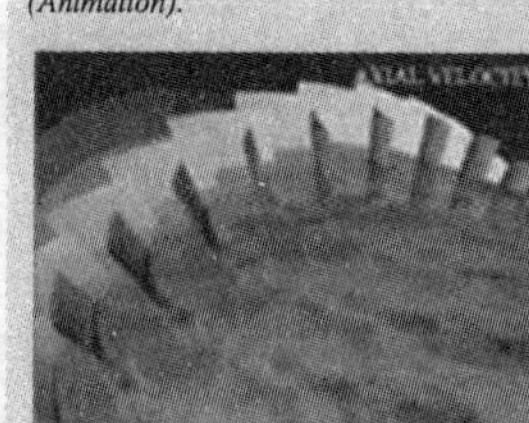

Three typical centrifugal machines are shown schematically in Fig. 10.1. The rotating element of a centrifugal pump or compressor is frequently called the impeller. Flow enters each machine nearly axially at small radius through the *eye* of the impeller, diagram (*a*), at radius r_1. Flow is turned and leaves through the impeller discharge at radius r_2, where the width is b_2. Diffusion of the flow is achieved in a centrifugal machine as it leaves the impeller and is collected in the *scroll* or *volute*, which gradually increases in area as it nears the outlet of the machine, diagram (*b*). The impeller usually has vanes; it may be *shrouded* (enclosed) as shown in diagram (*a*), or *open* as shown in diagram (*c*). The impeller vanes may be relatively straight, or they may curve to become nonradial at the outlet. Diagram (*c*) shows that the diffuser may have vanes to direct the flow between the impeller discharge and the volute; vanes allow for more efficient diffusion, but at increased fabrication cost. Centrifugal machines are capable of higher pressure ratios than axial machines, but they have a higher frontal area per unit mass flow.

Typical axial-flow and mixed-flow turbomachines are shown schematically in Fig. 10.2. Figure 10.2*a* shows a typical axial-flow compressor stage. In these machines the rotating element is referred to as the rotor, and flow diffusion is achieved in the stator. Flow enters nearly parallel to the rotor axis and maintains nearly the same radius through the stage. The mixed-flow pump in diagram (*b*) shows the flow being turned outward and moving to larger radius as it passes through the stage. Axial flow machines have higher efficiencies and less frontal area than centrifugal machines, but

Fig. 10.3 Photograph of a multiple-stage axial-flow compressor rotor for a gas turbine. (Photo courtesy of General Electric Company. © 2010 General Electric Company. All rights reserved.)

they cannot achieve as high pressure ratios. As a result, axial flow machines are more likely to consist of multiple stages, making them more complex than centrifugal machines. Figure 10.3 shows a multiple-stage axial flow compressor. In this photograph, the outer housing (to which the stator vanes are attached) has been removed, clearly showing the rows of rotor vanes.

The pressure rise that can be achieved efficiently in a single stage is limited, depending on the type of machine. The reason for this limitation can be understood based on the pressure gradients in these machines (see Section 9.6). In a pump or compressor, the boundary layer subjected to an adverse pressure gradient is not stable; so flow is more likely to encounter boundary-layer separation in a compressor or pump. Boundary-layer separation increases the drag on the impeller, resulting in a decrease in efficiency; therefore additional work is needed to compress the flow.

Fans, blowers, compressors, and pumps are found in many sizes and types, ranging from simple household units to complex industrial units of large capacity. Torque and power requirements for idealized pumps and turboblowers can be analyzed by applying the angular-momentum principle using a suitable control volume.

Propellers are essentially axial-flow devices that operate without an outer housing. Propellers may be designed to operate in gases or liquids. As you might expect, propellers designed for these very different applications are quite distinct. Marine propellers tend to have wide blades compared with their radii, giving high *solidity*. Aircraft propellers tend to have long, thin blades with relatively low solidity. These machines will be discussed in detail in Section 10.6.

Machines for Extracting Work (Power) from a Fluid

Machines that extract energy from a fluid in the form of work (or power) are called *turbines*. In *hydraulic turbines*, the working fluid is water, so the flow is incompressible. In *gas turbines* and *steam turbines*, the density of the working fluid may change significantly. In a turbine, a stage normally consists of an element to accelerate the flow, converting some of its pressure energy to kinetic energy, followed by a *rotor*, *wheel*, or *runner* extracts the kinetic energy from the flow via a set of *vanes*, *blades*, or *buckets* mounted on the wheel.

The two most general classifications of turbines are impulse and reaction turbines. *Impulse turbines* are driven by one or more high-speed free jets. The classic example of an impulse turbine is the waterwheel. In a waterwheel, the jets of water are driven by gravity; the kinetic energy of the water is transferred to the wheel, resulting in work. In more modern forms of impulse turbines, the jet is accelerated in a nozzle external to the turbine wheel. If friction and gravity are neglected, neither the fluid pressure nor speed relative to the runner changes as the fluid passes over the turbine buckets. Thus for an impulse turbine, the fluid acceleration and accompanying pressure drop take place in nozzles external to the blades, and the runner does not flow full of fluid; work is extracted as a result of the large momentum change of the fluid.

In *reaction turbines*, part of the pressure change takes place externally and part takes place within the moving blades. External acceleration occurs and the flow is turned to enter the runner in the proper direction as it passes through nozzles or stationary blades, called *guide vanes* or *wicket gates*. Additional fluid acceleration relative to the rotor occurs within the moving blades, so both the relative velocity and the pressure of the stream change across the runner. Because reaction turbines flow full of fluid, they generally can produce more power for a given overall size than impulse turbines.

Figure 10.4 shows turbines used for different applications. Figure 10.4*a* shows a Pelton wheel, a type of impulse turbine wheel used in hydroelectric power plants. Figure 10.4*b* is a photograph of an axial steam turbine rotor, an example of a reaction turbine. Figure 10.4*c* is a wind turbine farm. A wind turbine is another example of a reaction turbine, but, like a propeller, also operates without an outer housing. Modern wind turbines typically collect wind energy and convert it into electricity.

Several typical hydraulic turbines are shown schematically in Fig. 10.5. Figure 10.5*a* shows an impulse turbine driven by a single jet, which lies in the plane of the turbine runner. Water from the jet strikes each bucket in succession, is turned, and leaves the bucket with relative velocity nearly opposite to that with which it entered the bucket. Spent water falls into the *tailrace* (not shown).

A reaction turbine of the Francis type is shown in Fig. 10.5*b*. Incoming water flows circumferentially through the turbine casing. It enters the periphery of the stationary guide vanes and flows toward the runner. Water enters the runner nearly radially and is turned downward to leave nearly axially; the flow pattern may be thought of as a centrifugal pump in reverse. Water leaving the runner flows through a diffuser known as a *draft tube* before entering the tailrace. Figure 10.5*c* shows a propeller turbine of the Kaplan type. The water entry is similar to that in the Francis turbine, but it is

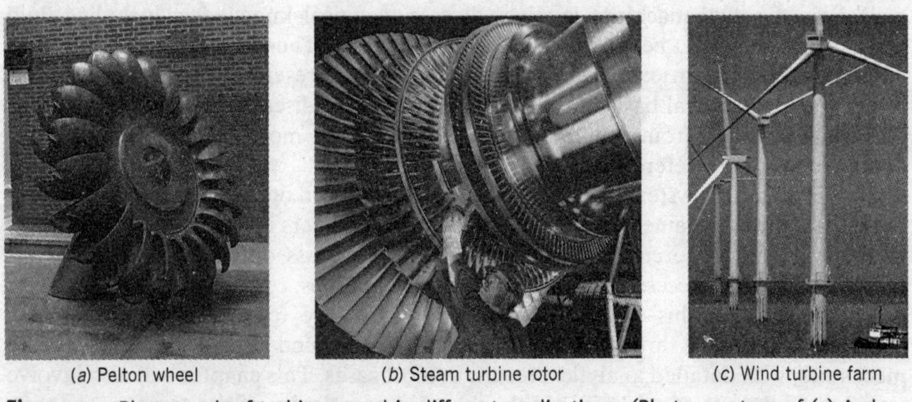

(*a*) Pelton wheel (*b*) Steam turbine rotor (*c*) Wind turbine farm

Fig. 10.4 Photograph of turbines used in different applications. (Photo courtesy of (*a*) Andy Dingley; (*b*) and (*c*) Siemens Energy © 2010.)

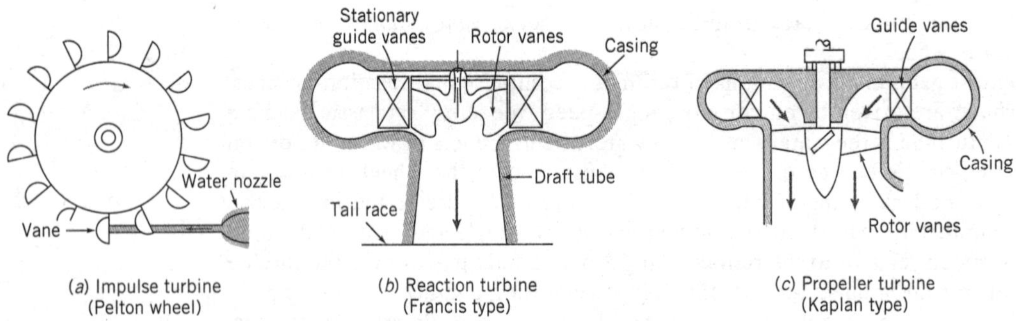

Fig. 10.5 Schematic diagrams of typical hydraulic turbines, adapted from [2].

turned to flow nearly axially before encountering the turbine runner. Flow leaving the runner may pass through a draft tube.

Thus turbines range from simple windmills to complex gas and steam turbines with many stages of carefully designed blading. These devices also can be analyzed in idealized form by applying the angular-momentum principle.

The allowable amount of pressure drop in a turbine stage is usually greater than the amount of pressure rise allowable in a compressor stage. The difference is due to the favorable pressure gradient (see Section 9.6), which makes boundary-layer separation much less likely than in the case of the compressor.

Dimensionless parameters, such as *specific speed, flow coefficient, torque coefficient, power coefficient*, and *pressure ratio*, frequently are used to characterize the performance of turbomachines. These parameters were introduced in Chapter 7; their development and use will be considered in more detail later in this chapter.

Scope of Coverage

According to Japikse [3], "Turbomachinery represents a $400 billion market (possibly much more) with enormous worldwide growth at this time. It is estimated that industrial centrifugal pumps alone consume 5 percent of all the energy produced in the USA." In addition, the demands for widely available, economical, green power will continue to drive research and development in the turbomachinery industry [4]. Therefore, proper design, construction, selection, and application of pumps and compressors are economically significant.

Design of actual machines involves diverse technical knowledge, including fluid mechanics, materials, bearings, seals, and vibrations. These topics are covered in numerous specialized texts. Our objective here is to present only enough detail to illustrate the analytical basis of fluid flow design and to discuss briefly the limitations on results obtained from simple analytical models. For more detailed design information, consult the references.

Applications or "system" engineering requires a wealth of experience. Much of this experience must be gained by working with other engineers in the field. Our coverage is not intended to be comprehensive; instead we discuss only the most important considerations for successful system application of pumps, compressors, and turbines.

The material in this chapter is of a different nature from that in the previous chapters. Chapters 1 through 9 covered much of the fundamental material of fluid mechanics, with detailed analytical results in most cases. This chapter will also involve significant amounts of analysis, but the inherent complexity of the topic means that, on many occasions, we need to resort to empirical results and correlations. To the student, this may appear as so much "hand-waving," but combining theory and experiment to deduce results is a very common approach in engineering science.

Turbomachinery Analysis 10.2

As in other analyses, the method of analysis used for turbomachinery is chosen according to the information sought. If overall information on flow rate, pressure change, torque, and power is desired, then a finite-control-volume analysis may be used. If detailed information is desired about blade angles or velocity profiles, then individual blade elements must be analyzed using an infinitesimal-control-volume or other detailed procedure. We consider only idealized flow processes in this book, so we concentrate on the approach using the finite control volume, applying the angular-momentum principle. The analysis that follows applies to machines both for doing work on, and extracting work from, a fluid flow.

The Angular-Momentum Principle: The Euler Turbomachine Equation

The angular-momentum principle was applied to finite control volumes in Chapter 4. The result was Eq. 4.46.

$$\vec{r}\times\vec{F}_s + \int_{CV}\vec{r}\times\vec{g}\rho d\forall + \vec{T}_{\text{shaft}} = \frac{\partial}{\partial t}\int_{CV}\vec{r}\times\vec{V}\rho d\forall + \int_{CS}\vec{r}\times\vec{V}\rho\vec{V}\cdot d\vec{A} \tag{4.46}$$

Equation 4.46 states that the moment of surface forces and body forces, plus the applied torque, lead to a change in the angular momentum of the flow. [The surface forces are due to friction and pressure, the body force is due to gravity, the applied torque could be positive or negative (depending on whether we are doing work on or extracting work from the fluid, respectively), and the angular-momentum change can arise as a change in angular momentum within the control volume, or flux of angular momentum across the control surface.]

We will now simplify Eq. 4.46 for analysis of turbomachinery. First, it is convenient to choose a fixed control volume enclosing the rotor to evaluate shaft torque. Because we are looking at control volumes for which we expect large shaft torques, as a first approximation torques due to surface forces may be ignored. This means we are neglecting friction and torque generated by pressure changes. The body force may be neglected by symmetry. Then, for steady flow, Eq. 4.46 becomes

$$\vec{T}_{\text{shaft}} = \int_{CS}\vec{r}\times\vec{V}\rho\vec{V}\cdot d\vec{A} \tag{10.1a}$$

Equation 10.1a states: For a turbomachine with work *input*, the torque *required* causes a change in the fluid angular momentum; for a turbomachine with work *output*, the torque *produced* is due to the change in fluid angular momentum. Let us write this equation in scalar form and illustrate its application to axial- and radial-flow machines.

As shown in Fig. 10.6, we select a *fixed* control volume enclosing a generalized turbomachine rotor. The fixed coordinate system is chosen with the z-axis aligned with the axis of rotation of the machine. The idealized velocity components are shown in the figure. The fluid enters the rotor at radial location, r_1, with uniform absolute velocity, $\vec{V}_1$; the fluid leaves the rotor at radial location, r_2, with uniform absolute velocity $\vec{V}_2$.

The integrand on the right side of Eq. 10.1a is the product of with the mass flow rate at each section. For *uniform flow* into the rotor at section 1, and out of the rotor at section 2, Eq. 10.1a becomes

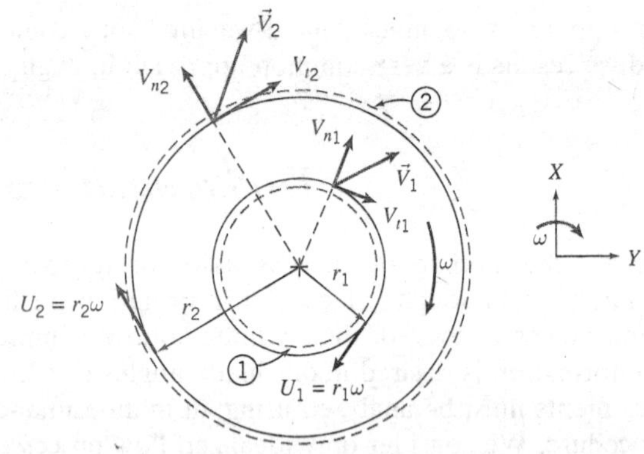

Fig. 10.6 Finite control volume and absolute velocity components for analysis of angular momentum.

$$T_{\text{shaft}}\hat{k} = (r_2 V_{t_2} - r_1 V_{t_1})\dot{m}\hat{k} \tag{10.1b}$$

(Note that in $\vec{r} \times \vec{V}$ the position vector $\vec{r}$ is purely radial; so only the tangential velocity component V_t counts.) In scalar form,

$$T_{\text{shaft}} = (r_2 V_{t_2} - r_1 V_{t_1})\dot{m} \tag{10.1c}$$

The assumptions we made in deriving this equation are *steady, frictionless flow; uniform flow* at inlet and exit; and *negligible pressure effects*. Equation 10.1c is the basic relationship between torque and angular momentum for all turbomachines. It often is called the *Euler turbomachine equation*.

Each velocity that appears in Eq. 10.1c is the tangential component of the absolute velocity of the fluid crossing the control surface. The tangential velocities are chosen positive when in the same direction as the blade speed, U. This sign convention gives $T_{\text{shaft}} > 0$ for pumps, fans, blowers, and compressors and $T_{\text{shaft}} < 0$ for turbines.

The rate of work done on a turbomachine rotor (the mechanical power, $\dot{W}_m$) is given by the dot product of rotor angular velocity, $\vec{\omega}$, and applied torque, $\vec{T}_{\text{shaft}}$. Using Eq. 10.1b, we obtain

$$\dot{W}_m = \vec{\omega}\cdot\vec{T}_{\text{shaft}} = \omega\hat{k}\cdot T_{\text{shaft}}\hat{k} = \omega\hat{k}\cdot(r_2 V_{t_2} - r_1 V_{t_1})\dot{m}\hat{k}$$

or

$$\dot{W}_m = \omega T_{\text{shaft}} = \omega(r_2 V_{t_2} - r_1 V_{t_1})\dot{m} \tag{10.2a}$$

According to Eq. 10.2a, the angular momentum of the fluid is increased by the addition of shaft work. For a pump, $\dot{W}_m > 0$ and the angular momentum of the fluid must increase. For a turbine, $\dot{W}_m < 0$ and the angular momentum of the fluid must decrease.

Equation 10.2a may be written in two other useful forms. Introducing $U = r\omega$, where U is the tangential speed of the rotor at radius r, we have

$$\dot{W}_m = (U_2 V_{t_2} - U_1 V_{t_1})\dot{m} \tag{10.2b}$$

Dividing Eq. 10.2b by $\dot{m}g$, we obtain a quantity with the dimensions of length, which may be viewed as the theoretical head added to the flow.[1]

[1]Since $\dot{W}_m$ has dimensions of energy per unit time and $\dot{m}g$ is weight flow per unit time, head, H, is actually energy per unit weight of flowing fluid.

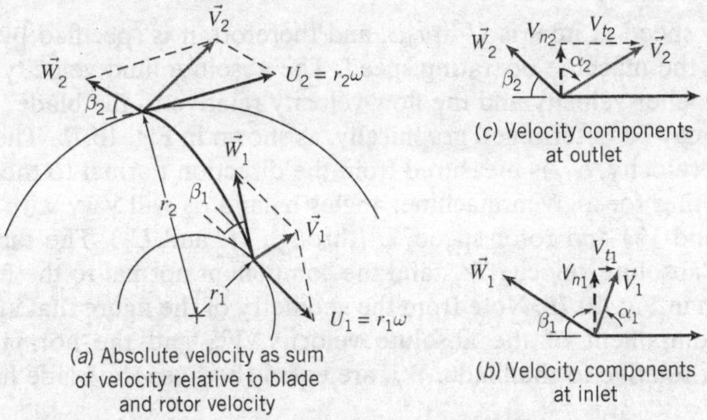

Fig. 10.7 Geometry and notation used to develop velocity diagrams for typical radial-flow machines.

$$H = \frac{\dot{W}_m}{\dot{m}g} = \frac{1}{g}(U_2 V_{t_2} - U_1 V_{t_1}) \tag{10.2c}$$

Equations 10.1 and 10.2 are simplified forms of the angular-momentum equation for a control volume. They all are written for a fixed control volume under the assumptions of steady, uniform flow at each section. The equations show that only the difference in the product rV_t or UV_t, between the outlet and inlet sections, is important in determining the torque applied to the rotor or the mechanical power. Although $r_2 > r_1$ in Fig. 10.6, no restriction has been made on geometry; the fluid may enter and leave at the same or different radii. Therefore, these equations may be used for axial, radial, or mixed-flow machines.

Velocity Diagrams

The equations that we have derived also suggest the importance of clearly defining the velocity components of the fluid and rotor at the inlet and outlet sections. For this purpose, it is useful to develop *velocity diagrams* (frequently called *velocity polygons*) for the inlet and outlet flows. Figure 10.7 shows the velocity diagrams and introduces the notation for blade and flow angles. The important notation to remember is that the variable V is typically used to indicate absolute velocity, that is, the velocity of the flow relative to a stationary observer, while the variable W is used to indicate flow velocity relative to the rotating blade.

Machines are designed such that at *design condition* the fluid moves smoothly (without disturbances) through the blades. In the idealized situation at the *design speed*, flow relative to the rotor is assumed to enter and leave tangent to the blade profile at each section. (This idealized inlet condition is sometimes called *shockless* entry flow.) At speeds other than design speed (and sometimes in reality, even at design speed!), the fluid may impact the blades at inlet, exit at an angle relative to the blade, or may have significant flow separation, leading to machine inefficiency. Figure 10.7 is representative of a typical radial flow machine. We assume the fluid is moving without major flow disturbances through the machine, as shown in Fig. 10.7*a*, with blade inlet and exit angles β_1 and β_2, respectively, relative to the circumferential direction. Note that although angles β_1 and β_2 are both less than 90° in Fig. 10.7, in general they can be less than, equal to, or greater than 90°, and the analysis that follows applies to all of these possibilities.

The runner speed at inlet is $U_1 = r_1\omega$, and therefore it is specified by the impeller geometry and the machine operating speed. The absolute fluid velocity is the vector sum of the impeller velocity and the flow velocity relative to the blade. The absolute inlet velocity may be determined graphically, as shown in Fig. 10.7*b*. The angle of the absolute fluid velocity, α_1, is measured from the direction normal to the flow area, as shown.[2] Note that for a given machine, angles α_1 and α_2 will vary with flow rate, Q, (through $\vec{V}_1$ and $\vec{V}_2$) and rotor speed, ω (through U_1 and U_2). The tangential component of the absolute velocity, V_{t_1}, and the component normal to the flow area, V_{n_1}, are also shown in Fig. 10.7*b*. Note from the geometry of the figure that at each section the normal component of the absolute velocity, V_n, and the normal component of the velocity relative to the blade, W_n, are equal (because the blade has no normal velocity).

To help determine the absolute velocity at the machine entrance, it is necessary to determine whether swirl exists at the entrance. Swirl, which may be present in the inlet flow or introduced by *inlet guide vanes*, is the presence of a circumferential velocity component. When the inlet flow is swirl free, the absolute inlet velocity will be purely radial. The inlet blade angle may be specified for the design flow rate and pump speed to provide a smooth entry flow relative to the orientation of the blades.

The velocity diagram is constructed similarly at the outlet section. The runner speed at the outlet is $U_2 = r_2\omega$, which again is known from the geometry and operating speed of the turbomachine. The relative flow is assumed to leave the impeller tangent to the blades, as shown in Fig. 10.7*c*. This idealizing assumption of perfect guidance fixes the direction of the relative outlet flow at design conditions.

For a centrifugal pump or reaction turbine, the velocity relative to the blade generally changes in magnitude from inlet to outlet. The continuity equation must be applied, using the impeller geometry, to determine the normal component of velocity at each section. The normal component, together with the outlet blade angle, is sufficient to establish the velocity relative to the blade at the impeller outlet for a radial-flow machine. The velocity diagram is completed by the vector addition of the velocity relative to the blade and the wheel velocity, as shown in Fig. 10.7*c*.

The inlet and outlet velocity diagrams provide all the information needed to calculate the ideal torque or power, absorbed or delivered by the impeller, using Eqs. 10.1 or 10.2. The results represent the performance of a turbomachine under idealized conditions at the design operating point, since we have assumed:

- Negligible torque due to surface forces (viscous and pressure).
- Inlet and exit flow tangent to blades.
- Uniform flow at inlet and exit.

An actual turbomachine is not likely to conform to all of these assumptions, so the results of our analysis represent the upper limit of the performance of actual machines.

Performance of an actual machine may be estimated using the same basic approach, but accounting for variations in flow properties across the blade span at the inlet and outlet sections, as well as for deviations between the blade angles and the flow directions. Such detailed calculations are beyond the scope of this book. The alternative is to measure the overall performance of a machine on a suitable test stand. Manufacturers' data are examples of measured performance information.

In Example 10.1 we will use the Euler Turbomachine Equation to analyze an idealized centrifugal pump.

[2]The notation varies from book to book, so be careful when comparing references.

Example 10.1 IDEALIZED CENTRIFUGAL PUMP

A centrifugal pump is used to pump 0.009 m^3/s of water. The water enters the impeller axially through a 32 mm diameter inlet. The inlet velocity is axial and uniform. The impeller outlet diameter is 100 mm. Flow leaves the impeller at 3 m/s relative to the blades, which are radial at the exit. The impeller speed is 3450 rpm. Determine the impeller exit width, b_2, the torque input, and the power predicted by the Euler turbine equation.

Given: Flow as shown in the figure:
$V_{r_2} = 3$ m/s, $Q = 0.009$ m^3/s.

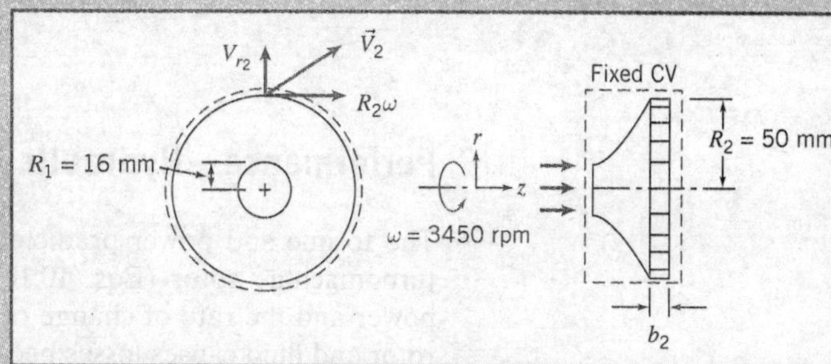

Find: (a) b_2.
(b) T_{shaft}.
(c) $\dot{W}_m$.

Solution:
Apply the Euler turbomachine equation to a fixed control volume.

Governing equations:

$$T_{shaft} = (r_2 V_{t_2} - r_1 V_{t_1})\dot{m} \qquad (10.1c)$$

$$\cancel{\frac{\partial}{\partial t}\int_{CV} \rho \, d\forall}^{\,=0(2)} + \int_{CS} \rho \vec{V} \cdot d\vec{A} = 0 \qquad (4.12)$$

Assumptions: (1) Neglect torques due to body and surface forces.
(2) Steady flow.
(3) Uniform flow at inlet and outlet sections.
(4) Incompressible flow.

Then, from continuity,

$$(-\rho V_1 \pi R_1^2) + (\rho V_{r_2} 2\pi R_2 b_2) = 0$$

or

$$\dot{m} = \rho Q = \rho V_{r_2} 2\pi R_2 b_2$$

so that

$$b_2 = \frac{Q}{2\pi R_2 V_{r_2}} = \frac{0.009 \text{ m}^3/\text{s}}{2\pi \times 50 \text{ mm} \times \frac{\text{m}}{1000 \text{ mm}} \times 3 \text{ m/s}}$$

$$b_2 = 9.5 \times 10^{-3} \text{m or } 9.5 \text{ mm}$$ b_2

For an axial inlet the tangential velocity $V_{t_1} = 0$, and for radial exit blades $V_{t_2} = R_2\omega$, so Eq. 10.1c reduces to

$$T_{shaft} = R_2^2 \omega \dot{m} = \omega R_2^2 \rho Q$$

where we have used continuity ($\dot{m} = \rho Q$).
Thus,

$$T_{shaft} = \omega R_2^2 \rho Q = 360 \frac{\text{rad}}{\text{s}} \times 8.1 \text{ N}\cdot\text{m} \times \frac{\text{J}}{\text{N}\cdot\text{m}} \times \frac{\text{W.s}}{\text{J}}$$

$$T_{shaft} = 8.1 \text{ N}\cdot\text{m}$$ T_{shaft}

and

$$\dot{W}_m = \omega T_{\text{shaft}} = 360\frac{\text{rad}}{\text{s}} \times (50\text{ mm})^2 \times \frac{\text{m}^2}{(1000\text{ mm})^2} \times 1000\frac{\text{kg}}{\text{m}^3} \times 0.009\frac{\text{m}^3}{\text{s}}$$

$$\dot{W}_m = 2.916 \times 10^3\text{ W} \longleftarrow \dot{W}_m$$

This problem illustrates the application of the Euler turbomachine equation for a fixed control volume to a centrifugal flow machine.

Performance—Hydraulic Power

The torque and power predicted by applying the angular-momentum equation to a turbomachine rotor (Eqs. 10.1c and 10.2a) are idealized values. In practice, rotor power and the rate of change of fluid energy are not equal. Energy *transfer* between rotor and fluid causes losses because of viscous effects, departures from uniform flow, and departures of flow direction from the blade angles. Kinetic energy *transformation* to pressure rise (which is ultimately the goal of these turbomachines) by diffusion in the fixed casing introduces more losses. Energy *dissipation* occurs in seals and bearings and in fluid friction between the rotor and housing of the machine ("windage" losses). Applying the first law of thermodynamics to a control volume surrounding the rotor shows that these "losses" in mechanical energy are irreversible conversions from mechanical energy to thermal energy. As was the case for the pipe flows discussed in Chapter 8, the thermal energy appears either as internal energy in the fluid stream or as heat transfer to the surroundings.

Because of these losses, in a pump the actual power delivered to the fluid is less than predicted by the angular-momentum equation. In the case of a turbine, the actual power delivered to the shaft is less than the power given up by the fluid stream.

We can define the power, head, and efficiency of a turbomachine based on whether the machine does work on the fluid or extracts work (or power) from the fluid.

For a pump, the *hydraulic power* is given by the rate of mechanical energy input to the fluid,

$$\dot{W}_h = \rho Q g H_p \tag{10.3a}$$

where

$$H_p = \left(\frac{p}{\rho g} + \frac{\overline{V}^2}{2g} + z\right)_{\text{discharge}} - \left(\frac{p}{\rho g} + \frac{\overline{V}^2}{2g} + z\right)_{\text{suction}} \tag{10.3b}$$

For a pump the *head* rise measured on a test stand is less than that produced by the impeller. The rate of mechanical energy input is greater than the rate of head rise produced by the impeller. The mechanical input power needed to drive the pump is related to the hydraulic power by defining *pump efficiency* as

$$\eta_p = \frac{\dot{W}_h}{\dot{W}_m} = \frac{\rho Q g H_p}{\omega T} \tag{10.3c}$$

To evaluate the actual change in head across a machine from Eq. 10.3b, we must know the pressure, fluid velocity, and elevation at two measurement sections. Fluid velocity can be calculated from the measured volume flow rate and passage diameters. (Suction and discharge lines for pumps usually have different inside diameters.)

Static pressure usually is measured in straight sections of pipe upstream from the pump inlet and downstream from the pump outlet, after diffusion has occurred within the pump casing. The elevation of each pressure gage may be recorded, or the static pressure readings may be corrected to the same elevation. (The pump centerline provides a convenient reference level.)

For a hydraulic turbine, the *hydraulic power* is defined as the rate of mechanical energy removal from the flowing fluid stream,

$$\dot{W}_h = \rho Q g H_t \tag{10.4a}$$

where

$$H_t = \left(\frac{p}{\rho g} + \frac{\overline{V}^2}{2g} + z\right)_{\text{inlet}} - \left(\frac{p}{\rho g} + \frac{\overline{V}^2}{2g} + z\right)_{\text{outlet}} \tag{10.4b}$$

For a hydraulic turbine, the power output obtained from the rotor (the mechanical power) is less than the rate of energy transfer from the fluid to the rotor, because the rotor must overcome friction and windage losses.

The mechanical power output obtained from the turbine is related to the hydraulic power by defining *turbine efficiency* as

$$\eta_t = \frac{\dot{W}_m}{\dot{W}_h} = \frac{\omega T}{\rho Q g H_t} \tag{10.4c}$$

Equations 10.4a and 10.4b show that *to obtain maximum power output from a hydraulic turbine, it is important to minimize the mechanical energy in the flow leaving the turbine*. This is accomplished by making the outlet pressure, flow speed, and elevation as small as practical. The turbine must be set as close to the tailwater level as possible, allowing for the level increase when the river floods. Tests to measure turbine efficiency may be performed at various output power levels and at different constant head conditions (see the discussion of Figs. 10.35 and 10.36).

Dimensional Analysis and Specific Speed

Dimensional analysis for turbomachines was introduced in Chapter 7, where dimensionless flow, head, and power coefficients were derived in generalized form. The independent parameters were the flow coefficient and a form of Reynolds number. The dependent parameters were the head and power coefficients.

Our objective here is to develop the forms of dimensionless coefficients in common use and to give examples illustrating their use in selecting a machine type, designing model tests, and scaling results. Since we developed an idealized theory for turbomachines, we can gain additional physical insight by developing dimensionless coefficients directly from the resulting computing equations. We will then apply these expressions to scaling of turbomachines through similarity rules in Section 10.3.

The dimensionless *flow coefficient*, Φ, is defined by normalizing the volume flow rate using the exit area and the wheel speed at the outlet. Thus

$$\Phi = \frac{Q}{A_2 U_2} = \frac{V_{n_2}}{U_2} \tag{10.5}$$

where V_{n_2} is the velocity component perpendicular to the exit area. This component is also referred to as the *meridional velocity* at the wheel exit plane. It appears in true projection in the *meridional plane*, which is any radial cross-section through the centerline of a machine.

A dimensionless head coefficient, Ψ, may be obtained by normalizing the head, H (Eq. 10.2c), with U_2^2/g. Thus

$$\Psi = \frac{gH}{U_2^2} \tag{10.6}$$

A dimensionless torque coefficient, τ, may be obtained by normalizing the torque, T (Eq. 10.1c), with $\rho A_2 U_2^2 R_2$. Thus

$$\tau = \frac{T}{\rho A_2 U_2^2 R_2} \tag{10.7}$$

Finally, the dimensionless power coefficient, Π, is obtained by normalizing the power, $\dot{W}$ (Eq. 10.2b), with $\dot{m}U_2^2 = \rho Q U_2^2$. Thus

$$\Pi = \frac{\dot{W}}{\rho Q U_2^2} = \frac{\dot{W}}{\rho \omega^2 Q R_2^2} \tag{10.8}$$

For pumps, mechanical input power exceeds hydraulic power, and the efficiency is defined as $\eta_p = \dot{W}_h/\dot{W}_m$ (Eq. 10.3c). Hence

$$\dot{W}_m = T\omega = \frac{1}{\eta_p}\dot{W}_h = \frac{\rho Q g H_p}{\eta_p} \tag{10.9}$$

Introducing dimensionless coefficients Φ (Eq. 10.5), Ψ (Eq. 10.6), and τ (Eq. 10.7) into Eq. 10.9, we obtain an analogous relation among the dimensionless coefficients as

$$\tau = \frac{\Psi\Phi}{\eta_p} \tag{10.10}$$

For turbines, mechanical output power is less than hydraulic power, and the efficiency is defined as $\eta_t = \dot{W}_m/\dot{W}_h$ (Eq. 10.4c). Hence

$$\dot{W}_m = T\omega = \eta_t \dot{W}_h = \eta_t \rho Q g H_p \tag{10.11}$$

Introducing dimensionless coefficients Φ, Ψ, and τ into Eq. 10.11, we obtain an analogous relation among the dimensionless coefficients as

$$\tau = \Psi\Phi\eta_t \tag{10.12}$$

The dimensionless coefficients form the basis for designing model tests and scaling the results. As shown in Chapter 7, the flow coefficient, Φ, is treated as the independent parameter. Then, if viscous effects are neglected, the head, torque, and power coefficients are treated as multiple dependent parameters. Under these assumptions, dynamic similarity is achieved when the flow coefficient is matched between model and prototype machines.

As discussed in Chapter 7, a useful parameter called *specific speed* can be obtained by combining the flow and head coefficients and eliminating the machine size. The result was

$$N_S = \frac{\omega Q^{1/2}}{h^{3/4}} \tag{7.22a}$$

When head is expressed as energy per unit mass (i.e., with dimensions equivalent to L^2/t^2, or g times head in height of liquid), and ω is expressed in radians per second, the specific speed defined by Eq. 7.22a is dimensionless.

Although specific speed is a dimensionless parameter, it is common practice to use an "engineering" equation form of Eq. 7.22a in which ω and Q are specified in units that are convenient but inconsistent, and energy per unit mass, h, is replaced with energy per unit weight of fluid, H. When this is done, the specific speed is not a unitless parameter and the magnitude of the specific speed depends on the units used to calculate it. Customary units used in U.S. engineering practice for pumps are rpm for ω, gpm for Q, and feet (energy per unit weight) for H. In practice, the symbol N is used to represent rate of rotation (ω) in rpm. Thus, the dimensional specific speed for pumps, expressed in U.S. customary units, as an "engineering" equation, becomes

$$N_{S_{cu}} = \frac{N(\text{rpm})[Q(\text{gpm})]^{1/2}}{[H(\text{ft})]^{3/4}} \tag{7.22b}$$

Values of the dimensionless specific speed, N_S (Eq. 7.22a), must be multiplied by 2733 to obtain the values of specific speed corresponding to this commonly used but inconsistent set of units (see Example 10.2).

For hydraulic turbines, we use the fact that power output is proportional to flow rate and head, $\mathcal{P} \propto \rho Q h$ in consistent units. Substituting $\mathcal{P}/\rho h$ for Q in Eq. 7.22a gives

$$N_S = \frac{\omega}{h^{3/4}}\left(\frac{\mathcal{P}}{\rho h}\right)^{1/2} = \frac{\omega \mathcal{P}^{1/2}}{\rho^{1/2} h^{5/4}} \tag{10.13a}$$

as the nondimensional form of the specific speed.

In U.S. engineering practice it is customary to drop the factor $\rho^{1/2}$ (water is invariably the working fluid in the turbines to which the specific speed is applied) and to use head H in place of energy per unit mass h. Customary units used in U.S. engineering practice for hydraulic turbines are rpm for ω, horsepower for $\mathcal{P}$, and feet for H. In practice, the symbol N is used to represent rate of rotation (ω) in rpm. Thus the dimensional specific speed for a hydraulic turbine, expressed in U.S. customary units, as an "engineering" equation, becomes

$$N_{S_{cu}} = \frac{N(\text{rpm})[\mathcal{P}(\text{hp})]^{1/2}}{[H(\text{ft})]^{5/4}} \tag{10.13b}$$

Values of the dimensionless specific speed for a hydraulic turbine, N_S (Eq. 10.13a), must be multiplied by 43.46 to obtain the values of specific speed corresponding to this commonly used but inconsistent set of units.

Specific speed may be thought of as the operating speed at which a pump produces unit head at unit volume flow rate (or, for a hydraulic turbine, unit power at unit head). To see this, solve for N in Eqs. 7.22b and 10.13b, respectively. For pumps

$$N(\text{rpm}) = N_{S_{cu}} \frac{[H(\text{ft})]^{3/4}}{[Q(\text{gpm})]^{1/2}}$$

and for hydraulic turbines

$$N(\text{rpm}) = N_{S_{cu}} \frac{[H(\text{ft})]^{5/4}}{[\mathcal{P}(\text{hp})]^{1/2}}$$

Holding specific speed constant describes all operating conditions of geometrically similar machines with similar flow conditions.

It is customary to characterize a machine by its specific speed at the design point. This specific speed has been found to characterize the hydraulic design features of a machine. Low specific speeds correspond to efficient operation of radial-flow machines. High specific speeds correspond to efficient operation of axial-flow machines. For a specified head and flow rate, one can choose either a low specific speed machine (which operates at low speed) or a high specific speed machine (which operates at higher speed).

Typical proportions for commercial pump designs and their variation with dimensionless specific speed are shown in Fig. 10.8. In this figure, the size of each machine has been adjusted to give the same head and flow rate for rotation at a speed corresponding to the specific speed. Thus it can be seen that if the machine's size and weight are critical, one should choose a higher specific speed. Figure 10.8 shows the trend from radial (purely centrifugal pumps), through mixed-flow, to axial-flow geometries as specific speed increases.

The corresponding efficiency trends for typical pumps are shown in Fig. 10.9, which shows that pump capacity generally increases as specific speed increases. The figure also shows that at any given specific speed, efficiency is higher for large pumps than for small ones. Physically this scale effect means that viscous losses become less important as the pump size is increased.

Characteristic proportions of hydraulic turbines also are correlated by specific speed, as shown in Fig. 10.10. As in Fig. 10.8, the machine size has been scaled in this illustration to deliver approximately the same power at unit head when rotating at a speed equal to the specific speed. The corresponding efficiency trends for typical turbine types are shown in Fig. 10.11.

Several variations of specific speed, calculated directly from engineering units, are widely used in practice. The most commonly used forms of specific speed for pumps are defined and compared in Example 10.2.

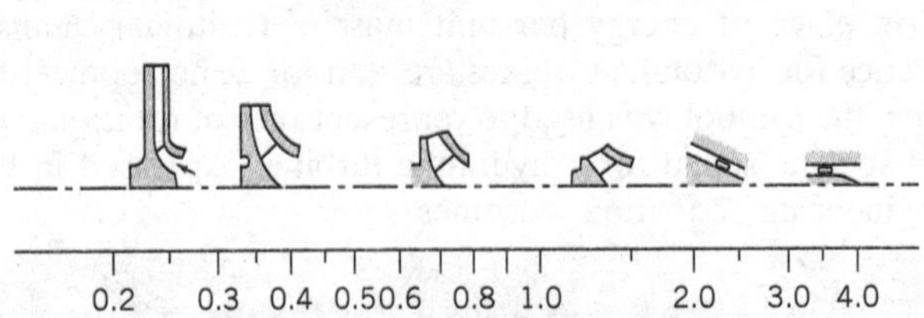

Fig. 10.8 Typical geometric proportions of commercial pumps as they vary with dimensionless specific speed [5].

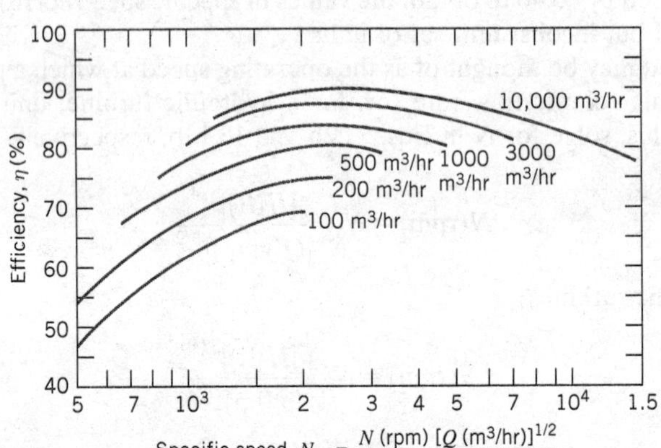

Fig. 10.9 Average efficiencies of commercial pumps as they vary with specific speed and pump size [6].

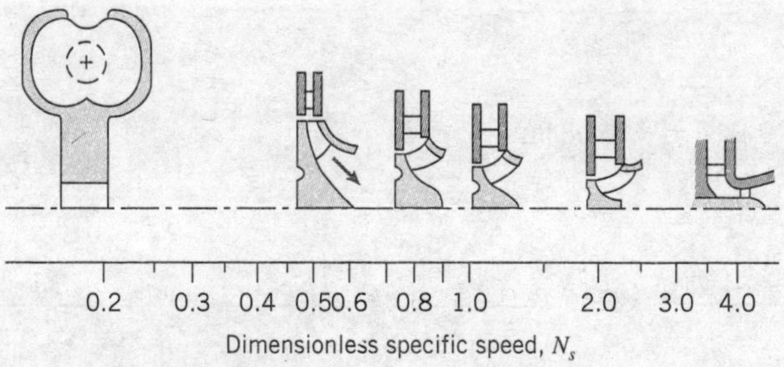

Fig. 10.10 Typical geometric proportions of commercial hydraulic turbines as they vary with dimensionless specific speed [5].

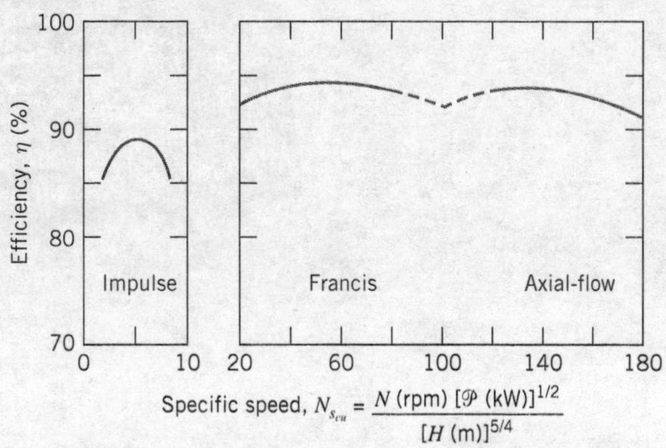

Fig. 10.11 Average efficiencies of commercial hydraulic turbines as they vary with specific speed [6].

Example 10.2 COMPARISON OF SPECIFIC SPEED DEFINITIONS

At the best efficiency point, a centrifugal pump, with impeller diameter D = 200 mm, produces H = 7 m at Q = 68 m^3/h with N = 1170 rpm. Compute the corresponding dimensionless and engineering specific speeds. Develop conversion factors to relate the specific speeds.

Given: Centrifugal pump at best efficiency point (BEP). Assume the pump characteristics are H = 7 m, Q = 68 m^3/h and N = 1170 rpm.

Find: (a) The specific speed in U.S. customary units.
(b) The specific speed in SI units.
(c) The specific speed in European units.
(d) Appropriate conversion factors to relate the specific speeds.

Solution:

Governing equations: $N_s = \dfrac{\omega Q^{1/2}}{h^{3/4}}$ and $N_{S_{cu}} = \dfrac{NQ^{1/2}}{H^{3/4}}$

From the given information, the specific speed in U.S. customary units is

$$N_{S_{cu}} = 1170\text{ rpm} \times (68)^{1/2}\left(\frac{\text{m}^3}{\text{h}}\right)^{1/2} \times \frac{1}{7^{3/4}\text{ m}^{3/4}} = 2242 \longleftarrow N_{S_{cu}}$$

The energy per unit mass is

$$h = gH = 9.81\,\frac{\text{m}}{\text{s}^2} \times 6.68\ \text{m} = 65.5\ \text{m}^2/\text{s}^2$$

The dimensionless specific speed is

$$N_s = 123\,\frac{\text{rad}}{\text{s}} \times (0.0190)^{1/2}\,\frac{\text{m}^{3/2}}{\text{s}^{1/2}} \times \frac{(\text{s}^2)^{3/4}}{(65.5)^{3/4}\,(\text{m}^2)^{3/4}} = 0.736 \qquad N_s$$

To relate the specific speeds, form the ratio:

$$\frac{N_{S_{cu}}}{N_s(\text{SI})} = \frac{2242}{0.736} = 3046$$

This problem demonstrates the use of "engineering" equations to calculate the dimensionless and engineering specific speed for pumps from each of three commonly used sets of units and to compare the results. (Three significant figures have been used for all calculations in this example. Slightly different results would be obtained if more significant figures were carried in intermediate calculations.)

10.3 Pumps, Fans, and Blowers

We will now look at the various types of fluid machines in greater detail. We will begin our discussion with rotating machines that perform work on an incompressible fluid, namely pumps, fans and blowers.

Application of Euler Turbomachine Equation to Centrifugal Pumps

As demonstrated in Example 10.1, the treatment from Section 10.2 may be applied directly to the analysis of centrifugal machines. Figure 10.7 in Section 10.2 represents the flow through a simple centrifugal pump impeller. If the fluid enters the impeller with a purely radial absolute velocity, then the fluid entering the impeller has no angular momentum and V_{t_1} is identically zero.

With $V_{t_1} = 0$, the increase in head (from Eq. 10.2c) is given by

$$H = \frac{U_2 V_{t_2}}{g} \tag{10.14}$$

From the exit velocity diagram of Fig. 10.7c,

$$V_{t_2} = U_2 - W_2 \cos \beta_2 = U_2 - \frac{V_{n_2}}{\sin \beta_2} \cos \beta_2 = U_2 - V_{n_2} \cot \beta_2 \tag{10.15}$$

Then

$$H = \frac{U_2^2 - U_2 V_{n_2} \cot \beta_2}{g} \tag{10.16}$$

For an impeller of width w, the volume flow rate is

$$Q = \pi D_2 w V_{n_2} \tag{10.17}$$

To express the increase in head in terms of volume flow rate, we substitute for V_{n_2} in terms of Q from Eq. 10.17. Thus

$$H = \frac{U_2^2}{g} - \frac{U_2 \cot \beta_2}{\pi D_2 w g} Q \tag{10.18a}$$

Equation 10.18a is of the form

$$H = C_1 - C_2 Q \tag{10.18b}$$

where constants C_1 and C_2 are functions of *machine geometry* and *speed*,

$$C_1 = \frac{U_2^2}{g} \quad \text{and} \quad C_2 = \frac{U_2 \cot \beta_2}{\pi D_2 w g}$$

Thus Eq. 10.18a predicts a linear variation of head, H, with volume flow rate, Q. Note that this linear relation is an idealized model; actual devices may have only an approximate linear variation and may be better modeled with a curve-fitting method based on measured data. (We will see an example of this in Example 10.5.)

Constant $C_1 = U_2^2/g$ represents the ideal head developed by the pump for zero flow rate; this is called the *shutoff head*. The slope of the curve of head versus flow rate (the $H - Q$ curve) depends on the sign and magnitude of C_2.

For radial outlet vanes, $\beta_2 = 90°$ and $C_2 = 0$. The tangential component of the absolute velocity at the outlet is equal to the wheel speed and is independent of flow rate. From Eq. 10.18a, the ideal head is independent of flow rate. This characteristic $H - Q$ curve is plotted in Fig. 10.12.

If the vanes are backward curved (as shown in Fig. 10.7*a*), $\beta_2 < 90°$ and $C_2 > 0$. Then the tangential component of the absolute outlet velocity is less than the wheel speed and it decreases in proportion to the flow rate. From Eq. 10.18a, the ideal head decreases linearly with increasing flow rate. The corresponding $H - Q$ curve is plotted in Fig. 10.12.

If the vanes are forward curved, then $\beta_2 > 90°$ and $C_2 < 0$. The tangential component of the absolute fluid velocity at the outlet is greater than the wheel speed, and it increases as the flow rate increases. From Eq. l0.7a, the ideal head increases linearly with increasing flow rate. The corresponding $H - Q$ curve is plotted in Fig. 10.12.

The characteristics of a radial-flow machine can be altered by changing the outlet vane angle; the idealized model developed above predicts the trends as the outlet vane angle is changed.

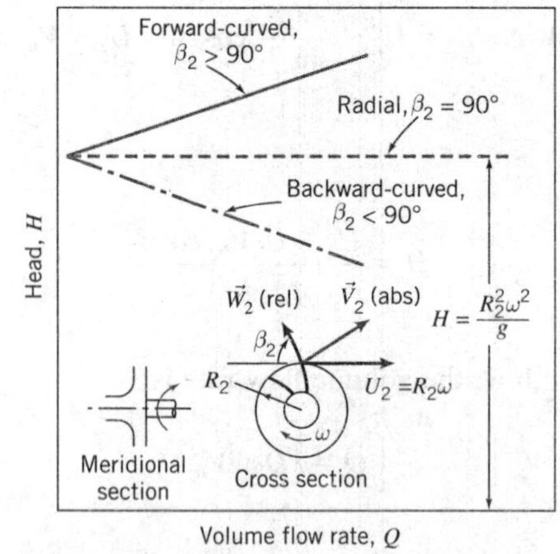

Fig. 10.12 Idealized relationship between head and volume flow rate for centrifugal pump with forward-curved, radial, and backward-curved impeller blades.

The predictions of the idealized angular-momentum theory for a centrifugal pump are summarized in Fig. 10.12. Forward-curved vanes are almost never used in practice because they tend to have an unstable operating point.

Application of the Euler Equation to Axial Flow Pumps and Fans

The Euler Turbomachine Equation developed in Section 10.2 can be used for axial-flow machines as well. However, in order to use this model, some assumptions need to be made. The most important assumption is that the flow properties at the mean radius (the midpoint of the rotor blades) fully represent the flow at all radii. This is a good assumption, provided the ratio of blade height to mean radius is approximately 0.2 or less [7]. At larger ratios a three-dimensional analysis will be necessary. Such an analysis is beyond the scope of this work, but other sources can provide information on this phenomenon, such as Dixon [7]. A second assumption is that there is no radial component to the flow velocity. This is a reasonable assumption, since many axial machines incorporate stators or sets of vanes which guide the flow into the machine, removing unwanted radial velocity components. The third assumption is that the flow only varies in the axial direction. *This is not the same as saying that there is only an axial component of velocity!* In fact, there will be a significant component of the velocity in the tangential direction as the flow passes through an axial-flow machine, i.e., the flow will have "swirl." The meaning of this assumption is that at a given axial location, the amount of swirl in the flow is constant, rather than varying between the blades of the machine [7].

The primary consequence of this model applied to axial-flow machines is that the radius used in Equations (10.1) is constant, i.e.,

$$r_1 = r_2 = R_m \tag{10.19a}$$

Since the angular velocity ω of the rotor is also constant, it follows that

$$U_1 = U_2 = U \tag{10.19b}$$

Therefore, Equations (10.1) and (10.2) reduce to:

$$T_{shaft} = R_m(V_{t_2} - V_{t_1})\dot{m} \tag{10.20}$$

$$\dot{W}_m = U(V_{t_2} - V_{t_1})\dot{m} \tag{10.21}$$

$$H = \frac{\dot{W}_m}{\dot{m}g} = \frac{U}{g}(V_{t_2} - V_{t_1}) \tag{10.22}$$

In Example 10.3 these special versions of the Euler turbomachine equation and velocity diagrams are utilized in the analysis of flow through an axial-flow fan.

Example 10.3 IDEALIZED AXIAL-FLOW FAN

An axial-flow fan operates at 1200 rpm. The blade tip diameter is 1.1 m and the hub diameter is 0.8 m. The inlet and exit angles at the mean blade radius are 30° and 60°, respectively. Inlet guide vanes give the absolute flow entering the first stage an angle of 30°. The fluid is air at standard conditions and the flow may be considered incompressible. There is no change in axial component of velocity across the rotor. Assume the relative flow enters and leaves the rotor at the geometric blade angles and use properties at the mean blade radius for calculations. For these idealized conditions, draw the inlet velocity diagram, determine the volume flow rate of the fan, and sketch the rotor blade shapes. Using the data so obtained, draw the outlet velocity diagram and calculate the minimum torque and power needed to drive the fan.

Given: Flow through rotor of axial-flow fan.

Tip diameter: 1.1 m
Hub diameter: 0.8 m
Operating speed: 1200 rpm
Absolute inlet angle: 30°
Blade inlet angle: 30°
Blade outlet angle: 60°

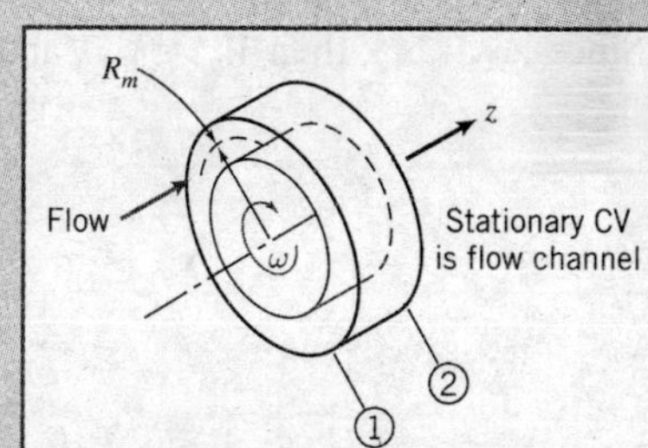

Fluid is air at standard conditions. Use properties at mean diameter of blades.

Find: (a) Inlet velocity diagram.
(b) Volume flow rate.
(c) Rotor blade shape.
(d) Outlet velocity diagram.
(e) Rotor torque.
(f) Power required.

Solution: Apply the Euler turbomachine equation to a fixed control volume.

Governing equation:

$$T_{shaft} = R_m(V_{t_2} - V_{t_1})\dot{m} = R_m(V_{t_2} - V_{t_1})\rho Q \tag{10.20}$$

Assumptions:
(1) Neglect torques due to body or surface forces.
(2) Steady flow.
(3) Uniform flow at inlet and outlet sections.
(4) Incompressible flow.
(5) No change in axial flow area.
(6) Use mean radius of rotor blades, R_m.

The blade shapes are

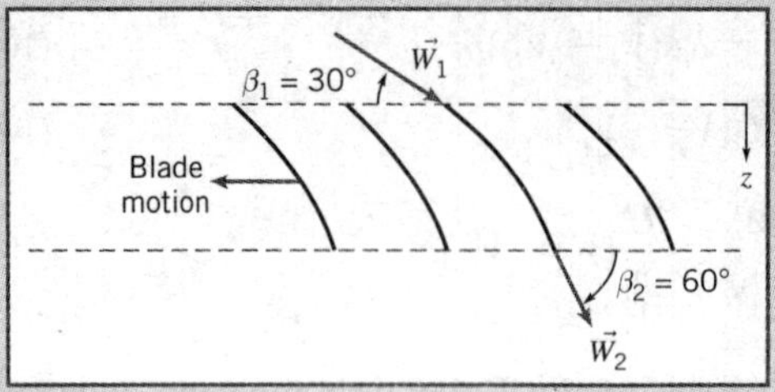

(Note that for an axial-flow machine the normal velocity components are parallel to the axis, not normal to the circumferential surface!)

The inlet velocity diagram is

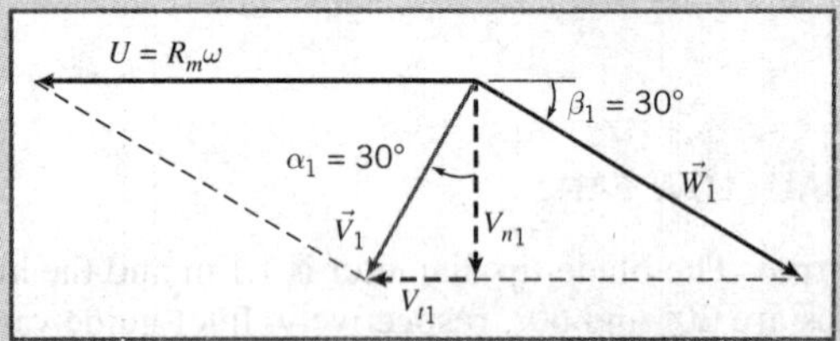

From continuity

$$(-\rho V_{n_1} A_1) + (\rho V_{n_2} A_2) = 0$$

or

$$Q = V_{n_1} A_1 = V_{n_2} A_2$$

Since $A_1 = A_2$, then $V_{n_1} = V_{n_2}$, and the outlet velocity diagram is as shown in the following figure:

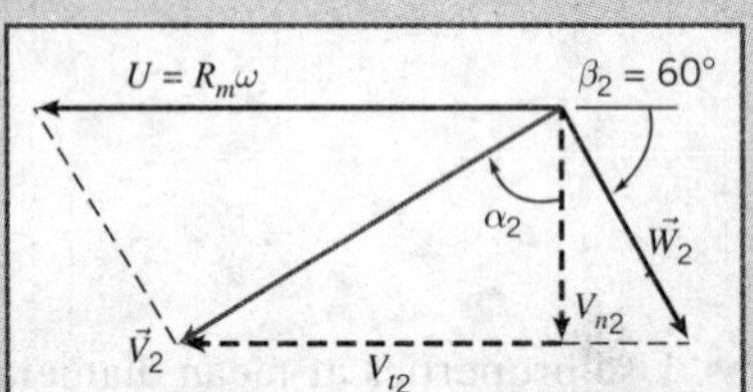

At the mean blade radius,

$$U = R_m \omega = \frac{D_m}{2}\omega$$

$$U = \frac{\frac{1}{2}(1.1 + 0.8)\text{m}}{2} \times 1200\frac{\text{rev}}{\text{min}} \times 2\pi\frac{\text{rad}}{\text{rev}} \times \frac{\text{min}}{60\text{ s}} = 59.7\text{ m/s}$$

From the geometry of the inlet velocity diagram,

$$U = V_{n_1}(\tan\alpha_1 + \cot\beta_1)$$

so that

$$V_{n_1} = \frac{U}{\tan\alpha_1 + \cot\beta_1} = 59.7\frac{\text{m}}{\text{s}} \times \frac{1}{\tan 30° + \cot 30°} = 25.9\text{ m/s}$$

Consequently,

$$V_1 = \frac{V_{n_1}}{\cos \alpha_1} = 25.9\,\frac{\text{m}}{\text{s}} \times \frac{1}{\cos 30^\circ} = 29.9\ \text{m/s}$$

$$V_{t_1} = V_1 \sin \alpha_1 = 29.9\,\frac{\text{m}}{\text{s}} \times \sin 30^\circ = 15.0\ \text{m/s}$$

and

$$W_1 = \frac{V_{n_1}}{\sin \beta_1} = 25.9\,\frac{\text{m}}{\text{s}} \times \frac{1}{\sin 30^\circ} = 51.8\ \text{m/s}$$

The volume flow rate is

$$Q = V_{n_1} A_1 = \frac{\pi}{4} V_{n_1}(D_t^2 - D_h^2) = \frac{\pi}{4} \times 25.9\,\frac{\text{m}}{\text{s}}\,[(1.1)^2 - (0.8)^2]\text{m}^2$$

$$Q = 11.6\ \text{m}^3/\text{s} \qquad Q$$

From the geometry of the outlet velocity diagram,

$$\tan \alpha_2 = \frac{V_{t_2}}{V_{n_2}} = \frac{U - V_{n_2} \cot \beta_2}{V_{n_2}} = \frac{U - V_{n_1} \cot \beta_2}{V_{n_1}}$$

or

$$\alpha_2 = \tan^{-1}\left[\frac{59.7\,\frac{\text{m}}{\text{s}} - 25.9\,\frac{\text{m}}{\text{s}} \times \cot 60^\circ}{25.9\,\frac{\text{m}}{\text{s}}}\right] = 59.9^\circ$$

and

$$V_2 = \frac{V_{n_2}}{\cos \alpha_2} = \frac{V_{n_1}}{\cos \alpha_2} = 25.9\,\frac{\text{m}}{\text{s}} \times \frac{1}{\cos 59.9^\circ} = 51.6\ \text{m/s}$$

Finally,

$$V_{t_2} = V_2 \sin \alpha_2 = 51.6\,\frac{\text{m}}{\text{s}} \times \sin 59.9^\circ = 44.6\ \text{m/s}$$

Applying Eq. 10.20

$$T_{\text{shaft}} = \rho Q R_{\text{m}} (V_{t_2} - V_{t_1})$$

$$= 1.23\,\frac{\text{kg}}{\text{m}^3} \times 11.6\,\frac{\text{m}^3}{\text{s}} \times \frac{0.95}{2}\,\text{m} \times (44.6 - 15.0)\,\frac{\text{m}}{\text{s}} \times \frac{\text{N}\cdot\text{s}^2}{\text{kg}\cdot\text{m}}$$

$$T_{\text{shaft}} = 201\ \text{N}\cdot\text{m} \qquad T_{\text{shaft}}$$

Thus the torque *on* the CV is in the same sense as $\vec{\omega}$. The power required is

$$\dot{W}_m = \vec{\omega}\cdot\vec{T} = \omega T_{\text{shaft}} = 1200\,\frac{\text{rev}}{\text{min}} \times 2\pi\,\frac{\text{rad}}{\text{rev}} \times \frac{\text{min}}{60\ \text{s}} \times 201\ \text{N}\cdot\text{m} \times \frac{\text{W}\cdot\text{s}}{\text{N}\cdot\text{m}}$$

$$\dot{W}_m = 25.3\ \text{kW} \qquad \dot{W}_m$$

This problem illustrates construction of velocity diagrams and application of the Euler turbomachine equation for a fixed control volume to an axial-flow machine under idealized conditions.

Performance Characteristics

To specify fluid machines for flow systems, the designer must know the pressure rise (or head), torque, power requirement, and efficiency of a machine. For a given machine, each of these characteristics is a function of flow rate; the characteristics for similar machines depend on size and operating speed. Here we define *performance characteristics* for pumps and turbines and review experimentally measured trends for typical machines.

The idealized analyses presented in Section 10.2 are useful to predict trends and to approximate the design-point performance of an energy-absorbing or an energy-producing machine. However, the complete performance of a real machine, including operation at off-design conditions, must be determined experimentally.

To determine performance, a pump, fan, blower, or compressor must be set up on an instrumented test stand with the capability of measuring flow rate, speed, input torque, and pressure rise. The test must be performed according to a standardized procedure corresponding to the machine being tested [8, 9]. Measurements are made as flow rate is varied from shutoff (zero flow) to maximum delivery by varying the load from maximum to minimum (by starting with a valve that is closed and opening it to fully open in stages). Power input to the machine is determined from a calibrated motor or calculated from measured speed and torque, and then efficiency is computed as illustrated in Example 10.4. Finally, the calculated characteristics are plotted in the desired engineering units or nondimensionally. If appropriate, smooth curves may be faired through the plotted points or curve-fits may be made to the results, as illustrated in Example 10.5.

Example 10.4 CALCULATION OF PUMP CHARACTERISTICS FROM TEST DATA

The flow system used to test a centrifugal pump at a nominal speed of 1750 rpm is shown. The liquid is water at 27°C, and the suction and discharge pipe diameters are 150 mm. Data measured during the test are given in the table. The electric motor is supplied at 460 V, 3-phase, and has a power factor of 0.875 and a constant efficiency of 90 percent.

Rate of Flow (m^3/h)	Suction Pressure (kPa-gage)	Discharge Pressure (kPa-gage)	Motor Current (amp)
0	−25	377	18.0
114	−29	324	25.1
182	−32	277	30.0
227	−39	230	32.6
250	−43	207	34.1
273	−46	179	35.4
318	−53	114	39.0
341	−58	69	40.9

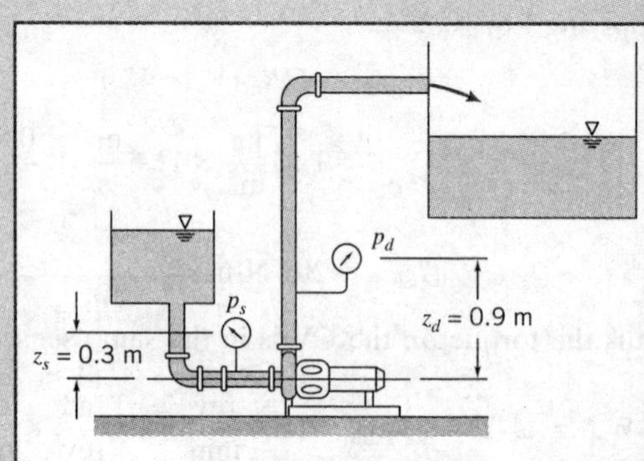

Calculate the net head delivered and the pump efficiency at a volume flow rate of 227 m^3/h. Plot the pump head, power input, and efficiency as functions of volume flow rate.

Given: Pump test flow system and data shown.

Find: (a) Pump head and efficiency at $Q = 227\ m^3/h$.
(b) Pump head, power input, and efficiency as a function of volume flow rate. Plot the results.

Solution:

Governing equations:

$$\dot{W}_h = \rho Q g H_p \qquad \eta_p = \frac{\dot{W}_h}{\dot{W}_m} = \frac{\rho Q g H_p}{\omega T}$$

$$H_p = \left(\frac{p}{\rho g} + \frac{\overline{V}^2}{2g} + z\right)_d - \left(\frac{p}{\rho g} + \frac{\overline{V}^2}{2g} + z\right)_s$$

Assumptions: (1) Steady flow.
(2) Uniform flow at each section.
(3) $\overline{V}_2 = \overline{V}_1$.
(4) Correct all heads to the same elevation.

Since $\overline{V}_1 = \overline{V}_2$, the pump head is

$$H_p = \frac{1}{g}\left[\left(\frac{p}{\rho} + gz\right)_d - \left(\frac{p}{\rho} + gz\right)_s\right] = \frac{p_2 - p_1}{\rho g}$$

where the discharge and suction pressures, *corrected to the same elevation*, are designated p_2 and p_1, respectively.
Correct measured static pressures to the pump centerline:

$$p_1 = p_s + \rho g z_s$$

$$p_1 = -39 \times 10^3 \text{ Pa} + 1000 \text{ kg/m}^3 \times 9.8 \text{ m/s}^2 \times 0.3 \text{ m} \times \frac{\text{N}\cdot\text{s}^2}{\text{kg}\cdot\text{m}} \times \frac{\text{Pa}\cdot\text{m}^2}{\text{N}} = -36.03 \times 10^3 \text{ Pa}$$

and

$$p_2 = p_d + \rho g z_d$$

$$p_2 = 230 \times 10^3 \text{ Pa} + 1000 \text{ kg/m}^3 \times 9.8 \text{m/s}^2 \times 0.9 \text{ m} \times \frac{\text{N}\cdot\text{s}^2}{\text{kg}\cdot\text{m}} \times \frac{\text{Pa}\cdot\text{m}^2}{\text{N}} = 238.82 \times 10^3 \text{ Pa}$$

Calculate the pump head:

$$H_p = (p_2 - p_1)/\rho g$$

$$H_p = \frac{[238.82 - (-36.06)] \times 10^3 \text{ Pa}}{1000 \text{ kg/m}^3 \times 9.8 \text{ m/s}^2} \times \frac{\text{kg}\cdot\text{m}}{\text{N}\cdot\text{s}^2} \times \frac{\text{N}}{\text{Pa}\cdot\text{m}^2} = 28.05 \text{ m} \longleftarrow H_p$$

Compute the hydraulic power delivered to the fluid:

$$\dot{W}_h = \rho Q g H_p = Q(p_2 - p_1)$$

$$= 227 \frac{\text{m}^3}{\text{h}} \times \frac{\text{h}}{3600\text{s}} \times [238.82 - (-36.06)] \times 10^3 \text{ Pa} \times \frac{\text{N}}{\text{Pa}\cdot\text{m}^2} \times \frac{\text{J}}{\text{N}\cdot\text{m}} \times \frac{\text{W}\cdot\text{s}}{\text{J}}$$

$$\dot{W}_h = 17.3 \times 10^3 \text{ W}$$

Calculate the motor power output (the mechanical power input to the pump) from electrical information:

$$\mathcal{P}_{\text{in}} = \eta\sqrt{3}(PF)EI$$

$$\mathcal{P}_{\text{in}} = 0.90 \times \sqrt{3} \times 0.875 \times 460 \text{ V} \times 32.6 \text{ A} \times \frac{\text{W}}{\text{VA}} = 20.5 \times 10^3 \text{ W}$$

The corresponding pump efficiency is

$$\eta_p = \frac{\dot{W}_h}{\dot{W}_m} = \frac{17.3 \times 10^3 \text{ W}}{28.5 \times 10^3 \text{ W}} = 0.844 \quad \text{or} \quad 84.4 \text{ percent} \longleftarrow \eta_p$$

Results from similar calculations at the other volume flow rates are plotted below:

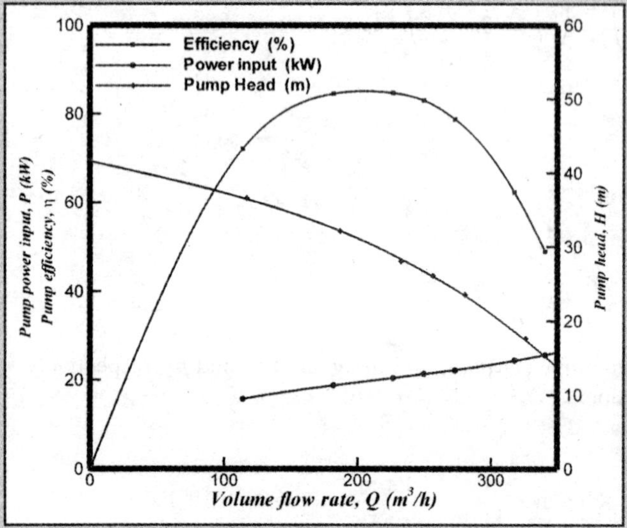

This problem illustrates the data reduction procedure used to obtain the performance curves for a pump from experimental data. The results calculated and plotted in this example are typical for a centrifugal pump driven at constant speed:

- ✓ The pressure rise is highest at shutoff (zero flow rate).
- ✓ Pressure rise decreases steadily as flow rate is increased; compare this typical experimental curve to the linear behavior predicted by Eq. 10.18b, and shown in Fig. 10.12, for idealized backward-curved impeller blades (used in most centrifugal pumps).
- ✓ Required power input increases with flow rate; the increase is generally nonlinear.
- ✓ Efficiency is zero at shutoff, rises to a peak as flow rate is increased, then drops off at larger flow rates; it stays near its maximum over a range of flow rates (in this example, from about 180 to 220 m^3/h).

This example is a little oversimplified because it is assumed that the electric motor efficiency is constant. In practice, motor efficiency varies with load, so must be either computed at each load from motor speed and torque measurements, or obtained from a calibration curve.

The *Excel* workbook for this Example was used for the calculations for each flow rate, and for generating the graph. It can be modified for use with other pump data.

Example 10.5 CURVE-FIT TO PUMP PERFORMANCE DATA

Pump test data were given and performance was calculated in Example 10.4. Fit a parabolic curve, $H = H_0 - AQ^2$, to these calculated pump performance results and compare the fitted curve with the measured data.

Given: Pump test data and performance calculated in Example 10.4.

Find: (a) Parabolic curve, $H = H_0 - AQ^2$, fitted to the pump performance data.
(b) Comparison of the curve-fit with the calculated performance.

Solution:
The curve-fit may be obtained by fitting a linear curve to H versus Q^2. Tabulating,

From calculated performance:			From the curve fit:	
Q (m^3/h)	Q^2 (m^6/h^2)	H (m)	H (m)	Error (%)
0	0	41.6	40.4	−3.0
114	1.3×10^4	36.6	37.4	2.1
182	3.3×10^4	32.1	32.8	2.0
227	5.2×10^4	28.0	28.6	1.8
250	6.3×10^4	26.1	26.0	−0.3
273	7.5×10^4	23.6	23.3	−1.2
318	10.1×10^4	17.6	17.2	−2.7
341	11.6×10^4	13.6	13.7	1.0
			Intercept =	40.4
			Slope =	-23×10^{-4}
			r^2 =	0.995

Using the method of least squares, the equation for the fitted curve is obtained as

$$H(\text{m}) = 40.4 - 2.3 \times 10^{-4}\ [Q(\text{m}^3/\text{h})]^2$$

with coefficient of determination $r^2 = 0.995$. (The closer r^2 is to unity, its maximum possible value, the better the fit.)

Always compare the results of a curve-fit with the data used to develop the fit. The figure shows the curve-fit (the solid line) and the experimental values (the points).

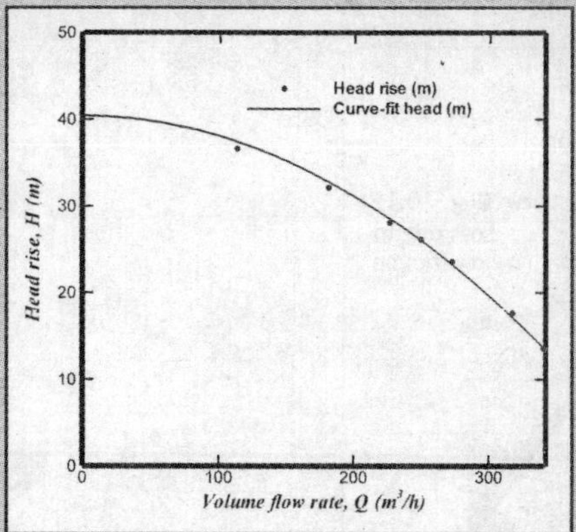

This problem illustrates that the pump test data for Example 10.4 can be fitted quite well to a parabolic curve. As with fitting a curve to any experimental data, our justifications for choosing a parabolic function in this case are:

✓ Experimental observation—the experimental data *looks* parabolic.

✓ Theory or concept—we will see later in this section that similarity rules suggest such a relation between head and flow rate.

The *Excel* workbook for this Example was used for the least-squares calculations, and for generating the graph. It can be modified for use with other pump data.

The basic procedure used to calculate machine performance was illustrated for a centrifugal pump in Example 10.4. The difference in static pressures between the pump suction and discharge was used to calculate the head rise produced by the pump. For pumps, dynamic pressure rise (or fluid kinetic energy change) typically is a small fraction of the head rise developed by the pump, so it may be neglected compared with the head rise.

Typical characteristic curves for a centrifugal pump tested at constant speed were shown qualitatively in Fig. 7.5;[3] the head versus capacity curve is reproduced in Fig. 10.13 to compare with characteristics predicted by the idealized analysis. Figure 10.13 shows that the head at any flow rate in the real machine may be significantly lower than is predicted by the idealized analysis. Some of the causes are:

1. At very low flow rate, some fluid recirculates in the impeller.
2. Friction loss and leakage loss both increase with flow rate.
3. "Shock loss" results from a mismatch between the direction of the relative velocity and the tangent to the impeller blade at the inlet.[4]

Curves such as those in Figs. 7.5 and 10.13 are measured at constant (design) speed with a single impeller diameter. It is common practice to vary pump capacity by changing the impeller size in a given casing. To present information compactly, data from tests of several impeller diameters may be plotted on a single graph, as shown in Fig. 10.14. As before, for each diameter, head is plotted versus flow rate; each curve is labeled with the corresponding diameter. Efficiency contours are plotted by joining points having the same constant efficiency. Power-requirement contours are also plotted. Finally, the *NPSH* requirements (which we have not yet defined; we will discuss its meaning later in this section) are shown for the extreme diameters; in Fig. 10.14, the curve for the 200 mm impeller lies between the curves for the 150 mm and 250 mm impellers.

With the advent of computer-aided analyses, the data of Fig. 10.14 are often tabulated for quick access by computer codes. Therefore, data are not always presented in the manner shown in this figure. Specifically, the data of Fig. 10.14 are simplified by reporting an average efficiency as a function of the flow rate only, as shown in Fig. 10.15, rather than as a function of flow rate and head. The figures in Appendix D display pump performance in this format.

For this typical machine, head is a maximum at shutoff and decreases continuously as flow rate increases. Input power is minimum at shutoff and increases as

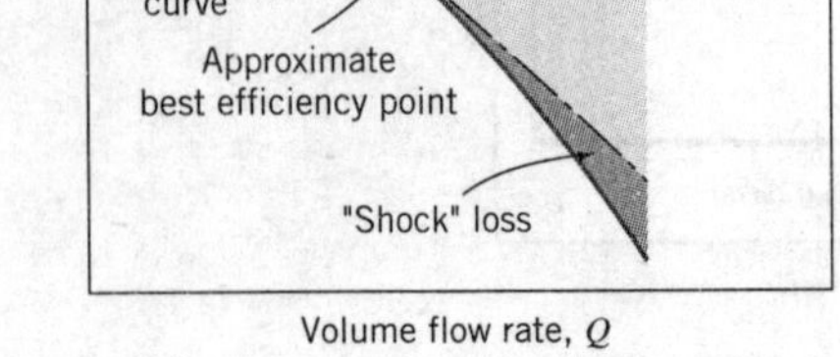

Fig. 10.13 Comparison of ideal and actual head-flow curves for a centrifugal pump with backward-curved impeller blades [10].

[3]The only important pump characteristic not shown in Fig. 7.5 is the net positive suction head (NPSH) required to prevent cavitation. Cavitation and NPSH will be treated later in this section.

[4]This loss is largest at high and low flow rates; it decreases essentially to zero as optimum operating conditions are approached [11].

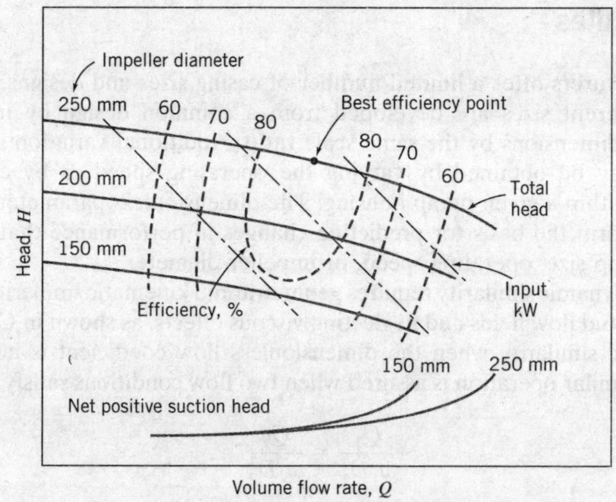

Fig. 10.14 Typical pump performance curves from tests with three impeller diameters at constant speed [10].

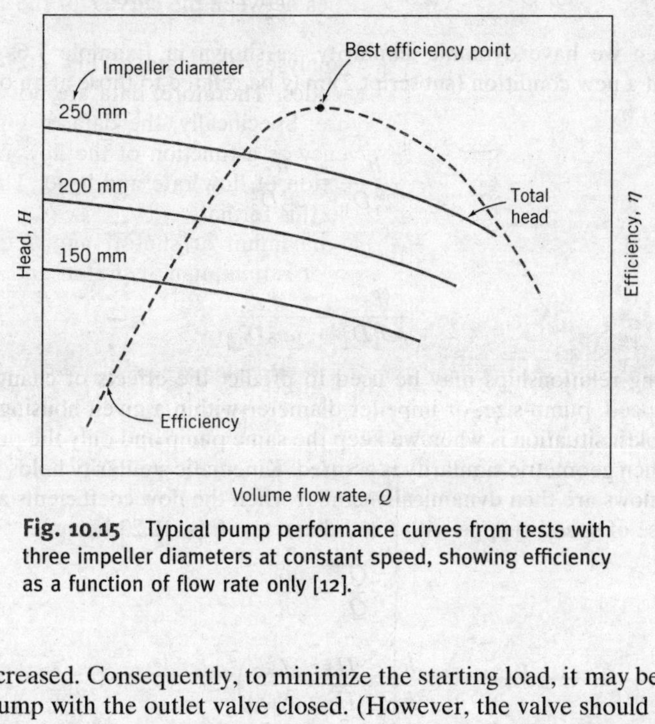

Fig. 10.15 Typical pump performance curves from tests with three impeller diameters at constant speed, showing efficiency as a function of flow rate only [12].

delivery is increased. Consequently, to minimize the starting load, it may be advisable to start the pump with the outlet valve closed. (However, the valve should not be left closed for long, lest the pump overheat as energy dissipated by friction is transferred to the water in the housing.) Pump efficiency increases with capacity until the *best efficiency point* (BEP) is reached, then decreases as flow rate is increased further. For minimum energy consumption, it is desirable to operate as close to BEP as possible.

Centrifugal pumps may be combined in parallel to deliver greater flow or in series to deliver greater head. A number of manufacturers build multistage pumps, which are essentially several pumps arranged in series within a single casing. Pumps and blowers are usually tested at several constant speeds. Common practice is to drive machines with electric motors at nearly constant speed, but in some system applications impressive energy savings can result from variable-speed operation. These pump application topics are discussed later in this section.

Similarity Rules

Pump manufacturers offer a limited number of casing sizes and designs. Frequently, casings of different sizes are developed from a common design by increasing or decreasing all dimensions by the same scale ratio. Additional variation in characteristic curves may be obtained by varying the operating speed or by changing the impeller size within a given pump housing. The dimensionless parameters developed in Chapter 7 form the basis for predicting changes in performance that result from changes in pump size, operating speed, or impeller diameter.

To achieve dynamic similarity requires geometric and kinematic similarity. Assuming similar pumps and flow fields and neglecting viscous effects, as shown in Chapter 7, we obtain dynamic similarity when the dimensionless flow coefficient is held constant. Dynamically similar operation is assured when two flow conditions satisfy the relation

$$\frac{Q_1}{\omega_1 D_1^3} = \frac{Q_2}{\omega_2 D_2^3} \tag{10.23a}$$

The dimensionless head and power coefficients depend only on the flow coefficient, i.e.,

$$\frac{h}{\omega^2 D^2} = f_1\left(\frac{Q}{\omega D^3}\right) \quad \text{and} \quad \frac{\mathcal{P}}{\rho \omega^3 D^5} = f_2\left(\frac{Q}{\omega D^3}\right)$$

Hence, when we have dynamic similarity, as shown in Example 7.6, pump characteristics at a new condition (subscript 2) may be related to those at an old condition (subscript 1) by

$$\frac{h_1}{\omega_1^2 D_1^2} = \frac{h_2}{\omega_2^2 D_2^2} \tag{10.23b}$$

and

$$\frac{\mathcal{P}_1}{\rho \omega_1^3 D_1^5} = \frac{\mathcal{P}_2}{\rho \omega_2^3 D_2^5} \tag{10.23c}$$

These scaling relationships may be used to predict the effects of changes in pump operating speed, pump size, or impeller diameter within a given housing.

The simplest situation is when we keep the same pump and only the pump speed is changed. Then geometric similarity is assured. Kinematic similarity holds if there is no cavitation; flows are then dynamically similar when the flow coefficients are matched. For this case of speed change with fixed diameter, Eqs. 10.23 become

$$\frac{Q_2}{Q_1} = \frac{\omega_2}{\omega_1} \tag{10.24a}$$

$$\frac{h_2}{h_1} = \frac{H_2}{H_1} = \left(\frac{\omega_2}{\omega_1}\right)^2 \tag{10.24b}$$

$$\frac{\mathcal{P}_2}{\mathcal{P}_1} = \left(\frac{\omega_2}{\omega_1}\right)^3 \tag{10.24c}$$

In Example 10.5, we showed that a pump performance curve may be modeled within engineering accuracy by the parabolic relationship,

$$H = H_0 - AQ^2 \tag{10.25a}$$

Since this representation contains two parameters, the pump curve for the new operating condition could be derived by scaling any two points from the performance curve measured at the original operating condition. Usually, the *shutoff condition* and

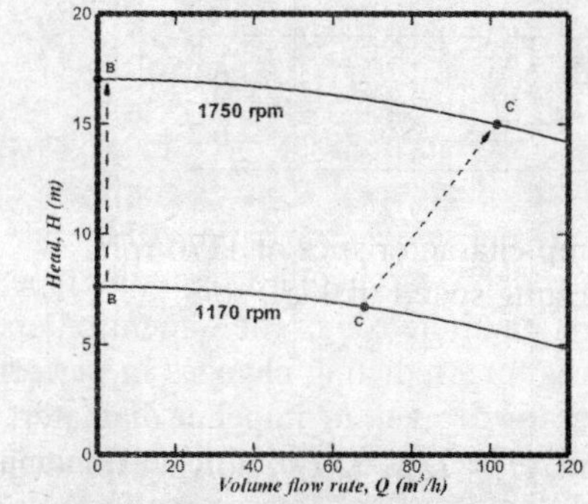

Fig. 10.16 Schematic of a pump performance curve, illustrating the effect of a change in pump operating speed.

the *best efficiency point* are chosen for scaling. These points are represented by points B and C in Fig. 10.16.

As shown by Eq. 10.24a, the flow rate increases by the ratio of operating speeds, so

$$Q_{B'} = \frac{\omega_2}{\omega_1} Q_B = 0 \quad \text{and} \quad Q_{C'} = \frac{\omega_2}{\omega_1} Q_C$$

Thus, point B' is located directly above point B, and point C' moves to the right of point C (in this example $\omega_2 > \omega_1$).

The head increases by the square of the speed ratio, so

$$H_{B'} = H_B \left(\frac{\omega_1}{\omega_2}\right)^2 \quad \text{and} \quad H_{C'} = H_C \left(\frac{\omega_2}{\omega_1}\right)^2$$

Points C and C', where dynamically similar flow conditions are present, are termed *homologous* points for the pump.

We can relate the old operating condition (e.g., running at speed $N_1 = 1170$ rpm, as shown in Fig. 10.16) to the new, primed one (e.g., running at speed $N_2 = 1750$ rpm in Fig. 10.16) using the parabolic relation and Eqs. 10.24a and 10.24b,

$$H = H'\left(\frac{\omega_1}{\omega_2}\right)^2 = H_0 - AQ^2 = H_0'\left(\frac{\omega_1}{\omega_2}\right)^2 - AQ'^2\left(\frac{\omega_1}{\omega_2}\right)^2$$

or

$$H' = H_0' - AQ'^2 \qquad (10.25b)$$

so that for a given pump the factor A remains unchanged as we change pump speed (as we will verify in Example 10.6).

Efficiency remains relatively constant between dynamically similar operating points when only the pump operating speed is changed. Application of these ideas is illustrated in Example 10.6.

Example 10.6 SCALING PUMP PERFORMANCE CURVES

When operated at $N = 1170$ rpm, a centrifugal pump, with impeller diameter $D = 200$ mm, has shutoff head $H_0 = 7.6$ m of water. At the same operating speed, best efficiency occurs at $Q = 68\ \text{m}^3/\text{h}$, where the head is $H = 6.7$ m of water. Fit these data at 1170 rpm with a parabola. Scale the results to a new operating speed of 1750 rpm. Plot and compare the results.

Given: Centrifugal pump (with $D = 200$ mm impeller) operated at $N = 1170$ rpm.

Q (m^3/hr)	0	68
H (m)	7.6	6.7

Find: (a) The equation of a parabola through the pump characteristics at 1170 rpm.
(b) The corresponding equation for a new operating speed of 1750 rpm.
(c) Comparison (plot) of the results.

Solution:
Assume a parabolic variation in pump head of the form, $H = H_0 - AQ^2$. Solving for A gives

$$A_1 = \frac{H_0 - H}{Q^2} = (7.6 - 6.7)\text{ m} \times \frac{1}{(68)^2(\text{m}^3/\text{h})^2} = 1.95 \times 10^{-4}\text{ m}/(\text{m}^3/\text{h})^2$$

The desired equation is

$$H(\text{m}) = 7.6 - 1.95 \times 10^{-4}\,[Q\ (\text{m}^3/\text{h})]^2$$

The pump remains the same, so the two flow conditions are geometrically similar. Assuming no cavitation occurs, the two flows also will be kinematically similar. Then dynamic similarity will be obtained when the two flow coefficients are matched. Denoting the 1170 rpm condition by subscript 1 and the 1750 rpm condition by subscript 2, we have

$$\frac{Q_2}{\omega_2 D_2^3} = \frac{Q_1}{\omega_1 D_1^3} \quad \text{or} \quad \frac{Q_2}{Q_1} = \frac{\omega_2}{\omega_1} = \frac{N_2}{N_1}$$

since $D_2 = D_1$. For the shutoff condition,

$$Q_2 = \frac{N_2}{N_1}Q_1 = \frac{1750\text{ rpm}}{1170\text{ rpm}} \times 0\text{ m}^3/\text{h} = 0\text{ m}^3/\text{h}$$

From the best efficiency point, the new flow rate is

$$Q_2 = \frac{N_2}{N_1}Q_1 = \frac{1750\text{ rpm}}{1170\text{ rpm}} \times 68\text{ m}^3/\text{h} = 101.7\text{ m}^3/\text{h}$$

The pump heads are related by

$$\frac{h_2}{h_1} = \frac{H_2}{H_1} = \frac{N_2^2 D_2^2}{N_1^2 D_1^2} \quad \text{or} \quad \frac{H_2}{H_1} = \frac{N_2^2}{N_1^2} = \left(\frac{N_2}{N_1}\right)^2$$

since $D_2 = D_1$. For the shutoff condition,

$$H_2 = \left(\frac{N_2}{N_1}\right)^2 H_1 = \left(\frac{1750\text{ rpm}}{1170\text{ rpm}}\right)^2 7.6\text{ m} = 17\text{ m}$$

At the best efficiency point,

$$H_2 = \left(\frac{N_2}{N_1}\right)^2 H_1 = \left(\frac{1750\text{ rpm}}{1170\text{ rpm}}\right)^2 6.7\text{ m} = 14.98\text{ m}$$

The curve parameter at 1750 rpm may now be found. Solving for A, we find

$$A_2 = \frac{H_{02} - H_2}{Q_2^2} = (17.0 - 14.98)\text{ m} \times \frac{1}{(101.7)^2(\text{m}^3/\text{h})^2} = 1.95 \times 10^{-4}\text{ m}/(\text{m}^3/\text{h})^2$$

Note that A_2 at 1750 rpm is the same as A_1 at 1170 rpm. Thus we have demonstrated that the coefficient A in the parabolic equation does not change when the pump speed is changed. The "engineering" equations for the two curves are

$$H_1 = 7.6 - 1.95 \times 10^{-4}[Q\ (\text{m}^3/\text{h})]^2 \text{ (at 1170 rpm)}$$

and

$$H_2 = 17.0 - 1.95 \times 10^{-4}[Q \ (\mathrm{m^3/h})]^2 \ (\text{at } 1750 \text{ rpm})$$

The pump curves are compared in the following plot:

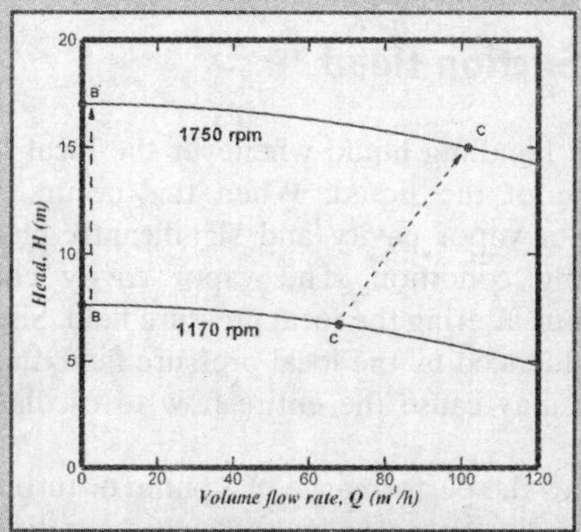

This Example illustrates the procedures for:

- ✓ Obtaining the parabolic "engineering" equation from shutoff head H_0 and best efficiency data on Q and H.
- ✓ Scaling pump curves from one speed to another.

The *Excel* workbook for this Example can be used to generate pump performance curves for a range of speeds.

In principle, geometric similarity would be maintained when pumps of the *same geometry*, differing in size only by a scale ratio, were tested at the *same operating speed*. The flow, head, and power would be predicted to vary with *pump size* as

$$Q_2 = Q_1\left(\frac{D_2}{D_1}\right)^3, \quad H_2 = H_1\left(\frac{D_2}{D_1}\right)^2, \quad \text{and} \quad \mathcal{P}_2 = \mathcal{P}_1\left(\frac{D_2}{D_1}\right)^5 \tag{10.26}$$

It is impractical to manufacture and test a series of pump models that differ in size by only a scale ratio. Instead it is common practice to test a given pump casing at a fixed speed with several impellers of different diameter [13]. Because pump casing width is the same for each test, impeller width also must be the same; only impeller diameter D is changed. As a result, volume flow rate scales in proportion to D^2, not to D^3. Pump input power at fixed speed scales as the product of flow rate and head, so it becomes proportional to D^4. Using this modified scaling method frequently gives results of acceptable accuracy, as demonstrated in several end-of-chapter problems where the method is checked against measured performance data from Appendix D.

It is not possible to compare the efficiencies at the two operating conditions directly. However, viscous effects should become relatively less important as the pump size increases. Thus efficiency should improve slightly as diameter is increased. Moody [14] suggested an empirical equation that may be used to estimate the maximum efficiency of a prototype pump based on test data from a geometrically similar model of the prototype pump. His equation is written

$$\frac{1-\eta_p}{1-\eta_m} = \left(\frac{D_m}{D_p}\right)^{1/5} \tag{10.27}$$

To develop Eq. 10.27, Moody assumed that only the surface resistance changes with model scale so that losses in passages of the same roughness vary as $1/D^5$. Unfortunately, it is difficult to maintain the same relative roughness between model and prototype pumps. Further, the Moody model does not account for any difference in mechanical losses between model and prototype, nor does it allow determination of off-peak efficiencies. Nevertheless, scaling of the maximum-efficiency point is useful to obtain a general estimate of the efficiency curve for the prototype pump.

Cavitation and Net Positive Suction Head

Cavitation can occur in any machine handling liquid whenever the local static pressure falls below the vapor pressure of the liquid. When this occurs, the liquid can locally flash to vapor, forming a vapor cavity and significantly changing the flow pattern from the noncavitating condition. The vapor cavity changes the effective shape of the flow passage, thus altering the local pressure field. Since the size and shape of the vapor cavity are influenced by the local pressure field, the flow may become unsteady. The unsteadiness may cause the entire flow to oscillate and the machine to vibrate.

CLASSIC VIDEO

Cavitation.

As cavitation commences, it reduces the performance of a pump or turbine rapidly. Thus cavitation must be avoided to maintain stable and efficient operation. In addition, local surface pressures may become high when the vapor cavity implodes or collapses, causing erosion damage or surface pitting. The damage may be severe enough to destroy a machine made from a brittle, low-strength material. Obviously cavitation also must be avoided to assure long machine life.

In a pump, cavitation tends to begin at the section where the flow is accelerated into the impeller. Cavitation in a turbine begins where pressure is lowest. The tendency to cavitate increases as local flow speeds increase; this occurs whenever flow rate or machine operating speed is increased.

Cavitation can be avoided if the pressure everywhere in the machine is kept above the vapor pressure of the operating liquid. At constant speed, this requires that a pressure somewhat greater than the vapor pressure of the liquid be maintained at a pump inlet (the *suction*). Because of pressure losses in the inlet piping, the suction pressure may be subatmospheric. Therefore it is important to carefully limit the pressure drop in the inlet piping system.

Net positive suction head (*NPSH*) is defined as the difference between the absolute stagnation pressure in the flow at the pump suction and the liquid vapor pressure, expressed as head of flowing liquid [15].[5] Hence the *NPSH* is a measure of the difference between the maximum possible pressure in the given flow and the pressure at which the liquid will start flashing over to a vapor; the larger the *NPSH*, the less likely cavitation is to occur. The *net positive suction head required* (*NPSHR*) by a specific pump to suppress cavitation varies with the liquid pumped, and with the liquid temperature and pump condition (e.g., as critical geometric features of the pump are affected by wear). *NPSHR* may be measured in a pump test facility by controlling the input pressure. The results are plotted on the pump performance curve. Typical pump characteristic curves for three impellers tested in the same housing were shown in Fig. 10.14. Experimentally determined *NPSHR* curves for the largest and smallest impeller diameters are plotted near the bottom of the figure.

The *net positive suction head available* (*NPSHA*) at the pump inlet must be greater than the *NPSHR* to suppress cavitation. Pressure drop in the inlet piping and pump entrance increases as volume flow rate increases. Thus for any system, the *NPSHA* decreases as flow rate is raised. The *NPSHR* of the pump increases as the flow rate is raised. Therefore, as the system flow rate is increased, the curves for *NPSHA* and *NPSHR* versus flow rate ultimately cross. Hence, for any inlet system, there is a flow

[5]*NPSH* may be expressed in any convenient units of measure, such as height of the flowing liquid, e.g., m of water (hence the term *suction head*), or kPa (abs). When expressed as *head*, *NPSH* is measured relative to the pump impeller centerline.

rate that cannot be exceeded if flow through the pump is to remain free from cavitation. Inlet pressure losses may be reduced by increasing the diameter of the inlet piping; for this reason, many centrifugal pumps have larger flanges or couplings at the inlet than at the outlet.

Example *10.7* CALCULATION OF NET POSITIVE SUCTION HEAD (*NPSH*)

A Peerless Type 4AE11 centrifugal pump (Fig. D.3, Appendix D) is tested at 1750 rpm using a flow system with the layout of Example 10.4. The water level in the inlet reservoir is 1 m above the pump centerline; the inlet line consists of 1.8 m of 125 mm diameter straight cast-iron pipe, a standard elbow, and a fully open gate valve. Calculate the net positive suction head available (*NPSHA*) at the pump inlet at a volume flow rate of 230 m^3/h of water at 30°C. Compare with the net positive suction head required (*NPSHR*) by the pump at this flow rate. Plot *NPSHA* and *NPSHR* for water at 30°C and 80°C versus volume flow rate.

Given: A Peerless Type 4AE11 centrifugal pump (Fig. D.3, Appendix D) is tested at 1750 rpm using a flow system with the layout of Example 10.4. The water level in the inlet reservoir is 1 m above the pump centerline; the inlet line has 1.8 m of 125 mm diameter straight cast-iron pipe, a standard elbow, and a fully open gate valve.

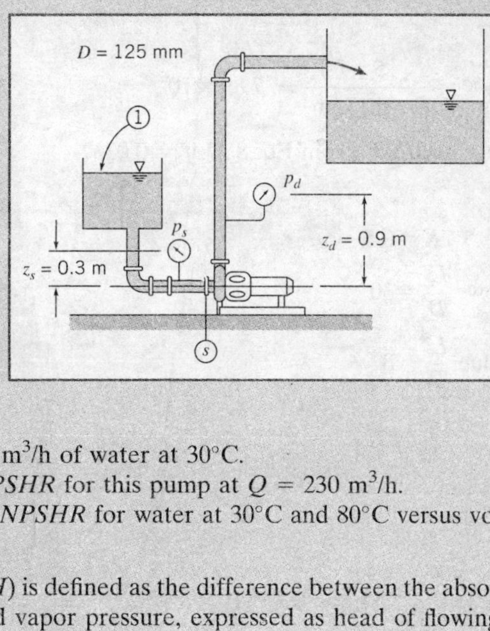

Find: (a) *NPSHA* at $Q = 230$ m^3/h of water at 30°C.
(b) Comparison with *NPSHR* for this pump at $Q = 230$ m^3/h.
(c) Plot of *NPSHA* and *NPSHR* for water at 30°C and 80°C versus volume flow rate.

Solution:

Net positive suction head (*NPSH*) is defined as the difference between the absolute stagnation pressure in the flow at the pump suction and the liquid vapor pressure, expressed as head of flowing liquid. Therefore it is necessary to calculate the head at the pump suction.

Apply the energy equation for steady, incompressible pipe flow to compute the pressure at the pump inlet and thus the *NPSHA*. Denote the reservoir level as ① and the pump suction as ⓢ, as shown above.

Governing equation:

$$p_1 + \frac{1}{2}\rho \overbrace{\bar{V}_1^2}^{\approx 0} + \rho g z_1 = p_s + \frac{1}{2}\rho \bar{V}_s^2 + \rho g_s + \rho h_{\ell_T}$$

Assumption: $\bar{V}_1$ is negligible. Thus

$$p_s = p_1 + \rho g(z_1 - z_s) - \frac{1}{2}\rho \bar{V}_s^2 - \rho h_{\ell_T} \tag{1}$$

The total head loss is

$$h_{\ell_T} = \left(\sum K + \sum f\frac{L_e}{D} + f\frac{L}{D}\right)\frac{1}{2}\rho\overline{V}_s^2 \tag{2}$$

Substituting Eq. 2 into Eq. 1 and dividing by ρg,

$$H_s = H_1 + z_1 - z_s - \left(\sum K + \sum f\frac{L_e}{D} + f\frac{L}{D} + 1\right)\frac{\overline{V}_s^2}{2g} \tag{3}$$

Evaluating the friction factor and head loss,

$$f = f(Re, e/D); \quad Re = \frac{\rho\overline{V}D}{\mu} = \frac{\overline{V}D}{\nu}; \quad \overline{V} = \frac{Q}{A}; \quad A = \frac{\pi D^2}{4}$$

For 125 mm (nominal) pipe, D = 128 mm

$$D = 128\text{ mm} \times \frac{\text{m}}{1000\text{ mm}} = 0.128\text{ m}, \quad A = \frac{\pi D^2}{4} = 0.0129\text{ m}^2$$

$$\overline{V} = 230\,\frac{\text{m}^3}{\text{h}} \times \frac{\text{h}}{3600\text{ s}} \times \frac{1}{0.0129\text{ m}^2} = 4.95\text{ m/s}$$

From Table A.8, for water at T = 30°C, $\nu = 8.03 \times 10^{-7}$ m²/s.

The Reynolds number is

$$Re = \frac{\overline{V}D}{\nu} = 4.95\frac{\text{m}}{\text{s}} \times 0.128\text{ m} \times \frac{\text{s}}{8.03 \times 10^{-7}\text{m}^2} = 7.89 \times 10^5$$

From Table 8.1, e = 0.26 mm, so e/D = 0.00203. From Eq. 8.37, f = 0.0237. The minor loss coefficients are

Entrance	$K = 0.5$
Standard elbow	$\frac{L_e}{D} = 30$
Open gate value	$\frac{L_e}{D} = 8$

Substituting,

$$\left(\sum K + \sum f\frac{L_e}{D} + f\frac{L}{D} + 1\right)$$

$$= 0.5 + 0.0237(30 + 8) + 0.0237\left(\frac{1.8}{0.128}\right) + 1 = 2.73$$

The heads are

$$H_1 = \frac{p_{\text{atm}}}{\rho g} = \frac{1.01325 \times 10^5\text{Pa}}{996\text{ kg/m}^3 \times 9.8\text{ m/s}^2} \times \frac{\text{N}}{\text{Pa.m}^2} \times \frac{\text{kg}\cdot\text{m}}{\text{N}\cdot\text{s}^2}$$

$$= 10.4\text{ m}$$

$$\frac{\overline{V}_s^2}{2g} = \frac{1}{2} \times (4.95)^2\frac{\text{m}^2}{\text{s}^2} \times \frac{\text{s}^2}{9.8\text{ m}} = 1.25\text{ m}$$

Thus,

$$H_s = 10.4\text{ m} + 1\text{ m} - 2.73 \times 1.25\text{ m} = 7.98\text{ m}$$

This problem illustrates the procedures used for checking whether a given pump is in danger of experiencing cavitation:

- ✓ Equation 3 and the plots show that the *NPSHA* decreases as flow rate Q (or $\overline{V}_s$) increases; on the other hand, the *NPSHR* increases with Q, so if the flow rate is high enough, a pump will likely experience cavitation (when *NPSHA* < *NPSHR*).
- ✓ The *NPSHR* for any pump increases with flow rate Q because local fluid velocities within the pump increase, causing locally reduced pressures and tending to promote cavitation.
- ✓ For this pump, at 30°C, the pump appears to have *NPSHA* > *NPSHR* at all flow rates, so it would never experience cavitation; at 80°C, cavitation would occur around 250 m³/h, but from Fig. D.3, the pump best efficiency is around 200 m³/h, so it would probably not be run at 250 m³/h—the pump would probably not cavitate even with the hotter water.

The *Excel* workbook for this Example can be used to generate the *NPSHA* and *NPSHR* curves for a variety of pumps and water temperatures.

To obtain *NPHSA*, add velocity head and subtract vapor head. Thus

$$NPHSA = H_s + \frac{\overline{V}_s^2}{2g} - H_v$$

The vapor pressure for water at 30°C is $p_v = 4.25$ kPa. The corresponding head is $H_v = 0.44$ m of water. Thus,

$$NPSHA = 7.98 + 1.25 - 0.44 = 8.79 \text{ m} \quad \longleftarrow \quad NPSHA$$

The pump curve (Fig. D.3, Appendix D) shows that at 230 m³/h the pump requires

$$NPSHR = 3.1 \text{ m} \quad \longleftarrow \quad NPSHR$$

Results of similar computations for water at 30°C are plotted in the figure on the left below. (*NPSHR* values are obtained from the pump curves in Fig. D.3, Appendix D.)

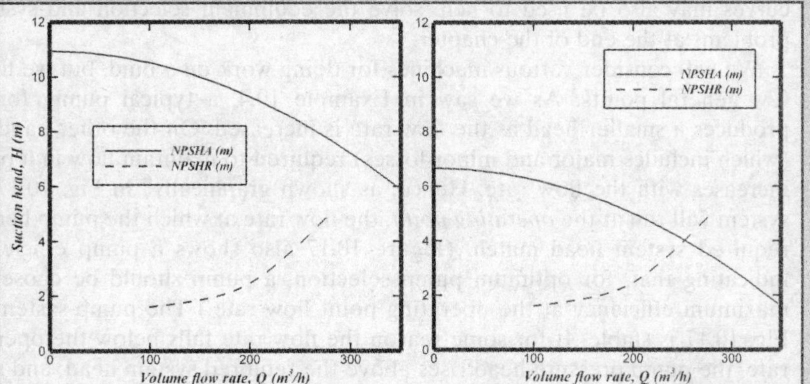

(*a*) Net positive suction head, water at 30 °C (*b*) Net positive suction head, water at 80 °C

Results of computation for water at 80°C are plotted in the figure on the right above. The vapor pressure for water at 80°C is $p_v = 47.4$ kPa. The corresponding head is $H_v = 4.98$ m of water. This high vapor pressure reduces the *NPSHA*, as shown in the plot.

Pump Selection: Applications to Fluid Systems

We define a *fluid system* as the combination of a fluid machine and a network of pipes or channels that convey fluid. The engineering application of fluid machines in an actual system requires matching the machine and system characteristics, while satisfying constraints of energy efficiency, capital economy, and durability. We have alluded to the vast assortment of hardware offered by competing suppliers; this variety verifies the commercial importance of fluid machinery in modern engineering systems.

Usually it is more economical to specify a production machine rather than a custom unit, because products of established vendors have known, published performance characteristics, and they must be durable to survive in the marketplace. Application engineering consists of making the best selection from catalogs of available products. In addition to machine characteristic curves, all manufacturers provide a wealth of dimensional data, alternative configuration and mounting schemes, and technical information bulletins to guide intelligent application of their products.

This section consists of a brief review of relevant theory, followed by example applications using data taken from manufacturer literature. Selected performance

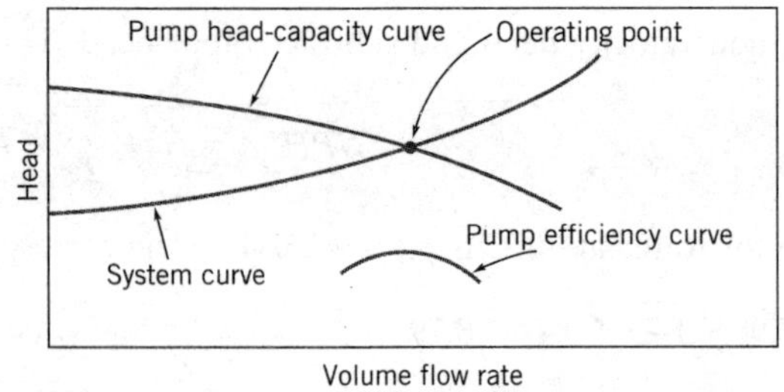

Fig. 10.17 Superimposed system head-flow and pump head-capacity curves.

curves for centrifugal pumps and fans are presented in Appendix D. These may be studied as typical examples of performance data supplied by manufacturers. The curves may also be used to help solve the equipment selection and system design problems at the end of the chapter.

We will consider various machines for doing work on a fluid, but we first make a few general points. As we saw in Example 10.4, a typical pump, for example, produces a smaller head as the flow rate is increased. On the other hand, the head (which includes major and minor losses) required to maintain flow in a pipe system increases with the flow rate. Hence, as shown graphically[6] in Fig. 10.17, a pump-system will run at the *operating point*, the flow rate at which the pump head rise and required system head match. (Figure 10.17 also shows a pump efficiency curve, indicating that, for optimum pump selection, a pump should be chosen that has maximum efficiency at the operating point flow rate.) The pump-system shown in Fig. 10.17 is stable. If for some reason the flow rate falls below the operating flow rate, the pump pressure head rises above the required system head, and so the flow rate increases back to the operating point. Conversely, if the flow rate momentarily increases, the required head exceeds the head provided by the pump, and the flow rate decreases back to the operating point. This notion of an operating point applies to each machine we will consider (although, as we will see, the operating points are not always stable).

The system pressure requirement at a given flow rate is composed of frictional pressure drop (major loss due to friction in straight sections of constant area and minor loss due to entrances, fittings, valves, and exits) and pressure changes due to gravity (static lift may be positive or negative). It is useful to discuss the two limiting cases of pure friction and pure lift before considering their combination.

The *all-friction* system head versus flow curve, with no static lift, starts at zero flow and head, as shown in Fig. 10.18*a*. For this system the total head required is the sum of major and minor losses,

$$h_{l_T} = \sum h_l + \sum h_{l_m} = \sum f \frac{L}{D}\frac{\bar{V}^2}{2} + \sum \left(f \frac{L_e}{D}\frac{\bar{V}^2}{2} + K\frac{\bar{V}^2}{2} \right)$$

For turbulent flow (the usual flow regime in engineering systems), as we learned in Chapter 8 (see Fig. 8.13), the friction factors approach constant and the minor loss coefficients K and equivalent lengths L_e are also constant. Hence $h_{l_T} \sim \bar{V}^2 \sim Q^2$ so that the system curve is approximately parabolic. (In reality, because the friction factors f only approach constants as the regime becomes fully turbulent, it turns out that $Q^{1.75} < h_{l_T} < Q^2$.) This means the system curve with pure friction becomes steeper as

[6]While a graphical representation is useful for visualizing the pump-system matching, we typically use analytical or numerical methods to determine the operating point (*Excel* is very useful for this).

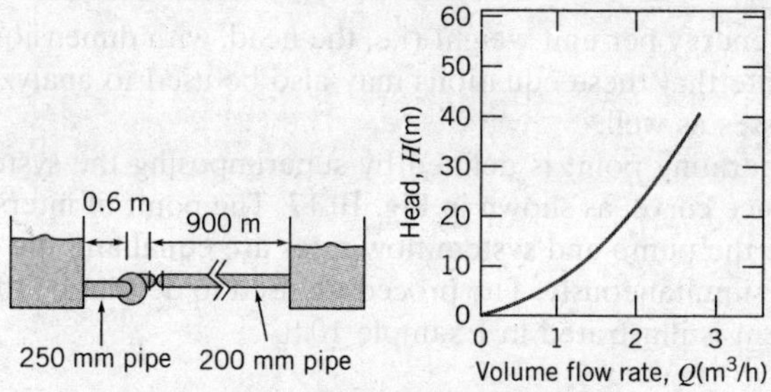

(*a*) Purely frictional pressure drop

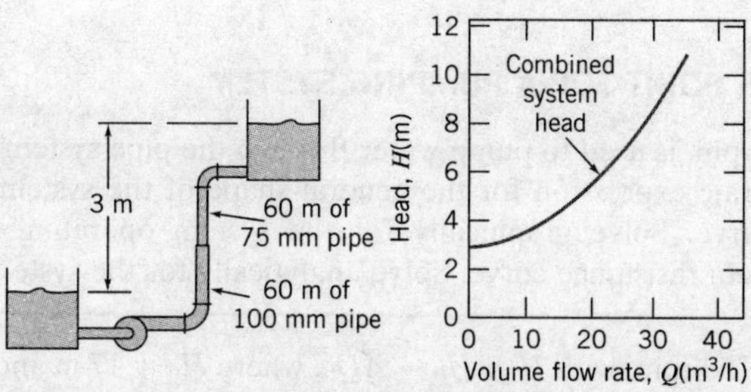

(*b*) Combination of frictional and graivty pressure changes

Fig. 10.18 Schematic diagrams illustrating basic types of system head-flow curves (adapted from [10]).

flow rate increases. To develop the friction curve, losses are computed at various flow rates and then plotted.

Pressure change due to elevation difference is independent of flow rate. Thus the *pure lift* system head-flow curve is a horizontal straight line. The gravity head is evaluated from the change in elevation in the system.

All actual flow systems have some frictional pressure drop and some elevation change. Thus all system head-flow curves may be treated as the sum of a frictional component and a static-lift component. The head for the complete system at any flow rate is the sum of the frictional and lift heads. The system head-flow curve is plotted in Fig. 10.18*b*.

Whether the resulting system curve is *steep* or *flat* depends on the relative importance of friction and gravity. Friction drop may be relatively unimportant in the water supply to a high-rise building (e.g., the Willis Tower, formerly the Sears Tower, in Chicago, which is nearly 400 m tall), and gravity lift may be negligible in an air-handling system for a one-story building.

In Section 8.7 we obtained a form of the energy equation for a control volume consisting of a pump-pipe system,

$$\left(\frac{p_1}{\rho} + \alpha_1 \frac{\overline{V}_1^2}{2} + gz_1\right) - \left(\frac{p_2}{\rho} + \alpha_2 \frac{\overline{V}_2^2}{2} + gz_2\right) = h_{l_T} - \Delta h_{\text{pump}} \tag{8.49}$$

Replacing Δh_{pump} with h_a, representing the head added by any machine (not only a pump) that does work on the fluid, and rearranging Eq. 8.49, we obtain a more general expression

$$\frac{p_1}{\rho} + \alpha_1 \frac{\overline{V}_1^2}{2} + gz_1 + h_a = \frac{p_2}{\rho} + \alpha_2 \frac{\overline{V}_2^2}{2} + gz_2 + h_{l_T} \tag{10.28a}$$

Dividing by g gives

$$\frac{p_1}{\rho g} + \alpha_1 \frac{\overline{V}_1^2}{2g} + z_1 + H_a = \frac{p_2}{\rho g} + \alpha_2 \frac{\overline{V}_2^2}{2g} + z_2 + \frac{h_{l_T}}{g} \tag{10.28b}$$

where H_a is the energy per unit weight (i.e, the head, with dimensions of L) added by the machine. Note that these equations may also be used to analyze a fluid machine with internal losses as well.

The pump operating point is defined by superimposing the system curve and the pump performance curve, as shown in Fig. 10.17. The point of intersection is the only condition where the pump and system flow rates are equal and the pump and system heads are equal simultaneously. The procedure used to determine the match point for a pumping system is illustrated in Example 10.8.

Example 10.8 FINDING THE OPERATING POINT FOR A PUMPING SYSTEM

The pump of Example 10.6 operating at 1750 rpm, is used to pump water through the pipe system (made from iron-cast tubes) of Fig. 10.18*a*. Develop an algebraic expression for the general shape of the system resistance curve. Calculate and plot the system resistance curve. Solve graphically for the system operating point. Obtain an approximate analytical expression for the system resistance curve. Solve analytically for the system operating point.

Given: Pump of Example 10.6 operating at 1750 rpm, with $H = H_0 - AQ^2$, where $H_0 = 17$ m and $A = 1.95 \times 10^{-4}$ m/(m^3/h)^2. System of Fig. 10.18*a*, where $L_1 = 0.6$ m of $D_1 = 250$ mm pipe and $L_2 = 900$ m of $D_2 = 200$ mm pipe, conveying water between two large reservoirs whose surfaces are at the same level.

Find: (a) A general algebraic expression for the system head curve.
(b) The system head curve by direct calculation.
(c) The system operating point using a graphical solution.
(d) An *approximate* analytical expression for the system head curve.
(e) The system operating point using the analytical expression of part (d).

Solution:

Apply the energy equation to the flow system of Fig. 10.18*a*.

Governing equation:

$$\frac{p_0}{\rho g} + \alpha_0 \frac{\bar{V}_0^2}{2g} + z_0 + H_a = \frac{p_3}{\rho g} + \alpha_3 \frac{\bar{V}_3^2}{2g} + z_3 + \frac{h_{l_T}}{g} \tag{10.24b}$$

where z_0 and z_3 are the surface elevations of the supply and discharge reservoirs, respectively.

Assumptions: (1) $p_0 = p_3 = p_{\text{atm}}$.
(2) $\bar{V}_0 = \bar{V}_3 = 0$.
(3) $z_0 = z_3$ (given).

Simplifying, we obtain

$$H_a = \frac{h_{l_T}}{g} = \frac{h_{l_{T_{01}}}}{g} + \frac{h_{l_{T_{23}}}}{g} = H_{l_T} \tag{1}$$

where sections ① and ② are located just upstream and downstream from the pump, respectively.

The total head losses are the sum of the major and minor losses, so

$$h_{l_{T_{01}}} = K_{\text{ent}} \frac{\bar{V}_1^2}{2} + f_1 \frac{L_1}{D_1} \frac{\bar{V}_1^2}{2} = \left(K_{\text{ent}} + f_1 \frac{L_1}{D_1} \right) \frac{\bar{V}_1^2}{2}$$

$$h_{l_{T_{23}}} = f_2 \frac{L_2}{D_2} \frac{\bar{V}_2^2}{2} + K_{\text{exit}} \frac{\bar{V}_2^2}{2} = \left(f_2 \frac{L_2}{D_2} + K_{\text{exit}} \right) \frac{\bar{V}_2^2}{2}$$

From continuity, $\bar{V}_1 A_1 = \bar{V}_2 A_2$, so $\bar{V}_1 = \bar{V}_2 \dfrac{A_2}{A_1} = \bar{V}_2 \left(\dfrac{D_2}{D_1} \right)^2$.

Hence

$$H_{l_T} = \frac{h_{l_T}}{g} = \left(K_{ent} + f_1\frac{L_1}{D_1}\right)\frac{\overline{V}_2^2}{2g}\left(\frac{D_2}{D_1}\right)^4 + \left(f_2\frac{L_2}{D_2} + K_{exit}\right)\frac{\overline{V}_2^2}{2g}$$

or, upon simplifying,

$$H_{l_T} = \left[\left(K_{ent} + f_1\frac{L_1}{D_1}\right)\left(\frac{D_2}{D_1}\right)^4 + f_2\frac{L_2}{D_2} + K_{exit}\right]\frac{\overline{V}_2^2}{2g} \qquad H_{l_T}$$

This is the head loss equation for the system. At the operating point, as indicated in Eq. 1, the head loss is equal to the head produced by the pump, given by

$$H_a = H_0 - AQ^2 \tag{2}$$

where $H_0 = 17$ m and $A = 1.95 \times 10^{-4}$ m/(m^3/h)^2.

The head loss in the system and head produced by the pump can be computed for a range of flow rates:

Q (m^3/h)	$\overline{V}_1$ (m/s)	Re_1 (1000)	f_1 (−)	$\overline{V}_2$ (m/s)	Re_2 (1000)	f_2 (−)	H_{l_T} (m)	H_a (m)
0	0.00	0	–	0.00	0	–	0.00	17.00
25	0.14	39	0.0249	0.22	49	0.0249	0.28	16.88
50	0.28	79	0.0228	0.44	99	0.0232	1.05	16.51
75	0.42	118	0.0220	0.66	148	0.0225	2.29	15.90
100	0.57	158	0.0215	0.88	197	0.0221	4.00	15.05
125	0.71	197	0.0212	1.11	247	0.0219	6.19	13.95
150	0.85	237	0.0210	1.33	296	0.0218	8.86	12.61
175	0.99	276	0.0208	1.55	345	0.0217	12.00	11.03
200	1.13	316	0.0207	1.77	394	0.0216	15.61	9.20
225	1.27	355	0.0206	1.99	444	0.0215	19.70	7.13
250	1.41	395	0.0206	2.21	493	0.0215	24.25	4.81

The pump curve and the system resistance curve are plotted below:

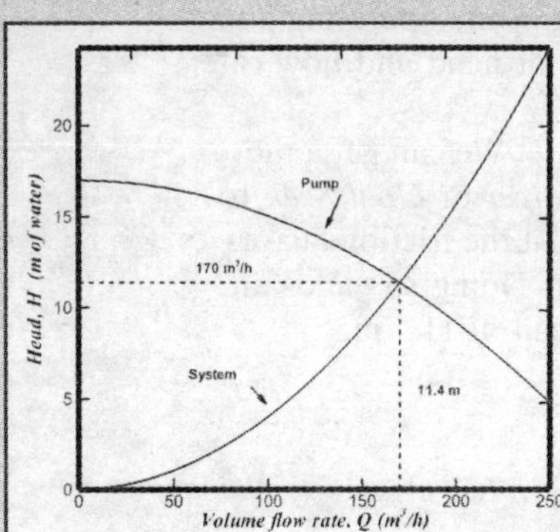

The graphical solution is shown on the plot. At the operating point, $H \approx 11.4$ m and $Q \approx 170.0$ m^3/h.

The approximate analytical expression for the system resistance curve is developed as follows. Because the Reynolds number corresponds to the fully turbulent regime, $f \approx$ const., we can simplify the equation for the head loss and write it in the form

$$H_{l_T} \approx CQ^2 \quad (3)$$

where $C = 8/\pi^2 D_2^4 g$ times the term in square brackets in the expression for H_{l_T}. We can obtain a value for C directly from Eq. 3 by using values for H_{l_T} and Q from the table at a point close to the anticipated operating point. For example, from the $Q = 150\ \text{m}^3/\text{h}$ data point,

$$C = \frac{H_{l_T}}{Q^2} = \frac{8.86\ \text{m}}{150^2\ (\text{m}^3/\text{h})^2} = 3.94 \times 10^{-4}\ \text{m}/(\text{m}^3/\text{h})^2$$

Hence, the approximate analytical expression for the system head curve is

$$H_{l_T} = 3.94 \times 10^{-4}\ \text{m}/(\text{m}^3/\text{h})^2 \times [Q(\text{m}^3/\text{h})]^2 \longleftarrow H_{l_T}$$

Using Eqs. 2 and 3 in Eq. 1, we obtain

$$H_0 - AQ^2 = CQ^2$$

Solving for Q, the volume flow rate at the operating point, gives

$$Q = \left[\frac{H_0}{A + C}\right]^{1/2}$$

For this case,

$$Q = \left[17\ \text{m} \times \frac{(\text{m}^3/\text{h})^2}{(1.95 \times 10^{-4} + 3.94 \times 10^{-4})\ \text{m}}\right]^{1/2} = 170.0\ \text{m}^3/\text{h} \longleftarrow Q$$

The volume flow rate may be substituted into either expression for head to calculate the head at the operating point as

$$H = CQ^2 = 3.94 \times 10^{-4} \frac{\text{m}}{(\text{m}^3/\text{h})^2} \times (170)^2 (\text{m}^3/\text{h})^2 = 11.4\ \text{m} \longleftarrow H$$

We can see that in this problem our reading of the operating point from the graph was pretty good: The reading of head and flow rate were in agreement with the calculated head.

Note that both sets of results are approximate. We can get a more accurate, and easier, result by using *Excel*'s *Solver* or *Goal Seek* to find the operating point, allowing for the fact that the friction factors vary, however slightly, with Reynolds number. Doing so yields an operating point flow rate of 170.2 m^3/h and head of 11.4 m.

This problem illustrates the procedures used to find the operating point of a pump and flow system.

- ✓ The approximate methods—graphical, and assuming friction losses are proportional to Q^2—yielded results close to the detailed computation using *Excel*. We conclude that since most pipe flow friction coefficients are accurate to only about ±10 percent anyway, the approximate methods are accurate enough. On the other hand, use of *Excel*, when available, is easier as well as being more accurate.
- ✓ Equation 3, for the head loss in the system, must be replaced with an equation of the form $H = Z_0 + CQ^2$ when the head H required by the system has a component Z_0 due to gravity as well as a component due to head losses.

The *Excel* workbook for this Example was used to generate the tabulated results as well as the most accurate solution. It can be adapted for use with other pump-pipe systems.

The shapes of both the pump curve and the system curve can be important to system stability in certain applications. The pump curve shown in Fig. 10.17 is typical of the curve for a new centrifugal pump of intermediate specific speed, for which the head decreases smoothly and monotonically as the flow rate increases from shutoff. Two effects take place gradually as the system ages: (1) The pump wears, and its performance decreases (it produces less pressure head; so the pump curve gradually moves downward toward lower head at each flow rate). (2) The system head increases (the system curve gradually moves toward higher head at each flow rate because of pipe aging[7]). The effect of these changes is to move the operating point toward lower flow rates over time. The magnitude of the change in flow rate depends on the shapes of the pump and system curves.

The capacity losses, as pump wear occurs, are compared for steep (friction dominated) and flat (gravity dominated) system curves in Fig. 10.19. The loss in capacity is greater for the flat system curve than for the steep system curve.

The pump efficiency curve is also plotted in Fig. 10.17. The original system operating point usually is chosen to coincide with the maximum efficiency by careful choice of pump size and operating speed. Pump wear increases internal leakage, thus reducing delivery and lowering peak efficiency. In addition, as shown in Fig. 10.19, the operating point moves toward lower flow rate, away from the best efficiency point. Thus the reduced system performance may not be accompanied by reduced energy usage.

Sometimes it is necessary to satisfy a high-head, low-flow requirement; this forces selection of a pump with low specific speed. Such a pump may have a performance curve with a slightly rising head near shutoff, as shown in Fig. 10.20. When the system curve is steep, the operating point is well-defined and no problems with system operation should result. However, use of the pump with a flat system curve could easily cause problems, especially if the actual system curve were slightly above the computed curve or the pump delivery were below the charted head capacity performance.

If there are two points of intersection of the pump and system curves, the system may operate at either point, depending on conditions at start-up; a disturbance could cause the system operating point to shift to the second point of intersection. Under certain conditions, the system operating point can alternate between the two points of intersection, causing unsteady flow and unsatisfactory performance.

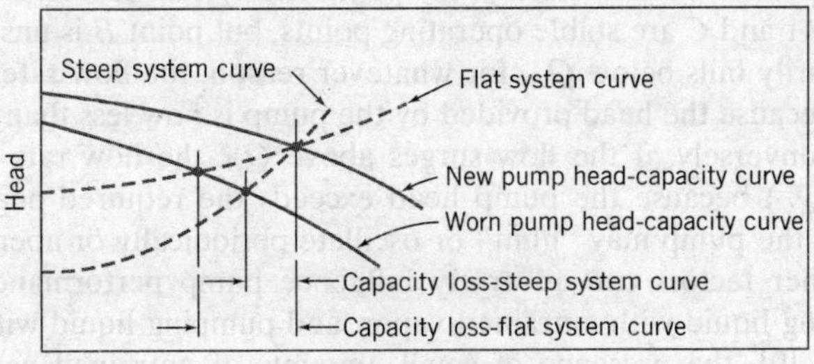

Fig. 10.19 Effect of pump wear on flow delivery to system.

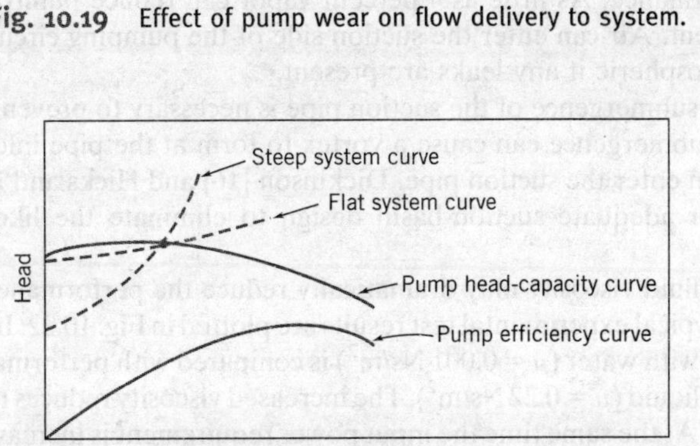

Fig. 10.20 Operation of low specific speed pump near shutoff.

[7]As the pipe ages, mineral deposits form on the wall (see Fig. 8.14), raising the relative roughness and reducing the pipe diameter compared with the as-new condition. See Problem 10.63 for typical friction factor data.

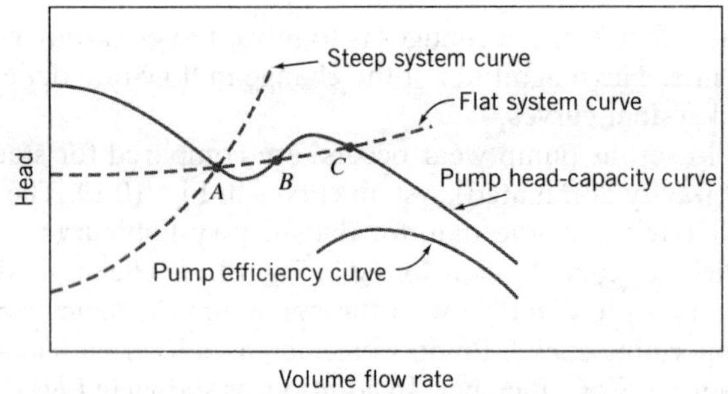

Fig. 10.21 Operation of high specific speed pump near the dip.

Instead of a single pump of low specific speed, a multistage pump may be used in this situation. Since the flow rate through all stages is the same, but the head per stage is less than in the single-stage unit, the specific speed of the multistage pump is higher (see Eq. 7.22a).

The head-flow characteristic curve of some high specific speed pumps shows a dip at capacities below the peak efficiency point, as shown in Fig. 10.21. Caution is needed in applying such pumps if it is ever necessary to operate the pump at or near the dip in the head-flow curve. No trouble should occur if the system characteristic is steep, for there will be only one point of intersection with the pump curve. Unless this intersection is near point *B*, the system should return to stable, steady-state operation following any transient disturbance.

Operation with a flat system curve is more problematic. It is possible to have one, two, or three points of intersection of the pump and system curves, as suggested in the figure. Points *A* and *C* are stable operating points, but point *B* is unstable: If the flow rate momentarily falls below Q_B, for whatever reason, the flow rate will continue to fall (to Q_A) because the head provided by the pump is now less than that required by the system; conversely, if the flow surges above Q_B, the flow rate will continue to increase (to Q_C) because the pump head exceeds the required head. With the flat system curve, the pump may "hunt" or oscillate periodically or aperiodically.

Several other factors can adversely influence pump performance: pumping hot liquid, pumping liquid with entrained vapor, and pumping liquid with high viscosity. According to [9], the presence of small amounts of entrained gas can drastically reduce performance. As little as 4 percent vapor can reduce pump capacity by more than 40 percent. Air can enter the suction side of the pumping circuit where pressure is below atmospheric if any leaks are present.

Adequate submergence of the suction pipe is necessary to prevent air entrainment. Insufficient submergence can cause a vortex to form at the pipe inlet. If the vortex is strong, air can enter the suction pipe. Dickinson [16] and Hicks and Edwards [17] give guidelines for adequate suction-basin design to eliminate the likelihood of vortex formation.

Increased fluid viscosity may dramatically reduce the performance of a centrifugal pump [17]. Typical experimental test results are plotted in Fig. 10.22. In the figure, pump performance with water ($\mu = 0.001\ \text{Ns/m}^2$) is compared with performance in pumping a more viscous liquid ($\mu = 0.22\ \text{Ns/m}^2$). The increased viscosity reduces the head produced by the pump. At the same time the input power requirement is increased. The result is a dramatic drop in pump efficiency at all flow rates.

Heating a liquid raises its vapor pressure. Thus to pump a hot liquid requires additional pressure at the pump inlet to prevent cavitation, as we saw in Example 10.7.

In some systems, such as city water supply or chilled-water circulation, there may be a wide range in demand with a relatively constant system resistance. In these cases, it may be possible to operate constant-speed pumps in series or parallel to supply the system requirements without excessive energy dissipation due to outlet throttling.

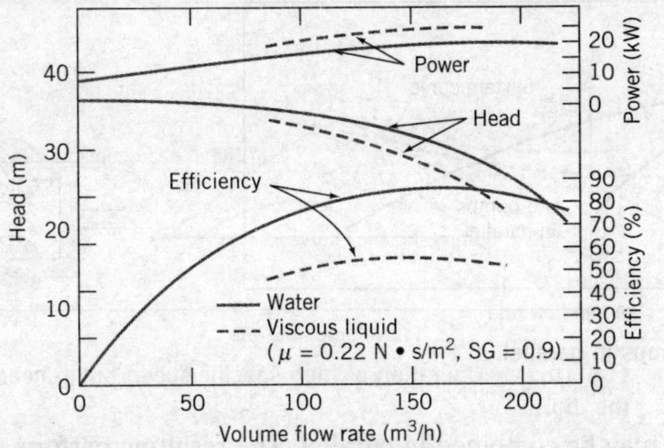

Fig. 10.22 Effect of liquid viscosity on performance of a centrifugal pump [9].

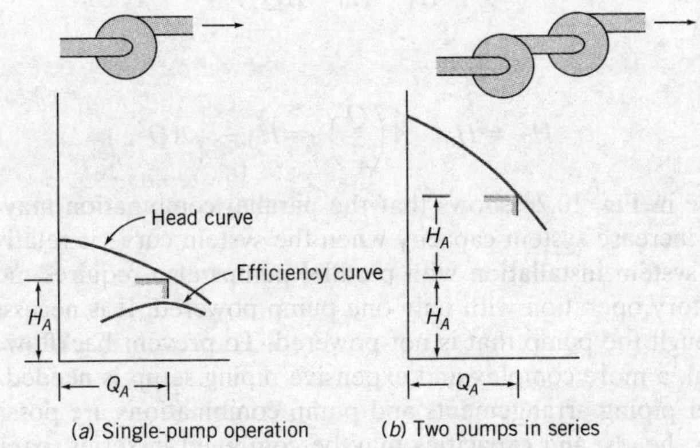

Fig. 10.23 Operation of two centrifugal pumps in series.

Two or more pumps may be operated in parallel or series to supply flow at high demand conditions, and fewer units can be used when demand is low.

For pumps in *series*, the combined performance curve is derived by adding the head rises at each flow rate (Fig. 10.23). The increase in flow rate gained by operating pumps in series depends on the resistance of the system being supplied. For two pumps in series, delivery will increase at any system head. The characteristic curves for one pump and for two identical pumps in series are

$$H_1 = H_0 - AQ^2$$

and

$$H_{2_s} = 2(H_0 - AQ^2) = 2H_0 - 2AQ^2$$

Figure 10.23 is a schematic illustrating the application of two identical pumps in series. A reasonable match to the system requirement is possible—while keeping efficiency high—if the system curve is relatively steep.

In an actual system, it is not appropriate simply to connect two pumps in series. If only one pump were powered, flow through the second, unpowered pump would cause additional losses, raising the system resistance. It also is desirable to arrange the pumps and piping so that each pump can be taken out of the pumping circuit for maintenance, repair, or replacement when needed. Thus a system of bypasses, valves, and check valves may be necessary in an actual installation [13, 17].

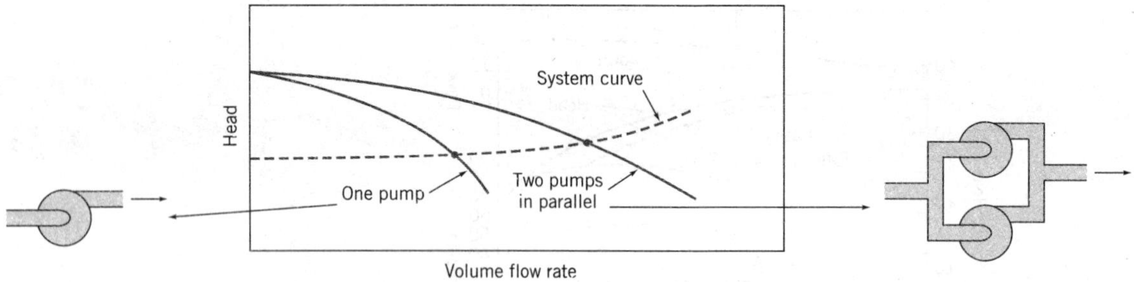

Fig. 10.24 Operation of two centrifugal pumps in parallel.

Pumps also may be combined in parallel. The resulting performance curve, shown in Fig. 10.24, is obtained by adding the pump capacities at each head. The characteristic curves for one pump and for two identical pumps in parallel are

$$H_1 = H_0 - AQ^2$$

and

$$H_{2_p} = H_0 - A\left(\frac{Q}{2}\right)^2 = H_0 - \frac{1}{4}AQ^2$$

The schematic in Fig. 10.24 shows that the parallel combination may be used most effectively to increase system capacity when the system curve is relatively flat.

An actual system installation with parallel pumps also requires more thought to allow satisfactory operation with only one pump powered. It is necessary to prevent backflow through the pump that is not powered. To prevent backflow and to permit pump removal, a more complex and expensive piping setup is needed.

Many other piping arrangements and pump combinations are possible. Pumps of different sizes, heads, and capacities may be combined in series, parallel, or series-parallel arrangements. Obviously the complexity of the piping and control system increases rapidly. In many applications the complexity is due to a requirement that the system handle a variety of flow rates—a range of flow rates can be generated by using pumps in parallel and in series and by using throttling valves. Throttling valves are usually necessary because constant-speed motors drive most pumps, so simply using a network of pumps (with some on and others off) without throttling valves allows the flow rate to be varied only in discrete steps. The disadvantage of throttling valves is that they can be a major loss of energy so that a given flow rate will require a larger power supply than would otherwise be the case. Some typical data for a throttling valve, given in Table 10.1 [18], show a decreasing valve efficiency (the percentage of pump pressure available that is not consumed by the valve) as the valve is used to reduce the flow rate.

Use of *variable-speed operation* allows infinitely variable control of system flow rate with high energy efficiency and without extra plumbing complexity. A further advantage is that a variable-speed drive system offers much simplified control of system flow rate. The cost of efficient variable-speed drive systems continues to decrease because of advances in power electronic components and circuits. The system flow rate can be controlled by varying pump operating speed with impressive savings in pumping power and energy usage. The input power reduction afforded by use of a variable-speed drive is illustrated in Table 10.1. At 250 m^3/h, the power input is cut almost 54 percent for the variable-speed system; the reduction at 136 m^3/h is more than 75 percent.

The reduction in input power requirement at reduced flow with the variable speed drive is impressive. The energy savings, and therefore the cost savings, depend on the specific duty cycle on which the machine operates. Armintor and Conners [18] present

Table 10.1

Power Requirements for Constant- and Variable-Speed Drive Pumps

Throttle Valve Control with Constant-Speed (1750 rpm) Motor

Flow Rate (m^3/h)	System Head (m)	Valve[a] Efficiency (%)	Pump Head (m)	Pump Efficiency (%)	Pump Power (kW)	Motor Efficiency (%)	Motor Input (kW)	Power Input[b] (kW)
386	54.9	100.0	54.9	80.0	72.1	90.8	79.4	79.6
341	45.7	78.1	58.5	78.4	69.3	90.7	76.4	76.5
309	39.9	66.2	60.4	76.8	66.1	90.7	72.9	73.0
250	31.1	49.5	62.8	72.4	59.0	90.6	65.1	65.2
204	25.3	39.5	64.0	67.0	53.2	90.3	58.9	59.0
136	18.9	29.0	65.2	54.0	44.8	90.0	49.8	49.9

Variable-Speed Drive with Energy-Efficient Motor

Flow Rate (m^3/h)	Pump/ System Head (m)	Pump Efficiency (%)	Pump Power (kW)	Motor Speed (rpm)	Motor Efficiency (%)	Motor Input (kW)	Control Efficiency (%)	Power Input (kW)
386	54.9	80.0	72.1	1750	93.7	77.0	97.0	79.3
341	45.7	79.6	53.3	1580	94.0	56.7	96.1	59.0
309	39.9	78.8	42.7	1470	93.9	45.4	95.0	47.8
250	31.1	78.4	27.0	1275	93.8	28.8	94.8	30.3
204	25.3	77.1	18.3	1140	92.3	19.8	92.8	21.3
136	18.9	72.0	9.8	960	90.0	10.8	89.1	12.2

Source: Based on Armintor and Conners [18].

[a]Valve efficiency is the ratio of system pressure to pump pressure.

[b]Power input is motor input divided by 0.998 starter efficiency.

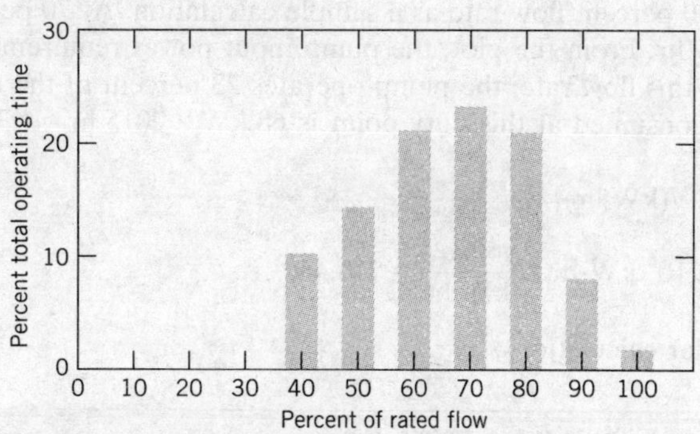

Fig. 10.25 Mean duty cycle for centrifugal pumps in the chemical and petroleum industries [18].

information on mean duty cycles for centrifugal pumps used in the chemical process industry; Fig. 10.25 is a plot showing the histogram of these data. The plot shows that although the system must be designed and installed to deliver full rated capacity, this condition seldom occurs. Instead, more than half the time, the system operates at 70 percent capacity or below. The energy savings that result from use of a variable speed drive for this duty cycle are estimated in Example 10.9.

Example 10.9 ENERGY SAVINGS WITH VARIABLE-SPEED CENTRIFUGAL PUMP DRIVE

Combine the information on mean duty cycle for centrifugal pumps given in Fig. 10.25 with the drive data in Table 10.1. Estimate the annual savings in pumping energy and cost that could be achieved by implementing a variable-speed drive system.

Given: Consider the variable-flow, variable-pressure pumping system of Table 10.1. Assume the system operates on the typical duty cycle shown in Fig. 10.25, 24 hours per day, year round.

Find: (a) An estimate of the reduction in annual energy usage obtained with the variable-speed drive.
(b) The energy costs and the cost saving due to variable-speed operation.

Solution:
Full-time operation involves 365 days × 24 hours per day, or 8760 hours per year. Thus the percentages in Fig. 10.27 may be multiplied by 8760 to give annual hours of operation.

First plot the pump input power versus flow rate using data from Table 10.1 to allow interpolation, as shown below.

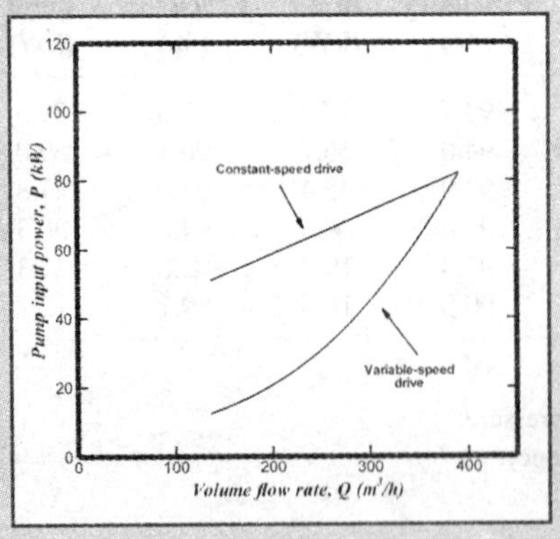

Illustrate the procedure using operation at 70 percent flow rate as a sample calculation. At 70 percent flow rate, the pump delivery is $0.7 \times 386\ \text{m}^3/\text{hr} = 270\ \text{m}^3/\text{hr}$. From the plot, the pump input power requirement at this flow rate is 68 kW for the constant-speed drive. At this flow rate, the pump operates 23 percent of the time, or $0.23 \times 8760 = 2015$ hours per year. The total energy consumed at this duty point is $68\ \text{kW} \times 2015\ \text{hr} = 1.37 \times 10^5\ \text{kW}\cdot\text{hr}$. The electrical energy consumed is

The corresponding cost of electricity [at \$0.12/(kW·hr)] is

$$C = 1.37 \times 10^5\ \text{kW}\cdot\text{hr} \times \frac{\$0.12}{\text{kW}\cdot\text{hr}} = \$16{,}440$$

The following tables were prepared using similar calculations:

Constant-Speed Drive, 8760 hr/yr

Flow (%)	Flow (m^3/h)	Time (%)	Time (hr)	Power (kW)	Energy (kW · hr)
100	386	2	175	80	1.4×10^4
90	348	8	701	77	5.4×10^4
80	309	21	1840	73	13.4×10^4
70	270	23	2010	68	13.7×10^4
60	232	21	1840	63	11.6×10^4
50	193	15	1310	57	7.5×10^4
40	154	10	876	52	4.6×10^4
				Total:	57.6×10^4

Summing the last column of the table shows that for the constant-speed drive system the annual energy consumption is 57.6×10^4 kW·hr.

At \$0.12 per kilowatt hour, the energy cost for the constant-speed drive system is

$$C = 57.6 \times 10^4 \text{ kW}\cdot\text{hr} \times \frac{\$0.12}{\text{kW}\cdot\text{hr}} = \$69{,}120 \leftarrow C_{CSD}$$

Variable-Speed Drive, 8760 hr/yr

Flow (%)	Flow (m^3/h)	Time (%)	Time (hr)	Power (kW)	Energy (kW·hr)
100	386	2	175	79	1.4×10^4
90	348	8	701	62	4.3×10^4
80	309	21	1840	48	8.8×10^4
70	270	23	2010	36	7.2×10^4
60	232	21	1840	27	5.0×10^4
50	193	15	1310	20	2.6×10^4
40	154	10	876	15	1.3×10^4
				Total:	30.6×10^4

Summing the last column of the table shows that for the variable-speed drive system, the annual energy consumption is 3.06×10^5 kW·hr. The electrical energy consumption is

At \$0.12 per kilowatt hour, the energy cost for the variable-speed drive system is only

$$C = 3.06 \times 10^5 \text{ kW}\cdot\text{hr} \times \frac{\$0.12}{\text{kW}\cdot\text{hr}} = \$36{,}720 \leftarrow C_{VSD}$$

Thus, in this application, the variable-speed drive reduces energy consumption by 270,000 kW·hr (47 percent). The cost saving is an impressive \$32,400 annually. One could afford to install a variable-speed drive even at considerable cost penalty. The savings in energy cost are appreciable each year and continue throughout the life of the system.

This problem illustrates the energy and cost savings that can be gained by the use of variable-speed pump drives. We see that the specific benefits depend on the system and its operating duty cycle. The *Excel* workbook for this Example was used for plotting the graph, for obtaining the interpolated data, and for performing all calculations. It can be easily modified for other such analyses. Note that results were rounded down to three significant figures *after* calculation.

Blowers and Fans

Fans are designed to handle air or vapor. Fan sizes range from that of the cooling fan in a notebook computer, which moves a cubic meter of air per hour and requires a few watts of power, to that of the ventilation fans for the Channel Tunnel, which move thousands of cubic meters of air per minute and require many hundreds of kilowatts of power. Fans are produced in varieties similar to those of pumps: They range from radial-flow (centrifugal) to axial-flow devices. As with pumps, the characteristic curve shapes for fans depend on the fan type. Some typical performance curves for centrifugal fans are presented in Appendix D. The curves may be used to choose fans to solve some of the equipment-selection and system design problems at the end of the chapter.

An exploded view of a medium-size centrifugal fan is shown in Fig. 10.26. Some commonly used terminology is shown on the figure. The pressure rise produced by fans is several orders of magnitude less than that for pumps. Another difference between fans and pumps is that measurement of flow rate is more difficult in gases and vapors than in liquids. There is no convenient analog to the catch-the-flow-in-a-bucket method of measuring liquid flow rates! Consequently, fan testing requires special facilities and procedures [20, 21]. Because the pressure rise produced by a fan is small, usually it is impractical to measure flow rate with a restriction flow meter such as an orifice, flow nozzle, or venturi. It may be necessary to use an auxiliary fan to develop enough pressure rise to permit measurement of flow rate with acceptable accuracy using a restriction flow meter. An alternative is to use an instrumented duct in which the flow rate is calculated from a pitot traverse. Appropriate standards may be consulted to obtain complete information on specific fan-test methods and data-reduction procedures for each application [20, 21].

The test and data reduction procedures for fans, blowers, and compressors are basically the same as for centrifugal pumps. However, blowers, and especially fans, add relatively small amounts of static head to gas or vapor flows. For these machines, the dynamic head may increase from inlet to discharge, and it may be appreciable compared with the static head rise. For these reasons, it is important to state clearly the basis on which performance calculations are made. Standard definitions are available for machine efficiency based on either the static-to-static pressure rise or the static-to-total pressure rise [20]. Data for both static and total pressure rise and for efficiency, based on both pressure rises, are frequently plotted on the same characteristic graph (Fig. 10.27).

The coordinates may be plotted in physical units (e.g., millimeters of water, kW, and m^3/min) or as dimensionless flow and pressure coefficients. The difference between the total and static pressures is the dynamic pressure, so the vertical distance between these two curves is proportional to Q^2.

Centrifugal fans are used frequently; we will use them as examples. The centrifugal fan developed from simple paddle-wheel designs, in which the wheel was a disk carrying radial flat plates. (This primitive form still is used in nonclogging fans such as in commercial clothes dryers.) Refinements have led to the three general types shown in Fig. 10.28*a–c*, with backward-curved, radial-tipped, and forward curved blades. All the fans illustrated have blades that are curved at their inlet edges to approximate shockless flow between the blade and the inlet flow direction. These three designs are typical of fans with sheet-metal blades, which are relatively simple to manufacture and thus relatively inexpensive. The forward-curved design illustrated in the figure has very closely spaced blades; it is frequently called a *squirrel-cage fan* because of its resemblance to the exercise wheels found in animal cages.

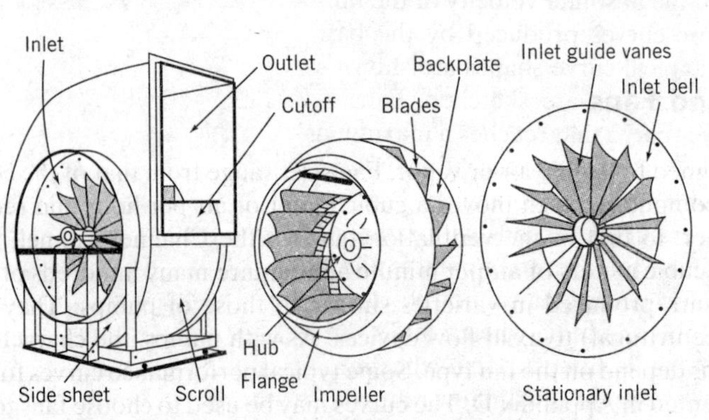

Fig. 10.26 Exploded view of typical centrifugal fan [19].

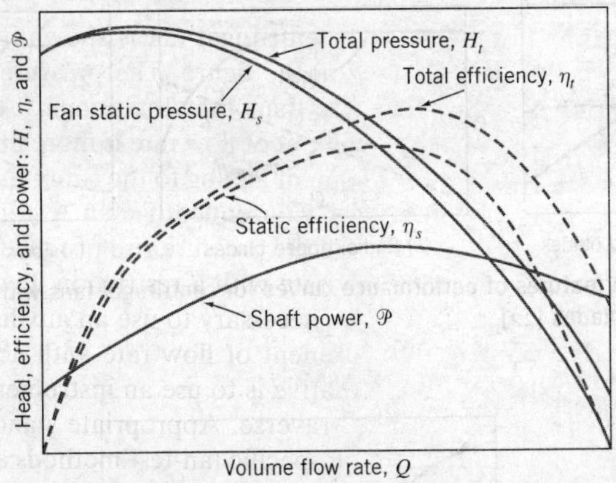

Fig. 10.27 Typical characteristic curves for fan with backward-curved blades [22].

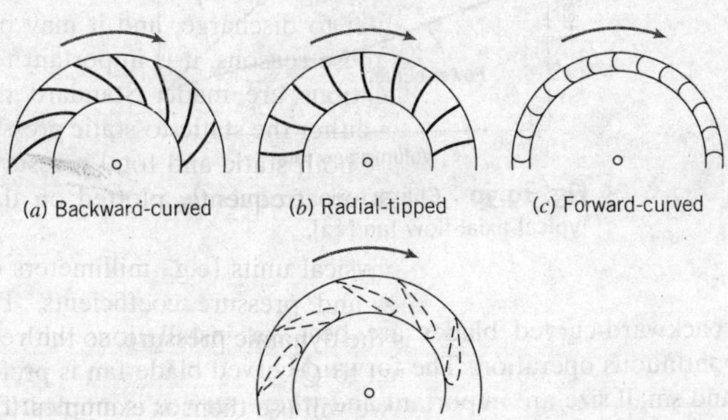

Fig. 10.28 Typical types of blading used for centrifugal fan wheels [22].

As fans become larger in size and power demand, efficiency becomes more important. The streamlined *airfoil blades* shown in Fig. 10.28*d* are much less sensitive to inlet flow direction and improve efficiency markedly compared with the thin blades shown in diagrams *a* through *c*. The added expense of airfoil blades for large metal fans may be life-cycle cost effective. Airfoil blades are being used more frequently on small fans as impellers molded from plastic become common.

As is true for pumps, the total pressure rise across a fan is approximately proportional to the absolute velocity of the fluid at the exit from the wheel. Therefore the characteristic curves produced by the basic blade shapes tend to differ from each other. The typical curve shapes are shown in Fig. 10.29, where both pressure rise and power requirements are sketched. Fans with backward-curved blade tips typically have a power curve that reaches a maximum and then decreases as flow rate increases. If the fan drive is sized properly to handle the peak power, it is impossible to overload the drive with this type of fan.

The power curves for fans with radial and forward-curved blades rise as flow rate increases. If the fan operating point is higher than the design flow rate, the motor may be overloaded. Such fans cannot be run for long periods at low back pressures. An example of this would be when a fan is run without a load to resist the flow—in other words, the fan is almost "free-wheeling." Because the power drawn by the fan monotonically increases with flow rate, the fan motor could eventually burn out under this free-wheeling condition.

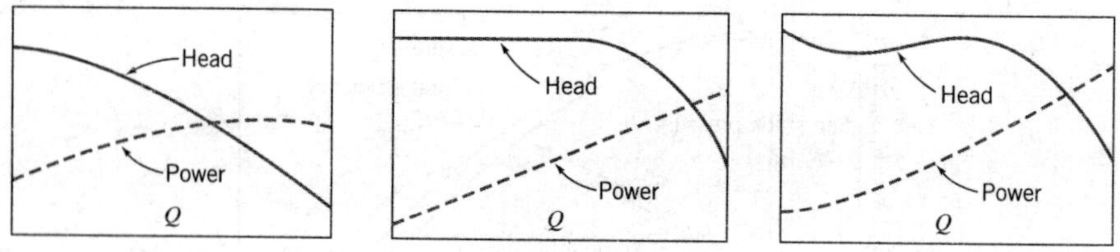

(*a*) Backward-curved blades (*b*) Radial-tipped blades (*c*) Forward-curved blades

Fig. 10.29 General features of performance curves for centrifugal fans with backward-, radial-, and forward-curved blades [22].

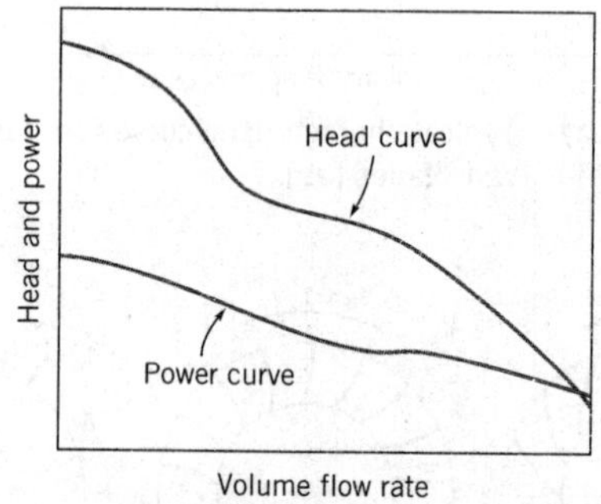

Fig. 10.30 Characteristic curves for a typical axial-flow fan [22].

Fans with backward-curved blades are best for installations with large power demand and continuous operation. The forward-curved blade fan is preferred where low first cost and small size are important and where service is intermittent. Forward curved blades require lower tip speed to produce a specified head; lower blade tip speed means reduced noise. Thus forward-curved blades may be specified for heating and air conditioning applications to minimize noise.

Characteristic curves for axial-flow (*propeller*) fans differ markedly from those for centrifugal fans. The power curve, Fig. 10.30, is especially different, as it tends to decrease continuously as flow rate increases. Thus it is impossible to overload a properly sized drive for an axial-flow fan. The simple propeller fan is often used for ventilation; it may be free-standing or mounted in an opening, as a window fan, with no inlet or outlet duct work. Ducted axial-flow fans have been studied extensively and developed to high efficiency [23]. Modern designs, with airfoil blades, mounted in ducts and often fitted with guide vanes, can deliver large volumes against high resistances with high efficiency. The primary deficiency of the axial-flow fan is the non-monotonic slope of the pressure characteristic: In certain ranges of flow rate the fan may pulsate. Because axial-flow fans tend to have high rotational speeds, they can be noisy.

Selection and installation of a fan always requires compromise. To minimize energy consumption, it is desirable to operate a fan at its highest efficiency point. To reduce the fan size for a given capacity, it is tempting to operate at higher flow rate than that at maximum efficiency. In an actual installation, this tradeoff must be made considering such factors as available space, initial cost, and annual hours of operation. It is not wise to operate a fan at a flow rate below maximum efficiency. Such a fan would be larger than necessary and some designs, particularly those with forward-curved blades, can be unstable and noisy when operated in this region.

It is necessary to consider the duct system at both the inlet and the outlet of the fan to develop a satisfactory installation. Anything that disrupts the uniform flow at the fan inlet is likely to impair performance. Nonuniform flow at the inlet causes the wheel to operate unsymmetrically and may decrease capacity dramatically. Swirling flow also adversely affects fan performance. Swirl in the direction of rotation reduces the pressure developed; swirl in the opposite direction can increase the power required to drive the fan.

The fan specialist may not be allowed total freedom in designing the best flow system for the fan. Sometimes a poor flow system can be improved without too much effort by adding splitters or straightening vanes to the inlet. Some fan manufacturers offer guide vanes that can be installed for this purpose.

Flow conditions at the fan discharge also affect installed performance. Every fan produces nonuniform outlet flow. When the fan is connected to a length of straight duct, the flow becomes more uniform and some excess kinetic energy is transformed to static pressure. If the fan discharges directly into a large space with no duct, the excess kinetic energy of the nonuniform flow is dissipated. A fan in a flow system with no discharge ducting may fall considerably short of the performance measured in a laboratory test setup.

The flow pattern at the fan outlet may be affected by the amount of resistance present downstream. The effect of the system on fan performance may be different at different points along the fan pressure-flow curve. Thus, it may not be possible to accurately predict the performance of a fan, *as installed*, on the basis of curves measured in the laboratory.

Fans may be scaled up or down in size or speed using the basic laws developed for fluid machines in Chapter 7. It is possible for two fans to operate with fluids of significantly different density,[8] so pressure is used instead of head (which uses density) as a dependent parameter and density must be retained in the dimensionless groups. The dimensionless groups appropriate for fan scaling are

$$\Pi_1 = \frac{Q}{\omega D^3}, \quad \Pi_2 = \frac{p}{\rho \omega^2 D^2}, \quad \text{and} \quad \Pi_3 = \frac{\mathcal{P}}{\rho \omega^3 D^5} \tag{10.29}$$

Once again, dynamic similarity is assured when the flow coefficients are matched. Thus when

$$Q' = Q\left(\frac{\omega'}{\omega}\right)\left(\frac{D'}{D}\right)^3 \tag{10.30a}$$

then

$$p' = p\left(\frac{\rho'}{\rho}\right)\left(\frac{\omega'}{\omega}\right)^2\left(\frac{D'}{D}\right)^2 \tag{10.30b}$$

and

$$\mathcal{P}' = \mathcal{P}\left(\frac{\rho'}{\rho}\right)\left(\frac{\omega'}{\omega}\right)^3\left(\frac{D'}{D}\right)^5 \tag{10.30c}$$

As a first approximation, the efficiency of the scaled fan is assumed to remain constant, so

$$\eta' = \eta \tag{10.30d}$$

When head is replaced by pressure, and density is included, the expression defining the specific speed of a fan becomes

$$N_S = \frac{\omega Q^{1/2} \rho^{3/4}}{p^{3/4}} \tag{10.31}$$

A fan scale-up with density variation is the subject of Example 10.10.

[8]Density of the flue gas handled by an induced-draft fan on a steam powerplant may be 40 percent less than the density of the air handled by the forced-draft fan in the same plant.

Example 10.10 SCALING OF FAN PERFORMANCE

Performance curves [20] are given below for a centrifugal fan with $D = 914$ mm and $N = 600$ rpm, as measured on a test stand using cool air ($\rho = 1.2\ \text{kg/m}^3$). Scale the data to predict the performance of a similar fan with $D' = 1070$ mm, $N' = 1150$ rpm, and $\rho' = 0.72\ \text{kg/m}^3$. Estimate the delivery and power of the larger fan when it operates at a system pressure equivalent to 190 mm of H_2O. Check the specific speed of the fan at the new operating point.

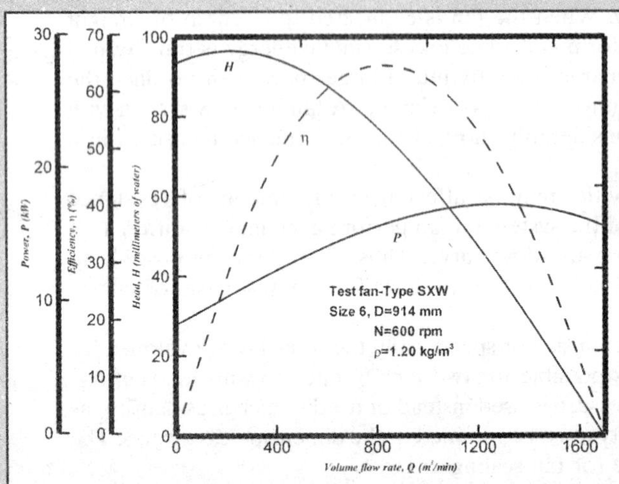

Given: Performance data as shown for centrifugal fan with $D = 914$ mm $N = 600$ rpm, and $\rho = 1.2\ \text{kg/m}^3$.

Find: (a) The predicted performance of a geometrically similar fan with $D' = 1070$ mm, at $N' = 1150$ rpm, with $\rho' = 0.72\ \text{kg/m}^3$.

(b) An estimate of the delivery and input power requirement if the larger fan operates against a system resistance of 190 mm H_2O.

(c) The specific speed of the larger fan at this operating point.

Solution:

Develop the performance curves at the new operating condition by scaling the test data point-by-point. Using Eqs. 10.30 and the data from the curves at $Q = 850\ \text{m}^3/\text{min}$, the new volume flow rate is

$$Q' = Q\left(\frac{N'}{N}\right)\left(\frac{D'}{D}\right)^3 = 850\ \text{m}^3/\text{min} \times \left(\frac{1150}{600}\right) \times \left(\frac{1070}{914}\right)^3 = 2614\ \text{m}^3/\text{min}$$

The fan pressure rise is

$$p' = p\frac{\rho'}{\rho}\left(\frac{N'}{N}\right)^2\left(\frac{D'}{D}\right)^2 = 75.2\ \text{mm H}_2\text{O} \times \left(\frac{0.72}{1.2}\right) \times \left(\frac{1150}{600}\right)^2 \times \left(\frac{1070}{914}\right)^2 = 227.2\ \text{mm H}_2\text{O}$$

and the new power input is

$$\mathcal{P}' = \mathcal{P}\left(\frac{\rho'}{\rho}\right)\left(\frac{N'}{N}\right)^3\left(\frac{D'}{D}\right)^5 = 16.0\ \text{kW} \times \left(\frac{0.72}{1.2}\right) \times \left(\frac{1150}{600}\right)^3 \times \left(\frac{1070}{914}\right)^5 = 148.6\ \text{kW}$$

We assume the efficiency remains constant between the two scaled points, so

$$\eta' = \eta = 64.8\%$$

Similar calculations at other operating points give the results tabulated on page 547:

Q (m^3/min)	p (mm H_2O)	$\mathcal{P}$ (kW)	η (%)	Q'(m^3/min)	p' (mm H_2O)	$\mathcal{P}'$ (kW)
0	93.5	8.3	0	0	282.4	77.1
283	95.3	11.3	37.4	870	287.9	105.0
566	88.9	13.9	59.2	1741	268.5	129.1
850	75.2	16.0	64.8	2614	227.2	148.6
1130	53.8	17.2	57.4	3475	162.5	159.8
1420	25.9	17.2	34.5	4367	78.2	159.8
1700	0	15.7	0	5228	0	145.8

To allow interpolation among the calculated points, it is convenient to plot the results:

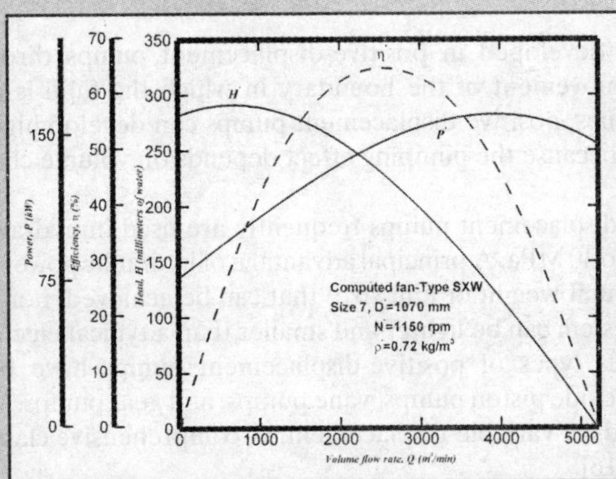

From the head-capacity curve, the larger fan should deliver 3107 m^3/min at 190 mm H_2O system head, with an efficiency of about 62.2 percent.

This operating point is only slightly to the right of peak efficiency for this fan, so it is a reasonable point at which to operate the fan. The engineering specific of the fan at this operating point is given by direct substitution into Eq. 10.31:

$$N_{s_{cu}} = \frac{\omega Q^{1/2}\rho^{3/4}}{p^{3/4}} = \frac{(1150 \text{ rpm}) \times (3107 \text{ m}^3/\text{min})^{1/2} \times (0.72 \text{ kg/m}^3)^{3/4}}{(190 \text{ mm H}_2\text{O})^{3/4}}$$

$$= 979 \longleftarrow N_{s_{cu}}$$

In nondimensional (SI) units,

$$N_s = \frac{(120 \text{ rad/s})(51.8 \text{ m}^3/\text{s})^{1/2}(0.72 \text{ kg/m}^3)^{3/4}}{(1.86 \times 10^3 \text{N/m}^2)^{3/4}} = 2.38 \longleftarrow N_s(\text{SI})$$

This problem illustrates the procedure for scaling performance of fans operating on gases with two different densities.

The *Excel* workbook for this Example was used for plotting the graphs, for obtaining the interpolated data, and for performing all calculations. It can be easily modified for other such analyses.

Three methods are available to control fan delivery: motor speed control, inlet dampers, and outlet throttling. Speed control was treated thoroughly in the section on pumps. The same benefits of reduced energy usage and noise are obtained with fans, and the cost of variable-speed drive systems continues to drop.

Inlet dampers may be used effectively on some large centrifugal fans. However, they decrease efficiency and cannot be used to reduce the fan flow rate below about 40 percent of rated capacity. Outlet throttling is cheap but wasteful of energy. For further details, consult either Jorgensen [19] or Berry [22]; both are particularly comprehensive. Osborne [24] also treats noise, vibration, and the mechanical design of fans.

Fans also may be combined in series, parallel, or more complex arrangements to match varying system resistance and flow needs. These combinations may be analyzed using the methods described for pumps. ASHRAE [25] and Idelchik [26] are excellent sources for loss data on air flow systems.

Blowers have performance characteristics similar to fans, but they operate (typically) at higher speeds and increase the fluid pressure more than do fans. Jorgensen [19] divides the territory between fans and compressors at an arbitrary pressure level that changes the air density by 5 percent; he does not demarcate between fans and blowers.

10.4 Positive Displacement Pumps

Pressure is developed in positive-displacement pumps through volume reductions caused by movement of the boundary in which the fluid is confined. In contrast to turbomachines, positive displacement pumps can develop high pressures at relatively low speeds because the pumping effect depends on volume change instead of dynamic action.

Positive-displacement pumps frequently are used in hydraulic systems at pressures ranging up to 40 MPa. A principal advantage of hydraulic power is the high *power density* (power per unit weight or unit size) that can be achieved: For a given power output, a hydraulic system can be lighter and smaller than a typical electric-drive system.

Numerous types of positive-displacement pumps have been developed. A few examples include piston pumps, vane pumps, and gear pumps. Within each type, pumps may be fixed- or variable-displacement. A comprehensive classification of pump types is given in [16].

The performance characteristics of most positive-displacement pumps are similar; in this section we shall focus on gear pumps. This pump type typically is used, for example, to supply pressurized lubricating oil in internal combustion engines. Figure 10.31 is a schematic diagram of a typical gear pump. Oil enters the space between the gears at the bottom of the pump cavity. Oil is carried outward and upward by the teeth of the rotating gears and exits through the outlet port at the top of the cavity. Pressure is generated as the oil is forced toward the pump outlet; leakage and backflow are prevented by the closely fitting gear teeth at the center of the pump, and by the small clearances maintained between the side faces of the gears and the pump housing. The close clearances require the hydraulic fluid to be kept extremely clean by full-flow filtration.

Figure 10.32 is a photo showing the parts of an actual gear pump; it gives a good idea of the robust housing and bearings needed to withstand the large pressure forces

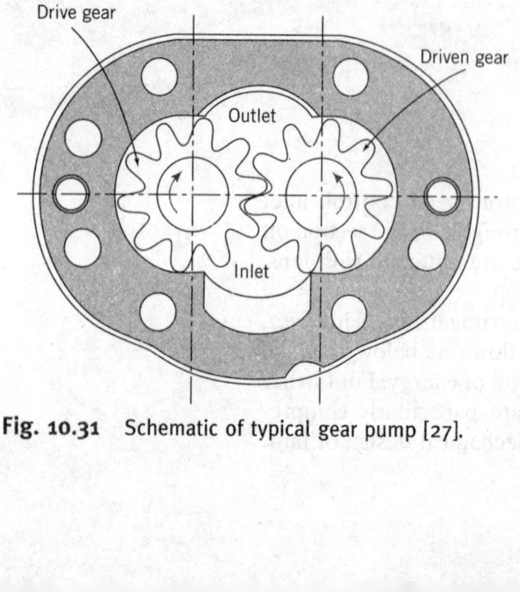

Fig. 10.31 Schematic of typical gear pump [27].

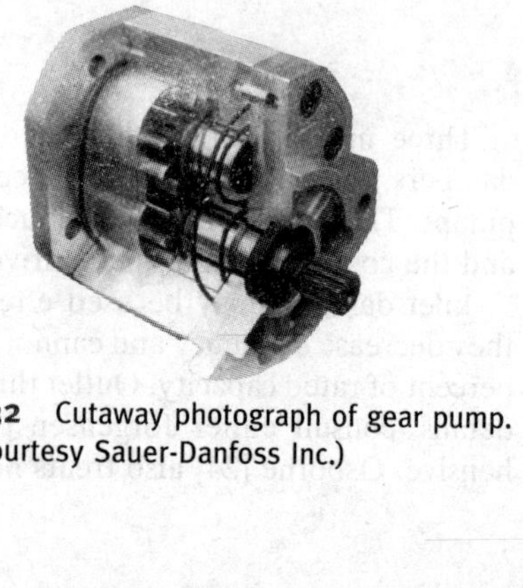

Fig. 10.32 Cutaway photograph of gear pump. (Photo courtesy Sauer-Danfoss Inc.)

developed within the pump. It also shows pressure-loaded side plates designed to "float"—to allow thermal expansion—while maintaining the smallest possible side clearance between gears and housing. Many ingenious designs have been developed for pumps; details are beyond the scope of our treatment here, which will focus on performance characteristics. For more details consult Lambeck [27] or Warring [28].

Typical performance curves of pressure versus delivery for a medium-duty gear pump are shown in Fig. 10.33. The pump size is specified by its displacement per revolution and the working fluid is characterized by its viscosity and temperature. Curves for tests at three constant speeds are presented in the diagram. At each speed, delivery decreases slightly as pressure is raised. The pump displaces the same volume, but as pressure is raised, both leakage and backflow increase; so delivery decreases slightly. Leakage fluid ends up in the pump housing; so a case drain must be provided to return this fluid to the system reservoir.

Volumetric efficiency—shown by the dashed curves—is defined as actual volumetric delivery divided by pump displacement. Volumetric efficiency decreases as pressure is raised or pump speed is reduced. *Overall efficiency*—shown by the solid curves—is defined as power delivered to the fluid divided by power input to the pump. Overall efficiency tends to rise (and reaches a maximum at intermediate pressure) as pump speed increases.

Thus far we have shown pumps of fixed displacement only. The extra cost and complication of variable-displacement pumps are motivated by the energy saving they permit during partial-flow operation. In a variable-displacement pump, delivery can be varied to accommodate the load. Load sensing can be used to reduce the delivery pressure and thus the energy expenditure still further during part-load operation. Some pump designs allow pressure relief to further reduce power loss during standby operation.

Figure 10.34 illustrates system losses with a fixed-displacement pump, compared with losses for variable-displacement and variable-pressure pumps. Assume the pressure and flow required by the load at partial-flow operation correspond to point L on the diagram. A fixed-displacement pump will operate along curve CD; its delivery will be at point A. Since the load requires only the flow at L, the remaining flow

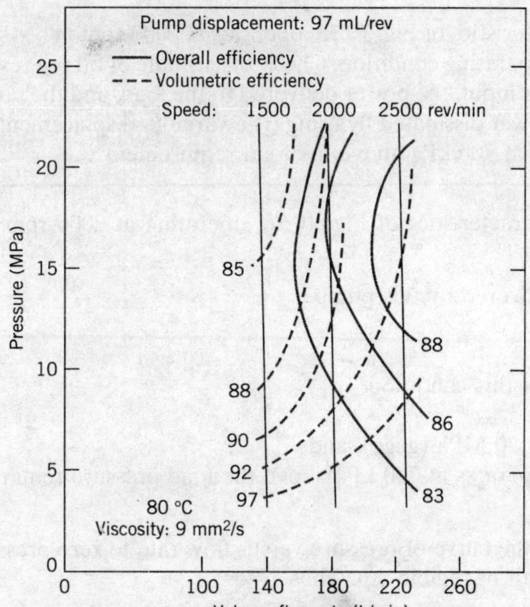

Fig. 10.33 Performance characteristics of typical gear pump [27].

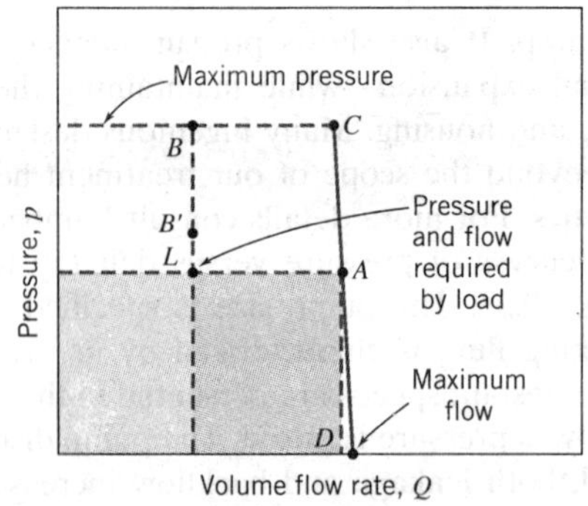

Fig. 10.34 Pressure-flow diagram illustrating system losses at part load [27].

(between L and A) must be bypassed back to the reservoir. Its pressure is dissipated by throttling. Consequently the system power loss will be the area beneath line LA.

A variable-displacement pump operating at constant pressure will deliver just enough flow to supply the load, but at a pressure represented by point B. The system power loss will be proportional to the area to the left of line BL. Control of delivery pressure using load sensing can be used to reduce power loss. With a load-sensing pump of variable displacement, the pressure supplied is only slightly higher than is needed to move the load. A pump with load sensing would operate at the flow and pressure of point B'. The system loss would be reduced significantly to the area to the left of line $B'L$.

The best system choice depends on the operating duty cycle. Complete details of these and other hydraulic power systems are presented in Lambeck [27].

Example 10.11 PERFORMANCE OF A POSITIVE-DISPLACEMENT PUMP

A hydraulic pump, with the performance characteristics of Fig. 10.33, operates at 2000 rpm in a system that requires $Q = 75$ L/min at $p = 10$ MPa to the load at one operating condition. Check the volume of oil per revolution delivered by this pump. Compute the required pump power input, the power delivered to the load, and the power dissipated by throttling at this condition. Compare with the power dissipated by using (i) a variable-displacement pump at 20 MPa and (ii) a pump with load sensing that operates at 700 kPa above the load requirement.

Given: Hydraulic pump, with performance characteristics of Fig. 10.36, operating at 2000 rpm. System requires $Q = 75$ L/min at $p = 10$ MPa (gage).

Find: (a) The volume of oil per revolution delivered by this pump.
(b) The required pump power input.
(c) The power delivered to the load.
(d) The power dissipated by throttling at this condition.
(e) The power dissipated using:
  (i) a variable-displacement pump at 20 MPa (gage), and
  (ii) a pump with load sensing that operates at 700 kPa above the load pressure requirement.

Solution:

To estimate the maximum delivery, extrapolate the curve of pressure versus flow rate to zero pressure. Under these conditions, $Q = 186$ L/min at $N = 2000$ rpm with negligible Δp. Thus

$$\forall = \frac{Q}{N} = 186 \frac{\text{L}}{\text{min}} \times \frac{\text{min}}{2000 \text{ rev}} \times \frac{1000 \text{ mL}^3}{\text{L}} = \frac{93 \text{ mL}}{\text{rev}} \longleftarrow \forall$$

The volumetric efficiency of the pump at maximum flow is

$$\eta_V = \frac{\forall_{calc}}{\forall_{pump}} = \frac{93}{97} = 0.959$$

The operating point of the pump may be found from Fig. 10.36. At 10 MPa (gage), it operates at $Q \approx 178$ L/min. The power delivered to the fluid is

$$\mathcal{P}_{fluid} = \rho Q g H_p = Q \Delta p_p$$

$$= 178 \frac{\text{L}}{\text{min}} \times \frac{\text{m}^3}{1000\ \text{L}} \times \frac{\text{min}}{60\ \text{s}} \times 10^7 \text{Pa} \times \frac{\text{N}}{\text{Pa} \cdot \text{m}^2} \times \frac{\text{J}}{\text{N} \cdot \text{m}} \times \frac{\text{W} \cdot \text{s}}{\text{J}}$$

$$\mathcal{P}_{fluid} = 29.7 \times 10^3\ \text{W}$$

From the graph, at this operating point, the pump efficiency is approximately $\eta = 0.84$. Therefore the required input power is

$$\mathcal{P}_{input} = \frac{\mathcal{P}_{fluid}}{\eta} = \frac{29.7 \times 10^3\ \text{W}}{0.84} = 35.4 \times 10^3\ \text{W} \longleftarrow \mathcal{P}_{input}$$

The power delivered to the load is

$$\mathcal{P}_{load} = Q_{load} \Delta p_{load}$$

$$= 75 \frac{\text{L}}{\text{min}} \times \frac{\text{m}^3}{1000\ \text{L}} \times \frac{\text{min}}{60\ \text{s}} \times 10^7\ \text{Pa} \times \frac{\text{N}}{\text{Pa} \cdot \text{m}^2} \times \frac{\text{J}}{\text{N} \cdot \text{m}} \times \frac{\text{W} \cdot \text{s}}{\text{J}}$$

$$\mathcal{P}_{load} = 12.5 \times 10^3\ \text{W} \longleftarrow \mathcal{P}_{load}$$

The power dissipated by throttling is

$$\mathcal{P}_{dissipated} = \mathcal{P}_{fluid} - \mathcal{P}_{load} = (29.7 - 12.5) \times 10^3\ \text{W} = 17.2 \times 10^3\ \text{W} \longleftarrow \mathcal{P}_{dissipated}$$

The dissipation with the variable-displacement pump is

$$\mathcal{P}_{var\text{-}disp} = Q_{load}(p_{oper} - p_{load})$$

$$= 75 \frac{\text{L}}{\text{min}} \times \frac{\text{m}^3}{1000\ \text{L}} \times \frac{\text{min}}{60\ \text{s}} \times (20 - 10) \times 10^6\ \text{Pa} \times \frac{\text{N}}{\text{Pa} \cdot \text{m}^2} \times \frac{\text{J}}{\text{N} \cdot \text{m}} \times \frac{\text{W} \cdot \text{s}}{\text{J}}$$

$$\mathcal{P}_{var\text{-}disp} = 12.5 \times 10^3\ \text{W} \longleftarrow \mathcal{P}_{var\text{-}disp}$$

The dissipation with the variable-displacement pump is therefore less than the 17.2×10^3 W dissipated with the constant-displacement pump and throttle. The saving is approximately 5×10^3 W.

The final computation is for the load-sensing pump. If the pump pressure is 700 kPa above that required by the load, the excess energy dissipation is

$$\mathcal{P}_{load\text{-}sense} = Q_{load}(p_{oper} - p_{load})$$

$$= 75 \frac{\text{L}}{\text{min}} \times \frac{\text{m}^3}{1000\ \text{L}} \times \frac{\text{min}}{60\ \text{s}} \times 700 \times 10^3\ \text{Pa} \times \frac{\text{N}}{\text{Pa} \cdot \text{m}^2} \times \frac{\text{J}}{\text{N} \cdot \text{m}} \times \frac{\text{W} \cdot \text{s}}{\text{J}}$$

$$\mathcal{P}_{load\text{-}sense} = 875\ \text{W} \longleftarrow \mathcal{P}_{load\text{-}sense}$$

This problem contrasts the performance of a system with a pump of constant displacement to that of a system with variable-displacement and load-sensing pumps. The specific savings depend on the system operating point and on the duty cycle of the system.

10.5 Hydraulic Turbines

Hydraulic Turbine Theory

The theory for machines doing work on a fluid, e.g., pumps, may be used for the analysis of machines extracting work from a fluid, namely turbines. The main difference is that the terms denoting torque, work, and power will be negative instead of positive. Example 10.12 below illustrates the application of the Euler turbomachine equation to a reaction turbine.

Example 10.12 IDEAL ANALYSIS OF A REACTION TURBINE

In a vertical-shaft Francis turbine the available head at the inlet flange of the turbine is 150 m and the vertical distance between the runner and the tailrace is 1.95 m. The runner tip speed is 34.5 m/s, the velocity of the water entering the runner is 39 m/s, and the velocity of the water exiting the runner is constant and equal to 10.5 m/s. The flow velocity at the exit of the draft tube is 3.45 m/s. The hydraulic energy losses estimated from the turbine are equal to 6 m at the volute, 1.05 m at the draft tube, and 9.9 m at the runner. Determine the pressure head (with respect to the tailrace) at the inlet and exit of the runner, the flow angle at the runner inlet, and the efficiency of the turbine.

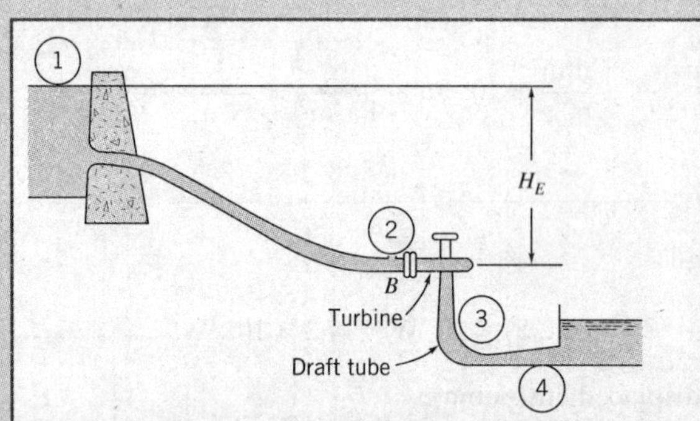

Given: Flow through a vertical shaft Francis turbine
Head at entrance: 150 m
Distance between runner and tailrace: 1.95 m
Runner tip speed: 34.5 m/s
Velocity at runner entrance: 39 m/s
Velocity at runner exit: 10.5 m/s
Flow velocity at draft tube exit: 3.45 m/s
Losses: 6 m at volute, 1.05 m at draft tube, 9.9 m at runner

Find: (a) Pressure head at inlet and exit of runner.
(b) Flow angle at runner inlet.
(c) Turbine efficiency.

Solution: Apply the energy and Euler turbomachine equations to the control volume.

Governing equations:

$$H = \frac{\dot{W}_m}{\dot{m}g} = \frac{1}{g}(U_2 V_{t_2} - U_1 V_{t_1}) \tag{10.2c}$$

$$\eta_t = \frac{\dot{W}_m}{\dot{W}_h} = \frac{\omega T}{\rho Q g H_t} \tag{10.4c}$$

$$\frac{p_1}{\rho g} + \alpha_1 \frac{\overline{V}_1^2}{2g} + z_1 + H_a = \frac{p_2}{\rho g} + \alpha_2 \frac{\overline{V}_2^2}{2g} + z_2 + \frac{h_{l_T}}{g} \tag{10.28b}$$

Assumptions: (1) Steady flow
(2) Uniform flow at each station
(3) Turbulent flow; $\alpha = 1$
(4) Reservoir and tailrace are at atmospheric pressure
(5) Reservoir is at stagnation condition; $\overline{V}_1 = 0$

(a) If we apply the energy equation between the runner exit and the tailrace:

$$H_3 = \frac{p_3 - p_{atm}}{\rho g} = \frac{\overline{V}_4^2 - \overline{V}_3^2}{2g} + \Delta H_{DT} + z_4$$

$$H_3 = \frac{1}{2} \times \left[\left(3.45 + \frac{\text{m}}{\text{s}}\right)^2 - \left(10.5\frac{\text{m}}{\text{s}}\right)^2\right] \times \frac{1\ \text{s}^2}{9.81\ \text{m}} + 1.05\ \text{m} - 1.95\ \text{m} = -5.91\ \text{m} \qquad H_3$$

(negative sign indicates suction)

Next we apply the energy equation between the runner entrance and the tailrace:

$$H_2 = \frac{p_2 - p_{atm}}{\rho g} = H_E - \Delta H_R - \frac{\overline{V}_2^2}{2g}$$

$$H_2 = 150\ \text{m} - 9.9\ \text{m} - \frac{1}{2} \times \left(39\frac{\text{m}}{\text{s}}\right)^2 \times \frac{1\text{s}^2}{9.81\ \text{m}} = 62.58\ \text{m} \qquad H_2$$

(b) Applying the energy equation across the entire system provides the work extraction through the turbine:

$$\frac{p_1}{\rho g} + \alpha_1 \frac{\overline{V}_1^2}{2g} + z_1 + H_a = \frac{p_4}{\rho g} + \alpha_4 \frac{\overline{V}_4^2}{2g} + z_4 + \frac{h_{l_T}}{g}$$

If we simplify the expression based on assumptions and solve for the head extracted at the turbine:

$$H_a = \frac{\overline{V}_4^2}{2g} - z_1 + z_4 + \sum \Delta H = \frac{\overline{V}_4^2}{2g} - (H_E + z) + (\Delta H_V + \Delta H_R + \Delta H_{DT})$$

Since station 1 is higher than station 4, we will take the negative of H_a and call that H_T, the head extracted at the turbine:

$$H_T = -\frac{\overline{V}_4^2}{2g} + (H_e + z) - (\Delta H_V + \Delta H_R + \Delta H_{DT})$$

$$= -\frac{1}{2} \times \left(3.45\frac{\text{m}}{\text{s}}\right)^2 \times \frac{1\text{s}^2}{9.81\ \text{m}} + (150\ \text{m} + 1.95\ \text{m}) - (6\ \text{m} + 9.9\ \text{m} + 1.05\ \text{m}) = 134.39\ \text{m}$$

Applying the Euler turbomachine equation to this system:

$$-H_T = \frac{U_3 V_{t_3} - U_2 V_{t_2}}{g}$$

Solving for the tangential velocity at 2:

$$V_{t_2} = \frac{g H_T}{U_2} = 9.81\frac{\text{m}}{\text{s}^2} \times 134.39\ \text{m} \times \frac{1}{34.5}\frac{\text{s}}{\text{m}} = 38.21\frac{\text{m}}{\text{s}}$$

Setting up the velocity triangle:

$$\beta_2 = \tan^{-1}\frac{V_{t_2} - U_2}{V_{n_2}} = \tan^{-1}\frac{38.21 - 34.5}{10.5} = 19.46° \longleftarrow \beta_2$$

$$\alpha_2 = \tan^{-1}\frac{V_{t_2}}{V_{n_2}} = \tan^{-1}\frac{38.21}{10.5} = 74.63 \longleftarrow \alpha_2$$

(c) To calculate the efficiency:

$$\eta_t = \frac{\dot{W}_m}{\dot{W}_h} = \frac{gH_T}{gH_E} = \frac{134.39}{150} = 89.59\% \longleftarrow \eta$$

This problem demonstrates the analysis of a hydraulic turbine with head losses and quantifies those effects in terms of a turbine efficiency. In addition, since the head at the turbine exit is below atmospheric, care must be taken to ensure that cavitation does not occur.

The trends predicted by the idealized angular-momentum theory, especially Eq. 10.18b and Fig. 10.12, are compared with experimental results in the next section.

Performance Characteristics for Hydraulic Turbines

The test procedure for turbines is similar to that for pumps, except that a dynamometer is used to absorb the turbine power output while speed and torque are measured. Turbines usually are intended to operate at a constant speed that is a fraction or multiple of the electric power frequency to be produced. Therefore turbine tests are run at constant speed under varying load, whereas water usage is measured and efficiency is calculated.

The impulse turbine is a relatively simple turbomachine, so we use it to illustrate typical test results. Impulse turbines are chosen when the head available exceeds about 300 m. Most impulse turbines used today are improved versions of the *Pelton wheel* developed in the 1880s by American mining engineer Lester Pelton [29]. An impulse turbine is supplied with water under high head through a long conduit called a *penstock*. The water is accelerated through a nozzle and discharges as a high-speed free jet at atmospheric pressure. The jet strikes deflecting buckets attached to the rim of a rotating wheel (Fig. 10.5*a*). Its kinetic energy is given up as it is turned by the buckets. Turbine output is controlled at essentially constant jet speed by changing the flow rate of water striking the buckets. A variable-area nozzle may be used to make small and gradual changes in turbine output. Larger or more rapid changes must be accomplished by means of jet deflectors, or auxiliary nozzles, to avoid sudden changes in flow speed and the resulting high pressures in the long water column in the penstock. Water discharged from the wheel at relatively low speed falls into the tailrace. The tailrace level is set to avoid submerging the wheel during flooded conditions. When large amounts of water are available, additional power can be obtained by connecting two wheels to a single shaft or by arranging two or more jets to strike a single wheel.

Figure 10.35 illustrates an impulse-turbine installation and the definitions of gross and net head [11]. The *gross head* available is the difference between the levels in the supply reservoir and the tailrace. The effective or *net head*, H, used to calculate efficiency, is the total head at the *entrance* to the nozzle, measured at the nozzle centerline [11]. Hence not all of the net head is converted into work at the turbine: Some is lost to turbine inefficiency, some is lost in the nozzle itself, and some is lost as residual kinetic energy in the exit flow. In practice, the penstock usually is sized so that at rated power the net head is 85–95 percent of the gross head.

In addition to nozzle loss, windage, bearing friction, and surface friction between the jet and bucket reduce performance compared with the ideal, frictionless case. Figure 10.36 shows typical results from tests performed at constant head.

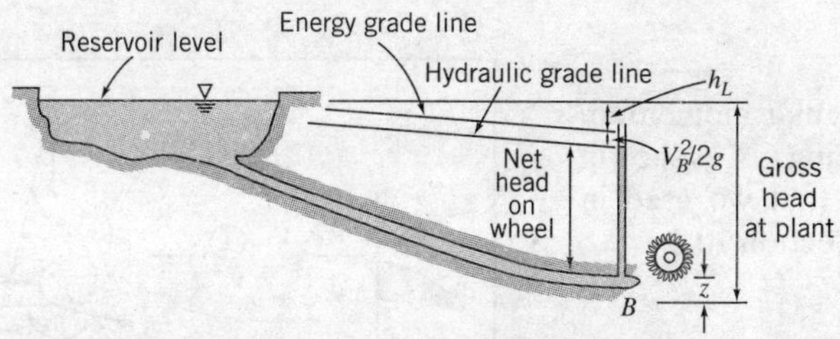

Fig. 10.35 Schematic of impulse-turbine installation, showing definitions of gross and net heads [11].

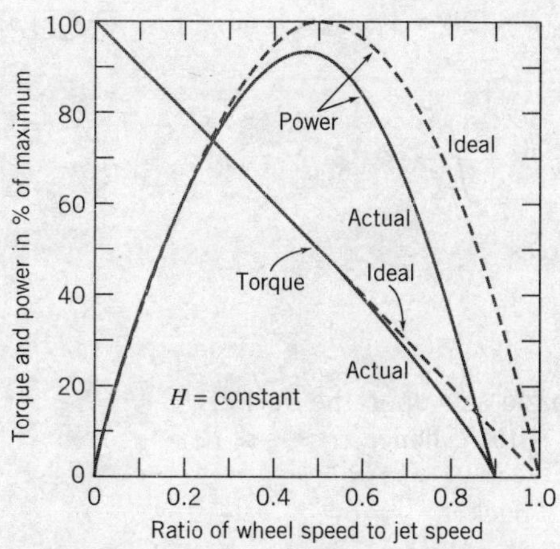

Fig. 10.36 Ideal and actual variable-speed performance for an impulse turbine [6].

The peak efficiency of the impulse turbine corresponds to the peak power, since the tests are performed at constant head and flow rate. For the ideal turbine, as we will see in Example 10.13, this occurs when the wheel speed is half the jet speed. As we will see, at this wheel speed the fluid exits the turbine at the lowest absolute velocity possible, hence minimizing the loss of kinetic energy at the exit. As indicated in Eq. 10.2a, if we minimize the exit velocity $\vec{V}_2$ we will maximize the turbine work $\dot{W}_m$, and hence the efficiency. In actual installations, peak efficiency occurs at a wheel speed only slightly less than half the jet speed. This condition fixes the wheel speed once the jet speed is determined for a given installation. For large units, overall efficiency may be as high as 88 percent [30].

Example 10.13 OPTIMUM SPEED FOR IMPULSE TURBINE

A Pelton wheel is a form of impulse turbine well adapted to situations of high head and low flow rate. Consider the Pelton wheel and single-jet arrangement shown, in which the jet stream strikes the bucket tangentially and is turned through angle θ. Obtain an expression for the torque exerted by the water stream on the wheel and the corresponding power output. Show that the power is a maximum when the bucket speed, $U = R\omega$, is half the jet speed, V.

Given: Pelton wheel and single jet shown.

Find: (a) Expression for torque exerted on the wheel.
(b) Expression for power output.
(c) Ratio of wheel speed U to jet speed V for maximum power.

Solution:

As an illustration of its use, we start with the angular-momentum equation, Eq. 4.52 (on the Web site), for a rotating CV as shown, rather than the inertial CV form, Eq. 4.46, that we used in deriving the Euler turbomachine equation in Section 10.2.

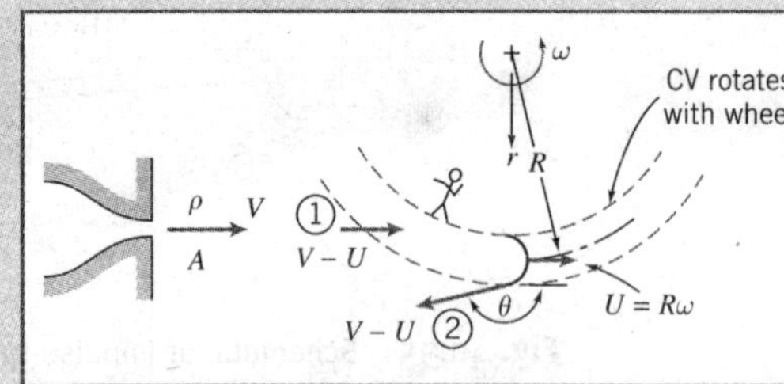

Governing equation:

$$\vec{r}\times\vec{F}_s + \int_{CV}\vec{r}\times\vec{g}\,\rho\,dV + \vec{T}_{\text{shaft}} - \int_{CV}\vec{r}\times[2\vec{\omega}\times\vec{V}_{xyz} + \vec{\omega}\times(\vec{\omega}\times\vec{r}) + \dot{\vec{\omega}}\times\vec{r}]\,\rho\,dV$$
$$= \frac{\partial}{\partial t}\int_{CV}\vec{r}\times\vec{V}_{xyz}\,\rho\,dV + \int_{CS}\vec{r}\times\vec{V}_{xyz}\,\rho\vec{V}_{xyz}\cdot d\vec{A} \tag{4.52}$$

(first term $= 0(1)$; second term $= 0(2)$; integral on the right of the first line $\approx 0(3)$; time-derivative term $= 0(4)$)

Assumptions:

(1) Neglect torque due to surface forces.
(2) Neglect torque due to body forces.
(3) Neglect mass of water on wheel.
(4) Steady flow with respect to wheel.
(5) All water that issues from the nozzle acts upon the buckets.
(6) Bucket height is small compared with R, hence $r_1 \approx r_2 \approx R$.
(7) Uniform flow at each section.
(8) No change in jet speed relative to bucket.

Then, since all water from the jet crosses the buckets,

$$\vec{T}_{\text{shaft}} = \vec{r}_1\times\vec{V}_1(-\rho VA) + \vec{r}_2\times\vec{V}_2(+\rho VA)$$

$$\vec{r}_1 = R\hat{e}_r \qquad \vec{r}_2 = R\hat{e}_r$$

$$\vec{V}_1 = (V-U)\hat{e}_\theta \qquad \vec{V}_2 = (V-U)\cos\theta\,\hat{e}_\theta + (V-U)\sin\theta\,\hat{e}_r$$

$$T_{\text{shaft}}\hat{k} = R(V-U)\hat{k}(-\rho VA) + R(V-U)\cos\theta\,\hat{k}(\rho VA)$$

so that finally

$$T_{\text{shaft}}\hat{k} = -R(1-\cos\theta)\rho VA(V-U)\hat{k}$$

This is the external torque of the shaft on the control volume, i.e., on the wheel. The torque exerted by the water on the wheel is equal and opposite,

$$\vec{T}_{\text{out}} = -\vec{T}_{\text{shaft}} = R(1-\cos\theta)\rho VA(V-U)\hat{k}$$

$$\vec{T}_{\text{out}} = \rho QR(V-U)\times(1-\cos\theta)\hat{k} \longleftarrow \underline{\vec{T}_{\text{out}}}$$

The corresponding power output is

$$\dot{W}_{\text{out}} = \vec{\omega}\cdot\vec{T}_{\text{out}} = R\omega(1-\cos\theta)\rho VA(V-U)$$

$$\dot{W}_{\text{out}} = \rho QU(V-U)\times(1-\cos\theta) \longleftarrow \underline{\dot{W}_{\text{out}}}$$

To find the condition for maximum power, differentiate the expression for power with respect to wheel speed U and set the result equal to zero. Thus

$$\frac{d\dot{W}}{dU} = \rho Q(V-U)(1-\cos\theta) + \rho QU(-1)(1-\cos\theta) = 0$$

$$\therefore (V-U) - U = V - 2U = 0$$

This problem illustrates the use of the angular-momentum equation for a rotating control volume, Eq. 4.52 (on the Web), to analyze flow through an ideal impulse turbine.

✓ The peak power occurs when the wheel speed is half the jet speed, which is a useful design criterion when selecting a turbine for a given available head.

✓ This problem also could be analyzed starting with an inertial control volume, i.e., using the Euler turbomachine equation (Problem 10.17).

Thus for maximum power, $U/V = \frac{1}{2}$ or $U = V/2$. ← U/V

Note: Turning the flow through $\theta = 180°$ would give maximum power with $U = V/2$. Under these conditions, theoretically the *absolute* velocity of the fluid at the exit (computed in the direction of U) would be $U - (V - U) = V/2 - (V - V/2) = 0$, so that the fluid would exit with zero kinetic energy, maximizing the power output. In practice, it is possible to deflect the jet stream through angles up to 165°. With $\theta = 165°$, $1 - \cos\theta \approx 1.97$, or about 1.5 percent below the value for maximum power.

In practice, hydraulic turbines usually are run at a constant speed, and output is varied by changing the opening area of the needle valve jet nozzle. Nozzle loss increases slightly and mechanical losses become a larger fraction of output as the valve is closed, so efficiency drops sharply at low load, as shown in Fig. 10.37. For this Pelton wheel, efficiency remains above 85 percent from 40 to 113 percent of full load.

At lower heads, reaction turbines provide better efficiency than impulse turbines. In contrast to flow in a centrifugal pump, used for doing work on a fluid, flow in a work-producing reaction turbine enters the rotor at the largest (outer) radial section and discharges at the smallest (inner) radial section after transferring most of its energy to the rotor. Reaction turbines tend to be high-flow, low-head machines. A typical reaction turbine installation is shown schematically in Fig. 10.38, where the terminology used to define the heads is indicated.

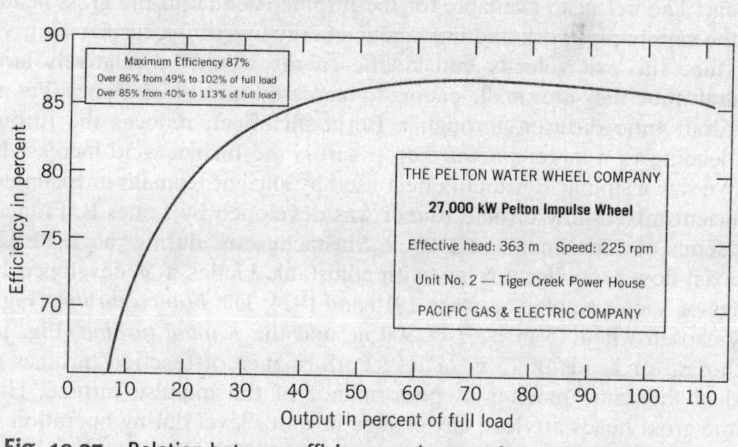

Fig. 10.37 Relation between efficiency and output for a typical Pelton water turbine (adapted from [30]).

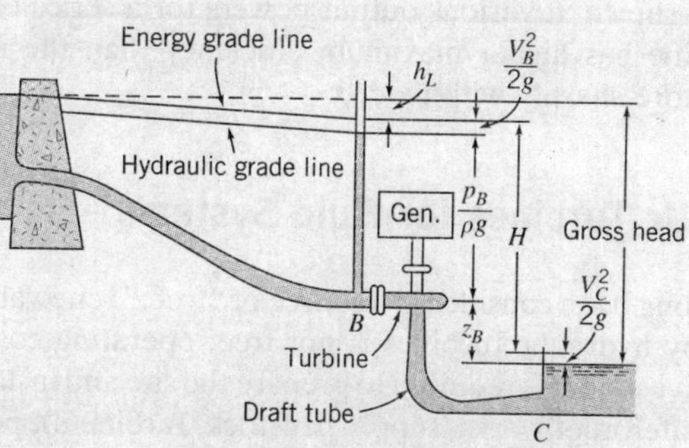

Fig. 10.38 Schematic of typical reaction turbine installation, showing definitions of head terminology [11].

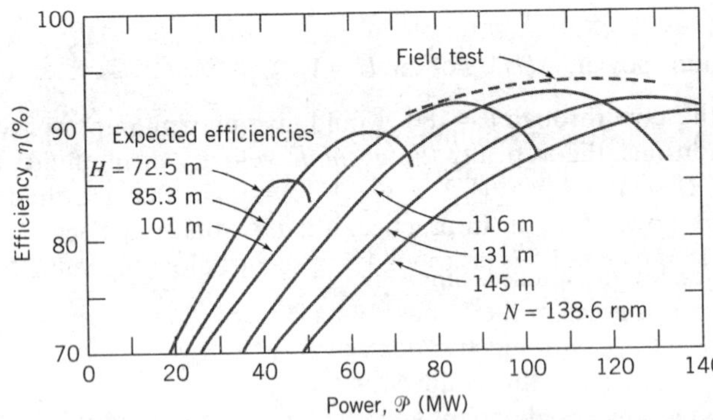

Fig. 10.39 Performance of typical reaction turbine as predicted by model tests (expected efficiencies) and confirmed by field test [6].

Reaction turbines flow full of water. Consequently, it is possible to use a diffuser or draft tube to regain a fraction of the kinetic energy that remains in water leaving the rotor. The draft tube forms an integral part of the installation design. As shown in Fig. 10.38, the *gross head* available is the difference between the supply reservoir head and the tailrace head. The *effective head* or *net head*, H, used to calculate efficiency, is the difference between the elevation of the energy grade line just upstream of the turbine and that of the draft tube discharge (section C). The benefit of the draft tube is clear: The net head available for the turbine is equal to the gross head minus losses in the supply pipework and the kinetic energy loss at the turbine exit. Without the draft tube the exit velocity and kinetic energy would be relatively large, but with the draft tube they are small, leading to increased turbine efficiency. Put another way, the draft tube diffuser, through a Bernoulli effect, reduces the turbine exit pressure, leading to a larger pressure drop across the turbine, and increased power output. (We saw a similar Bernoulli effect used by ancient Romans in Example 8.10.)

An efficient mixed-flow turbine runner was developed by James B. Francis using a careful series of experiments at Lowell, Massachusetts, during the 1840s [29]. An efficient axial-flow propeller turbine, with adjustable blades, was developed by German Professor Victor Kaplan between 1910 and 1924. The *Francis turbine* (Fig. 10.5*b*) is usually chosen when $15 \text{ m} \leq H \leq 300 \text{ m}$, and the *Kaplan turbine* (Fig. 10.5*c*) is usually chosen for heads of 15 m or less. Performance of reaction turbines may be measured in the same manner as performance of the impulse turbine. However, because the gross heads are less, any change in water level during operation is more significant. Consequently, measurements are made at a series of heads to completely define the performance of a reaction turbine.

An example of the data presentation for a reaction turbine is given in Fig. 10.39, where efficiency is shown at various output powers for a series of constant heads [6]. The reaction turbine has higher maximum efficiency than the impulse turbine, but efficiency varies more sharply with load.

Sizing Hydraulic Turbines for Fluid Systems

Falling water has long been considered a source of "free," renewable energy. In reality, power produced by hydraulic turbines is not free; operating costs are low, but considerable capital investment is required to prepare the site and install the equipment. At a minimum, the water inlet works, supply penstock, turbine(s), powerhouse, and controls must be provided. An economic analysis is necessary to determine the feasibility of developing any candidate site. In addition to economic factors, hydroelectric power plants must also be evaluated for their environmental impact; in recent years it has been found that such plants are not entirely benign and can be damaging, for example, to salmon runs.

Early in the industrial revolution, waterwheels were used to power grain mills and textile machinery. These plants had to be located at the site of the falling water, which limited use of water power to relatively small and local enterprises. The introduction of alternating current in the 1880s made it possible to transmit electrical energy efficiently over long distances. Since then nearly 40 percent of the available hydroelectric power resources in the United States have been developed and connected to the utility grid [31]. Hydroelectric power accounts for about 16 percent of the electrical energy produced in this country.

The United States has abundant and relatively cheap supplies of fossil fuels, mostly coal. Therefore at present the remaining hydropower resources in the United States are not considered economical compared to fossil-fired plants.

Worldwide, only about one-third of available hydropower resources have been developed commercially [32]. Considerably more hydropower will likely be developed in coming decades as countries become more industrialized. Many developing countries do not have their own supplies of fossil fuel. Hydropower may offer many such countries their only practical path to increased utility development. Consequently the design and installation of hydroelectric plants are likely to be important future engineering activities in developing countries.

To evaluate a candidate site for hydropower potential, one must know the average stream flow rate and the gross head available to make preliminary estimates of turbine type, number of turbines, and potential power production. Economic analyses are beyond the scope of this book, but we consider the fluids engineering fundamentals of impulse turbine performance to optimize the efficiency.

Hydraulic turbines convert the potential energy of stored water to mechanical work. To maximize turbine efficiency, it is always a design goal to discharge water from a turbine at ambient pressure, as close to the tailwater elevation as possible and with the minimum possible residual kinetic energy.

Conveying water flow into the turbine with minimum energy loss also is important. Numerous design details must be considered, such as inlet geometry, trash racks, etc. [31]. References 1, 6, 10, 31 and 33–38 contain a wealth of information about turbine siting, selection, hydraulic design, and optimization of hydropower plants. The number of large manufacturers has dwindled to just a few, but small-scale units are becoming plentiful [35]. The enormous cost of a commercial-scale hydro plant justifies the use of comprehensive scale-model testing to finalize design details. See [31] for a more detailed coverage of hydraulic power generation.

Hydraulic losses in long supply pipes (known as *penstocks*) must be considered when designing the installation for high-head machines such as impulse turbines; an optimum diameter for the inlet pipe that maximizes turbine output power can be determined for these units, as shown in Example 10.14.

Turbine power output is proportional to volume flow rate times the pressure difference across the nozzle. At zero flow, the full hydrostatic head is available but power is zero. As flow rate increases, the net head at the nozzle inlet decreases. Power first increases, reaches a maximum, then decreases again as flow rate increases. As we will see in Example 10.14, for a given penstock diameter, the theoretical maximum power is obtained when the system is designed so that one-third of the gross head is dissipated by friction losses in the penstock. In practice, penstock diameter is chosen larger than the theoretical minimum, and only 10–15 percent of the gross head is dissipated by friction [11].

A certain minimum penstock diameter is required to produce a given power output. The minimum diameter depends on the desired power output, the available head, and the penstock material and length. Some representative values are shown in Fig. 10.40.

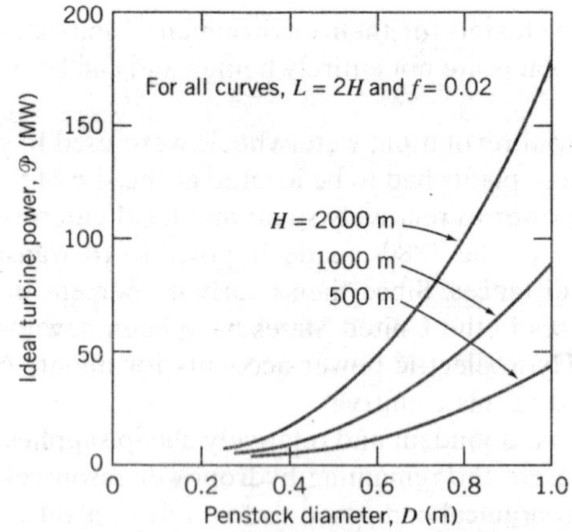

Fig. 10.40 Maximum hydraulic impulse turbine power output versus penstock diameter.

Example 10.14 PERFORMANCE AND OPTIMIZATION OF AN IMPULSE TURBINE

Consider the hypothetical impulse turbine installation shown. Analyze flow in the penstock to develop an expression for the optimum turbine output power as a function of jet diameter, D_j. Obtain an expression for the ratio of jet diameter, D_j, to penstock diameter, D, at which output power is maximized. Under conditions of maximum power output, show that the head loss in the penstock is one-third of the available head. Develop a parametric equation for the minimum penstock diameter needed to produce a specified power output, using gross head and penstock length as parameters.

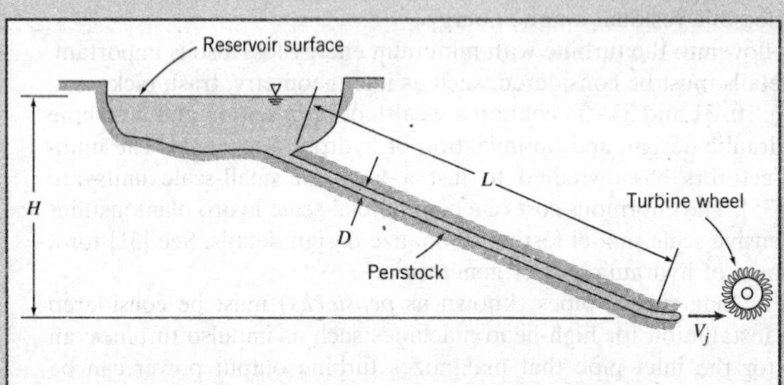

Given: Impulse turbine installation shown.

Find: (a) Expression for optimum turbine output power as a function of jet diameter.
(b) Expression for the ratio of jet diameter, D_j, to penstock diameter, D, at which output power is maximized.
(c) Ratio of head loss in penstock to available head for condition of maximum power.
(d) Parametric equation for the minimum penstock diameter needed to produce a specified output power, using gross head and penstock length as parameters.

Solution:

According to the results of Example 10.13, the output power of an idealized impulse turbine is given by $\mathcal{P}_{out} = \rho Q U (V - U)(1 - \cos\theta)$. For optimum power output, $U = V/2 = V_j/2$, and

$$\mathcal{P}_{\text{out}} = \rho Q \frac{V}{2}\left(V - \frac{V}{2}\right)(1 - \cos\theta) = \rho A_j V_j \frac{V_j}{2}\frac{V_j}{2}(1 - \cos\theta)$$

$$\mathcal{P}_{\text{out}} = \rho A_j \frac{V_j^3}{4}(1 - \cos\theta)$$

Thus output power is proportional to $A_j\, V_j^3$.

Apply the energy equation for steady incompressible pipe flow through the penstock to analyze V_j^2 at the nozzle outlet. Designate the free surface of the reservoir as section ①; there $\overline{V}_1 \approx 0$.

Governing equation:

$$\left(\frac{p_1}{\rho} + \alpha_1 \frac{\overset{\approx 0}{\overline{V}_1^2}}{2} + gz_1\right) - \left(\frac{p_j}{\rho} + \alpha_j \frac{\overline{V}_j^2}{2} + gz_j\right) = h_{l_T} = \left(K_{\text{ent}} + f\frac{L}{D}\right)\frac{\overline{V}_p^2}{2} + K_{\text{nozzle}}\frac{\overline{V}_j^2}{2}$$

Assumptions:
(1) Steady flow.
(2) Incompressible flow.
(3) Fully developed flow.
(4) Atmospheric pressure at jet exit.
(5) $\alpha_j = 1$, so $\overline{V}_j = V_j$.
(6) Uniform flow in penstock, so $\overline{V}_p = V$.
(7) $K_{\text{ent}} \ll f\dfrac{L}{D}$.
(8) $K_{\text{nozzle}} = 1$.

Then

$$g(z_1 - z_j) = gH = f\frac{L}{D}\frac{V^2}{2} + \frac{V_j^2}{2} \quad \text{or} \quad V_j^2 = 2gH - f\frac{L}{D}V^2 \tag{1}$$

Hence the available head is partly consumed in friction in the supply penstock, and the rest is available as kinetic energy at the jet exit—in other words, the jet kinetic energy is reduced by the loss in the penstock. However, this loss itself is a function of jet speed, as we see from continuity:

$$VA = V_j A_j, \text{ so } V = V_j \frac{A_j}{A} = V_j\left(\frac{D_j}{D}\right)^2 \quad \text{and} \quad V_j^2 = 2gH - f\frac{L}{D}V_j^2\left(\frac{D_j}{D}\right)^4$$

Solving for V_j, we obtain

$$V_j = \left[\frac{2gH}{\left\{1 + f\dfrac{L}{D}\left(\dfrac{D_j}{D}\right)^4\right\}}\right]^{1/2} \tag{2}$$

The turbine power can be written as

$$\mathcal{P} = \rho A_j \frac{V_j^3}{4}(1 - \cos\theta) = \rho\frac{\pi}{16}D_j^2\left[\frac{2gH}{\left\{1 + f\dfrac{L}{D}\left(\dfrac{D_j}{D}\right)^4\right\}}\right]^{3/2}(1 - \cos\theta)$$

$$\mathcal{P} = C_1 D_j^2\left[1 + f\frac{L}{D}\left(\frac{D_j}{D}\right)^4\right]^{-3/2} \longleftarrow \mathcal{P}$$

where $C_1 = \rho\pi(2gH)^{3/2}(1 - \cos\theta)/16 = \text{constant}$.

To find the condition for maximum power output, at fixed penstock diameter, D, differentiate with respect to D_j and set equal to zero,

$$\frac{d\mathcal{P}}{dD_j} = 2C_1 D_j\left[1 + f\frac{L}{D}\left(\frac{D_j}{D}\right)^4\right]^{-3/2} - \frac{3}{2}C_1 D_j^2\left[1 + f\frac{L}{D}\left(\frac{D_j}{D}\right)^4\right]^{-5/2} 4f\frac{L}{D}\frac{D_j^3}{D^4} = 0$$

Thus,

$$1 + f\frac{L}{D}\left(\frac{D_j}{D}\right)^4 = 3f\frac{L}{D}\left(\frac{D_j}{D}\right)^4$$

Solving for D_j/D, we obtain

$$\frac{D_j}{D} = \left[\frac{1}{2f\frac{L}{D}}\right]^{1/4}$$

$\dfrac{D_j}{D}$

At this optimum value of D_j/D, the jet speed is given by Eq. 2 as

$$V_j = \left[\frac{2gH}{\left\{1 + f\frac{L}{D}\left(\frac{D_j}{D}\right)^4\right\}}\right]^{1/2} = \sqrt{\frac{4}{3}gH}$$

The head loss at maximum power is then obtained from Eq. 1 after rearranging

$$h_l = f\frac{L}{D}\frac{V^2}{2} = gH - \frac{V_j^2}{2} = gH - \frac{2}{3}gH = \frac{1}{3}gH$$

and

$$\frac{h_l}{gH} = \frac{1}{3}$$

$\dfrac{h_l}{gH}$

Under the conditions of maximum power

$$\mathcal{P}_{max} = \rho V_j^3 \frac{A_j}{4}(1 - \cos\theta) = \rho\left(\frac{4}{3}gH\right)^{3/2}\frac{\pi}{16}\left[\frac{D^5}{2fL}\right]^{1/2}(1 - \cos\theta)$$

Finally, to solve for minimum penstock diameter for fixed output power, the equation may be written in the form

$$D \propto \left(\frac{L}{H}\right)^{1/5}\left(\frac{\mathcal{P}}{H}\right)^{2/5}$$

D

This problem illustrates the optimization of an idealized impulse turbine. The analysis determines the minimum penstock diameter needed to obtain a specified power output. In practice, larger diameters than this are used, reducing the frictional head loss below that computed here.

10.6 Propellers and Wind-Power Machines

As mentioned in Section 10.1, propellers and wind-power machines such as windmills and wind turbines may be considered axial-flow machines without housings [6]. Despite their long history (propellers were used on marine craft as early as 1776, and wind-power machines discovered in Persia date back to some time between the 6th and 10th centuries C.E. [39]), they have been proven to be efficient methods of propulsion and energy generation.

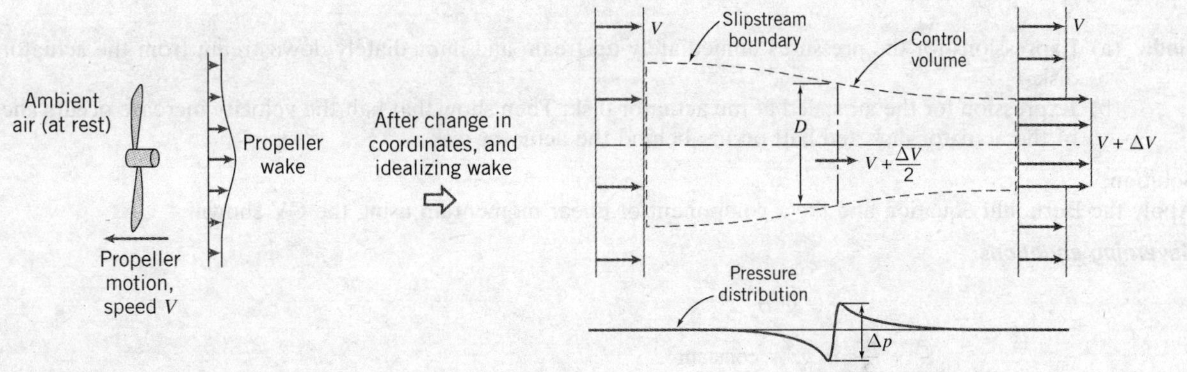

Fig. 10.41 One-dimensional flow model and control volume used to analyze an idealized propeller [6].

Propellers

In common with other propulsion devices, a propeller produces thrust by imparting linear momentum to a fluid. Thrust production always leaves the stream with some kinetic energy and angular momentum that are not recoverable, so the process is never 100 percent efficient.

The one-dimensional flow model shown schematically in Fig. 10.41 is drawn in absolute coordinates on the left and as seen by an observer moving with the propeller, at speed V, on the right. The wake is modeled as a uniform flow as shown, and in the new coordinates the flow is steady. The actual propeller is replaced conceptually by a thin *actuator disk*, across which flow speed is continuous but pressure rises abruptly. Relative to the propeller, the upstream flow is at speed V and ambient pressure. The axial speed at the actuator disk is $V + \Delta V/2$, with a corresponding reduction in pressure. Downstream, the speed is $V + \Delta V$ and the pressure returns to ambient. (Example 10.15 will show that half the speed increase occurs before and half after the actuator disk.) The contraction of the slipstream area to satisfy continuity and the pressure rise across the propeller disk are shown in the figure.

Not shown in the figure are the swirl velocities that result from the torque required to turn the propeller. The kinetic energy of the swirl in the slipstream also is lost unless it is removed by a counter-rotating propeller or partially recovered in stationary guide vanes.

As for all turbomachinery, propellers may be analyzed in two ways. Application of linear momentum in the axial direction, using a finite control volume, provides overall relations among slipstream speed, thrust, useful power output, and minimum residual kinetic energy in the slipstream. A more detailed *blade element* theory is needed to calculate the interaction between a propeller blade and the stream. A general relation for ideal propulsive efficiency can be derived using the control volume approach, as shown following Example 10.15.

Example 10.15 CONTROL VOLUME ANALYSIS OF IDEALIZED FLOW THROUGH A PROPELLER

Consider the one-dimensional model for the idealized flow through a propeller shown in Fig. 10.41. The propeller advances into still air at steady speed V_1. Obtain expressions for the pressure immediately upstream and the pressure immediately downstream from the actuator disk. Write the thrust on the propeller as the product of this pressure difference times the disk area. Equate this expression for thrust to one obtained by applying the linear momentum equation to the control volume. Show that half the velocity increase occurs ahead of and half behind the propeller disk.

Given: Propeller advancing into still air at speed V_1, as shown in Fig. 10.41.

Find: (a) Expressions for the pressures immediately upstream and immediately downstream from the actuator disk.

(b) Expression for the air speed at the actuator disk. Then show that half the velocity increase occurs ahead of the actuator disk and half occurs behind the actuator disk.

Solution:

Apply the Bernoulli equation and the x component of linear momentum using the CV shown.

Governing equations:

$$\frac{p}{\rho} + \frac{V^2}{2} + \overset{\approx 0(5)}{\cancel{gz}} = \text{constant}$$

$$F_{S_x} + \overset{=0(5)}{\cancel{F_{B_x}}} = \overset{=0(1)}{\cancel{\frac{\partial}{\partial t}\int_{\text{CV}} u_{xyz}\,\rho\, dV\!\!\!\!-}} + \int_{\text{CS}} u_{xyz}\,\rho \vec{V} \cdot d\vec{A}$$

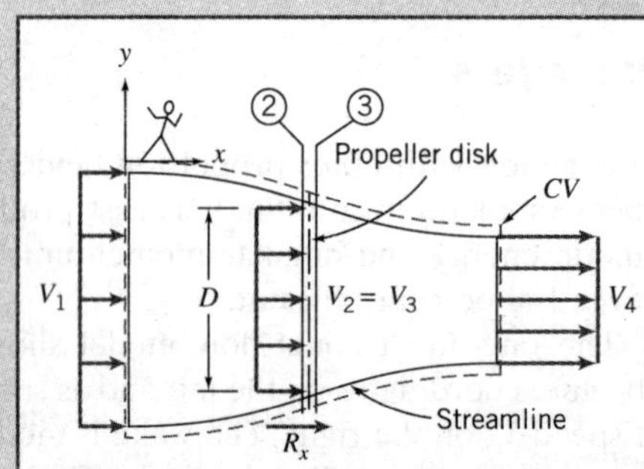

Assumptions:
(1) Steady flow relative to the CV.
(2) Incompressible flow.
(3) Flow along a streamline.
(4) Frictionless flow.
(5) Horizontal flow: neglect changes in z; $F_{B_x} = 0$.
(6) Uniform flow at each section.
(7) p_{atm} surrounds the CV.

Applying the Bernoulli equation from section ① to section ② gives

$$\frac{p_{\text{atm}}}{\rho} + \frac{V_1^2}{2} = \frac{p_2}{\rho} + \frac{V_2^2}{2}; \quad p_{2(\text{gage})} = \frac{1}{2}\rho\,(V_1^2 - V_2^2)$$

Applying Bernoulli from section ③ to section ④ gives

$$\frac{p_3}{\rho} + \frac{V_3^2}{2} = \frac{p_{\text{atm}}}{\rho} + \frac{V_4^2}{2}; \qquad p_{3(\text{gage})} = \frac{1}{2}\rho\,(V_4^2 - V_3^2)$$

The thrust on the propeller is given by

$$F_T = (p_3 - p_2)A = \frac{1}{2}\rho A(V_4^2 - V_1^2) \qquad (V_3 = V_2 = V)$$

From the momentum equation, using *relative* velocities,

$$R_x = F_T = u_1(-\dot{m}) + u_4(+\dot{m}) = \rho VA(V_4 - V_1) \qquad \{u_1 = V_1,\ u_4 = V_4\}$$

$$F_T = \rho VA(V_4 - V_1)$$

Equating these two expressions for F_T,

$$F_T = \frac{1}{2}\rho A(V_4^2 - V_1^2) = \rho VA(V_4 - V_1)$$

$$\text{or} \quad \frac{1}{2}(V_4 + V_1)(V_4 - V_1) = V(V_4 - V_1)$$

Thus, $V = \frac{1}{2}(V_1 + V_4)$, so

$$\Delta V_{12} = V - V_1 = \frac{1}{2}(V_1 + V_4) - V_1 = \frac{1}{2}(V_4 - V_1) = \frac{\Delta V}{2}$$

$$\Delta V_{34} = V_4 - V = V_4 - \frac{1}{2}(V_1 + V_4) = \frac{1}{2}(V_4 - V_1) = \frac{\Delta V}{2} \longleftarrow \text{Velocity Increase}$$

The purpose of this problem is to apply the continuity, momentum, and Bernoulli equations to an idealized flow model of a propeller, and to verify the Rankine theory of 1885 that half the velocity change occurs on either side of the propeller disk.

The continuity and momentum equations in control volume form were applied in Example 10.15 to the propeller flow shown in Fig. 10.41. The results obtained are discussed further below. The thrust produced is

$$F_T = \dot{m}\Delta V \tag{10.32}$$

For incompressible flow, in the absence of friction and heat transfer, the energy equation indicates that the minimum required input to the propeller is the power required to increase the kinetic energy of the flow, which may be expressed as

$$\mathcal{P}_{\text{input}} = \dot{m}\left[\frac{(V+\Delta V)^2}{2} - \frac{V^2}{2}\right] = \dot{m}\left[\frac{2V\Delta V + (\Delta V)^2}{2}\right] = \dot{m}V\Delta V\left[1 + \frac{\Delta V}{2V}\right] \tag{10.33}$$

The useful power produced is the product of thrust and speed of advance, V, of the propeller. Using Eq. 10.32, this may be written as

$$\mathcal{P}_{\text{useful}} = F_T V = \dot{m}V\Delta V \tag{10.34}$$

Combining Eqs. 10.33 and 10.34, and simplifying, gives the propulsive efficiency as

$$\eta = \frac{\mathcal{P}_{\text{useful}}}{\mathcal{P}_{\text{input}}} = \frac{1}{1 + \dfrac{\Delta V}{2V}} \tag{10.35}$$

Equations 10.32–10.35 are applicable to any device that creates thrust by increasing the speed of a fluid stream. Thus they apply equally well to propeller-driven or jet-propelled aircraft, boats, or ships.

Equation 10.35 for propulsive efficiency is of fundamental importance. It indicates that propulsive efficiency can be increased by reducing ΔV or by increasing V. At constant thrust, as shown by Eq. 10.32, ΔV can be reduced if $\dot{m}$ is increased, i.e., if more fluid is accelerated over a smaller speed increase. More mass flow can be handled if propeller diameter is increased, but overall size and tip speed ultimately limit this approach. The same principle is used to increase the propulsive efficiency of a turbofan engine by using a large fan to move additional air flow outside the engine core.

Propulsive efficiency also can be increased by increasing the speed of motion relative to the fluid. Speed of advance may be limited by cavitation in marine applications. Flight speed is limited for propeller-driven aircraft by compressibility effects at the propeller tips, but progress is being made in the design of propellers to maintain high efficiency with low noise levels while operating with transonic flow at the blade tips. Jet-propelled aircraft can fly much faster than propeller-driven craft, giving them superior propulsive efficiency.

The analysis provided does not reveal the length scale over which the axial velocity varies. Such an analysis is provided in [40]; the axial variation in velocity may be expressed as

$$V_{cl}(x) = V + \Delta V\left(1 - \frac{x}{\sqrt{x^2 + R^2}}\right) \tag{10.36}$$

In Eq. 10.36 $V_{cl}(x)$ is the centerline velocity at location x upstream of the disk, while V is the upstream velocity. This relationship is plotted in Fig. 10.42. The plot shows that the effects of the propeller are only felt at distances within two radii of the actuator disk.

A more detailed *blade element* theory may be used to calculate the interaction between a propeller blade and the stream and therefore to determine the effects of blade aerodynamic drag on the propeller efficiency. If the blade spacing is large and the

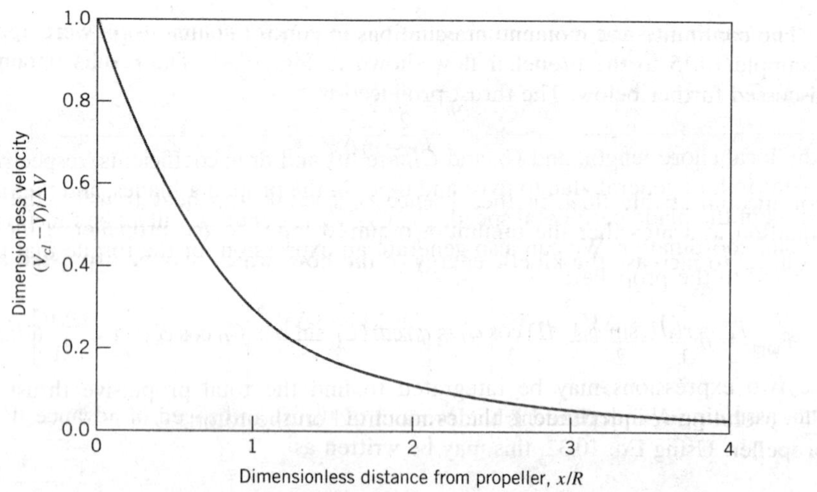

Fig. 10.42 Plot of velocity versus distance for flow of air near a propeller.

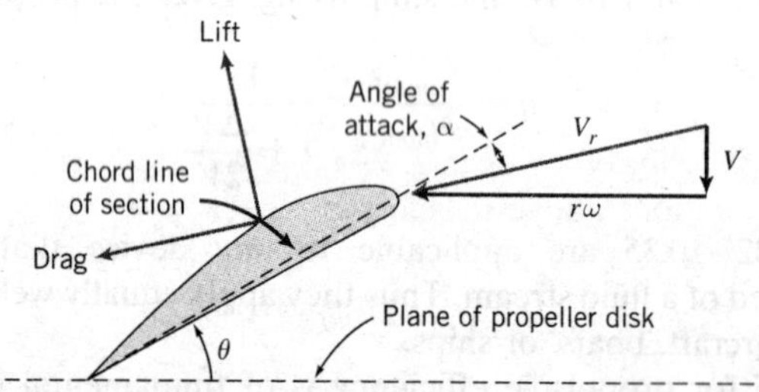

Fig. 10.43 Diagram of blade element and relative flow velocity vector.

disk loading[9] is light, blades can be considered independent, and relations can be derived for the torque required and the thrust produced by a propeller. These approximate relations are most accurate for low-solidity propellers.[10] Aircraft propellers typically are of fairly low solidity, having long, thin blades.

A schematic diagram of an element of a rotating propeller blade is shown in Fig. 10.43. The blade is set at angle θ to the plane of the propeller disk and has a thickness (into the plane of the page) of dr. Flow is shown as it would be seen by an observer on the propeller blade. Lift and drag forces are exerted on the blade perpendicular and parallel to the relative velocity vector V_r, respectively. We call the angle that V_r makes with the plane of the propeller disk the *effective pitch angle*, ϕ, and therefore the lift and drag forces are inclined at an angle to the propeller rotation axis and the plane of the propeller disk, respectively.

The relative speed of flow, V_r, passing over the blade element depends on both the blade peripheral speed, $r\omega$, and the *speed of advance*, V. Consequently, for a given blade setting, the angle of attack, α, depends on *both* V and $r\omega$. Thus, the performance of a propeller is influenced by both ω and V.

If we take a free-body diagram of the airfoil element of width dr in Fig. 10.43, we find that the magnitude of the resultant force dF_r parallel to the velocity vector $\vec{V}$:

$$dF_T = dL \cos\phi - dD \sin\phi = q_r c dr(C_L \cos\phi - C_D \sin\phi) \tag{10.37a}$$

[9]*Disk loading* is the propeller thrust divided by the swept area of the actuator disk.

[10]*Solidity* is defined as the ratio of projected blade area to the swept area of the actuator disk.

In this equation q_r is the dynamic pressure based on the relative velocity V_r,

$$q_r = \frac{1}{2}\rho V_r^2$$

c is the local chord length, and C_L and C_D are lift and drag coefficients, respectively, for the airfoil. In general, due to twist and taper in the propeller blades, and the radial variation of the blade peripheral speed, C_L, C_D, V_r, c, ϕ, and q_r will all be functions of the radial coordinate r. We can also generate an expression for the torque that must be applied to the propeller:

$$dT = r(dL \sin\phi + dD \cos\phi) = q_r rcdr(C_L \sin\phi + C_D \cos\phi) \qquad (10.37b)$$

These two expressions may be integrated to find the total propulsive thrust and torque, assuming N independent blades mounted on the rotor:

$$F_T = N\int_{r=R_{\text{hub}}}^{r=R} dF_T = qN\int_{R_{\text{hub}}}^{R} \frac{(C_L \cos\phi - C_D \sin\phi)}{\sin^2\phi} cdr \qquad (10.38a)$$

$$T = N\int_{r=R_{\text{hub}}}^{r=R} dT = qN\int_{R_{\text{hub}}}^{R} \frac{(C_L \sin\phi + C_D \cos\phi)}{\sin^2\phi} rcdr \qquad (10.38b)$$

In these equations, q_r is replaced by $q/\sin^2\phi$ based on the relationship between V and V_r. We will use the equations above to analyze the startup characteristics of a propeller in Example 10.16.

Example 10.16 PROPELLER STARTUP THRUST AND TORQUE

Use blade element theory to estimate the start-up thrust and torque for a propeller consisting of N independent blades with constant chord length, c, and at a constant angle, θ, with respect to the actuator disk plane.

Given: Propeller with N independent blades
Chord length c is constant
Angle with respect to actuator disk θ is constant

Find: Expressions for startup thrust and torque

Solution: Apply the equations presented above to the propeller:

Governing equations:

$$dF_T = dL \cos\phi - dD \sin\phi = q_r cdr(C_L \cos\phi - C_D \sin\phi) \qquad (10.37a)$$

$$dT = r(dL \sin\phi + dD \cos\phi) = q_r rcdr(C_L \sin\phi + C_D \cos\phi) \qquad (10.37b)$$

$$F_T = qN\int_{R_{\text{hub}}}^{R} \frac{(C_L \cos\phi - C_D \sin\phi)}{\sin^2\phi} cdr \qquad (10.38a)$$

$$T = qN\int_{R_{\text{hub}}}^{R} \frac{(C_L \sin\phi + C_D \cos\phi)}{\sin^2\phi} rcdr \qquad (10.38b)$$

Assumptions: Local wind velocity V is negligible.
Angular velocity ω is constant.

If at start-up we neglect the local wind velocity V, we find that the integrals in Eqs. 10.38 will be indeterminate since $q=0$ and $\phi=0$. Therefore, we will use the differential thrust and torque expressions given in Eqs. 10.37 and integrate them. At start-up, the relative velocity V_r is simply equal to the local blade element velocity $r\omega$. Therefore, the relative dynamic pressure q_r is equal to:

$$q_r = \frac{1}{2}\rho r^2\omega^2$$

When $\phi=0$, the differential thrust and torque expressions become

$$dF_T = \frac{1}{2}\rho r^2\omega^2 c\,dr(C_L\cos 0 - C_D\sin 0) = \frac{1}{2}\rho\omega^2 cC_L r^2 dr$$

$$dT = \frac{1}{2}\rho r^2\omega^2 rc\,dr(C_L\sin 0 + C_D\cos 0) = \frac{1}{2}\rho\omega^2 cC_D r^3 dr$$

We can then integrate the thrust and torque over the entire actuator disk:

$$F_T = N\int dF_T = \frac{1}{2}\rho\omega^2 cC_L \int_{R_{\text{hub}}}^{R} r^2 dr = \frac{1}{2}\rho\omega^2 cC_L \times \frac{1}{3}(R^3 - R_{\text{hub}}^3)$$

$$T = N\int dT = \frac{1}{2}\rho\omega^2 cC_D \int_{R_{\text{hub}}}^{R} r^3 dr = \frac{1}{2}\rho\omega^2 cC_D \times \frac{1}{4}(R^4 - R_{\text{hub}}^4)$$

When we collect terms and simplify we get the following expressions:

$$F_{T_{\text{startup}}} = \frac{\rho\omega^2 cC_L}{6}(R^3 - R_{\text{hub}}^3) \leftarrow F_{T_{\text{startup}}}$$

$$T_{\text{startup}} = \frac{\rho\omega^2 cC_D}{8}(R^4 - R_{\text{hub}}^4) \leftarrow T_{\text{startup}}$$

This problem demonstrates the analysis of a propeller using blade element theory. While the expressions here seem relatively simple, it is important to note that the lift and drag coefficients, C_L and C_D, are functions of the airfoil section being used, as well as the local angle of attack, α, which for $V=0$ is equal to the blade inclination angle θ. In addition, it should also be noted that when airfoil lift and drag coefficients are presented, such as in Figs. 9.17 or 9.19, they are typically given at high Reynolds numbers, where the flow is fully turbulent and the lift and drag are insensitive to changes in speed. Care needs to be taken to make sure that the lift and drag coefficients used are appropriate for the Reynolds number at startup.

While these expressions may be relatively simple to derive, they are difficult to evaluate. Even if the geometry of the propeller is adjusted to give constant geometric pitch,[11] the flow field in which it operates may not be uniform. Thus, the angle of attack across the blade elements may vary from the ideal, and it can be calculated only with the aid of a comprehensive computer code that can predict local flow directions and speeds. As a result, Eqs. 10.38 are not normally used, and propeller performance characteristics usually are measured experimentally.

Figure 10.44 shows typical measured characteristics for a marine propeller [6] and for an aircraft propeller [41]. The variables used to plot the characteristics are almost dimensionless: by convention, rotational speed, n, is expressed in revolutions per second (rather than as ω, in radians per second). The independent variable is the *speed of advance coefficient*, J,

$$J \equiv \frac{V}{nD} \tag{10.39}$$

[11] *Pitch* is defined as the distance a propeller would travel in still fluid per revolution if it advanced along the blade setting angle θ. The pitch, H, of this blade element is equal to $2\pi r\tan\theta$. To obtain constant pitch along the blade, θ must follow the relation, $\tan\theta = H/2\pi r$, from hub to tip. Thus the geometric blade angle is smallest at the tip and increases steadily toward the root.

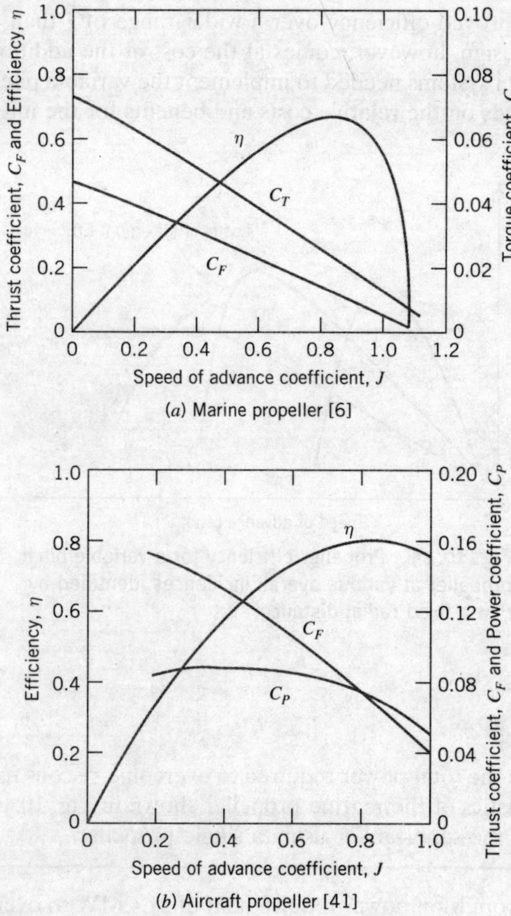

Fig. 10.44 Typical measured characteristics of two propellers.

Dependent variables are the *thrust coefficient*, C_F, the *torque coefficient*, C_T, the *power coefficient*, C_P, and the *propeller efficiency*, η, defined as

$$C_F = \frac{F_T}{\rho n^2 D^4}, \quad C_T = \frac{T}{\rho n^2 D^5}, \quad C_P = \frac{\mathcal{P}}{\rho n^3 D^5}, \quad \text{and} \quad \eta = \frac{F_T V}{\mathcal{P}_{input}} \tag{10.40}$$

The performance curves for both propellers show similar trends. Both thrust and torque coefficients are highest, and efficiency is zero, at zero speed of advance. This corresponds to the largest angle of attack for each blade element ($\alpha = \alpha_{max} = \theta$). Efficiency is zero because no useful work is being done by the stationary propeller. As advance speed increases, thrust and torque decrease smoothly. Efficiency increases to a maximum at an optimum advance speed and then decreases to zero as thrust tends to zero. (For example, if the blade element section is symmetric, this would theoretically occur when $\alpha = 0$, or when $\tan\theta = V/\omega r$.)

In order to improve performance, some propellers are designed with variable pitch. The performance of a variable-pitch propeller is shown in Fig. 10.45. This figure shows efficiency curves (solid curves) for a propeller set to different pitch angles. As we saw in Fig. 10.44, the propeller exhibits a maximum η at a certain value of J. However, the value of J needed for maximum η varies with θ. If we trace out all of the maxima, the result is the dashed curve in Fig. 10.45. Therefore, if we allow for the variation of θ, we may achieve improved efficiency over a wider range of J than with a fixed-pitch propeller. Such a design, however, comes at the cost of the additional complexity in actuators and control systems needed to implement the variable pitch, so the selection of this design depends on the relative costs and benefits for the intended application.

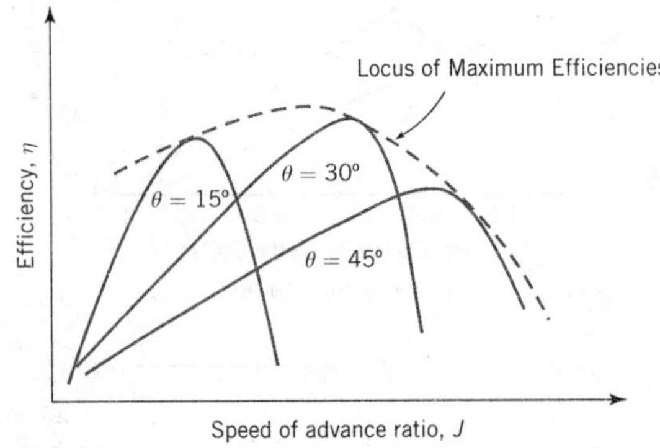

Fig. 10.45 Propeller efficiency for a variable-pitch propeller at various overall incidences identified by θ at a fixed radial distance.

Example 10.17 SIZING A MARINE PROPELLER

Consider the supertanker of Example 9.5. Assume the total power required to overcome viscous resistance and wave drag is 11.4 MW. Use the performance characteristics of the marine propeller shown in Fig. 10.44*a* to estimate the diameter and operating speed required to propel the supertanker using a single propeller.

Given: Supertanker of Example 9.5, with total propulsion power requirement of 11.4 MW to overcome viscous and wave drag, and performance data for the marine propeller shown in Fig. 10.44*a*.

Find: (a) An estimate of the diameter of a single propeller required to power the ship.
(b) The operating speed of this propeller.

Solution:
From the curves in Fig. 10.44*a*, at optimum propeller efficiency, the coefficients are

$$J = 0.85, \quad C_F = 0.10, \quad C_T = 0.020, \quad \text{and} \quad \eta = 0.66$$

The ship steams at $V = 6.69$ m/s and requires $\mathcal{P}_{\text{useful}} = 11.4$ MW. Therefore, the propeller thrust must be

$$F_T = \frac{\mathcal{P}_{\text{useful}}}{V} = 11.4 \times 10^6 \,\text{W} \times \frac{\text{s}}{6.69\,\text{m}} \times \frac{\text{N}\cdot\text{m}}{\text{W}\cdot\text{s}} = 1.70 \,\text{MN}$$

The required power input to the propeller is

$$\mathcal{P}_{\text{input}} = \frac{\mathcal{P}_{\text{useful}}}{\eta} = \frac{11.4\,\text{MW}}{0.66} = 17.3\,\text{MW}$$

From $J = \dfrac{V}{nD} = 0.85$, then

$$nD = \frac{V}{J} = 6.69\frac{\text{m}}{\text{s}} \times \frac{1}{0.85} = 7.87\,\text{m/s}$$

Since

$$C_F = \frac{F_T}{\rho n^2 D^4} = 0.10 = \frac{F_T}{\rho(n^2D^2)D^2} = \frac{F_T}{\rho(nD)^2 D^2}$$

solving for D gives

$$D = \left[\frac{F_T}{\rho (nD)^2 C_F}\right]^{1/2} = \left[1.70 \times 10^6\ \text{N} \times \frac{\text{m}^3}{1025\ \text{kg}} \times \frac{\text{s}^2}{(7.87)^2\ \text{m}^2} \times \frac{1}{0.10} \times \frac{\text{kg}\cdot\text{m}}{\text{N}\cdot\text{s}^2}\right]^{1/2}$$

$$D = 16.4\ \text{m} \qquad \longleftarrow \qquad D$$

This problem illustrates the use of normalized coefficient data for the preliminary sizing of a marine propeller. This preliminary design process would be repeated, using data for other propeller types, to find the optimum combination of propeller size, speed, and efficiency.

From $nD = \dfrac{V}{J} = 7.87\ \text{m/s}$, $n = \dfrac{nD}{D} = 7.87\dfrac{\text{m}}{\text{s}} \times \dfrac{1}{16.4\ \text{m}} = 0.480\ \text{rev/s}$

so that

$$n = \frac{0.480\ \text{rev}}{\text{s}} \times 60\frac{\text{s}}{\text{min}} = 28.8\ \text{rev/min} \qquad \longleftarrow \qquad n$$

The required propeller is quite large, but still smaller than the 25 m draft of the supertanker. The ship would need to take on seawater for ballast to keep the propeller submerged when not carrying a full cargo of petroleum.

Marine propellers tend to have high solidity. This packs a lot of lifting surface within the swept area of the disk to keep the pressure difference small across the propeller and to avoid cavitation. Cavitation tends to unload the blades of a marine propeller, reducing both the torque required and the thrust produced [6]. Cavitation becomes more prevalent along the blades as the cavitation number,

$$Ca = \frac{p - p_v}{\frac{1}{2}\rho V^2} \tag{10.41}$$

is reduced. Inspection of Eq. 10.41 shows that Ca decreases when p is reduced by operating near the free surface or by increasing V. Those who have operated motor boats also are aware that local cavitation can be caused by distorted flow approaching the propeller, e.g., from turning sharply.

Compressibility affects aircraft propellers when tip speeds approach the *critical Mach number*, at which the local Mach number approaches $M = 1$ at some point on the blade. Under these conditions, torque increases because of increased drag, thrust drops because of reduced section lift, and efficiency drops drastically.

If a propeller operates within the boundary layer of a propelled body, where the relative flow is slowed, its apparent thrust and torque may increase compared with those in a uniform freestream at the same rate of advance. The residual kinetic energy in the slipstream also may be reduced. The combination of these effects may increase the overall propulsive efficiency of the combined body and propeller. Advanced computer codes are used in the design of modern ships (and submarines, where noise may be an overriding consideration) to optimize performance of each propeller/hull combination.

For certain special applications, a propeller may be placed within a *shroud* or *duct*. Such configurations may be integrated into a hull (e.g., as a bow thruster to increase maneuverability), built into the wing of an aircraft, or placed on the deck of a hovercraft. Thrust may be improved by the favorable pressure forces on the duct lip, but efficiency may be reduced by the added skin-friction losses encountered in the duct.

Wind-Power Machines

Windmills (or more properly, wind turbines) have been used for centuries to harness the power of natural winds. Two well-known classical examples are shown in Fig. 10.46.

(a) Traditional Dutch Mill (b) American farm windmill

Fig. 10.46 Examples of well-known windmills [42]. (Photo courtesy of (a) Catharina van den Dikkenberg/iStockphoto, (b) U.S. Department of Agriculture.)

Dutch windmills (Fig. 10.46*a*) turned slowly so that the power could be used to turn stone wheels for milling grain; hence the name "windmill." They evolved into large structures; the practical maximum size was limited by the materials of the day. Calvert [43] reports that, based on his laboratory-scale tests, a traditional Dutch windmill of 26 m diameter produced 41 kW in a wind of 36 km/hr at an angular speed of 20 rpm. American multi-blade windmills (Fig. 10.46*b*) were found on many American farms between about 1850 and 1950. They performed valuable service in powering water pumps before rural electrification.

The recent emphasis on renewable resources has revived interest in windmill design and optimization. In 2008, the United States had over 25,000 MW of wind-based electric generation capacity, which generated 52 million MWh of electricity, representing 1.26 percent of the total electric energy consumption for that year [44]. In addition, in 2008 the United States overtook Germany to become the largest generator of wind-based electrical power in the world. Wind power accounts for 42 percent of *all* new generating capacity, up from only 2 percent in 2004. America's wind belt, which stretches across the Great Plains from Texas to the Dakotas, has been dubbed the "Saudi Arabia of wind" [45].

Schematics of wind turbine configurations are shown in Fig. 10.47. In general, wind turbines are classified in two ways. The first classification is the orientation of the turbine axis. Horizontal-axis wind turbine (HAWT) and vertical-axis wind turbine (VAWT) configurations have been studied extensively. Most HAWT designs feature two- or three-bladed propellers turning at high speed, mounted on a tower along with its electric generator. The large modern HAWT, shown in Fig. 10.48*a*, is capable of producing power in any wind above a light breeze. The wind turbine shown in Fig. 10.48*b* is a VAWT. This device uses a modern symmetric airfoil section for the rotor. Earlier designs of the VAWT, such as the *Darrieus troposkien* shape,[12] suffered from high bending stresses and pulsatory torques. More recent designs, such as the one shown in this figure, feature helical airfoils, which distribute the torques more evenly about the central axis. VAWTs feature a ground-mounted electric generator.

The second classification is how the wind energy is harnessed. The first group of turbines collects wind energy through drag forces; these wind turbines are typically of the vertical axis configuration only. The second group collects energy through lift forces. Lift-based wind turbines employ horizontal- or vertical-axis configurations. It is important to note that most of these designs are self-starting. The lift-type VAWT is

[12]This shape (which would be assumed by a flexible cord whirled about a vertical axis) minimizes bending stresses in the Darrieus turbine rotor.

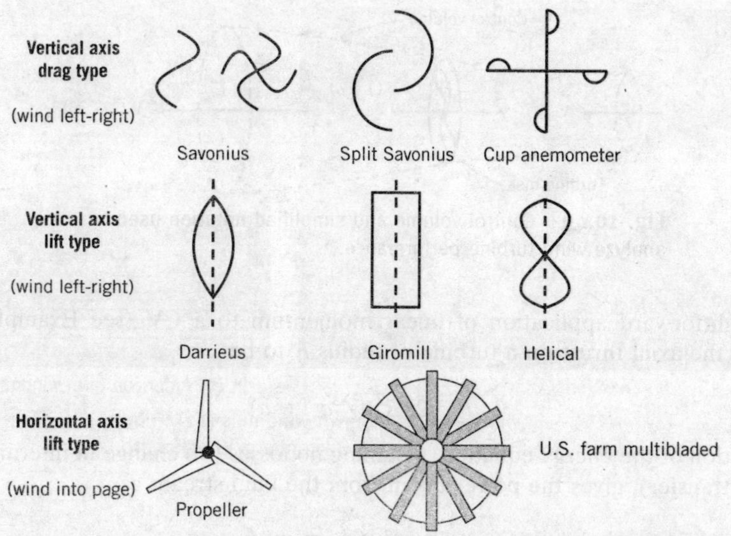

Fig. 10.47 Wind turbine configurations differentiated by axis orientation (horizontal versus vertical) and by nature of force exerted on the active element (lift versus drag).

(*a*) Horizontal-axis wind turbine (*b*) Vertical-axis wind turbine

Fig. 10.48 Examples of modern wind turbine designs. (Photos courtesy of (*a*) Siemens Energy, © 2010; (*b*) www.quietrevolution.co.uk.)

not capable of starting from rest; it can produce usable power only above a certain minimum angular speed. It is typically combined with a self-starting turbine, such as a Savonius rotor, to provide starting torque [40, 46].

A horizontal-axis wind turbine may be analyzed as a propeller operated in reverse. The Rankine model of one-dimensional flow incorporating an idealized actuator disk is shown in Fig. 10.49. The simplified notation of the figure is frequently used for analysis of wind turbines.

The wind speed far upstream is V. The stream is decelerated to $V(1 - a)$ at the turbine disk and to $V(1 - 2a)$ in the wake of the turbine (a is called the *interference factor*). Thus the stream tube of air captured by the wind turbine is small upstream and its diameter increases as it moves downstream.

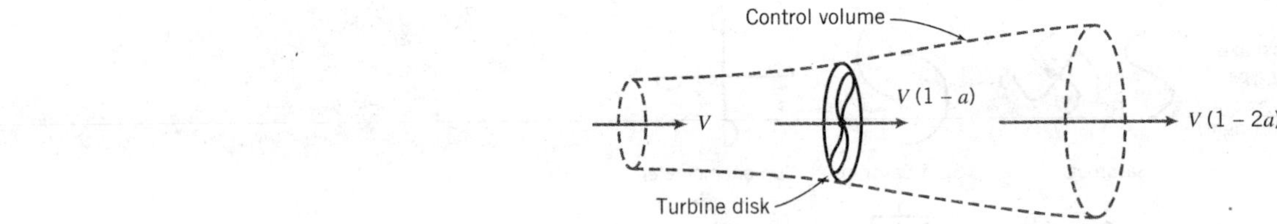

Fig. 10.49 Control volume and simplified notation used to analyze wind turbine performance.

Straightforward application of linear momentum to a CV (see Example 10.18) predicts the axial thrust on a turbine of radius R to be

$$F_T = 2\pi R^2 \rho V^2 a(1-a) \tag{10.42}$$

Application of the energy equation, assuming no losses (no change in internal energy or heat transfer), gives the power taken from the fluid stream as

$$\mathcal{P} = 2\pi R^2 \rho V^3 a(1-a)^2 \tag{10.43}$$

The efficiency of a wind turbine is most conveniently defined with reference to the kinetic energy flux contained within a stream tube the size of the actuator disk. This kinetic energy flux is

$$KEF = \frac{1}{2}\rho V^3 \pi R^2 \tag{10.44}$$

Combining Eqs. 10.43 and 10.44 gives the efficiency (or alternatively, the *power coefficient* [47]) as

$$\eta = \frac{\mathcal{P}}{KEF} = 4a(1-a)^2 \tag{10.45}$$

Betz [see 47] was the first to derive this result and to show that the theoretical efficiency is maximized when $a = 1/3$. The maximum theoretical efficiency is $\eta = 0.593$.

If the wind turbine is lightly loaded (a is small), it will affect a large mass of air per unit time, but the energy extracted per unit mass will be small and the efficiency low. Most of the kinetic energy in the initial air stream will be left in the wake and wasted. If the wind turbine is heavily loaded ($a \approx 1/2$), it will affect a much smaller mass of air per unit time. The energy removed per unit mass will be large, but the power produced will be small compared with the kinetic energy flux through the undisturbed area of the actuator disk. Thus a peak efficiency occurs at intermediate disk loadings.

The Rankine model includes some important assumptions that limit its applicability [47]. First, the wind turbine is assumed to affect only the air contained within the stream tube defined in Fig. 10.49. Second, the kinetic energy produced as swirl behind the turbine is not accounted for. Third, any radial pressure gradient is ignored. Glauert [see 41] partially accounted for the wake swirl to predict the dependence of ideal efficiency on tip-speed ratio, X,

$$X = \frac{R\omega}{V} \tag{10.46}$$

as shown in Fig. 10.50 (ω is the angular velocity of the turbine).

As the tip-speed ratio increases, ideal efficiency increases, approaching the peak value ($\eta = 0.593$) asymptotically. (Physically, the swirl left in the wake is reduced as the tip-speed ratio increases.) Avallone et al. [46] presents a summary of the detailed blade-element theory used to develop the limiting efficiency curve shown in Fig. 10.50.

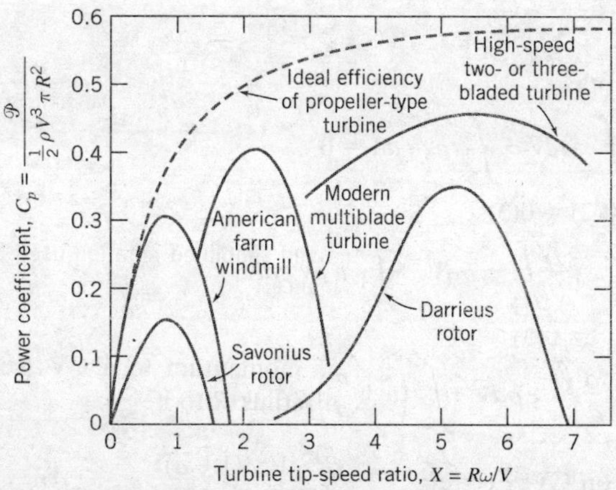

Fig. 10.50 Efficiency trends of wind turbine types versus tip-speed ratio [43].

Each type of wind turbine has its most favorable range of application. The traditional American multibladed windmill has a large number of blades and operates at relatively slow speed. Its solidity, σ (the ratio of blade area to the swept area of the turbine disk, πR^2), is high. Because of its relatively slow operating speed, its tip-speed ratio and theoretical performance limit are low. Its relatively poor performance, compared with its theoretical limit, is largely caused by use of crude blades, which are simply bent sheet metal surfaces rather than airfoil shapes.

It is necessary to increase the tip-speed ratio considerably to reach a more favorable operating range. Modern high-speed wind-turbine designs use carefully shaped airfoils and operate at tip-speed ratios up to 7 [48].

Example 10.18 PERFORMANCE OF AN IDEALIZED WINDMILL

Develop general expressions for thrust, power output, and efficiency of an idealized windmill, as shown in Fig. 10.49. Calculate the thrust, ideal efficiency, and actual efficiency for the Dutch windmill tested by Calvert ($D = 26$ m, $N = 20$ rpm, $V = 36$ km/hr, and $\mathcal{P}_{\text{output}} = 41$ kW).

Given: Idealized windmill, as shown in Fig. 10.49, and Dutch windmill tested by Calvert:

$$D = 26\text{ m} \quad N = 20\text{ rpm} \quad V = 36\text{ km/hr} \quad \mathcal{P}_{\text{output}} = 41\text{ kW}$$

Find: (a) General expressions for the ideal thrust, power output, and efficiency.
(b) The thrust, power output, and ideal and actual efficiencies for the Dutch windmill tested by Calvert.

Solution:
Apply the continuity, momentum (x component), and energy equations, using the CV and coordinates shown.

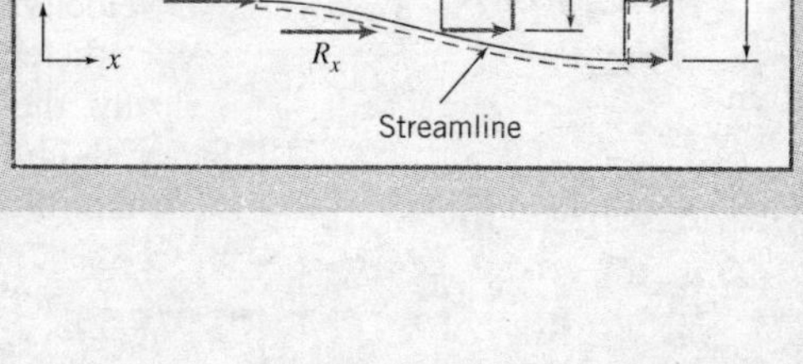

Governing equations:

$$\overset{=0(3)}{\cancel{\frac{\partial}{\partial t}}}\int_{CV} \rho\, d\mathcal{V} + \int_{CS} \rho \vec{V} \cdot d\vec{A} = 0$$

$$F_{S_x} + \overset{=0(2)}{\cancel{F_{B_x}}} = \overset{=0(3)}{\cancel{\frac{\partial}{\partial t}}}\int_{CS} u\, \rho\, d\mathcal{V} + \int_{CS} u\, \rho \vec{V} \cdot d\vec{A}$$

$$\overset{=0(7)}{\cancel{\dot{Q}}} - \dot{W}_s = \overset{=0(3)}{\cancel{\frac{\partial}{\partial t}}}\int_{CV} e\, \rho\, d\mathcal{V} + \int_{CS} \left(e + \frac{p}{\rho}\right) \rho \vec{V} \cdot d\vec{A}$$

Assumptions:
(1) Atmospheric pressure acts on CV; $F_{S_x} = R_x$.
(2) $F_{B_x} = 0$.
(3) Steady flow.
(4) Uniform flow at each section.
(5) Incompressible flow of standard air.
(6) $V_1 - V_2 = V_2 - V_3 = \frac{1}{2}(V_1 - V_3)$, as shown by Rankine.
(7) $Q = 0$.
(8) No change in internal energy for frictionless incompressible flow.

In terms of the interference factor, a, $V_1 = V$, $V_2 = (1-a)\,V$, and $V_3 = (1-2a)\,V$.

From continuity, for uniform flow at each cross section, $V_1A_1 = V_2A_2 = V_3A_3$.
From momentum,

$$R_x = u_1(-\rho V_1 A_1) + u_3(+\rho V_3 A_3) = (V_3 - V_1)\rho V_2 A_2 \qquad \{u_1 = V_1,\ u_3 = V_3\}$$

R_x is the external force acting *on* the control volume. The thrust force exerted *by* the CV *on* the surroundings is

$$K_x = -R_x = (V_1 - V_3)\rho V_2 A_2$$

In terms of the interference factor, the equation for thrust may be written in the general form,

$$K_x = \rho V^2\, \pi R^2 2a(1-a) \qquad K_x$$

(Set dK_x/da equal to zero to show that maximum thrust occurs when $a = \frac{1}{2}$.)
The energy equation becomes

$$-\dot{W}_s = \frac{V_1^2}{2}(-\rho V_1 A_1) + \frac{V_3^2}{2}(+\rho V_3 A_3) = \rho V_2 \pi R^2 \frac{1}{2}(V_3^2 - V_1^2)$$

The ideal output power, $\mathcal{P}$, is equal to $\dot{W}_s$. In terms of the interference factor,

$$\mathcal{P} = \dot{W}_s = \rho V(1-a)\pi R^2 \left[\frac{V^2}{2} - \frac{V^2}{2}(1-2a)^2\right] = \rho V^3 (1-a)\frac{\pi R^2}{2}\left[1 - (1-2a)^2\right]$$

After simplifying algebraically,

$$\mathcal{P}_{\text{ideal}} = 2\rho V^3 \pi R^2 a(1-a)^2 \qquad \mathcal{P}_{\text{ideal}}$$

The kinetic energy flux through a stream tube of undisturbed flow, equal in area to the actuator disk, is

$$KEF = \rho V \pi R^2 \frac{V^2}{2} = \frac{1}{2}\rho V^3 \pi R^2$$

Thus the ideal efficiency may be written

$$\eta = \frac{\mathcal{P}_{\text{ideal}}}{KEF} = \frac{2\rho V^3 \pi R^2 a(1-a)^2}{\frac{1}{2}\rho V^3 \pi R^2} = 4a(1-a)^2 \qquad \eta$$

To find the condition for maximum possible efficiency, set $d\eta/da$ equal to zero. The maximum efficiency is $\eta = 0.593$, which occurs when $a = 1/3$.

The Dutch windmill tested by Calvert had a tip-speed ratio of

$$X = \frac{NR}{V} = 20\frac{\text{rev}}{\text{min}} \times 2\pi\frac{\text{rad}}{\text{rev}} \times \frac{\text{min}}{60\text{ s}} \times 13\text{ m} \times \frac{\text{s}}{10\text{ m}} = 2.72 \quad X$$

The maximum theoretically attainable efficiency at this tip-speed ratio, accounting for swirl (Fig. 10.45), would be about 0.53.

The actual efficiency of the Dutch windmill is

$$\eta_{\text{actual}} = \frac{\mathcal{P}_{\text{actual}}}{KEF}$$

Based on Calvert's test data, the kinetic energy flux is

$$KEF = \frac{1}{2}\rho V^3 \pi R^2$$

$$= \frac{1}{2} \times 1.23\frac{\text{kg}}{\text{m}^3} \times (10)^3\frac{\text{m}^3}{\text{s}^3} \times \pi \times (13)^2\text{ m}^2 \times \frac{\text{N}\cdot\text{s}^2}{\text{kg}\cdot\text{m}} \times \frac{\text{W}\cdot\text{s}}{\text{N}\cdot\text{m}}$$

$$KEF = 3.27 \times 10^5\text{ W} \quad \text{or} \quad 327\text{ kW}$$

Substituting into the definition of actual efficiency,

$$\eta_{\text{actual}} = \frac{41\text{ kW}}{327\text{ kW}} = 0.125 \quad \eta_{\text{actual}}$$

Thus the actual efficiency of the Dutch windmill is about 24 percent of the maximum efficiency theoretically attainable at this tip-speed ratio.

The actual thrust on the Dutch windmill can only be estimated, because the interference factor, a, is not known. The maximum possible thrust would occur at $a = 1/2$, in which case,

$$K_x = \rho V^2 \pi R^2\, 2a(1-a)$$

$$= 1.23\frac{\text{kg}}{\text{m}^3} \times (10)^2\frac{\text{m}^2}{\text{s}^2} \times \pi \times (13)^2\text{ m}^2 \times 2\left(\frac{1}{2}\right)\left(1-\frac{1}{2}\right) \times \frac{\text{N}\cdot\text{s}^2}{\text{kg}\cdot\text{m}}$$

$$K_x = 3.27 \times 10^4\text{ N} \quad \text{or} \quad 32.7\text{ kN} \quad K_x$$

This does not sound like a large thrust force, considering the size ($D = 26$ m) of the windmill. However, $V = 36$ km/hr is only a moderate wind. The actual machine would have to withstand much more severe wind conditions during storms.

This problem illustrates application of the concepts of ideal thrust, power, and efficiency for a windmill, and calculation of these quantities for an actual machine.

The analysis of a VAWT is slightly different from that of a HAWT. The main reason for this difference can be seen in Fig. 10.51. In this figure, a cross section of one airfoil in a Darrieus turbine is shown rotating about the turbine axis. Assuming that the wind emanates from a constant direction, the airfoil angle of attack α will be a function of the azimuthal angle θ. The angle of attack is due to the relation between the effective velocity vector and the rotational direction. As θ varies, α will vary as well until it reaches a maximum value when θ is equal to 90°. In that configuration, the angle of attack is expressed by:

$$\alpha_m = \tan^{-1}\frac{V}{R\omega} \tag{10.47a}$$

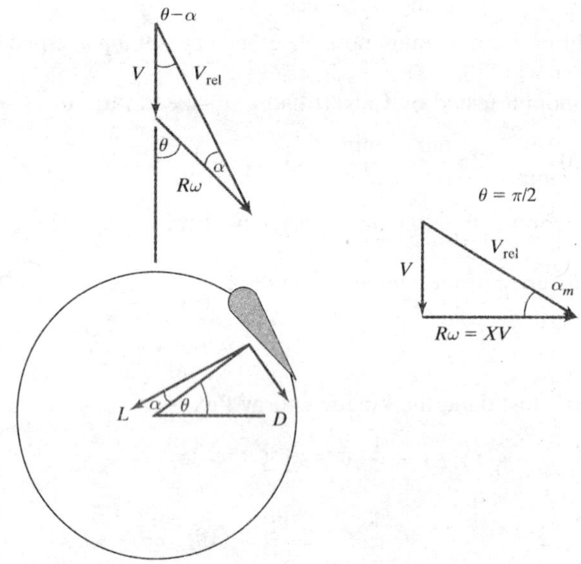

Fig. 10.51 Velocities around a Darrieus rotor blade element at a general azimuthal angle θ, as well as at $\theta = \pi/2$, where the airfoil angle of attack is maximized.

Equation 10.47a states that the maximum angle of attack is related to the wind velocity, the angular velocity of the rotor, and the local rotor radius. In terms of the tip speed ratio X from Eq. 10.46, Eq. 10.47a may then be rewritten as:

$$\alpha_m = \tan^{-1}\frac{1}{X} \tag{10.47b}$$

Since the maximum angle of attack must be less than that for stall (10°−15° for most typical airfoils), it follows that X should be a large number (at least on the order of 6). The lift and drag forces (L and D, respectively) acting on the airfoil can be seen in Fig. 10.51. These aerodynamic forces generate a torque on the rotor. The torque on the rotor at a given value of α is:

$$T = \omega R(L \sin\alpha - D\cos\alpha) \tag{10.48}$$

Now if the airfoil section being employed is symmetric (zero camber), then the lift coefficient is linearly proportional to the angle of attack [49]:

$$C_L = m\alpha \tag{10.49}$$

In Eq. 10.49, m is the slope of the lift curve, and is specific to the airfoil being used. In addition, the drag coefficient may be approximated by:

$$C_D = C_{D,0} + \frac{C_L^2}{\pi AR} \tag{9.43}$$

In this expression, $C_{D,0}$ is the drag coefficient at zero angle of attack, and AR is the aspect ratio of the airfoil. Now since the air velocity relative to the rotor is a function of α, which depends on θ, it follows that the lift and drag forces are functions of θ as well. Therefore, any quantification of rotor performance needs to be averaged over the entire range of θ. Decher [40] derived an expression for the efficiency of the rotor based on lift and drag effects, $\eta_{L/D}$. This expression is defined as the useful work out (the torque in Eq. 10.48) divided by the available power in the wind. In terms of the lift and drag, this expression is:

$$\eta_{L/D} = \frac{\overline{R\omega(L \sin\alpha - D\cos\alpha)}}{\overline{V(L\cos\alpha + D\sin\alpha)}}$$

The overbars in this equation indicate average values of those quantities. Since the lift and drag forces on the rotor change with θ, a time average of the forces needs to be calculated by integrating. Now once we substitute Eqs. 10.49 and 9.43 into this expression and average over a full revolution of the rotor ($0 \le \theta \le 2\pi$), the efficiency becomes:

$$\eta_{L/D} = \frac{1 - C_{D,0}\left(\dfrac{2}{C_{D,0}AR} + \dfrac{4X^3}{1+X^2}\right)}{1 + C_{D,0}\left(\dfrac{1}{2\pi} + \dfrac{3}{2C_{D,0}ARX^2}\right)} \tag{10.50}$$

This efficiency modifies the efficiency based on actuator disk theory (Eq. 10.45) for an estimate of the overall efficiency of the rotor:

$$\eta \approx \eta_{act\ disk}\eta_{L/D} \tag{10.51}$$

One must keep in mind, however, that in order to determine the efficiency of a complete rotor, one must add the contributions to the torque over the entire rotor. Since different parts of the rotor have different radii (different values of R), they will have different values of X. Based on Eq. 10.50, one might realize that the portions of the rotor with small radii will contribute very little to the torque compared to central portions of the rotor.

Example 10.19 ANALYSIS OF A GIROMILL

A Giromill wind turbine (see Fig. 10.47) has a height of 42 m and a diameter of 33 m. The airfoil section being used is a constant-width symmetric section with a stall angle of 12° and an aspect ratio of 50. Over the normal range of operation the airfoil lift coefficient can be described by the equation $C_L = 0.1097\alpha$, where α is the angle of attack in degrees. The drag coefficient at zero angle of attack is 0.006, and at other angles of attack the drag coefficient can be approximated by Eq. 9.43. If the Giromill rotates at 24 rpm, calculate the maximum allowable wind speed to avoid stall on the airfoil section. If the power generated at this minimum speed condition is 120 kW, what is the efficiency of the turbine?

Given: Giromill wind turbine
Height: 42 m
Diameter: 33 m
Minimum rotation speed: 24 rpm
Power: 120 kW
Airfoil is symmetric
Stall angle: 12°
Aspect ratio: 50
Lift coefficient is linear; $C_L = 0.1097\alpha$ (α is in degrees)
Drag coefficient is parabolic, $C_{D,0} = 0.006$

Find: (a) Maximum allowable wind speed to prevent stall.
(b) Turbine efficiency.

Solution:
Apply the equations presented above to the turbine:

Governing equations:

$$\alpha_m = \tan^{-1}\frac{V}{\omega R} = \tan^{-1}\frac{1}{X} \tag{10.47a, b}$$

$$KEF = \frac{1}{2}\rho V^3 \pi R^2 \tag{10.44}$$

$$\eta = \frac{\mathcal{P}}{KEF} \tag{10.45}$$

$$C_D = C_{D,0} + \frac{C_L^2}{\pi AR} \tag{9.43}$$

$$\eta_{L/D} = \frac{1 - C_{D,0}\left(\dfrac{2}{C_{D,0}AR} + \dfrac{4X^3}{1+X^2}\right)}{1 + C_{D,0}\left(\dfrac{1}{2\pi} + \dfrac{3}{2C_{D,0}ARX^2}\right)} \tag{10.50}$$

$$\eta \approx \eta_{act\ disk}\eta_{L/D} \tag{10.51}$$

Assumption: Standard atmosphere: $\rho = 1.23\ \text{kg/m}^3$

(a) To find the maximum speed, we solve Eq. 10.47a for the velocity:

$$V = R\omega \tan\alpha_m = 16.5 \times 24\frac{\text{rev}}{\text{min}} \times \frac{2\pi\ \text{rad}}{\text{rev}} \times \frac{\text{min}}{60\ \text{s}} \times \tan 12^\circ = 8.8\frac{\text{m}}{\text{s}}$$

$$V = 8.8\ \frac{\text{m}}{\text{s}} \longleftarrow V$$

(b) To determine the efficiency, we find the actuator disk efficiency and the lift/drag efficiency, per Eq. 10.51. To calculate the actuator disk efficiency, first we find the kinetic energy flux:

$$KEF = \frac{1}{2}\rho V^3 \pi R^2 = \frac{\pi}{2} \times 1.23\frac{\text{kg}}{\text{m}^3} \times \left(8.8\frac{\text{m}}{\text{s}}\right)^3 \times (16.5\ \text{m})^2 = 358.5\ \text{kW}$$

Therefore, the actuator disk efficiency is:

$$\eta = \frac{\mathcal{P}}{KEF} = \frac{120}{358.5} = 0.335$$

To find the lift/drag efficiency of the rotor, we need to find the tip speed ratio:

$$X = \frac{1}{\tan\alpha_m} = \frac{1}{\tan 12^\circ} = 4.705$$

This problem demonstrates the analysis of a VAWT, provided that the airfoil section used is below the stall angle. A more detailed analysis would be needed if a different type of section, such as the Darrieus turbine, were used, since the rotor radius is not constant.

Taking this value for X and the given data, we can calculate the lift/drag efficiency:

$$\eta_{L/D} = \frac{1 - C_{D,0}\left(\dfrac{2}{C_{D,0}AR} + \dfrac{4X^3}{1+X^2}\right)}{1 + C_{D,0}\left(\dfrac{1}{2\pi} + \dfrac{3}{2C_{D,0}ARX^2}\right)} = \frac{1 - 0.006 \times \left(\dfrac{2}{0.006 \times 50} + \dfrac{4 \times 4.705^3}{1 + 4.705^2}\right)}{1 + 0.006 \times \left(\dfrac{1}{2\pi} + \dfrac{3}{2 \times 0.006 \times 50 \times 4.705^2}\right)} = 0.850$$

So the overall efficiency is:

$$\eta \approx \eta_{act\ disk}\eta_{L/D} = 0.335 \times 0.850 = 0.285 \longleftarrow \eta$$

Compressible Flow Turbomachines 10.7

While the interaction of incompressible fluids with turbomachines is an important topic, both from a phenomenological and a practical point of view, there are many instances in which the flow through a turbomachine will experience significant changes in density. This is especially important in gas turbine (Brayton cycle) power generation (for both stationary and mobile power plants) and in steam turbine (Rankine cycle) power generation. We will investigate the modifications to the governing equations and dimensional analyses necessary in compressible flow applications. Where necessary, the reader is directed to the appropriate sections in Chapters 12 and 13 for further clarification.

Application of the Energy Equation to a Compressible Flow Machine

In Chapter 4 we looked at the first law of thermodynamics for an arbitrary control volume. The result was the energy equation, Eq. 4.56,

$$\dot{Q} - \dot{W}_s - \dot{W}_{\text{shear}} - \dot{W}_{\text{other}} = \frac{\partial}{\partial t}\int_{\text{CV}} e\rho d\forall + \int_{\text{CS}}\left(u + pv + \frac{V^2}{2} + gz\right)\rho\vec{V}\cdot d\vec{A} \tag{4.56}$$

Equation 4.56 states that the heat added to the system, minus the work done by the system results in an increase in energy for the system. In this equation, the work done by the system is assumed to consist of three parts. The first, known as "shaft work," is the useful work input/output we consider in the analysis of turbomachines. The second is the work done by fluid shear stresses at the control volume surface. The third, referred to as "other work," includes sources such as electromagnetic energy transfer.

We will now simplify Eq. 4.56 for compressible flow turbomachinery. First, turbomachines typically run at conditions such that heat transfer with the surroundings are minimized, and so the heat transfer term may be ignored. Second, work terms other than shaft work should be negligibly small, and so they can be ignored as well. Third, changes in gravitational potential energy should be small, and so that term can be dropped from the surface integral. Since enthalpy is defined as $h \equiv u + pv$, for steady flow, Eq. 4.56 becomes

$$\dot{W}_s = -\int_{\text{CS}}\left(h + \frac{V^2}{2}\right)\rho\vec{V}\cdot d\vec{A}$$

At this point, we introduce the *stagnation enthalpy*[13] as the sum of the fluid enthalpy and kinetic energy:

$$h_0 = h + \frac{V^2}{2}$$

Therefore, we may rewrite the energy equation as:

$$\dot{W}_s = -\int_{\text{CS}} h_0\rho\vec{V}\cdot d\vec{A} \tag{10.52a}$$

Equation 10.52a states that, for a turbomachine with work *input*, the power *required* causes an increase in the stagnation enthalpy in the fluid; for a turbomachine with work *output*, the power *produced* is due to a decrease in the stagnation enthalpy of the fluid.

[13]See Section 12.3 for a discussion of the stagnation state.

In this equation, $\dot{W}_s$ is positive when work is being done *by* the fluid (as in a turbine), while $\dot{W}_s$ is negative when work is being done *on* the fluid (as in a compressor).

It is important to note that the sign convention used in this equation appears to be contrary to that used in the Euler turbomachine equation, developed in Section 10.2. If you recall, in Eq. 10.2 a positive value of $\dot{W}_p$ indicated work done on the fluid, while a negative value indicated work done by the fluid. The difference to remember is that $\dot{W}_s$ is the mechanical power exerted *by* the working fluid on its surroundings, i.e., the rotor, whereas $\dot{W}_p$ is the mechanical power exerted *on* the working fluid by the rotor. Keeping this distinction in mind, it makes perfect sense that these two quantities would have equal magnitudes and opposite signs.

The integrand on the right side of Eq. 10.52a is the product of the stagnation enthalpy with the mass flow rate at each section. If we make the additional assumption of uniform flow into the machine at section 1, and out of the machine at section 2, Eq. 10.52a becomes

$$\dot{W}_s = -(h_{0_2} - h_{0_1})\dot{m} \tag{10.52b}$$

Compressors

Compressors may be centrifugal or axial, depending on specific speed. Automotive turbochargers, small gas-turbine engines, and natural-gas pipeline boosters usually are centrifugal. Large gas and steam turbines and jet aircraft engines (as seen in Figs. 10.3 and 10.4*b*) frequently are axial-flow machines.

Since the flow through a compressor will see a change in density, the dimensional analysis presented for incompressible flow is no longer appropriate. Rather, we quantify the performance of a compressor through Δh_{0_s}, the ideal rise in stagnation enthalpy of the flow,[14] the efficiency η, and the power $\mathcal{P}$. The functional relationship is:

$$\Delta h_{0_s}, \eta, \mathcal{P} = f(\mu, N, D, \dot{m}, \rho_{0_1}, c_{0_1}, k) \tag{10.53}$$

In this relation, the independent variables are, in order, viscosity, rotational speed, rotor diameter, mass flow rate, inlet stagnation density, inlet stagnation speed of sound, and ratio of specific heats.

If we apply the Buckingham Pi theorem to this system, the resulting dimensionless groups are:

$$\Pi_1 = \frac{\Delta h_{0_s}}{(ND)^2} \qquad \Pi_2 = \frac{\mathcal{P}}{\rho_{0_1} N^3 D^5}$$

$$\Pi_3 = \frac{\dot{m}}{\rho_{0_1} N D^3} \qquad \Pi_4 = \frac{\rho_{0_1} N D^2}{\mu}$$

$$\Pi_5 = \frac{ND}{c_{0_1}}$$

Since the efficiency η and ratio of specific heats k are dimensionless quantities, they can be thought of as Π-terms. The resulting functional relationships are:

[14] In Section 12.1, it is demonstrated that an adiabatic and reversible process is isentropic. It can be proven that an isentropic compression results in the minimum power input between two fixed pressures, and an isentropic expansion results in the maximum power output between two fixed pressures. Therefore, the isentropic compression/expansion process is considered the ideal for compressors and turbines, respectively. For more information, please consult Moran and Shapiro [50].

$$\frac{\Delta h_{0_s}}{(ND)^2}, \eta, \frac{\mathcal{P}}{\rho_{0_1} N^3 D^5} = f_1\left(\frac{\dot{m}}{\rho_{0_1} ND^3}, \frac{\rho_{0_1} ND^2}{\mu}, \frac{ND}{c_{0_1}}, k\right) \tag{10.54a}$$

This equation is actually an expression of three separate functions; that is, the terms $\Pi_1 = \Delta h_{0_s}/(ND)^2$, η, and $\Pi_2 = \mathcal{P}/\rho_{0_1} N^3 D^5$ are all functions of the other dimensionless quantities. $\Delta h_{0_s}/(ND)^2$ is a measure of the energy change in the flow and is the compressible analog to the head coefficient Ψ (Eq. 10.6). $\mathcal{P}/\rho_{0_1} N^3 D^5$ is a power coefficient, similar to that in Eq. 10.8. $\dot{m}/\rho_{0_1} ND^3$ is a mass flow coefficient, analogous to the incompressible flow coefficient Φ (Eq. 10.5). $\rho_{0_1} ND^2/\mu$ is a Reynolds number based on rotor tip speed, and ND/c_{0_1} is a Mach number based on rotor tip speed. Using the relationships for isentropic processes and for the compressible flow of a perfect gas, we can make some simplifications. As a result, Eq. 10.54a may be rewritten as:

$$\frac{p_{0_2}}{p_{0_1}}, \eta, \frac{\Delta T_0}{T_{0_1}} = f_2\left(\frac{\dot{m}\sqrt{RT_{0_1}}}{p_{0_1} D^2}, \text{Re}, \frac{ND}{\sqrt{RT_{0_1}}}, k\right) \tag{10.54b}$$

The functional relationships presented here can be used in the manner seen both in Chapter 7 and earlier in this chapter to investigate scaling the performance of similar flow machines. An example of this is presented in the following example.

Example 10.20 SCALING OF A COMPRESSOR

A 1/5 scale model of a prototype air compressor consuming 225 kW and running at a speed of 1000 rpm delivers a flow rate of 9 kg/s through a pressure ratio of 5. At dynamically and kinematically similar conditions, what would the operating speed, mass flow rate, and power consumption be for the full-scale prototype?

Given: 1/5 scale compressor model
Power: 225 kW
Speed: 1000 rpm
Pressure ratio: 5
Mass flow rate: 9 kg/s

Find: Prototype speed, mass flow rate, and power consumption at similar conditions.

Solution: Apply the equations presented above and the concepts presented in Chapter 7 on similitude to the compressor:

Governing equations:

$$\left(\frac{ND}{c_{0_1}}\right)_p = \left(\frac{ND}{c_{0_1}}\right)_m$$

$$\left(\frac{\dot{m}}{\rho_{0_1} ND^3}\right)_p = \left(\frac{\dot{m}}{\rho_{0_1} ND^3}\right)_m$$

$$\left(\frac{\mathcal{P}}{\rho_{0_1} N^3 D^5}\right)_p = \left(\frac{\mathcal{P}}{\rho_{0_1} N^3 D^5}\right)_m$$

Assumption: Similar entrance conditions for both model and prototype.

Similar entrance conditions would stipulate that the stagnation sound speed and density would be equal for both the model and the prototype. Solving the first equation for the prototype speed:

$$N_p = N_m \frac{D_m}{D_p}\frac{c_{0_{1_p}}}{c_{0_{1_m}}} = 1000\,\text{rpm} \times \frac{1}{5} \times 1 = 200\,\text{rpm}$$

$$N_p = 200\,\text{rpm} \longleftarrow N_p$$

Solving the second equation for the prototype mass flow rate:

$$\dot{m}_p = \dot{m}_m \frac{p_{0_{1_p}}}{p_{0_{1_m}}}\frac{N_p}{N_m}\left(\frac{D_p}{D_m}\right)^3 = 9\frac{\text{kg}}{\text{s}} \times \frac{200}{1000} \times \left(\frac{5}{1}\right)^3 = 225\frac{\text{kg}}{\text{s}}$$

$$\dot{m}_p = 225\,\text{kg/s} \longleftarrow \dot{m}_p$$

To calculate the power requirement for the prototype:

$$\mathcal{P}_p = \mathcal{P}_m \frac{p_{0_{1_p}}}{p_{0_{1_m}}}\left(\frac{N_p}{N_m}\right)^3\left(\frac{D_p}{D_m}\right)^5 = 225\,\text{kW} \times \left(\frac{200}{1000}\right)^3 \times \left(\frac{5}{1}\right)^5 = 5625\,\text{kW}$$

$$\mathcal{P}_p = 5625\,\text{kW} \longleftarrow P_p$$

This problem demonstrates the scaling of compressible flow machines. Note that if the working fluid for the two different scale machines were different, e.g., air versus helium, the effects of different gas constants and specific heat ratios would have to be taken into account.

Since most operability studies are performed on a single compressor design without scaling, and, using the same working fluid, all variables related to the scale and the fluid (specifically, D, R, and k) may be eliminated from the functional relationship. In addition, empirical studies have shown that, as in the case study of the centrifugal pump in Chapter 7, for sufficiently high values of Reynolds number the performance of the compressor is not dependent upon Reynolds number either; i.e., the flow is fully turbulent in the compressor. Once these variables are eliminated, Eq. 10.54b becomes

$$\frac{p_{0_2}}{p_{0_1}}, \eta, \frac{\Delta T_0}{T_{0_1}} = f_3\left(\frac{\dot{m}\sqrt{T_{0_1}}}{p_{0_1}}, \frac{N}{\sqrt{T_{0_1}}}\right) \tag{10.54c}$$

Note that this equation is no longer dimensionless. However, it is still useful in characterizing the performance of a compressor, provided the performance is assessed for a single machine using a single working fluid. The relationship portrayed in Eq. 10.54c is normally expressed in the form of a compressor operability map, as shown in Fig. 10.52. On this map we can see the compression ratio versus mass flow ratio ($\dot{m}\sqrt{T_{0_1}}/p_{0_1}$), with curves of constant normalized speed ($N/\sqrt{T_{0_1}}$) and efficiency. Often, the abscissa is a "corrected mass flow":

$$\dot{m}_{corr} = \frac{\dot{m}\sqrt{T_{0_1}/T_{\text{ref}}}}{p_{0_1}/p_{\text{ref}}}$$

and the lines of constant compressor speed are a "corrected speed":

$$N_{corr} = \frac{N}{\sqrt{T_{0_1}/T_{\text{ref}}}}$$

In these expressions, T_{ref} and p_{ref} are reference pressure and temperature (usually standard conditions one would expect at the entrance of such a machine). This allows the user to read the chart quickly in terms of "real" physical quantities and to be able to make adjustments for varying entrance conditions with a minimum of calculation. The operating line is the locus of points of maximum efficiency for a given mass flow.

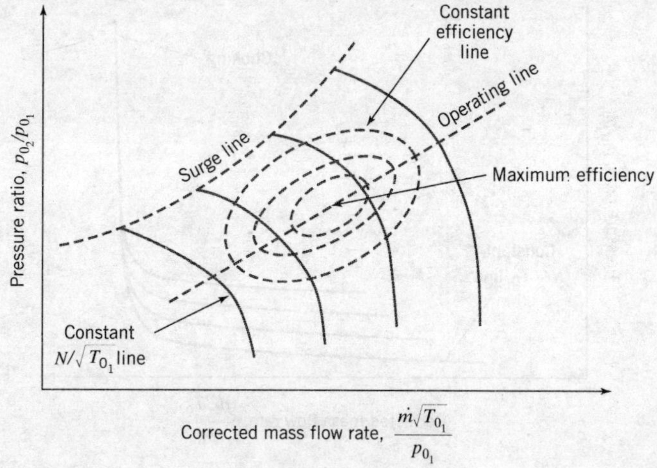

Fig. 10.52 Performance map for a compressor.

It is important to note that the compressor operability map of Fig. 10.52 bears a striking resemblance to the pump operability map of Fig. 10.14. Not only do both figures show the performance of a turbomachine performing work on a fluid, but the data are plotted in a similar fashion; level curves of constant efficiency are plotted on a plane of work output (head for the pump, pressure ratio for the compressor) versus flow input (volumetric flow rate for the pump, mass flow rate for the compressor).

This figure shows two of the phenomena that must be avoided in the operation of a compressor. The first is *choking*, which is experienced when the local Mach number at some point in the compressor reaches unity.[15] To explain choking in a physical sense, imagine that we run the compressor at constant speed and constant inlet pressure and that we can directly control the compressor exit pressure. On the compressor map, we would be traveling along a line of constant normalized speed. If we were to lower the exit pressure, the pressure ratio would decrease. If the compressor speed remains constant, the mass flow increases. However, we see that the lines of constant normalized speed turn downward if the mass flow rate is increased beyond a certain value, indicating a maximum possible flow rate for a given rotor speed, and the compressor is choked. When choking occurs, it is impossible to increase mass flow without increasing rotor speed.

The second phenomenon is *surge*, which is a cyclic pulsation phenomenon that causes the mass flow rate through the machine to vary, and can even reverse it! Surge occurs when the pressure ratio in the compressor is raised beyond a certain level for a given mass flow rate. As pressure ratio increases, the adverse pressure gradient across the compressor increases as well. This increase in pressure gradient can cause boundary-layer separation on the rotor surfaces and constrict flow through the space between two adjacent blades.[16] Therefore, the extra flow gets diverted to the next channel between blades. The separation is relieved in the previous channel and moves to the next channel, causing the cyclic pulsation mentioned above. Surge is accompanied by loud noises and can damage the compressor or related components; it too must be avoided. Fig. 10.52 shows the *surge line*, the locus of operating conditions beyond which surge will occur.

In general, as shown in Fig. 10.52, the higher the performance, the more narrow the range in which the compressor may be operated successfully. Thus a compressor must be carefully matched to its flow system to assure satisfactory operation. Compressor matching in natural gas pipeline applications is discussed by Vincent-Genod [51].

[15] Choking is also described from the standpoint of nozzle flow in Section 13.2.

[16] Boundary layer separation due to adverse pressure gradients is discussed in Section 9.6.

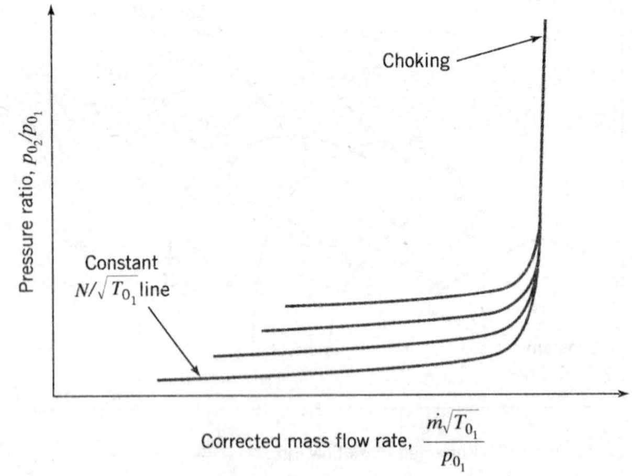

Fig. 10.53 Performance map for a compressible flow turbine.

Perhaps the most common application of high-speed fluid machinery today is in automotive turbochargers (worldwide many millions of cars are sold each year with turbochargers). Automotive turbocharger matching is described in manufacturers' literature [52].

Compressible-Flow Turbines

The flow through a gas turbine is governed by the same general relationship as the compressor, but the actual functional relationships are different. Figure 10.53 shows the performance map for a compressible flow turbine. As in the case for the compressor the turbine map shows lines of constant normalized speed on a graph of pressure ratio versus normalized mass flow rate. The most striking difference between this map and that for the compressor is that the performance is a very weak function of $N/\sqrt{T_{0_1}}$; the curves are set very close together. The choking of the turbine flow is well-defined on this map: There is a normalized mass flow that cannot be exceeded in the turbine, regardless of the pressure ratio.

10.8 Summary and Useful Equations

In this chapter, we:

✓ Defined the two major types of fluid machines: positive displacement machines and turbomachines.

✓ Defined, within the turbomachine category, radial, axial, and mixed-flow types, pumps, fans, blowers, compressors, and impulse and reaction turbines.

✓ Discussed various features of turbomachines, such as impellers, rotors, runners, scrolls (volutes), compressor stages, and draft tubes.

✓ Used the angular-momentum equation for a control volume to derive the Euler turbomachine equation.

✓ Drew velocity diagrams and applied the Euler turbomachine equation to the analysis of various idealized machines to derive ideal torque, head, and power.

✓ Evaluated the performance—head, power, and efficiency—of various actual machines from measured data.

✓ Defined and used dimensionless parameters to scale the performance of a fluid machine from one size, operating speed, and set of operating conditions to another.

✓ Discussed various defining parameters, such as pump efficiency, solidity, hydraulic power, mechanical power, turbine efficiency, shutoff head, shock loss, specific speed, cavitation, *NPSHR*, and *NPSHA*.

- ✓ Examined pumps for their compliance with the constraint that the net positive suction head available exceeds that required to avoid cavitation.
- ✓ Matched fluid machines for doing work on a fluid to pipe systems to obtain the operating point (flow rate and head).
- ✓ Predicted the effects of installing fluid machines in series and parallel on the operating point of a system.
- ✓ Discussed and analyzed turbomachines without housings, namely propellers and wind turbines.
- ✓ Discussed the use and performance of compressible flow turbomachines.

With these concepts and techniques, we learned how to use manufacturers' literature and other data to perform preliminary analyses and make appropriate selections of pumps, fans, hydraulic and wind turbines, and other fluid machines.

Note: Most of the Useful Equations in the table below have a number of constraints or limitations—*be sure to refer to their page numbers for details!*

Useful Equations

Euler turbomachine equation:	$T_{\text{shaft}} = (r_2 V_{t_2} - r_1 V_{t_1})\dot{m}$	**(10.1c)**	Page 500
Turbomachine theoretical power:	$\dot{W}_m = (U_2 V_{t_2} - U_1 V_{t_1})\dot{m}$	**(10.2b)**	Page 500
Turbomachine theoretical head:	$H = \dfrac{\dot{W}_m}{\dot{m}g} = \dfrac{1}{g}(U_2 V_{t_2} - U_1 V_{t_1})$	**(10.2c)**	Page 501
Pump power, head, and efficiency:	$\dot{W}_h = \rho Q g H_p$	**(10.3a)**	Page 504
	$H_p = \left(\dfrac{p}{\rho g} + \dfrac{\overline{V}^2}{2g} + z\right)_{\text{discharge}} - \left(\dfrac{p}{\rho g} + \dfrac{\overline{V}^2}{2g} + z\right)_{\text{suction}}$	**(10.3b)**	
	$\eta_p = \dfrac{\dot{W}_h}{\dot{W}_m} = \dfrac{\rho Q g H_p}{\omega T}$	**(10.3c)**	
Turbine power, head, and efficiency:	$\dot{W}_h = \rho Q g H_t$	**(10.4a)**	Page 505
	$H_t = \left(\dfrac{p}{\rho g} + \dfrac{\overline{V}^2}{2g} + z\right)_{\text{inlet}} - \left(\dfrac{p}{\rho g} + \dfrac{\overline{V}^2}{2g} + z\right)_{\text{outlet}}$	**(10.4b)**	
	$\eta_t = \dfrac{\dot{W}_m}{\dot{W}_h} = \dfrac{\omega T}{\rho Q g H_t}$	**(10.4c)**	
Dimensionless flow coefficient:	$\Phi = \dfrac{Q}{A_2 U_2} = \dfrac{V_{n_2}}{U_2}$	**(10.5)**	Page 505
Dimensionless head coefficient:	$\Psi = \dfrac{gH}{U_2^2}$	**(10.6)**	Page 506
Dimensionless torque coefficient:	$\tau = \dfrac{T}{\rho A_2 U_2^2 R_2}$	**(10.7)**	Page 506
Dimensionless power coefficient:	$\Pi = \dfrac{\dot{W}}{\rho Q U_2^2} = \dfrac{\dot{W}}{\rho \omega^2 Q R_2^2}$	**(10.8)**	Page 506

Centrifugal pump specific speed (in terms of head h):	$N_S = \frac{\omega Q^{1/2}}{h^{3/4}}$	**(7.22a)**	Page 506
Centrifugal pump specific speed (in terms of head H):	$N_{S_{cu}} = \frac{N(\text{rpm})[Q(\text{m}^3/\text{h})]^{1/2}}{[H(\text{m})]^{3/4}}$	**(7.22b)**	Page 507
Centrifugal turbine specific speed (in terms of head h):	$N_S = \frac{\omega}{h^{3/4}}\left(\frac{\mathcal{P}}{\rho h}\right)^{1/2} = \frac{\omega \mathcal{P}^{1/2}}{\rho^{1/2} h^{5/4}}$	**(10.13a)**	Page 507
Centrifugal turbine specific speed (in terms of head H):	$N_{S_{cu}} = \frac{N(\text{rpm})[\mathcal{P}(\text{kW})]^{1/2}}{[H(\text{m})]^{5/4}}$	**(10.13b)**	Page 507
Axial-flow turbomachine ideal performance:	$T_{\text{shaft}} = R_m(V_{t_2} - V_{t_1})\dot{m}$	**(10.20)**	Page 513
	$\dot{W}_m = U(V_{t_2} - V_{t_1})\dot{m}$	**(10.21)**	
	$H = \frac{\dot{W}_m}{\dot{m}g} = \frac{U}{g}(V_{t_2} - V_{t_1})$	**(10.22)**	
Propeller thrust:	$F_T = qN\int_{R_{hub}}^{R}\frac{(C_L\cos\phi - C_D\sin\phi)}{\sin^2\phi}cdr$	**(10.38a)**	Page 567
Propeller torque:	$T = qN\int_{R_{hub}}^{R}\frac{(C_L\sin\phi + C_D\cos\phi)}{\sin^2\phi}rcdr$	**(10.38b)**	Page 567
Propeller speed of advance coefficient:	$J \equiv \frac{V}{nD}$	**(10.39)**	Page 568
Propeller thrust, torque, power, and efficiency coefficients:	$C_F = \frac{F_T}{\rho n^2 D^4}, \quad C_T = \frac{T}{\rho n^2 D^5},$ $C_P = \frac{\mathcal{P}}{\rho n^3 D^5}, \quad \eta = \frac{F_T V}{\mathcal{P}_{input}}$	**(10.40)**	Page 569
Cavitation number:	$Ca = \frac{p - p_v}{\frac{1}{2}\rho V^2}$	**(10.41)**	Page 571
Actuator disk efficiency:	$\eta = \frac{\mathcal{P}}{KEF} = 4a(1-a)^2$	**(10.45)**	Page 574
Tip-speed ratio	$X = \frac{R\omega}{V}$	**(10.46)**	Page 574
VAWT efficiency:	$\eta_{L/D} = \frac{1 - C_{D,0}\left(\frac{2}{C_{D,0}AR} + \frac{4X^3}{1+X^2}\right)}{1 + C_{D,0}\left(\frac{1}{2\pi} + \frac{3}{2C_{D,0}ARX^2}\right)}$	**(10.50)**	Page 579
	$\eta \approx \eta_{act\ disk}\eta_{L/D}$	**(10.51)**	

Energy equation for compressible flow turbomachine:	$\dot{W}_s = -(h_{0_2} - h_{0_1})\dot{m}$	**(10.52b)**	Page 582
Performance parameters for compressible flow turbomachine:	$\frac{p_{0_2}}{p_{0_1}}, \eta, \frac{\Delta T_0}{T_{0_1}} = f_3\left(\frac{\dot{m}\sqrt{T_{0_1}}}{p_{0_1}}, \frac{N}{\sqrt{T_{0_1}}}\right)$	**(10.54c)**	Page 584

Case Study

The Little Engine That Could!

Silicon gas-turbine engines suitable for powering laptops or cell phones; a 6-mm-diameter turbine assembly. (Courtesy of Dr. Alan Epstein, M.I.T.)

Alan Epstein, a professor of aeronautics and astronautics at the Massachusetts Institute of Technology, and his team have done a lot of research on tiny gas-turbine engines made of silicon. They are about the size of a quarter (as shown in the figure) and can be easily mass produced. Unlike conventional large turbines that are assembled from many components, these turbines are built basically from a solid piece of silicon. Professor Epstein discovered that the basic concepts of turbine theory (discussed in this chapter) apply even to his microturbines; the fluid mechanics turns out to be the same as that for larger engines, as long as the passages made for gas flow are larger than about 1 μm in diameter (smaller than this and non-continuum molecular kinetics is needed).

The rotor and its airfoils are carved out of a single wafer, as shown in the figure. Additional "plumbing" and bearings are etched onto the wafers that are to sandwich the rotor. Combustion occurs just outside the rotor, at the same wafer level, spinning it by pushing on its airfoils from the outside. At more than a *million* rpm, these turbines make no audible noise (it's there, but not even your dog can hear it)! Electricity will then be generated using, for example, a tiny generator. The fuel source could be packaged with the engine or come as a replaceable cartridge like a cigarette lighter. In terms of power density, the little engine will easily beat batteries, with an output of somewhere between 50 and 100 watts!

References

1. Wilson, D. G., "Turbomachinery—From Paddle Wheels to Turbojets," *Mechanical Engineering*, *104*, 10, October 1982, pp. 28–40.
2. Logan, E. S., Jr., *Turbomachinery: Basic Theory and Applications*. New York: Dekker, 1981.
3. Japikse, D. "Teaching Design in an Engineering Education Curriculum: A Design Track Syllabus," TM-519, Concepts ETI Inc., White River Jct., VT 05001.
4. Postelwait, J., "Turbomachinery Industry Set for Growth," *Power Engineering*, http://pepei.pennnet.com/.
5. Sabersky, R. H., A. J. Acosta, E. G. Hauptmann, and E. M. Gates, *Fluid Flow: A First Course in Fluid Mechanics*, 4th ed. Englewood Cliffs, NJ: Prentice-Hall, 1999.
6. Daily, J. W., "Hydraulic Machinery," in Rouse, H., ed., *Engineering Hydraulics*. New York: Wiley, 1950.
7. Dixon, S. L., *Fluid Mechanics and Thermodynamics of Turbomachinery*, 5th ed. Amsterdam: Elsevier, 2005.
8. American Society of Mechanical Engineers, *Performance Test Codes: Centrifugal Pumps*, ASME PTC 8.2-1990. New York: ASME, 1990.
9. American Institute of Chemical Engineers, *Equipment Testing Procedure: Centrifugal Pumps (Newtonian Liquids)*. New York: AIChE, 1984.
10. Peerless Pump,* Brochure B-4003, "System Analysis for Pumping Equipment Selection," 1979.

*Peerless Pump Company, a member of the Sterling Group, P.O. Box 7026, Indianapolis, IN 46206-7026, U.S.A.

11. Daugherty, R. L., J. B. Franzini, and E. J. Finnemore, *Fluid Mechanics with Engineering Applications*, 8th ed. New York: McGraw-Hill, 1985.
12. Peerless Pump Company, *RAPID*, v 8.25.6, March 23, 2007.
13. Hodge, B. K., *Analysis and Design of Energy Systems*, 2nd ed. Englewood Cliffs, NJ: Prentice-Hall, 1990.
14. Moody, L. F., "Hydraulic Machinery," in *Handbook of Applied Hydraulics*, ed. by C. V. Davis. New York: McGraw-Hill, 1942.
15. Hydraulic Institute, *Hydraulic Institute Standards*. New York: Hydraulic Institute, 1969.
16. Dickinson, C., *Pumping Manual*, 8th ed. Surrey, England: Trade & Technical Press, Ltd., 1988.
17. Hicks, T. G., and T. W. Edwards, *Pump Application Engineering*. New York: McGraw-Hill, 1971.
18. Armintor, J. K., and D. P. Conners, "Pumping Applications in the Petroleum and Chemical Industries," *IEEE Transactions on Industry Applications, IA-23*, 1, January 1987.
19. Jorgensen, R., ed., *Fan Engineering*, 8th ed. Buffalo, NY: Buffalo Forge, 1983.
20. Air Movement and Control Association, *Laboratory Methods of Testing Fans for Rating*. AMCA Standard 210-74, ASHRAE Standard 51-75. Atlanta, GA: ASHRAE, 1975.
21. American Society of Mechanical Engineers, *Power Test Code for Fans*. New York: ASME, Power Test Codes, PTC 11-1946.
22. Berry, C. H., *Flow and Fan: Principles of Moving Air through Ducts*, 2nd ed. New York: Industrial Press, 1963.
23. Wallis, R. A., *Axial Flow Fans and Ducts*. New York: Wiley, 1983.
24. Osborne, W. C., *Fans*, 2nd ed. London: Pergamon Press, 1977.
25. American Society of Heating, Refrigeration, and Air Conditioning Engineers, *Handbook of Fundamentals*. Atlanta, GA: ASHRAE, 1980.
26. Idelchik, I. E., *Handbook of Hydraulic Resistance*, 2nd ed. New York: Hemisphere, 1986.
27. Lambeck, R. R., *Hydraulic Pumps and Motors: Selection and Application for Hydraulic Power Control Systems*. New York: Dekker, 1983.
28. Warring, R. H., ed., *Hydraulic Handbook*, 8th ed. Houston: Gulf Publishing Co., 1983.
29. Rouse, H., and S. Ince, *History of Hydraulics*. Iowa City, IA: Iowa University Press, 1957.
30. Russell, G. E., *Hydraulics*, 5th ed. New York: Henry Holt, 1942.
31. Gulliver, J. S., and R. E. A. Arndt, *Hydropower Engineering Handbook*. New York: McGraw-Hill, 1990.
32. World Energy Council, "2007 Survey of Energy Resources," World Energy Council, 2007.
33. Fritz, J. J., *Small and Mini Hydropower Systems: Resource Assessment and Project Feasibility*. New York: McGraw-Hill, 1984.
34. Gladwell, J. S., *Small Hydro: Some Practical Planning and Design Considerations*. Idaho Water Resources Institute. Moscow, ID: University of Idaho, April 1980.
35. McGuigan, D., *Small Scale Water Power*. Dorchester: Prism Press, 1978.
36. Olson, R. M., and S. J. Wright, *Essentials of Engineering Fluid Mechanics*, 5th ed. New York: Harper & Row, 1990.
37. Quick, R. S., "Problems Encountered in the Design and Operation of Impulse Turbines," *Transactions of the ASME*, *62*, 1940, pp. 15–27.
38. Warnick, C. C., *Hydropower Engineering*. Englewood Cliffs, NJ: Prentice-Hall, 1984.
39. Dodge, D. M., "Illustrated History of Wind Power Development," http://www.telosnet.com/wind/index.html.
40. Decher, R., *Energy Conversion: Systems, Flow Physics, and Engineering*. New York: Oxford University Press, 1994.
41. Durand, W. F., ed., *Aerodynamic Theory*, 6 Volumes. New York: Dover, 1963.
42. Putnam, P. C., *Power from the Wind*. New York: Van Nostrand, 1948.
43. Calvert, N. G., *Windpower Principles: Their Application on the Small Scale*. London: Griffin, 1978.
44. American Wind Energy Association, *Annual Wind Industry Report, Year Ending 2008*. Washington, DC: American Wind Energy Association, 2008.
45. "Wind Power in America: Becalmed," *The Economist*, *392*, 8642 (August 1, 2009).
46. Eldridge, F. R., *Wind Machines*, 2nd ed. New York: Van Nostrand Reinhold, 1980.
47. Avallone, E. A., T. Baumeister, III, and A. Sadegh, eds., *Marks' Standard Handbook for Mechanical Engineers*, 11th ed. New York: McGraw-Hill, 2007.
48. Migliore, P. G., "Comparison of NACA 6-Series and 4-Digit Airfoils for Darrieus Wind Turbines," *Journal of Energy*, *7*, 4, Jul–Aug 1983, pp. 291–292.
49. Anderson, J. D., *Introduction to Flight*, 4th ed. Boston: McGraw-Hill, 2000.
50. Moran, M. J., and H. N. Shapiro, *Fundamentals of Engineering Thermodynamics*, 6th ed. Hoboken, NJ: Wiley, 2007.
51. Vincent-Genod, J., *Fundamentals of Pipeline Engineering*. Houston: Gulf Publishing Co., 1984.
52. Warner-Ishi Turbocharger brochure. (Warner-Ishi, P.O. Box 580, Shelbyville, IL 62565-0580, U.S.A.).
53. White, F. M., *Fluid Mechanics*, 6th ed. New York: McGraw-Hill, 2007.
54. Sovern, D. T., and G. J. Poole, "Column Separation in Pumped Pipelines," in K. K. Kienow, ed., *Pipeline Design and Installation*, Proceedings of the International Conference on Pipeline Design and Installation, Las Vegas, Nevada, March 25–27, 1990. New York: American Society of Civil Engineers, 1990, pp. 230–243.
55. U.S. Department of the Interior, "Selecting Hydraulic Reaction Turbines," A Water Resources Technical Publication, *Engineering Monograph No. 20*. Denver, CO: U.S. Department of the Interior, Bureau of Reclamation, 1976.
56. Drella, M., "Aerodynamics of Human-Powered Flight," in *Annual Review of Fluid Mechanics*, *22*, pp. 93–110. Palo Alto, CA: Annual Reviews, 1990.
57. Abbott, I. H., and A. E. von Doenhoff, *Theory of Wing Sections, Including a Summary of Airfoil Data*. New York: Dover, 1959.

Problems

Introduction and Classification of Fluid Machines; Turbomachinery Analysis

10.1 Dimensions of a centrifugal pump impeller are

Parameter	Inlet, Section ①	Outlet, Section ②
Radius, r (mm)	175	500
Blade width, b (mm)	50	30
Blade angle, β (deg)	65	70

The pump handles water and is driven at 750 rpm. Calculate the theoretical head and mechanical power input if the flow rate is 0.75 m^3/s.

10.2 The geometry of a centrifugal water pump is $r_1 = 10$ cm, $r_2 = 20$ cm, $b_1 = b_2 = 4$ cm, $\beta_1 = 30°$, $\beta_2 = 15°$, and it runs at speed 1600 rpm. Estimate the discharge required for axial entry, the power generated in the water in watts, and the head produced.

10.3 A centrifugal pump running at 3000 rpm pumps water at a rate of 0.6 m^3/min. The water enters axially and leaves the impeller at 5.4 m/s relative to the blades, which are radial at the exit. If the pump requires 5 kW and is 72 percent efficient, estimate the basic dimensions (impeller exit diameter and width), using the Euler turbomachine equation.

10.4 Consider the centrifugal pump impeller dimensions given in Example 10.1. Estimate the ideal head rise and mechanical power input if the outlet blade angle is changed to 60°, 70°, 80°, or 85°.

10.5 Dimensions of a centrifugal pump impeller are

Parameter	Inlet, Section ①	Outlet, Section ②
Radius, r (mm)	380	1140
Blade width, b (mm)	120	80
Blade angle, β (deg)	40	60

The pump is driven at 575 rpm and the fluid is water. Calculate the theoretical head and mechanical power if the flow rate is 18,000.

10.6 Dimensions of a centrifugal pump impeller are

Parameter	Inlet, Section ①	Outlet, Section ②
Radius, r (mm)	75	250
Blade width, b (mm)	38	30
Blade angle, β (deg)	60	70

The pump is driven at 1250 rpm while pumping water. Calculate the theoretical head and mechanical power input if the flow rate is 340 m^3/h.

10.7 For the impeller of Problem 10.6, determine the rotational speed for which the tangential component of the inlet velocity is zero if the volume flow rate is 910 m^3/h. Calculate the theoretical head and mechanical power input.

10.8 A centrifugal water pump, with 15 cm diameter impeller and axial inlet flow, is driven at 1750 rpm. The impeller vanes are backward-curved (β_2=65°) and have axial width $b_2 = 2$ cm. For a volume flow rate of 225 m^3/hr determine the theoretical head rise and power input to the pump.

10.9 For the impeller of Problem 10.1, operating at 750 rpm, determine the volume flow rate for which the tangential component of the inlet velocity is zero. Calculate the theoretical head and mechanical power input.

10.10 Consider the geometry of the idealized centrifugal pump described in Problem 10.11. Draw inlet and outlet velocity diagrams assuming b = constant. Calculate the inlet blade angles required for "shockless" entry flow at the design flow rate. Evaluate the theoretical power input to the pump at the design flow rate.

10.11 Consider a centrifugal water pump whose geometry and flow conditions are as follows:

Impeller inlet radius, R_1	2.5 cm
Impeller outlet radius, R_2	18 cm
Impeller outlet width, b_2	1 cm
Design speed, N	1800 rpm
Design flow rate, Q	30 m^3/min
Backward-curved vanes (outlet blade angle), β_2	75°
Required flow rate range	50–150% of design

Assume ideal pump behavior with 100 percent efficiency. Find the shutoff head. Calculate the absolute and relative discharge velocities, the total head, and the theoretical power required at the design flow rate.

10.12 Consider the centrifugal pump impeller dimensions given in Example 10.1. Construct the velocity diagram for shockless flow at the impeller inlet, if b = constant. Calculate the effective flow angle with respect to the radial impeller blades for the case of no inlet swirl. Investigate the effects on flow angle of (a) variations in impeller width and (b) inlet swirl velocities.

10.13 For the impeller of Problem 10.5, determine the inlet blade angle for which the tangential component of the inlet velocity is zero if the volume flow rate is 28,000. Calculate the theoretical head and mechanical power input.

10.14 A centrifugal water pump designed to operate at 1300 rpm has dimensions

Parameter	Inlet	Outlet
Radius, r (mm)	100	175
Blade width, b (mm)	10	7.5
Blade angle, β (deg)		40

Draw the inlet velocity diagram for a volume flow rate of 35 L/s. Determine the inlet blade angle for which the entering velocity has no tangential component. Draw the outlet velocity diagram. Determine the outlet absolute flow angle (measured relative to the normal direction). Evaluate the hydraulic power delivered by the pump, if its efficiency is 75 percent. Determine the head developed by the pump.

10.15 A centrifugal pump runs at 1750 rpm while pumping water at a rate of 50 L/s. The water enters axially, and leaves tangential to the impeller blades. The impeller exit diameter and width are 300 mm and 10 mm, respectively. If the pump requires 45 kW, and is 75 percent efficient, estimate the exit angle of the impeller blades.

10.16 A centrifugal water pump designed to operate at 1200 rpm has dimensions

Parameter	Inlet	Outlet
Radius, r (mm)	90	150
Blade width, b (mm)	10	7.5
Blade angle, β (deg)	25	45

Determine the flow rate at which the entering velocity has no tangential component. Draw the outlet velocity diagram, and determine the outlet absolute flow angle (measured relative to the normal direction) at this flow rate. Evaluate the hydraulic power delivered by the pump if its efficiency is 70 percent. Determine the head developed by the pump.

10.17 Repeat the analysis for determining the optimum speed for an impulse turbine of Example 10.13, using the Euler turbomachine equation.

10.18 Kerosene is pumped by a centrifugal pump. When the flow rate is 0.025 m^3/s, the pump requires 15 kW input, and its efficiency is 82 percent. Calculate the pressure rise produced by the pump. Express this result as (a) meter of water and (b) meter of kerosene.

10.19 A centrifugal pump designed to deliver water at 30 L/s has dimensions

Parameter	Inlet	Outlet
Radius, r (mm)	75	150
Blade width, b (mm)	7.5	6.25
Blade angle, β (deg)	25	40

Draw the inlet velocity diagram. Determine the design speed if the entering velocity has no tangential component. Draw the outlet velocity diagram. Determine the outlet absolute flow angle (measured relative to the normal direction). Evaluate the theoretical head developed by the pump. Estimate the minimum mechanical power delivered to the pump.

Pumps, Fans, and Blowers

10.20 In the water pump of Problem 10.8, the pump casing acts as a diffuser, which converts 60 percent of the absolute velocity head at the impeller outlet to static pressure rise. The head loss through the pump suction and discharge channels is 0.75 times the radial component of velocity head leaving the impeller. Estimate the volume flow rate, head rise, power input, and pump efficiency at the maximum efficiency point. Assume the torque to overcome bearing, seal, and spin losses is 10 percent of the ideal torque at $Q = 0.065$ m^3/s.

10.21 The theoretical head delivered by a centrifugal pump at shutoff depends on the discharge radius and angular speed of the impeller. For preliminary design, it is useful to have a plot showing the theoretical shutoff characteristics and approximating the actual performance. Prepare a log-log plot of impeller radius versus theoretical head rise at shutoff with standard motor speeds as parameters. Assume the fluid is water and the actual head at the design flow rate is 70 percent of the theoretical shutoff head. (Show these as dashed lines on the plot.) Explain how this plot might be used for preliminary design.

10.22 Use data from Appendix D to choose points from the performance curves for a Peerless horizontal split case Type 16A18B pump at 705 and 880 nominal rpm. Obtain and plot curve-fits of total head versus delivery for this pump, with an 460 mm diameter impeller.

10.23 Use data from Appendix D to choose points from the performance curves for a Peerless horizontal split case Type 4AE12 pump at 1750 and 3550 nominal rpm. Obtain and plot curve-fits for total head versus delivery at each speed for this pump, with a 310 mm diameter impeller.

10.24 Data from tests of a water suction pump operated at 2000 rpm with a 35 cm diameter impeller are

Flow rate, Q ($m^3/s \times 10^3$)	17	26	38	45	63
Total head, H (m)	60	59	54	50	37
Power input, $\mathcal{P}$ (kW)	19	22	26	30	34

Plot the performance curves for this pump; include a curve of efficiency versus volume flow rate. Locate the best efficiency point and specify the pump rating at this point.

10.25 A 22.5 cm diameter centrifugal pump, running at 900 rpm with water at 20°C generates the following performance data:

Flow rate, Q (m^3/min)	0	5.7	11.3	17.0	22.6	28.3
Total head, H (m)	7	6.8	6.4	5.9	5.2	3.8
Power input, $\mathcal{P}$ (kW)	11.3	12.8	18.2	20.1	24.0	36.4

Plot the performance curves for this pump; include a curve of efficiency versus volume flow rate. Locate the best efficiency point. What is the specific speed for this pump?

10.26 An axial-flow fan operates in seal-level air at 1350 rpm and has a blade tip diameter of 1 m and a root diameter of 0.8 m. The inlet angles are $\alpha_1 = 55°$, $\beta_1 = 30°$, and at the exit $\beta_2 = 60°$. Estimate the flow volumetric flow rate, power, and the outlet angle, α_2.

10.27 Write the turbine specific speed in terms of the flow coefficient and the head coefficient.

10.28 Data measured during tests of a centrifugal pump driven at 3000 rpm are

Parameter	Inlet, Section ①	Outlet, Section ②
Gage pressure, p (kPa)	86	
Elevation above datum, z (m)	2.0	10
Average speed of flow, $\bar{V}$ (m/s)	2.0	4.6

The flow rate is 15 m^3/h and the torque applied to the pump shaft is 6.4 N·m. The pump efficiency is 75 percent, and the electric motor efficiency is 85 percent. Find the electric power required, and the gage pressure at section ②.

10.29 The *kilogram force* (kgf), defined as the force exerted by a kilogram mass in standard gravity, is commonly used in

European practice. The *metric horsepower* (hpm) is defined as 1 hpm ≡ 75 m·kgf/s. Develop a conversion relating metric horsepower to U.S. horsepower. Relate the specific speed for a hydraulic turbine—calculated in units of rpm, metric horsepower, and meters—to the specific speed calculated in U.S. customary units.

10.30 Write the pump specific speed in terms of the flow coefficient and the head coefficient.

10.31 A small centrifugal pump, when tested at $N=2875$ rpm with water, delivered $Q=0.016\ \text{m}^3/\text{s}$ and $H=40$ m at its best efficiency point ($\eta=0.70$). Determine the specific speed of the pump at this test condition. Sketch the impeller shape you expect. Compute the required power input to the pump.

10.32 Typical performance curves for a centrifugal pump, tested with three different impeller diameters in a single casing, are shown. Specify the flow rate and head produced by the pump at its best efficiency point with a 300 mm diameter impeller. Scale these data to predict the performance of this pump when tested with 275 mm and 325 mm impellers. Comment on the accuracy of the scaling procedure.

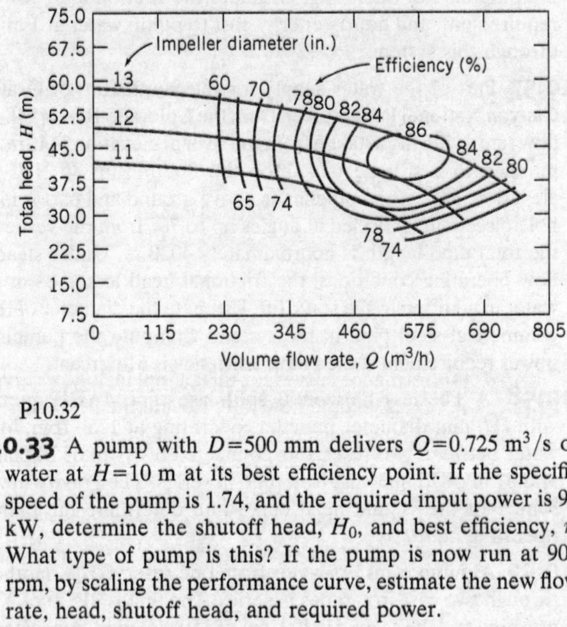

P10.32

10.33 A pump with $D=500$ mm delivers $Q=0.725\ \text{m}^3/\text{s}$ of water at $H=10$ m at its best efficiency point. If the specific speed of the pump is 1.74, and the required input power is 90 kW, determine the shutoff head, H_0, and best efficiency, η. What type of pump is this? If the pump is now run at 900 rpm, by scaling the performance curve, estimate the new flow rate, head, shutoff head, and required power.

10.34 At its best efficiency point ($\eta=0.85$), a mixed-flow pump, with $D=400$ mm, delivers $Q=1.2\ \text{m}^3/\text{s}$ of water at $H=50$ m when operating at $N=1500$ rpm. Calculate the specific speed of this pump. Estimate the required power input. Determine the curve-fit parameters of the pump performance curve based on the shutoff point and the best efficiency point. Scale the performance curve to estimate the flow, head, efficiency, and power input required to run the same pump at 750 rpm.

10.35 A pumping system must be specified for a lift station at a wastewater treatment facility. The average flow rate is 110 million liters per day and the required lift is 10 m. Non-clogging impellers must be used; about 65 percent efficiency is expected. For convenient installation, electric motors of 37.5 kW or less are desired. Determine the number of motor/pump units needed and recommend an appropriate operating speed.

10.36 A centrifugal water pump operates at 1750 rpm; the impeller has backward-curved vanes with $\beta_2=60°$ and $b_2=1.25$ cm. At a flow rate of $0.025\ \text{m}^3/\text{s}$, the radial outlet velocity is $V_{n_2}=3.5$ m/s. Estimate the head this pump could deliver at 1150 rpm.

10.37 A set of eight 30-kW motor-pump units is used to deliver water through an elevation of 30 m. The efficiency of the pumps is specified to be 65 percent. Estimate the delivery (liters per day) and select an appropriate operating speed.

10.38 Appendix D contains area bound curves for pump model selection and performance curves for individual pump models. Use these data to verify the similarity rules for a Peerless Type 4AE12 pump, with impeller diameter $D=11.0$ in., operated at 1750 and 3550 nominal rpm.

10.39 Appendix D contains area bound curves for pump model selection and performance curves for individual pump models. Use these data and the similarity rules to predict and plot the curves of head H (m) versus Q (m³/l) of a Peerless Type 10AE12 pump, with impeller diameter $D=305$ mm for nominal speeds of 1000, 1200, 1400, and 1600 rpm.

10.40 Consider the Peerless Type 16A18B horizontal split case centrifugal pump (Appendix D). Use these performance data to verify the similarity rules for (a) impeller diameter change and (b) operating speeds of 705 and 880 rpm (note the scale change between speeds).

10.41 Use data from Appendix D to verify the similarity rules for the effect of changing the impeller diameter of a Peerless Type 4AE12 pump operated at 1750 and 3550 nominal rpm.

10.42 Performance curves for Peerless horizontal split case pumps are presented in Appendix D. Develop and plot a curve-fit for a Type 10AE12 pump driven at 1150 nominal rpm using the procedure described in Example 10.6.

10.43 Performance curves for Peerless horizontal split case pumps are presented in Appendix D. Develop and plot curve-fits for a Type 16A18B pump, with impeller diameter $D=460$ mm, driven at 705 and 880 nominal rpm. Verify the effects of pump speed on scaling pump curves using the procedure described in Example 10.6.

10.44 Catalog data for a centrifugal water pump at design conditions are $Q=57\ \text{m}^3/\text{h}$ and $\Delta p=128$ kPa at 1750 rpm. A laboratory flume requires 45 m³/h at 9.8 m of head. The only motor available develops 2.2 kW at 1750 rpm. Is this motor suitable for the laboratory flume? How might the pump/motor match be improved?

10.45 Problem 10.21 suggests that pump head at best efficiency is typically about 70 percent of shutoff head. Use pump data from Appendix D to evaluate this suggestion. A further suggestion in Section 10.4 is that the appropriate scaling for tests of a pump casing with different impeller diameters is $Q \propto D^2$. Use pump data to evaluate this suggestion.

10.46 White [53] suggests modeling the efficiency for a centrifugal pump using the curve-fit, $\eta=aQ-bQ^3$, where a and b are constants. Describe a procedure to evaluate a and b from experimental data. Evaluate a and b using data for the Peerless Type 10AE12 pump, with impeller diameter D=305 mm, at 1760 rpm (Appendix D). Plot and illustrate the accuracy of the curve-fit by comparing measured and predicted efficiencies for this pump.

10.47 A fan operates at $Q = 6.3$ m^3/s, $H = 0.15$ m, and $N = 1440$ rpm. A smaller, geometrically similar fan is planned in a facility that will deliver the same head at the same efficiency as the larger fan, but at a speed of 1800 rpm. Determine the volumetric flow rate of the smaller fan.

10.48 A 1/3 scale model of a centrifugal water pump, when running at N_m=5100 rpm, produces a flow rate of Q_m=1 m^3/s with a head of H_m=5.4 m. Assuming the model and prototype efficiencies are comparable, estimate the flow rate, head, and power requirement if the design speed is 125 rpm.

10.49 Sometimes the variation of water viscosity with temperature can be used to achieve dynamic similarity. A model pump delivers 0.10 m^3/s of water at 15°C against a head of 27 m, when operating at 3600 rpm. Determine the water temperature that must be used to obtain dynamically similar operation at 1800 rpm. Estimate the volume flow rate and head produced by the pump at the lower-speed test condition. Comment on the *NPSH* requirements for the two tests.

10.50 A large deep fryer at a snack-food plant contains hot oil that is circulated through a heat exchanger by pumps. Solid particles and water droplets coming from the food product are observed in the flowing oil. What special factors must be considered in specifying the operating conditions for the pumps?

10.51 Data from tests of a pump operated at 1500 rpm, with a 30 cm. diameter impeller, are

Flow rate, Q (m^3/s × 10^3)	10	20	30	40	50	60	70
Net positive suction head required, *NPSR* (m)	2.2	2.4	2.6	3.1	3.6	4.1	5.1

Develop and plot a curve-fit equation for *NPSHR* versus volume flow rate in the form $NPSHR = a + bQ^2$, where a and b are constants. If the *NPSHA*=6 m, estimate the maximum allowable flow rate of this pump.

10.52 A four-stage boiler feed pump has suction and discharge lines of 10 cm and 7.5 cm inside diameter. At 3500 rpm, the pump is rated at 0.025 m^3/s against a head of 125 m while handling water at 115°C. The inlet pressure gage, located 50 cm below the impeller centerline, reads 150 kPa. The pump is to be factory certified by tests at the same flow rate, head rise, and speed, but using water at 27°C. Calculate the *NPSHA* at the pump inlet in the field installation. Evaluate the suction head that must be used in the factory test to duplicate field suction conditions.

10.53 The net positive suction head required (*NPSHR*) by a pump may be expressed approximately as a parabolic function of volume flow rate. The *NPSHR* for a particular pump operating at 1800 rpm is given as $H_r = H_0 = AQ^2$, where $H_0 = 3$ m of water and $A = 3000$ m/(m^3/s)^2. Assume the pipe system supplying the pump suction consists of a reservoir, whose surface is 6 m above the pump centerline, a square entrance, 6 m of 15 cm (nominal) cast-iron pipe, and a 90° elbow. Calculate the maximum volume flow rate at 20°C for which the suction head is sufficient to operate this pump without cavitation.

10.54 A centrifugal pump, operating at N=2265 rpm, lifts water between two reservoirs connected by 90 m of 150 mm and 30 m of 75 mm cast-iron pipe in series. The gravity lift is 7.6 m. Estimate the head requirement, power needed, and hourly cost of electrical energy to pump water at 45 m^3/h to the higher reservoir. Assume that electricity costs 12¢/kWhr, and that the electric motor efficiency is 85 percent.

10.55 For the pump and flow system of Problem 10.53, calculate the maximum flow rate for hot water at various temperatures and plot versus water temperature. (Be sure to consider the density variation as water temperature is varied.)

10.56 A centrifugal pump is installed in a piping system with L=300 m of D=40 cm cast-iron pipe. The downstream reservoir surface is 15 m lower than the upstream reservoir. Determine and plot the system head curve. Find the volume flow rate (magnitude and direction) through the system when the pump is not operating. Estimate the friction loss, power requirement, and hourly energy cost to pump water at 1 m^3/s through this system.

10.57 Part of the water supply for the South Rim of Grand Canyon National Park is taken from the Colorado River [54]. A flow rate of 136 m^3/h, taken from the river at elevation 1140 m, is pumped to a storage tank atop the South Rim at 2140 m elevation. Part of the pipeline is above ground and part is in a hole directionally drilled at angles up to 70° from the vertical; the total pipe length is approximately 4020 m. Under steady flow operating conditions, the frictional head loss is 88 m of water in addition to the static lift. Estimate the diameter of the commercial steel pipe in the system. Compute the pumping power requirement if the pump efficiency is 61 percent.

10.58 A Peerless horizontal split-case type 4AE12 pump with 280 mm diameter impeller, operating at 1750 rpm, lifts water between two reservoirs connected by 61 m of 100 mm and 61 m of 75 mm cast-iron pipe in series. The gravity lift is 3 m. Plot the system head curve and determine the pump operating point.

10.59 A pump transfers water from one reservoir to another through two cast-iron pipes in series. The first is 915 m of 230 mm pipe and the second is 300 mm of 150 mm pipe. A constant flow rate of 17 m^3/h is tapped off at the junction between the two pipes. Obtain and plot the system head versus flow rate curve. Find the delivery if the system is supplied by the pump of Example 10.6, operating at 1750 rpm.

10.60 Performance data for a pump are

H (m)	27.5	27	25	22	18	13	6.5
Q (m^3/s)	0	0.025	0.050	0.075	0.100	0.125	0.150

The pump is to be used to move water between two open reservoirs with an elevation increase of 7.5 m. The connecting pipe system consists of 500 m of commercial steel pipe containing two 90° elbows and an open gate valve. Find the flow rate if we use (a) 20 cm, (b) 30 cm, and (c) 40 cm (nominal) pipe.

10.61 Performance data for a pump are

H (m)	55	54	50	44	36	26	13
Q (m^3/hr)	0	105	210	315	420	525	630

Estimate the delivery when the pump is used to move water between two open reservoirs, through 365 m of 305 mm commercial steel pipe containing two 90° elbows and an open gate valve, if the elevation increase is 15 m. Determine the gate valve loss coefficient needed to reduce the volume flow rate by half.

10.62 Consider again the pump and piping system of Problem 10.61. Determine the volume flow rate and gate valve loss coefficient for the case of two identical pumps installed in *series*.

10.63 The resistance of a given pipe increases with age as deposits form, increasing the roughness and reducing the pipe diameter (see Fig. 8.14). Typical multipliers to be applied to the friction factor are given in [15]:

Pipe Age (years)	Small Pipes, 100–250 mm	Large Pipes, 300–1500 mm
New	1.00	1.00
10	2.20	1.60
20	5.00	2.00
30	7.25	2.20
40	8.75	2.40
50	9.60	2.86
60	10.0	3.70
70	10.1	4.70

Consider again the pump and piping system of Problem 10.61. Estimate the percentage reductions in volume flow rate that occur after (a) 20 years and (b) 40 years of use, if the pump characteristics remain constant. Repeat the calculation if the pump head is reduced 10 percent after 20 years of use and 25 percent after 40 years.

10.64 Consider again the pump and piping system of Problem 10.61. Determine the volume flow rate and gate valve loss coefficient for the case of two identical pumps installed in *parallel*.

10.65 Consider again the pump and piping system of Problem 10.64. Estimate the percentage reductions in volume flow rate that occur after (a) 20 years and (b) 40 years of use, if the pump characteristics remain constant. Repeat the calculation if the pump head is reduced 10 percent after 20 years of use and 25 percent after 40 years. (Use the data of Problem 10.63 for increase in pipe friction factor with age.)

10.66 Consider again the pump and piping system of Problem 10.62. Estimate the percentage reductions in volume flow rate that occur after (a) 20 years and (b) 40 years of use, if the pump characteristics remain constant. Repeat the calculation if the pump head is reduced 10 percent after 20 years of use and 25 percent after 40 years. (Use the data of Problem 10.63 for increase in pipe friction factor with age.)

10.67 The city of Englewood, Colorado, diverts water for municipal use from the South Platte River at elevation 1610 m [54]. The water is pumped to storage reservoirs at 1620-m elevation. The inside diameter of the steel water line is 68.5 cm; its length is 1770 m. The facility is designed for an initial capacity (flow rate) of 3200 m^3/hr, with an ultimate capacity of 3900 m^3/hr. Calculate and plot the system resistance curve. Ignore entrance losses. Specify an appropriate pumping system. Estimate the pumping power required for steady-state operation, at both the initial and ultimate flow rates.

10.68 A pump in the system shown draws water from a sump and delivers it to an open tank through 400 m of new, 10 cm diameter steel pipe. The vertical suction pipe is 2 m long and includes a foot valve with hinged disk and a 90° standard elbow. The discharge line includes two 90° standard elbows, an angle lift check valve, and a fully open gate valve. The design flow rate is 800 L/min. Find the head losses in the suction and discharge lines. Calculate the *NPSHA*. Select a pump suitable for this application.

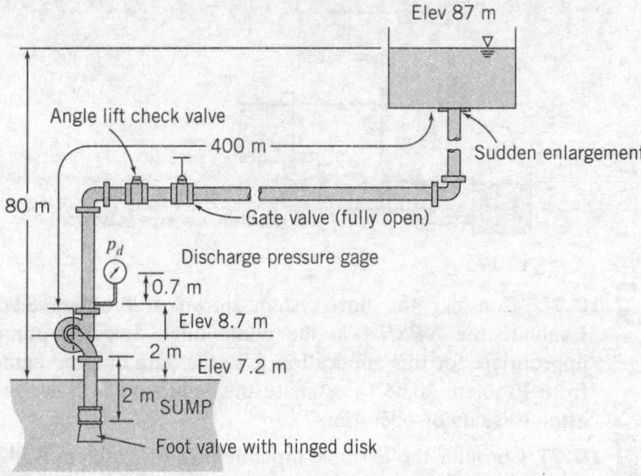

P10.68, P10.70

10.69 Consider the flow system described in Problem 8.175. Select a pump appropriate for this application. Check the *NPSHR* versus the *NPSHA* for this system.

10.70 Consider the flow system and data of Problem 10.68 and the data for pipe aging given in Problem 10.63. Select pump(s) that will maintain the system flow at the desired rate for (a) 10 years and (b) 20 years. Compare the delivery produced by these pumps with the delivery by the pump sized for new pipes only.

10.71 Consider the flow system shown in Problem 8.176. Select an appropriate pump for this application. Check the pump efficiency and power requirement compared with those in the problem statement.

10.72 Consider the flow system shown in Problem 8.124. Assume the minimum *NPSHR* at the pump inlet is 4.6 m of water. Select a pump appropriate for this application. Use the data for increase in friction factor with pipe age given in

Problem 10.65 to determine and compare the system flow rate after 10 years of operation.

10.73 Consider the pipe network of Problem 8.189. Select a pump suitable to deliver a total flow rate of 68 m^3/h through the pipe network.

10.74 A fire nozzle is supplied through 90 m of 75 mm diameter canvas hose (with $e = 0.3$ mm). Water from a hydrant is supplied at 345 kPa to a booster pump on board the pumper truck. At design operating conditions, the pressure at the nozzle inlet is 690 kPa, and the pressure drop along the hose is 7.5 kPa/m of length. Calculate the design flow rate and the maximum nozzle exit speed. Select a pump appropriate for this application, determine its efficiency at this operating condition, and calculate the power required to drive the pump.

10.75 A pumping system with two different static lifts is shown. Each reservoir is supplied by a line consisting of 300 m of 20 cm cast-iron pipe. Evaluate and plot the system head versus flow curve. Explain what happens when the pump head is less than the height of the upper reservoir. Calculate the flow rate delivered at a pump head of 26 m.

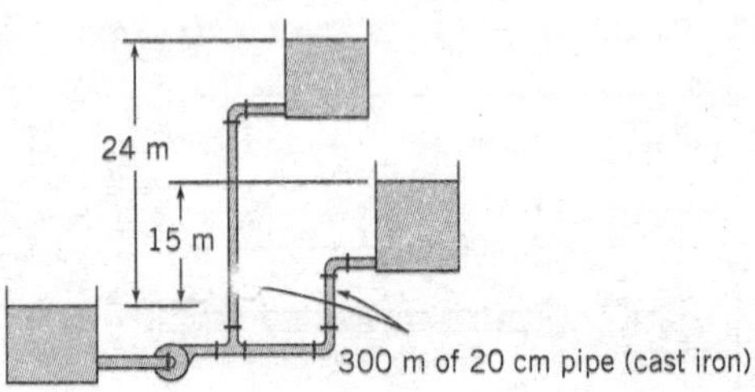

P10.75

10.76 Consider the flow system shown in Problem 8.90. Evaluate the *NPSHA* at the pump inlet. Select a pump appropriate for this application. Use the data on pipe aging from Problem 10.63 to estimate the reduction in flow rate after 10 years of operation.

10.77 Consider the gasoline pipeline flow of Problem 8.142. Select pumps that, combined in parallel, supply the total flow requirement. Calculate the power required for 4 pumps in parallel. Also calculate the volume flow rates and power required when only 1, 2, or 3 of these pumps operates.

10.78 Consider the chilled water circulation system of Problem 8.178. Select pumps that may be combined in parallel to supply the total flow requirement. Calculate the power required for 3 pumps in parallel. Also calculate the volume flow rates and power required when only 1 or 2 of these pumps operates.

10.79 Water for the sprinkler system at a lakeside summer home is to be drawn from the adjacent lake. The home is located on a bluff 33 m above the lake surface. The pump is located on level ground 3 m above the lake surface. The sprinkler system requires 40 L/min at 300 kPa (gage). The piping system is to be 2 cm diameter galvanized iron. The inlet section (between the lake and pump inlet) includes a reentrant inlet, one standard 45° elbow, one standard 90° elbow, and 20 m of pipe. The discharge section (between the pump outlet and the sprinkler connection) includes two standard 45° elbows and 45 m of pipe. Evaluate the head loss on the suction side of the pump. Calculate the gage pressure at the pump inlet. Determine the hydraulic power requirement of the pump. If the pipe diameter were increased to 4 cm., would the power requirement of the pump increase, decrease, or stay the same? What difference would it make if the pump were located halfway up the hill?

10.80 Consider the fire hose and nozzle of Problem 8.179. Specify an appropriate pump to supply four such hoses simultaneously. Calculate the power input to the pump.

10.81 Manufacturer's data for a submersible utility pump are

Discharge height (m)	0.3	0.7	1.5	3.0	4.5	6.0	8.0
Water flow rate (L/min)	77.2	75	71	61	51	26	0

The owner's manual also states, "Note: These ratings are based on discharge into 25-mm pipe with friction loss neglected. Using 20-mm garden hose adaptor, performance will be reduced approximately 15 percent." Plot a performance curve for the pump. Develop a curve-fit equation for the performance curve; show the curve-fit on the plot. Calculate and plot the pump delivery versus discharge height through a 15-m length of smooth 20-mm garden hose. Compare with the curve for delivery into 25-mm pipe.

10.82 Consider the swimming pool filtration system of Problem 8.190. Assume the pipe used is 20-mm PVC (smooth plastic). Specify the speed and impeller diameter and estimate the efficiency of a suitable pump.

10.83 Water is pumped from a lake (at $z = 0$) to a large storage tank located on a bluff above the lake. The pipe is 75 mm diameter galvanized iron. The inlet section (between the lake and the pump) includes one rounded inlet, one standard 90° elbow, and 15 m of pipe. The discharge section (between the pump outlet and the discharge to the open tank) includes two standard 90° elbows, one gate valve, and 46 m of pipe. The pipe discharge (into the side of the tank) is at $z = 21$ m. Calculate the system flow curve. Estimate the system operating point. Determine the power input to the pump if its efficiency at the operating point is 80 percent. Sketch the system curve when the water level in the upper tank reaches $z = 27$ m. If the water level in the upper tank is at $z = 23$ m and the valve is partially closed to reduce the flow rate to 10 m^3/h, sketch the system curve for this operating condition. Would you expect the pump efficiency to be higher for the first or second operating condition? Why?

10.84 Performance data for a centrifugal fan of 1 m diameter, tested at 650 rpm, are

Volume flow rate Q (m^3/s)	3	4	5	6	7	8
Static pressure rise, Δp (mm H_2O)	53	51	45	35	23	11
Power output $\mathcal{P}$ (kW)	2.05	2.37	2.60	2.62	2.61	2.4

Plot the performance data versus volume flow rate. Calculate static efficiency, and show the curve on the plot. Find the best efficiency point, and specify the fan rating at this point.

10.85 Using the fan of Problem 10.84, determine the minimum size square sheet-metal duct that will carry a flow of 15 m over a distance of 5.75 m^3/s. Estimate the increase in delivery if the fan speed is increased to 800 rpm.

10.86 The performance data of Problem 10.84 are for a 1 m diameter fan wheel. The fan also is manufactured with 1.025, 1.125, 1.250 and 1.375 m diameter wheels. Pick a standard fan to deliver 14 m^3/s against a 25 mm H_2O static pressure rise. Determine the required fan speed and input power required.

10.87 Consider the fan and performance data of Problem 10.84. At $Q = 5.75\ m^3/s$, the dynamic pressure is equal to 4 mm of water. Evaluate the fan outlet area. Plot total pressure rise and input power for this fan versus volume flow rate. Calculate the fan total efficiency, and show the curve on the plot. Find the best efficiency point, and specify the fan rating at this point.

10.88 Performance characteristics of a Howden Buffalo axial flow fan are presented below. The fan is used to power a wind tunnel with 0.3 m square test section. The tunnel consists of a smooth inlet contraction, two screens (each with loss coefficient $K = 0.12$), the test section, and a diffuser where the cross section is expanded to 610 mm diameter at the fan inlet. Flow from the fan is discharged back to the room. Calculate and plot the system characteristic curve of pressure loss versus volume flow rate. Estimate the maximum air flow speed available in this wind tunnel test section.

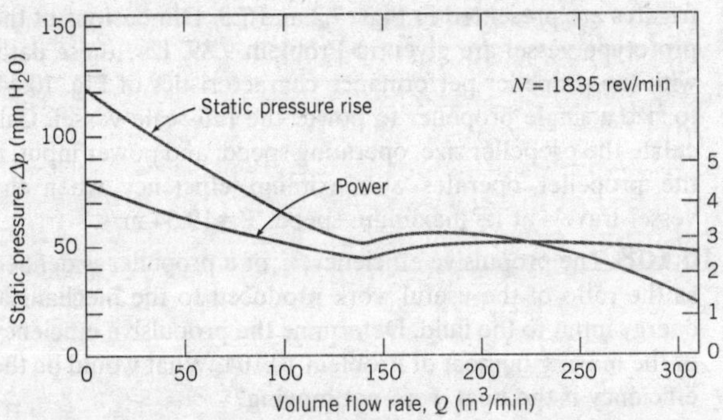

P10.88, P10.89

10.89 Consider again the axial-flow fan and wind tunnel of Problem 10.88. Scale the performance of the fan as it varies with operating speed. Develop and plot a "calibration curve" showing test section flow speed (in m/s) versus fan speed (in rpm).

10.90 Experimental test data for an aircraft engine fuel pump are presented below. This gear pump is required to supply jet fuel at 205 kg per hour and 1 MPa to the engine fuel controller. Tests were conducted at 10, 96, and 100 percent of the rated pump speed of 4536 rpm. At each constant speed, the back pressure on the pump was set, and the flow rate measured. On one graph, plot curves of pressure versus delivery at the three constant speeds. Estimate the pump displacement volume per revolution. Calculate the volumetric efficiency at each test point and sketch contours of constant η_v. Evaluate the energy loss caused by valve throttling at 100 percent speed and full delivery to the engine.

Pump Speed (rpm)	Back Pressure MPa (gage)	Fuel Flow (kg/hr)	Pump Speed (rpm)	Back Pressure MPa (gage)	Fuel Flow (kg/hr)	Pump Speed (rpm)	Back Pressure MPa (gage)	Fuel Flow (kg/hr)
	1.4	815		1.4	780		1.4	40
4536	2.1	815	4355	2.1	790	453	1.7	33
(100%)	2.8	815	(96%)	2.8	782	(10%)	2.1	26
	3.5	805		3.5	775		2.4	20
	6.3	775		6.3	775		2.8	14

* Fuel flow rate measured in pounds per hour (kg/h).

Hydraulic Turbines

10.91 A hydraulic turbine is designed to produce 26,800 kW at 95 rpm under 15 m of head. Laboratory facilities are available to provide 5 m of head and to absorb 35 kW from the model turbine. Determine (a) the appropriate model test speed and scale ratio and (b) volume flow rate, assuming a model efficiency of 86 percent.

10.92 Preliminary calculations for a hydroelectric power generation site show a net head of 715 m is available at a water flow rate of 2 m^3/s. Compare the geometry and efficiency of Pelton wheels designed to run at (a) 450 rpm and (b) 600 rpm.

10.93 Conditions at the inlet to the nozzle of a Pelton wheel are $p = 4.9$ MPa and $V = 24$ km/h. The jet diameter is $d = 190$ mm and the nozzle loss coefficient is $K_{nozzle} = 0.04$. The wheel diameter is $D = 2.4$ m. At this operating condition, $\eta = 0.86$. Calculate (a) the power output, (b) the normal operating speed, (c) the approximate runaway speed, (d) the torque at normal operating speed, and (e) the approximate torque at zero speed.

10.94 The reaction turbines at Niagara Falls are of the Francis type. The impeller outside diameter is 4.5 m. Each turbine produces 54 MW at 107 rpm, with 93.8 percent efficiency under 65 m of net head. Calculate the specific speed of these units. Evaluate the volume flow rate to each turbine. Estimate the penstock size if it is 400 m long and the net head is 83 percent of the gross head.

10.95 Francis turbine Units 19, 20, and 21, installed at the Grand Coulee Dam on the Columbia River, are *very* large [55]. Each runner is 830 mm in diameter and contains 550 tons of cast steel. At rated conditions, each turbine develops 610 MW at 72 rpm under 87 m of head. Efficiency is nearly 95 percent at rated conditions. The turbines operate at heads from 67 to 108 m. Calculate the specific speed at rated operating conditions. Estimate the maximum water flow rate through each turbine.

10.96 Measured data for performance of the reaction turbines at Shasta Dam near Redding, California, are shown in Fig. 10.39. Each turbine is rated at 77 × 10^3 kW when operating at 138.6 rpm under a net head of 115 m. Evaluate the specific speed and compute the shaft torque developed by each turbine at rated operating conditions. Calculate and plot the water flow rate per turbine required to produce rated output power as a function of head.

10.97 Figure 10.37 contains data for the efficiency of a large Pelton waterwheel installed in the Tiger Creek Power House of Pacific Gas & Electric Company near Jackson, California.

This unit is rated at 26.8 MW when operated at 225 rpm under a net head of 360 m of water. Assume reasonable flow angles and nozzle loss coefficient, and water at 15°C. Determine the rotor radius, and estimate the jet diameter and the mass flow rate of water.

10.98 An impulse turbine is to develop 15 MW from a single wheel at a location where the net head is 350 m. Determine the appropriate speed, wheel diameter, and jet diameter for single- and multiple-jet operation. Compare with a double-overhung wheel installation. Estimate the required water consumption.

10.99 An impulse turbine under a net head of 9.9 m was tested at a variety of speeds. The flow rate and the brake force needed to set the impeller speed were recorded:

Wheel Speed (rpm)	Flow rate (m^3/hr)	Brake Force (N) ($R = 0.15$ m)
0	13.15	11.70
1000	13.15	10.68
1500	13.15	9.88
1900	12.64	8.50
2200	11.93	6.45
2350	9.58	3.87
2600	7.85	1.51
2700	6.93	0.40

Calculate and plot the machine power output and efficiency as a function of water turbine speed.

10.100 In U.S. customary units, the common definition of specific speed for a hydraulic turbine is given by Eq. 10.13b. Develop a conversion between this definition and a truly dimensionless one in SI units. Evaluate the specific speed of an impulse turbine, operating at 400 rpm under a net head of 1190 ft with 86 percent efficiency, when supplied by a single 6-in.-diameter jet. Use both U.S. customary and SI units. Estimate the wheel diameter.

10.101 According to a spokesperson for Pacific Gas & Electric Company, the Tiger Creek plant, located east of Jackson, California, is one of 71 PG&E hydroelectric powerplants. The plant has 373 m of gross head, consumes 21 m^3/s of water, is rated at 60 MW, and operates at 58 MW. The plant is claimed to produce 0.785 kW·hr/(m^2·m) of water and 336.4×10^6 kW·hr/yr of operation. Estimate the net head at the site, the turbine specific speed, and its efficiency. Comment on the internal consistency of these data.

10.102 Design the piping system to supply a water turbine from a mountain reservoir. The reservoir surface is 320 m above the turbine site. The turbine efficiency is 83 percent, and it must produce 30 kW of mechanical power. Define the minimum standard-size pipe required to supply water to the turbine and the required volume flow rate of water. Discuss the effects of turbine efficiency, pipe roughness, and installing a diffuser at the turbine exit on the performance of the installation.

10.103 A small hydraulic impulse turbine is supplied with water through a penstock with diameter D and length L; the jet diameter is d. The elevation difference between the reservoir surface and nozzle centerline is Z. The nozzle head loss coefficient is K_{nozzle} and the loss coefficient from the reservoir to the penstock entrance is $K_{entrance}$. Determine the water jet speed, the volume flow rate, and the hydraulic power of the jet, for the case where $Z=90$ m, $L=300$ m, $D=150$ mm, $K_{entrance}=0.5$, $K_{nozzle}=0.04$, and $d=50$ mm, if the pipe is made from commercial steel. Plot the jet power as a function of jet diameter to determine the optimum jet diameter and the resulting hydraulic power of the jet. Comment on the effects of varying the loss coefficients and pipe roughness.

Propellers and Wind-Power Machines

10.104 The propeller on a fanboat used in the Florida Everglades moves air at the rate of 50 kg/s. When at rest, the speed of the slipstream behind the propeller is 45 m/s at a location where the pressure is atmospheric. Calculate (a) the propeller diameter, (b) the thrust produced at rest, and (c) the thrust produced when the fanboat is moving ahead at 15 m/s if the mass flow rate through the propeller remains constant.

10.105 A fanboat in the Florida Everglades is powered by a propeller, with $D=1.5$ m, driven at maximum speed, $N=1800$ rpm, by a 125 kW engine. Estimate the maximum thrust produced by the propeller at (a) standstill and (b) $V=12.5$ m/s.

10.106 A jet-propelled aircraft traveling at 225 m/s takes in 50 kg/s of air. If the propulsive efficiency (defined as the ratio of the useful work output to the mechanical energy input to the fluid) of the aircraft is 45 percent, determine the speed at which the exhaust is discharged relative to the aircraft.

10.107 Drag data for model and prototype guided missile frigates are presented in Figs. 7.2 and 7.3. Dimensions of the prototype vessel are given in Problem 9.89. Use these data, with the propeller performance characteristics of Fig. 10.44, to size a single propeller to power the full-scale vessel. Calculate the propeller size, operating speed, and power input, if the propeller operates at maximum efficiency when the vessel travels at its maximum speed, $V=19.34$ m/s.

10.108 The propulsive efficiency, η, of a propeller is defined as the ratio of the useful work produced to the mechanical energy input to the fluid. Determine the propulsive efficiency of the moving fanboat of Problem 10.104. What would be the efficiency if the boat were not moving?

10.109 The propeller for the Gossamer Condor human-powered aircraft has $D=3.6$ m and rotates at $N=107$ rpm. Additional details on the aircraft are given in Problem 9.174. Estimate the dimensionless performance characteristics and efficiency of this propeller at cruise conditions. Assume the pilot expends 70 percent of maximum power at cruise. (See Reference [56] for more information on human-powered flight.)

10.110 Equations for the thrust, power, and efficiency of propulsion devices were derived in Section 10.6. Show that these equations may be combined for the condition of constant thrust to obtain

$$\eta=\frac{2}{1+\left(1+\dfrac{F_T}{\dfrac{\rho V^2}{2}\dfrac{\pi D^2}{4}}\right)^{1/2}}$$

Interpret this result physically.

10.111 The National Aeronautics & Space Administration (NASA) and the U.S. Department of Energy (DOE) cosponsor a large demonstration wind turbine generator at Plum Brook, near Sandusky, Ohio [47]. The turbine has two blades, with a radius of 19 m, and delivers maximum power when the wind speed is above $V = 29$ km/h. It is designed to produce 100 kW with powertrain efficiency of 75 percent. The rotor is designed to operate at a constant speed of 45 rpm in winds over 2.6 m/s by controlling system load and adjusting blade angles. For the maximum power condition, estimate the rotor tip speed and power coefficient.

10.112 A typical American multiblade farm windmill has $D=2.1$ m and is designed to produce maximum power in winds with $V=24$ km/h. Estimate the rate of water delivery, as a function of the height to which the water is pumped, for this windmill.

10.113 A model of an American multiblade farm windmill is to be built for display. The model, with $D=1$ m, is to develop full power at $V=10$ m/s wind speed. Calculate the angular speed of the model for optimum power generation. Estimate the power output.

10.114 A large Darrieus vertical axis wind turbine was built by the U.S. Department of Energy near Sandia, New Mexico [48]. This machine is 18 m tall and has a 5-m radius; the area swept by the rotor is over 110 m^2. If the rotor is constrained to rotate at 70 rpm, plot the power this wind turbine can produce in kilowatts for wind speeds between 5 and 50 knots.

10.115 Lift and drag data for the NACA 23015 airfoil section are presented in Fig. 9.17. Consider a two-blade horizontal-axis propeller wind turbine with NACA 23015 blade section. Analyze the air flow relative to a blade element of the rotating wind turbine. Develop a numerical model for the blade element. Calculate the power coefficient developed by the blade element as a function of tip-speed ratio. Compare your result with the general trend of power output for high-speed two-bladed turbine rotors shown in Fig. 10.50.

10.116 Aluminum extrusions, patterned after NACA symmetric airfoil sections, frequently are used to form Darrieus wind turbine "blades." Below are section lift and drag coefficient data [57] for a NACA 0012 section, tested at $Re = 6 \times 10^6$ with standard roughness (the section stalled for $\alpha > 12°$):

Angle of attack, α (deg)	0	2	4	6	8	10	12
Lift coefficient, C_L (—)	0	0.23	0.45	0.68	0.82	0.94	1.02
Drag coefficient. C_D (—)	0.0098	0.0100	0.0119	0.0147	0.0194	—	—

Analyze the air flow relative to a blade element of a Darrieus wind turbine rotating about its troposkien axis. Develop a numerical model for the blade element. Calculate the power coefficient developed by the blade element as a function of tip-speed ratio. Compare your result with the general trend of power output for Darrieus rotors shown in Fig. 10.50.

Compressible-Flow Turbomachines

10.117 A prototype air compressor with a compression ratio of 7 is designed to take 8.9 kg/s air at 1 atmosphere and 20°C. The design point speed, power requirement, and efficiency are 600 rpm, 5.6 MW, and 80 percent, respectively. A 1:5-scale model of the prototype is built to help determine operability for the prototype. If the model takes in air at identical conditions to the prototype design point, what will the mass flow and power requirement be for operation at 80 percent efficiency?

10.118 A compressor has been designed for entrance conditions of 101.3 kPa and 21°C. To economize on the power required, it is being tested with a throttle in the entry duct to reduce the entry pressure. The characteristic curve for its normal design speed of 3200 rpm is being obtained on a day when the ambient temperature is 14.4°C. At what speed should the compressor be run? At the point on the characteristic curve at which the mass flow would normally be 57 kg/s, the entry pressure is 55.16 kPa. Calculate the actual mass flow rate during the test.

10.119 The turbine for a new jet engine was designed for entrance conditions of 1100 kPa and 925°C, ingesting 250 kg/s at a speed of 500 rpm, and exit conditions of 550 kPa and 730°C. If the altitude and fueling for the engine were changed such that the entrance conditions were now 965 kPa and 870°C, calculate the new operating speed, mass flow rate, and exit conditions for similar operation, i.e., equal efficiency.

10.120 We have seen many examples in Chapter 7 of replacing working fluids in order to more easily achieve similitude between models and prototypes. Describe the effects of testing an air compressor using helium as the working fluid on the dimensionless and dimensional parameters we have discussed for compressible flow machines.

11

Flow in Open Channels

Case Study in Energy and the Environment

Using a Reservoir as a Battery

We are all familiar with electric batteries; we have them in our cars, our laptops, our cell phones, and our MP3 players, to mention just a few of their uses. Batteries are energy storage devices that allow us to generate energy at one time and place and store it for use at a different time and in another place. The figure shows a rather mundane-looking dam (it's the Ffestiniog Dam in north Wales), but it's actually part of a pretty exciting development, the Ffestiniog Pumped Storage Scheme: it's a battery!

The idea of using reservoirs not only as a source of power but as a way to store power is not new; efforts were made in the 19th century. But it is becoming very important in optimizing power plant performance, as well as in storing renewable energy generated by wind,

The Ffestiniog dam.

wave, and ocean current farms, some of which we have reviewed in previous *Case Studies in Energy and the Environment.* Power companies have always had the problem that energy demand tends to have severe peaks and troughs; in the afternoon and evening there is high demand; in the middle of the night, low demand. However, for best efficiency, plants should operate at a steady energy output; in addition, the power company needs to have on hand extra power generation capability just for those peaks. On the other hand, renewable energy needs to be harvested when it's available—when the wind is blowing, when there are waves or decent currents flowing—and these times do not always correspond to the times when the energy is needed. With schemes like the one at Ffestiniog, at times of low electrical demand, excess generation capacity from the power company is used to pump water into an upper reservoir; when there is high demand, water is released back into a lower reservoir through a turbine, generating electricity. Reversible turbine/generator assemblies act as pump and turbine (usually of a Francis turbine design; see Chapter 10). The system's four water turbines can generate 360 MW of electricity within a minute of the need arising!

Some facilities worldwide are purely pumped-storage plants, which simply shift water between two reservoirs, but combined pump-storage plants that also generate their own electricity like conventional hydroelectric plants are becoming more common. The process is reasonably efficient and is the only way that huge amounts of energy can be stored (electric batteries are all relatively low capacity). Taking into account losses in the turbine/generator system and from evaporation loss at the exposed water surface, as well as the possibility of losses due to hydraulic jumps (discussed in this chapter) occurring at outlets, about 70 to 85 percent of the electrical energy used to pump the water into the elevated reservoir can be regained. In future years, increased effort will be placed on increasing the efficiency of these systems, and they will become much more common. The Ffestiniog system is for storing excess power plant energy, but in the future we may expect to see pumped-storage plants adjacent to a number of wind farms.

In this chapter we introduce some of the basic concepts in the study of open-channel flows. The topic of open-channel flow is covered in much more detail in a number of specialized texts [1–8].

Many flows in engineering and in nature occur with a free surface. An example of a human-made channel is shown in Fig. 11.1. This is a view of the 306 km-long Hayden-Rhodes Aqueduct, which is part of the Central Arizona Project (CAP). The CAP is a 541 km diversion canal in Arizona used to redirect water from

Fig. 11.1 Hayden-Rhodes Aqueduct, Central Arizona Project.
[Courtesy of the U.S. Bureau of Reclamation (1985), photograph by Joe Madrigal Jr.]

the Colorado River into central and southern Arizona, and it is the largest and most expensive aqueduct system ever constructed in the United States.

Because free surface flows differ in several important respects from the flows in closed conduits that we reviewed in Chapter 8, we treat them separately in this chapter. Familiar examples where the free surface is at atmospheric pressure include flows in rivers, aqueducts, and irrigation canals, flows in rooftop or street gutters, and drainage ditches. Human-made channels are given many different names, including canal, flume, or culvert. A *canal* usually is excavated below ground level and may be unlined or lined. Canals generally are long and of very mild slope; they are used to carry irrigation or storm water or for navigation. A *flume* usually is built above ground level to carry water across a depression. A *culvert*, which usually is designed to flow only part-full, is a short covered channel used to drain water under a highway or railroad embankment.

In this chapter we shall develop, using control volume concepts from Chapter 4, some basic theory for describing the behavior and classification of flows in natural and human-made channels. We shall consider:

- *Flows for which the local effects of area change predominate and frictional forces may be neglected.* An example is flow over a bump or depression, over the short length of which friction is negligible.
- *Flow with an abrupt change in depth.* This occurs during a hydraulic jump (see Fig. 11.12 for examples of hydraulic jumps).
- *Flow at what is called normal depth.* For this, the flow cross section does not vary in the flow direction; the liquid surface is parallel to the channel bed. This is analogous to fully developed flow in a pipe.
- *Gradually varied flow.* An example is flow in a channel in which the bed slope varies. The major objective in the analysis of gradually varied flow is to predict the shape of the free surface.

It is quite common to observe surface waves in flows with a free surface, the simplest example being when an object such as a pebble is thrown into the water. The propagation speed of a surface wave is analogous in many respects to the propagation of a sound wave in a compressible fluid medium (which we discuss in Chapter 12). We shall determine the factors that affect the speed of such surface waves. We will see that this is an important determinant in whether an open-channel flow is able to gradually adjust to changing conditions downstream or a hydraulic jump occurs.

This chapter also includes a brief discussion of flow measurement techniques for use in open channels.

Basic Concepts and Definitions 11.1

Before analyzing the different types of flows that may occur in an open channel, we will discuss some common concepts and state some simplifying assumptions. We are doing so explicitly, because there are some important differences between our previous studies of pipes and ducts in Chapter 8 and the study of open-channel flows.

One significant difference between flows in pipes and ducts is

- The driving force for open-channel flows is *gravity.*

(Note that some flows in pipes and ducts are also gravity driven (for example, flow down a full drainpipe), but typically flow is driven by a pressure difference generated by a device such as a pump.) The gravity force in open-channel flow is opposed by friction force on the solid boundaries of the channel.

Simplifying Assumptions

The flow in an open channel, especially in a natural one such as a river, is often very complex, three-dimensional, and unsteady. However, in most cases, we can obtain useful results by approximating such flows as being:

- *One-dimensional.*
- *Steady.*

A third simplifying assumption is:

- The flow at each section in an open-channel flow is approximated as a *uniform velocity*.

VIDEO

A Turbulent Channel (Animation).

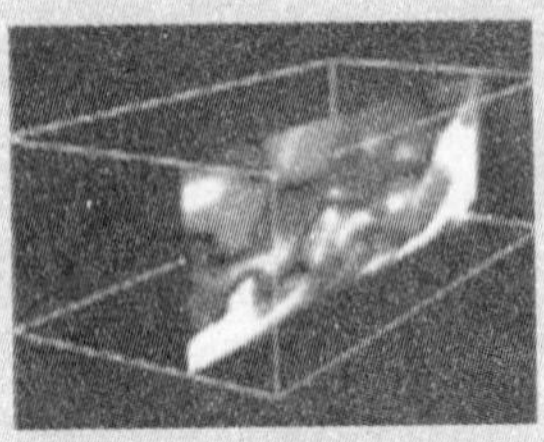

VIDEO

The Glen Canyon Dam: A Source of Turbulent Channel Flow.

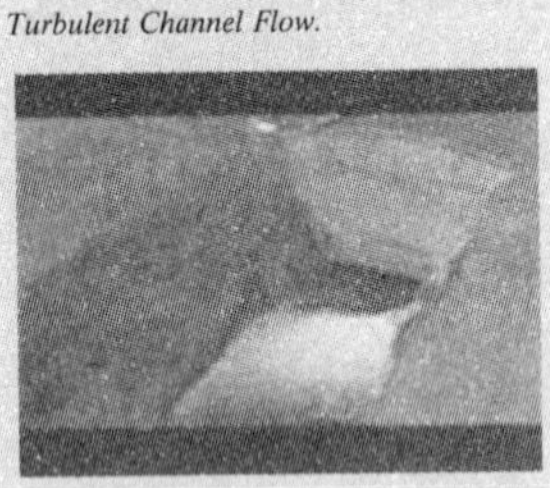

Typical contours of actual streamwise velocity for a number of open-channel sections are shown in Fig. 11.2. These would seem to indicate that the third assumption is invalid, but in fact it is a reasonable approximation, as we shall now justify. Most flows of interest are large in physical scale, so the Reynolds numbers generally are large. Consequently, open-channel flow seldom is laminar; in this chapter we will assume turbulent flow. As we saw in earlier chapters, turbulence tends to smooth out the velocity gradient (see Fig. 8.11 for turbulent pipe flow and Fig. 9.7*a* for turbulent boundary layers). Hence although the profiles, as shown in Fig. 11.2, are *not* uniform, as a reasonable approximation we will assume uniform velocity at each section, with the kinetic energy coefficient, α, taken to be unity (the kinetic energy coefficient is discussed in Section 8.6). This is illustrated in Fig. 11.3*a*.

Figure 11.2 shows that the measured maximum velocity occurs below the free surface, in spite of the fact that there is negligible shear stress due to air drag so one would expect the maximum velocity to occur at the free surface. Secondary flows are also responsible for distorting the axial velocity profile; examples of secondary flows are when a channel has a bend or curve or has an obstruction, such as a bridge pier. The high velocities that may be present in the vortices generated in such cases can seriously erode the bottom of a natural channel.

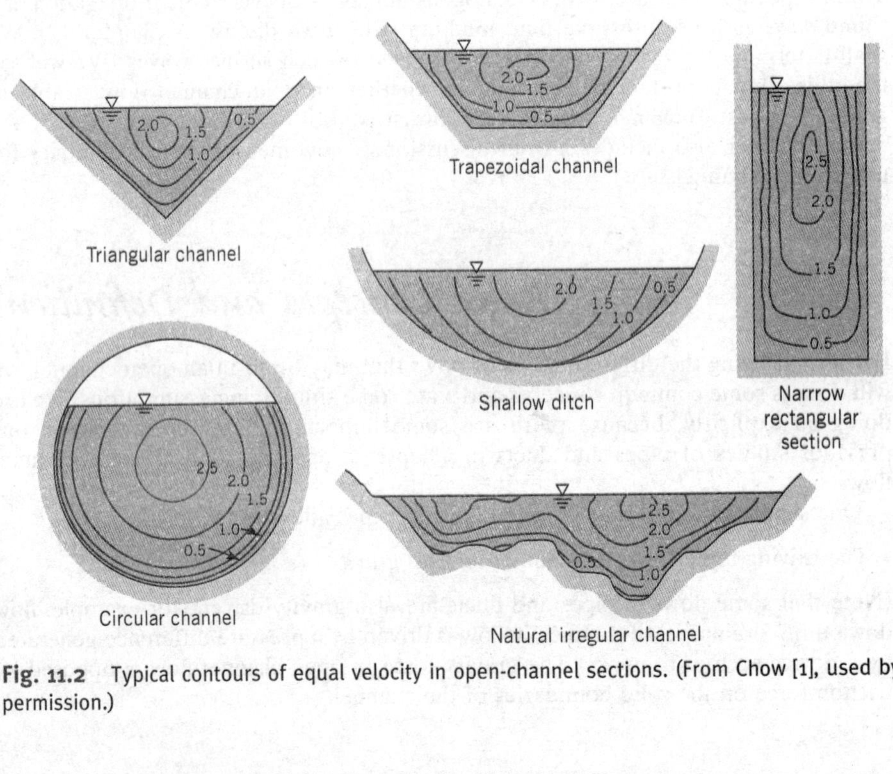

Fig. 11.2 Typical contours of equal velocity in open-channel sections. (From Chow [1], used by permission.)

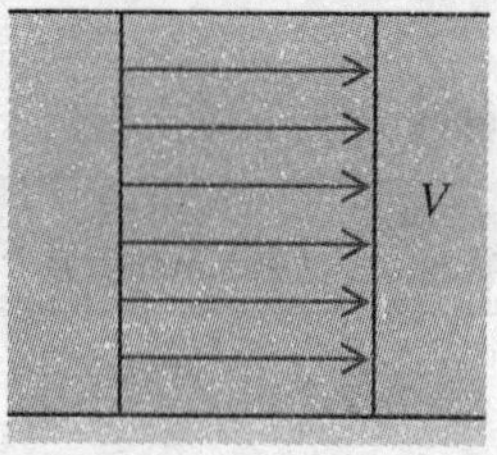

(a) Approximate velocity profile

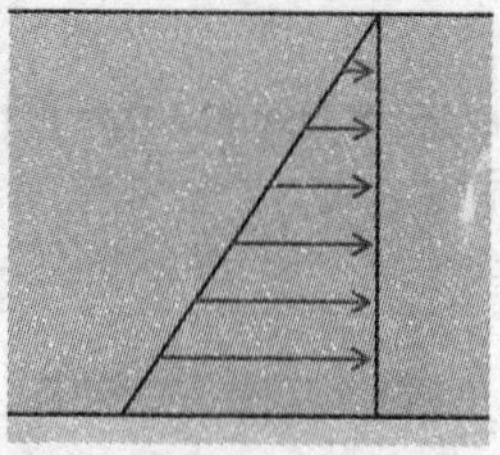

(b) Approximate pressure distribution (gage)

Fig. 11.3 Approximations for velocity profile and pressure distribution.

The next simplifying assumption we make is:

- The pressure distribution is approximated as *hydrostatic*.

This is illustrated in Fig. 11.3*b* and is a significant difference from the analysis of flows in pipes and ducts of Chapter 8; for these we found that the pressure was uniform at each axial location and varied in the streamwise direction. In open-channel flows, the free surface will be at atmospheric pressure (zero gage), so the pressure at the surface does not vary in the direction of flow. The major pressure variation occurs *across* each section; this will be exactly true if streamline curvature effects are negligible, which is often the case.

As in the case of turbulent flow in pipes, we must rely on empirical correlations to relate frictional effects to the average velocity of flow. The empirical correlation is included through a head loss term in the energy equation (Section 11.2). Additional complications in many practical cases include the presence of sediment or other particulate matter in the flow, as well as the erosion of earthen channels or structures by water action.

Channel Geometry

Channels may be constructed in a variety of cross-sectional shapes; in many cases regular geometric shapes are used. A channel with a constant slope and cross section is termed *prismatic*. Lined canals often are built with rectangular or trapezoidal sections; smaller troughs or ditches sometimes are triangular. Culverts and tunnels generally are circular or elliptical in section. Natural channels are highly irregular and nonprismatic, but often they are approximated using trapezoid or paraboloid sections. Geometric properties of common open-channel shapes are summarized in Table 11.1.

The *depth of flow*, y, is the perpendicular distance measured from the channel bed to the free surface. The *flow area*, A, is the cross section of the flow perpendicular to the flow direction. The *wetted perimeter*, P, is the length of the solid channel cross-section surface in contact with the liquid. The *hydraulic radius*, R_h, is defined as

$$R_h = \frac{A}{P} \tag{11.1}$$

For flow in noncircular closed conduits (Section 8.7), the hydraulic diameter was defined as

$$D_h = \frac{4A}{P} \tag{8.50}$$

Thus, for a circular pipe, the hydraulic diameter, from Eq. 8.50, is equal to the pipe diameter. From Eq. 11.1, the hydraulic radius for a circular pipe would then be *half* the actual pipe radius, which is a bit confusing! The hydraulic radius, as defined by Eq. 11.1, is commonly used in the analysis of open-channel flows, so it will be used

Table 11.1

Geometric Properties of Common Open-Channel Shapes

Shape	Section	Flow Area, A	Wetted Perimeter, P	Hydraulic Radius, R_h
Trapezoidal	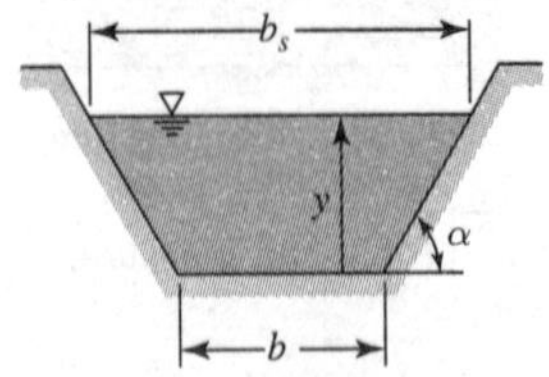	$y(b+y\cot\alpha)$	$b+\dfrac{2y}{\sin\alpha}$	$\dfrac{y(b+y\cot\alpha)}{b+\dfrac{2y}{\sin\alpha}}$
Triangular	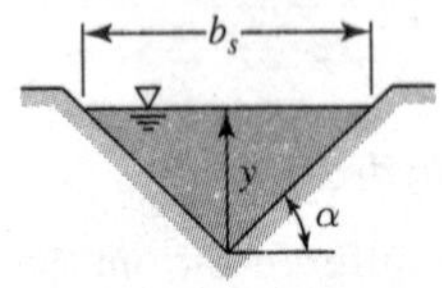	$y^2\cot\alpha$	$\dfrac{2y}{\sin\alpha}$	$\dfrac{y\cos\alpha}{2}$
Rectangular	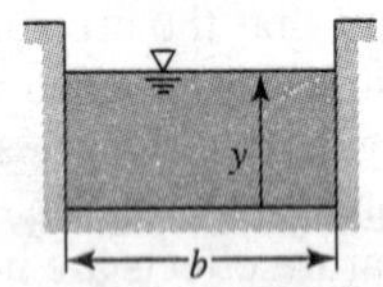	by	$b+2y$	$\dfrac{by}{b+2y}$
Wide Flat	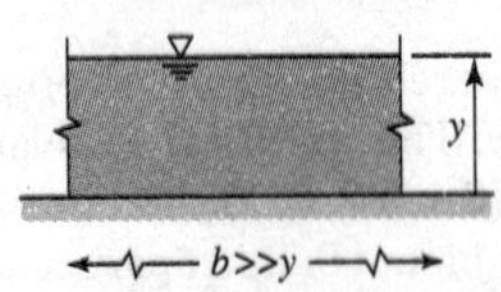	by	b	y
Circular	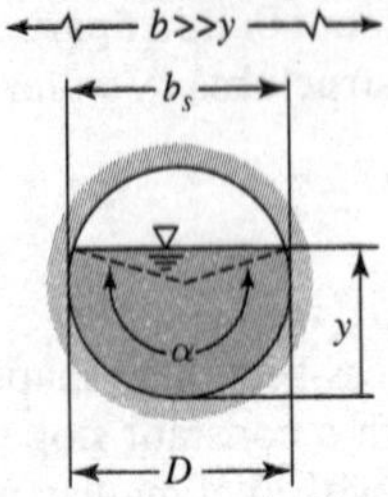	$(\alpha-\sin\alpha)\dfrac{D^2}{8}$	$\dfrac{\alpha D}{2}$	$\dfrac{D}{4}\left(1-\dfrac{\sin\alpha}{\alpha}\right)$

throughout this chapter. One reason for this usage is that the hydraulic radius of a wide channel, as seen in Table 11.1, is equal to the actual depth.

For nonrectangular channels, the *hydraulic depth* is defined as

$$y_h = \frac{A}{b_s} \tag{11.2}$$

where b_s is the width at the surface. Hence the hydraulic depth represents the *average depth* of the channel at any cross section. It gives the *depth of an equivalent rectangular channel.*

Speed of Surface Waves and the Froude Number

We will learn later in this chapter that the behavior of an open-channel flow as it encounters downstream changes (for example, a bump of the bed surface, a narrowing of the channel, or a change in slope) is strongly dependent on whether the flow is "slow" or "fast." A slow flow will have time to gradually adjust to changes downstream, whereas a fast flow will also sometimes gradually adjust but in some situations will do so "violently" (i.e., there will be a hydraulic jump; see Fig. 11.12*a* for an example). The question is what constitutes a slow or fast flow? These vague descriptions will be made more precise now. It turns out that the speed at which surface waves travel along the surface is key to defining more precisely the notions of slow and fast.

To determine the speed (or *celerity*) of surface waves, consider an open channel with movable end wall, containing a liquid initially at rest. If the end wall is given a sudden motion, as in Fig. 11.4*a*, a wave forms and travels down the channel at some speed, c (we assume a rectangular channel of width, b, for simplicity).

If we shift coordinates so that we are traveling with the wave speed, c, we obtain a steady control volume, as shown in Fig. 11.4*b* (where for now we assume $c > \Delta V$). To obtain an expression for c, we will use the continuity and momentum equations for this control volume. We also have the following assumptions:

1. Steady flow.
2. Incompressible flow.
3. Uniform velocity at each section.
4. Hydrostatic pressure distribution at each section.
5. Frictionless flow.

Assumption **1** is valid for the control volume in shifted coordinates. Assumption **2** is obviously valid for our liquid flow. Assumptions **3** and **4** are used for the entire chapter. Assumption **5** is valid in this case because we assume the area on which it acts, $b\Delta x$, is relatively small (the sketch is not to scale), so the total friction force is negligible.

For an *incompressible* flow with *uniform velocity* at each section, we can use the appropriate form of continuity from Chapter 4,

$$\sum_{CS} \vec{V} \cdot \vec{A} = 0 \tag{4.13b}$$

Applying Eq. 4.13b to the control volume, we obtain

$$(c - \Delta V)\{(y + \Delta y)b\} - cyb = 0 \tag{11.3}$$

or

$$cy - \Delta V y + c\Delta y - \Delta V \Delta y - cy = 0$$

Solving for ΔV,

$$\Delta V = c\frac{\Delta y}{y + \Delta y} \tag{11.4}$$

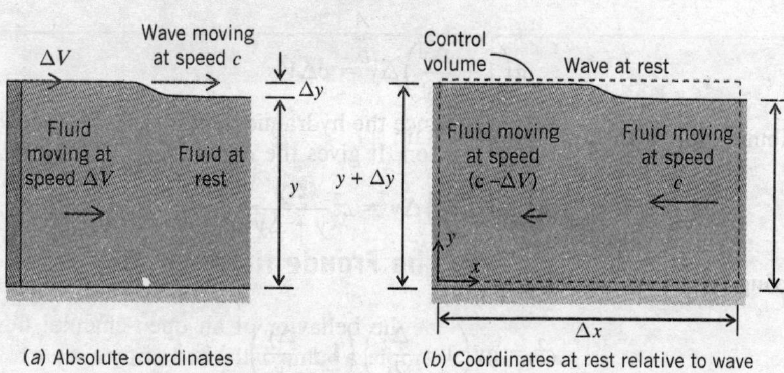

Fig. 11.4 Motion of a surface wave.

For the momentum equation, again with the assumption of uniform velocity at each section, we can use the following form of the x component of momentum

$$F_x = F_{S_x} + F_{B_x} = \frac{\partial}{\partial t}\int_{CV} u\rho\, d\forall + \sum_{CS} u\rho\vec{V}\cdot\vec{A} \tag{4.18d}$$

The unsteady term $\partial/\partial t$ disappears as the flow is *steady*, and the body force F_{B_x} is zero for *horizontal flow*. So we obtain

$$F_{S_x} = \sum_{CS} u\rho\vec{V}\cdot\vec{A} \tag{11.5}$$

The surface force consists of pressure forces on the two ends, and friction force on the bottom surface (the air at the free surface contributes negligible friction in open-channel flows). By assumption 5 we neglect friction. The gage pressure at the two ends is hydrostatic, as illustrated in Fig. 11.4*b*. We recall from our study of hydrostatics that the hydrostatic force F_R on a submerged vertical surface of area A is given by the simple result

$$F_R = p_c A \tag{3.10b}$$

where p_c is the pressure at the centroid of the vertical surface. For the two vertical surfaces of the control volume, then, we have

$$\begin{aligned} F_{S_x} &= F_{R_{\text{left}}} - F_{R_{\text{right}}} = (p_c A)_{\text{left}} - (p_c A)_{\text{right}} \\ &= \left\{\left(\rho g \frac{y+\Delta y}{2}\right)(y+\Delta y)b\right\} - \left\{\left(\rho g \frac{y}{2}\right)yb\right\} \\ &= \frac{\rho g b}{2}(y+\Delta y)^2 - \frac{\rho g b}{2}y^2 \end{aligned}$$

Using this result in Eq. 11.5 and evaluating the terms on the right,

$$\begin{aligned} F_{S_x} &= \frac{\rho g b}{2}(y+\Delta y)^2 - \frac{\rho g b}{2}y^2 = \sum_{CS} u\rho\vec{V}\cdot\vec{A} \\ &= -(c-\Delta V)\rho\{(c-\Delta V)(y+\Delta y)b\} - c\rho\{-cyb\} \end{aligned}$$

The two terms in braces are equal, from continuity as shown in Eq. 11.3, so the momentum equation simplifies to

$$gy\Delta y + \frac{g(\Delta y)^2}{2} = yc\Delta V$$

or

$$g\left(1 + \frac{\Delta y}{2y}\right)\Delta y = c\Delta V$$

Combining this with Eq. 11.4, we obtain

$$g\left(1 + \frac{\Delta y}{2y}\right)\Delta y = c^2\frac{\Delta y}{y+\Delta y}$$

and solving for c,

$$c^2 = gy\left(1 + \frac{\Delta y}{2y}\right)\left(1 + \frac{\Delta y}{y}\right)$$

For waves of relatively small amplitude ($\Delta y \ll y$), we can simplify this expression to

$$c = \sqrt{gy} \tag{11.6}$$

Hence the speed of a surface disturbance depends on the local fluid depth. For example, it explains why waves "crash" as they approach the beach. Out to sea, the water depths below wave crests and troughs are approximately the same, and hence so are their speeds. As the water depth decreases on the approach to the beach, the depth of crests start to become significantly larger than trough depths, causing crests to speed up and overtake the troughs.

Note that fluid properties do not enter into the speed: Viscosity is usually a minor factor, and it turns out that the disturbance or wave we have described is due to the interaction of gravitational and inertia forces, both of which are linear with density. Equation 11.6 was derived on the basis of one-dimensional motion (x direction); a more realistic model allowing two-dimensional fluid motion (x and y directions) shows that Eq. 11.6 applies for the limiting case of large wavelength waves (Problem 11.6 explores this). Also, there are other types of surface waves, such as capillary waves driven by surface tension, for which Eq. 11.6 does not apply (Problems 11.7 and 11.8 explore surface tension effects).

Example 11.1 SPEED OF FREE SURFACE WAVES

You are enjoying a summer's afternoon relaxing in a rowboat on a pond. You decide to find out how deep the water is by splashing your oar and timing how long it takes the wave you produce to reach the edge of the pond. (The pond is artificial; so it has approximately the same depth even to the shore.) From floats installed in the pond, you know you're 6 m from shore, and you measure the time for the wave to reach the edge to be 1.5 s. Estimate the pond depth. Does it matter if it's a freshwater pond or if it's filled with seawater?

Given: Time for a wave to reach the edge of a pond.

Find: Depth of the pond.

Solution: Use the wave speed equation, Eq. 11.6.

Governing equation: $c = \sqrt{gy}$

The time for a wave, speed c, to travel a distance L, is $\Delta t = \dfrac{L}{c}$, so $c = \dfrac{L}{\Delta t}$. Using this and Eq. 11.6,

$$\sqrt{gy} = \frac{L}{\Delta t}$$

where y is the depth, or

$$y = \frac{L^2}{g\Delta t^2}$$

Using the given data

$$y = 6^2\text{m}^2 \times \frac{1}{9.81}\frac{\text{s}^2}{\text{m}} \times \frac{1}{1.5^2\text{s}^2} = 1.63 \text{ m} \longleftarrow y$$

The pond depth is about 1.63 m.

The result obtained is independent of whether the water is fresh or saline, because the speed of these surface waves is independent of fluid properties.

The speed of surface disturbances given in Eq. 11.6 provides us with a more useful "litmus test" for categorizing the speed of a flow than the terms "slow" and "fast." To illustrate this, consider a flow moving at speed V, which experiences a disturbance at some point downstream. (The disturbance could be caused by a bump in the channel floor or by a barrier, for example.) The disturbance will travel upstream at speed c *relative to the fluid*. If the fluid speed is slow, $V < c$, and the disturbance will travel upstream at absolute speed $(c - V)$. However, if the fluid speed is fast, $V > c$, and the disturbance cannot travel upstream and instead is washed downstream at absolute speed $(V - c)$. This leads to radically different responses of slow and fast flows to a downstream disturbance. Hence, recalling Eq. 11.6 for the speed c, open-channel flows may be classified on the basis of Froude number first introduced in Chapter 7:

$$Fr = \frac{V}{\sqrt{gy}} \tag{11.7}$$

Instead of the rather loose terms "slow" and "fast," we now have the following criteria:

$Fr < 1$ Flow is *subcritical*, *tranquil*, or *streaming*. Disturbances can travel upstream; downstream conditions can affect the flow upstream. The flow can gradually adjust to the disturbance.

$Fr = 1$ Flow is *critical*.

$Fr > 1$ Flow is *supercritical*, *rapid*, or *shooting*. No disturbance can travel upstream; downstream conditions cannot be felt upstream. The flow may "violently" respond to the disturbance because the flow has no chance to adjust to the disturbance before encountering it.

Note that for nonrectangular channels we use the hydraulic depth y_h,

$$Fr = \frac{V}{\sqrt{gy_h}} \tag{11.8}$$

These regimes of flow behavior are qualitatively analogous to the subsonic, sonic, and supersonic regimes of gas flow that we will discuss in Chapter 12. (In that case we are also comparing a flow speed, V, to the speed of a wave, c, except that the wave is a sound wave rather than a surface wave.)

We will discuss the ramifications of these various Froude number regimes later in this chapter.

11.2 Energy Equation for Open-Channel Flows

In analyzing open-channel flows, we will use the continuity, momentum, and energy equations. Here we derive the appropriate form of the energy equation (we will use continuity and momentum when needed). As in the case of pipe flow, friction in open-channel flows results in a loss of mechanical energy; this can be characterized by a head loss. The temptation is to just use one of the forms of the energy equation for pipe flow we derived in Section 8.6, such as

$$\left(\frac{p_1}{\rho g} + \alpha_1 \frac{\bar{V}_2^2}{2g} + z_1\right) - \left(\frac{p_2}{\rho g} + \alpha_2 \frac{\bar{V}_2^2}{2g} + z_2\right) = \frac{h_{l_T}}{g} = H_{l_T} \tag{8.30}$$

The problem with this is that it was derived on the assumption of uniform pressure at each section, which is not the case in open-channel flow (we have a hydrostatic pressure variation at each location); we do not have a uniform p_1 at section ① and uniform p_2 at section ②!

Instead we need to derive an energy equation for open-channel flows from first principles. We will closely follow the steps outlined in Section 8.6 for pipe flows but use different assumptions. You are urged to review Section 8.6 in order to be aware of the similarities and differences between pipe flows and open-channel flows.

We will use the generic control volume shown in Fig. 11.5, with the following assumptions:

1. Steady flow
2. Incompressible flow
3. Uniform velocity at a section
4. Gradually varying depth so that pressure distribution is hydrostatic
5. Small bed slope
6. $\dot{W}_s = \dot{W}_{\text{shear}} = \dot{W}_{\text{other}} = 0$

We make a few comments here. We have seen assumptions **1–4** already; they will always apply in this chapter. Assumption **5** simplifies the analysis so that depth, y, is taken to be vertical and speed, V, is taken to be horizontal, rather than normal and parallel to the bed, respectively. Assumption **6** states that there is no shaft work, no work due to fluid shearing at the boundaries, and no other work. There is no shear work at the boundaries because on each part of the control surface the tangential velocity is zero (on the channel walls) or the shear stress is zero (the open surface), so no work can be done. Note that there can still be mechanical energy dissipation within the fluid due to friction.

We have chosen a generic control volume so that we can derive a generic energy equation for open-channel flows, that is, an equation that can be applied to a variety of flows such as ones with a variation in elevation, or a hydraulic jump, or a sluice gate, and so on, between sections ① and ②. Coordinate z indicates distances measured in the vertical direction; distances measured vertically from the channel bed are denoted by y. Note that y_1 and y_2 are the flow depths at sections ① and ②, respectively, and z_1 and z_2 are the corresponding channel elevations.

The energy equation for a control volume is

$$\dot{Q} - \underbrace{\dot{W}_s}_{=0(6)} - \underbrace{\dot{W}_{\text{shear}}}_{=0(6)} - \underbrace{\dot{W}_{\text{other}}}_{=0(6)} = \underbrace{\frac{\partial}{\partial t}\int_{\text{CV}} e\,\rho\,d\mathcal{V}}_{=0(1)} + \int_{\text{CS}} (e + pv)\rho\vec{V}\cdot d\vec{A} \tag{4.56}$$

$$e = u + \frac{V^2}{2} + gz$$

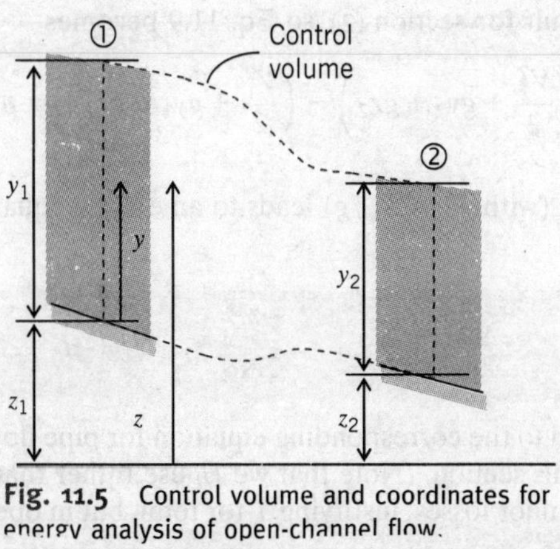

Fig. 11.5 Control volume and coordinates for energy analysis of open-channel flow.

Recall that u is the thermal specific energy and $v = 1/\rho$ is the specific volume. After using assumptions **1** and **6**, and rearranging, with $\dot{m} = \int \rho \vec{V} \cdot d\vec{A}$, and $dA = b\,dy$ where $b(y)$ is the channel width, we obtain

$$\dot{Q} = -\int_1 \left(\frac{p}{\rho} + \frac{V^2}{2} + gz\right)\rho V b\,dy - \int_1 u\rho V b\,dy + \int_2 \left(\frac{p}{\rho} + \frac{V^2}{2} + gz\right)\rho V b\,dy + \int_2 u\rho V b\,dy$$

$$= \int_1 \left(\frac{p}{\rho} + \frac{V^2}{2} + gz\right)\rho V b\,dy + \int_2 \left(\frac{p}{\rho} + \frac{V^2}{2} + gz\right)\rho V b\,dy + \dot{m}(u_2 - u_1)$$

or

$$\int_1 \left(\frac{p}{\rho} + \frac{V^2}{2} + gz\right)\rho V b\,dy - \int_2 \left(\frac{p}{\rho} + \frac{V^2}{2} + gz\right)\rho V b\,dy = \dot{m}(u_2 - u_1) - \dot{Q} = \dot{m}h_{l_T} \tag{11.9}$$

This states that the loss in mechanical energies ("pressure," kinetic and potential) through the control volume leads to a gain in the thermal energy and/or a loss of heat from the control volume. As in Section 8.6, these thermal effects are collected into the head loss term h_{l_T}.

The surface integrals in Eq. 11.9 can be simplified. The speed, V, is constant at each section by assumption **3**. The pressure, p, does vary across sections ① and ②, as does the potential, z. However, by assumption **4**, the pressure variation is hydrostatic. Hence, for section ①, using the notation of Fig. 11.5

$$p = \rho g(y_1 - y)$$

[so $p = \rho g y_1$ at the bed and $p = 0$ (gage) at the free surface] and

$$z = (z_1 + y)$$

Conveniently, we see that the pressure *decreases* linearly with y while z *increases* linearly with y, so the two terms together are constant,

$$\left(\frac{p}{\rho} + gz\right)_1 = g(y_1 - y) + g(z_1 + y) = g(y_1 + z_1)$$

Using these results in the first integral in Eq. 11.9,

$$\int_1 \left(\frac{p}{\rho} + \frac{V^2}{2} + gz\right)\rho V b\,dy = \int_1 \left(\frac{V^2}{2} + g(y_1 + z_1)\right)\rho V b\,dy = \left(\frac{V_1^2}{2} + gy_1 + gz_1\right)\dot{m}$$

We find a similar result for section ②, so Eq. 11.9 becomes

$$\left(\frac{V_2^2}{2} + gy_2 + gz_2\right) - \left(\frac{V_1^2}{2} + gy_1 + gz_1\right) = h_{l_T}$$

Finally, dividing by g (with $H_l = h_{l_T}/g$) leads to an energy equation for open-channel flow

$$\frac{V_1^2}{2g} + y_1 + z_1 = \frac{V_2^2}{2g} + y_2 + z_2 + H_l \tag{11.10}$$

This can be compared to the corresponding equation for pipe flow, Eq. 8.30, presented at the beginning of this section. (Note that we H_l use rather than H_{l_T}; in pipe flow we can have major and minor losses, justifying T for total, but in open-channel flow we do not make this distinction.) Equation 11.10 will prove useful to us for the remainder of the chapter and indicates that energy computations can be done simply from geometry (y and z) and velocity, V.

The *total head* or *energy head*, H, at any location in an open-channel flow can be defined from Eq. 11.10 as

$$H = \frac{V^2}{2g} + y + z \tag{11.11}$$

where y and z are the local *flow depth* and *channel bed elevation*, respectively (they no longer represent the coordinates shown in Fig. 11.5). This is a measure of the mechanical energy (kinetic and pressure/potential) of the flow. Using this in the energy equation, we obtain an alternative form

$$H_1 - H_2 = H_l \tag{11.12}$$

From this we see that the loss of total head depends on head loss due to friction.

Specific Energy

We can also define the *specific energy* (or *specific head*), denoted by the symbol E,

$$E = \frac{V^2}{2g} + y \tag{11.13}$$

This is a measure of the mechanical energy (kinetic and pressure/potential) of the flow above and beyond that due to channel bed elevation; it essentially indicates *the energy due to the flow's speed and depth*. Using Eq. 11.13 in Eq. 11.10, we obtain another form of the energy equation,

$$E_1 - E_2 + z_1 - z_2 = H_l \tag{11.14}$$

From this we see that the change in specific energy depends on friction and on channel elevation change. While the total head must decrease in the direction of flow (Eq. 11.12), the specific head may decrease, increase, or remain constant, depending on the bed elevation, z.

From continuity, $V = Q/A$, so the specific energy can be written

$$E = \frac{Q^2}{2gA^2} + y \tag{11.15}$$

For all channels A is a monotonically increasing function of flow depth (as Table 11.1 indicates); increasing the depth must lead to a larger flow area. Hence, Eq. 11.15 indicates that the specific energy is a combination of a hyperbolic-type decrease with depth and a linear increase with depth. This is illustrated in Fig. 11.6. We see that for a given flow rate, Q, there is a range of possible flow depths and energies, but one depth at which the specific energy is at a minimum. Instead of E versus y we typically plot y versus E so that the plot corresponds to the example flow section, as shown in Fig. 11.7.

Recalling that the specific energy, E, indicates actual energy (kinetic plus potential/pressure per unit mass flow rate) being carried by the flow, we see that a given flow, Q, can have a range of energies, E, and corresponding flow depths, y. Figure 11.7 also reveals some interesting flow phenomena. For a given flow, Q, and specific energy, E, there are two possible flow depths, y; these are called *alternate depths*. For example,

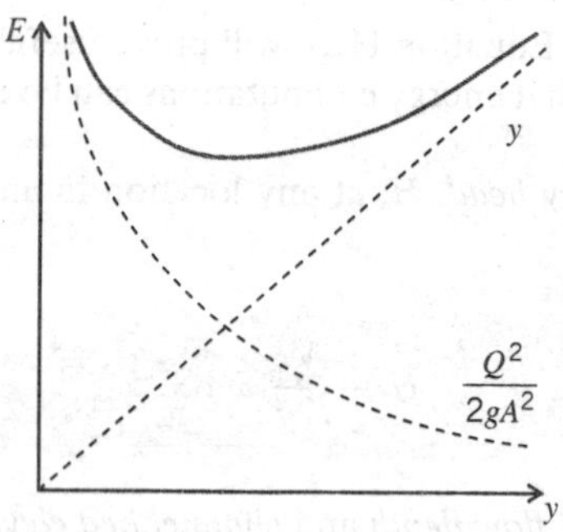

Fig. 11.6 Dependence of specific energy on flow depth for a given flow rate.

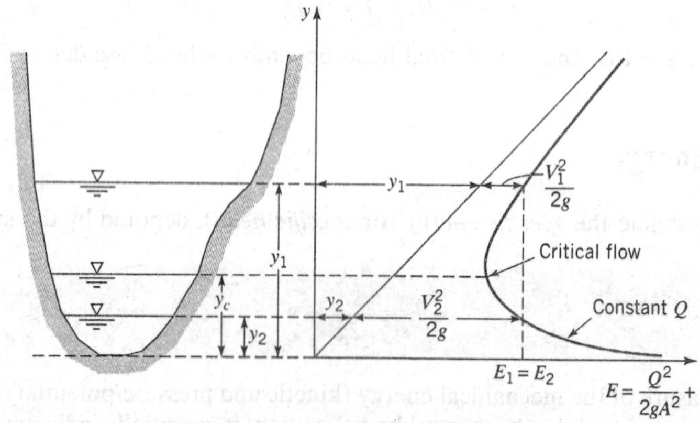

Fig. 11.7 Specific energy curve for a given flow rate.

we can have a flow at depth y_1 or depth y_2. The first flow has large depth and is moving slowly, and the second flow is shallow but fast moving. The plot graphically indicates this: For the first flow, E_1 is made up of a large y_1 and small $V_1^2/2g$; for the second flow, E_2 is made up of a small y_2 and large $V_2^2/2g$. We will see later that we can switch from one flow to another. We can also see (as we will demonstrate in Example 11.2 for a rectangular channel) that for a given Q, there is always one flow for which the specific energy is minimum, $E = E_{min}$; we will investigate this further after Example 11.2 and show that $E_{min} = E_{crit}$, where E_{crit} is the specific energy at critical conditions.

Example 11.2 SPECIFIC ENERGY CURVES FOR A RECTANGULAR CHANNEL

For a rectangular channel of width $b = 10$ m, construct a family of specific energy curves for $Q = 0$, 2, 5, and 10 m³/s. What are the minimum specific energies for these curves?

Given: Rectangular channel and range of flow rates.

Find: Curves of specific energy. For each flow rate, find the minimum specific energy.

Solution: Use the flow rate form of the specific energy equation (Eq. 11.15) for generating the curves.

Governing equation:

$$E = \frac{Q^2}{2gA^2} + y \qquad (11.15)$$

For the specific energy curves, express E as a function of depth, y.

$$E = \frac{Q^2}{2gA^2} + y = \frac{Q^2}{2g(by)^2} + y = \left(\frac{Q^2}{2gb^2}\right)\frac{1}{y^2} + y \tag{1}$$

The table and corresponding graph were generated from this equation using *Excel*.

y (m)	Specific Energy, E (m)			
	$Q=0$	$Q=2$	$Q=5$	$Q=10$
0.100	0.10	0.92	5.20	20.49
0.125	0.13	0.65	3.39	13.17
0.150	0.15	0.51	2.42	9.21
0.175	0.18	0.44	1.84	6.83
0.200	0.20	0.40	1.47	5.30
0.225	0.23	0.39	1.23	4.25
0.250	0.25	0.38	1.07	3.51
0.275	0.28	0.38	0.95	2.97
0.30	0.30	0.39	0.87	2.57
0.35	0.35	0.42	0.77	2.01
0.40	0.40	0.45	0.72	1.67
0.45	0.45	0.49	0.70	1.46
0.50	0.50	0.53	0.70	1.32
0.55	0.55	0.58	0.72	1.22
0.60	0.60	0.62	0.74	1.17
0.70	0.70	0.72	0.80	1.12
0.80	0.80	0.81	0.88	1.12
0.90	0.90	0.91	0.96	1.15
1.00	1.00	1.01	1.05	1.20
1.25	1.25	1.26	1.28	1.38
1.50	1.50	1.50	1.52	1.59
2.00	2.00	2.00	2.01	2.05
2.50	2.50	2.50	2.51	2.53

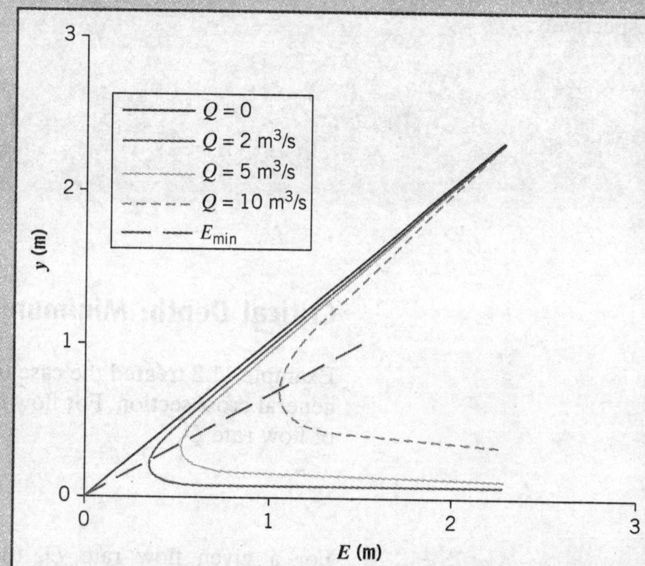

To find the minimum energy for a given Q, we differentiate Eq. 1,

$$\frac{dE}{dy} = \left(\frac{Q^2}{2gb^2}\right)\left(-\frac{2}{y^3}\right) + 1 = 0$$

Hence, the depth $y_{E_{min}}$ for minimum specific energy is

$$y_{E_{min}} = \left(\frac{Q^2}{gb^2}\right)^{\frac{1}{3}}$$

Using this in Eq. 11.15:

$$E_{min} = \frac{Q^2}{2gA^2} + y_{E_{min}} = \frac{Q^2}{2gb^2 y_{E_{min}}^2} + \left[\frac{Q^2}{gb^2}\right]^{\frac{1}{3}} = \frac{1}{2}\left[\frac{Q^2}{gb^2}\right]\left[\frac{gb^2}{Q^2}\right]^{\frac{2}{3}} + \left[\frac{Q^2}{gb^2}\right]^{\frac{1}{3}} = \frac{3}{2}\left[\frac{Q^2}{gb^2}\right]^{\frac{1}{3}}$$

$$E_{min} = \frac{3}{2}\left[\frac{Q^2}{gb^2}\right]^{\frac{1}{3}} = \frac{3}{2} y_{E_{min}} \tag{2}$$

Hence for a rectangular channel, we obtain a simple result for the minimum energy. Using Eq. 2 with the given data:

Q (m^3/s)	2	5	10
E_{min} (m)	0.302	0.755	1.51

The depths corresponding to these flows are 0.201 m, 0.503 m, and 1.01 m, respectively.

We will see in the next topic that the depth at which we have minimum energy is the critical depth, y_c, and $E_{min} = E_{crit}$.

The *Excel* workbook for this problem can be used for plotting specific energy curves for other rectangular channels. The depth for minimum energy is also obtained using *Solver*.

Critical Depth: Minimum Specific Energy

Example 11.2 treated the case of a rectangular channel. We now consider channels of general cross section. For flow in such a channel we have the specific energy in terms of flow rate Q,

$$E = \frac{Q^2}{2gA^2} + y \tag{11.15}$$

For a given flow rate Q, to find the depth for minimum specific energy, we differentiate:

$$\frac{dE}{dy} = 0 = -\frac{Q^2}{gA^3}\frac{dA}{dy} + 1 \tag{11.16}$$

To proceed further, it would seem we need $A(y)$; some examples of $A(y)$ are shown in Table 11.1. However, it turns out that for any given cross section we can write

$$dA = b_s dy \tag{11.17}$$

where, as we saw earlier, b_s is the width at the surface. This is indicated in Fig. 11.8; the incremental increase in area dA due to incremental depth change dy occurs at the free surface, where $b = b_s$.

Using Eq. 11.17 in Eq. 11.16 we find

$$-\frac{Q^2}{gA^3}\frac{dA}{dy} + 1 = -\frac{Q^2}{gA^3} b_s + 1 = 0$$

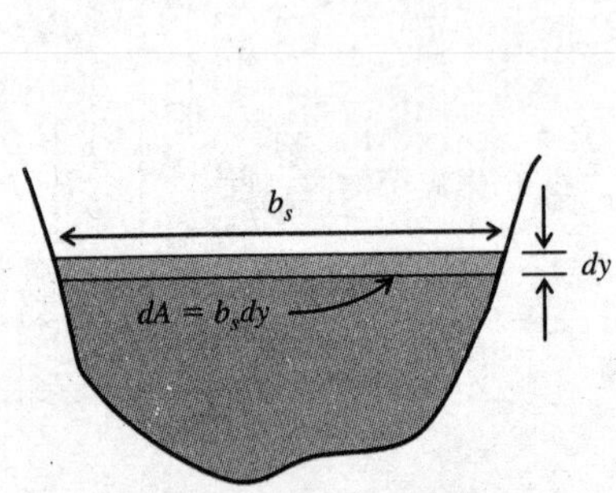

Fig. 11.8 Dependence of flow area change dA on depth change dy.

so

$$Q^2 = \frac{gA^3}{b_s} \tag{11.18}$$

for minimum specific energy. From continuity $V = Q/A$, so Eq. 11.18 leads to

$$V = \frac{Q}{A} = \frac{1}{A}\left[\frac{gA^3}{b_s}\right]^{1/2} = \sqrt{\frac{gA}{b_s}} \tag{11.19}$$

We have previously defined the hydraulic depth,

$$y_h = \frac{A}{b_s} \tag{11.2}$$

Hence, using Eq. 11.2 in Eq. 11.19, we obtain

$$V = \sqrt{gy_h} \tag{11.20}$$

But the Froude number is given by

$$Fr = \frac{V}{\sqrt{gy_h}} \tag{11.8}$$

Hence we see that, for minimum specific energy, $Fr = 1$, which corresponds to critical flow. We obtain the important result that, for flow in any open channel, *the specific energy is at its minimum at critical conditions.*

We collect Eqs. 11.18 and 11.20; for critical flow

$$Q^2 = \frac{gA_c^3}{b_{s_c}} \tag{11.21}$$

$$V_c = \sqrt{gy_{h_c}} \tag{11.22}$$

for $E = E_{\min}$. In these equations, A_c, V_c, b_{s_c} and y_{h_c} are the critical flow area, velocity, channel surface width, and hydraulic depth, respectively. Equation 11.21 can be used to find the critical depth, y_c, for a given channel cross-section shape, at a given flow rate. The equation is deceptively difficult: A_c and b_{s_c} each depend on flow depth y, often in a nonlinear fashion; so it must usually be iteratively solved for y. Once y_c is obtained, area, A_c, and surface width, b_{s_c}, can be computed, leading to y_{h_c} (using Eq. 11.2). This in turn is used in Eq. 11.22 to find the flow speed V_c (or $V_c = Q/A_c$ can be used). Finally, the minimum energy can be computed from Eq. 11.15.

For the particular case of a *rectangular channel*, we have $b_s = b = \text{constant}$ and $A = by$, so Eq. 11.21 becomes

$$Q^2 = \frac{gA_c^3}{b_{s_c}} = \frac{gb^3y_c^3}{b} = gb^2y_c^3$$

so

$$y_c = \left[\frac{Q^2}{gb^2}\right]^{1/3} \tag{11.23}$$

with

$$V_c = \sqrt{gy_c} = \left[\frac{gQ}{b}\right]^{1/3} \qquad (11.24)$$

For the rectangular channel, a particularly simple result for the minimum energy is obtained when Eq. 11.24 is used in Eq. 11.15,

$$E = E_{\text{min}} = \frac{V_c^2}{2g} + y_c = \frac{gy_c}{2g} + y_c$$

or

$$E_{\text{min}} = \frac{3}{2}y_c \qquad (11.25)$$

This is the same result we found in Example 11.2. The critical state is an important benchmark. It will be used in the next section to help determine what happens when a flow encounters an obstacle such as a bump. Also, near the minimum E, as Fig. 11.7 shows, the rate of change of y with E is nearly infinite. This means that for critical flow conditions, even small changes in E, due to channel irregularities or disturbances, can cause pronounced changes in fluid depth. Thus, surface waves, usually in an unstable manner, form when a flow is near critical conditions. Long runs of near-critical flow consequently are avoided in practice.

Example 11.3 CRITICAL DEPTH FOR TRIANGULAR SECTION

A steep-sided triangular section channel ($\alpha = 60°$) has a flow rate of 300 m³/s. Find the critical depth for this flow rate. Verify that the Froude number is unity.

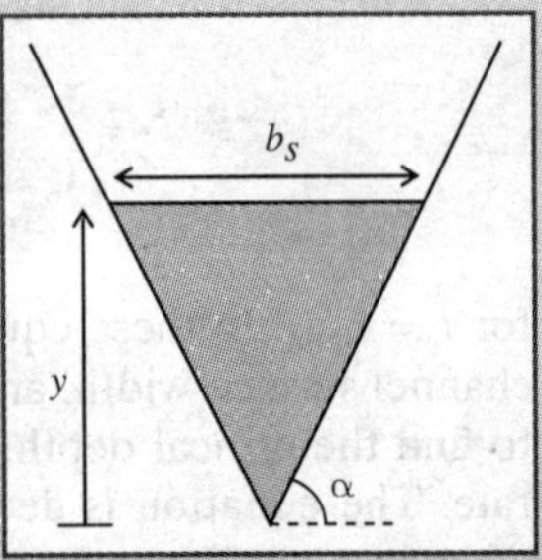

Given: Flow in a triangular section channel.

Find: Critical depth; verify that $Fr = 1$.

Solution:
Use the critical flow equation, Eq. 11.21.

Governing equations:

$$Q^2 = \frac{gA_c^3}{b_{s_c}} \qquad Fr = \frac{V}{\sqrt{gy_h}}$$

The given data is: $Q = 300\,\text{m}^3/\text{s} \qquad \alpha = 60°$

From Table 11.1 we have the following:

$$A = y^2 \cot \alpha$$

and from basic geometry

$$\tan \alpha = \frac{y}{b_s/2} \quad \text{so} \quad b_s = 2y \cot \alpha$$

Using these in Eq. 11.21

$$Q^2 = \frac{gA_c^3}{b_{s_c}} = \frac{g[y_c^2 \cot \alpha]^3}{2y_c \cot \alpha} = \frac{1}{2}gy_c^5 \cot^2\alpha$$

Hence

$$y_c = \left[\frac{2Q^2\tan^2\alpha}{g}\right]^{1/5}$$

Using the given data

$$y_c = \left[2\times 300^2\left(\frac{\text{m}^3}{\text{s}}\right)^2\times\tan^2\left(\frac{60\times\pi}{180}\right)\times\frac{\text{s}^2}{9.81\text{m}}\right]^{1/5} = [5.51\times 10^4\text{m}^5]^{1/5}$$

Finally

$$y_c = 8.88\text{ m}$$ ← y_c

To verify that $Fr = 1$, we need V and y_h.

From continuity

$$V_c = \frac{Q}{A_c} = \frac{Q}{y_c^2\cot\alpha} = 300\frac{\text{m}^3}{\text{s}}\times\frac{1}{8.88^2\text{m}^2}\times\frac{1}{\cot\left(\frac{60\times\pi}{180}\right)} = 6.60\text{ m/s}$$

and from the definition of hydraulic depth

$$y_{h_c} = \frac{A_c}{b_{s_c}} = \frac{y_c^2\cot\alpha}{2y_c\cot\alpha} = \frac{y_c}{2} = 4.44\text{ m}$$

Hence

$$Fr_c = \frac{V_c}{\sqrt{gy_{h_c}}} = \frac{6.60\frac{\text{m}}{\text{s}}}{\sqrt{9.81\frac{\text{m}}{\text{s}^2}\times 4.44\text{ m}}} = 1$$ ← $Fr_c = 1$

We have verified that at critical depth the Froude number is unity.

As with the rectangular channel, the triangular section channel analysis leads to an explicit equation for y_c from Eq. 11.21. Other more complicated channel cross sections often lead to an implicit equation that needs to be solved numerically.

Localized Effect of Area Change (Frictionless Flow) 11.3

We will next consider a simple flow case in which the channel bed is horizontal and for which the effects of channel cross section (area change) predominate: flow over a bump. Since this phenomenon is localized (it takes place over a short distance), the effects of friction (on either momentum or energy) may be reasonably neglected.

The energy equation, Eq. 11.10, with the assumption of no losses due to friction then becomes

$$\frac{V_1^2}{2g} + y_1 + z_1 = \frac{V_2^2}{2g} + y_2 + z_2 = \frac{V^2}{2g} + y + z = \text{const} \tag{11.26}$$

(Note that Eq. 11.26 could also have been obtained from by applying the Bernoulli equation between two points ① and ② on the surface, because all of the requirements of the Bernoulli equation are satisfied here.) Alternatively, using the definition of specific energy

$$E_1 + z_1 = E_2 + z_2 = E + z = \text{const}$$

We see that the specific energy of a frictionless flow will change only if there is a change in the elevation of the channel bed.

Flow over a Bump

Consider frictionless flow in a horizontal rectangular channel of constant width, b, with a bump in the channel bed, as illustrated in Fig. 11.9. (We choose a rectangular channel for simplicity, but the results we obtain will apply generally.) The bump height above the horizontal bed of the channel is $z = h(x)$; the water depth, $y(x)$, is measured from the local channel bottom surface.

Note that we have indicated two possibilities for the free surface behavior: Perhaps the flow gradually rises over the bump; perhaps it gradually dips over the bump. (There are other possibilities too!) One thing we can be sure of, however, is that if it rises, it will not have the same contour as the bump. (Can you explain why?) Applying the energy equation (Eq. 11.26) for frictionless flow between an upstream point ① and any point along the region of the bump,

$$\frac{V_1^2}{2g} + y_1 = E_1 = \frac{V^2}{2g} + y + h = E + h(x) = \text{const} \tag{11.27}$$

Equation 11.27 indicates that the specific energy must decrease through the bump, then increase back to its original value (of $E_1 = E_2$),

$$E(x) = E_1 - h(x) \tag{11.28}$$

From continuity

$$Q = bV_1 y_1 = bVy$$

Using this in Eq 11.27

$$\frac{Q^2}{2gb^2y_1^2} + y_1 = \frac{Q^2}{2gb^2y^2} + y + h = \text{const} \tag{11.29}$$

We can obtain an expression for the variation of the free surface depth by differentiating Eq. 11.29:

$$-\frac{Q^2}{gb^2y^3}\frac{dy}{dx} + \frac{dy}{dx} + \frac{dh}{dx} = 0$$

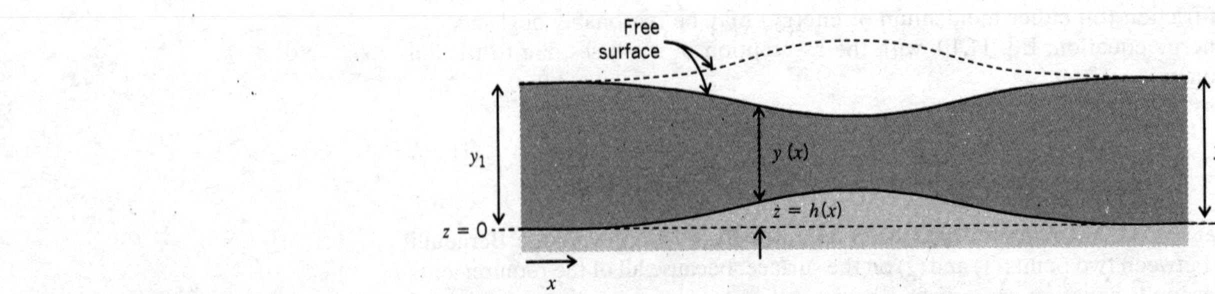

Fig. 11.9 Flow over a bump in a horizontal channel.

Solving for the slope of the free surface, we obtain

$$\frac{dy}{dx} = \frac{dh/dx}{\left[\dfrac{Q^2}{gb^2y^3} - 1\right]} = \frac{dh/dx}{\left[\dfrac{V^2}{gy} - 1\right]}$$

Finally,

$$\frac{dy}{dx} = \frac{1}{Fr^2 - 1}\frac{dh}{dx} \tag{11.30}$$

Equation 11.30 leads to the interesting conclusion that the response to a bump very much depends on the local Froude number, *Fr*.

$Fr < 1$ Flow is *subcritical*, *tranquil*, or *streaming*. When $Fr < 1$, $(Fr^2 - 1) < 1$ and the slope dy/dx of the free surface has the *opposite* sign to the slope dh/dx of the bump: When the bump elevation increases, the flow dips; when the bump elevation decreases, the flow depth increases. This is the solid free surface shown in Fig. 11.9.

$F_r = 1$ Flow is *critical*. When $Fr = 1$, $(Fr^2 - 1) = 0$. Eq. 11.30 predicts an infinite water surface slope, unless dh/dx equals zero at this instant. Since the free surface slope cannot be infinite, then dh/dx must be zero when $Fr = 1$; put another way, if we have $Fr = 1$ (we don't *have* to have $Fr = 1$ in a flow), it can *only* be at a location where $dh/dx = 0$ (at the crest of the bump, or where the channel is flat). If critical flow *is* attained, then downstream of the critical flow location the flow may be subcritical or supercritical, depending on downstream conditions. If critical flow does *not* occur where $dh/dx = 0$, then flow downstream from this location will be the same type as the flow upstream from the location.

$F_r > 1$ Flow is *supercritical*, *rapid*, or *shooting*. When $Fr > 1$, $(Fr^2 - 1) > 1$ and the slope dy/dx of the free surface has the same sign as the slope dh/dx of the bump: when the bump elevation increases, so does the flow depth; when the bump elevation decreases, so does the flow depth. This is the dashed free surface shown in Fig. 11.9.

The general trends for $Fr < 1$ and $Fr > 1$, for either an increasing or decreasing bed elevation, are illustrated in Fig. 11.10. The important point about critical flow ($Fr = 1$) is that, if it does occur, it can do so only where the bed elevation is constant.

An additional visual aid is provided by the specific energy graph of Fig. 11.11. This shows the specific energy curve for a given flow rate, Q. For a subcritical flow that is at state *a* before it encounters a bump, as the flow moves up the bump toward the bump peak, the specific energy must decrease (Eq. 11.28). Hence we move along the curve to point *b*. If point *b* corresponds to the bump peak, then we move back along the curve to *a* (note that this frictionless flow is reversible!) as the flow descends the bump. Alternatively, if the bump continues to increase beyond point *b*, we continue to move along the curve to the minimum energy point, point *e* where $E = E_{\text{min}} = E_{\text{crit}}$. As we have discussed, for frictionless flow to exist, point *e* can only be where $dh/dx = 0$ (the bump peak). For this case, something interesting happens as the flow descends down the bump: We can return along the curve to point *a*, or we can move along the curve to point *d*. This means that the surface of a subcritical flow that encounters a bump will dip and then *either* return to its original depth *or* (if the bump is high enough for the flow to reach critical conditions) may continue to accelerate and become shallower until it reaches the supercritical state corresponding to the original specific energy (point *d*). Which trend occurs depends on downstream conditions; for example, if there is

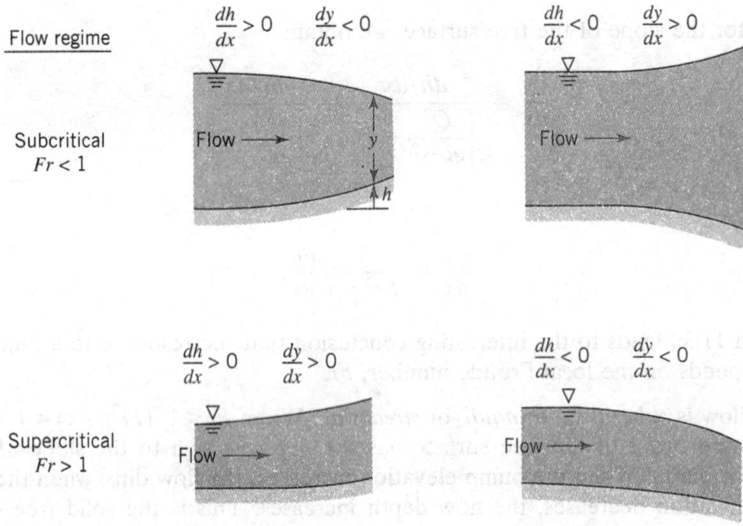

Fig. 11.10 Effects of bed elevation changes.

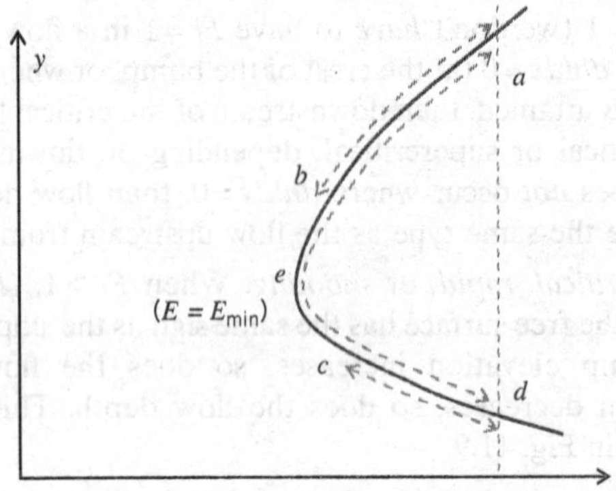

Fig. 11.11 Specific energy curve for flow over a bump.

some type of flow restriction, the flow downstream of the bump will return to its original subcritical state. Note that as we mentioned earlier, when a flow is at the critical state the surface behavior tends to display dramatic variations in behavior. Finally, Fig. 11.11 indicates that a supercritical flow (point *d*) that encounters a bump would increase in depth over the bump (to point *c* at the bump peak), and then return to its supercritical flow at point *d*. We also see that if the bump is high enough a supercritical flow could slow down to critical (point *e*) and then either return to supercritical (point *d*) or become subcritical (point *a*). Which of these possibilities actually occurs obviously depends on the bump shape, but also on upstream and downstream conditions (the last possibility is somewhat unlikely to occur in practice). The alert reader may ask, "What happens if the bump is so big that the specific energy wants to decrease below the minimum shown at point *e*?" The answer is that the flow will no longer conform to Eq. 11.26; the flow will no longer be frictionless, because a hydraulic jump will occur, consuming a significant amount of mechanical energy (see Section 11.4).

Example 11.4 FLOW IN A RECTANGULAR CHANNEL WITH A BUMP OR A NARROWING

A rectangular channel 2 m wide has a flow of 2.4 m^3/s at a depth of 1.0 m. Determine whether critical depth occurs at (a) a section where a bump of height $h = 0.20$ m is installed across the channel bed, (b) a side wall constriction (with no bumps) reducing the channel width to 1.7 m, and (c) both the bump and side wall constrictions combined. Neglect head losses of the bump and constriction caused by friction, expansion, and contraction.

Given: Rectangular channel with a bump, a side wall constriction, or both.

Find: Whether critical flow occurs.

Solution: Compare the specific energy to the minimum specific energy for the given flow rate in each case to establish whether critical depth occurs.

Governing equations:

$$E = \frac{Q^2}{2gA^2} + y \quad (11.15) \qquad y_c = \left[\frac{Q}{gb^2}\right]^{1/3} \quad (11.23)$$

$$E_{min} = \frac{3}{2}y_c \quad (11.25) \qquad E = E_1 - h \quad (11.28)$$

(a) Bump of height $h = 0.20$ m:

The initial specific energy, E_1, is

$$E_1 = y_1 + \frac{Q^2}{2gA^2} = y_1 + \frac{Q^2}{2gb^2y_1^2}$$

$$= 1.0\text{ m} + 2.4^2\left(\frac{\text{m}^3}{\text{s}}\right)^2 \times \frac{1}{2} \times \frac{\text{s}^2}{9.81\text{ m}} \times \frac{1}{2^2\text{ m}^2} \times \frac{1}{1^2\text{ m}^2}$$

$$E_1 = 1.073\text{ m}$$

Then the specific energy at the peak of the bump, E_{bump}, is obtained from Eq. 11.28

$$E_{bump} = E_1 - h = 1.073\text{ m} - 0.20\text{ m}$$
$$E_{bump} = 0.873\text{ m} \quad (1)$$

We must compare this to the minimum specific energy for the flow rate Q. First, the critical depth is

$$y_c = \left[\frac{Q^2}{gb^2}\right]^{1/3} = \left[2.4^2\left(\frac{\text{m}^3}{\text{s}}\right)^2 \times \frac{\text{s}^2}{9.81\text{ m}} \times \frac{1}{2^2\text{ m}^2}\right]^{1/3}$$

$$y_c = 0.528\text{ m}$$

(Note that we have $y_1 > y_c$, so we have a subcritical flow.)
Then the minimum specific energy is

$$E_{min} = \frac{3}{2}y_c = 0.791\text{ m} \quad (2)$$

Comparing Eqs. 1 and 2 we see that with the bump we do *not* attain critical conditions. ←

(b) A side wall constriction (with no bump) reducing the channel width to 1.7 m:

In this case the specific energy remains constant throughout ($h = 0$), even at the constriction; so

$$E_{constriction} = E_1 - h = E_1 = 1.073\text{ m} \quad (3)$$

However, at the constriction, we have a new value for b, ($b_{constriction} = 1.7$ m), and so a new critical depth

$$y_{c_{\text{constriction}}} = \left[\frac{Q^2}{gb^2_{\text{constriction}}}\right]^{1/3} = \left[2.4^2\left(\frac{\text{m}^3}{\text{s}}\right)^2 \times \frac{\text{s}^2}{9.81\text{ m}} \times \frac{1}{1.7^2\text{ m}^2}\right]^{1/3}$$

$$y_{c_{\text{constriction}}} = 0.588\text{ m}$$

Then the minimum specific energy *at the constriction* is

$$E_{\text{min}_{\text{constriction}}} = \frac{3}{2}y_{c_{\text{constriction}}} = 0.882\text{ m} \tag{4}$$

Comparing Eqs. 3 and 4 we see that with the constriction we do *not* attain critical conditions. ←

We might enquire as to what constriction *would* cause critical flow. To find this, solve

$$E = 1.073\text{ m} = E_{\text{min}} = \frac{3}{2}y_c = \frac{3}{2}\left[\frac{Q^2}{gb_c^2}\right]^{1/3}$$

for the critical channel width b_c.
Hence

$$\frac{Q^2}{gb_c^2} = \left[\frac{2}{3}E_{\text{min}}\right]^3$$

$$b_c = \frac{Q}{\sqrt{\frac{8}{27}gE^3_{\text{min}}}}$$

$$= \left(\frac{27}{8}\right)^{1/2} \times 2.4\left(\frac{\text{m}^3}{\text{s}}\right) \times \frac{\text{s}}{9.81^{1/2}\text{ m}^{1/2}} \times \frac{1}{1.073^{3/2}\text{ m}^{3/2}}$$

$$b_c = 1.27\text{ m}$$

To make the given flow attain critical conditions, the constriction should be 1.27 m; anything wider, and critical conditions are not reached.

(c) For a bump of $h = 0.20$ m *and* the constriction to $b = 1.7$ m:

We have already seen in case (a) that the bump ($h = 0.20$ m) was insufficient by itself to create critical conditions. From case (b) we saw that at the constriction the minimum specific energy is $E_{\text{min}} = 0.882$ m rather than $E_{\text{min}} = 0.791$ m in the main flow. When we have both factors present, we can compare the specific energy at the bump and constriction,

$$E_{\text{bump + constriction}} = E_{\text{bump}} = E_1 - h = 0.873\text{ m} \tag{5}$$

and the minimum specific energy for the flow at the bump and constriction,

$$E_{\text{min}_{\text{constriction}}} = \frac{3}{2}y_{c_{\text{constriction}}} = 0.882\text{ m} \tag{6}$$

From Eqs. 5 and 6 we see that with both factors the specific energy is actually *less* than the minimum. The fact that we must have a specific energy that is less than the minimum allowable means something has to give! What happens is that the flow assumptions become invalid; the flow may no longer be uniform or one-dimensional, or there may be a significant energy loss, for example due to a hydraulic jump occurring. (We will discuss hydraulic jumps in the next section.)

Hence the bump and constriction together *are* sufficient to make the flow reach critical state. ←

This Example illustrates how to determine whether a channel bump or constriction, or both, lead to critical flow conditions.

The Hydraulic Jump 11.4

We have shown that open-channel flow may be subcritical ($Fr < 1$) or supercritical ($Fr > 1$). For subcritical flow, disturbances caused by a change in bed slope or flow cross section may move upstream and downstream; the result is a smooth adjustment of the flow, as we have seen in the previous section. When flow at a section is supercritical, and downstream conditions will require a change to subcritical flow, the need for this change cannot be communicated upstream; the flow speed exceeds the speed of surface waves, which are the mechanism for transmitting changes. Thus a gradual change with a smooth transition through the critical point is not possible. The transition from supercritical to subcritical flow occurs abruptly through a *hydraulic jump*. Hydraulic jumps can occur in canals downstream of regulating sluices, at the foot of spillways (see Fig. 11.12*a*), where a steep channel slope suddenly becomes flat—and even in the home kitchen (see Fig. 11.12*b*)! The specific energy curve and general shape of a jump are shown in Fig. 11.13. We will see in this section that the jump always goes from a supercritical depth ($y_1 < y_c$) to a subcritical depth ($y_2 > y_c$) and that there will be a drop ΔE in the specific energy. Unlike the changes due to phenomena such as a bump, the abrupt change in depth involves a significant loss of mechanical energy through turbulent mixing.

The control volume for a hydraulic jump is sketched in Fig. 11.14.

We shall analyze the jump phenomenon by applying the basic equations to the control volume shown in the sketch. Experiments show that the jump occurs over a relatively short distance—at most, approximately six times the larger depth (y_2) [9]. In view of this short length, it is reasonable to assume that friction force F_f acting on the control volume is negligible compared to pressure forces. Note that we are therefore ignoring viscous effects for momentum considerations, but *not* for energy considerations (as we just mentioned, there is considerable turbulence in the jump). Although hydraulic jumps can occur on inclined surfaces, for simplicity we assume a horizontal bed, and rectangular channel of width b; the results we obtain will apply generally to hydraulic jumps.

Hence we have the following assumptions:

1. Steady flow
2. Incompressible flow
3. Uniform velocity at each section

VIDEO

A Laminar Hydraulic Jump.

(*a*) The Burdekin dam in Australia

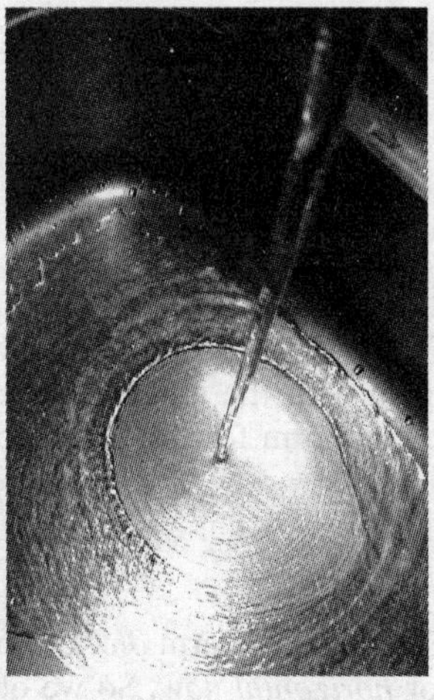

(*b*) The Kitchen Sink (James Kilfiger)

Fig. 11.12 Examples of a hydraulic jump.

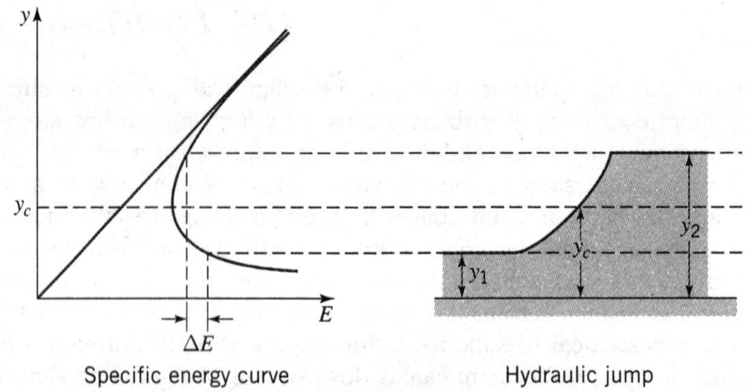

Fig. 11.13 Specific energy curve for flow through a hydraulic jump.

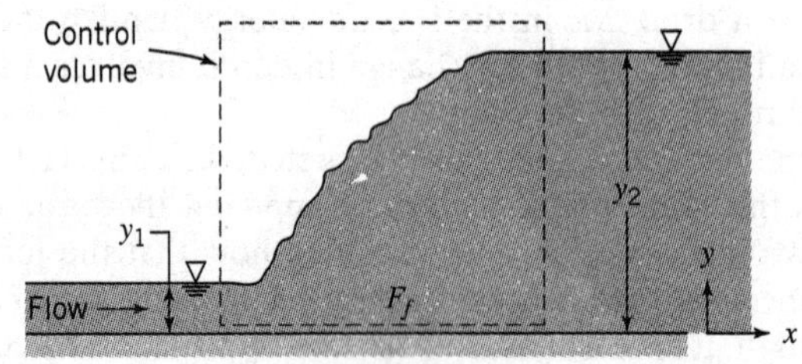

Fig. 11.14 Schematic of hydraulic jump, showing control volume used for analysis.

4. Hydrostatic pressure distribution at each section
5. Frictionless flow (for the momentum equation)

These assumptions are familiar from previous discussions in this chapter. For an *incompressible* flow with *uniform velocity* at each section, we can use the appropriate form of continuity from Chapter 4,

$$\sum_{CS} \vec{V} \cdot \vec{A} = 0 \tag{4.13b}$$

Applying Eq. 4.13b to the control volume we obtain

$$-V_1 b y_1 + V_2 b y_2 = 0$$

or

$$V_1 y_1 = V_2 y_2 \tag{11.31}$$

This is the continuity equation for the hydraulic jump. For the momentum equation, again with the assumption of uniform velocity at each section, we can use the following form for the x component of momentum

$$F_x = F_{S_x} + F_{B_x} = \frac{\partial}{\partial t}\int_{CV} u\rho \, d\forall \; + \sum_{CS} u\rho \vec{V} \cdot \vec{A} \tag{4.18d}$$

The unsteady term $\partial/\partial t$ disappears as the flow is *steady*, and the body force F_{B_x} is zero for *horizontal flow*. So we obtain

$$F_{S_x} = \sum_{CS} u\rho \vec{V} \cdot \vec{A} \tag{11.32}$$

The surface force consists of pressure forces on the two ends and friction force on the wetted surface. By assumption 5 we neglect friction. The gage pressure at the two ends is hydrostatic, as illustrated in Fig. 11.3*b*. We recall from our study of hydrostatics that the hydrostatic force, F_R, on a submerged vertical surface of area, A, is given by the simple result

$$F_R = p_c A \tag{3.10b}$$

where p_c is the pressure at the centroid of the vertical surface. For the two vertical surfaces of the control volume, then, we have

$$F_{S_x} = F_{R_1} - F_{R_2} = (p_c A)_1 - (p_c A)_2 = \{(\rho g y_1) y_1 b\} - \{(\rho g y_2) y_2 b\}$$
$$= \frac{\rho g b}{2}(y_1^2 - y_2^2)$$

Using this result in Eq. 11.32, and evaluating the terms on the right,

$$F_{S_x} = \frac{\rho g b}{2}(y_1^2 - y_2^2) = \sum_{CS} u \rho \vec{V} \cdot \vec{A} = V_1 \rho\{-V_1 y_1 b\} + V_2 \rho\{V_2 y_2 b\}$$

Rearranging and simplifying

$$\frac{V_1^2 y_1}{g} + \frac{y_1^2}{2} = \frac{V_2^2 y_2}{g} + \frac{y_2^2}{2} \tag{11.33}$$

This is the momentum equation for the hydraulic jump. We have already derived the energy equation for open-channel flows,

$$\frac{V_1^2}{2g} + y_1 + z_1 = \frac{V_2^2}{2g} + y_2 + z_2 + H_l \tag{11.10}$$

For our horizontal hydraulic jump, $z_1 = z_2$, so

$$E_1 = \frac{V_1^2}{2g} + y_1 = \frac{V_2^2}{2g} + y_2 + H_l = E_2 + H_l \tag{11.34}$$

This is the energy equation for the hydraulic jump; the loss of mechanical energy is

$$\Delta E = E_1 - E_2 = H_l$$

The continuity, momentum, and energy equations (Eqs. 11.31, 11.33, and 11.34, respectively) constitute a complete set for analyzing a hydraulic jump.

Depth Increase Across a Hydraulic Jump

To find the downstream or, as it is called, the *sequent* depth in terms of conditions upstream from the hydraulic jump, we begin by eliminating V_2 from the momentum equation. From continuity, $V_2 = V_1 y_1 / y_2$ (Eq. 11.31), so Eq. 11.33 can be written

$$\frac{V_1^2 y_1}{g} + \frac{y_1^2}{2} = \frac{V_1^2 y_1}{g}\left(\frac{y_1}{y_2}\right) + \frac{y_2^2}{2}$$

Rearranging

$$y_2^2 - y_1^2 = \frac{2V_1^2 y_1}{g}\left(1 - \frac{y_1}{y_2}\right) = \frac{2V_1^2 y_1}{g}\left(\frac{y_2 - y_1}{y_2}\right)$$

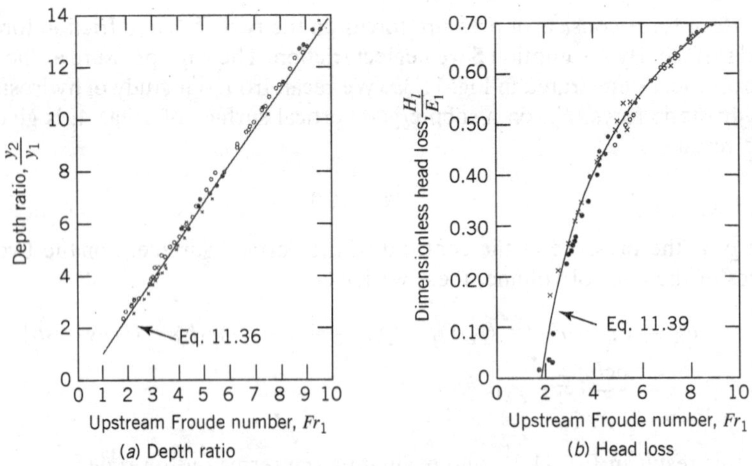

Fig. 11.15 Depth ratio and head loss for a hydraulic jump. (Data from Peterka [9].)

Dividing both sides by the common factor $(y_2 - y_1)$, we obtain

$$y_2 + y_1 = \frac{2V_1^2 y_1}{g y_2}$$

Next, multiplying by y_2 and dividing by y_1^2 gives

$$\left(\frac{y_2}{y_1}\right)^2 + \left(\frac{y_2}{y_1}\right) = \frac{2V_1^2}{g y_1} = 2Fr_1^2 \qquad (11.35)$$

Solving for y_2/y_1 using the quadratic formula (ignoring the physically meaningless negative root), we obtain

$$\frac{y_2}{y_1} = \frac{1}{2}\left[\sqrt{1 + 8Fr_1^2} - 1\right] \qquad (11.36)$$

Hence, the ratio of downstream to upstream depths across a hydraulic jump is only a function of the upstream Froude number. Equation 11.36 has been experimentally well verified, as can be seen in Fig. 11.15*a*. Depths y_1 and y_2 are referred to as *conjugate depths*. From Eq. 11.35, we see that an increase in depth $(y_2 > y_1)$ requires an upstream Froude number greater than one $(Fr_1 > 1)$. We have not yet established that we *must* have $Fr_1 > 1$, just that it must be for an increase in depth (theoretically we could have $Fr_1 < 1$ and $y_2 < y_1$); we will now consider the head loss to demonstrate that we *must* have $Fr_1 > 1$.

Head Loss Across a Hydraulic Jump

Hydraulic jumps often are used to dissipate energy below spillways as a means of preventing erosion of artificial or natural channel bottom or sides. It is therefore of interest to be able to determine the head loss due to a hydraulic jump.

From the energy equation for the jump, Eq. 11.34, we can solve for the head loss

$$H_l = E_1 - E_2 = \frac{V_1^2}{2g} + y_1 - \left(\frac{V_2^2}{2g} + y_2\right)$$

From continuity, $V_2 = V_1 y_1 / y_2$, so

$$H_l = \frac{V_1^2}{2g}\left[1 - \left(\frac{y_1}{y_2}\right)^2\right] + (y_1 - y_2)$$

or

$$\frac{H_l}{y_1} = \frac{Fr_1^2}{2}\left[1 - \left(\frac{y_1}{y_2}\right)^2\right] + \left[1 - \frac{y_2}{y_1}\right] \tag{11.37}$$

Solving Eq. 11.35 for Fr_1 in terms of y_2/y_1 and substituting into Eq.11.37, we obtain (after quite a bit of algebraic manipulation)

$$\frac{H_l}{y_1} = \frac{1}{4}\frac{\left[\frac{y_2}{y_1} - 1\right]^3}{\frac{y_2}{y_1}} \tag{11.38a}$$

Equation 11.38a is our proof that $y_2/y_1 > 1$; the left side is always positive (turbulence must lead to a loss of mechanical energy); so the cubed term must lead to a positive result. Then, from either Eq. 11.35 or Eq.11.36, we see that we must have $Fr_1 > 1$. An alternative form of this result is obtained after some minor rearranging,

$$H_l = \frac{[y_2 - y_1]^3}{4y_1y_2} \tag{11.38b}$$

which again shows that $y_2 > y_1$ for real flows ($H_l > 0$). Next, the specific energy, E_1, can be written as

$$E_1 = \frac{V_1^2}{2g} + y_1 = y_1\left[\frac{V_1^2}{2gy_1} + 1\right] = y_1\frac{(Fr_1^2 + 2)}{2}$$

Nondimensionalizing H_l using E_1,

$$\frac{H_l}{E_1} = \frac{1}{2}\frac{\left[\frac{y_2}{y_1} - 1\right]^3}{\frac{y_2}{y_1}[Fr_1^2 + 2]}$$

The depth ratio in terms of Fr_1 is given by Eq. 11.36. Hence H_l/E_1, can be written purely as a function of the upstream Froude number. The result, after some manipulation, is

$$\frac{H_l}{E_1} = \frac{\left[\sqrt{1 + 8Fr_1^2} - 3\right]^3}{8\left[\sqrt{1 + 8Fr_1^2} - 1\right][Fr_1^2 + 2]} \tag{11.39}$$

We see that the head loss, as a fraction of the original specific energy across a hydraulic jump, is only a function of the upstream Froude number. Equation 11.39 is experimentally well verified, as can be seen in Fig. 11.15*b*; the figure also shows that more than 70 percent of the mechanical energy of the entering stream is dissipated in jumps with $Fr_1 > 9$. Inspection of Eq. 11.39 also shows that if $Fr_1 = 1$, then $H_l = 0$, and that negative values are predicted for $Fr_1 < 1$. Since H_l must be positive in any real flow, this reconfirms that *a hydraulic jump can occur only in supercritical flow. Flow downstream from a jump always is subcritical.*

Example 11.5 HYDRAULIC JUMP IN A RECTANGULAR CHANNEL FLOW

A hydraulic jump occurs in a rectangular channel 3 m wide. The water depth before the jump is 0.6 m, and after the jump is 1.6 m. Compute (a) the flow rate in the channel (b) the critical depth (c) the head loss in the jump.

Given: Rectangular channel with hydraulic jump in which flow depth changes from 0.6 m to 1.6 m.

Find: Flow rate, critical depth, and head loss in the jump.

Solution: Use the equation that relates depths y_1 and y_2 in terms of the Froude number (Eq. 11.36); then use the Froude number (Eq. 11.7) to obtain the flow rate; use Eq. 11.23 to obtain the critical depth; and finally compute the head loss from Eq. 11.38b.

Governing equations:

$$\frac{y_2}{y_1} = \frac{1}{2}\left[-1 + \sqrt{1 + 8Fr_1^2}\right] \tag{11.36}$$

$$Fr = \frac{V}{\sqrt{gy}} \tag{11.7}$$

$$y_c = \left[\frac{Q^2}{gb^2}\right]^{1/3} \tag{11.23}$$

$$H_l = \frac{[y_2 - y_1]^3}{4y_1y_2} \tag{11.38b}$$

(a) From Eq. 11.36

$$Fr_1 = \sqrt{\frac{\left(1 + 2\frac{y_2}{y_1}\right)^2 - 1}{8}}$$

$$= \sqrt{\frac{\left(1 + 2 \times \frac{1.6\,\text{m}}{0.6\,\text{m}}\right)^2 - 1}{8}}$$

$$Fr_1 = 2.21$$

As expected, $Fr_1 > 1$ (supercritical flow). We can now use the definition of Froude number for open-channel flow to find V_1

$$Fr_1 = \frac{V_1}{\sqrt{gy_1}}$$

Hence

$$V_1 = Fr_1\sqrt{gy_1} = 2.21 \times \sqrt{\frac{9.81\,\text{m}}{\text{s}^2} \times 0.6\,\text{m}} = 5.36\,\text{m/s}$$

From this we can obtain the flow rate, Q.

$$Q = by_1V_1 = 3.0\,\text{m} \times 0.6\,\text{m} \times \frac{5.36\,\text{m}}{\text{s}}$$

$$Q = 9.65\,\text{m}^3/\text{s} \longleftarrow Q$$

(b) The critical depth can be obtained from Eq. 11.23.

$$y_c = \left[\frac{Q^2}{gb^2}\right]^{1/3}$$

$$= \left(9.65^2\,\frac{\text{m}^6}{\text{s}^2} \times \frac{\text{s}^2}{9.81\,\text{m}} \times \frac{1}{3.0^2\,\text{m}^2}\right)^{1/3}$$

$$y_c = 1.02\,\text{m} \longleftarrow y_c$$

Note that as illustrated in Fig. 11.13, $y_1 < y_c < y_2$.

(c) The head loss can be found from Eq. 11.38b.

$$H_l = \frac{[y_2 - y_1]^3}{4y_1y_2}$$

$$= \frac{1}{4}\frac{[1.6\,\text{m} - 0.6\,\text{m}]^3}{1.6\,\text{m} \times 0.6\,\text{m}} = 0.260\,\text{m} \longleftarrow H_l$$

As a verification of this result, we use the energy equation directly,

$$H_l = E_1 - E_2 = \left(y_1 + \frac{V_1^2}{2g}\right) - \left(y_2 + \frac{V_2^2}{2g}\right)$$

with $V_2 = Q/(by_2) = 2.01$ m/s,

$$H_l = \left(0.6\,\text{m} + 5.36^2\,\frac{\text{m}^2}{\text{s}^2} \times \frac{1}{2} \times \frac{\text{s}^2}{9.81\,\text{m}}\right)$$

$$- \left(1.6\,\text{m} + 2.01^2\,\frac{\text{m}^2}{\text{s}^2} \times \frac{1}{2} \times \frac{\text{s}^2}{9.81\,\text{m}}\right)$$

$$H_l = 0.258\,\text{m}$$

This Example illustrates computation of flow rate, critical depth, and head loss, for a hydraulic jump.

Steady Uniform Flow 11.5

After studying local effects such as bumps and hydraulic jumps, and defining some fundamental quantities such as the specific energy and critical velocity, we are ready to analyze flows in long stretches. Steady uniform flow is something that is to be expected to occur for channels of constant slope and cross section; Figs. 11.1 and 11.2 show examples of this kind of flow. Such flows are very common and important, and have been extensively studied.

The simplest such flow is *fully developed* flow; it is analogous to fully developed flow in pipes. A fully developed flow is one for which the channel is *prismatic*, that is, a channel with constant slope and cross section that flows at constant depth. This depth, y_n, is termed the *normal depth* and the flow is termed a *uniform flow*. Hence the expression *uniform flow* in this chapter has a different meaning than in earlier chapters. In earlier chapters it meant that the *velocity* was uniform *at a section* of the flow; in this chapter we use it to mean that, but in addition specifically that the *flow* is the same *at all sections*. Hence for the flow shown in Fig. 11.16, we have $A_1 = A_2 = A$ (cross-section areas), $Q_1 = Q_2 = Q$ (flow rates), $V_1 = V_2 = V$ (average velocity, $V = Q/A$), and $y_1 = y_2 = y_n$ (flow depth).

As before (Section 11.2), we use the following assumptions:

1. Steady flow
2. Incompressible flow
3. Uniform velocity at a section
4. Gradually varying depth so that pressure distribution is hydrostatic
5. Bed slope is small
6. $\dot{W}_s = \dot{W}_{\text{shear}} = \dot{W}_{\text{other}} = 0$

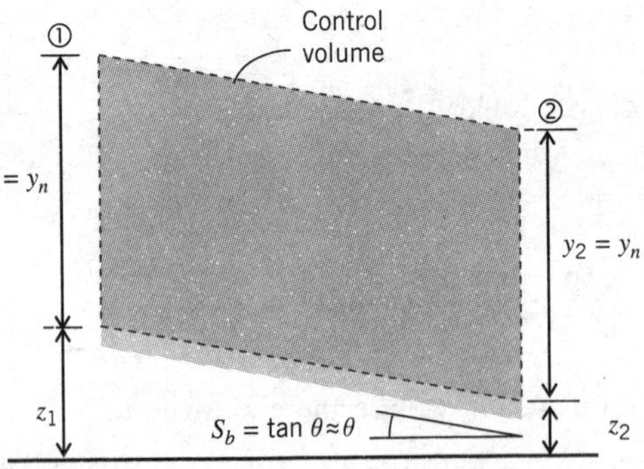

Fig. 11.16 Control volume for uniform channel flow.

Note that assumption **5** means that we can approximate the flow depth y to be vertical and flow speed horizontal. (Strictly speaking they should be normal and parallel to the channel bottom, respectively.)

The continuity equation is obvious for this case.

$$Q = V_1A_1 = V_2A_2 = VA$$

For the momentum equation, again with the assumption of uniform velocity at each section, we can use the following form for the x component of momentum

$$F_x = F_{S_x} + F_{B_x} = \frac{\partial}{\partial t}\int_{CV} u\rho\, d\forall + \sum_{CS} u\rho \vec{V}\cdot\vec{A} \tag{4.18d}$$

The unsteady term $\partial/\partial t$ disappears as the flow is steady, and the control surface summation is zero because $V_1 = V_2$; hence the right hand side is zero as there is no change of momentum for the control volume. The body force $F_{B_x} = W\sin\theta$ where W is the weight of fluid in the control volume; θ is the bed slope, as shown in Fig. 11.16. The surface force consists of the hydrostatic force on the two end surfaces at ① and ② and the friction force F_f on the wetted surface of the control volume; however, because we have the same pressure distributions at ① and ②, the net x component of pressure force is zero. Using all these results in Eq. 4.18d we obtain

$$-F_f + W\sin\theta = 0$$

or

$$F_f = W\sin\theta \tag{11.40}$$

We see that for flow at normal depth, the component of the gravity force driving the flow is just balanced by the friction force acting on the channel walls. This is in contrast to flow in a pipe or duct, for which (with the exception of pure gravity driven flow) we usually have a balance between an applied pressure gradient and the friction. The friction force may be expressed as the product of an average wall shear stress, τ_w, and the channel wetted surface area, PL (where L is the channel length), on which the stress acts

$$F_f = \tau_w PL \tag{11.41}$$

The component of gravity force can be written as

$$W \sin\theta = \rho g A L \sin\theta \approx \rho g A L \theta \approx \rho g A L S_b \tag{11.42}$$

where S_b is the channel bed slope. Using Eqs. 11.41 and 11.42 in Eq. 11.40,

$$\tau_w P L = \rho g A L S_b$$

or

$$\tau_w = \frac{\rho g A S_b}{P} = \rho g R_h S_b \tag{11.43}$$

where we have used the hydraulic radius, $R_h = A/P$ as defined in Eq. 11.1. In Chapter 9 we have previously introduced a skin friction coefficient,

$$C_f = \frac{\tau_w}{\frac{1}{2}\rho V^2} \tag{9.22}$$

Using this in Eq. 11.43

$$\frac{1}{2} C_f \rho V^2 = \rho g R_h S_b$$

so, solving for V

$$V = \sqrt{\frac{2g}{C_f}} \sqrt{R_h S_b} \tag{11.44}$$

The Manning Equation for Uniform Flow

Equation 11.44 gives the flow velocity V as a function of channel geometry, specifically the hydraulic radius, R_h and slope, S_b, but also the skin friction coefficient, C_f. This latter term is difficult to obtain experimentally or theoretically; it depends on a number of factors such as bed roughness and fluid properties, but also on the velocity itself (via the flow Reynolds number). Instead of this we define a new quantity,

$$C = \sqrt{\frac{2g}{C_f}}$$

so that Eq. 11.44 becomes

$$V = C\sqrt{R_h S_b} \tag{11.45}$$

Equation 11.45 is the well-known *Chezy equation*, and C is referred to as the *Chezy coefficient*. Experimental values of C were obtained by Manning [10]. He suggested that

$$C = \frac{1}{n} R_h^{1/6} \tag{11.46}$$

where n is a roughness coefficient having different values for different types of boundary roughness. Some representative values of n are listed in Table 11.2. Values of n for natural channels have also been published by the U.S. Geological Survey [13]. Substituting C from Eq. 11.46 into Eq. 11.45 results in the *Manning equation* for the velocity for flow at normal depth

$$V = \frac{1}{n} R_h^{2/3} S_b^{1/2} \tag{11.47}$$

which is valid for SI units. Manning's equation in SI units can also be expressed as

$$Q = \frac{1}{n} A R_h^{2/3} S_b^{1/2} \tag{11.48}$$

where A is in square meters. Note that a number of these equations, as well as many that follow, are "engineering" equations; that is, *the user needs to be aware of the required units of each term in the equation.* In Table 11.1 we have previously listed data on A and R_h for various channel geometries.

The relationship among variables in Eqs. 11.48 can be viewed in a number of ways. For example, it shows that the volume flow rate through a prismatic channel of given slope and roughness is a function of both channel size and channel shape. This is illustrated in Examples 11.6 and 11.7.

Table 11.2

A Selection of Manning's Roughness Coefficients

		Manning's n		
		Depth Ranges		
Lining Category	Lining Type	0–15 cm	15–60 cm	>60 cm
Rigid	Concrete	0.015	0.013	0.013
	Grouted riprap	0.040	0.030	0.028
	Stone masonry	0.042	0.032	0.030
	Soil cement	0.025	0.022	0.020
	Asphalt	0.018	0.016	0.016
Unlined	Bare soil	0.023	0.020	0.020
	Rock cut	0.045	0.035	0.025
Temporary	Woven paper net	0.016	0.015	0.015
	Jute net	0.028	0.022	0.019
	Fiberglass roving	0.028	0.021	0.019
	Straw with net	0.065	0.033	0.025
	Curled wood mat	0.066	0.035	0.028
	Synthetic mat	0.036	0.025	0.021
Gravel riprap	2.5 cm D_{50}	0.044	0.033	0.030
	5 cm D_{50}	0.066	0.041	0.034
Rock riprap	15 cm D_{50}	0.104	0.069	0.035
	12 in 30 cm D_{50}	—	0.078	0.040

Source: Linsley et al. [11] and Chen and Cotton [12].

Example 11.6 FLOW RATE IN A RECTANGULAR CHANNEL

An 2.4 m wide rectangular channel with a bed slope of 0.0004 m/m has a depth of flow of 0.6 m. Assuming steady uniform flow, determine the discharge in the channel. The Manning roughness coefficient is $n = 0.015$.

Given: Geometry of rectangular channel and flow depth.

Find: Flow rate Q.

Solution: Use the apprcpriate form of Manning's equation. For a problem in Engineering units, this is Eq. 11.48.

Governing equations:

$$Q = \frac{1}{n} A R_h^{2/3} S_b^{1/2} \qquad R_h = \frac{by}{b + 2y} \quad \text{(Table 11.1)}$$

Using this equation with the given data

$$Q = \frac{1}{n} A R_h^{2/3} S_b^{1/2}$$

$$= \frac{1}{0.015} \times (2.4\,\text{m} \times 0.6\,\text{m}) \times \left(\frac{2.4\,\text{m} \times 0.6\,\text{m}}{2.4\,\text{m} + 2 \times 0.6\,\text{m}} \right)^{2/3} \times \left(0.0004 \frac{\text{m}}{\text{m}} \right)^{1/2}$$

$$Q = 1.04\,\text{m}^3/\text{s} \longleftarrow Q$$

This Example demonstrates use of Manning's equation to solve for flow rate, Q. Note that because this is an "engineering" equation, the units do not cancel.

Example 11.7 FLOW VERSUS AREA THROUGH TWO CHANNEL SHAPES

Open channels, of square and semicircular shapes, are being considered for carrying flow on a slope of $S_b = 0.001$; the channel walls are to be poured concrete with $n = 0.015$. Evaluate the flow rate delivered by the channels for maximum dimensions between 0.5 and 2.0 m. Compare the channels on the basis of volume flow rate for given cross-sectional area.

Given: Square and semicircular channels; $S_b = 0.001$ and $n = 0.015$. Sizes between 0.5 and 2.0 m across.

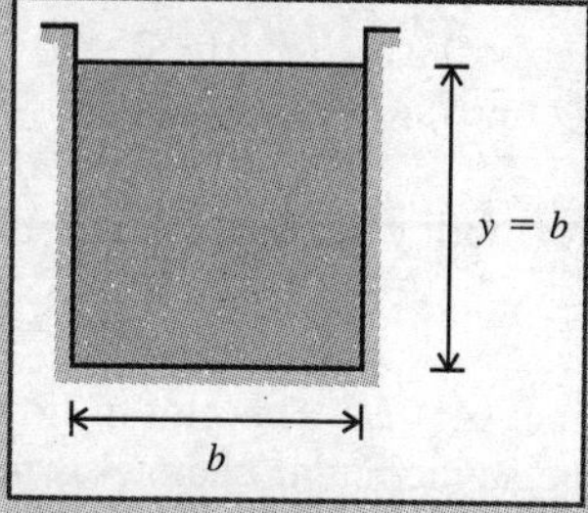

Find: Flow rate as a function of size. Compare channels on the basis of volume flow rate, Q, versus cross-sectional area, A.

Solution: Use the appropriate form of Manning's equation. For a problem in SI units, this is Eq. 11.48.

Governing equation:

$$Q = \frac{1}{n} A R_h^{2/3} S_b^{1/2} \tag{11.48}$$

Assumption: Flow at normal depth.

For the square channel,

$$P = 3b \quad \text{and} \quad A = b^2 \quad \text{so} \quad R_h = \frac{b}{3}$$

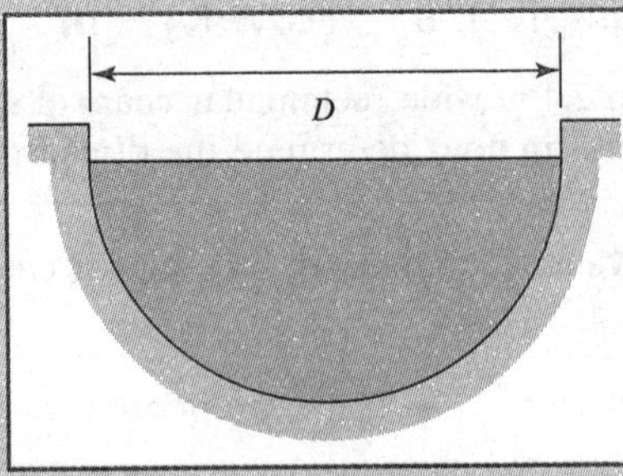

Using this in Eq. 11.48

$$Q = \frac{1}{n} A R_h^{2/3} S_b^{1/2} = \frac{1}{n} b^2 \left(\frac{b}{3}\right)^{2/3} S_b^{1/2} = \frac{1}{3^{2/3} n} S_b^{1/2} b^{8/3}$$

For $b = 1$ m,

$$Q = \frac{1}{3^{2/3}(0.015)} (0.001)^{1/2} (1)^{8/3} = 1.01 \text{ m}^3/\text{s} \longleftarrow Q$$

Tabulating for a range of sizes yields

b (m)	0.5	1.0	1.5	2.0
A (m^2)	0.25	1.00	2.25	4.00
Q (m^3/s)	0.160	1.01	2.99	6.44

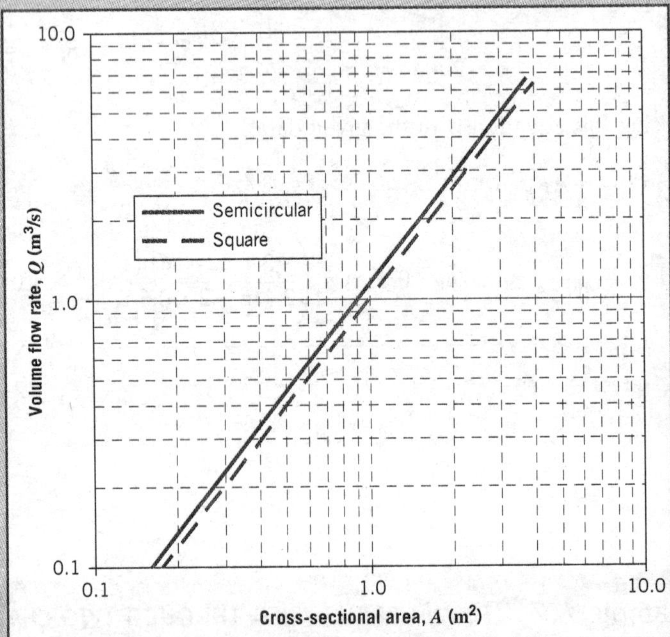

For the semicircular channel,

$$P = \frac{\pi D}{2} \quad \text{and} \quad A = \frac{\pi D^2}{8}$$

$$\text{so} \quad R_h = \frac{\pi D^2}{8} \frac{2}{\pi D} = \frac{D}{4}$$

Using this in Eq. 11.48

$$Q = \frac{1}{n} A R_h^{2/3} S_b^{1/2} = \frac{1}{n} \frac{\pi D^2}{8} \left(\frac{D}{4}\right)^{2/3} S_b^{1/2}$$

$$= \frac{\pi}{4^{5/3}(2)n} S_b^{1/2} D^{8/3}$$

For $D = 1$ m,

$$Q = \frac{\pi}{4^{5/3}(2)(0.015)} (0.001)^{1/2} (1)^{8/3} = 0.329 \text{ m}^3/\text{s} \longleftarrow Q$$

Tabulating for a range of sizes yields

D (m)	0.5	1.0	1.5	2.0
A (m^2)	0.0982	0.393	0.884	1.57
Q (m^3/s)	0.0517	0.329	0.969	2.09

For both channels, volume flow rate varies as

$$Q \sim L^{8/3} \quad \text{or} \quad Q \sim A^{4/3}$$

since $A \sim L^2$. The plot of flow rate versus cross-sectional area shows that the semicircular channel is more "efficient."

Performance of the two channels may be compared at any specified area. At $A = 1$ m^2, $Q/A = 1.01$ m/s for the square channel. For the semicircularchannel with $A = 1$ m^2, then $D = 1.60$ m, and $Q = 1.15$ m^3/s; so $Q/A = 1.15$ m/s. Thus the semicircular channel carries approximately 14 percent more flow per unit area than the square channel.

The comparison on cross-sectional area is important in determining the amount of excavation required to build the channel. The channel shapes also could be compared on the basis of perimeter, which would indicate the amount of concrete needed to finish the channel.

The *Excel* workbook for this problem can be used for computing data and plotting curves for other square and semicircular channels.

We have demonstrated that Eqs. 11.48 mean that, for normal flow, the flow rate depends on the channel size and shape. For a specified flow rate through a prismatic channel of given slope and roughness, Eqs. 11.48 also show that the depth of uniform flow is a function of both channel size and shape, as well as the slope. There is only one depth for uniform flow at a given flow rate; it may be greater than, less than, or equal to the critical depth. This is illustrated in Examples 11.8 and 11.9.

Example 11.8 NORMAL DEPTH IN A RECTANGULAR CHANNEL

Determine the normal depth (for uniform flow) if the channel described in Example 11.6 has a flow rate of 100 cfs.

Given: Geometric data on rectangular channel of Example 11.6.

Find: Normal depth for a flow rate $Q = 2.83\ \text{m}^3/\text{s}$.

Solution: Use the appropriate form of Manning's equation. For a problem in English Engineering units, this is Eq. 11.48.

Governing equations:

$$Q = \frac{1.49}{n} A R_h^{2/3} S_b^{1/2} \qquad R_h = \frac{b y_n}{b + 2y_n} \quad \text{(Table 11.1)}$$

Combining these equations

$$Q = \frac{1.49}{n} A R_h^{2/3} S_b^{1/2} = \frac{1.49}{n}(b y_n)\left(\frac{b y_n}{b + 2y_n}\right)^{2/3} S_b^{1/2}$$

Hence, after rearranging

$$\left(\frac{Qn}{1.49 b^{5/3} S_b^{1/2}}\right)^3 (b + 2y_n)^2 = y_n^5$$

Substituting $Q = 2.83\ \text{m}^3/\text{s}$, $n = 0.015$, $b = 2.4$ m, and $S_b = 0.0004$ and simplifying (remembering this is an "engineering" equation, in which we insert values without units),

$$0.48(8 + 2y_n)^2 = y_n^5$$

This nonlinear equation can be solved for y_n using a numerical method such as the Newton-Raphson method (or better yet using your calculator's solving feature or *Excel*'s *Goal Seek* or *Solver*!). We find

$$y_n = 1.232\ \text{m} \longleftarrow y_n$$

Note that there are five roots, but four of them are complex—mathematically correct but physically meaningless.

> ✓ This Example demonstrates the use of Manning's equation for finding the normal depth.
> ✓ This relatively simple physical problem still involved solving a nonlinear algebraic equation.
> The *Excel* workbook for this problem can be used for solving similar problems.

Example 11.9 DETERMINATION OF FLUME SIZE

An above-ground flume, built from timber, is to convey water from a mountain lake to a small hydroelectric plant. The flume is to deliver water at $Q = 2\ \text{m}^3/\text{s}$; the slope is $S_b = 0.002$ and $n = 0.013$. Evaluate the required flume size for (a) a rectangular section with $y/b = 0.5$ and (b) an equilateral triangular section.

Given: Flume to be built from timber, with $S_b = 0.002$, $n = 0.013$, and $Q = 2.00\ \text{m}^3/\text{s}$.

Find: Required flume size for:
(a) Rectangular section with $y/b = 0.5$.
(b) Equilateral triangular section.

Solution: Assume flume is long, so flow is uniform; it is at normal depth. Then Eq. 11.48 applies.

Governing equation:

$$Q = \frac{1}{n} A R_h^{2/3} S_b^{1/2} \tag{11.48}$$

The choice of channel shape fixes the relationship between R_h and A; so Eq. 11.48 may be solved for normal depth, y_n, thus determining the channel size required.

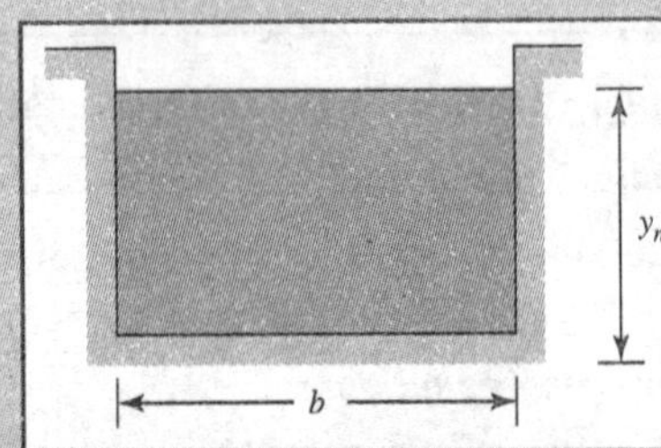

(a) Rectangular section

$$P = 2y_n + b; \; y_n/b = 0.5 \text{ so } b = 2y_n$$

$$P = 2y_n + 2y_n = 4y_n \qquad A = y_n b = y_n(2y_n) = 2y_n^2$$

$$\text{so } R_h = \frac{A}{P} = \frac{2y_n^2}{4y_n} = 0.5y_n$$

Using this in Eq. 11.48,

$$Q = \frac{1}{n} A R_h^{2/3} S_b^{1/2} = \frac{1}{n}(2y_n^2)(0.5y_n)^{2/3} S_b^{1/2} = \frac{2(0.5)^{2/3}}{n} y_n^{8/3} S_b^{1/2}$$

Solving for y_n

$$y_n = \left[\frac{nQ}{2(0.5)^{2/3} S_b^{1/2}}\right]^{3/8} = \left[\frac{0.013(2.00)}{2(0.5)^{2/3}(0.002)^{1/2}}\right]^{3/8} = 0.748 \text{ m}$$

The required dimensions for the rectangular channel are

$$y_n = 0.748 \text{ m} \qquad A = 1.12 \text{ m}^2$$
$$b = 1.50 \text{ m} \qquad p = 3.00 \text{ m}$$

← Flume size

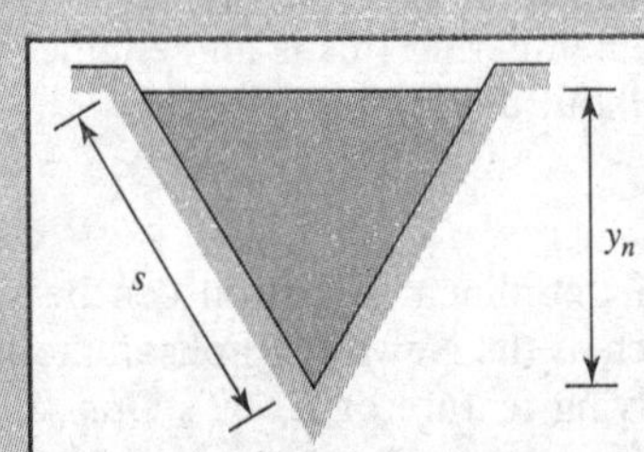

(b) Equilateral triangle section

$$P = 2s = \frac{2y_n}{\cos 30°} \qquad A = \frac{y_n s}{2} = \frac{y_n^2}{2\cos 30°}$$

$$\text{so } R_h = \frac{A}{P} = \frac{y_n}{4}$$

Using this in Eq. 11.48,

$$Q = \frac{1}{n} A R_h^{2/3} S_b^{1/2} = \frac{1}{n}\left(\frac{y_n^2}{2\cos 30°}\right)\left(\frac{y_n}{4}\right)^{2/3} S_b^{1/2} = \frac{1}{2\cos 30°(4)^{2/3} n} y_n^{8/3} S_b^{1/2}$$

Solving for y_n

$$y_n = \left[\frac{2\cos 30°(4)^{2/3} nQ}{S_b^{1/2}}\right]^{3/8} = \left[\frac{2\cos 30°(4)^{2/3}(0.013)(2.00)}{(0.002)^{1/2}}\right]^{3/8} = 1.42 \text{ m}$$

The required dimensions for the triangular channel are

$$y_n = 1.42 \text{ m} \qquad A = 1.16 \text{ m}^2$$
$$b_s = 1.64 \text{ m} \qquad p = 3.28 \text{ m}$$

← Flume size

Note that for the triangular channel

$$V = \frac{Q}{A} = 2.0\frac{\text{m}^3}{\text{s}} \times \frac{1}{1.16\,\text{m}^2} = 1.72\,\text{m/s}$$

and

$$Fr = = \frac{V}{\sqrt{gy_h}} = \frac{V}{\sqrt{gA/b_s}}$$

$$Fr = 1.72\frac{\text{m}}{\text{s}} \times \frac{1}{\left[9.81\frac{\text{m}}{\text{s}^2} \times 1.16\,\text{m}^2 \times \frac{1}{1.64\,\text{m}}\right]^{1/2}} = 0.653$$

This example shows the effect of channel shape on the size required to deliver a given flow at a specified bed slope and roughness coefficient. At specified S_b and n, flow may be subcritical, critical, or supercritical, depending on Q.

Hence this normal flow is subcritical (as is the flow in the rectangular channel).

Comparing results, we see that the rectangular flume would be cheaper to build; its perimeter is about 8.5 percent less than that of the triangular flume.

Energy Equation for Uniform Flow

To complete our discussion of normal flows, we consider the energy equation. The energy equation was already derived in Section 11.2.

$$\frac{V_1^2}{2g} + y_1 + z_1 = \frac{V_2^2}{2g} + y_2 + z_2 + H_l \tag{11.10}$$

In this case we obtain, with $V_1 = V_2 = V$, and $y_1 = y_2 = y_n$,

$$z_1 = z_2 + H_l$$

or

$$H_l = z_1 - z_2 = LS_b \tag{11.49}$$

where S_b is the slope of the bed and L is the distance between points ① and ②. Hence we see that for flow at normal depth, *the head loss due to friction is equal to the change in elevation of the bed.* The specific energy, E, is the same at all sections,

$$E = E_1 = \frac{V_1^2}{2g} + y_1 = E_2 = \frac{V_2^2}{2g} + y = \text{const}$$

For completeness we also compute the energy grade line EGL and hydraulic grade line HGL. From Section 6.5

$$EGL = \frac{p}{\rho g} + \frac{V^2}{2g} + z_{\text{total}} \tag{6.16b}$$

and

$$HGL = \frac{p}{\rho g} + z_{\text{total}} \tag{6.16c}$$

Note that we have used $z_{\text{total}} = z + y$ in Eqs. 6.16b and 6.16c (in Chapter 6, z is the total elevation of the free surface). Hence at any point on the free surface (recall that we are using gage pressures),

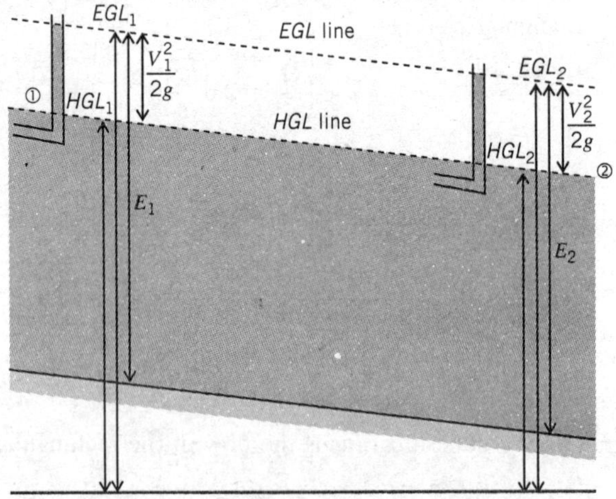

Fig. 11.17 Energy grade line, hydraulic grade line, and specific energy for uniform flow.

$$EGL = \frac{V^2}{2g} + z + y \tag{11.50}$$

and

$$HGL = z + y \tag{11.51}$$

Hence, using Eqs. 11.50 and 11.51 in Eqs. 11.10, between points ① and ② we obtain

$$EGL_1 - EGL_2 = H_l = z_1 - z_2$$

and (because $V_1 = V_2$)

$$HGL_1 - HGL_2 = H_l = z_1 - z_2$$

For normal flow, the energy grade line, the hydraulic grade line, and the channel bed are all parallel. The trends for the energy grade line, hydraulic grade line, and specific energy, are shown in Fig. 11.17.

Optimum Channel Cross Section

For given slope and roughness, the optimum channel cross section is that for which we need the smallest channel for a given flow rate; this is when Q/A is maximized. From Eq. 11.48 (using the SI version, although the results we obtain will apply generally)

$$\frac{Q}{A} = \frac{1}{n} R_h^{2/3} S_b^{1/2} \tag{11.52}$$

Thus the optimum cross section has maximum hydraulic radius, R_h. Since $R_h = A/P$, R_h is maximum when the wetted perimeter is minimum. Solving Eq. 11.52 for A (with $R_h = A/P$) then yields

$$A = \left[\frac{nQ}{S_b^{1/2}}\right]^{3/5} P^{2/5} \tag{11.53}$$

Table 11.3
Properties of Optimum Open-Channel Sections (SI Units)

Shape	Section	Optimum Geometry	Normal Depth, y_n	Cross-Sectional Area, A
Trapezoidal	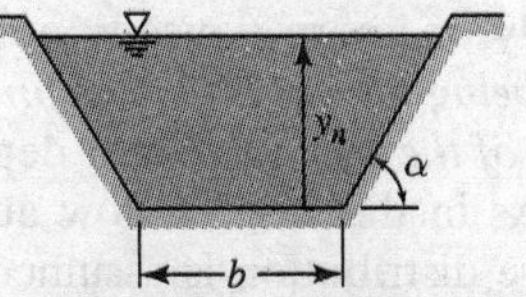	$\alpha = 60°$ $b = \frac{2}{\sqrt{3}} y_n$	$0.968 \left[\frac{Qn}{S_b^{1/2}}\right]^{3/8}$	$1.622 \left[\frac{Qn}{S_b^{1/2}}\right]^{3/4}$
Rectangular	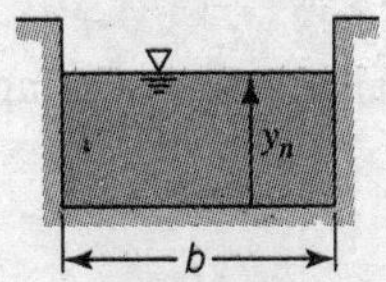	$b = 2y_n$	$0.917 \left[\frac{Qn}{S_b^{1/2}}\right]^{3/8}$	$1.682 \left[\frac{Qn}{S_b^{1/2}}\right]^{3/4}$
Triangular	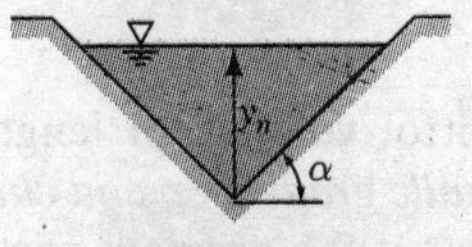	$\alpha = 45°$	$1.297 \left[\frac{Qn}{S_b^{1/2}}\right]^{3/8}$	$1.682 \left[\frac{Qn}{S_b^{1/2}}\right]^{3/4}$
Wide Flat	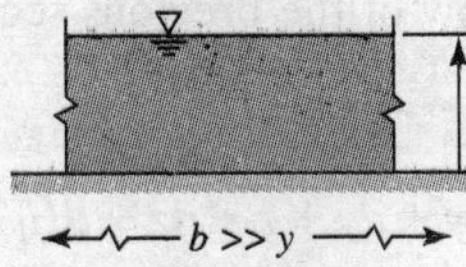	None	$1.00 \left[\frac{(Q/b)n}{S_b^{1/2}}\right]^{3/8}$	—
Circular	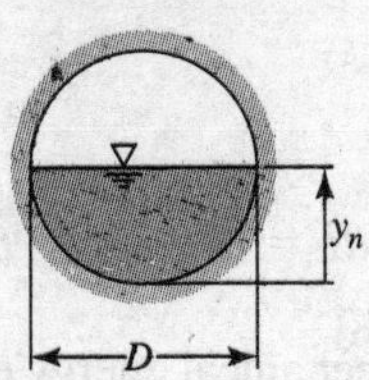	$D = 2y_n$	$1.00 \left[\frac{Qn}{S_b^{1/2}}\right]^{3/8}$	$1.583 \left[\frac{Qn}{S_b^{1/2}}\right]^{3/4}$

From Eq. 11.53, the flow area will be a minimum when the wetted perimeter is a minimum.

Wetted perimeter, P, is a function of channel shape. For any given prismatic channel shape (rectangular, trapezoidal, triangular, circular, etc.), the channel cross section can be optimized. Optimum cross sections for common channel shapes are given without proof in Table 11.3.

Once the optimum cross section for a given channel shape has been determined, expressions for normal depth, y_n, and area, A, as functions of flow rate can be obtained from Eq. 11.48. These expressions are included in Table 11.3.

Flow with Gradually Varying Depth 11.6

Most human-made channels are designed to have uniform flow (for example, see Fig. 11.1). However, this is not true in some situations. A channel can have nonuniform flow, that is, a flow for which the depth and hence speed, and so on vary along the channel for a number of reasons. Examples include when an open-channel flow encounters a change in bed slope, geometry, or roughness, or is adjusting itself back to normal depth after experiencing an upstream change (such as a sluice gate). We have already studied rapid, localized changes, such as that occurring in a hydraulic jump, but here we assume flow depth changes gradually. Flow with gradually varying depth is analyzed by applying the energy equation to a differential control volume; the result is a differential equation that relates changes in depth to distance along the flow. The resulting equation may be solved analytically or, more typically numerically, if *we approximate the head loss at each section as being the same as that for flow at normal depth, using the velocity and hydraulic radius of the section.* Water depth and channel bed height are assumed to change slowly. As in the case of flow at normal depth, velocity is assumed uniform, and the pressure distribution is assumed hydrostatic at each section.

The energy equation (Eq. 11.10) for open-channel flow was applied to a finite control volume in Section 11.2,

$$\frac{V_1^2}{2g} + y_1 + z_1 = \frac{V_2^2}{2g} + y_2 + z_2 + H_l \tag{11.10}$$

We apply this equation to the differential control volume, of length dx, shown in Fig. 11.18. Note that *the energy grade line, hydraulic grade line, and channel bottom all have different slopes*, unlike for the uniform flow of the previous section!

The energy equation becomes

$$\frac{V^2}{2g} + y + z = \frac{V^2}{2g} + d\left(\frac{V^2}{2g}\right) + y + dy + z + dz + dH_l$$

or after simplifying and rearranging

$$-d\left(\frac{V^2}{2g}\right) - dy - dz = dH_l \tag{11.54}$$

This is not surprising. The differential loss of mechanical energy equals the differential head loss. From channel geometry

$$dz = -S_b dx \tag{11.55}$$

We also have the approximation that the head loss in this differential nonuniform flow can be approximated by the head loss that uniform flow would have at the same flow rate, Q, at the section. Hence the differential head loss is approximated by

$$dH_l = S dx \tag{11.56}$$

where S is the slope of the EGL (see Fig. 11.18). Using Eqs. 11.55 and 11.56 in Eq. 11.54, dividing by dx, and rearranging, we obtain

$$\frac{d}{dx}\left(\frac{V^2}{2g}\right) + \frac{dy}{dx} = S_b - S \tag{11.57}$$

To eliminate the velocity derivative, we differentiate the continuity equation, $Q = VA = \text{const}$, to obtain

$$\frac{dQ}{dx} = 0 = A\frac{dV}{dx} + V\frac{dA}{dx}$$

or

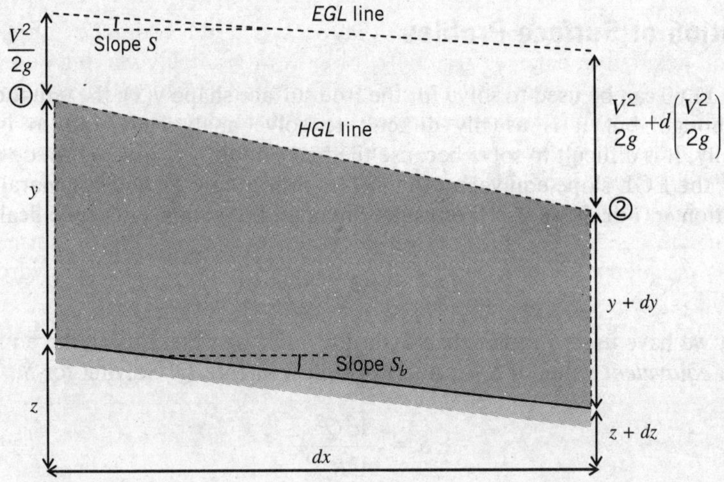

Fig. 11.18 Control volume for energy analysis of gradually varying flow.

$$\frac{dV}{dx} = -\frac{V}{A}\frac{dA}{dx} = -\frac{Vb_s}{A}\frac{dy}{dx} \tag{11.58}$$

where we have used $dA = b_s dy$ (Eq. 11.17), where b_s is the channel width at the free surface. Using Eq. 11.58 in Eq. 11.57, after rearranging

$$\frac{d}{dx}\left(\frac{V^2}{2g}\right) + \frac{dy}{dx} = \frac{V}{g}\frac{dV}{dx} + \frac{dy}{dx} = -\frac{V^2 b_s}{gA}\frac{dy}{dx} + \frac{dy}{dx} = S_b - S \tag{11.59}$$

Next, we recognize that

$$\frac{V^2 b_s}{gA} = \frac{V^2}{g\dfrac{A}{b_s}} = \frac{V^2}{gy_h} = Fr^2$$

where y_h is the hydraulic depth (Eq. 11.2). Using this in Eq. 11.59, we finally obtain our desired form of the *energy equation for gradually varying flow*

$$\frac{dy}{dx} = \frac{S_b - S}{1 - Fr^2} \tag{11.60}$$

This equation indicates how the depth y of the flow varies. Whether the flow becomes deeper ($dy/dx > 0$) or shallower ($dy/dx < 0$) depends on the sign of the right-hand side. For example, consider a channel that has a horizontal section ($S_b = 0$):

$$\frac{dy}{dx} = -\frac{S}{1 - Fr^2}$$

Because of friction the EGL always decreases, so $S > 0$. If the incoming flow is subcritical ($Fr < 1$), the flow depth will gradually decrease ($dy/dx < 0$); if the incoming flow is supercritical, ($Fr > 1$), the flow depth will gradually increase ($dy/dx > 0$). Note also that for critical flow ($Fr = 1$), the equation leads to a singularity, and gradually flow is no longer sustainable—something dramatic will happen (guess what).

Calculation of Surface Profiles

Equation 11.60 can be used to solve for the free surface shape $y(x)$; the equation looks simple enough, but it is usually difficult to solve analytically and so is solved numerically. It is difficult to solve because the bed slope, S_b, the local Froude number, Fr, and S, the EGL slope equivalent to uniform flow at rate Q, will in general all vary with location, x. For S, we use the results obtained in Section 11.5, specifically

$$Q = \frac{1}{n} A R_h^{2/3} S^{1/2} \tag{11.48}$$

Note that we have used S rather than S_b in Eq. 11.48 as we are using the equation to obtain an *equivalent* value of S for a uniform flow at rate Q! Solving for S,

$$S = \frac{n^2 Q^2}{A^2 R_h^{4/3}} \tag{11.61}$$

We can also express the Froude number as a function of Q,

$$Fr = \frac{V}{\sqrt{g y_h}} = \frac{Q}{A\sqrt{g y_h}} \tag{11.62}$$

Using Eqs. 11.61 and 11.62 in Eq. 11.60

$$\frac{dy}{dx} = \frac{S_b - S}{1 - Fr^2} = \frac{S_b - \dfrac{n^2 Q^2}{A^2 R_h^{4/3}}}{1 - \dfrac{Q^2}{A^2 g y_h}} \tag{11.63}$$

For a given channel (slope, S_b, and roughness coefficient, n, both of which may vary with x) and flow rate Q, the area A, hydraulic radius R_h, and hydraulic depth y_h are all functions of depth y (see Section 11.1). Hence Eqs. 11.63 are usually best solved using a suitable numerical integration scheme. Example 11.10 shows such a calculation for the simplest case, that of a rectangular channel.

Example 11.10 CALCULATION OF FREE SURFACE PROFILE

Water flows in a 5 m wide rectangular channel made from unfinished concrete with $n = 0.015$. The channel contains a long reach on which S_b is constant at $S_b = 0.020$. At one section, flow is at depth $y_1 = 1.5$ m, with speed $V_1 = 4.0$ m/s. Calculate and plot the free surface profile for the first 100 m of the channel, and find the final depth.

Given: Water flow in a rectangular channel.

Find: Plot of free surface profile; depth at 100 m.

Solution: Use the appropriate form of the equation for flow depth, Eq. 11.63a.

Governing equation:

$$\frac{dy}{dx} = \frac{S_b - S}{1 - Fr^2} = \frac{S_b - \dfrac{n^2 Q^2}{A^2 R_h^{4/3}}}{1 - \dfrac{Q^2}{A^2 g y_h}} \tag{11.63a}$$

We use Euler's method (see Section 5.5) to convert the differential equation to a difference equation. In this approach, the differential is converted to a difference,

$$\frac{dy}{dx} \approx \frac{\Delta y}{\Delta x} \tag{1}$$

where Δx and Δy are small but finite changes in x and y, respectively. Combining Eqs. 11.63a and 1, and rearranging,

$$\Delta y = \Delta x \left(\frac{S_b - \dfrac{n^2 Q^2}{A^2 R_h^{4/3}}}{1 - \dfrac{Q^2}{A^2 g y_h}} \right)$$

Finally, we let $\Delta y = y_{i+1} - y_i$, where y_i and y_{i+1} are the depths at point i and a point $(i+1)$ distance Δx further downstream,

$$y_{i+1} = y_i + \Delta x \left(\frac{S_{b_i} - \dfrac{n_i^2 Q^2}{A_i^2 R_{h_i}^{4/3}}}{1 - \dfrac{Q^2}{A_i^2 g y_{h_i}}} \right) \tag{2}$$

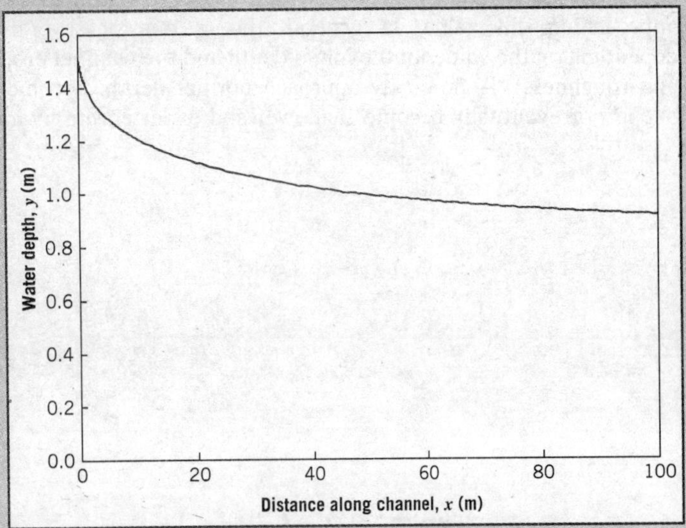

Equation 2 computes the depth, y_{i+1}, given data at point i. In the current application, S_b and n are constant, but A, R_h, and y_h will, of course, vary with x because they are functions of y. For a rectangular channel we have the following:

$$A_i = b y_i$$

$$R_{h_i} = \frac{b y_i}{b + 2 y_i}$$

$$y_{h_i} = \frac{A_i}{b_s} = \frac{A_i}{b} = \frac{b y_i}{b_s} = y_i$$

The calculations are conveniently performed and results plotted using *Excel*. Note that partial results are shown in the table, and that for the first meter, over which there is a rapid change in depth, the step size is $\Delta x = 0.05$.

i	x (m)	y (m)	A (m^2)	R_h (m)	y_h (m)
1	0.00	1.500	7.500	0.938	1.500
2	0.05	1.491	7.454	0.934	1.491
3	0.10	1.483	7.417	0.931	1.483
4	0.15	1.477	7.385	0.928	1.477
5	0.20	1.471	7.356	0.926	1.471
⋮	⋮	⋮	⋮	⋮	⋮
118	98	0.916	4.580	0.670	0.916
119	99	0.915	4.576	0.670	0.915
120	100	0.914	4.571	0.669	0.914

The depth at location $x = 100$ m is seen to be 0.914 m.

$$y(100\,\text{m}) = 0.914\,\text{m} \leftarrow \qquad y(100\,\text{m})$$

Note (following the solution procedure of Example 11.8) that the normal depth for this flow is $y_n = 0.858$ m; the flow depth is asymptotically approaching this value. In general, this is one of several possibilities, depending on the values of the initial depth and the channel properties (slope and roughness). A flow may approach normal depth, become deeper and deeper, or eventually become shallower and experience a hydraulic jump.

The accuracy of the results obtained obviously depends on the numerical model used; for example, a more accurate model is the RK4 method. Also, for the first meter or so, there are rapid changes in depth, bringing into question the validity of several assumptions, for example, uniform flow and hydrostatic pressure.

The *Excel* workbook for this problem can be modified for use in solving similar problems.

11.7 Discharge Measurement Using Weirs

A *weir* is a device (or overflow structure) that is placed normal to the direction of flow. The weir essentially backs up water so that, in flowing over the weir, the water goes through critical depth. Weirs have been used for the measurement of water flow in open channels for many years. Weirs can generally be classified as *sharp-crested weirs* and *broad-crested weirs.* Weirs are discussed in detail in Bos [14], Brater [15], and Replogle [16].

A *sharp-crested weir* is basically a thin plate mounted perpendicular to the flow with the top of the plate having a beveled, sharp edge, which makes the nappe spring clear from the plate (see Fig. 11.19).

The rate of flow is determined by measuring the head, typically in a stilling well (see Fig. 11.20) at a distance upstream from the crest. The head H is measured using a gage.

Suppressed Rectangular Weir

These sharp-crested weirs are as wide as the channel and the width of the nappe is the same length as the crest. Referring to Fig. 11.20, consider an elemental area $dA = bdh$ and assume the velocity is $V = \sqrt{2gh}$; then the elemental flow is

$$dQ = bdh\sqrt{2gh} = b\sqrt{2g}h^{1/2}dh$$

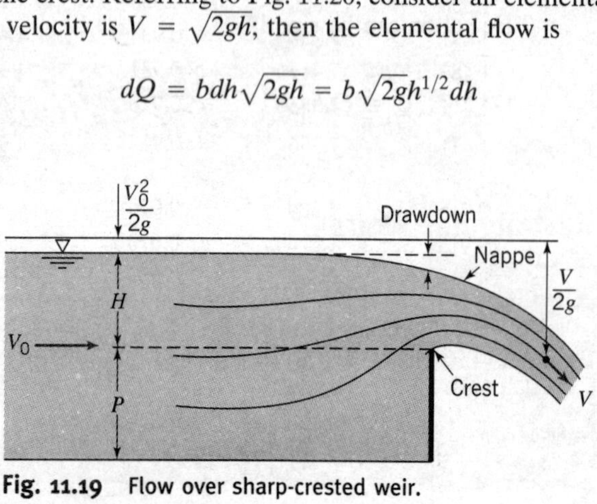

Fig. 11.19 Flow over sharp-crested weir.

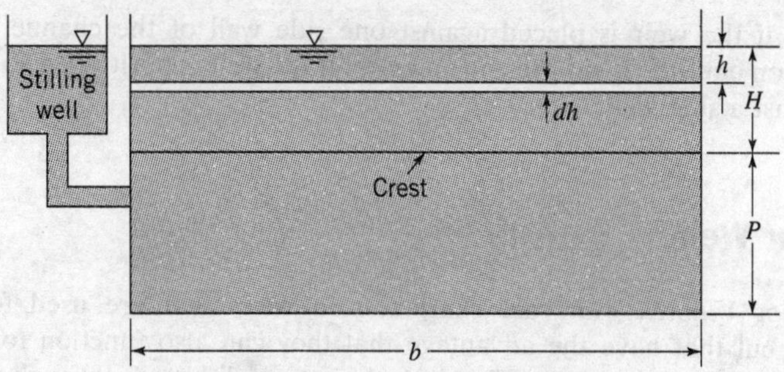

Fig. 11.20 Rectangular sharp-crested weir without end contraction.

The discharge is expressed by integrating this over the area above the top of the weir crest:

$$Q = \int_0^H dQ = \sqrt{2g}b \int_0^H h^{1/2} dh = \frac{2}{3}\sqrt{2g}bH^{3/2} \tag{11.64}$$

Friction effects have been neglected in the derivation of Eq. 11.64. The drawdown effect shown in Fig. 11.19 and the crest contraction indicate that the streamlines are not parallel or normal to the area in the plane. To account for these effects, a coefficient of discharge C_d is used, so that

$$Q = C_d \frac{2}{3}\sqrt{2g}bH^{3/2}$$

where C_d is approximately 0.62. This is the basic equation for a suppressed rectangular weir, which can be expressed more generally as

$$Q = C_w bH^{3/2} \tag{11.65}$$

where the C_w is the weir coefficient, $C_w = \frac{2}{3}C_d\sqrt{2g}$. For SI units, $C_w \approx 1.84$.

If the velocity of approach, V_a, where H is measured is appreciable, then the integration limits are

$$Q = \sqrt{2g}b \int_{V_a^2/2g}^{H + V_a^2/2g} h^{1/2} dh = C_w b \left[\left(H + \frac{V_a^2}{2g} \right)^{3/2} - \left(\frac{V_a^2}{2g} \right)^{3/2} \right] \tag{11.66}$$

When $(V_a^2/2g)^{3/2} \approx 0$ Eq. 11.66 can be simplified to

$$Q = C_w b \left(H + \frac{V_a^2}{2g} \right)^{3/2} \tag{11.67}$$

Contracted Rectangular Weirs

A *contracted horizontal weir* is another sharp-crested weir with a crest that is shorter than the width of the channel and one or two beveled end sections so that water contracts both horizontally and vertically. This forces the nappe width to be less than b. The effective crest length is

$$b' = b - 0.1\,nH$$

where $n = 1$ if the weir is placed against one side wall of the channel so that the contraction on one side is suppressed and $n = 2$ if the weir is positioned so that it is not placed against a side wall.

Triangular Weir

Triangular or *V-notch weirs* are sharp-crested weirs that are used for relatively small flows but that have the advantage that they can also function for reasonably large flows as well. Referring to Fig. 11.21, the rate of discharge through an elemental area, dA, is

$$dQ = C_d\sqrt{2gh}dA$$

where $dA = 2x dh$, and $x = (H - h)\tan(\theta/2)$; so $dA = 2(H - h)\tan(\theta/2)dh$. Then

$$dQ = C_d\sqrt{2gh}\left[2(H-h)\tan\left(\frac{\theta}{2}\right)dh\right]$$

and

$$Q = C_d 2\sqrt{2g}\tan\left(\frac{\theta}{2}\right)\int_0^H (H-h)h^{1/2}dh$$

$$= C_d\left(\frac{8}{15}\right)\sqrt{2g}\tan\left(\frac{\theta}{2}\right)H^{5/2}$$

$$Q = C_w H^{5/2}$$

The value of C_w for a value of $\theta = 90°$ (the most common) is $C_w = 1.38$ for SI units.

Broad-Crested Weir

Broad-crested weirs (Fig. 11.22) are essentially critical-depth weirs in that if the weirs are high enough, critical depth occurs on the crest of the weir. For critical flow conditions $y_c = (Q^2/gb^2)^{1/3}$ (Eq. 11.23) and $E = 3y_c/2$ (Eq. 11.25) for rectangular channels:

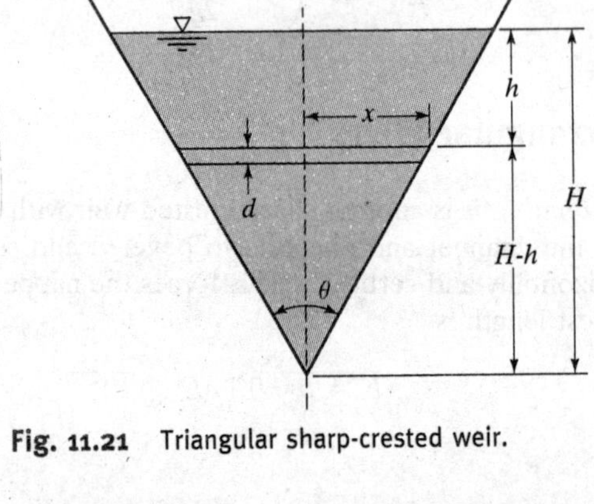

Fig. 11.21 Triangular sharp-crested weir.

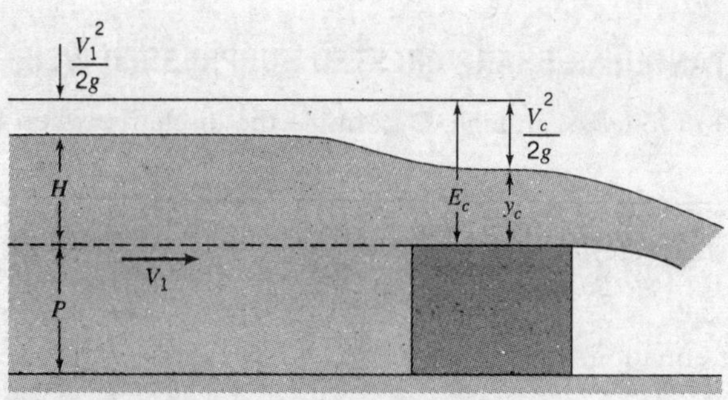

Fig. 11.22 Broad-crested weir.

$$Q = b\sqrt{g y_c^3} = b\sqrt{g\left(\frac{2}{3}E\right)^3} = b\left(\frac{2}{3}\right)^{3/2}\sqrt{g}E^{3/2}$$

or, assuming the approach velocity is negligible:

$$Q = b\left(\frac{2}{3}\right)^{3/2}\sqrt{g}H^{3/2}$$

$$Q = C_w b H^{3/2}$$

Figure 11.23 illustrates a broad-crested weir installation in a concrete-lined canal.

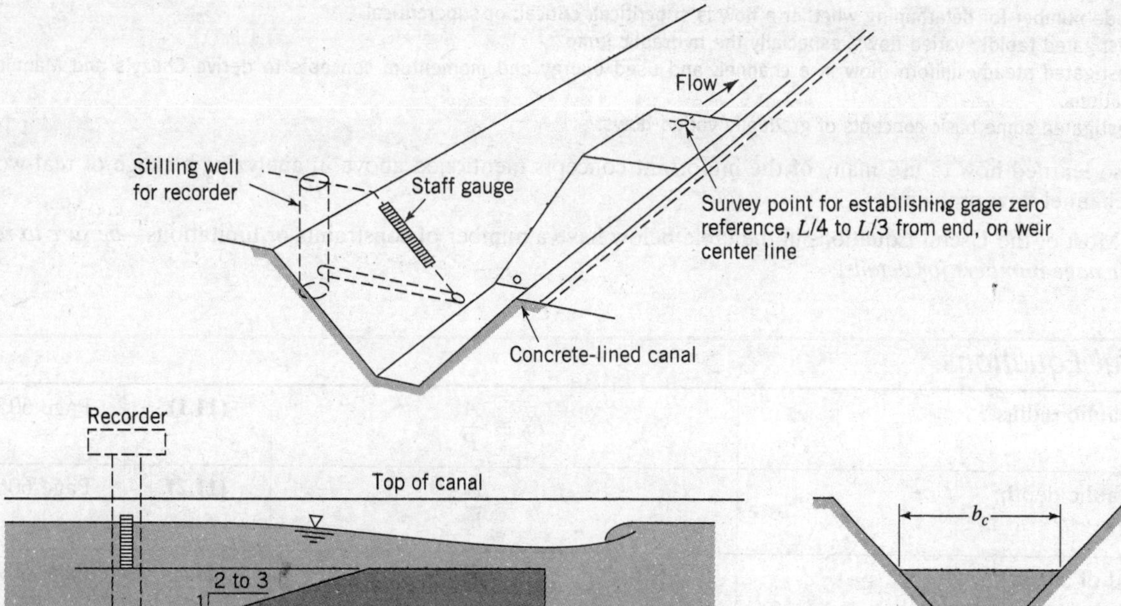

Fig. 11.23 Broad-crested weir in concrete-lined canal (from Bos et al. [14]).

Example 11.11 DISCHARGE FROM A RECTANGULAR SHARP-CRESTED SUPPRESSED WEIR

A rectangular, sharp-crested suppressed weir 3 m long is 1 m high. Determine the discharge when the head is 150 mm.

Given: Geometry and head of a rectangular sharp-crested suppressed weir.

Find: Discharge (flow rate), Q.

Solution: Use the appropriate weir discharge equation.

Governing equation:

$$Q = C_w b H^{3/2} \tag{11.65}$$

In Eq. 11.65 we use $C_w \approx 1.84$, and the given data, $b = 3$ m and $H = 150$ mm $= 0.15$ m, so

$$Q = 1.84 \times 3\,\text{m} \times (0.15\,\text{m})^{3/2}$$

$$Q = 0.321\,\text{m}^3/\text{s} \longleftarrow Q$$

Note that Eq. 11.65 is an "engineering" equation; so we do not expect the units to cancel.

This Example illustrates use of one of a number of weir-discharge equations.

11.8 Summary and Useful Equations

In this chapter, we:

- ✓ Derived an expression for the speed of surface waves and developed the notion of the specific energy of a flow, and derived the Froude number for determining whether a flow is subcritical, critical, or supercritical.
- ✓ Investigated rapidly varied flows, especially the hydraulic jump.
- ✓ Investigated steady uniform flow in a channel, and used energy and momentum concepts to derive Chezy's and Manning's equations.
- ✓ Investigated some basic concepts of gradually varied flows.

We also learned how to use many of the important concepts mentioned above in analyzing a range of real-world open-channel flow problems.

Note: Most of the Useful Equations in the table below have a number of constraints or limitations—*be sure to refer to their page numbers for details*!

Useful Equations

Hydraulic radius:	$R_h = \dfrac{A}{P}$	**(11.1)**	Page 605
Hydraulic depth:	$y_h = \dfrac{A}{b_s}$	**(11.2)**	Page 606
Speed of surface wave:	$c = \sqrt{gy}$	**(11.6)**	Page 609
Froude number:	$Fr = \dfrac{V}{\sqrt{gy}}$	**(11.7)**	Page 610

Energy equation for open-channel flow:	$\frac{V_1^2}{2g} + y_1 + z_1 = \frac{V_2^2}{2g} + y_2 + z_2 + H_l$	**(11.10)**	Page 612
Total head:	$H = \frac{V^2}{2g} + y + z$	**(11.11)**	Page 613
Specific energy:	$E = \frac{V^2}{2g} + y$	**(11.13)**	Page 613
Critical flow:	$Q^2 = \frac{gA_c^3}{b_{s_c}}$	**(11.21)**	Page 617
Critical velocity:	$V_c = \sqrt{gy_{h_c}}$	**(11.22)**	Page 617
Critical depth (rectangular channel):	$y_c = \left[\frac{Q^2}{gb^2}\right]^{1/3}$	**(11.23)**	Page 617
Critical velocity (rectangular channel):	$V_c = \sqrt{gy_c} = \left[\frac{gQ}{b}\right]^{1/3}$	**(11.24)**	Page 618
Minimum specific energy (rectangular channel):	$E_{\min} = \frac{3}{2}y_c$	**(11.25)**	Page 618
Hydraulic jump conjugate depths:	$\frac{y_2}{y_1} = \frac{1}{2}\left[\sqrt{1 + 8Fr_1^2} - 1\right]$	**(11.36)**	Page 628
Hydraulic jump head loss:	$H_l = \frac{[y_2 - y_1]^3}{4y_1y_2}$	**(11.38b)**	Page 629
Hydraulic jump head loss (in terms of Fr_1):	$\frac{H_l}{E_1} = \frac{\left[\sqrt{1 + 8Fr_1^2} - 3\right]^3}{8\left[\sqrt{1 + 8Fr_1^2} - 1\right][Fr_1^2 + 2]}$	**(11.39)**	Page 629
Chezy equation:	$V = C\sqrt{R_hS_b}$	**(11.45)**	Page 633
Chezy coefficient:	$C = \frac{1}{n}R_h^{1/6}$	**(11.46)**	Page 633
Manning equation for velocity (SI units)	$V = \frac{1}{n}R_h^{2/3}S_b^{1/2}$	**(11.47)**	Page 633
Manning equation for flow (SI units)	$Q = \frac{1}{n}AR_h^{2/3}S_b^{1/2}$	**(11.48)**	Page 634
Energy Grade Line	$EGL = \frac{V^2}{2g} + z + y$	**(11.50)**	Page 640
Hydraulic Grade Line	$HGL = z + y$	**(11.51)**	Page 640
Energy equation (gradually varying flow):	$\frac{dy}{dx} = \frac{S_b - S}{1 - Fr^2}$	**(11.60)**	Page 643

Case Study

The Three Gorges Dam

A model of the Three Gorges Dam.

This chapter provided an introduction to free surface flows, such as that at the exit from a dam. The Three Gorges Dam on the Yangtze River in China is the largest hydroelectric dam in the world. The generating capacity will eventually be 22,500 MW. The dam is more than 2 km wide and 185 m tall, and its reservoir will eventually stretch over 600 km upstream. The construction of the dam was very controversial: Millions of people had to be relocated, and we are still not sure of the long-term environmental consequences of this massive project.

The most significant function of the dam, apart from power generation, is to control flooding. The reservoir's flood storage capacity is 22 km^3; this will reduce the frequency of downstream flooding from once every 10 years to once every 100 years. Historically, a number of large cities and a lot of farmland have been vulnerable to flooding. For example, in 1954, almost 200,000 km^2 of land were flooded, killing over 30,000 people and forcing almost 20 million people to move; in 1998, a flood in the same area affected over 2 million people. With the dam, it is expected that major floods can be controlled.

Ship locks for river traffic to bypass the dam have been built so that shipping will become safer (the gorges were notoriously dangerous to navigate). Each ship lock is made up of five stages, taking around 4 hr in total to complete. In addition to the canal locks, the Three Gorges Dam is equipped with a ship lift capable of lifting ships of up to 3000 tons.

References

1. Chow, V. T., *Open-Channel Hydraulics*. New York: McGraw-Hill, 1959.
2. Henderson, F. M., *Open-Channel Flow*. New York: Macmillan, 1966.
3. French, R. H., *Open-Channel Hydraulics*. New York: McGraw-Hill, 1985.
4. Townson, J. M., *Free-Surface Hydraulics*. London: Unwin Hyman, 1991.
5. Chaudhry, M. H., *Open-Channel Flow*. Englewood Cliffs, NJ: Prentice Hall, 1993.
6. Jain, S. C., *Open Channel Flow*. New York: Wiley, 2001.
7. Sturm, T. W., *Open Channel Hydraulics*. New York: McGraw-Hill, 2001.
8. Mays, L. W., *Water Resources Engineering*, 2005 ed. New York: Wiley, 2005.
9. Peterka, A. J., "Hydraulic Design of Stilling Basins and Energy Dissipators," U.S. Department of the Interior, Bureau of Reclamation, Engineering Monograph No. 25 (Revised), July 1963.
10. Manning, R., "On the Flow of Water in Open Channels and Pipes." *Transactions Institute of Civil Engineers of Ireland*, vol. 20, pp. 161–209, Dublin, 1891; Supplement, vol. 24, pp. 179–207, 1895.
11. Linsley, R. K., J. B. Franzini, D. L. Freyberg, and G. Tchobanoglous, *Water Resources Engineering*. New York: McGraw-Hill, 1991.
12. Chen, Y. H., and G. K. Cotton, *Design of Roadside Channels with Flexible Linings*, Hydraulic Engineering Circular 15, FHWA-IP-87-7, Federal Highway Administration, McClean, VA, 1988.
13. Barnes, H. H., Jr., *Roughness Characteristics of Natural Channels*. U.S. Geological Survey, Water Supply Paper 1849. Washington, DC: U.S. Government Printing Office, 1962.
14. Bos, M. G., J. A. Replogle, and A. J. Clemmens, *Flow Measuring Flumes for Open Channel System*. New York: John Wiley & Sons, 1984.
15. Brater, E. F., H. W. King, J. E. Lindell, and C. Y. Wei, *Handbook of Hydraulics*, 7th ed. New York: McGraw-Hill, 1996.
16. Replogle, J. A., A. J. Clemmens, and C. A. Pugh, "Hydraulic Design of Flow Measuring Structures." *Hydraulic Design Handbook*, L. W. Mays, ed. New York: McGraw-Hill, 1999.

Problems

Basic Concepts and Definitions

11.1 Verify the equation given in Table 11.1 for the hydraulic radius of a trapezoidal channel. Plot the ratio R/y for $b = 2$ m with side slope angles of 30° and 60° for 0.5 m $< y <$ 3 m.

11.2 Verify the equation given in Table 11.1 for the hydraulic radius of a circular channel. Evaluate and plot the ratio R/D, for liquid depths between 0 and D.

11.3 A wave from a passing boat in a lake is observed to travel at 16 km/h. Determine the approximate water depth at this location.

11.4 A pebble is dropped into a stream of water that flows in a rectangular channel at 2 m depth. In one second, a ripple caused by the stone is carried 7 m downstream. What is the speed of the flowing water?

11.5 A pebble is dropped into a stream of water of uniform depth. A wave is observed to travel upstream 1.5 m in 1 s, and 3.9 m downstream in the same time. Determine the flow speed and depth.

11.6 Solution of the complete differential equations for wave motion without surface tension shows that wave speed is given by

$$c = \sqrt{\frac{g\lambda}{2\pi} \tanh\left(\frac{2\pi y}{\lambda}\right)}$$

where λ is the wave wavelength and y is the liquid depth. Show that when $\lambda/y \ll 1$, wave speed becomes proportional to $\sqrt{\lambda}$. In the limit as $\lambda/y \to \infty$, $c = \sqrt{gy}$. Determine the value of λ/y for which $c > 0.99\sqrt{gy}$.

11.7 Capillary waves (ripples) are small amplitude and wavelength waves, commonly seen, for example, when an insect or small particle hits the water surface. They are waves generated due to the interaction of the inertia force of the fluid ρ and the fluid surface tension σ. The wave length is

$$\lambda = 2\pi\sqrt{\frac{\sigma}{\rho g}}$$

Find the length of capillary waves in water and mercury.

11.8 Solution of the complete differential equations for wave motion in quiescent liquid, including the effects of surface tension, shows that wave speed is given by

$$c = \sqrt{\left(\frac{g\lambda}{2\pi} + \frac{2\pi\sigma}{\rho\lambda}\right) \tanh\left(\frac{2\pi y}{\lambda}\right)}$$

where λ is the wave wavelength, y is the liquid depth, and σ is the surface tension. Plot wave speed versus wavelength for the range 1 mm $< \lambda <$ 100 mm for (a) water and (b) mercury. Assume $y = 7$ mm for both liquids.

11.9 Surface waves are caused by a sharp object that just touches the free surface of a stream of flowing water, forming the wave pattern shown. The stream depth is 150 mm. Determine the flow speed and Froude number. Note that the wave travels at speed c (Eq. 11.6) normal to the wave front, as shown in the velocity diagram.

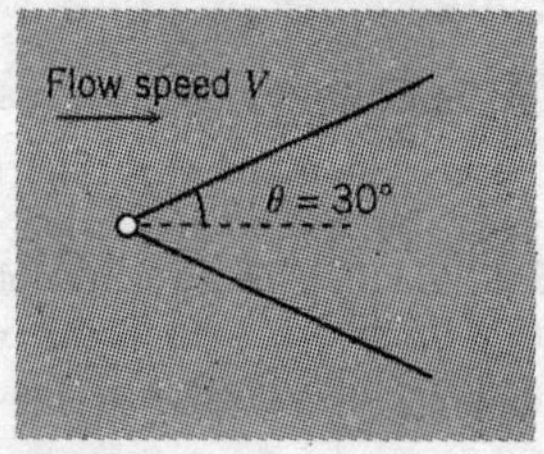

P11.9

11.10 The Froude number characterizes flow with a free surface. Plot on a log-log scale the speed versus depth for 0.1 m/s $< V <$ 3 m/s and 0.001 $< y <$ 1 m; plot the line $Fr = 1$, and indicate regions that correspond to tranquil and rapid flow.

11.11 A submerged body traveling horizontally beneath a liquid surface at a Froude number (based on body length) about 0.5 produces a strong surface wave pattern if submerged less than half its length. (The wave pattern of a surface ship also is pronounced at this Froude number.) On a log-log plot of speed versus body (or ship) length for 1 m/s $< V <$ 30 m/s and 1 m $< x <$ 300 m, plot the line $Fr = 0.5$.

11.12 Water flows in a rectangular channel at a depth of 750 mm. If the flow speed is (a) 1 m/s and (b) 4 m/s, compute the corresponding Froude numbers.

11.13 A long rectangular channel 3 m wide is observed to have a wavy surface at a depth of about 1.8 m. Estimate the rate of discharge.

Energy Equation for Open-Channel Flows

11.14 A partially open sluice gate in a 5 m wide rectangular channel carries water at 10 m^3/s. The upstream depth is 2.5 m. Find the downstream depth and Froude number.

11.15 For a rectangular channel of width $b = 20$ m, construct a family of specific energy curves for $Q = 0$, 25, 75, 125, and 200 m^3/s. What are the minimum specific energies for these curves?

11.16 Find the critical depth for flow at 3 m^3/s in a rectangular channel of width 2.5 m.

11.17 A trapezoidal channel with a bottom width of 6 m, side slopes of 1 to 2, channel bottom slope of 0.0016, and a Manning's n of 0.025 carries a discharge of 11.3 m^3/s. Compute the critical depth and velocity of this channel.

11.18 A rectangular channel carries a discharge of 0.93 m^3/s per meter of width. Determine the minimum specific energy possible for this flow. Compute the corresponding flow depth and speed.

11.19 Flow in the channel of Problem 11.18 ($E_{min} = 0.66$ m) is to be at twice the minimum specific energy. Compute the alternate depths for this E.

11.20 For a channel of nonrectangular cross section, critical depth occurs at minimum specific energy. Obtain a general equation for critical depth in a trapezoidal section in terms of Q, g, b, and α. It will be implicit in y_c!

11.21 Water flows at 8.5 m^3/s in a trapezoidal channel with bottom width of 2.4 m. The sides are sloped at 2:1. Find the critical depth for this channel.

Localized Effects of Area Change (Frictionless Flow)

11.22 Consider the Venturi flume shown. The bed is horizontal, and flow may be considered frictionless. The upstream depth is 0.3 m, and the downstream depth is 0.22 m. The upstream breadth is 0.6 m, and the breadth of the throat is 0.3 m. Estimate the flow rate through the flume.

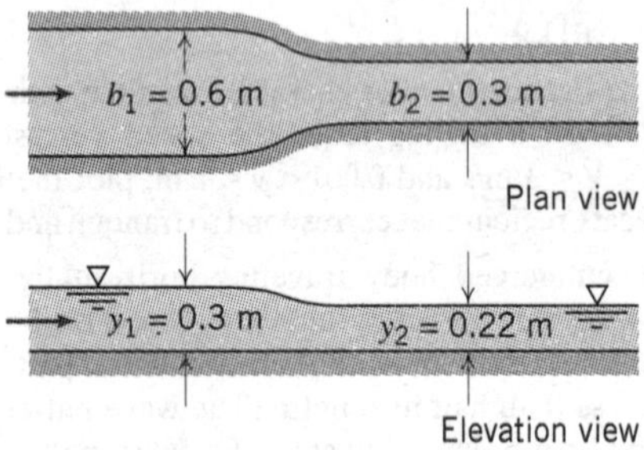

P11.22

11.23 A rectangular channel 3 m wide carries 2.83 m^3/s on a horizontal bed at 0.3 m depth. A smooth bump across the channel rises 10 cm. above the channel bottom. Find the elevation of the liquid free surface above the bump.

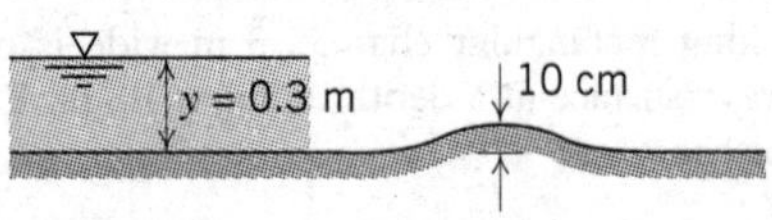

P11.23

11.24 A rectangular channel 3 m wide carries a discharge of 0.57 m^3/s at 0.27 m depth. A smooth bump 0.06 m high is placed on the floor of the channel. Estimate the local change in flow depth caused by the bump.

11.25 At a section of a 3 m-wide rectangular channel, the depth is 0.09 m for a discharge of 0.57 m^3/s. A smooth bump 0.03 m high is placed on the floor of the channel. Determine the local change in flow depth caused by the bump.

11.26 Water, at 0.9 m/s and 0.6 m depth, approaches a smooth rise in a wide channel. Estimate the stream depth after the 0.15 m rise.

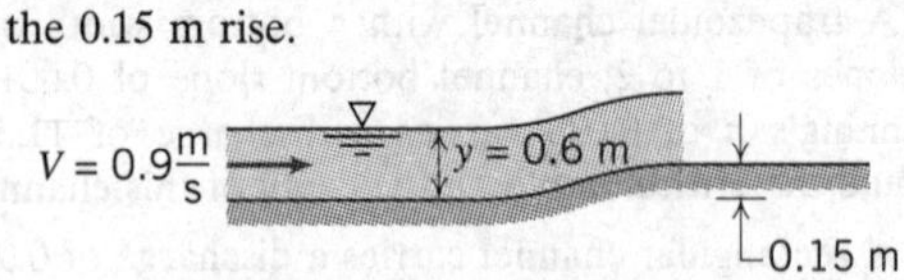

P11.26

11.27 Water issues from a sluice gate at 1.25 m depth. The discharge per unit width is 10 $m^3/s/m$. Estimate the water level far upstream where the flow speed is negligible. Calculate the maximum rate of flow per unit width that could be delivered through the sluice gate.

11.28 A horizontal rectangular channel 0.9 m wide contains a sluice gate. Upstream of the gate the depth is 1.8 m; the depth downstream is 0.27 m. Estimate the volume flow rate in the channel.

11.29 Flow through a sluice gate is shown. Estimate the water depth and velocity after the gate (well before the hydraulic jump).

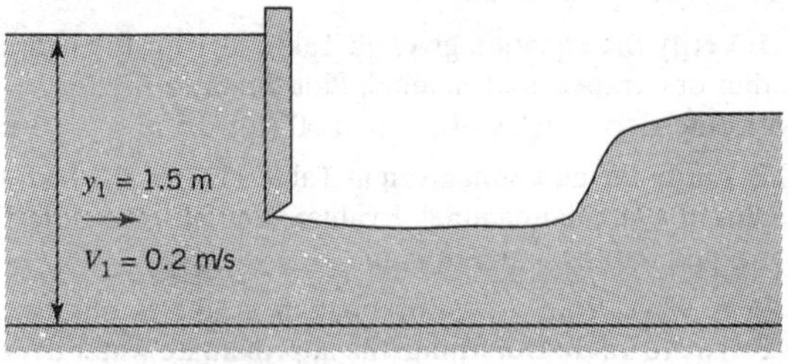

P11.29, P11.39

11.30 Rework Example 11.4 for a 350-mm-high bump and a side wall constriction that reduces the channel width to 1.5 m.

The Hydraulic Jump

11.31 Find the rate at which energy is being consumed (kW) by the hydraulic jump of Example 11.5. Is this sufficient to produce a significant temperature rise in the water?

11.32 A hydraulic jump occurs in a rectangular channel 4.0 m wide. The water depth before the jump is 0.4 m and 1.7 m after the jump. Compute the flow rate in the channel, the critical depth, and the head loss in the jump.

11.33 A wide channel carries 10 m^3/s per foot of width at a depth of 1 m at the toe of a hydraulic jump. Determine the depth of the jump and the head loss across it.

11.34 A hydraulic jump occurs in a wide horizontal channel. The discharge is 2 m^3/s per meter of width. The upstream depth is 500 mm. Determine the depth of the jump.

11.35 A hydraulic jump occurs in a rectangular channel. The flow rate is 6.5 m^3/s, and the depth before the jump is 0.4 m. Determine the depth behind the jump and the head loss, if the channel is 1 m wide.

11.36 The hydraulic jump may be used as a crude flow meter. Suppose that in a horizontal rectangular channel 1.5 m wide the observed depths before and after a hydraulic jump are 0.2 and 0.9 m. Find the rate of flow and the head loss.

11.37 A hydraulic jump occurs on a horizontal apron downstream from a wide spillway, at a location where depth is 0.9 m and speed is 25 m/s. Estimate the depth and speed downstream from the jump. Compare the specific energy downstream of the jump to that upstream.

11.38 A hydraulic jump occurs in a rectangular channel. The flow rate is 50 m^3/s, and the depth before the jump is 2 m. Determine the depth after the jump and the head loss, if the channel is 1 m wide.

11.39 Estimate the depth of water before and after the jump for the hydraulic jump downstream of the sluice gate of Fig. P11.29.

11.40 A positive surge wave, or moving hydraulic jump, can be produced in the laboratory by suddenly opening a sluice gate. Consider a surge of depth y_2 advancing into a quiescent channel of depth y_1. Obtain an expression for surge speed in terms of y_1 and y_2

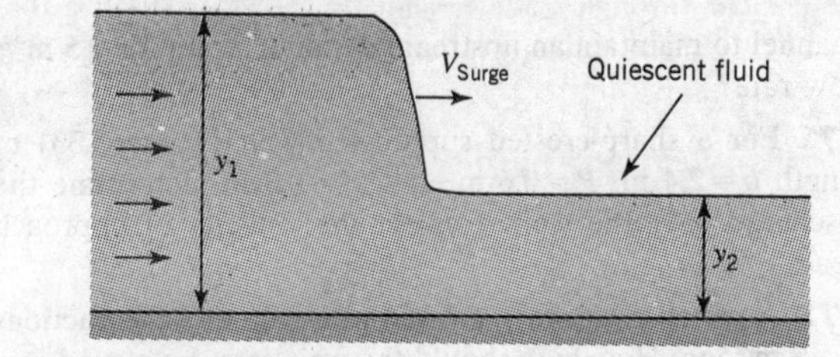

P11.40

11.41 A tidal bore (an abrupt translating wave or moving hydraulic jump) often forms when the tide flows into the wide estuary of a river. In one case, a bore is observed to have a height of 3.6 m above the undisturbed level of the river that is 2.4 m deep. The bore travels upstream at $V_{bore} = 28.97$ km/h. Determine the approximate speed V_r of the current of the undisturbed river.

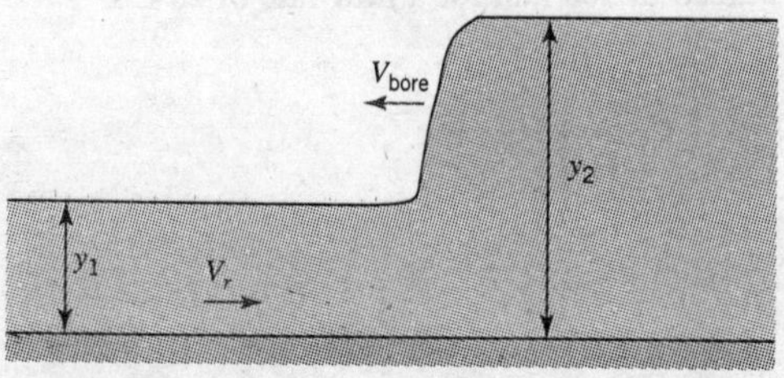

P11.41

Uniform Flow

11.42 A 2-m-wide rectangular channel with a bed slope of 0.0005 has a depth of flow of 1.5 m. Manning's roughness coefficient is 0.015. Determine the steady uniform discharge in the channel.

11.43 Determine the uniform flow depth in a rectangular channel 2.5 m wide with a discharge of 3 m^3/s. The slope is 0.0004 and Manning's roughness factor is 0.015.

11.44 Determine the uniform flow depth in a trapezoidal channel with a bottom width of 2.4 m and side slopes of 1 vertical to 2 horizontal. The discharge is 2.8 m^3/s. Manning's roughness factor is 0.015 and the channel bottom slope is 0.0004.

11.45 Determine the uniform flow depth in a trapezoidal channel with a bottom width of 2.5 m and side slopes of 1 vertical to 2 horizontal with a discharge of 3 m^3/s. The slope is 0.0004 and Manning's roughness factor is 0.015.

11.46 A rectangular flume built of concrete, with 1 m per 1000 m slope, is 1.8 m wide. Water flows at a normal depth of 0.9 m. Compute the discharge.

11.47 A rectangular flume built of timber is 0.9 m wide. The flume is to handle a flow of 2.55 m^3/s at a normal depth of 1.8 m. Determine the slope required.

11.48 A channel with square cross section is to carry 20 m^3/s of water at normal depth on a slope of 0.003. Compare the dimensions of the channel required for (a) concrete and (b) soil cement.

11.49 Water flows in a trapezoidal channel at a normal depth of 1.2 m. The bottom width is 2.4 m and the sides slope at 1:1 (45°). The flow rate is 7.1 m^3/s. The channel is excavated from bare soil. Find the bed slope.

11.50 A triangular channel with side angles of 45° is to carry 10 m^3/s at a slope of 0.001. The channel is concrete. Find the required dimensions.

11.51 A semicircular trough of corrugated steel, with diameter $D = 1$ m, carries water at depth $y = 0.25$ m. The slope is 0.01. Find the discharge.

11.52 Find the discharge at which the channel of Problem 11.51 flows full.

11.53 The flume of Problem 11.46 is fitted with a new plastic film liner ($n = 0.010$). Find the new depth of flow if the discharge remains constant at 2.42 m^3/s.

11.54 Discharge through the channel of Problem 11.49 is increased to 10 m^3/s. Find the corresponding normal depth if the bed slope is 0.00193.

11.55 The channel of Problem 11.49 has 0.00193 bed slope. Find the normal depth for the given discharge after a new plastic liner ($n = 0.010$) is installed.

11.56 Consider again the semicircular channel of Problem 11.51. Find the normal depth that corresponds to a discharge of 0.5 m^3/s.

11.57 Consider a symmetric open channel of triangular cross section. Show that for a given flow area, the wetted perimeter is minimized when the sides meet at a right angle.

11.58 Compute the normal depth and velocity of the channel of Problem 11.17.

11.59 Determine the cross section of the greatest hydraulic efficiency for a trapezoidal channel with side slope of 1 vertical to 2 horizontal if the design discharge is 250 m^3/s. The channel slope is 0.001 and Manning's roughness factor is 0.020.

11.60 For a trapezoidal shaped channel ($n = 0.014$ and slope $S_b = 0.0002$ with a 6 m bottom width and side slopes of 1 vertical to 1.5 horizontal), determine the normal depth for a discharge of 28.3 m^3/s.

11.61 Show that the best hydraulic trapezoidal section is one-half of a hexagon.

11.62 Compute the critical depth for the channel in Problem 11.42.

11.63 Consider a 2.45-m-wide rectangular channel with a bed slope of 0.0004 and a Manning's roughness factor of 0.015. A weir is placed in the channel, and the depth upstream of the weir is 1.52 m for a discharge of 5.66 m^3/s. Determine whether a hydraulic jump forms upstream of the weir.

11.64 An above-ground rectangular flume is to be constructed of timber. For a drop of 1.9 m/km, what will be the depth and width for the most economical flume if it is to discharge 1.1 m^3/s?

11.65 Consider flow in a rectangular channel. Show that, for flow at critical depth and optimum aspect ratio ($b = 2y$), the volume flow rate and bed slope are given by the expressions:

$$Q = 62.6 y_c^{5/2} \quad \text{and} \quad S_c = 24.7 \frac{n^2}{y_c^{1/3}}$$

11.66 A trapezoidal canal lined with brick has side slopes of 2:1 and bottom width of 3 m. It carries 17 m^3/s at critical speed. Determine the critical slope (the slope at which the depth is critical).

11.67 A wide flat unfinished concrete channel discharges water at 1.9 m^3/s per meter of width. Find the critical slope (the slope at which depth is critical).

11.68 An optimum rectangular storm sewer channel made of unfinished concrete is to be designed to carry a maximum flow rate of 2.83 m^3/s, at which the flow is at critical condition. Determine the channel width and slope.

Discharge Measurement

11.69 The crest of a broad-crested weir is 0.3 m below the level of an upstream reservoir, where the water depth is 2.4 m. For $C_w \approx 3.4$, what is the maximum flow rate per unit width that could pass over the weir?

11.70 A rectangular, sharp-crested weir with end contraction is 1.6 m long. How high should it be placed in a channel to maintain an upstream depth of 2.5 m for 0.5 m^3/s flow rate?

11.71 For a sharp-crested suppressed weir ($C_w \approx 3.33$) of length $b = 2.4$ m, $P = 0.6$ m, and $H = 0.3$ m, determine the discharge over the weir. Neglect the velocity of approach head.

11.72 A rectangular sharp-crested weir with end contractions is 1.5 m long. How high should the weir crest be placed in a channel to maintain an upstream depth of 2.5 m for 0.5 m^3/s flow rate?

11.73 Determine the head on a 60° V-notch weir for a discharge of 150 L/s. Take $C_d \approx 0.58$.

11.74 The head on a 90° V-notch weir is 0.45 m. Determine the discharge.

11.75 Determine the weir coefficient of a 90° V-notch weir for a head of 180 mm for a flow rate of 20 L/s.

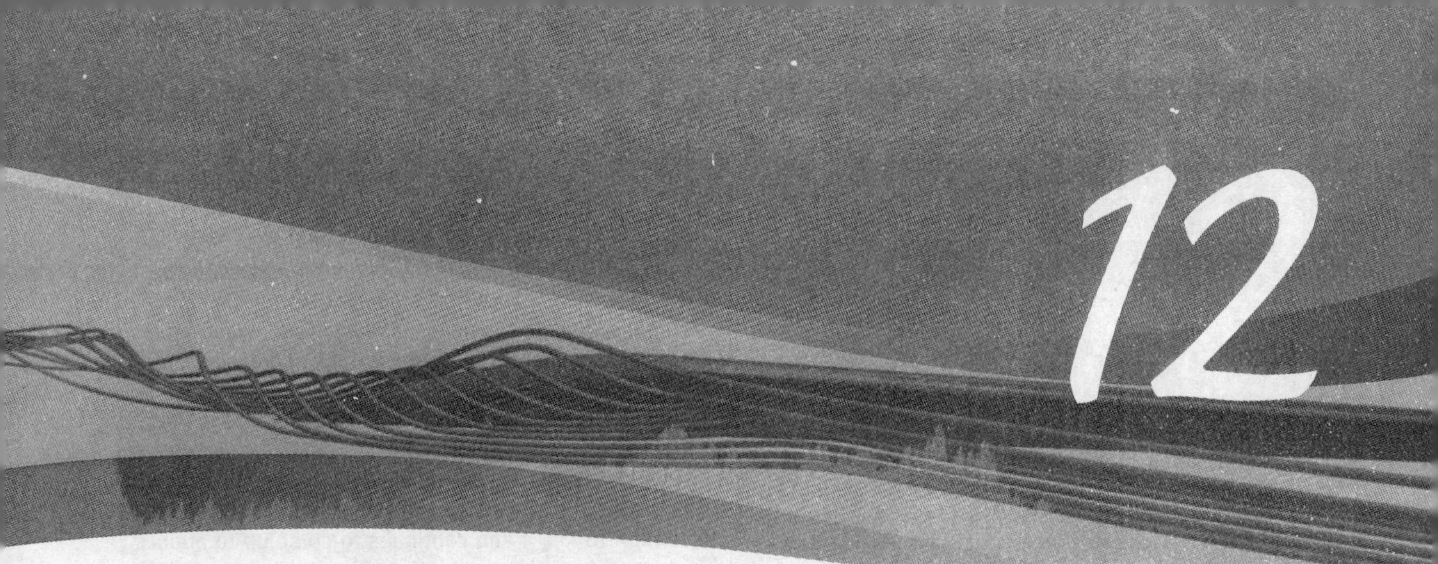

12 Introduction to Compressible Flow

Case Study in Energy and the Environment

Wind Power: *The Helix* Vertical Axis Wind Turbine

Most of the devices we have looked at in previous *Case Studies in Energy and the Environment* have been concerned with large-scale power production. However, a lot of work is being done on residential-scale devices. Scott Weinbrandt is the CEO of a company called *Helix Wind*. His background is in computer technology, and he has lived through the computer industry, moving from large central mainframes to distributed personal computers. Scott says he's seeing the same trend emerging in wind energy. His company is targeting urban residential (and commercial) customers with the company's range of small-scale helical-shaped turbines; one of the models is shown in the photograph. As can be seen, they are quite beautiful machines—excellent examples of how engineering at its best can create attractive, as well as functional, machines. *Helix Wind* is finding that some customers are even buying them just for the product's aesthetic value! The turbines are an elegant form of a Savonius turbine and so are generally considered less efficient at generating electricity than the common horizontal-axis propeller-driven turbines. (Figure P9.97 in the problem set for Chapter 9 shows a crude version of a Savonius turbine; such turbines are drag-based, as opposed to the lift-based propeller turbines.) An advantage of the helix design is that the helix shape generates a

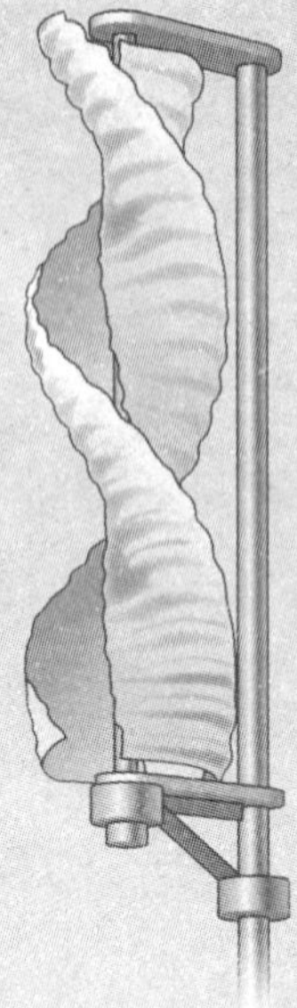

Helix S322 Turbine.

secondary upward flow that improves the overall aerodynamics, hence improving efficiency.

Helix Wind says a key benefit of the helical design is its ability to operate even in lower wind speeds, although high wind speeds will not present any difficulties. Recently an S322 model in California continued to operate in wind speeds in excess of 96 Km/h, whereas a nearby conventional wind farm had many damaged turbine blades.

The *Helix* turbines, being vertical-axis wind turbines (VAWTs), have a number of advantages: They are easy to maintain because most of the moving parts are located near the ground; no yaw device is needed to orient the device into the wind; they usually have a lower tip-speed ratio and so are less likely to break in high winds (see Fig. 10.50); they don't have to shut down in high wind speeds. On the other hand, a possible disadvantage of VAWTs is that they tend to be a little less efficient than horizontal-axis wind turbines. (See Fig. 10.50 for a comparison of Savonius turbine and conventional turbine efficiencies.) The *Helix* design has resulted in greatly increased efficiency, overcoming this disadvantage, and *Helix Wind* turbines are now only 6–7 percent less efficient than the most efficient HAWTs.

The low-speed 2 kW *Helix Wind* turbine (which is only about 2.7 m × 1.2 m) starts generating power at less than 16 km/h and continues to do so up to about 65 km/h; it continues to spin with no damage to the system in winds as high as 130 km/h, although no additional electricity is generated above the maximum output at 65 km/h. *Helix Wind* turbines are safe for wildlife because they spin at much lower speeds than horizontal turbines and appear as a solid mass rather than a blurry blade that a bird cannot see. At about 5 dB, the turbines are nearly silent because they operate with tip speeds close to the wind velocity (similar to the wind blowing around any stationary object such as a tree or house). Conventional wind turbines spin at up to 10 times the wind speed, which causes the whistling sound that can be heard around them. The turbine spins no matter what direction the wind comes from (including vertically if it's mounted on the side of a large building) and generates power even in turbulence-prone urban environments.

Helix Wind is confident that there will be a huge market for these residential and small commercial-scale turbines, including at such existing locations as cell phone towers, cruise ships, billboards, oil and gas pumping systems, and agricultural water pumping. The company says its wind turbines also are ideal for use in developing countries, for example, for pumping potable water and providing electricity.

In Chapter 2 we briefly discussed the two most important questions we must ask before analyzing a fluid flow: whether or not the flow is viscous, and whether or not the flow is compressible. We subsequently considered *incompressible, inviscid* flows (Chapter 6) and *incompressible, viscous* flows (Chapters 8 and 9). We are now ready to study flows that experience compressibility effects. Because this is an introductory text, our focus will be mainly on *one-dimensional compressible*, *inviscid* flows, although we will also review some important *compressible, viscous* flow phenomena. After our consideration of one-dimensional flows, we will introduce some basic concepts of *two-dimensional steady compressible flows*.

We first need to establish what we mean by a "compressible" flow. This is a flow in which there are significant or noticeable changes in fluid density. Just as inviscid fluids do not actually exist, so incompressible fluids do not actually exist. For example, in this text we have treated water as an incompressible fluid, although in fact the density of seawater increases by 1 percent for each mile or so of depth. Hence, whether or not a given flow can be treated as incompressible is a judgment call: Liquid flows will almost always be considered incompressible (exceptions include phenomena such as the "water hammer" effect in pipes), but gas flows could easily be either incompressible or compressible. We will learn in this chapter (in Example 12.5) that for Mach numbers M less than about 0.3, the change in gas density due to the flow will be less than 3 percent; this change is small enough in most engineering applications for the following rule: *A gas flow can be considered incompressible when* $M < 0.3$.

The consequences of compressibility are not limited simply to density changes. Density changes mean that we can have significant compression or expansion work on a gas, so the thermodynamic state of the fluid will change, meaning that in general *all* properties—temperature, internal energy, entropy, and so on—can change. In particular, density changes create a mechanism (just as viscosity did) for exchange of energy between "mechanical" energies (kinetic, potential, and "pressure") and the thermal internal energy. For this reason, we begin with a review of the thermodynamics needed to study compressible flow.

Review of Thermodynamics 12.1

The pressure, density, and temperature of a substance may be related by an equation of state. Although many substances are complex in behavior, experience shows that most gases of engineering interest, at moderate pressure and temperature, are well represented by the ideal gas equation of state,

$$p = \rho RT \tag{12.1}$$

where R is a unique constant for each gas;[1] R is given by

$$R = \frac{R_u}{M_m}$$

where R_u is the universal gas constant, $R_u = 8314$ N·m/(kgmole·K) and M_m is the molecular mass of the gas. Although the ideal gas equation is derived using a model that has the unrealistic assumptions that the gas molecules (a) take up zero volume (i.e., they are point masses) and (b) do not interact with one another, many real gases conform to Eq. 12.1, especially if the pressure is "low" enough and/or temperature "high" enough (see, e.g., [1–3]). For example, at room temperature, as long as the pressure is less than about 30 atm, Eq. 12.1 models the air density to better than 1 percent accuracy; similarly, Eq. 12.1 is accurate for air at 1 atm for temperatures that are greater than about −130°C (140 K).

The ideal gas has other features that are useful. In general, the *internal energy* of a simple substance may be expressed as a function of any two independent properties, e.g., $u = u(v, T)$, where $v \equiv 1/\rho$ is the *specific volume*. Then

$$du = \left(\frac{\partial u}{\partial T}\right)_v dT + \left(\frac{\partial u}{\partial v}\right)_T dv$$

The *specific heat at constant volume* is defined as $c_v \equiv (\partial u/\partial T)_v$, so that

$$du = c_v\, dT + \left(\frac{\partial u}{\partial v}\right)_T dv$$

[1]For air, $R = 287$ N·m/(kg·K)

In particular, for an ideal gas it can be shown (see, e.g., Chapter 11 of [1]) that the internal energy, u, is a function of temperature only, so $(\partial u/\partial v)_T = 0$, and

$$du = c_v\, dT \tag{12.2}$$

for an ideal gas. This means that internal energy and temperature changes may be related if c_v is known. Furthermore, since $u = u(T)$, then from Eq. 12.2, $c_v = c_v\,(T)$.

The *enthalpy* of any substance is defined as $h \equiv u + p/\rho$. For an ideal gas, $p = \rho RT$, and so $h = u + RT$. Since $u = u(T)$ for an ideal gas, h also must be a function of temperature alone.

We can obtain a relation between h and T by recalling once again that for a simple substance any property can be expressed as a function of any two other independent properties [1], e.g., $h = h(v, T)$ as we did for u, or $h = h(p, T)$. We choose the latter in order to develop a useful relation,

$$dh = \left(\frac{\partial h}{\partial T}\right)_p dT + \left(\frac{\partial h}{\partial p}\right)_T dp$$

Since the *specific heat at constant pressure* is defined as $c_p \equiv (\partial h/\partial T)_p$,

$$dh = c_p\, dT + \left(\frac{\partial h}{\partial p}\right)_T dp$$

We have shown that for an ideal gas h is a function of T only. Consequently, $(\partial h/\partial p)_T = 0$ and

$$dh = c_p\, dT \tag{12.3}$$

Since h is a function of T alone, Eq. 12.3 requires that c_p be a function of T only for an ideal gas.

Although specific heats for an ideal gas are functions of temperature, their difference is a constant for each gas. To see this, from

$$h = u + RT$$

we can write

$$dh = du + RdT$$

Combining this with Eq. 12.2 and Eq. 12.3, we can write

$$dh = c_p\, dT = du + RdT = c_v\, dT + R\, dT$$

Then

$$c_p - c_v = R \tag{12.4}$$

It may seem a bit odd that we have functions of temperature on the left of Eq. 12.4 but a constant on the right; it turns out that the specific heats of an ideal gas change with temperature at the same rate, so their *difference* is constant.

The *ratio of specific heats* is defined as

$$k \equiv \frac{c_p}{c_v} \tag{12.5}$$

Using the definition of k, we can solve Eq. 12.4 for either c_p or c_v in terms of k and R. Thus,

$$c_p = \frac{kR}{k-1} \tag{12.6a}$$

and

$$c_v = \frac{R}{k-1} \tag{12.6b}$$

Although the specific heats of an ideal gas may vary with temperature, for moderate temperature ranges they vary only slightly, and can be treated as constant, so

$$u_2 - u_1 = \int_{u_1}^{u_2} du = \int_{T_1}^{T_2} c_v \, dT = c_v(T_2 - T_1) \tag{12.7a}$$

$$h_2 - h_1 = \int_{h_1}^{h_2} dh = \int_{T_1}^{T_2} c_p \, dT = c_p(T_2 - T_1) \tag{12.7b}$$

Data for M_m, c_p, c_v, R, and k for common gases are given in Table A.6 of Appendix A.

We will find the property *entropy* to be extremely useful in analyzing compressible flows. State diagrams, particularly the temperature-entropy (Ts) diagram, are valuable aids in the physical interpretation of analytical results. Since we shall make extensive use of Ts diagrams in solving compressible flow problems, let us review briefly some useful relationships involving the property entropy [1–3].

Entropy is defined by the equation

$$\Delta S \equiv \int_{\text{rev}} \frac{\delta Q}{T} \quad \text{or} \quad dS = \left(\frac{\delta Q}{T}\right)_{\text{rev}} \tag{12.8}$$

where the subscript signifies *reversible*.

The inequality of Clausius, deduced from the second law, states that

$$\oint \frac{\delta Q}{T} \leq 0$$

As a consequence of the second law, we can write

$$dS \geq \frac{\delta Q}{T} \quad \text{or} \quad T\,dS \geq \delta Q \tag{12.9a}$$

For *reversible* processes, the equality holds, and

$$T\,ds = \frac{\delta Q}{m} \quad \text{(reversible process)} \tag{12.9b}$$

The inequality holds for *irreversible* processes, and

$$T\,ds > \frac{\delta Q}{m} \quad \text{(irreversible process)} \tag{12.9c}$$

For an *adiabatic* process, $\delta Q/m = 0$. Thus

$$ds = 0 \quad \text{(reversible adiabatic process)} \tag{12.9d}$$

and

$$ds > 0 \quad \text{(irreversible adiabatic process)} \tag{12.9e}$$

Thus a process that is *reversible and adiabatic* is also *isentropic*; the entropy remains constant during the process. Inequality 12.9e shows that entropy must *increase* for an adiabatic process that is irreversible.

Equations 12.9 show that any two of the restrictions—reversible, adiabatic, or isentropic—imply the third. For example, a process that is isentropic and reversible must also be adiabatic.

A useful relationship among properties (p, v, T, s, u) can be obtained by considering the first and second laws together. The result is the Gibbs, or $T\,ds$, equation

$$T\,ds = du + p\,dv \tag{12.10a}$$

This is a differential relationship among properties, valid for any process between any two equilibrium states. Although it is derived from the first and second laws, it is, in itself, a statement of neither.

An alternative form of Eq. 12.10a can be obtained by substituting

$$du = d(h - pv) = dh - p\,dv - v\,dp$$

to obtain

$$T\,ds = dh - v\,dp \tag{12.10b}$$

For an ideal gas, entropy change can be evaluated from the $T\,ds$ equations as

$$ds = \frac{du}{T} + \frac{p}{T}dv = c_v\frac{dT}{T} + R\frac{dv}{v}$$

$$ds = \frac{dh}{T} - \frac{v}{T}dp = c_p\frac{dT}{T} - R\frac{dp}{p}$$

For constant specific heats, these equations can be integrated to yield

$$s_2 - s_1 = c_v \ln\frac{T_2}{T_1} + R\ln\frac{v_2}{v_1} \tag{12.11a}$$

$$s_2 - s_1 = c_p \ln\frac{T_2}{T_1} - R\ln\frac{p_2}{p_1} \tag{12.11b}$$

and also

$$s_2 - s_1 = c_v \ln\frac{p_2}{p_1} + c_p\ln\frac{v_2}{v_1} \tag{12.11c}$$

(Equation 12.11c can be obtained from either Eq. 12.11a or 12.11b using Eq. 12.4 and the ideal gas equation, Eq. 12.1, written in the form $pv = RT$, to eliminate T.)

Example 12.1 shows use of the above governing equations (the $T\,ds$ equations) to evaluate property changes during a process.

For an ideal gas with constant specific heats, we can use Eqs. 12.11 to obtain relations valid for an isentropic process. From Eq. 12.11a

$$s_2 - s_1 = 0 = c_v \ln\frac{T_2}{T_1} + R\ln\frac{v_2}{v_1}$$

Then, using Eqs. 12.4 and 12.5,

$$\left(\frac{T_2}{T_1}\right)\left(\frac{v_2}{v_1}\right)^{R/c_v} = 0 \quad \text{or} \quad T_2v_2^{k-1} = T_1v_1^{k-1} = Tv^{k-1} = \text{constant}$$

where states 1 and 2 are arbitrary states of the isentropic process. Using $v = 1/\rho$,

$$Tv^{k-1} = \frac{T}{\rho^{k-1}} = \text{constant} \tag{12.12a}$$

We can apply a similar process to Eqs. 12.11b and 12.11c, respectively, and obtain the following useful relations:

$$Tp^{1-k/k} = \text{constant} \tag{12.12b}$$

$$pv^k = \frac{p}{\rho^k} = \text{constant} \qquad (12.12c)$$

Equations 12.12 are for an ideal gas undergoing an isentropic process.

Qualitative information that is useful in drawing state diagrams also can be obtained from the $T\,ds$ equations. To complete our review of the thermodynamic fundamentals, we evaluate the slopes of lines of constant pressure and of constant volume on the Ts diagram in Example 12.2.

Example 12.1 PROPERTY CHANGES IN COMPRESSIBLE DUCT FLOW

Air flows through a long duct of constant area at 0.15 kg/s. A short section of the duct is cooled by liquid nitrogen that surrounds the duct. The rate of heat loss in this section is 15.0 kJ/s from the air. The absolute pressure, temperature, and velocity entering the cooled section are 188 kPa, 440 K, and 210 m/s, respectively. At the outlet, the absolute pressure and temperature are 213 kPa and 351 K. Compute the duct cross-sectional area and the changes in enthalpy, internal energy, and entropy for this flow.

Given: Air flows steadily through a short section of constant-area duct that is cooled by liquid nitrogen.

$T_1 = 440$ K
$p_1 = 188$ kPa (abs)
$V_1 = 210$ m/s

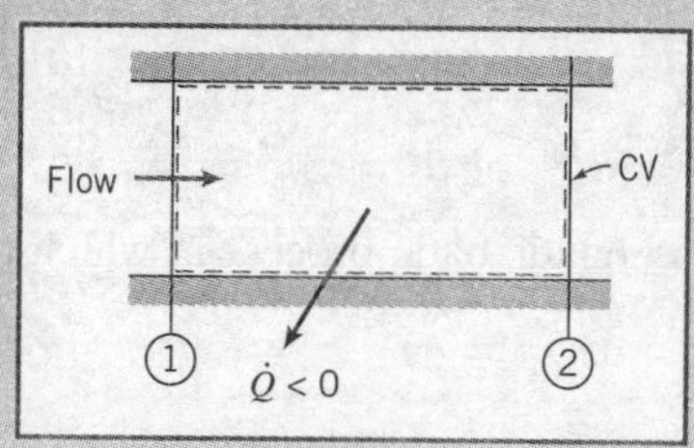

$T_2 = 351$ K
$p_2 = 213$ kPa (abs)

Find: (a) Duct area. (b) Δh. (c) Δu. (d) Δs.

Solution: The duct area may be found from the continuity equation.

Governing equation:

$$\overbrace{\frac{\partial}{\partial t}\int_{CV} \rho\, dV\!\!\!-}^{=0(1)} + \int_{CV} \rho \vec{V}\cdot d\vec{A} = 0 \qquad (4.12)$$

Assumptions: (1) Steady flow.
(2) Uniform flow at each section.
(3) Ideal gas with constant specific heats.

Then

$$(-\rho_1 V_1 A_1) + (\rho_2 V_2 A_2) = 0$$

or

$$\dot{m} = \rho_1 V_1 A = \rho_2 V_2 A$$

since $A = A_1 = A_2 =$ constant. Using the ideal gas relation, $p = \rho RT$, we find

$$\rho_1 = \frac{p_1}{RT_1} = 1.88 \times 10^5 \frac{\text{N}}{\text{m}^2} \times \frac{\text{kg}\cdot\text{K}}{287\,\text{N}\cdot\text{m}} \times \frac{1}{440\,\text{K}} = 1.49\ \text{kg/m}^3$$

From continuity,

$$A = \frac{\dot{m}}{\rho_1 V_1} = 0.15\frac{\text{kg}}{\text{s}} \times \frac{\text{m}^3}{1.49\,\text{kg}} \times \frac{\text{s}}{210\,\text{m}} = 4.79 \times 10^{-4}\,\text{m}^2 \qquad A$$

For an ideal gas, the change in enthalpy is

$$\Delta h = h_2 - h_1 = \int_{T_1}^{T_2} c_p\, dT = c_p(T_2 - T_1) \tag{12.7b}$$

$$\Delta h = 1.00 \frac{\text{kJ}}{\text{kg}\cdot\text{K}} \times (351 - 440)\text{K} = -89.0 \text{ kJ/kg}$$ ← Δh

Also, the change in internal energy is

$$\Delta u = u_2 - u_1 = \int_{T_1}^{T_2} c_v\, dT = c_v(T_2 - T_1) \tag{12.7a}$$

$$\Delta u = 0.717 \frac{\text{kJ}}{\text{kg}\cdot\text{K}} \times (351 - 440)\text{ K} = -63.8 \text{ kJ/kg}$$ ← Δu

The entropy change may be obtained from Eq. 12.11b,

$$\Delta s = s_2 - s_1 = c_p \ln \frac{T_2}{T_1} - R \ln \frac{p_2}{p_1}$$

$$= 1.00 \frac{\text{kJ}}{\text{kg}\cdot\text{K}} \times \ln\left(\frac{351}{440}\right) - 0.287 \frac{\text{kJ}}{\text{kg}\cdot\text{K}} \times \ln\left(\frac{2.13 \times 10^5}{1.88 \times 10^5}\right)$$

$$\Delta s = -0.262 \text{ kJ/(kg}\cdot\text{K)}$$ ← Δs

This Example illustrates the use of the governing equations for computing property changes of an ideal gas during a process.

We see that entropy may decrease for a nonadiabatic process in which the gas is cooled.

Example 12.2 CONSTANT-PROPERTY LINES ON A *Ts* DIAGRAM

For an ideal gas, find the equations for lines of (a) constant volume and (b) constant pressure in the *Ts* plane.

Find: Equations for lines of (a) constant volume and (b) constant pressure in the *Ts* plane for an ideal gas.

Solution:

(a) We are interested in the relation between T and s with the volume v held constant. This suggests use of Eq. 12.11a,

$$s_2 - s_1 = c_v \ln \frac{T_2}{T_1} + R \ln \overbrace{\frac{v_2}{v_1}}^{=0} \tag{12.8}$$

We relabel this equation so that state 1 is now reference state 0, and state 2 is an arbitrary state,

$$s - s_0 = c_v \ln \frac{T}{T_0} \quad \text{or} \quad T = T_0 e^{(s - s_0)/c_v} \tag{1}$$

Hence, we conclude that constant volume lines in the *Ts* plane are exponential.

(b) We are interested in the relation between T and s with the pressure p held constant. This suggests use of Eq. 12.11b, and using a similar approach to case (a), we find

$$T = T_0 e^{(s - s_0)/c_p} \tag{2}$$

Hence, we conclude that constant pressure lines in the Ts plane are also exponential.

What about the slope of these curves? Because $c_p > c_v$ for all gases, we can see that the exponential, and therefore the slope, of the constant pressure curve, Eq. 2, is smaller than that for the constant volume curve, Eq. 1.

This is shown in the sketch below:

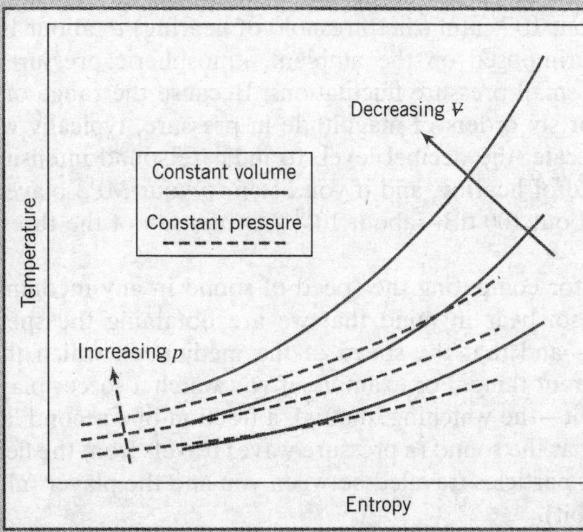

This Example illustrates use of governing equations to explore relations among properties.

Propagation of Sound Waves 12.2

Speed of Sound

A beginner to compressible flow studies might wonder what on earth sound has to do with the speeds present in a flow. We will see in this chapter and the next that the speed of sound, c, is an important marker in fluid mechanics: Flows with speeds less than the speed of sound are called *subsonic*; flows with speeds greater than the speed of sound are called *supersonic*; and we will learn that the behaviors of subsonic and supersonic flows are completely different. We have previously (in Chapter 2, and in Chapter 7 with Eq. 7.16) defined the Mach number M of a flow, and it is so important for our studies that we redefine it here,

VIDEO

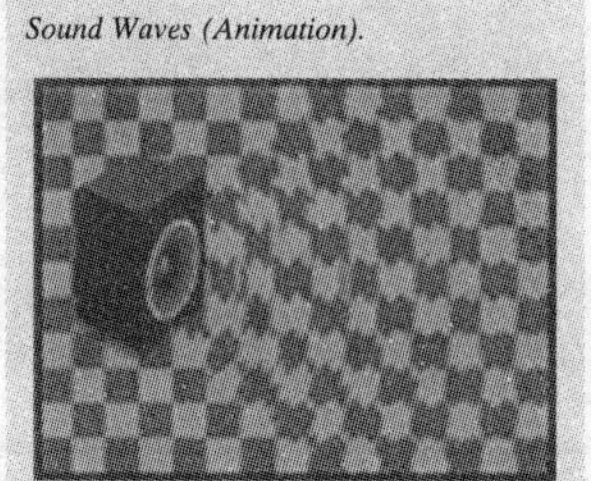

Sound Waves (Animation).

$$M \equiv \frac{V}{c} \tag{12.13}$$

where V is the speed (of the fluid, or in some cases of the aircraft), so that $M < 1$ and $M > 1$ correspond to subsonic and supersonic flow, respectively. In addition, we mentioned in Section 12.1 that we'll demonstrate in Example 12.5 that for $M < 0.3$, we can generally assume incompressible flow. Hence, knowledge of the Mach number value is important in fluid mechanics.

An answer to the question posed at the beginning of this section is that the speed of sound is important in fluid mechanics because this is the speed at which "signals" can travel through the medium. Consider, for example, an object such as an aircraft in motion—the air ultimately has to move out of its way. In Newton's day, it was thought that this happened when the (invisible) air particles literally bounced off the front of the object, like so many balls bouncing off a wall; now we know that in most instances the air starts moving out of the way well before encountering the object (this will *not* be true when we have supersonic flow!). How does the air "know" to move out of the way? It knows because as the object moves, it generates disturbances (infinitesimal pressure waves—sound waves) that emanate from the object in all directions. It is these waves that cumulatively "signal" the air and redirect it around the body as it approaches. These waves travel out at the speed of sound.

[Sound is a pressure wave of very low pressure magnitude, for human hearing typically in the range of about 10^{-9} atm (the threshold of hearing) to about 10^{-3} atm (you will feel pain!). Superimposed on the ambient atmospheric pressure, sound waves consist of extremely small pressure fluctuations. Because the range of human hearing covers about five or six orders of magnitude in pressure, typically we use a dimensionless logarithmic scale, the decibel level, to indicate sound intensity; 0 dB corresponds to the threshold of hearing, and if you listen to your MP3 player at full blast the sound will be at about 100 dB—about 10^{10} the intensity of the threshold of hearing!]

Let us derive a method for computing the speed of sound in any medium (solid, liquid, or gas). As we do so, bear in mind that we are obtaining the speed of a "signal"—a pressure wave—and that the speed of the medium in which the wave travels is a completely different thing. For example, if you watch a soccer player kick the ball (at the speed of light—the watching, that is), a fraction of a second later you will hear the thud of contact as the sound (a pressure wave) travels from the field up to you in the stands, but no air particles traveled between you and the player (all the air particles simply vibrated a bit).

Consider propagation of a sound wave of infinitesimal strength into an undisturbed medium, as shown in Fig. 12.1*a*. We are interested in relating the speed of wave propagation, c, to fluid property changes across the wave. If pressure and density in the undisturbed medium ahead of the wave are denoted by p and ρ, passage of the wave will cause them to undergo infinitesimal changes to become $p + dp$ and $\rho + d\rho$. Since the wave propagates into a stationary fluid, the velocity ahead of the wave, V_x, is zero. The magnitude of the velocity behind the wave, $V_x + dV_x$, then will be simply dV_x; in Fig. 12.1*a*, the direction of the motion behind the wave has been assumed to the left.[2]

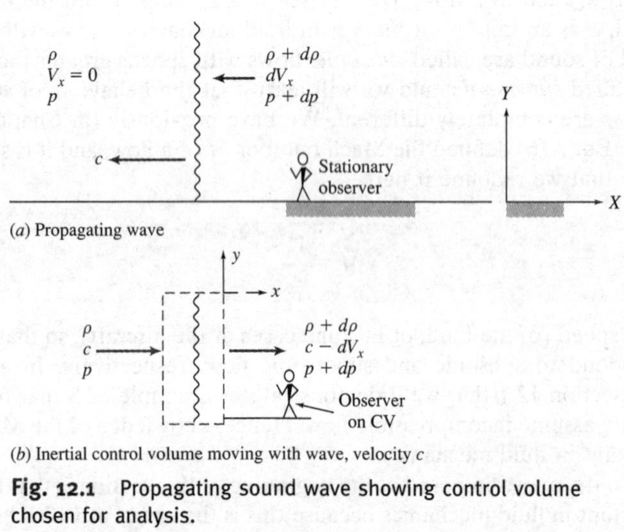

Fig. 12.1 Propagating sound wave showing control volume chosen for analysis.

[2]The same final result is obtained regardless of the direction initially assumed for motion behind the wave (see Problem 12.39).

The flow of Fig. 12.1*a* appears unsteady to a stationary observer, viewing the wave motion from a fixed point on the ground. However, the flow appears steady to an observer located *on* an inertial control volume moving with a segment of the wave, as shown in Fig. 12.1*b*. The velocity approaching the control volume is then c, and the velocity leaving is $c - dV_x$.

The basic equations may be applied to the differential control volume shown in Fig. 12.1*b* (we use V_x for the x component of velocity to avoid confusion with internal energy, u).

a. Continuity Equation

Governing equation:
$$\overbrace{\frac{\partial}{\partial t}\int_{CV} \rho \, d\forall}^{=0(1)} + \int_{CS} \rho \vec{V} \cdot d\vec{A} = 0 \qquad (4.12)$$

Assumptions: (1) Steady flow.
(2) Uniform flow at each section.

Then
$$(-\rho cA) + \{(\rho + d\rho)(c - dV_x)A\} = 0 \qquad (12.14a)$$

or
$$-\rho c A + \rho c A - \rho \, dV_x A + d\rho c A - \overbrace{d\rho \, dV_x A}^{\approx 0} = 0$$

or
$$dV_x = \frac{c}{\rho} d\rho \qquad (12.14b)$$

b. Momentum Equation

Governing equation:
$$F_{S_x} + \overbrace{F_{B_x}}^{=0(3)} = \overbrace{\frac{\partial}{\partial t}\int_{CV} V_x \, \rho \, d\forall}^{=0(1)} + \int_{CS} V_x \, \rho \vec{V} \cdot d\vec{A} \qquad (4.18a)$$

Assumption: (3) $F_{B_x} = 0$

The only surface forces acting in the x direction on the control volume of Fig. 12.1*b* are due to pressure (the infinitesimal upper and lower areas have zero friction because we assume the wave is one dimensional).

$$F_{S_x} = pA - (p + dp)A = -A \, dp$$

Substituting into the governing equation gives

$$-A \, dp = c(-\rho c A) + (c - dV_x)\{(\rho + d\rho)(c - dV_x)A\}$$

Using the continuity equation, (Eq. 12.14a), this reduces to

$$-A \, dp = c(-\rho c A) + (c - dV_x)(\rho c A) = (-c + c - dV_x)(\rho c A)$$
$$-A \, dp = -\rho c A \, dV_x$$

or
$$dV_x = \frac{1}{\rho c} dp \qquad (12.14c)$$

Combining Eqs. 12.14b and 12.14c, we obtain

$$dV_x = \frac{c}{\rho}d\rho = \frac{1}{\rho c}dp$$

from which

$$dp = c^2\, d\rho$$

or

$$c^2 = \frac{dp}{d\rho} \tag{12.15}$$

We have derived an expression for the speed of sound in any medium in terms of thermodynamic quantities! Equation 12.15 indicates that the speed of sound depends on how the pressure and density of the medium are related. To obtain the speed of sound in a medium we could measure the time a sound wave takes to travel a prescribed distance, or instead we could apply a small pressure change dp to a sample, measure the corresponding density change $d\rho$, and evaluate c from Eq. 12.15. For example, an *incompressible* medium would have $d\rho = 0$ for any dp, so $c \to \infty$. We can anticipate that solids and liquids (whose densities are difficult to change) will have relatively high c values, and gases (whose densities are easy to change) will have relatively low c values. There is only one problem with Eq. 12.15: For a simple substance, each property depends on any *two* independent properties [1]. For a sound wave, by definition we have an infinitesimal pressure change (i.e., it is *reversible*), and it occurs very quickly, so there is no time for any heat transfer to occur (i.e., it is *adiabatic*). Thus the sound wave propagates *isentropically*. Hence, if we express p as a function of density and entropy, $p = p(\rho, s)$, then

$$dp = \left(\frac{\partial p}{\partial \rho}\right)_s d\rho + \left(\frac{\partial p}{\partial s}\right)_\rho ds = \left(\frac{\partial p}{\partial \rho}\right)_s d\rho$$

so Eq. 12.15 becomes

$$c^2 = \frac{dp}{d\rho} = \left.\frac{\partial p}{\partial \rho}\right)_s$$

and

$$c = \sqrt{\left.\frac{\partial p}{\partial \rho}\right)_s} \tag{12.16}$$

We can now apply Eq. 12.16 to solids, liquids, and gases. For *solids* and *liquids* data are usually available on the bulk modulus E_v, which is a measure of how a pressure change affects a relative density change,

$$E_v = \frac{dp}{d\rho/\rho} = \rho\frac{dp}{d\rho}$$

For these media

$$c = \sqrt{E_v/\rho} \tag{12.17}$$

For an *ideal gas*, the pressure and density in isentropic flow are related by

$$\frac{p}{\rho^k} = \text{constant} \tag{12.12c}$$

Taking logarithms and differentiating, we obtain

$$\frac{dp}{p} - k\frac{d\rho}{\rho} = 0$$

Therefore,

$$\left.\frac{\partial p}{\partial \rho}\right)_s = k\frac{p}{\rho}$$

But $p/\rho = RT$, so finally

$$c = \sqrt{kRT} \tag{12.18}$$

for an ideal gas. The speed of sound in air has been measured precisely by numerous investigators [4]. The results agree closely with the theoretical prediction of Eq. 12.18.

The important feature of sound propagation in an ideal gas, as shown by Eq. 12.18, is that the *speed of sound is a function of temperature only*. The variation in atmospheric temperature with altitude on a standard day was discussed in Chapter 3; the properties are summarized in Table A.3. The corresponding variation in c is computed as an exercise in Problem 12.40 and plotted as a function of altitude.

Example 12.3 SPEED OF SOUND IN STEEL, WATER, SEAWATER, AND AIR

Find the speed of sound in (a) steel ($E_v \approx 200$ GN/m^2), (b) water (at 20°C), (c) seawater (at 20°C), and (d) air at sea level on a standard day.

Find: Speed of sound in (a) steel ($E_v = 200$ GN/m^2), (b) water (at 20°C), (c) seawater (at 20°C), and (d) air at sea level on a standard day.

Solution:

(a) For steel, a solid, we use Eq. 12.17, with ρ obtained from Table A.1(b),

$$c = \sqrt{E_v/\rho} = \sqrt{E_v/SG\rho_{H_2O}}$$

$$c = \sqrt{200 \times 10^9 \frac{N}{m^2} \times \frac{1}{7.83} \times \frac{1}{1000}\frac{m^3}{kg} \times \frac{kg\cdot m}{N\cdot s^2}} = 5050 \text{ m/s} \qquad c_{steel}$$

(b) For water we also use Eq. 12.17, with data obtained from Table A.2,

$$c = \sqrt{E_v/\rho} = \sqrt{E_v/SG\rho_{H_2O}}$$

$$c = \sqrt{2.24 \times 10^9 \frac{N}{m^2} \times \frac{1}{0.998} \times \frac{1}{1000}\frac{m^3}{kg} \times \frac{kg\cdot m}{N\cdot s^2}} = 1500 \text{ m/s} \qquad c_{water}$$

(c) For seawater we again use Eq. 12.17, with data obtained from Table A.2,

$$c = \sqrt{E_v/\rho} = \sqrt{E_v/SG\rho_{H_2O}}$$

$$c = \sqrt{2.42 \times 10^9 \frac{N}{m^2} \times \frac{1}{1.025} \times \frac{1}{1000}\frac{m^3}{kg} \times \frac{kg\cdot m}{N\cdot s^2}} = 1540 \text{ m/s} \qquad c_{seawater}$$

(d) For air we use Eq. 12.18, with the sea level temperature obtained from Table A.3,

$$c = \sqrt{kRT}$$

$$c = \sqrt{1.4 \times 287 \frac{\text{N}\cdot\text{m}}{\text{kg}\cdot\text{K}} \times 288\ \text{K} \times \frac{\text{kg}\cdot\text{m}}{\text{N}\cdot\text{s}^2}} = 340\ \text{m/s} \longleftarrow c_{\text{air (288 K)}}$$

This Example illustrates the relative magnitudes of the speed of sound in typical solids, liquids, and gases ($c_{solids} > c_{liquids} > c_{gases}$). Do not confuse the speed of sound with the *attenuation* of sound—the rate at which internal friction of the medium reduces the sound level—generally, solids and liquids attenuate sound much more rapidly than do gases.

Types of Flow—The Mach Cone

Flows for which $M < 1$ are *subsonic*, while those with $M > 1$ are *supersonic*. Flow fields that have both subsonic and supersonic regions are termed *transonic*. (The transonic regime occurs for Mach numbers between about 0.9 and 1.2.) Although most flows within our experience are subsonic, there are important practical cases where $M \geq 1$ occurs in a flow field. Perhaps the most obvious are supersonic aircraft and transonic flows in aircraft compressors and fans. Yet another flow regime, *hypersonic* flow ($M \gtrsim 5$), is of interest in missile and reentry-vehicle design. (The proposed National Aerospace Plane would have cruised at Mach numbers approaching 20.) Some important qualitative differences between subsonic and supersonic flows can be deduced from the properties of a simple moving sound source.

Consider a point source of sound that emits a pulse every Δt seconds. Each pulse expands outwards from its origination point at the speed of sound c, so at any instant t the pulse will be a sphere of radius ct centered at the pulse's origination point. We want to investigate what happens if the point source itself is moving. There are four possibilities, as shown in Fig. 12.2:

(a) $V = 0$. The point source is *stationary*. Figure 12.2*a* shows conditions after $3\Delta t$ seconds. The first pulse has expanded to a sphere of radius $c(3\Delta t)$, the second to a sphere of radius $c(2\Delta t)$, and the third to a sphere of radius $c(\Delta t)$; a new pulse is about to be emitted. The pulses constitute a set of ever-expanding concentric spheres.

(b) $0 < V < c$. The point source moves to the left at *subsonic* speed. Figure 12.2*b* shows conditions after $3\Delta t$ seconds. The source is shown at times $t = 0$, Δt, $2\Delta t$, and $3\Delta t$. The first pulse has expanded to a sphere of radius $c(3\Delta t)$ *centered where the source was originally*, the second to a sphere of radius $c(2\Delta t)$ *centered where the source was at time* Δt, and the third to a sphere of radius $c(\Delta t)$ *centered where the source was at time* $2\Delta t$; a new pulse is about to be emitted. The pulses again constitute a set of ever-expanding spheres, except now they are not concentric. The pulses are all expanding at constant speed c. We make two important notes: First, we can see that an observer who is ahead of the source (or whom the source is approaching) will hear the pulses at a higher frequency rate than will an observer who is behind the source (this is the Doppler effect that occurs when a vehicle approaches and passes); second, an observer ahead of the source hears the source *before* the source itself reaches the observer.

(c) $V = c$. The point source moves to the left at *sonic* speed. Figure 12.2*c* shows conditions after $3\Delta t$ seconds. The source is shown at times $t = 0$ (point 1), Δt (point 2),

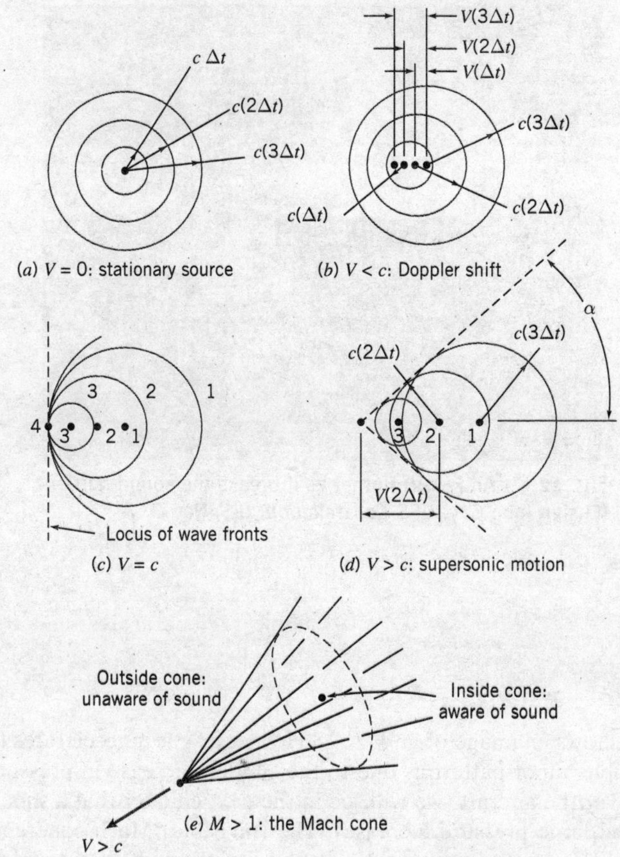

Fig. 12.2 Propagation of sound waves from a moving source: The Mach cone.

$2\Delta t$ (point 3), and $3\Delta t$ (point 4). The first pulse has expanded to sphere 1 of radius $c(3\Delta t)$ *centered at point 1*, the second to sphere 2 of radius $c(2\Delta t)$ *centered at point 2*, and the third to sphere 3 of radius $c(\Delta t)$ *centered around the source at point 3*. We can see once more that the pulses constitute a set of ever-expanding spheres, except now they are tangent to one another on the left! The pulses are all expanding at constant speed c, but the source is also moving at speed c, with the result that the source and all its pulses are traveling together to the left. We again make two important notes: First, we can see that an observer who is ahead of the source will *not* hear the pulses before the source reaches her; second, in theory, over time an unlimited number of pulses will accumulate at the front of the source, leading to a sound wave of unlimited amplitude (a source of concern to engineers trying to break the "sound barrier," which many people thought could not be broken—Chuck Yeager in a Bell X-1 was the first to do so in 1947).

(d) $V > c$. The point source moves to the left at *supersonic* speed. Figure 12.2*d* shows conditions after $3\Delta t$ seconds. By now it is clear how the spherical waves develop. We can see once more that the pulses constitute a set of ever-expanding spheres, except now the source is moving so fast it moves ahead of each sphere that it generates! For supersonic motion, the spheres generate what is called a *Mach cone* tangent to each sphere. The region inside the cone is called the *zone of action* and that outside the cone the *zone of silence*, for obvious reasons, as shown in Fig. 12.2*e*. From geometry, we see from Fig. 12.2*d* that

$$\sin\alpha = \frac{c}{V} = \frac{1}{M}$$

VIDEO

Shock Waves over a Supersonic Airplane.

Fig. 12.3 An F/A-18 Hornet as it breaks the sound barrier. (Ensign John Gay, USS *Constellation*, U.S. Navy.)

VIDEO

Shock Waves due to a Projectile.

or

$$\alpha = \sin^{-1}\left(\frac{1}{M}\right) \tag{12.19}$$

Figure 12.3 shows an image of an F/A-18 Hornet just as it accelerates to supersonic speed. The visible vapor pattern is due to the sudden increase in pressure as a shock wave washes over the aircraft (we will see in the next chapter that a shock wave leads to a sudden and large pressure increase). The (invisible) Mach cone emanates from the nose of the aircraft and passes through the periphery of the vapor disk.

Example 12.4 MACH CONE OF A BULLET

In tests of a protective material, we wish to photograph a bullet as it impacts a jacket made of the material. A camera is set up a perpendicular distance $h = 5$ m from the bullet trajectory. We wish to determine the perpendicular distance d from the target plane at which the camera must be placed such that the sound of the bullet will trigger the camera at the impact time. Note: The bullet speed is measured to be 550 m/s; the delay time of the camera is 0.005 s.

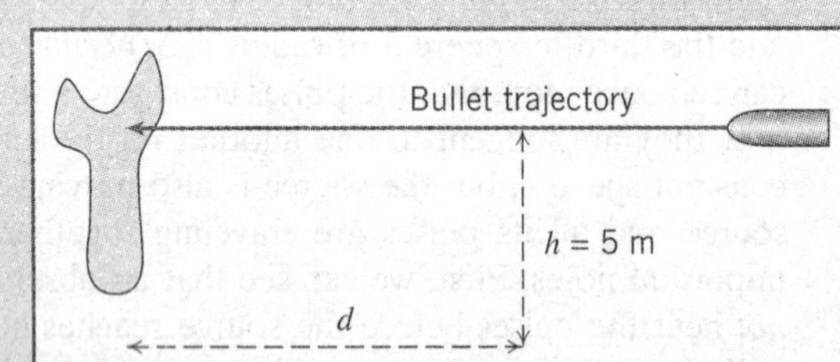

Find: Location of camera for capturing impact image.

Solution:

The correct value of d is that for which the bullet hits the target 0.005 s before the Mach wave reaches the camera. We must first find the Mach number of the bullet; then we can find the Mach angle; finally, we can use basic trigonometry to find d.

Assuming sea level conditions, from Table A.3 we have $T = 288$ K. Hence Eq. 12.18 yields

$$c = \sqrt{kRT}$$

$$c = \sqrt{1.4 \times 287\frac{\text{N}\cdot\text{m}}{\text{kg}\cdot\text{K}} \times 288\,\text{K} \times \frac{\text{kg}\cdot\text{m}}{\text{N}\cdot\text{s}^2}} = 340\ \text{m/s}$$

Then we can find the Mach number,

$$M = \frac{V}{c} = \frac{550\ \text{m/s}}{340\ \text{m/s}} = 1.62$$

From Eq. 12.19 we can next find the Mach angle,

$$\alpha = \sin^{-1}\left(\frac{1}{M}\right) = \sin^{-1}\left(\frac{1}{1.62}\right) = 38.2°$$

The distance x traveled by the bullet while the Mach wave reaches the camera is then

$$x = \frac{h}{\tan(\alpha)} = \frac{5\ \text{m}}{\tan(38.2°)} = 6.35\ \text{m}$$

Finally, we must add to this the time traveled by the bullet while the camera is operating, which is 0.005 s × 550 m/s,

$$d = 0.005\ \text{s} \times \frac{550\ \text{m}}{\text{s}} + 6.35\ \text{m} = 2.75\ \text{m} + 6.35\ \text{m}$$

$$d = 9.10\ \text{m} \qquad d$$

Reference State: Local Isentropic Stagnation Properties 12.3

In our study of compressible flow, we will discover that, in general, *all* properties (p, T, ρ, u, h, s, V) may be changing as the flow proceeds. We need to obtain reference conditions that we can use to relate conditions in a flow from point to point. For any flow, a reference condition is obtained when the fluid is (in reality or conceptually) brought to rest ($V = 0$). We will call this the *stagnation condition*, and the property values (p_0, T_0, ρ_0, u_0, h_0, s_0) at this state the *stagnation properties*. This process—of bringing the fluid to rest—is not as straightforward as it seems. For example, do we do so while there is friction, or while the fluid is being heated or cooled, or "violently," or in some other way? The simplest process to use is an isentropic process, in which there is no friction, no heat transfer, and no "violent" events. Hence, the properties we obtain will be the *local isentropic stagnation properties*. Why "local"? Because the actual flow can be any kind of flow, e.g., with friction, so it may or may not itself be isentropic. Hence, each point in the flow will have its own, or local, isentropic stagnation properties. This is illustrated in Fig. 12.4, showing a flow from some state ① to some new state ②. The local isentropic stagnation properties for each state, obtained by isentropically bringing the fluid to rest, are also shown. Hence, $s_{0_1} = s_1$ and $s_{0_2} = s_2$. The actual flow may or may not be isentropic. If it *is* isentropic, $s_1 = s_2 = s_{0_1} = s_{0_2}$, so the stagnation states are identical; if it is *not* isentropic, then $s_{0_1} \neq s_{0_2}$. We will see that changes in local isentropic stagnation properties will provide useful information about the flow.

We can obtain information on the reference isentropic stagnation state for *incompressible* flows by recalling the Bernoulli equation from Chapter 6

$$\frac{p}{\rho} + \frac{V^2}{2} + gz = \text{constant} \tag{6.8}$$

valid for a steady, incompressible, frictionless flow along a streamline. Equation 6.8 is valid for an incompressible isentropic process because it is reversible (frictionless

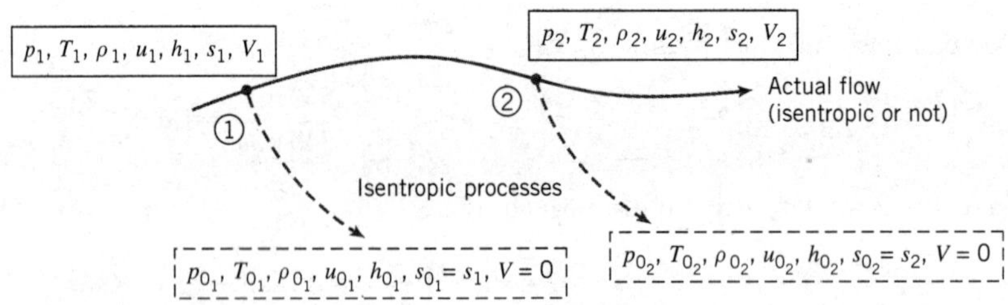

Fig. 12.4 Local isentropic stagnation properties.

and steady) and adiabatic (we did not include heat transfer considerations in its derivation). As we saw in Section 6.3, the Bernoulli equation leads to

$$p_0 = p + \frac{1}{2}\rho V^2 \tag{6.11}$$

(The gravity term drops out because we assume the reference state is at the same elevation as the actual state, and in any event in external flows it is usually much smaller than the other terms.) In Example 12.6 we compare isentropic stagnation conditions obtained assuming incompressibility (Eq. 6.11), and allowing for compressibility.

For *compressible* flows, we will focus on ideal gas behavior.

Local Isentropic Stagnation Properties for the Flow of an Ideal Gas

For a compressible flow we can derive the isentropic stagnation relations by applying the mass conservation (or continuity) and momentum equations to a differential control volume, and then integrating. For the process shown schematically in Fig. 12.4, we can depict the process from state ① to the corresponding stagnation state by imagining the control volume shown in Fig. 12.5. Consider first the continuity equation.

a. Continuity Equation

Governing equation:

$$\overbrace{\frac{\partial}{\partial t}\int_{\text{CV}} \rho \, d\mathcal{V}}^{=0(1)} + \int_{\text{CS}} \rho \vec{V} \cdot d\vec{A} = 0 \tag{4.12}$$

Assumptions: (1) Steady flow.
(2) Uniform flow at each section.

Then

$$(-\rho V_x A) + \{(\rho + d\rho)(V_x + dV_x)(A + dA)\} = 0$$

or

$$\rho V_x A = (\rho + d\rho)(V_x + dV_x)(A + dA) \tag{12.20a}$$

b. Momentum Equation

Governing equation:

$$F_{S_x} + \overbrace{F_{B_x}}^{=0(3)} = \overbrace{\frac{\partial}{\partial t}\int_{\text{CV}} V_x \, \rho \, d\mathcal{V}}^{=0(1)} + \int_{\text{CS}} V_x \, \rho \vec{V} \cdot d\vec{A} \tag{4.18a}$$

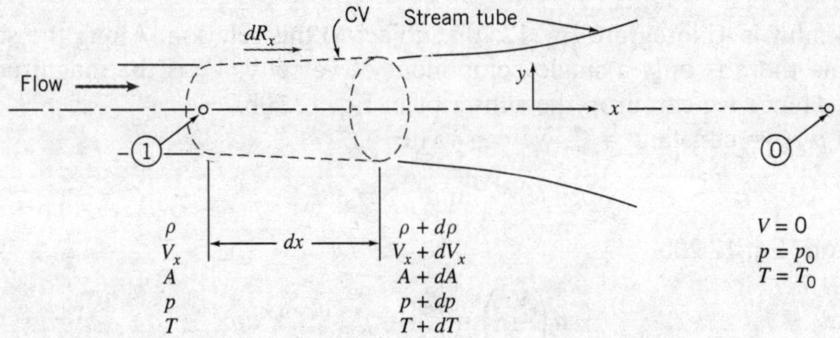

Fig. 12.5 Compressible flow in an infinitesimal stream tube.

Assumptions: (3) $F_{B_x} = 0$.
(4) Frictionless flow.

The surface forces acting on the infinitesimal control volume are

$$F_{S_x} = dR_x + pA - (p + dp)(A + dA)$$

The force dR_x is applied along the stream tube boundary, as shown in Fig. 12.5, where the average pressure is $p + dp/2$, and the area component in the x direction is dA. There is no friction. Thus,

$$F_{S_x} = \left(p + \frac{dp}{2}\right)dA + pA - (p + dp)(A + dA)$$

or

$$F_{S_x} = \cancel{p\,dA} + \overset{\approx 0}{\cancel{\frac{dp\,dA}{2}}} + \cancel{pA} - \cancel{pA} - dp\,A - \cancel{p\,dA} - \overset{\approx 0}{\cancel{dp\,dA}}$$

Substituting this result into the momentum equation gives

$$-dp\,A = V_x\{-\rho V_x A\} + (V_x + dV_x)\{(\rho + d\rho)(V_x + dV_x)(A + dA)\}$$

which may be simplified using Eq. 12.20a to obtain

$$-dp\,A = (-V_x + V_x + dV_x)(\rho V_x A)$$

Finally,

$$dp = -\rho V_x dV_x = -\rho\, d\left(\frac{V_x^2}{2}\right)$$

or

$$\frac{dp}{\rho} + d\left(\frac{V_x^2}{2}\right) = 0 \qquad (12.20b)$$

Equation 12.20b is a relation among properties during the deceleration process. (Note that for incompressible flow, it immediately leads to Eq. 6.11.) In developing this relation, we have specified a frictionless deceleration process. Before we can integrate between the initial and final (stagnation) states, we must specify the relation that exists between pressure, p, and density, ρ, along the process path.

Since the deceleration process is isentropic, then p and ρ for an ideal gas are related by the expression

$$\frac{p}{\rho^k} = \text{constant} \qquad (12.12c)$$

Our task now is to integrate Eq. 12.20b subject to this relation. Along the stagnation streamline there is only a single component of velocity; V_x is the magnitude of the velocity. Hence we can drop the subscript in Eq. 12.20b.

From p/ρ^k = constant = C, we can write

$$p = C\rho^k \quad \text{and} \quad \rho = p^{1/k}C^{-1/k}$$

Then, from Eq. 12.20b,

$$-d\left(\frac{V^2}{2}\right) = \frac{dp}{\rho} = p^{-1/k}C^{1/k}dp$$

We can integrate this equation between the initial state and the corresponding stagnation state

$$-\int_V^0 d\left(\frac{V^2}{2}\right) = C^{1/k}\int_p^{p_0} p^{-1/k}dp$$

to obtain

$$\frac{V^2}{2} = C^{1/k}\frac{k}{k-1}\left[p^{(k-1)/k}\right]_p^{p_0} = C^{1/k}\frac{k}{k-1}\left[p_0^{(k-1)/k} - p^{(k-1)/k}\right]$$

$$\frac{V^2}{2} = C^{1/k}\frac{k}{k-1}p^{(k-1)/k}\left[\left(\frac{p_0}{p}\right)^{(k-1)/k} - 1\right]$$

Since $C^{1/k} = p^{1/k}/\rho$,

$$\frac{V^2}{2} = \frac{k}{k-1}\frac{p^{1/k}}{\rho}p^{(k-1)/k}\left[\left(\frac{p_0}{p}\right)^{(k-1)/k} - 1\right]$$

$$\frac{V^2}{2} = \frac{k}{k-1}\frac{p}{\rho}\left[\left(\frac{p_0}{p}\right)^{(k-1)/k} - 1\right]$$

Since we seek an expression for stagnation pressure, we can rewrite this equation as

$$\left(\frac{p_0}{p}\right)^{(k-1)/k} = 1 + \frac{k-1}{k}\frac{\rho}{p}\frac{V^2}{2}$$

and

$$\frac{p_0}{p} = \left[1 + \frac{k-1}{k}\frac{\rho V^2}{2p}\right]^{k/(k-1)}$$

For an ideal gas, $p = \rho RT$, and hence

$$\frac{p_0}{p} = \left[1 + \frac{k-1}{2}\frac{V^2}{kRT}\right]^{k/(k-1)}$$

Also, for an ideal gas the sonic speed is $c = \sqrt{kRT}$, and thus

$$\frac{p_0}{p} = \left[1 + \frac{k-1}{2}\frac{V^2}{c^2}\right]^{k/(k-1)}$$

$$\frac{p_0}{p} = \left[1 + \frac{k-1}{2}M^2\right]^{k/(k-1)} \tag{12.21a}$$

Equation 12.21a enables us to calculate the local isentropic stagnation pressure at any point in a flow field of an ideal gas, provided that we know the static pressure and Mach number at that point.

We can readily obtain expressions for other isentropic stagnation properties by applying the relation

$$\frac{p}{\rho^k} = \text{constant}$$

between end states of the process. Thus

$$\frac{p_0}{p} = \left(\frac{\rho_0}{\rho}\right)^k \quad \text{and} \quad \frac{\rho_0}{\rho} = \left(\frac{p_0}{p}\right)^{1/k}$$

For an ideal gas, then,

$$\frac{T_0}{T} = \frac{p_0}{p}\frac{\rho}{\rho_0} = \frac{p_0}{p}\left(\frac{p_0}{p}\right)^{-1/k} = \left(\frac{p_0}{p}\right)^{(k-1)/k}$$

Using Eq. 12.21a, we can summarize the equations for determining local isentropic stagnation properties of an ideal gas as

$$\frac{p_0}{p} = \left[1 + \frac{k-1}{2}M^2\right]^{k/(k-1)} \tag{12.21a}$$

$$\frac{T_0}{T} = 1 + \frac{k-1}{2}M^2 \tag{12.21b}$$

$$\frac{\rho_0}{\rho} = \left[1 + \frac{k-1}{2}M^2\right]^{1/(k-1)} \tag{12.21c}$$

From Eqs. 12.21, the ratio of each local isentropic stagnation property to the corresponding static property at any point in a flow field for an ideal gas can be found if the local Mach number is known. We will usually use Eqs. 12.21 in lieu of the continuity and momentum equations for relating the properties at a state to that state's stagnation properties, but it is important to remember that we derived Eqs. 12.21 using these equations *and* the isentropic relation for an ideal gas. Appendix E.1 lists flow functions for property ratios T_0/T, p_0/p, and ρ_0/ρ, in terms of M for isentropic flow of an ideal gas. A table of values, as well as a plot of these property ratios is presented for air (k = 1.4) for a limited range of Mach numbers. The associated *Excel* workbook, *Isentropic Relations*, available on the Web site, can be used to print a larger table of values for air and other ideal gases. The calculation procedure is illustrated in Example 12.5.

The Mach number range for validity of the assumption of incompressible flow is investigated in Example 12.6.

Example 12.5 LOCAL ISENTROPIC STAGNATION CONDITIONS IN CHANNEL FLOW

Air flows steadily through the duct shown from 350 kPa (abs), 60°C, and 183 m/s at the inlet state to M = 1.3 at the outlet, where local isentropic stagnation conditions are known to be 385 kPa (abs) and 350 K. Compute the local isentropic stagnation pressure and temperature at the inlet and the static pressure and temperature at the duct outlet. Locate the inlet and outlet static state points on a Ts diagram, and indicate the stagnation processes.

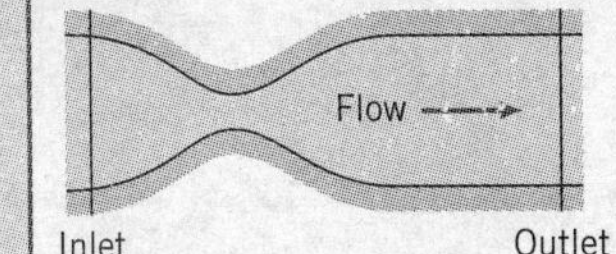

Given: Steady flow of air through a duct as shown in the sketch.

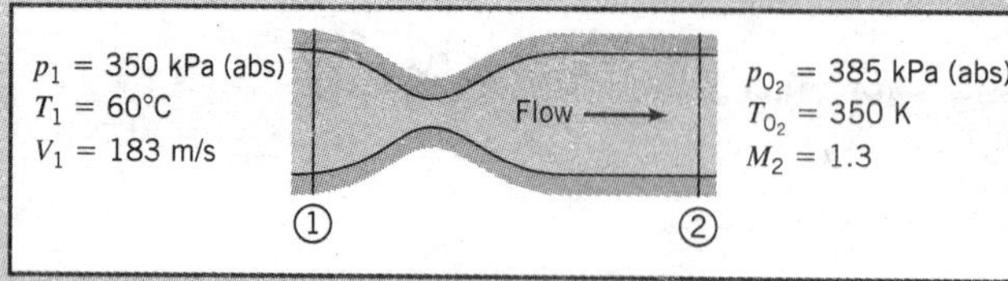

Find: (a) p_{0_1}.
(b) T_{0_1}.
(c) p_2.
(d) T_2.
(e) State points ① and ② on a Ts diagram; indicate the stagnation processes.

Solution: To evaluate local isentropic stagnation conditions at section ①, we must calculate the Mach number, $M_1 = V_1/c_1$. For an ideal gas, $c = \sqrt{kRT}$. Then

$$c_1 = \sqrt{kRT_1} = \left[1.4 \times 287\frac{\text{N}\cdot\text{m}}{\text{kg}\cdot\text{K}} \times (273+60)\text{ K} \times \frac{\text{kg}\cdot\text{m}}{\text{N}\cdot\text{s}^2}\right]^{1/2} = 366 \text{ m/s}$$

and

$$M_1 = \frac{V_1}{c_1} = \frac{183}{366} = 0.5$$

Local isentropic stagnation properties can be evaluated from Eqs. 12.21. Thus

$$p_{0_1} = p_1\left[1+\frac{k-1}{2}M_1^2\right]^{k(k-1)} = 350\text{ kPa }[1+0.2(0.5)^2]^{3.5} = 415\text{ kPa(abs)} \qquad p_{0_1}$$

$$T_{0_1} = T_1\left[1+\frac{k-1}{2}M_1^2\right] = 333\text{ K}[1+0.2(0.5)^2] = 350\text{ K} \qquad T_{0_1}$$

At section ②, Eqs. 12.21 can be applied again. Thus from Eq. 12.21a,

$$p_2 = \frac{p_{0_2}}{\left[1+\frac{k-1}{2}M_2^2\right]^{k/(k-1)}} = \frac{385\text{ kPa}}{[1+0.2(1.3)^2]^{3.5}} = 139\text{ kPa(abs)} \qquad p_2$$

From Eq. 12.21b,

$$T_2 = \frac{T_{0_2}}{1+\frac{k-1}{2}M_2^2} = \frac{350\text{ K}}{1+0.2(1.3)^2} = 262\text{ K} \qquad T_2$$

To locate states ① and ② in relation to one another, and sketch the stagnation processes on the Ts diagram, we need to find the change in entropy $s_2 - s_1$. At each state we have p and T, so it is convenient to use Eq. 12.11b,

$$s_2 - s_1 = c_p \ln\frac{T_2}{T_1} - R\ln\frac{p_2}{p_1}$$

$$= 1.00\frac{\text{kJ}}{\text{kg}\cdot\text{K}} \times \ln\left(\frac{262}{333}\right) - 0.287\frac{\text{kJ}}{\text{kg}\cdot\text{K}} \times \ln\left(\frac{139}{350}\right)$$

$$s_2 - s_1 = 0.0252 \text{ kJ/(kg}\cdot\text{K)}$$

Hence in this flow we have an increase in entropy. Perhaps there is irreversibility (e.g., friction), or heat is being added, or both. (We will see in Chapter 13 that the fact that $T_{0_1} = T_{0_2}$ for this particular flow means that actually we have an adiabatic flow.) We also found that $T_2 < T_1$ and that $p_2 < p_1$. We can now sketch the Ts diagram (and recall we saw in Example 12.2 that isobars (lines of constant pressure) in Ts space are exponential),

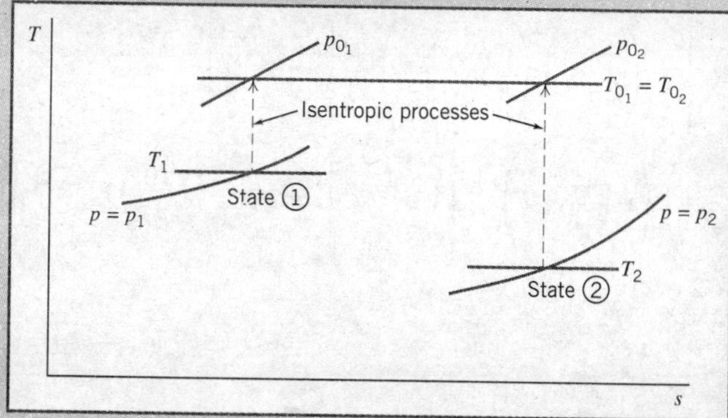

This problem illustrates use of the local isentropic stagnation properties (Eqs. 12.21) to relate different points in a flow.

The Excel workbook *Isentropic Relations*, available on the Web site, can be used for computing property ratios from the Mach number M, as well as for computing M from property ratios.

Example 12.6 MACH-NUMBER LIMIT FOR INCOMPRESSIBLE FLOW

We have derived equations for p_0/p for both compressible and "incompressible" flows. By writing both equations in terms of Mach number, compare their behavior. Find the Mach number below which the two equations agree within engineering accuracy.

Given: The incompressible and compressible forms of the equations for stagnation pressure, p_0.

Incompressible
$$p_0 = p + \frac{1}{2}\rho V^2 \tag{6.11}$$

Compressible
$$\frac{p_0}{p} = \left[1 + \frac{k-1}{2}M^2\right]^{k/(k-1)} \tag{12.21a}$$

Find: (a) Behavior of both equations as a function of Mach number.
(b) Mach number below which calculated values of p_0/p agree within engineering accuracy.

Solution: First, let us write Eq. 6.11 in terms of Mach number. Using the ideal gas equation of state and $c^2 = kRT$,

$$\frac{p_0}{p} = 1 + \frac{\rho V^2}{2p} = 1 + \frac{V^2}{2RT} = 1 + \frac{kV^2}{2kRT} = 1 + \frac{kV^2}{2c^2}$$

Thus,

$$\frac{p_0}{p} = 1 + \frac{k}{2}M^2 \tag{1}$$

for "incompressible" flow.

Equation 12.21a may be expanded using the binomial theorem,

$$(1+x)^n = 1 + nx + \frac{n(n-1)}{2!}x^2 + \cdots, |x| < 1$$

For Eq. 12.21a, $x = [(k-1)/2]M^2$, and $n = k/(k-1)$. Thus the series converges for $[(k-1)/2]M^2 < 1$, and for compressible flow,

$$\frac{p_0}{p} = 1 + \left(\frac{k}{k-1}\right)\left[\frac{k-1}{2}M^2\right] + \left(\frac{k}{k-1}\right)\left(\frac{k}{k-1}-1\right)\frac{1}{2!}\left[\frac{k-1}{2}M^2\right]^2$$

$$+ \left(\frac{k}{k-1}\right)\left(\frac{k}{k-1}-1\right)\left(\frac{k}{k-1}-2\right)\frac{1}{3!}\left[\frac{k-1}{2}M^2\right]^3 + \cdots \tag{2}$$

$$= 1 + \frac{k}{2}M^2 + \frac{k}{8}M^4 + \frac{k(2-k)}{48}M^6 + \cdots$$

$$\frac{p_0}{p} = 1 + \frac{k}{2}M^2\left[1 + \frac{1}{4}M^2 + \frac{(2-k)}{24}M^4 + \cdots\right]$$

In the limit, as $M \to 0$, the term in brackets in Eq. 2 approaches 1.0. Thus, for flow at low Mach number, the incompressible and compressible equations give the same result. The variation of p_0/p with Mach number is shown below. As Mach number is increased, the compressible equation gives a larger ratio, p_0/p.

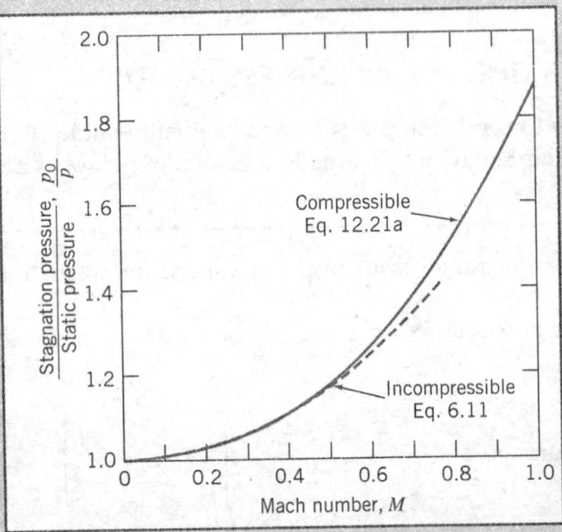

Equations 1 and 2 may be compared quantitatively most simply by writing

$$\frac{p_0}{p} - 1 = \frac{k}{2}M^2 \quad (\text{"incompressible"})$$

$$\frac{p_0}{p} - 1 = \frac{k}{2}M^2\left[1 + \frac{1}{4}M^2 + \frac{(2-k)}{24}M^4 + \cdots\right] \quad (\text{compressible})$$

The term in brackets is approximately equal to 1.02 at $M = 0.3$, and to 1.04 at $M = 0.4$. Thus, for calculations of engineering accuracy, *flow may be considered incompressible if $M < 0.3$*. The two agree within 5 percent for $M \lesssim 0.45$.

Critical Conditions 12.4

Stagnation conditions are extremely useful as reference conditions for thermodynamic properties; this is not true for velocity, since by definition $V = 0$ at stagnation. A useful reference value for velocity is the *critical speed*—the speed V we attain when a flow is either accelerated or decelerated (actually or conceptually) isentropically until we reach $M = 1$. Even if there is no point in a given flow field where the Mach number is equal to unity, such a hypothetical condition still is useful as a reference condition.

Using asterisks to denote conditions at $M = 1$, then by definition

$$V^* \equiv c^*$$

At critical conditions, Eqs. 12.21 for isentropic stagnation properties become

$$\frac{p_0}{p^*} = \left[\frac{k+1}{2}\right]^{k/(k-1)} \tag{12.22a}$$

$$\frac{T_0}{T^*} = \frac{k+1}{2} \tag{12.22b}$$

$$\frac{\rho_0}{\rho^*} = \left[\frac{k+1}{2}\right]^{1/(k-1)} \tag{12.22c}$$

The critical speed may be written in terms of either critical temperature, T^*, or isentropic stagnation temperature, T_0.

For an ideal gas, $c^* = \sqrt{kRT^*}$, and thus $V^* = \sqrt{kRT^*}$. Since, from Eq. 12.22b,

$$T^* = \frac{2}{k+1} T_0$$

we have

$$V^* = c^* = \sqrt{\frac{2k}{k+1} RT_0} \tag{12.23}$$

We shall use both stagnation conditions and critical conditions as reference conditions in the next chapter when we consider a variety of compressible flows.

12.5 Summary and Useful Equations

In this chapter, we:

- ✓ Reviewed the basic equations used in thermodynamics, including isentropic relations.
- ✓ Introduced some compressible flow terminology, such as definitions of the Mach number and subsonic, supersonic, transonic, and hypersonic flows.
- ✓ Learned about several phenomena having to do with sound, including that the speed of sound in an ideal gas is a function of temperature only ($c = \sqrt{kRT}$), and that the Mach cone and Mach angle determine when a supersonic vehicle is heard on the ground.
- ✓ Learned that there are two useful reference states for a compressible flow: the isentropic stagnation condition, and the isentropic critical condition.

Note: Most of the Useful Equations in the table below have a number of constraints or limitations—*be sure to refer to their page numbers for details!*

Useful Equations

Definition of Mach number M:	$M \equiv \frac{V}{c}$	**(12.13)**	Page 665
Speed of sound c:	$c = \sqrt{\left(\frac{\partial p}{\partial \rho}\right)_s}$	**(12.16)**	Page 668
Speed of sound c (solids and liquids):	$c = \sqrt{E_v/\rho}$	**(12.17)**	Page 668
Speed of sound c (ideal gas):	$c = \sqrt{kRT}$	**(12.18)**	Page 669
Mach cone angle α:	$\alpha = \sin^{-1}\left(\frac{1}{M}\right)$	**(12.19)**	Page 672
Isentropic pressure ratio (ideal gas, constant specific heats):	$\frac{p_0}{p} = \left[1 + \frac{k-1}{2}M^2\right]^{k/(k-1)}$	**(12.21a)**	Page 677
Isentropic temperature ratio (ideal gas, constant specific heats):	$\frac{T_0}{T} = 1 + \frac{k-1}{2}M^2$	**(12.21b)**	Page 677
Isentropic density ratio (ideal gas, constant specific heats):	$\frac{\rho_0}{\rho} = \left[1 + \frac{k-1}{2}M^2\right]^{1/(k-1)}$	**(12.21c)**	Page 677
Critical pressure ratio (ideal gas, constant specific heats):	$\frac{p_0}{p^*} = \left[\frac{k+1}{2}\right]^{k/(k-1)}$	**(12.22a)**	Page 681
Critical temperature ratio (ideal gas, constant specific heats):	$\frac{T_0}{T^*} = \frac{k+1}{2}$	**(12.22a)**	Page 681
Critical density ratio (ideal gas, constant specific heats):	$\frac{\rho_0}{\rho^*} = \left[\frac{k+1}{2}\right]^{1/(k-1)}$	**(12.22c)**	Page 681
Critical velocity V^* (ideal gas, constant specific heats):	$V^* = c^* = \sqrt{\frac{2k}{k+1}RT_0}$	**(12.23)**	Page 681

Case Study

Reducing the Sonic Boom

The SSBD aircraft trailing NASA's F-15B. (Courtesy of NASA.)

Most of us are familiar with the fact that supersonic aircraft generally have very sharp noses and wing leading edges, compared to subsonic aircraft (we will learn some reasons why supersonic aircraft are so shaped in the next chapter); compare, for example, the decommissioned supersonic Concorde to the subsonic Boeing 747 jumbo jet. We are also familiar with the notion of a sonic boom, a large window-rattling boom created when the remains of a supersonic jet's oblique shock wave wash over the ground. The sonic boom is an important reason that the Concorde was not allowed to fly supersonically over land, limiting its use and being one factor in the aircraft's limited commercial success.

The Defense Advanced Research Projects Agency (DARPA) and NASA have now shown that modifying an aircraft's shape can also change the shape of its sonic boom, thereby reducing the loudness developed. Their Shaped Sonic Boom Demonstration (SSBD) program found that by designing the aircraft to a specific shape, the pressure waves created by the aircraft can be kept from merging into a shock wave (see the discussion of the Mach cone in Section 12.2); when these weaker waves reach the ground, the loudness of the sonic boom is greatly reduced. For the demonstration, Northrop Grumman modified an F-5E fighter aircraft by installing a specially shaped "nose glove"; compare the nose of the SSBD (the lower aircraft in the figure) to the traditional supersonic nose of a supersonic F-15B.

It is possible that this new nose shape could eventually allow reintroduction of supersonic transports, even for the key routes across the continental United States.

References

1. Cengel, Y. A., and M. A. Boles, *Thermodynamics: An Engineering Approach*, 4th ed. New York: McGraw-Hill, 2002.
2. Borgnake, C., and R. E. Sonntag, *Fundamentals of Thermodynamics*, 7th ed. New York: Wiley, 2008.
3. Moran, M. J., and H. N. Shapiro, *Fundamentals of Engineering Thermodynamics*, 6th ed. New York: Wiley, 2007.
4. Wong, G. S. K.,"Speed of Sound in Standard Air," *J. Acoustical Society of America, 79*, 5, May 1986, pp. 1359–1366.

Problems

Review of Thermodynamics

12.1 An air flow in a duct passes through a thick filter. What happens to the pressure, temperature, and density of the air as it does so? *Hint*: This is a throttling process.

12.2 Air is expanded in a steady flow process through a turbine. Initial conditions are 1300°C and 2.0 MPa (abs). Final conditions are 500°C and atmospheric pressure. Show this process on a *Ts* diagram. Evaluate the changes in internal energy, enthalpy, and specific entropy for this process.

12.3 A vendor claims that an adiabatic air compressor takes in air at atmosphere pressure and 10°C and delivers the air at 1034 kPa and 93°C. Is this possible? Justify your answer by calculation. Sketch the process on a *Ts* diagram.

12.4 A turbine manufacturer claims that an adiabatic gas turbine can take flow at 1013.25 kPa and 1204°C and exhaust to atmospheric pressure at a temperature of 454°C. Sketch the process on a *Ts* diagram, and prove whether the manufacturer's claims are possible. Assume that the gas has the same properties as air.

12.5 Air initially at 345 kPa and 367 K expands to atmospheric pressure. The process by which this expansion occurs is defined by the expression $p\forall^{1.3}$ = constant. Calculate the final temperature and the change in entropy through this process.

12.6 What is the lowest possible delivery temperature generated by an adiabatic air compressor, starting with standard atmosphere conditions and delivering the air at 500 kPa (gage)? Sketch the process on a *Ts* diagram.

12.7 Air expands without heat transfer through a turbine from a pressure of 10 bars and a temperature of 1400 K to a pressure of 1 bar. If the turbine has an efficiency of 80 percent, determine the exit temperature, and the changes in enthalpy and entropy across the turbine. If the turbine is generating 1 MW of power, what is the mass flow rate of air through the turbine?

12.8 A test chamber is separated into two equal chambers by a rubber diaphragm. One contains air at 20°C and 200 kPa (absolute), and the other has a vacuum. If the diaphragm is punctured, find the pressure and temperature of the air after it expands to fill the chamber. *Hint*: This is a rapid, violent event, so is irreversible but adiabatic.

12.9 An automobile supercharger is a device that pressurizes the air that is used by the engine for combustion to increase the engine power (how does it differ from a turbocharger?). A supercharger takes in air at 21°C and atmospheric pressure and boosts it to 1.38 MPa, at an intake rate of 0.014 m^3/s. What are the pressure, temperature, and volume flow rate at the exit? (The relatively high exit temperature is the reason an intercooler is also used.) Assuming a 70 percent efficiency, what is the power drawn by the supercharger? *Hint*: The efficiency is defined as the ratio of the isentropic power to actual power.

12.10 Five kilograms of air is cooled in a closed tank from 250 to 50°C. The initial pressure is 3 MPa. Compute the changes in entropy, internal energy, and enthalpy. Show the process state points on a *Ts* diagram.

12.11 Air is contained in a piston-cylinder device. The temperature of the air is 100°C. Using the fact that for a reversible process the heat transfer $q = \int T ds$, compare the amount of heat (J/kg) required to raise the temperature of the air to 1200°C at (a) constant pressure and (b) constant volume. Verify your results using the first law of thermodynamics. Plot the processes on a *Ts* diagram.

12.12 The four-stroke Otto cycle of a typical automobile engine is sometimes modeled as an ideal air-standard closed system. In this simplified system the combustion process is modeled as a heating process, and the exhaust-intake process as a cooling process of the working fluid (air). The cycle consists of: isentropic compression from state ① ($p_1 = 100$ kPa (abs), $T_1 = 20°\text{C}$, $\mathcal{V}_1 = 500$ cc) to state ② ($\mathcal{V}_2 = \mathcal{V}_1/8.5$); isometric (constant volume) heat addition to state ③ ($T_3 = 2750°\text{C}$); isentropic expansion to state ④ ($\mathcal{V}_4 = \mathcal{V}_1$); and isometric cooling back to state ①. Plot the $p\mathcal{V}$ and Ts diagrams for this cycle, and find the efficiency, defined as the net work (the cycle area in $p\mathcal{V}$ space) divided by the heat added.

12.13 The four-stroke cycle of a typical diesel engine is sometimes modeled as an ideal air-standard closed system. In this simplified system the combustion process is modeled as a heating process, and the exhaust-intake process as a cooling process of the working fluid (air). The cycle consists of: isentropic compression from state ① ($p_1 = 100$ kPa (abs), $T_1 = 20°\text{C}$, $\mathcal{V}_1 = 500$ cc) to state ② ($\mathcal{V}_2 = \mathcal{V}_1/12.5$); isometric (constant volume) heat addition to state ③ ($T_3 = 3000°\text{C}$); isobaric heat addition to state ④ ($\mathcal{V}_4 = 1.75\mathcal{V}_3$); isentropic expansion to state ⑤; and isometric cooling back to state ①. Plot the $p\mathcal{V}$ and Ts diagrams for this cycle, and find the efficiency, defined as the net work (the cycle area in $p\mathcal{V}$ space) divided by the heat added.

12.14 A 1-m^3 tank contains air at 0.1 MPa (abs) and 20°C. The tank is pressurized to 2 MPa. Assuming that the tank is filled adiabatically and reversibly, calculate the final temperature of the air in the tank. Now assuming that the tank is filled isothermally, how much heat is lost by the air in the tank during filling? Which process (adiabatic or isothermal) results in a greater mass of air in the tank?

12.15 A tank of volume $\mathcal{V} = 10$ m^3 contains compressed air at 15°C. The gage pressure in the tank is 4.50 MPa. Evaluate the work required to fill the tank by compressing air from standard atmosphere conditions for (a) isothermal compression and (b) isentropic compression followed by cooling at constant pressure. What is the peak temperature of the isentropic compression process? Calculate the energy removed during cooling for process (b). Assume ideal gas behavior and reversible processes. Label state points on a Ts diagram and a $p\mathcal{V}$ diagram for each process.

12.16 Air enters a turbine in steady flow at 0.5 kg/s with negligible velocity. Inlet conditions are 1300°C and 2.0 MPa (abs). The air is expanded through the turbine to atmospheric pressure. If the actual temperature and velocity at the turbine exit are 500°C and 200 m/s, determine the power produced by the turbine. Label state points on a Ts diagram for this process.

12.17 Natural gas, with the thermodynamic properties of methane, flows in an underground pipeline of 0.6 m diameter. The gage pressure at the inlet to a compressor station is 0.5 MPa; outlet pressure is 8.0 MPa (gage). The gas temperature and speed at inlet are 13°C and 32 m/s, respectively. The compressor efficiency is $\eta = 0.85$. Calculate the mass flow rate of natural gas through the pipeline. Label state points on a Ts diagram for compressor inlet and outlet. Evaluate the gas temperature and speed at the compressor outlet and the power required to drive the compressor.

12.18 Over time the efficiency of the compressor of Problem 12.17 drops. At what efficiency will the power required to attain 8.0 MPa (gage) exceed 30 MW? Plot the required power and the gas exit temperature as functions of efficiency.

12.19 Improper maintenance on the turbine of Problem 12.7 has resulted in a gradual decrease in its efficiency over time. Assuming that the efficiency drops by 1 percent per year, how long would it take for the power output of the turbine to drop to 950 kW, assuming that entrance conditions, flow rate, and exhaust pressure were all kept constant?

12.20 In an isothermal process, 4.7×10^{-5}m^3 of standard air per second is pumped into a balloon. Tension in the rubber skin of the balloon is given by $\sigma = kA$, where $k = 31{,}416$ N/m^3, and A is the surface area of the balloon in m^2. Compute the time required to increase the balloon radius from 127 to 178 mm.

12.21 For the balloon process of Problem 12.20 we could define a "volumetric ratio" as the ratio of the volume of standard air supplied to the volume increase of the balloon, per unit time. Plot this ratio over time as the balloon radius is increased from 5 to 7 in.

Propagation of Sound Waves

12.22 A sound pulse level above about 20 Pa can cause permanent hearing damage. Assuming such a sound wave travels through air at 20°C and 100 kPa, estimate the density, temperature, and velocity change immediately after the sound wave passes.

12.23 Calculate the speed of sound at 20°C for (a) hydrogen, (b) helium, (c) methane, (d) nitrogen, and (e) carbon dioxide.

12.24 The bulk modulus E_v of a material indicates how hard it is to compress the material; a large E_v indicates the material requires a large pressure to compress. Is air "stiffer" when suddenly or slowly compressed? To answer this, find expressions in terms of instantaneous pressure p for the bulk modulus of air (kPa) when it is (a) rapidly compressed and (b) slowly compressed. *Hint*: Rapid compression is approximately isentropic (it is adiabatic because it is too quick for heat transfer to occur), and slow compression is isothermal (there is plenty of time for the air to equilibrate to ambient temperature).

12.25 You have designed a device for determining the bulk modulus, E_v, of a material. It works by measuring the time delay between sending a sound wave into a sample of the material and receiving the wave after it travels through the sample and bounces back. As a test, you use a 1-m rod of steel ($E_v \approx 200$ GN/m^2). What time delay should your device indicate? You now test a 1-m rod (1 cm diameter) of an unknown material and find a time delay of 0.5 ms. The mass of the rod is measured to be 0.25 kg. What is this material's bulk modulus?

12.26 Dolphins often hunt by listening for sounds made by their prey. They "hear" with the lower jaw, which conducts the sound vibrations to the middle ear via a fat-filled cavity in the lower jaw bone. If the prey is 1000 m away, how long after a sound is made does a dolphin hear it? Assume the seawater is at 20°.

12.27 A submarine sends a sonar signal to detect the enemy. The reflected wave returns after 3.25 s. Estimate the separation between the submarines. (As an approximation, assume the seawater is at 20°C.)

12.28 An airplane flies at 550 km/hr at 1500 m altitude on a standard day. The plane climbs to 15,000 m and flies at 1200 km/h. Calculate the Mach number of flight in both cases.

12.29 Next-generation missiles will use scramjet engines to travel at Mach numbers as high as 7. If a scramjet-powered missile travels at Mach 7 at an altitude of 25500 m, how long will it take for the missile to travel 1.095×10^6 m? Assume standard atmospheric conditions. (*Note:* This is the range for the Tomahawk missile, which uses a conventional propulsion system, but it takes 90 min to cover that same distance.)

12.30 Actual performance characteristics of the Lockheed SR-71 "Blackbird" reconnaissance aircraft never were released. However, it was thought to cruise at $M = 3.3$ at 25.9 km altitude. Evaluate the speed of sound and flight speed for these conditions. Compare to the muzzle speed of a 30-06 rifle bullet (700 m/s).

12.31 The Boeing 727 aircraft of Example 9.8 cruises at 835 km/h at 10.06 km altitude on a standard day. Calculate the cruise Mach number of the aircraft. If the maximum allowable operating Mach number for the aircraft is 0.9, what is the corresponding flight speed?

12.32 Investigate the effect of altitude on Mach number by plotting the Mach number of a 800 km/h airplane as it flies at altitudes ranging from sea level to 10 km.

12.33 You are watching a July 4th fireworks display from a distance of one mile. How long after you see an explosion do you hear it? You also watch New Year's fireworks (same place and distance). How long after you see an explosion do you hear it? Assume it's 24°C in July and −15°C in January.

12.34 The X-15 North American rocket plane held the record for the fastest manned flight. In 1967, the X-15 flew at a speed of 7270 km/h at an altitude of 58.4 km. At what Mach number did the X-15 fly?

12.35 You need to estimate the speed of a hypersonic aircraft traveling at Mach 7 and 36576 m. Not having a table of atmospheric properties handy, you remember that through the stratosphere (approximately 10800 to 21600 m) the temperature of air is nearly constant at 217 K, and you assume this temperature for your calculation. Later, you obtain the appropriate data and recalculate the speed. What was the percentage error? What would the percentage error have been if you used the air temperature at sea level?

12.36 The grandstand at the Kennedy Space Center is located 5.6 km away from the Space Shuttle Launch Pad. On a day when the air temperature is 27°C, how long does it take the sound from a blastoff to reach the spectators? If the launch was early on a winter morning, the temperature may be as low as 10°C. How long would the sound take to reach the spectators under those conditions?

12.37 While working on a pier on a mountain lake, you notice that the sounds of your hammering are echoing from the mountains in the distance. If the temperature is 25°C and the echoes reach you 3 seconds after the hammer strike, how far away are the mountains?

12.38 Use data for specific volume to calculate and plot the speed of sound in saturated liquid water over the temperature range from 0 to 200°C.

12.39 Re-derive the equation for sonic speed (Eq. 12.18) assuming that the direction of fluid motion behind the sound wave is dV_x to the *right*. Show that the result is identical to that given by Eq. 12.18.

12.40 Compute the speed of sound at sea level in standard air. By scanning data from Table A.3 into your PC (or using Fig. 3.3), evaluate the speed of sound and plot for altitudes to 90 km.

12.41 The temperature varies linearly from sea level to approximately 11 km altitude in the standard atmosphere. Evaluate the *lapse rate*—the rate of decrease of temperature with altitude—in the standard atmosphere. Derive an expression for the rate of change of sonic speed with altitude in an ideal gas under standard atmospheric conditions. Evaluate and plot from sea level to 10 km altitude.

12.42 Air at 25°C flows at $M = 1.9$. Determine the air speed and the Mach angle.

12.43 Consider the hypersonic aircraft of Problem 12.35. How long would it take for an observer to hear the aircraft after it flies over the observer? In that elapsed time, how far did the aircraft travel?

12.44 A projectile is fired into a gas (ratio of specific heats $k = 1.625$) in which the pressure is 450 kPa (abs) and the density is 4.5 kg/m^3. It is observed experimentally that a Mach cone emanates from the projectile with 25° total angle. What is the speed of the projectile with respect to the gas?

12.45 A photograph of a bullet shows a Mach angle of 32°. Determine the speed of the bullet for standard air.

12.46 The National Transonic Facility (NTF) is a high-speed wind tunnel designed to operate with air at cryogenic temperatures to reduce viscosity, thus raising the unit Reynolds number (Re/x) and reducing pumping power requirements. Operation is envisioned at temperatures of −168°C and below. A schlieren photograph taken in the NTF shows a Mach angle of 57° where $T = -168$°C and $p = 9$ kPa. Evaluate the local Mach number and flow speed. Calculate the unit Reynolds number for the flow.

12.47 An F-4 aircraft makes a high-speed pass over an airfield on a day when $T = 35$°C. The aircraft flies at $M = 1.4$ and 200 m altitude. Calculate the speed of the aircraft. How long after it passes directly over point A on the ground does its Mach cone pass over point A?

12.48 While jogging on the beach (it's a warm summer day, about 25°C) a high-speed jet flies overhead. You guesstimate that it's at an altitude of about 3000 m, and count off about 7.5 s before you hear it. Estimate the speed and Mach number of the jet.

12.49 An aircraft passes overhead at 3 km altitude. The aircraft flies at $M = 1.5$; assume air temperature is constant at 20°C. Find the air speed of the aircraft. A headwind blows at 30 m/s. How long after the aircraft passes directly overhead does its sound reach a point on the ground?

12.50 A supersonic aircraft flies at 3 km altitude at a speed of 1000 m/s on a standard day. How long after passing directly above a ground observer is the sound of the aircraft heard by the ground observer?

12.51 For the conditions of Problem 12.50, find the location at which the sound wave that first reaches the ground observer was emitted.

12.52 The Concorde supersonic transport cruised at $M = 2.2$ at 17 km altitude on a standard day. How long after the aircraft passed directly above a ground observer was the sound of the aircraft heard?

12.53 The airflow around an automobile is assumed to be incompressible. Investigate the validity of this assumption for an automobile traveling at 96 km/h. (Relative to the automobile the minimum air velocity is zero, and the maximum is approximately 190 km/h.)

12.54 Opponents of supersonic transport aircraft claim that sound waves can be refracted in the upper atmosphere and that, as a result, sonic booms can be heard several hundred miles away from the ground track of the aircraft. Explain the phenomenon of sound wave refraction.

Reference State: Local Isentropic Stagnation Properties

12.55 Plot the percentage discrepancy between the density at the stagnation point and the density at a location where the Mach number is M, of a compressible flow, for Mach numbers ranging from 0.05 to 0.95. Find the Mach numbers at which the discrepancy is 1 percent, 5 percent, and 10 percent.

12.56 Find the stagnation temperature at the nose of the missile described in Problem 12.29.

12.57 Find the stagnation temperature at the nose of the aircraft described in Problem 12.34.

12.58 Find the ratio of static to total pressure for a car moving at 88 km/h and for a Formula One race car traveling at 355 km/h at sea level. Do you expect the flow over either car to experience compressibility effects?

12.59 Find the dynamic and stagnation pressures for the missile described in Problem 12.29.

12.60 Find the dynamic and stagnation pressures for the aircraft described in Problem 12.34.

12.61 An aircraft flies at 250 m/s in air at 28 kPa and -50°C. Find the stagnation pressure at the nose of the aircraft.

12.62 Compute the air density in the undisturbed air, and at the stagnation point, of Problem 12.61. What is the percentage increase in density? Can we approximate this as an incompressible flow?

12.63 For an aircraft traveling at $M = 2$ at an elevation of 12 km, find the dynamic and stagnation pressures.

12.64 A body moves through standard air at 200 m/s. What is the stagnation pressure on the body? Assume (a) compressible flow and (b) incompressible flow.

12.65 Consider flow of standard air at 600 m/s. What is the local isentropic stagnation pressure? The stagnation enthalpy? The stagnation temperature?

12.66 A DC-10 aircraft cruises at 12 km altitude on a standard day. A pitot-static tube on the nose of the aircraft measures stagnation and static pressures of 29.6 kPa and 19.4 kPa. Calculate (a) the flight Mach number of the aircraft, (b) the speed of the aircraft, and (c) the stagnation temperature that would be sensed by a probe on the aircraft.

12.67 An aircraft cruises at $M = 0.65$ at 10 km altitude on a standard day. The aircraft speed is deduced from measurement of the difference between the stagnation and static pressures. What is the value of this difference? Compute the air speed from this actual difference assuming (a) compressibility and (b) incompressibility. Is the discrepancy in airspeed computations significant in this case?

12.68 The Anglo-French Concorde supersonic transport cruised at $M = 2.2$ at 20 km altitude. Evaluate the speed of sound, aircraft flight speed, and Mach angle. What was the maximum air temperature at stagnation points on the aircraft structure?

12.69 Modern high-speed aircraft use "air data computers" to compute air speed from measurement of the difference between the stagnation and static pressures. Plot, as a function of actual Mach number M, for $M = 0.1$ to $M = 0.9$, the percentage error in computing the Mach number assuming incompressibility (i.e., using the Bernoulli equation), from this pressure difference. Plot the percentage error in speed, as a function of speed, of an aircraft cruising at 12 km altitude, for a range of speeds corresponding to the actual Mach number ranging from $M = 0.1$ to $M = 0.9$.

12.70 A supersonic wind tunnel test section is designed to have $M = 2.5$ at 15°C and 35 kPa (abs). The fluid is air. Determine the required inlet stagnation conditions, T_0 and p_0. Calculate the required mass flow rate for a test section area of 0.175 m^2.

12.71 Air flows steadily through a length (① denotes inlet and ② denotes exit) of insulated constant-area duct. Properties change along the duct as a result of friction.

(a) Beginning with the control volume form of the first law of thermodynamics, show that the equation can be reduced to

$$h_1 + \frac{V_1^2}{2} = h_2 + \frac{V_2^2}{2} = \text{constant}$$

(b) Denoting the constant by h_0 (the stagnation enthalpy), show that for adiabatic flow of an ideal gas with friction

$$\frac{T_0}{T} = 1 + \frac{k-1}{2}M^2$$

(c) For this flow does $T_{0_1} = T_{0_2}$? $p_{0_1} = p_{0_2}$? Explain these results.

12.72 A new design for a supersonic transport is tested in a wind tunnel at $M = 1.8$. Air is the working fluid. The stagnation temperature and pressure for the wind tunnel are 1400 kPa and 260°C, respectively. The model wing area is 0.064 m^2. The measured lift and drag are 53,000 N and 7100 N, respectively. Find the lift and drag coefficients.

12.73 For aircraft flying at supersonic speeds, lift and drag coefficients are functions of Mach number only. A supersonic transport with wingspan of 75 m is to fly at 780 m/s at 20 km altitude on a standard day. Performance of the aircraft is to be measured from tests of a model with 0.9 m wingspan in a supersonic wind tunnel. The wind tunnel is to be supplied from a large reservoir of compressed air, which can be heated if desired. The static temperature of air in the test section is to be 10°C to avoid freezing of moisture. At what air speed should the wind tunnel tests be run to duplicate the Mach number of the prototype? What must be the stagnation temperature in the reservoir? What pressure is required in the reservoir if the test section pressure is to be 10 kPa (abs)?

12.74 Actual performance characteristics of the Lockheed SR-71 "Blackbird" reconnaissance aircraft were classified. However, it was thought to cruise at $M = 3.3$ at 26 km altitude. Calculate the aircraft flight speed for these conditions. Determine the local isentropic stagnation pressure. Because the aircraft speed is supersonic, a normal shock occurs in front of a total-head tube. The stagnation pressure decreases by 74.7 percent across the shock. Evaluate the stagnation pressure sensed by a probe on the aircraft. What is the maximum air temperature at stagnation points on the aircraft structure?

12.75 The NASA X-43A Hyper-X experimental vehicle traveled at $M = 9.68$ at an altitude of 33528 m. Calculate the flight speed for these conditions. Determine the local stagnation pressure. Because the aircraft speed is supersonic, a normal shock wave occurs in front of a total-head tube. However, the shock wave results in a stagnation pressure decrease of 99.6 percent. Evaluate the stagnation pressure sensed by a probe on the aircraft. What is the maximum air temperature at stagnation points on the aircraft structure?

12.76 Air flows in an insulated duct. At point ① the conditions are $M_1 = 0.1$, $T_1 = 20°C$, and $p_1 = 1.0$ MPa (abs). Downstream, at point ②, because of friction the conditions are $M_2 = 0.7$, $T_2 = -5.62°C$, and $p_2 = 136.5$ kPa (abs). (Four significant figures are given to minimize roundoff errors.) Compare the stagnation temperatures at points ① and ②, and explain the result. Compute the stagnation pressures at points ① and ②. Can you explain how it can be that the velocity *increases* for this frictional flow? Should this process be isentropic or not? Justify your answer by computing the change in entropy between points ① and ②. Plot static and stagnation state points on a *Ts* diagram.

12.77 Air is cooled as it flows without friction at a rate of 0.05 kg/s in a duct. At point ① the conditions are $M_1 = 0.5$, $T_1 = 500°C$, and $p_1 = 500$ kPa (abs). Downstream, at point ②, the conditions are $M_2 = 0.2$, $T_2 = -18.57°C$, and $p_2 = 639.2$ kPa (abs). (Four significant figures are given to minimize roundoff errors.) Compare the stagnation temperatures at points ① and ②, and explain the result. Compute the rate of cooling. Compute the stagnation pressures at points ① and ②. Should this process be isentropic or not? Justify your answer by computing the change in entropy between points ① and ②. Plot static and stagnation state points on a *Ts* diagram.

12.78 Consider steady, adiabatic flow of air through a long straight pipe with $A = 0.05$ m^2. At the inlet (section ①) the air is at 200 kPa (abs), 60°C, and 146 m/s. Downstream at section ②, the air is at 95.6 kPa (abs) and 280 m/s. Determine p_{0_1}, p_{0_2}, T_{0_1}, T_{0_2} and the entropy change for the flow. Show static and stagnation state points on a *Ts* diagram.

12.79 Air flows steadily through a constant-area duct. At section ①, the air is at 400 kPa (abs), 325 K, and 150 m/s. As a result of heat transfer and friction, the air at section ② downstream is at 275 kPa (abs), 450 K. Calculate the heat transfer per kilogram of air between sections ① and ②, and the stagnation pressure at section ②.

12.80 The combustion process in a ramjet engine is modeled as simple heat addition to air in a frictionless duct. Consider such a combustor, with air flowing at a rate of 0.045 kg/s. At point ① the conditions are M_1 0.2, $T_1 = 315.6°C$, and $p_1 = 48$ kPa. Downstream, at point ②, the conditions are $M_2 = 0.9$, $T_2 = 1032°C$, and $p_2 = 28.3$ kPa. Compare the stagnation temperatures at points ① and ②, and explain the result. Compute the rate of heat addition to the flow. Compute the stagnation pressures at points ① and ②. Should this process be isentropic or not? Justify your answer by computing the change in entropy between points ① and ②. Plot static and stagnation state points on a *Ts* diagram.

12.81 Let us revisit the ramjet combustor in Problem 12.80. To more accurately model the flow, we now include the effects of friction in the duct. Once the effects of friction have been included, we find that the conditions at state ② are now $M_2 = 0.9$, $T_2 = 904°C$, and $p_2 = 11.03$ kPa. Recalculate the heat transfer per pound of air between sections ① and ②, and the stagnation pressure at section ②.

12.82 Air passes through a normal shock in a supersonic wind tunnel. Upstream conditions are $M_1 = 1.8$, $T_1 = 270$ K, and $p_1 = 10.0$ kPa (abs). Downstream conditions are $M_2 = 0.6165$, $T_2 = 413.6$ K, and $p_2 = 36.13$ kPa (abs). (Four significant figures are given to minimize roundoff errors.) Evaluate local isentropic stagnation conditions (a) upstream from, and (b) downstream from, the normal shock. Calculate the change in specific entropy across the shock. Plot static and stagnation state points on a *Ts* diagram.

12.83 Air enters a turbine at $M_1 = 0.4$, $T_1 = 1250°C$, and $p_1 = 625$ kPa (abs). Conditions leaving the turbine are $M_2 = 0.8$, $T_2 = 650°C$, and $p_2 = 20$ kPa (abs). Evaluate local isentropic stagnation conditions (a) at the turbine inlet and (b) at the turbine outlet. Calculate the change in specific entropy across the turbine. Plot static and stagnation state points on a *Ts* diagram.

12.84 A Boeing 747 cruises at $M = 0.87$ at an altitude of 13 km on a standard day. A window in the cockpit is located where the external flow Mach number is 0.2 relative to the plane surface. The cabin is pressurized to an equivalent altitude of 2500 m in a standard atmosphere. Estimate the pressure difference across the window. Be sure to specify the direction of the net pressure force.

Critical Conditions

12.85 If a window of the cockpit in Problem 12.84 develops a tiny leak the air will start to rush out at critical speed. Find the mass flow rate if the leak area is 1 mm^2.

12.86 Space debris impact is a real concern for spacecraft. If a piece of space debris were to create a hole of 0.625 mm^2 area in the hull of the International Space Station (ISS), at what rate would air leak from the ISS? Assume that the atmosphere in the International Space Station (ISS) is air at a pressure of 101.3 kPa and a temperature of 18°C.

12.87 A CO_2 cartridge is used to propel a toy rocket. Gas in the cartridge is pressurized to 45 MPa (gage) and is at 25°C.

Calculate the critical conditions (temperature, pressure, and flow speed) that correspond to these stagnation conditions.

12.88 The gas storage reservoir for a high-speed wind tunnel contains helium at 2000 k and 5 MPa. Calculate the critical conditions (temperature, pressure, and flow speed) that correspond to these stagnation conditions.

12.89 Stagnation conditions in a solid propellant rocket motor are $T_0 = 3000$ K and $p_0 = 45$ MPa (gage). Critical conditions occur in the throat of the rocket nozzle where the Mach number is equal to one. Evaluate the temperature, pressure, and flow speed at the throat. Assume ideal gas behavior with $R = 323$ J/(kg·K) and $k = 1.2$.

12.90 The hot gas stream at the turbine inlet of a JT9-D jet engine is at 1500°C, 140 kPa (abs), and $M = 0.32$. Calculate the critical conditions (temperature, pressure, and flow speed) that correspond to these conditions. Assume the fluid properties of pure air.

12.91 Certain high-speed wind tunnels use combustion air heaters to generate the extreme pressures and temperatures required to accurately simulate flow at high Mach numbers. In one set of tests, a combustion air heater supplied stagnation conditions of 1.7 MPa and 1010 K. Calculate the critical pressure and temperature corresponding to these stagnation conditions.

12.92 The ramjet combustor exhaust from Problem 12.81 is accelerated through a nozzle to critical conditions. Calculate the temperature, pressure, and flow velocity at the nozzle exit. Assume fluid properties of pure air.

13

Compressible Flow

Case Study in Energy and the Environment

Wind Power: The *Windspire* Vertical Axis Wind Turbine

Wind turbine farms are now a common sight in many parts of the world. One of the earliest wind farms in the United States, the Altamont Pass Wind Farm in central California, has almost 5000 relatively small wind turbines of various types, making it at one time the largest farm in the world in terms of capacity. Altamont Pass is still the largest single concentration of wind turbines in the world, with a capacity of 576 MW, producing about 125 MW on average and 1.1 TWh annually. Even though the turbines are quite large, technology has improved since they were installed in the 1970s, and they are being gradually replaced with much larger and more cost-effective units. The smaller turbines are dangerous to various birds such as golden eagles (about 70 of these are killed each year). The new, larger units turn more slowly and, being so much larger and higher, are less hazardous to the local wildlife.

As we saw in the last *Case Study in Energy and the Environment*, a number of companies are developing small-scale alternatives to such wind farms. One such company is *Windspire Energy* in Nevada. Its wind turbines, as shown in the photograph, are low-cost, low-noise, attractive-looking wind power generators for use

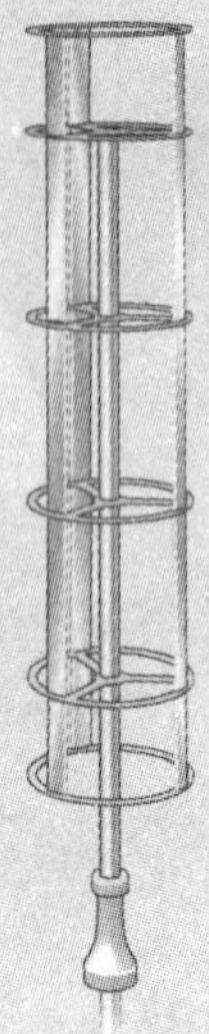

Windspire Turbines.

with residential, business, and commercial buildings. Each *Windspire* can generate about 1.2 kW of electricity; for comparison, a typical home solar array can generate up to 3 kW in full, direct sunlight. Manufactured in Michigan, the *Windspire* wind turbines are 10 m tall and 1 m wide, propeller-free, vertical-axis wind turbine (VAWTs, discussed in Chapter 10). The *Windspire* turbines, in contrast to the *Helix* VAWT in Chapter 12, are lift devices; that is, the vertical blades are essentially airfoils generating lift (see Chapter 9) and hence torque. *Windspires* are currently powering over 500 homes, small businesses, schools, museums, parks, vineyards, and commercial buildings.

Recently Adobe Systems Inc., makers of the commonly used Adobe Acrobat, installed 20 *Windspire* wind turbines at its San Jose, California, campus. This is in keeping with Adobe's leadership in green building efforts; the headquarters is the first commercial office building to receive the Leadership in Energy and Environmental Design (LEED-EB) Platinum certification for its headquarters. The new *Windspires* are located on Adobe's sixth-floor patio, which doubles as a rooftop garden and recreational area; the patio is located between three office towers, which create a wind tunnel effect from steady winds off the Pacific Ocean. Adobe selected the *Windspire* for its powerful, sleek, quiet, and aesthetically pleasing design. The tall, slender, delicate-looking cylinders look nothing like the giant turbine blades turning above the Altamont Pass, but they generate power all the same!

Sales of small turbines—those with a capacity of 100 kW or less—rose 78 percent in 2008; so the market is rapidly growing. Unlike the *Windspire*, most of the small turbines sold in the United States are horizontal-axis turbines (HAWTs). *Windspire Energy* believes that its VAWT *Windspire* has a number of advantages over HAWTs, including its much smaller footprint, lower noise level, and aesthetic appeal.

In Chapter 12 we reviewed some basic concepts of compressible flow. The main focus of this chapter is to discuss one-dimensional compressible flow in more detail. The first question we can ask is "What would cause the fluid properties to vary in a one-dimensional compressible flow?" The answer is that various phenomena can cause changes:

- Flow with varying area (causing the velocity to change, and hence other property changes).
- Normal shock (a "violent" adiabatic process that causes the entropy to increase, and hence other property changes).
- Flow in a channel with friction (causing the entropy to increase, and hence other property changes).
- Flow in a channel with heating or cooling (causing a change in fluid energy, and hence other property changes).

For simplicity, we will study each of these phenomena separately (bearing in mind that a real flow is likely to experience several of them simultaneously). After completing our treatment of one-dimensional flow, we will introduce some basic concepts of two-dimensional flows: oblique shocks and expansion waves.

Basic Equations for One-Dimensional Compressible Flow 13.1

Our first task is to develop general equations for a one-dimensional flow that express the basic laws from Chapter 4: *mass conservation* (continuity), *momentum*, the *first law of thermodynamics*, the *second law of thermodynamics*, and an *equation of state*. To do so, we will use the fixed control volume shown in Fig. 13.1. We initially assume that the flow is affected by *all* of the phenomena mentioned above (i.e., area change, friction, and heat transfer—even the normal shock will be described by this approach). Then, for each individual phenomenon we will simplify the equations to obtain useful results.

As shown in Fig. 13.1, the properties at sections ① and ② are labeled with corresponding subscripts. R_x is the x component of surface force from friction and pressure on the sides of the channel. There will also be surface forces from pressures at surfaces ① and ②. Note that the x component of body force is zero, so it is not shown. $\dot{Q}$ is the heat transfer.

a. Continuity Equation

Basic equation:

$$\overbrace{\frac{\partial}{\partial t}\int_{\mathrm{CV}} \rho \, d\forall}^{=0(1)} + \int_{\mathrm{CS}} \rho \vec{V} \cdot d\vec{A} = 0 \tag{4.12}$$

Assumptions: (1) Steady flow.
(2) One-dimensional flow.

Then

$$(-\rho_1 V_1 A_1) + (\rho_2 V_2 A_2) = 0$$

or

$$\rho_1 V_1 A_1 = \rho_2 V_2 A_2 = \rho V A = \dot{m} = \text{constant} \tag{13.1a}$$

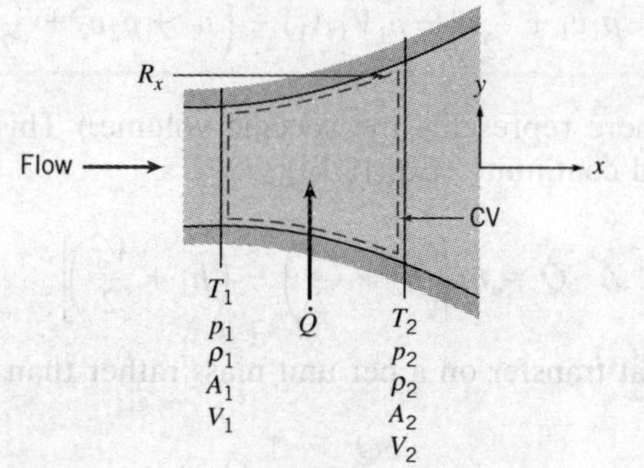

Fig. 13.1 Control volume for analysis of a general one-dimensional flow.

b. Momentum Equation

Basic equation:

$$F_{S_x} + \cancelto{0(3)}{F_{B_x}} = \cancelto{0(1)}{\frac{\partial}{\partial t}\int_{\text{CV}} V_x\,\rho\,d\forall} + \int_{\text{CS}} V_x\,\rho\vec{V}\cdot d\vec{A} \tag{4.18a}$$

Assumption: (3) $F_{B_x} = 0$

The surface force is caused by pressure forces at surfaces ① and ②, and by the friction and distributed pressure force, R_x, along the channel walls. Substituting gives

$$R_x + p_1A_1 - p_2A_2 = V_1(-\rho_1 V_1 A_1) + V_2(\rho_2 V_2 A_2)$$

Using continuity, we obtain

$$R_x + p_1A_1 - p_2A_2 = \dot{m}V_2 - \dot{m}V_1 \tag{13.1b}$$

c. First Law of Thermodynamics

Basic equation:

$$\dot{Q} - \cancel{\dot{W}_s} - \cancel{\dot{W}_{\text{shear}}} - \cancel{\dot{W}_{\text{other}}} = \cancel{\frac{\partial}{\partial t}\int_{\text{CV}} e\,\rho\,d\forall} + \int_{\text{CS}} (e + pv)\rho\,\vec{V}\cdot d\vec{A} \tag{4.56}$$

where

$$e = u + \frac{V^2}{2} + \cancelto{\simeq 0(6)}{gz}$$

Assumptions: (4) $\dot{W}_s = 0$.
(5) $\dot{W}_{\text{shear}} = \dot{W}_{\text{other}} = 0$.
(6) Effects of gravity are negligible.

(Note that even if we have friction, there is no friction *work* at the walls because with friction the velocity at the walls must be zero from the no-slip condition.) Under these assumptions, the first law reduces to

$$\dot{Q} = \left(u_1 + p_1v_1 + \frac{V_1^2}{2}\right)(-\rho_1 V_1 A_1) + \left(u_2 + p_2v_2 + \frac{V_2^2}{2}\right)(\rho_2 V_2 A_2)$$

(Remember that v here represents the specific volume.) This can be simplified by using $h \equiv u + pv$, and continuity (Eq. 13.1a),

$$\dot{Q} = \dot{m}\left[\left(h_2 + \frac{V_2^2}{2}\right) - \left(h_1 + \frac{V_1^2}{2}\right)\right]$$

We can write the heat transfer on a per unit mass rather than per unit time basis:

$$\frac{\delta Q}{dm} = \frac{1}{\dot{m}}\dot{Q}$$

so

$$\frac{\delta Q}{dm} + h_1 + \frac{V_1^2}{2} = h_2 + \frac{V_2^2}{2} \tag{13.1c}$$

Equation 13.1c expresses the fact that heat transfer changes the total energy (the sum of thermal energy h, and kinetic energy $V^2/2$) of the flowing fluid. This combination, $h + V^2/2$, occurs often in compressible flow, and is called the *stagnation enthalpy*, h_0. This is the enthalpy obtained if a flow is brought adiabatically to rest.

Hence, Eq. 13.1c can also be written

$$\frac{\delta Q}{dm} = h_{0_2} - h_{0_1}$$

We see that heat transfer causes the stagnation enthalpy, and hence, stagnation temperature, T_0, to change.

d. Second Law of Thermodynamics

Basic equation:

$$\overbrace{\frac{\partial}{\partial t}\int_{CV} s\rho d\forall}^{=0(1)} + \int_{CS} s\,\rho \vec{V}\cdot d\vec{A} \geq \int_{CS} \frac{1}{T}\left(\frac{\dot{Q}}{A}\right) dA \tag{4.58}$$

or

$$s_1(-\rho_1 V_1 A_1) + s_2(\rho_2 V_2 A_2) \geq \int_{CS} \frac{1}{T}\left(\frac{\dot{Q}}{A}\right) dA$$

and, again using continuity,

$$\dot{m}(s_2 - s_1) \geq \int_{CS} \frac{1}{T}\left(\frac{\dot{Q}}{A}\right) dA \tag{13.1d}$$

e. Equation of State

Equations of state are relations among intensive thermodynamic properties. These relations may be available as tabulated data or charts, or as algebraic equations. In general, regardless of the format of the data, as we discussed in Chapter 12 (see References [1–3] of that chapter), for a simple substance any property can be expressed as a function of any two other independent properties. For example, we could write $h = h(s, p)$, or $\rho = \rho(s; p)$, and so on.

We will primarily be concerned with ideal gases with constant specific heats, and for these we can write Eqs. 12.1 and 12.7b (renumbered for convenient use in this chapter),

$$p = \rho RT \tag{13.1e}$$

and

$$\Delta h = h_2 - h_1 = c_p \Delta T = c_p(T_2 - T_1) \tag{13.1f}$$

For ideal gases with constant specific heats, the change in entropy, $\Delta s = s_2 - s_1$, for any process can be computed from any of Eqs. 12.11. For example, Eq. 12.11b (renumbered for convenient use in this chapter) is

$$\Delta s = s_2 - s_1 = c_p \ln \frac{T_2}{T_1} - R \ln \frac{p_2}{p_1} \tag{13.1g}$$

We now have a basic set of equations for analyzing one-dimensional compressible flows of an ideal gas with constant specific heats:

$$\rho_1 V_1 A_1 = \rho_2 V_2 A_2 = \rho V A = \dot{m} = \text{constant} \tag{13.1a}$$

$$R_x + p_1 A_1 - p_2 A_2 = \dot{m} V_2 - \dot{m} V_1 \tag{13.1b}$$

$$\frac{\delta Q}{dm} + h_1 + \frac{V_1^2}{2} = h_2 + \frac{V_2^2}{2} \tag{13.1c}$$

$$\dot{m}(s_2 - s_1) \geq \int_{CS} \frac{1}{T}\left(\frac{\dot{Q}}{A}\right) dA \tag{13.1d}$$

$$p = \rho R T \tag{13.1e}$$

$$\Delta h = h_2 - h_1 = c_p \Delta T = c_p (T_2 - T_1) \tag{13.1f}$$

$$\Delta s = s_2 - s_1 = c_p \ln \frac{T_2}{T_1} - R \ln \frac{p_2}{p_1} \tag{13.1g}$$

Note that Eq. 13.1e applies only if we have an ideal gas; Equations 13.1f and 13.1g apply only if we have an ideal gas with constant specific heats. Our task is now to simplify this set of equations for each of the phenomena that can affect the flow:

- Flow with varying area.
- Normal shock.
- Flow in a channel with friction.
- Flow in a channel with heating or cooling.

13.2 Isentropic Flow of an Ideal Gas: Area Variation

The first phenomenon is one in which the flow is changed only by area variation—there is no heat transfer ($\delta Q/dm = 0$) or friction (so that R_x, the x component of surface force, results only from pressure on the sides of the channel), and there are no shocks. The absence of heat transfer, friction, and shocks (which are "violent" and therefore inherently irreversible) means the flow will be reversible and adiabatic, so Eq. 13.1d becomes

$$\dot{m}(s_2 - s_1) = \int_{CS} \frac{1}{T}\left(\frac{\dot{Q}}{A}\right) dA = 0$$

or

$$\Delta s = s_2 - s_1 = 0$$

so such a flow is *isentropic*. This means that Eq. 13.1g leads to the result we saw in Chapter 12,

$$T_1 p_1^{(1-k)/k} = T_2 p_2^{(1-k)/k} = T p^{(1-k)/k} = \text{constant} \tag{12.12b}$$

or its equivalent (which can be obtained by using the ideal gas equation of state in Eq. 12.12b to eliminate temperature),

$$\frac{p_1}{\rho_1^k} = \frac{p_2}{\rho_2^k} = \frac{p}{\rho^k} = \text{constant} \tag{12.12c}$$

Hence, the basic set of equations (Eqs. 13.1) becomes:

$$\rho_1 V_1 A_1 = \rho_2 V_2 A_2 = \rho V A = \dot{m} = \text{constant} \tag{13.2a}$$

$$R_x + p_1 A_1 - p_2 A_2 = \dot{m} V_2 - \dot{m} V_1 \tag{13.2b}$$

$$h_{0_1} = h_1 + \frac{V_1^2}{2} = h_2 + \frac{V_2^2}{2} = h_{0_2} = h_0 \tag{13.2c}$$

$$s_2 = s_1 = s \tag{13.2d}$$

$$p = \rho R T \tag{13.2e}$$

$$\Delta h = h_2 - h_1 = c_p \Delta T = c_p (T_2 - T_1) \tag{13.2f}$$

$$\frac{p_1}{\rho_1^k} = \frac{p_2}{\rho_2^k} = \frac{p}{\rho^k} = \text{constant} \tag{13.2g}$$

Note that Eqs. 13.2c, 13.2d, and 13.2f provide insight into how this process appears on an *hs* diagram and on a *Ts* diagram. From Eq. 13.2c, the total energy, or stagnation enthalpy h_0, of the fluid is constant; the enthalpy and kinetic energy may vary along the flow, but their sum is constant. This means that if the fluid accelerates, its temperature must decrease, and vice versa. Equation 13.2d indicates that the entropy remains constant. These results are shown for a typical process in Fig. 13.2.

Equation 13.2f indicates that the temperature and enthalpy are linearly related; hence, processes plotted on a *Ts* diagram will look very similar to that shown in Fig. 13.2 except for the vertical scale.

Equations 13.2 *could* be used to analyze isentropic flow in a channel of varying area. For example, if we know conditions at section ① (i.e., p_1, ρ_1, T_1, s_1, h_1, V_1, and A_1) we could use these equations to find conditions at some new section ② where the area is A_2: We would have seven equations and seven unknowns (p_2, ρ_2, T_2, s_2, h_2, V_2, and, if desired, the net pressure force on the walls R_x). We stress *could*, because in practice this procedure is unwieldy—we have a set of seven *nonlinear coupled algebraic* equations to solve (however, we will see, for example in Example 13.1, that *Excel* can be used to solve this set of equations). Instead we will usually use some of these equations as convenient but also take advantage of the results we obtained for isentropic flows in Chapter 12, and develop property relations in terms of the local Mach number, the stagnation conditions, and critical conditions.

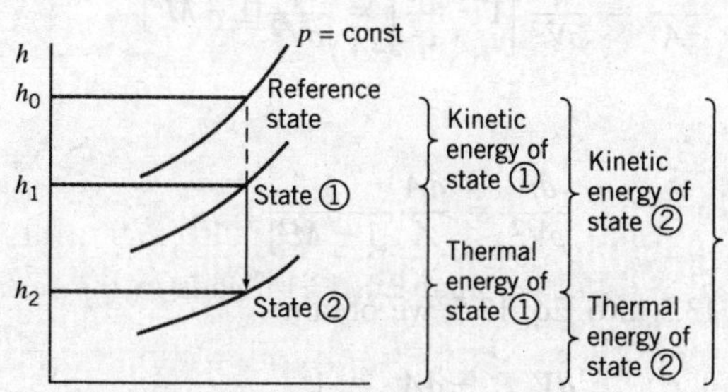

Fig. 13.2 Isentropic flow in the *hs* plane.

Before proceeding with this approach, we can gain insight into the isentropic process by reviewing the results we obtained in Chapter 12 when we analyzed a differential control volume (Fig. 12.5). The momentum equation for this was

$$\frac{dp}{\rho} + d\left(\frac{V^2}{2}\right) = 0 \tag{12.20b}$$

(Note that we could also have obtained this equation using our set of equations, Eqs. 13.1. If we applied Eqs. 13.1 to a differential control volume, we could replace ρ_1, V_1, and A_1 with ρ, V, and A, and ρ_2, V_2, and A_2 with $\rho + d\rho$, $V + dV$, and $A + dA$. Then Eq. 13.1a and 13.1b simplify to the above equation.) Then

$$dp = -\rho V\,dV$$

Dividing by ρV^2, we obtain

$$\frac{dp}{\rho V^2} = -\frac{dV}{V} \tag{13.3}$$

A convenient differential form of the continuity equation can be obtained from Eq. 13.2a, in the form

$$\rho A V = \text{constant}$$

Differentiating and dividing by $\rho A V$ yields

$$\frac{d\rho}{\rho} + \frac{dA}{A} + \frac{dV}{V} = 0 \tag{13.4}$$

Solving Eq. 13.4 for dA/A gives

$$\frac{dA}{A} = -\frac{dV}{V} - \frac{d\rho}{\rho}$$

Substituting from Eq. 13.3 gives

$$\frac{dA}{A} = \frac{dp}{\rho V^2} - \frac{d\rho}{\rho}$$

or

$$\frac{dA}{A} = \frac{dp}{\rho V^2}\left[1 - \frac{V^2}{dp/d\rho}\right]$$

Now recall that for an isentropic process, $dp/d\rho = \partial p/\partial\rho)_s = c^2$, so

$$\frac{dA}{A} = \frac{dp}{\rho V^2}\left[1 - \frac{V^2}{c^2}\right] = \frac{dp}{\rho V^2}[1 - M^2]$$

or

$$\frac{dp}{\rho V^2} = \frac{dA}{A}\frac{1}{[1 - M^2]} \tag{13.5}$$

Substituting from Eq. 13.3 into Eq. 13.5, we obtain

$$\frac{dV}{V} = -\frac{dA}{A}\frac{1}{[1 - M^2]} \tag{13.6}$$

Note that for an isentropic flow there can be no friction. Equations 13.5 and 13.6 confirm that for this case, from a momentum point of view we expect an increase in pressure to cause a decrease in speed, and vice versa. Although we cannot use them for computations (we have not so far determined how M varies with A), Eqs. 13.5 and 13.6 give us very interesting insights into how the pressure and velocity change as we change the area of the flow. Three possibilities are discussed below.

Subsonic Flow, $M < 1$

For $M < 1$, the factor $1/[1 - M^2]$ in Eqs. 13.5 and 13.6 is positive, so that a positive dA leads to a positive dp and a negative dV. These mathematical results mean that in a *divergent* section ($dA > 0$) the flow must experience an *increase* in pressure ($dp > 0$) and the velocity must *decrease* ($dV < 0$). Hence *a divergent channel is a subsonic diffuser* (a diffuser is a device that decelerates a flow).

On the other hand, a negative dA leads to a negative dp and a positive dV. These mathematical results mean that in a *convergent* section ($dA < 0$) the flow must experience a *decrease* in pressure ($dp < 0$) and the velocity must *increase* ($dV > 0$). Hence *a convergent channel is a subsonic nozzle* (a nozzle is a device that accelerates a flow).

These results are consistent with our everyday experience and are not surprising—for example, recall the venturi meter in Chapter 8, in which a reduction in area at the throat of the venturi led to a local increase in velocity, and because of the Bernoulli principle, to a pressure drop, and the divergent section led to pressure recovery and flow deceleration. (The Bernoulli principle assumes incompressible flow, which is the limiting case of subsonic flow.)

The subsonic diffuser and nozzle are also shown in Fig. 13.3.

Supersonic Flow, $M > 1$

For $M > 1$, the factor $1/[1 - M^2]$ in Eqs. 13.5 and 13.6 is negative, so that a positive dA leads to a negative dp and a positive dV. These mathematical results mean that in a *divergent* section ($dA > 0$) the flow must experience a *decrease* in pressure ($dp < 0$) and the velocity must *increase* ($dV > 0$). Hence *a divergent channel is a supersonic nozzle.*

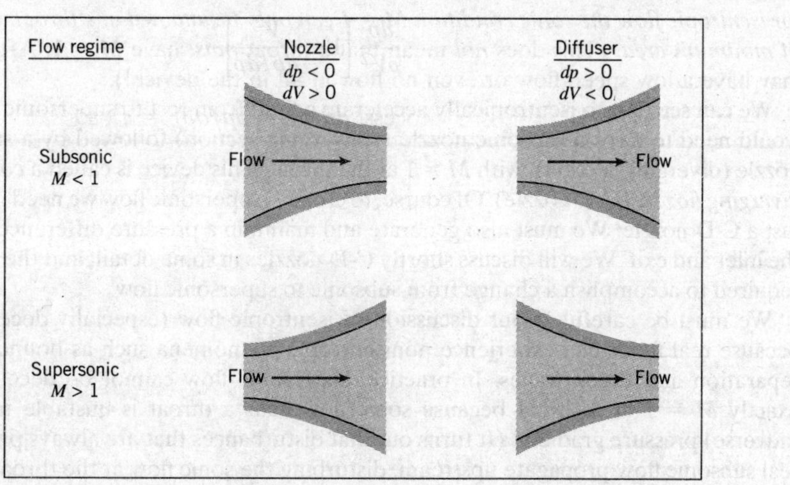

Fig. 13.3 Nozzle and diffuser shapes as a function of initial Mach number.

On the other hand, a negative dA leads to a positive dp and a negative dV. These mathematical results mean that in a *convergent* section ($dA < 0$) the flow must experience an *increase* in pressure ($dp > 0$) and the velocity must *decrease* ($dV < 0$). Hence *a convergent channel is a supersonic diffuser*.

These results are inconsistent with our everyday experience and are at first a bit surprising—they are the opposite of what we saw in the venturi meter! The results are consistent with the laws of physics; for example, an increase in pressure must lead to a flow deceleration because pressure forces are the only forces acting.

The supersonic nozzle and diffuser are shown in Fig. 13.3.

These somewhat counterintuitive results can be understood when we realize that we are used to assuming that ρ = constant, but we are now in a flow regime where the fluid density is a sensitive function of flow conditions. From Eq. 13.4,

$$\frac{dV}{V} = -\frac{dA}{A} - \frac{d\rho}{\rho}$$

For example, in a supersonic diverging flow (dA positive) the flow actually accelerates (dV also positive) because the density drops sharply ($d\rho$ is negative and large, with the net result that the right side of the equation is positive). We can see examples of supersonic diverging nozzles in the space shuttle main engines, each of which has a nozzle about 3 m long with a 2.4 m exit diameter. The maximum thrust is obtained from the engines when the combustion gases exit at the highest possible speed, which the nozzles achieve.

Sonic Flow, $M = 1$

As we approach $M = 1$, from either a subsonic or supersonic state, the factor $1/[1 - M^2]$ in Eqs. 13.5 and 13.6 approaches infinity, implying that the pressure and velocity changes also approach infinity. This is obviously unrealistic, so we must look for some other way for the equations to make physical sense. The only way we can avoid these singularities in pressure and velocity is if we require that $dA \to 0$ as $M \to 1$. Hence, for an isentropic flow, sonic conditions can only occur where the area is constant! We can be even more specific: We can imagine approaching $M = 1$ from either a subsonic or a supersonic state. A subsonic flow ($M < 1$) would need to be accelerated using a subsonic nozzle, which we have learned is a converging section; a supersonic flow ($M > 1$) would need to be decelerated using a supersonic diffuser, which is also a converging section. Hence, sonic conditions are limited not just to a location of constant area, but one that is a minimum area. The important result is that *for isentropic flow the sonic condition $M = 1$ can only be attained at a throat, or section of minimum area*. (This does *not* mean that a throat *must* have $M = 1$. After all, we may have a low speed flow or even no flow at all in the device!).

We can see that to isentropically accelerate a fluid from rest to supersonic speed we would need to have a subsonic nozzle (converging section) followed by a supersonic nozzle (diverging section), with $M = 1$ at the throat. This device is called a *converging-diverging nozzle* (C-D nozzle). Of course, to create a supersonic flow we need more than just a C-D nozzle: We must also generate and maintain a pressure difference between the inlet and exit. We will discuss shortly C-D nozzles in some detail, and the pressures required to accomplish a change from subsonic to supersonic flow.

We must be careful in our discussion of isentropic flow (especially deceleration), because real fluids can experience nonisentropic phenomena such as boundary-layer separation and shock waves. In practice, supersonic flow cannot be decelerated to exactly $M = 1$ at a throat because sonic flow near a throat is unstable in a rising (adverse) pressure gradient. (It turns out that disturbances that are always present in a real subsonic flow propagate upstream, disturbing the sonic flow at the throat, causing shock waves to form and travel upstream, where they may be disgorged from the inlet of the supersonic diffuser.)

The throat area of a real supersonic diffuser must be slightly larger than that required to reduce the flow to $M = 1$. Under the proper downstream conditions, a weak normal shock forms in the diverging channel just downstream from the throat. Flow leaving the shock is subsonic and decelerates in the diverging channel. Thus deceleration from supersonic to subsonic flow cannot occur isentropically in practice, since the weak normal shock causes an entropy increase. Normal shocks will be analyzed in Section 13.3.

For accelerating flows (favorable pressure gradients), the idealization of isentropic flow is generally a realistic model of the actual flow behavior. For decelerating flows, the idealization of isentropic flow may not be realistic because of the adverse pressure gradients and the attendant possibility of flow separation, as discussed for incompressible boundary-layer flow in Chapter 9.

Reference Stagnation and Critical Conditions for Isentropic Flow of an Ideal Gas

As we mentioned at the beginning of this section, in principle we could use Eqs. 13.2 to analyze one-dimensional isentropic flow of an ideal gas, but the computations would be somewhat tedious. Instead, because the flow is isentropic, we can use the results of Sections 12.3 (reference stagnation conditions) and 12.4 (reference critical conditions). The idea is illustrated in Fig. 13.4: Instead of using Eqs. 13.2 to compute, for example, properties at state ② from those at state ①, we can use state ① to determine two reference states (the stagnation state and the critical state), and then use these to obtain properties at state ②. We need two reference states because the reference stagnation state does not provide area information (mathematically the stagnation area is infinite).

We will use Eqs. 12.21 (renumbered for convenience),

$$\frac{p_0}{p} = \left[1 + \frac{k-1}{2}M^2\right]^{k/(k-1)} \tag{13.7a}$$

$$\frac{T_0}{T} = 1 + \frac{k-1}{2}M^2 \tag{13.7b}$$

$$\frac{\rho_0}{\rho} = \left[1 + \frac{k-1}{2}M^2\right]^{1/(k-1)} \tag{13.7c}$$

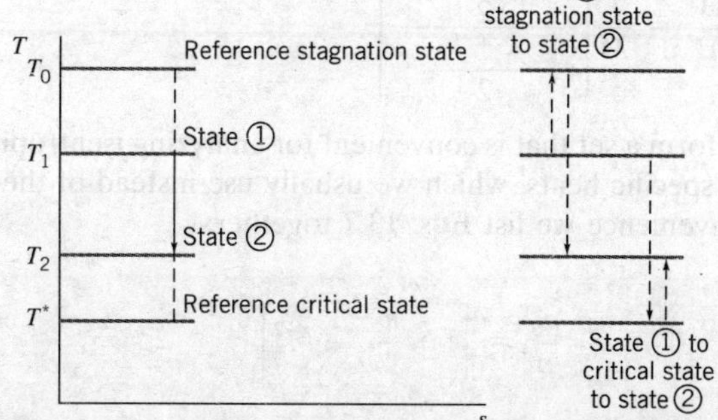

Fig. 13.4 Example of stagnation and critical reference states in the *Ts* plane.

We note that *the stagnation conditions are constant throughout the isentropic flow*. The critical conditions (when $M = 1$) were related to stagnation conditions in Section 12.4,

$$\frac{p_0}{p^*} = \left[\frac{k+1}{2}\right]^{k/(k-1)} \tag{12.22a}$$

$$\frac{T_0}{T^*} = \frac{k+1}{2} \tag{12.22b}$$

$$\frac{\rho_0}{\rho^*} = \left[\frac{k+1}{2}\right]^{1/(k-1)} \tag{12.22c}$$

$$V^* = c^* = \sqrt{\frac{2k}{k+1}RT_0} \tag{12.23}$$

Although a particular flow may never attain sonic conditions (as in the example in Fig. 13.4), we will still find the critical conditions useful as reference conditions. Equations 13.7a, 13.7b, and 13.7c relate local properties (p, ρ, T, and V) to stagnation properties (p_0, ρ_0, and T_0) via the Mach number M, and Eqs. 12.22 and 12.23 relate critical properties (p^*, ρ^*, T^*, and V^*) to stagnation properties (p_0, ρ_0, and T_0) respectively, but we have yet to obtain a relation between areas A and A^*. To do this we start with continuity (Eq. 13.2a) in the form

$$\rho AV = \text{constant} = \rho^* A^* V^*$$

Then

$$\frac{A}{A^*} = \frac{\rho^*}{\rho}\frac{V^*}{V} = \frac{\rho^*}{\rho}\frac{c^*}{Mc} = \frac{1}{M}\frac{\rho^*}{\rho}\sqrt{\frac{T^*}{T}}$$

$$\frac{A}{A^*} = \frac{1}{M}\frac{\rho^*}{\rho_0}\frac{\rho_0}{\rho}\sqrt{\frac{T^*/T_0}{T/T_0}}$$

$$\frac{A}{A^*} = \frac{1}{M}\frac{\left[1+\frac{k-1}{2}M^2\right]^{1/(k-1)}}{\left[\frac{k+1}{2}\right]^{1/(k-1)}}\left[\frac{1+\frac{k-1}{2}M^2}{\frac{k+1}{2}}\right]^{1/2}$$

$$\frac{A}{A^*} = \frac{1}{M}\left[\frac{1+\frac{k-1}{2}M^2}{\frac{k+1}{2}}\right]^{(k+1)/2(k-1)} \tag{13.7d}$$

Equations 13.7 form a set that is convenient for analyzing isentropic flow of an ideal gas with constant specific heats, which we usually use instead of the basic equations, Eqs. 13.2. For convenience we list Eqs. 13.7 together:

$$\frac{p_0}{p} = \left[1+\frac{k-1}{2}M^2\right]^{k/(k-1)} \tag{13.7a}$$

$$\frac{T_0}{T} = 1+\frac{k-1}{2}M^2 \tag{13.7b}$$

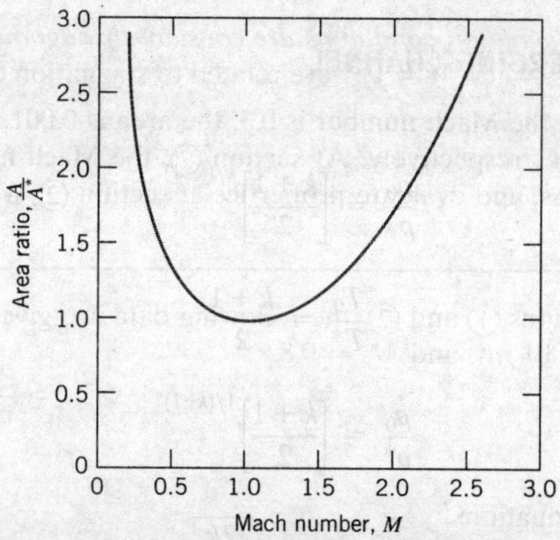

Fig. 13.5 Variation of A/A^* with Mach number for isentropic flow of an ideal gas with $k = 1.4$.

$$\frac{\rho_0}{\rho} = \left[1 + \frac{k-1}{2}M^2\right]^{1/(k-1)} \tag{13.7c}$$

$$\frac{A}{A^*} = \frac{1}{M}\left[\frac{1 + \frac{k-1}{2}M^2}{\frac{k+1}{2}}\right]^{(k+1)/2(k-1)} \tag{13.7d}$$

Equations 13.7 provide property relations in terms of the local Mach number, the stagnation conditions, and critical conditions; they are so useful that some calculators have some of them built in (for example, the *HP 48G* series [1]). It is a good idea to program them if your calculator does not already have them. There are even interactive Web sites that make them available (see, for example, [2]), and they are fairly easy to define in spreadsheets such as *Excel*. The reader is urged to download the *Excel* add-ins for these equations from the Web site; with the add-ins, functions are available for computing pressure, temperature, density or area ratios from M, and M from the ratios. While they are somewhat complicated algebraically, they have the advantage over the basic equations, Eq. 13.2, that they are not coupled. Each property can be found directly from its stagnation value and the Mach number. Equation 13.7d shows the relation between Mach number M and area A. The critical area A^* (defined whether or not a given flow ever attains sonic conditions) is used to normalize area A. For each Mach number M we obtain a unique area ratio, but as shown in Fig 13.5 each A/A^* ratio (except 1) has two possible Mach numbers—one subsonic, the other supersonic. The shape shown in Fig. 13.5 *looks* like a converging-diverging section for accelerating from a subsonic to a supersonic flow (with, as necessary, $M = 1$ only at the throat), but in practice this is not the shape to which such a passage would be built. For example, the diverging section usually will have a much less severe angle of divergence to reduce the chance of flow separation (in Fig. 13.5 the Mach number increases linearly, but this is not necessary).

Appendix E.1 lists flow functions for property ratios T_0/T, p_0/p, ρ_0/ρ, and A/A^* in terms of M for isentropic flow of an ideal gas. A table of values, as well as a plot of these property ratios, is presented for air ($k = 1.4$) for a limited range of Mach numbers. The associated *Excel* workbook, *Isentropic Relations*, can be used to print a larger table of values for air and other ideal gases.

Example 13.1 demonstrates use of some of the above equations. As shown in Fig. 13.4, we can use the equations to relate a property at one state to the stagnation value and then from the stagnation value to a second state, but note that we can accomplish this in one step—for example, p_2 can be obtained from p_1 by writing $p_2 = (p_2/p_0)(p_0/p_1)p_1$, where the pressure ratios come from Eq. 13.7a evaluated at the two Mach numbers.

Example 13.1 ISENTROPIC FLOW IN A CONVERGING CHANNEL

Air flows isentropically in a channel. At section ①, the Mach number is 0.3, the area is 0.001 m², and the absolute pressure and the temperature are 650 kPa and 62°C, respectively. At section ②, the Mach number is 0.8. Sketch the channel shape, plot a *Ts* diagram for the process, and evaluate properties at section ②. Verify that the results agree with the basic equations, Eqs. 13.2.

Given: Isentropic flow of air in a channel. At sections ① and ②, the following data are given: $M_1 = 0.3$, $T_1 = 62°\text{C}$, $p_1 = 650$ kPa (abs), $A_1 = 0.001\ \text{m}^2$, and $M_2 = 0.8$.

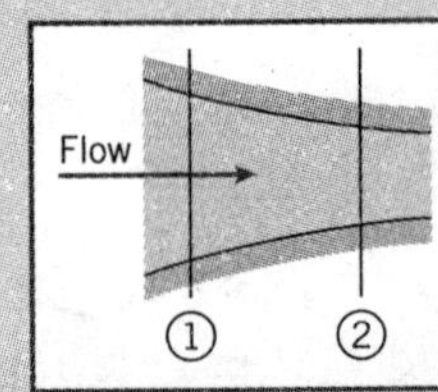

Find: (a) The channel shape.
(b) A *Ts* diagram for the process.
(c) Properties at section ②.
(d) Show that the results satisfy the basic equations.

Solution:

To accelerate a subsonic flow requires a converging nozzle. The channel shape must be as shown.

On the *Ts* plane, the process follows an s = constant line. Stagnation conditions remain fixed for isentropic flow. Consequently, the stagnation temperature at section ② can be calculated (for air, $k = 1.4$) from Eq. 13.7b.

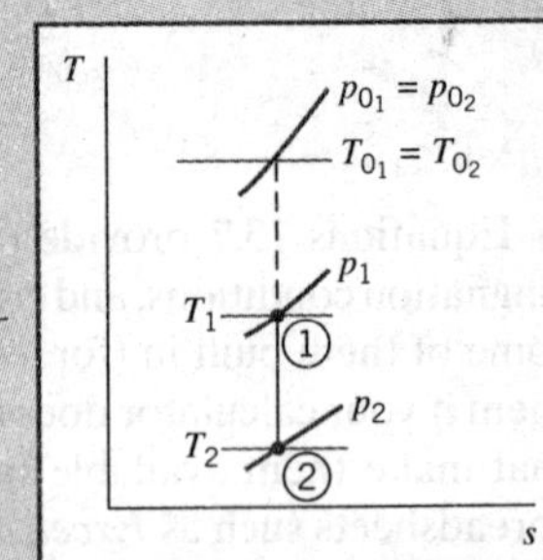

$$T_{0_2} = T_{0_1} = T_1\left[1 + \frac{k-1}{2}M_1^2\right]$$

$$= (62 + 273)\ \text{K}\left[1 + 0.2(0.3)^2\right]$$

$$T_{0_2} = T_{0_1} = 341\ \text{K}$$ T_{0_1}, T_{0_2}

For p_{0_2}, from Eq. 13.7a,

$$p_{0_2} = p_{0_1} = p_1\left[1 + \frac{k-1}{2}M_1^2\right]^{k/(k-1)} = 650\ \text{kPa}[1 + 0.2(0.3)^2]^{3.5}$$

$$p_{0_2} = 692\ \text{kPa (abs)}$$ p_{0_2}

For T_2, from Eq. 13.7b,

$$T_2 = T_{0_2}\Big/\left[1 + \frac{k-1}{2}M_2^2\right] = 341\ \text{K}\Big/\left[1 + 0.2(0.8)^2\right]$$

$$T_2 = 302\ \text{K}$$ T_2

For p_2, from Eq. 13.7a,

$$p_2 = p_{0_2}\Big/\left[1 + \frac{k-1}{2}M_2^2\right]^{k/k-1} = 692\ \text{kPa}\Big/\left[1 + 0.2(0.8)^2\right]^{3.5}$$

$$p_2 = 454\ \text{kPa}$$ p_2

Note that we could have directly computed T_2 from T_1 because T_0 = constant:

$$\frac{T_2}{T_1} = \frac{T_2}{T_0}\bigg/\frac{T_1}{T_0} = \left[1+\frac{k-1}{2}M_1^2\right]\bigg/\left[1+\frac{k-1}{2}M_2^2\right] = \left[1+0.2(0.3)^2\right]\big/\left[1+0.2(0.8)^2\right]$$

$$\frac{T_2}{T_1} = \frac{0.8865}{0.9823} = 0.9025$$

Hence,

$$T_2 = 0.9025\,T_1 = 0.9025(273+62)\,\text{K} = 302\,\text{K}$$

Similarly, for p_2,

$$\frac{p_2}{p_1} = \frac{p_2}{p_0}\bigg/\frac{p_1}{p_0} = 0.8865^{3.5}/0.9823^{3.5} = 0.6982$$

Hence,

$$p_2 = 0.6982\,p_1 = 0.6982(650\,\text{kPa}) = 454\,\text{kPa}$$

The density ρ_2 at section ② can be found from Eq. 13.7c using the same procedure we used for T_2 and p_2, or we can use the ideal gas equation of state, Eq. 13.2e,

$$\rho_2 = \frac{p_2}{RT_2} = 4.54\times10^5\frac{\text{N}}{\text{m}^2}\times\frac{\text{kg}\cdot\text{K}}{287\,\text{N}\cdot\text{m}}\times\frac{1}{302\,\text{K}} = 5.24\,\text{kg/m}^3 \qquad \rho_2$$

and the velocity at section ② is

$$V_2 = M_2c_2 = M_2\sqrt{kRT_2} = 0.8\times\sqrt{1.4\times287\frac{\text{N}\cdot\text{m}}{\text{kg}\cdot\text{K}}\times302\,\text{K}\times\frac{\text{kg}\cdot\text{m}}{\text{s}^2\cdot\text{N}}} = 279\,\text{m/s} \qquad V_2$$

The area A_2 can be computed from Eq. 13.7d, noting that A^* is constant for this flow,

$$\frac{A_2}{A_1} = \frac{A_2}{A^*}\frac{A^*}{A_1} = \frac{1}{M_2}\left[\frac{1+\frac{k-1}{2}M_2^2}{\frac{k+1}{2}}\right]^{(k+1)/2(k-1)}\Bigg/\frac{1}{M_1}\left[\frac{1+\frac{k-1}{2}M_1^2}{\frac{k+1}{2}}\right]^{(k+1)/2(k-1)}$$

$$= \frac{1}{0.8}\left[\frac{1+0.2(0.8)^2}{1.2}\right]^3\Bigg/\frac{1}{0.3}\left[\frac{1+0.2(0.3)^2}{1.2}\right]^3 = \frac{1.038}{2.035} = 0.5101$$

Hence,

$$A_2 = 0.5101A_1 = 0.5101(0.001\,\text{m}^2) = 5.10\times10^{-4}\,\text{m}^2 \qquad A_2$$

Note that $A_2 < A_1$ as expected.

Let us verify that these results satisfy the basic equations.

We first need to obtain ρ_1 and V_1:

$$\rho_1 = \frac{p_1}{RT_1} = 6.5\times10^5\frac{\text{N}}{\text{m}^2}\times\frac{\text{kg}\cdot\text{K}}{287\,\text{N}\cdot\text{m}}\times\frac{1}{335\,\text{K}} = 6.76\,\text{kg/m}^3$$

and

$$V_1 = M_1c_1 = M_1\sqrt{kRT_1} = 0.3\times\sqrt{1.4\times287\frac{\text{N}\cdot\text{m}}{\text{kg}\cdot\text{K}}\times335\,\text{K}\times\frac{\text{kg}\cdot\text{m}}{\text{s}^2\cdot\text{N}}} = 110\,\text{m/s}$$

The mass conservation equation is

$$\rho_1V_1A_1 = \rho_2V_2A_2 = \rho VA = \dot{m} = \text{constant} \qquad (13.2a)$$

$$\dot{m} = 6.76\frac{\text{kg}}{\text{m}^3} \times 110\frac{\text{m}}{\text{s}} \times 0.001\ \text{m}^2 = 5.24\frac{\text{kg}}{\text{m}^3} \times 279\frac{\text{m}}{\text{s}} \times 0.00051\ \text{m}^2 = 0.744\ \text{kg/s} \qquad \text{(Check!)}$$

We cannot check the momentum equation (Eq. 13.2b) because we do not know the force R_x produced by the walls of the device (we could use Eq. 13.2b to compute this if we wished). The energy equation is

$$h_{0_1} = h_1 + \frac{V_1^2}{2} = h_2 + \frac{V_2^2}{2} = h_{0_2} = h_0 \tag{13.2c}$$

We will check this by replacing enthalpy with temperature using Eq. 13.2f,

$$\Delta h = h_2 - h_1 = c_p \Delta T = c_p(T_2 - T_1) \tag{13.2f}$$

so the energy equation becomes

$$c_p T_1 + \frac{V_1^2}{2} = c_p T_2 + \frac{V_2^2}{2} = c_p T_0$$

Using c_p for air from Table A.6,

$$c_p T_1 + \frac{V_1^2}{2} = 1004\frac{\text{J}}{\text{kg}\cdot\text{K}} \times 335\ \text{K} + \frac{(110)^2}{2}\left(\frac{\text{m}}{\text{s}}\right)^2 \times \frac{\text{N}\cdot\text{s}^2}{\text{kg}\cdot\text{m}} \times \frac{\text{J}}{\text{N}\cdot\text{m}} = 342\ \text{kJ/kg}$$

$$c_p T_2 + \frac{V_2^2}{2} = 1004\frac{\text{J}}{\text{kg}\cdot\text{K}} \times 302\ \text{K} + \frac{(278)^2}{2}\left(\frac{\text{m}}{\text{s}}\right)^2 \times \frac{\text{N}\cdot\text{s}^2}{\text{kg}\cdot\text{m}} \times \frac{\text{J}}{\text{N}\cdot\text{m}} = 342\ \text{kJ/kg}$$

$$c_p T_0 = 1004\frac{\text{J}}{\text{kg}\cdot\text{K}} \times 341\ \text{K} = 342\ \text{kJ/kg} \qquad \text{(Check!)}$$

The final equation we can check is the relation between pressure and density for an isentropic process (Eq. 13.2g),

$$\frac{p_1}{\rho_1^k} = \frac{p_2}{\rho_2^k} = \frac{p}{\rho^k} = \text{constant} \qquad \text{(Check!)}$$

$$\frac{p_1}{\rho_1^{1.4}} = \frac{650\ \text{kPa}}{\left(6.76\ \dfrac{\text{kg}}{\text{m}^3}\right)^{1.4}} = \frac{p_2}{\rho_2^{1.4}} = \frac{454\ \text{kPa}}{\left(5.24\ \dfrac{\text{kg}}{\text{m}^3}\right)^{1.4}} = 44.7\frac{\text{kPa}}{\left(\dfrac{\text{kg}}{\text{m}^3}\right)^{1.4}} \qquad \text{(Check!)}$$

The basic equations are satisfied by our solution.

This Example illustrates:

- ✓ Use of the isentropic equations, Eqs. 13.7.
- ✓ That the isentropic equations are consistent with the basic equations, Eqs. 13.2.
- ✓ That the computations can be quite laborious without using preprogrammed isentropic relations (available, for example, in the *Excel* add-ins on the Web site)!

The *Excel* workbook for this Example is convenient for performing the calculations, using either the isentropic equations or the basic equations.

Isentropic Flow in a Converging Nozzle

Now that we have our computing equations (Eqs. 13.7) for analyzing isentropic flows, we are ready to see how we could obtain flow in a nozzle, starting from rest. We first look at the converging nozzle, and then the C-D nozzle. In either case, to produce a flow we must provide a pressure difference. For example, as illustrated in the converging nozzle shown in Fig. 13.6*a*, we can do this by providing the gas from a reservoir (or "plenum chamber") at p_0 and T_0, and using a vacuum pump/valve

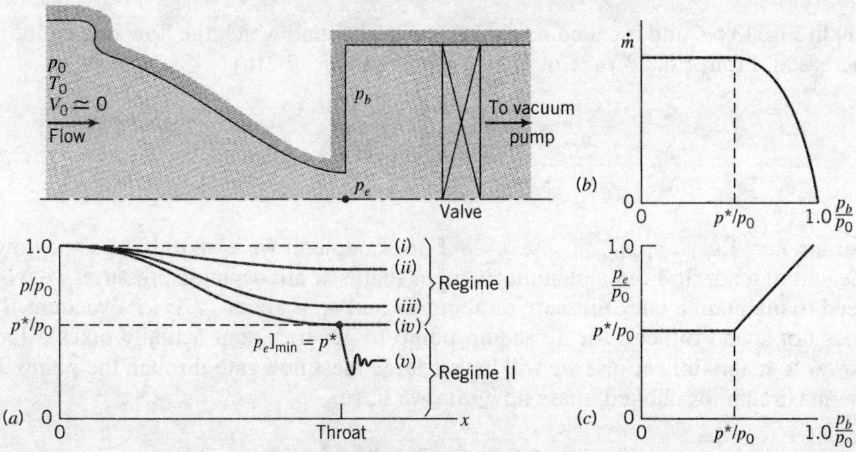

Fig. 13.6 Converging nozzle operating at various back pressures.

combination to create a low pressure, the "back pressure," p_b. We are interested in what happens to the gas properties as the gas flows through the nozzle, and also in knowing how the mass flow rate increases as we progressively lower the back pressure. (We could also produce a flow by maintaining a constant back pressure, e.g., atmospheric, and increasing the pressure in the plenum chamber.)

Let us call the pressure at the exit plane p_e. We will see that this will often be equal to the applied back pressure, p_b, but not always! The results we obtain as we progressively open the valve from a closed position are shown in Figs. 13.6*b* and 13.6*c*. We consider each of the cases shown.

When the valve is closed, there is no flow through the nozzle. The pressure is p_0 throughout, as shown by condition (*i*) in Fig. 13.6*a*.

If the back pressure, p_b, is now reduced to slightly less than p_0, there will be flow through the nozzle with a decrease in pressure in the direction of flow, as shown by condition (*ii*). Flow at the exit plane will be subsonic with the exit-plane pressure equal to the back pressure.

What happens as we continue to decrease the back pressure? As expected, the flow rate will continue to increase, and the exit-plane pressure will continue to decrease, as shown by condition (*iii*) in Fig. 13.6*a*. [Note that conditions (*ii*) and (*iii*) can be described using the Bernoulli equation (Eq. 6.8), as long as the maximum Mach number, at the exit plane, does not exceed $M = 0.3$.]

As we progressively lower the back pressure the flow rate increases, and hence, so do the velocity and Mach number at the exit plane. The question arises: "Is there a limit to the mass flow rate through the nozzle?" or, to put it another way, "Is there an upper iimit on the exit Mach number?" The answer to these questions is "Yes!" To see this, recall that for isentropic flow Eq. 13.6 applies:

$$\frac{dV}{V} = -\frac{dA}{A}\frac{1}{[1 - M^2]} \tag{13.6}$$

From this we learned that the *only* place we can have sonic conditions ($M = 1$) is where the change in area dA is zero. We *cannot* have sonic conditions anywhere in the converging section. Logically we can see that the *maximum exit Mach number is one*. Because the flow started from rest ($M = 0$), if we had $M > 1$ at the exit, we would have had to pass through $M = 1$ somewhere in the converging section, which would be a violation of Eq. 13.6.

Hence, the maximum flow rate occurs when we have sonic conditions at the exit plane, when $M_e = 1$, and $p_e = p_b = p^*$, the critical pressure. This is shown as condition (*iv*) in Fig. 13.6*a*, and is called a "choked flow," beyond which the flow rate cannot be increased. From Eq. 13.7a with $M = 1$ (or from Eq. 12.21a),

$$\left.\frac{p_e}{p_0}\right|_{\text{choked}} = \frac{p^*}{p_0} = \left(\frac{2}{k+1}\right)^{k/(k-1)} \tag{13.8}$$

For air, $k = 1.4$, so $p_e/p_0]_{\text{choked}} = 0.528$. For example, if we wish to have sonic flow at the exit of a nozzle from a plenum chamber that is at atmospheric pressure, we would need to maintain a back pressure of about 53.5 kPa, or about 47.85 kPa vacuum. This does not sound difficult for a vacuum pump to generate, but actually takes a lot of power to maintain, because we will have a large mass flow rate through the pump. For the maximum, or choked, mass flow rate we have

$$\dot{m}_{\text{choked}} = \rho^* V^* A^*$$

Using the ideal gas equation of state, Eq. 13.2e, and the stagnation to critical pressure and temperature ratios, Eqs. 13.7a and 13.7b respectively, with $M = 1$ (or Eqs. 12.21a and 12.21b, respectively), with $A^* = A_e$, it can be shown that this becomes

$$\dot{m}_{\text{choked}} = A_e p_0 \sqrt{\frac{k}{RT_0}\left(\frac{2}{k+1}\right)^{(k+1)/2(k-1)}} \tag{13.9a}$$

Note that for a given gas (k and R), the maximum flow rate in the converging nozzle depends *only* on the size of the exit area (A_e) and the conditions in the reservoir (p_0, T_0).

For air, for convenience we write an "engineering" form of Eq. 13.9a,

$$\dot{m}_{\text{choked}} = 0.04\frac{A_e p_0}{\sqrt{T_0}} \tag{13.9b}$$

with $\dot{m}_{\text{choked}}$ in kg/s, A_e in m^2, p_0 in Pa, and T_0 in K.

Suppose we now insist on lowering the back pressure below this "benchmark" level of p^*. Our next question is "What will happen to the flow in the nozzle?" The answer is "Nothing!" The flow remains choked: The mass flow rate does not increase, as shown in Fig. 13.6*b*, and the pressure distribution in the nozzle remains unchanged, with $p_e = p^* > p_b$, as shown in condition (*v*) in Figs. 13.6*a* and 13.6*c*. After exiting, the flow adjusts down to the applied back pressure, but does so in a nonisentropic, three-dimensional manner in a series of expansion waves and shocks, and for this part of the flow our one-dimensional, isentropic flow concepts no longer apply. We will return to this discussion in Section 13.4.

This idea of choked flow seems a bit strange, but can be explained in at least two ways. First, we have already discussed that to increase the mass flow rate beyond choked would require $M_e > 1$, which is not possible. Second, once the flow reaches sonic conditions, it becomes "deaf" to downstream conditions: Any change (i.e., a reduction) in the applied back pressure propagates in the fluid at the speed of sound in all directions, as we discussed in Chapter 12, so it gets "washed" downstream by the fluid which is moving at the speed of sound at the nozzle exit.

Flow through a converging nozzle may be divided into two regimes:

1. In Regime I, $1 \geq p_b/p_0 \geq p^*/p_0$. Flow to the throat is isentropic and $p_e = p_b$.
2. In Regime II, $p_b/p_0 < p^*/p_0$. Flow to the throat is isentropic, and $M_e = 1$. A nonisentropic expansion occurs in the flow leaving the nozzle and $p_e = p^* > p_b$ (entropy increases because this is adiabatic but irreversible).

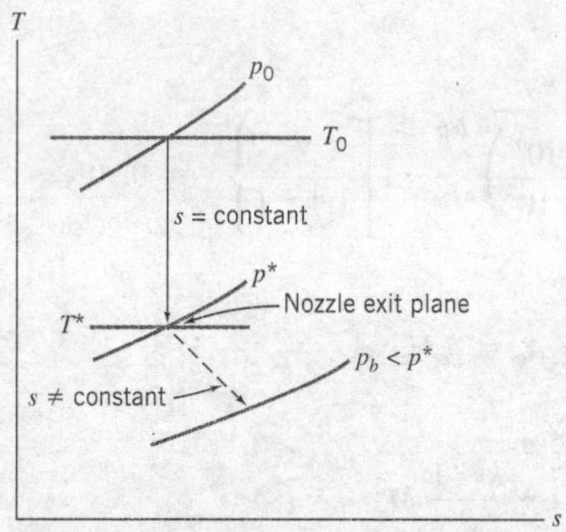

Fig. 13.7 Schematic *Ts* diagram for choked flow through a converging nozzle.

The flow processes corresponding to Regime II are shown on a *Ts* diagram in Fig. 13.7. Two problems involving converging nozzles are solved in Examples 13.2 and 13.3.

Although isentropic flow is an idealization, it often is a very good approximation for the actual behavior of nozzles. Since a nozzle is a device that accelerates a flow, the internal pressure gradient is favorable. This tends to keep the wall boundary layers thin and to minimize the effects of friction.

Example 13.2 ISENTROPIC FLOW IN A CONVERGING NOZZLE

A converging nozzle, with a throat area of 0.001 m^2, is operated with air at a back pressure of 591 kPa (abs). The nozzle is fed from a large plenum chamber where the absolute stagnation pressure and temperature are 1.0 MPa and 60°C. The exit Mach number and mass flow rate are to be determined.

Given: Air flow through a converging nozzle at the conditions shown: Flow is isentropic.

Find: (a) M_e.
(b) $\dot{m}$.

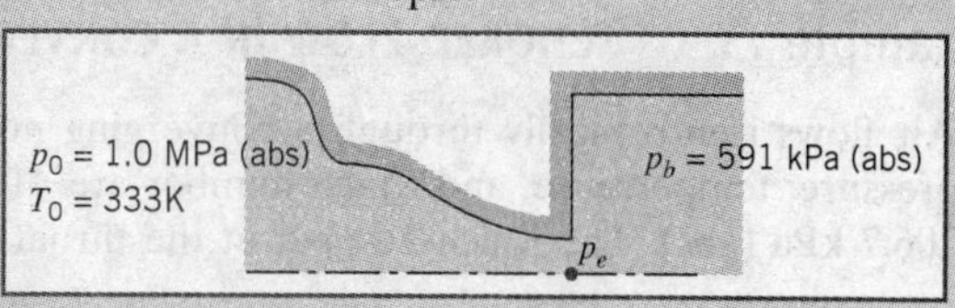

Solution:
The first step is to check for choking. The pressure ratio is

$$\frac{p_b}{p_0} = \frac{5.91 \times 10^5}{1.0 \times 10^6} = 0.591 > 0.528$$

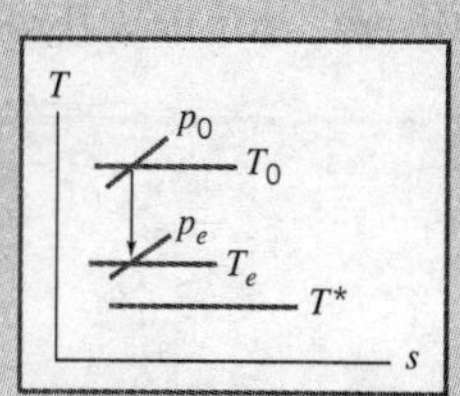

so the flow is *not* choked. Thus $p_b = p_e$, and the flow is isentropic, as sketched on the *Ts* diagram.

Since p_0 = constant, M_e may be found from the pressure ratio,

$$\frac{p_0}{p_e} = \left[1 + \frac{k-1}{2}M_e^2\right]^{k/(k-1)}$$

Solving for M_e, since $p_e = p_b$, we obtain

$$1 + \frac{k-1}{2}M_e^2 = \left(\frac{p_0}{p_b}\right)^{(k-1)/k}$$

and

$$M_e = \left\{\left[\left(\frac{p_0}{p_b}\right)^{(k-1)/k} - 1\right]\frac{2}{k-1}\right\}^{1/2} = \left\{\left[\left(\frac{1.0\times10^6}{5.91\times10^5}\right)^{0.286} - 1\right]\frac{2}{1.4-1}\right\}^{1/2} = 0.90 \longleftarrow \qquad M_e$$

The mass flow rate is

$$\dot{m} = \rho_e V_e A_e = \rho_e M_e c_e A_e$$

We need T_e to find ρ_e and c_e. Since T_0 = constant,

$$\frac{T_0}{T_e} = 1 + \frac{k-1}{2}M_e^2$$

or

$$T_e = \frac{T_0}{1+\frac{k-1}{2}M_e^2} = \frac{(273+60)\text{ K}}{1+0.2(0.9)^2} = 287\text{ K}$$

$$c_e = \sqrt{kRT_e} = \left[1.4\times287\frac{\text{N}\cdot\text{m}}{\text{kg}\cdot\text{K}}\times287\text{ K}\times\frac{\text{kg}\cdot\text{m}}{\text{N}\cdot\text{s}^2}\right]^{1/2} = 340\text{ m/s}$$

and

$$\rho_e = \frac{p_e}{RT_e} = 5.91\times10^5\frac{\text{N}}{\text{m}^2}\times\frac{\text{kg}\cdot\text{K}}{287\text{ N}\cdot\text{m}}\times\frac{1}{287\text{ K}} = 7.18\text{ kg/m}^3$$

Finally,

$$\dot{m} = \rho_e M_e c_e A_e = 7.18\frac{\text{kg}}{\text{m}^3}\times0.9\times340\frac{\text{m}}{\text{s}}\times0.001\text{ m}^2$$

$$= 2.20\text{ kg/s} \longleftarrow \qquad \dot{m}$$

This problem illustrates use of the isentropic equations, Eqs. 13.7, for a flow that is not choked.
The *Excel* workbook for this Example is convenient for performing the calculations (using either the isentropic equations or the basic equations). (The *Excel* add-ins for isentropic flow, on the Web site, also make calculations much easier.)

Example 13.3 CHOKED FLOW IN A CONVERGING NOZZLE

Air flows isentropically through a converging nozzle. At a section where the nozzle area is 0.0012 m^2, the local pressure, temperature, and Mach number are 413.4 kPa (abs), 4.5°C, and 0.52, respectively. The back pressure is 206.7 kPa (abs). The Mach number at the throat, the mass flow rate, and the throat area are to be determined.

Given: Air flow through a converging nozzle at the conditions shown:

$$M_1 = 0.52$$
$$T_1 = 4.5°\text{C}$$
$$p_1 = 413.4\text{ kPa(abs)}$$
$$A_1 = 0.0012\text{ m}^2$$

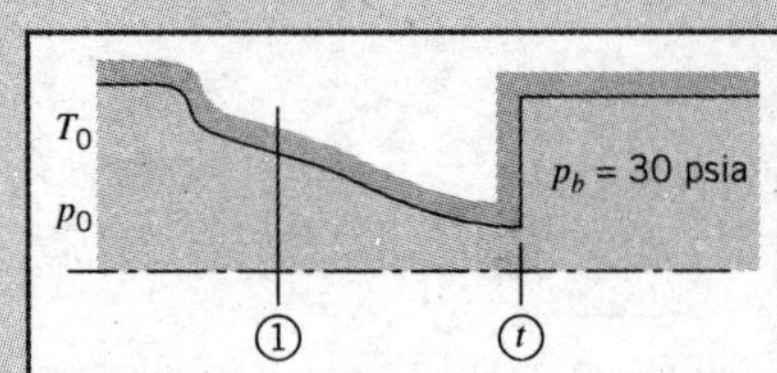

Find: (a) M_t. (b) $\dot{m}$. (c) A_t.

Solution:

First we check for choking, to determine if flow is isentropic down to p_b. To check, we evaluate the stagnation conditions.

$$p_0 = p_1\left[1+\frac{k-1}{2}M_1^2\right]^{k/(k-1)} = 413.4\text{ kPa (abs) }[1+0.2(0.52)^2]^{3.5} = 497.1\text{ kPa (abs)}$$

The back pressure ratio is

$$\frac{p_b}{p_0} = \frac{206.7}{497.1} = 0.416 < 0.528$$

so the flow is choked! For choked flow,

$$M_t = 1.0 \qquad M_t$$

The Ts diagram is

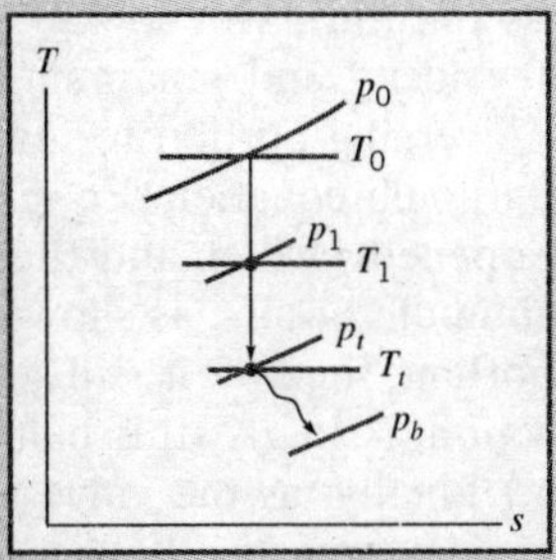

The mass flow rate may be found from conditions at section ①, using $\dot{m} = \rho_1 V_1 A_1$.

$$V_1 = M_1 c_1 = M_1\sqrt{kRT_1}$$

$$= 0.52\left[1.4 \times 287\frac{\text{N}\cdot\text{m}}{\text{kg}\cdot\text{k}} \times (273 + 4.5)\,\text{k}\right]^{1/2}$$

$$V_1 = 173.6\,\text{m/s}$$

$$\rho_1 = \frac{p_1}{RT_1} = 413.4 \times 1000\,\text{Pa} \times \frac{\text{kg}\cdot\text{k}}{287\,\text{N}\cdot\text{m}} \times \frac{1}{277.5\,\text{k}} \times \frac{\text{N}}{\text{Pa}\cdot\text{m}^2} = 5.19\,\text{kg/m}^3$$

$$\dot{m} = \rho_1 V_1 A_1 = 5.19\frac{\text{kg}}{\text{m}^3} \times 173.6\frac{\text{m}}{\text{s}} \times 0.0012\,\text{m}^2 = 1.08\,\text{kg/s} \qquad \dot{m}$$

From Eq. 13.6,

$$\frac{A_1}{A^*} = \frac{1}{M_1}\left[\frac{1 + \frac{k-1}{2}M_1^2}{\frac{k+1}{2}}\right]^{(k+1)/2(k-1)} = \frac{1}{0.52}\left[\frac{1 + 0.2(0.52)^2}{1.2}\right]^{3.00} = 1.303$$

For choked flow, $A_t = A^*$. Thus,

$$A_t = A^* = \frac{A_1}{1.303} = \frac{0.0012\,\text{m}^2}{1.303}$$

$$A_t = 9.21 \times 10^{-4}\,\text{m}^2 \qquad A_t$$

This problem illustrates use of the isentropic equations, Eqs. 13.7, for a flow that is choked.

✓ Because the flow is choked, we could also have used Eq. 13.9a for $\dot{m}$ (after finding T_0).

The *Excel* workbook for this Example is convenient for performing the calculations. (The *Excel* add-ins for isentropic flow, on the Web site, also make calculations much easier.)

Isentropic Flow in a Converging-Diverging Nozzle

Having considered isentropic flow in a converging nozzle, we turn now to isentropic flow in a converging-diverging (C-D) nozzle. As in the previous case, flow through the converging-diverging passage of Fig. 13.8 is induced by a vacuum pump downstream, and is controlled by the valve shown; upstream stagnation conditions are constant.

Pressure in the exit plane of the nozzle is p_e; the nozzle discharges to back pressure p_b. As for the converging nozzle, we wish to see, among other things, how the flow rate varies with the driving force, the applied pressure difference $(p_0 - p_b)$. Consider the effect of gradually reducing the back pressure. The results are illustrated graphically in Fig. 13.8. Let us consider each of the cases shown.

With the valve initially closed, there is no flow through the nozzle; the pressure is constant at p_0. Opening the valve slightly (p_b slightly less than p_0) produces pressure distribution curve (*i*). If the flow rate is low enough, the flow will be subsonic and essentially incompressible at all points on this curve. Under these conditions, the C-D nozzle will behave as a venturi, with flow accelerating in the converging portion until a point of maximum velocity and minimum pressure is reached at the throat, then decelerating in the diverging portion to the nozzle exit. (This behavior is described accurately by the Bernoulli equation, Eq. 6.8.)

As the valve is opened farther and the flow rate is increased, a more sharply defined pressure minimum occurs, as shown by curve (*ii*). Although compressibility effects become important, the flow is still subsonic everywhere, and flow decelerates in the diverging section. (Clearly, this behavior is *not* described accurately by the Bernoulli equation.) Finally, as the valve is opened farther, curve (*iii*) results. At the section of minimum area the flow finally reaches $M = 1$, and the nozzle is choked—the flow rate is the maximum possible for the given nozzle and stagnation conditions.

All flows with pressure distributions (*i*), (*ii*), and (*iii*) are isentropic; as we progress from (*i*) to (*ii*) to (*iii*) we are generating increasing mass flow rates. Finally, when curve (*iii*) is reached, critical conditions are present at the throat. For this flow rate, the flow is choked, and

$$\dot{m} = \rho^* V^* A^*$$

where $A^* = A_t$, just as it was for the converging nozzle, and for this maximum possible flow rate Eq. 13.9a applies (with A_e replaced with the throat area A_t),

$$\dot{m}_{\text{choked}} = A_t p_0 \sqrt{\frac{k}{RT_0}} \left(\frac{2}{k+1}\right)^{(k+1)/2(k-1)} \tag{13.10a}$$

Note that for a given gas (k and R), the maximum flow rate in the C-D nozzle depends *only* on the size of the throat area (A_t) and the conditions in the reservoir (p_0, T_0).

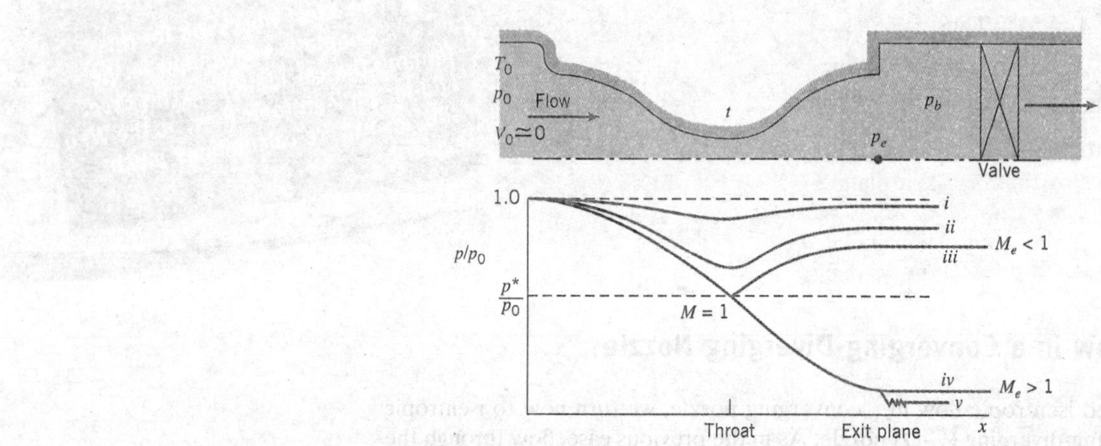

Fig. 13.8 Pressure distributions for isentropic flow in a converging-diverging nozzle.

For air, for convenience we write an "engineering" form of Eq. 13.10a,

$$\dot{m}_{\text{choked}} = 0.04 \frac{A_t p_0}{\sqrt{T_0}} \tag{13.10b}$$

with $\dot{m}_{\text{choked}}$ in kg/s, A_t in m^2, p_0 in Pa, and T_0 in K.

Any attempt to increase the flow rate by further lowering the back pressure will fail, for the two reasons we discussed earlier: once we attain sonic conditions, downstream changes can no longer be transmitted upstream; and we cannot exceed sonic conditions at the throat, because this would require passing through the sonic state somewhere in the converging section, which is not possible in isentropic flow. (Of course, we *could* increase the choked mass flow rate through a given C-D nozzle to any level desired by, for example, increasing the reservoir pressure.)

With sonic conditions at the throat, we consider what *can* happen to the flow in the diverging section. We have previously discussed (see Fig. 13.3) that a diverging section will decelerate a subsonic flow ($M < 1$) but will accelerate a supersonic flow ($M > 1$)—very different behaviors! The question arises: "Does a sonic flow behave as a subsonic or as a supersonic flow as it enters a diverging section?" The answer to this question is that it can behave like either one, depending on the downstream pressure! We have already seen subsonic flow behavior [curve (*iii*)]: the applied back pressure leads to a gradual downstream pressure increase, decelerating the flow. We now consider accelerating the choked flow.

To accelerate flow in the diverging section requires a pressure decrease. This condition is illustrated by curve (*iv*) in Fig. 13.8. The flow will accelerate isentropically in the nozzle provided the exit pressure is set at p_{iv}. Thus, we see that with a throat Mach number of unity, there are two possible isentropic flow conditions in the converging-diverging nozzle. This is consistent with the results of Fig. 13.5, where we found two Mach numbers for each A/A^* in isentropic flow.

Lowering the back pressure below condition (*iv*), say to condition (*v*), has no effect on flow in the nozzle. The flow is isentropic from the plenum chamber to the nozzle exit [as in condition (*iv*)] and then it undergoes a three-dimensional irreversible expansion to the lower back pressure. A nozzle operating under these conditions is said to be *underexpanded*, since additional expansion takes place outside the nozzle.

A converging-diverging nozzle generally is intended to produce supersonic flow at the exit plane. If the back pressure is set at p_{iv}, flow will be isentropic through the nozzle, and supersonic at the nozzle exit. Nozzles operating at $p_b = p_{iv}$ [corresponding to curve (*iv*) in Fig. 13.8] are said to operate at *design conditions*.

Flow leaving a C-D nozzle is supersonic when the back pressure is at or below nozzle design pressure. The exit Mach number is fixed once the area ratio, A_e/A^*, is specified. All other exit plane properties (for isentropic flow) are uniquely related to stagnation properties by the fixed exit plane Mach number.

The assumption of isentropic flow for a real nozzle at design conditions is a reasonable one. However, the one-dimensional flow model is inadequate for the design of relatively short nozzles to produce uniform supersonic exit flow.

Rocket-propelled vehicles use C-D nozzles to accelerate the exhaust gases to the maximum possible speed to produce high thrust. A propulsion nozzle is subject to varying ambient conditions during flight through the atmosphere, so it is impossible to attain the maximum theoretical thrust over the complete operating range. Because only a single supersonic Mach number can be obtained for each area ratio, nozzles for developing supersonic flow in wind tunnels often are built with interchangeable test sections, or with variable geometry.

You probably have noticed that nothing has been said about the operation of converging-diverging nozzles with back pressure in the range $p_{iii} > p_b > p_{iv}$. For such cases the flow cannot expand isentropically to p_b. Under these conditions a shock (which may be treated as an irreversible discontinuity involving entropy increase) occurs somewhere within the flow. Following a discussion of normal shocks in Section 13.3, we shall return to complete the discussion of converging-diverging nozzle flows in Section 13.4.

Nozzles operating with $p_{iii} > p_b > p_{iv}$ are said to be *overexpanded* because the pressure at some point in the nozzle is less than the back pressure. Obviously, an overexpanded nozzle could be made to operate at a new design condition by removing a portion of the diverging section.

In Example 13.4, we consider isentropic flow in a C-D nozzle; in Example 13.5, we consider choked flow in a C-D nozzle.

Example 13.4 ISENTROPIC FLOW IN A CONVERGING-DIVERGING NOZZLE

Air flows isentropically in a converging-diverging nozzle, with exit area of 0.001 m². The nozzle is fed from a large plenum where the stagnation conditions are 350K and 1.0 MPa (abs). The exit pressure is 954 kPa (abs) and the Mach number at the throat is 0.68. Fluid properties and area at the nozzle throat and the exit Mach number are to be determined.

Given: Isentropic flow of air in C-D nozzle as shown:

$$T_0 = 350\text{ K}$$
$$p_0 = 1.0\text{ MPa (abs)}$$
$$p_b = 954\text{ kPa (abs)}$$
$$M_t = 0.68 \quad A_e = 0.001\text{ m}^2$$

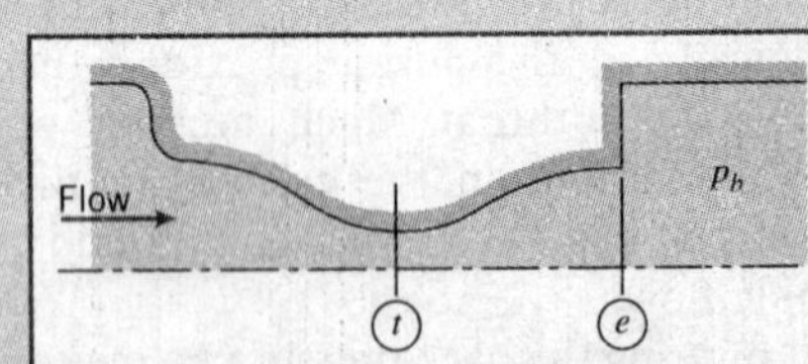

Find: (a) Properties and area at nozzle throat.
(b) M_e.

Solution:
Stagnation temperature is constant for isentropic flow. Thus, since

$$\frac{T_0}{T} = 1 + \frac{k-1}{2}M^2$$

then

$$T_t = \frac{T_0}{1 + \frac{k-1}{2}M_t^2} = \frac{350\text{ K}}{1 + 0.2(0.68)^2} = 320\text{ K} \longleftarrow \qquad T_t$$

Also, since p_0 is constant for isentropic flow, then

$$p_t = p_0\left(\frac{T_t}{T_0}\right)^{k/(k-1)} = p_0\left[\frac{1}{1 + \frac{k-1}{2}M_t^2}\right]^{k/(k-1)}$$

$$p_t = 1.0 \times 10^6\text{ Pa}\left[\frac{1}{1 + 0.2(0.68)^2}\right]^{3.5} = 734\text{ kPa (abs)} \longleftarrow \qquad p_t$$

so

$$\rho_t = \frac{p_t}{RT_t} = 7.34 \times 10^5 \frac{\text{N}}{\text{m}^2} \times \frac{\text{kg}\cdot\text{K}}{287\,\text{N}\cdot\text{m}} \times \frac{1}{320\,\text{K}} = 7.99\,\text{kg/m}^3 \longleftarrow \rho_t$$

and

$$V_t = M_t c_t = M_t\sqrt{kRT_t}$$

$$V_t = 0.68\left[14 \times 287 \frac{\text{N}\cdot\text{m}}{\text{kg}\cdot\text{K}} \times 320\,\text{K} \times \frac{\text{kg}\cdot\text{m}}{\text{N}\cdot\text{s}^2}\right]^{1/2} = 244\,\text{m/s} \longleftarrow V_t$$

From Eq. 13.7d we can obtain a value of A_t/A^*

$$\frac{A_t}{A^*} = \frac{1}{M_t}\left[\frac{1 + \frac{k-1}{2}M_t^2}{\frac{k+1}{2}}\right]^{(k+1)/2(k-1)} = \frac{1}{0.68}\left[\frac{1 + 0.2(0.68)^2}{1.2}\right]^{3.00} = 1.11$$

but at this point A^* is not known.

Since $M_t < 1$, flow at the exit must be subsonic. Therefore, $p_e = p_b$. Stagnation properties are constant, so

$$\frac{p_0}{p_e} = \left[1 + \frac{k-1}{2}M_e^2\right]^{k/(k-1)}$$

Solving for M_e gives

$$M_e = \left\{\left[\left(\frac{p_0}{p_e}\right)^{(k-1)/k} - 1\right]\frac{2}{k-1}\right\}^{1/2} = \left\{\left[\left(\frac{1.0 \times 10^6}{9.54 \times 10^5}\right)^{0.286} - 1\right](5)\right\}^{1/2} = 0.26 \longleftarrow M_e$$

The Ts diagram for this flow is

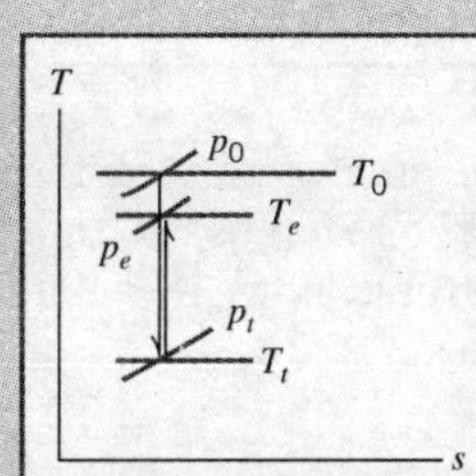

Since A_e and M_e are known, we can compute A^*. From Eq. 13.7d

$$\frac{A_e}{A^*} = \frac{1}{M_e}\left[\frac{1 + \frac{k-1}{2}M_e^2}{\frac{k+1}{2}}\right]^{(k+1)/2(k-1)} = \frac{1}{0.26}\left[\frac{1 + 0.2(0.26)^2}{1.2}\right]^{3.00} = 2.317$$

Thus,

$$A^* = \frac{A_e}{2.317} = \frac{0.001\,\text{m}^2}{2.317} = 4.32 \times 10^{-4}\,\text{m}^2$$

and

$$A_t = 1.110A^* = (1.110)(4.32 \times 10^{-4}\,\text{m}^2)$$
$$= 4.80 \times 10^{-4}\,\text{m}^2 \longleftarrow A_t$$

This problem illustrates use of the isentropic equations, Eqs. 13.7, for flow in a C-D nozzle that is not choked.

✓ Note that use of Eq. 13.7d allowed us to obtain the throat area without needing to first compute other properties.

The *Excel* workbook for this Example is convenient for performing the calculations (using either the isentropic equations or the basic equations). (The *Excel* add-ins for isentropic flow, on the Web site, also make calculations much easier.)

Example 13.5 ISENTROPIC FLOW IN A CONVERGING-DIVERGING NOZZLE: CHOKED FLOW

The nozzle of Example 13.4 has a design back pressure of 87.5 kPa (abs) but is operated at a back pressure of 50.0 kPa (abs). Assume flow within the nozzle is isentropic. Determine the exit Mach number and mass flow rate.

Given: Air flow through C-D nozzle as shown:

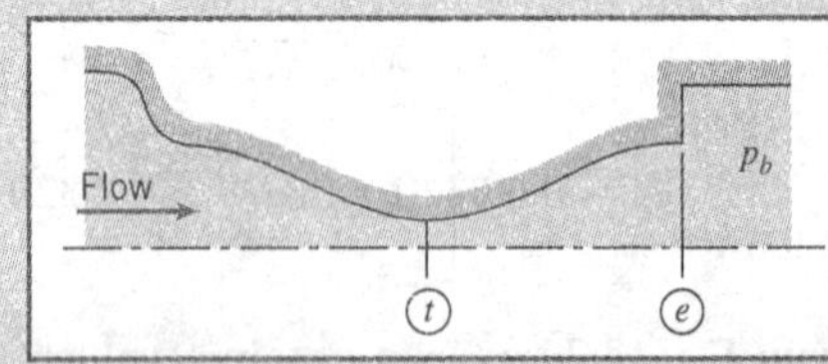

$$T_0 = 350\text{ K}$$
$$p_0 = 1.0\text{ MPa (abs)}$$
$$p_e(\text{design}) = 87.5\text{ kPa (abs)}$$
$$p_b = 50.0\text{ kPa (abs)}$$
$$A_e = 0.001\text{ m}^2$$
$$A_t = 4.8 \times 10^{-4}\text{m}^2 \text{ (Example 13.4)}$$

Find: (a) M_e.
(b) $\dot{m}$.

Solution:
The operating back pressure is *below* the design pressure. Consequently, the nozzle is underexpanded, and the *Ts* diagram and pressure distribution will be as shown:

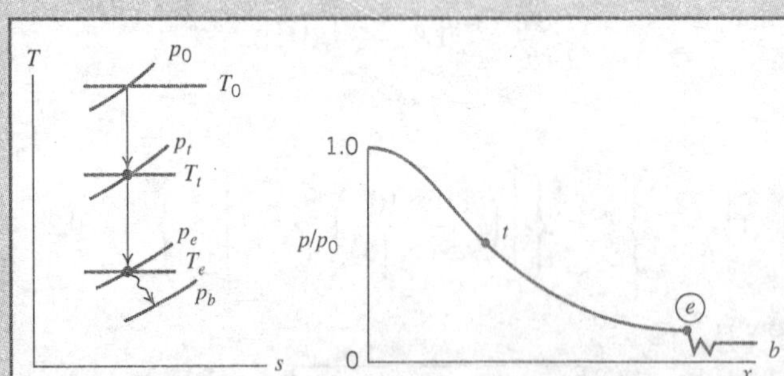

Flow *within* the nozzle will be isentropic, but the irreversible expansion from p_e to p_b will cause an entropy increase; $p_e = p_e(\text{design}) = 87.5$ kPa (abs).

Since stagnation properties are constant for isentropic flow, the exit Mach number can be computed from the pressure ratio. Thus

$$\frac{p_0}{p_e} = \left[1 + \frac{k-1}{2} M_e^2\right]^{k/(k-1)}$$

or

$$M_e = \left\{\left[\left(\frac{p_0}{p_e}\right)^{(k-1)/k} - 1\right]\frac{2}{k-1}\right\}^{1/2} = \left\{\left[\left(\frac{1.0 \times 10^6}{8.75 \times 10^4}\right)^{0.286} - 1\right]\frac{2}{0.4}\right\}^{1/2}$$

$$= 2.24 \qquad M_e$$

Because the flow is choked we can use Eq. 13.10b for the mass flow rate,

$$\dot{m}_{\text{chocked}} = 0.04\frac{A_t p_0}{\sqrt{T_0}} \qquad (13.10\text{b})$$

(with $\dot{m}_{\text{choked}}$ in kg/s, A_t in m^2, p_0 in Pa, and T_0 in K), so

$$\dot{m}_{\text{choked}} = 0.04 \times 4.8 \times 10^{-4} \times 1 \times 10^6/\sqrt{350}$$

$$\dot{m} = \dot{m}_{\text{choked}} = 1.04\text{ kg/s} \qquad \dot{m}$$

This problem illustrates use of the isentropic equations, Eqs. 13.7, for flow in a C-D nozzle that is choked.

✓ Note that we used Eq. 13.10b, an "engineering equation"—that is, an equation containing a coefficient that has units. While useful here, generally these equations are no longer used in engineering because their correct use depends on using input variable values in specific units.

The *Excel* workbook for this Example is convenient for performing the calculations (using either the isentropic equations or the basic equations). (The *Excel* add-ins for isentropic flow, on the Web site, also make calculations much easier.)

We have now completed our study of idealized one-dimensional isentropic flow in channels of varying area. In real channels, we will have friction and quite possibly heat transfer. We need to study the effects of these phenomena on a flow. Before this, in the next section, we study the effect of normal shocks (and see in Section 13.4 how these affect the C-D nozzle, in more detail than we did in this section).

Normal Shocks 13.3

We mentioned normal shocks in the previous section in the context of flow through a nozzle. In practice, these irreversible discontinuities can occur in any supersonic flow field, in either internal flow or external flow. Knowledge of property changes across shocks and of shock behavior is important in understanding the design of supersonic diffusers, e.g., for inlets on high performance aircraft, and supersonic wind tunnels. Accordingly, the purpose of this section is to analyze the normal shock process.

Before applying the basic equations to normal shocks, it is important to form a clear physical picture of the shock itself. Although it is physically impossible to have discontinuities in fluid properties, the normal shock is nearly discontinuous. The thickness of a shock is about 0.2 μm, or roughly 4 times the mean free path of the gas molecules [3]. Large changes in pressure, temperature, and other properties occur across this small distance. Fluid particle decelerations through the shock reach tens of millions of *g*s! These considerations justify treating the normal shock as an abrupt discontinuity; we are interested in changes occurring across the shock rather than in the details of its structure.

CLASSIC VIDEO

Channel Flow of a Compressible Fluid.

Consider the short control volume surrounding a normal shock standing in a passage of arbitrary shape shown in Fig. 13.9. As for isentropic flow with area variation (Section 13.2), our starting point in analyzing this normal shock is the set of basic equations (Eqs. 13.1), describing one-dimensional motion that may be affected by several phenomena: area change, friction, and heat transfer. These are

$$\rho_1 V_1 A_1 = \rho_2 V_2 A_2 = \rho V A = \dot{m} = \text{constant} \qquad (13.1a)$$

$$R_x + p_1 A_1 - p_2 A_2 = \dot{m} V_2 - \dot{m} V_1 \qquad (13.1b)$$

$$\frac{\delta Q}{dm} + h_1 + \frac{V_1^2}{2} = h_2 + \frac{V_2^2}{2} \qquad (13.1c)$$

$$\dot{m}(s_2 - s_1) \geq \int_{\text{CS}} \frac{1}{T}\left(\frac{\dot{Q}}{A}\right) dA \qquad (13.1d)$$

$$p = \rho R T \qquad (13.1e)$$

$$\Delta h = h_2 - h_1 = c_p \Delta T = c_p (T_2 - T_1) \qquad (13.1f)$$

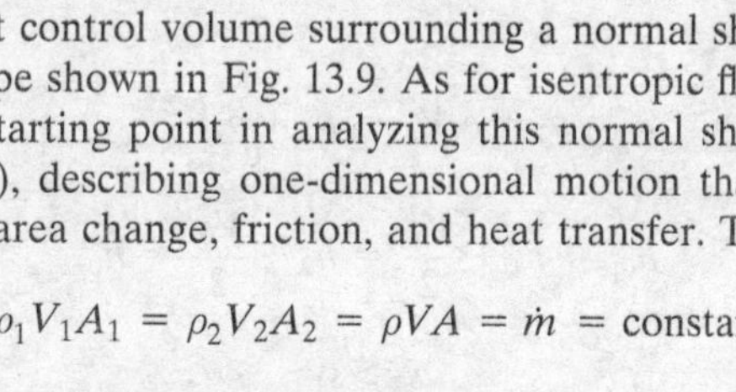

Fig. 13.9 Control volume used for analysis of normal shock.

$$\Delta s = s_2 - s_1 = c_p \ln \frac{T_2}{T_1} - R \ln \frac{p_2}{p_1} \tag{13.1g}$$

We recall that Equation 13.1a is *continuity*, Eq. 13.1b is a *momentum equation*, Eq. 13.1c is an *energy equation*, Eq. 13.1d is the *second law of thermodynamics*, and Eqs. 13.1e, 13.1f, and 13.1g are useful *property relations* for an ideal gas with constant specific heats.

We must simplify these equations for flow through a normal shock.

Basic Equations for a Normal Shock

We can now simplify Eqs. 13.1 for flow of an ideal gas with constant specific heats through a normal shock. The most important simplifying feature is that the width of the control volume is infinitesimal (in reality about 0.2 μm as we indicated), so $A_1 \approx A_2 \approx A$, the force due to the walls $R_x \approx 0$ (because the control volume wall surface area is infinitesimal), and the heat exchange with the walls $\delta Q/dm \approx 0$, for the same reason. Hence, for this flow our equations become

$$\rho_1 V_1 = \rho_2 V_2 = \frac{\dot{m}}{A} = \text{constant} \tag{13.11a}$$

$$p_1 A - p_2 A = \dot{m} V_2 - \dot{m} V_1$$

or, using Eq. 13.11a,

$$p_1 + \rho_1 V_1^2 = p_2 + \rho_2 V_2^2 \tag{13.11b}$$

$$h_{0_1} = h_1 + \frac{V_1^2}{2} = h_2 + \frac{V_2^2}{2} = h_{0_2} \tag{13.11c}$$

$$s_2 > s_1 \tag{13.11d}$$

$$p = \rho RT \tag{13.11e}$$

$$\Delta h = h_2 - h_1 = c_p \Delta T = c_p (T_2 - T_1) \tag{13.11f}$$

$$\Delta s = s_2 - s_1 = c_p \ln \frac{T_2}{T_1} - R \ln \frac{p_2}{p_1} \tag{13.11g}$$

Equations 13.11 can be used to analyze flow through a normal shock. For example, if we know conditions before the shock, at section ① (i.e., p_1, ρ_1, T_1, s_1, h_1, and V_1), we can use these equations to find conditions after the shock, at section ②. We have six equations (not including the constraint of Eq. 13.11d) and six unknowns (p_2, ρ_2, T_2, s_2, h_2, and V_2). Hence, for given upstream conditions there is a single unique downstream state. In practice this procedure is unwieldy—we have a set of *nonlinear coupled algebraic* equations to solve.

We can certainly use these equations for analyzing normal shocks, but we will usually find it more useful to develop normal shock functions based on M_1, the upstream Mach number. Before doing this, let us consider the set of equations. We have stated in this chapter that changes in a one-dimensional flow can be caused by area variation, friction, or heat transfer, but in deriving Eqs. 13.11 we have eliminated all three causes! In this case, then, what is causing the flow to change? Perhaps there are no changes through a normal shock! Indeed, if we examine each of these equations we see that each one is satisfied—has a possible "solution"—if all properties at location ② are equal to the corresponding properties at location ① (e.g., $p_2 = p_1$, $T_2 = T_1$) *except* for Eq. 13.11d, which expresses the second law of thermodynamics. Nature is telling us that in the absence of area change, friction, and heat transfer, flow properties will not change *except* in a very abrupt, irreversible manner, for which the entropy increases. In fact, all properties except T_0 *do* change through the shock. We must find a solution in which *all* of Eqs. 13.11 are satisfied. (Incidentally, because all the equations except Eq. 13.11d are satisfied by $p_2 = p_1$, $T_2 = T_1$, and so on, numerical searching methods such as *Excel*'s *Solver* have some difficulty in finding the correct solution!)

Because they are a set of nonlinear coupled equations, it is difficult to use Eqs. 13.11 to see exactly what happens through a normal shock. We will postpone formal proof of the results we are about to present until a subsequent subsection, where we recast the equations in terms of the incoming Mach number. This recasting is rather mathematical, so we present results of the analysis here for clarity. It turns out that a normal shock can occur only when the incoming flow is supersonic. Fluid flows will generally gradually adjust to downstream conditions (e.g., an obstacle in the flow) as the pressure field redirects the flow (e.g., around the object). However, if the flow is moving at such a speed that the pressure field cannot propagate upstream (when the flow speed, V, is greater than the local speed of sound, c, or in other words $M > 1$), then the fluid has to "violently" adjust to the downstream conditions. The shock that a supersonic flow may encounter is like a hammer blow that each fluid particle experiences; the pressure suddenly increases through the shock (as mentioned, over a distance $< 2\ \mu$m), so that, at the instant a particle is passing through the shock, there is a very large negative pressure gradient. This pressure gradient causes a dramatic reduction in speed, V, and hence a rapid rise in temperature, T, as kinetic energy is converted to internal thermal energy. We may wonder what happens to the density because both the temperature and pressure rise through the shock, leading to opposing changes in density; it turns out that the density, ρ, increases through the shock. Because the shock is adiabatic but highly irreversible, entropy, s, increases through the shock. Finally, we see that as speed, V, decreases and the speed of sound, c, increases (because temperature, T, increases) through the normal shock, the Mach number, M, decreases; in fact, we will see later that it always becomes subsonic. These results are shown graphically in Fig. 13.10 and in tabular form in Table 13.1.

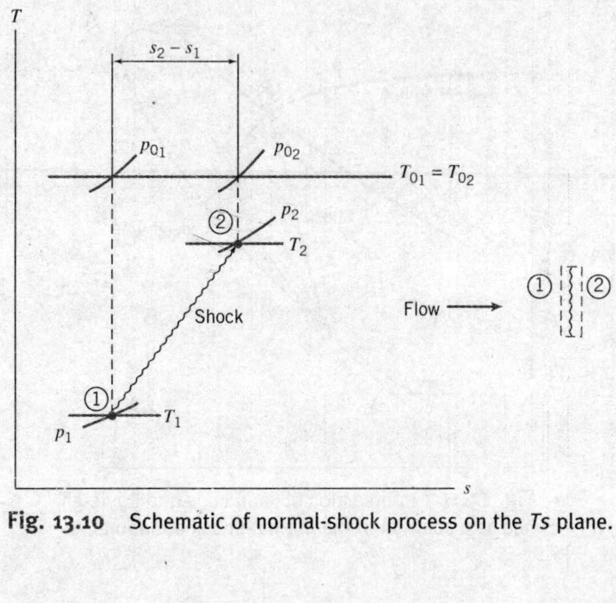

Fig. 13.10 Schematic of normal-shock process on the *Ts* plane.

Table 13.1

Summary of Property Changes Across a Normal Shock

Property	Effect	Obtained from:
Stagnation temperature	T_0 = Constant	Energy equation
Entropy	$s \Uparrow$	Second law
Stagnation pressure	$p_0 \Downarrow$	*Ts* diagram
Temperature	$T \Uparrow$	*Ts* diagram
Velocity	$V \Downarrow$	Energy equation, and effect on T
Density	$\rho \Uparrow$	Continuity equation, and effect on V
Pressure	$p \Uparrow$	Momentum equation, and effect on V
Mach number	$M \Downarrow$	$M = V/c$, and effects on V and T

Fanno and Rayleigh Interpretation of Normal Shock

A normal shock is a phenomenon in which fluid properties experience large changes over a very short distance and time; we are very far from equilibrium conditions! It is not easy to see what is happening in such a dramatic process. However, we can gain some insight by considering two processes in which flow properties are gradually changing: during a process involving friction and during a heat transfer process. Hence Eqs. 13.11 can be understood to some degree by considering *Fanno-line* (friction) and *Rayleigh-line* (heat transfer) curves. You may wish to postpone reading this subsection until these curves are discussed in much more detail in Sections 13.5 and 13.6, but the following discussion briefly and sufficiently describes them; they are shown schematically in Fig. 13.11. Both curves comply with our ubiquitous Eqs. 13.1.

In a *Ts* diagram, the Fanno line curve shows all possible states for a one-dimensional flow that is being changed *only* by friction (there is no area change and no heat transfer). The second law of thermodynamics requires that in this case entropy must increase, so that, as seen in Fig. 13.11, if the flow starts out subsonic, friction causes the flow to accelerate until it becomes sonic; a flow that starts out supersonic decelerates until it is sonic. We will see in Section 13.5 that the curve is generated from Eqs. 13.1. (It is counterintuitive that friction can accelerate a flow, but it can happen if the pressure

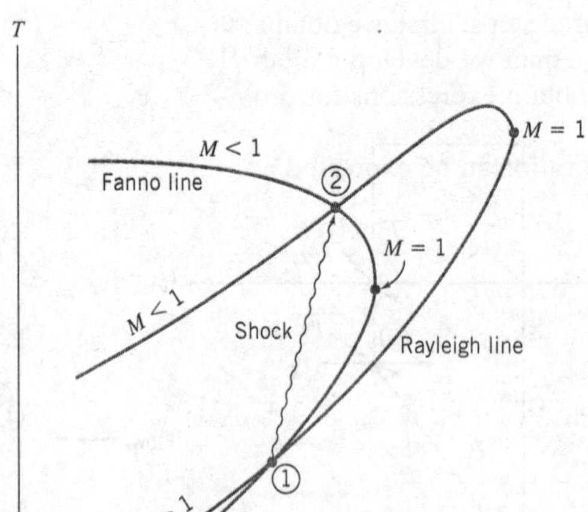

Fig. 13.11 Intersection of Fanno line and Rayleigh line as a solution of the normal-shock equations.

is falling sufficiently rapidly). We will wait until Section 13.5 to discuss what happens if friction continues after we reach sonic conditions! All such friction flows must move in the direction of increasing entropy.

The Rayleigh line shows all possible states for a one-dimensional flow that is experiencing *only* heat transfer (no area changes or friction). Heat addition corresponds to an increase in both entropy and temperature (except for a small region near $M = 1$), and both sub- and supersonic flows approach sonic conditions; a cooling process leads to reductions in temperature and entropy. We will see in Section 13.6 that the curve is generated from Eqs. 13.1.

The normal shock is obtained from a superposition of the two curves, as shown. States ① and ② are the beginning and end states of a flow that (following the Fanno line) has *only* friction present (no area changes or heat transfer); they are also the beginning and end states of a flow that (following the Rayleigh curve) has *only* heat transfer (no area changes or friction). This raises the question of how we can have a flow with simultaneously *only* friction and *only* heat transfer! The answer is that we mathematically can follow the Fanno line from ① to ②, although not in reality. We could actually have a flow in which we had friction from state ① to where $M = 1$, but from $M = 1$ to state ② we would have to have "negative friction" (recall that friction requires us to increase the entropy). Hence, from state ① to state ②, we have a fictitious process in which we have friction, then negative friction, ending in no net friction. A similar process applies to the Rayleigh line. To get from state ① to state ②, we would heat the flow and then cool it, with no net heat transfer. The conclusion we come to is that states ① and ② represent a change in a flow for which there is no heat transfer, no friction, and no area change; moreover, it is "violent" because the flow changes from state ① to state ② without following a process curve; so entropy must increase. Note that Fig. 13.11 shows some trends we have mentioned: The flow must go from super- to subsonic, and entropy and temperature must increase through a shock.

Normal-Shock Flow Functions for One-Dimensional Flow of an Ideal Gas

We have mentioned that the basic equations, Eqs. 13.11, can be used to analyze flows that experience a normal shock. As in isentropic flow, it is often more convenient to use Mach number-based equations, in this case based on the incoming Mach number, M_1. This involves three steps: First, we obtain property ratios (e.g., T_2/T_1 and p_2/p_1) in terms of M_1 and M_2, then we develop a relation between M_1 and M_2, and finally, we use this relation to obtain expressions for property ratios in terms of upstream Mach number, M_1.

The temperature ratio can be expressed as

$$\frac{T_2}{T_1} = \frac{T_2}{T_{0_2}} \frac{T_{0_2}}{T_{0_1}} \frac{T_{0_1}}{T_1}$$

Since stagnation temperature is constant across the shock, we have

$$\frac{T_2}{T_1} = \frac{1 + \dfrac{k-1}{2} M_1^2}{1 + \dfrac{k-1}{2} M_2^2} \tag{13.12}$$

A velocity ratio may be obtained by using

$$\frac{V_2}{V_1} = \frac{M_2 c_2}{M_1 c_1} = \frac{M_2}{M_1} \frac{\sqrt{kRT_2}}{\sqrt{kRT_1}} = \frac{M_2}{M_1} \sqrt{\frac{T_2}{T_1}}$$

or

$$\frac{V_2}{V_1} = \frac{M_2}{M_1}\left[\frac{1+\frac{k-1}{2}M_1^2}{1+\frac{k-1}{2}M_2^2}\right]^{1/2}$$

A ratio of densities may be obtained from the continuity equation

$$\rho_1 V_1 = \rho_2 V_2 \tag{13.11a}$$

so that

$$\frac{\rho_2}{\rho_1} = \frac{V_1}{V_2} = \frac{M_1}{M_2}\left[\frac{1+\frac{k-1}{2}M_2^2}{1+\frac{k-1}{2}M_1^2}\right]^{1/2} \tag{13.13}$$

Finally, we have the momentum equation,

$$p_1 + \rho_1 V_1^2 = p_2 + \rho_2 V_2^2 \tag{13.11b}$$

Substituting $\rho = p/RT$, and factoring out pressures, gives

$$p_1\left[1+\frac{V_1^2}{RT_1}\right] = p_2\left[1+\frac{V_2^2}{RT_2}\right]$$

Since

$$\frac{V^2}{RT} = k\frac{V^2}{kRT} = kM^2$$

then

$$p_1\left[1+kM_1^2\right] = p_2\left[1+kM_2^2\right]$$

Finally,

$$\frac{p_2}{p_1} = \frac{1+kM_1^2}{1+kM_2^2} \tag{13.14}$$

To solve for M_2 in terms of M_1, we must obtain another expression for one of the property ratios given by Eqs. 13.12 through 13.14.

From the ideal gas equation of state, the temperature ratio may be written as

$$\frac{T_2}{T_1} = \frac{p_2/\rho_2 R}{p_1/\rho_1 R} = \frac{p_2}{p_1}\frac{\rho_1}{\rho_2}$$

Substituting from Eqs. 13.13 and 13.14 yields

$$\frac{T_2}{T_1} = \left[\frac{1+kM_1^2}{1+kM_2^2}\right]\frac{M_2}{M_1}\left[\frac{1+\frac{k-1}{2}M_1^2}{1+\frac{k-1}{2}M_2^2}\right]^{1/2} \tag{13.15}$$

Equations 13.12 and 13.15 are two equations for T_2/T_1. We can combine them and solve for M_2 in terms of M_1. Combining and canceling gives

$$\left[\frac{1+\frac{k-1}{2}M_1^2}{1+\frac{k-1}{2}M_2^2}\right]^{1/2} = \frac{M_2}{M_1}\left[\frac{1+kM_1^2}{1+kM_2^2}\right]$$

Squaring, we obtain

$$\frac{1+\frac{k-1}{2}M_1^2}{1+\frac{k-1}{2}M_2^2} = \frac{M_2^2}{M_1^2}\left[\frac{1+2kM_1^2+k^2M_1^4}{1+2kM_2^2+k^2M_2^4}\right]$$

which may be solved explicitly for M_2^2. Two solutions are obtained:

$$M_2^2 = M_1^2 \tag{13.16a}$$

and

$$M_2^2 = \frac{M_1^2 + \frac{2}{k-1}}{\frac{2k}{k-1}M_1^2 - 1} \tag{13.16b}$$

Obviously, the first of these is trivial. The second expresses the unique dependence of M_2 on M_1.

Now, having a relationship between M_2 and M_1, we can solve for property ratios across a shock. Knowing M_1, we obtain M_2 from Eq. 13.16b; the property ratios can be determined subsequently from Eqs. 13.12 through 13.14.

Since the stagnation temperature remains constant, the stagnation temperature ratio across the shock is unity. The ratio of stagnation pressures is evaluated as

$$\frac{p_{0_2}}{p_{0_1}} = \frac{p_{0_2}}{p_2}\frac{p_2}{p_1}\frac{p_1}{p_{0_1}} = \frac{p_2}{p_1}\left[\frac{1+\frac{k-1}{2}M_2^2}{1+\frac{k-1}{2}M_1^2}\right]^{k/(k-1)} \tag{13.17}$$

Combining Eqs. 13.14 and 13.16b, we obtain (after considerable algebra)

$$\frac{p_2}{p_1} = \frac{1+kM_1^2}{1+kM_2^2} = \frac{2k}{k+1}M_1^2 - \frac{k-1}{k+1} \tag{13.18}$$

Using Eqs. 13.16b and 13.18, we find that Eq. 13.17 becomes

$$\frac{p_{0_2}}{p_{0_1}} = \frac{\left[\dfrac{\frac{k+1}{2}M_1^2}{1+\frac{k-1}{2}M_1^2}\right]^{k/(k-1)}}{\left[\frac{2k}{k+1}M_1^2 - \frac{k-1}{k+1}\right]^{1/(k-1)}} \tag{13.19}$$

After substituting for M_2^2 from Eq.13.16b into Eqs. 13.12 and 13.13, we summarize the set of Mach number-based equations (renumbered for convenience) for use with an ideal gas passing through a normal shock:

$$M_2^2 = \frac{M_1^2 + \frac{2}{k-1}}{\frac{2k}{k-1}M_1^2 - 1} \tag{13.20a}$$

$$\frac{p_{0_2}}{p_{0_1}} = \frac{\left[\dfrac{\dfrac{k+1}{2}M_1^2}{1+\dfrac{k-1}{2}M_1^2}\right]^{k/(k-1)}}{\left[\dfrac{2k}{k+1}M_1^2 - \dfrac{k-1}{k+1}\right]^{1/(k-1)}} \tag{13.20b}$$

$$\frac{T_2}{T_1} = \frac{\left(1+\dfrac{k-1}{2}M_1^2\right)\left(kM_1^2 - \dfrac{k-1}{2}\right)}{\left(\dfrac{k+1}{2}\right)^2 M_1^2} \tag{13.20c}$$

$$\frac{p_2}{p_1} = \frac{2k}{k+1}M_1^2 - \frac{k-1}{k+1} \tag{13.20d}$$

$$\frac{\rho_2}{\rho_1} = \frac{V_1}{V_2} = \frac{\dfrac{k+1}{2}M_1^2}{1+\dfrac{k-1}{2}M_1^2} \tag{13.20e}$$

Equations 13.20 are useful for analyzing flow through a normal shock. Note that all changes through a normal shock depend only on M_1, the incoming Mach number (as well as the fluid property, k, the ratio of specific heats). The equations are usually preferable to the original equations, Eq. 13.11, because they provide explicit, uncoupled expressions for property changes; Eqs. 13.11 are occasionally useful too. Note that Eq. 13.20d requires $M_1 > 1$ for $p_2 > p_1$, which agrees with our previous discussion. The ratio p_2/p_1 is known as the *strength* of the shock; the higher the incoming Mach number, the stronger (more violent) the shock.

Equations 13.20, while quite complex algebraically, provide explicit property relations in terms of the incoming Mach number, M_1. They are so useful that some calculators have some of them built in (e.g., the *HP 48G* series [1]); it is a good idea to program them if your calculator does not already have them. There are also interactive Web sites that make them available (see, e.g., [2]), and they are fairly easy to define in spreadsheets such as *Excel*. The reader is urged to download the *Excel* add-ins for these equations from the Web site; with the add-ins, functions are available for computing M_2, and the stagnation pressure, temperature, pressure, and density/velocity ratios, from M_1, and M_1 from these ratios. Appendix E.2 lists flow functions for M_2 and property ratios p_{0_2}/p_{0_1}, T_2/T_1, p_2/p_1, and ρ_2/ρ_1 (V_1/V_2) in terms of M_1 for normal-shock flow of an ideal gas. A table of values, as well as a plot of these property ratios, is presented for air ($k = 1.4$) for a limited range of Mach numbers. The associated *Excel* workbook, *Normal-Shock Relations*, can be used to print a larger table of values for air and other ideal gases.

A problem involving a normal shock is solved in Example 13.6.

Example 13.6 NORMAL SHOCK IN A DUCT

A normal shock stands in a duct. The fluid is air, which may be considered an ideal gas. Properties upstream from the shock are $T_1 = 5°\text{C}$, $p_1 = 65.0$ kPa (abs), and $V_1 = 668$ m/s. Determine properties downstream and $s_2 - s_1$. Sketch the process on a Ts diagram.

Given: Normal shock in a duct as shown:

$$T_1 = 5°\text{C}$$
$$P_1 = 65.0\ \text{kPa (abs)}$$
$$V_1 = 668\ \text{m/s}$$

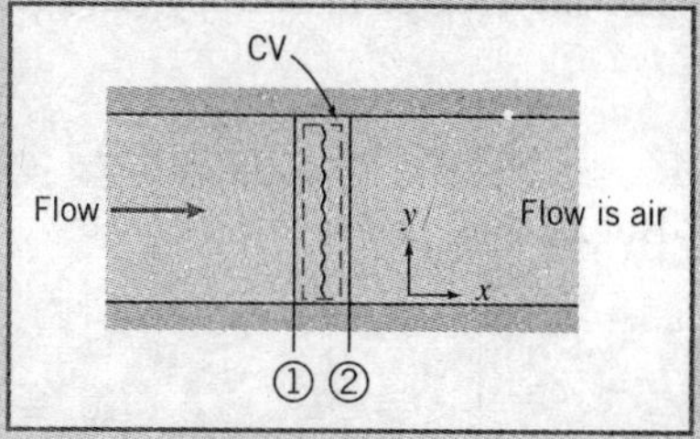

Find: (a) Properties at section ②.
(b) $s_2 - s_1$.
(c) Ts diagram.

Solution:
First compute the remaining properties at section ①. For an ideal gas,

$$\rho_1 = \frac{p_1}{RT_1} = 6.5 \times 10^4 \frac{\text{N}}{\text{m}^2} \times \frac{\text{kg}\cdot\text{K}}{287\ \text{N}\cdot\text{m}} \times \frac{1}{278\ \text{K}} = 0.815\ \text{kg/m}^3$$

$$c_1 = \sqrt{kRT_1} = \left[1.4 \times 287 \frac{\text{N}\cdot\text{m}}{\text{kg}\cdot\text{K}} \times 278\ \text{K} \times \frac{\text{kg}\cdot\text{m}}{\text{N}\cdot\text{s}^2}\right]^{1/2} = 334\ \text{m/s}$$

Then

$$M_1 = \frac{V_1}{c_1} = \frac{668}{334} = 2.00,$$ and (using isentropic stagnation relations, Eqs. 12.21b and 12.21a)

$$T_{0_1} = T_1\left(1 + \frac{k-1}{2}M_1^2\right) = 278\ \text{K}[1 + 0.2(2.0)^2] = 500\ \text{K}$$

$$p_{0_1} = p_1\left(1 + \frac{k-1}{2}M_1^2\right)^{k/(k-1)} = 65.0\ \text{kPa}[1 + 0.2(2.0)^2]^{3.5} = 509\ \text{kPa (abs)}$$

From the normal-shock flow functions, Eqs. 13.20, at $M_1 = 2.0$,

M_1	M_2	p_{0_2}/p_{0_1}	T_2/T_1	p_2/p_1	V_2/V_1
2.00	0.5774	0.7209	1.687	4.500	0.3750

From these data

$$T_2 = 1.687T_1 = (1.687)278\ \text{K} = 469\ \text{K} \quad \longleftarrow \quad T_2$$

$$p_2 = 4.500p_1 = (4.500)65.0\ \text{kPa} = 293\ \text{kPa (abs)} \quad \longleftarrow \quad p_2$$

$$V_2 = 0.3750V_1 = (0.3750)668\ \text{m/s} = 251\ \text{m/s} \quad \longleftarrow \quad V_2$$

For an ideal gas,

$$\rho_2 = \frac{p_2}{RT_2} = 2.93 \times 10^5 \frac{\text{N}}{\text{m}^2} \times \frac{\text{kg}\cdot\text{K}}{287\ \text{N}\cdot\text{m}} \times \frac{1}{469\ \text{K}} = 2.18\ \text{kg/m}^3 \quad \longleftarrow \quad \rho_2$$

Stagnation temperature is constant in adiabatic flow. Thus

$$T_{0_2} = T_{0_1} = 500\ \text{K} \quad \longleftarrow \quad T_{0_2}$$

Using the property ratios for a normal shock, we obtain

$$p_{0_2} = p_{0_1}\frac{p_{0_2}}{p_{0_1}} = 509\ \text{kPa}\ (0.7209) = 367\ \text{kPa (abs)} \quad \longleftarrow \quad p_{0_2}$$

For the change in entropy (Eq. 13.11g),

$$s_2 - s_1 = c_p \ln \frac{T_2}{T_1} - R \ln \frac{p_2}{p_1}$$

But $s_{0_2} - s_{0_1} = s_2 - s_1$, so

$$s_{0_2} - s_{0_1} = s_2 - s_1 = \overbrace{c_p \ln \frac{T_{0_2}}{T_{0_1}}}^{=0} - R \ln \frac{p_{0_2}}{p_{0_1}} = -0.287 \frac{\text{kJ}}{\text{kg} \cdot \text{K}} \times \ln(0.7209)$$

$$s_2 - s_1 = 0.0939 \text{ kJ}/(\text{kg} \cdot \text{K})$$ $\longleftarrow$ $s_2 - s_1$

The Ts diagram is

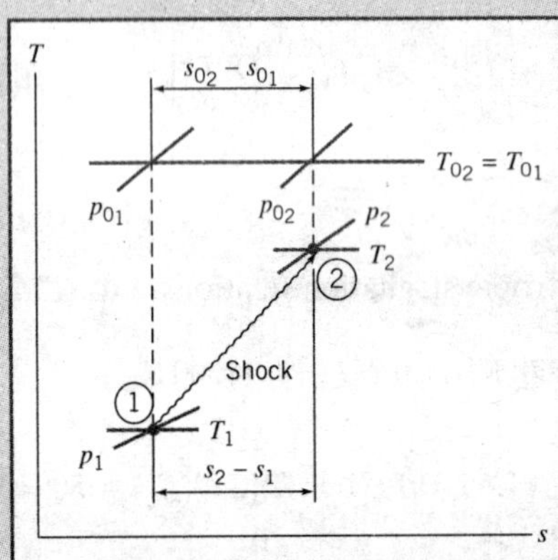

This problem illustrates the use of the normal shock relations, Eqs. 13.20, for analyzing flow of an ideal gas through a normal shock.

The *Excel* workbook for this Example is convenient for performing the calculations. (Alternatively, the normal shock relations *Excel* add-ins, available on the Web site, are useful for these calculations.)

13.4 Supersonic Channel Flow with Shocks

Supersonic flow is a necessary condition for a normal shock to occur. The possibility of a normal shock must be considered in any supersonic flow. Sometimes a shock *must* occur to match a downstream pressure condition; it is desirable to determine if a shock will occur and the shock location when it does occur.

In Section 13.3 we showed that stagnation pressure decreases dramatically across a shock: The stronger the shock, the larger the decrease in stagnation pressure. It is necessary to control shock position to obtain acceptable performance from a supersonic diffuser or supersonic wind tunnel.

In this section isentropic flow in a converging-diverging nozzle (Section 13.2) is extended to include shocks. Additional topics (on the Web site) include operation of supersonic diffusers and supersonic wind tunnels, flows with friction, and flows with heat addition.

Flow in a Converging-Diverging Nozzle

Since we have considered normal shocks, we now can complete our discussion of flow in a converging-diverging nozzle operating under varying back pressures, begun in

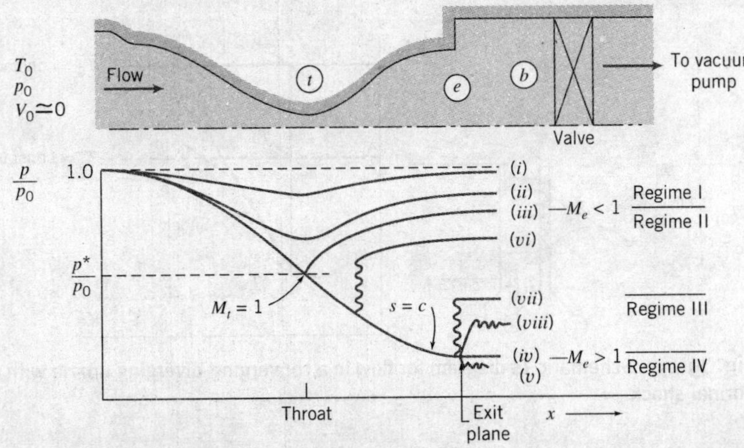

Fig. 13.12 Pressure distributions for flow in a converging-diverging nozzle for different back pressures.

Section 13.2. The pressure distribution through a nozzle for different back pressures is shown in Fig. 13.12.

Four flow regimes are possible. In Regime I the flow is subsonic throughout. The flow rate increases with decreasing back pressure. At condition (*iii*), which forms the dividing line between Regimes I and II, flow at the throat is sonic, and $M_t = 1$.

As the back pressure is lowered below condition (*iii*), a normal shock appears downstream from the throat, as shown by condition (*vi*). There is a pressure rise across the shock. Since the flow is subsonic ($M < 1$) after the shock, the flow decelerates, with an accompanying increase in pressure, through the diverging channel. As the back pressure is lowered further, the shock moves downstream until it appears at the exit plane (condition *vii*). In Regime II, as in Regime I, the exit flow is subsonic, and consequently $p_e = p_b$. Since flow properties at the throat are constant for all conditions in Regime II, the flow rate in Regime II does not vary with back pressure.

In Regime III, as exemplified by condition (*viii*), the back pressure is higher than the exit pressure, but not high enough to sustain a normal shock in the exit plane. The flow adjusts to the back pressure through a series of oblique compression shocks outside the nozzle; these oblique shocks cannot be treated by one-dimensional theory.

As previously noted in Section 13.2, condition (*iv*) represents the design condition. In Regime IV the flow adjusts to the lower back pressure through a series of oblique expansion waves outside the nozzle; these oblique expansion waves cannot be treated by one-dimensional theory.

The *Ts* diagram for converging-diverging nozzle flow with a normal shock is shown in Fig. 13.13; state ① is located immediately upstream from the shock and state ② is immediately downstream. The entropy increase across the shock moves the subsonic downstream flow to a new isentropic line. The critical temperature is constant, so p_2^* is lower than p_1^*. Since $\rho^* = p^*/RT^*$, the critical density downstream also is reduced. To carry the same mass flow rate, the downstream flow must have a larger critical area. From continuity (and the equation of state), the critical area ratio is the inverse of the critical pressure ratio, i.e., across a shock, $p^*A^* =$ constant.

If the Mach number (or position) of the normal shock in the nozzle is known, the exit-plane pressure can be calculated directly. In the more realistic situation, the exit-plane pressure is specified, and the position and strength of the shock are unknown. The subsonic flow downstream must leave the nozzle at the back pressure, so $p_b = p_e$. Then

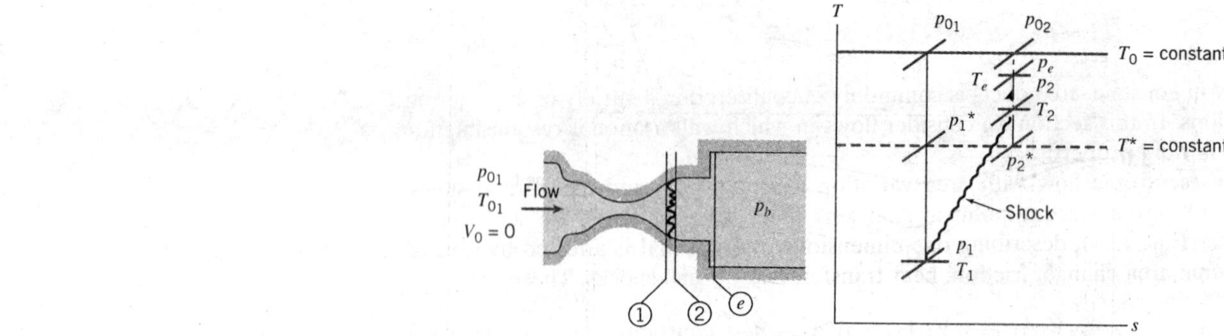

Fig. 13.13 Schematic *Ts* diagram for flow in a converging-diverging nozzle with a normal shock.

$$\frac{p_b}{p_{0_1}} = \frac{p_e}{p_{0_1}} = \frac{p_e}{p_{0_2}}\frac{p_{0_2}}{p_{0_1}} = \frac{p_e}{p_{0_2}}\frac{A_1^*}{A_2^*} = \frac{p_e}{p_{0_2}}\frac{A_t}{A_e}\frac{A_e}{A_2^*} \tag{13.21}$$

Because we have isentropic flow from state ② (after the shock) to the exit plane, $A_2^* = A_e^*$ and $p_{0_2} = p_{0_e}$. Then from Eq. 13.21 we can write

$$\frac{p_e}{p_{0_1}} = \frac{p_e}{p_{0_2}}\frac{A_t}{A_e}\frac{A_e}{A_2^*} = \frac{p_e}{p_{0_e}}\frac{A_t}{A_e}\frac{A_e}{A_e^*}$$

Rearranging,

$$\frac{p_e}{p_{0_1}}\frac{A_e}{A_t} = \frac{p_e}{p_{0_e}}\frac{A_e}{A_e^*} \tag{13.22}$$

In Eq. 13.22 the left side contains known quantities, and the right side is a function of the exit Mach number M_e only. The pressure ratio is obtained from the stagnation pressure relation (Eq. 12.21a); the area ratio is obtained from the isentropic area relation (Eq. 13.7d). Finding M_e from Eq. 13.43 usually requires iteration. (Problem 13.103 uses *Excel*'s *Solver* feature to perform the iteration.) The magnitude and location of the normal shock can be found once M_e is known by rearranging Eq. 13.43 (remembering that $p_{0_2} = p_{0_e}$),

$$\frac{p_{0_2}}{p_{0_1}} = \frac{A_t}{A_e}\frac{A_e}{A_e^*} \tag{13.23}$$

In Eq. 13.44 the right side is known (the first area ratio is given and the second is a function of M_e only), and the left side is a function of the Mach number before the shock, M_1, only (Eq. 13.41b). Hence, M_1 can be found. The area at which this shock occurs can then be found from the isentropic area relation (Eq. 13.7d, with $A^* = A_t$) for isentropic flow between the throat and state ①.

Supersonic Diffuser (on the Web)

Supersonic Wind Tunnel Operation (on the Web)

Supersonic Flow with Friction in a Constant-Area Channel (on the Web)

Supersonic Flow with Heat Addition in a Constant-Area Channel (on the Web)

Flow in a Constant-Area Duct with Friction 13.5

Gas flow in constant-area ducts is commonly encountered in a variety of engineering applications. In this section we consider flows in which wall friction is responsible for changes in fluid properties.

As for isentropic flow with area variation (Section 13.2) and the normal shock (Section 13.3), our starting point in analyzing flows with friction is the set of basic equations (Eqs. 13.1), describing one-dimensional motion that is affected by several phenomena: area change, friction, heat transfer, and normal shocks. These are

$$\rho_1 V_1 A_1 = \rho_2 V_2 A_2 = \rho V A = \dot{m} = \text{constant} \tag{13.1a}$$

$$R_x + p_1 A_1 - p_2 A_2 = \dot{m} V_2 - \dot{m} V_1 \tag{13.1b}$$

$$\frac{\delta Q}{dm} + h_1 + \frac{V_1^2}{2} = h_2 + \frac{V_2^2}{2} \tag{13.1c}$$

$$\dot{m}(s_2 - s_1) \geq \int_{CS} \frac{1}{T}\left(\frac{\dot{Q}}{A}\right) dA \tag{13.1d}$$

$$p = \rho R T \tag{13.1e}$$

$$\Delta h = h_2 - h_1 = c_p \Delta T = c_p (T_2 - T_1) \tag{13.1f}$$

$$\Delta s = s_2 - s_1 = c_p \ln \frac{T_2}{T_1} - R \ln \frac{p_2}{p_1} \tag{13.1g}$$

Equation 13.1a is *continuity*, Eq. 13.1b is a *momentum equation*, Eq. 13.1c is an *energy equation*, Eq. 13.1d is the *second law of thermodynamics*, and Eqs. 13.1e, 13.1f, and 13.1g are useful *property relations* for an ideal gas with constant specific heats.

We must simplify these equations for flow in a constant-area duct with friction. We must think about what happens to the heat that friction generates. There are two obvious cases we can consider: In the first we assume that the flow is *adiabatic*, so any heat generated remains in the fluid; in the second we assume that the flow remains *isothermal*, so the fluid either gives off heat or absorbs heat as necessary. While some flows may be neither adiabatic nor isothermal, many real-world flows are. Flow in a relatively short duct will be approximately adiabatic; flow in a very long duct (e.g., an uninsulated natural gas pipeline) will be approximately isothermal (the pipeline will be at the ambient temperature). We consider first adiabatic flow.

Basic Equations for Adiabatic Flow

We can simplify Eqs. 13.1 for frictional adiabatic flow in a constant-area duct of an ideal gas with constant specific heats, as shown in Fig. 13.18.

We now have $A_1 = A_2 = A$. In addition, for no heat transfer we have $\delta Q/dm = 0$. Finally, the force R_x is now due only to friction (no x component of surface force is caused by pressure on the parallel sides of the channel). Hence, for this flow our equations simplify to

$$\rho_1 V_1 = \rho_2 V_2 = \rho V \equiv G = \frac{\dot{m}}{A} = \text{constant} \tag{13.24a}$$

$$R_x + p_1 A - p_2 A = \dot{m} V_2 - \dot{m} V_1 \tag{13.24b}$$

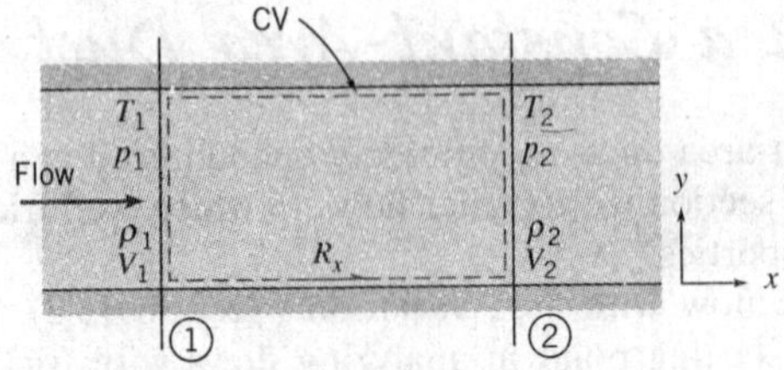

Fig. 13.18 Control volume used for integral analysis of frictional adiabatic flow.

$$h_{0_1} = h_1 + \frac{V_1^2}{2} = h_2 + \frac{V_2^2}{2} = h_{0_2} = h_0 \tag{13.24c}$$

$$s_2 > s_1 \tag{13.24d}$$

$$p = \rho RT \tag{13.24e}$$

$$\Delta h = h_2 - h_1 = c_p \Delta T = c_p(T_2 - T_1) \tag{13.24f}$$

$$\Delta s = s_2 - s_1 = c_p \ln \frac{T_2}{T_1} - R \ln \frac{p_2}{p_1} \tag{13.24g}$$

Equations 13.24 can be used to analyze frictional adiabatic flow in a channel of constant area. For example, if we know conditions at section ① (i.e., p_1, ρ_1, T_1, s_1, h_1, and V_1), we can use these equations to find conditions at some new section ② after the fluid has experienced a total friction force R_x. It is the effect of friction that causes fluid properties to change along the duct. For a known friction force we have six equations (not including the constraint of Eq. 13.24d) and six unknowns (p_2, ρ_2, T_2, s_2, h_2, and V_2). In practice this procedure is unwieldy—as for isentropic flow we have a set of *nonlinear coupled algebraic* equations to solve, and as for isentropic flow we will eventually develop alternative approaches. For now, let's see what Eqs. 13.24 indicate will happen to the flow.

Adiabatic Flow: The Fanno Line

If we were to attempt the calculations described above, as the flow progresses down the duct (i.e., for increasing values of R_x), we would develop a relationship between T and s shown qualitatively in Fig. 13.19 for two possibilities: a flow that was initially subsonic (starting at some point ①), and flow that was initially supersonic (starting at some point ①'). The locus of all possible downstream states is referred to as the *Fanno line*. Detailed calculations show some interesting features of Fanno-line flow. At the point of maximum entropy, the Mach number is unity. On the upper branch of the curve, the Mach number is always less than unity, and it increases monotonically as we proceed to the right along the curve. At every point on the lower portion of the curve, the Mach number is greater than unity; the Mach number decreases monotonically as we move to the right along the curve.

For any initial state on a Fanno line, each point on the Fanno line represents a mathematically possible downstream state. In Fig. 13.19 we generated the curves by repeatedly solving Eqs. 13.24 for increasing values of the friction force, R_x; the total friction force increases as we progress down the duct because we are including more and more surface area. Note the arrows in Fig. 13.19, indicating that, as required by Eq. 13.24d, the entropy must increase for this flow. In fact it is because we *do* have friction (an irreversibility) present in an adiabatic flow that this *must* happen.

Referring again to Fig. 13.19, we see that for an initially subsonic flow (state ①), the effect of friction is to increase the Mach number toward unity. For a flow that is initially supersonic (state ①′), the effect of friction is to decrease the Mach number toward unity.

In developing the simplified form of the first law for Fanno-line flow, Eq. 13.24c, we found that stagnation enthalpy remains constant. Consequently, when the fluid is an ideal gas with constant specific heats, stagnation temperature must also remain constant. What happens to stagnation pressure? Friction causes the local isentropic stagnation pressure to decrease for all Fanno-line flows, as shown in Fig. 13.20. Since entropy must increase in the direction of flow, the flow process must proceed to the right on the *Ts* diagram. In Fig. 13.20, a path from state ① to state ② is shown on the subsonic portion of the curve. The corresponding local isentropic stagnation pressures, p_{0_1} and p_{0_2}, clearly show that $p_{0_2} < p_{0_1}$. An identical result is obtained for flow on the supersonic branch of the curve from state ①′ to state ②′. Again $p_{0_{2'}} < p_{0_{1'}}$. Thus p_0 decreases for any Fanno-line flow.

The effects of friction on flow properties in Fanno-line flow are summarized in Table 13.2.

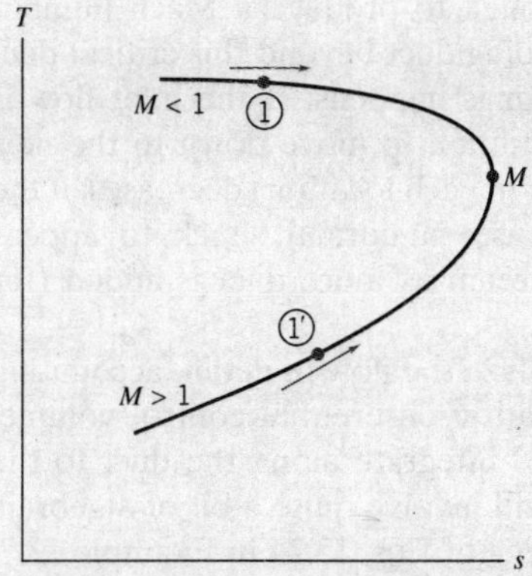

Fig. 13.19 Schematic *Ts* diagram for frictional adiabatic (Fanno-line) flow in a constant-area duct.

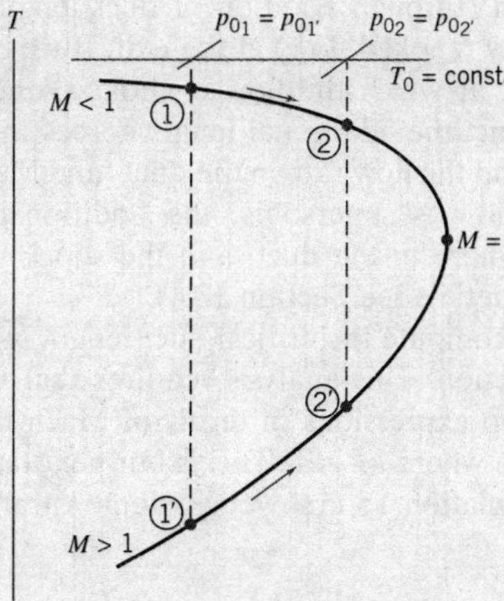

Fig. 13.20 Schematic of Fanno-line flow on *Ts* plane, showing reduction in local isentropic stagnation pressure caused by friction.

Table 13.2

Summary of Effects of Friction on Properties in Fanno-Line Flow

Property	Subsonic $M < 1$	Supersonic $M > 1$	Obtained from:
Stagnation temperature	T_0 = Constant	T_0 = Constant	Energy equation
Entropy	s⇧	s⇧	$T\,ds$ equation
Stagnation pressure	p_0⇩	p_0⇩	T_0 = constant; s⇧
Temperature	T⇩	T⇧	Shape of Fanno line
Velocity	V⇧	V⇩	Energy equation, and trend of T
Mach number	M⇧	M⇩	Trends of V, T, and definition of M
Density	ρ⇩	ρ⇧	Continuity equation, and effect on V
Pressure	p⇩	p⇧	Equation of state, and effects on ρ, T

In deducing the effect of friction on flow properties for Fanno-line flow, we used the shape of the Fanno line on the *Ts* diagram and the basic governing equations (Eqs. 13.24). You should follow through the logic indicated in the right column of the table. Note that the effect of friction is to accelerate a subsonic flow! This seems a real puzzle—a violation of Newton's second law—until we realize that the pressure is dropping quite rapidly, so the pressure gradient more than cancels the drag due to friction. We can also note that the density is decreasing in this flow (largely because of the pressure drop) mandating (from continuity) that the velocity must be increasing. All properties simultaneously affect one another (as expressed in the coupled set of equations, Eqs. 13.24), so it is not possible to conclude that the change in any one property is solely responsible for changes in any of the others. Note the parallel between normal shocks (Table 13.1) and supersonic flow with friction (Table 13.2). Both represent irreversible processes in supersonic flow, and all properties change in the same directions.

We have noted that entropy must increase in the direction of flow: It is the effect of friction that causes the change in flow properties along the Fanno-line curve. From Fig. 13.20, we see that there is a maximum entropy point corresponding to $M = 1$ for each Fanno line. The maximum entropy point is reached by increasing the amount of friction (through addition of duct length), just enough to produce a Mach number of unity (choked flow) at the exit. If we insist on adding duct beyond this critical duct length, at which the flow is choked, one of two things happens: If the inlet flow is subsonic, the additional length forces the sonic condition to move down to the new exit, and the flow rate in the duct (and Mach number at each location) decreases; if the inlet flow is supersonic, the additional length causes a normal shock to appear somewhere in the duct, and the shock moves upstream as more duct is added (for more details see Section 13.4).

To compute the critical duct length, we must analyze the flow in detail, accounting for friction. This analysis requires that we begin with a differential control volume, develop expressions in terms of Mach number, and integrate along the duct to the section where $M = 1$. This is our next task, and it will involve quite a bit of algebraic manipulation, so first we will demonstrate use of some of Eqs. 13.24 in Example 13.7.

Example 13.7 FRICTIONAL ADIABATIC FLOW IN A CONSTANT-AREA CHANNEL

Air flow is induced in an insulated tube of 7.16 mm diameter by a vacuum pump. The air is drawn from a room, where $p_0 = 101$ kPa (abs) and $T_0 = 23°$C, through a smoothly contoured converging nozzle. At section ①, where the nozzle joins the constant-area tube, the static pressure is 98.5 kPa (abs). At section ②, located some distance downstream in the constant-area tube, the air temperature is 14°C. Determine the mass flow rate, the local isentropic stagnation pressure at section ②, and the friction force on the duct wall between sections ① and ②.

Given: Air flow in insulated tube

Find: (a) $\dot{m}$.
(b) Stagnation pressure at section ②.
(c) Force on duct wall.

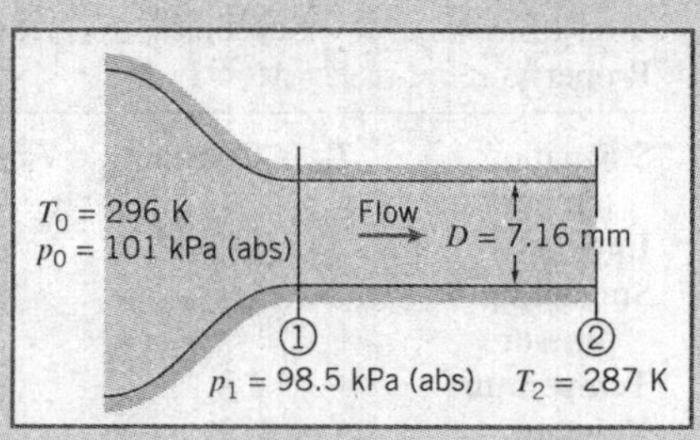

Solution:
The mass flow rate can be obtained from properties at section ①. For isentropic flow through the converging nozzle, local isentropic stagnation properties remain constant. Thus,

$$\frac{p_{0_1}}{p_1} = \left(1 + \frac{k-1}{2} M_1^2\right)^{k/(k-1)}$$

and

$$M_1 = \left\{\frac{2}{k-1}\left[\left(\frac{p_{0_1}}{p_1}\right)^{(k-1)/k} - 1\right]\right\}^{1/2} = \left\{\frac{2}{0.4}\left[\left(\frac{1.01\times10^5}{9.85\times10^4}\right)^{0.286} - 1\right]\right\}^{1/2} = 0.190$$

$$T_1 = \frac{T_{0_1}}{1+\frac{k-1}{2}M_1^2} = \frac{(273+23)\text{ K}}{1+0.2(0.190)^2} = 294\text{ K}$$

For an ideal gas,

$$\rho_1 = \frac{p_1}{RT_1} = 9.85\times10^4\frac{\text{N}}{\text{m}^2}\times\frac{\text{kg}\cdot\text{K}}{287\text{ N}\cdot\text{m}}\times\frac{1}{294\text{ K}} = 1.17\text{ kg/m}^3$$

$$V_1 = M_1c_1 = M_1\sqrt{kRT_1} = (0.190)\left[1.4\times287\frac{\text{N}\cdot\text{m}}{\text{kg}\cdot\text{K}}\times294\text{ K}\times\frac{\text{kg}\cdot\text{m}}{\text{N}\cdot\text{s}^2}\right]^{1/2}$$

$$V_1 = 65.3\text{ m/s}$$

The area, A_1, is

$$A_1 = A = \frac{\pi D^2}{4} = \frac{\pi}{4}(7.16\times10^{-3})^2\text{ m}^2 = 4.03\times10^{-5}\text{ m}^2$$

From continuity,

$$\dot{m} = \rho_1V_1A_1 = 1.17\frac{\text{kg}}{\text{m}^3}\times65.3\frac{\text{m}}{\text{s}}\times4.03\times10^{-5}\text{ m}^2$$

$$\dot{m} = 3.08\times10^{-3}\text{ kg/s} \longleftarrow \dot{m}$$

Flow is adiabatic, so T_0 is constant, and

$$T_{0_2} = T_{0_1} = 296\text{ K} \longleftarrow T_{0_2}$$

Then

$$\frac{T_{0_2}}{T_2} = 1 + \frac{k-1}{2}M_2^2$$

Solving for M_2 gives

$$M_2 = \left[\frac{2}{k-1}\left(\frac{T_{0_2}}{T_2} - 1\right)\right]^{1/2} = \left[\frac{2}{0.4}\left(\frac{296}{287} - 1\right)\right]^{1/2} = 0.396 \longleftarrow M_2$$

$$V_2 = M_2c_2 = M_2\sqrt{kRT_2} = (0.396)\left[1.4\times287\frac{\text{N}\cdot\text{m}}{\text{kg}\cdot\text{K}}\times287\text{ K}\times\frac{\text{kg}\cdot\text{m}}{\text{N}\cdot\text{s}^2}\right]^{1/2}$$

$$V_2 = 134\text{ m/s} \longleftarrow V_2$$

From continuity, Eq. 13.24a, $\rho_1V_1 = \rho_2V_2$, so

$$\rho_2 = \rho_1\frac{V_1}{V_2} = 1.17\frac{\text{kg}}{\text{m}^3}\times\frac{65.3}{134} = 0.570\text{ kg/m}^3 \longleftarrow \rho_2$$

and

$$p_2 = \rho_2RT_2 = 0.570\frac{\text{kg}}{\text{m}^3}\times287\frac{\text{N}\cdot\text{m}}{\text{kg}\cdot\text{K}}\times287\text{ K} = 47.0\text{ kPa(abs)} \longleftarrow p_2$$

The local isentropic stagnation pressure is

$$p_{0_2} = p_2\left(1 + \frac{k-1}{2}M_2^2\right)^{k/(k-1)} = 4.70 \times 10^4\ \text{Pa}[1 + 0.2(0.396)^2]^{3.5}$$

$$p_{0_2} = 52.4\ \text{kPa (abs)} \qquad\qquad p_{0_2}$$

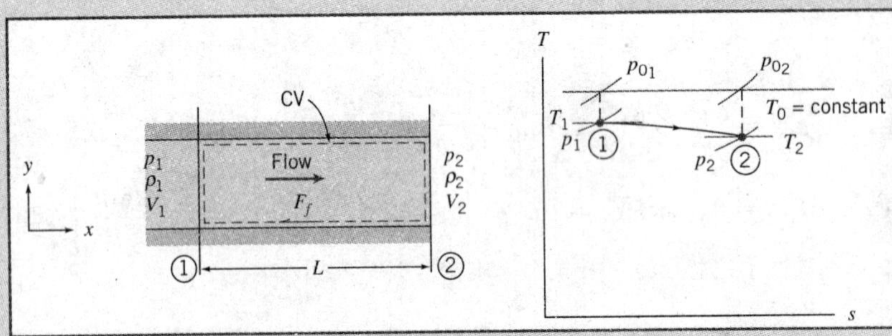

The friction force may be obtained using the momentum equation (Eq. 13.24b),

$$R_x + p_1 A - p_2 A = \dot{m}V_2 - \dot{m}V_2 \tag{13.24b}$$

which we apply to the control volume shown above (except we replace R_x from Fig. 13.18 with $-F_f$ because we know the friction force F_f on the fluid acts in the negative x direction).

$$-F_f = (p_2 - p_1)A + \dot{m}(V_2 - V_1)$$

$$-F_f = (4.70 - 9.85) \times 10^4 \frac{\text{N}}{\text{m}^2} \times 4.03 \times 10^{-5}\ \text{m}^2 + 3.08 \times 10^{-3}\frac{\text{kg}}{\text{s}} \times (134 - 65.3)\frac{\text{m}}{\text{s}} \times \frac{\text{N}\cdot\text{s}^2}{\text{kg}\cdot\text{m}}$$

or

$$F_f = 1.86\ \text{N} \quad \text{(to the left, as shown)}$$

This is the force exerted on the control volume by the duct wall. The force of the *fluid* on the *duct* is

$$K_x = -F_f = 1.86\ \text{N} \quad \text{(to the right)} \qquad\qquad K_x$$

This problem illustrates use of some of the basic equations, Eqs. 13.24, for flow in a duct with friction.
The *Excel* workbook for this Example is convenient for performing the calculations.

Fanno-Line Flow Functions for One-Dimensional Flow of an Ideal Gas

The primary independent variable in Fanno-line flow is the friction force, F_f. Knowledge of the total friction force between any two points in a Fanno-line flow enables us to predict downstream conditions from known upstream conditions. The total friction force is the integral of the wall shear stress over the duct surface area. Since wall shear stress varies along the duct, we must develop a differential equation and then integrate to find property variations. To set up the differential equation, we use the differential control volume shown in Fig. 13.21.

Comparing Fig. 13.21 to Fig. 13.19 we see that we can use the basic equations, Eqs. 13.24, for flow in a duct with friction, if we replace T_1, p_1, ρ_1, V_1, with T, p, ρ, V, and T_2, p_2, ρ_2, V_2, with $T + dT$, $p + dp$, $\rho + d\rho$, $V + dV$, and also R_x with $-dF_f$.

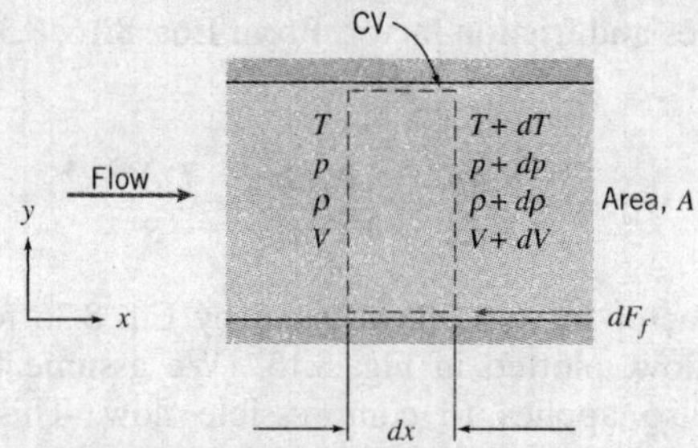

Fig. 13.21 Differential control volume used for analysis of Fanno-line flow.

The continuity equation (Eq. 13.24a) becomes

$$\rho V = (\rho + d\rho)(V + dV) = \frac{\dot{m}}{A}$$

so

$$\rho V = \rho V + \rho\, dV + d\rho\, V + d\rho\, dV$$

which reduces to

$$\rho\, dV + V\, d\rho = 0 \qquad (13.25a)$$

since products of differentials are negligible. The momentum equation (Eq. 13.24b) becomes

$$-dF_f + pA - (p + dp)A = \dot{m}(V + dV) - \dot{m}V$$

which reduces to

$$-\frac{dF_f}{A} - dp = \rho V\, dV \qquad (13.25b)$$

after using continuity ($\dot{m} = \rho AV$). The first law of thermodynamics (Eq. 13.24c) becomes

$$h + \frac{V^2}{2} = (h + dh) + \frac{(V + dV)^2}{2}$$

which reduces to

$$dh + d\left(\frac{V^2}{2}\right) = 0 \qquad (13.25c)$$

since products of differentials are negligible.

Equations 13.25 are differential equations that we can integrate to develop useful relations, but before doing so we need to see how we can relate the friction force F_f to other flow properties. First, we note that

$$dF_f = \tau_w\, dA_w = \tau_w P\, dx \qquad (13.26)$$

where P is the wetted perimeter of the duct. To obtain an expression for τ_w in terms of flow variables at each cross section, we assume changes in flow variables with x are gradual and use correlations developed in Chapter 8 for fully developed, incompressible duct flow. For incompressible flow, the local wall shear stress can be written in terms of flow properties and friction factor. From Eqs. 8.16, 8.32, and 8.34 we have, for incompressible flow,

$$\tau_w = -\frac{R}{2}\frac{dp}{dx} = \frac{\rho R}{2}\frac{dh_l}{dx} = \frac{f\rho V^2}{8} \tag{13.27}$$

where f is the friction factor for pipe flow, given by Eq. 8.36 for laminar flow and Eq. 8.37 for turbulent flow, plotted in Fig. 8.13. (We assume that this correlation of experimental data also applies to compressible flow. This assumption, when checked against experimental data, shows surprisingly good agreement for subsonic flows; data for supersonic flow are sparse.)

Ducts of other than circular shape can be included in our analysis by introducing the hydraulic diameter

$$D_h = \frac{4A}{P} \tag{8.50}$$

(Recall the factor of 4 was included in Eq. 8.50 so that D_h would reduce to diameter D for circular ducts.)

Combining Eqs. 8.50, 13.26, and 13.27, we obtain

$$dF_f = \tau_w P\,dx = f\frac{\rho V^2}{8}\frac{4A}{D_h}dx$$

or

$$dF_f = \frac{fA}{D_h}\frac{\rho V^2}{2}dx \tag{13.28}$$

Substituting this result into the momentum equation (Eq. 13.25b), we obtain

$$-\frac{f}{D_h}\frac{\rho V^2}{2}dx - dp = \rho V\,dV$$

or, after dividing by p,

$$\frac{dp}{p} = -\frac{f}{D_h}\frac{\rho V^2}{2p}dx - \frac{\rho V\,dV}{p}$$

Noting that $p/\rho = RT = c^2/k$, and $VdV = d(V^2/2)$, we obtain

$$\frac{dp}{p} = -\frac{f}{D_h}\frac{kM^2}{2}dx - \frac{k}{c^2}d\left(\frac{V^2}{2}\right)$$

and finally,

$$\frac{dp}{p} = -\frac{f}{D_h}\frac{kM^2}{2}dx - \frac{kM^2}{2}\frac{d(V^2)}{V^2} \tag{13.29}$$

To relate M and x, we must eliminate dp/p and $d(V^2)/V^2$ from Eq. 13.29. From the definition of Mach number, $M = V/c$, so $V^2 = M^2c^2 = M^2\,kRT$, and after differentiating this equation and dividing by the original equation,

$$\frac{d(V^2)}{V^2} = \frac{dT}{T} + \frac{d(M^2)}{M^2} \tag{13.30a}$$

From the continuity equation, Eq. 13.25a, $d\rho/\rho = -dV/V$ and so

$$\frac{d\rho}{\rho} = -\frac{1}{2}\frac{d(V^2)}{V^2}$$

From the ideal gas equation of state, $p = \rho RT$,

$$\frac{dp}{p} = \frac{d\rho}{\rho} + \frac{dT}{T}$$

Combining these three equations, we obtain

$$\frac{dp}{p} = \frac{1}{2}\frac{dT}{T} - \frac{1}{2}\frac{d(M^2)}{M^2} \tag{13.30b}$$

Substituting Eqs. 13.30 into Eq. 13.29 gives

$$\frac{1}{2}\frac{dT}{T} - \frac{1}{2}\frac{d(M^2)}{M^2} = -\frac{f}{D_h}\frac{kM^2}{2}dx - \frac{kM^2}{2}\frac{dT}{T} - \frac{kM^2}{2}\frac{d(M^2)}{M^2}$$

This equation can be simplified to

$$\left(\frac{1+kM^2}{2}\right)\frac{dT}{T} = -\frac{f}{D_h}\frac{kM^2}{2}dx + \left(\frac{1-kM^2}{2}\right)\frac{d(M^2)}{M^2} \tag{13.31}$$

We have been successful in reducing the number of variables somewhat. However, to relate M and x, we must obtain an expression for dT/T in terms of M. Such an expression can be obtained most readily from the stagnation temperature equation

$$\frac{T_0}{T} = 1 + \frac{k-1}{2}M^2 \tag{12.21b}$$

Since stagnation temperature is constant for Fanno-line flow,

$$T\left(1 + \frac{k-1}{2}M^2\right) = \text{constant}$$

and after differentiating this equation and dividing by the original equation,

$$\frac{dT}{T} + \frac{M^2\dfrac{(k-1)}{2}}{\left(1 + \dfrac{k-1}{2}M^2\right)}\frac{d(M^2)}{M^2} = 0$$

Substituting for dT/T into Eq. 13.31 yields

$$\frac{M^2\dfrac{(k-1)}{2}\left(\dfrac{1+kM^2}{2}\right)}{\left(1 + \dfrac{k-1}{2}M^2\right)}\frac{d(M^2)}{M^2} = \frac{f}{D_h}\frac{kM^2}{2}dx - \left(\frac{1-kM^2}{2}\right)\frac{d(M^2)}{M^2}$$

Combining terms, we obtain

$$\frac{(1-M^2)}{\left(1 + \dfrac{k-1}{2}M^2\right)}\frac{d(M^2)}{kM^4} = \frac{f}{D_h}dx \tag{13.32}$$

We have (finally!) obtained a differential equation that relates changes in M with x. Now we must integrate the equation to find M as a function of x.

Integrating Eq. 13.32 between states ① and ② would produce a complicated function of both M_1 and M_2. The function would have to be evaluated numerically for each new combination of M_1 and M_2 encountered in a problem. Calculations can be simplified considerably using critical conditions (where, by definition, $M = 1$). All Fanno-line flows tend toward $M = 1$, so integration is between a section where the

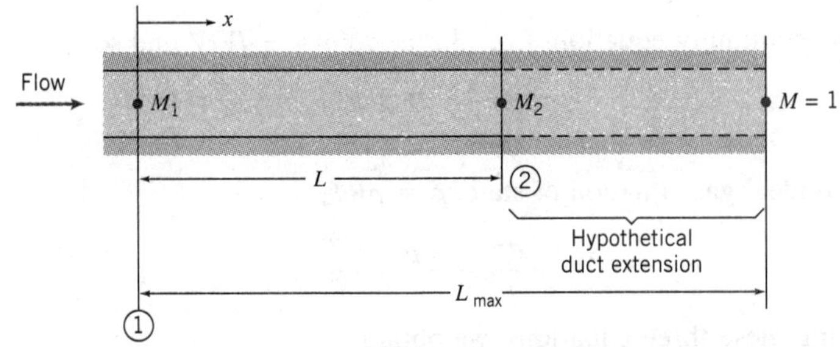

Fig. 13.22 Coordinates and notation used for analysis of Fanno-line flow.

Mach number is M and the section where sonic conditions occur (the critical conditions). Mach number will reach unity when the maximum possible length of duct is used, as shown schematically in Fig. 13.22.

The task is to perform the integration

$$\int_M^1 \frac{(1-M^2)}{kM^4\left(1+\frac{k-1}{2}M^2\right)}d(M^2) = \int_0^{L_{max}} \frac{f}{D_h}dx \tag{13.33}$$

The left side may be integrated by parts. On the right side, the friction factor, f, may vary with x, since Reynolds number will vary along the duct. Note, however, that since ρV is constant along the duct (from continuity), the variation in Reynolds number is caused solely by variations in fluid absolute viscosity.

For a mean friction factor, $\bar{f}$, defined over the duct length as

$$\bar{f} = \frac{1}{L_{max}}\int_0^{L_{max}} f\,dx$$

integration of Eq. 13.33 leads to

$$\frac{1-M^2}{kM^2} + \frac{k+1}{2k}\ln\left[\frac{(k+1)M^2}{2\left(1+\frac{k-1}{2}M^2\right)}\right] = \frac{\bar{f}L_{max}}{D_h} \tag{13.34a}$$

Equation 13.34a gives the maximum $\bar{f}L/D_h$ corresponding to any initial Mach number.

Since $\bar{f}L_{max}/D_h$ is a function of M, the duct length, L, required for the Mach number to change from M_1 to M_2 (as illustrated in Fig. 13.22) may be found from

$$\frac{\bar{f}L}{D_h} = \left(\frac{\bar{f}L_{max}}{D_h}\right)_{M_1} - \left(\frac{\bar{f}L_{max}}{D_h}\right)_{M_2}$$

Critical conditions are appropriate reference conditions to use in developing property ratio flow functions in terms of local Mach number. Thus, for example, since T_0 is constant, we can write

$$\frac{T}{T^*} = \frac{T/T_0}{T^*/T_0} = \frac{\left(\frac{k+1}{2}\right)}{\left(1+\frac{k-1}{2}M^2\right)} \tag{13.34b}$$

Similarly,

$$\frac{V}{V^*} = \frac{M\sqrt{kRT}}{\sqrt{kRT^*}} = M\sqrt{\frac{T}{T*}} = \left[\frac{\left(\frac{k+1}{2}\right)M^2}{1+\frac{k-1}{2}M^2}\right]^{1/2}$$

From continuity, $V/V^* = \rho^*/\rho$, so

$$\frac{V}{V^*} = \frac{\rho^*}{\rho} = \left[\frac{\left(\frac{k+1}{2}\right)M^2}{1+\frac{k-1}{2}M^2}\right]^{1/2} \tag{13.34c}$$

From the ideal gas equation of state,

$$\frac{p}{p^*} = \frac{\rho}{\rho^*}\frac{T}{T^*} = \frac{1}{M}\left[\frac{\left(\frac{k+1}{2}\right)}{1+\frac{k-1}{2}M^2}\right]^{1/2} \tag{13.34d}$$

The ratio of local stagnation pressure to the reference stagnation pressure is given by

$$\frac{p_0}{p_0^*} = \frac{p_0}{p}\frac{p}{p^*}\frac{p^*}{p_0^*}$$

$$\frac{p_0}{p_0^*} = \left(1+\frac{k-1}{2}M^2\right)^{k/(k-1)}\frac{1}{M}\left[\frac{\left(\frac{k+1}{2}\right)}{1+\frac{k-1}{2}M^2}\right]^{1/2}\frac{1}{\left(\frac{k+1}{2}\right)^{k/(k-1)}}$$

or

$$\frac{p_0}{p_0^*} = \frac{1}{M}\left[\left(\frac{2}{k+1}\right)\left(1+\frac{k-1}{2}M^2\right)\right]^{(k+1)/2(k-1)} \tag{13.34e}$$

Equations 13.34 form a complete set for analyzing flow of an ideal gas in a duct with friction, which we usually use instead of (or in addition to) the basic equations, Eqs. 13.24. For convenience we list them together:

$$\frac{\bar{f}L_{max}}{D_h} = \frac{1-M^2}{kM^2} + \frac{k+1}{2k}\ln\left[\frac{(k+1)M^2}{2\left(1+\frac{k-1}{2}M^2\right)}\right] \tag{13.34a}$$

$$\frac{T}{T^*} = \frac{\left(\frac{k+1}{2}\right)}{\left(1+\frac{k-1}{2}M^2\right)} \tag{13.34b}$$

$$\frac{V}{V^*} = \frac{\rho^*}{\rho} = \left[\frac{\left(\frac{k+1}{2}\right)M^2}{1+\frac{k-1}{2}M^2}\right]^{1/2} \tag{13.34c}$$

$$\frac{p}{p^*} = \frac{1}{M}\left[\frac{\left(\frac{k+1}{2}\right)}{1+\frac{k-1}{2}M^2}\right]^{1/2} \tag{13.34d}$$

$$\frac{p_0}{p_0^*} = \frac{1}{M}\left[\left(\frac{2}{k+1}\right)\left(1+\frac{k-1}{2}M^2\right)\right]^{(k+1)/2(k-1)} \tag{13.34e}$$

Equations 13.34, the Fanno-line relations, provide property relations in terms of the local Mach number and critical conditions. They are obviously quite algebraically complicated, but unlike Eqs. 13.24 are not coupled. It is a good idea to program them into your calculator. They are also fairly easy to define in spreadsheets such as *Excel*. The reader is urged to download the *Excel* add-ins for these equations from the Web site; with the add-ins, functions are available for computing the friction factor, temperature, velocity, pressure, and stagnation pressure ratios from M, and M from these ratios. It is important to remember that, as demonstrated in Fig. 12.4, the properties at a state, in any flow process, may be related to that state's isentropic stagnation properties through use of Eqs. 12.21. Appendix E.3 lists flow functions for property ratios p_0/p_0^*, T/T^*, p/p^*, ρ/ρ^*, (V^*/V), and $\bar{f}L_{max}/D_h$ in terms of M for Fanno-line flow of an ideal gas. A table of values, as well as a plot of these property ratios, is presented for air (k = 1.4) for a limited range of Mach numbers. The associated *Excel* workbook, *Fanno-Line Relations*, can be used to print a larger table of values for air and other ideal gases.

Example 13.8 FRICTIONAL ADIABATIC FLOW IN A CONSTANT-AREA CHANNEL: SOLUTION USING FANNO-LINE FLOW FUNCTIONS

Air flow is induced in a smooth insulated tube of 7.16 mm diameter by a vacuum pump. Air is drawn from a room, where $p_0 = 760$ mm Hg (abs) and $T_0 = 23°$C, through a smoothly contoured converging nozzle. At section ①, where the nozzle joins the constant-area tube, the static pressure is −18.9 mm Hg (gage). At section ②, located some distance downstream in the constant-area tube, the static pressure is −412 mm Hg (gage). The duct walls are smooth; assume the average friction factor, $\bar{f}$, is the value at section ①. Determine the length of duct required for choking from section ①, the Mach number at section ②, and the duct length, L_{12}, between sections ① and ②. Sketch the process on a Ts diagram.

Given:
Air flow (with friction) in an insulated constant-area tube.

Gage pressures: $p_1 = -18.9$ mm Hg, and $p_2 = -412$ mm Hg.
$M_3 = 1.0$

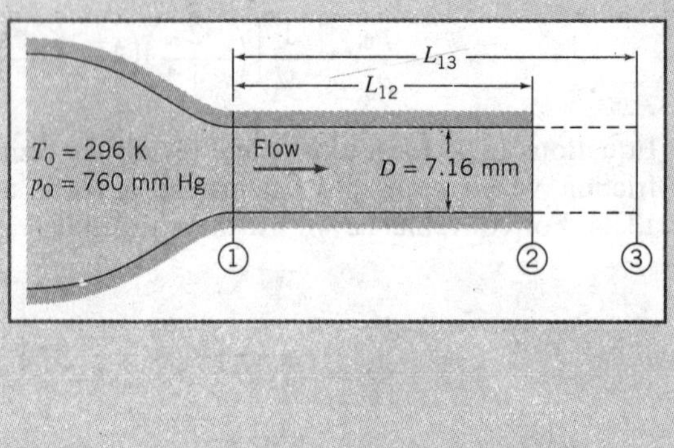

Find: (a) L_{13}. (b) M_2. (c) L_{12}. (d) Sketch the Ts diagram.

Solution:
Flow in the constant-area tube is frictional and adiabatic, a Fanno-line flow. To find the friction factor, we need to know the flow conditions at section ①. If we assume flow in the nozzle is isentropic, local properties at the nozzle exit may be computed using isentropic relations. Thus

$$\frac{p_{0_1}}{p_1} = \left(1 + \frac{k-1}{2}M_1^2\right)^{k/(k-1)}$$

Solving for M_1, we obtain

$$M_1 = \left\{\frac{2}{k-1}\left[\left(\frac{p_{0_1}}{p_1}\right)^{(k-1)/k} - 1\right]\right\}^{1/2} = \left\{\frac{2}{0.4}\left[\left(\frac{760}{760-18.9}\right)^{0.286} - 1\right]\right\}^{1/2} = 0.190$$

$$T_1 = \frac{T_{0_1}}{1 + \frac{k-1}{2}M_1^2} = \frac{296\text{ K}}{1 + 0.2(0.190)^2} = 294\text{ K}$$

$$V_1 = M_1 c_1 = M_1\sqrt{kRT_1} = 0.190\left[1.4 \times 287\frac{\text{N}\cdot\text{m}}{\text{kg}\cdot\text{K}} \times 294 \times \frac{\text{kg}\cdot\text{m}}{\text{N}\cdot\text{s}^2}\right]^{1/2}$$

$$V_1 = 65.3\text{ m/s}$$

Using the density of mercury at room temperature (23°C),

$$p_1 = g\rho_{Hg}h_1 = g\,\text{SG}\,\rho_{H_2O}h_1$$

$$= 9.81\frac{\text{m}}{\text{s}^2} \times 13.5 \times 1000\frac{\text{kg}}{\text{m}^3} \times (760 - 18.9) \times 10^{-3}\text{ m} \times \frac{\text{N}\cdot\text{s}^2}{\text{kg}\cdot\text{m}}$$

$$p_1 = 98.1\ \text{kPa(abs)}$$

$$\rho_1 = \frac{p_1}{RT_1} = 9.81\times10^4\frac{\text{N}}{\text{m}^2}\times\frac{\text{kg}\cdot\text{K}}{287\ \text{N}\cdot\text{m}}\times\frac{1}{294\ \text{K}} = 1.16\ \text{kg/m}^3$$

At $T = 294\ \text{K}\ (21°\text{C})$, $\mu = 1.82\times10^{-5}\ \text{kg/(m}\cdot\text{s)}$ from Table A.10, Appendix A. Thus

$$Re_1 = \frac{\rho_1 V_1 D_1}{\mu_1} = 1.16\frac{\text{kg}}{\text{m}^3}\times65.3\frac{\text{m}}{\text{s}}\times0.00716\ \text{m}\times\frac{\text{m}\cdot\text{s}}{1.82\times10^{-5}\ \text{kg}} = 2.98\times10^4$$

From Eq. 8.37 (turbulent flow), for smooth pipe, $f = 0.0235$.

From Appendix E.3 at $M_1 = 0.190$, $p/p^* = 5.745$ (Eq. 13.34d), and $\bar{f}L_{max}/D_h = 16.38$ (Eq. 13.34a). Thus, assuming $\bar{f} = f_1$,

$$L_{13} = (L_{max})_1 = \left(\frac{\bar{f}L_{max}}{D_h}\right)_1\frac{D_h}{f_1} = 16.38\times\frac{0.00716\ \text{m}}{0.0235} = 4.99\ \text{m} \qquad L_{13}$$

Since p^* is constant for all states on the same Fanno line, conditions at section ② can be determined from the pressure ratio, $(p/p^*)_2$. Thus

$$\left(\frac{p}{p^*}\right)_2 = \frac{p_2}{p^*} = \frac{p_2}{p_1}\frac{p_1}{p^*} = \frac{p_2}{p_1}\left(\frac{p}{p^*}\right)_1 = \left(\frac{760-412}{760-18.9}\right)5.745 = 2.698$$

where we used Eq. 13.34d to obtain the value of p/p^* at section ①. For $p/p^* = 2.698$ at section ②, Eq. 13.34d yields $M_2 = 0.400$ (after obtaining an initial guess value from the plot in Appendix E.3, and iterating).

$$M_2 = 0.400 \qquad M_2$$

The Ts diagram for this flow is

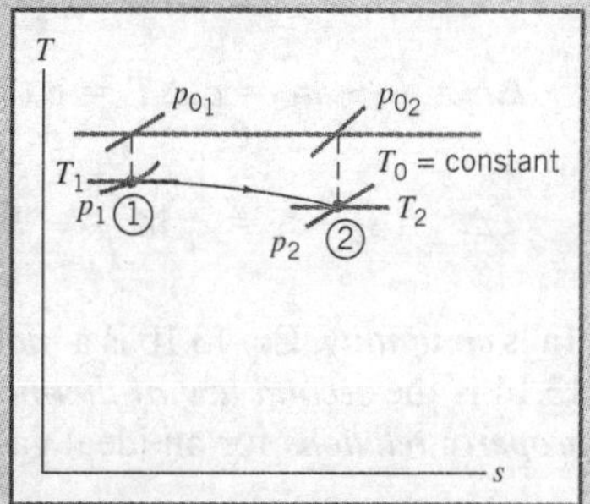

At $M_2 = 0.400$, $\bar{f}L_{max}/D_h = 2.309$ (Eq. 13.34a, Appendix E.3). Thus

$$L_{23} = (L_{max})_2 = \left(\frac{\bar{f}L_{max}}{D_h}\right)_2\frac{D_h}{f_1} = 2.309\times\frac{0.00716\ \text{m}}{0.0235} = 0.704\ \text{m}$$

Finally,

$$L_{12} = L_{13} - L_{23} = (4.99 - 0.704)\ \text{m} = 4.29\ \text{m} \qquad L_{12}$$

This problem illustrates use of the Fanno-line equations, Eqs. 13.34.

✓ These equations give the same results as the basic equations, Eqs. 13.24, as can be seen by comparing, for example, the value of M_2 obtained in this Example and in Example 13.7.

✓ The computations can be quite laborious without using preprogrammed Fanno-line relations (available, for example, in the *Excel* add-ins on the Web site)!

The *Excel* workbook for this Example is convenient for performing the calculations, either using the Fanno-line relations or the basic equations.

Isothermal Flow (on the Web)

13.6 Frictionless Flow in a Constant-Area Duct with Heat Exchange

To explore the effects of heat exchange on a compressible flow, we apply the basic equations to steady, one-dimensional, frictionless flow of an ideal gas with constant specific heats through the finite control volume shown in Fig. 13.23.

You may well be exhausted by the process by now, but as in Section 13.2 (effects of area variation only), Section 13.3 (the normal shock), and Section 13.5 (effects of friction only), our starting point in analyzing frictionless flows with heat exchange is the set of basic equations (Eqs. 13.1), describing one-dimensional motion that is affected by several phenomena: area change, friction, heat transfer, and normal shocks. These are

$$\rho_1 V_1 A_1 = \rho_2 V_2 A_2 = \rho V A = \dot{m} = \text{constant} \qquad (13.1a)$$

$$R_x + p_1 A_1 - p_2 A_2 = \dot{m} V_2 - \dot{m} V_1 \qquad (13.1b)$$

$$\frac{\delta Q}{dm} + h_1 + \frac{V_1^2}{2} = h_2 + \frac{V_2^2}{2} \qquad (13.1c)$$

$$\dot{m}(s_2 - s_1) \geq \int_{CS} \frac{1}{T}\left(\frac{\dot{Q}}{A}\right) dA \qquad (13.1d)$$

$$p = \rho R T \qquad (13.1e)$$

$$\Delta h = h_2 - h_1 = c_p \Delta T = c_p (T_2 - T_1) \qquad (13.1f)$$

$$\Delta s = s_2 - s_1 = c_p \ln \frac{T_2}{T_1} - R \ln \frac{p_2}{p_1} \qquad (13.1g)$$

We recall that Eq. 13.1a is *continuity*, Eq. 13.1b is a *momentum equation*, Eq. 13.1c is an *energy equation*, Eq. 13.1d is the *second law of thermodynamics*, and Eqs. 13.1e, 13.1f, and 13.1g are useful *property relations* for an ideal gas with constant specific heats.

Basic Equations for Flow with Heat Exchange

We simplify Eqs. 13.1 using the facts that $A_1 = A_2 = A$ and that $R_x = 0$. In addition we have the relation $h_0 = h + V^2/2$. Equations 13.1 become for this flow

$$\rho_1 V_1 = \rho_2 V_2 = \rho V A \equiv G = \frac{\dot{m}}{A} = \text{constant} \qquad (13.43a)$$

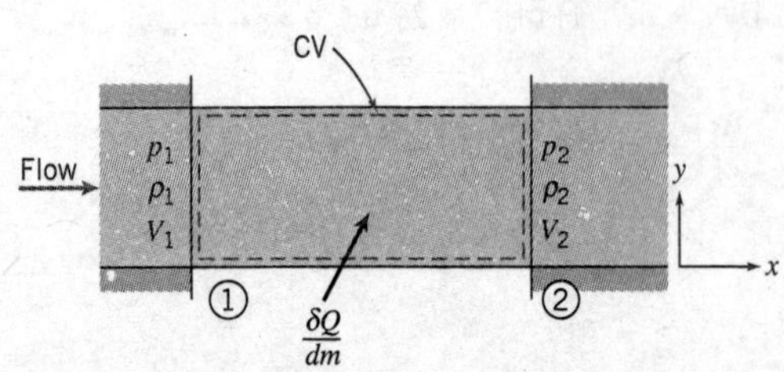

Fig. 13.23 Control volume used for integral analysis of frictionless flow with heat exchange.

$$p_1A - p_2A = \dot{m}V_2 - \dot{m}V_1 \tag{13.43b}$$

$$\frac{\delta Q}{dm} = \left(h_2 + \frac{V_2^2}{2}\right) - \left(h_1 + \frac{V_1^2}{2}\right) = h_{0_2} - h_{0_1} \tag{13.43c}$$

$$\dot{m}(s_2 - s_1) \geq \int_{CS} \frac{1}{T}\left(\frac{\dot{Q}}{A}\right) dA \tag{13.43d}$$

$$p = \rho RT \tag{13.43e}$$

$$\Delta h = h_2 - h_1 = c_p \Delta T = c_p(T_2 - T_1) \tag{13.43f}$$

$$\Delta s = s_2 - s_1 = c_p \ln \frac{T_2}{T_1} - R \ln \frac{p_2}{p_1} \tag{13.43g}$$

Note that Eq. 13.43c indicates that the heat exchange changes the total (kinetic plus internal) energy of the flow. Equation 13.43d is not very useful here. The inequality or equality may apply, depending on the nature of the heat exchange, but in any event we should *not* conclude that in this flow the entropy necessarily increases. For example, for a gradual cooling it will decrease!

Equations 13.43 can be used to analyze frictionless flow in a channel of constant area with heat exchange. For example, if we know conditions at section ① (i.e., p_1, ρ_1, T_1, s_1, h_1, and V_1) we can use these equations to find conditions at some new section ② after the fluid has experienced a total heat exchange $\delta Q/dm$. For a given heat exchange, we have six equations (not including the constraint of Eq. 13.43d) and six unknowns (p_2, ρ_2, T_2, s_2, h_2, and V_2). It is the effect of heat exchange that causes fluid properties to change along the duct. In practice, as we have seen for other flows, this procedure is unwieldy—we again have a set of *nonlinear coupled algebraic* equations to solve. We will use Eqs. 13.43 in Example 13.9. We will also develop some Mach number-based relations to supplement or replace the basic equations, and show how to use these in Example 13.10.

The Rayleigh Line

If we use Eqs. 13.43 to compute property values as a given flow proceeds with a prescribed heat exchange rate, we obtain a curve shown qualitatively in the *Ts* plane in Fig. 13.24. The locus of all possible downstream states is called the *Rayleigh line*. The calculations show some interesting features of Rayleigh-line flow. At the point of maximum temperature (point *a* of Fig. 13.24), the Mach number for an ideal gas is $1/\sqrt{k}$. At the point of maximum entropy (point *b* of Fig. 13.24), $M = 1$. On the upper branch of the curve, Mach number is always less than unity, and it increases monotonically as we proceed to the right along the curve. At every point on the lower portion of the curve, Mach number is greater than unity, and it decreases monotonically as we move to the right along the curve. Regardless of the initial Mach number, with heat addition the flow state proceeds to the right, and with heat rejection the flow state proceeds to the left along the Rayleigh line.

For any initial state in a Rayleigh-line flow, any other point on the Rayleigh line represents a mathematically possible downstream state. Although the Rayleigh line represents all mathematically possible states, are they all physically attainable downstream states? A moment's reflection will indicate that they are. Since we are considering a flow with heat exchange, the second law (Eq. 13.43d) does not impose any restrictions on the sign of the entropy change.

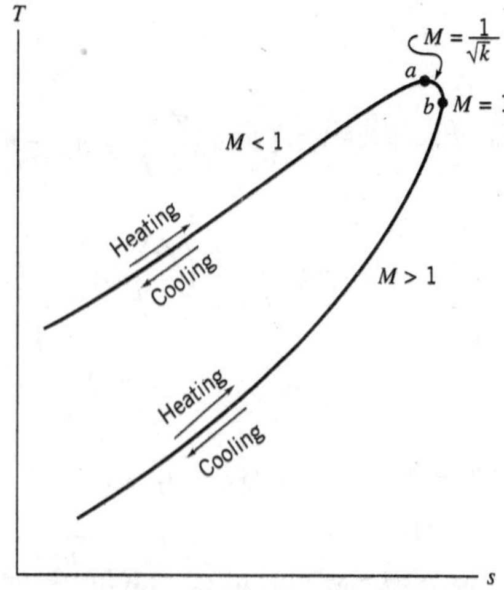

Fig. 13.24 Schematic *Ts* diagram for frictionless flow in a constant-area duct with heat exchange (Rayleigh-line flow).

The effects of heat exchange on properties in steady, frictionless, compressible flow of an ideal gas are summarized in Table 13.3; the basis of each indicated trend is discussed in the next few paragraphs.

The direction of entropy change is always determined by the heat exchange; entropy increases with heating and decreases with cooling. Similarly, the first law, Eq. 13.43c, shows that heating increases the stagnation enthalpy and cooling decreases it; since $\Delta h_0 = c_p \Delta T_0$, the effect on stagnation temperature is the same.

The effect of heating and cooling on temperature may be deduced from the shape of the Rayleigh line in Fig. 13.24. We see that except for the region $1/\sqrt{k} < M < 1$ (for air, $1/\sqrt{k} \approx 0.85$), heating causes T to increase, and cooling causes T to decrease. However, we also see the unexpected result that for $1/\sqrt{k} < M < 1$, *heat addition* causes the stream temperature to *decrease*, and *heat rejection* causes the stream temperature to *increase!*

For subsonic flow, the Mach number increases monotonically with heating, until $M = 1$ is reached. For given inlet conditions, all possible downstream states lie on a single Rayleigh line. Therefore, the point $M = 1$ determines the maximum possible heat addition without choking. If the flow is initially supersonic, heating will reduce the Mach number. Again, the maximum possible heat addition without choking is that which reduces the Mach number to $M = 1.0$.

The effect of heat exchange on static pressure is obtained from the shapes of the Rayleigh line and of constant-pressure lines on the *Ts* plane (see Fig. 13.25). For $M < 1$, pressure falls with heating, and for $M > 1$, pressure increases, as shown by the shapes of the constant-pressure lines. Once the pressure variation has been found, the effect on velocity may be found from the momentum equation,

$$p_1 A - p_2 A = \dot{m} V_2 - \dot{m} V_1 \tag{13.43b}$$

or

$$p + \left(\frac{\dot{m}}{A}\right) V = \text{constant}$$

Thus, since $\dot{m}/A$ is a positive constant, trends in p and V must be opposite. From the continuity equation, Eq. 13.43a, the trend in ρ is opposite to that in V.

Table 13.3

Summary of Effects of Heat Exchange on Fluid Properties

	Heating		Cooling		
Property	$M < 1$	$M > 1$	$M < 1$	$M > 1$	Obtained from:
Entropy	$s\Uparrow$	$s\Uparrow$	$s\Downarrow$	$s\Downarrow$	$T\,ds$ equation
Stagnation temperature	$T_0\Uparrow$	$T_0\Uparrow$	$T_0\Downarrow$	$T_0\Downarrow$	First law, and $\Delta h_0 = c_p\Delta T_0$
Temperature	$T\Uparrow\left(M < \frac{1}{\sqrt{k}}\right)$ $T\Downarrow\left(\frac{1}{\sqrt{k}} < M < 1\right)$	$T\Uparrow$	$T\Downarrow\left(M < \frac{1}{\sqrt{k}}\right)$ $T\Uparrow\left(\frac{1}{\sqrt{k}} < M < 1\right)$	$T\Downarrow$	Shape of Rayleigh line
Mach number	$M\Uparrow$	$M\Downarrow$	$M\Downarrow$	$M\Uparrow$	Trend on Rayleigh line
Pressure	$p\Downarrow$	$p\Uparrow$	$p\Uparrow$	$p\Downarrow$	Trend on Rayleigh line
Velocity	$V\Uparrow$	$V\Downarrow$	$V\Downarrow$	$V\Uparrow$	Momentum equation, and effect on p
Density	$\rho\Downarrow$	$\rho\Uparrow$	$\rho\Uparrow$	$\rho\Downarrow$	Continuity equation, and effect on V
Stagnation pressure	$p_0\Downarrow$	$p_0\Downarrow$	$p_0\Uparrow$	$p_0\Uparrow$	Fig. 13.25

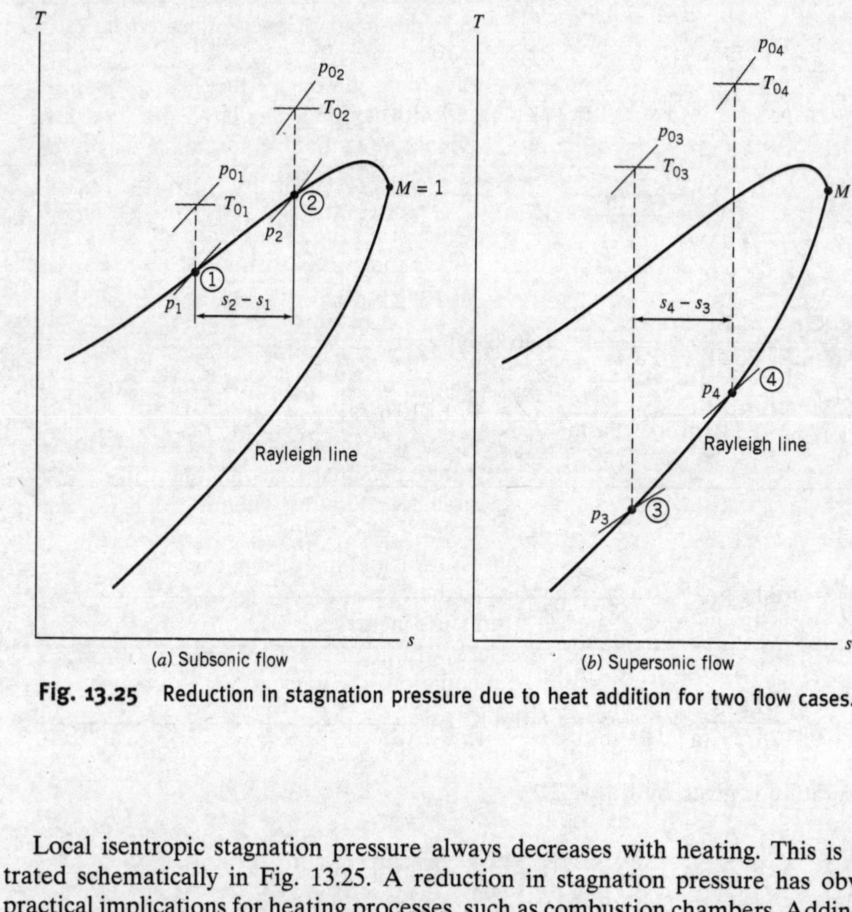

Fig. 13.25 Reduction in stagnation pressure due to heat addition for two flow cases.

Local isentropic stagnation pressure always decreases with heating. This is illustrated schematically in Fig. 13.25. A reduction in stagnation pressure has obvious practical implications for heating processes, such as combustion chambers. Adding the same amount of energy per unit mass (same change in T_0) causes a larger change in p_0 for supersonic flow; because heating occurs at a lower temperature in supersonic flow, the entropy increase is larger.

Example 13.9 FRICTIONLESS FLOW IN A CONSTANT-AREA DUCT WITH HEAT ADDITION

Air flows with negligible friction through a duct of area $A = 0.023\ \text{m}^2$. At section ①, flow properties are $T_1 = 330$ K, $p_1 = 138$ kPa (abs), and $V_1 = 110$ m/s. At section ②, $p_2 = 69$ kPa (abs). The flow is heated between sections ① and ②. Determine the properties at section ②, the energy added, and the entropy change. Finally, plot the process on a Ts diagram.

Given: Frictionless flow of air in duct shown:

$T_1 = 330$ K
$p_1 = 138$ kPa (abs) $\quad p_2 = 69$ kPa (abs)
$V_1 = 110$ m/s $\quad A_1 = A_2 = A = 0.023\ \text{m}^2$

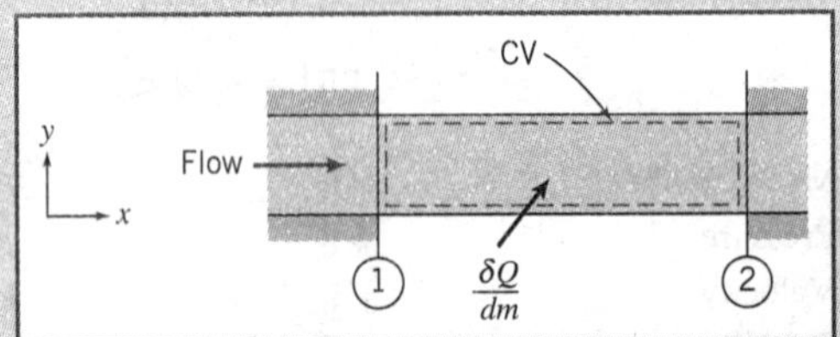

Find: (a) Properties at section ②.
(b) $\delta Q/dm$.
(c) $s_2 - s_1$.
(d) Ts diagram.

Solution:
The momentum equation (Eq. 13.43b) is

$$p_1A - p_2A = \dot{m}V_2 - \dot{m}V_1 \tag{13.43b}$$

or

$$p_1 - p_2 = \frac{\dot{m}}{A}(V_2 - V_1) = \rho_1 V_1(V_2 - V_1)$$

Solving for V_2 gives

$$V_2 = \frac{p_1 - p_2}{\rho_1 V_1} + V_1$$

For an ideal gas, Eq. 13.43e,

$$\rho_1 = \frac{p_1}{RT_1} = 138 \times 1000\ \text{Pa} \times \frac{\text{kg}\cdot\text{K}}{287\ \text{N}\cdot\text{m}} \times \frac{1}{330\ \text{K}} \times \frac{\text{N}}{\text{Pa}\cdot\text{m}^2} = 1.46\ \text{kg/m}^3$$

$$V_2 = (138 - 69) \times 1000\ \text{Pa} \times \frac{1}{1.46\ \text{kg/m}^3} \times \frac{1}{110\ \text{m/s}} \times \frac{\text{N}}{\text{Pa}\cdot\text{m}^2} \times \frac{\text{kg}\cdot\text{m}}{\text{N}\cdot\text{s}^2} + 110\ \text{m/s}$$

$$V_2 = 539.6\ \text{m/s} \qquad V_2$$

From continuity, Eq. 13.43a, $G = \rho_1 V_1 = \rho_2 V_2$, so

$$\rho_2 = \rho_1 \frac{V_1}{V_2} = 1.46 \frac{\text{kg}}{\text{m}^3}\left(\frac{110}{539.6}\right) = 0.3\ \text{kg/m}^3 \qquad \rho_2$$

Solving for T_2, we obtain

$$T_2 = \frac{p_2}{\rho_2 R} = 69 \times 1000\ \text{Pa} \times \frac{1}{0.3\ \text{kg/m}^3} \times \frac{\text{kg}\cdot\text{K}}{287\ \text{N}\cdot\text{m}} \times \frac{\text{N}}{\text{Pa}\cdot\text{m}^2} = 801.4\ \text{K} \qquad T_2$$

The local isentropic stagnation temperature is given by Eq. 12.21b,

$$T_{0_2} = T_2\left(1 + \frac{k-1}{2}M_2^2\right)$$

$$c_2 = \sqrt{kRT_2} = 567.5\ \text{m/s}; \qquad M_2 = \frac{V_2}{c_2} = \frac{539.6}{567.5} = 0.951$$

$$T_{0_2} = 801.4\ \text{K}[1 + 0.2(0.951)^2] = 946.4\ \text{R} \qquad T_{0_2}$$

and

$$p_{0_2} = p_2\left(\frac{T_{0_2}}{T_2}\right)^{k/(k-1)} = 69\,\text{kPa}\left(\frac{946.4}{801.4}\right)^{3.5} = 123.5\,\text{kPa} \longleftarrow p_{0_2}$$

The heat addition is obtained from the energy equation (Eq. 13.43c),

$$\frac{\delta Q}{dm} = \left(h_2 + \frac{V_2^2}{2}\right) - \left(h_1 + \frac{V_1^2}{2}\right) = h_{0_2} - h_{0_1} \tag{13.43c}$$

or

$$\frac{\delta Q}{dm} = h_{0_2} - h_{0_1} = c_p(T_{0_2} - T_{0_1})$$

We already obtained T_{0_2}. For T_{0_1} we have

$$T_{0_1} = T_1\left(1 + \frac{k-1}{2}M_1^2\right)$$

$$c_1 = \sqrt{kRT_1} = 364.1\,\text{m/s}; \qquad M_1 = \frac{V_1}{c_1} = \frac{110}{364.1} = 0.3$$

$$T_{0_1} = 330\,\text{K}[1 + 0.2(0.3)^2] = 336\,\text{K}$$

so

$$\frac{\delta Q}{dm} = 1.005\frac{\text{kJ}}{\text{kg}\cdot\text{K}}(946.4 - 336)\,\text{K} = 613.5\,\text{kJ/kg} \longleftarrow \delta Q/dm$$

For the change in entropy (Eq 13.43g),

$$\Delta s = s_2 - s_1 = c_p \ln\frac{T_2}{T_1} - R\ln\frac{p_2}{p_1} = c_p \ln\frac{T_2}{T_1} - (c_p - c_v)\ln\frac{p_2}{p_1} \tag{13.43g}$$

Then

$$s_2 - s_1 = 1.005\frac{\text{kJ}}{\text{kg}\cdot\text{K}} \times \ln\left(\frac{801.4}{330}\right) - (1.005 - 0.718)\frac{\text{kJ}}{\text{kg}\cdot\text{K}} \times \ln\left(\frac{69,000}{138,000}\right)$$

$$s_2 - s_1 = 1.091\,\text{kJ/kg}\cdot\text{K} \longleftarrow s_2 - s_1$$

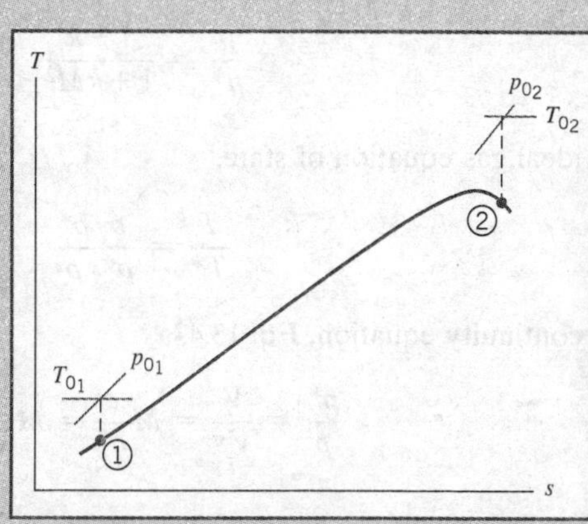

The process follows a Rayleigh line:
To complete our analysis, let us examine the change in p_0 by comparing p_{0_2} with p_{0_1}.

$$p_{0_1} = p_1\left(\frac{T_{0_1}}{T_1}\right)^{k/(k-1)} = 138\,\text{kPa}\left(\frac{336}{330}\right)^{3.5}$$

$$= 147\,\text{kPa} \qquad p_{0_1}$$

Comparing, we see that p_{0_2} is less than p_{0_1}.

This problem illustrates the use of the basic equations, Eqs. 13.43, for analyzing frictionless flow of an ideal gas in a duct with heat exchange.
The *Excel* workbook for this Example is convenient for performing the calculations.

Rayleigh-Line Flow Functions for One-Dimensional Flow of an Ideal Gas

Equations 13.43 are the basic equations for Rayleigh-line flow between two arbitrary states ① and ② in the flow. To reduce labor in solving problems, it is convenient to derive flow functions for property ratios in terms of local Mach number as we did for Fanno-line flow. The reference state is again taken as the critical condition where $M = 1$; properties at the critical condition are denoted by (*).

Dimensionless properties (such as p/p^* and T/T^*) may be obtained by writing the basic equations between a point in the flow where properties are M, T, p, etc., and the critical state ($M = 1$, with properties denoted as T^*, p^*, etc.).

The pressure ratio, p/p^*, may be obtained from the momentum equation, Eq. 13.43b,

$$pA - p^*A = \dot{m}V^* - \dot{m}V$$

or

$$p + \rho V^2 = p^* + \rho^* V^{*2}$$

Substituting $\rho = p/RT$, and factoring out pressures yields

$$p\left[1 + \frac{V^2}{RT}\right] = p^*\left[1 + \frac{V^{*2}}{RT^*}\right]$$

Noting that $V^2/RT = k(V^2/kRT) = kM^2$, we find

$$p\left[1 + kM^2\right] = p^*[1 + k]$$

and finally,

$$\frac{p}{p^*} = \frac{1+k}{1+kM^2} \tag{13.44a}$$

From the ideal gas equation of state,

$$\frac{T}{T^*} = \frac{p}{p^*}\frac{\rho^*}{\rho}$$

From the continuity equation, Eq. 13.43a,

$$\frac{\rho^*}{\rho} = \frac{V}{V^*} = M\frac{c}{c^*} = M\sqrt{\frac{T}{T^*}}$$

Then, substituting for ρ^*/ρ, we obtain

$$\frac{T}{T^*} = \frac{p}{p^*} M \sqrt{\frac{T}{T^*}}$$

Squaring and substituting from Eq. 13.44a gives

$$\frac{T}{T^*} = \left[\frac{p}{p^*} M\right]^2 = \left[M\left(\frac{1+k}{1+kM^2}\right)\right]^2 \tag{13.44b}$$

From continuity, using Eq. 13.44b,

$$\frac{\rho^*}{\rho} = \frac{V}{V^*} = \frac{M^2(1+k)}{1+kM^2} \tag{13.44c}$$

The dimensionless stagnation temperature, T_0/T_0^*, can be determined from

$$\frac{T_0}{T_0^*} = \frac{T_0}{T}\frac{T}{T^*}\frac{T^*}{T_0^*}$$

$$= \left(1 + \frac{k-1}{2}M^2\right)\left[M\left(\frac{1+k}{1+kM^2}\right)\right]^2 \frac{1}{\left(\frac{k+1}{2}\right)} \tag{13.44d}$$

$$\frac{T_0}{T_0^*} = \frac{2(k+1)M^2\left(1 + \frac{k-1}{2}M^2\right)}{(1+kM^2)^2}$$

Similarly,

$$\frac{p_0}{p_0^*} = \frac{p_0}{p}\frac{p}{p^*}\frac{p^*}{p_0^*}$$

$$= \left(1 + \frac{k-1}{2}M^2\right)^{k/(k-1)}\left(\frac{1+k}{1+kM^2}\right)\frac{1}{\left(\frac{k+1}{2}\right)^{k/(k-1)}}$$

$$\frac{p_0}{p_0^*} = \frac{1+k}{1+kM^2}\left[\left(\frac{2}{k+1}\right)\left(1 + \frac{k-1}{2}M^2\right)\right]^{k/(k-1)} \tag{13.44e}$$

For convenience we collect together the equations:

$$\frac{p}{p^*} = \frac{1+k}{1+kM^2} \tag{13.44a}$$

$$\frac{T}{T^*} = \left[M\left(\frac{1+k}{1+kM^2}\right)\right]^2 \tag{13.44b}$$

$$\frac{\rho^*}{\rho} = \frac{V}{V^*} = \frac{M^2(1+k)}{1+kM^2} \tag{13.44c}$$

$$\frac{T_0}{T_0^*} = \frac{2(k+1)M^2\left(1 + \frac{k-1}{2}M^2\right)}{(1+kM^2)^2} \tag{13.44d}$$

$$\frac{p_0}{p_0^*} = \frac{1+k}{1+kM^2}\left[\left(\frac{2}{k+1}\right)\left(1+\frac{k-1}{2}M^2\right)\right]^{k/k-1} \tag{13.44e}$$

Equations 13.44, the Rayleigh-line relations, provide property ratios in terms of the local Mach number and critical conditions. They are obviously complicated, but can be programmed into your calculator. They are also fairly easy to define in spreadsheets such as *Excel*. The reader is urged to download the *Excel* add-ins for these equations from the Web site; with the add-ins, functions are available for computing the pressure, temperature, density, and stagnation temperature and pressure ratios from M, and M from these ratios. In Example 13.10 we will explore their use. Appendix E.4 lists flow functions for property ratios T_0/T_0^*, p_0/p_0^*, T/T^*, p/p^*, and ρ^*/ρ (V/V^*), in terms of M for Rayleigh-line flow of an ideal gas. A table of values, as well as a plot of these property ratios, is presented for air ($k = 1.4$) for a limited range of Mach numbers. The associated *Excel* workbook, *Rayleigh-Line Relations*, can be used to print a larger table of values for air and other ideal gases.

Example 13.10 FRICTIONLESS FLOW IN A CONSTANT-AREA DUCT WITH HEAT ADDITION: SOLUTION USING RAYLEIGH-LINE FLOW FUNCTIONS

Air flows with negligible friction in a constant-area duct. At section ①, properties are $T_1 = 60°\text{C}$, $p_1 = 135$ kPa (abs), and $V_1 = 732$ m/s. Heat is added between section ① and section ②, where $M_2 = 1.2$. Determine the properties at section ②, the heat exchange per unit mass, and the entropy change, and sketch the process on a Ts diagram.

Given: Frictionless flow of air as shown:

$T_1 = 333$ K $M_2 = 1.2$
$p_1 = 135$ kPa (abs)
$V_1 = 732$ m/s

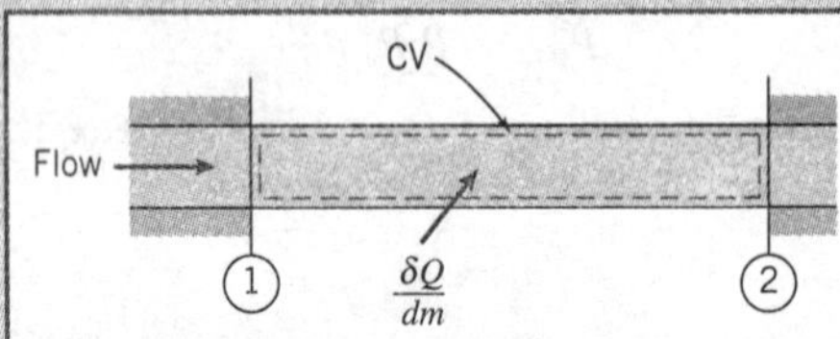

Find: (a) Properties at section ②.
(b) $\delta Q/dm$.
(c) $s_2 - s_1$.
(d) Ts diagram.

Solution:
To obtain property ratios, we need both Mach numbers.

$$c_1 = \sqrt{kRT_1} = \left[1.4 \times 287\frac{\text{N}\cdot\text{m}}{\text{kg}\cdot\text{K}} \times 333\text{ K} \times \frac{\text{kg}\cdot\text{m}}{\text{N}\cdot\text{s}^2}\right]^{1/2} = 366\text{ m/s}$$

$$M_1 = \frac{V_1}{c_1} = 732\frac{\text{m}}{\text{s}} \times \frac{\text{s}}{366\text{ m}} = 2.00$$

From the Rayleigh-line flow functions of Appendix E-4 we find the following:

M	T_0/T_0^*	p_0/p_0^*	T/T^*	p/p^*	V/V^*
2.00	0.7934	1.503	0.5289	0.3636	1.455
1.20	0.9787	1.019	0.9119	0.7958	1.146

Using these data and recognizing that critical properties are constant, we obtain

$$\frac{T_2}{T_1} = \frac{T_2/T^*}{T_1/T^*} = \frac{0.9119}{0.5289} = 1.72; \quad T_2 = 1.72T_1 = (1.72)333\,\text{K} = 573\,\text{K} \qquad T_2$$

$$\frac{p_2}{p_1} = \frac{p_2/p^*}{p_1/p^*} = \frac{0.7958}{0.3636} = 2.19; \quad p_2 = 2.19p_1 = (2.19)135\,\text{kPa}$$

$$p_2 = 296\,\text{kPa (abs)} \qquad p_2$$

$$\frac{V_2}{V_1} = \frac{V_2/V^*}{V_1/V^*} = \frac{1.146}{1.455} = 0.788; \quad V_2 = 0.788V_1 = (0.788)732\,\text{m/s}$$

$$V_2 = 577\,\text{m/s} \qquad V_2$$

$$\rho_2 = \frac{p_2}{RT_2} = 2.96\times10^5\frac{\text{N}}{\text{m}^2}\times\frac{\text{kg}\cdot\text{K}}{287\,\text{N}\cdot\text{m}}\times\frac{1}{573\,\text{K}} = 1.80\,\text{kg/m}^3 \qquad \rho_2$$

The heat addition may be determined from the energy equation, Eq. 13.43c,

$$\frac{\delta Q}{dm} = h_{0_2} - h_{0_1} = c_p(T_{0_2} - T_{0_1})$$

From the isentropic-stagnation functions (Eq. 12.21b) at $M = 2.0$,

$$\frac{T}{T_0} = \frac{T_1}{T_{0_1}} = 0.5556; \qquad T_{0_1} = \frac{T_1}{0.5556} = \frac{333\,\text{K}}{0.5556} = 599\,\text{K}$$

and at $M = 1.2$,

$$\frac{T}{T_0} = \frac{T_2}{T_{0_2}} = 0.7664; \qquad T_{0_2} = \frac{T_2}{0.7764} = \frac{573\,\text{K}}{0.7764} = 738\,\text{K} \qquad T_{0_2}$$

Substituting gives

$$\frac{\delta Q}{dm} = c_p(T_{0_2} - T_{0_1}) = 1.00\frac{\text{kJ}}{\text{kg}\cdot\text{K}}\times(738-599)\,\text{K} = 139\,\text{kJ/kg} \qquad \delta Q/dm$$

For the change in entropy (Eq. 13.43g),

$$s_2 - s_1 = c_p \ln\frac{T_2}{T_1} - R\ln\frac{p_2}{p_1}$$

$$= 1.00\frac{\text{kJ}}{\text{kg}\cdot\text{K}}\times\ln\left(\frac{573}{333}\right) - 287\frac{\text{N}\cdot\text{m}}{\text{kg}\cdot\text{K}}\times\ln\left(\frac{2.96\times10^5}{1.35\times10^5}\right)\times\frac{\text{kJ}}{1000\,\text{N}\cdot\text{m}}$$

$$s_2 - s_1 = 0.137\,\text{kJ/(kg}\cdot\text{K)} \qquad s_2 - s_1$$

Finally, check the effect on p_0. From the isentropic-stagnation function (Eq. 12.21a), at $M = 2.0$,

$$\frac{p}{p_0} = \frac{p_1}{p_{0_1}} = 0.1278; \qquad p_{0_1} = \frac{p_1}{0.1278} = \frac{135\,\text{kPa}}{0.1278} = 1.06\,\text{MPa (abs)}$$

and at $M = 1.2$,

$$\frac{p}{p_0} = \frac{p_2}{p_{0_2}} = 0.4124; \qquad p_{0_2} = \frac{p_2}{0.4124} = \frac{296\,\text{kPa}}{0.4124} = 718\,\text{kPa (abs)} \qquad p_{0_2}$$

Thus, $p_{0_2} < p_{0_1}$, as expected for a heating process.

The process follows the supersonic branch of a Rayleigh line:

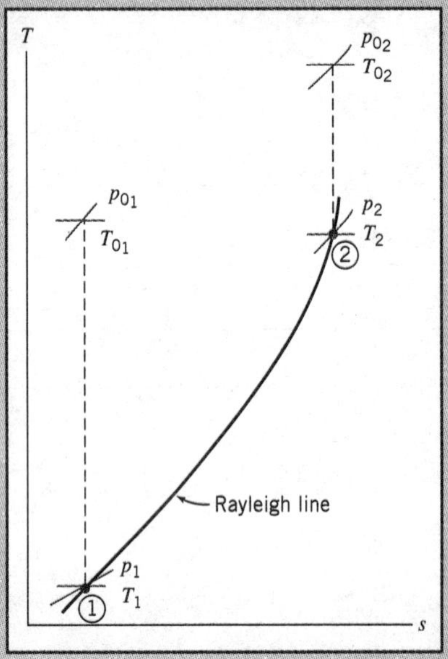

This problem illustrates the use of the Rayleigh-line equations, Eqs. 13.44, for analyzing frictionless flow of an ideal gas in a duct with heat exchange.
The *Excel* workbook for this Example is convenient for performing the calculations. (Alternatively, the Rayleigh-line *Excel* add-ins, available on the Web site, are useful for these calculations.)

13.7 Oblique Shocks and Expansion Waves

So far we have considered one-dimensional compressible flows. With the understanding we have developed, we are ready to introduce some basic concepts of two-dimensional flow: *oblique shocks* and *expansion waves.*

Oblique Shocks

VIDEO

Shock Waves over a Supersonic Airplane.

In Section 12.2, we discussed the Mach cone, with Mach angle α, that is generated by an airplane flying at $M > 1$, as shown (in airplane coordinates) in Fig. 13.26*a*. The Mach cone is a weak pressure (sound) wave, so weak that, as shown in Fig. 13.26*a*, it barely disturbs the streamlines—it is the limiting case of an oblique shock. If we zoom in on the airplane, we see that at the nose of the airplane we have an oblique shock—a shock wave that is aligned, at some angle $\beta < 90°$, to the flow. This oblique shock causes the streamlines to abruptly change direction (usually to follow the surface of the airplane or the airplane's airfoil). Further away from the airplane we still have an oblique shock, but it becomes progressively weaker (β decreases) and the streamlines experience smaller deflections until, far away from the airplane the oblique shock becomes a Mach cone ($\beta \rightarrow \alpha$) and the streamlines are essentially unaffected by the airplane.

A supersonic airplane does not necessarily generate an oblique shock that is attached to its nose—we may instead have a detached normal shock ahead of the airplane! In fact, as illustrated in Fig. 13.27, as an airplane accelerates to its supersonic cruising speed the flow will progress from subsonic, through supersonic with a detached normal shock, to attached oblique shocks that become increasingly "pressed" against the airplane's surface.

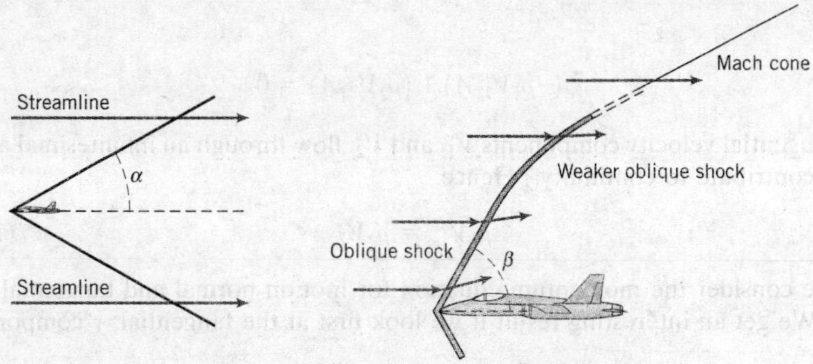

(*a*) Mach cone (*b*) Oblique shock

Fig. 13.26 Mach cone and oblique shock generated by airplane.

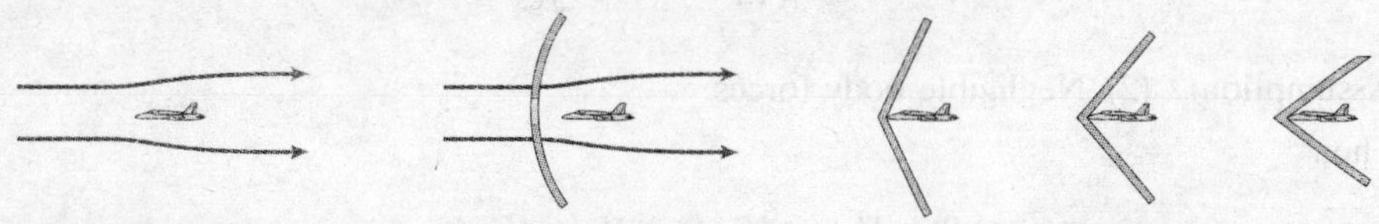

(*a*) Subsonic $M < 1$ (*b*) Low supersonic (*c*) Increasing M

Fig. 13.27 Airplane flow patterns as speed increases.

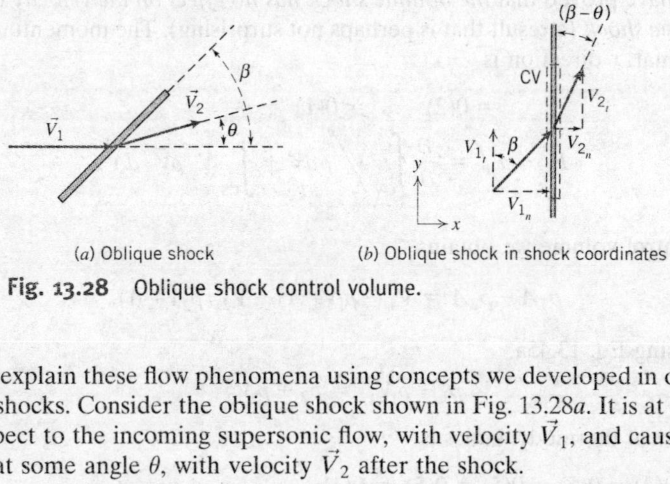

(*a*) Oblique shock (*b*) Oblique shock in shock coordinates

Fig. 13.28 Oblique shock control volume.

We can explain these flow phenomena using concepts we developed in our analysis of normal shocks. Consider the oblique shock shown in Fig. 13.28*a*. It is at some angle β with respect to the incoming supersonic flow, with velocity $\vec{V}_1$, and causes the flow to deflect at some angle θ, with velocity $\vec{V}_2$ after the shock.

VIDEO

Shock Waves due to a Projectile.

It is convenient to orient the xy coordinates orthogonal to the oblique shock, and decompose $\vec{V}_1$ and $\vec{V}_2$ into components normal and tangential to the shock, as shown in Fig. 13.28*b*, with appropriate subscripts. The control volume is assumed to have arbitrary area A before and after the shock, and infinitesimal thickness across the shock (the upper and lower surfaces in Fig. 13.28*b*). For this infinitesimal control volume, we can write the basic equations: continuity, momentum, and the first and second laws of thermodynamics.

The continuity equation is

$$\overbrace{\frac{\partial}{\partial t}\int_{\text{CV}} \rho \, dV}^{=0(1)} + \int_{\text{CS}} \rho \vec{V} \cdot d\vec{A} = 0 \qquad (4.12)$$

Assumption: (1) Steady flow.

Then

$$(-\rho_1 V_{1_n} A) + (\rho_2 V_{2_n} A) = 0$$

(The tangential velocity components V_{1_t} and V_{2_t} flow through an infinitesimal area, so do not contribute to continuity.) Hence,

$$\rho_1 V_{1_n} = \rho_2 V_{2_n} \tag{13.45a}$$

Next we consider the momentum equation for motion normal and tangential to the shock. We get an interesting result if we look first at the tangential, y component,

$$F_{S_y} + \underbrace{F_{B_y}}_{=0(2)} = \underbrace{\frac{\partial}{\partial t}\int_{CV} V_y \, \rho d\forall}_{=0(1)} + \int_{CS} V_y \, \rho \vec{V} \cdot d\vec{A} \tag{4.18b}$$

Assumption: (2) Negligible body forces.

Then

$$0 = V_{1_t}(-\rho_1 V_{1_n} A) + V_{2_t}(\rho_2 V_{2_n} A)$$

or, using Eq. 13.45a

$$V_{1_t} = V_{2_t} = V_t$$

Hence, we have proved that *the oblique shock has no effect on the velocity component parallel to the shock* (a result that is perhaps not surprising). The momentum equation for the normal, x direction is

$$F_{S_x} + \underbrace{F_{B_x}}_{=0(2)} = \underbrace{\frac{\partial}{\partial t}\int_{CV} V_x \, \rho d\forall}_{=0(1)} + \int_{CS} V_x \, \rho \vec{V} \cdot d\vec{A} \tag{4.18a}$$

For our control volume we obtain

$$p_1 A - p_2 A = V_{1_n}(-\rho_1 V_{1_n} A) + V_{2_n}(\rho_2 V_{2_n} A)$$

or, again using Eq. 13.45a,

$$p_1 + \rho_1 V_{1_n}^2 = p_2 + \rho_2 V_{2_n}^2 \tag{13.45b}$$

The first law of thermodynamics is

$$\underbrace{\dot{Q}}_{=0(4)} - \underbrace{\dot{W}_s}_{=0(5)} - \underbrace{\dot{W}_{\text{shear}}}_{=0(5)} - \underbrace{\dot{W}_{\text{other}}}_{=0(5)} = \underbrace{\frac{\partial}{\partial t}\int_{CV} e\,\rho \, d\forall}_{=0(1)} + \int_{CS} (e + pv)\, \rho \vec{V} \cdot d\vec{A} \tag{4.56}$$

where

$$e = u + \frac{V^2}{2} + \underbrace{gz}_{=0(6)}$$

Assumptions: (4) Adiabatic flow.
(5) No work terms.
(6) Negligible gravitational effect.

For our control volume we obtain

$$0 = \left(u_1 + p_1 v_1 + \frac{V_1^2}{2}\right)(-\rho_1 V_{1_n} A) + \left(u_2 + p_2 v_2 + \frac{V_2^2}{2}\right)(\rho_2 V_{2_n} A)$$

(Remember that v here represents the specific volume.) This can be simplified by using $h \equiv u + pv$, and continuity (Eq. 13.45a),

$$h_1 + \frac{V_1^2}{2} = h_2 + \frac{V_2^2}{2}$$

But each velocity can be replaced by its Pythagorean sum, so

$$h_1 + \frac{V_{1_n}^2 + V_{1_t}^2}{2} = h_2 + \frac{V_{2_n}^2 + V_{2_t}^2}{2}$$

We have already learned that the tangential velocity is constant, $V_{1_t} = V_{2_t} = V_t$, so the first law simplifies to

$$h_1 + \frac{V_{1_n}^2}{2} = h_2 + \frac{V_{2_n}^2}{2} \tag{13.45c}$$

Finally, the second law of thermodynamics is

$$\overbrace{\frac{\partial}{\partial t}\int_{CV} s\rho d\forall}^{=0(1)} + \int_{CS} s\rho \vec{V}\cdot d\vec{A} \geq \int_{CS} \overbrace{\frac{1}{T}\left(\frac{\dot{Q}}{A}\right)}^{=0(4)} dA \tag{4.58}$$

The shock is irreversible, so Eq. 4.58 for our control volume is

$$s_1(-\rho_1 V_{1_n} A) + s_2(\rho_2 V_{2_n} A) > 0$$

and, again using continuity,

$$s_2 > s_1 \tag{13.45d}$$

The continuity and momentum equations, and the first and second laws of thermodynamics, for an oblique shock, are given by Eqs. 13.45a through 13.45d, respectively. Examination of these equations shows that they are *identical* to the corresponding equations for a normal shock we derived is Section 13.3. Equations 13.11a through 13.11d, except V_1 and V_2, are replaced with normal velocity components V_{1_n} and V_{2_n}, respectively! Hence, we can use all of the concepts and equations of Section 13.3 for normal shocks, as long as we replace the velocities with their normal components only. The normal velocity components are given by

$$V_{1_n} = V_1 \sin\beta \tag{13.46a}$$

and

$$V_{2_n} = V_2 \sin(\beta - \theta) \tag{13.46b}$$

The corresponding Mach numbers are

$$M_{1_n} = \frac{V_{1_n}}{c_1} = M_1 \sin\beta \tag{13.47a}$$

and

$$M_{2_n} = \frac{V_{2_n}}{c_2} = M_2 \sin(\beta - \theta) \tag{13.47b}$$

The oblique shock equations for an ideal gas with constant specific heats are obtained directly from Eqs. 13.20:

$$M_{2_n}^2 = \frac{M_{1_n}^2 + \dfrac{2}{k-1}}{\dfrac{2k}{k-1} M_{1_n}^2 - 1} \tag{13.48a}$$

$$\frac{p_{0_2}}{p_{0_1}} = \frac{\left[\dfrac{\dfrac{k+1}{2} M_{1_n}^2}{1+\dfrac{k-1}{2} M_{1_n}^2}\right]^{k/(k-1)}}{\left[\dfrac{2k}{k+1} M_{1_n}^2 - \dfrac{k-1}{k+1}\right]^{1/(k-1)}} \tag{13.48b}$$

$$\frac{T_2}{T_1} = \frac{\left(1+\dfrac{k-1}{2} M_{1_n}^2\right)\left(k M_{1_n}^2 - \dfrac{k-1}{2}\right)}{\left(\dfrac{k+1}{2}\right)^2 M_{1_n}^2} \tag{13.48c}$$

$$\frac{p_2}{p_1} = \frac{2k}{k+1} M_{1_n}^2 - \frac{k-1}{k+1} \tag{13.48d}$$

$$\frac{\rho_2}{\rho_1} = \frac{V_{1_n}}{V_{2_n}} = \frac{\dfrac{k+1}{2} M_{1_n}^2}{1+\dfrac{k-1}{2} M_{1_n}^2} \tag{13.48e}$$

Equations 13.48, along with Eqs. 13.46 and 13.47, can be used to analyze oblique shock problems. Appendix E.5 lists flow functions for M_{2_n} and property ratios p_{0_2}/p_{0_1}, T_2/T_1, p_2/p_1, and ρ_2/ρ_1 (V_1/V_2) in terms of M_{1_n} for oblique-shock flow of an ideal gas. A table of values of these property ratios is presented for air ($k = 1.4$) for a limited range of Mach numbers. The associated *Excel* workbook, ***Oblique-Shock Relations***, can be used to print a larger table of values for air and other ideal gases. In essence, as demonstrated in Example 13.11, an oblique shock problem can be analyzed as an equivalent normal shock problem. The reader is urged to download the normal shock *Excel* add-ins from the Web site; they apply to these equations as well as Eqs. 13.20 for a normal shock.

Example 13.11 COMPARISON OF NORMAL AND OBLIQUE SHOCKS

Air at −2°C and 100 kPa is traveling at a speed of 1650 m/s. Find the pressure, temperature, and speed after the air experiences a normal shock. Compare with the pressure, temperature, and speed (and find the deflection angle θ) if the air instead experiences an oblique shock at angle $\beta = 30°$.

Given: Air flow with:

$$p_1 = 100\,\text{kPa}$$
$$T_1 = -2°\text{C}$$
$$V_1 = 1650\,\text{m/s}$$

Find: Downstream pressure, temperature, and speed if it experiences (a) a normal shock and (b) an oblique shock at angle $\beta = 30°$. Also find the deflection angle θ.

Solution:
(a) Normal shock

First compute the speed of sound,

$$c_1 = \sqrt{kRT_1} = \sqrt{1.4 \times 287 \frac{\text{N}\cdot\text{m}}{\text{kg}\cdot\text{K}} \times 271\ \text{K} \times \frac{\text{kg}\cdot\text{m}}{\text{s}^2\cdot\text{N}}} = 330\ \text{m/s}$$

Then the upstream Mach number is

$$M_1 = \frac{V_1}{c_1} = \frac{1650\ \text{m/s}}{330\ \text{m/s}} = 5.0$$

From the normal-shock flow functions, Eqs. 13.20, at $M_1 = 5.0$

M_1	M_2	p_2/p_1	T_2/T_1	V_2/V_1
5.0	0.4152	29.00	5.800	0.2000

From these data

$$T_2 = 5.800T_1 = (5.800)271\ \text{K} = 1572\ \text{K} = 1299°\text{C} \quad \longleftarrow T_2$$

$$p_2 = 29.00p_1 = (29.00)100\ \text{kPa} = 2.9\ \text{MPa} \quad \longleftarrow p_2$$

$$V_2 = 0.200V_1 = (0.200)1650\ \text{m/s} = 330\ \text{m/s} \quad \longleftarrow V_2$$

(b) Oblique shock

First compute the normal and tangential components of velocity,

$$V_{1_n} = V_1 \sin\beta = 1650\ \text{m/s} \times \sin 30° = 825\ \text{m/s}$$
$$V_{1_t} = V_1 \cos\beta = 1650\ \text{m/s} \times \cos 30° = 1429\ \text{m/s}$$

Then the upstream *normal* Mach number is

$$M_{1_n} = \frac{V_{1_n}}{c_1} = \frac{825\ \text{m/s}}{330\ \text{m/s}} = 2.5$$

From the oblique-shock flow functions, Eqs. 13.48, at $M_{1_n} = 2.5$

M_{1_n}	M_{2_n}	p_2/p_1	T_2/T_1	V_{2_n}/V_{1_n}
2.5	0.5130	7.125	2.138	0.300

From these data

$$T_2 = 2.138T_1 = (2.138)271\ \text{K} = 579\ \text{K} = 306°\text{C} \quad \longleftarrow T_2$$

$$p_2 = 7.125p_1 = (7.125)100\ \text{kPa} = 712.5\ \text{kPa} \quad \longleftarrow p_2$$

$$V_{2_n} = 0.300V_{1_n} = (0.300)825\ \text{m/s} = 247.5\ \text{m/s}$$

$$V_{2_t} = V_{1_t} = 1429\ \text{m/s} \quad \longleftarrow V_2$$

The downstream velocity is given by the Pythagorean sum of the velocity components,

$$V_2 = \sqrt{(V_{2_n}^2 + V_{2_t}^2)} = \sqrt{(247.5^2 + 1429^2)}\ \text{m/s} = 1450\ \text{m/s}$$

Note that

$$c_2 = \sqrt{kRT_2} = \sqrt{1.4 \times 287 \frac{\text{N}\cdot\text{m}}{\text{kg}\cdot\text{K}} 579\ \text{K} \times \frac{\text{kg}\cdot\text{m}}{\text{s}^2\cdot\text{N}}} = 482\ \text{m/s}$$

so that the downstream Mach number is

$$M_2 = \frac{V_2}{c_2} = \frac{1450\ \text{m/s}}{482\ \text{m/s}} = 3.01$$

Although the downstream normal Mach number must be subsonic, the actual downstream Mach number could be subsonic or supersonic (as in this case).

The deflection angle can be obtained from Eq. 13.46b

$$V_{2_n} = V_2 \sin(\beta - \theta)$$

or

$$\theta = \beta - \sin^{-1}\left(\frac{V_{2_n}}{V_2}\right) = 30° - \sin^{-1}\left(\frac{247.5}{1450}\right)$$

$$= 30° - 9.8° = 20.2° \longleftarrow \theta$$

This Example illustrates:

- ✓ That an oblique shock involves deflection of the flow through angle θ.
- ✓ Use of normal shock functions for solution of oblique shock problems.
- ✓ The important result that for a given supersonic flow an oblique shock will always be weaker than a normal shock, because $M_{1_n} < M_1$.
- ✓ That while $M_{2_n} < 1$ always, M_2 can be subsonic or supersonic (as in this case).

The *Excel* workbook for oblique shocks is convenient for performing these calculations. (Alternatively, the normal shock relations *Excel* add-ins, available on the Web site, are useful for these calculations.)

We can gain further insight into oblique shock behavior by combining some of our earlier equations to relate the deflection angle θ, the Mach number M_1, and the shock angle β. From the oblique shock geometry of Fig. 13.28*b*,

$$\frac{V_{1_n}}{V_{2_n}} = \frac{V_{1_t}\tan\beta}{V_{2_t}\tan(\beta - \theta)} = \frac{\tan\beta}{\tan(\beta - \theta)}$$

We can also relate the two normal velocities from Eq. 13.48e,

$$\frac{V_{1_n}}{V_{2_n}} = \frac{\frac{k+1}{2}M_{1_n}^2}{1 + \frac{k-1}{2}M_{1_n}^2} \tag{13.48e}$$

Equating the two expressions for the normal velocity ratio, we have

$$\frac{V_{1_n}}{V_{2_n}} = \frac{\tan\beta}{\tan(\beta - \theta)} = \frac{\frac{k+1}{2}M_{1_n}^2}{1 + \frac{k-1}{2}M_{1_n}^2}$$

and

$$\tan(\beta - \theta) = \frac{\tan\beta}{\frac{(k+1)}{2}M_{1_n}^2}\left(1 + \frac{k-1}{2}M_{1_n}^2\right)$$

Finally, if we use $M_{1_n} = M_1 \sin\beta$ in this expression and further simplify, we obtain (after using a trigonometric identity and more algebra)

$$\tan\theta = \frac{2\cot\beta(M_1^2\sin^2\beta - 1)}{M_1^2(k + \cos 2\beta) + 2} \tag{13.49}$$

Equation 13.49 relates the deflection angle θ to the incoming Mach number M_1 and the oblique shock angle β. For a given Mach number, we can compute θ as a function of β, as shown in Fig. 13.29 for air ($k = 1.4$).

Appendix E.5 presents a table of values of deflection angle θ as a function of Mach number M_1 and oblique shock angle β for air ($k = 1.4$) for a limited range of Mach numbers. The associated *Excel* workbook, *Oblique-Shock Relations*, can be used to print a larger table of values for air and other ideal gases. The reader is urged to download the oblique shock *Excel* add-in from the Web site; it can be used for solving Eq. 13.49 for the deflection angle θ, oblique shock β, or M_1.

We should note that we used M_1 and shock angle β to compute θ, but in reality the causality is the reverse: it is the deflection θ caused by an object such as the surface of an airplane wing that causes an oblique shock at angle β. We can draw some interesting conclusions from Fig. 13.29:

- For a given Mach number and deflection angle, there are generally *two* possible oblique shock angles—we could generate a weak shock (smaller β value, and hence, smaller normal Mach number, M_{1_n}) or a strong shock (larger β value, and hence, larger normal Mach number). In most cases the weak shock appears (exceptions include situations where the downstream pressure is forced to take on a large value as caused by, for example, an obstruction).
- For a given Mach number, there is a maximum deflection angle. For example, for air ($k = 1.4$), if $M_1 = 3$, the maximum deflection angle is $\theta_{max} \approx 34°$. Any attempt to deflect the flow at an angle $\theta > \theta_{max}$ would cause a detached normal shock to form instead of an oblique shock.
- For zero deflection ($\theta \to 0$), the weak shock becomes a Mach wave and $\beta \to \alpha = \sin^{-1}(1/M_1)$.

Figure 13.29 can be used to explain the phenomena shown in Fig. 13.27. If an airplane (or airplane wing), causing deflection θ, accelerates from subsonic through

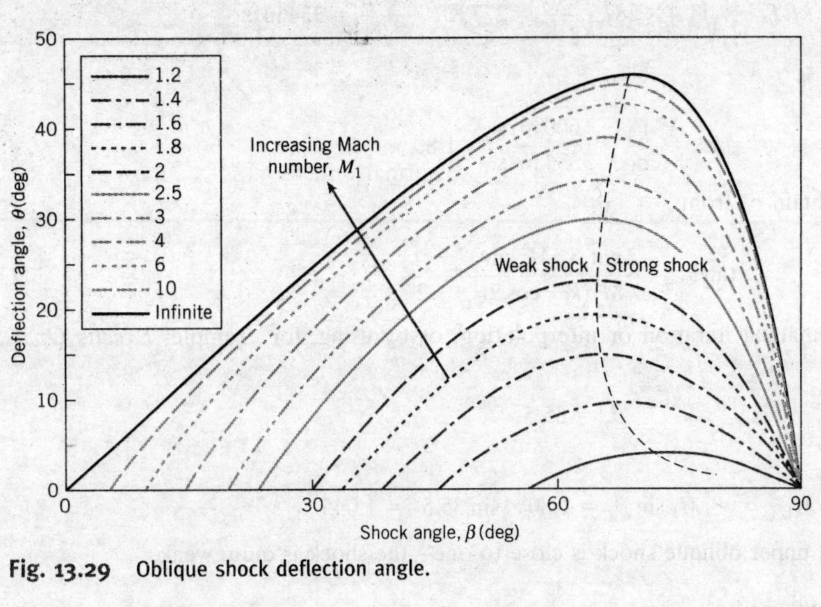

Fig. 13.29 Oblique shock deflection angle.

supersonic speed, we can plot the airplane's progress on Fig. 13.29 as a horizontal line from right to left, through lines of increasing Mach number. For example, for $\theta = 10°$, we obtain the following results: As M_1 increases from subsonic through about 1.4 there is no oblique shock solution—we instead either have no shock (subsonic flow) or a detached normal shock; at some Mach number the normal shock first attaches and becomes an oblique shock (Problem 13.187 shows that for $\theta = 10°$, the normal shock first attaches and becomes oblique at $M_1 \approx 1.42$, with $\beta \approx 67°$); as M_1 increases from 1.6 through 1.8, 2.0, 2.5, etc., toward infinity, from Fig. 13.29, $\beta \approx 51°, 44°, 39°, 32°$, toward 12°, respectively—the oblique shock angle progressively decreases, as we saw in Fig. 13.27.

A problem involving oblique shocks is solved in Example 13.12.

Example 13.12 OBLIQUE SHOCKS ON AN AIRFOIL

An airplane travels at a speed of 600 m/s in air at 4°C and 100 kPa. The airplane's airfoil has a sharp leading edge with included angle $\delta = 6°$, and an angle of attack $\alpha = 1°$. Find the pressures on the upper and lower surfaces of the airfoil immediately after the leading edge.

Given: Air flow over sharp leading edge with:

$p_1 = 100$ kPa $\qquad \delta = 6°$

$T_1 = 4°$C $\qquad \alpha = 1°$

$V_1 = 600$ m/s

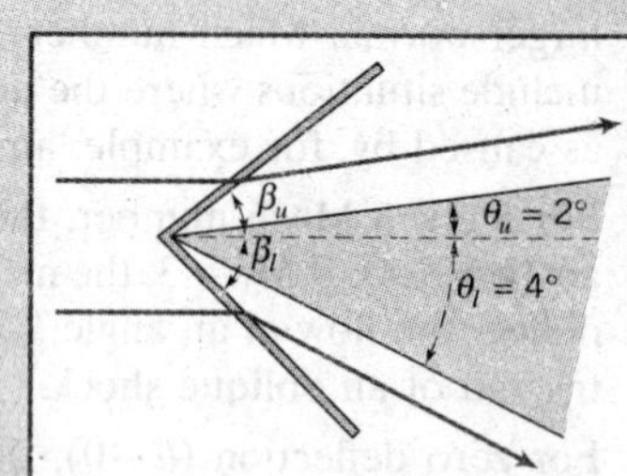

Find: Pressure on upper and lower surfaces.

Solution:

For an angle of attack of 1° of an airfoil with leading edge angle 6°, the deflection angles are $\theta_u = 2°$ and $\theta_l = 4°$ as shown.

(a) Upper surface

First compute the speed of sound,

$$c_1 = \sqrt{kRT_1} = \sqrt{1.4 \times 287 \frac{\text{N}\cdot\text{m}}{\text{kg}\cdot\text{K}} \times 277\,\text{K} \times \frac{\text{kg}\cdot\text{m}}{\text{s}^2\cdot\text{N}}} = 334\,\text{m/s}$$

Then the upstream Mach number is

$$M_1 = \frac{V_1}{c_1} = \frac{600\,\text{m/s}}{334\,\text{m/s}} = 1.80$$

For $M_1 = 1.80$ and $\theta_u = 2°$, we obtain β_u from

$$\tan\theta_u = \frac{2\cot\beta_u(M_1^2\sin^2\beta_u - 1)}{M_1^2(k + \cos 2\beta_u) + 2} \tag{13.49}$$

This can be solved for β_u using manual iteration or interpolation, or by using, for example, *Excel*'s *Goal Seek* function,

$$\beta_u = 35.5°$$

Then we can find $M_{1_{n(\text{upper})}}$,

$$M_{1_{n(\text{upper})}} = M_1 \sin\beta_u = 1.80 \times \sin 35.5° = 1.045$$

The normal Mach number for the upper oblique shock is close to one—the shock is quite weak.

From the oblique-shock pressure ratio, Eqs. 13.48d, at $M_{1_{n(\text{upper})}} = 1.045$,

$$\frac{p_{2(\text{upper})}}{p_1} = \frac{2k}{k+1} M_{1_{n(\text{upper})}}^2 - \frac{k-1}{k+1} = \frac{2 \times 1.4}{(1.4+1)}(1.045)^2 - \frac{(1.4-1)}{(1.4+1)} = 1.11$$

Hence,

$$p_{2_{(\text{upper})}} = 1.11\, p_1 = (1.11)100 \text{ kPa} = 111 \text{ kPa} \qquad \longleftarrow \qquad p_{2_{(\text{upper})}}$$

(b) Lower surface

For $M_1 = 1.80$ and $\theta_l = 4°$, we obtain β_l from

$$\tan \theta_l = \frac{2 \cot \beta_l (M_1^2 \sin^2 \beta_l - 1)}{M_1^2 (k + \cos 2\beta_l) + 2} \tag{13.49}$$

and find

$$\beta_l = 37.4°$$

Then we can find $M_{1_{n(\text{lower})}}$,

$$M_{1_{n(\text{lower})}} = M_1 \sin \beta_l = 1.80 \times \sin 37.4° = 1.093$$

The normal Mach number for the lower oblique shock is also close to one.

From the oblique-shock pressure ratio, Eq. 13.48d, at $M_{1_{n(\text{lower})}} = 1.093$,

$$\frac{p_{2(\text{lower})}}{p_1} = \frac{2k}{k+1} M_{1_{n(\text{lower})}}^2 - \frac{k-1}{k+1} = \frac{2 \times 1.4}{(1.4+1)}(1.093)^2 - \frac{(1.4-1)}{(1.4+1)} = 1.23$$

Hence,

$$p_{2_{(\text{lower})}} = 1.23\, p_1 = (1.23)100 \text{ kPa} = 123 \text{ kPa} \qquad \longleftarrow \qquad p_{2_{(\text{lower})}}$$

This Example illustrates the use of Eq. 13.49 for obtaining oblique shock data from the flow deflection. The *Excel* workbook for oblique shocks is convenient for performing these calculations. (Alternatively, the normal shock relations and oblique shock relations *Excel* add-ins, available on the Web site, are useful for these calculations.)

Isentropic Expansion Waves

Oblique shock waves occur when a flow is suddenly compressed as it is deflected. We can ask ourselves what happens if we *gradually* redirect a supersonic flow, for example, along a curved surface. The answer is that we may generate *isentropic compression or expansion waves*, as illustrated schematically in Figs. 13.30*a* and 13.30*b*, respectively. From Fig. 13.30*a* we see that a series of compression waves will eventually converge, and their cumulative effect will eventually generate an oblique shock not far from the curved surface. While compression waves do occur, they are not of great interest because the oblique shocks they lead to usually dominate the aerodynamics—at most the waves are a local phenomenon. On the other hand, as shown in Fig. 13.30*b*, expansion waves in series are divergent and so do not coalesce. Figure 13.30*c* shows expansion around a sharp-edged corner.

We wish to analyze these isentropic waves to obtain a relation between the deflection angle and the Mach number. First we note that each wave is a Mach wave, so is at angle $\alpha = \sin^{-1}(1/M)$, where M is the Mach number immediately before the wave. Compression waves are convergent because the wave angle α increases as the Mach number decreases. On the other hand, expansion waves are divergent because as the flow expands the Mach number increases, decreasing the Mach angle.

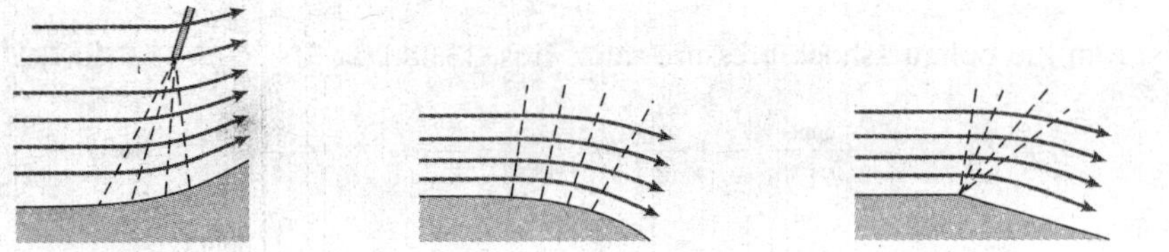

(a) Isentropic compression waves (b) Isentropic expansion waves (c) Isentropic expansion at a corner

Fig. 13.30 Isentropic compression and expansion waves.

Consider an isentropic wave as shown in Fig. 13.31*a*. (It happens to be a compression wave, but the analysis that follows applies to both expansion and compression waves.) The speed changes from V to $V+dV$, with deflection $d\theta$. As with the oblique shock analysis for Fig. 13.28*a*, it is convenient to orient the xy coordinates orthogonal to the wave, as shown in Fig. 13.31*b*. For this infinitesimal control volume, we can write the basic equations (continuity, momentum, and the first and second laws of thermodynamics). Comparing the isentropic wave control volume of Fig. 13.31*b* to the control volume for an oblique shock shown in Fig. 13.28, we see that the control volumes have similar features. However, an isentropic wave differs from an oblique shock wave in two important ways:

- The wave angle is $\alpha = \sin^{-1}(1/M)$, instead of angle β for the oblique shock.
- The changes in velocity and in density, pressure, etc., and the deflection angle, are all infinitesimals.

The second factor is the reason that the flow, which is adiabatic, is isentropic.

With these two differences in mind we repeat the analysis that we performed for the oblique shock. The continuity equation is

$$\overbrace{\frac{\partial}{\partial t}\int_{\mathrm{CV}} \rho\, d\forall}^{=0(1)} + \int_{\mathrm{CS}} \rho \vec{V}\cdot d\vec{A} = 0 \tag{4.12}$$

Assumption: (1) Steady flow.

Then

$$\{-\rho V \sin\alpha\, A\} + \{(\rho + d\rho)(V+dV)\sin(\alpha - d\theta)A\} = 0$$

or

$$\rho V \sin\alpha = (\rho + d\rho)(V + dV)\sin(\alpha - d\theta) \tag{13.50}$$

Next we consider the momentum equation for motion normal and tangent to the wave. We look first at the tangential, y component

$$F_{S_y} + \overbrace{F_{B_y}}^{=0(2)} = \overbrace{\frac{\partial}{\partial t}\int_{\mathrm{CV}} V_y\, \rho\, d\forall}^{=0(1)} + \int_{\mathrm{CS}} u\, V_y\, \rho \vec{V}\cdot d\vec{A} \tag{4.18b}$$

Assumption: (2) Negligible body forces.

Then

$$0 = V\cos\alpha\{-\rho V \sin\alpha\, A\} + (V+dV)\cos(\alpha - d\theta)\{(\rho + d\theta)(V+dV)\sin(\alpha - d\theta)A\}$$

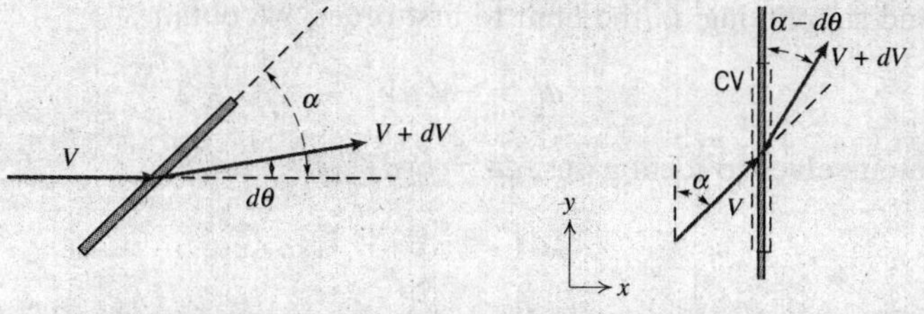

(a) Isentropic wave (b) Isentropic wave in wave coordinates

Fig. 13.31 Isentropic wave control volume.

or, using continuity (Eq. 13.50),

$$V\cos\alpha = (V + dV)\cos(\alpha - d\theta)$$

Expanding and simplifying [using the facts that, to first order, in the limit as $d\theta \to 0$, $\cos(d\theta) \to 1$, and $\sin(d\theta) \to d\theta$], we obtain

$$d\theta = -\frac{dV}{V\tan\alpha}$$

But $\sin\alpha = 1/M$, so $\tan\alpha = 1/\sqrt{M^2 - 1}$, and

$$d\theta = -\sqrt{M^2-1}\,\frac{dV}{V} \tag{13.51}$$

We skip the analysis of the normal, x component of momentum, and move on to the first law of thermodynamics, which is

$$\underbrace{\dot{Q}}_{=0(4)} - \underbrace{\dot{W}_s}_{=0(5)} - \underbrace{\dot{W}_{\text{shear}}}_{=0(5)} - \underbrace{\dot{W}_{\text{other}}}_{=0(5)} = \underbrace{\frac{\partial}{\partial t}\int_{\text{CV}} e\,\rho\,dV\!\!\!\!-}_{=0(1)} + \int_{\text{CS}} (e + pv)\,\rho\vec{V}\cdot d\vec{A} \tag{4.56}$$

where

$$e = u + \frac{V^2}{2} + \underbrace{gz}_{=0(6)}$$

Assumptions: (4) Adiabatic flow.
(5) No work terms.
(6) Negligible gravitational effect.

For our control volume we obtain (using $h \equiv u + pv$, where v represents the specific volume)

$$0 = \left\{h + \frac{V^2}{2}\right\}\{-\rho V\sin\alpha\, A\}$$
$$+\left\{(h + dh) + \frac{(V + dV)^2}{2}\right\}\{(\rho + d\rho)(V + dV)\sin(\alpha - d\theta)A\}$$

This can be simplified, using continuity (Eq. 13.50), to

$$h + \frac{V^2}{2} = (h + dh) + \frac{(V + dV)^2}{2}$$

Expanding and simplifying, in the limit to first order, we obtain

$$dh = -V\,dV$$

If we confine ourselves to ideal gases, $dh = c_p\,dT$, so

$$c_p\,dT = -V\,dV \tag{13.52}$$

Equation 13.52 relates differential changes in temperature and velocity. We can obtain a relation between M and V using $V = Mc = M\sqrt{kRT}$. Differentiating (and dividing the left side by V and the right by $M\sqrt{kRT}$),

$$\frac{dV}{V} = \frac{dM}{M} + \frac{1}{2}\frac{dT}{T}$$

Eliminating dT using Eq. 13.52,

$$\frac{dV}{V} = \frac{dM}{M} - \frac{1}{2}\frac{V\,dV}{c_pT} = \frac{dM}{M} - \frac{1}{2}\frac{dV}{V}\left(\frac{V^2}{c_pT}\right) = \frac{dM}{M} - \frac{1}{2}\frac{dV}{V}\left(\frac{M^2c^2}{c_pT}\right)$$

$$\frac{dV}{V} = \frac{dM}{M} - \frac{1}{2}\frac{dV}{V}\left(\frac{M^2\,kRT}{c_pT}\right) = \frac{dM}{M} - \frac{1}{2}\frac{dV}{V}M^2(k-1)$$

Hence,

$$\frac{dV}{V} = \frac{2}{2+M^2(k-1)}\frac{dM}{M} \tag{13.53}$$

Finally, combining Eqs. 13.51 and 13.53,

$$d\theta = -\frac{2\sqrt{M^2-1}}{2+M^2(k-1)}\frac{dM}{M} \tag{13.54}$$

We will generally apply Eq. 13.54 to expansion waves, for which $d\theta$ is negative, so it is convenient to change variables, $d\omega \equiv -d\theta$. Equation 13.54 relates the differential change in Mach number through an isentropic wave to the deflection angle. We can integrate this to obtain the deflection as a function of Mach number, to within a constant of integration. We *could* integrate Eq. 13.54 between the initial and final Mach numbers of a given flow, but it will be more convenient to integrate from a reference state, the critical speed ($M = 1$) to Mach number M, with ω arbitrarily set to zero at $M = 1$,

$$\int_0^{\omega} d\omega = \int_1^M \frac{2\sqrt{M^2-1}}{2+M^2(k-1)}\frac{dM}{M}$$

leading to the *Prandtl-Meyer supersonic expansion function*,

$$\omega = \sqrt{\frac{k+1}{k-1}}\tan^{-1}\left(\sqrt{\frac{k-1}{k+1}(M^2-1)}\right) - \tan^{-1}\left(\sqrt{M^2-1}\right) \tag{13.55}$$

We use Eq. 13.55 to relate the total deflection caused by isentropic expansion from M_1 to M_2,

$$\text{Deflection} = \omega_2 - \omega_1 = \omega(M_2) - \omega(M_1)$$

Appendix E.6 presents a table of values of the *Prandtl-Meyer supersonic expansion function*, ω, as a function of Mach number M for air ($k = 1.4$) for a limited range of Mach numbers. The associated *Excel* workbook, *Isentropic Expansion Wave*

Relations, can be used to print a larger table of values for air and other ideal gases. The reader is urged to download the isentropic expansion wave relations *Excel* add-in from the Web site; it can be used for solving Eq. 13.55 for the Prandtl-Meyer expansion function ω, or M.

We have already indicated that the flow is isentropic. We can verify this by using the second law of thermodynamics,

$$\overbrace{\frac{\partial}{\partial t}\int_{CV} s\,\rho\,dV}^{=0(1)} + \int_{CS} s\,\rho\,\vec{V}\cdot d\vec{A} \geq \int_{CS} \frac{1}{T}\overbrace{\left(\frac{\dot{Q}}{A}\right)}^{=0(4)} dA \tag{4.58}$$

The wave is reversible, so Eq. 4.58 for our control volume is

$$s\{-\rho V \sin\alpha\, A\} + (s+ds)\{(\rho+d\rho)(V+dV)\sin(\alpha-d\theta)A\} = 0$$

and using continuity (Eq. 13.50), this reduces to

$$ds = 0$$

The flow is demonstrated to be isentropic. Hence, stagnation properties are constant and the local isentropic stagnation property equations (Section 12.3) will be useful here.

$$\frac{p_0}{p} = \left[1 + \frac{k-1}{2}M^2\right]^{k/(k-1)} \tag{12.21a}$$

$$\frac{T_0}{T} = 1 + \frac{k-1}{2}M^2 \tag{12.21b}$$

$$\frac{\rho_0}{\rho} = \left[1 + \frac{k-1}{2}M^2\right]^{1/(k-1)} \tag{12.21c}$$

Equation 13.55, together with Eqs. 12.21a through 12.21c, can be used for analyzing isentropic expansion or compression waves. (We never got around to deriving the normal component of momentum—the above analysis provides a complete set of equations.) A problem involving expansion waves is solved in Example 13.13.

Example 13.13 EXPANSION WAVE ON AN AIRFOIL

The airplane of Example 13.12 (speed of 600 m/s in air at 4°C and 100 kPa, with a sharp leading edge with included angle $\delta = 6°$) now has an angle of attack $\alpha = 6°$. Find the pressures on the upper and lower surfaces of the airfoil immediately after the leading edge.

Given: Air flow over sharp leading edge with:

$p_1 = 100$ kPa $\quad \delta = 6°$

$T_1 = 4°$C $\quad \alpha = 6°$

$V_1 = 600$ m/s

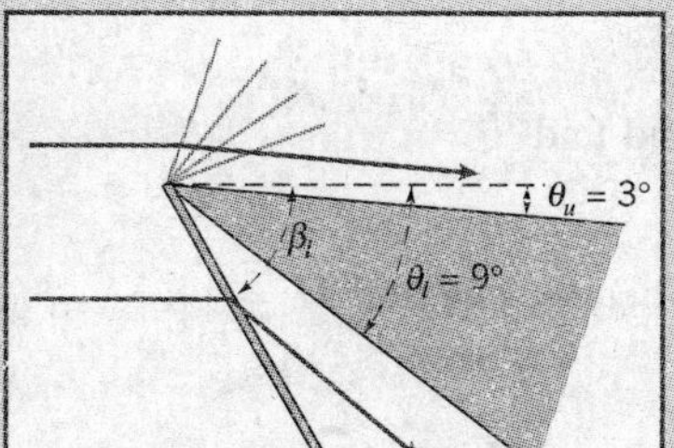

Find: Pressures on upper and lower surfaces.

Solution:

For an angle of attack of 6° of an airfoil with leading edge angle 6°, the deflection angles are $\theta_u = 3°$ and $\theta_l = 9°$ as shown.

(a) Upper surface—isentropic expansion

First compute the speed of sound,

$$c_1 = \sqrt{kRT_1} = \sqrt{1.4 \times 287 \frac{\text{N}\cdot\text{m}}{\text{kg}\cdot\text{K}} \times 277\,\text{K} \times \frac{\text{kg}\cdot\text{m}}{\text{s}^2\cdot\text{N}}} = 334\,\text{m/s}$$

Then the upstream Mach number is

$$M_1 = \frac{V_1}{c_1} = \frac{600\,\text{m/s}}{334\,\text{m/s}} = 1.80$$

For $M_1 = 1.80$, the Prandtl-Meyer function ω_1 is obtained from

$$\omega_1 = \sqrt{\frac{k+1}{k-1}} \tan^{-1}\left(\sqrt{\frac{k-1}{k+1}(M_1^2 - 1)}\right) - \tan^{-1}\left(\sqrt{M_1^2 - 1}\right) \tag{13.55}$$

so

$$\omega_1 = \sqrt{\frac{1.4+1}{1.4-1}} \tan^{-1}\left(\sqrt{\frac{1.4-1}{1.4+1}(1.80^2 - 1)}\right) - \tan^{-1}\left(\sqrt{1.80^2 - 1}\right) = 20.7^\circ$$

The Prandtl-Meyer function value on the upper surface, ω_u, is then

$$\omega_u = \omega_1 + \theta_u = 20.7^\circ + 3^\circ = 23.7^\circ$$

For this Prandtl-Meyer function value, $M_{2_{(\text{upper})}}$ is obtained from Eq. 13.55:

$$\omega_u = \sqrt{\frac{k+1}{k-1}} \tan^{-1}\left(\sqrt{\frac{k-1}{k+1}\left(M^2_{2_{(\text{upper})}} - 1\right)}\right) - \tan^{-1}\left(\sqrt{\left(M^2_{2_{(\text{upper})}} - 1\right)}\right)$$

This can be solved using manual iteration or interpolation, or by using, for example, *Excel*'s *Goal Seek* function,

$$M_{2_{(\text{upper})}} = 1.90$$

Finally, we can find $p_{2_{(\text{upper})}}$ from repeated use of Eq. 12.21a,

$$p_{2_{(\text{upper})}} = \frac{p_{2_{(\text{upper})}}}{p_0}\frac{p_0}{p_1}p_1 = \left\{\left[1 + \frac{k-1}{2}M_1^2\right]^{k/(k-1)} \Big/ \left[1 + \frac{k-1}{2}M^2_{2(\text{upper})}\right]^{k/(k-1)}\right\} p_1$$

$$= \{[1 + (0.2)1.80^2]^{3.5}/[1 + (0.2)1.90^2]^{3.5}\} \times 100\,\text{kPa}$$

so

$$p_{2_{(\text{upper})}} = 85.8\,\text{kPa} \quad \longleftarrow \quad p_{2_{(\text{upper})}}$$

(b) Lower surface—oblique shock

For $M_1 = 1.80$ and $\theta_1 = 9^\circ$, we obtain β_l from

$$\tan\theta_l = \frac{2\cot\beta_l(M_1^2\sin^2\beta_l - 1)}{M_1^2(k + \cos 2\beta_l) + 2} \tag{13.49}$$

and find

$$\beta_l = 42.8^\circ$$

Then we can find $M_{1_{n(\text{lower})}}$,

$$M_{1_{n(\text{lower})}} = M_1 \sin\beta_l = 1.80 \times \sin 42.8^\circ = 1.223$$

From the oblique-shock pressure ratio, Eq. 13.48d, at $M_{1_{n(\text{lower})}} = 1.223$,

$$\frac{p_{2_{(\text{lower})}}}{p_1} = \frac{2k}{k+1}M^2_{1_{n(\text{lower})}} - \frac{k-1}{k+1} = \frac{2\times 1.4}{(1.4+1)}(1.223)^2 - \frac{(1.4-1)}{(1.4+1)} = 1.58$$

Hence,

$$p_{2_{(\text{lower})}} = 1.58p_1 = (1.58)100\text{ kPa}$$
$$= 158\text{ kPa} \longleftarrow p_{2_{(\text{lower})}}$$

This Example illustrates the use of Eq. 13.55 and the isentropic stagnation relations for analysis of isentropic expansion waves and the use of Eq. 13.49 for an oblique shock.
The *Excel* workbooks for isentropic expansion waves and oblique shocks are convenient for performing these calculations. (Alternatively, the normal and oblique shock relations, and the isentropic expansion wave *Excel* add-ins, available on the Web site, are useful for these calculations.)

Example 13.13 hints at the approach we can use to obtain lift and drag coefficients of a supersonic wing, illustrated in Example 13.14.

Example 13.14 LIFT AND DRAG COEFFICIENTS OF A SUPERSONIC AIRFOIL

The airplane of Example 13.13 has a symmetric diamond cross section (sharp leading and trailing edges of angle $\delta = 6°$). For a speed of 600 m/s in air at 4°C and 100 kPa, find the pressure distribution on the upper and lower surfaces, and the lift and drag coefficients for an angle of attack of $\alpha = 6°$.

Given: Air flow over symmetric section shown with:

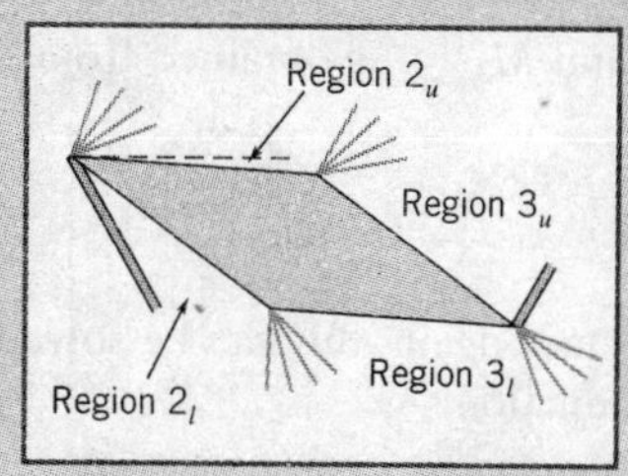

$$p_1 = 100\text{ kPa} \qquad \delta = 6°$$
$$T_1 = 4°\text{C} \qquad \alpha = 6°$$
$$V_1 = 600\text{ m/s}$$

Find: Pressure distribution, and lift and drag coefficients.

Solution:

We first need to obtain the pressures on the four surfaces of the airfoil. We have already obtained in Example 13.13 the data for Region 2_u and Region 2_l:

$$M_{2_{(\text{upper})}} = 1.90 \qquad p_{2_{(\text{upper})}} = 85.8\text{ kPa} \longleftarrow p_{2_{(\text{upper})}}$$

$$M_{2_{(\text{lower})}} = 1.489 \qquad p_{2_{(\text{lower})}} = 158\text{ kPa} \longleftarrow p_{2_{(\text{lower})}}$$

(Note that $M_{2_{(\text{lower})}} = 1.489$ is obtained from $M_{1_n} = 1.223$ in Example 13.13 by direct use of Eqs. 13.48a and 13.47b.) In addition, for Region 2_u, we found the Prandtl-Meyer function to be 23.7°. Hence, for Region 3_u, we can find the value of the Prandtl-Meyer function from the deflection angle. For 6° leading and trailing edges, the airfoil angles at the upper and lower surfaces are each 174°. Hence, at the upper and lower surfaces the deflections are each 6°.

For Region 3_u,

$$\omega_{3_{(\text{upper})}} = \omega_{2_{(\text{upper})}} + \theta = 23.7° + 6° = 29.7°$$

For this Prandtl-Meyer function value, $M_{3_{(upper)}}$ is obtained from Eq. 13.55,

$$\omega_{3_{(upper)}} = \sqrt{\frac{k+1}{k-1}}\tan^{-1}\left(\sqrt{\frac{k-1}{k+1}\left(M_{3_{(upper)}}^2-1\right)}\right) - \tan^{-1}\left(\sqrt{M_{3_{(upper)}}^2-1}\right)$$

This can be solved using manual iteration or interpolation, or by using, for example, *Excel*'s *Goal Seek* function,

$$M_{3(upper)} = 2.12$$

Finally, we can find $p_{3_{(upper)}}$ from repeated use of Eq. 12.21a,

$$p_{3_{(upper)}} = \frac{p_{3_{(upper)}}}{p_0}\frac{p_0}{p_1}p_1 = \left\{\left[1+\frac{k-1}{2}M_1^2\right]^{k/(k-1)} \Big/ \left[1+\frac{k-1}{2}M_{3_{(upper)}}^2\right]^{k/(k-1)}\right\}p_1$$

$$= \left\{\left[1+(0.2)1.80^2\right]^{3.5}/\left[1+(0.2)2.12^2\right]^{3.5}\right\} \times 100\,\text{kPa}$$

so

$$p_{3_{(upper)}} = 60.9\,\text{kPa} \quad \longleftarrow \quad p_{3_{(upper)}}$$

For Region 3_l, we first need to find the Prandtl-Meyer function in the previous region, Region 2_l. For $M_{2_{(lower)}} = 1.489$, we find from Eq. 13.55,

$$\omega_{2_{(lower)}} = \sqrt{\frac{k+1}{k-1}}\tan^{-1}\left(\sqrt{\frac{k-1}{k+1}\left(M_{2_{(lower)}}^2-1\right)}\right) - \tan^{-1}\left(\sqrt{M_{2_{(lower)}}^2-1}\right)$$

$$= \sqrt{\frac{1.4+1}{1.4-1}}\tan^{-1}\left(\sqrt{\frac{1.4-1}{1.4+1}(1.489^2-1)}\right) - \tan^{-1}\left(\sqrt{1.489^2-1}\right)$$

so

$$\omega_{2_{(lower)}} = 11.58°$$

Hence, for Region 3_l,

$$\omega_{3_{(lower)}} = \omega_{2_{(lower)}} + \theta = 11.58° + 6° = 17.6°$$

and $M_{3_{(lower)}}$ is obtained from Eq. 13.55,

$$\omega_{3_{(lower)}} = \sqrt{\frac{k+1}{k-1}}\tan^{-1}\left(\sqrt{\frac{k-1}{k+1}\left(M_{3_{(lower)}}^2-1\right)}\right) - \tan^{-1}\left(\sqrt{M_{3_{(lower)}}^2-1}\right)$$

Once again, this can be solved using manual iteration or interpolation, or by using, for example, *Excel*'s *Goal Seek* function,

$$M_{3_{(lower)}} = 1.693$$

Finally, we can find $p_{3_{(lower)}}$ from repeated use of Eq. 12.21a,

$$p_{3_{(lower)}} = \frac{p_{3_{(lower)}}}{p_{0_2}}\frac{p_{0_2}}{p_{2_{(lower)}}}p_{2_{(lower)}} = \left\{\left[1+\frac{k-1}{2}M_{2_{(lower)}}^2\right]^{k/(k-1)} \Big/ \left[1+\frac{k-1}{2}M_{3_{(lower)}}^2\right]^{k/(k-1)}\right\}p_{2_{(lower)}}$$

$$= \left\{\left[1+(0.2)1.489^2\right]^{3.5}/\left[1+(0.2)1.693^2\right]^{3.5}\right\} \times 158\,\text{kPa}$$

Hence,

$$p_{3_{(lower)}} = 117\,\text{kPa} \quad \longleftarrow \quad p_{3_{(lower)}}$$

(Note that we cannot use p_0, the stagnation pressure of the incoming flow for computing this pressure, because the flow experienced a shock before reaching the lower surface.)

To compute the lift and drag coefficients, we need the lift and drag force.

First we find the vertical and horizontal forces with respect to coordinates orthogonal to the airfoil.

The vertical force (assuming the chord c and span s are in meters) is given by

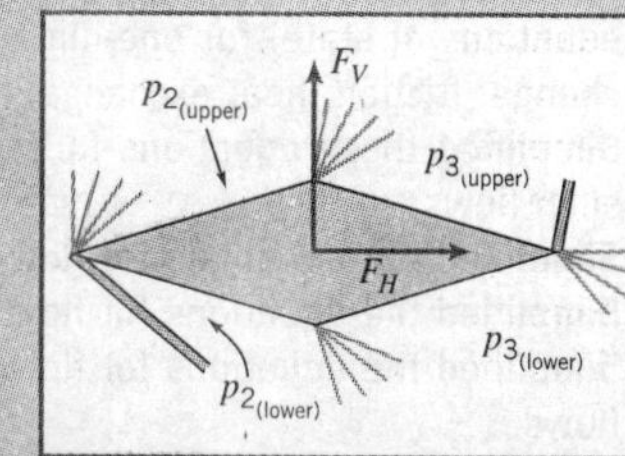

$$F_V = s\frac{c}{2}\{(p_{2_{(lower)}} + p_{3_{(lower)}}) - (p_{2_{(lower)}} + p_{3_{(upper)}})\}$$

$$F_V = s(\text{m})\frac{c(\text{m})}{2}\{(158 + 117) - (85.8 + 60.9)\}(\text{kPa}) = 64.2\, sc \text{ kN}$$

and the horizontal force by

$$F_H = s\frac{c}{2}\tan 3^\circ\{(p_{2_{(upper)}} + p_{2_{(lower)}}) - (p_{3_{(upper)}} + p_{3_{(lower)}})\}$$

$$F_H = s(\text{m})\frac{c(\text{m})}{2}\tan 3^\circ\{(85.8 + 158) - (60.9 + 117)\}(\text{kPa}) = 1.73\, sc \text{ kN}$$

The lift and drag forces (per unit plan area) are then

$$F_L = F_V \cos 6^\circ - F_H \sin 6^\circ = 63.6\, sc \text{ kN}$$

and

$$F_D = F_V \sin 6^\circ + F_H \cos 6^\circ = 8.42\, sc \text{ kN}$$

The lift and drag coefficients require the air density,

$$\rho = \frac{p}{RT} = 100 \times 10^3 \frac{\text{N}}{\text{m}^2} \times \frac{1}{287}\frac{\text{kg}\cdot\text{K}}{\text{N}\cdot\text{m}} \times \frac{1}{277\text{ K}} = 1.258 \text{ kg/m}^3$$

The lift coefficient is then

$$C_L = \frac{F_L}{\frac{1}{2}\rho V^2 sc} = 2 \times 63.6 \times 10^3 \text{ N} \times \frac{1}{1.258}\frac{\text{m}^3}{\text{kg}} \times \frac{1}{(600)^2}\frac{\text{s}^2}{\text{m}^2} \times \frac{\text{kg}\cdot\text{m}}{\text{N}\cdot\text{s}^2} = 0.281 \qquad C_L$$

and the drag coefficient is

$$C_D = \frac{F_D}{\frac{1}{2}\rho V^2 sc} = 2 \times 8.42 \times 10^3 \text{ N} \times \frac{1}{1.258}\frac{\text{m}^3}{\text{kg}}\frac{1}{(600)^2}\frac{\text{s}^2}{\text{m}^2} \times \frac{\text{kg}\cdot\text{m}}{\text{N}\cdot\text{s}^2} = 0.037 \qquad C_D$$

(Note that instead of using $\frac{1}{2}\rho V^2$ in the denominator of the coefficients, we could have used $\frac{1}{2}kp\,M^2$.) The lift-drag ratio is approximately 7.6.

This Example illustrates the use of oblique shock and isentropic expansion wave equations to determine the pressure distribution on an airfoil.

✓ We did not need to analyze the flow after the trailing expansion waves and oblique shock—unlike subsonic flow, the downstream condition has no effect on the airfoil.

✓ Unlike a subsonic flow, a supersonic flow can generate drag even in the absence of boundary layers and flow separation.

✓ Note that, unlike a subsonic flow, a supersonic flow *can* negotiate a sharp corner, even if we include the effect of a viscous boundary layer (as we have not done here). This is because an expanding supersonic flow has a negative pressure gradient, i.e., it is *not* adverse!

✓ An actual airfoil is not likely to have planar surfaces, so more sophisticated techniques than we can cover here are needed. However, this example illustrates the kind of results to be expected when analyzing a supersonic airfoil.

The *Excel* workbooks for oblique shocks and isentropic expansion waves are convenient for performing these calculations. (Alternatively, the normal and oblique shock relations, and the isentropic expansion wave *Excel* add-ins, available on the Web site, are useful for these calculations.)

13.8 Summary and Useful Equations

In this chapter, we:

- ✓ Developed a set of governing equations (continuity, the momentum equation, the first and second laws of thermodynamics, and equations of state) for one-dimensional flow of a compressible fluid (in particular an ideal gas) as it may be affected by area change, friction, heat exchange, and normal shocks.
- ✓ Simplified these equations for isentropic flow affected only by area change, and developed isentropic relations for analyzing such flows.
- ✓ Simplified the equations for flow through a normal shock, and developed normal-shock relations for analyzing such flows.
- ✓ Simplified the equations for flow affected only by friction, and developed the Fanno-line relations for analyzing such flows.
- ✓ Simplified the equations for flow affected only by heat exchange, and developed the Rayleigh-line relations for analyzing such flows.
- ✓ Introduced some basic concepts of two-dimensional flow: oblique shocks and expansion waves.

While investigating the above flows we developed insight into some interesting compressible flow phenomena, including:

- ✓ Use of *Ts* plots in visualizing flow behavior.
- ✓ Flow through, and necessary shape of, subsonic and supersonic nozzles and diffusers.
- ✓ The phenomenon of choked flow in converging nozzles and C-D nozzles, and the circumstances under which shock waves develop in C-D nozzles.
- ✓ *The phenomena of choked flow in flows with friction and flows with heat exchange.
- ✓ Computation of pressures and lift and drag coefficients for a supersonic airfoil.

Note: Most of the Useful Equations in the table below have a number of constraints or limitations—*be sure to refer to their page numbers for details*! In particular, most of them assume an ideal gas, with constant specific heats.

Useful Equations

One-dimensional flow equations:	$\rho_1 V_1 A_1 = \rho_2 V_2 A_2 = \rho V A = \dot{m} = \text{constant}$	**(13.1a)**	Page 694
	$R_x + p_1 A_1 - p_2 A_2 = \dot{m} V_2 - \dot{m} V_1$	**(13.1b)**	
	$\frac{\delta Q}{dm} + h_1 + \frac{V_1^2}{2} = h_2 + \frac{V_2^2}{2}$	**(13.1c)**	
	$\dot{m}(s_2 - s_1) \geq \int_{CS} \frac{1}{T}\left(\frac{\dot{Q}}{A}\right) dA$	**(13.1d)**	
	$p = \rho R T$	**(13.1e)**	
	$\Delta h = h_2 - h_1 = c_p \Delta T = c_p(T_2 - T_1)$	**(13.1f)**	
	$\Delta s = s_2 - s_1 = c_p \ln \frac{T_2}{T_1} - R \ln \frac{p_2}{p_1}$	**(13.1g)**	
Isentropic relations: [Note: These equations are a little cumbersome for practical use by hand. They are listed (and tabulated and plotted for air) in Appendix E. You are urged to download the *Excel* add-ins from the Web site for use in computing with these equations.]	$\frac{p_0}{p} = f(M)$	**(13.7a)**	Page 700
	$\frac{T_0}{T} = f(M)$	**(13.7b)**	
	$\frac{\rho_0}{\rho} = f(M)$	**(13.7c)**	
	$\frac{A}{A^*} = f(M)$	**(13.7d)**	Page 701

*This topic applies to sections that may be omitted without loss of continuity in the text material.

Pressure ratio for choked converging nozzle, $p_e/p_0\|_{\text{choked}}$:	$\left.\frac{p_e}{p_0}\right\|_{\text{choked}} = \frac{p^*}{p_0} = \left(\frac{2}{k+1}\right)^{k/(k-1)}$	**(13.8)**	Page 706
Mass flow rate for choked converging nozzle:	$\dot{m}_{\text{choked}} = A_e p_0 \sqrt{\frac{k}{RT_0}}\left(\frac{2}{k+1}\right)^{(k+1)/2(k-1)}$	**(13.9a)**	Page 706
Mass flow rate for choked converging nozzle (SI units):	$\dot{m}_{\text{choked}} = 0.04\frac{A_e p_0}{\sqrt{T_0}}$	**(13.9b)**	Page 706
Mass flow rate for choked converging nozzle (English Engineering units):	$\dot{m}_{\text{choked}} = 76.6\frac{A_e p_0}{\sqrt{T_0}}$	**(13.9c)**	Page 706
Mass flow rate for choked converging-diverging nozzle:	$\dot{m}_{\text{choked}} = A_t p_0 \sqrt{\frac{k}{RT_0}}\left(\frac{2}{k+1}\right)^{(k+1)/2(k-1)}$	**(13.10a)**	Page 710
Mass flow rate for choked converging-diverging nozzle (SI units):	$\dot{m}_{\text{choked}} = 0.04\frac{A_t p_0}{\sqrt{T_0}}$	**(13.10b)**	Page 711
Mass flow rate for choked converging-diverging nozzle (English Engineering units):	$\dot{m}_{\text{choked}} = 76.6\frac{A_t p_0}{\sqrt{T_0}}$	**(13.10c)**	Page 711
Normal shock relations: [Note: These equations are too cumbersome for practical use by hand. They are listed (and tabulated and plotted for air) in Appendix E. You are urged to download the *Excel* add-ins from the Web site for use in computing with these equations.]	$M_2 = f(M_1)$	**(13.20a)**	Page 721
	$\frac{p_{0_2}}{p_{0_1}} = f(M_1)$	**(13.20b)**	Page 722
	$\frac{T_2}{T_1} = f(M_1)$	**(13.20c)**	
	$\frac{p_2}{p_1} = f(M_1)$	**(13.20d)**	
	$\frac{\rho_2}{\rho_1} = \frac{V_1}{V_2} = f(M_1)$	**(13.20e)**	
Useful relations for determining the normal shock location in converging-diverging nozzle:	$\frac{p_e}{p_{0_1}}\frac{A_e}{A_t} = \frac{p_e}{p_{0_e}}\frac{A_e}{A_e^*}$	**(13.22)**	Page 726
	$\frac{p_{0_2}}{p_{0_1}} = \frac{A_t}{A_e}\frac{A_e}{A_e^*}$	**(13.23)**	
Fanno-line relations (friction): [Note: These equations are too cumbersome for practical use by hand. They are listed (and tabulated and plotted for air) in Appendix E. You are urged to download the *Excel* add-ins from the Web site for use in computing with these equations.]	$\frac{\bar{f}L_{\max}}{D_h} = f(M)$	**(13.34a)**	Page 737
	$\frac{T}{T^*} = f(M)$	**(13.34b)**	
	$\frac{V}{V^*} = \frac{\rho^*}{\rho} = f(M)$	**(13.34c)**	

	$\frac{p}{p^*} = f(M)$	**(13.34d)**	
	$\frac{p_0}{p_0^*} = f(M)$	**(13.34e)**	
Isothermal flow relations (friction): [Note: These equations are too cumbersome for practical use by hand. They are listed (and tabulated and plotted for air) in Appendix E. You are urged to download the *Excel* add-ins from the Web site for use in computing with these equations.]	$\frac{p_2}{p_1} = \frac{\rho_2}{\rho_1} = \frac{V_1}{V_2} = \frac{M_1}{M_2}$	**(13.42a)**	Page W-26
	$\frac{T_{0_2}}{T_{0_1}} = f\left(\frac{M_1}{M_2}\right)$	**(13.42b)**	
	$\frac{fL}{D_h} = f\left(\frac{M_1}{M_2}\right)$	**(13.42c)**	
Rayleigh-line relations (heat transfer): [Note: These equations are too cumbersome for practical use by hand. They are listed (and tabulated and plotted for air) in Appendix E. You are urged to download the *Excel* add-ins from the Web site for use in computing with these equations.]	$\frac{p}{p^*} = f(M)$	**(13.44a)**	Page 747
	$\frac{T}{T^*} = f(M)$	**(13.44b)**	
	$\frac{\rho^*}{\rho} = \frac{V}{V^*} = f(M)$	**(13.44c)**	
	$\frac{T_0}{T_0^*} = f(M)$	**(13.44d)**	
	$\frac{p_0}{p_0^*} = f(M)$	**(13.44e)**	Page 748
Geometric relations, oblique shock:	$M_{1_n} = \frac{V_{1_n}}{c_1} = M_1 \sin\beta$	**(13.47a)**	Page 753
	$M_{2_n} = \frac{V_{2_n}}{c_2} = M_2 \sin(\beta - \theta)$	**(13.47b)**	
Oblique shock relations: [Note: These equations are too cumbersome for practical use by hand. They are listed (and tabulated and plotted for air) in Appendix E. You are urged to download the *Excel* add-ins from the Web site for use in computing with these equations.]	$M_{2_n} = f(M_{1_n})$	**(13.48a)**	Page 754
	$\frac{p_{0_2}}{p_{0_1}} = f(M_{1_n})$	**(13.48b)**	
	$\frac{T_2}{T_1} = f(M_{1_n})$	**(13.48c)**	
	$\frac{p_2}{p_1} = f(M_{1_n})$	**(13.48d)**	
	$\frac{\rho_2}{\rho_1} = \frac{V_{1_n}}{V_{2_n}} = f(M_{1_n})$	**(13.48e)**	
Relation of β, θ, and M_1 for oblique shock:	$\tan\theta = \frac{2\cot\beta(M_1^2\sin^2\beta - 1)}{M_1^2(k + \cos 2\beta) + 2}$	**(13.49)**	Page 757
Prandtl-Meyer expansion function, ω:	$\omega = \sqrt{\frac{k+1}{k-1}}\tan^{-1}\left(\sqrt{\frac{k-1}{k+1}(M^2-1)}\right) - \tan^{-1}(\sqrt{M^2-1})$	**(13.55)**	Page 762

Case Study

The X-43A/Hyper-X Airplane

The X-43A/Hyper-X at $M = 7$ (CFD image showing pressure contours). (Courtesy of NASA.)

Superman is faster than a speeding bullet. So how fast is that? It turns out that the highest speed of a bullet is about 1500 m/s, or about Mach 4.5 at sea level. Can humans keep up with Superman? If we're in orbit we can (What is the Mach number of the Space Shuttle in orbit?—it's a trick question!), because there's no drag—once we get up to speed, we can stay there—but flying at hypersonic speeds (i.e., above about $M = 5$) in the atmosphere requires tremendous engine thrust and an engine that can function at all at such speeds. In 2004, an air-breathing X-43A managed to fly at almost Mach 10, or about 11,200 km/h. The hypersonic scramjet engine in this airplane is actually integrated into the airframe, and the entire lower surface of the vehicle is shaped to make the engine work. The bulge on the underside in the figure is the engine. Unlike the turbojet engines used in many jet aircraft, which have fans and compressors as major components, the scramjet, amazingly, has no moving parts, so if you were to look inside it there wouldn't be much to see! Instead it uses geometry to develop a shock train that reduces the speed of the airflow from hypersonic to supersonic velocities. The scramjet, which is essentially a ramjet with supersonic combustion, doesn't need to slow the flow down to subsonic speeds. The compression ram on the undersurface of the aircraft slows the flow down from hypersonic to supersonic speed before it reaches the engine. It does this by causing a sequence of oblique shocks (which we discussed in this chapter) that successively slow the flow down and also increase the air density. As the supersonic, relatively high-density air passes through the engine, hydrogen fuel is injected and combusts, creating tremendous thrust at the exhaust. Once at hypersonic speed, the combustion process is self-sustaining.

One of the problems the engineers faced was how to start the engine. First, the airplane has to be accelerated above Mach 4 by conventional means (by a jet engine or rocket, or by piggy-backing another aircraft), and then the scramjet fuel can be started and ignited. This sounds simple enough, but the ignition process has been compared to "lighting a match in a hurricane"! The solution was to ignite using a mixture of pyrophoric silane (which auto-ignites when exposed to air) and hydrogen, then switch to pure hydrogen.

The X-43A/Hyper-X is experimental, but in future we may expect to see scramjets in military applications (aircraft and missiles), then possibly in commercial aircraft. Conceivably, you could live in New York, go to a meeting in Los Angeles, and be back in New York for dinner!

References

1. *HP 48G Series User's Guide*, Hewlett-Packard Company, Corvallis Division, 1000 N.E. Circle Blvd., Corvallis, OR 97330.
2. *Isentropic Calculator* (http://www.aoe.vt.edu/aoe3114/calc.html), William Devenport, Aerospace and Ocean Engineering, Virginia Polytechnic Institute and State University.
3. Hermann, R., *Supersonic Inlet Diffusers*. Minneapolis, MN: Minneapolis-Honeywell Regulator Co., Aeronautical Division, 1956.
4. Runstadler, P.W., Jr., "Diffuser Data Book," Creare, Inc., Hanover, NH, Technical Note 186, 1975.
5. Seddon, J., and E. L. Goldsmith, *Intake Aerodynamics*. New York: American Institute of Aeronautics and Astronautics, 1985.
6. Shapiro, A. H., *The Dynamics and Thermodynamics of Compressible Fluid Flow*, Vol. *1*. New York: Ronald Press, 1953.
7. Zucrow, M. J., and J. D. Hoffman, *Compressible Flow*, Vol. *1*. New York: Wiley, 1976.
8. Baals, D. W., and W. R. Corliss, *Wind Tunnels of NASA*. Washington, D.C.: National Aeronautics and Space Administration, SP-440, 1981.
9. Pope, A., and K. L. Goin, *High-Speed Wind Tunnel Testing*. New York: Krieger. 1978.
10. Glass, I.I., "Some Aspects of Shock-Wave Research," *AIAA J., 25*, 2, February 1987, pp. 214–229.

Problems

Isentropic Flow—Area Variation

Most of the problems in this chapter involve computation of isentropic, Fanno, Rayleigh, normal shock, oblique shock, or isentropic expansion wave effects. The Web site for the text has associated *Excel* workbooks for each of these phenomena, and these are recommended for use while solving the problems (the Web site also has *Excel* add-in functions, which the reader is urged to download and install). To avoid needless duplication, the computer symbol will only be used next to problems when *Excel* has an *additional* benefit (e.g., for graphing).

13.1 Air is extracted from a large tank in which the temperature and pressure are 70°C and 101 kPa (abs), respectively, through a nozzle. At one location in the nozzle the static pressure is 25 kPa and the diameter is 15 cm. What is the mass flow rate? Assume isentropic flow.

13.2 Steam flows steadily and isentropically through a nozzle. At an upstream section where the speed is negligible, the temperature and pressure are 48°C and 6.2 MPa (abs). At a section where the nozzle diameter is 4.8 mm, the steam pressure is 4.1 MPa (abs). Determine the speed and Mach number at this section and the mass flow rate of steam. Sketch the passage shape.

13.3 Steam flows steadily and isentropically through a nozzle. At an upstream section where the speed is negligible, the temperature and pressure are 450°C and 6 MPa (abs). At a section where the nozzle diameter is 2 cm, the steam pressure is 2 MPa (abs). Determine the speed and Mach number at this section and the mass flow rate of steam. Sketch the passage shape.

13.4 Nitrogen flows through a diverging section of duct with $A_1 = 0.15$ m^2 and $A_2 = 0.45$ m^2. If $M_1 = 0.7$ and $p_1 = 450$ kPa, find M_2 and p_2.

13.5 Nitrogen flows through a diverging section of duct with $A_1 = 0.15$ m^2 and $A_2 = 0.45$ m^2. If $M_1 = 1.7$ and $T_1 = 30$°C, find M_2 and T_2.

13.6 At a section in a passage, the pressure is 150 kPa (abs), the temperature is 10°C, and the speed is 120 m/s. For isentropic flow of air, determine the Mach number at the point where the pressure is 50 kPa (abs). Sketch the passage shape.

13.7 At a section in a passage, the pressure is 207 kPa (abs), the temperature is 38°C, and the speed is 533 m/s. At a section downstream the Mach number is 2.5. Determine the pressure at this downstream location for isentropic flow of air. Sketch the passage shape.

13.8 Oxygen flows into an insulated duct with initial conditions of 200 kPa, 420 K, and 200 m/s. The area changes from $A_1 = 0.6$ m^2 to $A_2 = 0.5$ m^2. Compute M_1, p_{0_1}, and T_{0_1}. Is this duct a nozzle or a diffuser? Calculate the exit conditions (pressure, temperature, and Mach number) provided that there are no losses.

13.9 Air is flowing in an adiabatic system at 9.07 kg/s. At one section, the pressure is 206.8 kPa, the temperature is 648.9°C, and the area is 51.61 cm^2. At a downstream section, $M_2 = 1.2$. Sketch the flow passage. Find the exit area provided the flow is reversible.

13.10 Air flows isentropically through a converging-diverging nozzle from a large tank containing air at 250°C. At two locations where the area is 1 cm^2, the static pressures are 200 kPa and 50 kPa. Find the mass flow rate, the throat area, and the Mach numbers at the two locations.

13.11 Air flows steadily and isentropically through a passage. At section ①, where the cross-sectional area is 0.02 m^2, the air is at 40.0 kPa (abs), 60°C, and $M = 2.0$. At section ② downstream, the speed is 519 m/s. Calculate the Mach number at section ②. Sketch the shape of the passage between sections ① and ②.

13.12 Air flows steadily and isentropically through a passage at 68 kg/s. At the section where the diameter is $D = 0.9$ m, $M = 1.75$, $T = 0$°C, and $p = 172.2$ kPa. Determine the speed and cross-sectional area downstream where $T = 107$°C. Sketch the flow passage.

13.13 Air, at an absolute pressure of 60.0 kPa and 27°C, enters a passage at 486 m/s, where $A = 0.02$ m^2. At section ② downstream, $p = 78.8$ kPa (abs). Assuming isentropic flow, calculate the Mach number at section ②. Sketch the flow passage.

13.14 Air flows adiabatically through a duct. At the entrance, the static temperature and pressure are 310 K and 200 kPa, respectively. At the exit, the static and stagnation temperatures are 294 K and 316 K, respectively, and the static pressure is 125 kPa. Find (a) the Mach numbers of the flow at the entrance and exit and (b) the area ratio A_2/A_1.

13.15 Atmospheric air (101 kPa and 20°C) is drawn into a receiving pipe via a converging nozzle. The throat cross-section diameter is 1 cm. Plot the mass flow rate delivered for the receiving pipe pressure ranging from 100 kPa down to 5 kPa.

13.16 For isentropic flow of air, at a section in a passage, $A = 0.25$ m^2, $p = 15$ kPa (abs), $T = 10$°C, and $V = 590$ m/s. Find the Mach number and the mass flow rate. At a section downstream the temperature is 137°C and the Mach number is 0.75. Determine the cross-sectional area and pressure at this downstream location. Sketch the passage shape.

13.17 A passage is designed to expand air isentropically to atmospheric pressure from a large tank in which properties are held constant at 5.3°C and 310 kPa. The desired flow rate is 1 kg/s. Assuming the passage is 6.1 m long, and that the Mach number increases linearly with position in the passage, plot the cross-sectional area and pressure as functions of position.

13.18 Repeat Problem 13.15 if the converging nozzle is replaced with a converging-diverging nozzle with an exit diameter of 2.5 cm (same throat area).

13.19 Air flows isentropically through a converging nozzle into a receiver where the pressure is 250 kPa (abs). If the pressure is 350 kPa (abs) and the speed is 150 m/s at the nozzle location where the Mach number is 0.5, determine the pressure, speed, and Mach number at the nozzle throat.

13.20 Air flows isentropically through a converging nozzle into a receiver in which the absolute pressure is 241 kPa (abs). The air enters the nozzle with negligible speed at a pressure of 413 kPa (abs) and a temperature of 93°C. Determine the mass flow rate through the nozzle for a throat diameter of 100 mm.

13.21 Air flows through a diverging duct. At the entrance to the duct, the Mach number is 1 and the area is 0.2 m^2. At the exit to the duct, the area is 0.5 m^2. What are the two possible exit Mach numbers for this duct?

13.22 Air is flowing steadily through a series of three tanks. The first very large tank contains air at 650 kPa and 35°C. Air flows from it to a second tank through a converging nozzle with exit area 1 cm^2. Finally the air flows from the second tank to a third very large tank through an identical nozzle. The flow rate through the two nozzles is the same, and the flow in them is isentropic. The pressure in the third tank is 65 kPa. Find the mass flow rate, and the pressure in the second tank.

13.23 Air flowing isentropically through a converging nozzle discharges to the atmosphere. At the section where the absolute pressure is 250 kPa, the temperature is 20°C and the air speed is 200 m/s. Determine the nozzle throat pressure.

13.24 Air flows from a large tank (p = 650 kPa (abs), T = 550°C) through a converging nozzle, with a throat area of 600 mm^2, and discharges to the atmosphere. Determine the mass rate of flow for isentropic flow through the nozzle.

13.25 Air flowing isentropically through a converging nozzle discharges to the atmosphere. At a section the area is $A = 0.05\ m^2$, T = 3.3°C, and V = 200 m/s. If the flow is just choked, find the pressure and the Mach number at this location. What is the throat area? What is the mass flow rate?

13.26 A converging nozzle is connected to a large tank that contains compressed air at 15°C. The nozzle exit area is 0.001 m^2. The exhaust is discharged to the atmosphere. To obtain a satisfactory shadow photograph of the flow pattern leaving the nozzle exit, the pressure in the exit plane must be greater than 325 kPa (gage). What pressure is required in the tank? What mass flow rate of air must be supplied if the system is to run continuously? Show static and stagnation state points on a Ts diagram.

13.27 Air, with p_0 = 650 kPa (abs) and T_0 = 350 K, flows isentropically through a converging nozzle. At the section in the nozzle where the area is $2.6 \times 10^{-3}\ m^2$, the Mach number is 0.5. The nozzle discharges to a back pressure of 270 kPa (abs). Determine the exit area of the nozzle.

13.28 Air flows through a converging duct. At the entrance, the static temperature is 250 K, the static pressure is 310 kPa, the stagnation pressure is 350 kPa, and the area is 0.36 m^2. At the exit, the area is 0.27 m^2. Assuming isentropic flow through the duct, what are the exit temperature and the mass flow rate of air through the duct?

13.29 Air at 0°C is contained in a large tank on the space shuttle. A converging section with exit area $1 \times 10^{-3}\ m^2$ is attached to the tank, through which the air exits to space at a rate of 2 kg/s. What are the pressure in the tank, and the pressure, temperature, and speed at the exit?

13.30 A large tank supplies air to a converging nozzle that discharges to atmospheric pressure. Assume the flow is reversible and adiabatic. For what range of tank pressures will the flow at the nozzle exit be sonic? If the tank pressure is 600 kPa (abs) and the temperature is 600 K, determine the mass flow rate through the nozzle, if the exit area is $1.29 \times 10^{-3}\ m^2$.

13.31 Nitrogen is stored in a large chamber at 450 K and 150 kPa. The gas leaves the chamber through a converging-only nozzle with an outlet area of 30 cm^2. The ambient room pressure is 100 kPa, and the flow through the nozzle is isentropic. What is the mass flow rate of the nitrogen? If the room pressure could be lowered, what is the maximum possible mass flow rate for the nitrogen?

13.32 A large tank initially is evacuated to −10 kPa (gage). (Ambient conditions are 101 kPa at 20°C.) At t = 0, an orifice of 5 mm diameter is opened in the tank wall; the vena contracta area is 65 percent of the geometric area. Calculate the mass flow rate at which air initially enters the tank. Show the process on a Ts diagram. Make a schematic plot of mass flow rate as a function of time. Explain why the plot is nonlinear.

13.33 A 50 cm diameter spherical cavity initially is evacuated. The cavity is to be filled with air for a combustion experiment. The pressure is to be 45 kPa (abs), measured after its temperature reaches T_{atm}. Assume the valve on the cavity is a converging nozzle with throat diameter of 1 mm, and the surrounding air is at standard conditions. For how long should the valve be opened to achieve the desired final pressure in the cavity? Calculate the entropy change for the air in the cavity.

13.34 Air flows isentropically through a converging nozzle attached to a large tank, where the absolute pressure is 171 kPa and the temperature is 27°C. At the inlet section the Mach number is 0.2. The nozzle discharges to the atmosphere; the discharge area is 0.015 m^2. Determine the magnitude and direction of the force that must be applied to hold the nozzle in place.

13.35 Consider a "rocket cart" propelled by a jet supplied from a tank of compressed air on the cart. Initially, air in the tank is at 1.3 MPa (abs) and 20°C, and the mass of the cart and tank is M_0 = 25 kg. The air exhausts through a converging nozzle with exit area A_e = 30 mm^2. Rolling resistance of the cart is F_R = 6 N; aerodynamic resistance is negligible. For the instant after air begins to flow through the nozzle: (a) compute the pressure in the nozzle exit plane, (b) evaluate the mass flow rate of air through the nozzle, and (c) calculate the acceleration of the tank and cart assembly.

13.36 A stream of air flowing in a duct ($A = 5 \times 10^{-4}\ m^2$) is at p = 300 kPa (abs), has M = 0.5, and flows at $\dot{m}$ = 0.25 kg/s. Determine the local isentropic stagnation temperature. If the cross-sectional area of the passage were reduced downstream, determine the maximum percentage reduction of area allowable without reducing the flow rate (assume isentropic flow). Determine the speed and pressure at the minimum area location.

13.37 An air-jet-driven experimental rocket of 25 kg mass is to be launched from the space shuttle into space. The temperature of the air in the rocket's tank is 125°C. A converging section with exit area 25 mm^2 is attached to the tank, through which the air exits to space at a rate of 0.05 kg/s. What is the pressure in the tank, and the pressure, temperature, and air speed at the exit when the rocket is first released? What is the initial acceleration of the rocket?

13.38 Air enters a converging-diverging nozzle at 2 MPa (abs) and 313 K. At the exit of the nozzle, the pressure is 200 kPa (abs). Assume adiabatic, frictionless flow through the nozzle. The throat area is 20 cm^2. What is the area at the nozzle exit? What is the mass flow rate of the air?

13.39 Hydrogen is expanded adiabatically, without friction from 345 kPa, at 282.2°C, and at negligible velocity to 137.9 kPa via a converging-diverging nozzle. What is the exit Mach number?

13.40 A cylinder of gas used for welding contains helium at 20 MPa (gage) and room temperature. The cylinder is knocked over, its valve is broken off, and gas escapes through a converging passage. The minimum flow diameter is 10 mm at the outlet section where the gas flow is uniform. Find (a) the mass flow rate at which gas leaves the cylinder and (b) the instantaneous acceleration of the cylinder (assume the cylinder axis is horizontal and its mass is 65 kg). Show static and stagnation states and the process path on a *Ts* diagram.

13.41 A converging nozzle is bolted to the side of a large tank. Air inside the tank is maintained at a constant 345 kPa (abs) and 38°C. The inlet area of the nozzle is 6450 mm^2 and the exit area is 645 mm^2 The nozzle discharges to the atmosphere. For isentropic flow in the nozzle, determine the total force on the bolts, and indicate whether the bolts are in tension or compression.

13.42 An insulated spherical air tank with diameter $D = 2$ m is used in a blowdown installation. Initially the tank is charged to 2.75 MPa (abs) at 450 K. The mass flow rate of air from the tank is a function of time; during the first 30 s of blowdown 30 kg of air leaves the tank. Determine the air temperature in the tank after 30 s of blowdown. Estimate the nozzle throat area.

13.43 An ideal gas, with $k = 1.25$, flows isentropically through the converging nozzle shown and discharges into a large duct where the pressure is $p_2 = 172$ kPa (abs). The gas is *not* air and the gas constant, R, is unknown. Flow is steady and uniform at all cross sections. Find the exit area of the nozzle, A_2, and the exit speed, V_2.

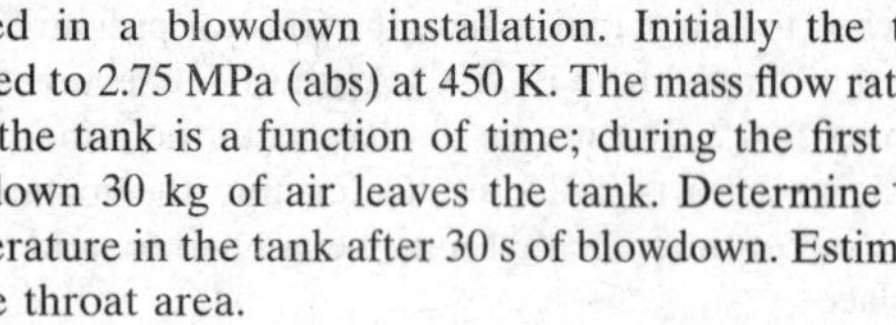

P13.43

13.44 A jet transport aircraft, with pressurized cabin, cruises at 11 km altitude. The cabin temperature and pressure initially are at 25°C and equivalent to 2.5 km altitude. The interior volume of the cabin is 25 m^3. Air escapes through a small hole with effective flow area of 0.002 m^2. Calculate the time required for the cabin pressure to decrease by 40 percent. Plot the cabin pressure as a function of time.

13.45 At some point upstream of the throat of a converging-diverging duct, air flows at a speed of 15.24 m/s, with pressure and temperature of 103.4 kPa and 21°C, respectively. If the throat area is 0.09 m^2, and the discharge from the duct is supersonic, find the mass flow rate of air, assuming frictionless, adiabatic flow.

13.46 A converging-diverging nozzle is attached to a very large tank of air in which the pressure is 150 kPa and the temperature is 35°C. The nozzle exhausts to the atmosphere where the pressure is 101 kPa. The exit diameter of the nozzle is 2.75 cm. What is the flow rate through the nozzle? Assume the flow is isentropic.

13.47 A large insulated tank, pressurized to 620 kPa (gage), supplies air to a converging nozzle which discharges to atmosphere. The initial temperature in the tank is 127°C. When flow through the nozzle is initiated, what is the Mach number in the exit plane of the nozzle? What is the pressure in the exit plane when the flow is initiated? At what condition will the exit-plane Mach number change? How will the exit-plane pressure vary with time? How will flow rate through the nozzle vary with time? What would you estimate the air temperature in the tank to be when flow through the nozzle approaches zero?

13.48 Air escapes from a high-pressure bicycle tire through a hole with diameter $d = 0.254$ mm. The initial pressure in the tire is $p_1 = 620$ kPa (gage). (Assume the temperature remains constant at 27°C.) The internal volume of the tire is approximately 4.26×10^{-4} m^3, and is constant. Estimate the time needed for the pressure in the tire to drop to 310 kPa (gage). Compute the change in specific entropy of the air in the tire during this process. Plot the tire pressure as a function of time.

13.49 At the design condition of the system of Problem 13.46, the exit Mach number is $M_e = 2.0$. Find the pressure in the tank of Problem 13.46 (keeping the temperature constant) for this condition. What is the flow rate? What is the throat area?

13.50 When performing tests in a wind tunnel at conditions near Mach 1, the effects of model blockage become very important. Consider a wind tunnel with a test section of 0.09 m^2 cross section. If the test section conditions are $M = 1.20$ and $T = 21$°C, how much area blockage could be tolerated before the flow choked in the test section? If a model with 18.75 cm^2 projected frontal area were inserted in the tunnel, what would the air velocity be in the test section?

13.51 A pitot static probe is placed in a converging-diverging duct through which air flows. The duct is fed by a reservoir kept at 20°C. If the probe reads a static pressure of 75 kPa and a stagnation pressure of 100 kPa at a location where the area is 0.00645 m^2, what is the local velocity and the mass flow rate of air?

13.52 A converging-diverging nozzle, with a throat area of 1290 mm^2, is connected to a large tank in which air is kept at a pressure of 550 kPa and a temperature of 15°C. If the nozzle is to operate at design conditions (flow is isentropic) and the ambient pressure outside the nozzle is 89 kPa, calculate the exit area of the nozzle and the mass flow rate.

13.53 A converging-diverging nozzle, designed to expand air to $M = 3.0$, has a 250 mm^2 exit area. The nozzle is bolted to the side of a large tank and discharges to standard atmosphere. Air in the tank is pressurized to 4.5 MPa (gage) at

750 K. Assume flow within the nozzle is isentropic. Evaluate the pressure in the nozzle exit plane. Calculate the mass flow rate of air through the nozzle.

13.54 Methane is stored in a tank at 520 kPa and 27°C. It discharges to another tank via a converging-only nozzle, with exit area 625 mm^2. What is the initial mass flow rate of methane when the discharge tank is at a pressure of (a) 103.4 kPa, and (b) 413.69 kPa?

13.55 Air, at a stagnation pressure of 7.20 MPa (abs) and a stagnation temperature of 1100 K, flows isentropically through a converging-diverging nozzle having a throat area of 0.01 m^2. Determine the speed and the mass flow rate at the downstream section where the Mach number is 4.0.

13.56 Air is to be expanded through a converging-diverging nozzle by a frictionless adiabatic process, from a pressure of 1.10 MPa (abs) and a temperature of 115°C, to a pressure of 141 kPa (abs). Determine the throat and exit areas for a well-designed shockless nozzle, if the mass flow rate is 2 kg/s.

13.57 Air flows isentropically through a converging-diverging nozzle attached to a large tank, in which the pressure is 1.73 MPa and the temperature is 280 K. The nozzle is operating at design conditions for which the nozzle exit pressure, p_e, is equal to the surrounding atmospheric pressure, p_a. The exit area of the nozzle is A_e = 1016 mm^2. Calculate the flow rate through the nozzle. Plot the mass flow rate as the temperature of the tank is progressively increased to 1110 K (all pressures remaining the same). Explain this result (e.g., compare the mass flow rates at 280 K and 1110 K).

13.58 A small, solid fuel rocket motor is tested on a thrust stand. The chamber pressure and temperature are 4 MPa and 3250 K. The propulsion nozzle is designed to expand the exhaust gases isentropically to a pressure of 75 kPa. The nozzle exit diameter is 25 cm. Treat the gas as ideal with k = 1.25 and R = 300 J/(kg·K). Determine the mass flow rate of propellant gas and the thrust force exerted against the test stand.

13.59 Nitrogen, at a pressure and temperature of 371 kPa (abs) and 400 K, enters a nozzle with negligible speed. The exhaust jet is directed against a large flat plate that is perpendicular to the jet axis. The flow leaves the nozzle at atmospheric pressure. The exit area is 0.003 m^2. Find the force required to hold the plate.

13.60 A liquid rocket motor is fueled with hydrogen and oxygen. The chamber temperature and absolute pressure are 3300 K and 6.90 MPa. The nozzle is designed to expand the exhaust gases isentropically to a design back pressure corresponding to an altitude of 10 km on a standard day. The thrust produced by the motor is to be 100 kN at design conditions. Treat the exhaust gases as water vapor and assume ideal gas behavior. Determine the propellant mass flow rate needed to produce the desired thrust, the nozzle exit area, and the area ratio, A_e/A_t.

13.61 A small rocket motor, fueled with hydrogen and oxygen, is tested on a thrust stand at a simulated altitude of 10 km. The motor is operated at chamber stagnation conditions of 1500 K and 8.0 MPa (gage). The combustion product is water vapor, which may be treated as an ideal gas. Expansion occurs through a converging-diverging nozzle with design Mach number of 3.5 and exit area of 700 mm^2. Evaluate the pressure at the nozzle exit plane. Calculate the mass flow rate of exhaust gas. Determine the force exerted by the rocket motor on the thrust stand.

13.62 A CO_2 cartridge is used to propel a small rocket cart. Compressed gas, stored at 35 MPa and 20°C, is expanded through a smoothly contoured converging nozzle with 0.5 mm throat diameter. The back pressure is atmospheric. Calculate the pressure at the nozzle throat. Evaluate the mass flow rate of carbon dioxide through the nozzle. Determine the thrust available to propel the cart. How much would the thrust increase if a diverging section were added to the nozzle to expand the gas to atmospheric pressure? What is the exit area? Show stagnation states, static states, and the processes on a Ts diagram.

13.63 A rocket motor is being tested at sea level where the pressure is 101.3 kPa. The chamber pressure is 1206.6 kPa, the chamber temperature is 3000 K, and the nozzle has a throat area of 645 mm^2. The exhaust gas has a ratio of specific heats of k = 1.25 and a gas constant R = 376.96 N·m/kg·K. Assuming adiabatic, frictionless flow in the nozzle, determine (a) the nozzle exit area and velocity and (b) the thrust generated.

13.64 If the rocket motor of Problem 13.63 is modified by cutting off the diverging portion of the nozzle, what will be the exit pressure and thrust?

13.65 Consider the converging-diverging option of Problem 13.62. To what pressure would the compressed gas need to be raised (keeping the temperature at 20°C) to develop a thrust of 15N? (Assume isentropic flow.)

Normal Shocks

13.66 Testing of a demolition explosion is to be evaluated. Sensors indicate that the shock wave generated at the instant of explosion is 30 MPa (abs). If the explosion occurs in air at 20°C and 101 kPa, find the speed of the shock wave, and the temperature and speed of the air just after the shock passes. As an approximation assume k = 1.4. (Why is this an approximation?)

13.67 A standing normal shock occurs in air which is flowing at a Mach number of 1.75. What are the pressure and temperature ratios across the shock? What is the increase in entropy across the shock?

13.68 Air flows into a converging duct, and a normal shock stands at the exit of the duct. Downstream of the shock, the Mach number is 0.54. If $p_2/p_1 = 2$, compute the Mach number at the entrance of the duct and the area ratio A_1/A_2.

13.69 A normal shock occurs when a pitot-static tube is inserted into a supersonic wind tunnel. Pressures measured by the tube are p_{0_2} = 69 kPa (abs) and p_2 = 55 kPa (abs). Before the shock, T_1 = 158 K and p_1 = 10 kPa (abs). Calculate the air speed in the wind tunnel.

13.70 A large tank containing air at 860 kPa (abs) and 79°C is attached to a converging-diverging nozzle that has a throat area of 967 mm^2 through which the air is exiting. A normal shock sits at a point in the nozzle where the area is 1613 mm^2. The nozzle exit area is 2258 mm^2. What are the Mach numbers just after the shock and at the exit? What are the stagnation and static pressures before and after the shock?

13.71 A total-pressure probe is placed in a supersonic wind tunnel where $T = 294$ K and $M = 2.0$. A normal shock stands in front of the probe. Behind the shock, $M_2 = 0.577$ and $p_2 =$ 40 kPa (abs). Find (a) the downstream stagnation pressure and stagnation temperature and (b) all fluid properties upstream from the shock. Show static and stagnation state points and the process path on a Ts diagram.

13.72 Air flows steadily through a long, insulated constant-area pipe. At section ①, $M_1 = 2.0$, $T_1 = 60°$C, and $p_1 = 247$ kPa (abs). At section ②, downstream from a normal shock, $V_2 = 330$ m/s. Determine the density and Mach number at section ②. Make a qualitative sketch of the pressure distribution along the pipe.

13.73 A wind tunnel nozzle is designed to operate at a Mach number of 5. To check the flow velocity, a pitot probe is placed at the nozzle exit. Since the probe tip is blunt, a normal shock stands off the tip of the probe. If the nozzle exit static pressure is 10 kPa, what absolute pressure should the pitot probe measure? If the stagnation temperature before the nozzle is 1450 K, what is the nozzle exit velocity?

13.74 Air approaches a normal shock at $V_1 = 900$ m/s, $p_1 = 50$ kPa, and $T_1 = 220$ K. What are the velocity and pressure after the shock? What would the velocity and pressure be if the flow were decelerated isentropically to the same Mach number?

13.75 Air with stagnation conditions of 1034.21 kPa and 204.44°C accelerates through a converging-diverging nozzle with throat area 0.0019 m^2. A normal shock is located where the area is 0.0038 m^2. What is the Mach number before and after the shock? What is the rate of entropy generation through the nozzle, if there is negligible friction between the flow and the nozzle walls?

13.76 Air approaches a normal shock at $M_1 = 2.5$, with $T_{0_1} = 694$ K and $p_1 = 138$ kPa (abs). Determine the speed and temperature of the air leaving the shock and the entropy change across the shock.

13.77 Air undergoes a normal shock. Upstream, $T_1 = 35°$C, $p_1 = 229$ kPa (abs), and $V_1 = 704$ m/s. Determine the temperature and stagnation pressure of the air stream leaving the shock.

13.78 A normal shock stands in a constant-area duct. Air approaches the shock with $T_{0_1} = 550$ K, $p_{0_1} = 650$ kPa (abs), and $M_1 = 2.5$. Determine the static pressure downstream from the shock. Compare the downstream pressure with that reached by decelerating isentropically to the same subsonic Mach number.

13.79 A normal shock occurs in air at a section where $V_1 =$ 3200 km/h, $T_1 = -26°$C, and $p_1 = 34$ kPa (abs). Determine the speed and Mach number downstream from the shock, and the change in stagnation pressure across the shock.

13.80 Air approaches a normal shock with $T_1 = -22°$C, $p_1 =$ 101 kPa (abs), and $V_1 = 2800$ km/h. Determine the speed immediately downstream from the shock and the pressure change across the shock. Calculate the corresponding pressure change for a frictionless, shockless deceleration between the same speeds.

13.81 A supersonic aircraft cruises at $M = 2.2$ at 12 km altitude. A pitot tube is used to sense pressure for calculating air speed. A normal shock stands in front of the tube. Evaluate the local isentropic stagnation conditions in front of the shock. Estimate the stagnation pressure sensed by the pitot tube. Show static and stagnation state points and the process path on a Ts diagram.

13.82 The Concorde supersonic transport flew at $M = 2.2$ at 20 km altitude. Air is decelerated isentropically by the engine inlet system to a local Mach number of 1.3. The air passed through a normal shock and was decelerated further to $M = 0.4$ at the engine compressor section. Assume, as a first approximation, that this subsonic diffusion process was isentropic and use standard atmosphere data for freestream conditions. Determine the temperature, pressure, and stagnation pressure of the air entering the engine compressor.

13.83 Stagnation pressure and temperature probes are located on the nose of a supersonic aircraft. At 10,700 m altitude a normal shock stands in front of the probes. The temperature probe indicates $T_0 = 215°$C behind the shock. Calculate the Mach number and air speed of the plane. Find the static and stagnation pressures behind the shock. Show the process and the static and stagnation state points on a Ts diagram.

13.84 The NASA X-43A *Hyper-X* experimental hypersonic vehicle flew at Mach 9.68 at an altitude of 33,528 m. Stagnation pressure and temperature probes were located on the nose of the aircraft. A normal shock wave stood in front of these probes. Estimate the stagnation pressure and temperature measured by the probes.

13.85 Equations 13.20 are a useful set of equations for analyzing flow through a normal shock. Derive another useful equation, the Rankine-Hugoniot relation,

$$\frac{p_2}{p_1} = \frac{(k+1)\frac{\rho_2}{\rho_1} - (k-1)}{(k+1) - (k-1)\frac{\rho_2}{\rho_1}}$$

and use it to find the density ratio for air as $p_2/p_1 \to \infty$.

13.86 A supersonic aircraft cruises at $M = 2.7$ at 18,000 m altitude. A normal shock stands in front of a pitot tube on the aircraft; the tube senses a stagnation pressure of 72 kPa (abs). Calculate the static pressure and temperature behind the shock. Evaluate the loss in stagnation pressure through the shock. Determine the change in specific entropy across the shock. Show static and stagnation states and the process path on a Ts diagram.

13.87 An aircraft is in supersonic flight at 10 km altitude on a standard day. The true air speed of the plane is 659 m/s. Calculate the flight Mach number of the aircraft. A total-head tube attached to the plane is used to sense stagnation pressure which is converted to flight Mach number by an on-board computer. However, the computer programmer has ignored the normal shock that stands in front of the total-head tube and has assumed isentropic flow. Evaluate the pressure sensed by the total-head tube. Determine the erroneous air speed calculated by the computer program.

13.88 A supersonic aircraft flies at $M_1 = 2.7$ at 20 km altitude on a standard day. Air enters the engine inlet system, where it is slowed isentropically to $M_2 = 1.3$. A normal shock occurs at that location. The resulting subsonic flow is decelerated further to $M_4 = 0.40$. The subsonic diffusion is adiabatic but not isentropic; the final pressure is 104 kPa (abs). Evaluate (a) the stagnation temperature for the flow, (b) the pressure change across the shock, (c) the entropy change, $s_4 - s_1$, and (d) the final stagnation pressure. Sketch the process path on a Ts diagram, indicating all static and stagnation states.

13.89 A blast wave propagates outward from an explosion. At large radii, curvature is small and the wave may be treated as a strong normal shock. (The pressure and temperature rise associated with the blast wave decrease as the wave travels outward.) At one instant, a blast wave front travels at $M = 1.60$ with respect to undisturbed air at standard conditions. Find (a) the speed of the air behind the blast wave with respect to the wave and (b) the speed of the air behind the blast wave as seen by an observer on the ground. Draw a Ts diagram for the process as seen by an observer on the wave, indicating static and stagnation state points and property values.

Supersonic Channel Flow with Shocks

13.90 Consider a supersonic wind tunnel starting as shown. The nozzle throat area is 0.12 m^2, and the test section design Mach number is 2.50. As the tunnel starts, a normal shock stands in the divergence of the nozzle where the area is 0.28 m^2. Upstream stagnation conditions are $T_0 = 600$ K and $p_0 = 790$ kPa (abs). Find the minimum theoretically possible diffuser throat area at this instant. Calculate the entropy increase across the shock.

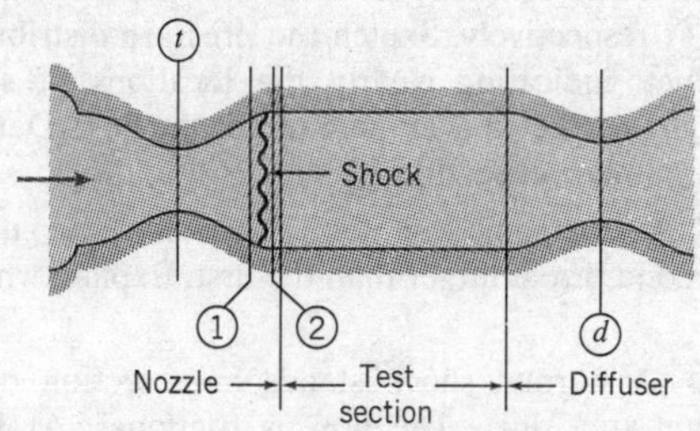

P13.90

13.91 Air enters a wind tunnel with stagnation conditions of 101.35 kPa and 23.89°C. The test section has a cross-sectional area of 0.0929 m^2 and a Mach number of 2.3. Find (a) the throat area of the nozzle, (b) the mass flow rate, (c) the pressure and temperature in the test section, and (d) the minimum possible throat area for the diffuser to ensure starting.

13.92 Air flows through a converging-diverging nozzle with $A_e/A_t = 3.5$. The upstream stagnation conditions are atmospheric; the back pressure is maintained by a vacuum pump. Determine the back pressure required to cause a normal shock to stand in the nozzle exit plane and the flow speed leaving the shock.

13.93 A supersonic wind tunnel is to be operated at $M = 2.2$ in the test section. Upstream from the test section, the nozzle throat area is 0.07 m^2. Air is supplied at stagnation conditions of 500 K and 1.0 MPa (abs). At one flow condition, while the tunnel is being brought up to speed, a normal shock stands at the nozzle exit plane. The flow is steady. For this *starting* condition, immediately downstream from the shock find (a) the Mach number, (b) the static pressure, (c) the stagnation pressure, and (d) the minimum area theoretically possible for the second throat downstream from the test section. On a Ts diagram show static and stagnation state points and the process path.

13.94 A converging-diverging nozzle is attached to a large tank of air, in which $T_{0_1} = 300$ K and $p_{0_1} = 250$ kPa (abs). At the nozzle throat the pressure is 132 kPa (abs). In the diverging section, the pressure falls to 68.1 kPa before rising suddenly across a normal shock. At the nozzle exit the pressure is 180 kPa. Find the Mach number immediately behind the shock. Determine the pressure immediately downstream from the shock. Calculate the entropy change across the shock. Sketch the Ts diagram for this flow, indicating static and stagnation state points for conditions at the nozzle throat, both sides of the shock, and the exit plane.

13.95 A converging-diverging nozzle expands air from 120°C and 348 kPa (abs) to 101 kPa (abs). The throat and exit plane areas are 517 and 592 mm^2, respectively. Calculate the exit Mach number. Evaluate the mass flow rate through the nozzle.

13.96 A converging-diverging nozzle, with throat area A_t = 645 mm^2, is attached to a large tank in which the pressure and temperature are maintained at 690 kPa (abs) and 330 K. The nozzle exit area is 1020 mm^2 Determine the exit Mach number at design conditions. Referring to Fig. 13.12, determine the back pressures corresponding to the boundaries of Regimes I, II, III, and IV. Sketch the corresponding plot for this nozzle.

13.97 A converging-diverging nozzle is designed to produce a Mach number of 2.5 with air. What operating pressure ratios ($p_b/p_{t\ inlet}$) will cause this nozzle to operate with isentropic flow throughout and supersonic flow at the exit (the so-called "third critical point"), with isentropic flow throughout and subsonic flow at the exit (the "first critical point"), and with a normal shock at the nozzle exit (the "second critical point")?

13.98 Oxygen flows through a converging-diverging nozzle with a exit-to-throat area ratio of 3.0. The stagnation pressure at the inlet is 827.37 kPa, and the back pressure is 344.74 kPa. Compute the pressure ratios for the nozzle and demonstrate that a normal shock wave should be located within the diverging portion of the nozzle. Compute the area ratio at which the shock occurs, the pre- and post-shock Mach numbers, and the Mach number at the nozzle exit.

13.99 A converging-diverging nozzle, with $A_e/A_t = 4.0$, is designed to expand air isentropically to atmospheric pressure. Determine the exit Mach number at design conditions and the required inlet stagnation pressure. Referring to Fig. 13.12, determine the back pressures that correspond to the boundaries of Regimes I, II, III, and IV. Sketch the plot of pressure ratio versus axial distance for this nozzle.

13.100 A normal shock occurs in the diverging section of a converging-diverging nozzle where $A = 25$ cm^2 and $M = 2.75$. Upstream, $T_0 = 550$ K and $p_0 = 700$ kPa (abs). The nozzle exit area is 40 cm^2. Assume the flow is isentropic except across the shock. Determine the nozzle exit pressure, throat area, and mass flow rate.

13.101 Air flows adiabatically from a reservoir, where T = 60°C and p = 600 kPa (abs), through a converging-diverging nozzle with A_e/A_t = 4.0. A normal shock occurs where M = 2.42. Assuming isentropic flow before and after the shock, determine the back pressure downstream from the nozzle. Sketch the pressure distribution.

13.102 A converging-diverging nozzle is designed to expand air isentropically to atmospheric pressure from a large tank, where T_0 = 150°C and p_0 = 790 kPa (abs). A normal shock stands in the diverging section, where p = 160 kPa (abs) and A = 600 mm^2. Determine the nozzle back pressure, exit area, and throat area.

13.103 A converging-diverging nozzle, with design pressure ratio p_e/p_0 = 0.128, is operated with a back pressure condition such that p_b/p_0 = 0.830, causing a normal shock to stand in the diverging section. Determine the Mach number at which the shock occurs.

13.104 Air flows through a converging-diverging nozzle, with A_e/A_t = 3.5. The upstream stagnation conditions are atmospheric; the back pressure is maintained by a vacuum system. Determine the range of back pressures for which a normal shock will occur within the nozzle and the corresponding mass flow rate, if A_t = 500 mm^2.

13.105 A converging-diverging nozzle, with A_e/A_t = 1.633, is designed to operate with atmospheric pressure at the exit plane. Determine the range(s) of inlet stagnation pressures for which the nozzle will be free from normal shocks.

13.106 Air flows through a converging-diverging nozzle with A_e/A_t = 1.87. Upstream, T_{0_1} = 115°C and p_{0_1} = 690 kPa (abs). The back pressure is maintained at 275 kPa. Determine the Mach number and flow speed in the nozzle exit plane.

13.107 A normal shock occurs in the diverging section of a converging-diverging nozzle where A = 2580 mm^2 and M = 2.00. Upstream, T_{0_1} = 555 K and p_{0_1} = 690 kPa. The nozzle exit area is 3870 mm^2 Assume that flow is isentropic except across the shock. Find the nozzle exit pressure. Show the processes on a Ts diagram, and indicate the static and stagnation state points.

13.108 Consider flow of air through a converging-diverging nozzle. Sketch the approximate behavior of the mass flow rate versus back pressure ratio, p_b/p_0. Sketch the variation of pressure with distance along the nozzle, and the Ts diagram for the nozzle flow, when the back pressure is p^*.

13.109 Air enters a converging-diverging nozzle with an area ratio of 1.76. Entrance stagnation conditions are 1034 kPa and 93.3°C. A normal shock stands at a location where the area is 1.2 times the throat area. Determine the exit Mach number and static pressure. What is the design point exit pressure?

13.110 A stationary normal shock stands in the diverging section of a converging-diverging nozzle. The Mach number ahead of the shock is 3.0. The nozzle area at the shock is 500 mm^2. The nozzle is fed from a large tank where the pressure is 1000 kPa (gage) and the temperature is 400 K. Find the Mach number, stagnation pressure, and static pressure after the shock. Calculate the nozzle throat area. Evaluate the entropy change across the shock. Finally, if the nozzle exit area is 600 mm^2, estimate the exit Mach number. Would the actual exit Mach number be higher, lower, or the same as your estimate? Why?

13.111 Air flows adiabatically from a reservoir, where T_{0_1} = 60°C and p_{0_1} = 600 kPa (abs), through a converging-diverging nozzle. The design Mach number of the nozzle is 2.94. A normal shock occurs at the location in the nozzle where M = 2.42. Assuming isentropic flow before and after the shock, determine the back pressure downstream from the nozzle. Sketch the pressure distribution.

***13.112** Air flows through a converging-diverging nozzle with an area ratio of 2.5. Stagnation conditions at the inlet are 1 MPa and 320 K. A constant-area, adiabatic duct with $L/D = 10$ and $f = 0.03$ is attached to the nozzle outlet. (a) Compute the back pressure that would place a normal shock at the nozzle exit. (b) What back pressure would place the normal shock at the duct exit? (c) What back pressure would result in shock-free flow?

***13.113** Consider the setup of Problem 13.112, except that the constant-area duct is frictionless and no longer adiabatic. A normal shock stands at the duct exit, after which the temperature is 350 K. Calculate the Mach number after the shock wave, and the heat addition in the constant-area duct.

***13.114** A normal shock stands in a section of insulated constant-area duct. The flow is frictional. At section ①, some distance upstream from the shock, T_1 = 260 K. At section ④, some distance downstream from the shock, T_4 = 417 K and M_4 = 1.0. Denote conditions immediately upstream and downstream from the shock by subscripts ② and ③, respectively. Sketch the pressure distribution along the duct, indicating clearly the locations of sections ① through ④. Sketch a Ts diagram for the flow. Determine the Mach number at section ①.

***13.115** A supersonic wind tunnel must have two throats, with the second throat larger than the first. Explain why this must be so.

***13.116** A normal shock stands in a section of insulated constant-area duct. The flow is frictional. At section ①, some distance upstream from the shock, T_1 = 370 K, p_{0_1} = 540 kPa (abs), and M_1 = 2.05. At section ④, some distance downstream from the shock, M_4 = 1.00. Calculate the air speed, V_2, immediately ahead of the shock, where T_2 = 198°C. Evaluate the entropy change, $s_4 - s_1$.

Flow with Friction

13.117 Nitrogen is discharged from a 30-cm-diameter duct at $M_2 = 0.85$, $T_2 = 300$ K, and $p_2 = 200$ kPa. The temperature at the inlet of the duct is $T_1 = 330$ K. Compute the pressure at the inlet and the mass flow rate.

13.118 Room air is drawn into an insulated duct of constant area through a smoothly contoured converging nozzle. Room conditions are T = 27°C and p = 101.3 kPa (abs). The duct diameter is D = 25 mm. The pressure at the duct inlet (nozzle outlet) is p_1 = 90 kPa (abs). Find (a) the mass flow rate in the duct and (b) the range of exit pressures for which the duct exit flow is choked.

*These problems require material from sections that may be omitted without loss of continuity in the text material.

13.119 Air from a large reservoir at 170 kPa (abs) and 120°C flows isentropically through a converging nozzle into an insulated pipe at 165 kPa (abs). The pipe flow experiences friction effects. Obtain a plot of the *Ts* diagram for this flow, until $M = 1$. Also plot the pressure and speed distributions from the entrance to the location at which $M = 1$.

13.120 Repeat Problem 13.119 except the nozzle is now a converging-diverging nozzle delivering the air to the pipe at 17 kPa (abs).

13.121 A 5 m duct 35 cm in diameter contains oxygen flowing at the rate of 40 kg/s. The inlet conditions are $p_1 = 200$ kPa and $T_1 = 450$ K. The exit pressure is $p_2 = 160$ kPa. Calculate the inlet and exit Mach number, and the exit stagnation pressure and temperature. Determine the friction factor, and estimate the absolute roughness of the duct material.

13.122 Air flows steadily and adiabatically from a large tank through a converging nozzle connected to an insulated constant-area duct. The nozzle may be considered frictionless. Air in the tank is at $p = 1$ MPa (abs) and $T = 120$°C. The absolute pressure at the nozzle exit (duct inlet) is 860 kPa (abs). Determine the pressure at the end of the duct, if the temperature there is 65°C. Find the entropy increase.

13.123 A Fanno-line flow apparatus in an undergraduate fluid mechanics laboratory consists of a smooth brass tube of 7.16 mm inside diameter, fed by a converging nozzle. The lab temperature and uncorrected barometer reading are 23.5°C and 755.1 mm of mercury. The pressure at the exit from the converging nozzle (entrance to the constant-area duct) is −20.8 mm of mercury (gage). Compute the Mach number at the entrance to the constant-area tube. Calculate the mass flow rate in the tube. Evaluate the pressure at the location in the tube where the Mach number is 0.4.

13.124 Measurements are made of compressible flow in a long smooth 7.16 mm i.d. tube. Air is drawn from the surroundings (20°C and 101 kPa) by a vacuum pump downstream. Pressure readings along the tube become steady when the downstream pressure is reduced to 626 mm Hg (vacuum) or below. For these conditions, determine (a) the maximum mass flow rate possible through the tube, (b) the stagnation pressure of the air leaving the tube, and (c) the entropy change of the air in the tube. Show static and stagnation state points and the process path on a *Ts* diagram.

13.125 Air flows through a smooth well-insulated 100 mm diameter pipe at 0.76 kg/s. At one section the air is at 690 kPa (abs) and 27°C. Determine the minimum pressure and the maximum speed that can occur in the pipe.

13.126 Nitrogen at stagnation conditions of 724 kPa and 38°C flows through an insulated converging-diverging nozzle without friction. The nozzle, which has an exit-to-throat area ratio of 4, discharges supersonically into a constant area duct, which has a friction length $fL/D = 0.355$. Determine the temperature and pressure at the exit of the duct.

13.127 A converging-diverging nozzle discharges air into an insulated pipe with area $A = 645$ mm^2. At the pipe inlet, $p = 127$ kPa , $T = 38$°C, and $M = 2.0$. For shockless flow to a Mach number of unity at the pipe exit, calculate the exit temperature, the net force of the fluid on the pipe, and the entropy change.

13.128 Air is drawn from the atmosphere (20°C and 101 kPa) through a converging nozzle into a long insulated 20-mm-diameter tube of constant area. Flow in the nozzle is isentropic. The pressure at the inlet to the constant-area tube is $p_i = 99.4$ kPa. Evaluate the mass flow rate through the tube. Calculate T^* and p^* for the isentropic process. Calculate T^* and p^* for flow leaving the constant-area tube. Show the corresponding static and stagnation state points on a *Ts* diagram.

13.129 Air flows through a converging nozzle and then a length of insulated duct. The air is supplied from a tank where the temperature is constant at 15°C and the pressure is variable. The outlet end of the duct exhausts to atmosphere. When the exit flow is just choked, pressure measurements show the duct inlet pressure and Mach number are 367 kPa and 0.30. Determine the pressure in the tank and the temperature, stagnation pressure, and mass flow rate of the outlet flow, if the tube diameter is 6.3 mm. Show on a *Ts* diagram the effect of raising the tank pressure to 690 kPa. Sketch the pressure distribution versus distance along the channel for this new flow condition.

13.130 A constant-area duct is fed by a converging-only nozzle. The nozzle receives air from a large chamber at $p_1 = 600$ kPa and $T_1 = 550$ K. The duct has a friction length of 5.3, and it is choked at the exit. What is the pressure at the end of the duct? If 80 percent of the duct is removed, and the conditions at station 1 and the friction factor remain constant, what is the new exit pressure and Mach number? Sketch both of these processes on a *Ts* diagram.

13.131 We wish to build a supersonic wind tunnel using an insulated nozzle and constant-area duct assembly. Shock-free operation is desired, with $M_1 = 2.1$ at the test section inlet and $M_2 = 1.1$ at the test section outlet. Stagnation conditions are $T_0 = 295$ K and $p_0 = 101$ kPa (abs). Calculate the outlet pressure and temperature and the entropy change through the test section.

13.132 Consider adiabatic flow of air in a constant-area pipe with friction. At one section of the pipe, $p_0 = 690$ kPa, $T_0 = 280$ K, and $M = 0.70$. If the cross-sectional area is 0.09 m^2 and the Mach number at the exit is $M_2 = 1$, find the friction force exerted on the fluid by the pipe.

13.133 For the conditions of Problem 13.122, find the length, L, of commercial steel pipe of 50 mm diameter between sections ① and ②.

13.134 Consider the laboratory Fanno-line flow channel of Problem 13.123. Assume laboratory conditions are 22.5°C and 760 mm of mercury (uncorrected). The manometer reading at a pressure tap at the end of the converging nozzle is −11.8 mm of mercury (gage). Calculate the Mach number at this location. Determine the duct length required to attain choked flow. Calculate the temperature and stagnation pressure at the choked state in the constant-area duct.

13.135 A 0.6096 m × 0.6096 m duct is 12.19 m long. Air enters at $M_1 = 3.0$ and leaves at $M_2 = 1.7$, with $T_2 = 280$ K and $p_2 = 758.73$ kPa. Find the static and stagnation conditions at the entrance. What is the friction factor for the duct?

13.136 Air flows in a 0.0762 m (nominal) i.d. pipe that is 3.048 m long. The air enters with a Mach number of 0.5 and a temperature of 21.11°C. What friction factor would cause the flow to be sonic at the exit? If the exit pressure is 101.39 kPa and the pipe is made of cast iron, estimate the inlet pressure.

13.137 For the conditions of Problem 13.132, determine the duct length. Assume the duct is circular and made from commercial steel. Plot the variations of pressure and Mach number versus distance along the duct.

13.138 Using coordinates T/T_0 and $(s - s*)/c_p$, where $s*$ is the entropy at $M = 1$, plot the Fanno line starting from the inlet conditions specified in Example 13.8. Proceed to $M = 1$.

13.139 Consider the flow described in Example 13.8. Using the flow functions for Fanno-line flow of an ideal gas, plot static pressure, temperature, and Mach number versus L/D measured from the tube inlet; continue until the choked state is reached.

13.140 Using coordinates $T/T*$ and $(s - s*)/c_p$, where $s*$ is the entropy at $M = 1$, plot the Fanno line for air flow for $0.1 < M < 3.0$.

13.141 Air flows through a 12 m length of insulated constant-area duct with $D = 0.65$ m. The relative roughness is $e/D = 0.002$. At the duct inlet, $T_1 = 38°\text{C}$ and $p_1 = 117$ kPa. At a location downstream, $p_2 = 101$ kPa, and the flow is subsonic. Is sufficient information given to solve for M_1 and M_2? Prove your answer graphically. Find the mass flow rate in the duct and T_2.

13.142 Air brought into a tube through a converging-diverging nozzle initially has stagnation temperature and pressure of 550 K and 1.35 MPa (abs). Flow in the nozzle is isentropic; flow in the tube is adiabatic. At the junction between the nozzle and tube the pressure is 15 kPa. The tube is 1.5 m long and 2.5 cm in diameter. If the outlet Mach number is unity, find the average friction factor over the tube length. Calculate the change in pressure between the tube inlet and discharge.

13.143 For the conditions of Problem 13.127, determine the duct length. Assume the duct is circular and made from commercial steel. Plot the variations of pressure and Mach number versus distance along the duct.

13.144 A smooth constant-area duct assembly ($D = 150$ mm) is to be fed by a converging-diverging nozzle from a tank containing air at 295 K and 1.0 MPa (abs). Shock-free operation is desired. The Mach number at the duct inlet is to be 2.1 and the Mach number at the duct outlet is to be 1.4. The entire assembly will be insulated. Find (a) the pressure required at the duct outlet, (b) the duct length required, and (c) the change in specific entropy. Show the static and stagnation state points and the process path on a Ts diagram.

***13.145** Natural gas is to be pumped through 96 km of 76 cm-diameter pipe with an average friction factor of 0.025. The temperature of the gas remains constant at 60°C, and the mass flow rate is 18.14 kg/s. The downstream pressure of the gas is 150 kPa. Estimate the required entrance pressure, and the power needed to pump the gas through the pipe.

***13.146** Air flows through a 25.4 mm diameter, 3.048 m long tube. The friction factor of the tube is 0.03. If the entrance conditions are 103.4 kPa and 297 K, calculate the mass flow rate for (a) incompressible flow (using the methods of Chapter 8), (b) adiabatic (Fanno) flow, and (c) isothermal flow. Assume for parts (b) and (c) that the exit pressure is 101.3 kPa.

***13.147** A 15 m umbilical line for an astronaut on a space walk is held at a constant temperature of 20°C. Oxygen is supplied to the astronaut at a rate of 10 L/min, through a 1 cm tube in the umbilical line with an average friction factor of 0.01. If the oxygen pressure at the downstream end is 30 kPa, what does the upstream pressure need to be? How much power is needed to feed the oxygen to the astronaut?

***13.148** Air enters a 15 cm diameter pipe at 15°C, 1.5 MPa, and 60 m/s. The average friction factor is 0.013. Flow is isothermal. Calculate the local Mach number and the distance from the entrance of the channel, at the point where the pressure reaches 500 kPa.

***13.149** In long, constant-area pipelines, as used for natural gas, temperature is constant. Assume gas leaves a pumping station at 350 kPa and 20°C at $M = 0.10$. At the section along the pipe where the pressure has dropped to 150 kPa, calculate the Mach number of the flow. Is heat added to or removed from the gas over the length between the pressure taps? Justify your answer: Sketch the process on a Ts diagram. Indicate (qualitatively) T_{0_1}, T_{0_2}, and p_{0_2}.

***13.150** A clean steel pipe is 290 m long and 133 mm inside diameter. Air at 27°C, 827 kPa, and 24 m/s enters the pipe. Calculate and compare the pressure drops through the pipe for (a) incompressible, (b) isothermal, and (c) adiabatic flows.

***13.151** Air enters a horizontal channel of constant area at 93°C, 4.1 MPa, and 107 m/s. Determine the limiting pressure for isothermal flow. Compare with the limiting pressure for frictional adiabatic flow.

***13.152** Natural gas (molecular mass $M_m = 18$ and $k = 1.3$) is to be pumped through a 914 mm i.d. pipe connecting two compressor stations 64 km apart. At the upstream station the pressure is not to exceed 620 kPa, and at the downstream station it is to be at least 69 kPa. Calculate the maximum allowable rate of flow (m^3/day at 21°C and 101.3 kPa) assuming sufficient heat exchange through the pipe to maintain the gas at 21°C.

Flow with Heat Exchange

13.153 Air from a large reservoir at 172 kPa and 120°C flows isentropically through a converging nozzle into a frictionless pipe at 165 kPa. The flow is heated as it flows along the pipe. Obtain a plot of the Ts diagram for this flow, until $M = 1$. Also plot the pressure and speed distributions from the entrance to the location at which $M = 1$.

13.154 Air enters a constant-area duct with $M_1 = 3.0$ and $T_1 = 250$ K. Heat transfer decreases the outlet Mach number to $M_2 = 1.60$. Compute the exit static and stagnation temperatures, and find the magnitude and direction of the heat transfer.

13.155 Repeat Problem 13.153 except the nozzle is now a converging-diverging nozzle delivering the air to the pipe at 2.5 psia.

13.156 Consider frictionless flow of air in a constant-area duct. At section ①, $M_1 = 0.50$, $p_1 = 1.10$ MPa (abs), and

*These problems require material from sections that may be omitted without loss of continuity in the text material

$T_{0_1} = 333$ K. Through the effect of heat exchange, the Mach number at section ② is $M_2 = 0.90$ and the stagnation temperature is $T_{0_2} = 478$ K. Determine the amount of heat exchange per unit mass to or from the fluid between sections ① and ② and the pressure difference, $p_1 - p_2$.

13.157 Air flows without friction through a short duct of constant area. At the duct entrance, $M_1 = 0.30$, $T_1 = 50°\text{C}$, and $\rho_1 = 2.16\ \text{kg/m}^3$. As a result of heating, the Mach number and pressure at the tube outlet are $M_2 = 0.60$ and $p_2 = 150$ kPa. Determine the heat addition per unit mass and the entropy change for the process.

13.158 Air enters a 150 mm diameter duct with a velocity of 90 m/s. The entrance conditions area 101.3 kPa and 93°C. How much heat must be added to the flow to yield (a) maximum static temperature at the exit, and (b) sonic flow at the exit?

13.159 Liquid Freon, used to cool electronic components, flows steadily into a horizontal tube of constant diameter, $D = 16.5$ mm. Heat is transferred to the flow, and the liquid boils and leaves the tube as vapor. The effects of friction are negligible compared with the effects of heat addition. Flow conditions are shown. Find (a) the rate of heat transfer and (b) the pressure difference, $p_1 - p_2$.

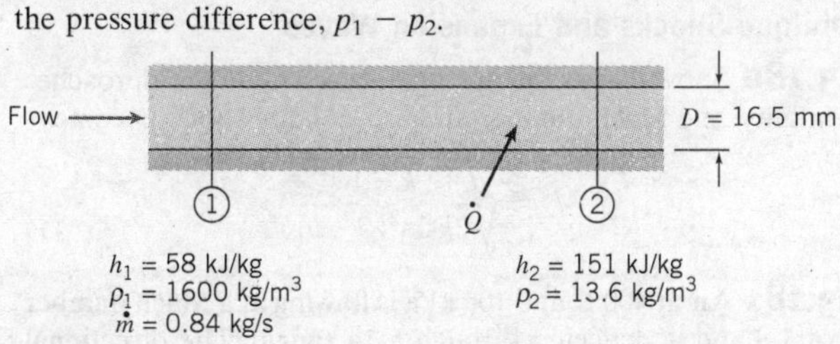

P13.159

13.160 Air flows through a 5 cm inside diameter pipe with negligible friction. Inlet conditions are $T_1 = 15°\text{C}$, $p_1 = 1$ MPa (abs), and $M_1 = 0.35$. Determine the heat exchange per kg of air required to produce $M_2 = 1.0$ at the pipe exit, where $p_2 = 500$ kPa.

13.161 Air flows at 1.42 kg/s through a 100 mm diameter duct. At the inlet section, the temperature and absolute pressure are 52°C and 60.0 kPa. At the section downstream where the flow is choked, $T_2 = 45°\text{C}$. Determine the heat addition per unit mass, the entropy change, and the change in stagnation pressure for the process, assuming frictionless flow.

13.162 Consider frictionless flow of air in a duct of constant area, $A = 0.008\ \text{m}^2$. At one section, the static properties are 278 K and 103 kPa and the Mach number is 0.2. At a section downstream, the static pressure is 69 kPa. Draw a Ts diagram showing the static and stagnation states. Calculate the flow speed and temperature at the downstream location. Evaluate the rate of heat exchange for the process.

13.163 Nitrogen flows through a frictionless duct. At the entrance of the duct, the conditions are $M_1 = 0.75$, $T_{0_1} = 280$ K, and $p_1 = 165.47$ kPa. At the exit of the duct the pressure is $p_2 = 275.79$ kPa. Determine the direction and the amount of the heat transfer with the nitrogen.

13.164 A combustor from a JT8D jet engine (as used on the Douglas DC-9 aircraft) has an air flow rate of 6.8 kg/s. The area is constant and frictional effects are negligible. Properties at the combustor inlet are 700 K, 1.62 MPa, and 185 m/s. At the combustor outlet, $T = 1022$ K and $M = 0.476$. The heating value of the fuel is 41,900 kJ/kg; the air-fuel ratio is large enough so properties are those of air. Calculate the pressure at the combustor outlet. Determine the rate of energy addition to the air stream. Find the mass flow rate of fuel required; compare it to the air flow rate. Show the process on a Ts diagram, indicating static and stagnation states and the process path.

13.165 Consider frictionless flow of air in a duct with $D = 10$ cm. At section ①, the temperature and pressure are 0°C and 70 kPa; the mass flow rate is 0.5 kg/s. How much heat may be added without choking the flow? Evaluate the resulting change in stagnation pressure.

13.166 A constant-area duct is fed with air from a converging-diverging nozzle. At the entrance to the duct, the following properties are known: $p_{0_1} = 800$ kPa (abs), $T_{0_1} = 700$ K, and $M_1 = 3.0$. A short distance down the duct (at section ②) $p_2 = 46.4$ kPa. Assuming frictionless flow, determine the speed and Mach number at section ②, and the heat exchange between the inlet and section ②.

13.167 Air flows steadily and without friction at 1.83 kg/s through a duct with cross-sectional area of 0.02 m². At the duct inlet, the temperature and absolute pressure are 260°C and 126 kPa. The exit flow discharges subsonically to atmospheric pressure. Determine the Mach number, temperature, and stagnation temperature at the duct outlet and the heat exchange rate.

13.168 20 kg/s of air enters a 0.06 m² duct at a pressure of 320 kPa, and a temperature of 350 K. Find the exit conditions (pressure, temperature, and Mach number) if heat is added to the duct at a rate of 650 kJ/kg of air.

13.169 Air enters a frictionless, constant-area duct with $p_1 = 135$ kPa, $T_1 = 500$ K, and $V_1 = 540$ m/s. How much heat transfer is needed to choke the flow? Is the heat transfer into or out of the duct?

13.170 In the frictionless flow of air through a 100 mm diameter duct, 1.42 kg/s enters at 52°C and 60.0 kPa (abs). Determine the amount of heat that must be added to choke the flow, and the fluid properties at the choked state.

13.171 Air flows without friction in a short section of constant-area duct. At the duct inlet, $M_1 = 0.30$, $T_1 = 50°\text{C}$, and $\rho_1 = 2.16\ \text{kg/m}^3$. At the duct outlet, $M_2 = 0.60$. Determine the heat addition per unit mass, the entropy change, and the change in stagnation pressure for the process.

13.172 Air, from an aircraft inlet system, enters the engine combustion chamber, where heat is added during a frictionless process in a tube with constant area of 0.01 m². The local isentropic stagnation temperature and Mach number entering the combustor are 427 K and 0.3. The mass flow rate is 0.5 kg/s. When the rate of heat addition is set at 404 kW, flow leaves the combustor at 1026 K and 22.9 kPa (abs). Determine for this process (a) the Mach number at the combustor outlet, (b) the static pressure at the combustor inlet, and (c) the change in local isentropic stagnation pressure during the heat addition process. Show static and stagnation state points and indicate the process path on a Ts diagram.

13.173 Air enters a frictionless, constant-area duct with $M_1 = 2.0$, $T_1 = 168$ K, and $p_1 = 482.63$ kPa. Heat transfer occurs as the air travels down the duct. A converging section ($A_2/A_3 = 1.5$) is placed at the end of the constant area duct and $M_3 = 1.0$. Assuming isentropic flow (aside from the heat transfer through the duct), calculate the amount and direction of heat transfer.

13.174 Consider steady, one-dimensional flow of air in a combustor with constant area of 0.05 m^2, where hydrocarbon fuel, added to the air stream, burns. The process is equivalent to simple heating because the amount of fuel is small compared to the amount of air; heating occurs over a short distance so that friction is negligible. Properties at the combustor inlet are 450 K, 1.38 MPa, and $M = 0.3$. The speed at the combustor outlet must not exceed 610 m/s. Find the properties at the combustor outlet and the heat addition rate. Show the process path on a Ts diagram, indicating static and stagnation state points before and after the heat addition.

13.175 Flow in a gas turbine combustor is modeled as steady, one-dimensional, frictionless heating of air in a channel of constant area. For a certain process, the inlet conditions are 500°C, 1.5 MPa (abs), and $M = 0.5$. Calculate the maximum possible heat addition. Find all fluid properties at the outlet section and the reduction in stagnation pressure. Show the process path on a Ts diagram, indicating all static and stagnation state points.

13.176 A supersonic wind tunnel is supplied from a high-pressure tank of air at 25°C. The test section temperature is to be maintained above 0°C to prevent formation of ice particles. To accomplish this, air from the tank is heated before it flows into a converging-diverging nozzle which feeds the test section. The heating is done in a short section with constant area. The heater output is $\dot{Q} = 10$ kW. The design Mach number in the wind tunnel test section is to be 3.0. Evaluate the stagnation temperature required at the heater exit. Calculate the maximum mass flow rate at which air can be supplied to the wind tunnel test section. Determine the area ratio, A_e/A_t.

13.177 Consider steady flow of air in a combustor where thermal energy is added by burning fuel. Neglect friction. Assume thermodynamic properties are constant and equal to those of pure air. Calculate the stagnation temperature at the burner exit. Compute the Mach number at the burner exit. Evaluate the heat addition per unit mass and the heat exchange rate. Express the rate of heat addition as a fraction of the maximum rate of heat addition possible with this inlet Mach number.

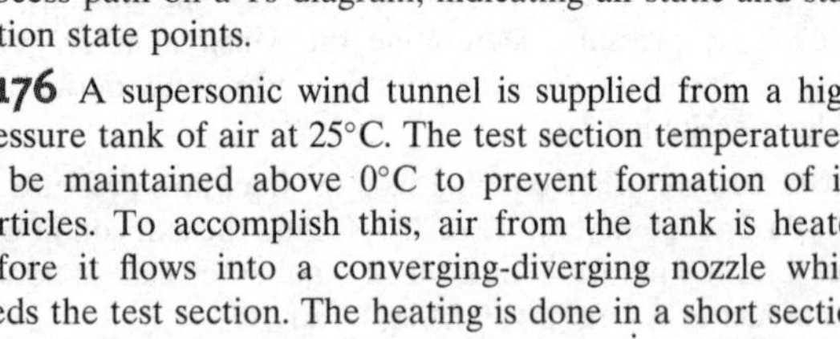

P13.177

13.178 A jet transport aircraft cruises at $M = 0.85$ at an altitude of 12,000 m. Air for the cabin pressurization system is taken aboard through an inlet duct and slowed isentropically to 30 m/s relative to the aircraft. Then it enters a compressor, where its pressure is raised adiabatically to provide a cabin pressure equivalent to 2400 m altitude. The air temperature increase across the compressor is 77°C. Finally, the air is cooled to 21°C (in a heat exchanger with negligible friction) before it is added to the cabin air. Sketch a diagram of the system, labeling all components and numbering appropriate cross sections. Determine the stagnation and static temperature and pressure at each cross section. Sketch to scale and label a Ts diagram showing the static and stagnation state points and indicating the process paths. Evaluate the work added in the compressor and the energy rejected in the heat exchanger.

13.179 Frictionless flow of air in a constant-area duct discharges to atmospheric pressure at section ②. Upstream at section ①, $M_1 = 3.0$, $T_1 = 120$ K, and $p_1 = 12$ kPa. Between sections ① and ②, 113 kJ/kg of air is added to the flow. Determine M_2 and p_2. In addition to a Ts diagram, sketch the pressure distribution versus distance along the channel, labeling sections ① and ②.

Oblique Shocks and Expansion Waves

13.180 Show that as the upstream Mach number approaches infinity, the Mach number after an oblique shock becomes

$$M_2 \approx \sqrt{\frac{k-1}{2k\sin^2(\beta-\theta)}}$$

13.181 Air at 400 K and 100 kPa is flowing at a Mach number of 1.8 and is deflected through a 14° angle. The directional change is accompanied by an oblique shock. What are the possible shock angles? For each of these shock angles, what is the pressure and temperature after the shock?

13.182 Consider supersonic flow of air at $M_1 = 3.0$. What is the range of possible values of the oblique shock angle β? For this range of β, plot the pressure ratio across the shock.

13.183 Supersonic air flow at $M_1 = 2.5$ and 80 kPa (abs) is deflected by an oblique shock with angle $\beta = 35°$. Find the Mach number and pressure after the shock, and the deflection angle. Compare these results to those obtained if instead the flow had experienced a normal shock. What is the smallest possible value of angle β for this upstream Mach number?

13.184 The temperature and Mach number before an oblique shock are $T_1 = 10$°C and $M_1 = 3.25$, respectively, and the pressure ratio across the shock is 5. Find the deflection angle, θ, the shock angle, β, and the Mach number after the shock, M_2.

13.185 The air velocities before and after an oblique shock are 1250 m/s and 650 m/s, respectively, and the deflection angle is $\theta = 35°$. Find the oblique shock angle β, and the pressure ratio across the shock.

13.186 An airfoil has a sharp leading edge with an included angle of $\delta = 60°$. It is being tested in a wind tunnel running at 1200 m/s (the air pressure and temperature upstream are 75 kPa and 3.5°C). Plot the pressure and temperature in the region adjacent to the upper surface as functions of angle of attack, α, ranging from $\alpha = 0°$ to 30°. What are the maximum pressure and temperature? (Ignore the possibility of a detached shock developing if α is too large; see Problem 13.189.)

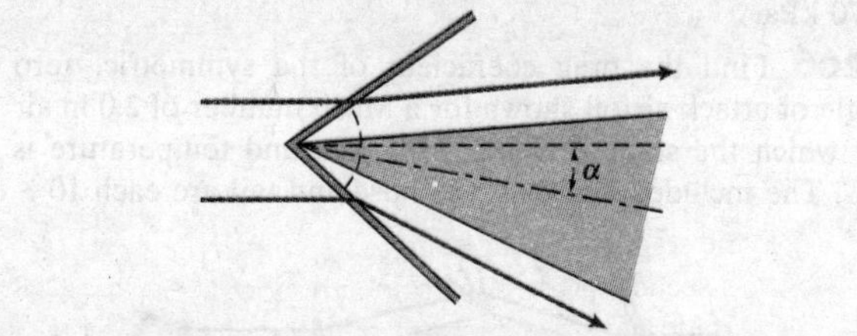

P13.186, P13.189

13.187 An airfoil at zero angle of attack has a sharp leading edge with an included angle of 20°. It is being tested over a range of speeds in a wind tunnel. The air temperature upstream is maintained at 15°C. Determine the Mach number and corresponding air speed at which a detached normal shock first attaches to the leading edge, and the angle of the resulting oblique shock. Plot the oblique shock angle β as a function of upstream Mach number M_1, from the minimum attached-shock value through $M_1 = 7$.

13.188 The wedge-shaped airfoil shown has chord $c = 1.5$ m and included angle $\delta = 7°$. Find the lift per unit span at a Mach number of 2.75 in air for which the static pressure is 70 kPa.

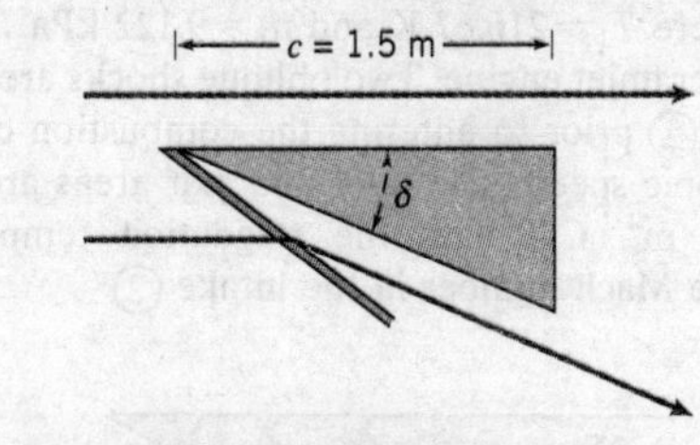

P13.188, P13.205

13.189 The airfoil of Problem 13.186 will develop a detached shock on the lower surface if the angle of attack, α, exceeds a certain value. What is this angle of attack? Plot the pressure and temperature in the region adjacent to the lower surface as functions of angle of attack, α, ranging from $\alpha =$ 0° to the angle at which the shock becomes detached. What are the maximum pressure and temperature?

13.190 An oblique shock causes a flow that was at $M = 4$ and a static pressure of 75 kPa to slow down to $M = 2.5$. Find the deflection angle and the static pressure after the shock.

13.191 The wedge-shaped airfoil shown has chord $c = 2$ m and angles $\delta_{lower} = 15°$ and $\delta_{upper} = 5°$. Find the lift per unit span at a Mach number of 2.75 in air at a static pressure of 75 kPa.

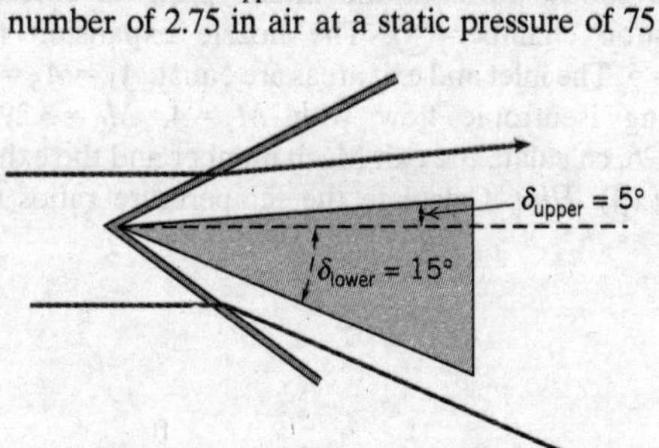

P13.191

13.192 Air flows at a Mach number of 3.3, with static conditions of 37.78°C and 137.9 kPa. An oblique shock is observed at an angle of 45° relative to the flow. Calculate the post-shock conditions (pressure, temperature, Mach number). What is the deflection angle for the flow? Is this a strong or a weak shock?

13.193 Air entering the inlet of a jet engine is turned through an angle of 8°, creating an oblique shock. If the freestream flow of air is at Mach 4 and 55.16 kPa, what is the pressure after the oblique shock? What would the pressure be if the flow were through two separate 4° wedges instead of a single 8° wedge?

13.194 Air having an initial Mach number of 2.3 and static conditions of 101.35 kPa and 26.67°C is turned through an angle of 10°. The resulting shock at the corner is reflected from the opposite wall, turning the flow back 10° to its original direction. Calculate the pressure, temperature, and Mach number after the initial and reflected shock waves.

13.195 A wedge-shaped projectile (half angle is 10°) is launched through air at 6.89 kPa and −12.22°C. If the static pressure measurement on the surface of the wedge is 20.68 kPa, calculate the speed at which the projectile is moving through the air.

13.196 Air at Mach 2 and 1 atmosphere is turned through an expansion of 16°, followed by another turn of 16°, causing an oblique shock wave. Calculate the Mach number and pressure downstream of the oblique shock.

13.197 Air at Mach 2.0 and 34.47 kPa static pressure is turned through an angle of 20°. Determine the resulting static pressure and stagnation pressure when the turning is achieved through (a) a single oblique shock, (b) two oblique shocks, each turning the flow 10°, and (c) an isentropic compression wave system.

13.198 Air flows isentropically at $M = 2.5$ in a duct. There is a 7.5° contraction that triggers an oblique shock, which in turn reflects off a wall generating a second oblique shock. This second shock is necessary so the flow ends up flowing parallel to the channel walls after the two shocks. Find the Mach number and pressure in the contraction and downstream of the contraction. (Note that the convex corner will have expansion waves to redirect the flow along the upper wall.)

P13.198

13.199 The geometry of the fuselage and engine cowling near the inlet to the engine of a supersonic fighter aircraft is designed so that the incoming air at $M = 3$ is deflected 7.5 degrees, and then experiences a normal shock at the engine entrance. If the incoming air is at 50 kPa, what is the pressure of the air entering the engine? What would be the pressure if the incoming air was slowed down by only a normal shock?

13.200 Air flows at Mach number of 1.5, static pressure 95 kPa, and is expanded by angles $\theta_1 = 15°$ and $\theta_2 = 15°$, as shown. Find the pressure changes.

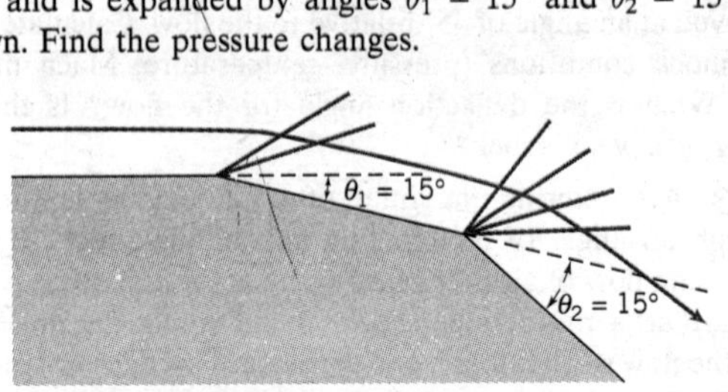

P13.200, P13.203

13.201 A flow at $M = 2.5$ is deflected by a combination of interacting oblique shocks as shown. The first shock pair is aligned at 30° to the flow. A second oblique shock pair deflects the flow again so it ends up parallel to the original flow. If the pressure before any deflections is 50 kPa, find the pressure after two deflections.

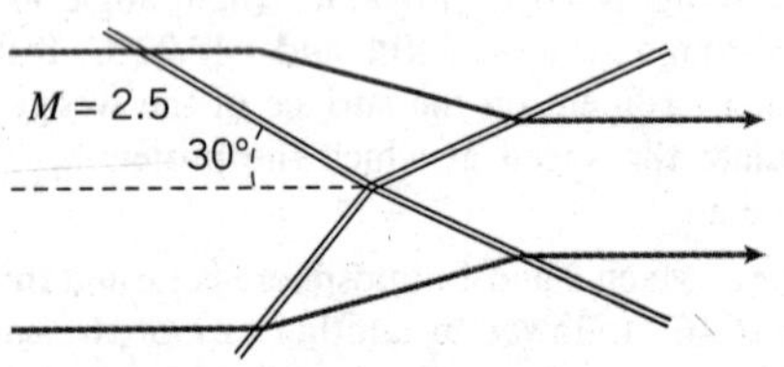

P13.201

13.202 Compare the static and stagnation pressures produced by (a) an oblique shock and (b) isentropic *compression* waves as they each deflect a flow at a Mach number of 3.5 through a deflection angle of 35° in air for which the static pressure is 50 kPa.

13.203 Find the incoming and intermediate Mach numbers and static pressures if, after two expansions of $\theta_1 = 15°$ and $\theta_2 = 15°$, the Mach number is 4, and static pressure is 10 kPa.

13.204 Find the lift and drag per unit span on the airfoil shown for flight at a Mach number of 1.75 in air for which the static pressure is 50 kPa. The chord length is 1 m.

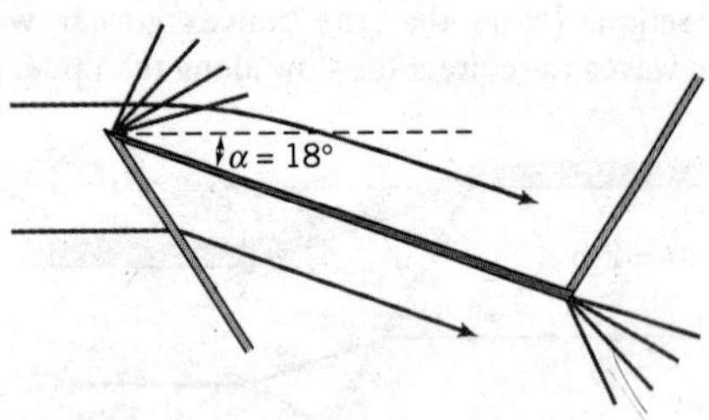

P13.204, P13.207

13.205 Consider the wedge-shaped airfoil of Problem 13.188. Suppose the oblique shock could be replaced by isentropic *compression* waves. Find the lift per unit span at the Mach number of 2.75 in air for which the static pressure is 70 kPa.

13.206 Find the drag coefficient of the symmetric, zero angle of attack airfoil shown for a Mach number of 2.0 in air for which the static pressure is 95 kPa and temperature is 0°C. The included angles at the nose and tail are each 10°.

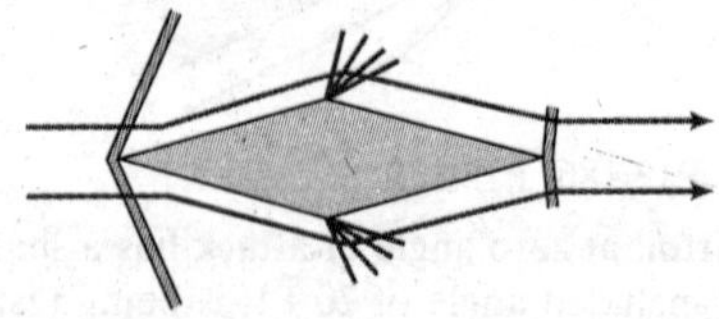

P13.206, P13.208

13.207 Plot the lift and drag per unit span, and the lift/drag ratio, as functions of angle of attack for $\alpha = 0°$ to 18°, for the airfoil shown, for flight at a Mach number of 1.75 in air for which the static pressure is 50 kPa. The chord length is 1 m.

13.208 Find the lift and drag coefficients of the airfoil of Problem 13.206 if the airfoil now has an angle of attack of 12°.

13.209 An airplane is flying at Mach 5 at an altitude of 16,764 m, where $T_1 = 216.67$ K and $p_1 = 9.122$ kPa. The airplane uses a scramjet engine. Two oblique shocks are formed in the intake ② prior to entering the combustion chamber ③ at supersonic speed. The inlet and exit areas are equal, $A_1 = A_5 = 0.2$ m^2. Calculate the stagnation temperature, T_2/T_1, and the Mach number in the intake ②.

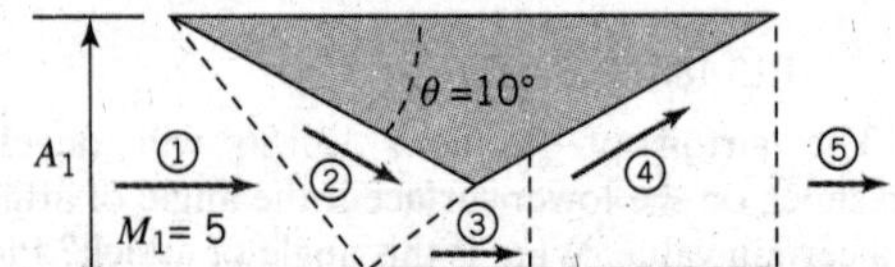

P13.209, P13.210, P13.211

13.210 Two oblique shocks are formed in a scramjet engine intake prior to entering the combustion chamber. The inlet Mach number is $M_1 = 5$, the incoming air temperature is $T_1 = 216.67$ K, $p_1 = 9.122$ kPa, and $A_1 = 0.2\,\text{m}^2$. Calculate M_3 in the combustion chamber ③ if $M_2 = 4.0$.

13.211 An airplane is flying at Mach 5, where $T_1 = 216.67$ K. Oblique shocks form in the intake prior to entering the combustion chamber ③. The nozzle expansion ratio is $A_5/A_4 = 5$. The inlet and exit areas are equal, $A_1 = A_5 = 0.2\,\text{m}^2$. Assuming isentropic flow with $M_2 = 4$, $M_3 = 3.295$, and $M_4 = 1.26$, calculate the exit Mach number and the exhaust jet velocity ⑤. *Hint:* Calculate the temperature ratios in each section.

Appendix A

Fluid Property Data

Specific Gravity A.1

Specific gravity data for several common liquids and solids are presented in Figs. A.1*a* and A.1*b* and in Tables A.1 and A.2. For liquids specific gravity is a function of temperature. (Density data for water and air are given as functions of temperature in Tables A.7 through A.10.) For most liquids specific gravity decreases as temperature increases. Water is unique: It displays a maximum density of 1000 kg/m^3 at 4°C. The maximum density of water is used as a reference value to calculate specific gravity. Thus

$$SG \equiv \frac{\rho}{\rho_{H_2O}\ (\text{at } 4^\circ\text{C})}$$

Consequently the maximum SG of water is exactly unity.

Specific gravities for solids are relatively insensitive to temperature; values given in Table A.1 were measured at 20°C.

The specific gravity of seawater depends on both its temperature and salinity. A representative value for ocean water is SG = 1.025, as given in Table A.2.

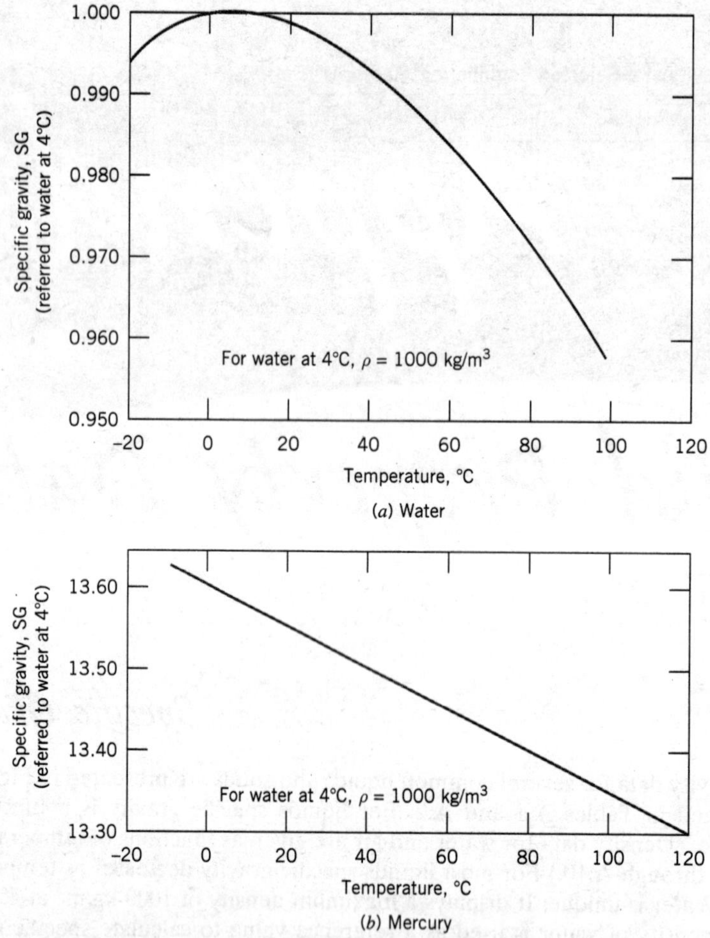

Fig. A.1 **Specific gravity of water and mercury as functions of temperature. (Data from Reference [1].) (The specific gravity of mercury varies linearly with temperature. The variation is given by $SG = 13.60 - 0.00240\,T$ when T is measured in degrees C.)**

Table A.1

Specific Gravities of Selected Engineering Materials

(a) Common Manometer Liquids at 20°C

Liquid	Specific Gravity
E.V. Hill blue oil	0.797
Meriam red oil	0.827
Benzene	0.879
Dibutyl phthalate	1.04
Monochloronaphthalene	1.20
Carbon tetrachloride	1.595
Bromoethylbenzene (Meriam blue)	1.75
Tetrabromoethane	2.95
Mercury	13.55

Source: Data from References [1–3].

Table A.1

Specific Gravities of Selected Engineering Materials *(continued)*

(b) Common Materials

Material	Specific Gravity (—)
Aluminum	2.64
Balsa wood	0.14
Brass	8.55
Cast Iron	7.08
Concrete (cured)	2.4[a]
Concrete (liquid)	2.5[a]
Copper	8.91
Ice (0°C)	0.917
Lead	11.4
Oak	0.77
Steel	7.83
Styrofoam (1 kg/m^3)	0.0160
Styrofoam (3 kg/m^3)	0.0481
Uranium (depleted)	18.7
White pine	0.43

Source: Data from Reference [4].
[a]depending on aggregate.

Table A.2

Physical Properties of Common Liquids at 20°C

Liquid	Isentropic Bulk Modulus[a] (GN/m^2)	Specific Gravity (—)
Benzene	1.48	0.879
Carbon tetrachloride	1.36	1.595
Castor oil	2.11	0.969
Crude oil	—	0.82–0.92
Ethanol	—	0.789
Gasoline	—	0.72
Glycerin	4.59	1.26
Heptane	0.886	0.684
Kerosene	1.43	0.82
Lubricating oil	1.44	0.88
Methanol	—	0.796
Mercury	28.5	13.55
Octane	0.963	0.702
Seawater[b]	2.42	1.025
SAE 10W oil	—	0.92
Water	2.24	0.998

Source: Data from References [1, 5, 6].
[a]Calculated from speed of sound; 1 GN/m^2 = 10^9 N/m^2.
[b]Dynamic viscosity of seawater at 20°C is $\mu = 1.08 \times 10^{-3}$ N · s/m^2. (Thus, the kinematic viscosity of seawater is about 5 percent higher than that of freshwater.)

Table A.3

Properties of the U.S. Standard Atmosphere

Geometric Altitude (m)	Temperature (K)	p/p_{SL} (—)	ρ/ρ_{SL} (—)
−500	291.4	1.061	1.049
0	288.2	1.000[a]	1.000[b]
500	284.9	0.9421	0.9529
1,000	281.7	0.8870	0.9075
1,500	278.4	0.8345	0.8638
2,000	275.2	0.7846	0.8217
2,500	271.9	0.7372	0.7812
3,000	268.7	0.6920	0.7423
3,500	265.4	0.6492	0.7048
4,000	262.2	0.6085	0.6689
4,500	258.9	0.5700	0.6343
5,000	255.7	0.5334	0.6012
6,000	249.2	0.4660	0.5389
7,000	242.7	0.4057	0.4817
8,000	236.2	0.3519	0.4292
9,000	229.7	0.3040	0.3813
10,000	223.3	0.2615	0.3376
11,000	216.8	0.2240	0.2978
12,000	216.7	0.1915	0.2546
13,000	216.7	0.1636	0.2176
14,000	216.7	0.1399	0.1860
15,000	216.7	0.1195	0.1590
16,000	216.7	0.1022	0.1359
17,000	216.7	0.08734	0.1162
18,000	216.7	0.07466	0.09930
19,000	216.7	0.06383	0.08489
20,000	216.7	0.05457	0.07258
22,000	218.6	0.03995	0.05266
24,000	220.6	0.02933	0.03832
26,000	222.5	0.02160	0.02797
28,000	224.5	0.01595	0.02047
30,000	226.5	0.01181	0.01503
40,000	250.4	0.002834	0.003262
50,000	270.7	0.0007874	0.0008383
60,000	255.8	0.0002217	0.0002497
70,000	219.7	0.00005448	0.00007146
80,000	180.7	0.00001023	0.00001632
90,000	180.7	0.000001622	0.000002588

Source: Data from Reference [7].

[a] $p_{SL} = 1.01325 \times 10^5$ N/m² (abs).

[b] $\rho_{SL} = 1.2250$ kg/m³.

A.2 Surface Tension

The values of surface tension, σ, for most organic compounds are remarkably similar at room temperature; the typical range is 25 to 40 mN/m. Water is higher, at about 73 mN/m at 20°C. Liquid metals have values in the range between 300 and 600 mN/m; mercury has a value of about 480 mN/m at 20°C. Surface tension decreases with temperature; the decrease is nearly linear with absolute temperature. Surface tension at the critical temperature is zero.

Values of σ are usually reported for surfaces in contact with the pure vapor of the liquid being studied or with air. At low pressures both values are about the same.

Table A.4

Surface Tension of Common Liquids at 20°C

Liquid	Surface Tension, σ (mN/m)[a]	Contact Angle, θ (degrees)
(a) In contact with air		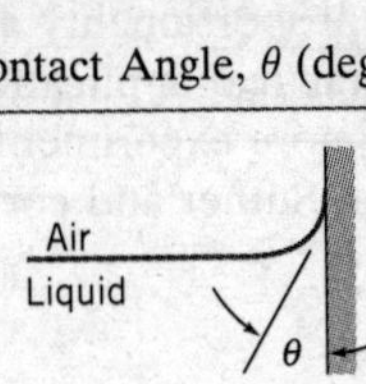
Benzene	28.9	
Carbon tetrachloride	27.0	
Ethanol	22.3	
Glycerin	63.0	
Hexane	18.4	
Kerosene	26.8	
Lube oil	25–35	
Mercury	484	140
Methanol	22.6	
Octane	21.8	
Water	72.8	~0

Source: Data from References [1, 5, 8, 9].

Liquid	Surface Tension, σ (mN/m)[a]	Contact Angle, θ (degrees)
(b) In contact with water		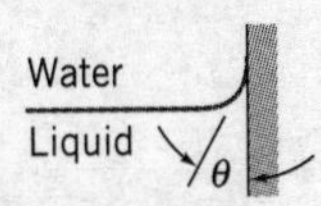
Benzene	35.0	
Carbon tetrachloride	45.0	
Hexane	51.1	
Mercury	375	140
Methanol	22.7	
Octane	50.8	

Source: Data from References [1, 5, 8, 9].

[a] $1\ \text{mN/m} = 10^{-3}\ \text{N/m}$.

The Physical Nature of Viscosity A.3

Viscosity is a measure of internal fluid friction, i.e., resistance to deformation. The mechanism of gas viscosity is reasonably well understood, but the theory is poorly developed for liquids. We can gain some insight into the physical nature of viscous flow by discussing these mechanisms briefly.

The viscosity of a Newtonian fluid is fixed by the state of the material. Thus $\mu = \mu(T, p)$. Temperature is the more important variable, so let us consider it first. Excellent empirical equations for viscosity as a function of temperature are available.

Effect of Temperature on Viscosity

a. Gases

All gas molecules are in continuous random motion. When there is bulk motion due to flow, the bulk motion is superimposed on the random motions. It is then distributed throughout the fluid by molecular collisions. Analyses based on kinetic theory predict

$$\mu \propto \sqrt{T}$$

The kinetic theory prediction is in fair agreement with experimental trends, but the constant of proportionality and one or more correction factors must be determined; this limits practical application of this simple equation.

If two or more experimental points are available, the data may be correlated using the empirical Sutherland correlation [7]

$$\mu = \frac{bT^{1/2}}{1 + S/T} \tag{A.1}$$

Constants b and S may be determined most simply by writing

$$\mu = \frac{bT^{3/2}}{S + T}$$

or

$$\frac{T^{3/2}}{\mu} = \left(\frac{1}{b}\right)T + \frac{S}{b}$$

(Compare this with $y = mx + c$.) From a plot of $T^{3/2}/\mu$ versus T, one obtains the slope, $1/b$, and the intercept, S/b. For air,

$$b = 1.458 \times 10^{-6} \frac{\text{kg}}{\text{m} \cdot \text{s} \cdot \text{K}^{1/2}}$$

$$S = 110.4 \text{ K}$$

These constants were used with Eq. A.1 to compute viscosities for the standard atmosphere in [7], the air viscosity values at various temperatures shown in Table A.10, and using appropriate conversion factors, the values shown in Table A.9.

b. Liquids

Viscosities for liquids cannot be estimated well theoretically. The phenomenon of momentum transfer by molecular collisions is overshadowed in liquids by the effects of interacting force fields among the closely packed liquid molecules.

Liquid viscosities are affected drastically by temperature. This dependence on absolute temperature may be represented by the empirical equation

$$\mu = Ae^{B/(T-C)} \tag{A.2}$$

or the equivalent form

$$\mu = A10^{B/(T-C)} \tag{A.3}$$

where T is absolute temperature.

Equation A.3 requires at least three points to fit constants A, B, and C. In theory it is possible to determine the constants from measurements of viscosity at just three temperatures. It is better practice to use more data and to obtain the constants from a statistical fit to the data.

However a curve-fit is developed, always compare the resulting line or curve with the available data. The best way is to critically inspect a plot of the curve-fit compared with the data. In general, curve-fit results will be satisfactory only when the quality of the available data and that of the empirical relation are known to be excellent.

Data for the dynamic viscosity of water are fitted well using constant values $A = 2.414 \times 10^{-5}$ N·s/m², $B = 247.8$ K, and $C = 140$ K. Reference [10] states that using these constants in Eq. A.3 predicts water viscosity within ±2.5 percent over the temperature range from 0°C to 370°C. Equation A.3 and *Excel* were used to compute

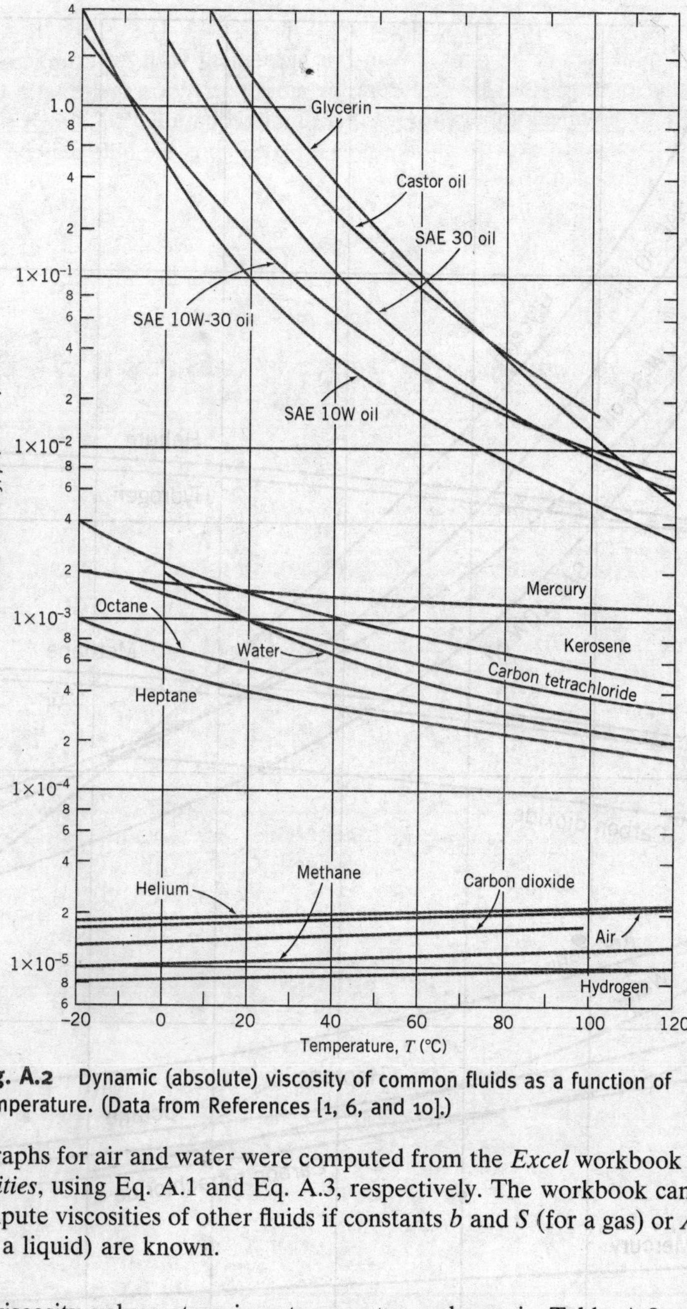

Fig. A.2 Dynamic (absolute) viscosity of common fluids as a function of temperature. (Data from References [1, 6, and 10].)

The graphs for air and water were computed from the *Excel* workbook *Absolute Viscosities*, using Eq. A.1 and Eq. A.3, respectively. The workbook can be used to compute viscosities of other fluids if constants b and S (for a gas) or A, B, and C (for a liquid) are known.

the water viscosity values at various temperatures shown in Table A.8, and using appropriate conversion factors, the values shown in Table A.7.

Note that the viscosity of a liquid decreases with temperature, while that of a gas increases with temperature.

Effect of Pressure on Viscosity

a. Gases

The viscosity of gases is essentially independent of pressure between a few hundredths of an atmosphere and a few atmospheres. However, viscosity at high pressures increases with pressure (or density).

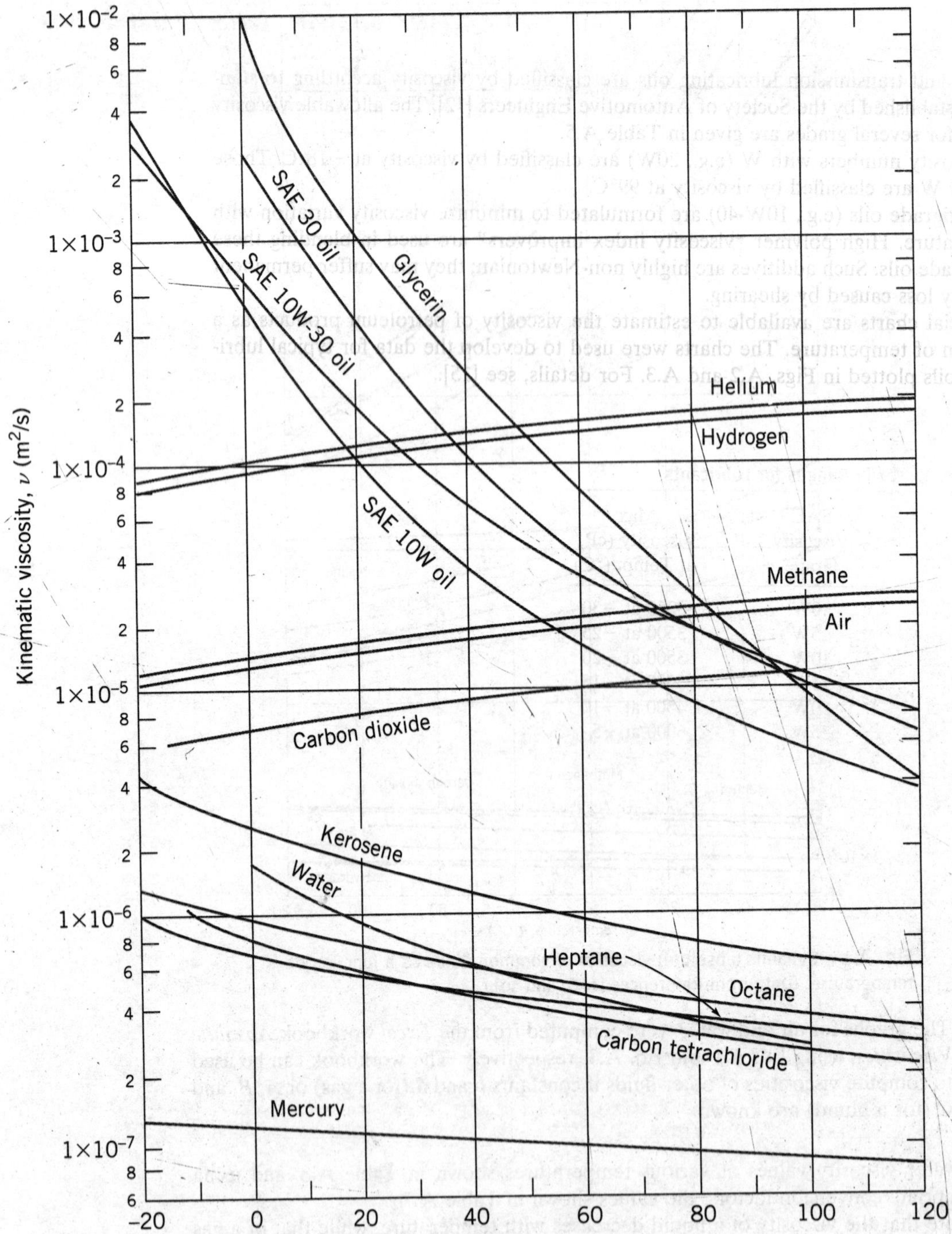

Fig. A.3 Kinematic viscosity of common fluids (at atmospheric pressure) as a function of temperature. (Data from References [1, 6, and 10].)

b. Liquids

The viscosities of most liquids are not affected by moderate pressures, but large increases have been found at very high pressures. For example, the viscosity of water at 10,000 atm is twice that at 1 atm. More complex compounds show a viscosity increase of several orders of magnitude over the same pressure range.

More information may be found in Reid and Sherwood [11].

Lubricating Oils A.4

Engine and transmission lubricating oils are classified by viscosity according to standards established by the Society of Automotive Engineers [12]. The allowable viscosity ranges for several grades are given in Table A.5.

Viscosity numbers with W (e.g., 20W) are classified by viscosity at −18°C. Those without W are classified by viscosity at 99°C.

Multigrade oils (e.g., 10W-40) are formulated to minimize viscosity variation with temperature. High polymer "viscosity index improvers" are used in blending these multigrade oils. Such additives are highly non-Newtonian; they may suffer permanent viscosity loss caused by shearing.

Special charts are available to estimate the viscosity of petroleum products as a function of temperature. The charts were used to develop the data for typical lubricating oils plotted in Figs. A.2 and A.3. For details, see [15].

Table A.5

Allowable Viscosity Ranges for Lubricants

Engine Oil	SAE Viscosity Grade	Max. Viscosity (cP)[a] at Temp. (°C)	Viscosity (cSt)[b] at 100°C Min	Max
	0W	3250 at −30	3.8	—
	5W	3500 at −25	3.8	—
	10W	3500 at −20	4.1	—
	15W	3500 at −15	5.6	—
	20W	4500 at −10	5.6	—
	25W	6000 at −5	9.3	—
	20	—	5.6	<9.3
	30	—	9.3	<12.5
	40	—	12.5	<16.3
	50	—	16.3	<21.9
Axle and Manual Transmission Lubricant	**SAE Viscosity Grade**	**Max. Temp. (°C) for Viscosity of 150,000 cP**	**Viscosity (cSt) at 100°C Min**	**Max**
	70W	−55	4.1	—
	75W	−40	4.1	—
	80W	−26	7.0	—
	85W	−12	11.0	—
	90	—	13.5	<24.0
	140	—	24.0	<41.0
	250	—	41.0	—
Automatic Transmission Fluid (Typical)	**Maximum Viscosity (cP)**	**Temperature (°C)**	**Viscosity (cSt) at 100°C Min**	**Max**
	50000	−40	6.5	8.5
	4000	−23.3	6.5	8.5
	1700	−18	6.5	8.5

Source: Data from References [12–14].

[a]centipoise = 1 cP = 1 mPa · s = 10^{-3} Pa · s.

[b]centistoke = 10^{-6} m²/s.

Properties of Common Gases, Air and Water A.5

Table A.6

Thermodynamic Properties of Common Gases at STP[a]

Gas	Chemical Symbol	Molecular Mass, M_m	R^b $\left(\frac{J}{kg \cdot K}\right)$	c_p $\left(\frac{J}{kg \cdot K}\right)$	c_v $\left(\frac{J}{kg \cdot K}\right)$	$k = \frac{c_p}{c_v}$ (−)
Air	—	28.98	286.9	1004	717.4	1.40
Carbon dioxide	CO_2	44.01	188.9	840.4	651.4	1.29
Carbon monoxide	CO	28.01	296.8	1039	742.1	1.40
Helium	He	4.003	2077	5225	3147	1.66
Hydrogen	H_2	2.016	4124	14,180	10,060	1.41
Methane	CH_4	16.04	518.3	2190	1672	1.31
Nitrogen	N_2	28.01	296.8	1039	742.0	1.40
Oxygen	O_2	32.00	259.8	909.4	649.6	1.40
Steam[c]	H_2O	18.02	461.4	~2000	~1540	~1.30

Source: Data from References [7, 16, 17].

[a]STP = standard temperature and pressure, $T = 15°C$ and $p = 101.325$ kPa (abs).

[b]$R \equiv R_u/M_m$; $R_u = 8314.3$ J/(kgmol · K); 1J = 1N · m.

[c]Water vapor behaves as an ideal gas when superheated by 55°C or more.

Table A.7

Properties of Water (SI Units)

Temperature, T (°C)	Density, ρ (kg/m^3)	Dynamic Viscosity, μ (N · s/m^2)	Kinematic Viscosity, ν (m^2/s)	Surface Tension, σ (N/m)	Vapor Pressure, p_v (kPa)	Bulk Modulus, E_v (GPa)
0	1000	1.76E-03	1.76E-06	0.0757	0.661	2.01
5	1000	1.51E-03	1.51E-06	0.0749	0.872	
10	1000	1.30E-03	1.30E-06	0.0742	1.23	
15	999	1.14E-03	1.14E-06	0.0735	1.71	
20	998	1.01E-03	1.01E-06	0.0727	2.34	2.21
25	997	8.93E-04	8.96E-07	0.0720	3.17	
30	996	8.00E-04	8.03E-07	0.0712	4.25	
35	994	7.21E-04	7.25E-07	0.0704	5.63	
40	992	6.53E-04	6.59E-07	0.0696	7.38	
45	990	5.95E-04	6.02E-07	0.0688	9.59	
50	988	5.46E-04	5.52E-07	0.0679	12.4	2.29
55	986	5.02E-04	5.09E-07	0.0671	15.8	
60	983	4.64E-04	4.72E-07	0.0662	19.9	
65	980	4.31E-04	4.40E-07	0.0654	25.0	
70	978	4.01E-04	4.10E-07	0.0645	31.2	
75	975	3.75E-04	3.85E-07	0.0636	38.6	
80	972	3.52E-04	3.62E-07	0.0627	47.4	
85	969	3.31E-04	3.41E-07	0.0618	57.8	
90	965	3.12E-04	3.23E-07	0.0608	70.1	2.12
95	962	2.95E-04	3.06E-07	0.0599	84.6	
100	958	2.79E-04	2.92E-07	0.0589	101	

Table A.8

Properties of Air at Atmospheric Pressure (SI Units)

Temperature, T (°C)	Density, ρ (kg/m^3)	Dynamic Viscosity, μ (N·s/m^2)	Kinematic Viscosity, ν (m^2/s)
0	1.29	1.72E-05	1.33E-05
5	1.27	1.74E-05	1.37E-05
10	1.25	1.76E-05	1.41E-05
15	1.23	1.79E-05	1.45E-05
20	1.21	1.81E-05	1.50E-05
25	1.19	1.84E-05	1.54E-05
30	1.17	1.86E-05	1.59E-05
35	1.15	1.88E-05	1.64E-05
40	1.13	1.91E-05	1.69E-05
45	1.11	1.93E-05	1.74E-05
50	1.09	1.95E-05	1.79E-05
55	1.08	1.98E-05	1.83E-05
60	1.06	2.00E-05	1.89E-05
65	1.04	2.02E-05	1.94E-05
70	1.03	2.04E-05	1.98E-05
75	1.01	2.06E-05	2.04E-05
80	1.00	2.09E-05	2.09E-05
85	0.987	2.11E-05	2.14E-05
90	0.973	2.13E-05	2.19E-05
95	0.960	2.15E-05	2.24E-05
100	0.947	2.17E-05	2.29E-05

References

1. *Handbook of Chemistry and Physics*, 62nd ed. Cleveland, OH: Chemical Rubber Publishing Co., 1981–1982.
2. "Meriam Standard Indicating Fluids," Pamphlet No. 920GEN: 430-1, The Meriam Instrument Co., 10920 Madison Avenue, Cleveland, OH 44102.
3. E. Vernon Hill, Inc., P.O. Box 7053, Corte Madera, CA 94925.
4. Avallone, E. A., and T. Baumeister, III, eds., *Marks' Standard Handbook for Mechanical Engineers*, 11th ed. New York: McGraw-Hill, 2007.
5. *Handbook of Tables for Applied Engineering Science*. Cleveland, OH: Chemical Rubber Publishing Co., 1970.
6. Vargaftik, N. B., *Tables on the Thermophysical Properties of Liquids and Gases*, 2nd ed. Washington, DC: Hemisphere Publishing Corp., 1975.
7. *The U.S. Standard Atmosphere (1976)*. Washington, DC: U. S. Government Printing Office, 1976.
8. Trefethen, L., "Surface Tension in Fluid Mechanics," in *Illustrated Experiments in Fluid Mechanics*. Cambridge, MA: The M.I.T. Press, 1972.
9. Streeter, V. L., ed., *Handbook of Fluid Dynamics*. New York: McGraw-Hill, 1961.
10. Touloukian, Y. S., S. C. Saxena, and P. Hestermans, *Thermophysical Properties of Matter, the TPRC Data Series. Vol. 11—Viscosity*. New York: Plenum Publishing Corp., 1975.
11. Reid, R. C., and T. K. Sherwood, *The Properties of Gases and Liquids*, 2nd ed. New York: McGraw-Hill, 1966.
12. "Engine Oil Viscosity Classification—SAE Standard J300 Jun86," *SAE Handbook*, 1987 ed. Warrendale, PA: Society of Automotive Engineers, 1987.
13. "Axle and Manual Transmission Lubricant Viscosity Classification—SAE Standard J306 Mar85," *SAE Handbook*, 1987 ed. Warrendale, PA: Society of Automotive Engineers, 1987.
14. "Fluid for Passenger Car Type Automatic Transmissions—SAE Information Report J311 Apr86," *SAE Handbook*, 1987 ed. Warrendale, PA: Society of Automotive Engineers, 1987.
15. ASTM Standard D 341–77, "Viscosity-Temperature Charts for Liquid Petroleum Products," American Society for Testing and Materials, 1916 Race Street, Philadelphia, PA 19103.
16. NASA, *Compressed Gas Handbook* (Revised). Washington, DC: National Aeronautics and Space Administration, SP-3045, 1970.
17. ASME, *Thermodynamic and Transport Properties of Steam*. New York: American Society of Mechanical Engineers, 1967.

Appendix B

Equations of Motion in Cylindrical Coordinates

The continuity equation in cylindrical coordinates for constant density is

$$\frac{1}{r}\frac{\partial}{\partial r}(rv_r) + \frac{1}{r}\frac{\partial}{\partial \theta}(v_\theta) + \frac{\partial}{\partial z}(v_z) = 0 \tag{B.1}$$

Normal and shear stresses in cylindrical coordinates for constant density and viscosity are

$$\sigma_{rr} = -p + 2\mu\frac{\partial v_r}{\partial r} \qquad \tau_{r\theta} = \mu\left[r\frac{\partial}{\partial r}\left(\frac{v_\theta}{r}\right) + \frac{1}{r}\frac{\partial v_r}{\partial \theta}\right]$$

$$\sigma_{\theta\theta} = -p + 2\mu\left(\frac{1}{r}\frac{\partial v_\theta}{\partial \theta} + \frac{v_r}{r}\right) \qquad \tau_{\theta z} = \mu\left(\frac{\partial v_\theta}{\partial z} + \frac{1}{r}\frac{\partial v_z}{\partial \theta}\right)$$

$$\sigma_{zz} = -p + 2\mu\frac{\partial v_z}{\partial z} \qquad \tau_{zr} = \mu\left(\frac{\partial v_r}{\partial z} + \frac{\partial v_z}{\partial r}\right) \tag{B.2}$$

The Navier–Stokes equations in cylindrical coordinates for constant density and viscosity are

r component:

$$\rho\left(\frac{\partial v_r}{\partial t} + v_r\frac{\partial v_r}{\partial r} + \frac{v_\theta}{r}\frac{\partial v_r}{\partial \theta} - \frac{v_\theta^2}{r} + v_z\frac{\partial v_r}{\partial z}\right)$$

$$= \rho g_r - \frac{\partial p}{\partial r} + \mu\left\{\frac{\partial}{\partial r}\left(\frac{1}{r}\frac{\partial}{\partial r}[rv_r]\right) + \frac{1}{r^2}\frac{\partial^2 v_r}{\partial \theta^2} - \frac{2}{r^2}\frac{\partial v_\theta}{\partial \theta} + \frac{\partial^2 v_r}{\partial z^2}\right\} \tag{B.3a}$$

θ component:

$$\rho\left(\frac{\partial v_\theta}{\partial t} + v_r\frac{\partial v_\theta}{\partial r} + \frac{v_\theta}{r}\frac{\partial v_\theta}{\partial \theta} + \frac{v_r v_\theta}{r} + v_z\frac{\partial v_\theta}{\partial z}\right)$$

$$= \rho g_\theta - \frac{1}{r}\frac{\partial p}{\partial \theta} + \mu\left\{\frac{\partial}{\partial r}\left(\frac{1}{r}\frac{\partial}{\partial r}[r v_\theta]\right) + \frac{1}{r^2}\frac{\partial^2 v_\theta}{\partial \theta^2} + \frac{2}{r^2}\frac{\partial v_r}{\partial \theta} + \frac{\partial^2 v_\theta}{\partial z^2}\right\} \qquad \text{(B.3b)}$$

z component:

$$\rho\left(\frac{\partial v_z}{\partial t} + v_r\frac{\partial v_z}{\partial r} + \frac{v_\theta}{r}\frac{\partial v_z}{\partial \theta} + v_z\frac{\partial v_z}{\partial z}\right)$$

$$= \rho g_z - \frac{\partial p}{\partial z} + \mu\left\{\frac{1}{r}\frac{\partial}{\partial r}\left(r\frac{\partial v_z}{\partial r}\right) + \frac{1}{r^2}\frac{\partial^2 v_z}{\partial \theta^2} + \frac{\partial^2 v_z}{\partial z^2}\right\} \qquad \text{(B.3c)}$$

Appendix C

Videos for Fluid Mechanics

Referenced in the text are the following videos available at www.wiley.com/go/global/fox.

Chapter 2

 Fluid as a Continuum

 Streaklines

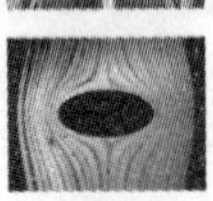 Streamlines

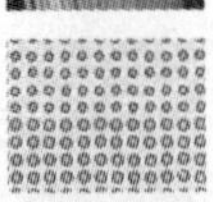 Molecular Interactions at the Interface

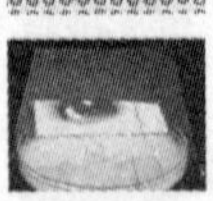 Shrinking Soap Film

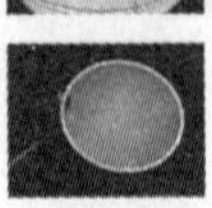 Soap Film Burst

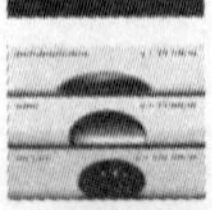 Wetted and Non-wetted Surfaces

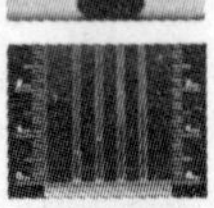 Capillary Rise

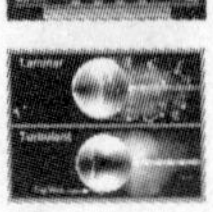 Examples of Flow over a Sphere

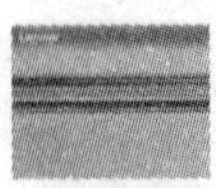 Internal Laminar Flow in a Tube

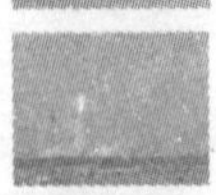 The Space Shuttle: An External Turbulent Flow

 Boundary Layer Flow

 Streamlined Flow over an Airfoil

 Streamlines around a Car

 Laminar and Turbulent Flow

 Compressible Flow: Shock Waves

Chapter 3

 Hydraulic Force Amplification

Chapter 4

 Mass Conservation: Filling a Tank

 Momentum Effect: A Jet Impacting a Surface

Chapter 5

 An Example of Streamlines/Streaklines

 Particle Motion in a Channel

 Particle Motion over a Cylinder

 Linear Deformation

 CFD: Turbulent Flow in a Channel

- CFD: Flow Past a Cylinder
- CFD: Fully Turbulent Duct Flow

Chapter 6

- An Example of Irrotational Flow

Chapter 7

- Flow Around a Sphere 1
- Flow Around a Sphere 2
- Geometric, Not Dynamic, Similarity: Flow Past a Block 1
- Geometric, Not Dynamic, Similarity: Flow Past a Block 2

Chapter 8

- The Reynolds Transition Experiment
- Variable Viscosity Experiment (Animation)
- Variable Viscosity Experiment: Pressure Drop
- Laminar Pipe Flow: Velocity Profile
- Pipe Flow: Laminar
- Pipe Flow: Transitional
- Pipe Flow: Turbulent
- Fully-Developed Pipe Flow
- Laminar Flow Exiting from a Tube
- The Glen Canyon Dam: A Turbulent Pipe Flow
- Computer Simulation: Turbulent Channel Flow 1
- Computer Simulation: Turbulent Channel Flow 2
- Computer Simulation: Turbulent Channel Flow 3
- Flow Visualization: Laser Induced Fluorescence
- Laser Doppler Anemometry (Animation)

Chapter 9

- Flow around an Airfoil
- Flow Separation on an Airfoil
- Laminar and Turbulent Boundary Layers
- Growth of the Boundary Layer
- Effect of Viscosity on Boundary Layer Growth
- Examples of Boundary Layer Growth
- Flow Separation: Sudden Expansion
- Flow Separation: Airfoil
- Flow about a Sports Car
- Plate Normal to the Flow
- An Object with a High Drag Coefficient
- Examples of Flow around a Sphere
- Laminar and Turbulent Flow Past a Sphere
- Flow Separation on a Cylinder
- Vortex Trail behind a Cylinder
- Low Reynolds Number Flow over a Cylinder
- Flow Separation behind a Cylinder
- Flow past a Model A and a Sports Car
- Flow Past an Airfoil ($a = 0^o$)
- Flow Past an Airfoil ($a = 10^o$)
- Flow Past an Airfoil ($a = 20^o$)
- Wing Tip Vortices
- Leading Edge Slats

Chapter 10

- Flow in an Axial Flow Compressor (Animation)

Chapter 11

- A Turbulent Channel (Animation)
- The Glen Canyon Dam: A Source of Turbulent Channel Flow
- A Laminar Hydraulic Jump

Chapter 12

Sound Waves (Animation)

Shock Waves over a Supersonic Airplane

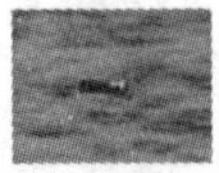
Shock Waves due to a Projectile

Chapter 13

Shock Waves over a Supersonic Airplane

Shock Waves due to a Projectile

The following videos were developed by the National Committee for Fluid Mechanics Films (NCFMF) and are referenced as "Classic Videos" in the text. They may all be viewed for free at http://web.mit.edu/hml/ncfmf.html or linked to from www.wiley.com/go/gobal/fox.

These videos are supplied by:

Encyclopaedia Britannica Educational Corporation
331 North La Salle Street
Chicago, IL 60654

Aerodynamic Generation of Sound (44 min, principals: M. J. Lighthill, J. E. Ffowcs-Williams)
Boundary Layer Control (25 min, principal: D. C. Hazen)
Cavitation (31 min, principal: P. Eisenberg)
Channel Flow of a Compressible Fluid (29 min, principal: D. E. Coles)
Deformation of Continuous Media (38 min, principal: J. L. Lumley)
Eulerian and Lagrangian Descriptions in Fluid Mechanics (27 min, principal: J. L. Lumley)
Flow Instabilities (27 min, principal: E. L. Mollo-Christensen)
Flow Visualization (31 min, principal: S. J. Kline)
The Fluid Dynamics of Drag (4 parts, 120 min, principal: A. H. Shapiro)
Fundamentals of Boundary Layers (24 min, principal: F. H. Abernathy)
Low-Reynolds-Number Flows (33 min, principal: Sir G. I. Taylor)
Magnetohydrodynamics (27 min, principal: J. A. Shercliff)
Pressure Fields and Fluid Acceleration (30 min, principal: A. H. Shapiro)
Rarefied Gas Dynamics (33 min, principals: F. C. Hurlbut, F. S. Sherman)
Rheological Behavior of Fluids (22 min, principal: H. Markovitz)
Rotating Flows (29 min, principal: D. Fultz)
Secondary Flow (30 min, principal: E. S. Taylor)
Stratified Flow (26 min, principal: R. R. Long)
Surface Tension in Fluid Mechanics (29 min, principal: L. M. Trefethen)
Turbulence (29 min, principal: R. W. Stewart)
Vorticity (2 parts, 44 min, principal: A. H. Shapiro)
Waves in Fluids (33 min, principal: A. E. Bryson)

For a list of additional videos on fluid mechanics, visit www.wiley.com/go/gobal/fox.

Appendix D

Selected Performance Curves for Pumps and Fans

Introduction D.1

Many firms, worldwide, manufacture fluid machines in numerous standard types and sizes. Each manufacturer publishes complete performance data to allow application of its machines in systems. This Appendix contains selected performance data for use in solving pump and fan system problems. Two pump types and one fan type are included.

Choice of a manufacturer may be based on established practice, location, or cost. Once a manufacturer is chosen, machine selection is a three-step process:

1. Select a machine type, suited to the application, from a manufacturer's full-line catalog, which gives the ranges of pressure rise (head) and flow rate for each machine type.
2. Choose an appropriate machine model and driver speed from a master selector chart, which superposes the head and flow rate ranges of a series of machines on one graph.
3. Verify that the candidate machine is satisfactory for the intended application, using a detailed performance curve for the specific machine.

It is wise to consult with experienced system engineers, either employed by the machine manufacturer or in your own organization, before making a final purchase decision.

Many manufacturers currently use computerized procedures to select a machine that is most suitable for each given application. Such procedures are simply automated versions of the traditional selection method. Use of the master selector chart and the detailed performance curves is illustrated below for pumps and fans, using data from one manufacturer of each type of machine. Literature of other manufacturers differs in detail but contains the necessary information for machine selection.

D.2 Pump Selection

Representative data are shown in Figs. D.1 through D.10 for Peerless[1] horizontal split case single-stage (series AE) pumps and in Figs. D.11 and D.12 for Peerless multi-stage (series TU and TUT) pumps.

Figures D.1 and D.2 are master pump selector charts for series AE pumps at 3500 and 1750 nominal rpm. On these charts, the model number (e.g., 6AE14) indicates the discharge line size (6 in. or about 150 mm nominal pipe), the pump series (AE), and the maximum impeller diameter (approximately 14 in. or 350 mm).

Figures D.3 through D.10 are detailed performance charts for individual pump models in the AE series.

Figures D.11 and D.12 are master pump selector charts for series TU and TUT pumps at 1750 nominal rpm. Data for two-stage pumps are presented in Fig. D.11, while Fig. D.12 contains data for pumps with three, four, and five stages.

Each pump performance chart contains curves of total head versus volume flow rate; curves for several impeller diameters—tested in the same casing—are presented on a single graph. Each performance chart also contains curves showing pump efficiency and driver power; the net positive suction head (*NPSH*) requirement, as it varies with flow rate, is shown by the curve at the bottom of each chart. The best efficiency point (BEP) for each impeller may be found using the efficiency curves.

Use of the master pump selector chart and detailed performance curves is illustrated in Example D.1.

Example D.1 PUMP SELECTION PROCEDURE

Select a pump to deliver 400 m^3/h of water at 36 m total head. Choose the appropriate pump model and driver speed. Specify the pump efficiency, driver power, and *NPSH* requirement.

Given: Select a pump to deliver 400 m^3/h of water at 36 m total head.

Find: (a) Pump model and driver speed.
(b) Pump efficiency.
(c) Driver power.
(d) *NPSH* requirement.

Solution:
Use the pump selection procedure described in Section D-1. (The numbers below correspond to the numbered steps given in the procedure.)

1. Select a machine type suited to the application. (This step actually requires a manufacturer's full-line catalog, which is not reproduced here. The Peerless product line catalog specifies a maximum delivery and head of 570 m^3/h and 200 m for series AE pumps. Therefore the required performance can be obtained; assume the selection is to be made from this series.)

[1]Peerless Pump Company, P.O. Box 7026, Indianapolis, IN 46207-7026.

2. Consult the master pump selector chart. The desired operating point is not within any pump contour on the 3500 rpm selector chart (Fig. D.1). From the 1750 rpm chart (Fig. D.2), select a model 6AE14 pump. From the performance curve for the 6AE14 pump (Fig. D.6), choose a 325 mm impeller.
3. Verify the performance of the machine using the detailed performance chart. On the performance chart for the 6AE14 pump, project up from the abscissa at $Q = 400\ \text{m}^3/\text{h}$. Project across from $H = 36$ m on the ordinate. The intersection is the pump performance at the desired operating point:

$$\eta \approx 85.8 \text{ percent} \qquad \mathcal{P} \approx 48 \text{ kW}$$

From the operating point, project down to the *NPSH* requirement curve. At the intersection, read $NPSH \approx 5$ m.

This completes the selection process for this pump. One should consult with experienced system engineers to verify that the system operating condition has been predicted accurately and the pump has been selected correctly.

Fan Selection D.3

Fan selection is similar to pump selection. A representative master fan selection chart is shown in Fig. D.13 for a series of Howden Buffalo[2] axial-flow fans. The chart shows the efficiency of the entire series of fans as a function of total pressure rise and flow rate. The series of numbers for each fan indicates the fan diameter in inches, the hub diameter in inches, and the fan speed in revolutions per minute. For instance, a 54-26-870 fan has a fan diameter of 54 in. or 1350 mm, a hub diameter of 26 in. or 650 mm, and should be operated at 870 rpm.

Normally, final evaluation of suitability of the fan model for the application would be done using detailed performance charts for the specific model. Instead, we use the efficiencies from Fig D.13, which are indicated by the shading of the different zones on the map. To calculate the power requirement for the fan motor, we use the following equation:

$$\mathcal{P}(\text{kW}) = \frac{Q(\text{m}^3/s) \times \Delta p(\text{mm} \cdot \text{H}_2\text{O})}{\eta}$$

A sample fan selection is presented in Example D.2.

Example D.2 FAN SELECTION PROCEDURE

Select an axial flow fan to deliver 850 m^3/min of standard air at 32 mm. H_2O total pressure. Choose the appropriate fan model and driver speed. Specify the fan efficiency and driver power.

Given: Select an axial flow fan to deliver 850 m^3/min of standard air at 32 mm H_2O total head.

Find: (a) Fan size and driver speed.
(b) Fan efficiency.
(c) Driver power.

Solution:
Use the fan selection procedure described in Section D-1. (The numbers below correspond to the numbered steps given in the procedure.)

[2]Howden Buffalo Inc., 2029 W. DeKalb St., Camden, SC 29020.

1. Select a machine type suited to the application. (This step actually requires a manufacturer's full-line catalog, which is not reproduced here. Assume the fan selection is to be made from the axial fan data presented in Fig. D.13.)
2. Consult the master fan selector chart. The desired operating point is within the contour for the 48-21-860 fan on the selector chart (Fig. D.13). To achieve the desired performance requires driving the fan at 860 rpm.
3. Verify the performance of the machine using a detailed performance chart. To determine the efficiency, we consult Fig D.13 again. We estimate an efficiency of 85 percent. To determine the motor power requirement, we use the equation given above:

$$\mathscr{P} = \frac{Q\Delta p}{\eta} = 8.50\frac{\text{m}^3}{\text{min}} \times \frac{\text{min}}{60\text{s}} \times 32 \text{ mm H}_2\text{O}$$

$$\times 999\frac{\text{kg}}{\text{m}^3} \times 9.81\frac{\text{m}}{\text{s}^2} \times \frac{1}{0.85} \times \frac{\text{m}}{1000 \text{ mm}} = 5.23 \text{ kW}$$

This completes the fan selection process. Again, one should consult with experienced system engineers to verify that the system operating condition has been predicted accurately and the fan has been selected correctly.

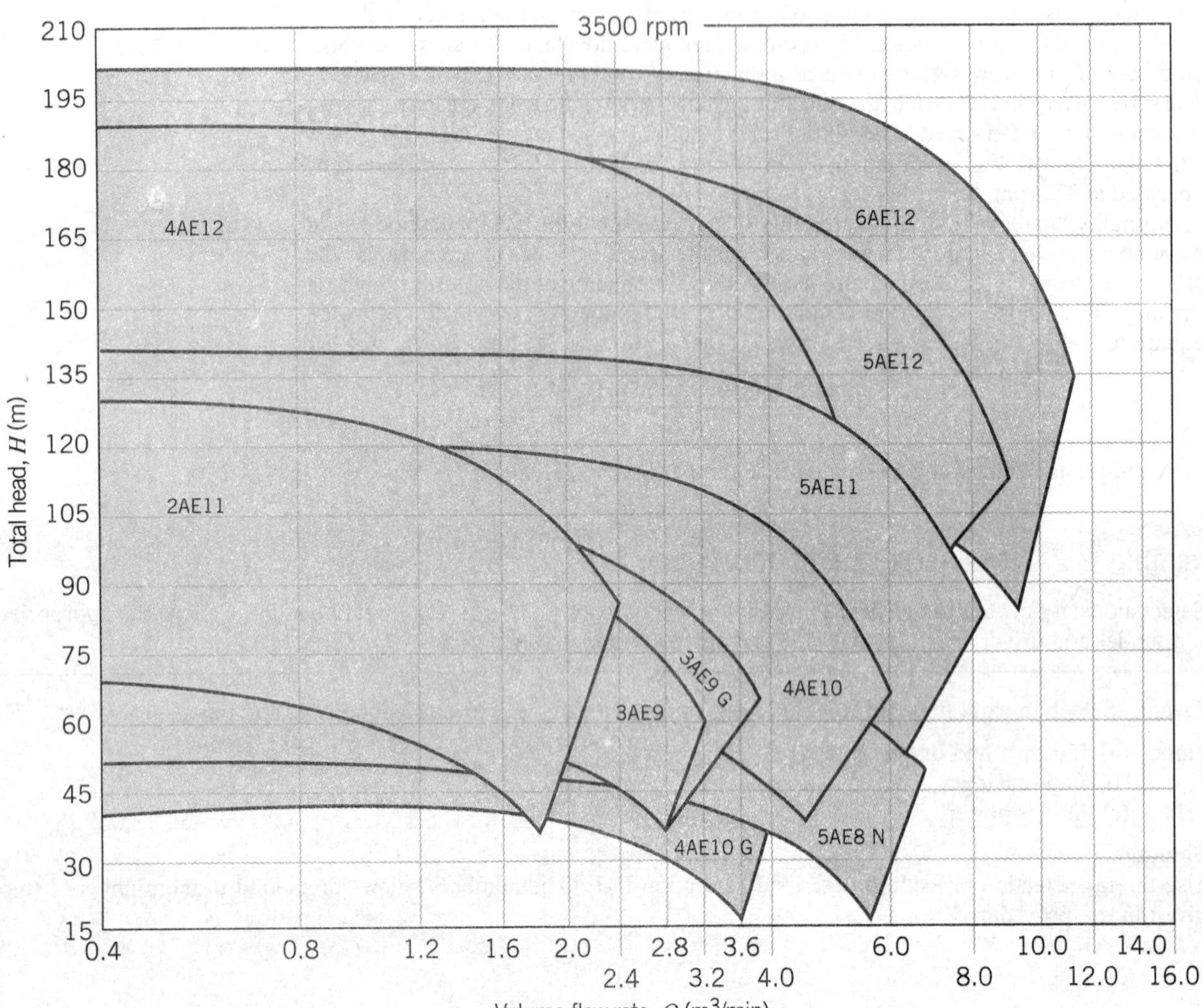

Fig. D.1 Selector chart for Peerless horizontal split case (series AE) pumps at 3500 nominal rpm.

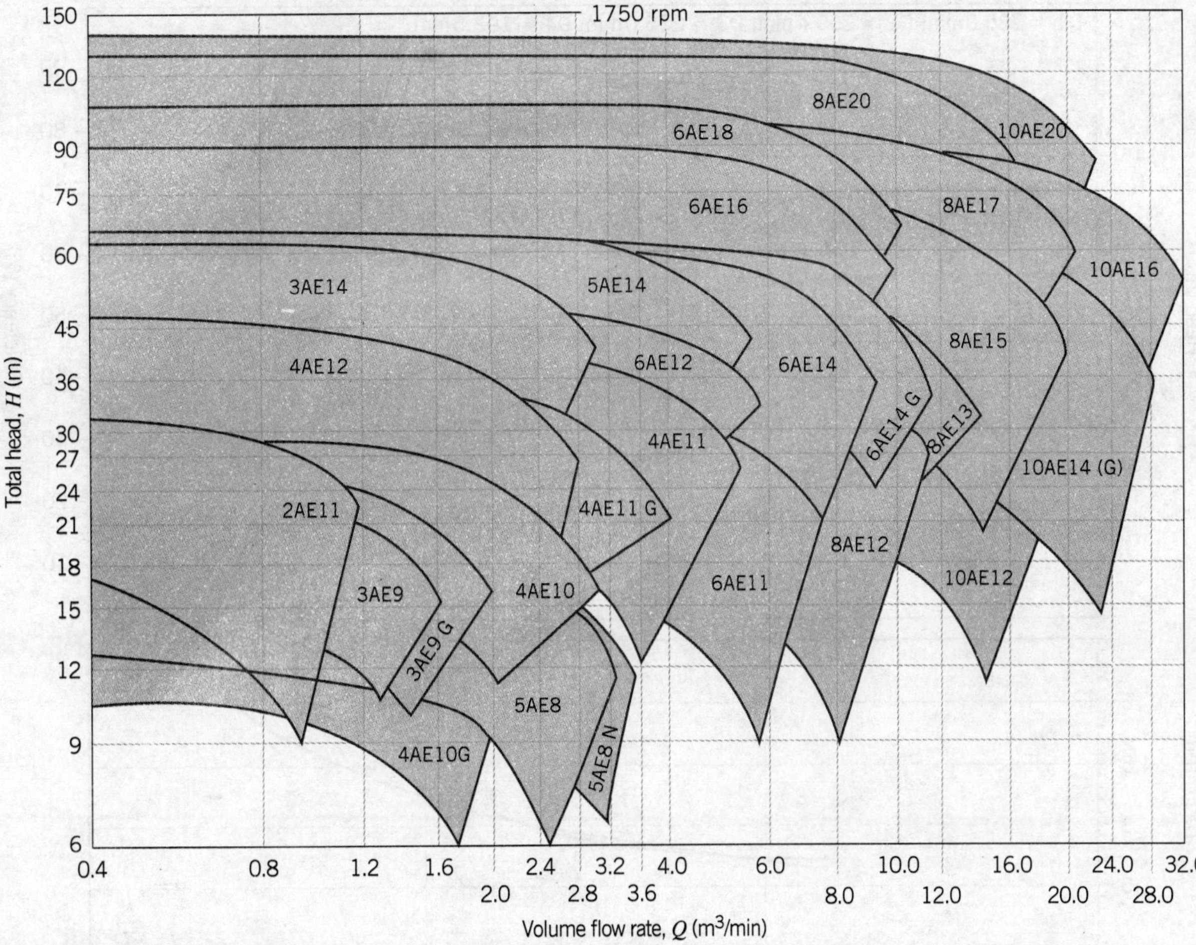

Fig. D.2 Selector chart for Peerless horizontal split case (series AE) pumps at 1750 nominal rpm.

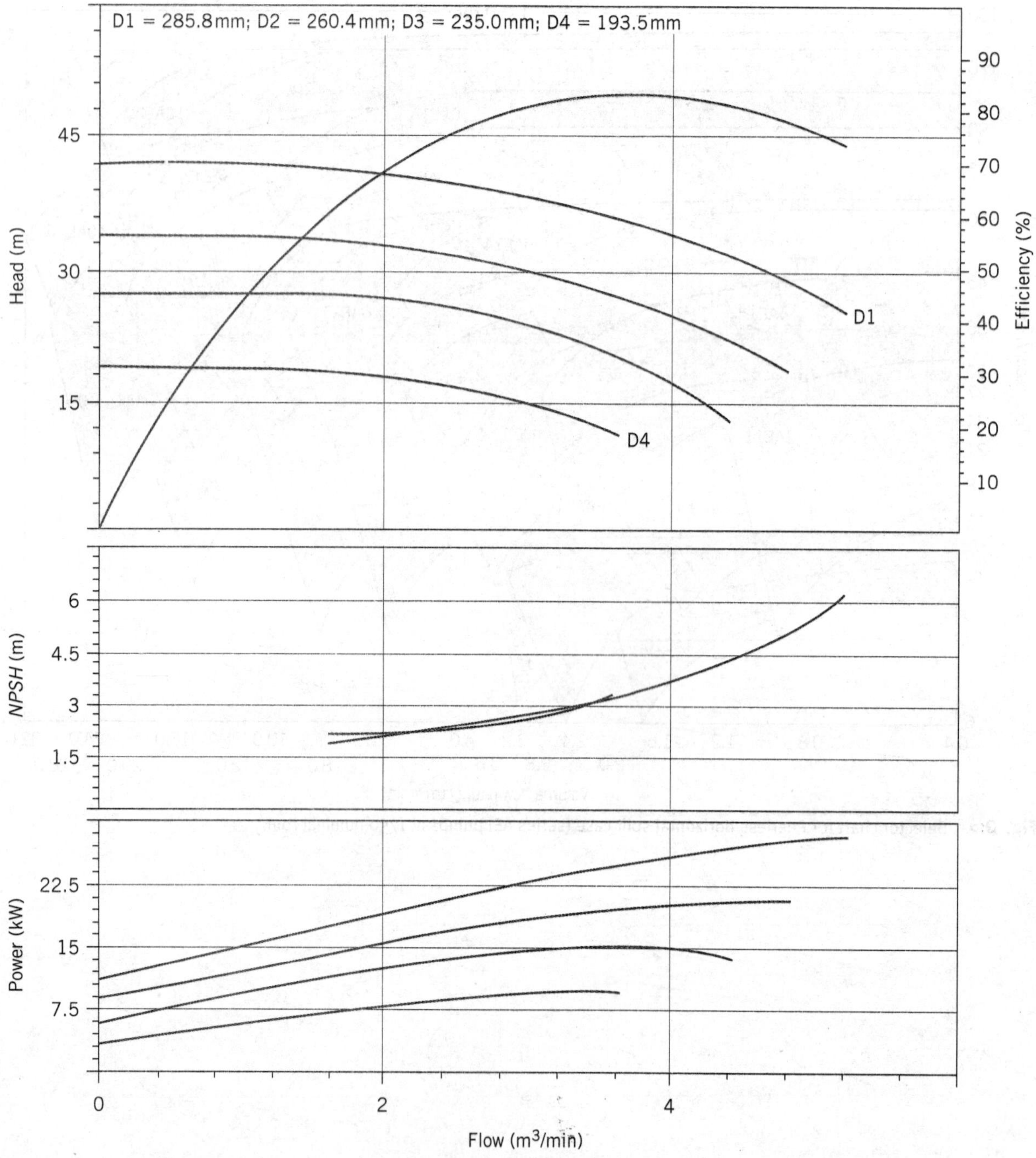

Fig. D.3 Performance curve for Peerless 4AE11 pump at 1750 rpm.

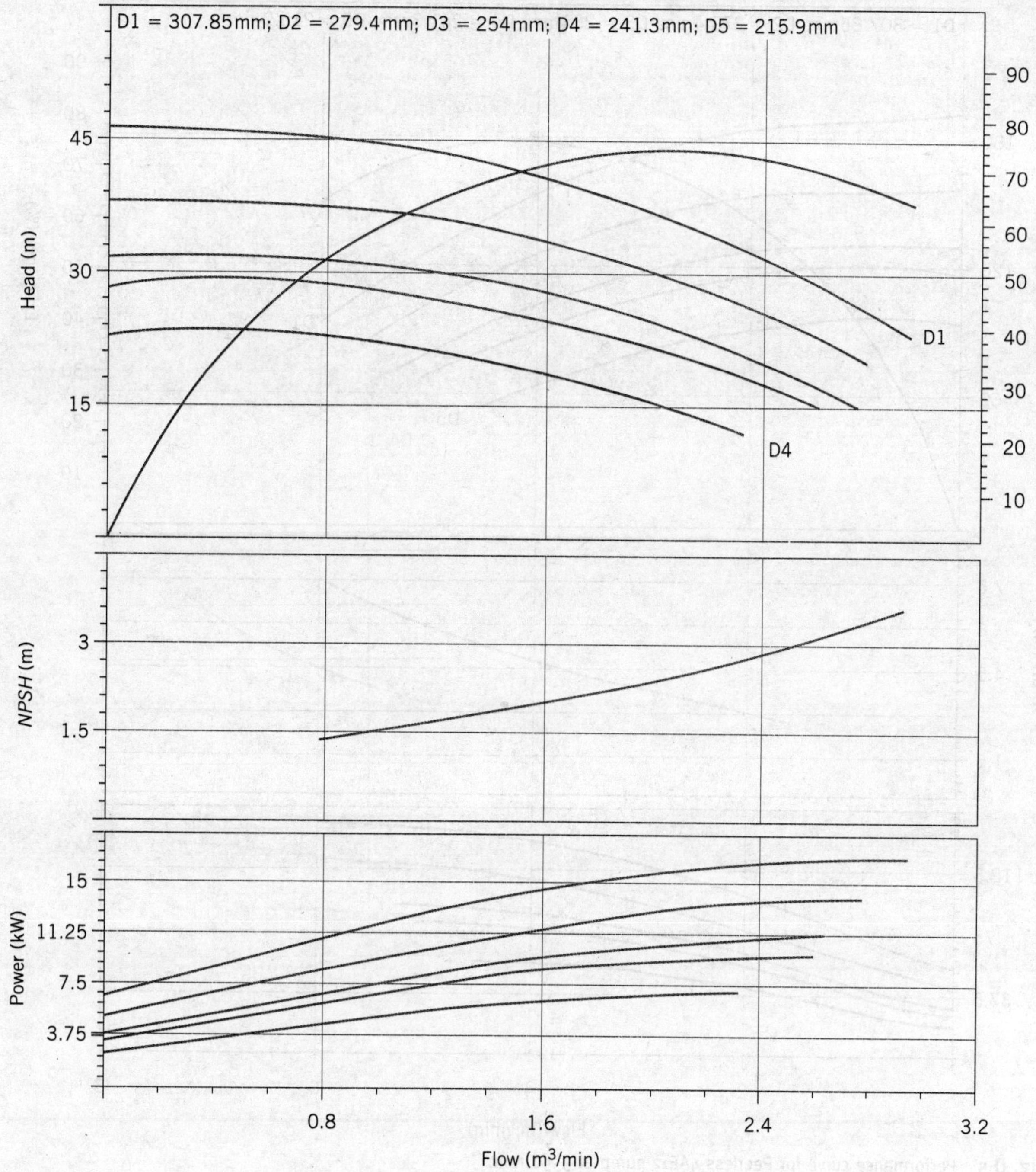

Fig. D.4 Performance curve for Peerless 4AE12 pump at 1750 rpm.

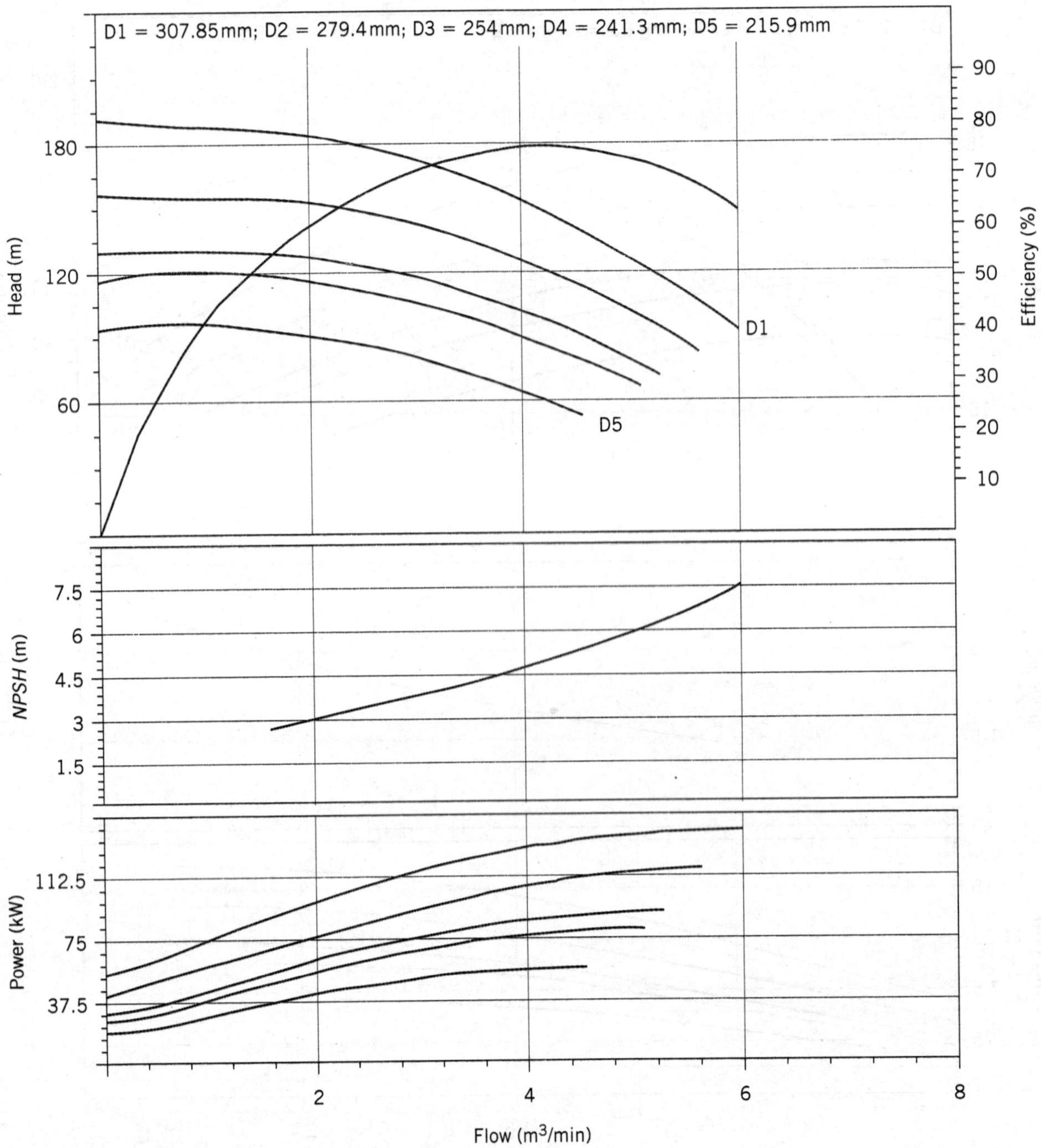

Fig. D.5 Performance curve for Peerless 4AE12 pump at 3550 rpm.

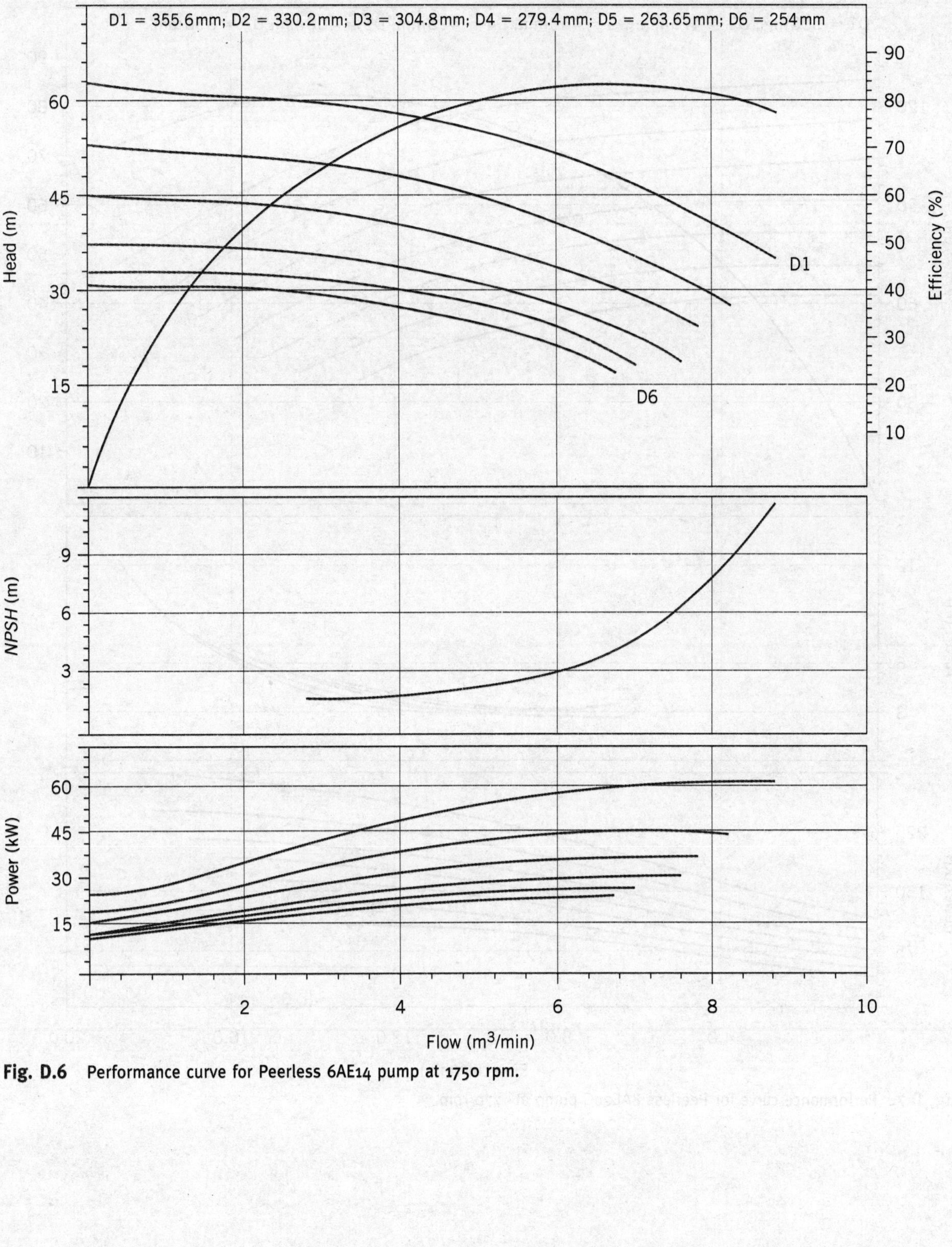

Fig. D.6 Performance curve for Peerless 6AE14 pump at 1750 rpm.

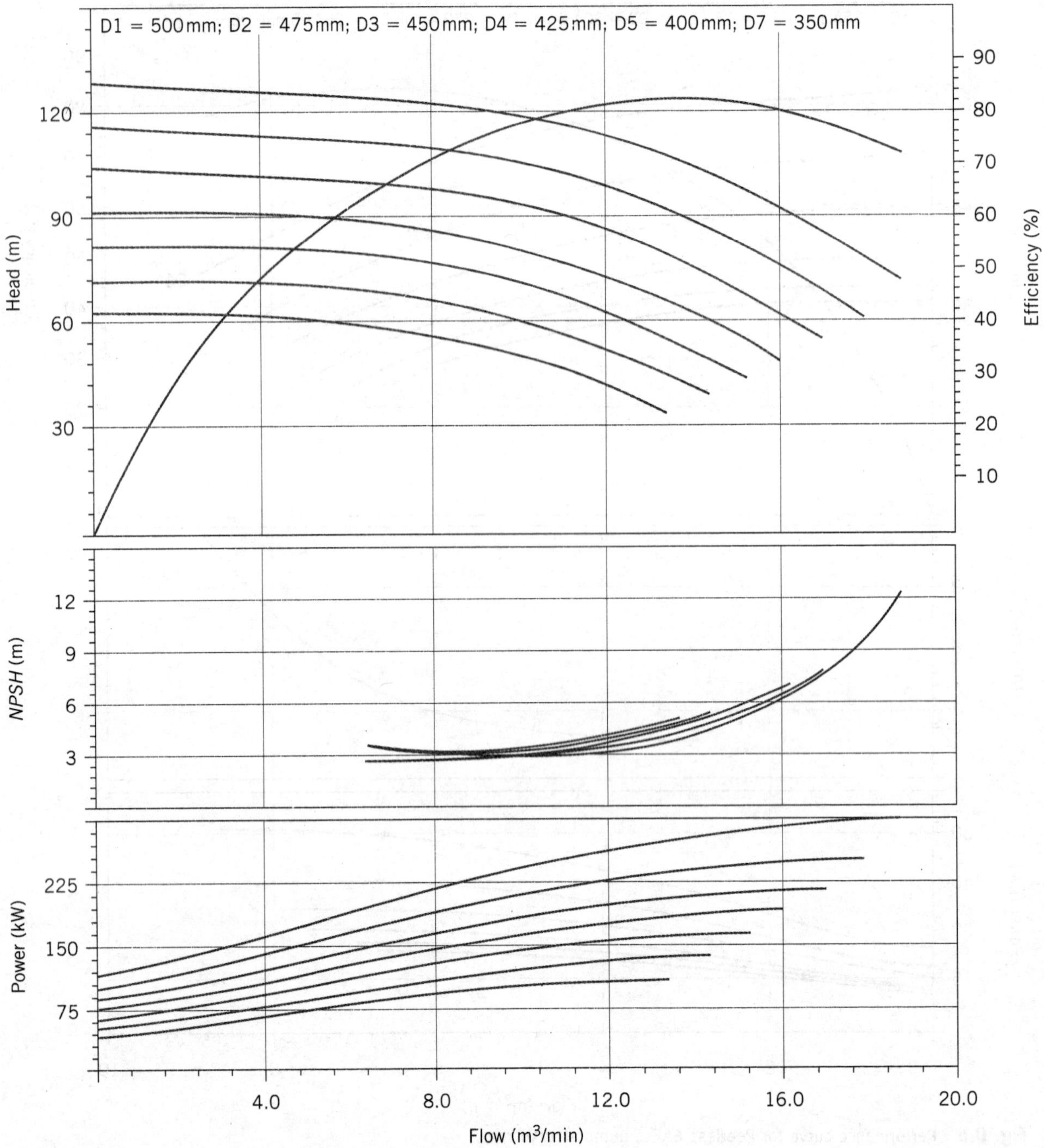

Fig. D.7 Performance curve for Peerless 8AE20G pump at 1770 rpm.

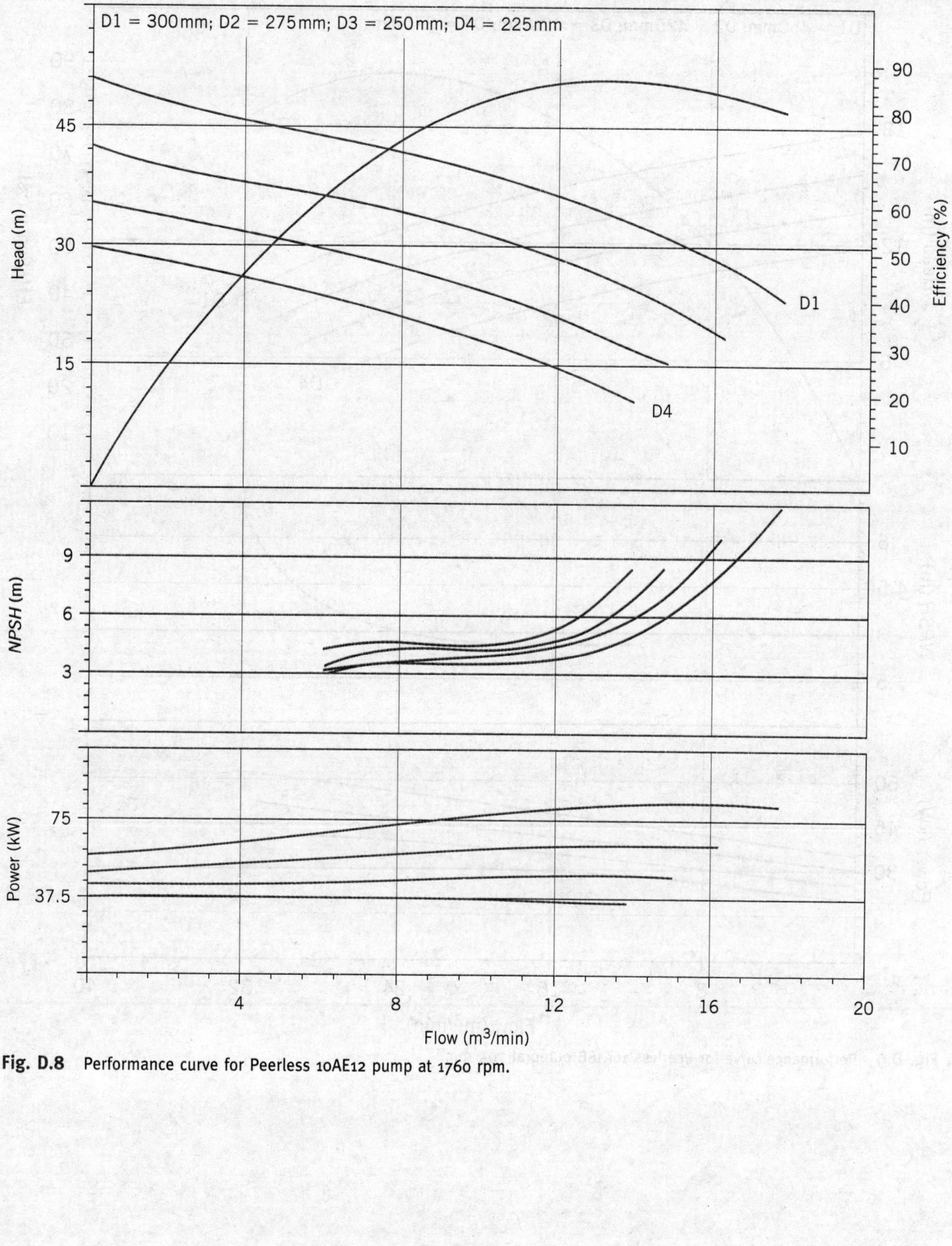

Fig. D.8 Performance curve for Peerless 10AE12 pump at 1760 rpm.

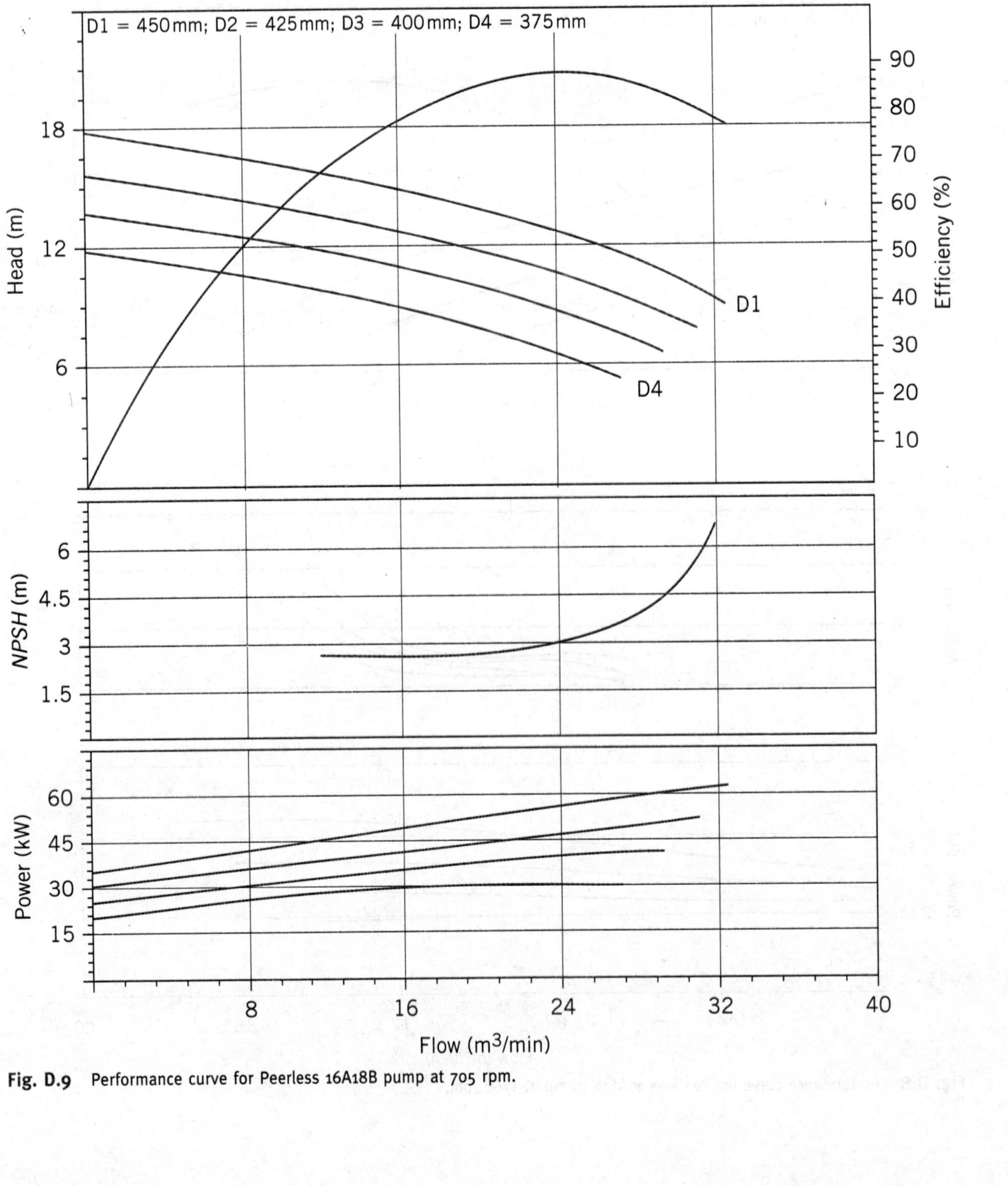

Fig. D.9 Performance curve for Peerless 16A18B pump at 705 rpm.

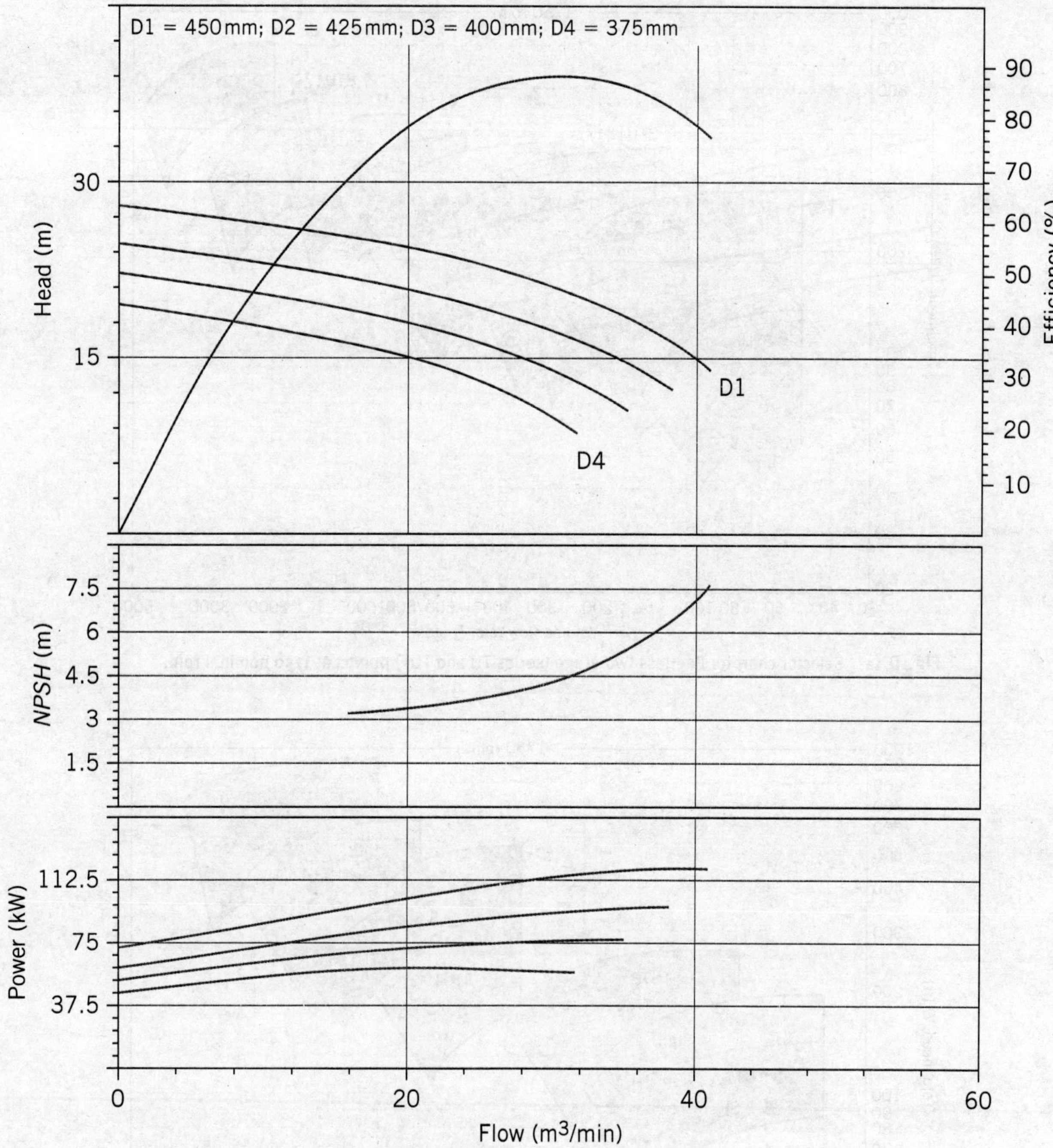

Fig. D.10 Performance curve for Peerless 16A18B pump at 880 rpm.

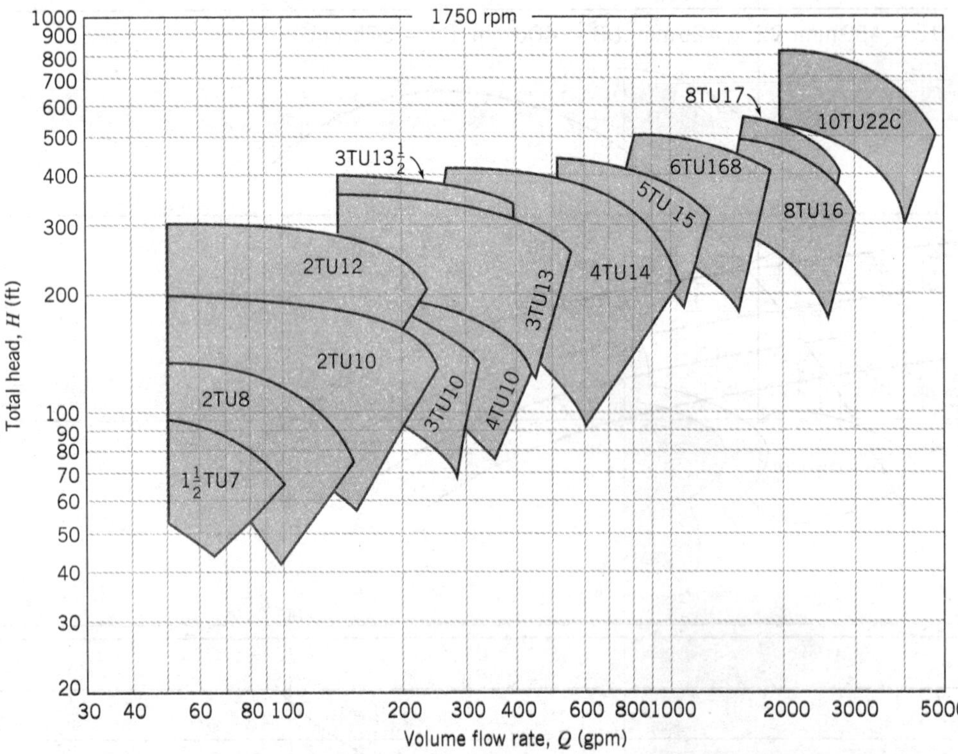

Fig. D.11 Selector chart for Peerless two-stage (series TU and TUT) pumps at 1750 nominal rpm.

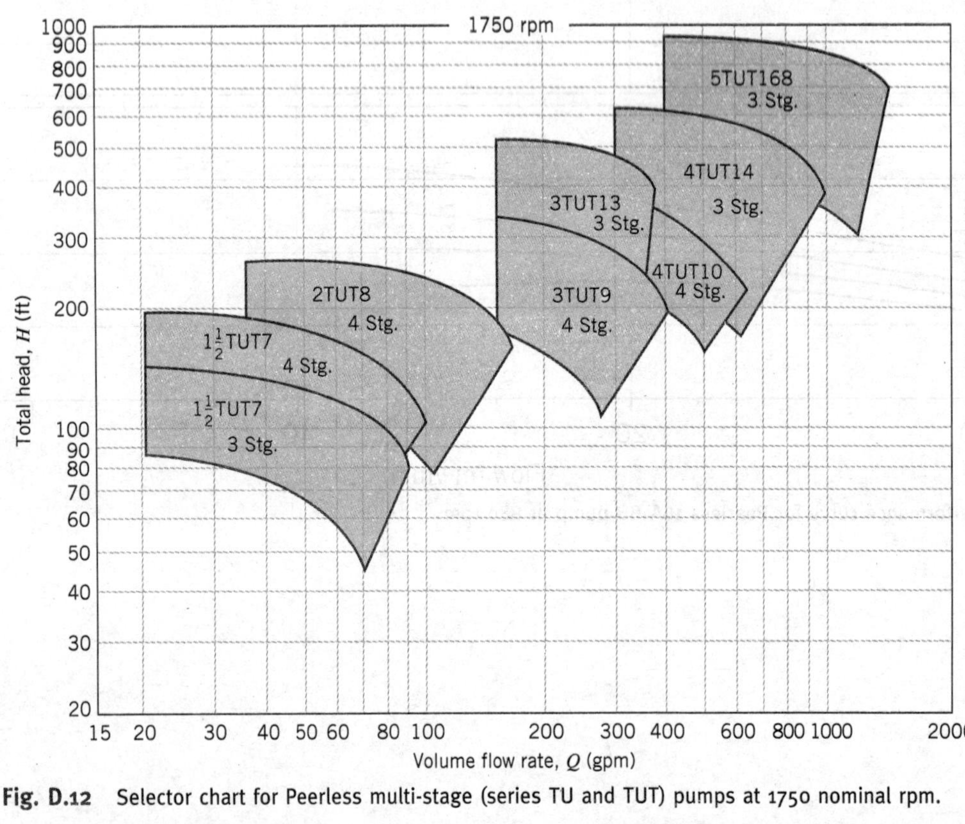

Fig. D.12 Selector chart for Peerless multi-stage (series TU and TUT) pumps at 1750 nominal rpm.

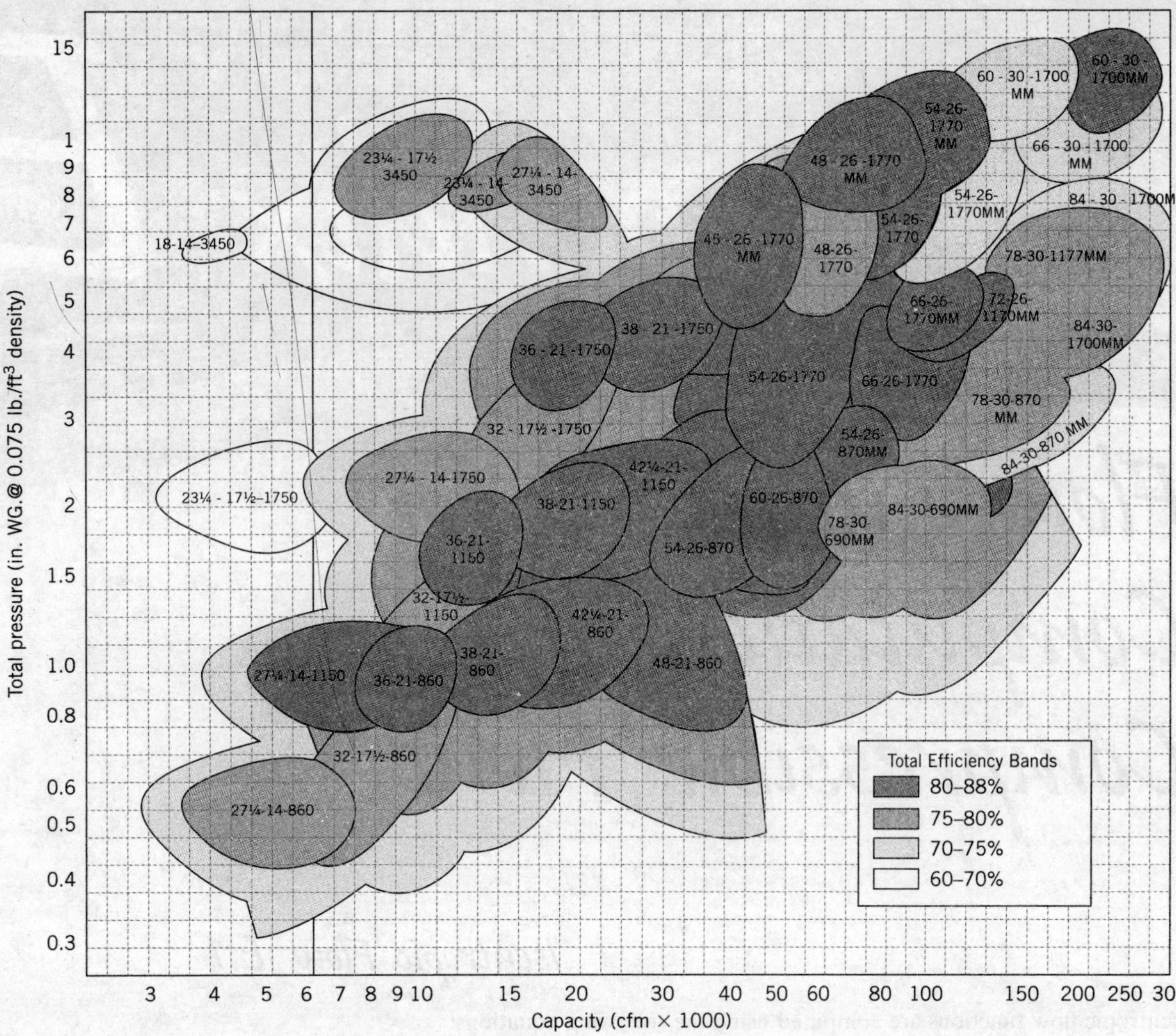

Fig. D.13 Master fan selection chart for Howden Buffalo axial fans.

References

1. Peerless Pump literature:
   - Horizontal Split Case Single Stage Double Suction Pumps, Series AE, Brochure B-1200, 2003.
   - Horizontal Split Case, Multistage Single Suction Pumps, Types TU, TUT, 60 Hertz, Performance Curves, Brochure B-1440, 2003.
   - RAPID v8.25.6, March 2007.
2. Buffalo Forge literature:
   - Axivane Axial Fan Optimum Efficiency Selection Chart, n.d.

Appendix E

Flow Functions for Computation of Compressible Flow

Isentropic Flow E.1

Isentropic flow functions are computed using the following equations:

$$\frac{T_0}{T} = 1 + \frac{k-1}{2}M^2 \qquad (12.21b)/(13.7b)$$

$$\frac{p_0}{p} = \left[1 + \frac{k-1}{2}M^2\right]^{k/(k-1)} \qquad (12.21a)/(13.7a)$$

$$\frac{\rho_0}{\rho} = \left[1 + \frac{k-1}{2}M^2\right]^{1/(k-1)} \qquad (12.21c)/(13.7c)$$

$$\frac{A}{A^*} = \frac{1}{M}\left[\frac{1 + \frac{k-1}{2}M^2}{\frac{k+1}{2}}\right]^{(k+1)/2(k-1)} \qquad (13.7d)$$

Representative values of the isentropic flow functions for $k = 1.4$ are presented in Table E.1 and plotted in Fig. E.1.

Table E.1

Isentropic Flow Functions (one-dimensional flow, ideal gas, k = 1.4)

M	T/T_0	p/p_0	ρ/ρ_0	A/A^*
0.00	1.0000	1.0000	1.0000	∞
0.50	0.9524	0.8430	0.8852	1.340
1.00	0.8333	0.5283	0.6339	1.000
1.50	0.6897	0.2724	0.3950	1.176
2.00	0.5556	0.1278	0.2301	1.688
2.50	0.4444	0.05853	0.1317	2.637
3.00	0.3571	0.02722	0.07623	4.235
3.50	0.2899	0.01311	0.04523	6.790
4.00	0.2381	0.006586	0.02766	10.72
4.50	0.1980	0.003455	0.01745	16.56
5.00	0.1667	0.001890	0.01134	25.00

This table was computed from the *Excel* workbook *Isentropic Relations*. The workbook contains a more detailed, printable version of the table and can be easily modified to generate data for a different Mach number range, or for a different gas.

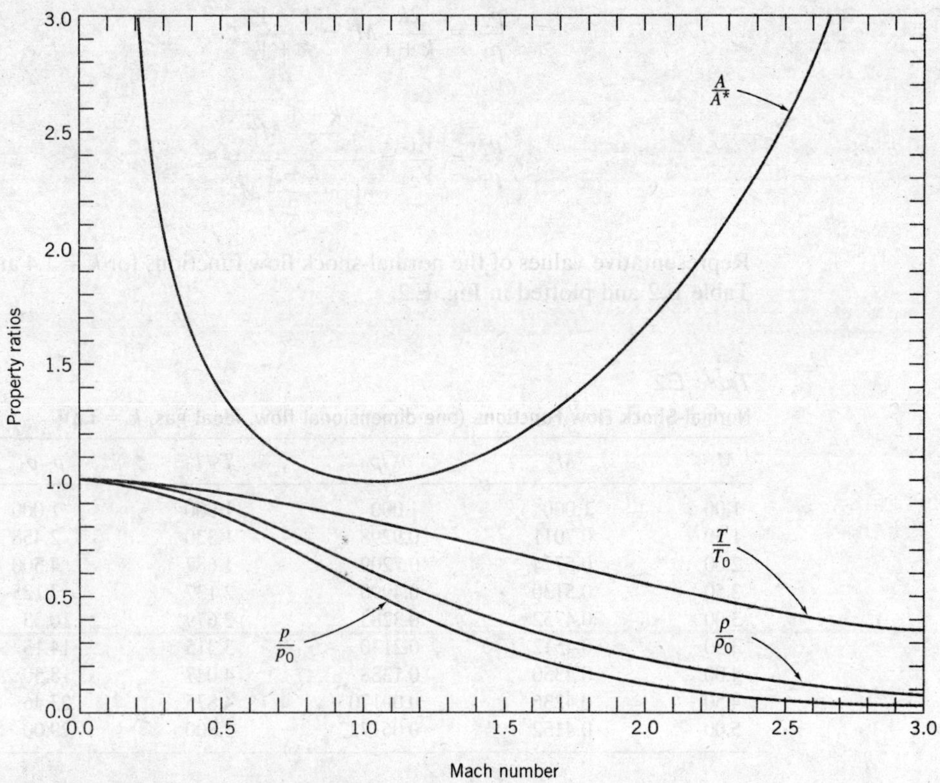

Fig. E.1 Isentropic flow functions.

This graph was generated from the *Excel* workbook. The workbook can be modified easily to generate curves for a different gas.

E.2 Normal Shock

Normal-shock flow functions are computed using the following equations:

$$M_2^2 = \frac{M_1^2 + \frac{2}{k-1}}{\frac{2k}{k-1}M_1^2 - 1} \tag{13.20a}$$

$$\frac{p_{0_2}}{p_{0_1}} = \frac{\left[\dfrac{\frac{k+1}{2}M_1^2}{1+\frac{k-1}{2}M_1^2}\right]^{k/(k-1)}}{\left[\frac{2k}{k+1}M_1^2 - \frac{k-1}{k+1}\right]^{1/(k-1)}} \tag{13.20b}$$

$$\frac{T_2}{T_1} = \frac{\left(1+\frac{k-1}{2}M_1^2\right)\left(kM_1^2 - \frac{k-1}{2}\right)}{\left(\frac{k+1}{2}\right)^2 M_1^2} \tag{13.20c}$$

$$\frac{p_2}{p_1} = \frac{2k}{k+1}M_1^2 - \frac{k-1}{k+1} \tag{13.20d}$$

$$\frac{\rho_2}{\rho_1} = \frac{V_1}{V_2} = \frac{\frac{k+1}{2}M_1^2}{1+\frac{k-1}{2}M_1^2} \tag{13.20e}$$

Representative values of the normal-shock flow functions for $k = 1.4$ are presented in Table E.2 and plotted in Fig. E.2.

Table E.2

Normal-Shock Flow Functions (one-dimensional flow, ideal gas, $k = 1.4$)

M_1	M_2	p_{0_2}/p_{0_1}	T_2/T_1	p_2/p_1	ρ_2/ρ_1
1.00	1.000	1.000	1.000	1.000	1.000
1.50	0.7011	0.9298	1.320	2.458	1.862
2.00	0.5774	0.7209	1.687	4.500	2.667
2.50	0.5130	0.4990	2.137	7.125	3.333
3.00	0.4752	0.3283	2.679	10.33	3.857
3.50	0.4512	0.2130	3.315	14.13	4.261
4.00	0.4350	0.1388	4.047	18.50	4.571
4.50	0.4236	0.09170	4.875	23.46	4.812
5.00	0.4152	0.06172	5.800	29.00	5.000

This table was computed from the *Excel* workbook *Normal-Shock Relations*. The workbook contains a more detailed, printable version of the table and can be modified easily to generate data for a different Mach number range, or for a different gas.

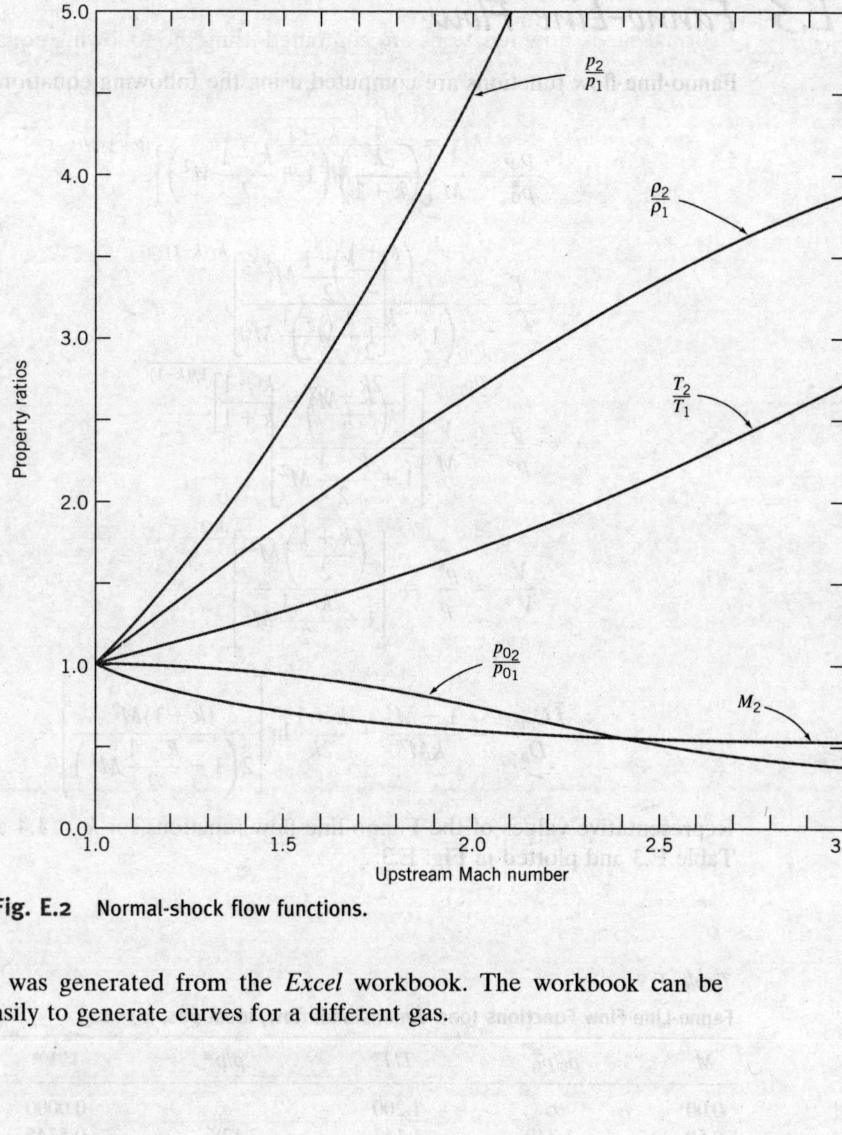

Fig. E.2 Normal-shock flow functions.

This graph was generated from the *Excel* workbook. The workbook can be modified easily to generate curves for a different gas.

E.3 Fanno-Line Flow

Fanno-line flow functions are computed using the following equations:

$$\frac{p_0}{p_0^*} = \frac{1}{M}\left[\left(\frac{2}{k+1}\right)\left(1+\frac{k-1}{2}M^2\right)\right]^{(k+1)/2(k-1)} \tag{13.34e}$$

$$\frac{T}{T^*} = \frac{\left(\frac{k+1}{2}\right)}{\left(1+\frac{k-1}{2}M^2\right)} \tag{13.34b}$$

$$\frac{p}{p^*} = \frac{1}{M}\left[\frac{\left(\frac{k+1}{2}\right)}{1+\frac{k-1}{2}M^2}\right]^{1/2} \tag{13.34d}$$

$$\frac{V}{V^*} = \frac{\rho^*}{\rho} = \left[\frac{\left(\frac{k+1}{2}\right)M^2}{1+\frac{k-1}{2}M^2}\right]^{1/2} \tag{13.34c}$$

$$\frac{\bar{f}L_{max}}{D_h} = \frac{1-M^2}{kM^2} + \frac{k+1}{2k}\ln\left[\frac{(k+1)M^2}{2\left(1+\frac{k-1}{2}M^2\right)}\right] \tag{13.34a}$$

Representative values of the Fanno-line flow functions for $k = 1.4$ are presented in Table E.3 and plotted in Fig. E.3.

Table E.3

Fanno-Line Flow Functions (one-dimensional flow, ideal gas, $k = 1.4$)

M	p_0/p_0^*	T/T^*	p/p^*	V/V^*	$\bar{f}L_{max}/D_h$
0.00	∞	1.200	∞	0.0000	∞
0.50	1.340	1.143	2.138	0.5345	1.069
1.00	1.000	1.000	1.000	1.000	0.0000
1.50	1.176	0.8276	0.6065	1.365	0.1361
2.00	1.688	0.6667	0.4083	1.633	0.3050
2.50	2.637	0.5333	0.2921	1.826	0.4320
3.00	4.235	0.4286	0.2182	1.964	0.5222
3.50	6.790	0.3478	0.1685	2.064	0.5864
4.00	10.72	0.2857	0.1336	2.138	0.6331
4.50	16.56	0.2376	0.1083	2.194	0.6676
5.00	25.00	0.2000	0.08944	2.236	0.6938

This table was computed from the *Excel* workbook *Fanno-Line Relations*. The workbook contains a more detailed, printable version of the table and can be modified easily to generate data for a different Mach number range, or for a different gas.

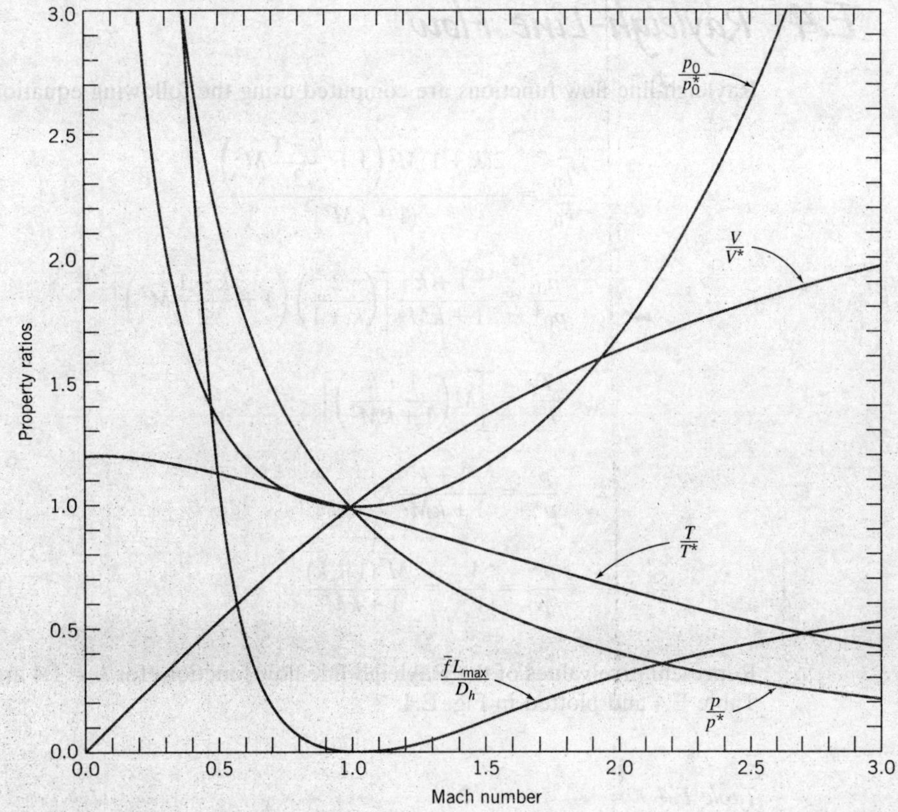

Fig. E.3 Fanno-line flow functions.

This graph was generated from the *Excel* workbook. The workbook can be modified easily to generate curves for a different gas.

E.4 Rayleigh-Line Flow

Rayleigh-line flow functions are computed using the following equations:

$$\frac{T_0}{T_0^*} = \frac{2(k+1)M^2\left(1+\frac{k-1}{2}M^2\right)}{(1+kM^2)^2} \tag{13.44d}$$

$$\frac{p_0}{p_0{}^*} = \frac{1+k}{1+kM^2}\left[\left(\frac{2}{k+1}\right)\left(1+\frac{k-1}{2}M^2\right)\right]^{k/(k-1)} \tag{13.44e}$$

$$\frac{T}{T^*} = \left[M\left(\frac{1+k}{1+kM^2}\right)\right]^2 \tag{13.44b}$$

$$\frac{p}{p^*} = \frac{1+k}{1+kM^2} \tag{13.44a}$$

$$\frac{\rho^*}{\rho} = \frac{V}{V^*} = \frac{M^2(1+k)}{1+kM^2} \tag{13.44c}$$

Representative values of the Rayleigh-line flow functions for $k = 1.4$ are presented in Table E.4 and plotted in Fig. E.4.

Table E.4

Rayleigh-Line Flow Functions (one-dimensional flow, ideal gas, $k = 1.4$)

M	T_0/T_0^*	p_0/p_0^*	T/T^*	p/p^*	V/V^*
0.00	0.0000	1.268	0.0000	2.400	0.0000
0.50	0.6914	1.114	0.7901	1.778	0.4444
1.00	1.000	1.000	1.000	1.000	1.000
1.50	0.9093	1.122	0.7525	0.5783	1.301
2.00	0.7934	1.503	0.5289	0.3636	1.455
2.50	0.7101	2.222	0.3787	0.2462	1.539
3.00	0.6540	3.424	0.2803	0.1765	1.588
3.50	0.6158	5.328	0.2142	0.1322	1.620
4.00	0.5891	8.227	0.1683	0.1026	1.641
4.50	0.5698	12.50	0.1354	0.08177	1.656
5.00	0.5556	18.63	0.1111	0.06667	1.667

This table was computed from the *Excel* workbook *Rayleigh-Line Relations*. The workbook contains a more detailed, printable version of the table and can be easily modified to generate data for a different Mach number range, or for a different gas.

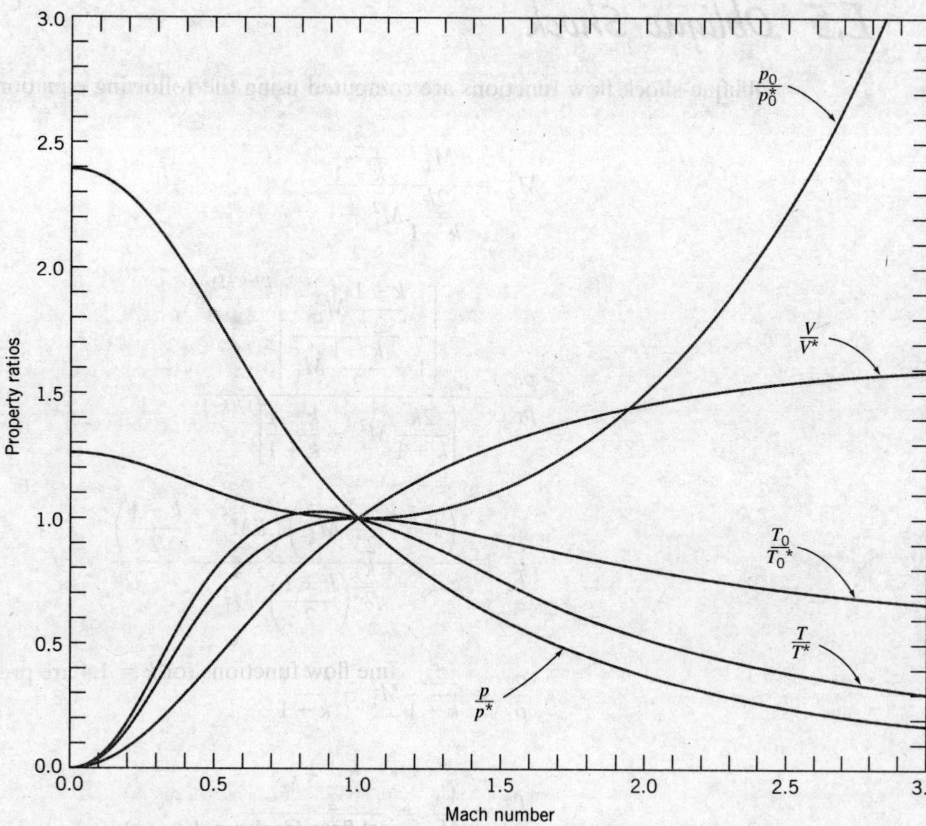

Fig. E.4 Rayleigh-line flow functions.

This graph was generated from the *Excel* workbook. The workbook can be modified easily to generate curves for a different gas.

E.5 Oblique Shock

Oblique-shock flow functions are computed using the following equations:

$$M_{2_n}^2 = \frac{M_{1_n}^2 + \dfrac{2}{k-1}}{\dfrac{2k}{k-1} M_{1_n}^2 - 1} \tag{13.48a}$$

$$\frac{p_{0_2}}{p_{0_1}} = \frac{\left[\dfrac{\dfrac{k+1}{2} M_{1_n}^2}{1 + \dfrac{k-1}{2} M_{1_n}^2}\right]^{k/(k-1)}}{\left[\dfrac{2k}{k+1} M_{1_n}^2 - \dfrac{k-1}{k+1}\right]^{1/(k-1)}} \tag{13.48b}$$

$$\frac{T_2}{T_1} = \frac{\left(1 + \dfrac{k-1}{2} M_{1_n}^2\right)\left(k M_{1_n}^2 - \dfrac{k-1}{2}\right)}{\left(\dfrac{k+1}{2}\right)^2 M_{1_n}^2} \tag{13.48c}$$

$$\frac{p_2}{p_1} = \frac{2k}{k+1} M_{1_n}^2 - \frac{k-1}{k+1} \tag{13.48d}$$

$$\frac{\rho_2}{\rho_1} = \frac{V_{1_n}}{V_{2_n}} = \frac{\dfrac{k+1}{2} M_{1_n}^2}{1 + \dfrac{k-1}{2} M_{1_n}^2} \tag{13.48e}$$

Representative values of the oblique-shock flow functions for $k = 1.4$ are presented in Table E.5 (identical to Table E.2 except for the Mach number notations).

Table E.5
Oblique-Shock Flow Functions (ideal gas, k = 1.4)

M_{1_n}	M_{2_n}	p_{0_2}/p_{0_1}	T_2/T_1	p_2/p_1	ρ_2/ρ_1
1.00	1.000	1.0000	1.000	1.000	1.000
1.50	0.7011	0.9298	1.320	2.458	1.862
2.00	0.5774	0.7209	1.687	4.500	2.667
2.50	0.5130	0.4990	2.137	7.125	3.333
3.00	0.4752	0.3283	2.679	10.33	3.857
3.50	0.4512	0.2130	3.315	14.13	4.261
4.00	0.4350	0.1388	4.047	18.50	4.571
4.50	0.4236	0.09170	4.875	23.46	4.812
5.00	0.4152	0.06172	5.800	29.00	5.000

The deflection angle θ, oblique-shock angle β, and Mach number M_1 are related using the following equation:

$$\tan\theta = \frac{2\cot\beta(M_1^2\sin^2\beta - 1)}{M_1^2(k + \cos 2\beta) + 2} \tag{13.49}$$

Representative values of angle θ are presented in Table E.6.

Table E.6

Oblique-Shock Deflection Angle θ (deg) (ideal gas, k = 1.4)

Shock angle β (deg)	Mach number M_1										
	1.2	1.4	1.6	1.8	2	2.5	3	4	6	10	∞
0	—	—	—	—	—	—	—	—	—	—	—
5	—	—	—	—	—	—	—	—	—	—	4.16
10	—	—	—	—	—	—	—	—	0.64	5.53	8.32
15	—	—	—	—	—	—	—	0.80	7.18	10.5	12.4
20	—	—	—	—	—	—	0.77	7.44	12.4	15.1	16.5
25	—	—	—	—	—	1.93	7.28	12.9	17.1	19.3	20.6
30	—	—	—	—	—	7.99	12.8	17.8	21.5	23.4	24.5
35	—	—	—	1.41	5.75	13.2	17.6	22.2	25.6	27.3	28.3
40	—	—	1.31	6.49	10.6	17.7	21.8	26.2	29.4	31.1	32.0
45	—	—	5.73	10.7	14.7	21.6	25.6	29.8	33.0	34.6	35.5
50	—	3.28	9.31	14.2	18.1	24.9	28.9	33.1	36.2	37.8	38.8
55	—	6.18	12.1	16.9	20.7	27.4	31.5	35.8	39.0	40.7	41.6
60	1.61	8.20	13.9	18.6	22.4	29.2	33.3	37.8	41.1	42.9	43.9
65	3.16	9.27	14.6	19.2	23.0	29.8	34.1	38.7	42.3	44.2	45.3
70	3.88	9.32	14.2	18.5	22.1	28.9	33.3	38.2	42.1	44.2	45.4
75	3.80	8.29	12.5	16.2	19.5	25.9	30.2	35.3	39.5	41.8	43.1
80	3.01	6.25	9.34	12.2	14.8	20.1	23.9	28.7	32.8	35.2	36.6
85	1.66	3.36	5.03	6.61	8.08	11.2	13.6	16.8	19.7	21.6	22.7
90	0	0	0	0	0	0	0	0	0	0	0

Tables E.5 and E.6 were computed from the *Excel* workbook *Oblique-Shock Relations*. The workbook contains a more detailed, printable version of the tables and can be modified easily to generate data for a different Mach number range, or for a different gas.

13.6 *Isentropic Expansion Wave Relations*

The Prandtl-Meyer supersonic expansion function, ω, is

$$\omega = \sqrt{\frac{k+1}{k-1}} \tan^{-1}\left(\sqrt{\frac{k-1}{k+1}(M^2-1)}\right) - \tan^{-1}\left(\sqrt{M^2-1}\right) \qquad (13.55)$$

Representative values of angle ω are presented in Table E.7.

Table E.7

Prandtl-Meyer Supersonic Expansion Function ω (deg) (ideal gas, $k = 1.4$)

M	ω (deg)	M	ω (deg)	M	ω (deg)	M	ω (deg)
1.00	0.00	2.00	26.4	3.00	49.8	4.00	65.8
1.05	0.49	2.05	27.7	3.05	50.7	4.05	66.4
1.10	1.34	2.10	29.1	3.10	51.6	4.10	67.1
1.15	2.38	2.15	30.4	3.15	52.6	4.15	67.7
1.20	3.56	2.20	31.7	3.20	53.5	4.20	68.3
1.25	4.83	2.25	33.0	3.25	54.4	4.25	68.9
1.30	6.17	2.30	34.3	3.30	55.2	4.30	69.5
1.35	7.56	2.35	35.5	3.35	56.1	4.35	70.1
1.40	8.99	2.40	36.7	3.40	56.9	4.40	70.7
1.45	10.4	2.45	37.9	3.45	57.7	4.45	71.3
1.50	11.9	2.50	39.1	3.50	58.5	4.50	71.8
1.55	13.4	2.55	40.3	3.55	59.3	4.55	72.4
1.60	14.9	2.60	41.4	3.60	60.1	4.60	72.9
1.65	16.3	2.65	42.5	3.65	60.9	4.65	73.4
1.70	17.8	2.70	43.6	3.70	61.6	4.70	74.0
1.75	19.3	2.75	44.7	3.75	62.3	4.75	74.5
1.80	20.7	2.80	45.7	3.80	63.0	4.80	75.0
1.85	22.2	2.85	46.8	3.85	63.7	4.85	75.5
1.90	23.6	2.90	47.8	3.90	64.4	4.90	76.0
1.95	25.0	2.95	48.8	3.95	65.1	4.95	76.4
2.00	26.4	3.00	49.8	4.00	65.8	5.00	76.9

This table was computed from the *Excel* workbook *Isentropic Expansion Wave Relations*. The workbook contains a more detailed, printable version of the table and can be easily modified to generate data for a different Mach number range, or for a different gas.

Appendix F

Analysis of Experimental Uncertainty

Introduction F.1

Experimental data often are used to supplement engineering analysis as a basis for design. Not all data are equally good; the validity of data should be documented before test results are used for design. Uncertainty analysis is the procedure used to quantify data validity and accuracy.

Analysis of uncertainty also is useful during experiment design. Careful study may indicate potential sources of unacceptable error and suggest improved measurement methods.

Types of Error F.2

Errors always are present when experimental measurements are made. Aside from gross blunders by the experimenter, experimental error may be of two types. Fixed (or systematic) error causes repeated measurements to be in error by the same amount for each trial. Fixed error is the same for each reading and can be removed by proper calibration or correction. Random error (nonrepeatability) is different for every reading and hence cannot be removed. The factors that introduce random error are uncertain by their nature. The objective of uncertainty analysis is to estimate the probable random error in experimental results.

We assume that equipment has been constructed correctly and calibrated properly to eliminate fixed errors. We assume that instrumentation has adequate resolution and that fluctuations in readings are not excessive. We assume also that care is used in making and recording observations so that only random errors remain.

F.3 Estimation of Uncertainty

Our goal is to estimate the uncertainty of experimental measurements and calculated results due to random errors. The procedure has three steps:

1. Estimate the uncertainty interval for each measured quantity.
2. State the confidence limit on each measurement.
3. Analyze the propagation of uncertainty into results calculated from experimental data.

Below we outline the procedure for each step and illustrate applications with examples.

Step 1 *Estimate the measurement uncertainty interval.* Designate the measured variables in an experiment as $x_1, x_2, \ldots, x_n$. One possible way to find the uncertainty interval for each variable would be to repeat each measurement many times. The result would be a distribution of data for each variable. Random errors in measurement usually produce a *normal* (*Gaussian*) frequency distribution of measured values. The data scatter for a normal distribution is characterized by the standard deviation, σ. The uncertainty interval for each measured variable, x_i, may be stated as $\pm n\sigma_i$, where $n = 1, 2,$ or 3.

However, the most typical situation in engineering work is a "single-sample" experiment, where only one measurement is made for each point [1]. A reasonable estimate of the measurement uncertainty due to random error in a single-sample experiment usually is plus or minus half the smallest scale division (the *least count*) of the instrument. However, this approach also must be used with caution, as illustrated in the following example.

Example F.1 UNCERTAINTY IN BAROMETER READING

The observed height of the mercury barometer column is $h = 752.6$ mm. The least count on the vernier scale is 0.1 mm, so one might estimate the probable measurement error as ± 0.05 mm.

A measurement probably could not be made this precisely. The barometer sliders and meniscus must be aligned by eye. The slider has a least count of 1 mm. As a conservative estimate, a measurement could be made to the nearest millimeter. The probable value of a single measurement then would be expressed as 752.6 ± 0.5 mm. The relative uncertainty in barometric height would be stated as

$$u_h = \pm \frac{0.5 \text{ mm}}{752.6 \text{ mm}} = \pm 0.000664 \quad \text{or} \quad \pm 0.0664 \text{ percent}$$

Comments:

1. An uncertainty interval of ±0.1 percent corresponds to a result specified to three significant figures; this precision is sufficient for most engineering work.
2. The measurement of barometer height was precise, as shown by the uncertainty estimate. But was it accurate? At typical room temperatures, the observed barometer reading must be reduced by a temperature correction of nearly 3 mm! This is an example of a fixed error that requires a correction factor.

When repeated measurements of a variable are available, they are usually normally distributed data, for which over 99 percent of measured values of x_i lie within $\pm 3\sigma_i$ of the mean value, 95 percent lie within $\pm 2\sigma_i$, and 68 percent lie within $\pm \sigma_i$ of the mean value of the data set [2]. Thus it would be possible to quantify expected errors within any desired *confidence limit* if a statistically significant set of data were available.

The method of repeated measurements usually is impractical. In most applications it is impossible to obtain enough data for a statistically significant sample owing to the excessive time and cost involved. However, the normal distribution suggests several important concepts:

1. Small errors are more likely than large ones.
2. Plus and minus errors are about equally likely.
3. No finite maximum error can be specified.

Step 2 *State the confidence limit on each measurement.* The uncertainty interval of a measurement should be stated at specified odds. For example, one may write $h = 752.6 \pm 0.5$ mm (20 to 1). This means that one is willing to bet 20 to 1 that the height of the mercury column actually is within ± 0.5 mm of the stated value. It should be obvious [3] that "... the specification of such odds can only be made by the experimenter based on ... total laboratory experience. There is no substitute for sound engineering judgment in estimating the uncertainty of a measured variable."

The confidence interval statement is based on the concept of standard deviation for a normal distribution. Odds of about 370 to 1 correspond to $\pm 3\sigma$; 99.7 percent of all future readings are expected to fall within the interval. Odds of about 20 to 1 correspond to $\pm 2\sigma$ and odds of 3 to 1 correspond to $\pm \sigma$ confidence limits. Odds of 20 to 1 typically are used for engineering work.

Step 3 *Analyze the propagation of uncertainty in calculations.* Suppose that measurements of independent variables, $x_1, x_2, \ldots, x_n$, are made in the laboratory. The relative uncertainty of each independently measured quantity is estimated as u_i. The measurements are used to calculate some result, R, for the experiment. We wish to analyze how errors in the x_is *propagate* into the calculation of R from measured values.

In general, R may be expressed mathematically as $R = R(x_1, x_2, \ldots, x_n)$. The effect on R of an error in measuring an individual x_i may be estimated by analogy to the derivative of a function [4]. A variation, δx_i, in x_i would cause variation δR_i in R,

$$\delta R_i = \frac{\partial R}{\partial x_i} \delta x_i$$

The relative variation in R is

$$\frac{\delta R_i}{R} = \frac{1}{R}\frac{\partial R}{\partial x_i} \delta x_i = \frac{x_i}{R}\frac{\partial R}{\partial x_i}\frac{\delta x_i}{x_i} \tag{F.1}$$

Equation F.1 may be used to estimate the relative uncertainty in the result due to uncertainty in x_i. Introducing the notation for relative uncertainty, we obtain

$$u_{R_i} = \frac{x_i}{R}\frac{\partial R}{\partial x_i} u_{x_i} \tag{F.2}$$

How do we estimate the relative uncertainty in R caused by the combined effects of the relative uncertainties in all the x_is? The random error in each variable has a range of values within the uncertainty interval. It is unlikely that all errors will have adverse values at the same time. It can be shown [1] that the best representation for the relative uncertainty of the result is

$$u_R = \pm\left[\left(\frac{x_1}{R}\frac{\partial R}{\partial x_1} u_1\right)^2 + \left(\frac{x_2}{R}\frac{\partial R}{\partial x_2} u_2\right)^2 + \cdots + \left(\frac{x_n}{R}\frac{\partial R}{\partial x_n} u_n\right)^2\right]^{1/2} \tag{F.3}$$

Example F.2 UNCERTAINTY IN VOLUME OF CYLINDER

Obtain an expression for the uncertainty in determining the volume of a cylinder from measurements of its radius and height. The volume of a cylinder in terms of radius and height is

$$\mathcal{V} = \mathcal{V}(r, h) = \pi r^2 h$$

Differentiating, we obtain

$$d\mathcal{V} = \frac{\partial \mathcal{V}}{\partial r} dr + \frac{\partial \mathcal{V}}{\partial h} dh = 2\pi rh\, dr + \pi r^2\, dh$$

since

$$\frac{\partial \mathcal{V}}{\partial r} = 2\pi rh \quad \text{and} \quad \frac{\partial \mathcal{V}}{\partial h} = \pi r^2$$

From Eq. F.2, the relative uncertainty due to radius is

$$u_{\mathcal{V},r} = \frac{\delta \mathcal{V}_r}{\mathcal{V}} = \frac{r}{\mathcal{V}} \frac{\partial \mathcal{V}}{\partial r} u_r = \frac{r}{\pi r^2 h}(2\pi rh)u_r = 2u_r$$

and the relative uncertainty due to height is

$$u_{\mathcal{V},h} = \frac{\delta \mathcal{V}_h}{\mathcal{V}} = \frac{h}{\mathcal{V}} \frac{\partial \mathcal{V}}{\partial h} u_h = \frac{h}{\pi r^2 h}(\pi r^2)u_h = u_h$$

The relative uncertainty in volume is then

$$u_{\mathcal{V}} = \pm[(2u_r)^2 + (u_h)^2]^{1/2} \tag{F.4}$$

Comment: The coefficient 2, in Eq. F.4, shows that the uncertainty in measuring cylinder radius has a larger effect than the uncertainty in measuring height. This is true because the radius is squared in the equation for volume.

F.4 Applications to Data

Applications to data obtained from laboratory measurements are illustrated in the following examples.

Example F.3 UNCERTAINTY IN LIQUID MASS FLOW RATE

The mass flow rate of water through a tube is to be determined by collecting water in a beaker. The mass flow rate is calculated from the net mass of water collected divided by the time interval,

$$\dot{m} = \frac{\Delta m}{\Delta t} \tag{F.5}$$

where $\Delta m = m_f - m_e$. Error estimates for the measured quantities are

Mass of full beaker, $m_f = 400 \pm 2$ g (20 to 1)
Mass of empty beaker, $m_e = 200 \pm 2$ g (20 to 1)
Collection time interval, $\Delta t = 10 \pm 0.2$ s (20 to 1)

The relative uncertainties in measured quantities are

$$u_{m_f} = \pm \frac{2\text{ g}}{400\text{ g}} = \pm 0.005$$

$$u_{m_e} = \pm\frac{2\text{ g}}{200\text{ g}} = \pm 0.01$$

$$u_{\Delta t} = \pm\frac{0.2\text{ s}}{10\text{ s}} = \pm 0.02$$

The relative uncertainty in the measured value of net mass is calculated from Eq. F.3 as

$$u_{\Delta m} = \pm\left[\left(\frac{m_f}{\Delta m}\frac{\partial \Delta m}{\partial m_f}u_{m_f}\right)^2 + \left(\frac{m_e}{\Delta m}\frac{\partial \Delta m}{\partial m_e}u_{m_e}\right)^2\right]^{1/2}$$

$$= \pm\{[(2)(1)(\pm 0.005)]^2 + [(1)(-1)(\pm 0.01)]^2\}^{1/2}$$

$$u_{\Delta m} = \pm 0.0141$$

Because $\dot{m} = \dot{m}(\Delta m, \Delta t)$, we may write Eq. F.3 as

$$u_{\dot{m}} = \pm\left[\left(\frac{\Delta m}{\dot{m}}\frac{\partial \dot{m}}{\partial \Delta m}u_{\Delta m}\right)^2 + \left(\frac{\Delta t}{\dot{m}}\frac{\partial \dot{m}}{\partial \Delta t}u_{\Delta t}\right)^2\right]^{1/2} \tag{F.6}$$

The required partial derivative terms are

$$\frac{\Delta m}{\dot{m}}\frac{\partial \dot{m}}{\partial \Delta m} = 1 \quad \text{and} \quad \frac{\Delta t}{\dot{m}}\frac{\partial \dot{m}}{\partial \Delta t} = -1$$

Substituting into Eq. F.6 gives

$$u_{\dot{m}} = \pm\{[(1)(\pm 0.0141)]^2 + [(-1)(\pm 0.02)]^2\}^{1/2}$$

$$u_{\dot{m}} = \pm 0.0245 \quad \text{or} \quad \pm 2.45 \text{ percent (20 to 1)}$$

Comment:
The 2 percent uncertainty interval in time measurement makes the most important contribution to the uncertainty interval in the result.

Example F.4 UNCERTAINTY IN THE REYNOLDS NUMBER FOR WATER FLOW

The Reynolds number is to be calculated for flow of water in a tube. The computing equation for the Reynolds number is

$$Re = \frac{4\dot{m}}{\pi\mu D} = Re(\dot{m}, D, \mu) \tag{F.7}$$

We have considered the uncertainty interval in calculating the mass flow rate. What about uncertainties in μ and D? The tube diameter is given as $D = 6.35$ mm. Do we assume that it is exact? The diameter might be measured to the nearest 0.1 mm. If so, the relative uncertainty in diameter would be estimated as

$$u_D = \pm\frac{0.05\text{ mm}}{6.35\text{ mm}} = \pm 0.00787 \quad \text{or} \quad \pm 0.787 \text{ percent}$$

The viscosity of water depends on temperature. The temperature is estimated as $T = 24 \pm 0.5°\text{C}$. How will the uncertainty in temperature affect the uncertainty in μ? One way to estimate this is to write

$$u_{\mu(T)} = \pm\frac{\delta\mu}{\mu} = \frac{1}{\mu}\frac{d\mu}{dT}(\pm\delta T) \tag{F.8}$$

The derivative can be estimated from tabulated viscosity data near the nominal temperature of 24°C. Thus

$$\frac{d\mu}{dT} \approx \frac{\Delta\mu}{\Delta T} = \frac{\mu(25^\circ\text{C}) - \mu(23^\circ\text{C})}{(25-23)^\circ\text{C}} = (0.000890 - 0.000933)\frac{\text{N}\cdot\text{s}}{\text{m}^2} \times \frac{1}{2^\circ\text{C}}$$

$$\frac{d\mu}{dT} = -2.15 \times 10^{-5}\ \text{N}\cdot\text{s}/(\text{m}^2\cdot{}^\circ\text{C})$$

It follows from Eq. F.8 that the relative uncertainty in viscosity due to temperature is

$$u_{\mu(T)} = \frac{1}{0.000911}\frac{\text{m}^2}{\text{N}\cdot\text{s}} \times -2.15 \times 10^{-5}\frac{\text{N}\cdot\text{s}}{\text{m}^2\cdot{}^\circ\text{C}} \times (\pm 0.5^\circ\text{C})$$

$$u_{\mu(T)} = \pm 0.0118 \quad \text{or} \quad \pm 1.18 \text{ percent}$$

Tabulated viscosity data themselves also have some uncertainty. If this is ±1.0 percent, an estimate for the resulting relative uncertainty in viscosity is

$$u_\mu = \pm[(\pm 0.01)^2 + (\pm 0.0118)^2]^{1/2} = \pm 0.0155 \quad \text{or} \quad \pm 1.55 \text{ percent}$$

The uncertainties in mass flow rate, tube diameter, and viscosity needed to compute the uncertainty interval for the calculated Reynolds number now are known. The required partial derivatives, determined from Eq. F.7, are

$$\frac{\dot{m}}{Re}\frac{\partial Re}{\partial \dot{m}} = \frac{\dot{m}}{Re}\frac{4}{\pi\mu D} = \frac{Re}{Re} = 1$$

$$\frac{\mu}{Re}\frac{\partial Re}{\partial \mu} = \frac{\mu}{Re}(-1)\frac{4\dot{m}}{\pi\mu^2 D} = -\frac{Re}{Re} = -1$$

$$\frac{D}{Re}\frac{\partial Re}{\partial D} = \frac{D}{Re}(-1)\frac{4\dot{m}}{\pi\mu D^2} = -\frac{Re}{Re} = -1$$

Substituting into Eq. F.3 gives

$$u_{Re} = \pm\left\{\left[\frac{\dot{m}}{Re}\frac{\partial Re}{\partial \dot{m}}u_{\dot{m}}\right]^2 + \left[\frac{\mu}{Re}\frac{\partial Re}{\partial \mu}u_\mu\right]^2 + \left[\frac{D}{Re}\frac{\partial Re}{\partial D}u_D\right]^2\right\}^{1/2}$$

$$u_{Re} = \pm\left\{[(1)(\pm 0.0245)]^2 + [(-1)(\pm 0.0155)]^2 + [(-1)(\pm 0.00787)]^2\right\}^{1/2}$$

$$u_{Re} = \pm 0.0300 \quad \text{or} \quad \pm 3.00 \text{ percent}$$

Comment:
Examples F.3 and F.4 illustrate two points important for experiment design. First, the mass of water collected, Δm, is calculated from two measured quantities, m_f and m_e. For any stated uncertainty interval in the measurements of m_f and m_e, the *relative* uncertainty in Δm can be decreased by making Δm larger. This might be accomplished by using larger containers or a longer measuring interval, Δt, which also would reduce the relative uncertainty in the measured Δt. Second, the uncertainty in tabulated property data may be significant. The data uncertainty also is increased by the uncertainty in measurement of fluid temperature.

Example F.5 UNCERTAINTY IN AIR SPEED

Air speed is calculated from pitot tube measurements in a wind tunnel. From the Bernoulli equation,

$$V = \left(\frac{2gh\rho_{\text{water}}}{\rho_{\text{air}}}\right)^{1/2} \tag{F.9}$$

where h is the observed height of the manometer column.

The only new element in this example is the square root. The variation in V due to the uncertainty interval in h is

$$\frac{h}{V}\frac{\partial V}{\partial h} = \frac{h}{V}\frac{1}{2}\left(\frac{2gh\rho_{\text{water}}}{\rho_{\text{air}}}\right)^{-1/2}\frac{2g\rho_{\text{water}}}{\rho_{\text{air}}}$$

$$\frac{h}{V}\frac{\partial V}{\partial h} = \frac{h}{V}\frac{1}{2}\frac{1}{V}\frac{2g\rho_{\text{water}}}{\rho_{\text{air}}} = \frac{1}{2}\frac{V^2}{V^2} = \frac{1}{2}$$

Using Eq. F.3, we calculate the relative uncertainty in V as

$$u_V = \pm\left[\left(\frac{1}{2}u_h\right)^2 + \left(\frac{1}{2}u_{\rho_{\text{water}}}\right)^2 + \left(-\frac{1}{2}u_{\rho_{\text{air}}}\right)^2\right]^{1/2}$$

If $u_h = \pm 0.01$ and the other uncertainties are negligible,

$$u_V = \pm\left\{\left[\frac{1}{2}(\pm 0.01)\right]^2\right\}^{1/2}$$

$$u_V = \pm 0.00500 \quad \text{or} \quad \pm 0.500 \text{ percent}$$

Comment:
The square root reduces the relative uncertainty in the calculated velocity to half that of u_h.

F.5 Summary

A statement of the probable uncertainty of data is an important part of reporting experimental results completely and clearly. The American Society of Mechanical Engineers requires that all manuscripts submitted for journal publication include an adequate statement of uncertainty of experimental data [5]. Estimating uncertainty in experimental results requires care, experience, and judgment, in common with many endeavors in engineering. We have emphasized the need to quantify the uncertainty of measurements, but space allows including only a few examples. Much more information is available in the references that follow (e.g., [4, 6, 7]). We urge you to consult them when designing experiments or analyzing data.

References

1. Kline, S. J., and F. A. McClintock, "Describing Uncertainties in Single-Sample Experiments," *Mechanical Engineering, 75*, 1, January 1953, pp. 3–9.
2. Pugh, E. M., and G. H. Winslow, *The Analysis of Physical Measurements*. Reading, MA: Addison-Wesley, 1966.
3. Doebelin, E. O., *Measurement Systems*, 4th ed. New York: McGraw-Hill, 1990.
4. Young, H. D., *Statistical Treatment of Experimental Data*. New York: McGraw-Hill, 1962.
5. Rood, E. P., and D. P. Telionis, "JFE Policy on Reporting Uncertainties in Experimental Measurements and Results," *Transactions of ASME, Journal of Fluids Engineering, 113*, 3, September 1991, pp. 313–314.
6. Coleman, H. W., and W. G. Steele, *Experimentation and Uncertainty Analysis for Engineers*. New York: Wiley, 1989.
7. Holman, J. P., *Experimental Methods for Engineers*, 5th ed. New York: McGraw-Hill, 1989.

Appendix G

SI Units, Prefixes, and Conversion Factors

Table G.1

SI Units and Prefixes[a]

SI Units	Quantity	Unit	SI Symbol	Formula
SI base units:	Length	meter	m	—
	Mass	kilogram	kg	—
	Time	second	s	—
	Temperature	kelvin	K	—
SI supplementary unit:	Plane angle	radian	rad	—
SI derived units:	Energy	joule	J	$N \cdot m$
	Force	newton	N	$kg \cdot m/s^2$
	Power	watt	W	J/s
	Pressure	pascal	Pa	N/m^2
	Work	joule	J	$N \cdot m$
SI prefixes	**Multiplication Factor**		**Prefix**	**SI Symbol**
	$1\,000\,000\,000\,000 = 10^{12}$		tera	T
	$1\,000\,000\,000 = 10^{9}$		giga	G
	$1\,000\,000 = 10^{6}$		mega	M
	$1\,000 = 10^{3}$		kilo	k
	$0.01 = 10^{-2}$		centi[b]	c
	$0.001 = 10^{-3}$		milli	m
	$0.000\,001 = 10^{-6}$		micro	μ
	$0.000\,000\,001 = 10^{-9}$		nano	n
	$0.000\,000\,000\,001 = 10^{-12}$		pico	p

[a]Source: ASTM Standard for Metric Practice E 380-97, 1997.

[b]To be avoided where possible.

Unit Conversions G.1

The data needed to solve problems are not always available in consistent units. Thus it often is necessary to convert from one system of units to another.

In principle, all derived units can be expressed in terms of basic units. Then, only conversion factors for basic units would be required.

In practice, many engineering quantities are expressed in terms of defined units, for example, the horsepower, British thermal unit (Btu), quart, or nautical mile. Definitions for such quantities are necessary, and additional conversion factors are useful in calculations.

Basic SI units and necessary conversion factors, plus a few definitions and convenient conversion factors are given in Table G.2.

Table G.2

Conversion Factors and Definitions

Fundamental Dimension	English Unit	Exact SI Value	Approximate SI Value
Length	1 in.	0.0254 m	—
Mass	1 lbm	0.453 592 37 kg	0.454 kg
Temperature	1°F	5/9 K	—

Definitions:

Acceleration of gravity:	$g = 9.8066$ m/s^2 (= 32.174 ft/s^2)
Energy:	Btu (British thermal unit) ≡ amount of energy required to raise the temperature of 1 lbm of water 1°F (1 Btu = 778.2 ft · lbf) kilocalorie ≡ amount of energy required to raise the temperature of 1 kg of water 1 K(1 kcal = 4187 J)
Length:	1 mile = 5280 ft; 1 nautical mile = 6076.1 ft = 1852 m (exact)
Power:	1 horsepower ≡ 550 ft · lbf/s
Pressure:	1 bar ≡ 10^5 Pa
Temperature:	degree Fahrenheit, $T_F = \frac{9}{5}T_C + 32$ (where T_C is degrees Celsius) degree Rankine, $T_R = T_F + 459.67$ Kelvin, $T_K = T_C + 273.15$ (exact)
Viscosity:	1 Poise ≡ 0.1 kg/(m · s) 1 Stoke ≡ 0.0001 m^2/s
Volume:	1 gal ≡ 231 in.^{3} (1 ft^3 = 7.48 gal)

Useful Conversion Factors:

Length:	1 ft = 0.3048 m	Power:	1 hp = 745.7 W
	1 in. = 25.4 mm		1 ft · lbf/s = 1.356 W
Mass:	1 lbm = 0.4536 kg		1 Btu/hr = 0.2931 W
	1 slug = 14.59 kg	Area	1 ft^2 = 0.0929 m^2
Force:	1 lbf = 4.448 N		1 acre = 4047 m^2
	1 kgf = 9.807 N	Volume:	1 ft^3 = 0.02832 m^3
Velocity:	1 ft/s = 0.3048 m/s		1 gal (US) = 0.003785 m^3
	1 ft/s = 15/22 mph		1 gal (US) = 3.785 L
	1 mph = 0.447 m/s	Volume flow rate:	1 ft^3/s = 0.02832 m^3/s
Pressure:	1 psi = 6.895 kPa		1 gpm = 6.309 × 10^{-5} m^3/s
	1 lbf/ft^2 = 47.88 Pa	Viscosity (dynamic)	1 lbf · s/ft^2 = 47.88 N · s/m^2
	1 atm = 101.3 kPa		1 g/(cm · s) = 0.1 N · s/m^2
	1 atm = 14.7 psi		1 Poise = 0.1 N · s/m^2
	1 in. Hg = 3.386 kPa	Viscosity (kinematic)	1 ft^2/s = 0.0929 m^2/s
	1 mm Hg = 133.3 Pa		1 Stoke = 0.0001 m^2/s
Energy:	1 Btu = 1.055 kJ		
	1 ft · lbf = 1.356 J		
	1 cal = 4.187 J		

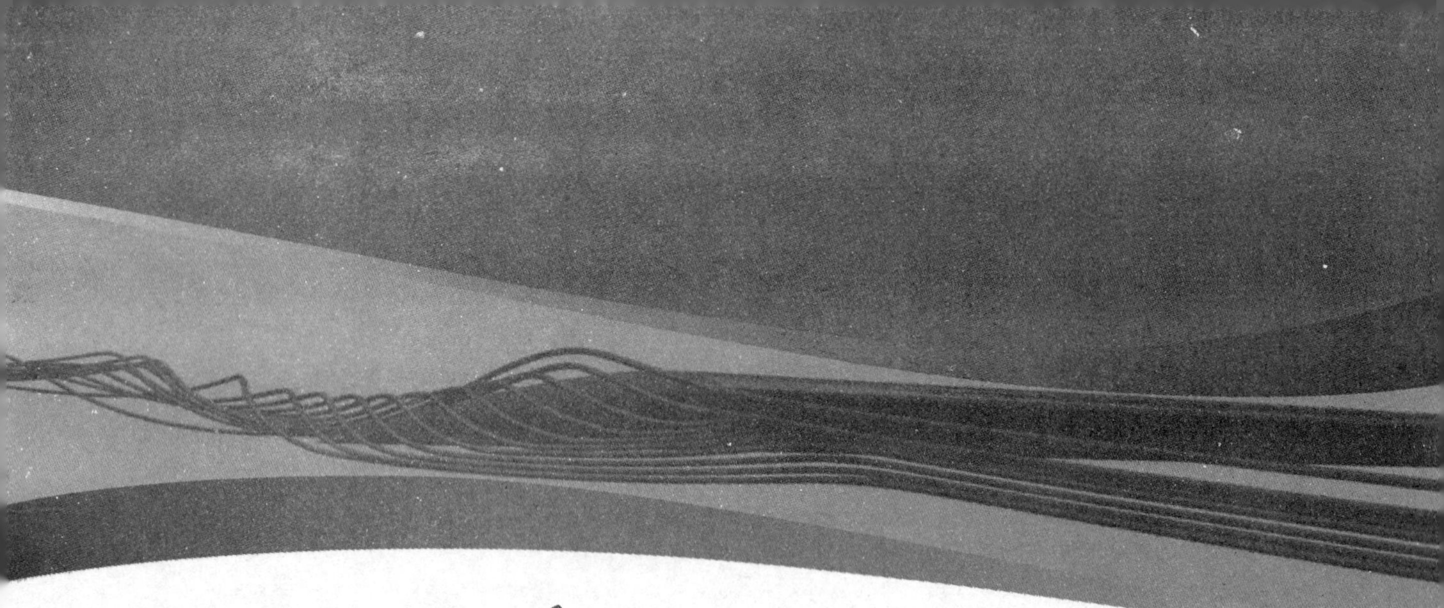

Answers to Selected Problems

1.5 $M = 26.6$ kg

1.7 $t = 3W/gk$

1.9 $L = 0.249$ m $D = 0.487$ m

1.11 $V_{max} = 0.0435$ m/s $t = 0.0133$ s

1.13 $y = 0.922$ mm

1.17 a) $kg \cdot m^2/s^3$ b) $kg/m \cdot s^2$ c) $kg/m \cdot s^2$ d) 1/s e) $kg \cdot m^2/s^2$
f) $kg \cdot m^2/s^2$ g) $kg \cdot m/s$ h) $kg/m \cdot s^2$ i) dimensionless
j) $kg \cdot m^2/s$

1.19 a) 10.76 ft^2/s b) 0.134 hp c) 0.43 Btu/lbm

1.21 a) 0.998 Btu/lbm · °R b) 67.1 mi/hr c) 305 in^3

1.23 a) 0.0472 m^3/s b) 0.0189 m^3 c) 29.1 m/s d) 2.19×10^4 m^2

1.25 a) 6.36×10^{-3} ft^3 b) 402 hp c) 1.044 $lbf \cdot s/ft^2$ d) 431 $ft^2 \cdot hr$

1.27 $Q = 397$ L/min

1.29 SG = 13.6 $v = 7.38 \times 10^{-5}$ m^3/kg $\gamma_E = 1.33 \times 10^5 \dfrac{N}{m^3}$

$\gamma_M = 2.26 \times 10^4$ N/m^3

1.31 $V = 86.5$ m/s $V = 58.2$ m/s using wrong units

1.33 $c = 2.38$ $K^{1/2} \cdot s/m$

1.35 C_D is dimensionless

1.37 c: N · s/m, k: N/m, f: N,

1.39 m, $m/(L/min)^2$

1.41 $\rho = 1.06 \pm 3.47 \times 10^{-3}$ kg/m^3 ($\pm 0.328\%$)

1.43 $\rho = 930 \pm 27.2$ kg/m^3

1.45 $t = 1, 5, 5$ s Flow rate uncertainty = ±5.0, 1.0, 1.0%

1.47 $\forall = 350 \times 10^{-6}$ m^3 $L = 102 \pm 0.0153$ mm ($\pm 1.53\%$)

1.49 $\delta_x = \pm 0.158$ mm

1.51 $H = 17.3 \pm 0.164$ m $\theta_{min} = 31.4°$

1.53 $V = 6.34 \dfrac{\text{cm}}{\text{min}}$ $D = 4.48$ cm

2.1 1) 2D, Unsteady 2) 2D, Steady 3) 1D, Steady 4) 1D, Steady
5) 1D, Unsteady 6) 2D, Steady 7) 2D, Unsteady
8) 3D, Steady

2.3 2D $\vec{V} = 0$ (lower disk) $\vec{V} = \hat{e}_\theta r\omega$ (upper disk)

2.5 A is irrelevant for streamline shapes; determines velocity magnitudes.

2.7 Streamlines: $y = c\,x^{-\frac{b}{a}t}$

2.9 Streamlines: $y = \dfrac{\text{const}}{x + \frac{B}{A}}$

2.11 Streamlines: $x^2 + y^2 = c$ Vortex model of center of a tornado.

2.13 Streamlines: $y = cx$ Models a sink (see Chapter 6).

2.15 Streamline & pathline (steady flow): $y = \dfrac{1}{2}x^2$

2.17 $\omega = \dfrac{K}{2\pi a^2}$

2.21 Lagrangian: $x(t) = 2t + 1$ $y(t) = 1 - t^2$

Pathlines: $y(x) = 1 - \dfrac{(x-1)^2}{4}$ Streamlines: $y(x,t) = 1 - t(x-1)$

2.23 Pathlines: $x(t) = e^{\frac{t}{5}}$ $y(t) = e^{\frac{t^2}{50}}$ Streamlines: $y(x,t) = x^{t/5}$

2.25 Pathlines: $y = 4t + 1,\ x = 3e^{0.05t^2}$

Streamlines: $y = 1 + \dfrac{40}{t}\ln\left(\dfrac{x}{3}\right)$

2.27 Streaklines: $y(t_0) = v_0\sin(\omega t_0)(t - t_0),\ x(t_0) = u_0(t - t_0)$

2.29 Streaklines: $y = e^{(t-\tau)},\ x = e^{(t-\tau)+0.1(t^2-\tau^2)}$

Streamlines: $y = x^{\frac{1}{(1+0.2t)}}$

2.31 Streamlines: $y^3 = 6x + 180$ (26.3 m, 6 m) (31.7 m, 4 m)

2.33 Streamline: $y(x) = 5\ln(x) + 1$ At 5 s: (em, 6 m) At 10 s: (e^2m, 11 m)

2.35 At 2 s: (1.91 m, 2.8 m) At 3 s: (1.49 m, 3.0 m)

2.37 $\nu = \dfrac{b' T^{3/2}}{1 + S'/T}$ $b' = 4.13 \times 10^{-9}$ m^2/s·K$^{3/2}$ $S' = 110.4$ K

2.39 $b = 1.52 \times 10^{-6}$ kg/m·s·K$^{1/2}$ $S = 102$ K

2.41 $F = 2.28$ N

2.43 $\tau_{yx} = 14.74 \dfrac{\text{N}}{\text{m}^2}$ (in positive x direction)

2.45 $F = 74.52$ N

2.47 $L = 0.66$ m

2.49 $t = 1.93$ s

2.51 $V = \dfrac{Mgd\sin\theta}{\mu A}\left(1 - e^{-\frac{\mu A}{Md}t}\right)$ $V = 0.404$ m/s $\mu = 1.08$ N · s/m²

2.53 $F = 2.83$ N

2.55 $r_{\tau=0} = 2.25$ mm $\tau_{rx_{\text{tube}}} = -2.37$ Pa $\tau_{rx_{\text{filament}}} = 2.52$ kPa

2.57 $V = 0.483$ m/s $V_{12} = 0.333$ m/s $V_{23} = 0.383$ m/s

2.59 $\mu = 8.07 \times 10^{-4}$ N · s/m²

2.61 $t = 4.00$ s

2.63 $T = \dfrac{2\pi\mu\omega h R^3}{a}$ $\omega = \dfrac{mga}{2\pi R^2 \mu h}\left[1 - e^{-\frac{2\pi R\mu h}{a(m_1+m_2)}t}\right]$ $\omega_{\max} = \dfrac{mga}{2\pi R^2\mu h}$

2.65 $\omega = \dfrac{A}{B}\left(1 - e^{-\frac{B}{C}t}\right)$ $\omega_{\max} = 25.1$ rpm $t = 0.671$ s

2.67 $\dot{\gamma} = \dfrac{\omega}{\theta}$ $T = \dfrac{2}{3}\pi R^3 V \tau_{yx}$

2.69 Bingham plastic $\mu_p = 0.652$ N · s/m²

2.71 $T = \dfrac{\pi\mu\Delta\omega R^4}{2a}$ $P_0 = \dfrac{\pi\mu\omega_0\Delta\omega R^4}{2a}$ $s = \dfrac{2aT}{\pi\mu R^4\omega_i}$

2.73 $\mu = \dfrac{2a\cos(\theta)T}{\pi\omega\tan^3(\theta)H^4}$ $\mu = 1.01$ N · s/m² Castor Oil

2.75 $T = \dfrac{2\pi\mu\omega R^4}{h}\left(\dfrac{\cos^3(\alpha)}{3} - \cos(\alpha) + \dfrac{2}{3}\right)$

2.77 $\Delta p = 2.91$ kPa

2.79 $A = 0.400$ mm $b = 4.37$ mm^{-1}

2.83 $a = 0$ $b = 2U$ $c = -U$

2.85 $V = 368$ km/h

2.87 $Re = 1389$ $T_{\text{turb}} = 52$°C

2.89 SG = 0.9 $\gamma = 8830$ N/m³ Laminar flow

2.91 $V = 667$ km/hr

3.1 $p = 25.44$ MPa $t = 2.27$ cm

3.3 $\Delta p = 6.72$ mm Hg $\Delta z = 173$ m

3.5 $F = 270$ N $T = 0.282$ N

3.9 $p = 316$ kPa (gage) $p_{SL} = 253$ kPa (gage)

3.11 SG = 1.77 $p_{\text{upper}} = 3.525$ kPa $p_{\text{lower}} = 6.13$ kPa

3.13 $\Delta\rho/\rho = 4.55\%$ $\Delta p/p = 2.24\%$

3.15 $p = -1.471$ kPa (gage)

3.17 $p = 6.39$ kPa (gage) $h = 39.3$ mm

3.19 $p = 128$ kPa (gage)

3.21 $H = 17.75$ mm

3.23 $h = 42.3$ mm

3.25 Amplification factor = 5.78

3.27 $p = 24.7$ kPa (gage) $h = 0.116$ m

3.29 $l = 1.600$ m

3.31 $l = 0.549$ m

3.33 $s = 6$

3.35 $h = 7.85$ mm $s = 0.308$

3.37 $l = 0.407$ m

3.39 $x = 0.2531$ cm

3.43 $\Delta z = 270$ m for 3% pressure drop $\Delta z = 455$ m for 5% density drop

3.45 $\rho = 3.32 \times 10^{-3}$ kg/m^3

3.47 $F_{1atm} = 14.7$ kN $F_{0.5atm} = 52.6$ kN

3.49 $p_A = 1.96$ kPa $p_B = 2820.375$ Pa $p_C = 7112.25$ Pa $p_{air1} = 245.25$ Pa
$p_{air2} = 4537.125$ Pa

3.51 $F_R = 354$ N $y' = 0.285$ m

3.55 $F_R = 2.04$ N

3.57 $F_R = 552$ kN $y' = 2.00$ m $x' = 2.50$ m

3.59 $F = 2627.8$ N

3.63 $D = 2.6$ m

3.65 $d = 2.66$ m

3.67 SG = 0.542

3.69 $F_V = 831$ kN

3.71 $F_V = 7.62$ kN $x'F_V = 3.76$ kN·m $F_A = -5.71$ kN

3.73 $F_V = 2.19$ kN $x' = 0.243$ m

3.75 $F_V = -\rho g w R^2\pi/4$ $x' = 4R/3\pi$

3.77 $F_V = 1.83 \times 10^7$ N $\alpha = 19.9°$

3.79 $F_V = 416$ kN $F_H = 370$ kN $\alpha = 48.3°$ $F = 557$ kN

3.81 $\dfrac{F_H}{w} = \dfrac{\rho g R^2}{2}$ $\dfrac{m}{w} = \rho R^2\left(1 + \dfrac{3\pi}{4}\right)$

3.85 $M = 734$ kg

3.87 $M = 631$ kg

3.89 $h = 177$ mm

3.91 $SG = SG_w \dfrac{F_{air}}{F_{air} - F_{net}}$

3.93 $\forall = 2.52\text{L}$ six weights needed

3.97 $F_B = 8.03 \times 10^{-11}$ N $V = 3.4 \times 10^{-4}$ m/s $\left(20.4 \frac{\text{mm}}{\text{min}}\right)$

3.99 $h = 48.3$ km

3.101 $M_B = 29.1$ kg

3.105 $D = 0.737$ m

3.107 $f = 0.286$ cycle/s ($\omega = 1.79$ rad/s)

3.109 $F = 159.4$ N

3.115 $h = aL/g$

3.117 Cavitation does not occur.

3.119 $\Delta p = \rho\omega^2 R^2/2$ $\omega = 7.16$ rad/s

3.121 $\alpha = 13.30°$

3.123 $dy/dx = -0.25$ $p = 105 - 1.96x$ (p: kPa, x: m)

3.125 $T = 405$ N $\Delta p = 3.03$ kPa

4.1 $x = 0.49$ m $x = 0.43$ m $x = 0.122$ m

4.3 $V = 0.577$ m/s $\theta = 48.2°$

4.5 $V = 87.5 \frac{\text{km}}{\text{h}}$ $t = 4.13$ s

4.7 $V_{dry} = 51.67$ km/h $V_{wet} = 33.82$ km/h

4.9 $t = 1.72$ hr

4.11 $\Delta U = 459$ MJ $\Delta U_{System} = 0$ $\Delta T/\Delta t = 6.09°$C/hr

4.13 $\int \vec{V}\cdot d\vec{A} = 30\,\text{m}^3/\text{s}$ $\rho \int \vec{V}(\vec{V}\cdot d\vec{A}) = (80\hat{i} + 75\hat{j})\,\text{kg}\cdot\text{m/s}^2$

4.15 $Q = -\frac{1}{2}Vhw$ m.f. $= -\frac{1}{3}\rho V^2 wh\hat{i}$

4.17 $Q = \frac{1}{2}u_{max}\pi R^2$ m.f. $= \frac{1}{3}\rho u_{max}^2 \pi R^2 \hat{i}$

4.19 $V_{jet} = 5.7$ m/s $V_{pipe} = 0.49$ m/s

4.21 $t_{exit} = 126$ s $t_{drain} = 506$ s $Q_{drain} = 0.0242$ m^3/s

4.23 $Q_{cool} = 1800 \frac{\text{L}}{\text{min}}$ $\dot{m}_{cool} = 1.08 \times 10^5$ kg/hr

$\dot{m}_{moist} = 5.86$ kg/s $\dot{m}_{air} = 4.36$ kg/s

4.25 $Q = -0.2$ m^3/s (inwards flow)

4.27 $t = 5.83$ min

4.29 $Q = 160$ L/s $V = 1.68$ m/s $w = 1.10$ m

4.31 $\rho = 0.267$ kg/m^3

4.33 $\frac{\dot{m}}{w} = \frac{\rho^2 g \sin(\theta) h^3}{3\mu}$

4.35 $U = 1.5$ m/s

4.37 $Q = 1.05 \times 10^{-5}$ m^3/s (10.45 mL/s) $V_{ave} = 0.139$ m/s

$u_{max} = 0.213$ m/s

4.39 $V_{min} = 5.0$ m/s

4.41 $\dot{m} = 16.2$ kg/s

4.43 $\partial M_{CV}/\partial t = -0.91$ kg/s

$\partial \forall_{oil}/\partial t = -1.036 \times 10^{-3}$ m^3/s $\left(1.036 \dfrac{\text{L}}{\text{S}}\right)$

4.45 $d\rho_{tank}/dt = -0.2582$ kg/m^3/s

4.47 $Q = 57$ m^3/s $\quad A = 0.47 \times 10^7$ m^2

4.49 $t_{3\text{-}2} = 3.83$ s $\quad t_{2\text{-}1} = 4.53$ s

4.51 $dy/dt = -9.01$ mm/s

4.53 $Q_{cd} = 4.50 \times 10^{-3}$ m^3/s $\quad Q_{ad} = 6.0 \times 10^{-4}$ m^3/s

$Q_{bc} = 1.65 \times 10^{-3}$ m^3/s

4.55 $t = \dfrac{8}{5}\dfrac{\tan^2(\theta) y_0^{5/2}}{\sqrt{2g}d^2}$ $\quad t_{drain} = 2.53$ min $\quad t_{12\text{-}6} = 2.10$ min

$t_{6\text{-}0} = 0.448$ min

4.57 $t_{500\text{ kPa}} = 42.2$ days $\quad p_{30\text{ day (Exact)}} = 639$ kPa $\quad p_{30\text{ day (Saying)}} = 493$ kPa

$\Delta p = 51$ kPa

4.59 $mf_2/mf_1 = 1.2$

4.61 $mf_x = 840$ N $\quad mf_y = -2075$ N

4.63 $V = 1243$ m/s $\quad F = 78.3$ N

4.65 $F = 90.4$ kN

4.67 $T = 3.12$ N

4.69 $F = 156$ N

4.71 Block slides $\quad M_{min} = 7.14$ kg

4.73 $M_{payload} = 671$ kg

4.75 $R_x = -\rho V^2 \dfrac{\pi D^2}{4}(1 + \sin\theta)\left[1 - \left(\dfrac{d}{D}\right)^2\right]$ $\quad R_x = -314$ N

4.77 $F = 11.6$ kN

4.79 $R_x = -668$ N

4.81 $F_y = 7.16$ N

4.83 $Q = 0.424$ m^3/s $\quad R_y = 4.05$ kN

4.85 $\dot{m}_{air} = 928$ kg/s $\quad T = 229$ kN

4.87 $V = 21.8$ m/s

4.89 $R_x = -4.68$ kN $\quad R_y = 1.66$ kN

4.91 $V_2 = 6.5$ m/s $\quad \Delta p = 85.1$ kPa

4.93 $F = 2456$ N

4.95 $R_x = 1760$ N

4.97 $F_x = 779$ N $\quad F_y = -387$ N

4.99 $\dot{m}_{air} = 6.63$ kg/s $\quad V_{max} = 18.8$ m/s $\quad F_{drag} = 54.1$ N

4.101 $U_1 = 10$ m/s $\quad u_{max} = 15$ m/s $\quad \Delta p = 15$ kPa

4.103 $R_x = -7.90 \times 10^{-4}$ N

4.105 $F = 52.1$ N

4.107 $\theta = \sin^{-1}\left(\frac{\alpha}{1-\alpha}\right) \qquad R_x = -\rho V^2 wh[1-(1-2\alpha)^{1/2}]$

4.109 $h_2/h = (1+\sin\theta)/2$

4.111 $h = 0.17$ m (170 mm) $\qquad F = 0.78$ N

4.113 $V = \sqrt{V_0^2 + 2gh} \qquad R_z = \rho V_0 A_0 \sqrt{V_0^2 + 2gh} \qquad R_z = 3.56$ N (upwards)

4.115 $M = \dfrac{(V_0 - V_2\cos\theta)\rho V_0 A_1}{g} \qquad M = 4.46$ kg $\qquad M_w = 2.06$ kg

4.117 $F = 1.14$ kN

4.119 $V(z) = \sqrt{V_0^2 + 2gz} \qquad D(z) = \dfrac{D_0}{\left(1 + \dfrac{2gz}{V_0^2}\right)^{1/4}} \qquad V = 1.03$ m/s

$D = 2.62$ mm

4.121 $V(z) = \sqrt{V_0^2 + 2gz} \qquad A(z) = \dfrac{A_0}{\sqrt{1 + \dfrac{2gz}{V_0^2}}} \qquad z_{1/2} = \dfrac{3V_0^2}{2g}$

4.123 $p(r) - p_{\text{atm}} = \dfrac{\rho}{2}\left(\dfrac{Q}{2\pi Rh}\right)^2\left[1 - \left(\dfrac{R}{r}\right)^2\right]$

4.125 $h_1 = \sqrt{h_2^2 + \dfrac{2Q^2}{gb^2 h_2}}$

4.127 $R_x = -940$ N $\qquad R_y = 252$ N

4.129 $R_x = -167$ N

4.131 $R_x = -1.73$ kN

4.133 $F_x = \rho(V-U)^2 A(1-\cos\theta) \qquad P = \rho U(V-U)^2 A(1-\cos\theta)$

4.135 $R_x = 4.24$ kN $\qquad t = 4.17$ mm

4.137 $U = V/2$

4.139 $\dfrac{\dot{m}_2}{\dot{m}_3} = \dfrac{1}{2} \qquad U = 3.03$ m/s

4.141 $U_t = 15.8$ m/s

4.143 $U(t) = \dfrac{\sqrt{\dfrac{g\mu_k M}{\rho A}}}{\tanh\left[\sqrt{\dfrac{g\mu_k \rho A}{M}}t + \tanh^{-1}\left(\sqrt{\dfrac{g\mu_k M}{\rho A}}\dfrac{1}{V}\right)\right]}$

$a(t) = \dfrac{\rho\left(V - U(t)\right)^2 A}{M} - g\mu_k$

4.145 $\theta = 19.7°$

4.147 $t = 23.5$ s

4.149 $U = 22.5$ m/s

4.151 $V(1\text{ s}) = 5.13$ m/s $\quad x(1\text{ s}) = 1.94$ m
$V(2\text{ s}) = 3.18$ m/s $\quad x(2\text{ s}) = 3.47$ m

4.153 $a = \dfrac{\rho(V-U)^2 A}{M} - \dfrac{kU}{M}$

4.155 $U(t) = U_0 e^{-\frac{4\rho VA}{M}t} \qquad a(t) = -\dfrac{4\rho VA}{M} U_0 e^{-\frac{4\rho VA}{M}t}$

4.157 $t = 0.867$ s $\quad x_{rest} = 6.26$ m

4.159 $Q = 0.0469\ \text{m}^3/\text{s}$

4.161 $t = 126$ s

4.163 $U_{max} = 354$ m/s $\quad \Delta U/U = 1.08\%$

4.165 $U = U_0 - V_e \ln\left(1 - \dfrac{\dot{m}t}{M_0}\right)$ Mass fraction = 38.3%

4.167 $m_{fuel} = 38.1$ kg

4.169 $a_{rfy} = 169\ \text{m/s}^2 \qquad U = -\left[V_e + \dfrac{(p_e - p_{atm})A_e}{\dot{m}}\right]\ln\left(\dfrac{M_0 - \dot{m}t}{M_0}\right) - gt$

4.173 $a_{rfx} = \dfrac{2\rho V^2 A}{M}\left[\dfrac{1}{1+\dfrac{2\rho VA}{M}t}\right]^2 \qquad t = \dfrac{M}{2\rho VA}$

4.175 $\dfrac{U}{V} = 1 - \dfrac{1}{\sqrt{1+\dfrac{2\rho VA}{M_0}t}}$

4.177 $\dot{m}_{init} = 0.111$ kg/s $\quad \dot{m}_{final} = 0.0556$ kg/s $\quad t = 20.8$ min

4.179 $\dfrac{d^2h}{dt^2} = \rho\left(\sqrt{V_0^2 - 2gh} - \dfrac{dh}{dt}\right)^2 \dfrac{A_0 V_0}{M\sqrt{V_0^2 - 2gh}} - g$

4.181 $\partial M_{CV}/\partial t = -0.165$ kg/s $\quad \partial P_{xCV}/\partial t = -2.1$ mN
Ratio $= -4.62 \times 10^{-4}\%$

4.185 Moment = 6.98 kN · m $\quad V = 24.3$ m/s

4.187 $F_x = 23.4$ kN/−22.8 kN $\quad$ Moment = −468 kN · m

4.189 $\dot{\omega} = \dfrac{3}{AR^2}\left(-\omega VAR - \dfrac{QV}{2}\right) \qquad \omega_{max} = -28.9$ rad/s (−270 rpm)

4.191 $T = 1.62$ N · m $\quad \omega = 113$ rpm

4.193 $T = 0.0722$ N · m

4.195 $T = -0.0161$ N · m

4.197 $\omega = 6.04$ rad/s (57.7 rpm) $\quad A = 1720\ \text{m}^2$

4.199 $T_{shaft} = 29.4$ N · m $\quad M_x = 51.0$ N · m $\quad M_y = 1.4$ N · m

4.201 $R_y = \rho V^2 wh\cos\theta$ (applied below Point O) $\quad$ Equilibrium when $\theta = 0°$

4.203 $\dot{W}_m = 80.0$ kW

4.205 $\eta = 79.0\%$

4.207 $\delta Q/dm = -21$ kJ/kg $\quad \dot{Q} = -189$ kJ/s

4.209 $Q = 1032$ L/min $\quad z_{max} = 65.2$ m $\quad R_x = 602$ N

4.211 V = 70.3 m/s $\quad \dot{W}_{min} = 360$ kW

5.1 a) Possible b) Not possible c) Not possible d) Not possible

5.3 Equation valid for steady and unsteady flow

Infinite number of solutions $\quad v(x,y) = 1.8y - 3.3\dfrac{y^2}{2}$

5.5 Equation valid for steady and unsteady flow

Infinite number of solutions $\quad v(x,y) = -3xy^2$

5.7 Equation valid for steady and unsteady flow $\quad u(x,y) = \dfrac{9}{2}x^2y^2 - \dfrac{3}{4}x^4$

5.9 Equation valid for steady and unsteady flow $\quad v(x,y) = -10e^{x/5}\sin\left(\dfrac{y}{5}\right)$

5.11 $\left.\dfrac{v}{U}\right)_{max} = 0.00167 \quad (0.167\%)$

5.13 $\left.\dfrac{v}{U}\right)_{max} = \dfrac{3}{8}\dfrac{\delta}{x}\left[\left(\dfrac{y}{\delta}\right)^2 - \dfrac{1}{2}\left(\dfrac{y}{\delta}\right)^4\right] \quad \dfrac{y}{\delta} = 1 \quad \left.\dfrac{v}{U}\right)_{max} = 0.00188$

5.15 $u(x,y) = \dfrac{3}{2}Bx^2y^2 \quad xy^{3/2} = C$

5.19 a) Possible b) Possible c) Possible

5.21 $V_\theta = -\dfrac{\Lambda \sin\theta}{r^2} + f(r)$

5.25 $\psi = \dfrac{Uy^2}{2h} \quad y = \dfrac{h}{\sqrt{2}}$

5.27 Incompressible flow $\quad \psi = A\theta - B\ln r$

5.29 Two-dimensional, incompressible $\quad \psi = -\dfrac{x^4}{4} + \dfrac{3}{2}x^2z^2 - \dfrac{z^4}{4}$

5.31 $\psi = \dfrac{Uy^2}{2\delta} \quad \dfrac{y}{\delta} = \dfrac{1}{2}$ at one-quarter flow rate

$\dfrac{y}{\delta} = \dfrac{1}{\sqrt{2}}$ at one-half flow rate

5.33 $\psi = -\dfrac{2U\delta}{\pi}\cos\left(\dfrac{\pi y}{2\delta}\right) \quad \dfrac{y}{\delta} = 0.460$ at one-quarter flow rate

$\dfrac{y}{\delta} = 0.667$ at one-half flow rate

5.35 $\psi = -\dfrac{\omega r^2}{2} + C \quad Q = -0.001100\ \text{m}^3/\text{s}\cdot\text{m} \quad Q = 0.001100\ \text{m}^3/\text{s}\cdot\text{m}$

5.37 $\psi = -C\ln r + C_1 \quad Q = -0.0547\ \text{m}^3/\text{s}\cdot\text{m} \quad Q = 0.0547\ \text{m}^3/\text{s}\cdot\text{m}$

5.39 Possible flow field $\quad a = 69.9\ \text{m/s}^2$

5.41 $v(x,y) = -A(5x^4y - 10x^2y^3 + y^5) \quad a = 5.41 \times 10^4\ \text{m/s}^2$

5.43 $v = Ay\sin\left(\dfrac{2\pi t}{T}\right)$

$\vec{a}_{p,conv} = A^2\sin^2\left(\dfrac{2\pi t}{T}\right) \times (x\hat{i} + y\hat{j}) \quad \vec{a}_{p,local} = \dfrac{2\pi A}{T}\cos\left(\dfrac{2\pi t}{T}\right)$

5.45 Incompressible $a_x = -\dfrac{A^2x}{(x^2+y^2)^2}$ $a_y = -\dfrac{A^2y}{(x^2+y^2)^2}$

$a = -\dfrac{100}{r^3}$

5.49 $a = -\left(\dfrac{Q}{2\pi h}\right)^2 \dfrac{1}{r^3}$ (Radial)

5.51 $\left.\dfrac{Dc}{Dt}\right)_{upstream} = 0$ $\left.\dfrac{Dc}{Dt}\right)_{drift} = \dfrac{125\times10^{-6}}{\text{hr}}$

$\left.\dfrac{Dc}{Dt}\right)_{downstream} = \dfrac{250\times10^{-6}}{\text{hr}}$

5.53 $\dfrac{DT}{Dt} = -7.8^\circ\text{C/min}$

5.59 $a_{px} = -\dfrac{U^2}{4x}\left(\dfrac{y}{\delta}\right)^2$ $a_{py} = -\dfrac{U^2}{4x}\left(\dfrac{y}{\delta}\right)^2\dfrac{y}{x}$ $a_{px\max} = -\dfrac{U^2}{4x}$

$a_{py\max} = -\dfrac{U^2}{4x}\dfrac{\delta}{x}$ ratio = 100

5.61 $xy = 8$ $\vec{V} = 10\hat{i} - 20\hat{j}$ $\vec{a_p} = 0\hat{i} - 0\hat{j}$ (Local)

$\vec{a_p} = 50\hat{i} + 100\hat{j}$ (Convective) $\vec{a_p} = 50\hat{i} + 100\hat{j}$ (Total)

5.63 $V_z = v_0\left(1 - \dfrac{z}{h}\right)$ $a_{pr} = \dfrac{v_0^2 r}{4h^2}$ $a_{pz} = \dfrac{v_0^2}{h}\left(\dfrac{z}{h} - 1\right)$

5.65 $v = v_0\left(1 - \dfrac{y}{h}\right)$ $\vec{a} = \dfrac{v_0^2 x}{h^2}\hat{i} + \dfrac{v_0^2}{h}\left(\dfrac{y}{h} - 1\right)\hat{j}$

5.67 $x_p = x_0 e^{At}$ $y_p = y_0 e^{-At}$

5.69 a) Not irrotational b) Not irrotational c) Not irrotational
d) Not irrotational

5.71 $\Gamma = -0.1\ \text{m}^2/\text{s}$ $\Gamma = -0.1\ \text{m}^2/\text{s}$

5.73 Not incompressible Not irrotational

5.75 Incompressible $\vec{\omega} = -0.5\hat{k}\dfrac{\text{rad}}{\text{s}}$ $\Gamma = -0.5\ \text{m}^2/\text{s}$

5.77 Incompressible Irrotational

5.79 Incompressible Not irrotational

5.81 $\vec{V} = -2y\hat{i} - 2x\hat{j}$

5.83 $\psi = \dfrac{A}{2}(y^2 - x^2) + By$ $\Gamma = 0$

5.85 $\Gamma = -UL\dfrac{h}{b}\left(1 - \dfrac{h}{b}\right)$ $\Gamma = -UL/4\ (h = b/2)$ $\Gamma = 0\ (h = b)$

5.87 Zero linear deformation $\dfrac{\partial V_r}{\partial z} + \dfrac{\partial V_z}{\partial r} = -\dfrac{2V_{\max} r}{R^2}$ $\zeta = -V_{\max}\dfrac{2r}{R^2}\hat{e}_\theta$

5.91 $\dfrac{dF_{\max}}{d\forall} = -\mu U\left(\dfrac{\pi}{2\delta}\right)^2$ $\dfrac{dF_{\max}}{d\forall} = -1.85\ \dfrac{\text{kN}}{\text{m}^3}$

5.93 $\dfrac{dF_{\max}}{d\forall} = -\dfrac{4\mu u_{\max}}{R^2}$ $-2.144\ \dfrac{\text{N}}{\text{m}^3}$

5.95 $u(y) = \frac{JB}{8\mu}(h^2 - 4y^2)$

5.97 $u(y) = -\frac{\varepsilon\zeta}{\mu}E \qquad V = 70.8 \times 10^{-6}$ m/s

6.1 $\vec{a} = (2.8\hat{i} + 6.3\hat{j})\frac{\text{m}}{\text{s}^2} \qquad \nabla p = (-2884\hat{i} - 16{,}593\hat{j})\frac{\text{Pa}}{\text{m}}$

6.3 $\vec{a}_{\text{local}} = B(\hat{i} + \hat{j})$
$\vec{a}_{\text{conv}} = A(Ax - Bt)\hat{i} + A(Ay + Bt)\hat{j}$
$\vec{a}_{\text{total}} = (A^2x - ABt + B)\hat{i} + (A^2y + ABt + B)\hat{j}$
$\nabla p = 6.99\hat{i} - 14.0\hat{j} - 9.80\hat{k}$ kPa/m

6.5 $\vec{a} = (12.7\hat{i} + 32.4\hat{j})\frac{\text{m}}{\text{s}^2} \qquad \nabla p = (-13.1\hat{i} - 43.5\hat{j})\frac{\text{kPa}}{\text{m}}$

6.7 $\vec{a}_{\text{local}} = 2\pi A\omega\cos(2\pi\omega t)(x\hat{i} - y\hat{j})$
$\vec{a}_{\text{conv}} = A^2\sin^2(2\pi\omega t)(x\hat{i} + y\hat{j})$
$\vec{a}_{\text{total}} = \vec{a}_{\text{local}} + \vec{a}_{\text{conv}}$
$\vec{a}_{\text{local}}(0) = 12.6(\hat{i} - \hat{j})\ \text{m/s}^2, \quad \vec{a}_{\text{conv}}(0) = 0\ \text{m/s}^2$
$\vec{a}_{\text{local}}(0.5\ \text{s}) = 12.6(-\hat{i} + \hat{j})\ \text{m/s}^2, \quad \vec{a}_{\text{conv}}(0.5\ \text{s}) = 0\ \text{m/s}^2$
$\vec{a}_{\text{local}}(1\ \text{s}) = 12.6(\hat{i} - \hat{j})\ \text{m/s}^2, \quad \vec{a}_{\text{conv}}(1\ \text{s}) = 0\ \text{m/s}^2$
$\nabla p(0) = -25.1\hat{i} + 25.1\hat{j}$ Pa/m
$\nabla p(0.5\ \text{s}) = 25.1\hat{i} - 25.1\hat{j}$ Pa/m $\quad \nabla p(1\ \text{s}) = -25.1\hat{i} + 25.1\hat{j}$ Pa/m

6.9 Incompressible Stagnation point: (2.5, 1.5)
$\nabla p = -\rho[(4x - 10)\hat{i} + (4y - 6)\hat{j} + g\hat{k}] \qquad \Delta p = 9.6$ Pa

6.11 $\frac{dp}{dx} = \rho\frac{U^2}{L}\left(1 - \frac{x}{L}\right) \qquad p_{\text{out}} = 241$ kPa (gage)

6.13 $a_r = -\frac{V_r^2}{r} \qquad \frac{\partial p}{\partial r} = \rho\frac{V_r^2}{r}$

6.15 $a_{px} = \frac{16v_0^2x}{D^2} \qquad \frac{\partial p}{\partial x} = -\frac{16\rho v_0^2x}{D^2} \qquad p(0) = 8\rho v_0^2\left(\frac{L}{D}\right)^2$
$\vec{a} = -0.127\hat{e}_r + 0\hat{e}_\theta\ \text{m/s}^2 \qquad \vec{a} = -0.127\hat{e}_r + 0\hat{e}_\theta\ \text{m/s}^2$
$\vec{a} = -0.0158\hat{e}_r + 0\hat{e}_\theta\ \text{m/s}^2 \qquad \nabla p = 127\hat{e}_r + 0\hat{e}_\theta$ Pa/m
$\nabla p = 127\hat{e}_r + 0\hat{e}_\theta$ Pa/m $\qquad \nabla p = 15.8\hat{e}_r + 0\hat{e}_\theta$ Pa/m

6.17 $\frac{\partial p}{\partial x} = -\frac{\rho u_0^2 e^{-\frac{x}{2a}}(2e^{-\frac{x}{2a}} - 1)}{2a(e^{-\frac{x}{a}} - e^{-\frac{x}{2a}} + 1)^3} \qquad p = p_0 + \frac{\rho u_0^2}{2}\left[1 - 1/(1 + e^{-\frac{x}{a}} - e^{-\frac{x}{2a}})^2\right]$

6.19 $a_x = -\frac{2V_i^2(D_o - D_i)}{D_iL\left[1 + \frac{(D_o - D_i)}{D_iL}x\right]^5} \qquad \left.\frac{\partial p}{\partial x}\right|_{\text{max}} = 100$ kPa/m $\qquad L \geq 4$ m

6.21 $\nabla p = -4.23\hat{i} - 12.1\hat{j}\ \text{N/m}^3$ Streamlines: $\frac{x}{h}\left(1 - \frac{y}{h}\right) = \text{const}$

6.23 $\vec{a}_p = \frac{q^2}{h}\left[\frac{x}{h}\hat{i} + \left(\frac{y}{h} - 1\right)\hat{j}\right] \qquad \frac{\partial p}{\partial x} = -\frac{\rho q^2x}{h^2} \qquad F_{\text{net}} = \frac{\rho q^2b^3L}{12h^2}$
$q = 0.0432\ \text{m}^3/\text{s/m}^2 \qquad U_{\text{max}} = 1.73$ m/s

6.25 $B = -0.6\ \text{m}^{-2}\cdot\text{s}^{-1} \qquad \vec{a}_p = 6\hat{i} + 3\hat{j}\ \text{m/s}^2 \qquad a_n = 6.45\ \text{m/s}^2$

6.27 $B = -8 \text{ m}^{-3} \cdot \text{s}^{-1}$ Streamline: $y^5 - 10y^3x^2 + 5yx^4 = -38$

$\vec{a}_p = 4A^2(x^2 + y^2)^3(x\hat{i} + y\hat{j})$ $R = 0.822$ m

6.29 $\nabla p = \dfrac{4\rho U^2}{a} \sin\theta(\sin\theta\,\hat{e}_r - \cos\theta\,\hat{e}_\theta)$ $p(\theta) = -2U^2\rho\sin^2\theta$

$p_{\min} = -13.8$ kPa

6.31 $a_r/g = -2876$ $\partial p/\partial r = 43{,}731$ Pa/m

6.33 $\vec{a}_p = 3\hat{i} + 2\hat{j}$ m/s^2 $\vec{V} = 3\hat{i} - 2\hat{j}$ m/s

$\vec{a}_t = 1.16\hat{i} - 0.771\hat{j}$ m/s^2 $\partial p/\partial s = -1.71$ N/m^2/m

6.35 $\vec{a}_p = (0.6\hat{i} + 1.2\hat{j})\dfrac{\text{m}}{\text{s}^2}$ $R = 1.76$ m

6.37 $\vec{a}_p = 4\hat{i} + 2\hat{j}$ m/s^2 $R = 5.84$ m

6.39 $p_{\text{dyn}} = 475$ Pa $h_{\text{dyn}} = 48.4$ mm

6.41 $F = 1.70$ N $F = 6.80$ N

6.43 $h = 628$ mm

6.47 $p_{0,j} = 799$ kPa (gage) $p_{0,rel} = 312$ kPa (gage)

$\vec{V}_{abs} = 2.5\hat{i} + 21.7\hat{j}$ m/s $p_{0,fixed} = 237$ kPa (gage)

6.49 $p_2 = 291$ kPa (gage)

6.51 $Q(h) = \dfrac{\pi D^2}{4}\sqrt{2gh}$ $h = 147$ mm

6.53 $p_{\text{Diet}} = 4.90$ kPa (gage) $p_{\text{Regular}} = 5.44$ kPa (gage)

6.55 $$A = A_0\sqrt{\frac{1}{1 + \dfrac{2g(z_1 - z)}{V_0^2}}}$$

6.57 $p_{r=50\text{ mm}} = -404$ Pa (gage)

6.59 $p_1 = 29.4$ kPa (gage) $p_0 = 229$ kPa $V_{rel} = 24.7$ m/s $V_{abs} = 44.7$ m/s

6.61 $Q = 1104\,\dfrac{\text{L}}{\text{min}}$ (0.0184 m^3/s)

6.63 $p = p_\infty + \dfrac{1}{2}\rho U^2(1 - 4\sin^2\theta)$ $\theta = 30°, 150°, 210°, 330°$

6.67 $Q = 18.5$ L/s $R_x = -2.42$ kN

6.69 $p_1 = 11.7$ kPa (gage) $R_x = -22.6$ N

6.71 $p_2 = 17.6$ kPa (gage) (132 mm Hg) $p_3 = 1.75$ kPa (gage) (13.2 mm Hg)

$R_x = 0.156$ N $R_y = -0.957$ N

6.73 $V_2 = 3.05$ m/s $p_{0,2} = 4.65$ kPa (gage) $F_y = 11.5$ N

6.77 $$\frac{h}{h_0} = \left[1 - \sqrt{\frac{g}{2h_0\left\{\left(\dfrac{D}{d}\right)^4 - 1\right\}}}\,t\right]^2$$

6.79 $F = 83.3$ kN

6.81 $\dot{m} = A\sqrt{2p\rho}$ $\quad \frac{dM}{dt} = -\rho_w \frac{d\forall_{air}}{dt}$ $\quad M_w = \rho_w \forall_0 \left[\frac{V_t}{\forall_0} - \left(1 + 1.5\sqrt{\frac{2p_0}{\rho_w}\frac{At}{\forall_0}}\right)^{\frac{2}{3}} \right]$

6.83 $\dot{m} = A\sqrt{2p\rho}$ $\quad \frac{dM}{dt} = -\rho_w \frac{d\forall_{air}}{dt}$

$$M_w = \rho_w \forall_0 \left\{ \frac{V_t}{\forall_0} - \left[1 + 1.70\sqrt{\frac{2p_0}{\rho_w}\frac{At}{\forall_0}}\right]^{0.588} \right\}$$

6.87 $C_c = 1/2$

6.89 $p = 12.5$ kPa

6.91 $dQ/dt = 0.0516\ \text{m}^3/\text{s/s}$

6.93 $D_j/D_1 = 0.32$

6.97 Bernoulli can be applied

6.99 Rotational flow Points on same streamline, so $\Delta p = -126$ kPa

6.101 $\psi = \frac{q}{2\pi}\left[\tan^{-1}\left(\frac{y-h}{x-h}\right) + \tan^{-1}\left(\frac{y+h}{x-h}\right) + \tan^{-1}\left(\frac{y+h}{x+h}\right) + \tan^{-1}\left(\frac{y-h}{x+h}\right)\right]$

$$\phi = -\frac{q}{4\pi}\ln[\{(x-h)^2 + (y-h)^2\}\{(x-h)^2 + (y+h)^2\}\{(x+h)^2 + (y-h)^2\}$$
$$\{(x+h)^2 + (y+h)^2\}]$$

$$u(x) = \frac{q}{\pi}\left[\frac{x-h}{(x-h)^2 + h^2} + \frac{x+h}{(x+h)^2 + h^2}\right]$$

6.103 $\psi = -\frac{K}{4\pi}\ln\left[\frac{\{(x-h)^2 + (y-h)^2\}\ \{(x+h)^2 + (y+h)^2\}}{\{(x-h)^2 + (y+h)^2\}\ \{(x+h)^2 + (y-h)^2\}}\right]$

$$\phi = \frac{K}{2\pi}\left[\tan^{-1}\left(\frac{y-h}{x-h}\right) - \tan^{-1}\left(\frac{y+h}{x-h}\right) + \tan^{-1}\left(\frac{y+h}{x+h}\right) - \tan^{-1}\left(\frac{y-h}{x+h}\right)\right]$$

$$u(x) = \frac{Kh}{\pi}\left[\frac{1}{(x-h)^2 + h^2} - \frac{1}{(x+h)^2 + h^2}\right]$$

6.105 $u(x,y) = 20xy^3 - 20x^3y$ $\quad v(x,y) = 30x^2y^2 - 5x^4 - 5y^4$
$\phi(x,y) = 5x^4y - 10x^2y^3 + y^5$

6.107 $B = -3A$ $\quad \phi(x,y) = -6Axy + 3Ax^2y - Ay^3$

6.109 $\psi = \frac{B}{2}(x^2 - y^2) - 2Axy$

6.111 $\psi(x,y) = 20x^3y^3 - 6x^5y - 6xy^5$

6.115 $|V| = x^2 + y^2$ $\quad \psi = 3Ax^2y - \frac{B}{3}x^3$

6.117 $Q = 1.25\ \text{m}^3/\text{s/m}$ $\quad \phi = \frac{B}{2}(y^2 - x^2)$

6.121 $\psi = -\frac{q}{2\pi}\theta - \frac{K}{2\pi}\ln r$ $\quad \psi = -\frac{q}{2\pi}\ln r - \frac{K}{2\pi}\theta$ $\quad r > 9.77$ m
$p = -6.37$ kPa (gage)

6.123 $h = 0.162$ m $\quad V = 44.3$ m/s $\quad p = -957$ Pa (gage)

6.125 $R_x = -5.51$ kN/m

6.127 Stagnations points: $\theta = 63°, 297°$ $r = 1.82$ m $\Delta p = 317$ Pa

7.1 $\dfrac{gL}{V_0^2}, \dfrac{\sigma}{\rho L V_0^2}$

7.3 $\dfrac{V_0^2}{gL}$

7.5 $\dfrac{gL}{V_0^2}, \dfrac{\mu}{\rho V_0 L}$

7.7 $\dfrac{\nu}{V_0 L}\left(=\dfrac{1}{\text{Re}}\right)$

7.9 $\dfrac{\Delta p}{\rho \overline{V}^2}, \dfrac{\nu}{D\overline{V}}, \dfrac{L}{D}$

7.11 $F \propto V^2$

7.13 $D = \rho L^2 c^2 f\left(\dfrac{\lambda}{L}\right)$

7.15 $\dfrac{F}{V^2 \rho A} = f\left(\dfrac{c}{V}\right)$

7.17 $\dfrac{\tau_w}{\rho U^2} = f\left(\dfrac{\mu}{\rho U X}\right)$

7.19 $\dfrac{W}{g\rho p^3}, \dfrac{\sigma}{g\rho p^2}$

7.21 $\dfrac{\overline{u}}{u_*}, \dfrac{\nu}{y u_*}$

7.23 $V = C\sqrt{\dfrac{\sigma}{\lambda\rho}}$

7.25 $\dfrac{T}{Fe}, \dfrac{\mu e^2 \omega}{F}, \dfrac{\sigma e}{F}$

7.27 $\dfrac{W}{D^2 \omega \mu} = f\left(\dfrac{l}{D}, \dfrac{c}{D}\right)$

7.29 $E = \rho V^3 f\left(\dfrac{nr}{V}\right)$

7.31 $\dfrac{Q}{Vh^2} = f\left(\dfrac{\rho V h}{\mu}, \dfrac{V^2}{gh}\right)$

7.33 Four dimensionless groups, three repeating parameters, $\dfrac{\mu}{\rho\sqrt{d^3 g}}$

7.35 $\dfrac{d}{D}, \dfrac{\mu}{\rho V D}, \dfrac{\sigma}{\rho D V^2}$

7.37 $\dfrac{d}{D}, \dfrac{\mu}{\rho V D}, \dfrac{\sigma}{\rho D V^2}, \dfrac{L}{D}$

7.39 $\dfrac{V^2}{g\delta}, \dfrac{\mu^2 \delta^3}{m^2 g}, \theta, \dfrac{A}{\delta^2}$

7.41 $\dfrac{\mathcal{P}}{\rho D^5 \omega_{\max}^3}, \dfrac{\mu}{\rho D^2 \omega_{\max}}, \dfrac{H}{D}, \dfrac{h}{D}, \dfrac{f}{\omega_{\max}}$

7.43 $\frac{\delta}{D}, \frac{L}{D}, \frac{\mu\omega D^3}{T}, \frac{I\omega^2}{T}$

7.45 $\frac{T}{\rho V^2 D^3}, \frac{\mu}{\rho VD}, \frac{\omega D}{V}, \frac{d}{D}$

7.47 $\frac{\dot{m}}{\delta\rho\alpha} = f\left(\frac{D}{\alpha}\right)$

7.49 Three dimensionless groups $\dot{Q} = \rho V^3 L^2 f\left(\frac{c_p\Theta}{V^2}, \frac{\mu}{\rho VL}\right)$

7.51 $\frac{dT}{dt}\frac{Lc_p}{V^3} = f\left(\frac{c}{c_p}, \frac{k}{\rho L^2 c_p}, \frac{\mu}{\rho LV}\right)$

7.53 $\Pi_1 = \frac{u}{U} \quad \Pi_2 = \frac{y}{\delta} \quad \Pi_3 = \frac{(dU/dy)\delta}{U} \quad \Pi_4 = \frac{\nu}{\delta U}$

7.55 $V_w = 6.90$ m/s $\quad F_{air} = 522$ N

7.57 $V_{air} > V_{water} \quad V_{air} = 15.1 V_{water}$

7.59 $V_m/V_p = 0.339 \quad F_p = 213$ N

7.61 $p_m = 1.934$ MPa $\quad F_p = 43.4$ kN

7.63 $V_m = 6$ m/s $\quad F_p = 1.03$ N

7.65 $D_m = 12.81$ cm $\quad \omega_m = 900$ rpm

7.67 $V_p = 6$ m/s $\quad \omega_p = 102$ rpm

7.69 $V_{H_2O} = 0.0142\ \frac{m}{s} \quad \Delta p_{H_2O} = 98.6$ Pa

7.71 $C_{D_m} = 0.0970 \quad Re_m = Re_p \quad F_{d_p} = 460$ N

7.73 $V_m = 0.1875$ m/s $\quad \omega_m = 0.9375$ Hz at standard conditions
$V_m = 0.286$ m/s $\quad \omega_m = 1.43$ Hz in hot air
$V_m = 0.01262$ m/s $\quad \omega_m = 0.0631$ Hz in water

7.75 $\nu_m = 4.11 \times 10^{-8}$ m²/s

7.77 $V_m = 9.02$ m/s $\quad F_p/F_m = 0.265 \quad p_{min} = 22.6$ kPa $\quad p_{tank} = 79.4$ kPa

7.79 $V_R = 175$ km/h @ 5°C $\quad V_R = 123.7$ km/h @ 65°C
$V_R = 289$ km/h using CO_2

7.81 $\frac{V_\theta}{\omega r} = g\left(\frac{\mu}{\rho\omega r^2}, \omega\tau\right)$ honey takes less time than water to reach steady state motion

7.83 Model $= \frac{1}{50} \times$ Prototype $\quad$ Adequate Reynolds number not achievable

7.87 $D = 245$ N at 15 knots $\quad D = 435$ N at 20 knots

7.89 $h_m = 13.8$ J/kg $\quad Q_m = 0.166$ m³/s $\quad D_m = 0.120$ m

7.91 $\frac{F_t}{\rho\omega^2 D^4} = f_1\left(\frac{V}{\omega D}, \frac{g}{\omega^2 D}\right)$

$\frac{T}{\rho\omega^2 D^5} = f_2\left(\frac{V}{\omega D}, \frac{g}{\omega^2 D}\right) \quad \frac{\mathcal{P}}{\rho\omega^3 D^5} = f_3\left(\frac{V}{\omega D}, \frac{g}{\omega^2 D}\right)$

7.93 K.E. ratio = 7.22

8.1 $Q = 5.17 \times 10^{-3}$ m^3/s $L = 3.12 - 5.00$ m (turbulent)
$L = 17.3$ m (laminar)

8.3 Smallest turbulent first

$Q_{\text{large}} = 7.63 \times 10^{-4}$ m^3/s Smallest, middle fully developed; largest only fully developed if turbulent

$Q_{\text{mid}} = 4.58 \times 10^{-4}$ m^3 /s Smallest fully developed; middle only fully developed if turbulent

$Q_{\text{small}} = 3.05 \times 10^{-4}$ m^3/s All fully developed

8.7 $\overline{V}/u_{\max} = 2/3$

8.9 $Q = 3.47 \times 10^{-7}$ m^3/s (347 mm^3/s)

8.11 $\tau_{yx} = -1.88$ Pa $Q/b = -5.63 \times 10^{-6}$ m^2/s

8.13 $Q = 3.97 \times 10^{-9}$ m^3/s (3.97×10^{-6} L/s)

8.15 $W = 0.143$ m $dp/dx = -9.79 \dfrac{\text{MPa}}{\text{m}}$ $h = 1.452 \times 10^{-5}$ m

8.19 $\overline{V} = \dfrac{Q}{2\pi rh}$ $\dfrac{dp}{dr} = -\dfrac{6\mu Q}{\pi rh^3}$ $p = p_{\text{atm}} - \dfrac{6\mu Q}{\pi rh^3}\ln\left(\dfrac{r}{R}\right)$
$(p = p_0,\ r < R_0)$

8.21 $n = 1.48$ (dilatant)

8.23 $\partial p/\partial x = -92.6$ Pa/m

8.25 $u_{\text{interface}} = 4.6$ m/s

8.27 $\partial p/\partial x = -2U\mu/a^2$ $\partial p/\partial x = 2U\mu/a^2$

8.29 $\nu = 1.00 \times 10^{-4}$ m^2/s

8.31 $\tau = \rho g \sin(\theta)(h - y)$ $Q/w = 217$ mm^3/s/mm $Re = 0.163$

8.33 $y(u_{\max}) = 2.08$ mm $u_{\max} = 0.625$ m/s $Q/w = 0.011$ m^2

8.35 $U = 0.504 \dfrac{\text{m}}{\text{s}}$ $\tau_{yx} = 0.4032$ Pa $\partial p/\partial x = 1.29 \dfrac{\text{kPa}}{\text{m}}$

8.37 $\dfrac{dV}{dt} = -\dfrac{\mu wL}{mh}V$ $t = 1.06$ s

8.39 $\mathcal{P}_{\text{v}} = \dfrac{\pi\mu\omega^2 D^3 L}{4a}$ $\mathcal{P}_{\text{p}} = \dfrac{\pi D a^3 \Delta p^2}{12\mu L}$ $\mathcal{P}_{\text{v}} = 3\mathcal{P}_{\text{p}}$

8.43 $t = 19.9$ min. $\mu = 0.199$ kg/m · s

8.45 B.C.: $y = 0,\ u = U_0;\ y = h,\ \tau = 0$ $u = \dfrac{\rho g}{\mu}\left(\dfrac{y^2}{2} - hy\right) + U_0$

8.49 $r = 0.707R$

8.51 $Q = 21.5$ mm^3/s (1290 mm^3/min)

8.53 $\tau = c_1/r$ $u = \dfrac{c_1}{\mu}\ln r + c_2$ $c_1 = \dfrac{\mu V_0}{\ln(r_i/r_o)}$ $c_2 = -\dfrac{V_0 \ln r_o}{\ln(r_i/r_o)}$

8.55 $r = R\left[\frac{1}{2}\frac{(1-k^2)}{\ln(1/k)}\right]^{1/2}$

8.57 % change $= -100/(1 + \ln k)$

8.59 $R_{hyd} = -\frac{8\mu}{3\pi\alpha}\left[\frac{1}{(r_0+\alpha z)^3} - \frac{1}{r_0^3}\right]$

8.61 $u(r) = \frac{n}{(n+1)}\left(\frac{1}{2k}\frac{\partial p}{\partial x}\right)^{\frac{1}{n}}\left(r^{\frac{n+1}{n}} - R^{\frac{n+1}{n}}\right)$

$$Q = -\frac{n\pi}{(3n+1)}\left(\frac{1}{2k}\frac{\partial p}{\partial x}\right)^{\frac{1}{n}} R^{\frac{3n+1}{n}}$$

$$\overline{V} = -\frac{n}{(3n+1)}\left(\frac{1}{2k}\frac{\partial p}{\partial x}\right)^{\frac{1}{n}} R^{\frac{n+1}{n}}$$

$\partial p/\partial x = -325k$ Pa/m ($n = 0.5$), $-42.4k$ Pa/m ($n = 1$), $-5.42k$ Pa/m ($n = 1$)

8.63 $u_0 = -\frac{a^2b^2}{2\mu(a^2+b^2)}\frac{\partial p}{\partial x}$ $\quad Q = -\frac{\pi a^3 b^3}{4\mu(a^2+b^2)}\frac{\partial p}{\partial x}$

$$Q_{pipe} = -\frac{\pi R^4}{8\mu}\frac{\partial p}{\partial x} \qquad Q_{elliptic} = -\frac{29\pi R^4}{104\mu}\frac{\partial p}{\partial x}$$

8.67 $F = 123$ N (in both cases)

8.69 $\tau_w = -9.33$ Pa

8.71 $\mu = 2.413 \times 10^8 \frac{\text{N}\cdot\text{s}}{\text{m}^2}$

8.73 $n = 6.21$ $\quad n = 8.55$

8.75 $\beta_{lam} = 4/3$ $\quad \beta_{turb} = 1.02$

8.77 $\alpha = 2$

8.79 $H_{lT} = 1.33$ m $\quad h_{lT} = 13.0$ J/kg

8.81 $V_1 = 3.70$ m/s

8.83 $\Delta Q = 3 \times 10^{-3}$ m³/s ($Q = 0.019$ m³/s)

8.85 $\overline{V}_1 = 2$ m/s

8.87 $p_2 = 1.68$ MPa

8.89 $\frac{d\bar{u}}{dy} = 971\ \text{s}^{-1}$ $\quad \tau_w = 1.73 \times 10^{-2}\ \frac{\text{N}}{\text{m}^2}$ $\quad \tau_w = 0.02$ N/m²

8.91 $f = 0.0390$ $\quad Re = 3183$ $\quad$ Turbulent

8.93 Maximum = 2.12% at $Re = 10000$ and $e/D = 0.01$

8.97 $p_2 = 171$ kPa $\quad p_2 = 155$ kPa

8.99 $Q = 1.114 \times 10^{-3}$ m³/s (0.067 m³/min, 67 L/min)

8.105 $K = 9.38 \times 10^{-4}$

8.107 $Q = 3.97$ L/s $\quad Q = 3.64$ L/s $\quad (\Delta Q = -0.33$ L/s)

$Q = 4.77$ L/s ($\Delta Q = 0.80$ L/s, a gain)

8.109 $\Delta p = 172.8$ kPa $\quad K = 0.3$

8.111 $h_{l_m} = (1-AR)^2 \frac{V_1^2}{2}$

8.113 $\overline{V}_1 = \sqrt{\frac{2\Delta p}{\rho(1-AR^2-K)}}$ Inviscid assumption: Lower indicated flow/larger Δp

8.115 $Q = 0.345$ L/min $d = 3.65$ m

8.117 $d = 6.13$ m (or 6.16 m if $\alpha = 2$, laminar)

8.119 $Q = 7.66 \times 10^{-5}$ m^3/s (0.0766 L/s) $h = 545$ mm $h = 475$ mm

8.123 $h = 79.6$ m $\Delta p = 781$ kPa

8.127 $\Delta p = 0.001$ m H_2O

8.129 $V_B = 4.04$ m/s $L_A = 12.8$ m (Not feasible!) $\Delta p = 29.9$ kPa

8.131 $\Delta p/L = 1.32$ N/m^3 (round) $\Delta p/L = 1.53$ N/m^3 (1:1) (+15.9%)
$\Delta p/L = 1.65$ N/m^3 (2:1) (+25%)
$\Delta p/L = 1.82$ N/m^3 (3:1) (+37.9%)

8.133 $p_1 = 1290$ kPa

8.135 $h = 1.51$ m $V = 9.41$ m/s

8.137 $V = 1.38$ m/s $Q = 6.80 \times 10^{-4}$ m^3/s (0.680 L/s)

8.139 $L = 26.5$ m

8.141 $t = 16.7$ min

8.143 $Q = 0.0395$ m^3/s

8.145 $dh/dt = 42.3$ mm/s

8.147 Rate of downpour = 0.143 cm/min

8.153 $Q = 6.68 \times 10^{-3}$ m^3/s $p_{min} = -20.0$ kPa (gage)

8.155 $Q = 26.7$ gpm $Q = 28.2$ gpm $Q = 31.3$ gpm

8.157 $Q = 5.30 \times 10^{-4}$ m^3/s $Q = 5.35 \times 10^{-4}$ m^3/s (diffuser)

8.159 $L = 0.296$ m

8.161 $D = 56.0$ mm

8.163 $D = 5.0 - 5.1$ cm (corresponds to standard 50 mm pipe)

8.165 $D = 150$ mm (nominal)

8.169 $\overline{V} = 6.46$ m/s $p_F = 705$ kPa (gage) $\mathcal{P} = 832$ kW $\tau_w = 88.6$ Pa

8.171 $dQ/dt = -0.524$ m^3/s/min

8.173 $\mathcal{P} = 6.074$ kW

8.175 $\Delta p = 150$ kPa

8.177 $D = 48$ mm $\Delta p = 3840$ kPa $\mathcal{P}_{pump} = 24.3$ kW

8.179 $Q = 5.58 \times 10^{-3}$ m^3/s (0.335 m^3/min) $V = 37.9$ m/s $\mathcal{P} = 8.77$ kW

8.181 Cost = \$34.17/day

8.183 $Q = 0.0419$ m^3/s $\Delta p = 487$ kPa $\mathcal{P} = 29.1$ kW

8.185 $Q = 2.31$ m^3/s

8.187 $Q_0 = 0.00812\ m^3/s$ $Q_1 = 0.00286\ m^3/s$ $Q_2 = 0.00379\ m^3/s$
$Q_3 = 0.00147\ m^3/s$ $Q_4 = 0.00526\ m^3/s$

8.189 $Q_0 = 0.0029\ m^3/s$ $Q_1 = 0.0022\ m^3/s$
$Q_2 = 0.007\ m^3/s$ $Q_3 = 0.0029\ m^3/s$

8.193 $\Delta p = 25.8$ kPa

8.195 $Q = 0.042\ m^3/s$

8.197 $Q = 0.00611\ m^3/s$

8.199 $\Delta t = 40.8$ mm $\dot{m}_{min} = 0.0220$ kg/s

8.203 $Re_d = 1800$ $f = 0.0356$ $p_2 = -290$ Pa (gage) (29.6 mm Hg)

9.3 $x_p = 10.4$ cm at takeoff $x_p = 7.47$ cm at cruise

9.5 $U = 87.2$ m/s for American golf ball $U = 91.2$ m/s for British golf ball
$U = 16.9$ m/s for soccer ball

9.9 $A = U$ $B = \pi/2\delta$ $C = 0$

9.11 $\dfrac{\delta^*}{\delta} = 0.375$ $\dfrac{\theta}{\delta} = 0.139$

9.13 $\dfrac{\delta^*}{\delta} = 0.396$ $\dfrac{\theta}{\delta} = 0.152$

9.15 Linear: $\dfrac{\theta}{\delta} = 0.167$ Sinusoidal: $\dfrac{\theta}{\delta} = 0.137$ Parabolic: $\dfrac{\theta}{\delta} = 0.133$

9.17 Power: $\dfrac{\delta^*}{\delta} = 0.125,\ \dfrac{\theta}{\delta} = 0.0972$ Parabolic: $\dfrac{\delta^*}{\delta} = 0.333,\ \dfrac{\theta}{\delta} = 0.133$

9.19 $\dot{m} = 50.4$ kg/s $D = 50.4$ N (more than Problem 9.18)

9.21 $\delta_2^* = 3.125$ mm $\Delta U/U = 3.91\%$

9.23 $U_2 = 25.5$ m/s $\Delta p = -15.8$ Pa

9.25 $\Delta p = -56.7\ N/m^2$

9.27 $U_2 = 24.6$ m/s $p_2 = -44.5$ mm H_2O

9.29 $\delta_2^* = 2.54$ mm $\Delta p = -107$ Pa $F_D = 2.04$ N

9.35 $y = 0.305$ cm $\dfrac{dy}{dx} = \dfrac{1}{2\sqrt{Re_x}}\dfrac{\eta f' - f}{f'}$ $\tau_w = 0.332\dfrac{\rho U^2}{\sqrt{Re_x}}$
$Re_L = 3.33 \times 10^5$ $\theta_L = 0.115$ cm

9.39 $\theta_L = 0.278$ mm $F_D = 0.850$ N

9.41 $F_D = 26.3$ N (long way) $F_D = 45.5$ N (short way)

9.43 $F_D = 1.68 \times 10^{-2}$ N (both sides) (twice as much as Problem 9.42)

9.45 $F_D = 6.9 \times 10^{-3}$ N (both sides) (higher than Problem 9.44)

9.47 $\dfrac{\delta}{x} = \dfrac{3.46}{\sqrt{Re_x}}$ $C_f = \dfrac{0.577}{\sqrt{Re_x}}$

9.49 $F_D = 0.557$ N

9.51 $\tau_w = 0.289\dfrac{\mu U}{x}\sqrt{Re_x}$ $F_D = \rho U^2 b\theta_L$ $F_D = 0.0563$ N

9.53 $\tau_w = 0.0297\dfrac{\rho U^2}{Re_x^{1/5}}$ $F_D = 0.0360\dfrac{\rho U^2 bL}{Re_x^{1/5}}$
$F_D = 2.27$ N

9.55 $U = 1.81$, 2.42, 3.63, and 7.25 m/s

9.57 $F_D = 0.3638$ N (both sides)

9.59 $F_D = 11.12$ N (separate, both sides)
$F_D = 8.46$ N (composite, both sides)

9.61 $\dfrac{\delta}{x} = \dfrac{0.353}{Re_x^{1/5}}$ $c_f = \dfrac{0.0605}{Re_x^{1/5}}$ $F_D = 2.38$ N

9.63 $\delta_L = 31.3$ mm $\tau_{w_L} = 0.798$ Pa $F_D = 0.700$ N

9.65 $w_2 = 80.3$ mm

9.67 $\Delta p = -6.16$ Pa $L = 0.233$ m

9.69 $mf = \dfrac{1}{3}\rho U^2 \delta W$ (linear) $mf = \dfrac{1}{2}\rho U^2 \delta W$ (sinusoid)
$mf = \dfrac{8}{15}\rho U^2 \delta W$ (parabolic) Linear profile separates first

9.73 $U_2 = 2.79$ m/s $\Delta p = 0.0940$ mm H_2O

9.79 $F_D = 11.12$ N (separate, both sides)
$F_D = 8.46$ N (composite, both sides)

9.81 $F_D = 7190$ N $\mathcal{P} = 1.598$ MW

9.83 $Re_L = 1.547 \times 10^7$ $x_t = 53.3$ mm $F_D = 98.0$ N $\mathcal{P} = 15.3$ kW

9.85 $x_t = 0.0745$ m $\delta = 0.0810$ m $F_D = 278$ N

9.87 $F_D = 3.02 \times 10^4$ N Savings of \$20,644/year assuming fuel costs \$0.26 per Kg

9.91 $F_D = 13.71$ kN $\mathcal{P} = 42.3$ kW

9.93 $d_i = 96.5$ mm

9.95 $D = 3.80$ m (single), 2.20 m (three chutes) $-1.01\ g$ maximum acceleration

9.97 B is 20.8% better than A ($H > D$)

9.99 She can cycle with the headwind, but she cannot reach the top speed with the tailwind.

9.101 Uphill: $V_{max} = 9.47$ km/hr without wind, $V_{max} = 8.94$ km/hr with headwind
Downhill: $V_{max} = 63.6$ km/hr without wind, $V_{max} = 73.0$ km/hr with tailwind
Coasting downhill: $V_{max} = 58.1$ km/hr without wind, $V_{max} = 68.1$ km/hr with tailwind

9.105 $V = \sqrt{\left[\dfrac{2mg\sin\theta}{C_D A\rho\cos^2\theta}\right]}$ $t = 1.30$ mm

9.107 $M = 0.0451$ kg

9.111 $FE = 2.89$ km/m^3 $\Delta Q = 6.85$ m^3/yr (9.34)

9.113 $V = 76.12$ km/h (1970's car) $V = 96$ km/h (current car)

9.115 $\vec{U} = \dfrac{Q_s\vec{E}_\infty}{6\pi\mu a}$ $u = -0.053$ m/s when $a = 1\ \mu$m $u = -0.0053$ m/s when $a = 10\ \mu$m

9.117 $C_D = 1.17$

9.119 $V = \sqrt{\left[\dfrac{2mg}{\rho A}\dfrac{1}{\left(\frac{A_1}{A_2}\right)^2 - 2\left(\frac{A_1}{A_2}\right) + 1}\right]}$

9.121 $F_D = C_D A \frac{1}{2}\rho(V-U)^2$ $T = C_D A \frac{1}{2}\rho(V-U)^2 R$

$\mathcal{P} = C_D A \frac{1}{2}\rho(V-U)^2 U$ $\omega_{opt} = \dfrac{V}{3R}$

9.123 $M = 1067$ N · m

9.125 $E_{calm} = 8.86$ kcal $E_{wind} = 12.79$ kcal

9.127 $V = 23.3$ m/s $Re = 48{,}200$ $F_D = 0.111$ N

9.129 $x = 13.9$ m

9.131 $C_D = 61.9$ $\rho_s = 3720$ kg/m^3 $V = 0.731$ m/s

9.133 $M = 0.0471$ kg

9.135 $F_D = \frac{7}{9} C_D \frac{1}{2}\rho U^2 DH$ $M = \frac{7}{16} C_D \frac{1}{2}\rho U^2 DH^2$

$\dfrac{F_D}{F_{D_{uniform}}} = \dfrac{7}{9}$ $\dfrac{M}{M_{uniform}} = \dfrac{7}{8}$

9.137 $C_L = 1.01$ $C_D = 0.0654$

9.139 $D = 7.99$ mm $y = 121$ mm

9.141 $t = 4.69$ s $x = 70.9$ m

9.143 $x_{max} = 48.7$ m (both methods)

9.145 $F_D = 59.1$ N $\Delta FC = 9.65 \times 10^{-3}$ kg/min
$FE = 27.2$ mpg (original design) $F_D = 213$ N
$\Delta FC = 3.48 \times 10^{-2}$ kg/min $FE = 21.9$ mpg (cheaper design)
Rounded corner design rental is \$9.69 cheaper, including cost of gasoline

9.147 $C_D = 0.610$ $V = 60$ km/h

9.149 $V_b = V_w - \sqrt{\dfrac{2F_R}{\rho(C_{Du}A_u + C_{Db}A_b)}}$ $V_b = 4.56$ m/s (16.4 km/hr)

9.151 $t = 4.93$ s $h = 30.0$ m

9.153 $x \approx 203$ m

9.157 $\Delta\mathcal{P} = 16.3$ kW (94%)

9.161 $A = 8.30$ m^2 $T = 1275$ N $\mathcal{P} = 79.7$ kW

9.163 $M = 19.5$ kg $\mathcal{P} = 542$ W

9.165 $M = 37.9$ kg $\mathcal{P} = 3.02$ kW

9.167 $V = 289$ km/h

9.169 $T = 75.6$ kN

9.171 $V = 156$ km/h $F_D = 4.06$ kN $\mathcal{P} = 175$ kW (minimum speed)
$V = 303$ km/h $F_D = 2.01$ kN $\mathcal{P} = 169$ kW (maximum speed)

9.173 $\theta = 3.42°$ $L = 168$ km

9.175 For a race car, effective; for a passenger car, not effective

9.181 $F_L = 0.0822$ N $= 0.175\ mg$ $F_D = 0.111$ N $= 0.236\ mg$

9.183 $\omega = 14{,}000 - 17{,}000$ rpm $x = 1.21$ m

9.185 $\omega = 3090$ rpm

10.1 $H = 135$ m $\dot{W} = 994$ kW

10.3 $r_2 = 6.04$ cm $b_2 = 0.488$ cm

10.5 $\dot{W} = 2.15 \times 10^7$ W $H = 439.2$ m

10.7 $\dot{W} = 161$ kW $H = 65$ m

10.9 $Q = 1.62$ m^3/s $H = 132$ m $\dot{W} = 2100$ kW

10.11 $H_0 = 117$ m $w_2 = 45.78$ m/s $V_2 = 49$ m/s $\dot{W} = 374$ kW
$H = 76.4$ m

10.13 $\beta_1 = 50°$ $\dot{W} = 3.24 \times 10^4$ kW $H = 425.4$ m

10.15 $\beta_1 = 61.3°$

10.19 $\alpha_2 = 79.3°$ $\dot{W} = 33.1$ kW $H = 113$ m

10.25 $Q = 19.4$ m^3/min $H = 5.52$ m $\eta = 82.6\%$ $N_s = 2.64$

10.27 $N_s = \Pi_3^{\frac{1}{2}}/\Pi_2^{\frac{5}{4}}$

10.29 1 hp = 1.01 hpm $N_{s_{cu}} = 0.228\ N_s$(rpm, hpm, m)

10.31 $\dot{W} = 8.97$ kW

10.33 $H_0 = 25.8$ m $\eta = 78.9\%$ $Q' = 1.07$ m^3/s $H' = 21.9$ m
$H_0' = 56.6$ m $\dot{W}' = 292$ kW

10.35 at least 6 pumps $N \approx 473$ rpm

10.37 $Q = 4.58 \times 10^7$ L/day $N \approx 2410$ rpm

10.47 $D_2/D_1 = 0.8$ $Q_2 = 4.03$ m^3/s

10.49 $T_2 = 48°$C $Q_2 = 0.0500$ m^3/s $H_2 = 6.75$ m
Inlet pressure must be increased 9.57 kPa to avoid cavitation

10.51 $Q = 81.2 \times 10^{-3}$ m^3/s

10.53 $Q = 0.0625$ m^3/s

10.57 $D = 0.15$ m $\dot{W} = 660.3$ kW

10.59 $Q_1 = 142$ m^3/h

10.61 $Q = 614$ m^3/h $L_e/D_{valve} = 26900$

10.63 $Q_{loss} = 37\ m^3/h$ (6.0% loss in 20 years)
$Q_{loss} = 50\ m^3/h$ (8.2% loss in 40 years)
$Q_{loss} = 57\ m^3/h$ (9.3% loss in 20 years)
$Q_{loss} = 111\ m^3/h$ (18.1% loss in 40 years)

10.65 $Q_{loss} = 150\ m^3/h$ (14.4% loss in 20 years)
$Q_{loss} = 194\ m^3/h$ (18.7% loss in 40 years)
$Q_{loss} = 195\ m^3/h$ (18.8% loss in 20 years)
$Q_{loss} = 322\ m^3/h$ (31.0% loss in 40 years)

10.67 $\dot{W}_{initial} = 191.2\ kW$ $\dot{W}_{full} = 286\ kW$

10.69 $H_p = 37.124\ m$ An 275 mm *4AE12* pump would work
$NPSHA = 24.99\ m > NPSHR \approx 1.5\ m$

10.71 A *5TUT168* would work $\eta \approx (0.86)^3 = 0.636 = 63.6\%$

10.75 $Q = 0.13\ m^3/s$

10.77 *10TU22C* pumps would work $Q = 0.95\ m^3/s$ $\dot{W} \approx 2.14\ MW$
1 pump: $Q \approx 0.42\ m^3/s$ $\dot{W} \approx 0.17\ MW$
2 pumps: $Q \approx 0.72\ m^3/s$ $\dot{W} \approx 0.82\ MW$
3 pumps: $Q \approx 0.90\ m^3/s$ $\dot{W} \approx 1.57\ MW$
4 pumps: $Q \approx 0.95\ m^3/s$ $\dot{W} \approx 2.14\ MW$

10.79 $H_{lt12} = 8.84\ m$ $p_2 = -54.9\ kPa$ (gage) $\dot{W} = 556\ W$
31% decrease for 4 cm pipe

10.83 $H = 36.6\ m$ $\dot{W} = 7.88\ kW$

10.85 $H = 1.284\ m$ $H = 1.703\ m$ at higher speed

10.87 $A_e = 0.72\ m^2$ $Q = 4.98\ m^3/s$ $h_t = 4.73\ cm$
$\dot{W} = 2.30\ kW$ $\eta = 90.8\%$

10.91 $N = 566\ rpm$ $D_m/D_p = 0.092$ $Q = 0.83\ m^3/s$

10.93 $\dot{W} = 11.7\ MW$ $N = 356\ rpm$ $N_{run} = 756\ rpm$
$T = 2.09 \times 10^5\ N \cdot m$ $T_{stall} = 5.45 \times 10^5\ N \cdot m$

10.95 $N_{s_{cu}} = 212$ $Q = 978\ m^3/s$

10.97 $R = 1.643\ m$ $D_j = 37.0\ cm$ $\dot{m} = 8830\ kg/s$

10.101 $H_{net} \approx 324\ m$ $N_s \approx 0.115$ $\eta \approx 87\%$

10.103 $V_j = 35.3\ m/s$ $Q = 0.069\ m^3/s$ $\dot{W} = 42.2\ kW$
Optimum $d_j \approx 55 - 56\ mm$

10.105 $F_T = 893\ N$ (at rest) $F_T = 809\ N$ (at speed)

10.107 $D = 5.66\ m$ $n = 241\ rpm$ (4.02 rev/s) $\dot{W} = 54\ MW$

10.109 $J = 0.745$ $C_F = 0.0452$ $\eta = 77.7\%$ $C_T = 0.00689$
$C_P = 0.0039$

10.111 $U = 79.6\ \frac{m}{s}$ $C_P = 0.364$

10.113 $N = 153\ rpm$ $\dot{W} = 144\ W$

10.117 $\dot{m} = 0.356\ \text{kg/s}$ $\dot{W} = 0.244\ \text{MW}$

10.119 $N = 488$ rpm $\dot{m} = 226\ \text{kg/s}$ $T_{02} = 686.59°\text{C}$ $p_{02} = 482.5$ kPa

11.3 $y = 1.97$ m

11.5 $V_{stream} = 1.2$ m/s $y = 0.74$ m

11.7 $\lambda_{Hg} = 12$ mm, 0.472 in. $\lambda_w = 17.1$ mm, 0.675 in.

11.9 $V_{stream} = 2.43$ m/s $Fr = 2$

11.13 $Q = 22.68\ \text{m}^3/\text{s}$

11.15 $E_c = NA$, 0.547 m, 1.14 m, 1.60 m, 2.19 m

11.17 $y_c = 0.681$ m $V_c = 2.61$ m/s

11.19 $y = 0.198$ m, 1.3 m

11.21 $y_c = 1$ m

11.23 $y = 0.4$ m

11.25 $\Delta y = 0.015$ m ($y_2 = 0.105$ m)

11.27 $q_{max} = 16.3\ \text{m}^3/\text{s/m}$

11.29 $y_2 = 4.45$ m $V_2 = 9.31$ m

11.31 $\dot{Q} = 24.4$ kW $\Delta T = 6.04 \times 10^{-4}$ °C

11.33 $y_2 = 4.04$ m $H_l = 1.74$ m

11.35 $y_2 = 4.45$ m/s $H_l = 9.31$ m/s

11.37 $y_2 = 10.3$ m $V_2 = 2.19$ m/s $E_1 = 32.8$ m $E_2 = 10.5$ m

11.39 $y_{before} = 0.0563$ m $y_{after} = 0.543$ m

11.41 $V_r = 2.1$ m/s (7.56 km/h)

11.43 $y = 1.24$ m

11.45 $y = 0.815$ m

11.47 $S_b = 2.03 \times 10^{-3}$

11.49 $S_b = 1.60 \times 10^{-3}$

11.51 $Q = 0.194\ \text{m}^3/\text{s}$

11.53 $y = 0.752$ m

11.55 $y = 0.775$ m

11.59 $y = 5.66$ m $b = 2.67$ m

11.63 There is no jump

11.67 $S_c = 2.46 \times 10^{-3}$

11.69 $q = 0.302\ \text{m}^3/\text{s/m}$

11.71 $Q = 0.73\ \text{m}^3/\text{s}$

11.73 $H = 0.514$ m

11.75 $C_w = 1.45$

12.1 T = const. p decreases ρ decreases
(Irreversible adiabatic process)

12.3 $\Delta s = -434.86 \dfrac{J}{kg \cdot K}$ so not feasible!
(violates 2nd law of thermodynamics)

12.5 $T_2 = 276.6$ K $\Delta s = 67.54 \dfrac{J}{kg \cdot K}$

12.7 $T_2 = 860$ K $\Delta h = 542$ kJ/kg $\Delta s = 171.7$ J/kg · K $\dot{m} = 1.845$ kg/s

12.9 $\dot{W} = 7.86$ kW

12.11 $\delta Q/dm = 1104$ kJ/kg at constant pressure
$\delta Q/dm = 789$ kJ/kg at constant volume

12.13 $\eta = 58.8\%$

12.15 $W = 176$ MJ $W_s = 228$ MJ T_s (max) = 858 K $Q_s = -317$ MJ

12.17 $\dot{m} = 36.7$ kg/s $T_2 = 572$ K $V_2 = 4.75$ m/s $\dot{W} = 23$ MW

12.19 $\Delta t = 4$ years

12.23 $c_{H_2} = 1305$ m/s $c_{He} = 1005$ m/s $c_{CH_4} = 446$ m/s $c_{N_2} = 349$ m/s
$c_{CO_2} = 267$ m/s

12.25 $\Delta t = 198$ μs $E_v = 12.7$ GN/m^2

12.27 $x = 2.5$ km

12.29 $\Delta t = 523$ s (8.72 min)

12.31 $M = 0.776$ $V = 269$ m/s

12.33 $\Delta t = 4.66$ s (July) $\Delta t = 5.00$ s (January)

12.35 −5.42% (assuming stratospheric temperature)
+9.08% (assuming sea level temperature)

12.37 $x = 519$ m

12.43 $\Delta t = 116.1$ s

12.45 $V = 642$ m/s

12.47 $V = 493$ m/s $\Delta t = 0.398$ s

12.49 $V = 515$ m/s $t = 6.92$ s

12.51 $\Delta x \approx 1043 - 1085$ m

12.53 Density change < 1.15%, so incompressible

12.55 $M = 0.142$ (1%) $M = 0.322$ (5%) $M = 0.464$ (10%)

12.57 $T_0 = 2290$ K

12.59 $p_0 = 9.1$ MPa $p_{dyn} = 7.6$ MPa

12.61 $p_0 = 44.2$ kPa

12.63 $p_{dyn} = 54.3$ kPa $p_0 = 152$ kPa

12.65 $p_0 = 546$ kPa $h_0 - h = 178$ kJ/kg $T_0 = 466$ K

12.67 $p_0 - p = 8.67$ kPa $V = 195$ m/s $V = 205$ m/s Error using
Bernoulli = 5.13%

12.71 T_0 = const (isoenergetic) p_0 decreases (irreversible adiabatic)

12.73 $V = 890$ m/s $T_0 = 677$ K $p_0 = 212$ kPa

12.75 $p_0 = 119.7$ kPa $T_0 = 4638$ K

12.77 $T_{0_1} = 812$ K (539°C) $T_{0_2} = 257$ K (−16.5°C) $\dot{Q} = -27.9$ kW
$\Delta s = -1186$ J/kg-K

12.79 $\delta Q/dm = 160$ kJ/kg $p_{0_2} = 385$ kPa

12.81 $\dot{Q} = 34.99$ kW $p_{0_2} = 18.66$ kPa

12.83 $p_{0_1} = 698$ kPa $T_{0_1} = 1572$ K (1299°C) $p_{0_2} = 30$ kPa
$T_{0_2} = 1041$ K (768°C) $\Delta s = 485$ J/kg-K

12.85 $\dot{m} = 1.83 \times 10^{-4}$ kg/s

12.87 $T^* = 260$ K $p^* = 24.7$ MPa $V^* = 252$ m/s

12.89 $T_t = 2730$ K $p_t = 25.5$ MPa $V_t = 1030$ m/s

13.1 $\dot{m} = 3.18$ kg/s

13.3 $V = 781$ m/s $M = 1.35$ $\dot{m} = 2.00$ kg/s

13.5 $M = 2.94$ $T = -98$°C

13.7 $p_2 = 45$ kPa

13.9 Converging duct $A = 6.5734$ cm^2

13.11 $M_2 = 1.20$ Supersonic diffuser

13.13 $M_2 = 1.20$ Supersonic diffuser

13.19 $p_t = 250$ kPa $V_t = 252$ m/s $M_t = 0.883$

13.21 $M = 0.240$ $M = 2.44$

13.23 $p_t = 166$ kPa

13.25 $p = 150$ kPa $M = 0.60$ $A_t = 0.0421$ m^2 $\dot{m} = 18.9$ kg/s

13.27 $A_t = 1.94 \times 10^{-3}$ m^2

13.29 $p_0 = 817$ kPa $p_e = 432$ kPa $T_e = 288$ K (−45.5°C) $V_e = 302$ m/s

13.31 $\dot{m} = 0.807$ kg/s $\dot{m}_{max} = 0.843$ kg/s

13.33 $\Delta t = 374$ s (6.23 min) $\Delta s = 232$ J/kg · K

13.35 $p_e = 687$ kPa $\dot{m} = 0.0921$ kg/s $a_{rfx} = 1.62$ m/s^2

13.37 $p_0 = 9.87$ kPa (abs) $p_e = 5.21$ kPa (abs) $T_e = 332$ K (58.7°C)
$V_e = 365$ m/s $a_x = 1.25$ m/s^2

13.39 $M = 1.706$

13.41 $R_x = 1.36$ kN (Tension)

13.43 $A_2 = 0.058$ m^2 $V_2 = 200.5$ m/s

13.45 $\dot{m} = 21.94$ kg/s

13.47 $M_e = 1$ $p_e = 381$ kPa Pressure and flow decrease asymptotically
$T_f = 228$ K (−45°C)

13.49 $p_0 = 793$ kPa $\dot{m} = 0.67$ kg/s $A_t = 3.704$ cm^2

13.51 $V = 225$ m/s $\dot{m} = 1.292$ kg/s

13.53 $p_e = 125$ kPa (abs) $\dot{m} = 0.401$ kg/s

13.55 $V_1 = 1300$ m/s $\dot{m} = 87.4$ kg/s

13.57 $\dot{m} = 1.59$ kg/s Mass flow rate decreases by a factor of 2

13.59 $R_x = 950$ N

13.61 $p_e = 88.3$ kPa $\dot{m} = 0.499$ kg/s $R_x = -1026$ N (to left)

13.63 $A_e = 1557.68$ mm^2 $V_e = 2101.9$ m/s $R_x = 1014$ N

13.65 $p_0 = 44.6$ MPa

13.67 $p_2/p_1 = 3.41$ $T_2/T_1 = 1.50$ $\Delta s = 51.8$ J/kg · K

13.69 $V = 509$ m/s

13.71 $p_1 = 8.89$ kPa $\rho_1 = 0.105$ kg/m^3 $V_1 = 687$ m/s $T_{0_1} = 529$ K
$p_{0_1} = 69.6$ kPa $T_{0_2} = 529$ K $p_{0_2} = 49.4$ kPa

13.73 $p_{0_2} = 327$ kPa $V_2 = 1558$ m/s

13.75 $M_1 = 2.20$ $M_2 = 0.547$ $\Delta s = 131.9 \dfrac{\text{N} \cdot \text{m}}{\text{kg} \cdot \text{K}}$

13.77 $T_2 = 520$ K $p_{0_2} = 1.29$ MPa (abs)

13.79 $M_2 = 0.486$ $V_2 = 865$ km/h (240 m/s) $\Delta p_0 = 607$ kPa

13.81 $T_{0_1} = 426$ K $p_{0_1} = 207$ kPa (abs) $p_{0_2} = 130$ kPa (abs)

13.83 $M_1 = 2.48$ $V_1 = 736$ m/s $p_{0_2} = 201$ kPa $p_2 = 167.5$ kPa

13.87 $M_1 = 2.20$ $p_{0_2} = 178$ kPa $V_1 = 568$ m/s ("Isentropic")

13.89 $V_2 = 268$ m/s (Relative to wave), $= -276$ m/s (Relative to ground)

13.91 $A_t = 0.424$ m^2 $\dot{m} = 10.07$ kg/s $p_1 = 8.105$ kPa $T_1 = 144.33$ K
$A_s^* = 0.0726$ m^2

13.93 $M_{2d} = 0.547$ $p_{2d} = 512$ kPa $p_{0_{2d}} = 628$ kPa $A_s^* = 0.111$ m^2

13.95 $M_e = 1.452$ $\dot{m} = 0.34$ kg/s

13.97 $p_{b1}/p_0 = 0.965$ $p_{b2}/p_0 = 0.417$ $p_{b3}/p_0 = 0.0585$

13.99 $M_e = 2.94$ $p_0 = 3.39$ MPa $p_{b1} = 3.35$ MPa $p_{b2} = 1.00$ MPa
$p_{b3} = 101$ kPa

13.101 $p_b = 301$ kPa

13.103 $M_1 = 1.50$

13.105 $p_{atm} < p_0 < 112$ kPa (abs); $p_0 > 743$ kPa (abs)

13.107 $p_3 = 459$ kPa

13.109 $M_e = 0.392$ $p_e = 854$ kPa $p_d = 122.22$ kPa

13.111 $p_b = 301$ kPa

13.113 $M_e = 0.627$ $\delta Q/dm = 57.8$ kJ/kg

13.117 $p_1 = 396$ kPa (abs) $\dot{m} = 47.6$ kg/s

13.121 $M_1 = 0.601$ $M_2 = 0.738$ $p_{0_2} = 230$ kPa (abs) $T_{0_2} = 482$ K
$f = 0.0241$ $e = 0.0776$ cm

13.123 $M_1 = 0.200$ $\dot{m} = 3.19 \times 10^{-3}$ kg/s $p_2 = 47.9$ kPa (abs)

13.125 $p_{min} = 130.14$ kPa $V_{max} = 318$ m/s

13.127 $T_e = 467$ K (194°C) $R_x = 59.4$ N (to right) $\Delta s = 151.2$ J/kg · K

13.129 $p_{0_1} = 390.6$ kPa $T_2 = 240$ K $p_{0_2} = 191.8$ kPa $\dot{m} = 0.014$ kg/s

13.131 $T_2 = 238$ K $p_2 = 26.1$ kPa (abs) $\Delta s = 172$ J/kg · K

13.133 $L = 3.6$ m

13.135 $p_1 = 322.72$ kPa $T_1 = 157.8$ K $p_{0_1} = 11.85$ MPa $T_{0_1} = 441.84$ K
$f = 0.01572$

13.137 $L = 5.66$ m

13.141 $T_2 = 306$ K $\dot{m} = 78.25$ kg/s

13.143 $L = 394$ mm

13.145 $p_1 = 943.2$ kPa $\dot{W} = 18.6$ kW

13.147 $p_1 = 30$ kPa $\dot{W} = 0.131$ μW

13.149 $M_2 = 0.233$ Heat added

13.151 $p_2 = 1.35$ MPa (Isothermal) $p_2 = 1.24$ MPa (Adiabatic)

13.157 $\delta Q/dm = 449$ kJ/kg $\Delta s = 0.892$ kJ/kg · K

13.159 Note: $\rho_2 = 13.6$ kg/m^3 $\dot{Q} = 113$ kJ/s $\Delta p = 1.12$ MPa
With wrong $\rho_2 = 1600$ kg/m^3: $\dot{Q} = 78$ kJ/s $\Delta p = -6.9$ kPa

13.161 $\delta Q/dm = 18$ kJ/kg $\Delta s = 0.0532$ kJ/kg · K $\Delta p_0 = 2.0$ kPa

13.163 $\delta Q/dm = -221.43$ J/kg (negative sign indicates heat lost)

13.165 $\delta Q/dm = 1.12$ MJ/kg $\Delta p_0 = -13.5$ kPa

13.167 $M_2 = 0.50$ $T_{0_2} = 1556$ K $T_2 = 1480$ K $\dot{Q} = 1.86$ MJ/s

13.169 $\delta Q/dm = 14.71$ kJ/kg (heat added)

13.171 $\delta Q/dm = 447$ kJ/kg $\Delta s = 0.889$ kJ/kg · K $\Delta p_0 = 22$ kPa

13.173 $\delta Q/dm = 11.68$ kJ/kg (heat added)

13.175 $\delta Q/dm = 364$ kJ/kg $\Delta p_0 = -182$ kPa $T_{0_2} = 1174$ K
$p_{0_2} = 1.60$ MPa $T_2 = 978$ K $p_2 = 0.844$ MPa $\rho_2 = 3.01$ kg/m^3

13.177 $M_2 = 0.60$ $T_{0_2} = 966$ K
$\delta Q/dm = 343$ kJ/kg (61.6% of max) $\dot{Q} = 4010$ kW

13.179 $M_2 = 1.74$ $p_2 = 31$ kPa

13.181 $\beta = 49.7°$ $p_2 = 203$ kPa $T_{0_2} = 495$ K (weak wave) $\beta = 78.0°$
$p_2 = 345$ kPa $T_{0_2} = 601$ K (strong wave)

13.183 $M_2 = 1.95$ $p_2 = 179$ kPa $M_2 = 0.513$ (Normal shock)
$p_2 = 570$ kPa (Normal shock) $\beta_{min} = 23.6°$

13.185 $\beta = 62.5°$ $p_2/p_1 = 9.15$

13.187 $M_1 = 1.42$ $V_1 = 483$ m/s $\beta = 67.4°$

13.189 $\alpha = 7.31°$ $p_{max} = 931$ kPa $T_{max} = 564$°C

13.191 $L/w = 183$ kN/m

13.193 $p = 116.76$ kPa (one shock) $p = 117.76$ kPa (two shocks)

13.195 $V_1 = 1593.7$ m/s

13.197 $p = 97.99$ kPa (one shock) $p = 96.63$ kPa (two shocks)
$p = 96.33$ kPa (isentropic compression)

13.199 $p = 690$ kPa $p = 517$ kPa (Normal shock only)

13.201 $p = 130$ kPa

13.203 $M_1 = 3.05$ $p_1 = 38.1$ kPa $M = 2.36$ $p = 110$ kPa

13.205 $L/w = 64.7$ kN/m

13.209 $T_2/T_1 = 1.429$ $M_2 = 4.00$

13.211 $M_5 = 3.23$ $V_5 = 1656$ m/s

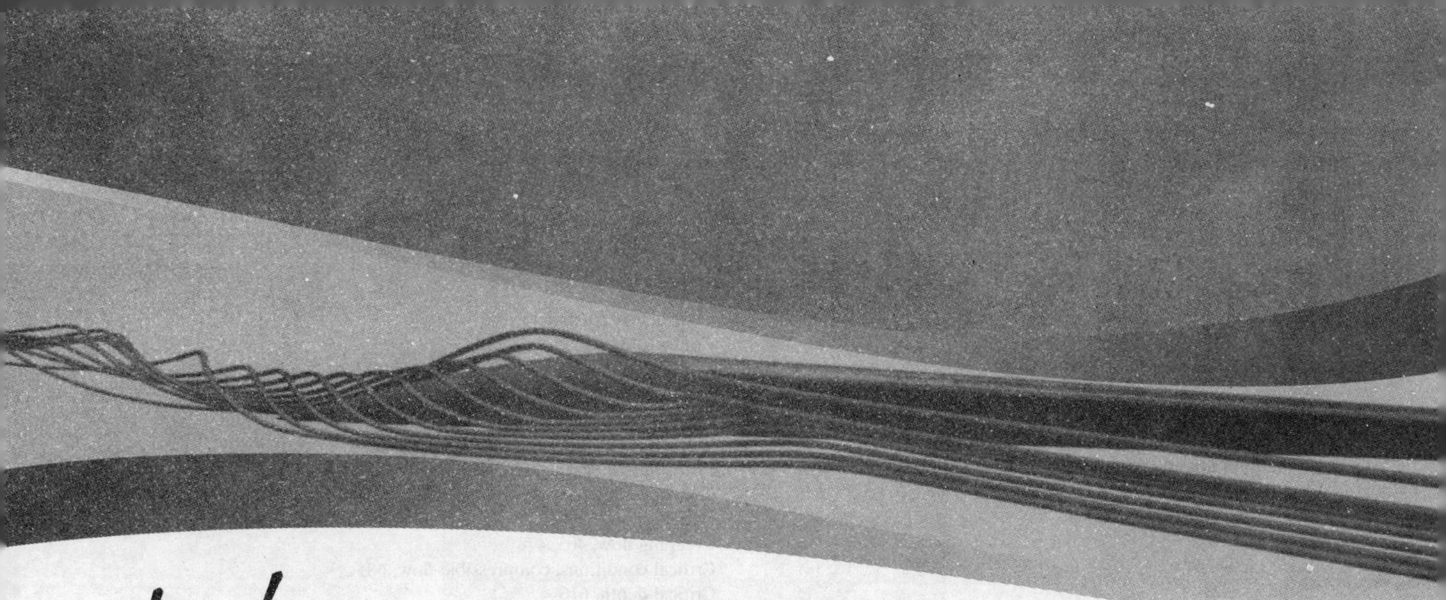

Index

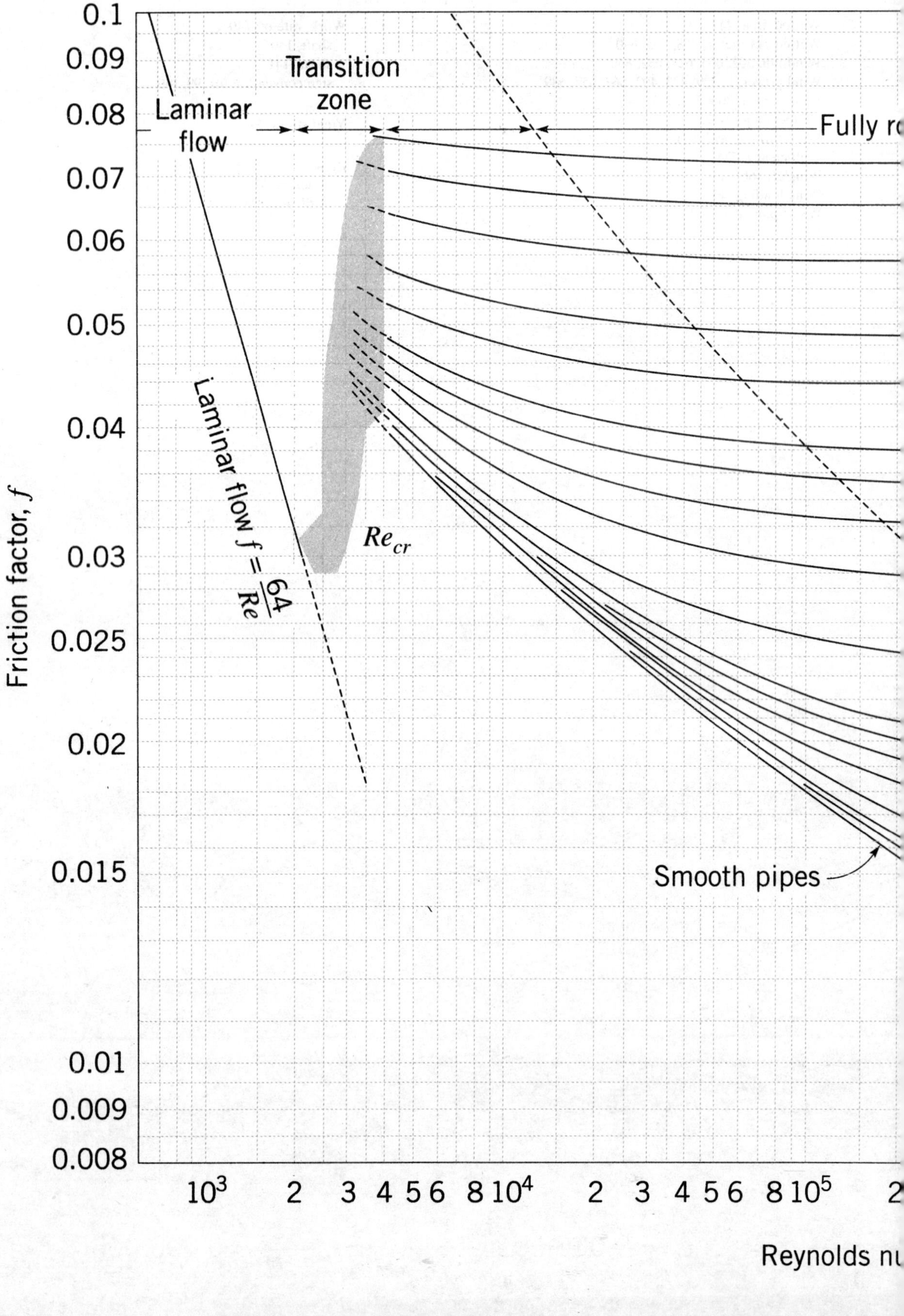
0.1
0.09
0.08
0.07
0.06
0.05
0.04
0.03
0.025
0.02
0.015
0.01
0.009
0.008
Friction factor, f
Laminar flow
Transition zone
Fully r
Laminar flow $f = \frac{64}{Re}$
Re_{cr}
Smooth pipes
10^3 2 3 4 5 6 8 10^4 2 3 4 5 6 8 10^5 2
Reynolds nu

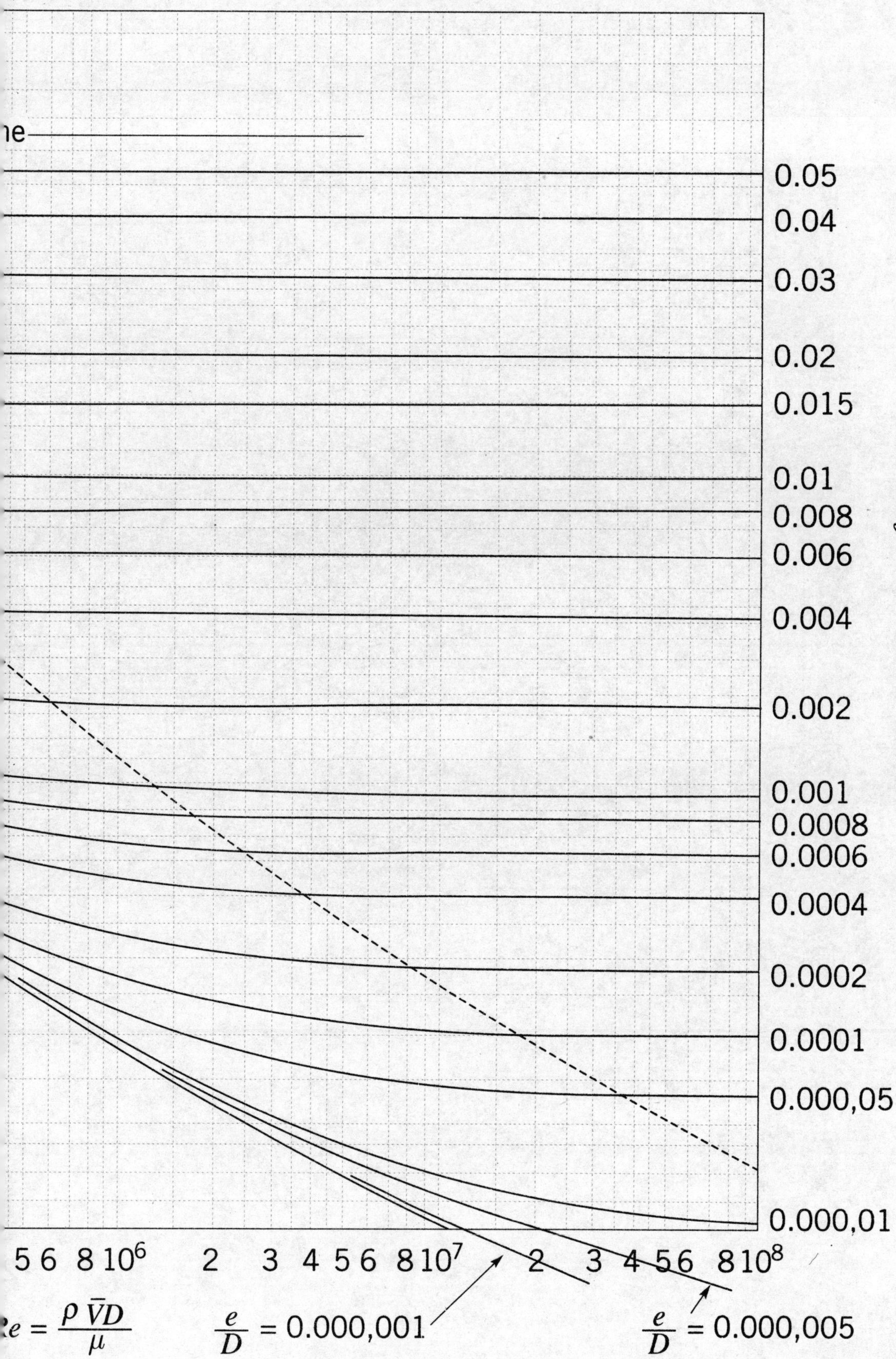
ne
0.05
0.04
0.03
0.02
0.015
0.01
0.008
0.006
0.004
0.002
0.001
0.0008
0.0006
0.0004
0.0002
0.0001
0.000,05
0.000,01
Relative roughness, $\frac{e}{D}$
5 6 8 10^6 2 3 4 5 6 8 10^7 2 3 4 5 6 8 10^8
$Re = \frac{\rho \bar{V} D}{\mu}$
$\frac{e}{D} = 0.000,001$
$\frac{e}{D} = 0.000,005$